REACTIVE INTERMEDIATE CHEMISTRY

REACTIVE INTERMEDIATE CHEMISTRY

Edited by

Robert A. Moss
Department of Chemistry
Rutgers University
New Brunswick, NJ

Matthew S. Platz
Department of Chemistry
Ohio State University
Columbus, OH

Maitland Jones, Jr.
Department of Chemistry
Princeton University
Princeton, NJ

WILEY-INTERSCIENCE

A John Wiley & Sons, Inc., Publication

Published by John Wiley & Sons, Inc., Hoboken, New Jersey.
Published simultaneously in Canada.

For general information on our other products and services please contact our Customer Care Department within the U.S. at 877-762-2974, outside the U.S. at 317-572-3993 or fax 317-572-4002.

Wiley also publishes its books in a variety of electronic formats. Some content that appears in print, however, may not be available in electronic format.

Library of Congress Cataloging-in-Publication Data:

Reactive intermediate chemistry / edited by Robert A. Moss, Matthew S. Platz, Maitland Jones, Jr.
    p. cm.
Includes index.
ISBN 0-471-23324-2 (cloth : acid-free paper)
  1. Chemical reaction, Conditions and laws of. 2. Intermediates (Chemistry)
3. Organic reaction mechanisms   I. Moss, Robert A. II. Platz, Matthew,
III. Jones, Maitland, 1937-
  QD502.R44   2004
  541'.394—dc22   2003021188

Printed in the United States of America

10 9 8 7 6 5 4 3 2

CONTENTS

The chemistry of reactive intermediates is central to a modern mechanistic and quantitative understanding of organic chemistry. Moreover, it underlies a significant portion of modern synthetic chemistry, and is frequently integral to a molecular view of biological chemistry. In the early 1980s, several books or series focused on the area, including three volumes of *Reactive Intermediates* by Abramovitch, three volumes of the same title by Jones and Moss, and *Reactive Molecules* by Wentrup. These authors and editors attempted to survey the entire panoply of reactive intermediates and to highlight their centrality to mechanistic organic chemistry.

Over the past 15 years, however, there has been no major attempt to integrate the subject further. Meanwhile, the field has evolved from a *product-driven* enterprise, in which the primary information was the analysis of reaction products followed by the deduction of probable mechanisms and intermediates. Today, *direct observation* of the intermediates themselves is more common. In other words, chemists can now look at the beasts themselves, rather than make inferences based upon their footprints.

Two main driving forces were at work in directing this change: advances in spectroscopic techniques, including matrix isolation and laser flash photolytic spectroscopy, and rapid progress in theory and computational methods, including the introduction of powerful ab initio and density functional theory. Spectroscopic tools now permit the visualization of reactive intermediates on the microsecond to femtosecond time scales, while theoretical methods enable us to calculate the energies and spectra of the intermediates, the activation energies of their reactions, and many of the features of the energy surfaces upon which those reactions play out. Meanwhile, the emerging field of reaction dynamics promises to create new paradigms for reactions that occur on relatively flat surfaces. Theory has also evolved from being limited to triatomics in the gas phase to a point at which it can accurately simulate the IR and UV–vis spectra of polyatomic intermediates in the gas phase and solution, affording corresponding and useful interaction with the new time-resolved and matrix isolation spectroscopic methods. In fact, "evolution" might be too timid a term for the past two decades of reactive intermediate chemistry; "revolution" might be more apt.

Reactive Intermediate Chemistry is an attempt to provide an updated survey and analysis of the field. We have adopted a "three-dimensional" approach. Reactive Intermediates are considered by type (e.g., carbocations, radicals, carbanions, carbenes, nitrenes, arynes, etc.); they are examined according to the kinetic realms that they inhabit, with special emphasis on rapid processes occurring on the nanosecond, picosecond, and femtosecond time scales; and they are described in historical context, connecting the older, product-based, deductive analyses of their chemistry with the contemporary, direct observational and computational approaches. At the same time, the emerging methods involving dynamics and theory are highlighted.

We intend *Reactive Intermediate Chemistry* to serve as a free-standing resource to be used by the entire chemical community. It should be especially useful for first or second year graduate students, for whom it could form the basis of a course in reactive intermediate chemistry. To that end, each chapter features a concluding section devoted to a summary of the current situation, as well as a roll call of near-term problems and probable research directions. A list of key reviews and suggestions for additional reading accompanies each chapter.

We offer this book in the hope that it will soon be outdated, with confidence that progress in the field will shortly require a new, or very different edition.

ROBERT A. MOSS
New Brunswick, NJ

MATTHEW S. PLATZ
Columbus, OH

MAITLAND JONES, JR.
Princeton, NJ

PART 1

REACTIVE INTERMEDIATES

CHAPTER 1

Carbocations

ROBERT A. McCLELLAND

Department of Chemistry, University of Toronto, Toronto, Ontario M5S 3H6, Canada

Reactive Intermediate Chemistry, edited by Robert A. Moss, Matthew S. Platz, and Maitland Jones, Jr.
ISBN 0-471-23324-2   Copyright © 2004 John Wiley & Sons, Inc.

1. HISTORICAL PERSPECTIVE

1.1. Definitions

Carbocations are a class of reactive intermediates that have been studied for 100 years, since the colored solution formed when triphenylmethanol was dissolved in sulfuric acid was characterized as containing the triphenylmethyl cation.[1] In the early literature, cations such as Ph_3C^+ and the *tert*-butyl cation were referred to as carbonium ions. Following suggestions of Olah,[2] such cations where the positive carbon has a coordination number of 3 are now termed *carbenium ions* with *carbonium ions* reserved for cases such as nonclassical ions where the coordination number is 5 or greater. *Carbocation* is the generic name for an ion with a positive charge on carbon.

1.2. Early Studies

Examples of highly stabilized carbocations as persistent ions in solution have been known for some time. Triarylmethyl halides readily ionize to triarylmethyl cations in weakly nucleophilic solvents such as sulfur dioxide.[3] Tri- and diarylmethyl alcohols undergo equilibrium ionization to cations in aqueous solutions of sulfuric acid and other strong acids (Eq. 1). Equilibrium constants K_R for this pseudo-acid–base equilibrium (Eq. 2), were determined by acidity function techniques.[4,5] Values of pK_R depend strongly on structure and substituents.[6] The parent triphenylmethyl cation has a pK_R of -6.6 and forms only in solutions with 50% or greater H_2SO_4. Derivatives with three *p*-MeO and *p*-Me$_2$N groups have pK_R values of -0.8 and 9.4, respectively. The latter cation, crystal violet, is stable in neutral aqueous solution. The aromatic tropylium ion $C_7H_7^+$ has a pK_R of 4.8, and forms from the alcohol in weakly acidic aqueous solutions.

$$R^+ + H_2O \rightleftharpoons ROH + H^+ \qquad (1)$$
$$K_R = [ROH][H^+]/[R^+] \qquad (2)$$

1.3. Carbocations as Reactive Intermediates

Simple carbocations are initially encountered in the chemical literature of the 1920s and 1930s, through proposals of their occurrence as short-lived intermediates of organic reactions such as S_N1 solvolysis, electrophilic alkene addition, and electrophilic aromatic substitution. Three chemists, Hans Meerwein of Germany, Sir Christopher Ingold of England, and Frank Whitmore of the United States, were the pioneers in this field. A number of experimental observations led to the proposals of the intermediate cations. These included substituent effects, orientation in electrophilic addition and substitution reactions, solvent effects especially on the rates of solvolysis, rearrangements, and common ion inhibition. Such criteria are discussed in standard textbooks[7,8] and the reader is referred to these for a detailed discussion. It was also recognized that in many of these reactions the carbocation intermediate formed and reacted at the stage of an ion pair. Winstein proposed the involvement of two ion pairs, a contact or intimate ion pair and a solvent separated ion pair.[8,9] The reader is referred to Chapter 19 in this volume for a discussion of recent studies of such ion pairs involving modern laser methods.[10] Reactions that are on the borderline of S_N1 and S_N2 solvolysis are discussed in Chapter 2 in this volume.[11]

2. PERSISTENT CARBOCATIONS UNDER STABLE ION CONDITIONS

2.1. Superacids

The modern era of carbocation chemistry began in the 1960s with the discovery by George Olah that simple alkyl carbenium ions could be observed at low temperatures in solutions now termed superacids. While a number of these superacids now exist, the most common are mixtures of a strong protic acid such as HSO_3F or HF to which has been added SbF_5. Substances of low basicity such as SO_2 and SO_2F_2 are often added as a diluent. The initial investigations of Olah employed SbF_5/SO_2. In addition to being sufficiently polar to support the formation of charged species, superacids are very strongly acidic, being orders of magnitude more acidic than sulfuric acid and perchloric acid. Thus, they are capable of protonating very weakly basic carbon bases to form carbocations. They are also very weakly basic and nucleophilic, so that the carbocations, once formed do not react. The combination of high acidity and weak basicity also means that equilibria such as $R_2C{=}CH_2 + H^+ \rightleftharpoons R_2C^+{-}Me$ lie far to the right. This finding suppresses a problem common to less acidic solutions, a reaction of the carbocation with the olefin resulting in complex oligomeric mixtures.

2.2. Stable Ion Chemistry

Olah's initial studies focused on the *tert*-butyl cation and related simple alkyl systems, as obtained, for example, by dissolving *tert*-butyl fluoride in SbF_5/SO_2.[12] The hydrogen nuclear magnetic resonance (¹H NMR) spectrum of this solution showed considerable deshielding of the methyl protons, from ~1.5 to 4.3 ppm. These protons also appeared as a singlet, in contrast to the doublet seen in the starting fluoride associated with H—F coupling. While this result was consistent with the formation of the *tert*-butyl cation, also possible was a polarized donor–acceptor complex that undergoes rapid fluoride exchange.

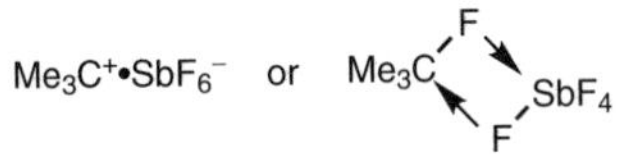

$$Me_3C^+{\cdot}SbF_6^- \quad \text{or} \quad Me_3C{\Big\langle}{\overset{F}{\underset{F}{}}}SbF_4$$

Evidence that the cation was the product accumulated quickly.[1] A ¹³C NMR spectrum, only obtained with difficulty in 1960, revealed that the chemical shift of the tertiary carbon was 335.2 ppm, which was at that time the record for the most deshielded ^{13}C signal. The ^{1}H NMR of the isopropyl cation obtained with Me$_2$CHF and SbF$_5$ showed 13.0 ppm for the CH proton, as well as a ^{13}C chemical shift for the central carbon at 320.6 ppm. Such highly deshielded chemical shifts can only reasonably be explained by the formation of a full cation. In addition, the ^{1}H NMR of the cation EtC$^+$Me$_2$ showed a large coupling between the methyl protons next to the C$^+$ center and the CH$_2$ protons. This coupling is consistent with transmission through an sp^2 carbon, but not through an sp^3 carbon, as would be the case for the donor–acceptor complex. Finally, the planar nature of the cation was firmly established by comparing the infrared (IR) and Raman spectra with those obtained for neutral, isoelectronic boron derivatives, trimethylboron, for example, being the model for the *tert*-butyl cation.

In the ~40 years since Olah's original publications, an impressive body of work has appeared studying carbocations under what are frequently termed "stable ion conditions." Problems such as local overheating and polymerization that were encountered in some of the initial studies were eliminated by improvements introduced by Ahlberg and Ek[13] and Saunders et al.[14] In addition to the solution-phase studies in superacids, Myhre and Yannoni[15] have been able to obtain ^{13}C NMR spectra of carbocations at very low temperatures (down to 5 K) in solid-state matrices of antimony pentafluoride. Sunko et al.[16] employed a similar matrix deposition technique to obtain low-temperature IR spectra. It is probably fair to say that nowadays most common carbocations that one could imagine have been studied. The structures shown below are a limited set of examples. Included are aromatically stabilized cations, vinyl cations, acylium ions, halonium ions, and dications. There is even a recent report of the very unstable phenyl cation ($C_6H_5^+$) generated in an argon matrix at 8 K.[17]

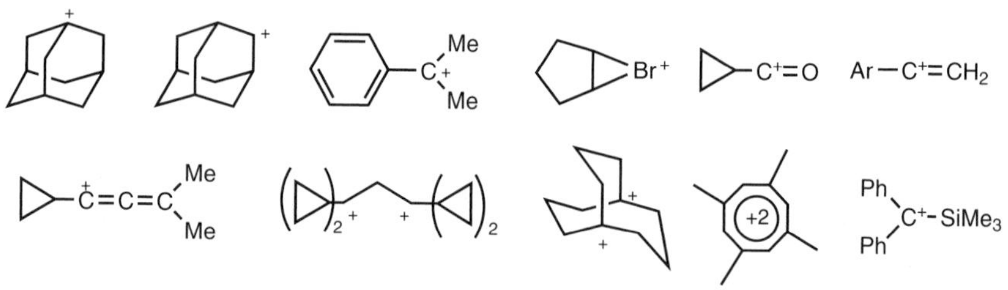

2.3. Theory

In the early days of stable ion chemistry, the experimental measurements of parameters such as NMR chemical shifts and IR frequencies were mainly descriptive, with the structures of the carbocations being inferred from such measurements. While in cases such as the *tert*-butyl cation there could be no doubt of the nature of the intermediate, in many cases, such as the 2-butyl cation and the nonclassical ions, ambiguity existed. A major advance in reliably resolving such uncertainties has been the application of theory. The high-speed computers and sophisticated ab initio programs now available allow properties such as relative energies, geometries, IR frequencies, and NMR chemical shifts to be computed within experimental accuracy. Theoretical methods distinguish transition states, that is, saddle points on a potential energy surface from true ground-state structures that represent real reaction intermediates. Experimental observations such as IR frequencies and NMR chemical shifts can be compared with values computed for the ground-state structures. Where these match, there is reasonable confidence that the structure being observed in the experiment corresponds to the computed structure.

An example of this approach is the 2-butyl cation, an ion that was first observed in 1968.[18] After some debate as to the structure of the ion being observed, a high-level computational search of the $C_4H_9^+$ potential energy surface revealed only two minima corresponding to the 2-butyl derivative, a partially bridged 2-butyl cation (**1**) and the hydrogen-bridged structure (**2**), with the hydrogen-bridged structure more stable by 0.4 kcal/mol.[19] Comparison of the IR frequencies computed for the two structures with an experimental spectrum obtained for the cation in a low-temperature matrix, lead to the conclusion that the experimental cation was principally the hydrogen-bridged ion (**2**).[20]

A second example is the 1-methyl-1-cyclohexyl cation, a tertiary ion where experimental studies had led to the suggestion that there was actually a pair of structures undergoing rapid equilibration even at very low temperatures. These two structures were conformational isomers: **3**, where the ring is distorted to maximize C_β—C_γ hyperconjugation, and **4** with C_β—H_{axial} hyperconjugation. The molecular orbital (MO) calculations performed at the time of the initial experimental observations could only be done using semiempirical methods and were not consistent with this idea. More recent high level ab initio calculations, however, provide full support.[21] The two structures, termed hyperconjomers, are ground states. Moreover, theory provides a close match with three different experimental observations—energy differences between the two, α-deuterium isotope effects, and ^{13}C NMR spectra.

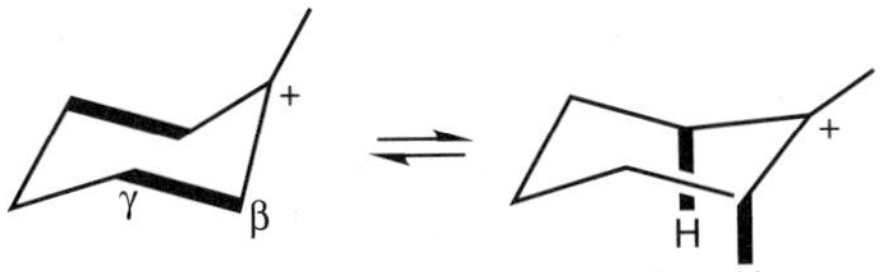

High-level computations are now routine, and failure to match computed and experimental parameters is more likely due to a flaw in the experiment (except for very large structures). The ab initio methods of course provide the full geometry of the ground state. Thus, correspondence of the experimental observations with those computed for a particular ground state means that, within a reasonable confidence level, the detailed geometry of the cation being observed experimentally is known.

2.4. Carbocation Rearrangements

A long-established feature of the carbocation intermediates of reactions, such as S_N1 solvolysis and electrophilic aromatic alkylation, is a skeletal rearrangement involving a 1,2-shift of a hydrogen atom, or an alkyl, or aryl group. The stable ion studies revealed just how facile these rearrangements were. Systems where a more stable cation could form by a simple 1,2-shift did indeed produce only that more stable ion even at very low temperatures (see, e.g., Eq. 3).

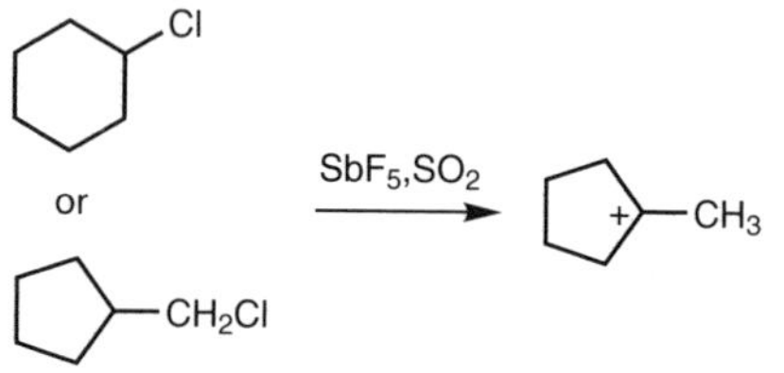

$$\text{(3)}$$

Exceptions to this rule were cations such as the secondary 2-adamantyl cation **5**, which does not rearrange to the more stable tertiary 1-adamantyl cation **6** by an intramolecular 1,2-hydride shift (Eq. 4).[22] The migrating group in a 1,2-shift migrates to the adjacent empty p orbital, so that there is a stereochemical requirement that the migrating group and the p orbital have a dihedral angle close to zero degrees. The 2-adamantyl cation, where the angle is 90°, is the worst possible case.

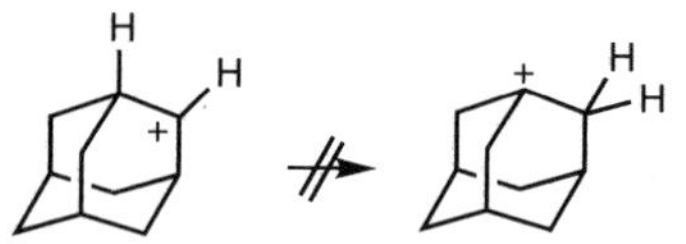

$$\text{(4)}$$

5   **6**

For ions of similar or identical stability, 1,2-shifts are also extremely rapid.[15] Even at very low temperatures, the ^1H NMR spectrum of the 2,3,3-trimethyl-2-butyl cation **7** has only one signal, while the ^{13}C NMR shows two signals, one for the methyl carbons and one for the quaternary carbons. As shown in Scheme 1.1, these lie almost exactly midway between those calculated for the static ion. A similar result is obtained for the 1,2-dimethylcyclopentyl cation **8**, and other tertiary cations related by 1,2-shifts. Thus, even at very low temperatures, 1,2-shifts

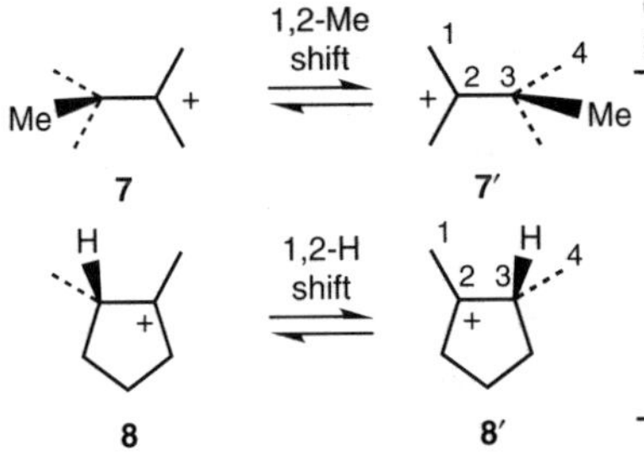

	^{13}C NMR Chemical Shifts		
	Carbon Number	Calc.	Obs.
1,2-Me shift	C2 - 336		
	C3 - 59		198
	C1 - 43		
	C4 - 22		32
1,2-H shift	C2 - 329		
	C3 - 75		203
	C1 - 33		
	C4 - 12		23

Scheme 1.1

between equilibrating degenerate ions are extremely rapid. The activation barriers associated with these rearrangements are <5 kcal/mol. (In some cases these have been measured.)

Static forms of cations such as **7** and **8** have been observed by ^{13}C NMR in solid SbF_5 matrices, with chemical shifts for the static ions matching computed values.[15] The signals for the static ions could be observed even at temperatures above the solution-phase limits, which suggests there is a variation in the lattice sites in the solid, and the cations find themselves in different environments. A broad distribution of rearrangement barriers results.

Some systems show NMR coalescence consistent with a rearrangement to an intermediate cation that is less stable.[23] The ^1H NMR of the isopropyl cation **9**, for example, shows a coalescence of the two proton signals between 0 and 40 °C. The activation barrier is 16 kcal/mol, close to the difference between a secondary and a primary cation. Calculations show that a corner protonated cyclopropane (**11**) is the only other minimum on the $C_3H_7^+$ energy surface.[24] The 1-propyl cation **10** is a transition state.

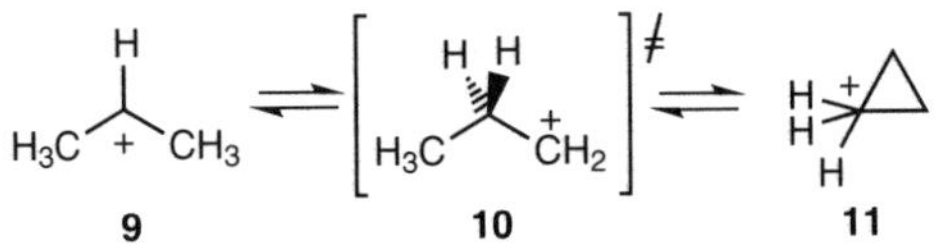

$$(5)$$

2.5. Nonclassical Ions

Nonclassical carbocations (or carbonium ions employing Olah's terminology)[2] are cations in which there is delocalization of σ electrons. These are distinguished from carbenium ions in terms of the following definitions.[25] Carbenium ions (or classical carbonium ions) "can be represented by a single Lewis structure involving only two-electron, two-center bonds. Traditionally, π-conjugated cations such as allyl are included in the category." A carbonium ion "cannot be represented adequately by a single Lewis structure. Such a cation contains one or more carbon or hydrogen bridges joining two-electron deficient centers. The bridging atoms have coordination numbers higher than usual, typically five or more for carbon and two or more for hydrogen. Such ions contain two-electron, three center bonds."

The most extensively studied, and at one time highly controversial, system is the 2-norbornyl cation.[26] Initial suggestions of a nonclassical structure came out of studies of solvolysis reactions, principally by the group of Winstein et al.[27] Observations included: (a) exo derivatives (e.g., **12**) solvolyzed over two orders of magnitude more rapidly than endo derivatives **13**; (b) the product from both was exclusively exo (**16**); (c) **16** was 100% racemized from chiral **12**, while their was some retention of chirality from endo **13**; (d) isotopic tracer studies with an exo reagent showed that 50% of the product was derived from solvent addition at C2 where the leaving group was originally attached, but 50% derived from addition at C1.[28] These observations (and others) led to the proposal that the exo substrate ionizes with anchimeric assistance from the C1—C6 bond forming the symmetrically bridged **14** with a plane of symmetry. The carbons that were C1 and C2 in the starting material are equivalent in this cation. Solvent (Sol) adds equally to these two carbons, from the exo face as shown in Scheme 1.2. This results in 100% racemic exo product. The C1—C6 bond in **13** is not aligned correctly to participate in the ionization. Thus, the endo reagent ionizes initially (and more slowly) to a classical structure (**15**), probably as an ion pair. This cation rearranges to the nonclassical **14**, but, especially in nucleophilic solvents, is partly trapped by solvent from the exo face before rearrangement. The latter reaction accounts for the observation that racemization is not complete in the endo system.

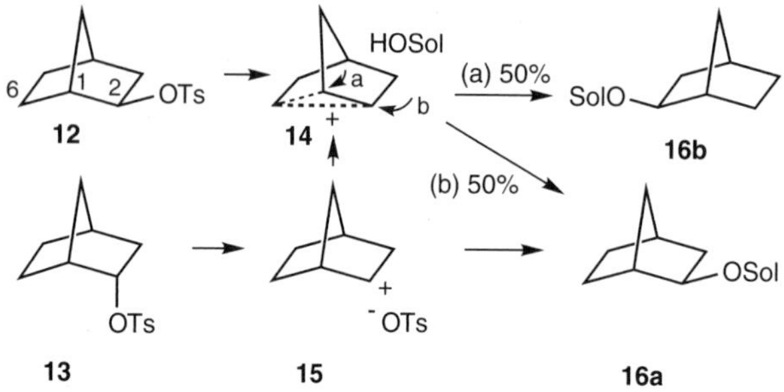

Scheme 1.2

For a number of years, a storm of controversy raged over this proposal, with H. C. Brown as the chief opponent.[26a] Brown ruled out anchimeric assistance as an explanation for the rate acceleration of the exo derivative, arguing that exo was normal, but that endo was unusually slow because of a steric effect. The racemization and isotopic tracer results, he proposed, could be explained by a rapid equilibrium between the classical ions **15** and **17** (see Scheme 1.3), with a steric effect responsible for the exo addition of nucleophiles. In terms of the cation, the question revolves around the issue whether the classical ions **15** and **17** should be joined by the equilibrium depiction (the rapidly rearranging scenario) or with a resonance depiction (the nonclassical scenario). The former, which is supported by the rapid migrations in cations such as **7**, has the bridged structure **14** (depicted in two ways in Scheme 1.3) as the transition state in the rearrangement. In the latter interpretation, **14** is the ground state, with the classical structures representing resonance contributors according to the resonance convention.

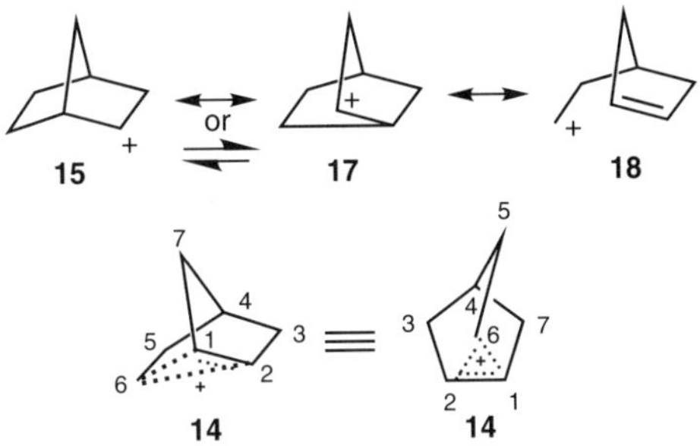

Scheme 1.3

The 2-norbornyl cation has nowadays been extensively studied under stable ion conditions, and there can be no doubt that this ion is the symmetrically bridged **14**. At low temperatures (where hydride shifts are frozen out), both the ^{13}C and ^{1}H NMR point to a symmetrical structure, with, for example, two equivalent strongly deshielded protons or carbons for H—C1 and H—C2. Moreover, the solid-state ^{13}C NMR spectrum obtained in amorphous SbF$_5$ is unchanged from $-144\,°$C down to $-268\,°$C (5 K!).[29] The result at 5 K means that if the symmetry is due to rapidly equilibrating classical ions, the activation barrier for the rearrangement is <0.2 kcal/mol. It has been argued that such a barrier is too low for the extensive bonding and geometrical reorganization that would have to occur. Although initial computations were equivocal because of the limitations of the low level of theory, more recent high-level calculations provide full support for the single minimum potential of **14**.[30] The classical structure is not even a minimum. Moreover there is good agreement in both ^{13}C NMR chemical shifts[30] and IR frequencies[31] for values calculated for the bridged structure and for those observed experimentally. A comparison of the carbon 1s photoelectron spectrum of the *tert*-butyl cation and the norbornyl cation shows the expected difference between classical and nonclassical structures.[32]

Using methods such as those discussed for the norbornyl cation, nonclassical structures have now been established for a number of carbocations.[15,16,24,33] Representative examples are shown below. The 7-phenyl-7-norbornenyl cation **19** exists as a bridged structure **20**, in which the formally empty *p* orbital at C7 overlaps with the C2—C3 double bond. This example is of a homoallylic cation. The cyclopropyl-carbinyl cation **21**, historically one of the first systems where nonclassical ions were proposed, has been shown to exist in superacids mainly as the nonclassical bicyclo-butonium ion **22**, although it appears as if there is a small amount of the classical **21** present in a rapid equilibrium. Cations **23** and **24** are examples of μ-hydridobridged structures.[34] Solvolysis reactions generating carbocations on medium ring carbocycles often show transannular shifts in the products. Initially, it was proposed that structures such as **23** were transition states in such rearrangements. Stable ion studies accompanied by computations have established that the bridged structures are actually minima in a number of such examples.

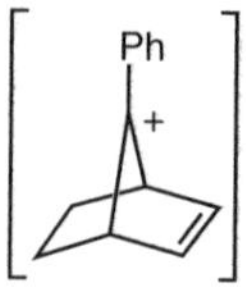

19 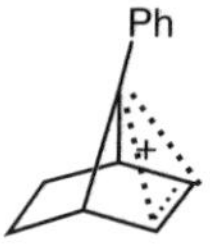 20 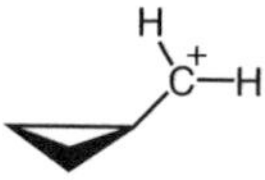 21 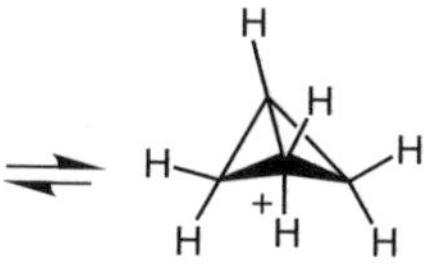 22

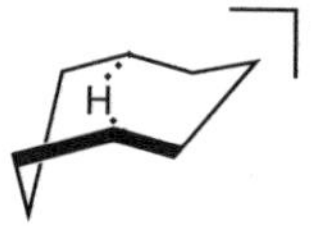

23 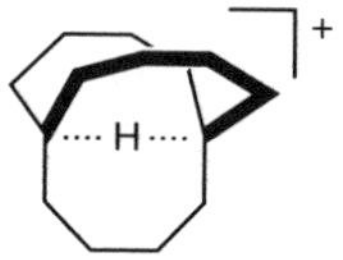 24

2.6. Isotopic Perturbation of Symmetry

The question whether symmetry in the NMR spectrum is due to rapid interconversion of 2 equivalent structures or to a single intermediate structure can be resolved using the tool of isotopic perturbation.[33] Cation **25**, with L = H, shows one signal in the ^{13}C NMR for C_x and C_y because of the symmetry of the delocalized allylic cation. Substituting L = D (*isotopic perturbation of resonance*) has little effect since the π delocalization is not significantly perturbed. The signal for C_x and C_y are split in **25**, L = D, but only by 0.5 ppm (C_x downfield).

25a 25b

(6)

This situation is quite different in the system of Eq. 7. As discussed previously (see Scheme 1.1) there is only one ^{13}C signal for C_x and C_y when L = H because of a rapid 1,2-hydride shift. Substituting L = D (*isotopic perturbation of equilibrium*) now has a large effect. The cation **8a**, with a Me group next to the positively charged carbon, is energetically more stable than **8b** where there is a CD_3 group. This is a β-deuterium equilibrium isotope effect; C—H hyperconjugation stabilizes an adjacent positive charge more than C—D hyperconjugation. The hydride shift is still occurring very rapidly, but the ^{13}C signals are now a weighted average. With $K < 1, C_x$ spends more of its time as the cationic carbon than C_y. Thus, compared to the single chemical shift observed where L = H, C_x is shifted downfield and C_y, which spends more of its time as an sp^3 carbon, upfield. The difference between the signals for C^+ and $C(sp^3)$ is large, $\sim$250 ppm (see Scheme 1.1). Thus the splitting of the signals for C_x and C_y in the deuterated cation is large. For example, C_x is 81.8 ppm downfield of C_y at -142 °C.

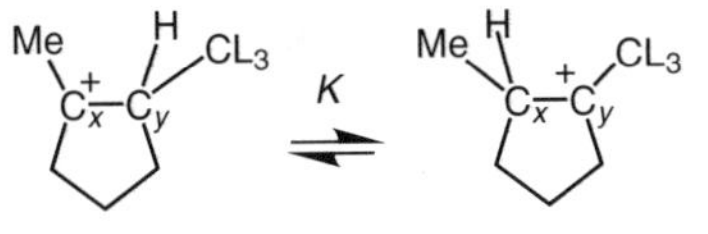

(7)

8a   **8b**

This approach has been applied to the norbornyl cation, where the NMR even at very low temperatures showed equivalent ^{13}C NMR signals for C1 and C2. The monodeuterated **26** was prepared, which showed a difference of only 2 ppm in the signals for C1 and C2.[35] This small splitting is inconsistent with rapidly equilibrating ions but is expected for the symmetrical nonclassical cation.

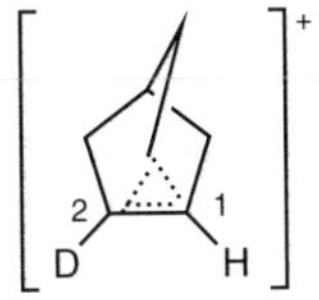

26

2.7. Crystal Structures

The first X-ray crystal structure of a carbocation salt was reported in 1965.[36] Triphenylmethyl perchlorate (**27**) has a planar central carbon. The three phenyl rings are each twisted $\sim$30°, so that overall the cation has a propellor shape. Disordered perchlorate anions sit above and below the central carbon, with a C—Cl separation of 4.09 Å.

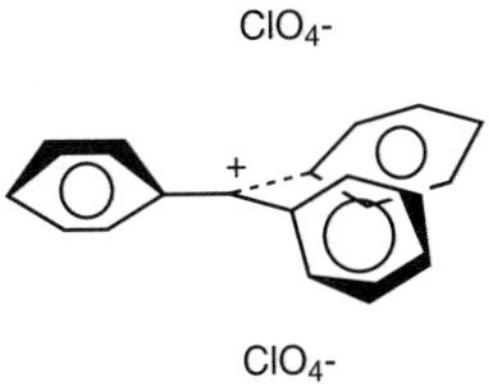

ClO_4^-

ClO_4^-

27

Since that time, a number of X-ray structures have been reported, although many of these, especially the earlier ones,[37] have represented salts of highly conjugatively stabilized cations. Problems such as the unstable nature of the salts and disordered crystals have plagued many attempts. No crystal structure of the parent 2-norbornyl cation, for example, has been reported. Through great persistence Laube[38] has managed to obtain crystals structures of some important cations. Representative examples follow.

The *tert*-butyl cation, as its $Sb_2F_{11}^-$ salt, has the expected planar central carbon, with relatively short C^+–C_α bonds (1.44 Å).[39] The structure is in agreement with the C—H hyperconjugation model of Eq. 8.

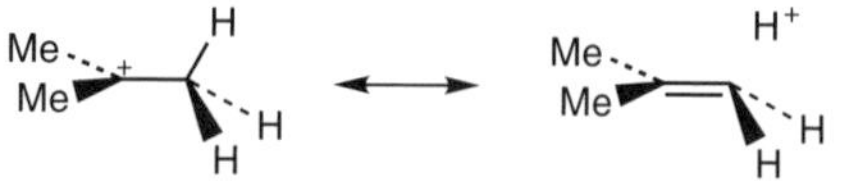

$$(8)$$

The remarkable effect of hyperconjugation is seen in the structure of the 3,5,7-trimethyl-1-adamantyl cation (**28**), also obtained as its $Sb_2F_{11}^-$ salt.[40] The cation shows a significant flattening at C1, with short C^+–C_α bonds (1.44 Å) and relatively long C_α–C_β bonds (1.61 Å). These point to significant C—C hyperconjugation, as shown in Eq. 9.

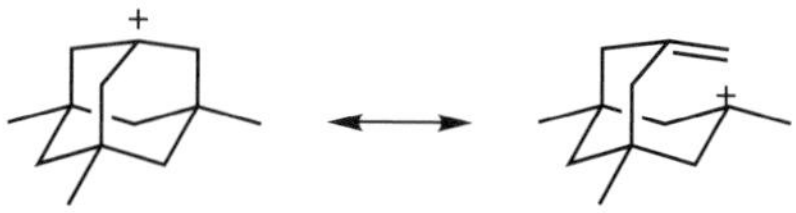

$$(9)$$

28

As shown below, the 2,3-dimethyl-7-phenyl-7-norbornenyl cation (**29**) shows significant distortion of the bridging C7 toward the C2—C3 double bond.[41] The C7–C3 distance is only 1.87 Å, and the C2—C3 bond has elongated to 1.41 Å. This structure is in agreement with the nonclassical (bis-homoaromatic) character of the non-methylated analogue **20**.

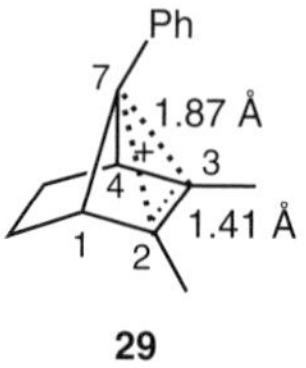

29

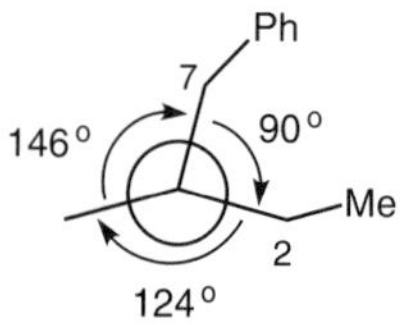

Newman projection
along C1—C4

As a further comment, the structure (i.e., bond lengths, bond angles) computed by high-level methods generally shows good agreement with the X-ray structure. Thus, the counterions in the crystal do not distort the structure to a significant extent. Laube has in fact suggested that the locations of the anions approximate the position of a leaving group or nucleophile in the transition state for the ionization or cation–anion combination. As a final comment, the Reed group at Riverside has recently reported some interesting solid superacids with very weakly nucleophilic carborane anions.[42,43] The crystal structure of the parent benzenium ion (protonated benzene) has been obtained from such a superacid.

3. REACTIVITY OF CARBOCATIONS

3.1. Introduction

While providing structural details and rates of intramolecular rearrangements–fragmentations, stable ion studies do not address questions relating to the reactivity of a carbocation with a second reagent, either with a solvent such as water or with some added nucleophile (Nu). Shown in Scheme 1.4 is the kinetic scheme for a limiting S_N1 solvolysis, that is, for a solvolysis that proceeds by way of a free carbocation. As is true for any reaction in which a reactive intermediate is formed as a stationary-state species, studies of this system cannot provide absolute rate constants. However, as recognized in early studies,[44] relative rate constants or selectivities can be obtained.

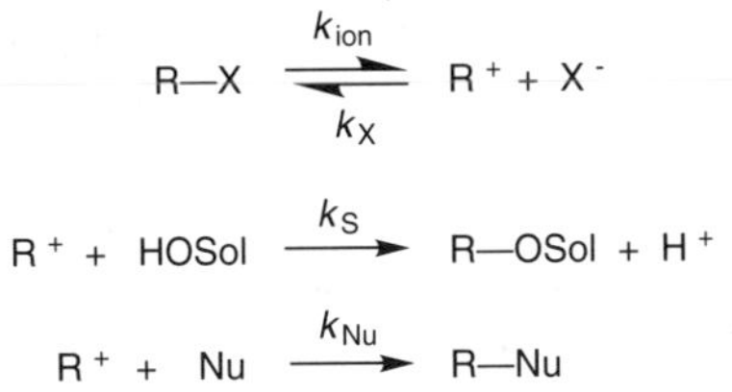

$$R\text{—}X \underset{k_X}{\overset{k_{ion}}{\rightleftharpoons}} R^+ + X^-$$

$$R^+ + HOSol \xrightarrow{k_S} R\text{—}OSol + H^+$$

$$R^+ + Nu \xrightarrow{k_{Nu}} R\text{—}Nu$$

Scheme 1.4

Using the *common ion effect*, first-order rate constants k_{obs} are obtained for the solvolysis to ROSol in the presence of an excess of the leaving group X^-. With the steady-state assumption for the cation R^+ and recognizing that $[X^-] \gg [RX]$,

$$\text{Rate} = -\left(\frac{d[RX]}{dt}\right) = k_{obs}[RX] = \left(\frac{k_{ion}k_s}{k_x[X^-] + k_s}\right)[RX] \tag{10}$$

where, as will be the case throughout this chapter, the rate constant k_s is defined as the first-order rate constant for the decay of the cation in solvent. Rearranging Eq. 10,

$$\frac{1}{k_{obs}} = \left(\frac{1}{k_{ion}}\right)\left(\frac{k_x}{k_s}\right)[X^-] + \left(\frac{1}{k_{ion}}\right) \tag{11}$$

and the relative rate constants for return to starting material (k_x) and capture by solvent (k_s) are obtained as the ratio of the slope to the intercept of a plot of $1/k_{obs}$ versus $[X^-]$.

In the method of *competition kinetics*, the intermediate cation is partitioned between two competing nucleophiles whose concentration is in excess over that of the cation precursor. For Scheme 1.4, where one of the nucleophiles is the solvent,

$$\frac{[RNu]}{[ROSol]} = \left(\frac{k_{Nu}}{k_s}\right) \cdot [Nu] \tag{12}$$

and, because the concentration of the nucleophile is known, the product ratio can be converted into a rate constant ratio.

The latter method, with azide ion and water as the competing nucleophiles in 80% acetone, was found to lead to an apparent reactivity–selectivity relation. The latter, in general terms, states that the selectivity of a reactive intermediate toward two competing reagents will decrease as the reactivity of the intermediate increases (or its stability decreases). The argument is that a highly reactive unstable intermediate will have an early transition state in its reactions, and thus show little discrimination. A less reactive and more stable intermediate on the other hand will have a later transition state, and thus show greater selectivity. Raber et al.[45] took the rate constant for the solvolysis (k_{ion} of Scheme 1.4) as a measure of the reactivity of the cation intermediates of a series of solvolysis reactions. Their argument was that k_{ion} measures cation stability (and hence reactivity), the more stable ion forming more easily ($> k_{ion}$) but, once formed, reacting more slowly. For 16 systems, a plot of log k_{ion} versus log($k_{az} : k_w$) (az = azide, w = water) was found to be approximately linear. In other words, the more easily formed and more stable cations showed a greater selectivity toward azide ion.

3.2. Ritchie's N$_+$ Scale

Ritchie was the first to directly measure the absolute reactivity of cations toward solvent and added nucleophiles.[6] The cations were highly stabilized examples, triarylmethyl cations bearing stabilizing substituents such as **30** and **31**, xanthylium ions (e.g., **32**) and tropylium ions (e.g., **33**). The feature (and requirement) of these cations was that they had a lifetime in water such that kinetics could be followed by conventional or stopped-flow spectroscopy whereby one solution containing the pre-formed cation was added to a second solution. The time required to mix these solutions was the important factor and limited measurements to cations with lifetimes longer than several milliseconds. The lifetimes in water for **30–33** are provided below. Lifetime is defined as the reciprocal of the first-order rate constant for the decay of the cation in solvent.

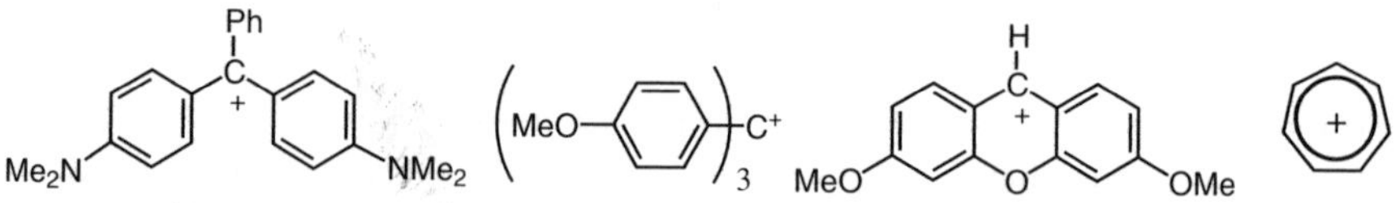

30 (5000 s)   **31** (0.07 s)   **32** (0.4 s)   **33** (0.07 s)

TABLE 1.1. Selected Values for N$_+$

Nuc(solv)	N$_+$	Nuc(solv)	N$_+$
H$_2$O(H$_2$O)	0.0	HOO$^-$(H$_2$O)	8.5
MeOH(MeOH)	0.5	CN$^-$(DMSO)^a	8.6
CN$^-$ (H$_2$O)	3.8	N$_3^-$(MeOH)	8.8
NH$_3$(H$_2$O)	3.9	PhS$^-$(H$_2$O)	9.1
OH$^-$(H$_2$O)	4.5	N$_3^-$(DMSO)^a	10.7
EtNH$_2$(H$_2$O)	5.3	PhS$^-$(MeOH)	10.4
N$_3^-$(H$_2$O)	7.5	PhS$^-$(DMSO)^a	13.1

^{a} Dimethyl sulfoxide = DMSO.

One of the surprising features of Ritchie's studies was a constant selectivity. While the absolute rate constants differed considerably over the series of cations, the selectivities toward pairs of nucleophiles were constant and independent of the reactivities of R$^+$. Rate constants were correlated by a simple two parameter equation

$$\log k_{Nu}(R_n^+) = \log k_0 + N_+ \tag{13}$$

where $\log k_{Nu}$ is the rate constant for cation R_n^+ reacting with Nu, $\log k_0$ is a parameter that depends only on the cation, and N$_+$ is a parameter that depends only on the nucleophile and the solvent. The latter was defined relative to k_w, the first-order rate constant for decay in water, as N$_+$ = $\log(k_{Nu} : k_w)$. Selected values are provided in Table 1.1. According to Eq. 13, plots of $\log k_{Nu}$ versus N$_+$ are parallel, differing only in the reactivity of the cation as defined by $\log k_0$. Figure 1.1 depicts this relationship in schematic terms.

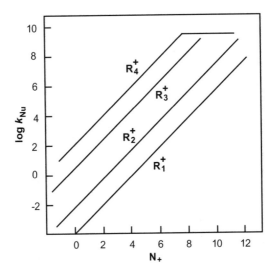

Figure 1.1. Constant selectivity with changing reactivity.

3.3. Azide Clock

There is a dichotomy between the constant selectivity observed with the very stable cations by Ritchie, and the changing selectivity for the cations studied by Raber et al.[45] The cations in the latter series are, however, much more reactive, and this led Rappoport[46] and Richard and Jencks[47] to provide an explanation. They suggested that for reactive carbocations powerful nucleophiles such as azide reacted on every encounter, that is, at the diffusion-controlled rate. Changes to encounter control had been suggested by Ritchie through the observation of sharp breaks in plots of $\log k_{Nu}$ versus N_+ when k_{Nu} surpassed $\sim 10^9\,M^{-1}s^{-1}$ (see R_4^+ in Fig. 1.1). According to Rappoport and Richard and Jencks, k_{az} was constant with changing cation reactivity since this nucleophile was already reacting as fast as it could. The apparent linear correlation of $\log k_{ion}$ and $\log (k_{az} : k_w)$ therefore was just a correlation of $\log k_{ion}$ and $-\log k_w$.

Recognizing this, Richard and Jencks, proposed using azide ion as a clock for obtaining absolute reactivities of less stable cations. The basic assumption is that azide ion is reacting at the diffusion limit with the cation. Taking $5 \times 10^9\,M^{-1}s^{-1}$ as the second-order rate constant for this reaction, measurement of the selectivity $k_{az} : k_{Nu}$ for the competition between azide ion and a second nucleophile then provides the absolute rate constant k_{Nu}, since k_{az} is "known." The "clock" approach has now been applied to a number of cations, with measurements of selectivities by both competition kinetics and common ion inhibition. Other nucleophiles have been employed as the clock. The laser flash photolysis (LFP) experiments to be discussed later have verified the azide clock assumption. Cations with lifetimes in water less than about 100 μs do react with azide ion with a rate constant in the range $5-10\times 10^9\,M^{-1}s^{-1}$,[48] which means that rate constants obtained by a "clock" method can be viewed with reasonable confidence.

3.4. Flash Photolytic Generation of Carbocations

Laser flash photolysis provides access to the absolute rate constants for more reactive cations. To apply this technique, a photochemical reaction must be available that generates the desired intermediate. A number of such reactions have now been found,[48] and are briefly summarized below.

The most common and widely studied photochemical reaction is *photoheterolysis*, the heterolytic cleavage of a C–X bond in the excited state. Like its ground-state analogue, the S_N1 reaction, polar solvents such as water, alcohols, and acetonitrile are required. Unlike the ground-state reaction, a competing photohomolysis is usually observed, which can be recognized in the products through the observation of both cation- and radical-derived products (Scheme 1.5). In LFP experiments, the cation and radical transients are easily distinguished by their kinetics. Cations decay exponentially by reacting with the solvent. The decay is not affected by oxygen, but is accelerated by added nucleophiles, azide ion being a particularly good indicator of a carbocation. Radicals decay by second-order kinetics (usually), and are quenched by oxygen but not azide.

X = —Cl, —Br, —OAc, —OC$_6$H$_4$-4-CN, —PAr$_3$$^+$, etc.

$$Ar_2CH\text{—}X \xrightarrow{h\nu} {}^1(Ar_2CH\text{—}X) \longrightarrow \begin{cases} \overset{+}{Ar_2CH} \xrightarrow{HOS} Ar_2CHOS \\ \\ \overset{\bullet}{Ar_2CH} \longrightarrow Ar_2CHCHAr_2 \\ \qquad\qquad (\text{+ other products}) \end{cases}$$

Scheme 1.5

A variety of leaving groups can be employed, as illustrated by studies involving diarylmethyl cations.[49,50] There is a general trend that better leaving groups yield more cation, although there are exceptions. Overall the effect of leaving group is highly attenuated relative to the situation in ground-state solvolysis. With respect to the cationic component there are some fascinating differences between the ground- and excited-state reactions. As one example, observed in one of the early studies of photosolvolysis,[51] *m*-methoxy groups promote photoheterolysis of a benzylic—X bond much better than *p*-methoxy. A second example is shown in Eq. 14. 9-Fluorenol **34**, when irradiated in water and alcohols, cleaves the poor leaving group hydroxide with a high quantum efficiency to give the 9-fluorenyl cation **35**. Under the same conditions, diphenylmethanol yields no cation. Wan and Krogh attributed the remarkable result in the fluorenyl system to the 4*n* π cyclic system of the 9-fluorenyl cation, which in the excited state is expected to have aromatic character.[52]

$$\text{(14)}$$

34   **35**

Alkenes, alkynes, and arenes become stronger bases in the singlet excited state. As a result, *photoprotonation* can occur under much more weakly acidic conditions than required in the ground state. Excited styrenes and phenylacetylenes, for example, are protonated by the solvent in 2,2,2-trifluoroethanol (TFE) and 1,1,1,3,3,3-hexafluoroisopropanol (HFIP), giving rise to phenethyl and α-arylvinylcations that can be observed by LFP (see Eq. 15).[53] In a similar manner, benzenium ions can be observed by photoprotonation of electron-rich aromatics in HFIP.[54] Equation 16 provides an example where the orientation of the protonation is different in the excited state from that of the ground state.

$$ArCH=CH_2 \xrightarrow{h\nu} {}^1(ArCH=CH_2) \xrightarrow{HOS} Ar\overset{+}{C}H\text{–}CH_3 \qquad \text{(15)}$$

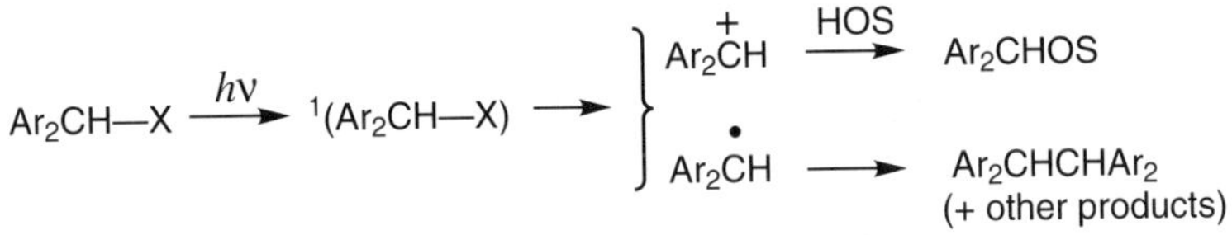

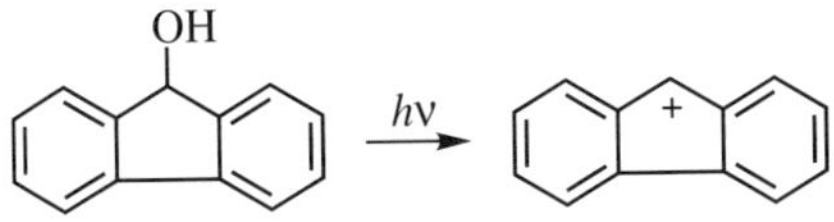

36   **37**   **38**

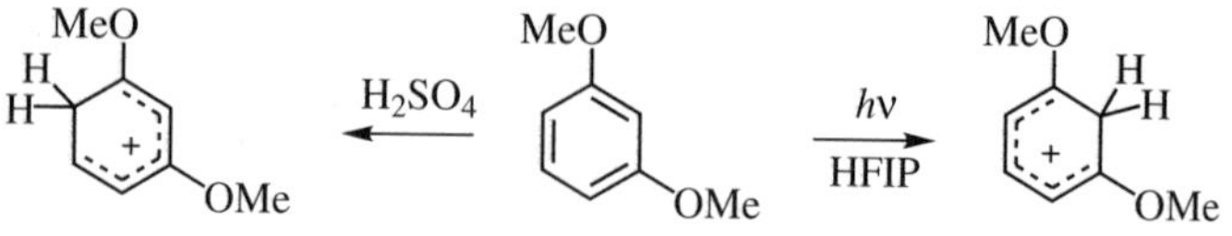

$$\text{(16)}$$

Carbocations have also been obtained by protonation of photochemically generated carbenes (see Eq. 17),[55,56] by the fragmentation of photochemically generated cation radicals (see Eq. 18),[57] and by the addition of one photochemically generated cation to an arene (or alkene) to generate a second cation.[58-60] As illustrated in Eq. 19, the last method has been employed to convert invisible carbocations into visible ones. Short-lived aryl cations[59] and secondary alkyl cations[60] are quenched by electron-rich aromatics such as mesitylene and 1,3,5-trimethoxybenzene in HFIP to give benzenium ions that can be observed by LFP in this solvent.

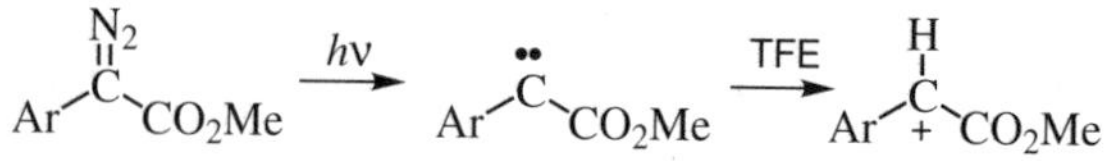

$$\text{(17)}$$

$(\text{Ar} = 4\text{-MeOC}_6\text{H}_4)$

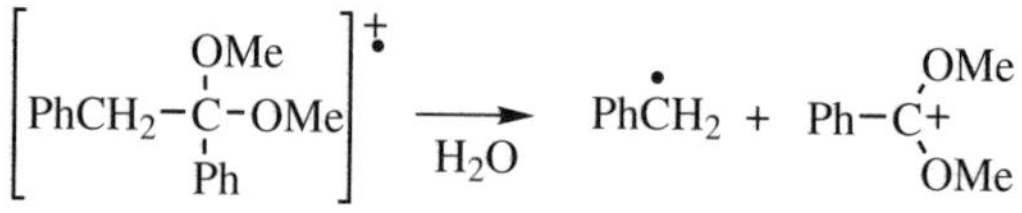

$$\text{(18)}$$

$$\text{(19)}$$

As in any LFP experiment the cation must be observable, and it must have a lifetime longer than the pulse of light employed to generate it. Since absorption spectroscopy has been the standard detection method to date, the majority of cations that have been studied are conjugated, usually benzylic-like or benzenium ions. The trick of turning invisible cations into visible ones has thus far seen limited application. With respect to lifetime, protic solvents react with a given cation in the order MeOH > EtOH > H$_2$O ≫ TFE ≫ HFIP (Table 1.2), which can be exploited to increase lifetime into a range where a particular cation becomes detectable. The parent diphenylmethyl cation, for example, requires picosecond LFP to be observed in methanol, ethanol, and water,[61] but can be studied by nanosecond LFP in TFE and in HFIP. In fact, it is quite long lived in HFIP. The 9-fluorenyl cation requires picosecond LFP even in TFE but is observed with nanosecond LFP in HFIP.

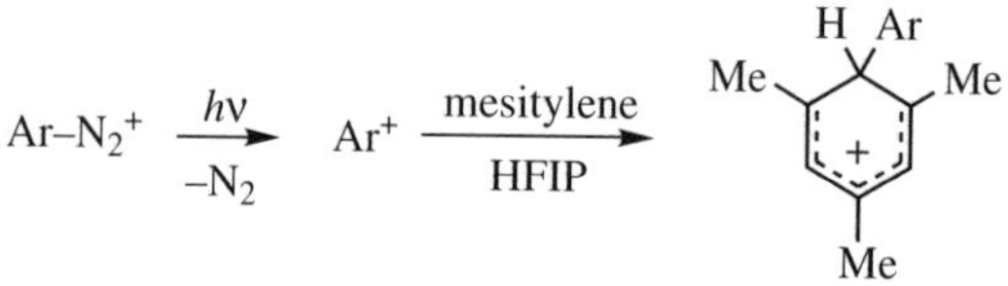

TABLE 1.2. Effect of Solvent on Rate Constants k_s (s^{-1}, 20 °C) for Decay of Cations[a]

Cation	MeOH	EtOH	H_2O	TFE	HFIP
$(An)_2CH^+$	**8.6×10^6**	**5.5×10^6**	**1.0×10^5**	**1.4×10^1**	
Ph_2CH^+	2.5×10^{10}	1.4×10^{10}	1.3×10^9	3.2×10^6	$\sim 1 \times 10^1$
9-Fluorenyl			$> 4 \times 10^{10}$	8×10^8	2×10^4

[a] Rate constants in bold have been measured using nanosecond LFP. Others require picosecond LFP. (An = 4-methoxyphenyl.)

As a final comment there is the question whether the cation observed with LFP is the same as the one produced in a ground-state reaction. With nanosecond LFP, the time interval from excitation to observation is likely sufficient to ensure that this is the case. Two pieces of evidence can be cited. (1) There is a good correspondence of ultraviolet–visible (UV–vis) spectra for the transient cations with ones obtained for solutions of ground-state cations under strongly acidic solutions. (2) Ratios of rate constant obtained directly by LFP agree with selectivities measured for the ground-state reactions. Diffusional separation of ion pairs is complete within 1–10 ns,[62] so that a transient cation observed with nanosecond LFP is a free ion. At shorter times, that is, in picosecond LFP, ion pairs can be observed and their dynamics studied.[62]

3.5. Lifetimes of Carbocations in Protic Solvents

Table 1.3 provides rate constants for the decay of selected carbocations and oxocarbocations in H_2O, TFE, and HFIP. As a general comment, water, methanol, and ethanol are highly reactive solvents where many carbocations that are written as free cations in standard textbooks have very short lifetimes. The diphenylmethyl cation, with two conjugating phenyl groups, has a lifetime in water of only 1 ns. Cations such as the benzyl cation,[63] simple tertiary alkyl cations such as tert-butyl,[64] and oxocarbocations derived from aldehydes[65] and simple glycosides,[66] if they exist at all, have aqueous lifetimes in the picosecond range, and do not form and react in water as free ions. This topic is discussed in more detail in Chapter 2 in this volume.

Lifetimes are longer in the more weakly nucleophilic TFE and HFIP, and cations whose existence is on the borderline in water and simple alkanols can become quite long lived, especially in HFIP. Benzenium ions such as protonated mesitylene can also be observed in HFIP, and there is an estimate for a simple secondary cation **77**.[60] There is one estimate of the lifetime of an acylium ion (**72**), based upon the clock approach.[67] Even with the powerful electron-donor 4-Me_2N on the aromatic ring, this cation appears to be very short lived in water.

The σ^+ substituent constant scale was developed based upon the solvolysis of substituted cumyl chlorides, where cumyl cations are formed in the rate-determining ionization. For the carbocation-forming reaction of Eq. 20, there is a good correlation of log k_H versus the sum of the σ^+ constants for the substituents directly on the carbon.[68] Such a correlation clearly does not exist for the reaction where the

TABLE 1.3. Solvent Reactivities of Selected Carbocations at 20 °C

Cation	k_w (s^{-1})	k_{TFE} (s^{-1})	k_{HFIP} (s^{-1})
(MeO)$_3$C$^+$ (**52**)	1.4×10^3		
PhC$^+$(OMe)$_2$ (**53**)	1.3×10^5		
MeC$^+$(OMe)$_2$ (**54**)	1.3×10^5		
HC$^+$(OEt)$_2$ (**55**)	2.0×10^7		
PhC$^+$(Me)(OMe) (**56**)	5×10^7		
PhCH$^+$(OMe) (**57**)	2×10^9		
Me$_2$C$^+$(OMe) (**58**)	1×10^9		
EtCH$^+$(OEt) (**59**)	6×10^{10}		
Ph$_3$C$^+$ (**60**)	1.5×10^5		
(4-MeOC$_6$H$_4$)$_2$CH$^+$ (**61**)	1.0×10^5	1.4×10^1	
(4-MeOC$_6$H$_4$)PhCH$^+$ (**62**)	2.0×10^6	1.2×10^3	
Ph$_2$CH$^+$ (**63**)	1.3×10^9	3.2×10^6	$\sim 1 \times 10^1$
9-Xanthylium (**64**)	2.3×10^4		
9-Phenyl-9-fluorenyl (**65**)	1.5×10^7	1.5×10^4	
9-Fluorenyl (**35**)	$>4 \times 10^{10}$	8×10^8	2×10^4
4-MeOC$_6$H$_4$C$^+$Me$_2$ (**66**)	$\sim 4 \times 10^7$	1.6×10^4	
4-MeOC$_6$H$_4$C$^+$HMe (**67**)		3.9×10^5	
4-MeOC$_6$H$_4$CH$_2^+$ (**68**)		4.3×10^6	3×10^2
4-MeOC$_6$H$_4$C$^+$=CH$_2$ (**69**)		1.3×10^6	
PhC$^+$Me$_2$ (**70**)			9×10^3
PhC$^+$HMe (**71**)			6×10^5
4-Me$_2$NC$_6$H$_4$C$^+$=O (**72**)	1×10^{10}		
2,4,6-Trimethoxybenzenium (**73**)	5.8×10^5	7×10^2	
2,4-Dimethoxybenzenium (**74**)			2×10^3
2-Methoxybenzenium (**75**)			7×10^5
2,4,6-Trimethylbenzenium (**76**)			1×10^5
Me$_2$CH$^+$ (**77**)			7×10^9

cations react with solvent.[69] This finding is dramatically demonstrated by comparing the triphenylmethyl cation **60**, $\Sigma\sigma^+ = -0.51$, with the dialkoxycarbocation **55**, which is two orders of magnitude more reactive despite having $\Sigma\sigma^+ = -1.6$.

$$\begin{array}{c} R \\ \diagdown \\ \diagup \\ R \end{array} C=CH_2 + H^+ \quad \xrightarrow{k_H} \quad \begin{array}{c} R \\ \diagdown \\ \diagup \\ R \end{array} \overset{+}{C}-CH_3 \qquad (20)$$

Even for a series with varying aromatic substituents, the correlations with σ^+ deviate significantly from linearity. Typical behavior is illustrated with data for monosubstituted triarylmethyl cations in Figure 1.2. Significant deviations are observed in the points for the para π donors. Moreover, these deviations are in the direction that indicates that these substituents have kinetic stabilizing effects greater than indicated by σ^+. In fact, there are good correlations with σ^{C+}, a parameter based on ^{13}C NMR chemical shifts of benzylic-type cations obtained under stable ion conditions.[70]

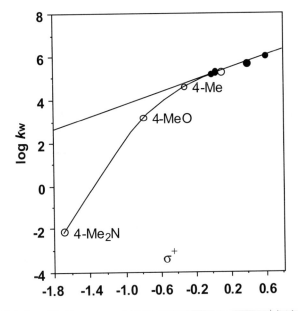

Figure 1.2. Correlation of $\log k_w$ and σ^+ for $X-C_6H_4C^+(Ph)_2$.

3.6. Rate-Equilibrium Correlation

A plot of $\log k_w$ versus pK_R (see Eqs. 1 and 2) represents a linear free energy correlation, where the rate constant and the equilibrium constant refer to the same reaction. With flash photolysis providing k_w for more reactive cations, a plot covering a range of 23 pK_R units can be constructed for carbocation hydration (Fig. 1.3). It can be seen that the data separate into families comprising structurally related ions, with different intrinsic reactivities. For the cations of Figure 1.3, the decreasing order of intrinsic reactivity is 9-xanthylium $\sim$ cyclic dialkoxycarbocation >phenyltropylium >diarylmethyl>triarylmethyl $\sim$9-aryl-9-fluorenyl.

Lines in Figure 1.3 have been drawn with the assumption that there is a linear relation. Scatter, however, may mask a gentle downward curvature consistent with the transition state becoming more cation-like as the reactivity of the cation increases. However, if there is curvature, it is slight. The quantity $d(\log k_w)/dpK_R$ is remarkably unchanged (-0.6 ± 0.1). For a very large change in thermodynamic driving force, there is remarkably little change in the position of the transition state. This point has recently been emphasized through the measurement of secondary kinetic and equilibrium isotope effects for the water addition to Ar_2CL^+ (L = H, D).[71] These also show a transition state 50–60% developed along the reaction coordinate, with no significant change over a wide range of reactivity. This is clearly consistent with the slopes in Figure 1.3.

Further insight into the nature of the transition state for carbocation hydration comes from application of the two parameter Yukawa-Tsuno equation (Eq. 21).[72]

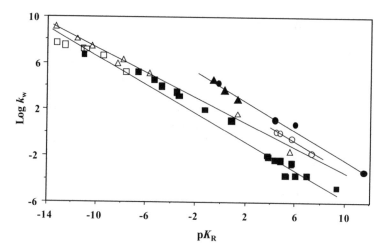

Figure 1.3. Rate-equilibrium correlation for hydration of carbocations: triarylmethyl and 9-aryl-9-fluorenyl (■ and □, respectively, slope = −0.60); diarylmethyl (△, −0.54); aryltropylium (○, −0.68); 9-xanthylium and cyclic phenyldialkoxycarbocations (● and ▲, respectively, slope = −0.63).

This equation was originally introduced for systems where the interactions lie between those of the two reactions that define the σ and the σ^+ scales with the parameter r^+ intended to lie between 0 and 1. However, $\log k_w$ (and pK_R) require r^+ values significantly >1 (Table 1.4), which is consistent with the idea that σ^+ underestimates the effect of para π donors on reactions of fully formed cations. The rate constants for the cation-forming reaction of Eq. 1 ($\log k_H$) on the other hand correlate with r^+ slightly <1. In other words, these show a reasonable correlation with σ^+, which is not surprising since both k_H and σ^+ refer to rate constants of reactions forming carbocations.

$$\log(k/k_0) = \rho\{\sigma + r^+(\sigma^+ - \sigma)\} \tag{21}$$

TABLE 1.4. Results of Fitting to Two Parameter Eqs. 21 and 22

Parameter	Ar—$^+$CHMe	Ph$_2$C$^+$—Ar
$r^+(k_w)$	2.3	3.6
$r^+(k_H)$	0.9	0.7
$r^+(K_R)$	1.4	1.6
$\rho_n(\text{norm}) = \rho_n(k_w)/\rho_n(K_R)$	0.36	0.33
$\rho_r(\text{norm}) = \rho_r(k_w)/\rho_r(K_R)$	0.61	0.72

The two-parameter equation suggested by Jencks and co-workers[47b] (Eq. 22) separates substituent effects into a polar or inductive contribution described by ρ_n and a resonance contribution described by ρ_r.

$$\log(k/k_0) = \rho_n\sigma + \rho_r(\sigma^+ - \sigma) \tag{22}$$

Where both rate and equilibrium constants are known, "normalized" ρ values (ρ_{norm}) are defined as the ratios $\rho_n(k_w)/\rho_n(K_R)$ and $\rho_r(k_w)/\rho_r(K_R)$. Since the value of $\rho(K_R)$ measures the overall substituent effect on proceeding from reagent to product, ρ_{norm} measures the fraction that has occurred at the transition state. As seen in Table 1.4, $\rho_n(norm)$ in the direction of cation hydration, is significantly <0.5, whereas $\rho_r(norm)$ is >0.5. The inequality means that there is imbalance in the transition state. When comparing the transition state to starting carbocation, the substituent "sees" a much larger fractional decrease in positive charge in its resonance interaction as compared to the decrease that it "sees" through the polar interaction. This imbalance is shown schematically in Scheme 1.6 where the numbers represent the charges experienced by the substituent normalized to +1 in the cation and 0 in the alcohol. A simple explanation is that the incoming water nucleophile has pyramidalized the alpha carbon, so that the resonance interaction has considerably diminished by the time the transition state is reached. There is still substantial positive charge at the alpha carbon, however, so that the polar interaction is still quite strong. For the reverse direction where the cation is forming, the transition state is well developed with respect to the polar interaction, but now the resonance interaction lags behind. For the latter reason, it is not surprising that the σ^+ substituent scale based upon rate constants for carbocation formation considerably underestimate the full conjugative effect of the aryl substituents.

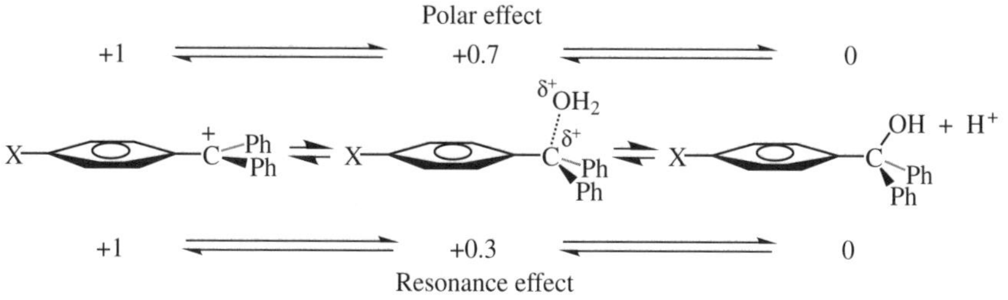

Polar effect

$$+1 \rightleftharpoons +0.7 \rightleftharpoons 0$$

$$+1 \rightleftharpoons +0.3 \rightleftharpoons 0$$

Resonance effect

Scheme 1.6

3.7. Reactivity with Added Nucleophiles

Rate constants for the reactions of carbocations with added nucleophiles are obtained in LFP experiments as the slopes of linear plots of first-order rate constants for cation decay against the concentrations of added nucleophile. One of the first detailed studies using LFP showed that rate constants for the parent triphenylmethyl cation did not adhere to the simple Ritchie N_+ relation of Eq. 13, but that the slope of a plot of $\log k_{Nu}$ versus N_+ was significantly <1.[73] This finding has been verified

with other cations.[74] It is now clear that the stable cations with lifetimes of the orders of magnitude of that employed by Ritchie do show an approximately constant selectivity as expressed by Eq. 13, but as the cations become more reactive there is a changeover. The more reactive cations seem to show a decreasing selectivity with increasing reactivity, in accordance with the reactivity–selectivity principle.

The extent to which the latter conclusion is influenced by reactions occurring at close to the diffusion limit is still not clear. It has also been established that nucleophiles can reach a reactivity limit well below diffusion because of a requirement for desolvation,[75–77] which is seen, for example, in a LFP study of the reaction of primary amines and diarylmethyl cations.[75] In 100% acetonitrile, these reactions are fast, with $k(RNH_2)$ approaching the encounter limit. There is a small dependence on amine structure, in the expected direction of $k(RNH_2)$ increasing with increasing amine basicity. As water is added to the solvent, $k(RNH_2)$ decreases regardless of the nature of RNH_2, but the effect is more pronounced the more basic the amine. By the time highly aqueous solvents are reached, reactivity no longer parallels basicity. For example, for some cations $k(n$-propylamine) is smaller than $k(2$-cyanoethylamine) despite the weaker basicity of the cyano-substituted amine. Scheme 1.7 provides an explanation. For an amine (or indeed any nucleophile) to react with a carbocation in water, the hydrogen bond to the nucleophile must be broken. One can envisage two ways that this occurs. One involves forming the reactive complex **81** from the small concentration of free, non-hydrogen-bonded amine **79** in equilibrium with the hydrogen-bonded **78**. The second involves the sequence **78** ⇌ **80** ⇌ **81**, in which the hydrated amine encounters the cation, and the water molecule is then moved out of the way. Either model explains the inversion of nucleophilic reactivity and basicity for the more reactive cations. For such cations, the actual barrier to formation of the nitrogen–carbon bond is low. What determines the reactivity is the ease of forming the reactive complex **81**, which depends inversely on the strength of the hydrogen bond. Because this hydrogen bond is stronger for more basic amines, these react more slowly.

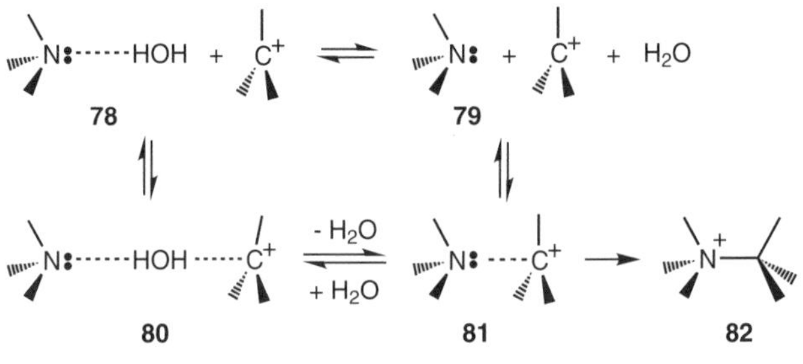

Scheme 1.7

Rate constants for cations reacting with carbon nucleophiles have also been obtained by LFP. HFIP is a good solvent for such studies, because cations can be generated photochemically in this solvent, but it is quite weakly nucleophilic, allowing the carbon nucleophile to compete. Equation 23 (**Fl** = 9-fluorenyl) describes one study in which the fast electrophilic addition step of a Friedel–Crafts alkylation was directly studied.[78,79] In fact, with electron-rich aromatics the cyclohexadienyl cation could be seen to grow as the fluorenyl cation decayed, so that both cationic intermediates of the alkylation were observed in the same experiment. The second-order rate constants for the addition of **Fl**$^+$ shows a strong dependence on the substituent for the less electron-rich aromatics, with log k_2 correlating with σ^+ with a slope ρ^+ of -8. This result is obviously indicative of a transition state with considerable cyclohexadienyl cation character. Anisole shows a negative deviation from the correlation line. In fact, anisole and other electron-rich aromatics all react with very similar rate constants in the range 1–$2 \times 10^9 \, M^{-1}s^{-1}$, indicating that these reactions have become diffusion controlled. The same rate constants mean that there is no intermolecular selectivity in a competition involving two electron-rich systems. Interestingly, the products still display large positional preferences, that is, intramolecular selectivity. Thus, although the encounter complex reacts to form products rather than separate them (to account for diffusion control), it still determines the carbon to which it attaches the fluorenyl group based upon the stabilities of the cyclohexadienyl cations.

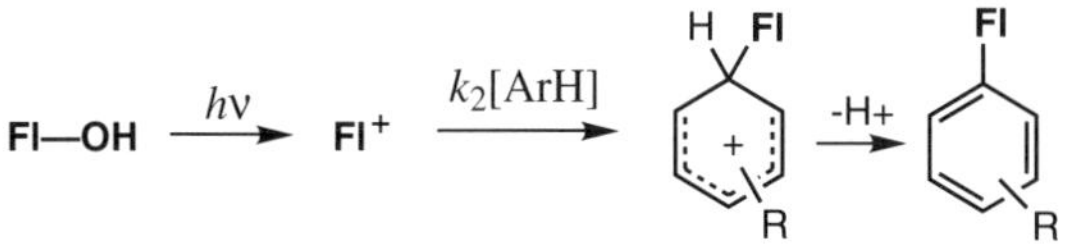

$$\text{Fl—OH} \xrightarrow{h\nu} \text{Fl}^+ \xrightarrow{k_2[\text{ArH}]} \quad \xrightarrow{-\text{H}+} \quad \tag{23}$$

The 9-fluorenyl cation also reacts at the diffusion limit with substituted styrenes, as indicated by the observation that the rate constants are $3 \times 10^9 \, M^{-1}s^{-1}$ independent of the styrene substituent. The reaction of styrene with the phenethyl cation (**71**), the first step in the polymerization of styrene (Sty) catalyzed by Brønsted acids, has also been directly observed (Scheme 1.8).[53] Photoprotonation of styrene

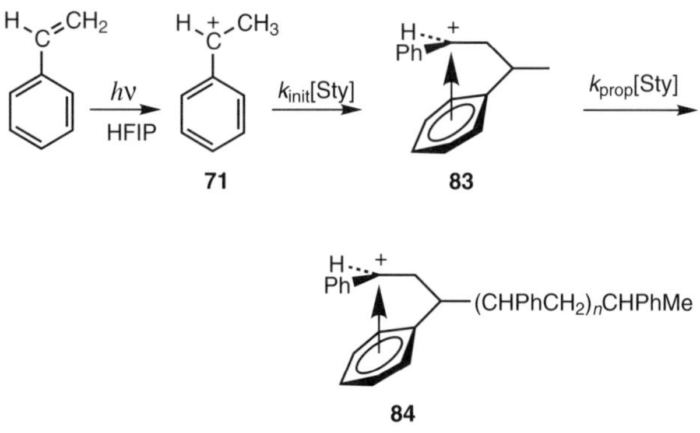

Scheme 1.8

in HFIP generates **71**, observed with LFP at $\lambda_{max} = 325$ nm. The cation reacts with the styrene not consumed in the photolysis with $k_{init} = 1 \times 10^9\,M^{-1}s^{-1}$. As the absorbance due to **71** decreases at 325 nm, the absorbance of the dimer cation **83**, $\lambda_{max} = 345$ nm, is observed to grow in at higher wavelengths. This absorbance is considerably more stable, decaying only on the milliseconds time scale. In both reactivity and λ_{max}, the dimer **83** is very similar to the propagating cation **84** observed in stopped-flow experiments, where k_{prop} has been measured as $1 \times 10^5\,M^{-1}s^{-1}$.[80] Thus, the monomer **71** has a lower λ_{max} and reacts four orders of magnitude more quickly than the dimer and oligomer cations **83** and **84**. One possible explanation is that the latter cations are stabilized by a charge-transfer interaction with the nearby phenyl ring, as depicted in Scheme 1.8.

TABLE 1.5. Electrophilicity and Nucleophilicity Parameters[a]

Electrophile	E	Nucleophile	$N(s)$
Ph_2CH^+	5.90	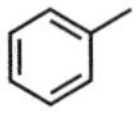	-4.47 (1.32)
4-MeC$_6$H$_4$CH$^+$(Ph)	4.59		-2.44 (1.09)
(4-MeC$_6$H$_4$)$_2$CH$^+$	3.63	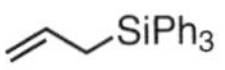 $\diagdown$SiPh$_3$	-0.13 (1.21)
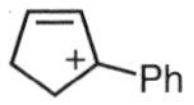 (Ph-cyclopentenyl cation)	2.93	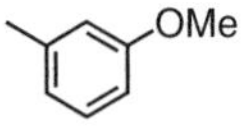 OMe	0.13 (1.27)
4-MeOC$_6$H$_4$CH$^+$(Ph)	2.11	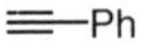 $\equiv$—Ph	0.34 (0.68)
4-MeC$_6$H$_4$CH$^+$(OMe)	1.95	 Ph	0.78 (0.95)
4-MeOC$_6$H$_4$CH$^+$(OMe)	0.14	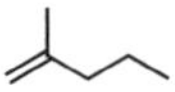	0.96 (1.0)
(4-MeOC$_6$H$_4$)$_2$CH$^+$	0.00		
Co$_2$(CO)$_6$ alkyne cation	-0.83	$\diagdown$SiMe$_3$	1.79 (0.96)
dithiane cation -H	-2.14	HSiPh$_3$[b]	2.06 (0.68)
tropylium cation	-3.72	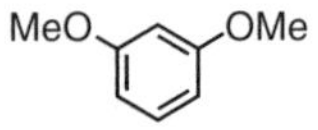 MeO OMe	2.48 (1.09)
		 $\diagdown$SnPh$_3$	3.09 (0.90)
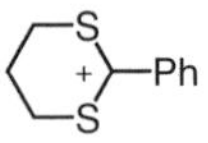 dithiane-Ph cation	-5.88	HSiEt$_3$[b]	3.64 (0.65)
(4-Me$_2$NC$_6$H$_4$)$_2$CH$^+$	-7.02	HGePh$_3$[b]	3.99 (0.62)
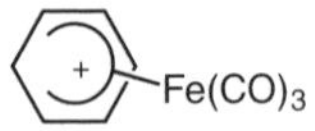 Fe(CO)$_3$ diene cation	-7.78	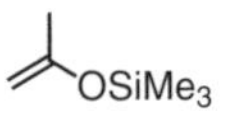 OSiMe$_3$	5.41 (0.91)
(Ferrocenyl)$_2$CH$^+$	-8.54	 $\diagdown$SnBu$_3$	5.46 (0.89)
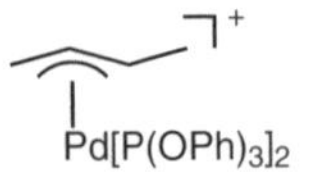		HSnPh$_3$[b]	5.64 (0.59)
Pd[P(OPh)$_3$]$_2$ allyl cation	-10.33	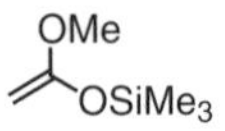 OMe OSiMe$_3$	8.23 (0.96)
		piperidine/morpholine-N-aryl	11.40 (0.83) X = O
			13.36 (0.81) X = CH$_2$

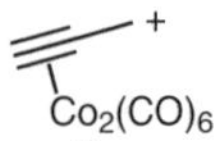

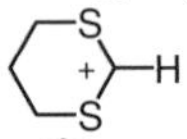

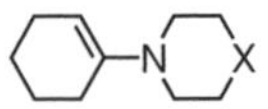

[a] Data from Ref. 81c.
[b] Reacts by hydride transfer.

3.8. Mayr's Scales of Electrophilicity and Nucleophilicity

In a series of reports published over the last 10–15 years, Mayr and co-workers[81] obtained second-order rate constants for reactions of carbocations and other electrophiles such as metal–π complexes with a series of nucleophiles, especially π-nucleophiles where a C—C bond is formed. An impressive body of reactivity data has been accumulated, and, including data from other groups, correlated by the following equation,

$$\log k(20\,^{\circ}\mathrm{C}) = s(N + E) \qquad (24)$$

where k is the second-order rate constant for the electrophile–nucleophile combination, E is the electrophilicity parameter, N is the nucleophilicity parameter, and s is the nucleophile-dependent slope parameter. Values of these parameters for selected electrophiles and nucleophiles are provided in Table 1.5. Equation 24 takes the same form as the Ritchie equation, Eq. 13, with the addition of the s parameter. It can be seen that s lies in a narrow range around unity; a value of 1 converts Eq. 24 into Eq. 13, and indeed many of the systems investigated by the Mayr group display a constant selectivity with changing reactivity. The Mayr correlation now covers a reactivity range of 30 orders of magnitude in both electrophiles and nucleophiles. The equation reproduces rate constants within a factor of 10–100, excluding reagents of steric bulk such as trityl cations. It has the power of predicting with reasonable precision the rate constant for a particular combination of cation and electrophile, and thus is of value in organic synthesis.

4. MISCELLANEOUS TOPICS

4.1. Carbocations with Electron-Withdrawing Substituents

There has been considerable interest in benzylic cations in which the α-carbon bears an electron-withdrawing group such as CN, CF_3, or COR. Originally, these cations were investigated in the context of solvolysis reactions. Despite large rate decreases associated with the substitution of the electron-withdrawing group, a cationic intermediate was still proposed.[82] This result has been verified by azide-clock experiments, which have suggested that the cation has a similar lifetime to analogues without the electron-withdrawing group.[83] The LFP studies have provided absolute rate constants.[84] Representative values are given in Table 1.6.

Taking the comparison of **67** and its α-CF_3 analogue **85** as the example (Eq. 25), a 10^7–10^9 difference is observed in the rate constants k_{for} for formation of the cations, but there is virtually no difference in the rate constants k_{rev} for their reactions with nucleophiles. In other words, the α-CF_3 cation is considerably less stable in the thermodynamic sense but of comparable stability in the kinetic sense. This finding has been explained[83b,85] by a model separating the substituent effects into polar and resonance contributions. The polar effect, which substantially destabilizes

TABLE 1.6. Rate Constants for Decay of Cations with Electron-Withdrawing Substituents

4-MeOC$_6$H$_4$CH$^+$R	R = Me (**67**)	$k_w = 5 \times 10^7 \, s^{-1}$
	R = CF$_3$ (**85**)	$k_w = 5 \times 10^7 \, s^{-1}$
	R = COOMe (**86**)	$k_w = 5 \times 10^7 \, s^{-1}$
	R = CONMe$_2$ (**87**)	$k_w = 1.6 \times 10^6 \, s^{-1}$
	R = CSNMe$_2$ (**88**)	$k_w = 1.5 \times 10^2 \, s^{-1}$
4-MeC$_6$H$_4$CH$^+$R	R = Me (**89**)	$k_{HFIP} = 5 \times 10^4 \, s^{-1}$
	R = COOMe (**90**)	$k_{HFIP} = 3 \times 10^5 \, s^{-1}$
4-MeOC$_6$H$_4$C$^+$R$_2$	R = Me (**66**)	$k_w = 4 \times 10^7 \, s^{-1}$
	R = CF$_3$ (**91**)	$k_w = 1.0 \times 10^7 \, s^{-1}$

the CF$_3$-substituted cation, dominates the equilibrium and is responsible for the thermodynamic destabilization. The polar effect, however, is partially compensated by a increased resonance stabilization in the CF$_3$-substituted cation due to the greater importance of the quinonoid resonance contributor **85b** as the positive charge is moved away from the electron-withdrawing CF$_3$ group. Evidence for the importance of this resonance contributor is the observation that α-CF$_3$-substituted cations show the highly unusual feature of reacting with nucleophiles at the para position of the benzene ring. As has been discussed in Section 3.6, there is an imbalance in the transition state for a cation–nucleophile combination, with a substantial loss of the resonance interaction at the transition state, but less of a change in the polar interaction. With the greater importance of the latter on the overall equilibrium, it is argued that the opposing changes in resonance and polar effects almost cancel, with the result that the effects on k_{rev} are small.

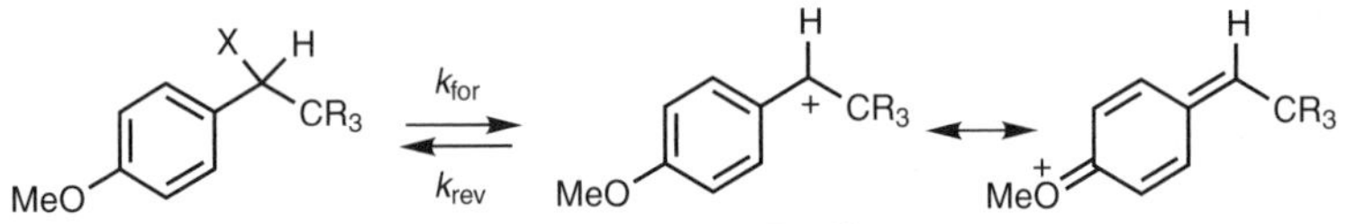

$$(25)$$

67a, R = H
85a, R = F

67b
85b

In Table 1.6, it can be seen that the relative kinetic stability of the cation substituted with an electron-withdrawing group is also found in comparing cations **89** and **90** where the para position is substituted by a methyl group. Other cations of interest are those substituted with an α-thioamido group such as **88**. These exhibit a substantial kinetic stabilization. The reasons for this are not entirely clear at this time. One suggestion is that there is stabilization by a bridging interaction with the C=S group.

4.2 Silyl Cations

This section provides a discussion of R$_3$Si$^+$, the silicon analogue of the carbenium ion. While such species do exist in the gas phase, their existence in condensed phases has proved controversial. The high bond dissociation energy of silicon–halogen and silicon–oxygen bonds has ruled out traditional S_N1 solvolyses as a route to these intermediates. The reactive intermediates **92** and **93** decay by heterolytic cleavage of a silicon–carbon bond. The LFP studies of these reactions, however, have shown that even in the weakly nucleophilic solvents acetonitrile (**92**)[86] and HFIP (**93**)[87] the C–Si cleavage occurs with solvent participation at the silicon so that no free silyl cation is formed. Mayr et al.[88], on the other hand, have concluded that hydride transfer from R_3SiH to diarylcarbenium ions in CH_2Cl_2 does proceed to give R_3Si^+ as an intermediate.

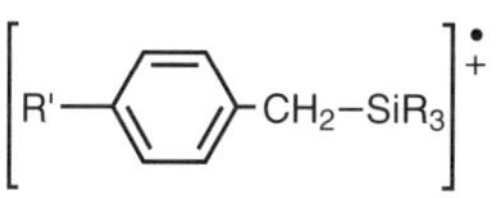

Lambert et al.[89] summarized the history of the attempts by his group and others to observe R_3Si^+ under stable ion conditions. Hydride transfer from R_3SiH to salts of carbocations such as the trityl cation does occur and has been used extensively. Initial studies involved silanes such as Ph_2MeSiH and the perchlorate salt of the trityl cation. The products were found to be covalent, that is, $R_3Si–OClO_3$. Switching to alkylthiosilanes such as $(i\text{-}PrS)_3SiH$ to take advantage of stabilization by the sulfur led to conducting solutions, but the products were complexes with the solvent. This conclusion was based upon ^{29}Si NMR chemical shifts that were typically in the range of 30–50 ppm, far upfield from a value $\sim$300 ppm calculated for the free silyl cation. The trityl counterion was then changed to $(C_6F_5)_4B^-$ and the hydride transfer performed in aromatic solvents such as benzene and toluene. Once again, the value of the ^{29}Si resonance, 80–100 ppm, led to the conclusion that there was coordination with the solvent, which was verified by a crystal structure of the toluene complex of the $Et_3Si^+ \cdot (C_6F_5)_4B^-$ salt **94**. This revealed a toluene ring geometry that was essentially unperturbed and a C–Si bond distance of 2.18 Å, considerably lengthened from the normal distance of 1.85 Å. The silicon portion, however, was not planar, which is consistent with tetracoordination. Thus, while the borate counterion was well removed, there was loose coordination to the toluene. Studies by the Reed[90] group with halocarboranes as the anion produced similar materials described as "closely approaching a silyl cation." Here there was coordination between the silicon and halogen atom of the counteranion. Although the silicon–halogen distance was long, the geometry around the silicon was still not planar, as required for the free cation.

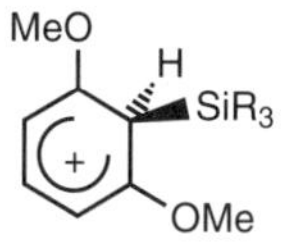

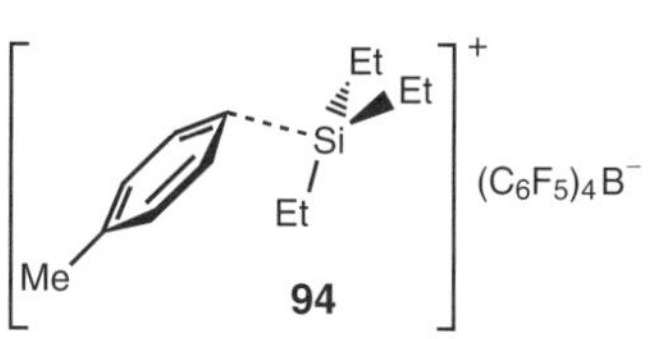

In addition to its interesting structure, the triethylsilylium–aromatic complex has proved useful in preparing other cations. Reaction with 1,1-diphenylethylene, for example, provided the cation **95**, the first example of a persistent β-silyl substituted carbocation (i.e., where decomposition by loss of the silyl group did not occur).[91]

$$[Et_3Si^+ \cdot (C_6F_5)_4B^-]_{ArH} + CH_2{=}CPh_2 \rightarrow Et_3SiCH_2CPh_2^+ \cdot (C_6F_5)_4B^- \qquad (26)$$
$$\underset{\textbf{95}}{}$$

Lambert then turned to the mesityl (2,4,6-trimethylphenyl) group Mes, in which the two ortho methyl groups provide steric hindrance that might prevent coordination. Trimesitylsilane, (Mes)$_3$SiH, did not transfer its hydride, however, presumably because the carbocation could not approach close enough. In what has been termed the "allyl leaving-group approach," trimesitylallylsilane was prepared. Its reaction in C$_6$D$_6$ with the β-silyl carbenium salt **95** led to the first free silyl cation **96**,[92] with a chemical shift of 225 ppm in good agreement with the computed value of 230 ppm. This reaction presumably proceeds via electrophilic addition to the terminal end of the allyl group to give an intermediate **97**, followed by cleavage of the silyl cation. Other examples of silyl cations have more recently been reported.[93] A crystal structure of a carborane salt of (Mes)$_3$Si$^+$ has confirmed that it has a free three-coordinate silyl cation.[94]

$$(Mes)_3Si{-}CH_2CH{=}CH_2 + Et_3SiCH_2CPh_2{}^{+\bullet}(C_6F_5)_4B^- \longrightarrow (Mes)_3Si^{+\bullet}(C_6F_5)_4B^-$$
$$\underset{\textbf{95}}{} \qquad\qquad\qquad\qquad\qquad\qquad\qquad\qquad \underset{\textbf{96}}{}$$

$$\text{via } (Mes)_3Si{-}CH_2{-}CH^+{-}CH_2{-}E$$
$$\underset{\textbf{97}}{} \qquad\qquad\qquad\qquad (27)$$

4.3. Carbocations in Zeolites[95]

Zeolites are porous aluminosilicate caged structures that can have both Brønsted and Lewis acid sites approaching superacid strengths. These structures play an important role in the petroleum industry because of their ability to catalyze transformations of hydrocarbons in reactions that proceed by way of carbocation intermediates. This has led to a considerable interest in observing such species within the zeolite framework. Relatively stabilized cations such as triarylmethyl, xanthylium, dibenzotropylium, and cyclopentenyl cations can be observed as persistent ions in acidic zeolites, where they have been characterized in similar manners as in superacids, that is, by solid-state NMR, IR, and UV–vis spectroscopy. Less stable cations such as cumyl cations and the 9-fluorenyl cation do not persist. However, such ions have recently been studied as transient intermediates using LFP with diffuse reflectance detection.[96] These studies provide information about the reactivity of carbocations generated within the zeolite cavities. There are some interesting differences from reactions in homogeneous solution associated with both the active and passive influences of the zeolite environment. In the former sense, the zeolite can directly participate in cation decay by direct participation as a nucleophile, leading to framework-bound products. In the latter sense, the zeolite can have significant effects on the reactions with added nucleophiles, slowing down diffusional encounter but enhancing the reactivity once the nucleophile is in the same cage in the zeolite.

4.4. Carbocations in Carcinogenesis

Most chemical carcinogens share the property of forming DNA adducts, a lesion, which if not repaired, can lead to mutagens and cancer. While some carcinogens form these adducts via radical chemistry, the more common mechanism is one where DNA reacts as a nucleophile.[97] The carcinogen is either an electrophile or is converted by metabolism into an electrophile. Of these, there are several important systems where the electrophile is a delocalized carbocation. Scheme 1.9 summarizes two of the widely studied examples. The polycyclic aromatic hydrocarbon (PAH) carcinogens, typified by benzo[a]pyrene (**98**),[98] undergo metabolic activation to a diol epoxide (**99**), that (possibly in the presence of DNA)[99] undergoes acid-catalyzed ring opening to a delocalized benzylic-type cation (**100**). Tamoxifen (**101**) (Ar $= -C_6H_4$-4-OCH$_2$CH$_2$NMe$_2$), an antiestrogen employed in the treatment of breast cancer, causes a small number of endometrial cancers. A mechanism has been implicated involving allylic hydroxylation, followed by sulphation to the ester **102**. This ester is short lived, undergoing S_N1 ionization to the allylic cation **103**.[100] As is typical of delocalized carbocations, **103** and the benzylic cations derived from the PAHs alkylate deoxyribonucleic acid (DNA) at the exocyclic amino groups of adenine and guanine to give relatively stable adducts such as **104**. This behavior contrasts dramatically with that of S_N2 alkylating agents, which react principally at N_7.[97]

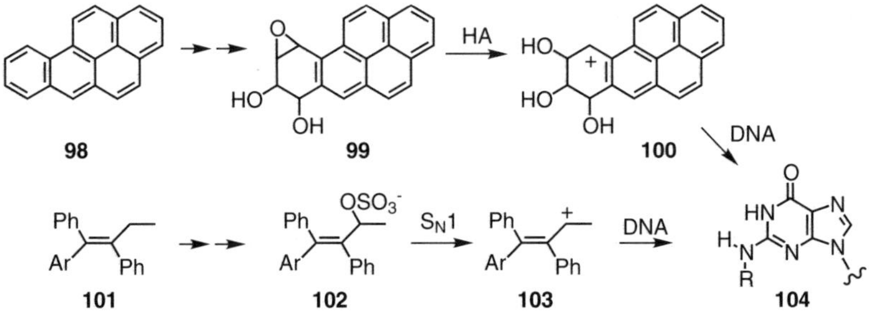

Scheme 1.9

Benzylic-type cations derived from PAHs have been studied under superacid conditions, where, not surprisingly, they are relatively stable.[101] Lifetimes in water of diastereomeric forms of the benzo[a]pyrene derivative (**100**) have been determined by the azide clock approach to be 50 ns.[98] The tamoxifen cation **103** has been studied directly by LFP; this has a lifetime at pH 7 of 25 μs.[100] As nucleophiles, guanine and adenine in monomeric forms do not compete effectively with water for such cations.[102] There are indications, however, that when incorporated in DNA they are significantly more nucleophilic.[97] The reasons for this are a subject of current investigation. It has been suggested that in the double helix the hydrogen-bonded partner functions as a general base catalyst removing one of the NH protons at the same time as the cation reacts.[103] There are also indications that delocalized cations preassociate in some manner with DNA before reaction occurs.[104]

4.5. Carbocations in Biosynthesis

The isoprene pathway produces a diverse range of natural products such as terpenes and steroids. A number of complex biochemical transformations are involved, many of which have been proposed to involve short-lived carbocation intermediates. Two recent studies provide a brief introduction.

The enzyme isopentenyl diphosphate: dimethylallyl diphosphate isomerase catalyzes a key early step whereby **105** is isomerized to **107** (Eq. 28, OPP = diphosphate). A number of studies, including a recent crystal structure,[105] have led to a mechanism whereby an intermediate tertiary carbocation **106** is formed by protonation by a cysteine residue C_{67}. Deprotonation is proposed to involve a metal bound glutamate E_{116} facing the cysteine in the active site. An interesting aspect of the enzyme is the requirement for a tryptophan residue W_{161}, which is suggested to stabilize the cationic intermediate through quadrupole-charge interaction with the indole π electrons.

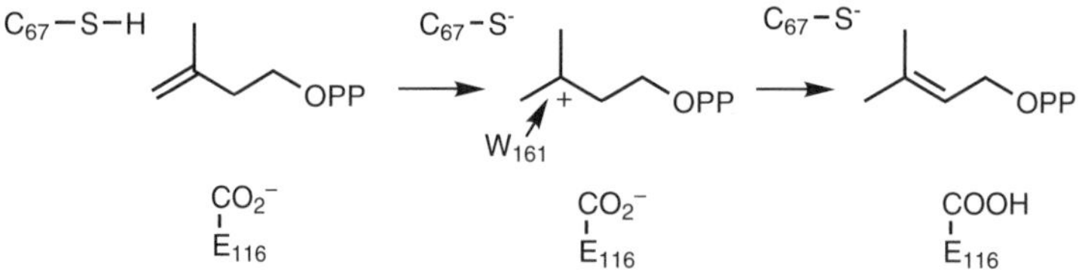

$$(28)$$

Squalene synthase catalyzes the first committed step in cholesterol biosynthesis, condensation of two farnesyl diphosphates to form presqualene diphosphate (**108**), with subsequent rearrangement and reduction to squalene (**112**) (Eq. 29). It has been proposed that **108** ionizes to the cyclopropylcarbinyl cation **109**, which rearranges via a cyclobutyl cation **110** (which may be a transition state) to a second more stable cyclopropylcarbinyl cation **111**.[106] Reductive ring opening produces squalene. The enzyme controls both the regio- and stereochemistry of the reaction. Of interest are model studies suggesting that the rearrangement of **109** to **111** is a minor reaction channel without the enzyme, with the preferred reaction being rearrangement to the allylic cation **113**. Within the active site of the enzyme, the transition state for the cyclopropylcarbinyl–cyclopropylcarbinyl rearrangement must be stabilized by >7.9 kcal/mol relative to that of the competing rearrangement.

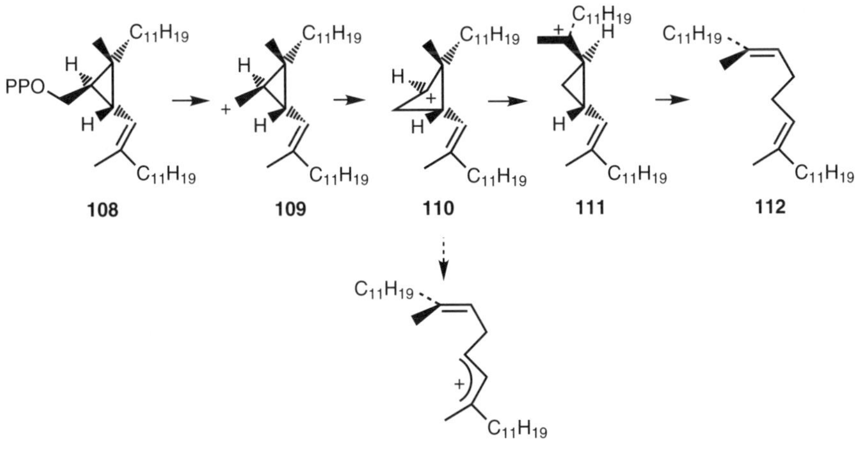

108   **109**   **110**   **111**   **112**

113   (29)

   As a general comment, the cations that have been implicated in such biosyntheses are of the type for which analogues have been observed in superacids. However, many of these cations, (e.g., **106** and **109**) would have a questionable existence as a free cation in an aqueous solution. This finding raises an interesting question whether they do have more than a fleeting existence within the active site of the enzyme. Does the enzyme provide some form of stabilization, such as that suggested when **106** is formed in the active site of isopentenyl diphosphate: dimethylallyl diphosphate isomerase?

5. CONCLUSION AND OUTLOOK

The study of carbocations has now passed its centenary since the observation and assignment of the triphenylmethyl cation. Their existence as reactive intermediates in a number of important organic and biological reactions is well established. In some respects, the field is quite mature. Exhaustive studies of solvolysis and electrophilic addition and substitution reactions have been performed, and the role of carbocations, where they are intermediates, is delineated. The stable ion observations have provided important information about their structure, and the rapid rates of their intramolecular rearrangements. Modern computational methods, often in combination with stable ion experiments, provide details of the structure of the cations with reasonable precision. The controversial issue of nonclassical ions has more or less been resolved. A significant amount of reactivity data also now exists, in particular reactivity data for carbocations obtained using time-resolved methods under conditions where the cation is normally found as a reactive intermediate. Having said this, there is still an enormous amount of activity in the field.

The roles of carbocations in commercially important hydrocarbon transformations are still not perfectly understood. The same can be said for carbocations in biological systems. Significant questions concerning reactivity still need to be explained. Why do so many reactions of carbocations show constant selectivity, in violation of the reactivity–selectivity principle? Is it possible to develop a unified scale of electrophilicity–nucleophilicity, in particular one that incorporates these parameters into the general framework of Lewis acidity and basicity. Finally, quite sophisticated synthetic transformations are being developed that employ carbocations, based upon insights revealed by the mechanistic studies.

SUGGESTED READING

G. K. S. Prakash and P. v. R. Schleyer, Eds., *Stable Ion Chemistry*, John Wiley & Sons, Inc., New York, 1997.

G. A. Olah, "100 Years of Carbocations and their Significance in Chemistry," *J. Org. Chem.* **2001**, *18*, 5943.

Number 12 of *Acc. Chem Res.* **1983**, where three papers discuss norbornyl cations.

C. D. Ritchie, "Cation–Anion Combination Reactions," *Can. J. Chem.* **1986**, *64*, 2239.

R. A. McClelland, "Flash Photolysis Generation and Reactivities of Carbenium Ions and Nitrenium Ions," *Tetrahedron* **1996**, *52*, 6823.

J. P. Richard, "A Consideration of the Barrier for Carbocation–Nucleophile Combination Reactions," *Tetrahedron* **1994**, *51*, 7981.

H. Mayr, T. Bug, M. F. Gotta, N. Hering, B. Irrgang, B. Janker, B. Kempf, R. Loos, A. R. Offial, G. Remennikov, and H. Schimmel, "Reference Scales for the Characterization of Cationic Electrophiles and Neutral Nucleophiles," *J. Am. Chem. Soc.* **2001**, *123*, 9500.

J. Lambert, "Preparation of the First Tricoordinate Silyl Cation" *J. Phys. Org. Chem.* **2001**, *14*, 370.

REFERENCES

1. See G. A. Olah, *J. Org. Chem.* **2001**, *18*, 5943.

2. G. A. Olah, *J. Am. Chem. Soc.* **1972**, *94*, 808.

3. N. N. Lichtin, *Prog. Phys. Org. Chem.* **1963**, *1*, 75.

4. N. C. Deno, J. J. Jaruzelski, and A. Schriesheim, *J. Am. Chem. Soc.* **1955**, *77*, 3044.

5. C. H. Rochester, *Acidity Functions*, Academic Press, London, **1970**.

6. C. D. Ritchie, (a) *Acc. Chem. Res.* **1972**, *5*, 348. (b) *Can. J. Chem.* **1986**, *64*, 2239.

7. C. K. Ingold, *Structure and Mechanism in Organic Chemistry*, 2nd Ed., Cornell University Press, Ithaca, NY, **1969**.

8. T. H. Lowry and K. S. Richardson, *Mechanism and Theory in Organic Chemistry*, 3rd Ed., Harper and Row, New York, **1986**, Chapters 4 and 5.

9. J. M. Harris, *Prog. Phys. Org. Chem.* **1974**, *11*, 89.

10. E. Hilinski, in *Reactive Intermediate Chemistry*, R. A. Moss, M. S. Platz, and M. Jones, Jr., Eds., John Wiley & Sons, Inc., New York, 2004, Chapter 19.

11. T. L. Amyes, M. M. Toteva, and J. Richard, in *Reactive Intermediate Chemistry*, R. A. Moss, M. S. Platz, and M. Jones, Jr., Eds., John Wiley & Sons, Inc., New York, 2004, Chapter 2.

12. G. A. Olah, E. B. Baker, J. C. Evans, W. S. Tolgyesi, J. S. McIntyre, and I. J. Bastien, *J. Am. Chem. Soc.* **1964**, *86*, 1360.

13. P. Ahlberg and M. Ek. *J. Chem. Soc., Chem. Commun.* **1979**, 624.

14. M. Saunders, D. Cox, and J. R. Lloyd, *J. Am. Chem. Soc.* **1979**, *101*, 6656.

15. P. C. Myrhe and C. S. Yannoni, in *Stable Ion Chemistry*, G. K. S. Prakash and P. v. R. Schleyer, Eds., John Wiley & Sons, Inc., New York, 1997, pp. 389–432.

16. D. E. Sunko, in *Stable Ion Chemistry*, G. K. S. Prakash and P. v. R. Schleyer, Eds., John Wiley & Sons, Inc., New York, 1997, pp. 349–388.

17. M. Winkler and W. Sander, *Angew. Chem. Int. Ed. Engl.* **2000**, *39*, 2014.

18. M. Saunders and E. L. Hagen, *J. Am. Chem. Soc.* **1968**, *90*, 6881.

19. J. W. de. M. Carneiro, P. v. R. Schleyer, W. Koch, and K. Raghavachari, *J. Am. Chem. Soc.* **1990**, *112*, 4064.

20. P. Buzek, P. v. R. Schleyer, S. Sieber, W. Koch, J. W. de M. Carneiro, H. Vancik, and D. E. Sunko, *J. Chem. Soc., Chem. Commun.* **1991**, 671.

21. A. Rauk, T. S. Sorensen, and P. v. R. Schleyer, *J. Chem. Soc., Perkin Trans. 2* **2001**, 869.

22. P. v. R. Schleyer, L. K. M. Lam, D. J. Raber, J. L. Fry, M. A. McKervey, J. R. Alford, B. D. Cuddy, V. G. Keizer, H. W. Geluk, and J. L. M. A. Schlatman, *J. Am. Chem. Soc.* **1970**, *92*, 5246.

23. M. Saunders, P. Vogel, E. L. Hagen, and J. Rosenfield, *Acc. Chem. Res.* **1973**, *6*, 53.

24. P. v. R. Schleyer, C. Maerker, P. Buzek, and S. Sieber, in *Stable Ion Chemistry*, G. K. S. Prakash and P. v. R. Schleyer, Eds., John Wiley & Sons, Inc., New York, **1997**, pp. 19–74.

25. H. C. Brown, in *The Nonclassical Ion Problem*, Plenum, New York, **1977**, p. 49.

26. (a) H. C. Brown, *Acc. Chem. Res.* **1983**, *16*, 432. (b) G. A. Olah, *Acc. Chem. Res.* **1983**, *16*, 440. (c) C. Walling, *Acc. Chem. Res.* **1983**, *16*, 448.

27. S. Winstein and D. S. Trifan, *J. Am. Chem. Soc.* **1949**, *71*, 2963; **1952**, *74*, 1147, 1154.

28. J. D. Roberts and C. C. Lee, *J. Am. Chem. Soc.* **1951**, *73*, 5009.

29. C. S. Yannoni, V. Macho, and P. C. Myrhe, *J. Am. Chem. Soc.* **1982**, *104*, 7380.

30. P. v. R. Schleyer and S. Sieber, *Angew. Chem. Int. Ed. Engl.* **1993**, *32*, 1606.

31. W. Koch, B. Liu, D. J. DeFrees, D. E. Sunko, and H. Vancik, *Angew. Chem.* **1990**, *102*, 198.

32. G. A. Olah, *Angew. Chem. Int. Ed. Engl.* **1973**, *12*, 173.

33. M. Saunders, H. A. Jiminez-Vazquez, and O. Kronja, in *Stable Ion Chemistry*, G. K. S. Prakash and P. v. R. Schleyer, Eds., John Wiley & Sons, Inc., New York, **1997**, pp. 297–322.

34. T. S. Sorensen, in *Stable Ion Chemistry*, G. K. S. Prakash and P. v. R. Schleyer, Eds., John Wiley & Sons, Inc., New York, **1997**, pp. 75–136.

35. M. Saunders and M. R. Kates, *J. Am. Chem. Soc.* **1980**, *102*, 6867; **1983**, *105*, 3571.

36. A. H. de M. Gomes, C. H. MacGillvary, and K. Eriks, *Acta Crystallogr.* **1965**, *18*, 437.

37. M. Sundaralingham and A. K. Chwang, in *Carbonium Ions*, Vol. 5, G. A. Olah and P. v. R. Schleyer, Eds., John Wiley & Sons, Inc., New York, **1976**, p. 2427.

38. T. Laube, in *Stable Ion Chemistry*, G. K. S. Prakash and P. v. R. Schleyer, Eds., John Wiley & Sons, Inc., New York, **1997**, pp. 453–496.

39. S. Hollenstein and T. Laube, *J. Am. Chem. Soc.* **1993**, *115*, 7240.

40. T. Laube, *Angew. Chem.* **1986**, *98*, 368.

41. T. Laube and C. Lohse, *J. Am. Chem. Soc.* **1994**, *166*, 2001.

42. D. Stasko and C. A. Reed, *J. Am. Chem. Soc.* **2002**, *124*, 1148.

43. C. A. Reed, N. L. P. Fackler, K.-C. Kim, D. Stasko, D. R. Evans, P. D. W. Boyd, and C. E. F. Richard, *J. Am. Chem. Soc.* **1999**, *121*, 6314.

44. L. C. Bateman, E. D. Hughes, and C. K. Ingold, *J. Chem. Soc.* **1940**, 1017.

45. D. J. Raber, J. M. Harris, R. E. Hall, and P. v. R. Schleyer, *J. Am. Chem. Soc.* **1971**, *93*, 4821.

46. Z. Rappoport, *Tetrahedron Lett.* **1979**, 2559; R. Ta-Shma and Z. Rappoport, *J. Am. Chem. Soc.* **1983**, *105*, 6082.

47. (a) J. P. Richard and W. P. Jencks, *J. Am. Chem. Soc.* **1982**, *104*, 4689; **1982**, *104*, 4691; (b) J. P. Richard, M. E. Rothenburg, and W. P. Jencks, *J. Am. Chem. Soc.* **1984**, *106*, 1373.

48. R. A. McClelland, *Tetrahedron* **1996**, *52*, 6823.

49. J. Bartl, S. Steenken, H. Mayr, and R. A. McClelland, *J. Am. Chem. Soc.* **1990**, *112*, 6918.

50. E. O. Alonso, L. J. Johnston, J. C. Scaiano, and V. G. Toscano, *J. Am. Chem. Soc.* **1990**, *112*, 1270.

51. H. E. Zimmerman and S. Somasekhara, *J. Am. Chem. Soc.* **1963**, *85*, 922.

52. P. Wan and E. Krogh, *J. Chem. Soc., Chem. Commun*, **1985**, 1027; *J. Am. Chem. Soc.* **1989**, *111*, 4887.

53. F. L. Cozens, V. M. Kanagasabapathy, R. A. McClelland, and S. Steenken, *Can. J. Chem.* **1999**, *77*, 2069.

54. S. Steenken and R. A. McClelland, *J. Am. Chem. Soc.* **1990**, *112*, 9648.

55. W. Kirmse, J. Kilian, and S. Steenken, *J. Am. Chem. Soc.* **1990**, *112*, 6399.

56. N. P. Schepp and J. Wirz, *J. Am. Chem. Soc.* **1994**, *116*, 11749.

57. S. Steenken and R. A. McClelland, *J. Am. Chem. Soc.* **1989**, *111*, 4967.

58. R. A. McClelland, F. L. Cozens, J. Li, and S. Steenken, *J. Chem. Soc., Perkin Trans. 2* **1996**, 1531.

59. S. Steenken, M. Ashokkumar, P. Maruthamuthu, and R. A. McClelland, *J. Am. Chem. Soc.* **1998**, *120*, 11925.

60. J. P. Pezacki, D. Shukla, J. Lustyk, and J. Warkentin, *J. Am. Chem. Soc.* **1999**, *121*, 6589.

61. J. Chateauneuf, *J. Chem. Soc., Chem. Commun.* **1991**, 1437.

62. (a) T. Yabe and J. K. Kochi, *J. Am. Chem. Soc.* **1992**, *114*, 4491; (b) K. S. Peters and B. Li, *J. Phys. Chem.* **1994**, *98*, 401.

63. J. I. Finneman and J. C. Fishbein, *J. Am. Chem. Soc.* **1995**, *117*, 4228.

64. (a) M. M. Toteva and J. P. Richard, *J. Am. Chem. Soc.* **1996**, *118*, 11434. (b) O. Kronja, M. Birus, and M. Saunders, *J. Chem. Soc., Perkin Trans. 2*, **1999**, 1375.

65. T. L. Amyes and W. P. Jencks, *J. Am. Chem. Soc.* **1989**, *111*, 7888.

66. (a) N. S. Banait and W. P. Jencks, *J. Am. Chem. Soc.* **1991**, *113*, 7951. (b) J. Zhu and A. J. Bennet, *J. Org. Chem.* **2000**, *65*, 4423.

67. B. D. Song and W. P. Jencks, *J. Am. Chem. Soc.* **1989**, *111*, 8470.

68. K. M. Koshy, D. Roy, and T. T. Tidwell, *J. Am. Chem. Soc.* **1979**, 101, 357.

69. See Figure 2 of Ref. 48.

70. H. C. Brown, D. P. Kelley, and M. Periasamy, *Proc. Natl. Acad. Sci. U.S.A.* **1980**, *77*, 6956.

71. T. V. Pham and R. A. McClelland, *Can. J. Chem.*, **2001**, *79*, 1887.

72. Y. Yukawa, Y. Tsuno, and M. Sawada, *Bull. Chem. Soc. Jpn.* **1966**, *39*, 2274.

73. R. A. McClelland, N. Banait, and S. Steenken, *J. Am. Chem. Soc.* **1986**, *108*, 7023.

74. J. P. Richard, T. L. Amyes, and T. Vontor, *J. Am. Chem. Soc.* **1992**, *114*, 5626.

75. R. A. McClelland, V. M. Kanagasabapathy, N. S. Banait, and S. Steenken, *J. Am. Chem. Soc.* **1992**, *114*, 1816.

76. J. P. Richard, *J. Chem. Soc., Chem. Commun.* **1987**, 1768.

77. W. P. Jencks, M. T. Haber, D. Herschlag, and K. Nazaretian, *J. Am. Chem. Soc.* **1986**, *108*, 479.

78. F. Cozens, J. Li, R. A. McClelland, and S. Steenken, *Angew. Chem., Int. Ed. Engl.* **1992**, *31*, 743.

79. R. A. McClelland, F. L. Cozens, J. Li, and S. Steenken, *J. Chem. Soc., Perkin Trans. 2*, **1996**, 1531.

80. (a) M. Villesange, A. Rives, C. Bunel, J.-P. Vairon, M. Froeyen, M. Van Beylen, and A. Persoons, *Makromol. Chem., Makromol. Symp.* **1991**, *47*, 271. (b) T. Kunitake and K. Takarabe, *Macromolecules* **1979**, *12*, 1061.

81. (a) H. Mayr and M. Patz, *Angew. Chem. Int. Ed. Engl.* **1994**, *33*, 938; (b) H. Mayr, M. Patz, M. F. Gotta, and A. R. Ofial, *Pure Appl. Chem.* **1998**, *70*, 1993; (c) H. Mayr, T. Bug, M. F. Gotta, N. Hering, B. Irrgang, B. Janker, B. Kempf, R. Loos, A. R. Offial, G. Remennikov, and H. Schimmel, *J. Am. Chem. Soc.* **2001**, *123*, 9500.

82. (a) A. D. Allen and T. T. Tidwell, in *Advances in Carbocation Chemistry*, X. Creary, Ed., JAI Press, Greenwich, CT. **1989**, Vol. 1, p. 1; (b) X. Creary, A. C. Hopkinson, and E. Lee-Ruff, in *Advances in Carbocation Chemistry*, Vol. 1, X. Creary, Ed., JAI Press, Greenwich, CT. **1989**, p. 45; (c) P. G. Gassman and T. T. Tidwell, *Acc. Chem. Res.* **1983**, *16*, 279.

83. (a) J. P. Richard, T. L. Amyes, L. Bei, and V. Stubblefield, *J. Am. Chem. Soc.* **1990**, *112*, 9513. (b) T. L. Amyes, I. W. Stevens, and J. P. Richard, *J. Org. Chem.* **1993**, *58*, 6057.

84. (a) R. A. McClelland, F. L. Cozens, S. Steenken, T. L. Amyes, and J. P. Richard, *J. Chem. Soc., Perkin Trans. 2*, **1993**, 1717; (b) N. P. Schepp and J. Wirz, *J. Am. Chem. Soc.* **1994**, *116*, 11749; (c) R. A. McClelland, V. E. Licence, and J. P. Richard, *Can. J. Chem.* **1998**, *76*, 1910.

85. J. P. Richard, *Tetrahedron* **1994**, *51*, 7981.

86. K. P. Dockery, J. P. Dinnocenzo, S. Farid, J. L. Goodman, I. R. Gould, and W. P. Todd, *J. Am. Chem. Soc.* **1997**, *119*, 1876.

87. C. S. Q. Lew and R. A. McClelland, *J. Am. Chem. Soc.* **1993**, *115*, 11516.

88. H. Mayr, N. Basso, and G. Hagen, *J. Am. Chem. Soc.* **1992**, *114*, 3060.

89. J. B. Lambert, Y. Zhao, and S. M. Zhang, *J. Phys. Org. Chem.* **2001**, *14*, 370.

90. C. A. Reed, *Acc. Chem. Res.* **1995**, *31*, 325.

91. J. B. Lambert and Y. Zhao, *J. Am. Chem. Soc.* **1996**, *118*, 7867.

92. J. B. Lambert, Y. Zhao, H. Wu, W. C. Tse, and B. Kuhlman, *J. Am. Chem. Soc.* **1999**, *121*, 5003.

93. (a) T. Nishinaga, Y. Izukawa, and K. Komatsu, *J. Am. Chem. Soc.* **2000**, *122*, 9312; (b) A. Sekiguchi, T. Matsuno, and M. Ichinohe, *J. Am. Chem. Soc.* **2000**, *122*, 11250.

94. K.-C. Kim, C. A. Reed, D. W. Elliott, L. J. Mueller, F. Tham, L. Lin., and J. B. Lambert, *Science* **2002**, *297*, 825.

95. For reviews see A. Corma, *Chem. Rev.* **1995**, *95*, 559. J. F. Haw, J. B. Nicholas, T. Xu, L. W. Beck, and D. B. Ferguson, *Acc. Chem. Res.* **1996**, *29*, 259. I. Kiricisi, H. Forster, G. Tasi, and J. B. Nagy, *Chem. Rev.* **1999**, *99*, 2085.

96. See M. A. O'Neill, F. L. Cozens, and N. P. Schepp, *J. Am. Chem. Soc.* **2000**, *122*, 6017; *Tetrahedron* **2000**, *56*, 6969.

97. A. Dipple, *Carcinogenesis* **1995**, *16*, 437.

98. (a) H. B. Islam, S. C. Gupta, H. Yagi, D. M. Jerina, and D. L. Whalen, *J. Am. Chem. Soc.* **1990**, *112*, 6363; (b) B. Lin, N. Islam, S. Friedman, H. Yagi, D. M. Jerina, and D. L. Whalen, *J. Am. Chem. Soc.* **1998**, *120*, 4327.

99. N. B. Islam, D. L. Whalen, H. Yagi, and D. M. Jerina, *J. Am. Chem. Soc.* **1987**, *109*, 2108.

100. C. Sanchez, S. Shibutani, L. Dasaradhi, J. L. Bolton, P. W. Fan, and R. A. McClelland, *J. Am. Chem. Soc.* **1998**, *120*, 13513.

101. See K. K. Laali and P. E. Hansen, *J. Org. Chem.* **1997**, *62*, 5804.

102. R. A. McClelland, C. Sanchez, E. Sauer, and S. Vukovic, *Can. J. Chem.* **2002**, *80*, 281.

103. J. J. Dannenberg and M. Tomasz, *J. Am. Chem. Soc.* **2000**, *122*, 2062.

104. R. A. McClelland and G. Marji, unpublished work.

105. V. Durbecq, G. Sainz, Y. Oudjama, B. Clantin, C. Bompard-Gilles, C. Tricot, J. Caillet, V. Stalon, L. Droogmans, and V. Villeret, *EMBO J.* **2001**, *20*, 1530.

106. B. S. J. Blagg, M. B. Jarstfer, D. H. Rogers, and C. D. Poulter, *J. Am. Chem. Soc.* **2002**, *124*, 8846.

CHAPTER 2

Crossing the Borderline Between S_N1 and S_N2 Nucleophilic Substitution at Aliphatic Carbon

TINA L. AMYES, MARIA M. TOTEVA, and JOHN P. RICHARD

Department of Chemistry, University at Buffalo, SUNY, Buffalo, NY

1. INTRODUCTION

Organic reactions that involve the cleavage and/or formation of two or more bonds may proceed by either a stepwise mechanism, in which bond cleavage and

Reactive Intermediate Chemistry, edited by Robert A. Moss, Matthew S. Platz, and Maitland Jones, Jr.
ISBN 0-471-23324-2   Copyright © 2004 John Wiley & Sons, Inc.

formation occur as separate events, or by a concerted mechanism in which they take place in a single reaction stage.[1] Hughes and Ingold first showed that aliphatic nucleophilic substitution with a zero-order dependence of rate on nucleophile (Nu) concentration (S_N1) is favored when the carbocation intermediate is strongly stabilized by electron-donating substituents.[2] Conversely, nucleophilic substitution with a first-order dependence on nucleophile concentration (S_N2) is favored when this mechanism avoids the formation of the unstable intermediate of the stepwise reaction[2] (see Scheme 2.1).

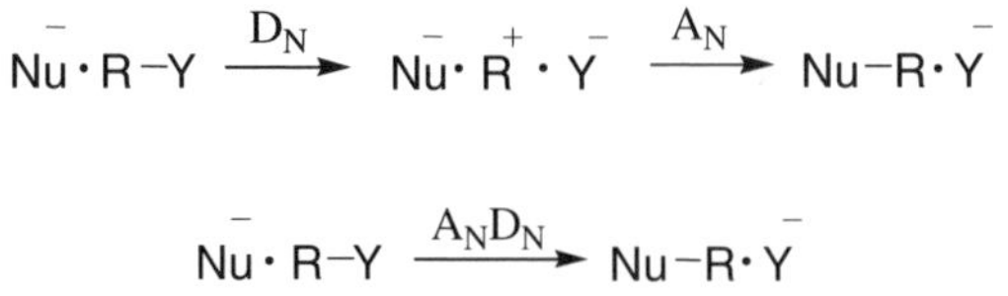

$$ \overset{-}{Nu} \cdot R{-}Y \xrightarrow{D_N} \overset{-}{Nu} \cdot \overset{+}{R} \cdot \overset{-}{Y} \xrightarrow{A_N} Nu{-}R \cdot \overset{-}{Y} \qquad (A) $$

$$ \overset{-}{Nu} \cdot R{-}Y \xrightarrow{A_N D_N} Nu{-}R \cdot \overset{-}{Y} \qquad (B) $$

Scheme 2.1

Similar qualitative relationships between reaction mechanism and the stability of the putative reactive intermediates have been observed for a variety of organic reactions, including alkene-forming elimination reactions,[3] and nucleophilic substitution at vinylic[4] and at carbonyl carbon.[5] The nomenclature for reaction mechanisms has evolved through the years and we will adopt the International Union of Pure and Applied Chemistry (IUPAC) nomenclature[6] and refer to stepwise substitution (S_N1) as $D_N + A_N$ (Scheme 2.1A) and concerted bimolecular substitution (S_N2) as $A_N D_N$ (Scheme 2.1B), except when we want to emphasize that the distinction in reaction mechanism is based solely upon the experimentally determined kinetic order of the reaction with respect to the nucleophile.

While the mechanistic imperatives are clear for the "limiting" cases of stepwise nucleophilic substitution through stable carbocation intermediates and concerted bimolecular substitution that avoids the formation of unstable intermediates, a more difficult question is how to model the transition across the borderline between mechanisms where the carbocation intermediate is unstable, but the advantage to a concerted substitution is only beginning to become significant. The term "borderline" implies the existence of a line separating $D_N + A_N$ from $A_N D_N$ nucleophilic substitution reactions, whose position depends, in some sense, on the lifetime of the reaction intermediate. It has been argued that nucleophilic aliphatic substitution generally occurs by a stepwise ($D_N + A_N$) reaction mechanism when the carbocation intermediate exists in an energy well for at least the time of a bond vibration ($\sim 10^{-13}$ s); and, that the change to an $A_N D_N$ mechanism is "enforced" by the disappearance of the energy well for the reaction intermediate (Fig. 2.1). This crossover between mechanisms places the borderline at the point where the lifetime of the intermediate in the presence of the incoming nucleophile is close to the vibrational limit of $\sim 10^{-13}$ s.[1]

This notion that reaction mechanisms are strictly enforced by the intermediate lifetime implies the existence of a narrow borderline region and a sharp change in reaction mechanism with changing lifetime of the carbocation intermediate. However, a narrow borderline region is not observed in all cases. The problem is

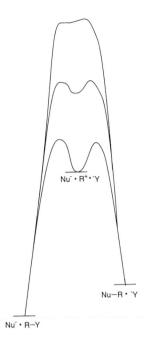

Figure 2.1. One-dimensional (1D) free energy reaction coordinate profiles that show the $D_N + A_N$ reaction mechanism through a carbocation intermediate and the change to an $A_N D_N$ reaction in which the intermediate is too unstable to exist in an energy well for the time of a bond vibration.

in part semantic. A borderline between reaction mechanisms may be defined with some rigor by the intermediate lifetime, which determines whether its existence in an energy well is possible. On the other hand, the term "borderline reaction" is less rigorous and generally refers to reactions whose observable properties are in some way intermediate between stepwise and concerted. We will define a borderline reaction as one that apparently proceeds through a "hot", unstable, carbocation "intermediate" that shows a small selectivity for reaction with added nucleophiles, so that the yield of the nucleophilic substitution product is very low and it is not possible to determine whether the substitution reaction is kinetically zero-order (S_N1) or first-order (S_N2) in nucleophile concentration. Such reactions generally show other borderline properties, including internal return of ion pair reaction intermediates,[7] and reaction with an excess of inversion or retention of configuration at carbon.[8–10] The intermediates of borderline reactions are clearly unstable and their lifetimes may lie close to the vibrational limit ($\sim 10^{-13}$ s), which marks the borderline between $D_N + A_N$ and $A_N D_N$. However, there is no strict requirement that crossing this rigorously defined borderline will result in the appearance of a *kinetically* significant bimolecular reaction of the nucleophile. Sometimes this is observed, but in other cases it is not, for reasons that will be discussed in this chapter.

2. NUCLEOPHILIC SUBSTITUTION OF AZIDE ION AT BENZYLIC CARBON

2.1. Introduction

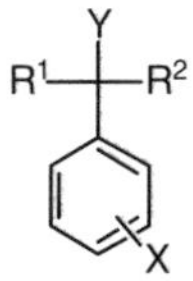

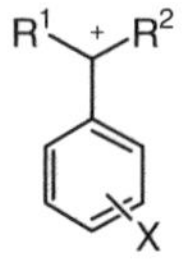

$$
\begin{array}{ll}
\text{X-1-Y} & R^1 = H, R^2 = Me \\
\text{X-2-Y} & R^1 = Me, R^2 = Me
\end{array}
\qquad
\begin{array}{ll}
\text{X-1}^+ & R^1 = H, R^2 = Me \\
\text{X-2}^+ & R^1 = Me, R^2 = Me
\end{array}
$$

The benzylic substrates X-1-Y and X-2-Y have provided a useful platform for examining the changes in reaction mechanism for nucleophilic substitution that occur as the lifetime of the carbocation intermediate is decreased systematically by varying the meta- and para- aromatic ring substituents. When X is strongly resonance electron-donating, X-1-Y and X-2-Y react by a stepwise $D_N + A_N$ mechanism through carbocation intermediates with significant and known lifetimes. The change from a strongly electron-donating (4-Me$_2$N) to an electron-withdrawing (3,5-di-CF$_3$) ring substituent X results in a $\sim$30 kcal/mol increase in the thermodynamic barrier to substrate ionization to form the corresponding benzylic carbocation reaction intermediate, through a combination of resonance and polar substituent effects. This series allows for a thorough examination of the changes in reaction mechanism that occur as the lifetime of the putative carbocation intermediate reaches the limiting value of the time for a bond vibration.[11–16]

2.1.1. Ring-Substituted 1-Phenylethyl Derivatives. Figure 2.2 ($\bullet$ and $\bigcirc$) shows that there is a sharp change from a stepwise to a concerted mechanism for nucleophilic substitution of azide ion at 1-phenylethyl derivatives (X-**1**-Y) in 50:50 (v/v) water/trifluoroethanol as X is changed from strongly electron donating to electron withdrawing.[12,15] This change in mechanism gives rise to a narrow minimum in the V-shaped plot

$$
\left(\frac{k_{az}}{k_s}\right)_{obsd} = \frac{[\text{X-1-N}_3]}{[\text{X-1-OSolv}][\text{N}_3^-]} \tag{1}
$$

of the nucleophile selectivity for azide ion and solvent determined from product analysis $[(k_{az}/k_s)_{obsd}, M^{-1}, \text{Eq. 1}]$, against the Hammett ring substituent constant for X, σ^+, or σ (Fig. 2.2).[12,15] When X is strongly electron donating ($\sigma^+ \leq -0.32$, $\bullet$) the formation of X-**1**-N$_3$ from reaction of X-**1**-Y is zero order in [N$_3^-$], and the product selectivity $(k_{az}/k_s)_{obsd}$ (M^{-1}) is independent of the leaving group Y$^-$. By contrast, the product X-**1**-N$_3$ of the reaction of azide ion with X-**1**-Cl forms *exclusively* by a reaction that is first order in [N$_3^-$] when X is electron withdrawing

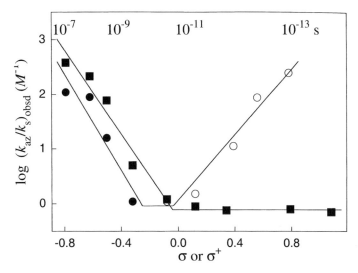

Figure 2.2. The change with changing aromatic ring substituent X in the azide (az) ion selectivities $(k_{az}/k_s)_{obsd}$ (M^{-1}) determined by analysis of the products of the reactions of ring-substituted 1-phenylethyl derivatives (X-**1**-Y) and ring-substituted cumyl derivatives (X-**2**-Y) in 50/50 (v/v) water/trifluoroethanol at 25 °C.[15,16] The selectivities are plotted against the appropriate Hammett substituent constant σ^+ or σ for the aromatic ring substituent. The lifetimes $1/k_s$ (s) of the carbocations X-**1**$^+$, determined either directly using the azide ion clock or estimated by extrapolation of a linear free energy Hammett relationship, are given along the top of this figure. The lifetimes of the carbocations X-**2**$^+$ are about fourfold longer than those of the corresponding X-**1**$^+$. Key: (●) reactions of X-**1**-Y (Y = Cl or ring-substituted benzoate) that are kinetically zero order with respect to azide ion; (○) reactions of X-**1**-Cl that are kinetically first order with respect to azide ion; (■) reactions of X-**2**-Y (Y = Cl or ring-substituted benzoate) that are kinetically zero order with respect to azide ion.

($\sigma^+ \geq -0.08$, Fig. 2.2, ○),[13] which shows that there is a single rate- and product-determining step for these reactions (see Eq. 1). The possibility that this rate-determining step involves collapse of an ion pair or triple ion intermediate to product was considered, but was excluded by several lines of experimental evidence that showed that these kinetically bimolecular reactions of X-**1**-Y proceed by a concerted mechanism.[13]

The descending product selectivities $(k_{az}/k_s)_{obsd}$ (M^{-1}) on the left-hand limb of Figure 2.2 (●) for the stepwise reactions of X-**1**-Y(Scheme 2.2) are due to the increase in the value of k_s (s^{-1}), with decreasing stability of the carbocation intermediate X-**1**$^+$, relative to the constant value of k_{az} $(M^{-1} s^{-1})$ for the diffusion-limited reaction of azide ion.[12,15] The lifetimes $1/k_s$ in the top left half of Figure 2.2 for addition of solvent to X-**1**$^+$ ($\sigma^+ \leq -0.32$) were calculated from $(k_{az}/k_s)_{obsd}$ (M^{-1}) for partitioning of X-**1**$^+$ and $k_{az} = 5 \times 10^9$ $M^{-1} s^{-1}$ for the diffusion-limited reaction of azide ion.[15] There is good agreement between

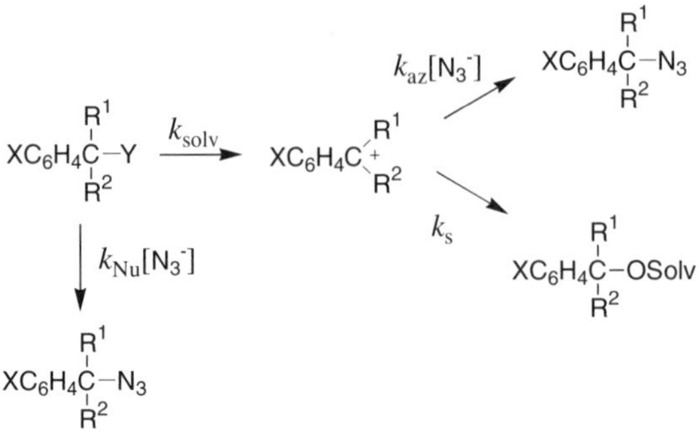

Scheme 2.2

k_s (s^{-1}) for (4-MeO)-$\mathbf{1}^+$ calculated by the "azide ion clock",[15] and the value determined directly by monitoring the disappearance of (4-MeO)-$\mathbf{1}^+$ generated by laser flash photolysis.[17] The lifetimes of X-$\mathbf{1}^+$ in the top right half of Figure 2.2 ($\sigma^+ \geq -0.08$) were estimated by extrapolation of the excellent linear logarithmic correlation of $1/k_s$ with the Hammett substituent constant for X for $\sigma^+ \leq -0.32$.[15]

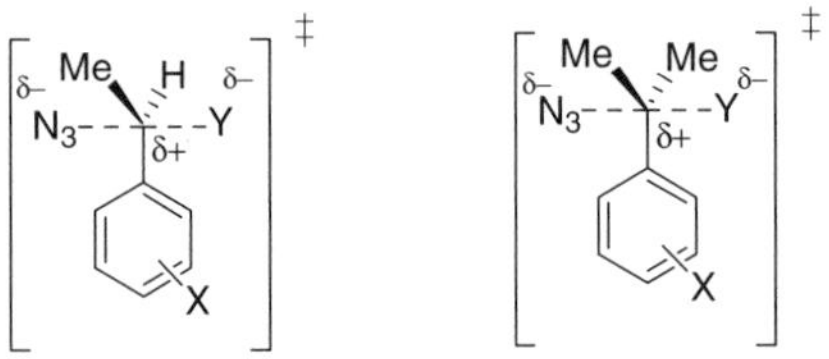

$[\text{X-}\mathbf{1}\text{-N}_3]^{\ddagger}$ $[\text{X-}\mathbf{2}\text{-N}_3]^{\ddagger}$

The break to the ascending right-hand limb in Figure 2.2 (O, $\sigma^+ \geq -0.08$) marks the appearance of a concerted bimolecular substitution reaction of azide ion with X-$\mathbf{1}$-Y, with a second-order rate constant k_{Nu} (Scheme 2.2). There is a very narrow borderline between the kinetically S_N1 and S_N2 nucleophilic substitution reactions of X-$\mathbf{1}$-Cl ($-0.32 \leq \sigma^+ \leq -0.08$, Fig. 2.2). The increasing product selectivities $(k_{az}/k_s)_{obsd}$ observed as the transition states for nucleophilic substitution of solvent and azide ion at X-$\mathbf{1}$-Cl are destabilized by electron-withdrawing X ($\sigma^+ \geq -0.08$) reflect the partial neutralization by azide ion of the developing positive charge in the transition state for bimolecular nucleophilic substitution, $[\text{X-}\mathbf{1}\text{-N}_3]^{\ddagger}$. This results in a 2.7-unit more positive Hammett reaction constant for concerted bimolecular nucleophilic substitution by azide ion, $\rho_{Nu} = -2.9$, compared with $\rho_{solv} = -5.6$ for the stepwise solvolysis of X-$\mathbf{1}$-Cl.[13]

In the presence of solvent alone, the lifetime of the intermediate of the stepwise reaction of X-$\mathbf{1}$-Y in the narrow borderline between the S_N1 and S_N2 substitution reactions of azide ion ($-0.32 \leq \sigma^+ \leq -0.08$, Fig. 2.2) is $\sim 1/k_s = 10^{-10}$ s.[15] Azide ion is $\sim 10^6$–10^7-fold more reactive than water toward triarylmethyl carbocations and related electrophiles, and this selectivity is independent of carbocation reactivity, so long as the reactions of both azide ion and solvent are limited by

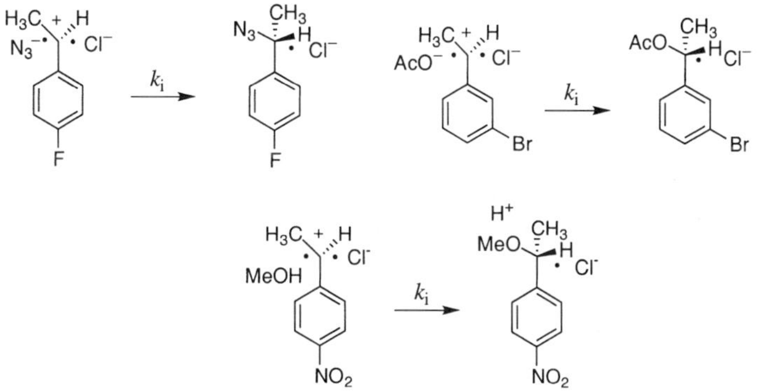

Scheme 2.3

the rate constant for chemical bond formation.[18,19] However, an azide ion selectivity of $k_{az}/k_s = 10^6 \ M^{-1}$ cannot be maintained across this sharp borderline, because this would require an impossibly large rate constant of $k_i \approx 10^{16} \ s^{-1}$ for collapse of the triple ion complex $N_3^- \bullet (4\text{-F})\text{-}\mathbf{1}^+ \bullet Cl^-$ ($\sigma^+ = -0.08$ for 4-F) to product (Scheme 2.3). In fact, $(4\text{-F})\text{-}\mathbf{1}^+$ is unlikely to exist for the time of a bond vibration (10^{-13} s) in the presence of azide ion because this would require a difference in the chemical barriers for the reaction of azide ion and solvent of only $\sim$4 kcal/mol, which is smaller than the $\sim$6 kcal/mol difference observed for a wide variety of carbocation-azide ion addition reactions. It was concluded that the concerted mechanism for the reaction of azide ion with X-**1**-Y ($\sigma^+ \geq -0.08$) is probably "enforced" because, in the presence of azide ion, the putative carbocation intermediate X-**1**$^+$ cannot exist in an energy well.[13] We will return to Scheme 2.3 in Section 2.3.1.

2.1.2. Ring-Substituted Cumyl Derivatives.
In contrast with the sharp change in the kinetic order for nucleophilic substitution of azide ion at 1-phenylethyl derivatives with changing aromatic ring substituent X (Fig. 2.2), there is no corresponding change from S_N1 to S_N2 nucleophilic substitution of azide ion at ring-substituted cumyl derivatives (X-**2**-Y) (Fig. 2.2, ∎).[16] The product selectivities $(k_{az}/k_s)_{obsd}$, $(M^{-1}$, Eq. 1) for reactions of X-**2**-Y in 50:50 (v/v) water/trifluoroethanol decrease from $(k_{az}/k_s)_{obsd} = 380$ to $0.7 \ M^{-1}$ as the ring substituent X is changed from electron donating (4-MeO, $\sigma^+ = -0.79$) to weakly electron withdrawing (3-MeO, $\sigma = 0.12$). This selectivity then remains almost constant as X is changed to strongly electron withdrawing (3,5-di-CF$_3$, $\Sigma\sigma = 1.08$). In all cases, the observed first-order rate constant, k_{obsd} (s^{-1}), for the reaction of X-**2**-Y is independent of [N$_3^-$]. However, throughout the broad region of minimum selectivity, where $(k_{az}/k_s)_{obsd} = 0.7 \ M^{-1}$ (Fig. 2.2, ∎), it is possible that small increases in k_{obsd} due to concerted nucleophilic substitution by azide ion are masked by compensating decreases due to a specific azide ion salt effect.[16] The reaction mechanism in this region is borderline, because it is not known whether the formation of X-**2**-N$_3$ is kinetically zero-order (S_N1) or first-order (S_N2) with respect to azide ion.

The addition of an α-Me group to X-**1**$^+$ to give X-**2**$^+$ results in a three- to four-fold increase in the selectivity $(k_{az}/k_s)_{obsd}$ (M^{-1}) for reaction of the carbocation with azide ion and a solvent of 50:50 (v/v) water/trifluoroethanol.[16] This is due to the decrease in k_s (s^{-1}) relative to the constant value of $k_{az} = 5 \times 10^9$ M^{-1} s^{-1} for the diffusion-limited trapping of both X-**1**$^+$ and X-**2**$^+$ by azide ion.[16] The lifetimes, $1/k_s$, for X-**2**$^+$ are only three- to fourfold longer than those of the corresponding 1-phenylethyl carbocations;[15–17] and an extrapolation of a good linear relationship between log $1/k_s$ for addition of solvent to X-**2**$^+$ and σ^+ for X $(\sigma^+ \leq -0.08)$ gives a lifetime of $1/k_s \approx 10^{-13}$ s for (3,5-di-CF$_3$)-**2**$^+$. The lifetimes of X-**2**$^+$ in the presence of azide ion must be even shorter than the lifetime in the presence of solvent alone, and they should reach the vibrational limit of $\sim 10^{-13}$ s as X-**2**$^+$ is destabilized by strongly electron-withdrawing X. However, in this case there is no break to an ascending limb in Figure 2.2 due to the appearance of a concerted bimolecular nucleophilic substitution reaction of azide ion with X-**2**-Y as is observed for X-**1**-Y. Rather, a broad borderline region for the reaction of azide ion with X-**2**-Y is observed.

Students of reaction mechanism will recognize intuitively that the difference between the narrow and broad borderline regions observed for nucleophilic substitution of azide ion at secondary and tertiary carbon (Fig. 2.2) is due to the greater steric hindrance to bimolecular nucleophilic substitution at the tertiary carbon.[2] This leads to a large difference in the effects of an α-Me group on k_{solv} (s^{-1}) for the stepwise solvolysis and k_{Nu} $(M^{-1}$ s$^{-1})$ for concerted bimolecular nucleophilic substitution reactions of X-**2**-Y. The barriers to solvolysis and the reaction of 1 M azide ion with (4-F)-**1**-Cl $(\sigma^+ = -0.08$ for 4-F) are similar, because the addition of 1 M azide ion results in about a doubling of k_{obsd} (s^{-1}) in 50:50 (v/v) water/trifluoroethanol and the reaction gives a $\sim$50% yield of (4-F)-**1**-N$_3$.[13] Now, the addition of an α-Me group to (4-F)-**1**-Cl to give (4-F)-**2**-Cl will stabilize the cationic transition state for the stepwise solvolysis reaction (k_{solv}), but will lead to a *relative* destabilization of the transition state for bimolecular nucleophilic substitution of azide ion (k_{Nu}), due to the increased steric crowding. The result is that $k_{solv} \gg k_{Nu}[N_3^-]$ for the reaction of (4-F)-**2**-Cl, and the formation of (4-F)-**2**-N$_3$ occurs mainly by nucleophile trapping of the carbocation intermediate of a stepwise reaction.[16] A change to a ring substituent X that is more electron withdrawing than 4-F will further destabilize (4-F)-**2**$^+$, and will favor concerted nucleophilic substitution that avoids formation of the unstable putative intermediate X-**2**$^+$. However, the resistance of tertiary carbon to formation of the pentavalent transition state for concerted nucleophilic substitution is so great that k_{Nu} for the concerted reaction remains much smaller than k_{solv}, even when X is strongly electron withdrawing and X-**2**$^+$ is highly unstable.

2.2. More O'Ferrall Diagrams

Two-dimensional (2D) More O'Ferrall reaction coordinate diagrams (Fig. 2.3) are helpful in rationalizing the differences in the borderline regions for nucleophilic substitution at X-**1**-Y and X-**2**-Y that are illustrated by the data in Figure 2.2. These

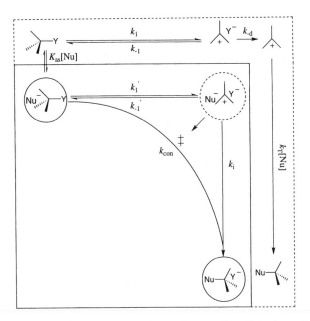

Figure 2.3. More O'Ferrall reaction coordinate diagram for aliphatic nucleophilic substitution, constructed according to Scheme 2.4 of the text. The fully stepwise reaction through a carbocation intermediate that is sufficiently long lived to diffuse through solvent and undergo trapping by nucleophiles (k_T[Nu]) is shown in the "wings" around this diagram (dotted lines). The stepwise preassociation mechanism (K_{as} and k_1') runs through the triple ion intermediate in the inner upper right hand corner, and the concerted pathway (k_{con}) runs through the interior of the diagram.

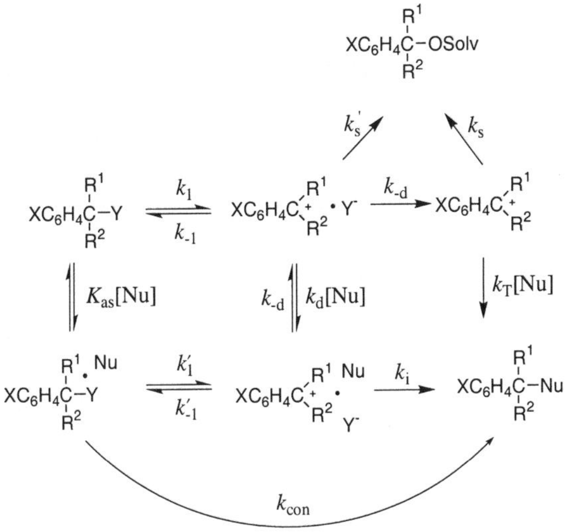

Scheme 2.4

diagrams utilize separate coordinates to show cleavage of the bond to the leaving group Y (x axis, Fig. 2.3) and formation of the bond to the incoming nucleophilic reagent Nu (y axis, Fig. 2.3). Figure 2.3 also includes the microscopic rate and association constants defined in Scheme 2.4. Note that the encounter between $R^+ \bullet Y^-$ and Nu to form a triple ion intermediate (k_d, Scheme 2.4) is not shown on this diagram. This reaction is a minor pathway for formation of the nucleophilic substitution product in the polar solvent water,[20] because diffusion-controlled trapping ($k_d \approx 5 \times 10^9\ M^{-1}\ s^{-1}$) is unable to compete effectively with the fast diffusional separation of the ion pair to the free ions ($k_{-d} \approx 1.6 \times 10^{10}\ s^{-1}$).[14]

Figure 2.3 may be used to illustrate the following progression in pathways for nucleophilic substitution at X-**1**-Y and X-**2**-Y with changes in the lifetime of the respective carbocation reaction intermediates.[13,16,20]

2.2.1. Stepwise Ionization and Trapping of a Liberated Reaction Intermediate.

Nucleophilic substitution of azide ion at X-**1**-Y and X-**2**-Y by a stepwise reaction mechanism around the "wings" of Figure 2.3 (k_1, k_{-d}, k_T) is strongly favored over the concerted reaction (k_{con}) when these reactions give carbocation intermediates that are stabilized by electron-donating ring substituents X.[14,16] These moderately stable carbocations exhibit a marked selectivity $(k_{az}/k_s)_{obsd} = k_T/k_s\ (M^{-1})$ for reaction with azide ion and solvent, which decreases as $k_s\ (s^{-1})$ for the reaction of the carbocation with the solvent increases relative to $k_T = k_d = 5 \times 10^9\ M^{-1}\ s^{-1}$ for the diffusion-limited trapping of the carbocation by azide ion.

2.2.2. Stepwise Preassociation Reactions.

The selectivity $k_{az}/k_s\ (M^{-1})$ for trapping of X-**1**$^+$ by azide ion and solvent decreases with increasing carbocation reactivity (k_s), until k_s is so large that the direct addition of solvent to the carbocation-leaving group ion pair (k'_s) is faster than its diffusional separation to the free ions: $k'_s \gg k_{-d} \approx 1.6 \times 10^{10}\ s^{-1}$ (Scheme 2.4).[14] [The rate constant k'_s for the addition of aqueous solvents to ion pairs is similar to k_s for the addition of solvent to the free carbocation.[21,22]] At this point there will be essentially *no* formation of the nucleophilic substitution product R—Nu by trapping of the free carbocation (k_T, Scheme 2.4) because the carbocation-leaving group ion pair intermediate is too short lived to undergo separation to the free ions. The formation of R—Nu can occur only when the nucleophile is present in a "preassociation complex" with the substrate R—Y at the time of its ionization (K_{as}, Scheme 2.4). This corresponds to a stepwise preassociation reaction of the nucleophile that runs through the inner upper right hand corner of the diagram in Figure 2.3 (K_{as}, k'_1, and k_i, Scheme 2.4).[16,23]

The yield of the nucleophilic substitution product from the stepwise preassociation reaction mechanism must be small in water, because association complexes between anions and neutral molecules in this solvent are weak (K_{as} is small). The dominance of this unfavorable reaction pathway is *enforced* when the competing formation of the nucleophilic substitution product by trapping of the free carbocation intermediate of the stepwise reaction is prevented by the *faster* direct addition of solvent to the carbocation-leaving group ion pair ($k'_s \gg k_{-d} \approx$ 1.6×10^{10} s^{-1}).

Minimum azide ion selectivities of $(k_{az}/k_s)_{obsd} = 1.1$ M^{-1} for reaction of the secondary substrate (4-Me)-**1**-Y[15] and $(k_{az}/k_s)_{obsd} = 0.7$ M^{-1} for reaction of the tertiary substrates X-**2**-Y ($\sigma \geq 0.12$)[16] are observed (Fig. 2.2). Around 70% of (4-Me)-**1**-N$_3$ obtained from the reaction of azide ion with (4-Me)-**1**-Cl in 50:50 (v/v) water/trifluoroethanol is formed by trapping of a liberated carbocation intermediate, and the remaining 30% is formed by a preassociation reaction mechanism.[14] For the tertiary substrates X-**2**-Cl, values of $k'_s \gg k_{-d}$ are observed when X is strongly electron withdrawing ($\sigma \geq 0.34$). These substrates react with azide ion to form X-**2**-N$_3$ exclusively by a preassociation reaction mechanism.[16] The minimum observed selectivity of $(k_{az}/k_s)_{obsd} = 0.7$ M^{-1} is consistent either with $K_{as} = 0.7$ M^{-1} for formation of an encounter complex between azide ion and substrate, which then undergoes unassisted ionization to form a triple ion intermediate ($k'_1 = k_1$, Scheme 2.4); or, with a smaller association constant and a small compensating rate increase from a formally bimolecular substitution reaction ($k'_1 > k_1$, Scheme 2.4).

The yield of the nucleophilic substitution product from the stepwise preassociation mechanism ($k'_1 = k_1$, Scheme 2.4) is small, because of the low concentration of the preassociation complex ($K_{as} \approx 0.7$ M^{-1} for the reaction of X-**2**-Y). Formally, the stepwise preassociation reaction is kinetically bimolecular, because both the nucleophile and the substrate are present in the rate-determining step (k'_1). In fact, these reactions are *borderline* between S$_N$1 and S$_N$2 because the kinetic order with respect to the nucleophile cannot be rigorously determined. A small rate increase may be due to either formation of nucleophile adduct by bimolecular nucleophilic substitution or a positive specific salt effect, while a formally bimolecular reaction may *appear* unimolecular due to an offsetting negative specific salt effect on the reaction rate.

2.2.3. Coupling and the Change to a Concerted Reaction Mechanism.

Further destabilization of the already unstable intermediate of a stepwise preassociation reaction (triple ion intermediate in the inner upper right hand corner of Fig. 2.3) will eventually result in the disappearance of the energy well for this intermediate. The question then is whether crossing the relatively sharp borderline that marks the disappearance of the energy well for the reaction intermediate is accompanied by the appearance of a concerted *bimolecular* reaction of the nucleophile with the substrate (k_{con}, Fig. 2.3 and Scheme 2.4). The factors that control the relative barriers to passage through the inner upper right hand corner of Figure 2.3 (k'_1) and through the interior of the diagram (k_{con}) are most appropriate for consideration by theoretical and computational chemists. However, the following simple model to describe these relative barrier heights provides a useful point of departure for more advanced discussion.[24]

The barrier to reactions in which two processes occur in a concerted (synchronous) fashion has been proposed to be approximately equal to the sum of the barriers to these individual steps in a fully stepwise reaction [(A + B), Fig. 2.4(I)]

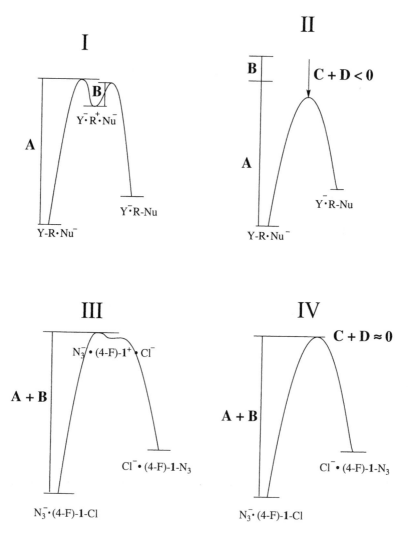

Figure 2.4. One-dimensional cross-sections of the reaction coordinate profiles for stepwise nucleophilic substitution along the inside upper and right-hand edges of the 2D diagram in Figure 2.3 and for concerted substitution through the interior of Figure 2.3. Barrier A is the barrier to formation and B is the barrier to breakdown of the carbocation "intermediate" of the "stepwise" reaction. The barrier to the concerted reaction is equal to (A + B) plus any stabilization that results from the coupling of bond formation and cleavage (C), along with the contribution of intramolecular interactions in the transition state (D), which may be either stabilizing or destabilizing.[24] Key: (I) Profile for the case where the intermediate of the stepwise reaction lies in a free energy well. (II) A competing concerted reaction will be observed when (C + D) < 0. (III) Profile for the nucleophilic substitution reaction of azide ion with 1-(4-fluorophenyl)ethyl chloride [(4-F)-**1**-Cl] for which there is no significant free energy well for the triple ion "intermediate" containing azide ion (B ≈ 0). (IV) The barrier to concerted substitution by azide ion at (4-F)-**1**-Cl is similar to that for formation of (4-F)-**1**$^+$ so that (C + D) ≈ 0.

minus the sum of (a) Transition state *stabilization* that results from the coupling of these two processes in a concerted reaction [C, Fig. 2.4(II)]; and (b) Intramolecular interactions in the transition state for the concerted reaction, which may be either stabilizing or destabilizing [D, Fig. 2.4(II)].[24] The advantage to coupling of two or more processes in a concerted reaction is normally small, with the exception of pericyclic reactions. Therefore, such coupled concerted reactions are usually observed only when the barrier to conversion of the reaction intermediate to product [B, Fig. 2.4(I)] is small or absent, so that any advantage to coupling results directly in a lowering of the barrier to the coupled concerted reaction relative to that for formation of the intermediate of the competing stepwise reaction.[24]

2.3. Crossing the Borderline between Mechanisms

2.3.1. Nucleophilic Substitution at X-1-Y.

The sharp transition from a stepwise to a concerted nucleophilic substitution reaction of azide ion with X-1-Y occurs at about the point of the disappearance of the energy well for the carbocation in the triple ion intermediate. This transition is illustrated by Figure 2.4(III), which shows the reaction coordinate profile for the reaction of azide ion with (4-F)-1-Cl through a carbocation "intermediate" that is too unstable to exist in an energy well, and Figure 2.4(IV), which shows the profile for the coupled concerted reaction (k_{con}, Fig. 2.3), which passes over an energy barrier similar to that for the "stepwise" reaction. The observed barrier to concerted nucleophilic substitution of 1 *M* azide ion at (4-F)-1-Cl is roughly equal to that for ionization of this substrate,[13] so that, for this substrate, any stabilization of the pentavalent transition state for the concerted reaction that results from the coupling of the two processes is offset by unfavorable steric interactions [(C + D) ≈ 0, Fig. 2.4(IV)].

The advantage to the coupling of bond formation and bond cleavage in nucleophilic substitution at X-1-Y depends on the ring substituent X, the leaving group Y, and the incoming nucleophile Nu.

1. The partial neutralization of positive charge at the benzylic carbon from partial bonding to the incoming nucleophilic reagent in the transition state for concerted bimolecular substitution at X-1-Cl (k_{con}, Fig. 2.3) decreases the barrier to formation of this transition state compared to that for the more cationic transition state for the stepwise reaction (k_1, Fig. 2.3). The relative stabilization of the transition state for the concerted reaction increases as the ring substituent X is made more electron withdrawing.[13] This corresponds to an increase in the advantage to coupling for the concerted reaction (C, Fig. 2.4) with decreasing stability of the avoided carbocation intermediate of the stepwise reaction.

2. Nucleophilic substitution of azide ion at (4-Me)-1-Cl is zero order in the concentration of azide ion [N_3^-]; but, there is a strong bimolecular substitution reaction of azide ion with (4-Me)-1-S(Me)$_2^+$.[13] This change in the kinetic order for the reaction of azide ion shows that the pentavalent transition state

for bimolecular nucleophilic substitution is stabilized relative to that for the competing stepwise reaction by "synergistic" interactions between the azide ion nucleophile and the dimethyl sulfide leaving group at (4-Me)-**1**-$S(Me)_2^+$.[25,26] Similarly, the change from a chloride to a bromide ion leaving group at X-**1**-Y (X = 4-F, 3-MeO, 3-Br) results in a small increase in the rate constant for bimolecular substitution of azide ion relative to that for the competing stepwise reaction, but the change from a chloride to a tosylate leaving group at $(4-NO_2)$-**1**-Y leads to the opposite change in the relative magnitude of these rate constants (Fig. 2.5).

3. The point of the change from a stepwise to a concerted mechanism for nucleophilic substitution at X-**1**-Cl may be detected as an upward break in the observed nucleophile selectivity k_{Nu}/k_s with decreasing stability of the putative intermediate X-**1**$^+$ (Fig. 2.2). Figure 2.5 shows that the position of this break and the change in mechanism shifts to more electron-withdrawing X as the reactivity of the nucleophile is decreased, from X = 4-F for

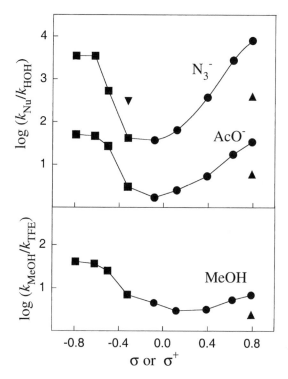

Figure 2.5. Nucleophile selectivities determined from product analysis for the reactions of ring-substituted 1-phenylethyl derivatives (X-**1**-Y) with azide ion, acetate ion and methanol in 50:50 (v/v) water/trifluoroethanol. The selectivities are plotted against the appropriate Hammett substituent constant σ^+ or σ. Leaving group Y: (■) ring-substituted benzoates; (●) chloride; (▼) dimethyl sulfide; (▲) tosylate.

$Nu = N_3^-$, to $X = 3$-Br for $Nu = AcO^-$, and $X = 4$-NO_2 for $Nu = MeOH$ (Scheme 2.3).[13] In all cases, the transition from stepwise to concerted bimolecular substitution occurs close to the point where the estimated rate constant for collapse to product of the triple ion intermediate containing the carbocation-nucleophile ion pair just reaches the vibrational limit (k_i, Scheme 2.3).[13] The breadth (range of X) of the region of minimum observed nucleophile selectivity for borderline nucleophilic substitution increases with decreasing nucleophile reactivity (Fig. 2.5). This "delay" in the appearance of concerted bimolecular substitution reflects the decreasing advantage to the coupling of bond formation and bond cleavage in the transition state for the concerted reaction with decreasing nucleophile reactivity.

2.3.2. Nucleophilic Substitution at X-2-Y.
The small azide ion product selectivity $(k_{az}/k_s)_{obsd} = 0.7\,M^{-1}$ observed for the reaction of azide ion and solvent with X-**2**-Y when X is electron-withdrawing (Fig. 2.2) is consistent with formation of an association complex between the substrate and azide ion ($K_{as} \approx 0.7\,M^{-1}$) that is converted to the product X-**2**-N_3 with little or no nucleophilic assistance. A change to strongly electron-withdrawing X results in the disappearance of the barrier to collapse of the triple ion intermediate to the azide ion adduct. However, a pentavalent transition state in the interior of Figure 2.3 fails to appear, even when the carbocation in the triple ion intermediate of the stepwise reaction no longer exists for the time of a bond vibration. Therefore, the barrier to nucleophilic substitution at X-**2**-Y by a coupled concerted reaction mechanism (k_{con}, Fig. 2.3) remains much higher than that for reaction through the inner upper right hand corner, even as the "corner" changes from a "well" for a reaction intermediate to a "flat plateau". This is because the transition state for the fully concerted reaction is destabilized by steric crowding ($C + D \gg 0$, Fig. 2.6).

These data and conclusions for the reactions of X-**2**-Y pose a problem in nomenclature. Stepwise reactions may be defined as those that proceed through an intermediate, and concerted reactions as those that proceed in a single stage that avoids formation of an intermediate. A simple division into just these two classes is possible, provided nucleophilic substitution reactions that proceed *without formation of a discrete intermediate* are defined as concerted, whether or not they exhibit bimolecular kinetics. We favor this inclusive definition of a concerted reaction, even though it leads to ambiguity about the concerted nature for some reactions. Consider the nucleophilic substitution of azide ion with the tertiary substrates X-**2**-Y in the case where the carbocation X-**2**$^+$ cannot exist in an energy well in the presence of azide ion for the time of a bond vibration. The coordinate for this reaction runs along the inner upper edge of Figure 2.3, but the energy well for the triple ion intermediate of the stepwise reaction is now replaced by a "flat plateau", so that there is first full cleavage of the bond to leaving group and *then* formation of the bond to nucleophile, but *no transition state stabilization from the coupling of these two processes* that is implied by the term *concerted*. Jencks proposed that such reactions be referred to as "uncoupled concerted", which indicates that there is no reaction

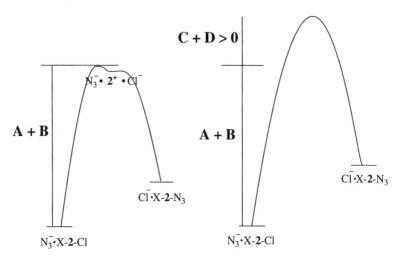

Figure 2.6. One-dimensional cross-sections of the 2D diagram in Figure 2.3 for the stepwise and concerted nucleophilic substitution reactions of azide ion with ring-substituted cumyl chlorides X-2-Cl, which take place in the inner part of the diagram. There is a large steric hindrance to formation of the transition state for the concerted reaction $[(C + D) \gg 0]$ and the barrier to this reaction (right-hand diagram) remains higher than that for the stepwise substitution (left-hand diagram), even when there is no significant free energy well for the reaction intermediate (B $\approx$ 0).

intermediate but that energetically favorable coupling of the bond cleavage and bond formation processes is not observed.[1]

The change from a stepwise preassociation mechanism through a triple ion intermediate to an uncoupled concerted reaction occurs as the triple ion becomes too unstable to exist in an energy well for the time of a bond vibration ($\sim 10^{-13}$ s). The borderline between these two reaction mechanisms is poorly marked, and there are no clear experimental protocols for its detection. These two reaction mechanisms cannot be distinguished by experiments designed to characterize their transition states, which lie at essentially the same position in the inner upper right hand corner of Figure 2.3. Only low yields of the nucleophilic substitution product are obtained from both stepwise preassociation and uncoupled concerted reactions, because K_{as} (M^{-1}) for formation of the preassociation complex in water is small and there is no significant nucleophilic assistance for either reaction. Degenerate reactions detected using isotopically labeled substrates, such as the exchange of ^{18}O between bridging and nonbridging positions during the solvolysis of alkyl sulfonates, are sometimes thought to provide evidence for formation of an ion pair intermediate in which the oxygens of the sulfonate leaving group become equivalent.[7,27] However, such exchange reactions may also proceed by an uncoupled concerted mechanism for which there is no energy well for the putative ion pair intermediate.[1,28] Picosecond and femtosecond kinetic methods have the potential to directly characterize carbocation intermediates that react faster than the reorganization of the surrounding solvation shell, but this has not yet been realized in practice.

There are several related examples of reactions for which there is no intermediate with a vibrational lifetime, but no significant energetic advantage to formation of the transition state for a coupled concerted mechanism that avoids formation of the intermediate. These include the stereomutation of cis and trans disubstituted cyclopropanes,[29,30] the thermal deazatization of 2,3-diazabicyclo[2.2.1]hept-2-ene,[31] and the thermal interconversion of bicyclo[3.2.0]hept-2-ene and bicyclo[2.2.1]hept-2-ene.[32] (See Chapter 21 by B. K. Carpenter and Chapter 22 by W. T. Borden.) These reactions proceed through unstable diradical species that lie at flat plateaus on energy landscapes in which substrate bonds undergo rotation to give a second diradical species that collapses to products. Each of these radicals must be fully formed in order for their interconversion through bond rotation to occur, but there is no lowering of the activation barrier to the reaction through the coupling of bond cleavage and bond rotation. These systems are discussed in Chapters 21 and 22 in this volume.

2.3.3. Nucleophilic Substitution at Benzyl Derivatives.

The sharp break from a stepwise to a concerted mechanism that is observed for nucleophilic substitution of azide ion at X-**1**-Y (Figs. 2.2 and 2.5) is blurred for nucleophilic substitution at the primary 4-methoxybenzyl derivatives (4-MeO,H)-**3**-Y. For example, the secondary substrate (4-MeO)-**1**-Cl reacts exclusively by a stepwise mechanism through the liberated carbocation intermediate (4-MeO)-**1**$^+$, which shows a moderately large selectivity toward azide ion ($k_{az}/k_s = 100\ M^{-1}$ in 50:50 (v/v) water/trifluoroethanol).[15] The removal of an α-Me group from (4-MeO)-**1**-Cl to give (4-MeO,H)-**3**-Cl increases the barrier to ionization of the substrate in the stepwise reaction relative to that for the concerted bimolecular substitution of azide ion. The result is that *both* of these mechanisms are observed *concurrently* for nucleophilic substitution of azide ion at (4-MeO,H)-**3**-Cl in water/acetone solvents.[33] These concurrent stepwise and concerted nucleophilic substitution reactions of azide ion with (4-MeO,H)-**3**-Cl show that there is no sharp borderline between mechanisms for substitution at primary benzylic carbon, but instead a region of "overlap" where both mechanisms are observed.

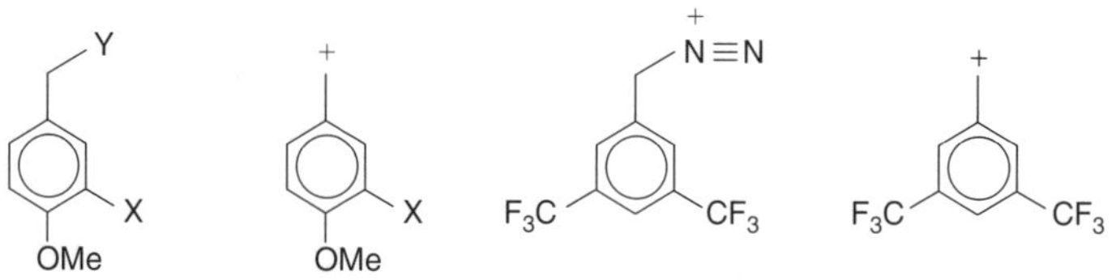

(4-MeO,X)-**3**-Y   (4-MeO,X)-**3**$^+$ 4-N$_2^+$ **4**$^+$

There is minimal steric hindrance in the transition state for coupled concerted bimolecular nucleophilic substitution at primary carbon [D $\approx$ 0, Fig. 2.4(II)] to offset the stabilization of this transition state that results from the coupling of bond breaking and bond formation [C < 0, Fig. 2.4(II)]. The result is significant net stabilization of this transition state from coupling, so that the concerted reaction is kinetically significant even when there is a substantial barrier to the addition of solvent (and possibly azide ion) to the carbocation intermediate of the stepwise reaction.

The transition states for the stepwise (k_1, Fig. 2.3) and concerted (k_{con}) reactions of (4-MeO,X)-**3**-Y lie at distinct well-separated positions on the More O'Ferrall diagram and show different sensitivities to changes in solvent polarity, meta substituents X at the aromatic ring, and the leaving group Y. For example, in 50:50 (v/v) water/trifluoroethanol (4-MeO,H)-**3**-Cl reacts with azide ion *exclusively* by a stepwise mechanism through the primary carbocation intermediate (4-MeO,H)-**3**$^+$ with a selectivity for reaction with azide ion and solvent of $k_{az}/k_s = 25$ M^{-1}. However, two-thirds of the azide ion substitution product obtained from the reaction of (4-MeO,H)-**3**-Cl in the less polar solvent 80:20 acetone/water forms by concerted bimolecular substitution and only one-third forms by trapping of the carbocation intermediate (4-MeO,H)-**3**$^+$ with a selectivity of $k_{az}/k_s = 8$ M^{-1}.[33] The preferred pathway for the reaction of azide ion with (4-MeO,X)-**3**-Y in 50:50 (v/v) water/trifluoroethanol changes from $D_N + A_N$ to A_ND_N as the meta aromatic substituent X is changed from H to NO_2.[34] Finally, concerted bimolecular substitution of azide ion at (4-MeO,H)-**3**-SMe_2^+ is favored by "synergistic" interactions between the nucleophile and the leaving group,[25,26] and this reaction is important even when the transition state for the stepwise reaction is stabilized by the strongly ionizing solvent water.[35–37]

Destabilization of primary benzylic carbocations by electron-withdrawing aromatic ring substituents must eventually result in the disappearance of the energy well for this intermediate, in which case concerted nucleophilic substitution will be "enforced" in the sense that no stepwise reaction is possible. These concerted bimolecular nucleophilic substitution reactions are normally strongly coupled.[38] However, the nucleophile selectivity calculated from the low yields of the azide ion substitution product obtained from the reaction of **4**-N_2^+ in the presence of 2 M azide ion in water that contains 4% 2-propanol, $k_{az}/k_s = 0.21$ M^{-1}, shows that there is little or no stabilization of the transition state for the concerted reaction from the coupling of bond formation and bond cleavage at this substrate,[23] even though the 3,5-(bis)trifluoromethylbenzyl carbocation **4**$^+$ cannot have a significant lifetime in the presence of azide ion. Therefore, despite the minimal steric hindrance to substitution at the primary benzylic carbon and the instability of **4**$^+$, both of which strongly favor the coupling of bond formation and bond cleavage and reaction of **4**-N_2^+ by a concerted mechanism, no such coupling is observed. Here the bond to the weakly basic nitrogen leaving group is so labile[39] that its cleavage occurs without any assistance from added nucleophiles. This reaction is another example of an uncoupled concerted mechanism that proceeds via the inner upper right hand corner of the reaction coordinate diagram in Figure 2.3.

3. ALIPHATIC NUCLEOPHILIC SUBSTITUTION AT TERTIARY CARBON

3.1. Reaction Mechanism

The essential features of the mechanism for aliphatic nucleophilic substitution at tertiary carbon were established in studies by Hughes and Ingold.[40] However, as chemists probed more deeply, the problems associated with the characterization of borderline reaction mechanisms were encountered, and controversy remains to this day about whether these problems have been entirely solved.[41] What is generally accepted is that *tert*-butyl derivatives undergo borderline solvolysis reactions through a *tert*-butyl carbocation intermediate that is too unstable to diffuse freely through nucleophilic solvents such as methanol and water. The borderline nature of substitution reactions at tertiary carbon is exemplified by the following observations.

1. The rate constant for solvolysis of the model tertiary substrate **5-Cl** is independent of the concentration of added azide ion, and the reaction gives only a low yield of the azide ion adduct (e.g., 16% in the presence of 0.50 *M* NaN$_3$ in 50:50 (v/v) water/trifluoroethanol].[42] Therefore, this is a borderline reaction for which it is not possible to determine the kinetic order with respect to azide ion, because of uncertainties about specific salt effects on the reaction.[42]

2. There is no detectable common bromide ion inhibition of the reaction of *tert*-butyl bromide in 90% acetone in water,[40,43] or common chloride ion inhibition of the reaction of **5-Cl** in 50:50 (v/v) water/trifluoroethanol or methanol/water.[42] There is also little exchange of ^{36}Cl from Na^{36}Cl into unreacted *tert*-butyl chloride during its reaction in methanol.[44]

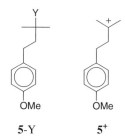

5-Y   5$^+$

3. There is little scrambling of ^{18}O between the bridging and nonbridging positions during the reaction of ^{18}O-labeled *tert*-butyl *p*-nitrobenzoate in 80% acetone in water.[45]

4. Stepwise solvolysis of chiral substrates through a planar achiral carbocation reaction intermediate normally results in the formation of racemic products. However, the solvolysis of chiral tertiary derivatives **6-Y** proceeds with either

partial inversion or retention of configuration, depending upon the solvent and the leaving group.[46]

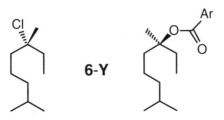

6-Y

Ar = 4-nitrobenzoate, phthalate

These results show that there is no detectable nucleophilic assistance to substitution at tertiary carbon by the very good nucleophile azide ion. In addition, the minimal trapping of simple tertiary carbocations by nucleophilic anions, the dramatic effects of the leaving group anion on the reaction stereochemistry, and the observation of limited ^{18}O-scrambling during the solvolysis of labeled *tert*-butyl *p*-nitrobenzoate[45] require that the addition of solvent to the carbocation occurs before separation and/or reorganization of the first-formed intimate ion pair intermediate. These data show that the formation of nucleophilic substitution products proceeds by moving from the weak preassociation complex in the inner upper left hand corner of Figure 2.3, through the inner upper right hand corner, and on to product.[16,42]

An important question is whether nucleophilic substitution at tertiary carbon proceeds though a carbocation intermediate that shows a significant chemical barrier to the addition of solvent and other nucleophiles. The yield of the azide ion substitution product from the reaction of **5-Cl** is similar to that observed for the reactions of X-**2**-Y when this product forms exclusively by conversion of the preassociation complex to product.[16] Therefore the carbocation **5**$^+$ is too unstable to escape from an aqueous solvation shell and undergo diffusion-controlled trapping by azide ion. This result sets a lower limit of $k'_s \approx k_s \geq k_{-d} \approx 1.6 \times 10^{10}$ s^{-1} (Scheme 2.4)[14] for addition of solvent to the ion pair intermediate **5**$^+ \bullet$Cl$^-$.

A more precise value of k_s (s^{-1}) for addition of solvent to **5**$^+$ was estimated as follows:[42]

1. Values of k_s (s^{-1}) for reaction of the more stable tertiary carbocations X-**2**$^+$ with a solvent of 50:50 (v/v) water/trifluoroethanol, determined using the azide ion clock,[16] were used to establish a good linear logarithmic relationship between k_s (s^{-1}) for reaction of X-**2**$^+$ and the first-order rate constants k_{obsd} (s^{-1}) for their formation as intermediates in the stepwise reaction of X-**2**-Cl in the same solvent (Eq. 2 and Scheme 2.5).[42]

$$\log k_s = -0.53 \log k_{obsd} + 10.6 \tag{2}$$

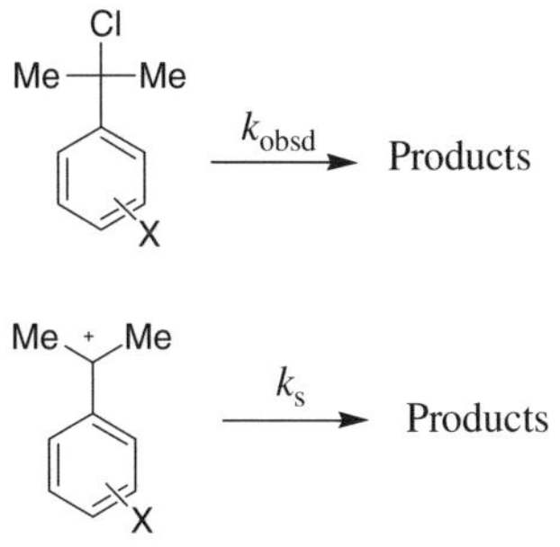

Scheme 2.5

2. This linear correlation was then assumed to hold for the formation and re-action of aliphatic tertiary carbocations, and values of $k_s = 3.5$ and 1.6×10^{12} s^{-1} for addition of 50:50 (v/v) water/trifluoroethanol to $\mathbf{5}^+$ and the *tert*-butyl carbocation, respectively, were estimated from the values of k_{obsd} for reaction of the corresponding tertiary aliphatic chlorides using Eq. 2.[42]

The value of $k_s \approx 10^{12}$ s^{-1} estimated for capture of the *tert*-butyl carbocation and $\mathbf{5}^+$ by water/trifluoroethanol is larger than both the rate constant for their diffusional encounter with external nucleophilic reagents and that for reorganization of the carbocation solvation shell by the dielectric relaxation of solvent ($k_{reorg} \approx 10^{11}$ s^{-1}).[47,48] This result suggests that the ion pair or ion molecule intermediates of the solvolysis of precursors to these carbocations undergo direct reaction with a molecule of solvent that is present within the solvation shell at the time of carbocation formation, with $k'_s \approx k_s = k_{reorg} \approx 10^{11}$ s^{-1}.[42]

The fleeting lifetimes of tertiary carbocations, and the difficulties in describing the energy profiles for their formation and reaction, are highlighted by the results of attempts to determine rate constant ratios for the partitioning of $\mathbf{5}^+$ and the *tert*-butyl carbocation between capture by nucleophilic solvents and deprotonation that are independent of the pathway for formation of the carbocation.[42,49] A telling result is the observation of different yields of the trisubstituted alkene **8** and the nucleophilic substitution products **5-OTFE** and **5-OMe** from the acid-catalyzed reactions of **5-OH** and the disubstituted alkene **7** in 50:45:5 (v/v/v) water/trifluoroethanol/methanol (Scheme 2.6).[42] The data require either that the carbocation intermediates of these two acid-catalyzed reactions are too short lived to come to equilibrium with respect to their respective solvation shells, or that the formation of these intermediates is avoided in concerted reactions in which there is minimal coupling of cleavage of the bond to leaving group and formation of the bond to nucleophile/and or proton loss.[42]

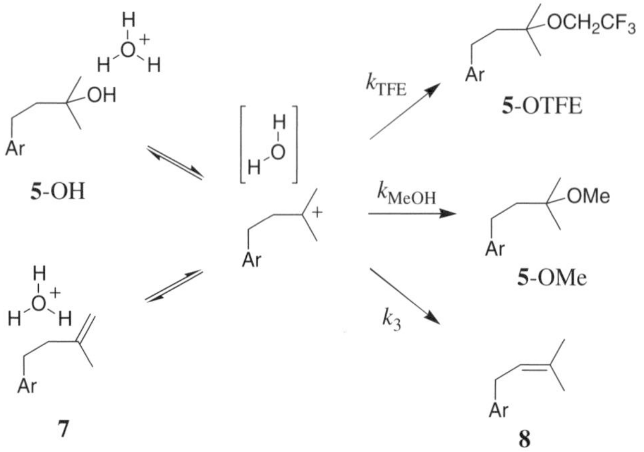

Scheme 2.6

3.2. Nucleophilic Solvent Participation and Nucleophilic Solvation

There is an ongoing controversy about whether there is any stabilization of the transition state for nucleophilic substitution at tertiary aliphatic carbon from interaction with nucleophilic solvent.[41,50] This controversy has developed with the increasing sophistication of experiments to characterize solvent effects on the rate constants for solvolysis reactions. Grunwald and Winstein[51] determined rate constants for solvolysis of *tert*-butyl chloride in a wide variety of solvents and used these data to define the solvent ionizing parameter Y (Eq. 3). They next found that rate constants for solvolysis of primary and secondary aliphatic carbon show a smaller sensitivity (m) to changes in Y than those for the parent solvolysis reaction of *tert*-butyl chloride (for which $m = 1$ by definition).[52] A second term was added (ℓN) to account for the effect of changes in solvent nucleophilicity on k_{obsd} that result from transition state stabilization by a nucleophilic interaction between solvent and substrate.[52,53] It was first assumed that there is no significant stabilization of the transition state for solvolysis of *tert*-butyl chloride from such a nucleophilic interaction. However, a close examination of extensive rate data revealed, in some cases, a correlation between rate constants for solvolysis of *tert*-butyl derivatives and solvent nucleophicity.[54–56]

$$Y = \log(k_{obsd}/k_0) \quad (Y = 0 \text{ for reaction in 80\% ethanol in water}) \quad (3)$$

Many other solvent parameters have been defined in an attempt to model as thoroughly as possible solvent effects on the rate constants for solvolysis. These include: (a) Several scales of solvent ionizing power Y_X developed for different substrates R—X that are thought to undergo limiting stepwise solvolysis.[57] (b) Several different scales of solvent nucleophilicity developed for substrates of different charge type that undergo concerted bimolecular substitution by solvent.[53] (c) An *aromatic ring parameter (I)*, which was added to account for differences in the interactions of solvent with aromatic rings in the ground and transition states for solvolysis of ring-substituted benzyl and benzhydryl derivatives.[58]

The development of these various solvent parameters and scales has been accompanied by the realization that there are uncertainties in the physical property of the solvent that is correlated by a particular parameter in cases where systematic changes in solvent structure affect several solvent properties. Consider a reaction that shows no rate dependence on the basicity of hydroxylic solvents, and a second reaction that proceeds through a transition state in which there is a small transition state stabilization from a nucleophilic interaction with the hydroxyl group. The rate constants for the latter reaction will increase more sharply with changing solvent nucleophilicity than those for the former, and they should show a correlation with some solvent nucleophilicity parameter. This trend was observed in a comparison of the effects of solvent on the rate constants for solvolysis of 1-adamantyl and *tert*-butyl halides, and is consistent with a greater stabilization of the transition state for reaction of the latter by interaction with nucleophilic solvents.[59,60]

The problems with this interpretation are (a) There is a strong correlation between the nucleophilicity of hydroxylic solvents and the acidity of their hydroxylic proton; and, (b) The transition state for solvolysis is stabilized by an electrophilic (hydrogen-bonding) interaction between this acidic proton and the leaving group anion. Therefore, a greater electrophilic stabilization of the transition state for solvolysis of a 1-adamantyl halide than for the corresponding *tert*-butyl halide would result in a relative *decrease* in the rate constant for reaction of the former with decreasing solvent acidity (*increasing* nucleophilicity), which is not easily distinguished from a relative *increase* in the rate constant for solvolysis of the *tert*-butyl halide due to a nucleophilic interaction.[41,61]

Results such as these have led to recurring discussions about the extent of stabilization of the transition state for heterolytic cleavage at tertiary carbon by "nucleophilic assistance" from solvent. The difficulty lies in reconciling studies that suggest that there is a small dependence of k_{obsd} (s^{-1}) for solvolysis of *tert*-butyl chloride,[59,60] and some cumyl derivatives,[62,63] on solvent nucleophilicity with other work that shows there is no detectable stabilization of the transition state for these reactions by interaction with the strongly nucleophilic azide and hydroxide ions, and the strong neutral nucleophile propanethiol.[42]

Experiments to estimate directly the difference in the free energy of solvation of the transition states for solvolysis of tertiary derivatives in the nucleophilic solvent ethanol and the non-nucleophilic solvent hexafluoroisopropanol suggest that any controversy about the role of solvent in transition state stabilization may be resolved by careful use of the term "solvation". There is good linear correlation, with slope < 1, between the *difference* in activation barriers for solvolysis of 1-adamantyl chloride and a series of caged and bridgehead tertiary alkyl chlorides RCl [$\Delta\Delta G^{\ddagger} = (\Delta G^{\ddagger})_{AdCl} - (\Delta G^{\ddagger})_{RCl}$] and the *relative* Gibbs free energy change for transfer of chloride ion between RCl and the 1-adamantyl carbocation in the gas phase (ΔG°, Scheme 2.7). This correlation is observed for solvolysis in both the weakly nucleophilic solvent 97% HFIP/H$_2$O (HFIP = hexafluoroisopropanol) and

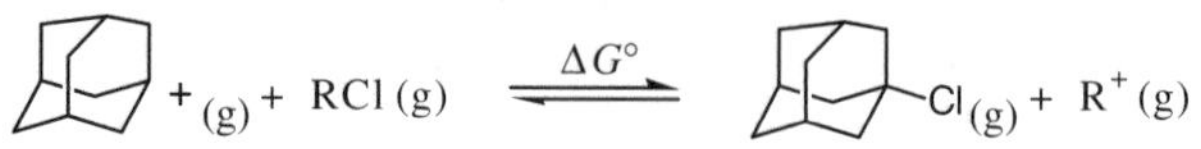

Scheme 2.7

the strongly nucleophilic solvent 80% EtOH/H_2O.[64] However, the values of $\Delta\Delta G^{\ddagger}$ for the acyclic *tert*-butyl chloride showed significant positive deviations from the linear correlation that was established using the data for caged and bridgehead substrates, for which only "frontside solvation" of the carbocation is important, because "backside solvation" is restricted by the hydrocarbon framework. This positive deviation was larger for solvolysis of *tert*-butyl chloride in 80% EtOH/ H_2O (7.4 kcal/mol) than for its solvolysis in 97% HFIP/H_2O (3.1 kcal/mol). The difference in positive deviations shows that there is a larger stabilization of the transition state for solvolysis of *tert*-butyl chloride by backside solvation by the strongly nucleophilic 80% EtOH/H_2O compared with the weakly nucleophilic 97% HFIP/H_2O.

These data show that backside interactions between solvent and caged and bridgehead carbocations are minimized by the steric "shielding" by the hydrocarbon framework of these tertiary carbocations. This may be thought of as steric hindrance to the *nucleophilic solvation* that arises from charge–dipole interactions between solvent and the cationic center (Scheme 2.8A). The 4.3 kcal/mol greater stabilization of the transition state for solvolysis of *tert*-butyl chloride by 80% EtOH/H_2O than by 97% HFIP/H_2O may reflect the reduced steric bulk of the former solvent, which should allow for interaction of a larger number of solvent molecules with the developing cationic center in the transition state. By comparison, the total solvation energy of carbocations in water is $\sim$50 kcal/mol.[65]

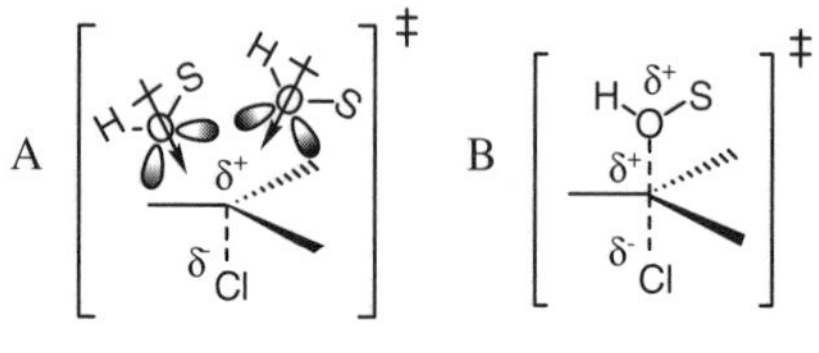

Scheme 2.8

Similar changes in *nucleophilic* (or *dipole*) *solvation* (Scheme 2.8A) provide a simple explanation for the observation of other correlations between rate constants for solvolysis and solvent nucleophilicity. This interpretation does not require that there be stabilization of the transition state for solvolysis of tertiary derivatives by a partial covalent interaction between nucleophile and electrophile.[50] We have defined this latter interaction as *nucleophilic solvent participation* (Scheme 2.8B) and have argued that the results of simple and direct experiments to detect stabilization of the transition state for reaction of simple tertiary derivatives by good nucleophiles such as propanethiol and azide ion effectively exclude this possibility.[50]

In summary, controversy concerning the mechanism for solvolysis at tertiary carbon is semantic and can be avoided by making a clear distinction between (a) *nucleophilic solvation*, which is stabilization of the transition state for *stepwise* solvolysis through carbocation or ion pair intermediates by charge–dipole interactions with nucleophilic solvents (Scheme 2.8A); and, (b) *nucleophilic solvent participation*, which is stabilization of the transition state for *concerted* solvolysis by formation of a partial covalent bond to the solvent nucleophile (Scheme 2.8B).

4. CONCLUSION AND OUTLOOK

Generally, only a single stepwise or concerted pathway for aliphatic nucleophilic substitution is detected by experiment because of the very different activation barriers for formation of the respective reaction transition states for these reactions. The description of the borderline between stepwise and concerted nucleophilic substitution reactions presented in this chapter has been obtained through a search for those rare substrates that show comparable barriers to these two reactions; and through the characterization of the barrier for nucleophile addition to the putative carbocation intermediate of the stepwise reaction in the region of this change in mechanism.

No break to a fully concerted reaction mechanism is observed for nucleophilic substitution of azide ion at the tertiary benzylic substrates X-**2**-Y, even as the energy well for the tertiary carbocation is transformed into a "flat plateau." Removal of a single α-methyl group from X-**2**-Y to give X-**1**-Y reduces the barrier to concerted bimolecular substitution of azide ion relative to that for the stepwise solvolysis, and similar barriers are observed for these reactions of the secondary substrates X-**1**-Cl at the point where the energy well for the intermediate is transformed into a "flat plateau." There is minimal steric hindrance to concerted bimolecular nucleophilic substitution at the primary benzylic substrates (4-MeO,X)-**3**-Y, and concerted bimolecular substitution by azide ion is observed even when there is a substantial barrier for addition of solvent to the carbocation intermediate of the stepwise reaction. The simple conclusion from these results is that the advantage to the coupling of bond formation and bond cleavage in concerted nucleophilic substitution at benzylic carbon is relatively small, but that the advantage changes significantly and systematically with changing structure of the nucleophile, electrophilic carbon, and leaving group.

The description of the borderline between stepwise and "concerted" nucleophilic substitution remains murky in cases where there is no significant stabilization of the transition state for the concerted reaction through the coupling of bond cleavage and formation. The reason is that there are no simple experimental protocols to detect the point at which the energy well for the carbocation intermediate of the stepwise reaction in the upper right hand corner of Figure 2.3 is transformed into a "flat plateau." This question might be addressed experimentally, through rapidly improving fast kinetic methods to characterize reactive intermediates.

Much additional experimental work might be done to further characterize the borderline between stepwise and concerted nucleophilic substitution, but in our opinion the results of such work would not affect the most important features of the description of the borderline region presented in this chapter. An important limitation of kinetic experiments is that they provide information only about saddle points on energy surfaces such as that represented by Figure 2.3, and about the curvature of the energy surface in the region of these saddle points.[66] A more detailed description of these energy surfaces will become available through computational studies once it has been demonstrated, in cases where a comparison is possible, that there is good agreement between experiment and calculation.

ACKNOWLEDGMENT

We thank the National Institutes of Health (GM 39754) for support of the work from our laboratory described in this chapter.

SUGGESTED READING

A. Williams, *Concerted Organic and Bioorganic Mechanisms*, CRC Press, New York, **2000**.

W. P. Jencks, "How Does a Reaction Choose Its Mechanism?," *Chem. Soc. Rev.* **1981**, *10*, 345.

J. P. Richard, "Simple Relationships between Carbocation Lifetime and the Mechanism for Nucleophilic Substitution at Saturated Carbon," *Adv. Carbocation Chem.* **1989**, *1*, 122.

T. W. Bentley and G. Llewellyn, "Y_x Scales of Solvent Ionizing Power," *Prog. Phys. Org. Chem.* **1990**, *17*, 121.

M. J. S. Dewar, "Multibond Reactions Cannot Normally be Synchronous," *J. Am. Chem. Soc.* **1984**, *106*, 209.

J. P. Richard, M. E. Rothenberg, and W. P. Jencks, "Formation and Stability of Ring-Substituted 1-Phenylethyl Carbocations," *J. Am. Chem. Soc.* **1984**, *106*, 1361.

T. L. Amyes and J. P. Richard, "Concurrent Stepwise and Concerted Substitution Reactions of 4-Methoxybenzyl Derivatives and the Lifetime of the 4-Methoxybenzyl Carbocation," *J. Am. Chem. Soc.* **1990**, *112*, 9507.

J. P. Richard, T. L. Amyes, and T. Vontor, "Absence of Nucleophilic Assistance by Solvent and Azide Ion to the Reaction of Cumyl Derivatives: Mechanism of Nucleophilic Substitution at Tertiary Carbon," *J. Am. Chem. Soc.* **1991**, *113*, 5871.

M. M. Toteva and J. P. Richard, "Mechanism for Nucleophilic Substitution and Elimination Reactions at Tertiary Carbon in Largely Aqueous Solutions: Lifetime of a Simple Tertiary Carbocation," *J. Am. Chem. Soc.* **1996**, *118*, 11434.

J. P. Richard, M. M. Toteva, and T. L. Amyes, "What is the Stabilizing Interaction with Nucleophilic Solvents in the Transition State for Solvolysis of Tertiary Derivatives: Nucleophilic Solvent Participation or Nucleophilic Solvation?" *Org. Lett.* **2001**, *3*, 2225.

REFERENCES

1. W. P. Jencks, *Chem. Soc. Rev.* **1981**, *10*, 345.
2. C. K. Ingold, *Structure and Mechanism in Organic Chemistry*; 2nd ed., Cornell University Press, Ithaca, NY, **1969**.
3. J. R. Keeffe and W. P. Jencks, *J. Am. Chem. Soc.* **1983**, *105*, 265.
4. T. Okuyama and G. Lodder, *Adv. Phys. Org. Chem.* **2002**, *37*, 1.
5. W. P. Jencks, *Acc. Chem. Res.* **1976**, *9*, 425.
6. R. D. Guthrie and W. P. Jencks, *Acc. Chem. Res.* **1989**, *22*, 343.
7. J. M. Harris, *Prog. Phys. Org. Chem.* **1974**, *11*, 89.
8. M. L. Sinnott and W. P. Jencks, *J. Am. Chem. Soc.* **1980**, *102*, 2026.
9. E. D. Hughes, C. K. Ingold, R. J. L. Martin, and D. F. Meigh, *Nature (London)* **1950**, *166*, 679.
10. W. v. E. Doering and H. H. Zeiss, *J. Am. Chem. Soc.* **1953**, *75*, 4733.
11. J. P. Richard and W. P. Jencks, *J. Am. Chem. Soc.* **1982**, *104*, 4691.
12. J. P. Richard and W. P. Jencks, *J. Am. Chem. Soc.* **1982**, *104*, 4689.
13. J. P. Richard and W. P. Jencks, *J. Am. Chem. Soc.* **1984**, *106*, 1383.
14. J. P. Richard and W. P. Jencks, *J. Am. Chem. Soc.* **1984**, *106*, 1373.
15. J. P. Richard, M. E. Rothenberg, and W. P. Jencks, *J. Am. Chem. Soc.* **1984**, *106*, 1361.
16. J. P. Richard, T. L. Amyes, and T. Vontor, *J. Am. Chem. Soc.* **1991**, *113*, 5871.
17. R. A. McClelland, F. L. Cozens, S. Steenken, T. L. Amyes, and J. P. Richard, *J. Chem. Soc., Perkin Trans. 2* **1993**, 1717.
18. R. Ta-Shma and Z. Rappoport, *J. Am. Chem. Soc.* **1983**, *105*, 6082.
19. C. D. Ritchie, *Acc. Chem. Res.* **1972**, *5*, 348.
20. T. L. Amyes and W. P. Jencks, *J. Am. Chem. Soc.* **1989**, *111*, 7900.
21. C. D. Ritchie and T. C. Hofelich, *J. Am. Chem. Soc.* **1980**, *102*, 7039.
22. J. P. Richard, *J. Org. Chem.* **1992**, *57*, 625.
23. J. I. Finnemann and J. C. Fishbein, *J. Am. Chem. Soc.* **1995**, *117*, 4228.
24. M. J. S. Dewar, *J. Am. Chem. Soc.* **1984**, *106*, 209.
25. R. G. Pearson and J. Songstad, *J. Org. Chem.* **1967**, *89*, 2899.
26. R. G. Pearson and J. Songstad, *J. Am. Chem. Soc.* **1967**, *89*, 1827.
27. C. Paradisi and J. F. Bunnett, *J. Am. Chem. Soc.* **1985**, *107*, 8223.
28. P. E. Dietze and M. Wojciechowski, *J. Am. Chem. Soc.* **1990**, *112*, 5240.
29. C. Doubleday, Jr., K. Bolton, and W. L. Hase, *J. Am. Chem. Soc.* **1997**, *119*, 5251.
30. D. A. Hrovat, S. Fang, W. T. Borden, and B. K. Carpenter, *J. Am. Chem. Soc.* **1997**, *119*, 5253.
31. M. B. Reyes and B. K. Carpenter, *J. Am. Chem. Soc.* **2000**, *122*, 10163.
32. B. K. Carpenter, *J. Am. Chem. Soc.* **1996**, *118*, 10329.
33. T. L. Amyes and J. P. Richard, *J. Am. Chem. Soc.* **1990**, *112*, 9507.
34. P. E. Yeary, Ph.D. Thesis, University of Kentucky, Lexington, KY, 1993.
35. N. Buckley and N. J. Oppenheimer, *J. Org. Chem.* **1997**, *62*, 540.
36. N. Buckley and N. J. Oppenheimer, *J. Org. Chem.* **1994**, *59*, 5717.

37. D. N. Kevill, N. H. J. Ismail, and M. J. D'Souza, *J. Org. Chem.* **1994**, *59*, 6303.

38. P. R. Young and W. P. Jencks, *J. Am. Chem. Soc.* **1979**, *101*, 3288.

39. R. Glaser, G. S.-C. Choy, and M. K. Hall, *J. Am. Chem. Soc.* **1991**, *113*, 1109.

40. L. C. Bateman, M. G. Church, E. D. Hughes, C. K. Ingold, and N. A. Taher, *J. Chem. Soc.* **1940**, 979.

41. J. J. Gajewski, *J. Am. Chem. Soc.* **2001**, *123*, 10877.

42. M. M. Toteva and J. P. Richard, *J. Am. Chem. Soc.* **1996**, *118*, 11434.

43. L. C. Bateman, E. D. Hughes, and C. K. Ingold, *J. Chem. Soc.* **1940**, 960.

44. C. A. Bunton and B. Nayak, *J. Chem. Soc.* **1959**, 3854.

45. H. L. Goering and J. F. Levy, *J. Am. Chem. Soc.* **1962**, *84*, 3853.

46. P. Muller and J.-C. Rossier, *J. Chem. Soc., Perkin Trans. 2* **2000**, 2232.

47. U. Kaatze, *J. Chem. Eng. Data* **1989**, *34*, 371.

48. U. Kaatze, R. Pottel, and A. Schumacher, *J. Phys. Chem.* **1992**, *96*, 6017.

49. M. Cocivera and S. Winstein, *J. Am. Chem. Soc.* **1963**, *85*, 1702.

50. J. P. Richard, M. M. Toteva, and T. L. Amyes, *Org. Lett.* **2001**, *3*, 2225.

51. E. Grunwald and S. Winstein, *J. Am. Chem. Soc.* **1948**, *70*, 846.

52. S. Winstein, E. Grunwald, and H. W. Jones, *J. Am. Chem. Soc.* **1951**, *73*, 2700.

53. D. N. Kevill, *Advances in Quantitative Structure–Property Relationships* **1996**, *1*, 81.

54. T. W. Bentley, C. T. Bowen, W. Parker, and C. I. F. Watt, *J. Am. Chem. Soc.* **1979**, *101*, 2486.

55. T. W. Bentley and G. E. Carter, *J. Am. Chem. Soc.* **1982**, *104*, 5741.

56. D. J. Raber, W. C. Neal, M. D. Dukes, J. M. Harris, and D. L. Mount, *J. Am. Chem. Soc.* **1978**, *100*, 8137.

57. T. W. Bentley and G. Llewellyn, *Prog. Phys. Org. Chem.* **1990**, *17*, 121.

58. D. N. Kevill, S. W. Anderson, and N. H. J. Ismail, *J. Org. Chem.* **1996**, *61*, 7256.

59. T. W. Bentley and G. E. Carter, *J. Am. Chem. Soc.* **1982**, *104*, 5741.

60. D. N. Kevill and S. W. Anderson, *J. Am. Chem. Soc.* **1986**, *108*, 1579.

61. J. M. Harris, S. P. McManus, M. R. Sedaghat-Herati, N. Neamati-Mazraeh, M. J. Kamlet, R. M. Doherty, R. W. Taft, and M. H. Abraham, in *Nucleophilicity (Advances in Chemistry Series)*, Vol. 215, J. M. Harris and S. P. McManus, Eds., American Chemical Society, Washington, DC, **1987**, p. 247*f*.

62. K.-T. Liu, P. S. Chen, C. R. Hu, and H. C. Sheu, *J. Phys. Org. Chem.* **1993**, *6*, 122.

63. K.-T. Liu, L.-W. Chang, and P.-S. Chen, *J. Org. Chem.* **1992**, *57*, 4791.

64. K. Takeuchi, M. Takasuka, E. Shiba, T. Kinoshita, T. Okazaki, J.-L. M. Abboud, R. Notario, and O. Castano, *J. Am. Chem. Soc.* **2000**, *122*, 7351.

65. M. H. Abraham, *J. Chem. Soc., Perkin Trans. 2* **1973**, 1893.

66. W. P. Jencks, *Bull. Soc. Chim. Fr.* **1988**, 218.

CHAPTER 3

Carbanions

SCOTT GRONERT

Department of Chemistry and Biochemistry, San Francisco State University,
San Francisco, CA

Reactive Intermediate Chemistry, edited by Robert A. Moss, Matthew S. Platz, and Maitland Jones, Jr.
ISBN 0-471-23324-2   Copyright © 2004 John Wiley & Sons, Inc.

69

1. INTRODUCTION

Carbanions provide a rich chemistry and have been the subject of intense study over the past 100 years, partly because of their central role in important synthetic reaction schemes. Given the low electronegativity of carbon, it is not surprising that carbon-centered anions can be highly reactive and appear as transient intermediates in reaction mechanisms; however, when certain conditions are met, they can be prepared and studied as stable species. For example, the addition of multiple electron-withdrawing substituents can lead to a carbanion that is stable in aqueous solution [e.g., $^-C(CN)_3$] or bonding with a cation can render a carbanion that is stable in a weakly acidic solvent such as diethyl ether (e.g., MeMgBr). As both of these examples suggest, the stability of a carbanion is intimately associated with its basicity, which in turn, is generally measured by the acidity (i.e., pK_a) of the parent carbon acid.

This chapter will begin with a brief overview of the development of carbanion chemistry followed by a section devoted to the structure and stability of carbanions. Methods of measuring carbon acidity and systematic trends in carbanion stability will be key elements in this chapter. Next, processes in which carbanions appear as transient, reactive intermediates will be presented and typical carbanion mechanisms will be outlined. Finally, some new developments in the field will be described. Although the synthetic utility of carbanions will be alluded to many times in this chapter, specific uses of carbanion-like reagents in synthesis will not be explored. This topic is exceptionally broad and well beyond the scope of this chapter.

2. DEVELOPMENT OF CARBANION CHEMISTRY

The origins of carbanion chemistry are deeply rooted in synthesis given the great utility of these species in forming new carbon–carbon bonds. Carbanions are highly nucleophilic and rarely undergo rearrangement reactions so they are very attractive reactive intermediates from a synthetic point of view. As a result, metal salts of carbanions have been widely used in synthesis for >100 years.[1–8] Certainly the best known example is the Grignard reagent[9,10] where bonding of magnesium to a carbanion center leads to a species that is relatively easy to prepare and handle. However, the initial evidence of organomagnesium compounds actually predates Grignard and goes back to the middle of the nineteenth century.[11,12] Organolithium compounds were investigated later[13] and are more difficult to handle because the

C–Li bond has greater ionic character, yielding a more carbanion-like salt with higher reactivity.[14] On the other hand, carbanions derived from highly acidic species such as malonate esters are readily formed under mild conditions and have long been used in the generation of new carbon–carbon bonds.[15–17]

The modern study of carbanions as reactive intermediates and the investigation of their physical properties begins with early efforts to gauge their basicity. In their classic paper, "The Study of Extremely Weak Acids," Conant and Wheland[18] qualitatively explored the basicity of a range of stabilized carbanions such as poly-arylmethides by allowing their sodium or potassium salts to react with a carbon acid and using color changes to indicate whether a proton transfer had occurred or not. Although crude in design relative to later studies, the results, which were later elaborated by McEwen,[19] are in surprisingly good accord with modern values when the same reference points are used and the comparison is limited to moderately acidic hydrocarbons such as cyclopentadiene derivatives and polyarylmethanes. For the most basic carbanions such as those derived from simple alkanes, equilibrium methods proved to be unsuitable because of the slow kinetics of proton transfer. One alternative was to explore related equilibria such as transmetalation reactions of organolithium salts with alkyl iodides.[20,21] At about the same time, kinetic methods were developed to gauge the basicity of these carbanions, and took advantage of the Brønsted relationship between proton-transfer rate and ΔpK_a.[22] Pioneering work by Shatenshtein[23] and Streitwieser and co-workers[24–36] using isotope-exchange rates in ammonia and cyclohexylamine, respectively, provided the first quantitative measures of basicity in carbanions derived from saturated hydrocarbons. These studies also lead to a much more detailed understanding of the proton-transfer process in solution. Cram[37] brought these data together in his seminal book on carbanion chemistry and referred to it as the MSAD scale, named for the four important contributors: McEwen, Streitwieser, Applequist, and Dessy. Over time, both the equilibrium and kinetic methods evolved and yielded very accurate measures of carbanion stability in a range of solvents including dimethyl sulfoxide (DMSO),[38,39] cyclohexylamine,[40–47] and ethers such as tetrahydrofuran (THF)[48–51] and 1,2-dimethoxyethane (DME).[52,53] These studies highlighted the variety of ion-pairing motifs that are available to carbanion salts and provided important insights into the factors that influence condensed-phase ion-pairing behavior. Most recently, carbanions have been studied in the gas phase via mass spectrometry (MS). This work has lead to accurate data for a wide range of bare carbanions[54] and allowed for an analysis of the fundamental factors that affect carbanion stability in the absence of solvation and ion-pairing effects.

3. STRUCTURE OF CARBANIONS

3.1. Geometries

For an sp^3 hybridized carbanion, one expects that it will prefer a structure where the central carbon adopts a trigonal-pyramidal geometry with a lone-pair occupying one of the typical tetrahedral valencies. This result is in direct analogy to a nitrogen-containing species like NH_3, a compound that is isoelectronic with CH_3^-. This assumption has been confirmed experimentally for the methyl anion by photoelectron spectroscopy (PES) measurements that suggest a barrier to inversion and therefore a pyramidal structure.[55] Although it is experimentally difficult to explore the structural details of bare carbanions, a great deal has been discovered through computational work.[56] High-level work on the methyl anion indicates a pyramidal structure with H–C–H angles (109.4°) very much like those found in CH_4 (i.e., a tetrahedral arrangement).[57] The C–H bond lengths are stretched to 1.101 Å in CH_3^- (cf. 1.077 Å for the methyl radical) most likely as a mechanism for dispersing the charge. The calculated barrier to inversion is 2.1 kcal/mol, which is about 4 kcal/mol smaller than that seen in NH_3.[58] The smaller barrier to inversion in the methyl anion can be attributed to the more diffuse nature of the carbon lone pair that gains less stabilization from introduction of s character into the orbital. In other words, the orbital contraction that occurs in going from a pure p (planar carbanion) to an sp^3 hybridized lone pair (pyramidal) has a smaller energetic effect because the lone-pair electrons are more dispersed in the carbanion. The low electronegativity of carbon also is evident in the electron binding energy of CH_3^-, a mere 1.8 kcal/mol.[55] In fact, computational work indicates that other simple alkyl anions like ethyl and isopropyl are unbound (i.e., negative electron binding energy) and would spontaneously eject an electron if they were formed in the gas phase.[59] Of course, these species can be stable in the condensed phase because bonding to metals stabilizes the carbanion and effectively increases the electron binding energy.

Analogously, carbanions located at sp^2 hybridized centers are expected to have bent geometries with a lone pair occupying the site of the missing proton on the previously trigonal-planar carbon. High-level computations confirm this structure and indicate a barrier to inversion of the carbanion center of ~31 kcal/mol.[56] Photoelectron spectroscopy on the vinyl anion gives an electron binding energy of 15.4 kcal/mol.[60] The much higher electron affinity of vinyl compared to methyl can be attributed to the greater amount of s character in the orbital that forms the lone pair. This difference in hybridization also plays a role in the inversion barrier because when the carbanion adopts a linear geometry ($H–C=CH_2$), the carbanion lone pair must reside in a purely p orbital and loses the advantage of sp^2 hybridization. The effect of hybridization on carbanion basicity will be discussed in detail in Section 4.2.2.

The ethynyl anion provides an example of an sp hybridized carbanion. The high degree of s character in the lone-pair orbital leads to a very large electron affinity for the ethynyl group. Photoelectron spectroscopy indicates an electron binding energy of nearly 70 kcal/mol for the $HC{\equiv}C^-$ anion.[60]

3.2. Stereochemistry and Racemization

For a carbanion derived from a tetrahedral carbon (i.e., sp^3 hybridized) to exhibit chirality, it must either remain pyramidal or if inversion occurs, one face must

remain distinct from the other. The latter requirement can be attained by the presence of chiral auxiliaries apart from the carbanion center, or from a barrier to rotation in the carbanion that locks the system into a single, asymmetric conformation.
The first question to be addressed is the barrier to inversion. As noted above, the
inversion barrier in the bare methyl anion is exceedingly small, ~2 kcal/mol, so no
stereochemical integrity is expected.[56] In fact, most studies have shown that simple
sp^3 hybridized carbanions generally are unable to retain their asymmetry even
under conditions where they react soon after being formed.[61,62] Coordination of
a carbanion to a metal cation such as lithium should increase the barrier to inversion
because as the carbon lone pair shifts to the opposing side of the carbon, significant
electrostatic attraction is lost until the cation is able to reorient itself on the opposite
side. As the carbon–metal bond becomes more covalent, this effect becomes more
important and chiral mercury salts of carbanions are easily prepared and retain their
asymmetry under most conditions.[63] For lithium salts, a few structural motifs have
been identified that lead to sp^3 hybridized carbanions with appreciable stereochemical integrity. One of the earliest to be identified was the cyclopropyl carbanion.
In this case, inversion is hindered by ring strain (in the inversion, the carbanion
must become sp^2 hybridized) and a large inversion barrier, 16.3 kcal/mol, has
been predicted by high-level ab initio calculations.[64] Not surprisingly, the isoelectronic, nitrogen analogue of the cyclopropyl anion, aziridine, has been shown
experimentally to have an unusually large barrier to inversion (19.1 kcal/mol).[65]
As a result of the significant inversion barrier, the lithium salts of cyclopropyl
anions can be generated in enantiomerically pure form at moderate temperatures
from the action of an alkyllithium on an appropriate chiral precursor such as a
halide or a stannane (direct reduction of chiral alkyl halides by lithium metal generally leads to racemization presumably due to the presence of radical intermediates[66]). For example, when (S) 1-methyl-2,2-diphenylcyclopropylbromide is treated
with butyllithium followed by CO_2 in ether or THF, there is complete retention of
stereochemistry in the resulting carboxylic acid (Eq. 1).[67]

$$(1)$$

More recently, Kass and co-workers[68] showed that bare cyclopropyl carbanions can
be configurationally stable in the gas phase. After fluorodesilylation of the cis and
trans isomers of 2-trimethylsilylcyclopropanecarboxaldehyde, they found that the
isomers did not interconvert at room temperature and give different product distributions in gas-phase reactions (Scheme 3.1). High level ab initio calculations suggest an inversion barrier (15.9 kcal/mol) that is close to that of the parent
cyclopropyl carbanion and consistent with configurational stability at 25 °C.
Although these species are diastereomers, the absence of inversion indicates that
with the proper substrates (i.e., enantiomerically pure), asymmetric gas-phase
cyclopropyl carbanions can be generated.

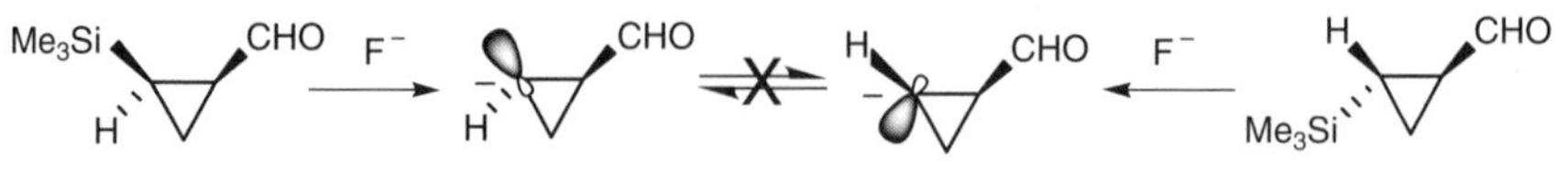

Scheme 3.1

Cram et al.[69] and Corey et al.[70] both demonstrated that salts of carbanions stabilized by an adjacent sulfone or related group can also retain their stereochemistry. For example, Cram found that a labeled, chiral 2-octyl phenyl sulfone underwent hydrogen/deuterium (H/D) exchange with the solvent >100 times faster than the optical activity was lost (Eq. 2).[71] This provides strong evidence that the carbanion, once generated, maintained its stereochemical integrity long enough for solvent to protonate it and complete the H/D exchange process.

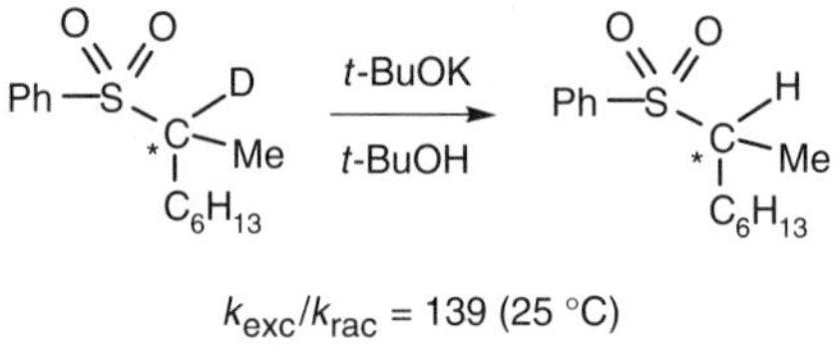

$$k_{exc}/k_{rac} = 139 \ (25 \ °C)$$

In this case, computational evidence does not point to a large barrier to inversion,[72,73] so other factors are at play. A likely explanation is that the sulfone group prevents rotation around the C–S bond. This has been confirmed by ab initio calculations that indicate a barrier of over 14 kcal/mol to C–S rotation in the $MeSO_2CH_2^-$ anion.[72] Although the participation of d orbitals in the bonding to sulfur had been implicated in the large rotational barrier,[74,75] it appears that it is simply the result of a strong electrostatic interaction with the polar SO_2 group that is maximized by aligning the carbon lone pair such that it is gauche to both sulfonyl oxygens.[72] One mechanism for these carbanions to remain asymmetric is illustrated in Scheme 3.2. It is based on the following assumptions: (a) The carbanion is effectively planar (i.e., the inversion barrier, if any, allows for rapid interconversion); (b) the C–S rotational barrier locks the carbanion into one conformation; and (c) the departing and returning group (deuteron and proton in this example) must exit and enter from the same face (gauche to both oxygens in this drawing). This analysis is generally supported by the available evidence.[76]

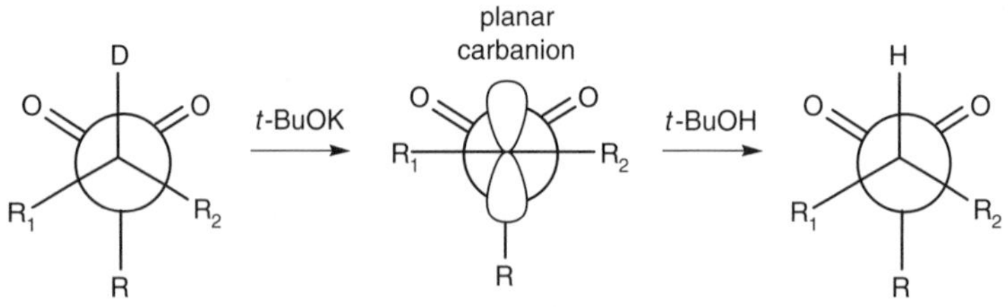

Scheme 3.2

3.3. Magnetic Properties and Nuclear Magnetic Resonance

Nuclear magnetic resonance (NMR) has proven to be a very powerful technique for probing the structures of carbanions in the condensed phase.[77–82] In particular, much work has been completed on the ion-pairing behavior of carbanions with lithium cations as well as the formation of aggregates of these lithium salts. A full discussion of this topic, particularly the methodology, is beyond the scope of this chapter, but a brief overview is appropriate.

The carbanion center is naturally electron-rich so NMR signals associated with it will appear unusually far upfield. For example, isopropyllithium gives an hydrogen NMR (^{1}H NMR) around -0.7 ppm for the proton on the secondary carbon.[77] Similarly, the ^{13}C NMR signal for a carbanion is also upfield from tetramethylsilane (TMS) and methyllithium gives a carbon resonance in the range of -10 to -20 ppm depending on the solvent and temperature. The other unusual feature that NMR reveals about the lithium salts of carbanions is the formation of aggregates (oligomers) in solution. The aggregation of lithium salts has also been confirmed by colligative measurements,[83,84] crystallography,[85,86] and computational modeling.[87] The high charge densities of the lithium cation and carbanion combined with the low dielectric constants of the appropriate NMR solvents (i.e., hydrocarbons and ethers) lead to a situation where electrostatic forces favor the formation of dimers, tetramers, and higher aggregates. A classic example is methyllithium, which has been shown by a number of techniques to be a tetramer in solvents like diethyl ether, **1**.[88,89] This arrangement allows each lithium cation to interact with three methyl anions and vice versa. The resulting, strong electrostatic stabilizations outweigh the obvious entropic disadvantage of a polymeric structure. The equilibria involving the lithium salts can be dynamic with mixtures of monomers, dimers, and higher aggregates coexisting. The position of the equilibria as well as the rate constants for interconversion between aggregates are naturally dependent on the nature of the solvent, the temperature, and any added ligands. For example, phenyllithium is mainly a dimer, **2**, in a mixture of diethyl ether and tetramethylethylenediamine ($Me_2NCH_2CH_2NMe_2$, TMEDA), but is monomeric, **3**, when dissolved in THF containing pentamethyldiethylenetriamine ($Me_2NCH_2CH_2NMeCH_2CH_2NMe_2$).[77,90]

| 1 | 2 | 3 |

The situation becomes even more complex when other lithium salts are present in the solution. For example, when a lithium halide such as LiBr is added to solutions of alkyllithiums, mixed aggregates are formed where a bromide anion takes the place of a carbanion.[91–94] These mixed aggregates may play important roles in synthetic chemistry because alkyllithiums are often formed by the action of lithium metal on an alkyl bromide so LiBr is a common contaminant. Kinetic studies have shown that the presence of organolithium aggregates or halide-containing mixed aggregates has a significant impact on reaction rates.[95–97]

4. BASICITY OF CARBANIONS–ACIDITY OF CARBON ACIDS

The most widely studied physical property of carbanions is their basicity, which of course is a direct measure of the acidity of the parent carbon acid. Carbon acidity measurements date back to the early part of the twentieth century and a myriad of techniques have been employed for the measurements. Although early measurements were only able to provide semiquantitative data, more recent ones have resulted in accurate acidity measurements across a vast range of effective acid dissociation constants, K_a values. This section will begin with a brief description of definitions and methodologies followed by representative data as well as applications of those data.

4.1. Definitions and Methodologies

Acidity has been most widely studied in aqueous solution and is easily defined in this medium by the acid dissociation equilibrium constant, K_a (Eq. 3) where S represents the solvent employed in the measurement. Given the magnitudes of K_a values, they are often expressed as pK_a values for convenience (Eq. 4).

$$AH + S \overset{K_a}{\rightleftharpoons} A^- + SH^+ \tag{3}$$

$$pK_a = -\log K_a \tag{4}$$

One assumption of this treatment is that the role of the solvent is limited to acting as a solvating agent for the proton and the other species in the equilibrium. In other words, the solvent cannot react with either the parent acid (AH) or its conjugate base (A$^-$). Because carbanions are often exceptionally strong bases, the requirement that they do not react with the solvent is a major limitation. This limitation is known as the *leveling effect*[98] and simply stated, it is not possible to use equilibrium techniques to measure the acidity of a species that is a weaker acid than the solvent because the resulting carbanion would be able to deprotonate the solvent. Very few carbon acids are more acidic than water, so measurements involving these species most often have been completed in other, less acidic solvents.

The next issue that arises from the weak acidity of carbon acids involves the degree of self-dissociation. In Eq. 3, the equilibrium constant is determined by measuring the concentration of the four species in the equation, but this requires that the carbon acid self-dissociates to an extent that a measurable quantity of the carbanion is formed. Again, because carbon acids are generally weak, this requirement often is not met and therefore another type of equilibrium measurement must be employed. In this case, a measurable concentration of the carbanion is generated in solution by the addition of a very strong base, and the acidity is measured relative to another species of similar acidity (Eq. 5). The solvent no longer appears in this equilibrium expression, but it does play a role through its solvation of the other species in the equilibrium, especially the anions. If one of the species in the equilibrium measurement has a known K_a value, then the relative measurement from Eq. 5 can be converted to an absolute value (i.e., pK_a).

$$A_1H + A_2^- \overset{K_{rel}}{\rightleftharpoons} A_1^- + A_2H \tag{5}$$

A common technique for measuring the K_{rel} values has been to employ species that produce anions with useful ultraviolet (UV) or visible (vis) absorbances and then determine the concentrations of these species spectrophotometrically. Alternatively, NMR measurements could be employed, but generally they require higher concentrations than the spectrophotometric methods. A hidden assumption in Eq. 5 is that the carbanion is fully dissociated in solution to give a free anion. Of course, most simple salts do fully dissociate in aqueous solution, but this is not necessarily true in the less polar solvents that are typical employed with carbanion salts. For example, dissociation is commonly observed for potassium salts of carbanions in DMSO because the solvent has an exceptionally large dielectric constant ($\varepsilon = 46.7$) and solvates cations very well, whereas dissociation occurs to a small extent in common solvents such as DME and THF (dielectric constants of 7.2 and 7.6, respectively). In these situations, the counterion, M^+, plays a role in the measurements because it is the relative stability of the ion pairs that determines the position of the equilibrium constant (Eq. 6).

$$A_1H + A_2^-M^+ \overset{K_{rel}}{\rightleftharpoons} A_1^-M^+ + A_2H \tag{6}$$

Here, the measurements lead to values that are best described as "ion-pair" acidities. Although it is not possible to directly relate these values to true pK_a values because they do not reflect the stability of the bare carbanion, they do provide useful data, particularly for very weak acids, and are relevant to many synthetic applications of carbanions because the solvents and concentrations employed in synthetic work generally lead to ion pairing.

In the case of very weak acids, it has not been possible to find suitable conditions for establishing and monitoring equilibria such as those shown in Eqs. 5 and 6. The carbanions react too readily with the solvent or equilibrium with the reference acid (i.e., A_1H) is reached too slowly for practical measurements. Here, kinetic measurements have been employed. These studies rely on the presence of a Brønsted relationship[22] between the rate of deprotonation of a carbon acid and the stability of the resulting carbanion. This correlation has proven to be viable in many instances and kinetic acidities are available in a range of solvents.[23,99,100] In general, these methods have used isotope exchange to monitor the rate of deprotonation. For example, when a mixture of α-deuterated and tritiated toluene ($PhCH_2D$ or $PhCH_2T$) is treated with lithium cyclohexylamide ($c\text{-}C_6H_{11}NHLi$) in cyclohexylamine for an extended time period, the disappearance of the isotopic labels gives a measure of both the deprotonation rate as well as the primary isotope effect in the deprotonation.[99,101] When applied to a set of compounds of unknown acidity, the measurements provide a qualitative ranking of their acidities. However, if the acidities of some of the species are already known, then these can be used to establish a linear Brønsted relationship, and the acidities of the unknown species can be determined by interpolation or extrapolation from the data in the Brønsted plot. One complication in kinetic measurements of acidity is the possibility of *internal return*. In this case, the isotopic label is returned to the acid rather than being replaced by a proton.

$$AD + B^- \underset{k_{-1}}{\overset{k_1}{\rightleftharpoons}} A^- \cdots DB \tag{7a}$$

$$A^- \cdots DB + HB \xrightarrow{k_2} A^- \cdots HB + DB \tag{7b}$$

$$A^- \cdots HB \rightleftharpoons AH + B^- \tag{7c}$$

This situation is illustrated in Eq. 7 where B^- represents the conjugate base of the solvent, BH, used in the kinetic experiment. After the base, B^-, removes the deuteron from the acid donor, A, (Eq. 7a) it is still complexed to it. At this point, the deuterated base, DB, may diffuse away and be replaced by a proton-bearing analogue, HB (Eq. 7b), or it can return the deuteron to the conjugate base of the acid. Under the typical conditions, the step in Eq. 7b is irreversible (the concentration of HB is always much greater than DB) and the rate of isotope exchange can be expressed in terms of the following elementary rate constants.

$$k_{obs} = \frac{k_1 k_2}{k_2 + k_{-1}} \tag{8}$$

If DB quickly diffuses away (k_2 is large compared to k_{-1}), the expression simplifies to $k_{obs} = k_1$ and is solely dependent on the rate of deuteron abstraction. In other words, the isotopic exchange rate truly reflects the rate of deuteron abstraction. If k_2 is relatively small, then the expression simplifies to $k_{obs} = k_1 k_2 / k_{-1}$ and the isotope exchange kinetics are also dependent on the competition between diffusion (k_2) and internal return (k_{-1}). Internal return can be identified by the absence of an isotope effect in the exchange process. This occurs because k_2 is not expected to have a significant isotope effect and the primary isotope effects inherent to k_1 and k_{-1} are expected to nearly cancel. Streitwieser's use of dual labeled systems (deuterium and tritium) allows isotope effects to be measured in the exchange experiment (i.e., k_D/k_T) and internal return can be identified directly.[26]

Carbanion stability has also been determined in the gas phase using mass spectrometric techniques. This allows for a measure of the intrinsic stability in the absence of any solvation effects. Given the low dielectric of a vacuum, self-ionization is not energetically viable. For example, dissociation of H_2O in the gas phase is endoergic by 384 kcal/mol,[54] corresponding to a pK_a value of >250! Consequently, acidity measurements have generally relied on measuring relative stabilities (Eq. 5) and then linking these measurements to a compound whose absolute acidity has been measured independently (usually indirectly via a combination of spectroscopic and kinetic measurements, see Section 4.4). Given that self-ionization is so endothermic in a gas-phase environment, acidities under these conditions are more conveniently characterized by energy changes rather than equilibrium constants (i.e., K_a). Either the free energy (ΔG_{acid}) or enthalpy (ΔH_{acid}) of deprotonation can be used as a measure of the acidity (Eq. 9). Proton affinity (PA) is often used to describe the stability of an anion in the gas phase and is equivalent to the ΔH_{acid} value for the parent acid (Eq. 10).

$$AH \xrightarrow[\text{or } \Delta H_{acid}]{\Delta G_{acid}} A^- + H^+ \tag{9}$$

$$PA(A^-) = \Delta H_{acid}(AH) \tag{10}$$

Measurements have been completed in the gas phase for a very wide range of organic species and for carbon acids, ΔG_{acid} values have been measured that range from a low of 289 kcal/mol for a highly fluorinated bis(sulfone), $(CF_3CF_2CF_2CF_2SO_2)_2CH_2$, to 409 kcal/mol for methane.[54] This represents an astounding, 80 orders-of-magnitude variation in the effective equilibrium constant for deprotonation. Kinetic methods have also been developed for estimating gas-phase acidities and they will be discussed in Section 4.2.1.

4.2. Structural Effects on Carbanion Basicity–Carbon Acidity

4.2.1. Carbanions Derived from sp³ Hybridized C–H Bonds.

There are very limited data in the gas phase or in solution on the equilibrium acidity of saturated hydrocarbons. Instead, the most comprehensive set of data comes from gas-phase kinetic measurements by DePuy et al.[102] Although it is not clear whether the values are quantitatively accurate across the whole series of compounds,[103] the work provides a good qualitative picture of structural effects in the simplest and weakest of the carbon acids. The method relies on measuring branching ratios in the reaction of HO^- with a trimethylsilyl derivative of the alkane. In the process, either the unique alkane (Eq. 11a) or methane (Eq. 11b) is expelled to give a siloxide anion. Relying on the fact that the transition states for these reactions have a high degree of carbanion character (Eq. 12), it is possible to relate the branching ratio to the relative acidities of the expelled groups and develop a correlation based on species of known acidity. This approach is a variation of the kinetic acidity methods described above and is closely related to a common gas-phase methodology, the Cooks kinetic method.[104]

$$Me_3Si-R \xrightarrow{HO^-} \begin{cases} Me_3SiO^- + RH & (11a) \\ RMe_2SiO^- + CH_4 & (11b) \end{cases}$$

$$Me_3SiR \xrightarrow{HO^-} Me_3\overset{-}{Si}\overset{OH}{\underset{R}{\diagup}} \longrightarrow \left[\begin{array}{c} O^{\cdots H} \\ \underset{Me_3Si----R}{|} \end{array}\right]^{\ddagger} \longrightarrow Me_3SiO^- + RH \quad (12)$$

TABLE 3.1. Gas-Phase ΔH_{acid} Values of Simple Alkanes[a]

Alkane[b]	ΔH_{acid} (exp)[c]	ΔH_{acid} (comp)[d]
CH$_4$	417	418
CH$_3$CH$_3$	420	421
CH$_3$**CH$_2$**CH$_3$	416	416
CH$_3$**CH**(CH$_3$)$_2$	413	
CH$_3$**C**(CH$_3$)$_3$	409	413[e]
Me**CH$_2$**Me	419	418
Me$_3$**CH**	413	412
c-C$_3$H$_6$	412	416[e]
c-C$_4$H$_8$	417	415[e]
c-C$_5$H$_{10}$	416	414[e]

[a] In kilocalories per mole (kcal/mol).
[b] Site of deprotonation is in bold.
[c] Values from gas-phase kinetic experiments (see Ref. 102).
[d] Values from G2 calculations (see Ref. 105).
[e] Values from MP2/6-31 + G** calculations (see Ref. 106).

Data for a representative set of simple alkanes are given in Table 3.1 along with values calculated at the G2[105] or MP2[106] levels of theory. There is a good correspondence in these systems between the kinetic and computed acidities, but it should be noted that DFT calculations[106] at the B3LYP/6-311++G(2d,p) level on these systems give ΔH_{acid} values that tend to be ~3–5 kcal/mol smaller than the MP2, G2, or experimental values. Starting with methane, the effect of adding methyl groups to the α-carbon can be evaluated with the acidities of ethane, propane (2° position), and isobutane (3° position). One might expect two effects from the added methyl groups. First, they are polarizable and therefore could stabilize the resulting carbanions by effectively dispersing the charge. Second, methyl groups are viewed as electron donors in Hammett-type correlations[107,108] (they have negative σ-values) so they should destabilize an adjacent carbanion. It can be seen in Table 3.1 that adding the first methyl group (i.e., ethane) reduces the acidity (i.e., increases ΔH_{acid}). However, it is not a simple trend and the addition of more methyl groups to the carbanion center (i.e., the isopropyl and *tert*-butyl carbanions) eventually leads to an increase in acidity. DePuy et al.[102] has argued that the trend can be explained by competing effects where destabilization from the electron-donating characteristics of the methyl group quickly is saturated and then overwhelmed by stabilization gained from the polarizability of the additional methyl groups. A simpler trend is observed in the addition of methyl groups at the β-carbon. Here, polarizability is the main factor and the acidity increases with each added methyl group. Across the series from ethane to neopentane, there is a smooth drop in the ΔH_{acid} values from 420 to 409 kcal/mol with each methyl group causing about a 3-kcal/mol increase in acidity. Data in Table 3.1 also indicate that ring strain can increase acidity. Specifically, cyclopropane is much more acidic than other monocyclic hydrocarbons or the secondary position of a simple alkane (i.e., propane). It is well known that ring strain causes an increase in the amount of s character in the C—H bonds attached to the ring in order to accommodate the greater p character needed for the small angle C—C bonds. This enhances their acidity because the carbanion's lone pair uses an orbital that is effectively more electronegative—more s character leads to a more compact orbital that places electron density nearer to the nucleus. To illustrate this point, Natural Bond Order (NBO) analysis from MP2 calculations indicates that the lone pair of the carbanion from cyclopropane has 34% s character, as opposed to 25% in a perfect sp^3 hybrid.[106] This effect does not play a significant role in cyclobutane (only 24% s character in the carbanion lone pair) and the four-membered ring compound exhibits an acidity that is consistent with a system containing secondary carbons. A similar argument can be developed using the NMR ^{13}C–^{1}H coupling constants of the ring C—H bonds to estimate the %s character of the bonding hybrid (32% for cyclopropane and 27% for cyclobutane).[109] Moreover, Streitwieser et al.[110] showed that the ^{13}C–^{1}H coupling constants correlate with the kinetic acidities of a series of cycloalkanes.

When assessing the effect of structure on the stability of more complex carbanions derived from carbon centers that are initially sp^3 hybridized, there are two general classes of species to consider. In the first case (**4**), the carbanion can be viewed (at least to a rough approximation) as a localized charge center that is stabilized by a substituent that acts a dipole, a polarizable group, an electron-withdrawing group, or some combination of these roles. These systems can be analyzed within the fundamental framework of substituent effects. In the second case (**5**), the carbanion is part of a cyclic π system and the charge is dispersed equally or nearly equally over several carbons. Here, the nature of the π system plays a critical role in the stability of the carbanion though substituents attached to the π-system can also play important roles.

$$Z—CH_2^-$$

4   **5**

We will begin with the first group of species and then move to the second. To initially avoid a discussion of solvation effects on carbanion stability, we will begin with gas-phase data for simple representatives of model classes of compounds (Table 3.2). Later, the gas-phase data will be compared to data that have been obtained in solution (Section 4.3.3).

When a substituent, Z, is added to a carbanion, it can have three important modes of stabilizing the anion. First, if it is more electronegative than carbon, it can inductively stabilize the carbanion by shifting electron density along the C—Z σ bond thereby effectively reducing the charge on the carbon (**6**). Second, if the substituent has low-lying π orbitals it can conjugate with the carbon lone pair and disperse the charge through a π-bonded resonance form (**7**). Third, the substituent can polarize its own electron density thereby creating an effective dipole

TABLE 3.2. Gas-Phase ΔH_{acid} Values for Me–Z Compounds^a

Z	ΔH_{acid}
H	417
F	409
Cl	400
Br	397
OMe	407
SMe	390
PMe$_2$	385
SiMe$_3$	388
S(O)Me	374
S(O)$_2$Me	366
CH=CH$_2$	389
Ph	381
C≡CH	381
C(O)Me	369
C≡N	373
NO$_2$	357

^{a} In kilocalories per mole (kcal/mol.) (see Ref. 54).

(or enhancing one that is already present in the neutral carbon acid) to stabilize the carbanion center (**8**).

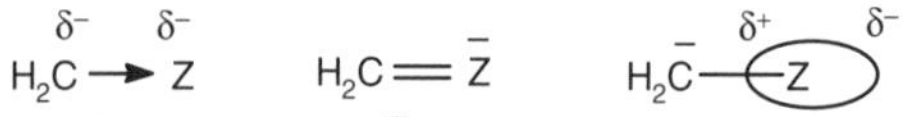

In Table 3.2, gas-phase ΔH_{acid} values are given for a group of substituted methane derivatives. The first few entries involve the addition of a halogen to the carbon acid. As expected, these electron-withdrawing groups inductively stabilize the resulting carbanion and therefore increase the acidity. Counter to the variation in electronegativities, the stabilizing effect on the carbanion increases along the series F < Cl < Br. This trend highlights the fact that the larger halogens are more polarizable. A similar, but more dramatic trend is seen in the data for oxygen and sulfur substituents (OMe and SMe) where the latter provides 17 kcal/mol more stabilization than the former. Obviously, polarizability is playing a much greater role than electronegativity in these gas-phase stabilizations. Another factor that Wiberg and Castejon[111] have highlighted in this system is *negative hyperconjugation*.[37,112] Here, conjugation of the carbon lone pair with a σ* orbital on sulfur leads to charge dispersal and stabilization (**10**).

Using computational methods, Wiberg showed that this effect is only significant with sulfur, not oxygen substitution. This mode of stabilization is in effect, a resonance description of a type of polarization in that electron density is shifting in the substituent group in response to the carbanion. However, negative hyperconjugation also indicates the formation of a partial π bond to the sulfur. The entries for the other third period elements (phosphorous and silicon) also highlight the importance of polarizability in the gas phase. Although there is a significant variation in electronegativity across the series Si < P < S < Cl, there is a relatively small variation in the stabilizing effects with the most electronegative element, chlorine, providing the least stabilization to the carbanion. Here, it appears that the additional Me groups used to fill-out the valencies of the third period elements in moving from right to left across the periodic table (i.e., MeS vs. Me$_2$P vs. Me$_3$Si) may provide additional polarizability that counters the effects of electronegativity. Of course, negative hyperconjugation could also be playing a role with phosphorous and silicon. The next two entries in the table involve sulfoxide and sulfone substituent groups. Both are highly polar with large internal dipole moments and therefore act as powerful stabilizing groups via inductive and through-space field effects (i.e., the interaction of the S—O dipole with the carbanion center). These groups lead to much more stable carbanions than those stabilized by the halogens or a simple sulfur group (SMe).

The rest of the entries in the table have π systems to conjugate with the carbon lone pair and therefore offer additional resonance forms for the carbanion. The vinyl and phenyl substituents are weakly polar and their effect is mainly from resonance delocalization along with polarizability. The greater stabilizing effect of the phenyl group is probably a result of its larger size and therefore greater polarizability. The ethynyl substituent is about as stabilizing as a phenyl group and this is probably related to the unusually high electronegativity of an sp-hybridized carbon (see Section 4.2.2 for more discussion of this effect). The final entries in the table are for species that allow for resonance forms where the carbanion's charge is carried by a more electronegative atom, oxygen or nitrogen (evidence for resonance delocalization onto nitrogen is not conclusive, see below).

$$H_2C=\overset{\overset{\displaystyle O^-}{\displaystyle |}}{C}CH_3 \qquad H_2C=C=N^- \qquad H_2C=\overset{+}{N}\overset{\displaystyle O^-}{\underset{\displaystyle O^-}{\diagup}}$$

As a result, these anions can gain exceptional stability and their carbon acids are unusually acidic. Because the above resonance forms probably play very important roles in stabilizing the anion, these species are not purely carbanions. However, thermodynamics favors reactions (protonation or alkylation) at carbon so their behavior is generally characteristic of carbanions. It also should be noted that kinetics often favor reaction at the heteroatom in these anions though leading to the less stable product.[113–115]

In Table 3.3, data are given for some polysubstituted methane derivatives. Here it is possible to judge the effects of sequential additions of the same substituent. The

TABLE 3.3. Gas-Phase ΔH_{acid} Values for Polysubstituted Methanes[a]

Species	ΔH_{acid}	$\Delta(Subst)$[b]
H_3CCl	400	−17
H_2CCl_2	376	−24
$HCCl_3$	358	−18
H_3CPh	381	−36
H_2CPh_2	364	−17
$HCPh_3$	359	−5
$CH_3C(O)Me$	369	−48
$H_2C[C(O)Me]_2$	344	−25
$HC[C(O)Me]_3$	329	−15
$CH_3C{\equiv}N$	373	−44
$H_2C(C{\equiv}N)_2$	336	−37

[a] In kilocalories per mole (kcal/mol) (see Ref. 54).
[b] Effect of additional substituent. In monosubstituted system, difference from methane.

first entries involve the addition of chlorines. It can be seen that to a rough approximation, each chlorine has a similar acidifying effect on the carbon acid. In contrast, the acidifying effect of additional phenyl groups drops off dramatically and the third one in the triphenylmethyl anion has only a marginal effect. The difference between these two systems can be linked to how they stabilize the carbanion. It appears that inductive stabilization (i.e., the Cl system) is not affected by the presence of additional substituents (however, this pattern was not observed in the inductive *destabilization* of carbanions by methyl groups, Section 4.2.1). In contrast, resonance stabilization (i.e., the phenyl system) is deeply attenuated by other conjugating substituents. A rational explanation is that each conjugating substituent reduces the charge density on the carbanion, leaving less density to be delocalized by the next conjugating substituent.[116] Of course, steric effects also play a role as the poly-substituted systems become crowded making conjugation and optimum orbital overlap difficult. A similar trend is seen with the ketone substituent, which is also expected to be dominated by resonance effects. However, the drop-off is not as dramatic as with the phenyl group hinting that dipole effects play a significant role with this substituent. The final entries refer to the cyano group. Data are available for mono- and dicyanomethane and the acidifying effect of the second group nearly matches the first. This suggests that although resonance stabilization is possible with the cyano group, it acts mainly through inductive and field effects, which are diminished to a lesser extent by polysubstitution. Data from aqueous solution also support this conclusion.[117]

Conjugated ring systems offer an alternative mode for the stabilization of a carbanion center. The most common situation is where deprotonation completes a cyclic π system leading to a highly stabilized, aromatic anion. The best known example is cyclopentadiene, which leads to a six-electron, aromatic ring after deprotonation. The ΔH_{acid} value for cyclopentadiene, 354 kcal/mol, is much lower than expected for a carbanion conjugated to a simple polyene system (Fig. 3.1). For example, pentadiene has a ΔH_{acid} value of 369 kcal/mol indicating that aromaticity provides roughly 15 kcal/mol of stabilization to the carbanion. To put this energy difference in perspective, it would translate to a difference of >10 units in terms of pK_a values at 300 K. Indene and fluorene represent larger analogues and also display unusually high acidity with ΔH_{acid} values near that of cyclopentadiene. Here, larger size and greater polarizability do not lead to a significant acidity enhancement suggesting that fusing the additional aromatic rings onto the cyclopentadiene framework has a mild acid weakening effect that balances the increased polarizability. This topic will be discussed in more detail in Section 4.3.3. Phenalene offers another example of a system where deprotonation completes a π system and in this case, produces a 14-electron, aromatic carbanion. Unlike the cyclopentadiene derivatives, it is not possible to draw out resonance forms for the phenalenide anion where each of the ring carbons bears the charge. Nonetheless, phenalene has a ΔH_{acid} value that is comparable to cyclopentadiene and much smaller than that

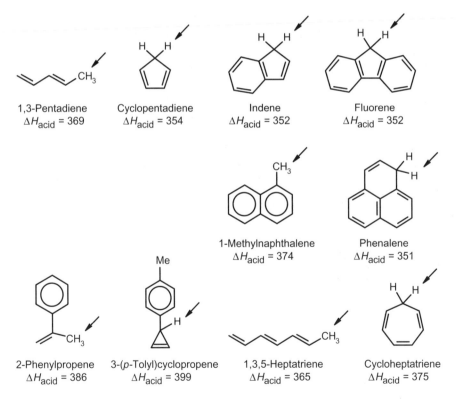

Figure 3.1. Gas-phase ΔH_{acid} values (kcal/mol) for carbon acids that give aromatic or antiaromatic carbanions.[54] Acyclic analogues are included for comparison. Arrow indicates acidic hydrogen.

of 1-methylnaphthalene, an analogue that contains some, but not all of the stabilizing features of phenalene (i.e., it lacks one double bond).

Antiaromaticity also plays a role in the stability of carbanions. In Figure 3.1, examples of four- and eight-electron antiaromatic systems are provided. Kass and co-workers[118] determined the gas-phase acidity of a 3-arylcyclopropene derivative and its ΔH_{acid} value is much higher than that for a typical allylic anion. Compared to 2-phenylpropene, the ring system destabilizes the carbanion by 13 kcal/mol. One could also compare the cyclopropene system to a saturated analogue, phenylcyclopropane, and that suggests that antiaromaticity destabilizes the anion by 9 kcal/mol. Neither of these is a perfect analogue because the former lacks the acidifying effect of a three-membered ring structure and the latter lacks allylic conjugation to the carbanion. In any case, these analogies suggest that the cyclopropenyl anion's stability is reduced due to antiaromaticity. The effects of antiaromaticity are also seen in the deprotonation of cycloheptatriene, which leads to an eight-electron, antiaromatic system and has a ΔH_{acid} value that is considerably higher than that of an acyclic analogue, 1,3,5-heptatriene. Both of these examples highlight the fact that antiaromaticity as well as aromaticity play significant roles in the stability of carbanions.

4.2.2. Carbanions Derived from sp^2 and sp Hybridized C–H Bonds.

In the discussion of carbanions derived from cycloalkanes, it was noted that increased s character in the carbon lone pair leads to a more stable carbanion. A similar effect is expected in carbanions derived from sp^2 and sp hybridized C–H bonds. In Table 3.4, gas-phase data are given for a variety of unsaturated systems. The experimental values for these species are generally very accurate and it can be seen that in most cases, the computed values match very well in these systems. One exception is ethyne, where a range of computed values has been reported, both higher and lower.[119] However, high-level calculations on ethyne do give a value close to experiment.[120] It can be seen that in going from ethane (Table 3.1) to ethene, the ΔH_{acid} value drops 11 kcal/mol indicating significant stabilization is gained in the

TABLE 3.4. Gas-Phase ΔH_{acid} Values of sp and sp^2 Hybridized Carbon Acids[a]

Carbon Acid	ΔH_{acid} (exp)[b]	ΔH_{acid} (comp)[c]
$H_2C{=}CH_2$	409	408
$c\text{-}C_6H_6$	402	402[d]
$HC{\equiv}CH$	378	372[e]
$H_2C{=}C{=}CH_2$	381	383
$H_2C{=}C{=}O$	365	366

[a] In kilocalories per mole (kcal/mol).
[b] See Ref. 54.
[c] Values from CCSD(T)/6-311 + + G(d,p) calculations (see Ref. 121).
[d] Values from MP2/6-31 + G(d,p) calculations (see Ref. 122).
[e] Values from MP2/aug-cc-pVDZ calculations (see Ref. 123).

sp^2 hybridized carbanion. For comparison, the effect of the hybridization change in going from propane (2° carbon) to cyclopropane leads to a 7 kcal/mol increase in acidity (Table 3.1). Given that the NBO calculations suggested that the cyclopropyl anion had nearly sp^2 hybridization in its lone pair (34% s character),[106] it is not surprising that a similar acidifying effect is seen in going from ethane to ethene. Benzene provides another example of an sp^2 hybridized carbon acid and its ΔH_{acid} value is somewhat smaller than ethene because it is a larger, more polarizable system. The shift to an sp hybridized system leads to a dramatic increase in acidity. Ethyne's ΔH_{acid} value is 31 kcal/mol below that of ethene. A large effect is expected because the system is going from roughly 33% s character in ethene to 50% in ethyne, a much larger change than in going from ethane to ethene. This trend is also seen in the C–H homolytic bond energies and bond distances across the ethane, ethene, ethyne series. Allene presents a special case because the resulting carbanion has a resonance form that is best described as an ethynyl stabilized carbanion (propargyl system). This resonance form and the resulting delocalization explain the relatively high acidity of allene compared to ethene. As expected, ketene is also unusually acidic for an sp^2 hybridized carbon acid and again, this results from the presence of a highly stable resonance form in the anion, an oxide in this case.

$$H\bar{C}=C=CH_2 \longleftrightarrow HC\equiv C-\bar{C}H_2$$

$$H\bar{C}=C=O \longleftrightarrow HC\equiv C-\bar{O}$$

4.3. Condensed-Phase Carbon Acidity Measurements

Carbon acidities have been measured in several solvents, but the greatest amount of data has been accumulated in aqueous mixtures,[124,125] DMSO,[38,39] cyclohexylamine,[40–47] and various ethers such as DME[52,53] and THF.[48–51] These solvents offer different limitations and environments for the resulting carbanions. The high dielectric of water (78.5) leads to the formation of free-ions, but its relatively low pK_a (15.5) limits its use to exceptionally strong carbon acids due to the leveling effect (see above). DMSO also has a high dielectric (46.7) so free ions are common, but it has a much higher pK_a (35.1) than water so a wider range of species may be studied in it. Cyclohexylamine is not nearly as polar as the former solvents and leads to ion-pair formation for carbanion salts; however, its low acidity (cesium ion pair $pK_a = 41.6$) allows for the study of weaker acids than DMSO. Finally, the ethers also generally lead to ion-paired carbanion salts and have high effective pK_a values. The ethers do not yield stable carbanions when deprotonated, but instead, decompose. Therefore, the solvent cannot be used as an absolute standard and the effective pK_a is essentially a measure of the kinetic stability of the solvent towards highly basic carbanions. In this section, no effort will be made to present a comprehensive listing of all the measurements completed in solution. Many workers have made important contributions in a variety of solvents and the data can be found in the original references and reviews.[23,37,38,125] Instead, general trends will be presented, free ion and ion-pair acidities will be contrasted, and comparisons to the gas phase will be made.

4.3.1. Carbon Acidity in DMSO.

Although DMSO has a high dielectric constant, it is unable to stabilize anions via hydrogen bonding. Consequently, anions that are capable of strong hydrogen-bonding interactions are much more difficult to form in DMSO than water. On the other hand, anions that are diffuse or highly delocalized, a common situation with carbanions, are not as sensitive to the hydrogen bonding ability of the solvent and are nearly as stable in DMSO as in water. This relationship is illustrated in Table 3.5 with data from Bordwell's group for measurements made in DMSO at 25 °C. For compounds that give small, highly localized anions like HCl and MeOH, there is a very large difference between the water and DMSO values (10–15 units), but for more delocalized systems the difference is much smaller. For example, the pK_a of malononitrile [$H_2C(C{\equiv}N)_2$] is the same in water and DMSO, which suggests that for carbanions in general, the DMSO values are reasonably close to what might be expected in an aqueous-like environment.

Table 3.6 contains a sample of pK_a values determined in DMSO for a variety of common functional groups. These were determined in Bordwell's group by measuring equilibria between carbon acid pairs (i.e., Eq. 5) and then linking the resulting scale to the absolute pK_a of the solvent. It can be seen that highly polar substituents capable of conjugation such as the nitro, carbonyl, or cyano group lead to pK_a values in the range from 17–31. In other words, carbanions stabilized by a single substituent, no matter how potent, are very basic in DMSO solution. Multiple acidifying groups are needed to bring carbanions into the pK_a range where any significant self-ionization might occur. For example, malononitrile and 1,3-pentadiene have pK_a values in DMSO that are <16 (11.0 and 13.3, respectively). In the absence of polar substituents, resonance stabilized carbanions have pK_a values that are >30. For example, the series from toluene to triphenylmethane have pK_a values from 30.6–43. In contrast, systems that lead to aromatic anions have significantly lower pK_a values. Cyclopentadiene has a pK_a of 18 and its larger derivatives, indene and fluorene, have pK_a values of 20.1 and 22.6, respectively. Finally, sp hybridization leads to a significant increase in acidity with phenylacetylene exhibiting a pK_a of

TABLE 3.5. The pK_a Values Measured in Water and DMSO[a]

Species[b]	pK_a(water)	pK_a(DMSO)
HCl	−8.0	1.8
MeSO$_3$H	−0.6	1.6
MeCO$_2$H	4.8	12.3
MeC(O)CH$_2$C(O)Me	8.9	13.3
H$_2$C(C≡N)$_2$	11.0	11.0
MeOH	15.5	29.0

[a] See Ref. 38.

TABLE 3.6. Measured pK_a Values in DMSO for Representative Functional Groups[a]

Species[b]	pK_a
CH_3NO_2	17.2
$CH_3C(O)CH_3$	26.5
$CH_3C(O)OEt^c$	27.4
$CH_3SO_2CH_3$	31.1
$CH_3C{\equiv}N$	31.3
$CH_3S(O)CH_3$	35.1
Cyclopentadiene	18.0
Indene[d]	20.1
Fluorene[d]	22.6
$PhC{\equiv}CH$	28.7
Ph_3CH	30.6
Ph_2CH_2	32.2
$PhCH_3$	43[e]

[a] In kilocalories per mole (kcal/mol) (see Ref. 38).
[b] Site of deprotonation is in bold.
[c] See Ref. 126.
[d] See Figure 3.1 for structures.
[e] Estimated value.

28.7, considerably lower than the resonance stabilized, sp^3 hybridized hydrocarbon systems like toluene.

The data in DMSO also allow for an evaluation of the effect of α-alkyl groups on the stability of a carbanion. Because alkyl groups are generally categorized as electron donors, one might expect that they would destabilize a carbanion, but the data indicate that the situation is not so simple. In Table 3.7, pK_a and ΔH_{acid} values are

TABLE 3.7. Effect of α-Methyl Groups on the Acidity of Carbon Acids in DMSO and the Gas Phase

Species[a]	$pK_a(DMSO)^b$	$\Delta H_{acid}{}^c$
$PhSO_2CH_3$	29.5	362.7
$PhSO_2CH_2Me$	31.3	365.0
$CH_2(C{\equiv}N)_2$	11.4	
$MeCH(C{\equiv}N)_2$	12.4	
CH_3NO_2	17.7	356.4
$MeCH_2NO_2$	17.0	355.9
Me_2CHNO_2	16.9	356.0
Fluorene	22.9	351.7
9-Methylfluorene[d]	22.3	350.8

[a] Site of deprotonation is in bold.
[b] See Ref. 127. Reported on a per acidic hydrogen basis in this table.
[c] In kilocalories per mole (kcal/mol) (see Ref. 54).
[d] The 9-position is the point of deprotonation.

given for three sets of homologues that differ only by the number of methyl groups attached to the carbanion center. In the first two sets, phenylsulfone and malononitrile derivatives, the addition of an α-methyl group causes a significant decrease in the acidity (destabilizes the carbanion). This result is most likely from the aforementioned electron-donating capabilities of the methyl groups. Because each of these systems is expected to give a fairly localized carbanion, the effect is maximized. The destabilizing effect of a methyl group was also seen in the gas phase in going from methane to ethane (Section 4.2.1). For nitro-stabilized carbanions, one sees the opposite effect, a weak, but general increase in acidity as methyl groups are added to the α carbon. This effect is also observed with fluorene (see Fig. 3.1 for the structure) and its 9-methyl derivative. In both of these systems, deprotonation leads to an sp^2 hybridized carbanion with extensive charge delocalization into other parts of the molecule. It is well known that alkyl substituents stabilize sp^2 hybridized centers (i.e., more highly substituted alkenes are more stable). Bordwell et al.[127] argued that this effect counters the inductive effect of the methyl group and renders it mildly stabilizing for a carbanion. From the data in Table 3.7 as well as the data presented earlier on the gas-phase acidity of alkanes (Section 4.2.1 and Table 3.1), it is clear that alkyl groups do not have a uniform effect on the stability of a carbanion center and systems have to be analyzed on an individual basis taking into account polarizability as well as inductive, hybridization, and steric effects.

Finally, Arnett et al.[128–130] developed a different approach to determining the stability of carbanions in DMSO. By means of solution calorimetry, the heat of protonation of a carbanion can be assessed in this solvent. As one might expect, there is a good correlation between the relative heats of protonation measured in this way and relative free energies of protonation measured by Bordwell using the methods described above.[130]

4.3.2. Ion-Pairing and Carbon Acidity.

Many studies of carbon acidity have been completed in solvents such as ethers that favor the formation of ion-paired carbanions. In these cases, the equilibrium measurements (Eq. 6) reveal the relative stabilities of the metal salts of the carbanions. If the interaction of the metal cation with the carbanion is weak or at least uniformly strong across a group of compounds, then the ion-pair acidities will parallel those observed in true pK_a measurements. On the other hand, if there are strong, specific interactions in the salts, then the carbanion may be preferentially stabilized causing its parent compound to appear unusually acidic. In Figure 3.2, the ion-pair acidities in THF and cyclohexylamine (CHA) of a series of hydrocarbons (e.g., cyclopentadiene and fluorene derivatives as well as di- and triphenylmethane) are plotted against the pK_a values in DMSO. Independent of the solvent (THF or CHA) or counterion (Li^+ or Cs^+), there are good correlations between the ion-pair and free-ion values. Clearly, the ion-pairing energies in the hydrocarbon derived carbanions are relatively constant and therefore do not affect the acidity measurements. The point that falls far below the correlation line is not for a hydrocarbon, but instead is for a ketone, p-biphenyl isopropyl ketone. Streitwieser and co-workers[131,132] determined lithium and cesium

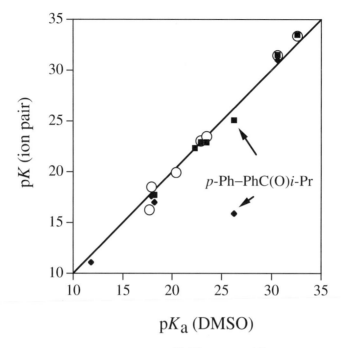

Figure 3.2. Plot of ion-pair acidities in THF[131–133] and CHA[125] versus free ion values in DMSO.[38] Filled squares are for cesium ion pairs and filled diamonds are for lithium ion pairs in THF. Open circles are for cesium ion pairs in CHA. For convenience, a line with slope = 1 and intercept = 0 has been placed on the plot.

ion-pair acidities for this compound in its monomeric form. The cesium ion-pair acidity [pK_{Cs}(THF) = 25.08] matches the DMSO pK_a value of a simpler analogue, phenyl isopropyl ketone (pK_a = 26.25), but the lithium ion-pair acidity is much lower [pK_{Li}(THF) = 15.88]. The difference between the ion-pair acidities in THF and free ion acidities in DMSO can be analyzed with the following set of linked equilibria suggested by Streitwieser and co-workers[133] (Scheme 3.3).

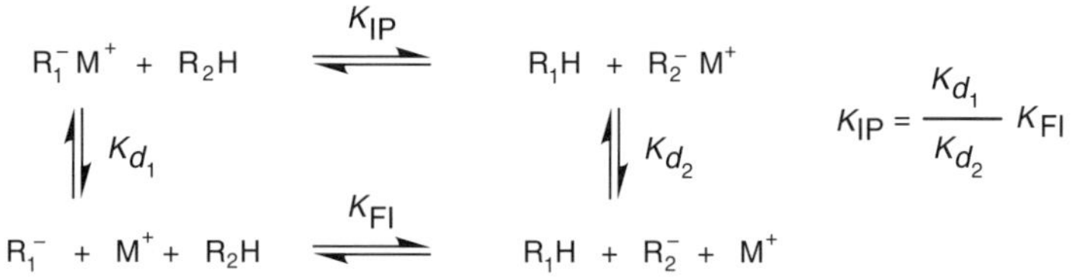

Scheme 3.3

If ion-pair dissociation constants, K_{d1} and K_{d2}, are roughly the same, which tends to be true for hydrocarbons, then the ion-pair K_{IP} and true, free-ion K_{FI} values will be approximately equal. This factor is why such a good correlation is seen in Figure 3.2 for the hydrocarbon species. Although the K_d values for the lithium and cesium salts of hydrocarbon-derived carbanions are different in THF,[133] the lithium values are uniformly ~100 times larger due to the stronger interaction of the cation with THF. This effect cancels out when both scales are arbitrarily referenced to the DMSO value for fluorene (22.9 on a per hydrogen basis). Note that the situation is somewhat more complicated in THF because two types of ion pairs are formed, contact ion and solvent-separated ion pairs. In the former, the cation directly interacts with the carbanion whereas in the later, the cation is fully solvated and it is the cation–solvent complex that interacts with the carbanion. Cesium generally prefers contact ion pairs, but lithium prefers solvent-separated ion pairs with delocalized hydrocarbon carbanions because it allows for strong electrostatic interactions of the lithium cation with the oxygens of THF.

For an ion-pair acidity value to differ significantly from the free-ion value, Scheme 3.3 indicates that there must be a large difference between K_{d1} and K_{d2}. In other words, the species of interest must have an ion-pair dissociation constant that is very different from the reference hydrocarbon used in the measurement (R_1H). For example, when R_2^- is an enolate, K_{d2} will be very small for lithium because there will be a strong electrostatic interaction with the highly charged oxygen of the enolate and the salt will resist dissociation. As a result, K_{IP} will be unusually large and the parent acid, R_2H, will appear to be unusually acidic under the ion-pairing conditions. The 9.2 pK difference between the lithium and cesium ion-pair acidities of p-biphenyl isopropyl ketone indicates that the formation of the lithium ion pair (relative to the cesium ion pair) provides roughly nine orders of magnitude of stabilization to the enolate. This is an overstatement of the true effect because lithium has a higher dissociation constant than cesium with the hydrocarbon reference species, but it still implies that K_d is nearly 10 million times smaller for lithium cations than cesium cations with this enolate. Obviously, there is a very strong electrostatic interaction between the small, densely charged lithium cation and the oxygen of the enolate (in this case, the lithium must be adopting a contact ion pair). The large cesium cation doesn't offer the high charge density of the lithium cation and therefore ion-pairing has little effect on the stability of the enolate relative to the delocalized carbanions used to develop the acidity scale (i.e., those that fit the correlation line in Figure 3.2), and hence the cesium ion-pair pK is close to the DMSO value.

Finally, it is worth noting that large differences in acidity can occur in switching to solvents capable of hydrogen bonding. As expected, the effect is closely related to nature of the atom bearing the bulk of the charge in the anion. Cox and Stewart[134] developed an acidity function that uses data in aqueous DMSO mixtures to extrapolate to a dilute aqueous reference state. With this approach, the pK_a of fluorene is nearly the same as in pure DMSO (22.1 vs. 22.6). However, acetone's pK value is >6 units lower in aqueous solution[100] than DMSO. Clearly, the localization of charge on the oxygen is allowing the acetone enolate to benefit more from the access to hydrogen bonding and it is greatly stabilized by the solvent relative to a delocalized carbanion derived from a hydrocarbon. Nitroalkanes also exhibit large pK_a shifts in going to hydrogen-bonding solvents due to the importance of a resonance form where the anion's charge is carried by the oxygens.[38]

4.3.3. Gas-Phase versus Condensed-Phase Acidities.

As noted above, in the absence of strong interactions with a counterion or hydrogen bonding, carbon acidities are relatively insensitive to the solvent. As a result, one expects, at least for hydrocarbons, that there would be a good correlation between gas- and condensed-phase acidities. Taft and Bordwell[135] compared data from the gas phase and DMSO and presented several conclusions about solvation effects. First, for hydrocarbons that give delocalized carbanions, there is an excellent correlation between the gas-phase and DMSO data, with nearly a 1:1 correspondence in the relative free energies of deprotonation—of course, the absolute free energies of deprotonation are much higher in the gas phase. A sample of such data is presented in Figure 3.3. Here, the linear fit has a slope of 1.026 and a correlation coefficient r^2 value of 0.980. Second, the addition of fused benzene rings to a cyclopentadiene core clearly reduces acidity in solution, but has little effect in the gas phase (see Fig. 3.1 for structures). In terms of relative free energies of deprotonation, DMSO yields the ordering cyclopentadiene (0.0 kcal mol^{-1}) < indene (2.9 kcal/mol) < fluorene (6.3 kcal/mol). In contrast, the gas-phase values are reversed and

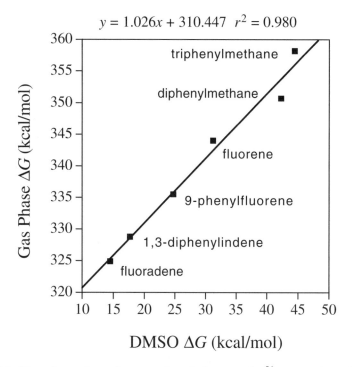

$$y = 1.026x + 310.447 \quad r^2 = 0.980$$

Figure 3.3. Plot of gas-phase free energies of deprotonation[54] versus those found in DMSO.[135]

closer together with cyclopentadiene (0.0 kcal/mol) > indene (−3.1 kcal/mol) > fluorene (−3.7 kcal/mol).[135] Despite all of them yielding highly delocalized hydrocarbon carbanions, the change in the size across this series has a marked effect on their solvation energies in DMSO. The effect is not limited to this solvent and a similar trend is observed for the cesium ion pairs of these carbanions in cyclohexylamine. Clearly, the larger carbanions benefit from greater charge delocalization in the gas phase, but this also reduces their solvation energy significantly and effectively destabilizes them in solution. Third, field effects caused by the interaction of an internal dipole with a carbanion are generally attenuated in solution. For example, perfluoro substitution on fluorene causes a 26.7 kcal/mol decrease in the gas-phase free energy of deprotonation, but only a 16.1 kcal/mol decrease in DMSO. The internal dipoles effectively provide another mode of charge delocalization, and therefore reduce the solvation energy of the carbanion. However, there are examples where the opposite is true. A para nitro group enhances the gas-phase acidity of toluene by 28.4 kcal/mol, but the effect in DMSO is 31.2 kcal/mol. In other words, the nitro group increases the solvation energy of the resulting benzylic carbanion. This increase suggests that the solvent must have favorable electrostatic interactions with the charge-bearing oxygens of the nitro group that can overcome the effects of delocalizing the charge (Scheme 3.4).

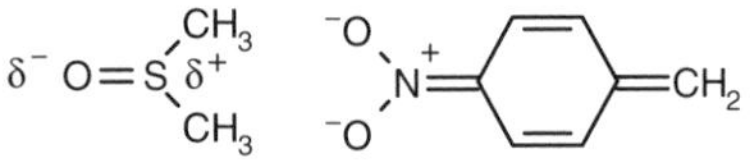

Scheme 3.4

4.3.4. Kinetic Acidities in the Condensed Phase.

For very weak acids, it is not always possible to establish proton-transfer equilibria in solution because the carbanions are too basic to be stable in the solvent system or the rate of establishing the equilibrium is too slow. In these cases, workers have turned to kinetic methods that rely on the assumption of a Brønsted correlation[22] between the rate of proton transfer and the acidity of the hydrocarbon. In other words, $\log k$ for isotope exchange is linearly related to the pK of the hydrocarbon (Eq. 13). The α value takes into account the fact that factors that stabilize a carbanion generally are only partially realized at the transition state for proton transfer (there is only partial charge development at that point) so the rate is less sensitive to structural effects than the pK. As a result, α values are expected to be between zero and one. Once the correlation in Eq. 13 is established for species of known pK, the relationship can be used with kinetic data to extrapolate to values for species of unknown pK.

$$\log k = -\alpha pK + \text{constant} \qquad (13)$$

As noted above, internal return can be a problem in kinetic acidity measurements, but Streitwieser et al.[136] showed that this is not a major impediment in cyclohexyl-

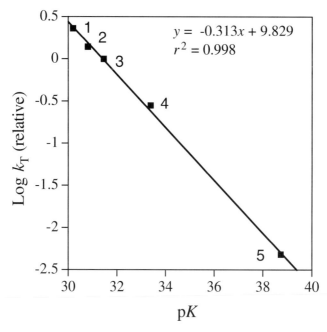

Figure 3.4. Plot of the rate of tritium exchange in lithium cyclohexylamide/cyclohexylamine versus cesium ion pair acidity in cyclohexylamine.[137] The following species are included: (1) p-biphenyldiphenylmethane; (2) di-p-biphenylmethane; (3) triphenylmethane; (4) diphenylmethane; (5) p-methylbiphenyl.

amine. The assumption presented in Eq. 13 appears to be valid as long as structurally similar species are compared. For example, there is an excellent linear relationship between the tritium isotope exchange rates of polyaryl methanes (e.g., triphenylmethane, diphenylmethane) in lithium cyclohexylamide/cyclohexylamine solution and their cesium ion-pair pK in cyclohexylamine (Fig. 3.4).[137] An α value of 0.31 is obtained for this set. When data for cyclopentadiene derivatives (e.g., indene, fluorene) are treated in a similar way, an α value of 0.37 is obtained.[138] In sharp contrast are highly localized carbanions such as polyhalogenated cycloalkanes and benzenes where α values near unity are obtained.[139] These results suggest that the transition state for proton transfer closely resembles the carbanion in localized systems like the benzenes and cycloalkanes, but when resonance delocalization is important in the carbanion, that stabilizing factor is not fully realized at the transition state and as a result has a lesser effect on the rate than the equilibrium acidity. This situation is easily rationalized by realizing that the transferring proton, with its significant positive charge, electrostatically localizes the charge at the site of deprotonation and limits the impact of resonance delocalization on the transition state's stability. This finding is illustrated in Scheme 3.5 for cyclopentadiene reacting with a base, B^-. The positively charged proton in the transition state forces the deprotonating carbon to remain pyramidal and retain the bulk of the developing negative charge, limiting the effects of resonance. In the product, the charge is fully delocalized and the effects of resonance stabilization are fully realized.

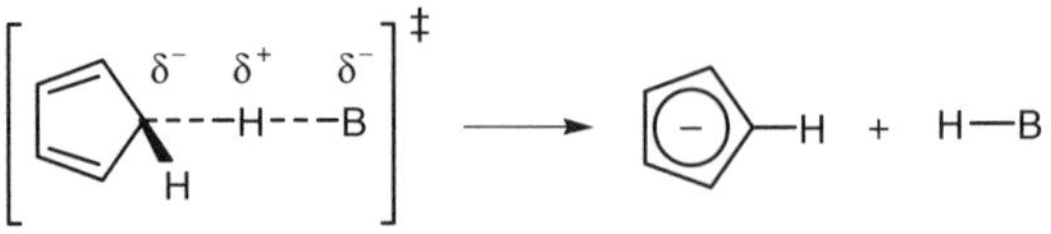

Scheme 3.5

In some cases, anomalous values have been obtained in rate–equilibrium relationships involving proton transfers. The most famous one involves nitroalkanes where substituents cause a greater effect on the relative rate of proton transfer than the position of the equilibrium.[140–144] In other words, a plot of log k versus pK gives a slope >1. Bordwell and Hughes[141] suggested an explanation much like the situation shown in Scheme 3.5. In the rate-determining step, hydrogen-bonding causes charge to be localized at the carbanion center rather than on the nitro group's oxygens. If the substituents are better able to interact with the carbanion center than the nitro group, their impact is greatest at the transition state (i.e., more charge is localized at the carbanion in the transition state). After the proton transfer is complete, the charge relocates onto the oxygens of the nitro group and the substituent's effect is diminished. Recent computational work supports this view.[145] However, it appears that nitroalkanes do not give anomalous proton-transfer behavior in the gas phase so the importance of a high degree of charge delocalization onto the nitro's oxygens seems to be a solvation phenomenon.[146] As noted in Section 4.3.3, solvation has marked effects on the stability of nitro-substituted carbanions.

In a novel kinetic approach, Dorfman et al.[147] developed methods for rapidly generating very reactive carbanions such as the benzyl anion in solvent mixtures containing water and alcohols. With pulsed radiolysis techniques, they have been able to study the fast and very exothermic reactions of carbanions with these solvents. The studies have shown that despite the high exothermicity, the protonation is not diffusion controlled and depends on the nature of the carbanion's counterion.

4.4. Carbon Acidities and Bond Strengths

A very useful thermodynamic cycle links three important physical properties: homolytic bond dissociation energies (BDE), electron affinities (EA), and acidities. It has been used in the gas phase and solution to determine, sometimes with high accuracy, carbon acidities (Scheme 3.6).[148] For example, the BDE of methane has been established as 104.9 ± 0.1 kcal/mol[149] and the EA of the methyl radical, 1.8 ± 0.7 kcal/mol, has been determined with high accuracy by photoelectron spectroscopy (PES) on the methyl anion (i.e., electron binding energy measurements).[55] Of course, the ionization potential of the hydrogen atom is well established, 313.6 kcal/mol, and as a result, a gas-phase acidity (ΔH_{acid}) of 416.7 ± 0.7 kcal/mol has been

$$BDE$$
$$AH \longrightarrow A\cdot + H\cdot$$

$$-EA(A\cdot)$$
$$A\cdot \longrightarrow A^-$$

$$IP(H\cdot)$$
$$H\cdot \longrightarrow H^+$$

$$\Delta H_{RXN}$$
$$AH \longrightarrow A^- + H^+$$

$$\Delta H_{RXN} = BDE - EA(A\cdot) + IP(H\cdot)$$

Scheme 3.6

determined for methane.[54] The cycle can also be used to calculate BDEs using acidity and EA data. Accurate bond strengths in fundamental species such as acetylene[150] and benzene[151] have been determined in this way. Finally, the same set of equations can be applied to condensed-phase data with the terms taking slightly different meanings (e.g., instead of EA, the reduction potential of the radical is used). This approach has been widely exploited by Bordwell and co-workers[152] who have used their extensive set of acidity measurements in DMSO as a foundation for determining a wide range of C–H BDE values. Note that earlier, Breslow[153] applied a related set of equations and used electrochemical data to estimate the acidity of very weak carbon acids.

5. REACTIVITY

Carbanions play critical roles in a wide variety of reaction pathways. As stated in the Introduction, this chapter will not focus on the synthetic utility of carbanions, but will instead focus on their mechanistic significance. In this section, a sample of important reaction mechanisms that involve transient or relatively short-lived carbanion intermediates will be introduced. As you will see, the key element in these mechanisms is the ability to form a carbanion that is reasonably stable, and often the kinetics of the reactions are dominated by carbanion stability. The role of carbanion intermediates in elimination reactions will be presented in some detail as a way to illustrate some of the methods that have been developed to probe for carbanion intermediates in reaction mechanisms. Other processes including additions and rearrangement reactions will be presented in less detail, but the role of carbanion stability in these reactions will be outlined.

5.1. Carbanion Intermediates in Elimination Reactions

It is well known that base-induced elimination reactions can proceed either by a single, concerted step (E2), or by two steps, proton transfer and leaving group expulsion, with a carbanion intermediate (E1cB) to yield an alkene.[154–156] The deciding factor between the two pathways is whether the carbanion intermediate is stable enough such that the leaving group expulsion step has a barrier on the potential energy surface. In simple alkyl halide systems (e.g., ethyl bromide), the carbanion is a high-energy species and the leaving group expulsion step does not have a barrier so the two steps, merge into one and an E2 mechanism is observed. However, if powerful carbanion stabilizing groups are introduced at the β-carbon, it is possible to shift the system to an E1cB pathway.

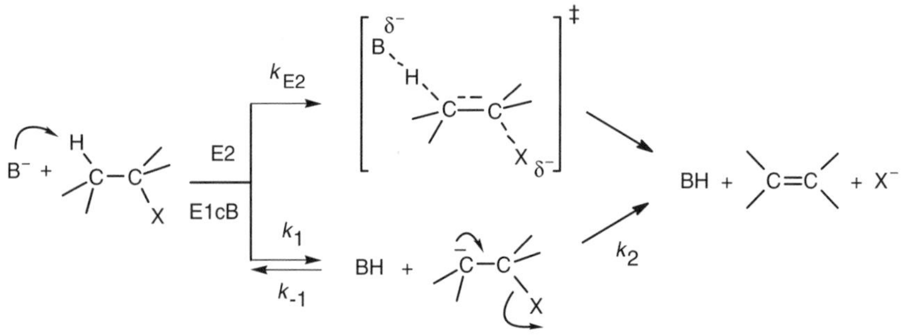

Scheme 3.7

As indicated in Scheme 3.7, the first step of an E1cB mechanism can be reversible and therefore deprotonation at the β-carbon does not always lead to product formation. By applying a steady-state approximation to the carbanion concentration, the following rate law is obtained for an E1cB reaction:

$$\text{rate} = k_{\text{obs}}[\text{RX}][\text{B}^-] \qquad k_{\text{obs}} = \frac{k_2 k_1}{k_{-1}[\text{BH}] + k_2} \qquad (14)$$

The carbanion can be destroyed in two ways, k_2 or k_{-1}, and two limiting types of behavior are expected for E1cB mechanisms. If $k_2 \gg k_{-1}[\text{BH}]$, then the carbanion always decomposes to the alkene product and the rate law simplifies to $k_{\text{obs}} = k_1$. In other words, the rate is only dependent on carbanion formation and the rate law has a form that is identical to what would be obtained in a concerted E2 reaction. This has been referred to as an E1cB$_{\text{IRR}}$ mechanism, where IRR indicates that deprotonation is irreversible. On the other hand, if $k_2 \ll k_{-1}[\text{BH}]$, then carbanion formation rarely leads to the product and the rate law simplifies to the following:

$$k_{\text{obs}} = \frac{k_2 k_1}{k_{-1}[\text{BH}]} \qquad (15)$$

These reactions often have been run in buffer solutions where [B$^-$] and [BH] represent the concentrations of the buffer's components. In Eq. 15, it can be seen that the rate depends on the concentration of the proton donor in the buffer solution. The factor k_1/k_{-1} also appears in the expression and is a measure of the relative acidities of the buffer acid [BH] and the elimination substrate [RX] (i.e., the more acidic the substrate, the higher k_1/k_{-1} and k_{obs}). The limiting behavior defined by Eq. 15 has been referred to as E1cB$_R$, where R indicates reversible deprotonation. Establishing experimental evidence for the E1cB mechanisms has proven to be challenging because their rate laws are nearly indistinguishable under many circumstances. As noted above, the E1cB$_{IRR}$ mechanism shares the same rate law as an E2 mechanism. Moreover, the rate law for an E1cB$_R$ mechanism takes the same form under common experimental conditions. Often in these elimination reactions, the solvent is the conjugate acid of the base (e.g., MeO$^-$/MeOH solutions). In this case, BH becomes a constant and Eq. 15 simplifies to $k_{obs} = k_2 k_1/k_{-1}$, the same form that is expected for an E2 or an E1cB$_{IRR}$ mechanism (i.e., k_{obs} is independent of any concentrations). Generally, it has taken the assembled data from a set of experiments to elucidate the mechanisms of E1cB processes.

There are a few obvious differences in the mechanisms that can be exploited in an effort to distinguish between them. First, the reversible nature of the first step in an E1cB$_R$ mechanism allows for the possibility of hydrogen/deuterium (H/D) exchange on the β-carbon in competition with elimination. Second, large primary deuterium isotope effects are expected for E2 and E1cB$_{IRR}$ mechanisms because C$_\beta$–H cleavage is part of the rate-determining step, whereas the preequilibrium nature of an E1cB$_R$ mechanism should lead to a much smaller deuterium isotope effect. Third, isotope or substituent effects associated with the leaving group should be most evident in E2 or E1cB$_R$ mechanisms because leaving group cleavage is part of the rate-determining step. Finally, the concentration of the conjugate acid of the base only appears in the rate law of an E1cB$_R$ mechanism. The work of Keeffe and Jencks[157,158] on the eliminations of p-nitrophenylethylammonium ions provides a good illustration of the strategies needed to solve this complex mechanistic problem. In this case, isotope exchange offers one clue to the mechanistic pathway. If an E1cB$_R$ mechanism is active, then isotope exchange at the β-carbon should be possible because in this mechanism, the β-carbon should be deprotonated/reprotonated multiple times before the elimination occurs (i.e., the first step is fast relative to the second one in the E1cB$_R$ mechanism). For example, when the substrate in Scheme 3.8 was treated with a basic buffer in D$_2$O, the ^{1}H NMR spectrum indicates that the β-hydrogens in the substrate are almost completely replaced by deuterium by the time that the elimination reaction has reached 50% completion.

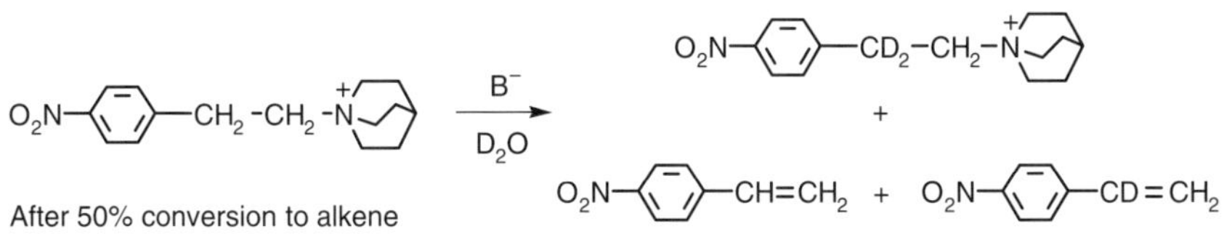

Scheme 3.8

It has been argued that the occurrence of isotope exchange at the β-carbon does not prove an $E1cB_R$ mechanism, only that a carbanion is formed under the reaction conditions—it is possible that an E2 process is responsible for product formation (i.e., $k_1 > k_{E2} > k_2 k_1 / k_{-1}[BH]$).[159] In other words, the carbanion is formed and exchanges quickly, but the elimination from the carbanion intermediate (k_2) is slow relative to a concerted E2 reaction. To further probe the mechanism, Keeffe and Jencks[157,158] examined the effect of buffer concentration and solvent isotope (H_2O vs. D_2O) on the rate. In their system, they found that increasing $[B^-]$ while keeping $[B^-]/[BH]$ constant led to a curved plot of rate versus $[B^-]$. This result is consistent with Eq. 14 (i.e., k_{obs} is dependent on [BH] and changes with increasing buffer concentration) and suggests a situation where $k_{-1}[BH]$ must at least be of comparable magnitude to k_2 (i.e., not an $E1cB_{IRR}$ mechanism). Here, increasing $[B^-]$ and [BH] causes the rate of deprotonation and reprotonation to increase, shifting the system toward the $E1cB_R$ extreme (Eq. 15). They also found a large, inverse solvent isotope effect, which again is consistent with Eq. 14. In D_2O, k_{-1} is diminished by a primary isotope effect, and therefore the carbanion intermediate has a longer lifetime and a greater propensity for forming the alkene product. Taken together, the data clearly point to an E1cB mechanism where k_2 is either partly or entirely rate determining (i.e., $E1cB_R$).

Identifying $E1cB_{IRR}$ reactions presents special problems because the rate law takes the same form as an E2 reaction and, like an E2 reaction, it should display a large primary kinetic isotope effect when the β hydrogens are substituted with deuterium. Distinguishing between them requires evidence concerning the cleavage of the C—X bond in the rate-determining step. For an E2 reaction, there should be extensive cleavage of this bond, but in an $E1cB_{IRR}$ reaction, it is only moderately weakened by a hyperconjugative interaction of the carbanion lone pair with the C—X bond. Curvature in linear free-energy relationships based on substituents attached to the leaving group has been used as evidence for the emergence of an $E1cB_{IRR}$ mechanism. Because the leaving group will bear a smaller negative charge in an $E1cB_{IRR}$ than an E2 mechanism, it should be less susceptible to the effects of electron-donating or -withdrawing groups. In the case of fluorenylmethyl benzenesulfonates (**11**) reacting with amine bases,[160] distinct curvature and a shift to a shallower slope (i.e., substituent has smaller effect on rate) is observed in a Hammett-type[107] substituent effect plot when electron-donating groups are added to the benzenesulfonate. Of course, these groups make the benzenesulfonate a weaker leaving group (i.e., stronger base) and apparently provide marginal stability to the carbanion intermediate and shift the mechanism from E2 to $E1cB_{IRR}$.

11   **12**

Another method of seeking evidence of the E1cB$_{IRR}$ mechanism is to exam heavy-atom isotope effects in the leaving group. Of course, these should be much more significant in an E2 process because the bond is breaking in the transition state. For example, Thibblin and co-workers[161] found that in the base-induced elimination of an alkyl halide in which the β-carbon is unusually acidic (indene derivative, **12**), moderately strong bases (triethylamine and methoxide) lead to a significant $^{35}Cl/^{37}Cl$ isotope effect ($k^{35}/k^{37} = 1.010 - 1.009$, where a maximum effect of 1.014 is expected for complete cleavage). When combined with large primary deuterium isotope effects on the β-carbon (7.1–8.4), these results indicate that both the C$_\alpha$—H and C$_\beta$—Cl bonds are breaking at the transition state, and hence it is a concerted E2 transition state. However, Saunders[162] used computational approaches to show that hyperconjugation to the leaving group in the proton-transfer transition state leading to an E1cB intermediate can be extensive and therefore large leaving group isotope effects are also possible in an E1cB$_{IRR}$ mechanism and are not necessarily reliable evidence of an E2 reaction.

5.2. Carbanion Intermediates in Addition Reactions

5.2.1. Nucleophilic Additions to Alkenes.
Nucleophilic additions to alkenes (Scheme 3.9) are mechanistically very closely related to an E1cB process. In fact, the addition process simply involves a reversal of the steps in response to an equilibrium constant that favors the addition product over the alkene. A notable example is the Michael addition of an enolate to an alkene bearing a strong electron-withdrawing group (EWG).

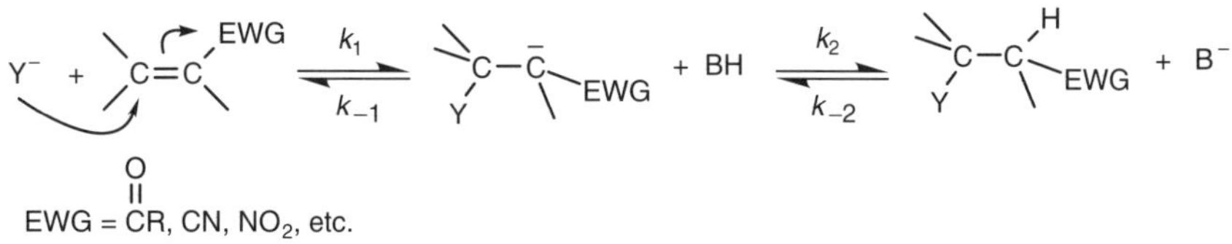

Scheme 3.9

In most practical applications, the rate-determining step is addition of the nucleophile (k_1) to give the carbanion intermediate, which is followed by a relatively fast protonation step (k_2). This situation represents the reverse of an E1cB$_R$ mechanism in that both fundamentally share the same rate-determining step, the step linking the alkene, and the carbanion intermediate. As expected, the rate constants of alkene additions are very sensitive to the nature of the EWG and their ability to stabilize the carbanion intermediate. An interesting example that points out the importance of carbanion stability involves additions to 9-methylenefluorene systems that bear strong electron-withdrawing substituents on the exocyclic methylene group (Scheme 3.10).[163] Here, addition could possibly occur either at the methylene group to give an aromatic fluorenyl carbanion or at the 9-position to give a

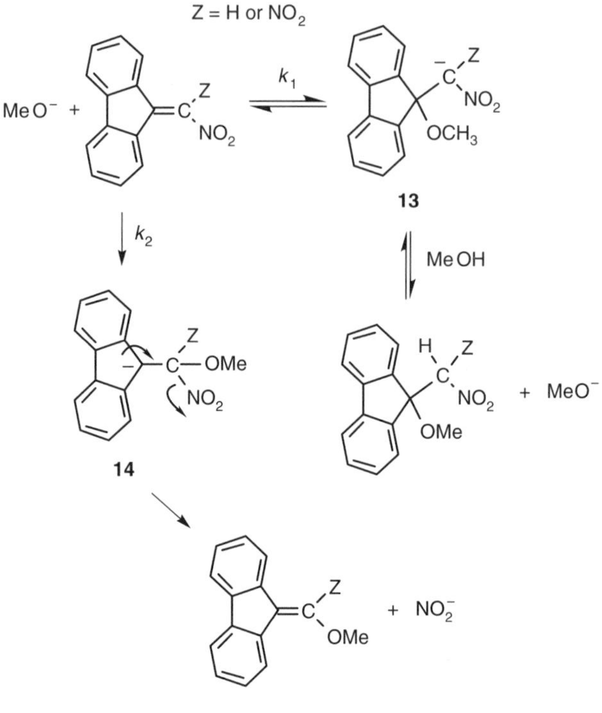

Scheme 3.10

carbanion that is stabilized by the attached EWGs. With a single nitro group on the methylene (Z = H), both pathways are active and addition at the 9-position is favored by a factor of ~4:1 for $Y^- = MeO^-$ in MeOH solvent. Attack at the methylene carbon does not lead to a stable addition product, but instead the carbanion eliminates nitrite to complete a vinylic substitution process, another example of a mechanism that incorporates a carbanion intermediate. This pathway becomes available because carbanion **14** has a good leaving group, NO_2^-, at the β-position and the overall process is exothermic (NO_2^- is a better leaving group than MeO^-). This result again points to the similarities between these carbanion intermediates and those of an E1cB process. In a solvent such as methanol, one expects that the nitro stabilized anion, **13**, would be more stable than the fluorenyl anion, **14**, because in this type of solvent nitro-substituted carbons have pK_a values near 10, whereas fluorenes typically are in the low 20's. Given this large difference in carbanion stability, one might expect addition to give **13** would be preferred by a much larger factor, but as noted earlier (Section 4.3.4), delocalization of carbanion lone pairs into nitro substituents is not fully realized at the transition states for their formation and therefore kinetically, nitro groups do not appear to be as stabilizing.[140–144,164] In fact, there is a good correlation between the kinetics of forming stabilized carbanions either via proton transfer (i.e., kinetic acidities) or by alkene additions.[163] When Z = NO₂, addition occurs exclusively at the

9-position to give carbanion **13** because the two nitro groups lead to a carbanion that is much more stable than a fluorenyl anion.[100] The substitution pathway that appears in Scheme 3.10 is particularly common in reactions of vinylic halides that bear good EWGs on the β-carbon. The reaction in Eq. 16 illustrates the importance of carbanion stability on the reaction rates of these processes.[165]

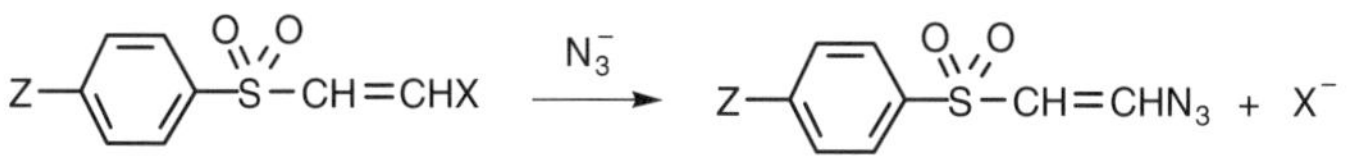

$$
Z-\underset{\substack{}}{\underset{\overset{O_\backslash\backslash{/}O}{\overset{\|}{S}}-CH=CHX}{\bigcirc}} \quad \xrightarrow{N_3^-} \quad Z-\underset{\substack{}}{\underset{\overset{O_\backslash\backslash{/}O}{\overset{\|}{S}}-CH=CHN_3}{\bigcirc}} + X^- \quad (16)
$$

Z = Me or NO$_2$; X = Cl or Br

Changing the leaving group from Cl$^-$ to Br$^-$ causes a rate increase of only about a factor of 2, whereas changing Z from Me to NO$_2$, increases the rate by roughly a factor of 50. Clearly, stabilization of the carbanion is dominating in this system and leaving group ability only has a minor effect on the transition state for addition.

Alkene addition processes have also been characterized in the gas phase via MS and carbanion intermediates have been identified. For example, Bernasconi et al.[166] showed that nucleophiles like acetone enolate readily add to alkenes with good EWGs such as a nitrile (i.e., acrylonitrile). One aspect of the gas-phase work is that localized nucleophiles commonly used in solution such as MeO$^-$ are strong bases in the gas phase, so proton abstraction competes with addition. In the case of acrylonitrile, MeO$^-$ gives 100% proton abstraction whereas acetone enolate gives 100% addition (Scheme 3.11). The absence of proton transfer with the enolate is understandable because it is an endothermic process in the gas phase. The preference for proton abstraction with MeO$^-$ is a result of the process being significantly exothermic ($\Delta H° = -10$ kcal/mol) and entropically much more favorable than the addition process. Deprotonated nitromethane illustrates a third variation in that after the addition process, a subsequent intramolecular substitution process occurs that leads to the formation of a cyclopropane derivative. A related process, the Favorskii rearrangement, is presented in Section 5.3.3.

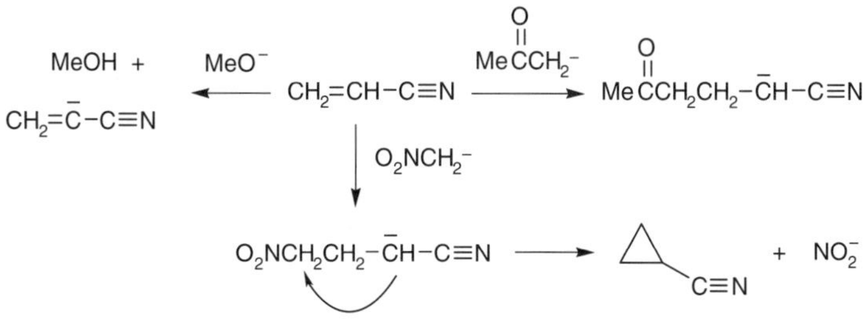

Scheme 3.11

5.2.2. Nucleophilic Aromatic Substitution.

A natural extension of alkene addition processes is aromatic nucleophilic substitution. Again, the ease of the process is highly dependent on the stability of the intermediate carbanion and strong EWGs are needed to facilitate these reactions in solution. The classic example is the reaction of a picryl (2,4,6-trinitrophenyl) ether with an alkoxide (Eq. 17). This reaction leads to a stable carbanion complex whose structure was initially confirmed by Meisenheimer, and they have come to be known as Meisenheimer complexes.[167]

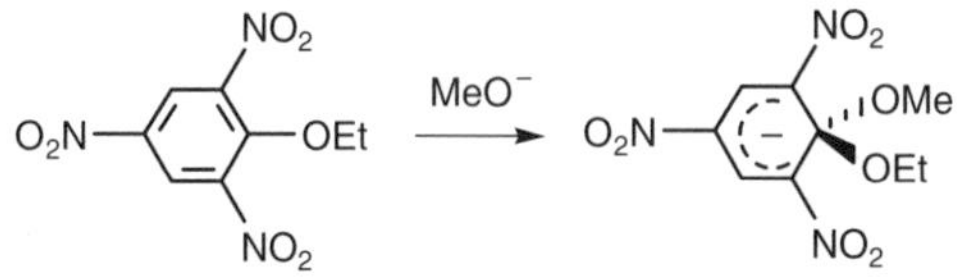

$$\text{(17)}$$

The high stability of the carbanion intermediate in this case is not surprising because it is delocalized (pentadienyl) and bears three strong EWGs. The intermediate carbanion need not be stable and if it is not, a substitution process occurs. In Eq. 18, this is illustrated for the gas-phase reaction of MeO^- with fluorobenzene.[168]

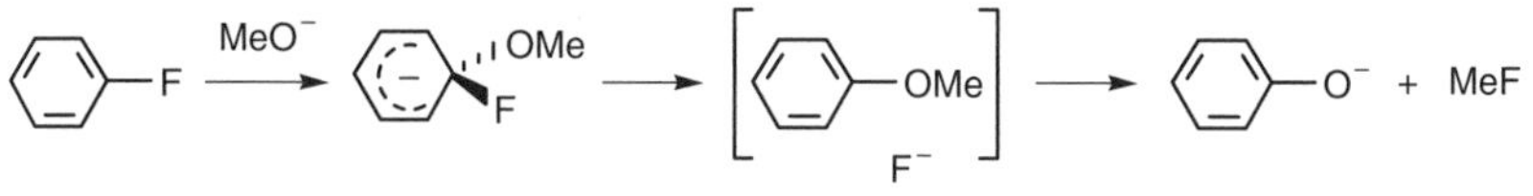

$$\text{(18)}$$

Localized anions such as methoxide are very strong bases in the gas phase, so the addition to give the pentadienyl carbanion intermediate is favorable under these circumstances. This outcome can be rationalized in the following way. The proton affinity of methoxide is 382 kcal/mol in the gas phase, whereas the proton affinity of the pentadienyl anion is only 369 kcal/mol. Of course, the carbanion intermediate will have an even lower proton affinity because it is stabilized by the presence of a pair of EWGs, the fluoro and methoxy groups. The conversion to a much weaker base as well as the formation of a new, strong C—O bond can overcome the disadvantage of the loss of aromaticity in forming the Meisenheimer complex in this case. In solution, species like methoxide are much weaker bases, so EWGs are needed to make the Meisenheimer complex a viable intermediate. In the gas-phase reaction (Eq. 18), the overall process can be completed by an S_N2 reaction between the departing fluoride and the anisole. Again, this can be understood in terms of proton affinities. In the product complex, fluoride with a proton affinity of 371 kcal/mol is a stronger base than phenoxide, whose proton affinity is 350 kcal/mol, and therefore, an exothermic S_N2 reaction is possible before the products separate. The mechanisms of nucleophilic aromatic substitution have been widely studied and the results point to a delicate balance with variations in the rate-determining step depending on the nature of the nucelophile, leaving group, and solvent. Descriptions of these studies have appeared in various reviews.[169–171]

5.3. Carbanion Intermediates in Rearrangements

Carbanions also appear as intermediates in rearrangement processes. In some cases, this involves the rearrangement from one carbanion to another, but in other cases, carbanions are involved in the conversion to a heteroatom-centered anion such as an alkoxide.

5.3.1. Wittig Rearrangement.

5.3.1. Wittig Rearrangement. There is a continuing controversy over the role of carbanion intermediates in this process. The reaction involves the formation of a carbanion at the α-carbon of an ether leading to a rearrangement that produces an alkoxide (Eq. 19).[172]

$$R_1{-}\overset{\bar{}}{C}H{-}O{-}R_2 \quad \underset{R_1{-}\overset{\cdot}{C}H{-}O^-}{\overset{R_1{-}CH{=}O}{\rightleftharpoons}} \quad \overset{R_1}{\underset{R_2}{\diagdown}}CH{-}O^- \tag{19}$$

It is formally a 1,2 migration of the alkyl group, R_2, from oxygen to carbon and in solution, generally requires a strong base such as an alkyllithium salt to form the initial carbanion. The issue is the nature of the intermediate, which could either be a carbanion–carbonyl or radical anion–radical complex pair (Eq. 19). The real question is whether the transferring alkyl group or the carbonyl has a higher effective electron affinity in the intermediate complex. Both carbonyl compounds and simple alkyl groups are known to have low electron affinities, so the question is not straightforward to answer. Often R_1 is an aromatic group that helps to stabilize the initial carbanion and would also stabilize a potential carbonyl radical anion. On the other hand, the dipole of the carbonyl in the complex would electrostatically stabilize a carbanion intermediate and increase the apparent electron affinity of the transferring alkyl group. Data in solution are not conclusive, but suggest a radical anion mechanism. For example, the migratory aptitude of the alkyl groups (e.g., $R_2 = $ benzyl > methyl, ethyl > phenyl) parallels radical stability more closely than the expected carbanion stability.[173] In addition, Garst and Smith[174] showed that when the migrating alkyl group was 5-hexenyl, it undergoes, to some extent, a cyclization process well known for the 5-hexenyl radical (Eq. 20). However, the reaction has also been shown to produce intermolecular products that are most consistent with a true carbonyl intermediate that escapes the complex with the transferring carbanion. For example, the reaction of dibenzyl ether with methyllithium produces varying yields of 1-phenylethanol, the condensation product of benzaldehyde and methyllithium (Eq. 21) suggesting that the aldehyde escaped from the benzyl anion in the initial complex and was trapped by excess methyllithium rather than the benzyl anion.[175]

$$Ph_2\overset{\bar{}}{C}{-}O{-}(CH_2)_4CH{=}CH_2 \longrightarrow$$

$$\underset{Ph_2\overset{O^-}{\underset{|}{C}}{-}CH_2{-}\!\!\bigtriangleup}{} \; + \; Ph_2\overset{O^-}{\underset{|}{C}}{-}(CH_2)_4CH{=}CH_2 \; + \; \text{other products} \tag{20}$$

$$\bigcirc\!\!{-}CH_2{-}O{-}CH_2{-}\!\!\bigcirc \; \xrightarrow{\text{MeLi}} \; \bigcirc\!\!{-}\overset{O^-}{\underset{|}{C}}H{-}CH_2{-}\!\!\bigcirc \; + \; \bigcirc\!\!{-}\overset{O^-}{\underset{|}{C}}H{-}Me \tag{21}$$

Gas-phase Wittig rearrangements are also known and similar migratory propensities have been observed, but the data have been interpreted in terms of a carbanion intermediate mechanism.[176] Computational modeling offers an explanation for the seemingly conflicting results on this mechanism.[177] For a bare carbanion (i.e., gas phase), high-level calculations favor a heterolytic cleavage of the C—O bond to give a carbonyl–carbanion pair. In contrast, complexation of a lithium cation to the system preferentially stabilizes the carbonyl radical anion via a strong O^-/Li^+ electrostatic interaction and favors a pathway involving radical migration. When solvation is included in the computational model, the effect of the lithium is weakened and leads to a situation where competition between the mechanisms might occur.

With allylic alcohols, there is the possibility of a [2,3] variant of the Wittig rearrangement that can compete with the [1,2] rearrangement described above (Eq. 22, the indicated electron flow is for the [2,3] rearrangement).[178] The reaction is expected to be a one-step, pericyclic process without a distinct carbanion intermediate. This rearrangement has proven to be useful synthetically because its concerted nature can lead to high stereoselectivity.[179]

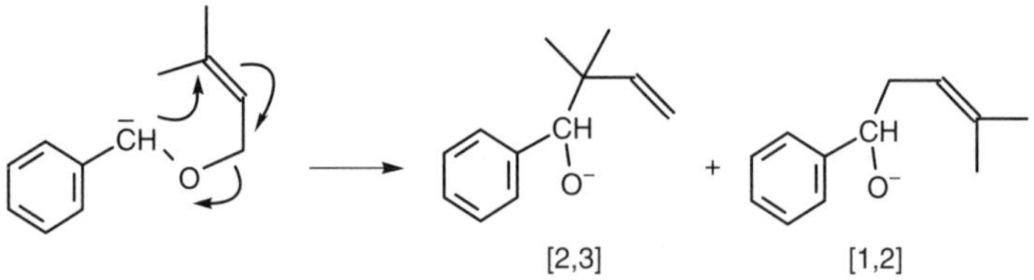

$$[2,3] \qquad\qquad [1,2] \qquad\qquad (22)$$

5.3.2. 1,2 Phenyl Migrations.
Phenyl migrations of carbanion intermediates are an intramolecular example of nucleophilic aromatic substitution. They often occur without the need of an activating group and are facilitated by an entropically favorable transition state because the nucleophile and phenyl ring are always in close proximity. A classic example is provided by Zimmerman and Zweig[180] who found that 2,2-diphenyl-1-propyllithium derivatives rearrange to give more stable 1,2-diphenyl-2-propyllithiums (Eq. 23).

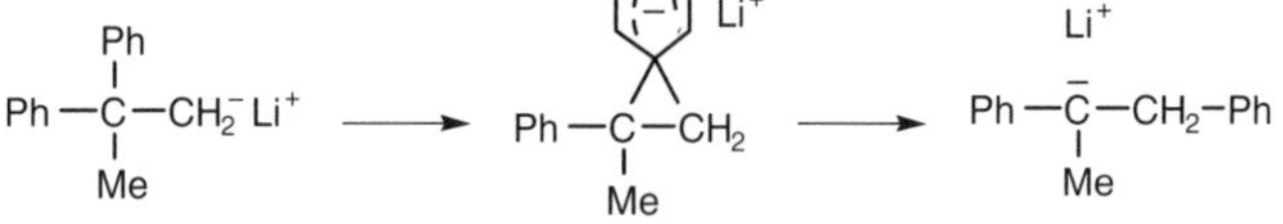

$$\qquad\qquad\qquad\qquad\qquad\qquad\qquad\qquad (23)$$

15

The spirocyclic species, **15**, on the reaction path could either be an intermediate or a transition state, but low-temperature NMR studies have identified a closely related species as a stable compound and proven its structure via reaction chemistry.[181] However, there is evidence that this type of reaction could also proceed via a dissociative mechanism where a carbanion is eliminated to form an alkene intermediate followed by recombination. Of course, this is simply the carbon-centered equivalent of a Wittig rearrangement. Evidence for this pathway can be found in the rearrangement of 2,2,3-triphenyl-1-propyllithium in the presence of carbon-13 labeled benzyllithium (Scheme 3.12) When the rearrangement product is trapped by reaction with CO_2, the label is statistically incorporated in the resulting carboxylic acid (i.e., in proportion to the amount of the labeled benzyllithium). Grovenstein and co-worker[182] completed a wide range of studies on this rearrangement and found that the mechanism is rather sensitive to the reaction conditions, especially the nature of the ion-pairing interaction with the metal counterion. Analogous migrations involving larger spirocyclic systems (i.e., four- and five-membered rings fused to the migrating phenyl) are also known.[183]

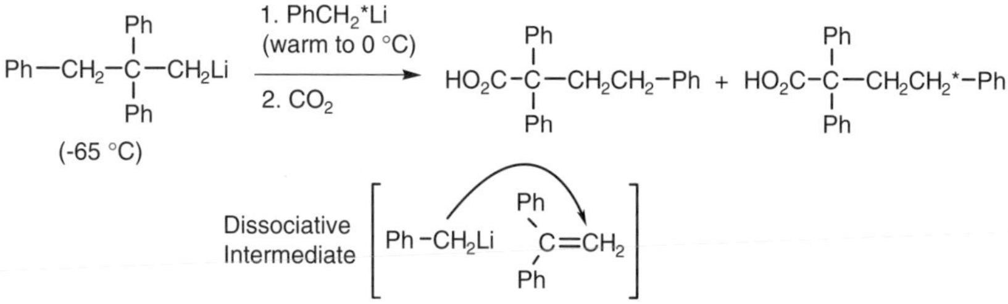

Scheme 3.12

5.3.3. Favorskii Rearrangement.

Like a 1,2 phenyl migration, the Favorskii[184] rearrangement involves a three-membered ring intermediate. When an α-halo ketone is treated with a base, the resulting enolate can cyclize to yield a cyclopropanone derivative by displacing the halide. (Scheme 3.13). In the presence of an alcoholic base, addition occurs at the carbonyl carbon causing fragmentation

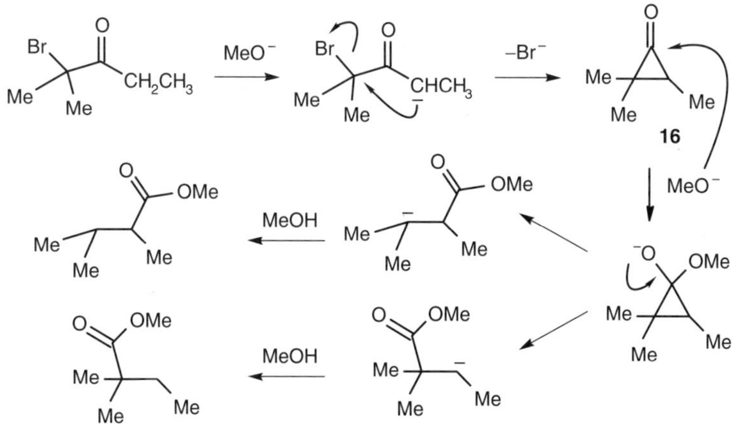

Scheme 3.13

of the ring and the production of a localized carbanion that is trapped by solvent. The driving force for forming the highly basic carbanion is undoubtedly the release of ring strain. In acyclic ketones, the process results in the formation of an ester with the transfer of one of the ketone's alkyl groups to the α position of the opposing alkyl group. In cyclic ketones, a ring contraction occurs (e.g., see Eq. 24). Support for the mechanism in Scheme 3.13 comes from the fact that when the cyclopropanone intermediate, **16**, is prepared independently and treated with MeO⁻, a similar mixture of esters results.[185,186] As is the case in many of these carbanion systems, other pathways are possible and the mechanism can shift to alternative routes with variations in the substituent patterns. For example, Bordwell and Carlson[187] showed that in some cases, the addition of methyl substituents on the halogen-bearing carbon can alter the rate-determining step and lead to substitution rather than rearrangement products (i.e., the halogen undergoes a solvolysis reaction). In addition, computational work suggests that in the reaction of 2-chlorocyclobutanone with hydroxide, the Favorskii rearrangement involves direct addition of the nucleophile at the carbonyl followed by ring-opening–ring-closing to give the expected ring contraction product (Eq. 24).[188] Apparently the high strain of a bicyclobutanone inhibits the conventional pathway (i.e., analogous to that in Scheme 3.13), so addition at the carbonyl becomes the more favorable process.

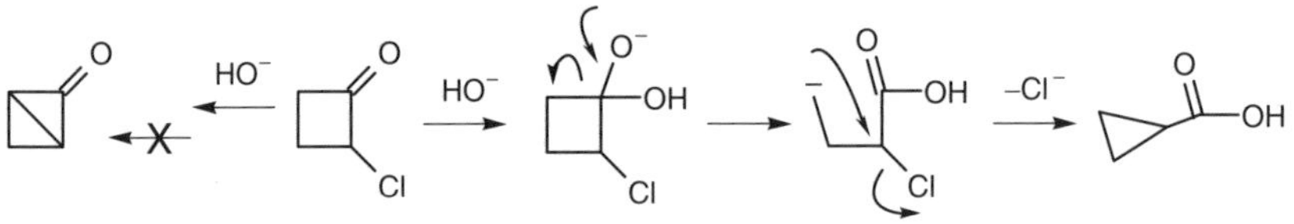

$$(24)$$

5.4. Carbanion Reactions in the Gas Phase

As noted in Section 4.2.1, the gas phase has proven to be a useful medium for probing the physical properties of carbanions, specifically, their basicity. In addition, the gas phase allows chemists to study organic reaction mechanisms in the absence of solvation and ion-pairing effects. This environment provides valuable data on the intrinsic, or baseline, reactivity of these systems and gives useful clues as to the roles that solvent and counterions play in the mechanisms. Although a variety of carbanion reactions have been explored in the gas phase,[189] two will be considered here (1) S_N2 substitutions and (2) nucleophilic acyl substitutions. Both of these reactions highlight some of the characteristic features of gas-phase carbanion chemistry.

5.4.1. S_N2 Reactions. Given its central role in the development of modern physical organic chemistry, it is no surprise that the S_N2 substitution reaction is the most widely studied of all gas-phase anionic processes. In a classic study, Olmstead and Brauman[190] used an ion cyclotron resonance (ICR) spectrometer to obtain

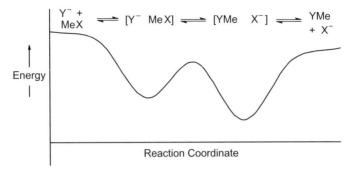

Figure 3.5. Diagram of a double-well potential energy surface for an S_N2 reaction.[190]

data on gas-phase S_N2 reactions and suggested that these reactions involve a double-well potential energy surface (Fig. 3.5). The reactions, although exothermic, do not occur at the collision controlled rate so a barrier must exist on the surface. On the other hand, there is always a long-range attractive potential between an ion and a neutral molecule (i.e., either ion–dipole or ion–induced-dipole), so the initial part of the surface must be attractive. The simplest way to account for both of these effects is a double-well potential where the initial attraction leads to a minimum on the surface for a loosely bound ion–molecule complex. From the complex, the energy rises as the system passes through the S_N2 transition state and finally drops as an ion–molecule complex is formed between the products. Given the low pressures generally employed in mass spectrometers, no collisions occur during the reaction process so the system retains all its energy (i.e., it is still at the energy level of the separated reactants) and the final complex possesses enough energy to dissociate to give the separated products. This type of surface is not unique to S_N2 reactions and is observed in many ionic gas-phase reactions.

In Table 3.8, rate constants are given for the reactions of several carbanions with methyl chloride. Most of the data was obtained in a flowing-after glow instrument.

TABLE 3.8. Rate Constants for Reactions with Methyl Chloride[a]

Nucleophile	PA (kcal/mol)[b]	Rate[c]
Ph^-	402	8.7
$MeSCH_2^-$	390	3.0
$CH_2CHCH_2^-$	389	2.9
$PhCH_2^-$	381	0.15
HCC^-	378	1.3
CF_3^-	377	0.56

[a] Data from Ref. 189.
[b] Proton affinity data from Ref. 54.
[c] Rates in units of 10^{-10} cm^3 molecule^{-1} s^{-1}.

In this method, the reactant ions are entrained in a fast flow of helium containing a small amount of a reagent gas. Reactions occur in the flowing helium on a time scale of ~10 ms and the products are detected at the end of the flow tube by a mass analyzer, usually a quadrupole. The data indicate that the S_N2 reactivity of a carbanion is closely related to its basicity, but not perfectly. Specifically, the benzyl anion is more basic than acetylide, but reacts nearly 10 times slower. Benzyl also has a lower rate constant than CF_3^-, another less basic carbanion. However, a low S_N2 rate constant for the benzyl anion is not unexpected. This carbanion derives much of its stability from its ability to delocalize charge into the benzene ring. It is logical that with reduced charge density at the benzylic carbon, the carbanion is less nucleophilic than a highly localized carbanion such as acetylide. Similar effects were noted in Section 4.3.4 when Brønsted correlations were discussed with respect to proton-transfer rates involving delocalized carbanions.

When a ketone or aldehyde enolate is allowed to react with an alkyl halide, there are two possible modes of attack because the anion can act as a carbon or oxygen nucleophile. Determining the site of alkylation has proven to be a challenging task in gas-phase studies because mass spectrometers are only able to identify charged products, but the characteristic ones in this case are neutral. The most direct evidence comes from a very difficult experiment involving the collection of the neutral products in a flowing afterglow mass spectrometer. In this way, Ellison and co-workers[114] showed that for the reaction of cyclohexanone enolate with MeBr, the sole product is the result of oxygen alkylation (Scheme 3.14). Of course, alkylation at carbon would be more exothermic, but the greater charge density on the oxygen makes it the more nucleophilic site. This type of behavior has also been seen in condensed-phase reactions of enolates.[113,115]

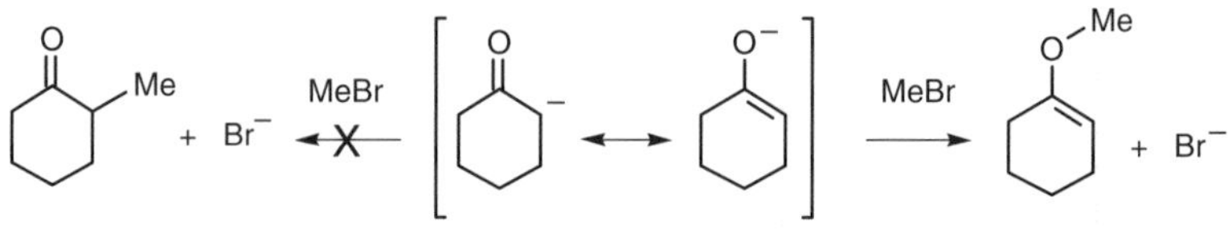

Scheme 3.14

5.4.2. Nucleophilic Acyl Substitution.
Studies of nucleophilic acyl substitution have also been completed in the gas phase. As in the condensed phase, esters offer a range of reactivity with strong bases. Two modes of attack are common: (1) the nucleophile can attack the carbonyl carbon in a nucleophilic acyl substitution process or (2) it can act as a base and deprotonate the ester at the α-carbon. In addition, the nucleophile can attack the alkyl group of the ester in either an S_N2 substitution or E2 elimination process. An interesting comparison can be made between the reactivity of CF_3^- and $N{\equiv}CCH_2^-$ in their reactions with methyl acetate. The latter ion undergoes a Claissen-type reaction with the ester to produce a stabilized enolate (Scheme 3.15).[191] After the nucleophilic acyl substitution process is complete, the methoxide is left in an ion–molecule complex with cyanoacetone, and a proton transfer occurs before the complex separates giving the enolate. The tetrahedral addition species is shown as an intermediate in this case (i.e., it is a

minimum and is formed without crossing a barrier), but in acyl substitutions on other substrates, such as acid chlorides, Brauman has shown that the tetrahedral species can be an energy maximum (i.e., transition state) on the potential energy surface.[192,193] With CF_3^-, an addition product is formed, but it does not continue with the substitution process.[194] The difference in reactivity between the two anions can be understood from the thermodynamics of the reactions. The cyanoacetone that is formed in the reaction with $^-CH_2C{\equiv}N$ is expected to be very acidic given that ketones are reasonably acidic in the gas phase and cyano is a powerful acidifying group. The enhanced acidity allows the overall substitution–methanal expulsion process to be exothermic. With CF_3^-, the resulting ketone, $MeC(O)CF_3$, is not quite acidic enough to allow for an exothermic substitution process. Of course, it is more acidic than MeOH in the gas phase (i.e., MeO^- can deprotonate it), but not by enough to overcome the inherent endothermicity of the conversion of an ester to a ketone. With both nucleophiles, proton transfer also occurs and produces the ester enolate of methyl acetate.

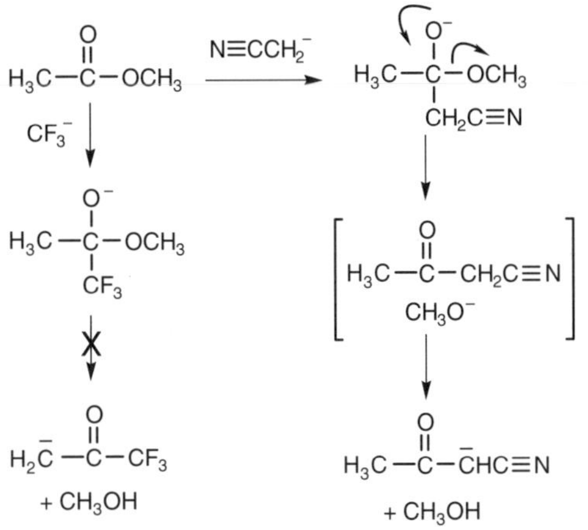

Scheme 3.15

When CF_3^- is allowed to react with methyl benzoate, a new reaction channel appears along with the formation of the carbonyl addition product (Eq. 25).[194] The nucleophile attacks at the methyl group to give an S_N2 substitution with the formation of the benzoate ion. In this case, proton transfer is not possible (no α hydrogens), but benzoate is a better leaving group than acetate and substitution at the methyl group becomes viable.

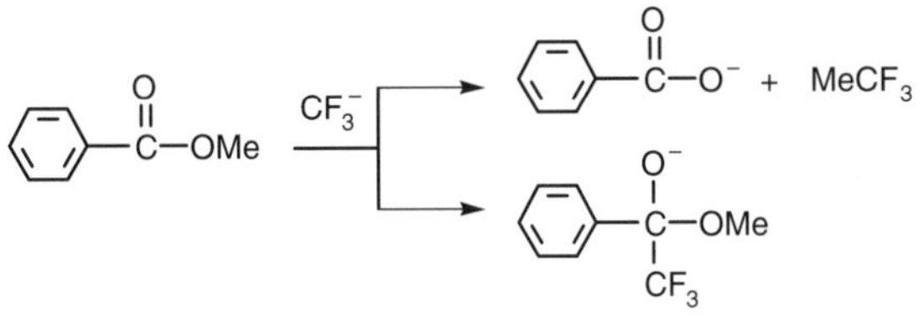

(25)

6. CONCLUSION AND OUTLOOK

Although past studies have laid out an impressive foundation for our understanding of carbanions as reactive intermediates, there are still many facets of the subject to be explored. Three areas seem particularly rich for future studies. First, work in the 1990s began the process of exploring carbanion salts under conditions that are similar to those employed in organic synthesis. The solvents, concentrations, and cosalts that are involved in synthesis offer a range of added complications in the study of carbanion reactivity, particularly in terms of the ion pairing of the carbanion. Work by Collum,[195–197] Streitwieser,[95,131,132,198] their co-workers, and others[199–202] has pointed to a wide variety of aggregates that may be simultaneously present in solution and have differing reactivities. A key issue is unraveling which of these aggregate structures is responsible for the observed chemistry and requires a knowledge of the concentrations of the various aggregates as well as their relative reactivities. The situation is complicated by the fact that the ionic products of the reactions can join into the mix of aggregates and alter the kinetics of the reaction as it progresses. This area is barely tapped and future experimental and computational work on the problem should offer important insights of synthetic utility.

Second, although a considerable body of work on carbanions already has been completed in the gas phase, much remains to be done.[189] The gas phase offers two attractive features in studying reactive intermediates. For one, by eliminating solvent and ion-pairing effects, it provides a baseline for judging the intrinsic factors that control carbanion stability and reactivity. Comparisons between gas-phase and condensed-phase data give clear clues as to the roles that solvent and counterions play in stabilizing carbanions and directing their reactivity. In the next few years, much will be learned as chemists explore a wider range of carbanion reactions in the gas phase. In addition, gas-phase studies, particularly those that also incorporate spectroscopy (e.g., PES), offer the potential of gaining exquisitely accurate data on the basicity of carbanions as well as their structural properties.[203,204]

Finally, carbanions play important roles in a number of biological processes and many enzyme-catalyzed reactions involve carbanions as reactive intermediates.[205–209] These systems can be exceptionally large and complicated, and offer new challenges in terms of methodologies. Problems involving carbanion intermediates in biology will be addressed by not only the standard approaches of physical organic chemistry, but also the emerging technologies of molecular biology and computational modeling. One already is seeing powerful collaborations of these methods in pursuit of detailed mechanisms for biologically relevant transformations and undoubtedly this area will see much growth in the next decade.[210,211]

SUGGESTED READING

E. Buncel and J. M. Dust, *Carbanion Chemistry: Structures and Mechanisms*, American Chemical Society, Washington, DC, **2002**.

D. J. Cram, *Fundamentals of Carbanion Chemistry*, Academic Press, New York, **1965**.

E. Buncel and T. Durst, Eds., *Comprehensive Carbanion Chemistry, Part A: Structure and Reactivity*, Elsevier, New York, **1980**.

E. Buncel and T. Durst, Eds., *Comprehensive Carbanion Chemistry, Part B: Selectivity in Carbon–Carbon Bond Forming Reactions*, Elsevier, New York, **1983**.

E. Buncel and T. Durst, Eds., *Comprehensive Carbanion Chemistry, Part C: Ground and Excited State Reactivity*, Elsevier, New York, **1987**.

R. B. Bates and C. A. Ogle, *Carbanion Chemistry*, Springer-Verlag, Berlin, 1983.

E. M. Kaiser and D. W. Slocum, in *Organic Reactive Intermediates*, S. P. McManus, Ed., Academic Press, New York, **1973**.

T. H. Lowry and K. S. Richardson, *Mechanism and Theory in Organic Chemistry*, 3rd ed., Harper Collins, New York, **1987**, pp. 284–308 and 517–540.

R. R. Squires, "Gas Phase Carbanion Chemistry," *Acc. Chem Res.* **1992**, *25*, 461.

W. B. Farnham, "Fluorinated Carbanions," *Chem Rev.* **1996**, *96*, 1633.

REFERENCES

1. M. Schlosser, *Organometallics in Synthesis : A Manual*; 2nd ed., John Wiley & Sons, Inc., New York, **2002**.

2. J. H. Bateson and M. B. Mitchell, *Organometallic Reagents in Organic Synthesis*, Academic Press, London, **1994**.

3. F. R. Hartley, *Organometallic Compounds in Organic and Biological Syntheses*, John Wiley & Sons, Inc., New York, **1989**.

4. S. Patai, Ed., *The Chemistry of the Carbonyl Group*, Vol. 1, Wiley-Interscience, London, **1966**.

5. R. E. Gawley, *Principles of Asymmetric Synthesis*, Elsevier, Oxford, **1996**.

6. B. J. Wakefield, *Organolithium Methods*, Academic Press, London, **1988**.

7. B. J. Wakefield, *Organomagnesium Methods in Organic Synthesis*, Academic Press, London, **1995**.

8. J. C. Stowell, *Carbanions in Organic Synthesis*, Wiley-Interscience, New York, **1979**.

9. V. Grignard, *Ann. Chim.* **1901**, *24*, 433.

10. M. S. Kharasch and O. Reinmuth, *Grignard Reactions of Nonmetallic Substances*, Prentice Hall, New York, **1954**.

11. W. Hallwachs and F. Schafarik, *Annals* **1859**, *109*, 206.

12. A. Cahours, *Annals* **1859**, *114*, 227.

13. W. Schlenk and J. Holtz, *Berchte* **1917**, *50*, 271.

14. A. Streitwieser, Jr., S. M. Bachrach, A. Dorigo, and P. v. R. Schleyer, in *Lithium Chemistry: A Theoretical and Experimental Overview*, A.-M. Sapse, and P. v. R. Schleyer, Eds., Wiley-Interscience, New York, **1995**.

15. G. Jones, *Org. React.* **1967**, *15*, 204.

16. E. Knoevenagel, *Berchte* **1898**, *31*, 2596.

17. L. Claisen, *Berchte* **1890**, *23*, 977.

18. J. B. Conant and G. W. Wheland, *J. Am. Chem. Soc.* **1932**, *54*, 1212.

19. W. K. McEwen, *J. Am. Chem. Soc.* **1936**, *58*, 1124.

20. D. E. Applequist and D. F. O'Brien, *J. Am. Chem. Soc.* **1963**, *85*, 743.

21. R. M. Salinger and R. E. Dessey, *Tetrahedron Lett.* **1963**, *11*, 729.

22. J. N. Brønsted, *Chem. Rev.* **1928**, *5*, 231.

23. A. I. Shatenshtein, *Adv. Phys. Org. Chem.* **1963**, *1*, 155.

24. A. Streitwieser, Jr., R. A. Caldwell, and M. R. Granger, *J. Am. Chem. Soc.* **1964**, *86*, 3578.

25. A. Streitwieser, Jr., R. A. Caldwell, R. G. Lawler, and G. R. Ziegler, *J. Am. Chem. Soc.* **1965**, *87*, 5399.

26. A. Streitwieser, Jr., W. B. Hollyhead, G. Sonnichsen, A. H. Pudjaatmaka, C. J. Chang, and T. L. Kruger, *J. Ame. Chem. Soc.* **1971**, *93*, 5096.

27. A. Streitwieser, Jr. and R. G. Lawler, *J. Am. Chem. Soc.* **1963**, *85*, 2854.

28. A. Streitwieser, Jr., D. E. Van Sickle, and W. C. Langworthy, *J. Am. Chem. Soc.* **1962**, *84*, 244.

29. A. Streitwieser, Jr. and H. F. Koch, *J. Am. Chem. Soc.* **1964**, *86*, 404.

30. A. Streitwieser, Jr. and D. Holtz, *J. Am. Chem. Soc.* **1967**, *89*, 692.

31. A. Streitwieser, Jr. and J. S. Humphrey, Jr., *J. Am. Chem. Soc.* **1967**, *89*, 3767.

32. A. Streitwieser, Jr. and G. R. Ziegler, *Tetrahedron Lett.* **1971**, 415.

33. A. Streitwieser, Jr. W. B. Hollyhead, and A. H. Pudjaatmaka, *J. Am. Chem. Soc.* **1971**, *93*, 5088.

34. A. Streitwieser, Jr., P. C. Mowery, and W. R. Young, *Tetrahedron Lett.* **1971**, 3931.

35. A. Streitwieser, Jr. and D. W. Boerth, *J. Am. Chem. Soc.* **1978**, *100*, 755.

36. A. Streitwieser, Jr. and D. E. Van Sickle, *J. Am. Chem. Soc.* **1962**, *84*, 249.

37. D. J. Cram, *Fundamentals of Carbanion Chemistry*, Academic Press, New York, 1965.

38. F. G. Bordwell, *Acc. Chem. Res.* **1988**, *21*, 456.

39. M. I. Terekhova, E. S. Petrov, S. P. Mesyats, and A. I. Shatenshtein, *Zh. Obshch. Khim.* **1975**, *45*, 1529.

40. A. Streitwieser, Jr., J. H. Hammons, and E. Ciuffarin, *J. Am. Chem. Soc.* **1967**, *89*, 59.

41. A. Streitwieser, Jr. and D. M. E. Reuben, *J. Am. Chem. Soc.* **1971**, *93*, 1794.

42. A. Streitwieser, Jr., P. J. Scannon, and H. M. Niemeyer, *J. Am. Chem. Soc.* **1972**, *94*, 7936.

43. A. Streitwieser, Jr., C. J. Chang, and D. M. E. Reuben, *J. Am. Chem. Soc.* **1972**, *94*, 5730.

44. A. Streitwieser, Jr., C. J. Chang, and W. B. Hollyhead, *J. Am. Chem. Soc.* **1972**, *94*, 5292.

45. A. Streitwieser, Jr. and P. J. Scannon, *J. Am. Chem. Soc.* **1973**, *95*, 6273.

46. A. Streitwieser, Jr., E. A. Vorpagel, and C. C. Chen, *J. Am. Chem. Soc.* **1985**, *107*, 6970.

47. A. Streitwieser, Jr. and G. W. Schriver, *Hetero. Chem.* **1997**, *8*, 533.

48. S. Gronert and A. Streitwieser, Jr., *J. Am. Chem. Soc.* **1986**, *108*, 7016.

49. D. A. Bors, M. J. Kaufman, and A. Streitwieser, Jr., *J. Am. Chem. Soc.* **1985**, *107*, 6975.

50. A. Streitwieser, Jr., J. C. Ciula, and J. A. Krom, Thiele *J. Org. Chem.* **1991**, *56*, 1074.

51. A. Streitwieser, Jr., D. Z. Wang, and M. Stratakis, *Can. J. Chem.* **1998**, *76*, 765.

52. E. S. Petrov, M. I. Terekhova, and A. I. Shatenshtein, *Zh. Obshch. Khim.* **1974**, *44*, 1118.

53. E. S. Petrov, M. I. Terekhova, A. I. Shahtenstein, B. A. Trofimov, R. G. Mirskov, and M. G. Voronkov, *Dokl. Akad. Nauk. SSSR* **1973**, *211*, 1393.

54. J. E. Bartmess, in *NIST Standard Reference Database Number 69*, W. G. Mallard, and P. J. Linstrom, Eds., National Institute of Standards and Technology (http://webbook. nist.gov): Gaithersburg, MD, 2002.

55. G. B. Ellison, P. C. Engelking, and W. C. Lineberger, *J. Am. Chem. Soc.* **1978**, *100*, 2556.

56. R. H. Nobes, D. Poppinger, W.-K. Li, and L. Radom, in *Comprehensive Carbanion Chemistry, Part C*, E. Buncel, and T. Durst, Eds., Elsevier, New York, **1987**; p. 1.

57. D. A. Dixon, D. Feller and K. A. Peterson, *J. Phys. Chem. A* **1997**, *101*, 9405.

58. A. Rauk, L. C. Allen, and K. Mislow, *Angew. Chem. Int. Ed. Engl.* **1970**, *9*, 400.

59. P. v. R. Schleyer and G. W. Spitznagel, *Tetrahedron Lett.* **1986**, *27*, 4411.

60. K. M. Ervin, S. Gronert, S. E. Barlow, M. K. Gilles, A. G. Harrison, V. M. Bierbaum, C. H. DePuy, W. C. Lineberger, and G. B. Ellison, *J. Am. Chem. Soc.* **1990**, *112*, 5750.

61. R. L. Letsinger, *J. Am. Chem. Soc.* **1950**, *72*, 4842.

62. D. Y. Curtin and W. J. Koehl, *J. Am. Chem. Soc.* **1962**, *84*, 1967.

63. F. R. Jensen, L. D. Whipple, D. K. Wedegaertner, and J. A. Landgrebe, *J. Am. Chem. Soc.* **1960**, *82*, 2466.

64. P. K. Chou, G. D. Dahlke, and S. R. Kass, *J. Am. Chem. Soc.* **1993**, *115*, 315.

65. R. E. Carter, T. Drakenberg, and N.-A. Bergman, *J. Am. Chem. Soc.* **1975**, *97*, 6990.

66. M. J. S. Dewar and J. M. Harris, *J. Am. Chem. Soc.* **1969**, *91*, 3652.

67. H. M. Walborsky, F. J. Impastato, and A. E. Young, *J. Am. Chem. Soc.* **1964**, *86*, 3283.

68. M. C. Baschky, K. C. Peterson, and S. R. Kass, *J. Am. Chem. Soc.* **1994**, *116*, 7218.

69. D. J. Cram, W. D. Nielsen, and B. Rickborn, *J. Am. Chem. Soc.* **1960**, *82*, 6415.

70. E. J. Corey and E. T. Kaiser, *J. Am. Chem. Soc.* **1961**, *83*, 490.

71. D. J. Cram, D. A. Scott, and W. D. Nielsen, *J. Am. Chem. Soc.* **1961**, *83*, 3696.

72. D. A. Bors and A. Streitwieser, Jr., *J. Am. Chem. Soc.* **1986**, *108*, 1397.

73. S. Wolfe, A. Stolow, and L. A. LaJohn, *Tetrahedron Lett.* **1983**, *24*, 4071.

74. K. A. R. Mitchell, *Chem. Rev.* **1969**, *69*, 157.

75. H. E. Zimmerman and B. S. Thyagarajan, *J. Am. Chem. Soc.* **1960**, *82*, 2505.

76. H.-J. Gais and G. Hellman, *J. Am. Chem. Soc.* **1992**, *114*, 4439.

77. H. Gunther, *J. Braz. Chem. Soc.* **1999**, *10*, 241.

78. W. Bauer, in *Lithium Chemistry: A Theoretical and Experimental Overview*, A.-M. Sapse, and P. v. R. Schleyer, Eds., Wiley-Interscience, New York, **1995**.

79. D. O'Brien, in *Comprehensive Carbanion Chemistry, Part A: Structure and Reactivity*; E. Buncel, and T. Durst, Eds., Elsevier, Amsterdam, The Netherlands, 1980, p. 271.

80. L. D. McKeever, in *Ions and Ion Pairs in Organic Reactions*, M. Szwarc, Ed., Wiley-Interscience, New York, **1972**.

81. G. Fraenkel, H. Hsu, and B. M. Su, in *Lithium: Current Applications in Science and Medical Technology*, R. Bach, Ed., John Wiley & Sons, Inc., New York, 1985.

82. L. M. Jackman and J. Bortiatynski, *Adv. in Carbanion Chem.* **1992**, *1*, 45.

83. T. L. Brown, *Acc. Chem. Res.* **1968**, *1*, 23.

84. S. Bywater and D. J. Worsfold, *Adv. Organomet. Chem.* **1967**, *10*, 1.

85. E. Weiss and G. Hencken, *J. Organomet. Chem.* **1970**, *21*, 265.

86. P. G. Willard and M. J. Hintze, *J. Am. Chem. Soc.* **1987**, *109*, 5539.

87. A.-M. Sapse, D. C. Jain, and K. Raghavachari, in *Lithium Chemistry: A Theoretical and Experimental Overview*, A.-M. Sapse, and P. v. R. Schleyer, Eds., Wiley-Interscience, New York, **1995**.

88. L. M. Seitz and T. L. Brown, *J. Am. Chem. Soc.* **1966**, *88*, 2174.

89. R. Waack and P. West, *J. Am. Chem. Soc.* **1967**, *89*, 4395.

90. H. J. Reich, D. P. Green, M. A. Medina, W. S. Goldenberg, B. O. Gudmundsson, R. R. Dykstra, and N. H. Phillips, *J. Am. Chem. Soc.* **1998**, *120*, 7201.

91. O. Eppers and H. Gunther, *Helv. Chim. Acta* **1990**, *73*, 2071.

92. D. P. Novak and D. L. Brown, *J. Am. Chem. Soc.* **1972**, *94*, 3793.

93. R. Waak, M. A. Doran, and E. B. Baker, *Chem. Cummun.* **1967**, 1291.

94. P. L. Hall, J. H. Gilchrist, A. T. Harrison, D. J. Fuller, and D. B. Collum, *J. Am. Chem. Soc.* **1991**, *113*, 9575.

95. F. Abu-Hasanayn and A. Streitwieser, Jr., *J. Am. Chem. Soc.* **1996**, *118*, 8136.

96. A. Streitwieser, Jr., E. Juaristi, Y.-J. Kim, and J. K. Pugh, *Org. Lett.* **2000**, *2*, 3739.

97. X. Sun and D. B. Collum, *J. Am. Chem. Soc.* **2000**, *122*, 2459.

98. T. H. Lowry and K. S. Richardson, *Mechanism and Theory in Organic Chemistry*; 3rd ed., Harper Collins, New York, 1987.

99. A. Streitwieser, Jr., M. R. Granger, F. Mares, and R. A. Wolf, *J. Am. Chem. Soc.* **1973**, *95*, 4257.

100. R. G. Pearson and R. L. Dillon, *J. Am. Chem. Soc* **1953**, *75*, 24393.

101. A. Streitwieser, Jr., P. H. Owens, G. Sonnichsen, and W. Smith, *J. Am. Chem. Soc.* **1973**, *95*, 4254.

102. C. H. DePuy, S. Gronert, S. E. Barlow, V. M. Bierbaum, and R. Damrauer, *J. Am. Chem. Soc.* **1989**, *111*, 1968.

103. J. Wong, K. A. Sannes, C. E. Johnson, and J. I. Brauman, *J. Am. Chem. Soc.* **2000**, *122*, 10878.

104. R. G. Cooks, J. S. Patrick, T. Kotiaho, and S. A. McLuckey, *Mass Spectrum. Rev.* **1994**, *13*, 287.

105. P. Burk and K. Sillar, *J. Mol. Struct. (Theochem.)* **2001**, *535*, 49.

106. R. R. Sauers, *Tetrahedron* **1999**, *55*, 10013.

107. L. P. Hammett, *J. Am. Chem. Soc.* **1937**, *59*, 96.

108. C. Hansch, A. Leo, and R. W. Taft, *Chem. Rev.* **1991**, *91*, 165.

109. K. Mislow, *Introduction to Stereochemistry*, W. A. Benjamin, Inc., New York, **1966**.

110. A. Streitwieser, Jr., R. A. Caldwell, and W. R. Young, *J. Am. Chem. Soc.* **1969**, *91*, 529.

111. K. B. Wiberg and H. Castejon, *J. Am. Chem. Soc.* **1994**, *116*, 10489.

112. A. E. Reed and P. v. R. Schleyer, *J. Am. Chem. Soc.* **1990**, *112*, 1434.

113. O. A. Reutov, I. P. Beletskaya, and A. L. Kurts, *Ambident Anions*, Plenum, New York, 1983.

114. M. E. Jones, S. R. Kass, J. Filley, R. M. Barkley, and G. B. Ellison, *J. Am. Chem. Soc.* **1985**, *107*, 109.

115. Y. Chiang, A. J. Kresge, and P. A. Walsh, *J. Am. Chem. Soc.* **1986**, *108*, 6314.

116. F. G. Bordwell, J. E. Barres, J. E. Bartmess, G. J. McCollum, M. Van Der Puy, N. R. Vanier, and W. S. Matthews, *J. Org. Chem.* **1977**, *42*, 321.

117. R. Stewart, *The Proton: Applications to Organic Chemistry*, Academic Press, Orlando, FL, **1985**.

118. K. M. Broadus, S. D. Han, and S. R. Kass, *J. Org. Chem.* **2001**, *66*, 99.

119. U. Molder, P. Burk, and I. A. Koppel, *Int. J. Quant. Chem.* **2001**, *82*, 73.

120. S. Gronert, unpublished results.

121. C. F. Bernasconi and P. J. Wenzel, *J. Am. Chem. Soc.* **2001**, *123*, 7146.

122. Z. Glasovac, M. Eckert-Maksic, K. M. Broadus, M. C. Hare, and S. R. Kass, *J. Org. Chem.* **2000**, *65*, 1818.

123. U. Molder, P. Burk, and I. A. Koppel, *Int. J. Quantum Chem.* **2001**, *82*, 73.

124. R. A. Cox and R. Stewart, *J. Am. Chem. Soc.* **1976**, *98*, 488.

125. A. Streitwieser, Jr., E. Juaristi, and L. L. Nebenzahl, in *Comprehensive Carbanion Chemistry, Part A: Structure and Reactivity*, E. Buncel, and T. Durst, Eds., Elsevier, Amsterdam, The Netherlands, **1980**, p. 323.

126. E. M. Arnett and J. A. Harrelson, Jr., *J. Am. Chem. Soc.* **1987**, *109*, 809.

127. F. G. Bordwell, J. E. Bartmess, and J. A. Hautala, *J. Org. Chem.* **1978**, *43*, 3095.

128. E. M. Arnett, T. C. Moriarty, L. E. Small, J. P. Rudolph, and R. P. Quirk, *J. Am. Chem. Soc.* **1973**, *95*, 1492.

129. E. M. Arnett and K. G. Ventatasubramaniam, *J. Org. Chem.* **1983**, *48*, 1569.

130. E. M. Arnett and L. E. Small, *J. Am. Chem. Soc.* **1977**, *99*, 808.

131. A. Streitwieser, Jr., J. A. Krom, and K. V. Kilway, *J. Am. Chem. Soc.* **1998**, *120*, 10801.

132. A. Abbotto and A. Streitwieser, Jr., *J. Am. Chem. Soc.* **1995**, *117*, 6358.

133. M. J. Kaufman, S. Gronert, and A. Streitwieser, Jr., *J. Am. Chem. Soc.* **1988**, *110*, 2829.

134. R. A. Cox and R. Stewart, *J. Am. Chem. Soc.* **1976**, *98*, 488.

135. R. W. Taft and F. G. Bordwell, *Acc. Chem. Res.* **1988**, *21*, 463.

136. A. Streitwieser, Jr., W. B. Hollyhead, G. Sonnichsen, A. H. Pudjaatmaka, C. J. Chang, and T. L. Kruger, *J. Am. Chem. Soc.* **1971**, *93*, 5096.

137. A. Streitwieser, Jr., *J. Am. Chem. Soc.* **1973**, *95*, 4257.

138. A. Streitwieser, Jr., M. J. Kaufman, D. A. Bors, J. R. Murdoch, C. A. MacArthur, J. T. Murphy, and C. C. Shen, *J. Am. Chem. Soc.* **1985**, *107*, 6983.

139. A. Streitwieser, Jr., D. Holtz, G. R. Ziegler, J. O. Stoffer, M. L. Brokaw, and F. Guibe, *J. Am. Chem. Soc.* **1976**, *98*, 5229.

140. F. G. Bordwell, W. J. Boyle, Jr., and K. C. Yee, *J. Am. Chem. Soc.* **1970**, *92*, 5926.

141. F. G. Bordwell and D. L. Hughes, *J. Am. Chem. Soc.* **1985**, *107*, 4737.

142. J. R. Keeffe, J. Morey, C. A. Palmer, and J. C. Lee, *J. Am. Chem. Soc.* **1979**, *101*, 1295.

143. A. J. Kresge, *Can. J. Chem.* **1974**, *52*, 1897.

144. A. Pross, *Adv. Phys. Org. Chem.* **1985**, *21*, 166.

145. D. Beksic, J. Bertran, J. M. Lluch, and J. T. Hynes, *J. Phys. Chem. A* **1998**, *102*, 3977.

146. H. Yamataka, Mustanir, and M. Mishima, *J. Am. Chem. Soc.* **1999**, *121*, 10223.

147. L. M. Dorfman, R. J. Sujdak, and B. Bockrath, *Acc. Chem. Res.* **1976**, *9*, 352.

148. J. Berkowitz, G. B. Ellison, and D. Gutman, *J. Phys. Chem.* **1994**, *98*, 2744.

149. K. E. McCulloh and V. H. Dibelerr, *J. Chem. Phys.* **1976**, *64*, 4445.

150. K. M. Ervin, S. Gronert, S. E. Barlow, M. K. Gilles, A. G. Harrison, V. M. Bierbaum, C. H. Depuy, W. C. Lineberger, and G. B. Ellison, *J. Am. Chem. Soc.* **1990**, *112*, 5750.

151. G. E. Davico, V. M. Bierbaum, C. H. DePuy, G. B. Ellison, and R. R. Squires, *J. Am. Chem. Soc.* **1995**, *117*, 2590.

152. F. G. Bordwell and X. M. Zhang, *Acc. Chem. Res.* **1993**, *26*, 510.

153. R. Breslow, *Pure Appl. Chem.* **1974**, *40*, 493.

154. W. H. Saunders, *Acc. Chem. Res.* **1976**, *9*, 19.

155. W. P. Jencks, *Acc. Chem. Res.* **1980**, *13*, 161.

156. F. G. Bordwell, *Acc. Chem. Res.* **1972**, *5*, 374.

157. J. R. Keeffe and W. P. Jencks, *J. Am. Chem. Soc.* **1981**, *103*, 2547.

158. J. R. Keeffe and W. P. Jencks, *J. Am. Chem. Soc.* **1983**, *105*, 265.

159. R. Breslow, *Tetrahedron Lett.* **1964**, 399.

160. F. G. Larkin, R. A. More O'Farrall, and D. G. Murphy, *Collect. Czech. Chem. Commun.* **1999**, *64*, 1833.

161. Z. S. Jia, J. Rudzinski, P. Paneth, and A. Thibblin, *J. Org. Chem.* **2002**, *67*, 177.

162. W. H. Saunders, *J. Org. Chem.* **1999**, *64*, 861.

163. S. Hoz, *J. Org. Chem.* **1983**, *48*, 2904.

164. C. F. Bernasconi, *Acc. Chem. Res.* **1987**, *20*, 301.

165. A. Campagni, G. Modena, and P. E. Todesco, *Gazz. Chim. Ital.* **1960**, *90*, 694.

166. C. F. Bernasconi, M. W. Stronach, C. H. DePuy, and S. Gronert, *J. Am. Chem. Soc.* **1990**, *112*, 9044.

167. J. Meisenheimer, *Justus Liebigs Ann. Chem.* **1902**, *323*, 205.

168. S. M. Briscese and J. M. Riveros, *J. Am. Chem. Soc.* **1975**, *97*, 230.

169. C. F. Bernasconi, *Acc. Chem. Res.* **1978**, *11*, 147.

170. E. Buncel, J. M. Dust, and F. Terrier, *Chem. Rev.* **1995**, *95*, 2261.

171. J. F. Bunnett and R. E. Zahler, *Chem. Rev.* **1951**, *49*, 273.

172. G. Wittig and L. Lohman, *Ann.* **1942**, *550*, 260.

173. G. Wittig, *Angew. Chem.* **1954**, *66*, 10.

174. J. F. Garst and C. D. Smith, *J. Am. Chem. Soc.* **1976**, *98*, 1526.

175. P. T. Lansbury and V. A. Pattison, *J. Org. Chem.* **1962**, *27*, 1933.

176. J. C. Sheldon, M. S. Taylor, J. H. Bowie, S. Dua, C. S. B. Chia, and P. C. H. Eichinger, *J. Chem. Soc., Perkin 2* **1999**, 333.

177. P. Antoniotti and G. Tonachini, *J. Org. Chem.* **1998**, *63*, 9756.

178. T. Nakai and K. Mikami, *Chem. Rev.* **1986**, *86*, 885.

179. E. J. Verner and T. Cohen, *J. Am. Chem. Soc.* **1992**, *114*, 375.

180. H. E. Zimmerman and A. Zwieg, *J. Am. Chem. Soc.* **1961**, *83*, 1196.

181. J. A. Bertrand and E. Grovenstein, Jr., *J. Am. Chem. Soc.* **1976**, *98*, 7835.

182. E. Grovenstein, Jr. and R. E. Williamson, *J. Am. Chem. Soc.* **1975**, *97*, 646.

183. E. Grovenstein, Jr. and P.-C. Lu, *J. Org. Chem.* **1982**, *47*, 2928.

184. A. S. Kende, *Org. React.* **1960**, *11*, 261.

185. C. Rappe, L. Knutsson, N. J. Turro, and R. B. Gagosian, *J. Am. Chem. Soc.* **1970**, *92*, 2032.

186. W. B. Hanmmond and N. J. Turro, *J. Am. Chem. Soc.* **1966**, *88*, 2880.

187. F. G. Bordwell and M. W. Carlson, *J. Am. Chem. Soc.* **1969**, *92*, 3370.

188. R. Castillo, J. Andres, and V. Moliner, *J. Phys. Chem. B* **2001**, *105*, 2453.

189. S. Gronert, *Chem. Rev.* **2001**, *101*, 329.

190. W. N. Olmstead and J. I. Brauman, *J. Am. Chem. Soc.* **1977**, *99*, 4219.

191. B. T. Frink and C. M. Hadad, *J. Chem Soc., Perkin Trans. 2* **1999**, 2397.

192. J. L. Wilbur and J. I. Brauman, *J. Am. Chem. Soc.* **1994**, *116*, 9216.

193. S. Baer, E. B. Brinkman, and J. I. Brauman, *J. Am. Chem. Soc.* **1991**, *113*, 805.

194. R. N. McDonald and A. K. Chowdhury, *J. Am. Chem. Soc.* **1983**, *105*, 7267.

195. X. F. Sun and D. B. Collum, *J. Am. Chem. Soc.* **2000**, *122*, 2452.

196. X. F. Sun and D. B. Collum, *J. Am. Chem. Soc.* **2000**, *122*, 2459.

197. J. L. Rutherford, D. Hoffmann, and D. B. Collum, *J. Am. Chem. Soc.* **2002**, *124*, 264.

198. F. Abu-Hasanayn and A. Streitwieser, Jr., *J. Org. Chem.* **1998**, *63*, 2954.

199. A. A. Arest-Yakubovich, *J. Polym. Sci. Pol. Chem.* **1997**, *35*, 3613.

200. S. E. Denmark and C. T. Chen, *J. Am. Chem. Soc.* **1995**, *117*, 11879.

201. D. Baskaran, A. H. E. Muller, and S. Sivaram, *Macromol. Chem. Phys.* **2000**, *201*, 1901.

202. E. M. Arnett, V. DePalma, S. Maroldo, and L. S. Small, *Pure Appl. Chem.* **1979**, *51*, 131.

203. P. G. Wenthold and W. C. Lineberger, *Acc. Chem. Res.* **1999**, *32*, 597.

204. K. M. Ervin, *Chem. Rev.* **2001**, *101*, 391.

205. L. Feng, Y. Li, and J. F. Kirsch, *J. Phys. Org. Chem.* **1998**, *11*, 536.

206. J. P. Richard and T. L. Amyes, *Curr. Opin. Chem. Bio.* **2001**, *5*, 626.

207. K. S. Kim, K. S. Oh, and J. Y. Lee, *Proc. Natl. Acad. Sci.* **2000**, *97*, 6373.

208. P. F. Fitzpatrick, K. Kurtz, G. Gadda, J. Denu, M. Rishavy, and W. W. Cleland, *Biomed. Health Res.* **1999**, *27*, 176.

209. A. Argyrou and M. W. Washabaugh, *J. Am. Chem. Soc.* **1999**, *121*, 12054.

210. M. Gondry, J. Dubois, M. Terrier, and F. Lederer, *Eur. J. Biochem.* **2001**, *268*, 4918.

211. A. Witkowski, A. K. Joshi, Y. Lindqvist, and S. Smith, *Biochemistry* **1999**, *38*, 11643.

CHAPTER 4

Radicals

MARTIN NEWCOMB

Department of Chemistry, University of Illinois at Chicago, Chicago, IL

Reactive Intermediate Chemistry, edited by Robert A. Moss, Matthew S. Platz, and Maitland Jones, Jr.
ISBN 0-471-23324-2 Copyright © 2004 John Wiley & Sons, Inc.

1. STRUCTURE AND REACTIVITY

1.1. Structure

Organic radicals are species that contain an unpaired electron in the highest energy (singly) occupied molecular orbital, referred to as SOMO for semioccupied MOs. The structures of organic radicals are familiar to most chemists, who were introduced to radicals in their first organic chemistry course. As with closed-shell molecules, conformational interconversions of radicals are low-energy processes. Unlike closed-shell molecules, however, configurational interconversions of radicals are also often low-energy processes. Radical configurations are described according to whether the SOMO is a *p*-type orbital (π radical) or a hybrid orbital (σ radical). Examples of organic radicals are shown in Figure 4.1.

A trivalent π radical is planar, and a trivalent σ radical is pyramidyl. Both types of trivalent radicals have staggered and eclipsed conformations. For a planar radical, staggered and eclipsed refer to the substituents at the radical center and not to the *p* orbital. A divalent π radical is linear, and a divalent σ radical is bent. In the case of heteroatom radicals, such as alkoxyl and aminyl radicals, two low-energy electronic states exist, and the odd electron can be in a *p* orbital with the lone pair in a hybrid orbital (π radical) or vice versa (σ radical).

Interconversions of acyclic carbon-centered radicals between π and σ types are low-energy processes. The methyl radical is planar, but increasing alkyl substitution at the radical center results in an increasing preference for pyramidalization. The *tert*-butyl radical is pyramidalized with the methyl groups ~10° from planarity (the deviation from planarity for a tetrahedral atom is 19°) and a barrier to inversion of ~0.5 kcal/mol.[1] When a radical center is in a carbocycle, a planar radical is favored for all cases except the cyclopropyl radical, and the barrier for inversion in cyclopropyl is only ~3 kcal/mol.[2]

Substitution of electronegative groups for hydrogen in the methyl radical favors pyramidalization, and the trifluoromethyl radical is tetrahedral.[3] This effect is a result of interaction of the SOMO with the lowest unoccupied molecular orbital (LUMO). The SOMO in the methyl radical is a carbon 2*p* orbital, and it is orthogonal to the LUMO, whereas, in a pyramidal radical, the SOMO is a hybrid orbital that can interact with the LUMO. Electronegative groups increase the energy of the SOMO by π donation and decrease the energy of the LUMO by σ withdrawal. As the energy levels of the SOMO and the LUMO approach one another, interaction

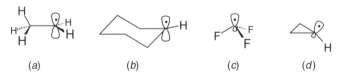

(a)   (b)   (c)   (d)

Figure 4.1. The (*a*) ethyl (shown in a staggered conformation) and (*b*) cyclohexyl radical are π radicals, and the (*c*) trifluoromethyl radical and (*d*) cyclopropyl radical are σ radicals.

between them is increasingly favored, and a σ radical becomes the low-energy configuration. On the other hand, conjugation of the trivalent radical center with a π system favors a planar structure. For example, in the case of the benzyl radical, the energy required for deformation from the planar structure in the transition states for reactions results in rates of additions to alkenes that are slower than expected from thermodynamic considerations.[4]

Divalent radicals are usually σ radicals. The vinyl radical is bent, and the barrier for inversion through the linear form is 3 kcal/mol.[5] Vinyl radicals with sigma substituents also are bent, but π substituents give linear vinyl radicals.[6] The formyl radical and acyl radicals are bent.[7]

The conformational barriers in acyclic radicals are smaller than those in closed-shell acycles, with the barrier to rotation in the ethyl radical on the order of tenths of a kilocalorie per mole.[1,8] The barriers increase for heteroatom-substituted radicals, such as the hydroxymethyl radical, which has a rotational barrier of ∼5 kcal/mol. Radicals that are conjugated with a π system, such as allyl, benzyl, and radicals adjacent to a carbonyl group, have barriers to rotation on the order of 10 kcal/mol. Such barriers can lead to rotational rate constants that are smaller than the rate constants of competing radical reactions, as was demonstrated with α-amide radicals,[9] and this type of effect permits acyclic stereocontrol in some cases.[10]

1.2. Radical Stabilities and C–H Bond Dissociation Energies

Radical stabilization energies (RSEs) and bond dissociation energies (BDEs) are critically important values for understanding radical reactivities. Homolytic BDEs are strongly related to the stabilities of the radicals formed by homolysis reactions, although BDEs and RSEs are not exactly equivalent.[11] Among the most precisely measured BDE values are those for cleavage of bonds to hydrogen atoms, and C–H BDEs for small organic compounds are determined with precisions of 0.5–2 kcal/mol. Several compilations of BDEs are available,[12–15] and these works summarize the methods used to obtain the thermochemical data. When using BDE values, one should ensure that the scales from different sources are equivalent by referring to the BDEs of simple molecules such as methane and water. Table 4.1 contains representative BDE values for small compounds.

The X–H BDEs increase across a row of the periodic table, reflecting increasing nuclear charge on atom X, and decrease as one moves down a column in the periodic table. The nuclear charge effect in the second row is dramatic with BDEs increasing by 30 kcal/mol from methane to HF, following the increase in the electronegativity of atom X. The same pattern is observed for hybridization of carbon with BDEs increasing as the amount of s character in the C–H bond increases. For example, the BDEs for ethane, ethylene, and acetylene are 101, 111, and 133 kcal/mol, respectively, and the high degree of s character in the C–H bonds of cyclopropane results in a C–H BDE that is greater than that of methane.

Any substituent on a trivalent C-centered radical is stabilizing. This phenomenon is a result of interactions resulting from mixing the SOMO containing the odd

TABLE 4.1. Bond Dissociation Energies (in kcal/mol)[a]

H_2	104.2	c-C_3H_5–H	106
HF	136.4 ± 0.2	H_2CCH–H	111.2 ± 0.8
HCl	103.15 ± 0.03	Ph–H	111.2 ± 0.8
HBr	87.54 ± 0.05	HCC–H	132.8 ± 0.7
HI	71.32 ± 0.06	CH_2CHCH_2–H	88.2 ± 2.1
H_2O	119.30 ± 0.05	$PhCH_2$–H	88.5 ± 1.5
CH_3O–H	104.2 ± 0.9	$HOCH_2$–H	96.06 ± 0.15
H_2S	91.2 ± 0.7	$NCCH_2$–H	94.8 ± 2.1
CH_3S–H	87.4 ± 0.5	$HC(O)CH_2$–H	94.3 ± 2.2
NH_3	108.2 ± 0.3	$RC(O)O$–H	105
CH_4	104.9 ± 0.1	$CH_3C(O)$–H	89.4 ± 0.3
CH_3CH_2–H	101.1 ± 0.4	$HOC(O)$–H	>89.5
$(CH_3)_2CH$–H	98.6 ± 0.4	$[(CH_3)_3Si]_3Si$–H	84
$(CH_3)_3C$–H	96.5 ± 0.4	Bu_3Sn–H	79

[a] Enthalpies for bond homolysis at 25 °C. Data taken from (12–15).

electron with either filled or empty MOs. In all cases, π-type interactions give new MOs of lower and higher energies. When the SOMO mixes with a filled MO, such as the C–H σ-bonding orbital of the methyl group in the ethyl radical or the lone pair on an adjacent oxygen atom, two electrons occupy the lower energy orbital and one occupies the higher energy orbital. When the SOMO mixes with an empty orbital, such as the LUMO of an electron-withdrawing group, the odd electron resides in the new low-energy combination MO.

Delocalization of the odd electron into extended π systems results in considerable radical stabilization. The C–H BDE at C3 of propene is reduced by 13 kcal/mol relative to that of ethane. That the stabilization effect in the allyl radical is due primarily to delocalization in the π system is shown by the fact that the rotational barrier for allyl is ~9 kcal/mol greater than that for ethyl.[16] Extending the conjugated system has a nearly additive effect, and the C–H BDE at C3 of 1,4-pentadiene is 10 kcal/mol smaller than that of propene.[12] Delocalization of the odd electron in the benzyl radical results in about one-half of the electron density residing at the benzylic carbon, and the C–H BDE of the methyl group in toluene is the same as that in propene.

Computational determinations of radical stabilities and BDE values have advanced to the point that computed and experimental values are in very good-to-excellent agreement, and it is noteworthy that relatively economical hybrid density functional theory computations can give excellent BDE results.[17] An advantage of computations is that radical stabilities can be determined for cases that would be difficult to measure; for example, the BDE value for the O–H bond in mercaptoethanol can be computed even though the S–H bond is weaker. Reliable estimations of BDE values for large molecules that contain isolated groups can be made from the BDE values for small molecules, but computations are recommended for large molecules when functional groups interact with the radical center.

In addition, computed BDE values are valuable for bonds that do not involve a hydrogen atom because experimental BDE values are mainly for X—H and X—C bonds.

1.3. Stable and Persistent Radicals

Most radicals are highly reactive and short lived, often reacting with one another with diffusion-controlled rates. Examples of thermodynamically stable radicals are known, however (Fig. 4.2). Nitroxyl radicals lacking β-hydrogen atoms, such as 2,2,6,6-tetramethylpiperidine-*N*-oxyl (TEMPO), or with bridgehead β-hydrogen atoms equilibrate with their dimers. The nitroxyl radicals are favored at equilibrium because the O—O bond formed by coupling is weak. The triphenylmethyl (or trityl) radical also equilibrates with its dimer, a quinoid species, with the dimer favored at equilibrium. An adequate concentration of trityl exists in solution at room temperature such that the solution has a yellow color. This color was observed by Gomberg[18] upon reaction of triphenylmethyl bromide with metals and reported in 1900 as evidence for formation of the trityl radical; Gomberg's work is often considered to be the beginning of organic radical chemistry. In the case of trityl, delocalization of the odd electron in the radical and strain in the dimer are the origins of the stability. Some radicals are thermodynamically disfavored with respect to dimerization but have relatively long lifetimes due to steric hindrance that slows radical–radical reactions.

Although persistent radicals can be thermodynamically favored with respect to their dimers, they often react rapidly with other molecules and radicals. For example, TEMPO couples with alkyl radicals with rate constants that are nearly as large as diffusional rate constants[19] to give oxime ethers that are stable

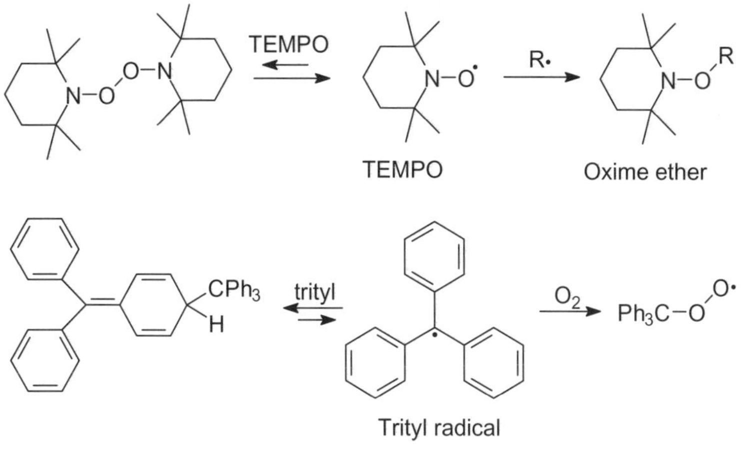

Figure 4.2. The persistent radicals TEMPO and trityl react rapidly with alkyl radicals and oxygen.

at room temperature, and Gomberg noted that the yellow solutions containing trityl lost their color when oxygen was admitted to the reaction mixture. This property can be exploited in synthesis and polymerization chemistry in radical nonchain reactions as discussed in Section 3.3, and it is a feature of many constructive enzyme-catalyzed radical reactions in Nature.

2. IDENTIFICATION AND CHARACTERIZATION OF RADICALS

Despite short lifetimes, radicals can be observed directly by several spectroscopic techniques. A transient radical also can be implicated by the products of its reactions, either closed-shell molecules formed in a specific sequence of radical reactions or "stable" radicals formed by reaction of a short-lived radical with a closed-shell molecule. Several characterization methods for radicals are unique, and they are discussed briefly in this section.

2.1. Inference from Products

As with any intermediate, a transient radical can be implicated from products formed in a reaction specific to the radical of interest. Experimentally, this is the basis of so-called *mechanistic probe* studies. An application of this method might employ, for example, 6-bromo-1-hexene as a probe for a radical intermediate as shown in Figure 4.3. If the 5-hexenyl radical is formed as a transient with an adequate lifetime, then cyclization of this radical to the cyclopentylmethyl radical could eventually give the cyclic product, and detection of the cyclic product provides evidence that a radical was formed. The mechanistic probe approach is deceptively simple, however. To be useful, one must exclude other possibilities for formation of the rearranged product *and* demonstrate that the transient was formed in the reaction of interest and not in a side reaction. The latter is especially difficult to demonstrate, and, unfortunately, some mechanistic probe studies that seemingly provided "proof" of radical intermediates were later found to be complicated by radical-forming side reactions.

Somewhat more definitive evidence that a transient radical was produced is provided when the radical reacts with a trapping agent to give a stable radical product

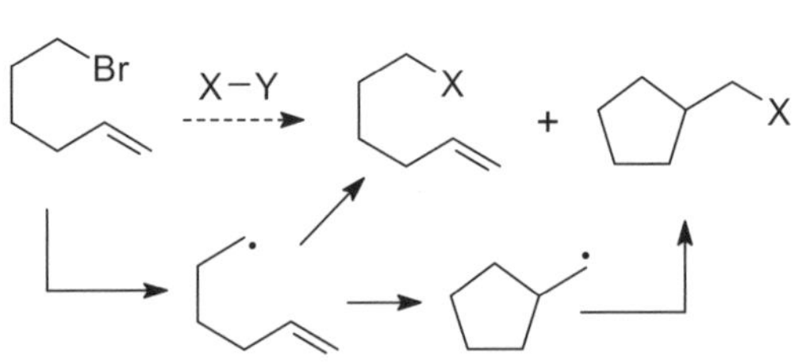

Figure 4.3. Design of a radical probe mechanistic study. Formation of the rearranged product implicates the intermediate 5-hexenyl radical that cyclized to cyclopentylmethyl.

because these reactions are specific for radicals. Examples include additions of radicals to nitroso compounds or nitrones, both of which give N-oxyl radical products that can be detected by electron spin resonance (ESR) spectroscopy. The technique is known as *spin trapping*, and can provide structural information about the initial radical that is trapped and quantitative information about the concentrations of radicals formed in addition to the qualitative confirmation that a radical was present.[20] One common spin trap is C-phenyl-N-*tert*-butylnitrone (PBN), which reacts as shown. Nitrones react with oxygen-centered radicals also, and water-soluble nitrones have been used to quantitate oxygen radicals in biological systems, so-called "active oxygen" intermediates, by quantitative ESR methods.[21]

PBN   A stable N-oxyl radical

2.2. Indirect Kinetic Determinations

The kinetics of radical reactions can be studied by direct methods (discussed in Chapter 18 of this volume) or by indirect methods. Indirect kinetic studies[22] require no special instrumentation, and are popular for obtaining relative or absolute rate constants for an intermediate that might be formed in a specific conversion. The radical of interest is generated in a reaction where two pathways compete, the reaction of interest and the basis reaction, for which a rate constant is known. Relative rate constants are then determined from the product mixture by spectroscopy or chromatography and used with the basis rate constant to calculate the absolute rate constant for the reaction of interest. The method is most easily applied when the competing reactions are first order or pseudo-first order, but that is not a requirement.[22]

When the basis reaction in the competition kinetic scheme is a calibrated first-order rearrangement, a cyclization, ring opening, or rearrangement reaction, then the radical is called a *radical clock*.[22–24] Calibrated alkyl radical clocks that cover a kinetic range >12 orders of magnitude are available.[22,24] In the example in Figure 4.3, the 5-hexenyl radical cyclizes with a known rate constant and can serve as a radical clock. The product ratio from the competition reaction is given by Eq. 1, where k_{XY} is the second-order rate constant for reaction of X–Y, k_C is the first-order rate constant for cyclization of the 5-hexenyl radical, [XY] is the concentration of reactant X–Y, [**R**] is the concentration of the 5-hexenyl radical, and [**C**] and [**A**] are the concentrations of cyclic and acylic products found at the end of the reaction. The concentration of 5-hexenyl radical in Eq. 1 cancels, and rearrangement gives Eq. 2. Note that Eq. 2 pertains to the situation where the concentration of X–Y was great enough such that it effectively does not vary during the reaction (i.e., pseudo-first-order conditions). Note also that one needs to determine from independent studies that the products actually were formed in the radical reactions of interest and not by sequences that involved side reactions.

$$[\mathbf{A}]/[\mathbf{C}] = (k_{XY}[\mathbf{XY}][\mathbf{R}])/(k_C[\mathbf{R}]) \qquad (1)$$

$$k_{XY} = k_C([\mathbf{A}]/[\mathbf{C}])[\mathbf{XY}]^{-1} \qquad (2)$$

2.3. Electron Spin Resonance Spectroscopy

Electron spin resonance spectroscopy, also known as electron paramagnetic resonance or EPR spectroscopy, permits direct detection of radicals and has been employed extensively to determine structural features of radicals. Electron spin resonance is similar to nuclear magnetic resonance (NMR) spectroscopy, with which most chemists are quite familiar. In ESR, the spin of the electron is the counterpart to the odd nuclear spin in NMR. When the sample is in a magnetic field, the spin states of the odd electron differ in energy, and one measures energy absorbed by the sample when the low-energy-spin electrons are promoted to the higher energy level. The energy gaps in ESR are much larger than those in NMR, and high sensitivity is possible without the use of superconducting magnets. In relatively small magnetic fields of 3000–12,000 G, electrons absorb in the microwave region, 9000–34,000 MHz.

Much of the nomenclature of ESR spectroscopy is analogous to that in NMR spectroscopy. The *g factor* in ESR is similar to the chemical shift in NMR. Couplings in ESR spectroscopy refer to electron–electron couplings in polyradicals and are not important for most organic radicals. *Hyperfine couplings* in ESR are couplings of the electron with nuclear spins on atoms, and have direct analogy to nuclear spin couplings in NMR spectroscopy. For example, the ESR spectrum of the methyl radical appears as a 1:3:3:1 quartet due to hyperfine coupling of the electron with the three equivalent protons. The ESR spectrum[25] of a triarylmethyl radical in Figure 4.4 demonstrates both the high resolution and hyperfine couplings. It is a well-resolved septet (6 ortho protons) of quartets (3 hydroxy protons).

ESR hyperfine couplings are listed as a values in analogy to the J values of NMR couplings. They are usually given in gauss (G), but they also can be given in millitesla (mT) or MHz (1 G = 0.1 mT = 2.80 MHz, when $g = 2.0023$). They are quite large with the result that couplings of remote nuclei can be observed giving rich structural information about the radical. Absolute values of the hyperfine couplings in the ethyl radical, for example, are $a = 22.4$ G for the α-protons and $a = 26.9$ G for the β-protons where instrumental resolution is in the tenths of gauss, and hyperfine couplings to the γ-protons in the cyclohexyl radical of $a = 0.5$ and 0.7 G are readily observed.

2.3.1. Structural Information from ESR Spectroscopy. ESR spectroscopy provides considerable information about the structures of radicals. The identity of

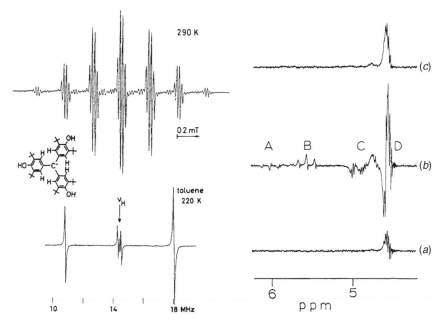

Figure 4.4. Spectra of radicals. **Left:** ESR (top) and ENDOR (bottom) spectra of a triarylmethyl radical. [Reprinted with permission from B. Kirste, W. Harrer, and H. Kurreck, *J. Am. Chem. Soc.* **1985**, *107*, 20–28. Copyright © 1985 American Chemical Society.] **Right:** CIDNP effects in the vinyl proton absorbance region in a 60 MHz NMR spectrum from reaction of BuLi and BuBr before mixing (*a*), 30 s after mixing (*b*) and after completion of the reaction (*c*). [Reprinted with permission from H. R. Ward, and R. G. Lawler, *J. Am. Chem. Soc.* **1967**, *89*, 5518–5519. Copyright © 1967 American Chemical Society.]

the atom with the odd electron can be obtained from the g factor. For organic radicals, the g factor is close to that of a free electron. Structural information is obtained from hyperfine coupling constants.[26]

The hyperfine coupling between a nuclear spin and the electron spin at a radical center is a function of the amount of odd electron spin at the nucleus, which in turn is a function of the degree of s character in the orbital containing the odd electron. In the hydrogen atom, the electron is in the $1s$ orbital resulting in a maximum hyperfine interaction, and $a = 507$ G. For the planar methyl radical, with the electron nominally residing in a $2p$ orbital, one might expect no ^{13}C hyperfine coupling because the p orbital has a node at the nucleus. However, a coupling of $a = 39$ G is observed for ^{13}C in methyl due to orbital mixing that gives some s character to the orbital containing the odd electron. The maximum coupling possible for ^{13}C is 1100 G, and the observed 39 g indicates somewhat $<4\%$ s character in the hybrid orbital. The formalism used is $a = A_0(C_{ns})$, where A_0 is the maximum coupling and C_{ns} is the coefficient of the s orbital in the hybrid orbital.

Proton hyperfine couplings in the ESR spectrum of the methyl radical are -23 G, where the sign of the coupling is not important except for CIDNP effects. The

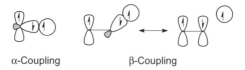

α-Coupling   β-Coupling

Figure 4.5. Origin of α- and β-proton hyperfine couplings.

electron spin density residing at the proton nuclei arises from spin polarization of the electrons in the CH bonds (Fig. 4.5). The electron in the CH bond that has spin parallel to that of the unpaired electron has increased electron density adjacent to the proton. The a value for the α protons on the carbon containing the odd electron in a p orbital are given by the McConnell equation (Eq. 3), where Q^H_{CH} is a constant approximately equal to -28 G, and ρ_C is the spin density on the carbon atom. In the methyl radical, ρ_C is unity, and $|a_{(obs)}| = 23$ G. In the benzene radical anion, spin density is equally distributed on the six carbon atoms, ρ_C is 0.17, and $|a_{(obs)}| = 3.1$ G for each proton.

$$a^H_\alpha = Q^H_{CH}\rho_C \tag{3}$$

The protons on atoms adjacent to a radical center also couple with the unpaired electron, giving rise to β-proton hyperfine couplings (Fig. 4.5). These interactions are from orbital mixings that are equivalent to hyperconjugation of the CH bonding electrons with the radical center. The extent of mixing is a function of the dihedral angle (θ) defined by the orbital of the radical, the atom of the radical center, the adjacent atom, and the hydrogen atom. A maximum in the β-proton hyperfine coupling in an alkyl radical will arise when the CH bond is parallel to the p orbital, and the coupling will be zero when the CH bond is perpendicular to the orbital.

Various qualitative equations have been used for β-proton hyperfine interactions, but they all include a $\cos^2 \theta$ term. In Eq. 4, the value of Q^H_{CHH} is 50 G, and ρ_C is again the spin density on the carbon atom. The Heller–McConnell equation (Eq. 5) contains two constants; A is ~ 5 G, and B is 40–50 G. The ESR spectrum is time averaged on the time scale of most conformational changes in acyclic radicals, and the value of a^H_β is a weighted average of the conformational populations. For a freely rotating methyl group in the position adjacent to the radical center, as in the ethyl radical, the integral of $\cos^2\theta$ from 0 to 2π is 0.5, and the value of a^H_β predicted from Eq. 4 is 25 G; the observed value is $a^H_\beta = 27$ G.

$$a^H_\beta = Q^H_{CCH}\rho_C(\cos^2 \theta) \tag{4}$$

$$a^H_\beta = (A + B\cos^2 \theta)\rho_C \tag{5}$$

γ-Proton hyperfine couplings also have a $\cos^2 \theta$ function, where the dihedral angle is defined by the orbital containing the odd electron and the CH bond. In the 1-adamantyl radical, the γ protons with 180° dihedral angles and $\cos^2 \theta = 1$

have $a = 3.08$, and the γ protons with $60°$ dihedral angles and $\cos^2 \theta = 0.25$ have $a = 0.80$ G.

The utility of ESR spectra for determining the structures of radicals is demonstrated by considering some examples. Methyl group substitution for hydrogen in the methyl radical ultimately results in slight deviation from planarity with a low inversion barrier. The ^{13}C a values for methyl, ethyl, isopropyl, and *tert*-butyl are 38.3, 39.1, 41.3, and 45.2 G, respectively. The *tert*-butyl radical is indicated to have $\sim 10°$ deviation from planarity, which is confirmed by infrared (IR) and Raman spectroscopy.[1]

Substituting fluorine atoms for hydrogen in the methyl radical results in increasing pyramidalization with increasing numbers of fluorine atoms as discussed in Section 1.1. For the series of radicals CH_3, CH_2F, CHF_2, and CF_3, the ^{13}C hyperfine couplings are 38, 55, 146, and 272 G, respectively.[27] Using the value of 1100 G for the coupling of ^{13}C to an electron in a $2s$ orbital, one concludes that the odd electron in the trifluoromethyl radical resides in an orbital that has 25% s character (i.e., the radical is tetrahedral).[3]

For the vinyl radical, the ^{13}C hyperfine coupling for the α-carbon is 107.6 G, which would suggest $\sim 10\%$ s character in the hybrid orbital. The vinyl radical clearly has the odd electron in a σ-type orbital because the β protons of the vinyl radical have distinct hyperfine couplings with $a = 37$ G for the *cis*-H, and $a = 65$ G for the *trans*-H. The cyclopropyl radical also is a σ radical with an α-^{13}C a value of 98 G, whereas the cyclohexyl radical, which has a nearly planar radical center, has an α-^{13}C a value of 41 G.

2.4. Electron Nuclear Double Resonance Spectroscopy

Electron Nuclear Double Resonance (ENDOR) spectroscopy is a technique similar to double resonance NMR spectroscopic methods. In an ENDOR experiment, the sample is irradiated with a constant microwave frequency at the resonance energy of the unpaired electron. The intensity of the microwave beam is adjusted such that the ESR signal is partially saturated. The sample is then irradiated with radio frequency in the region of the NMR absorbances of the coupled nuclei in the magnetic field employed. As the radio frequency (rf) is varied, the energy absorbed in the microwave region is monitored. When the rf matches the resonance condition of a nucleus coupled to the unpaired electron, an Overhauser relaxation of the partially saturated electron spin occurs, and an increase in absorbance of microwave energy is observed. One obtains an output of ESR signal intensity as a function of rf (Fig. 4.4).

The hyperfine interactions of proton nuclei with the unpaired electron in a radical splits the proton signals into doublets with large coupling constants (up to 100 MHz) in comparison to the J values of ~ 10 Hz observed in NMR spectroscopy. For the small magnetic field employed in an ESR spectrometer, all uncoupled protons absorb at effectively the same rf. The ENDOR spectrum appears as a series of doublets with lines spaced equally upfield and downfield from the expected absorbance of an uncoupled proton.

ENDOR spectroscopy has two advantages over ESR spectroscopy. The simplicity of the ENDOR spectrum is obvious. For the triarylmethyl radical in Figure 4.4, the ENDOR spectrum[25] consists of six lines (the central line in the figure from coupling to the *tert*-butyl protons is actually two closely spaced lines), whereas the ESR spectrum has 28 well-resolved lines. A second advantage of ENDOR spectroscopy is that identification of the nuclei coupled to the unpaired electron is simple. For example, the *a* values for a ^{13}C atom at a carbon radical center and an α proton are similar, but the radio wave frequencies at which ^{13}C and ^{1}H absorb in a given magnetic field are tremendously different (e.g., 75 vs. 300 MHz). This feature makes ENDOR spectroscopy especially attractive for identifying the types of the atoms coupled to the unpaired electron, which, for example, is useful in characterizing large, complex biological radicals.

2.5. Chemically Induced Dynamic Nuclear Polarization Effects

Chemically induced dynamic nuclear polarization (CIDNP; pronounced sid-nip or kid-nip) is a phenomenon first reported in the 1960s, which involves the observation of strongly enhanced absorbances and strong emissions in NMR spectra obtained from solutions in which a radical reaction is occurring. The CIDNP is a misnomer based on the apparent similarity of the effect to those seen in dynamic nuclear polarizations, where one pumps electron spins and observes NMR signals. An operational theory explaining CIDNP was expounded soon after early reports of the effect,[28,29] and it is known as the *radical pair theory*. The effect involves the phenomena described below in the context of a reaction that initially gives a singlet radical pair, but triplet radical pairs formed in photolysis reactions also can be explained.

Assume that a homolysis reaction is being conducted in a tube in an NMR spectrometer. A singlet radical pair is formed by homolysis, and this pair can recombine, diffuse apart, or undergo intersystem crossing (ISC, an electron spin flip) to give a triplet pair. The triplet pair thus formed can diffuse apart or return to the singlet pair by ISC, but it cannot recombine because that would produce a high-energy triplet state of a closed-shell molecule. Because diffusional processes have the same rates for both types of pairs, the relative amount of escape from the triplet pair is greater than that from the singlet pair.

The radicals formed in the homolysis maintain the nuclear spin character of their protons during the lifetime of the radical pair. Differences in proton nuclear spin give slightly different energy levels for the radicals in a magnetic field. This phenomenon is the same effect as seen in hyperfine coupling that gives multiplets in the ESR spectrum, but the magnetic field in this experiment is that of the NMR spectrometer. The rates of ISC differ slightly for different energy sublevels of the radical pairs and are thus a function of the nuclear spins of the protons. Therefore, the singlet and triplet radical pairs develop non-Boltzmann proton nuclear spin distributions. Further, diffusive free radicals that escape from the radical pairs also will have non-Boltzmann proton nuclear spin distributions.

A radical that escapes from a radical pair can react with a molecule to give a closed-shell product and a new radical, or it can react with another radical to give a new radical pair. A closed-shell product also can be formed by recombination of radicals in the original or new singlet radical pair. In both types of reactions, the closed-shell products will have non-Boltzmann proton nuclear spin populations, and it is these products that produce the CIDNP effects in NMR spectra. The nuclear spins will relax thermally, but, if the NMR spectrum is obtained before thermal relaxation is complete, one will observe large enhanced absorption or emission from the product molecules.

CIDNP effects can be seen when a reaction that gives a radical pair is reasonably fast, with a lifetime on the order of minutes. The small differences in high- and low-energy nuclear spin states that exist in the magnetic field of the NMR spectrometer, giving nearly equal populations of the spin states when thermally equilibrated, and the extremely high sensitivity of an NMR instrument are the features that combine to make CIDNP effects observable. The NMR spectrometer reports on differences in nuclear spin state populations of parts per million, and, for example, only a small fraction of nuclear-spin excited molecules are needed to obtain an emission signal. The high sensitivity of the method can be a problem, however, because it is possible to observe CIDNP signals originating from minor byproducts of the reaction. This is seen in the example in Figure 4.4 where CIDNP signals from 1-butene (at δ 5.6 and 6.1) are apparent, but those NMR signals are not observed in the spectrum of the product.

The spectra shown in Figure 4.4 were obtained from a reaction of *tert*-butyl-lithium with 1-bromobutane.[30] One deduces that both the *tert*-butyl and the butyl radical were produced and that they reacted in disproportionation reactions to give in part isobutylene and 1-butene, respectively. In the spectrum recorded at 30 s, vinyl proton signals from 1-butene are in emission (δ 5.0 and δ 6.1) and enhanced absorption (δ 5.6). The isobutylene vinyl proton signal at δ 4.6 is in emission on the left side and enhanced absorption is on the right side. This phenomenon is known as a *multiplet effect*, and it is due to differences in ISC rates for radical pairs containing *tert*-butyl radicals with different proton nuclear spins. Note that the *tert*-butyl-lithium sample contained an impurity of isobutylene before the reaction, and the amount of isobutylene was increased after the reaction.

2.6. Absorption Spectroscopy

In principle, absorption spectroscopy techniques can be used to characterize radicals. The key issues are the sensitivity of the method, the concentrations of radicals that are produced, and the molar absorptivities of the radicals. High-energy electron beams in pulse radiolysis and ultraviolet–visible (UV–vis) light from lasers can produce relatively high radical concentrations in the $1-10 \times 10^{-5}$ M range, and UV–vis spectroscopy is possible with sensitive photomultipliers. A compilation of absorption spectra for radicals contains many examples.[31] Infrared (IR) spectroscopy can be used for select cases, such as carbonyl-containing radicals, but it is less useful than UV–vis spectroscopy. Time-resolved absorption spectroscopy is used for direct kinetic studies. Dynamic ESR spectroscopy also can be employed for kinetic studies,[32,33] and this was the most important kinetic method available for reactions in organic solvents before the introduction of lasers. Chapter 18 in this volume[21b] on kinetics in the nanosecond realm discusses the methods.

3. MULTISTEP RADICAL REACTIONS

Most radicals are highly reactive, and there are few examples where one would produce a stable radical product in a reaction. Reference to a "radical reaction" in synthesis or in Nature, almost always concerns a sequence of elementary reactions that give a composite reaction. Multistep radical sequences are discussed in general terms in this section so that the elementary radical reactions presented later can be viewed in the context of real conversions. The sequences can be either radical chain reactions or radical nonchain reactions. Most synthetic applications involve radical chain reactions, and these comprise the bulk of organic synthetic sequences and commercial applications. Nonchain reaction sequences are largely involved in radical reactions in biology. Some synthetic radical conversions are nonchain processes, and some recent advances in commercial polymerization reactions involve nonchain sequences.

3.1. Radical Chain Reactions

Radical chain reactions are comprised of three distinct parts: initiation, propagation steps, and termination. The initiation portion involves one or more elementary reactions that produce a radical that can participate in one of the propagation steps. The propagation sequence is where the desired products are formed; it consists of two or more reactions in which one product of each elementary reaction is a radical that serves as a reactant in another step of the sequence. Radicals are destroyed in termination steps that give nonradical products by radical–radical coupling and disproportionation reactions.

A simple radical chain reaction sequence is shown in Figure 4.6. This reaction sequence is an example of a "tin hydride reaction," where an alkyl halide is reduced to an alkane. The initiation steps involve thermolysis of azobisisobutyryl-nitrile (AIBN) to give the 1-cyano-1-methylethyl radical that reacts with tributyltin hydride (the most commonly used tin hydride) to give the tributylstannyl radical. In one propagation step, the tin radical reacts with the alkyl halide to give an alkyl radical and tributyltin bromide. In the second propagation step, the alkyl radical reacts with the tin hydride to give an alkane and another tributylstannyl radical. In this example, three types of radical–radical reactions can serve as termination steps in principle, but the most important termination reaction will involve self-reactions of the radical with the largest concentration as discussed in Section 3.2. The key to a successful chain reaction sequence is that the velocities of the propagation reactions are much greater than those of the termination reactions such that many cycles of the propagation sequence are completed for each radical initiation and termination reaction. To obtain the high velocities necessary in the propagation steps, the rate constants for these reactions must be reasonably large. In practice, it

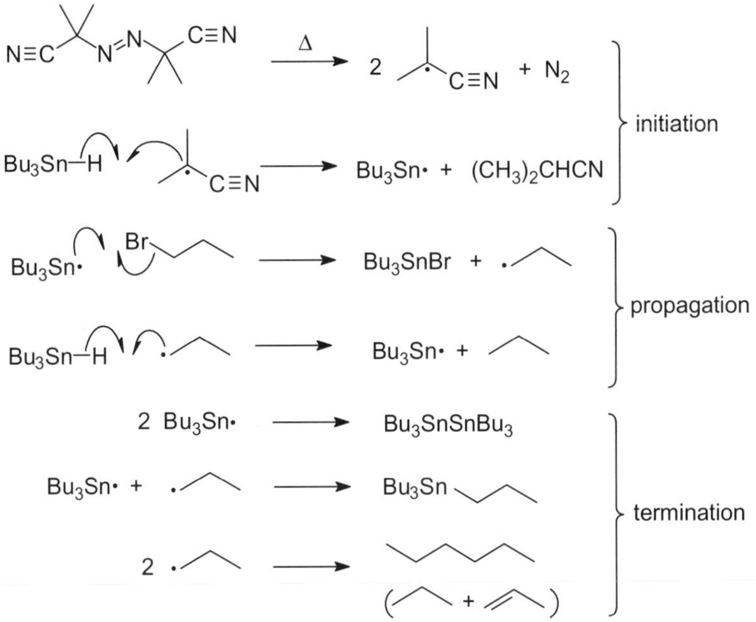

Figure 4.6. Elementary processes in the radical chain reduction of an alkyl bromide.

is difficult to use a propagation reaction that has a pseudo-first-order rate constant $< 1 \times 10^4$ s^{-1} at ambient temperature.

Additional elementary reactions can be incorporated into the propagation sequence to give more sophisticated synthetic conversions, but additional propagation steps inevitably produce increasing numbers of undesired reactions. For example, assume one wanted to prepare the addition product from the *tert*-butyl radical and methyl acrylate (Fig. 4.7). The desired propagation sequence now consists of three elementary reactions: bromine abstraction by the stannyl radical, addition of *tert*-butyl radical to the acrylate ester, and tin hydride trapping of the adduct radical. Two serious side reactions are now possible, reduction of the *tert*-butyl radical in competition with the addition to methyl acrylate, and addition of the adduct radical to a second molecule of acrylate in competition with the desired tin hydride reaction. Altering the concentration of one reagent to maximize a reaction in the desired sequence, such as decreasing the tin hydride concentration to avoid reduction of the alkyl radical, will reduce the efficiency of the other desired reaction. In practice, the conversion shown in Figure 4.7 cannot be made highly efficient because the rate constants for reaction of tin hydride with both types of radicals are essentially the same as are the rate constants for reaction of methyl acrylate with both types of radicals. If the newly added propagation reaction was a unimolecular reaction, such as a 5-*exo* radical cyclization, however, high yields of product could be obtained from the sequence halide reaction with stannyl radical, radical cyclization, and reaction of the cyclic radical with tin hydride.

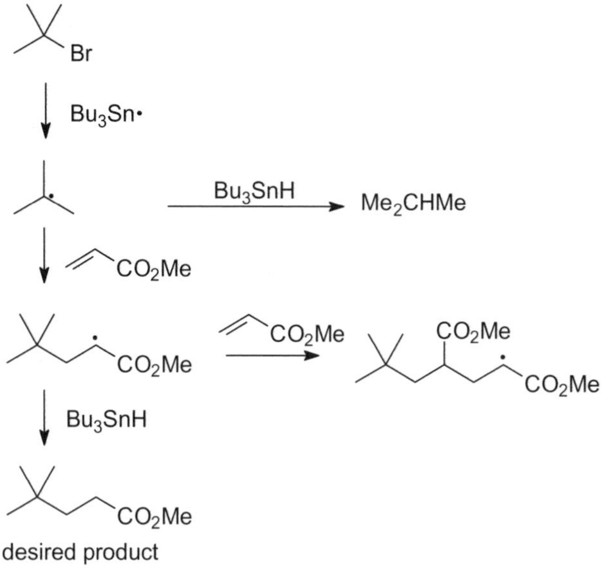

Figure 4.7. A three-step radical propagation sequence (top to bottom) that has two serious possible side reactions.

The number of undesired reactions increases rapidly as more steps are added to the propagation sequence with the result that achieving high-yield multistep conversions might seem unlikely. Nonetheless, many complex sequences can be achieved by controlling concentrations of reagents, incorporating unimolecular processes, and matching the polarities of radicals and radicophiles in the desired reaction sequence. For planning complex sequences, one needs absolute or relative kinetic information.

In radical chain polymerization reactions, the propagation sequence involves additions of a growing polymer chain to monomer molecules. In principle, each step is distinct, but, after the first addition step of the polymerization, subsequently formed polymeric radicals are similar in reactivity. To a good approximation, the polymerization reaction will behave as if there is a single propagation step. However, when the polymeric radicals grow large enough such that the reaction solution becomes viscous, termination reactions slow dramatically, and the concentration of radicals increases resulting in accelerated polymerization in the gel solution.[34]

3.2. Velocities of Radical Chain Reactions

Radical chain reactions are complicated because multiple reactions occur, but the overall velocity of the sequence can be given in simplified form by applying steady-state approximations. An important feature of any chain reaction is that the velocities of all propagation steps must be identical because the radicals formed as products in each elementary reaction are the reactants in another elementary reaction. In addition, under steady-state conditions, the total velocity for radical termination reactions is equal to the velocity of radical initiation reactions.

The equality in the velocities of the chain propagation steps results in a relatively high concentration of the radical that reacts in the slowest chain propagation step. For example, for the tin hydride reduction of an alkyl bromide shown in Figure 4.6, the reaction of Bu_3SnH with an alkyl radical has a rate constant of $2 \times 10^6 \, M^{-1} \, s^{-1}$ at ambient temperature, and the rate constant for reaction of the tributylstannyl radical with 1-bromopropane is $\sim 3 \times 10^7 \, M^{-1} \, s^{-1}$. If the initial concentrations of Bu_3SnH and 1-bromopropane were equal, then the concentration of the propyl radical must be 15 times as great as the concentration of the stannyl radical for the velocities of the propagation steps to be equal. Because the termination steps in this radical sequence occur with similar (diffusion-controlled) rate constants, the most significant termination step will be the self-reaction of propyl radicals. Accordingly, the square of the propyl radical concentration is the concentration term that appears in the rate law for the termination reaction. The rate expression for the chain reaction is derived by setting the equality for velocities of initiation and termination and using the required equalities in the velocities of the propagation reactions. The derived rate law contains only one propagation rate constant, that for the "slow" propagation step. Specifically, for the tin hydride reduction in Figure 4.6, the rate law is

$$d[\mathrm{RH}]/dt = k_{\mathrm{SnH}}[\mathrm{Bu_3SnH}](Fk_{\mathrm{In}}[\mathrm{AIBN}]/2k_{\mathrm{T}})^{1/2}$$

where k_{SnH} is the rate constant for the tin hydride reaction with the alkyl radical, F is the fraction of free radicals that escape from the solvent cage (between 0 and 2) when AIBN is thermolyzed, k_{In} is the rate constant for AIBN thermolysis, and k_{T} is the diffusional rate constant for radical–radical reactions (k_{T} is discussed in Section 4.3).

The slow radical propagation step in the chain sequence is termed the *rate-controlling step* because its rate constant appears in the rate law for the chain reaction, and its importance is apparent. If one substituted 1-iodopropane for 1-bromopropane in the reaction in Figure 4.6 and the concentrations and reaction conditions were otherwise unchanged, then the velocity of the chain reaction would not be altered. The steady-state concentration of the stannyl radical would be considerably reduced, however, from 7% of the concentration of the propyl radical in the bromopropane reduction to 0.1% of the concentration of the propyl radical in the iodopropane reduction. Alternatively, if the metal hydride reducing agent were changed to *tris*-(trimethylsilyl)silane, $(Me_3Si)_3SiH$, which reacts with the propyl radical with a rate constant ~ 0.2 times that of tin hydride, then the velocity of the chain reaction would be reduced by a factor of 5.

Rate laws for radical chain reactions initiated by thermolysis are 1.5 order, first order in the component reacting in the rate-controlling step, and 0.5 order in the initiator. When the initiator is the same component as that reacting in the rate-controlling step, the reaction will be 1.5 order in this reagent. When chain reactions are initiated by photolysis instead of thermolysis, the rate constant for initiation, the initiator concentration, and the efficiency term F are replaced by the quantum efficiency of the photochemical reaction and either the photon flux, for solutions that absorb all light, or the initiator's molar extinction coefficient and concentration, for solutions that transmit light.

3.3. Radical Nonchain Reaction Sequences

Radical chain reactions are by far the major class of radical reactions encountered in organic chemistry journal articles and are involved in the bulk of industrial applications of radical chemistry, but radical nonchain sequences are ubiquitous in Nature and arguably provide the largest number of types of radical conversions. In a radical nonchain reaction sequence, the velocity of a termination step is greater than the velocity of at least one potential propagation step. In a perfect nonchain reaction sequence, each initiation reaction would result in a single molecule of product, but that is difficult to achieve in condensed-phase reactions. Nature can accomplish highly efficient radical nonchain reactions by isolating the radical in an enzyme, and there are many such enzyme-catalyzed reactions.

Radical nonchain reaction sequences can result in high yields of a specific product when one of the radicals is persistent in solution, and the description of the phenomenon is known as the *persistent radical effect*.[35,36] The model of the radical chain reaction sequence is changed such that one radical self-termination reaction does not occur or has a slow rate. The radical that does not self-terminate efficiently is the persistent radical, and it accumulates to a relatively high concentration early in the reaction. As the concentration of the persistent radical increases, the velocities of cross-termination reactions involving the persistent radical, which still have diffusion-controlled rate constants, increase such that these are the only important termination reactions. Moreover, the velocity of the cross-termination reaction will eventually exceed the velocity of a propagation step in a normal chain sequence. At this point, the chain sequence is aborted, and the overall process can become a highly efficient nonchain reaction with products formed only from the cross-termination reaction. The negative side of radical nonchain sequences is that a stoichiometric amount of initiator is needed, which can result in high yields of an undesired byproduct. If the initiator is the precursor to the radical that will be functionalized, however, this is not an added problem.

Radical chain processes "break down" whenever the velocity of a termination reaction is comparable to the velocity of the rate-controlling step in a chain reaction. This situation would occur, for example, if one attempted to use Et_3SiH as the hydrogen atom donor in the alkyl halide reduction sequence in Figure 4.6 and employed typical "tin-hydride" reaction conditions because the rate constant for reaction of the silane with an alkyl radical is ~ 4 orders of magnitude smaller than that for reaction of Bu_3SnH. Such a slow reaction would not lead to a synthetically useful nonchain sequence, however, because no radical is persistent in this case. In fact, a silane-based radical chain reduction of an alkyl halide could be accomplished successfully if the velocity of the initiation reaction was reduced enough such that it (and, hence, also the velocity of alkyl radical termination reactions) was much less than the velocity of the silane reaction with the alkyl radical. However, this would require inert solvents and an unacceptable increase in the total reaction time.

Some examples of radical nonchain reactions are shown in Figure 4.8. In the Barton reaction (first example in Fig. 4.8),[37] photolysis of a nitrite ester gives an

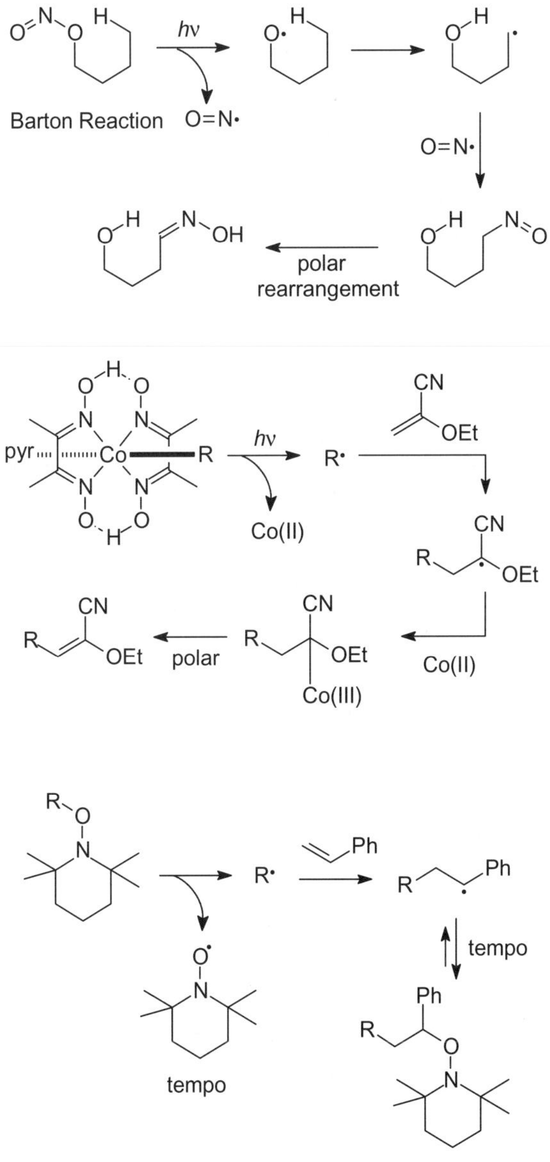

Figure 4.8. Some radical nonchain reactions.

alkoxyl radical and nitric oxide, which is the persistent radical, the alkoxyl radical abstracts hydrogen from a remote carbon, and the carbon radical thus formed reacts with nitric oxide. The reaction of the cobalt complex in the second example is modeled in part after reactions of enzymes that employ coenzyme B_{12} for radical production; in this case, the Co(II) species formed by photolysis is the persistent radical.[38] The third example shows an application of the persistent radical effect in a recently developed polymerization sequence that is one type of *living radical polymerization*.[39,40] Hydroxylamine ethers dissociate thermally to give carbon radicals and nitroxyl radicals, which are persistent. Following a radical addition step, the nitroxyl radical traps the extended polymeric radical. Subsequent dissociation of the new hydroxylamine ether regenerates the polymeric radical that adds to another monomer, and so on. Living radical polymerizations achieve narrow molecular weight distributions because, effectively, no radical termination reactions occur other than the cross-termination reaction. In these examples of radical non-chain reactions, recombination of the persistent radical with the initially formed reactive radical is thermodynamically favored, but such recombination only regenerates the starting materials.

4. ELEMENTARY RADICAL REACTIONS

4.1. Initiations: Radicals from Closed-Shell Compounds

Radicals are formed in thermal and photochemical reactions that homolyze bonds and in electron-transfer processes. In synthetic applications of radicals, a thermal initiator is most often employed to generate radicals. Thermal initiators are also widely used in industrial polymerization applications, but many commercial products employ photoinitiators that can be used to initiate a polymerization after a formulation of monomers is applied. Electron-transfer reactions, either reductions or oxidations, can produce radicals, but strong reducing and oxidizing agents will further reduce or oxidize the radical; therefore, this method requires that the reducing or oxidizing agent reacts relatively slowly with the initially formed radical.

4.1.1. Thermolysis Reactions. Any organic compound will react to give radicals if heated to a high enough temperature, but relatively low-temperature homolysis reactions are possible with azo and peroxy compounds, which contain weak heteroatom–heteroatom bonds.[41] Figure 4.9 shows the more common thermal radical initiators. Reactive species such as di-*tert*-butyl peroxyoxalate[42] and di-*tert*-butyl hyponitrite[43] are too unstable to be stored for long periods and are prepared soon before use, but AIBN, diacyl peroxides, peroxy esters, and dialkyl peroxides are commercially available. In synthetic applications of radical chain reactions, it is common to employ a reaction temperature that will result in about a 1-h half-life for the initiator so that low concentrations of radicals will be maintained. For some of the initiators in Figure 4.9, the temperatures that give 1-h half-lives are as follows:[44] di-*tert*-butyl peroxyoxalate, 45 °C; di-*tert*-butyl hyponitrite, 55 °C; AIBN, 81 °C;

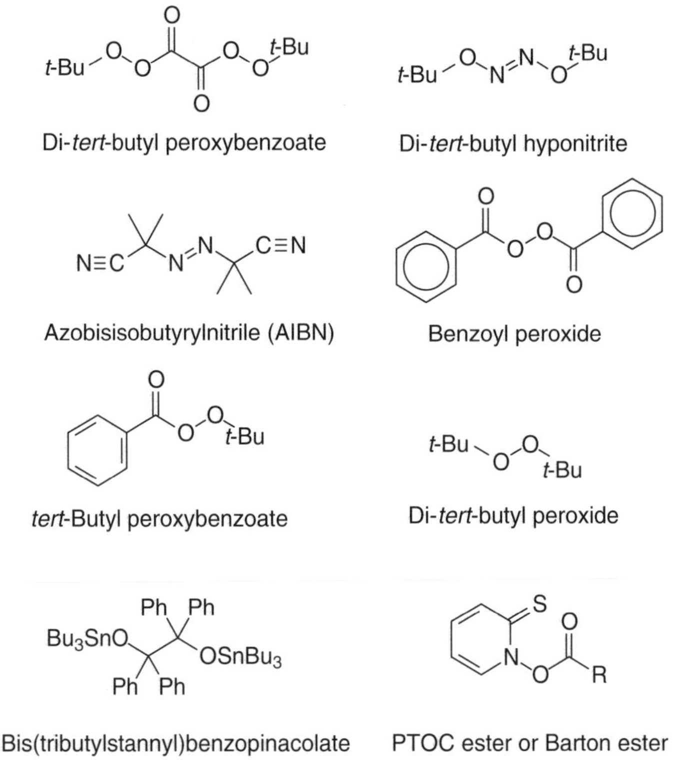

Di-*tert*-butyl peroxybenzoate   Di-*tert*-butyl hyponitrite

Azobisisobutyrylnitrile (AIBN)   Benzoyl peroxide

tert-Butyl peroxybenzoate   Di-*tert*-butyl peroxide

Bis(tributylstannyl)benzopinacolate   PTOC ester or Barton ester

Figure 4.9. Common thermal initiators. The acronym for **p**yridine-2-**t**hi**o**ne**o**xy**c**arbonyl is PTOC.

benzoyl peroxide, 91 °C; *tert*-butyl peroxybenzoate, 125 °C; di-*tert*-butyl peroxide, 150 °C. Many other thermal radical initiators can be prepared; for example, bis-(tributylstannyl)benzopinacolate, prepared from benzopinacol and tributylstannyl-dimethylamine,[45] decomposes at ~80 °C to give benzophenone and tributylstannyl radicals. Diacylperoxides from alkanoic acids decompose at lower temperatures than does benzoyl peroxide.

The initiation reactions for AIBN and for a diacylperoxide are shown in Figure 4.10. Initial homolysis gives radicals that further react by loss of nitrogen

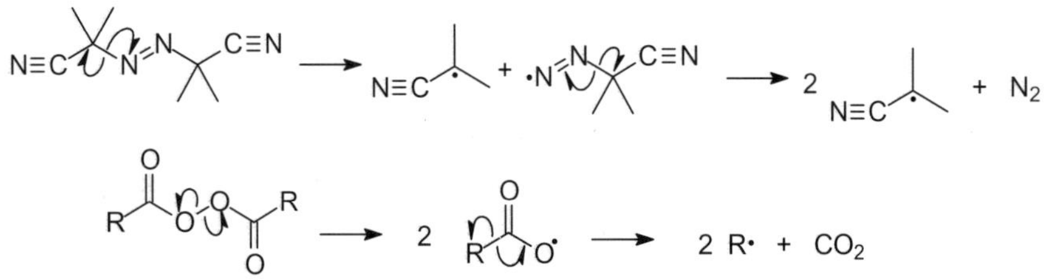

Figure 4.10. Thermal initiation reaction sequences for AIBN and for a diacyl peroxide.

and carbon dioxide, respectively. The decarboxylation reaction of an alkyl acyloxyl radical is comparable to or faster than diffusional processes, but aryl acyloxyl radicals decarboxylate less rapidly[46] due to the instability of an aryl radical. The 2-cyano-2-methylethyl radical from AIBN is a relatively stable radical; it will react with tin hydride or (Me$_3$Si)$_3$SiH in an initiation sequence such as that shown in Figure 4.6, but it is not reactive enough for many radical reactions, such as addition to an acrylate ester, and it adds reversibly to the thione group of a xanthate. When a highly reactive radical from the initiator is necessary, diacyl peroxides are preferred over AIBN.

A relatively new method for thermal radical initiation has gained popularity in synthetic applications. A small amount of Et$_3$B is added to the reaction mixture that has not been rigorously deoxygenated.[47,48] The borane reacts with oxygen, apparently generating ethyl radicals. A major advantage for this mode of radical initiation is the low reaction temperatures (as low as $-78\,^{\circ}$C) that can be achieved when stereoselective reactions are desired.[49]

Radical initiation in synthetic applications can be achieved without a special initiator when one employs members of the PTOC family of radical precursors or related thione compounds (see Fig. 4.9). The PTOC precursors were developed mainly by Barton's group[50,51] in the 1980s and comprise mixed anhydrides of a thiohydroxamic acid and a carboxylic acid or other acid. The commonly used thiohydroxamic acid is N-hydroxypyridine-2-thione. The PTOC derivatives are typically generated *in situ*, and the reaction mixture is heated or photolyzed with visible light to initiate radical reactions. Rate constants for thermal decomposition of PTOC derivatives have not been determined, but PTOC esters appear to be unstable >50–60 $^{\circ}$C, and radical initiation can be accomplished in refluxing benzene or toluene. Related oxygen-centered radical precursors react thermally and in chain reactions to give alkoxyl radicals.[52] A family of nitrogen-centered radical precursors is known for aminyl and amidyl radicals and aminium radical cations.[53,54]

4.1.2. Photolysis.

One method of photochemical initiation involves photochemically activated homolysis reactions. The common thermal initiators can be cleaved photochemically, but the molar extinction coefficients of some of these species are small leading to poor efficiency. For example, laser generation of the *tert*-butoxyl radical in laser flash photolysis studies involves irradiation of a solution containing 50% di-*tert*-butyl peroxide with a high-energy eximer laser.[55] Photolysis of AIBN with light from a mercury lamp is relatively efficient. The PTOC derivatives are yellow-colored compounds with a strong long wavelength absorbance centered at $\sim\lambda_{\max} = 360$ nm. These species are cleaved readily with visible light irradiation from a conventional tungsten-filament bulb. Many thermally stable compounds with relatively weak bonds can be cleaved photochemically; thus, an alkyl iodide, an alkyl phenylselenide, or a ditin compound such as hexabutyldistannane can be irradiated to initiate radical reactions.

Another method of photochemical initiation involves electron-transfer reactions of electronically excited states that are produced photochemically. The process is known as *photoinduced electron transfer* (PET).[56,57] The excited states are strong oxidants or reductants, and they oxidize or reduce substrates to give radical ions. Much of the chemistry of PET involves reactions of the radical ions and is discussed in Chapter 6, but cleavage reactions of the radical ions will produce radicals.

4.1.3. Electron Transfer. Both oxidative and reductive electron-transfer reactions will initiate radical chemistry. Chemical redox is discussed here, but the principles also apply to electrochemical reactions.[58] When an alkyl halide is reduced, the radical ions lose halide in a concomitant or subsequent heterolytic reaction that gives a radical. In reactions of halides with metals such as lithium or magnesium, the organic radical is further reduced to give an organometal species. With mild reducing agents such as samarium diiodide (SmI$_2$),[59] reduction of an organic halide gives a radical that is subsequently reduced with a half-life in the microsecond range,[60] and some radical reactions compete favorably with the second reduction step. A simple example is shown in Figure 4.11; the radical cyclization step competes effectively with the reduction of the initial radical, and the product radical is ultimately reduced to an organosamarium reagent that is protonated by alcohol.[60]

Radicals also can be produced by oxidation with the same constraints as in reductive entries to radicals. That is, the oxidizing agent must react relatively slowly with the radical. Manganese(III) acetate[61] has been widely used for entries to enol radicals that react by radical addition reactions to give products that are further oxidized. An example is shown in Figure 4.11.[62] Cerium(IV) reagents can be used in similar oxidation-initiated radical reactions.[63] Hypervalent iodine species oxidize a number of compounds including alcohols to alkoxyl radicals,[64] and anilides to *N*-arylamidyl radicals (Fig. 4.11).[65]

Reversible redox reactions can initiate radical chemistry without a follow-up reduction or oxidation reaction. In successful reactions of this type, the redox step that produces the radical is thermodynamically disfavored. For example, Cu(I) complexes react reversibly with alkyl halides to give Cu(II) halide complexes and an alkyl radical. The alkyl radical can react in, for example, an addition reaction, and the product radical will react with the Cu(II) halide to give a new alkyl halide. This type of reaction sequence, which has been applied in living radical polymerizations, is in the general family of nonchain radical reactions discussed earlier.[66,67]

$$Cu(I) \text{ complex} + R-X \rightleftharpoons Cu(II)X \text{ complex} + R^{\bullet}$$

4.2. Elementary Propagation Reactions

Radicals undergo both homolytic and heterolytic versions of many reactions, whereas heterolytic processes predominate in reactions of closed-shell species. A major difference between radical and closed-shell molecule reactions is the rates of reactions. In general, rate constants for radical reactions are much larger than the rate constants for equivalent reactions in closed-shell molecules, in part due

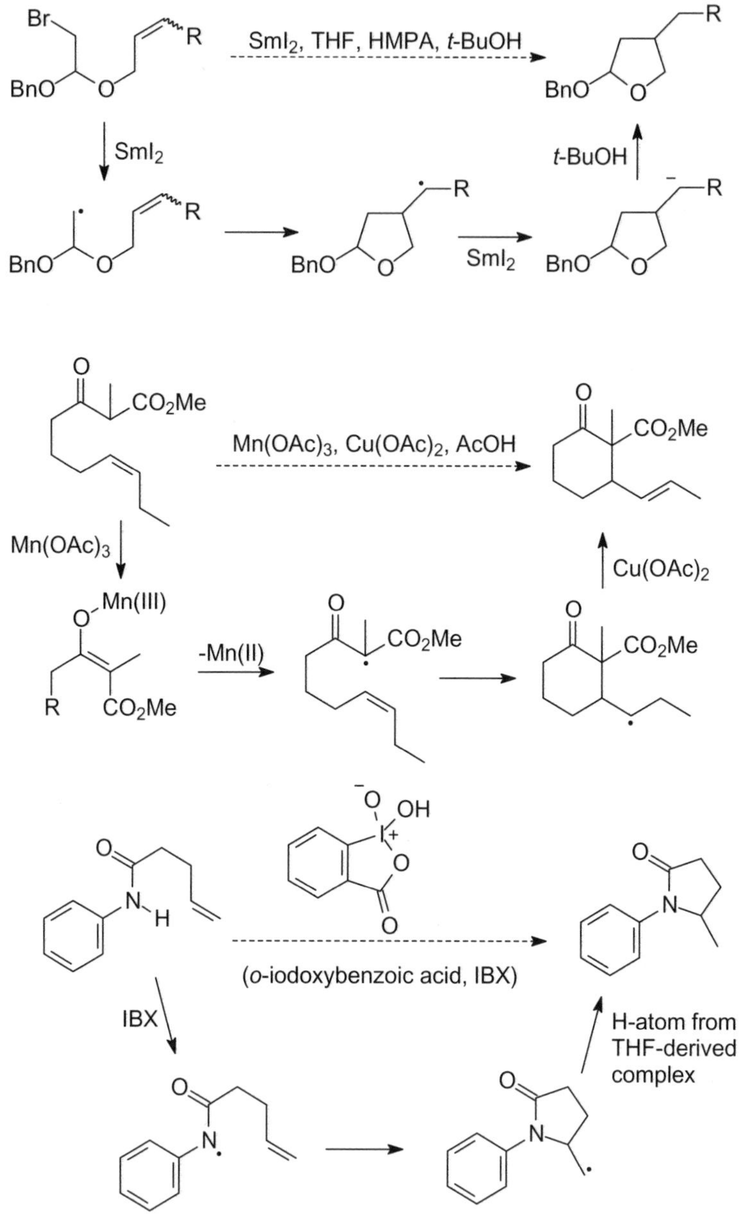

Figure 4.11. Examples of redox-initiated radical reactions. Samarium diiodide reduction of the bromide gives a radical that cyclizes faster than the second reduction reaction. Manganese triacetate oxidation of the β-keto ester gives an enol radical that is not further oxidized by the manganese reagent. The IBX oxidizes anilides to the corresponding radicals. Hexamethylphosphoramide = HMPA and Tetrahydrofuran = THF.

to the absence of one electron that would occupy a high energy orbital in the transition state in a radical reaction.

4.2.1. Homolytic Atom- and Group-Transfer Reactions.

Atom- and group-transfer reactions are among the most important radical propagation reactions in fine chemical synthesis, but they usually are undesired side reactions in radical polymerizations because they terminate polymer chain growth even though they are not chain-termination reactions *per se*. Atom-transfer reactions include hydrogen and halogen transfers. Group transfers involve the exchange of pseudo-halogens, such as a phenylselenyl group transfer. Relatively large multiatom groups, such as the allyl group, transfer by a composite addition–elimination sequence discussed in Section 4.2.4.

4.2.1.1. Hydrogen Atom Transfers.

Hydrogen atom transfers occur in a single-step process that can be termed S_H2 (substitution, homolytic, bimolecular) in analogy to the well-known S_N2 substitution reaction, and many of these reactions are kinetically well characterized. The reaction of Bu_3SnH with the propyl radical shown in Figure 4.6 is an example. In chain reactions for synthesis, group 14 (IV A) metal hydrides (mainly stannanes and silanes) are commonly employed because the metal-centered radical thus formed will react with an organic halide, pseudo-halide, thione, or other radical precursor to propagate the chain reaction, and extensive tables of rate constants for hydrogen atom transfer reactions of group 14 (IV A) hydrides are available.[15,68] Commonly employed group 14 (IV A) hydrides are tributyltin hydride, triphenyltin hydride, and tris(trimethylsilyl)silane, $(TMS)_3SiH$. Tributylgermanium hydride can be used, but it seldom has been, and triethylsilane reacts too slowly with carbon radicals to be useful in chain reactions. Thiols and selenols react rapidly with carbon radials and can be employed in some chain reactions, but the thiyl and selenyl radicals will not react rapidly with an organic halide. Other hydrogen atom donors that have been employed in synthetic applications are phosphines and reactive hydrocarbons such as 1,4-cyclohexadiene. Hydrogen atom transfer reactions from organic solvents are possible, and oxygen-centered radicals and aryl radicals abstract hydrogen from ethers rapidly, but alkyl radicals react slowly enough with ethers such as THF that the reactions are not of consequence.

Figure 4.12 shows rate constants at ambient temperature for reactions of group 14 (IV A) and group (VI A) 16 metal hydrides with several types of organic radicals.[15,69–72] Within each group, the rate constants increase with decreasing M–H bond strength, reflecting the thermodynamics of the hydrogen atom transfer reaction. In addition, the kinetics reflect a polar contribution. As the nucleophilicity of the radical increases, the rate constants for reactions with electron-rich hydrogen atom donors from Group 14 (IV A) decrease. The group 16 (VI A) hydrides are electron-poor hydrogen atom donors, and increasing nucleophilicity in the radical results in faster hydrogen atom transfer reactions in this group. Note that, whereas the S–H BDE for a thiol is greater than the Sn–H BDE for a stannane, an alkyl radical reacts faster with t-BuSH than with Bu_3SnH.

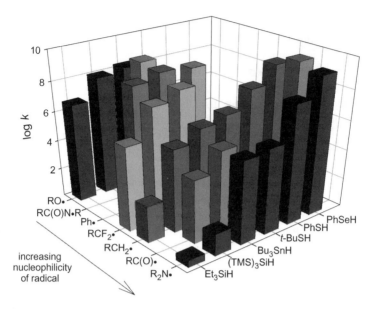

Figure 4.12. Second-order rate constants for reactions of hydrogen atom donors with various radical types at ambient temperature. Data sources: group 14 (IV A) hydrides (15); aminyl radicals (69); amidyl radicals (70); alkyl radials with group 16 (VI A) hydrides (71); acyl radical with PhSeH (72).

The polar effect on the kinetics of atom-transfer reactions permits the use of combinations of a group 14 (IV A) and a group 16 (VI A) metal hydride when neither reagent can be used independently. The compound Et_3SiH reacts too slowly with alkyl radicals to support a chain reaction, and a thiol cannot support a radical chain reduction of an alkyl halide because the thiyl radical will not abstract the halogen atom from the alkyl halide. However, the two reagents can be used in combination for the alkyl halide reduction chain reaction.[73] An alkyl radical reacts rapidly with t-BuSH, the thiyl radical thus formed reacts rapidly with Et_3SiH, and the silyl radical abstracts a halogen atom efficiently from an alkyl halide.[74] In a similar manner, a combination of Bu_3SnH and a catalytic amount of PhSeH, conveniently generated by reaction of PhSeSePh with the tin hydride *in situ*, can be employed when one wishes to take advantage of the very fast trapping of alkyl radicals by PhSeH.[71,75]

4.2.1.2. Halogen and Chalcogen Transfers. Halogen and chalcogen transfers to carbon radicals[76] are synthetically important for radical functionalization reactions, and organic halides and, less commonly, alkyl phenyl chalcogenides are useful radical precursors. Halogen atom transfer reactions and pseudo-halogen group transfers most likely occur by S_H2 mechanisms. It is possible that intermediates are formed with the larger atoms such as iodine and tellurium, but a large leaving-group effect was found in reactions of arene-substituted arylselenides with alkyl radicals, suggesting that an intermediate adduct was not formed in these reactions.[77] Group 14 (IV A) radicals react rapidly with alkyl halides and pseudo-halides to generate alkyl radicals, but group 16 (VI A) radicals (oxyl, thiyl, selenyl) do not abstract halogen atoms.

The rate constants for reactions of alkyl radicals with various organic halides demonstrate a clear dependence on the thermodynamics of the reactions. Iodides react faster than bromides, which react faster than chlorides, and the rate constants for reactions for a series of bromides or iodides correlate with the stability of the radical product.[78,79] The reactions of a primary alkyl radical with an iodomalonate and with a bromomalonate are quite fast[78] ($k = 2 \times 10^9\,M^{-1}\,s^{-1}$ and $k = 1 \times 10^6\,M^{-1}\,s^{-1}$, respectively, at 50 °C). To a good approximation, the rate constants for reactions of RSePh are the same as those for reactions of RBr, and the rate constants for reactions of RTePh are about the same as those for reactions of RI.[77] Dichalcogenides are useful for radical functionalization reactions; they react with primary alkyl radicals at ambient temperature with the following rate constants: MeSSMe, $6 \times 10^4\,M^{-1}\,s^{-1}$; PhSSPh, $2 \times 10^5\,M^{-1}\,s^{-1}$; PhSeSePh, $2.6 \times 10^7\,M^{-1}\,s^{-1}$; PhTeTePh, $1.1 \times 10^8\,M^{-1}\,s^{-1}$.[22]

Abstractions of halogen atoms from alkyl halides by silyl, germanyl, and stannyl radicals are among the most important reactions for producing an organic radical in chain-reaction sequences. These reactions are very fast as shown in Figure 4.13.[74,80] Both the tris(trimethylsilyl)silyl radical[81] and the tributylgermanyl radical[80] react with alkyl halides with rate constants similar to those of the tributylstannyl radical. Reactions of the group 14 (IV A) metal-centered radicals with alkyl phenyl selenides have rate constants comparable to those for the analogous bromides.

Intramolecular group-transfer reactions[82,83] can involve either direct displacements or addition–fragmentation reactions that are discussed in Section 4.2.4. A direct displacement is shown in the reaction of aryl radical **1**, produced from the corresponding bromide. Cyclization of **1** is fast enough to compete efficiently

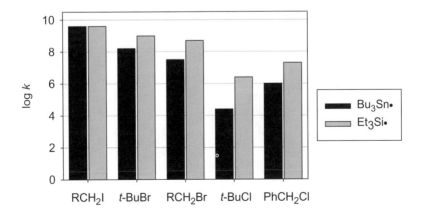

Figure 4.13. Second-order rate constants for reactions of tributylstannyl and triethylsilyl radicals with various alkyl halides at ambient temperature. [Data from (74) and (80).]

with reduction of the aryl radical by dilute tin hydride.[84] As in other types of reactions, the intramolecular aspect of the reaction permits processes that might be too slow to observe for the analogous bimolecular reaction.

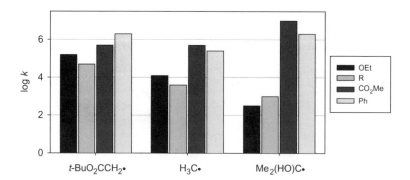

4.2.2. Homolytic Radical Addition and Elimination Reactions

4.2.2.1. Additions. Homolytic bimolecular addition reactions of carbon-centered radicals to unsaturated groups have been studied in detail because these are the reactions of synthesis and polymerization. Within this group, radical additions to substituted alkenes are by far the best understood. An excellent compilation of rate constants for carbon radical additions to alkenes is recommended for many specific kinetic values.[85]

Rates of radical additions to alkenes are controlled mainly by the enthalpy of the reaction, which is the origin of regioselectivity in additions to unsymmetrical systems, with polar effects superimposed when there is a favorable match between the electrophilic or nucleophilic character of the radical and that of the radicophile.[85] For example, in the addition of an alkyl radical to methyl acrylate (**2**), the nucleophilic alkyl radical interacts favorably with the resonance structure **3**. Polar effects are apparent in the representative rate constants shown in Figure 4.14 for additions of carbon radicals to terminal alkenes.[86–88] Addition of the electron-deficient or electrophilic *tert*-butoxycarbonylmethyl radical to the electron-deficient molecule methyl acrylate is ~10 times as fast as addition of

Figure 4.14. Second-order rate constants for additions of radicals to terminal alkenes at ambient temperature. The substituents on the alkenes within each group are shown in the legend. [Data from (86–88).]

this radical to an electron-rich enol ether and slower than addition to styrene. On the other hand, the nucleophilic 1-hydroxy-1-methylethyl radical adds to methyl acrylate >4 orders of magnitude faster than it reacts with an enol ether and faster than it adds to styrene.

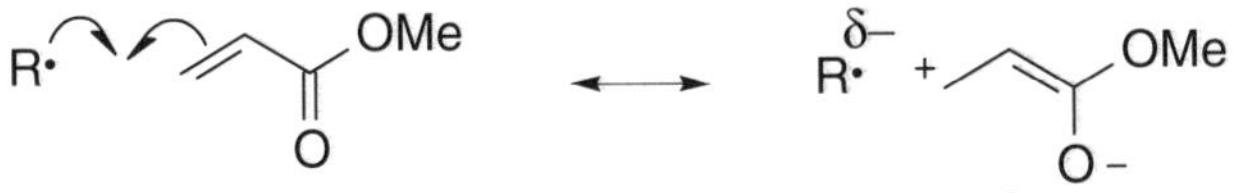

Intramolecular homolytic additions of radicals to alkenes, or radical cyclizations, are among the historically important reactions that led organic synthetic chemists to radical methodology. Cyclizations of carbon-centered radicals onto unactivated alkene moieties in five-membered ring-forming reactions are fast enough to be accomplished with no special techniques due to the acceleration afforded by the intramolecular nature of the reaction. Thus, five-membered carbocycles, which have limited entries from other synthetic approaches, are readily available by radical methods. In unsubstituted systems, cyclization of the 5-hexenyl radical ($k = 2 \times 10^5 \text{ s}^{-1}$) and the 6-heptenyl radical ($k = 6000 \text{ s}^{-1}$) are fast enough to be used in synthetic sequences. When an accelerating group such as an ester in an acrylate is present, macrocycles with 10–18-membered rings can be prepared in radical cyclizations.[89]

The regioselectivity of radical cyclization reactions will be biased by a radical stabilizing group on an unsaturated moiety as in intermolecular additions, but there is a strong preference for exo cyclizations over endo cyclizations in small ring-forming reactions when no stabilizing group is present. Thus, the 5-exo cyclization of the 5-hexenyl radical (**4**) is 50 times faster than 6-endo cyclization even though cyclizations onto the terminal carbon gives the thermodynamically favored radical.[90] The explanation of this kinetic preference with force field calculations was an early success of computational methods applied to study radical reactions.[91–93] Stereoselectivity in 5-exo cyclizations that give disubstituted cyclopentyl products is modest, but, again, the product ratios are well modeled in the force field studies.[93]

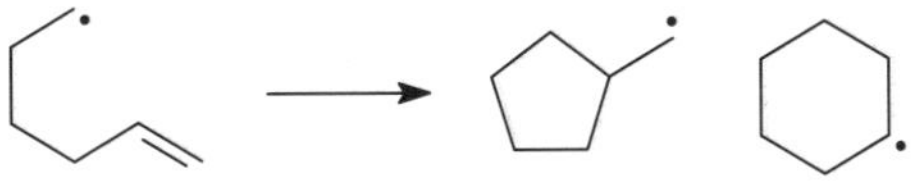

50 : 1

4

Rate constants for several 5-exo radical cyclizations are shown in Figure 4.15.[71,90,94–96] The general trends in the kinetics are similar to those found in intermolecular additions. It is tempting to assume that the large rate enhancements found when a radical stabilizing group is bonded to the incipient radical center will be paralleled by large rate retardations when the same group is present on the original radical center. That is not the general case, however, because the kinetic

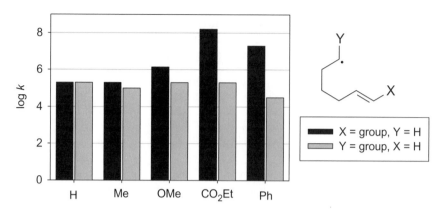

Figure 4.15. First-order rate constants for 5-exo cyclizations at ambient temperature. [Data from (71, 90, 94–96).]

acceleration for the substituent on the alkene can derive from a polarized transition state. The rate constant for 5-exo cyclization of the 1-methyl-1-ethoxycarbonyl-5-hexenyl radical (**5**) is, in fact, an order of magnitude smaller than that for cyclization of the analogous tertiary radical 1,1-dimethyl-5-hexenyl, but this is due to a steric effect in the transition state for cyclization caused by enforced planarity of the radical center by the ester group and not due to electronic stabilization in **5** per se.[94]

$$\text{CO}_2\text{Et} \longrightarrow \text{EtO}_2\text{C}$$

5

Whereas additions of carbon radicals to alkene moieties are the best character-ized homolytic additions, carbon radicals are known to add to a wide range of unsa-turated systems.[97] These include polyenes,[85] alkynes,[85] arenes,[87] heteroarenes,[98,99] carbon monoxide,[7,100,101] isonitriles,[100,102] nitriles,[103] imines and derivatives,[104,105] aldehydes,[106,107] nitrones,[107] and thiones.[108] Many of these reactions, such as addi-tion of an alkyl radical to a carbonyl group,[106] are thermodynamically unfavorable and readily reversible, and they form the basis of composite group-transfer reac-tions discussed below.

In a similar manner, many additions of heteroatom radicals to unsaturated posi-tions have been studied. In many cases, addition reactions of heteroatom radicals to alkenes are reversible and thermodynamically disfavored, but their occurrence is apparent. For example, the rapid addition and elimination of thiyl radicals to unsa-turated fatty acid methyl esters results in isomerization reactions from which kinetic parameters can be obtained.[109] Additions of group 14 (IV A) metal-centered radicals to activated alkenes at ambient temperature have rate constants in the 10^6–$10^8\ M^{-1}\,s^{-1}$ range.[80] Highly reactive alkoxyl radicals add rapidly to the alkene moiety to give thermodynamically favored carbon radical products.[110] For example, cyclization of the 4-pentenyl-1-oxyl radical (**6**) is ∼3 orders of magnitude faster than cyclization of the 5-hexenyl radical and has the same 50:1 preference for 5-exo cyclization over 6-endo cyclization as observed with the carbon radical analogue.[111] Cyclization of the dialkylaminyl radical **7** is slower than cyclization of 5-hexenyl and approximately theromoneutral,[112] whereas cyclization of amidyl radical (**8**) is very fast and strongly favored thermodynamically.[70]

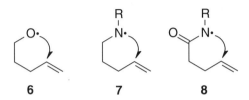

6   **7**   **8**

4.2.2.2. Fragmentations. Homolytic fragmentations of carbon radicals to give carbon radical products mainly involve ring openings of strained three- and four-membered rings. The simplest representative, the cyclopropylcarbinyl radical (**9**) ring opening to the 3-butenyl radical, is one of the most well-studied radical reactions and is the archetypal fast radical reaction with a rate constant for ring opening at ambient temperature of $7 \times 10^7\ s^{-1}$.[71,113] Cyclopropylcarbinyl compounds have been used as probes to test for radical intermediates in a wide range of reactions. The cyclobutylcarbinyl radical opens to the 4-pentenyl radical with a rate constant that is smaller by >4 orders of magnitude.[114]

9

Rate constants for ring openings of substituted cyclopropylcarbinyl radicals show the same types of kinetic effects as seen in radical additions to substituted alkenes with an obvious relationship to the thermodynamics of the reaction (Fig. 4.16).[71,113,115–121] Minor steric effects influence the kinetics, however, and these are apparent in computational results that accurately reproduce the barriers for reaction.[115] When the substituent is at the radical center, the kinetics of ring openings are only slightly affected unless a conjugating group such as phenyl is the substituent. As with 5-hexenyl radical cyclizations, this might appear counterintuitive, but the only group that is clearly out of line with the thermochemistry is the ester group, and ring openings of cyclopropylcarbinyl radicals with an ester at the radical center are accelerated by polarized transition states.[121]

A large number of homolytic fragmentation reactions of carbon radicals with β-leaving groups are known from studies in the gas and condensed phase.[122,123]

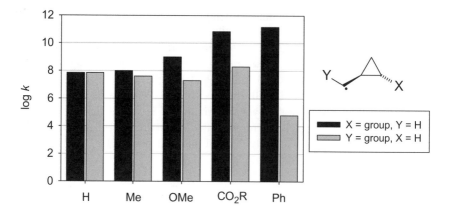

Figure 4.16. First-order rate constants for ring openings of substituted cyclopropylcarbinyl radicals at ambient temperature. [Data from (71, 113, 115–121).]

In many cases, rate constants for the fragmentation reactions are not available, but the reactions can be deduced to be quite fast because intermediate radicals are not trapped efficiently by reactive hydrogen atom donors.

Homolytic cleavage reactions of alkoxyl radicals are quite common, and the reactions have good synthetic utility.[52,124] Cleavage of the *tert*-butoxyl radical (**10**) gives methyl radical and acetone with a rate constant of 2×10^4 s^{-1} at ambient temperature in benzene.[125] This fragmentation represents the low end of the kinetic range for alkoxyl radical cleavages that give alkyl radicals because other alkyl radicals are more stable than methyl. The rate of homolysis is also strongly influenced by the stability of the carbonyl product formed in the elimination. For example, when the same alkyl radical was expelled from a series of alkoxyl radicals, the rate constants for elimination were a function of the carbonyl compound formed with the following relative values:[126] 0.013 (formaldehyde), 0.19 (acetaldehyde), 1 (acetone), 8.8 (benzaldehyde), and ≥ 22 (benzophenone).

10

Some homolytic fragmentation reactions are driven by formation of small, stable molecules. Alkyl acyloxyl radicals (RCO$_2^{\bullet}$) decarboxylate rapidly ($k > 1 \times 10^9$ s^{-1}) to give alkyl radicals, and even aryl acyloxyl radicals (ArCO$_2^{\bullet}$) decarboxylate to aryl radicals with rate constants in the 10^6 s^{-1} range.[46] Azo radicals produced in the homolysis of azo initiators eliminate nitrogen rapidly. Elimination of carbon monoxide from acyl radicals occurs but is slow enough ($k \approx 10^4$–10^5 s^{-1})[7,33] such that the acyl radical can be trapped in a bimolecular process, and addition of a radical to carbon monoxide followed by acyl radical trapping is a useful synthetic carbonylation sequence.[7]

4.2.3. Heterolytic Radical Addition and Fragmentation Reactions

4.2.3.1. Additions. Radicals can react with anionic species to give radical anion adducts as shown for radical **11**. Such addition reactions are steps in chain reaction processes described as $S_{RN}1$ (unimolecular radical nucleophilic substitution) reactions.[127,128] The $S_{RN}1$ reactions typically involve aryl radicals that are produced by reduction of an aryl halide to give the corresponding radical anion that expels halide. Therefore, these reactions provide an important method for aromatic substitution.[129] In a typical reaction sequence, the radical anion adduct **12** would transfer an electron to a molecule of reactant (ArX) in a chain propagation step to give the neutral product of an overall substitution reaction sequence.

Whether the addition step in an $S_{RN}1$ sequence is labeled as heterolytic or homolytic is a semantic question that depends on the MO energy levels of the initially formed adduct. Alkyl radicals also can react in $S_{RN}1$ sequences. If a simple alkyl radical reacted with an anion, the partially occupied MO in the radical anion adduct would reside on the original anion moiety, and the reaction would have the appearance of a homolytic addition. In many $S_{RN}1$ examples involving alkyl radical reactions, however, the radical is stabilized by strong electron-withdrawing groups such as nitro, and charge in the radical anion would be carried largely on the original radical part of the adduct.

Aryl radical additions to anions are generally very fast, with many reactions occurring at or near the diffusion limit. For example, competition studies involving mixtures of nucleophiles competing for the phenyl radical showed that the relative reactivities were within a factor of 10, suggesting encounter control,[130] and absolute rate constants for additions of cyanophenyl and 1-naphthyl radicals to thiophenoxide, diethyl phosphite anion, and the enolate of acetone are within an order of magnitude of the diffusional rate constant.[131]

4.2.3.2. Fragmentations. Heterolytic fragmentation reactions of radicals with leaving groups β to the radical center produce a radical cation and an anion. Such reactions might seem rare, but in actuality they are likely to be relatively common. The reaction sequence is described in the same context as heterolysis of a closed-shell species (Fig. 4.17). Dissociation of the radical gives an intimate or contact pair comprised of the radical cation and anion. The pair can recombine to give the original radical or a rearranged radical, or it can react in, for example, a proton transfer reaction. The contact ion pair can become solvated, and the solvent-separated ion

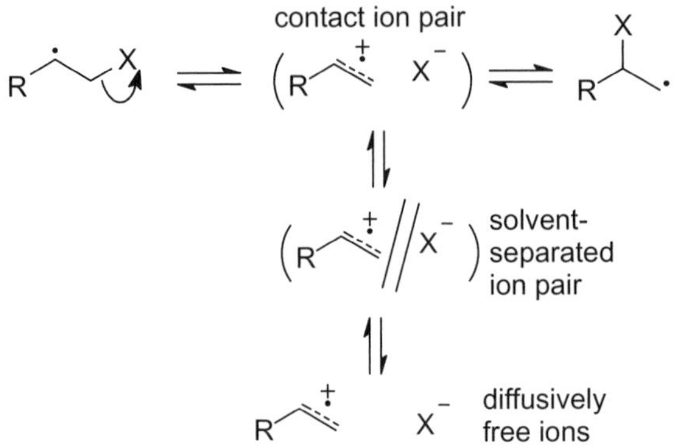

Figure 4.17. Ion-pair formation in a heterolytic radical fragmentation.

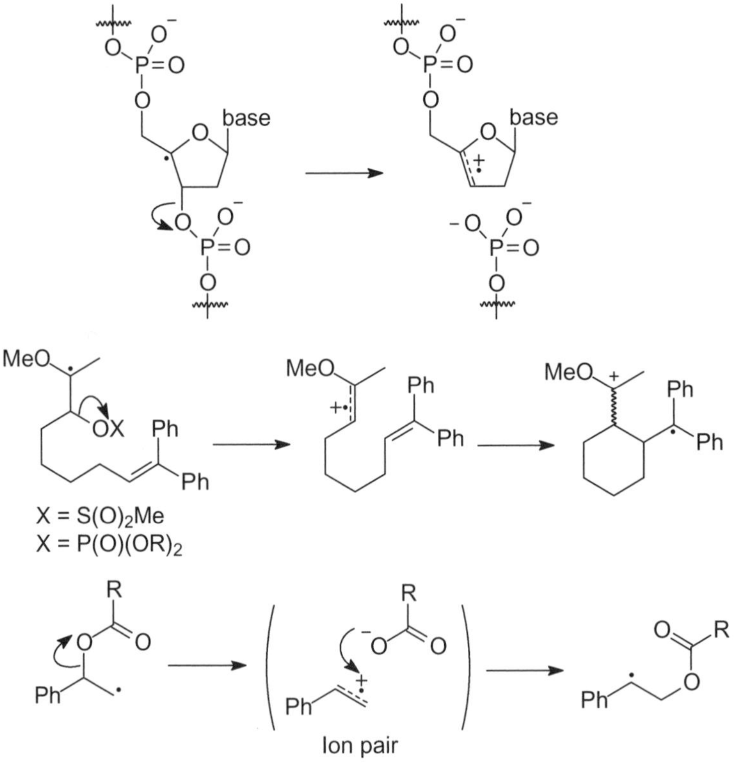

Figure 4.18. Radical heterolysis reactions. The third example is a possible reaction pathway for 1,2-migrations of ester groups.

pair can diffuse apart to give free ions. In highly polar media, the heterolysis reaction might be observed directly by the formation of free radical cations and anions. In low-polarity solvents, the ions can react with one another without diffusing apart, and the mechanism might be inferred.

Examples of radical heterolysis reactions are shown in Figure 4.18. Heterolysis of C4′ deoxyribonucleic acid (DNA) radicals is thought to be the anaerobic pathway for strand cleavage induced by DNA-damaging drugs such as iron-bleomycin. The pathway was supported in an important early study that showed that various alkyl radicals with β-leaving groups (halide, phosphate, sulfonate) reacted in water to give acid, presumably by heterolysis followed by nucleophilic capture or deprotonation of the radical cation by water.[132] Heterolysis of phosphate groups and the mesylate group in β-substituted α-methoxy radicals gives diffusively free radical cations in moderately polar solvents such as acetonitrile.[133,134] 1,2-Migrations of carboxylate and phosphate groups observed when β-ester radicals react in low polarity solvents are well-known reactions[135] that might proceed via ion pairs produced in the dissociative pathway.[136] The radical heterolysis reaction provides a nonoxidative entry to radical cations and could become an important synthetic method.[137]

4.2.4. Composite Group-Transfer Reactions.

Intermolecular group transfers of large organic moieties are composite reactions involving formation of an intermediate by homolytic addition followed by homolytic fragmentation. In organic synthetic conversions, this type of reaction provides a means for maintaining functionality at a radical position as opposed to completing a chain sequence with a hydrogen atom transfer reaction. The sequence is employed often in radical polymerizations to control the growth of polymer chains without terminating radical chain reactions and to add functional groups into polymer chains.[122]

Some examples of addition–elimination reactions are shown in Figure 4.19. The allylation reaction shown[138,139] is typical of a sequence often employed in organic

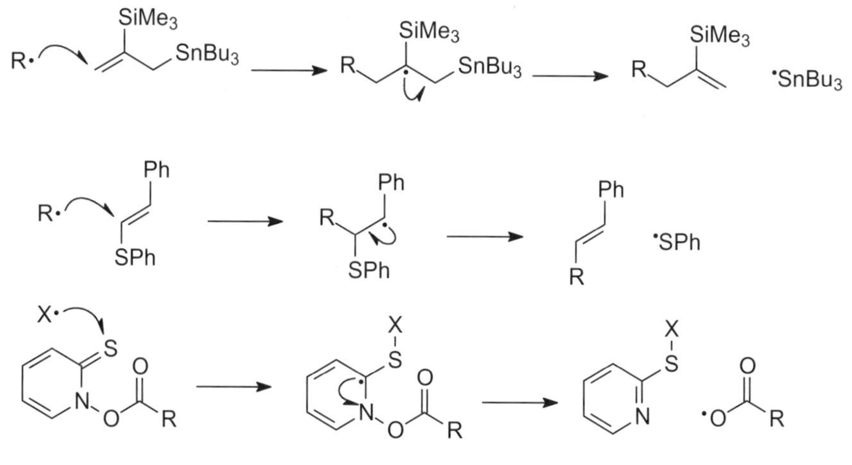

Figure 4.19. Examples of group transfers by addition–elimination.

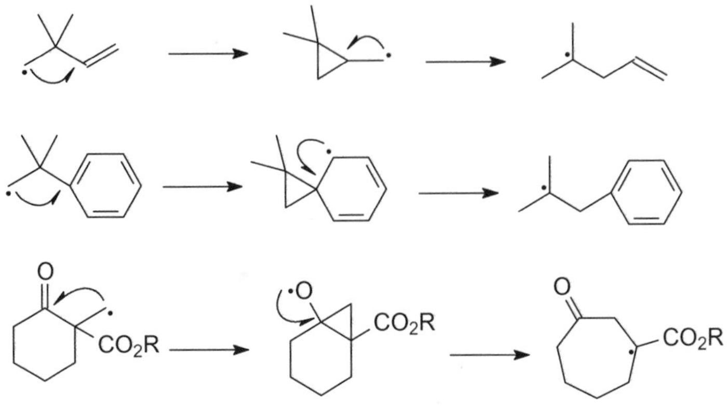

Figure 4.20. Migrations involving addition–elimination.

synthesis. Vinylation of a radical center is possible when the radicophile is a β-substituted styrene[140] or acrylate. The third example in Figure 4.19 is a propagation step when Barton's PTOC esters are used as radical precursors.[50] Addition to the thione moiety occurs at sulfur to give an intermediate that fragments. Xanthate esters and other thiones react in a similar manner as PTOC esters,[108,141] and the thione addition–fragmentation sequence is the basis of the Barton–McCombie deoxygenation reaction.[142,143]

Intramolecular homolytic addition–elimination reactions result in group migrations (Fig. 4.20). Most commonly, 1,2-migrations are involved because formation of a small ring has only a slight entropy penalty even if the enthalpy is unfavorable due to ring strain in the intermediate. The migration of the vinyl group in the 2,2-dimethyl-3-butenyl radical is representative; a small amount of cyclic product is obtained when the reaction is conducted in the presence of thiophenol.[144] Other groups that migrate in a similar manner include carbonyls and their derivatives, acetylenes, aryl groups, and the cyano group.[103,145–147] The second example in the figure is the migration of the phenyl group in the neophyl rearrangement, a well-calibrated[125] reaction that has been known for many years. The cyclization onto a carbonyl group in the third example is the heart of an important ring-expansion sequence for synthesis.[148] As one would expect, migration reactions with heteroatom-centered radicals also occur; for example, phenyl migration in 1-phenylalkoxyl radicals[149] is analogous to the neophyl rearrangement.

4.3. Radical Termination Reactions

Termination reactions convert radicals to closed-shell compounds. Radical–radical coupling reactions are the reverse of homolytic cleavage reactions and are common, but radicals with β-hydrogen atoms also react in disproportionation reactions as shown for **13**. The selectivity of radical–radical terminations is low because the reactions are highly exothermic, and activation barriers for radical–radical reactions are typically smaller than the apparent activation barriers for diffusion.

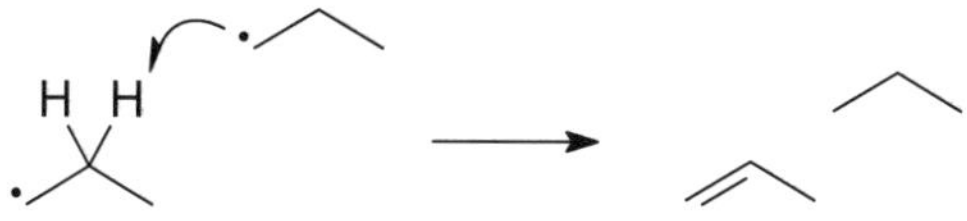

13

The termination rate constant k_T appeared in the rate law for a radical chain reaction in Section 3.2. It is typically a diffusional rate constant, on the order of 10^{10} M^{-1} s^{-1} at ambient temperature. Diffusional rate constants can be predicted from diffusion theory or determined experimentally using, for example, the hydrocarbon analogue of a radical as a model. The rate constant for termination is often less than the rate constant for diffusion, however, due to a phenomenon known as *spin statistical selection*.[150] When two mono-radicals form an encounter complex, the ensemble has a 25% probability of being in a singlet state and a 75% probability of being in a triplet state. If intersystem crossing (or spin flip) is slow on the scale of the lifetime of the encounter complex, which is common for a pair of organic radicals, then the triplet ensemble cannot react because it would give a high-energy excited state product. Thus, the termination rate constant in this case would be 25% of the diffusional rate constant. When heavy atoms are present in the radicals, intersystem crossing rates can increase, and the termination rate constants can approach the diffusional limit.

Knowledge of diffusional rate constants might seem arcane, but the diffusional values were critically important for establishing absolute rate constants for radical reactions. Before lasers were available, kinetic ESR studies were typically performed at low temperatures with continuous irradiation from a UV source.[32] In this experimental design, a reaction is followed that is in competition with radical–radical termination reactions, and one obtains a ratio of the rate constant for the reaction of interest to the (diffusional) rate constant for termination. Increasing understanding of diffusion resulted in adjustments in diffusional rate constants to larger numbers and, accordingly, larger derived rate constants for the radical reactions. In reading the literature on radical chemistry, one should be aware of the possibility that currently accepted values for radical reaction rate constants can be a factor of 2–4 larger than values appearing in articles that are 30–40 years old.

5. CONCLUSION AND OUTLOOK

Organic radical chemistry in synthesis has truly blossomed in the past two decades, and more dramatic advances are expected given the vast scope of synthesis. Radical methodology is attractive because typical conditions for carbon-radical generation and functionalization do not require protection of alcohol and carbonyl functionality and because cyclic compounds can be produced and further functionalized in cascade sequences. Significant progress in, for example, stereocontrol,[10,151,152] asymmetric induction,[153,154] and kinetic information changed somewhat mysterious reactions into powerful synthetic methods. Another driving force for further developing radical-based synthetic methodology is "green chemistry." Virtually all radicals are stable in water, and radical-based functionalization reactions can, in principle, allow one to avoid organic solvents.

Computational methods[93] have advanced to the point that quite accurate bond energies in general and activation barriers for reactions of "apolar" radicals can be computed with good accuracy. This finding in part reflects the fact that homolytic reactions do not create charge, and computations in the gas phase provide good approximations for neutral ground and transition states in condensed phase. Further advances in computational methods are expected in reactions of polarized radicals and reactions that involve heterolytic pathways, and this should follow general advances in handling solvent interactions and charge effects in polar reactions.

Radicals in biological reactions have only become well-appreciated recently. Although alluded to in this chapter, biological radical reactions have not been a focus, which is not a reflection of a lack of examples, however. In addition to destructive reactions, such as the DNA fragmentation in Figure 4.18, a wide range of constructive biological radical reactions are known. Perhaps the most impressive is the reduction of ribonucleotides to $2'$-deoxyribonucelotides by ribonucleotide reductase, a radical reaction involved in the production of all DNA.[155] Many radical reactions in biology are known to be effected by coenzyme B_{12}-dependent and S-adenosylmethionine-dependent enzymes, both of which give the $5'$-deoxyadenosin-$5'$-yl radical that abstracts a hydrogen atom from substrate.[155] The mechanisms of many of the biological radical reactions are poorly understood, however, in part because organic radical analogues are not known. It is possible that much new radical chemistry remains to be uncovered by studying biological radical reactions and that this will involve "polar radical" reaction pathways where enzymes catalyze reactions of radicals with polar groups and regions, much like they catalyze polar reactions of closed-shell molecules.

SUGGESTED READINGS

P. Renaud and M. P. Sibi, Eds., *Radicals in Organic Synthesis*, Wiley-VCH, Weinheim, **2001** (organic synthesis).

K. Matyjaszewski and T. P. Davis Eds., *Handbook of Radical Polymerization*, Wiley-Interscience, Hoboken, NJ, **2002** (polymer chemistry).

H. Fischer and L. Radom, "Factors controlling the addition of carbon-centered radicals to alkenes—an experimental and theoritical perspective," *Angew. Chem., Int. Ed. Engl.* **2001**, *40*, 1340 (rate constants for radical additions to alkenes).

C. Chatgilialoglu and M. Newcomb, "Hydrogen donor abilities of the group 14 hydrides," *Adv. Organometal Chem.* **1999**, *44*, 67 (rate constants for radical reactions with Group 14 (IV A) metal hydrides).

J. Berkowitz, G. B. Ellison, and D. Gutman, "Three Methods to measure RH Bond Energies," *J. Phys. Chem.* **1994**, *98*, 2744 (bond dissociation energies).

REFERENCES

1. J. Pacansky, W. Koch, and M. D. Miller, *J. Am. Chem. Soc.* **1991**, *113*, 317.

2. L. J. Johnston and K. U. Ingold, *J. Am. Chem. Soc.* **1986**, *108*, 2343.

3. W. R. Dolbier, *Chem. Rev.* **1996**, *96*, 1557.

4. K. Heberger, M. Walbiner, and H. Fischer, *Angew. Chem., Int. Ed. Engl.* **1992**, *31*, 635.

5. R. W. Fessenden and R. H. Schuler, *J. Chem. Phys.* **1963**, *39*, 2147.

6. C. Galli, A. Guarnieri, H. Koch, P. Mencarelli, and Z. Rappoport, *J. Org. Chem.* **1997**, *62*, 4072.

7. H. Chatgilialoglu, D. Crich, M. Komatsu, and I. Ryu, *Chem. Rev.* **1999**, *99*, 1991.

8. T. J. Sears, P. M. Johnson, P. Jin, and S. Oatis, *J. Chem. Phys.* **1996**, *104*, 781.

9. O. M. Musa, J. H. Horner, and M. Newcomb, *J. Org. Chem.* **1999**, *64*, 1022.

10. D. P. Curran, N. A. Porter, and B. Giese, *Stereochemistry of Radical Reactions: Concepts, Guidelines, and Synthetic Applications*, VCH, Weinheim, **1995**.

11. C. Rüchardt, *Angew. Chem., Int. Ed. Engl.* **1970**, *9*, 830.

12. D. F. McMillen and D. M. Golden, *Ann. Rev. Phys. Chem.* **1982**, *33*, 493.

13. D. Griller, J. M. Kanabus-Kaminska, and A. Maccoll, *Theochem* **1988**, *40*, 125.

14. J. Berkowitz, G. B. Ellison, and D. Gutman, *J. Phys. Chem.* **1994**, *98*, 2744.

15. C. Chatgilialoglu and M. Newcomb, *Adv. Organometal. Chem.* **1999**, *44*, 67.

16. R. Sustmann and H. Trill, *J. Am. Chem. Soc.* **1974**, *96*, 4343.

17. D. J. Henry, C. J. Parkinson, P. M. Mayer, and L. Radom, *J. Phys. Chem. A* **2001**, *105*, 6750.

18. M. Gomberg, *J. Am. Chem. Soc.* **1900**, *22*, 757.

19. A. L. J. Beckwith, V. W. Bowry, and K. U. Ingold, *J. Am. Chem. Soc.* **1992**, *114*, 4983.

20. M. J. Davies, *Res. Chem. Intermed.* **1993**, *19*, 669.

21. C. Frejaville, H. Karoui, B. Tuccio, F. Lemoigne, M. Culcasi, S. Pietri, R. Lauricella, and P. Tordo, *J. Chem. Soc., Chem. Commun.* **1994**, 1793.

22. M. Newcomb, *Tetrahedron* **1993**, *49*, 1151.

23. D. Griller and K. U. Ingold, *Acc. Chem. Res.* **1980**, *13*, 317.

24. M. Newcomb, in *Radicals in Organic Synthesis*, Vol. 1, P. Renaud and M. P. Sibi, Eds., Wiley-VCH, Weinheim, **2001**, pp. 318–336.

25. B. Kirste, W. Harrer, and H. Kurreck, *J. Am. Chem. Soc.* **1985**, *107*, 20.

26. B. J. Tabner, *Electron Spin Reson.* **1984**, *8*, 243.

27. R. W. Fessenden and R. H. Schuler, *J. Chem. Phys.* **1965**, *43*, 2704.

28. R. G. Lawler, *J. Am. Chem. Soc.* **1967**, *67*, 5519.

29. J. Bargon and H. Fischer, *Z. Naturforsch. A* **1967**, *22*, 1556.

30. H. R. Ward and R. G. Lawler, *J. Am. Chem. Soc.* **1967**, *89*, 5518.

31. C. Chatgilialoglu, in *Handbook of Organic Photochemistry*, Vol. 2, J. C. Scaiano, Ed., CRC Press, Boca Raton, FL, **1989**, pp. 3–11.

32. D. Griller and K. U. Ingold, *Acc. Chem. Res.* **1980**, *13*, 193.

33. H. Fischer and H. Paul, *Acc. Chem. Res.* **1987**, *20*, 200.

34. B. O'Shaughnessy and J. Yu, *Phys. Rev. Lett.* **1994**, *73*, 1723.

35. D. Griller and K. U. Ingold, *Acc. Chem. Res.* **1976**, *9*, 13.

36. H. Fischer, *J. Am. Chem. Soc.* **1986**, *108*, 3925.

37. D. H. R. Barton, J. M. Beaton, L. E. Geller, and M. M. Pechet, *J. Am. Chem. Soc.* **1960**, *82*, 2640.

38. J. Iqbal, R. Sanghi, and J. P. Nandy, in *Radicals in Organic Synthesis*, Vol. 1, P. Renaud, M. P. Sibi, Eds., Wiley-VCH, Weinheim, **2001**, pp. 135–151.

39. C. J. Hawker, in *Handbook of Radical Polymerization*, K. Matyjaszewsi and T. P. Davis, Eds., Wiley, Hoboken, NJ, **2002**, pp. 463–521.

40. M. Georges, in *Radicals in Organic Synthesis*, Vol. 1, P. Renaud and M. P. Sibi, Eds., Wiley-VCH, Weinheim, **2001**, pp. 479–488.

41. Y. Kita and M. Matsugi, in *Radicals in Organic Synthesis*, Vol. 1, P. Renaud and M. P. Sibi, Eds., Wiley-VCH, Weinheim, **2001**, pp. 1–10.

42. P. G. Bartlett, E. P. Benzing, and R. E. Pincock, *J. Am. Chem. Soc.* **1960**, *82*, 1762.

43. G. D. Mendenhall, *Tetrahedron Lett.* **1983**, *24*, 451.

44. C. Walling, *Tetrahedron* **1985**, *41*, 3887.

45. D. J. Hart, R. Krishnamurthy, L. M. Pook, and F. L. Seely, *Tetrahedron Lett.* **1993**, *34*, 7819.

46. J. Chateauneuf, J. Lusztyk, and K. U. Ingold, *J. Am. Chem. Soc.* **1988**, *110*, 2886.

47. K. Nozaki, K. Oshima, and K. Utimoto, *J. Am. Chem. Soc.* **1987**, *109*, 2547.

48. H. Yorimitsu and K. Oshima, in *Radicals in Organic Synthesis*, Vol. 1, P. Renaud and M. P. Sibi, Eds., Wiley-VCH, Weinheim, **2001**, pp. 11–27.

49. H. Yorimitsu and K. Oshima, in *Radicals in Organic Synthesis*, Vol. 1, M. P. Sibi, Ed., Wiley-VCH, Weinheim, **2002**, pp. 11–27.

50. D. H. R. Barton, D. Crich, and W. B. Motherwell, *Tetrahedron* **1985**, *41*, 3901–3924.

51. W. B. Motherwell and C. Imboden, in *Radicals in Organic Synthesis*, Vol. 1, P. Renaud and M. P. Sibi, Eds., Wiley-VCH, Weinheim, **2001**, pp. 109–134.

52. J. Hartung, T. Gottwald, and K. Spehar, *Synthesis-Stuttgart* **2002**, 1469–1498.

53. J. L. Esker and M. Newcomb, in *Advances in Heterocyclic Chemistry*, Vol. 58, A. R. Katritzky, Ed., Academic Press, **1900**, San Diego, CA, **1993**, pp. 1–45.

54. L. Stella, in *Radicals in Organic Synthesis*, Vol. 2, P. Renaud, M. P. Sibi, Eds., Wiley-VCH, Weinheim, **2001**, pp. 407–426.

55. C. Chatgilialoglu, K. U. Ingold, and, J. C. Scaiano, *J. Am. Chem. Soc.* **1981**, *103*, 7739.

56. P. Maslak, *Top. Curr. Chem.* **1993**, *168*, 1.

57. J. Cossy, in *Radicals in Organic Synthesis*, Vol. 1, P. Renaud and M. P. Sibi, Eds., Wiley-VCH, Weinheim, **2001**, pp. 229–249.

58. H. J. Schäfer, in *Radicals in Organic Synthesis*, Vol. 1, P. Renaud and M. P. Sibi, Eds., Wiley-VCH, Weinheim, **2001**, pp. 250–297.

59. G. A. Molander, in *Radicals in Organic Synthesis*, Vol. 1, P. Renaud and M. P. Sibi, Eds., Wiley-VCH, Weinheim, **2001**, pp. 153–182.

60. G. A. Molander and C. R. Harris, *Chem. Rev.* **1996**, *96*, 307–338.

61. B. B. Snider, in *Radicals in Organic Synthesis*, Vol. 1, P. Renaud and M. P. Sibi, Eds., Wiley-VCH, Weinheim, **2001**, pp. 198–218.

62. B. B. Snider, *Chem. Rev.* **1996**, *96*, 339.

63. T. Linker, in *Radicals in Organic Synthesis*, Vol. 1, P. Renaud and M. P. Sibi, Eds., Wiley-VCH, Weinheim, **2001**, pp. 219–228.

64. A. Boto, R. Hernandez, E. Suarez, C. Betancor, and M. S. Rodriguez, *J. Org. Chem.* **1995**, *60*, 8209.

65. K. C. Nicolaou, P. S. Baran, R. Kranich, Y. L. Zhong, K. Sugita, and N. Zou, *Angew. Chem., Int. Ed. Engl.* **2001**, *40*, 202.

66. K. Matyjaszewski, *J. Macromol. Sci. Pure Appl. Chem.* **1997**, *A34*, 1785.

67. K. Matyjaszewski, T. E. Patten, and J. H. Xia, *J. Am. Chem. Soc.* **1997**, *119*, 674.

68. C. Chatgilialoglu, in *Radicals in Organic Synthesis*, Vol. 1, P. Renaud, and M. P. Sibi, Eds., Wiley-VCH, Weinheim, **2001**, pp. 28–49.

69. O. M. Musa, J. H. Horner, H. Shahin, and M. Newcomb, *J. Am. Chem. Soc.* **1996**, *118*, 3862.

70. J. H. Horner, O. M. Musa, A. Bouvier, and M. Newcomb, *J. Am. Chem. Soc.* **1998**, *120*, 7738.

71. M. Newcomb, S. Y. Choi, and J. H. Horner, *J. Org. Chem.* **1999**, *64*, 1225.

72. P. A. Simakov, F. N. Martinez, J. H. Horner, and M. Newcomb, *J. Org. Chem.* **1998**, *63*, 1226.

73. R. P. Allen, B. P. Roberts, and C. R. Willis, *J. Chem. Soc., Chem. Commun.* **1989**, 1387.

74. C. Chatgilialoglu, K. U. Ingold, and J. C. Scaiano, *J. Am. Chem. Soc.* **1982**, *104*, 5123.

75. D. Crich and Q. W. Yao, *J. Org. Chem.* **1995**, *60*, 84.

76. J. Byers, in *Radicals in Organic Synthesis*, Vol. 1, P. Renaud and M. P. Sibi, Eds., Wiley-VCH, Weinheim, **2001**, pp. 72–89.

77. D. P. Curran, A. A. Martin-Esker, S. B. Ko, and M. Newcomb, *J. Org. Chem.* **1993**, *58*, 4691.

78. D. P. Curran, E. Bosch, J. Kaplan, and M. Newcomb, *J. Org. Chem.* **1989**, *54*, 1826.

79. M. Newcomb, R. M. Sanchez, and J. Kaplan, *J. Am. Chem. Soc.* **1987**, *109*, 1195.

80. K. U. Ingold, J. Lusztyk, and J. C. Scaiano, *J. Am. Chem. Soc.* **1984**, *106*, 343.

81. C. Chatgilialoglu, D. Griller, and M. Lesage, *J. Org. Chem.* **1989**, *54*, 2492.

82. K. U. Ingold and B. P. Roberts, *Free Radical Substitution Reactions*, Wiley-Interscience, New York, 1971.

83. C. H. Schiesser and L. M. Wild, *Tetrahedron* **1996**, *52*, 13265.

84. A. L. J. Beckwith and D. R. Boate, *J. Org. Chem.* **1988**, *53*, 4339.

85. H. Fischer and L. Radom, *Angew. Chem., Int. Ed. Engl.* **2001**, *40*, 1340.

86. J. Q. Wu, I. Beranek, and H. Fischer, *Helv. Chim. Acta* **1995**, *78*, 194.

87. T. Zytowski and H. Fischer, *J. Am. Chem. Soc.* **1997**, *119*, 12869.

88. K. Heberger and H. Fischer, *Int. J. Chem. Kinet.* **1993**, *25*, 913.

89. N. A. Porter, D. R. Magnin, and B. T. Wright, *J. Am. Chem. Soc.* **1986**, *108*, 2787.

90. A. L. J. Beckwith and G. Moad, *J. Chem. Soc., Chem. Commun.* **1974**, 472.

91. A. L. J. Beckwith and C. H. Schiesser, *Tetrahedron* **1985**, *41*, 3925.

92. D. C. Spellmeyer and K. N. Houk, *J. Org. Chem.* **1987**, *52*, 959.

93. C. H. Schiesser and M. A. Skidmore, in *Radicals in Organic Synthesis*, Vol. 1, P. Renaud and M. P. Sibi, Eds., Wiley-VCH, Weinheim, **2001**, pp. 337–359.

94. M. Newcomb, M. A. Filipkowski, and C. C. Johnson, *Tetrahedron Lett.* **1995**, *36*, 3643.

95. S.-U. Park, S.-K. Chung, and M. Newcomb, *J. Am. Chem. Soc.* **1986**, *108*, 240.

96. C. Walling and A. Cioffari, *J. Am. Chem. Soc.* **1972**, *94*, 6064.

97. S. Kim and J. Y. Yoon, in *Radicals in Organic Synthesis*, Vol. 2, P. Renaud and M. P. Sibi, Eds., Wiley-VCH, Weinheim, **2001**, pp. 1–21.

98. D. C. Harrowven, B. J. Sutton, and S. Coulton, *Tetrahedron Lett.* **2001**, *42*, 9061.

99. A. Demircan and P. J. Parsons, *Synlett* **1998**, 1215.

100. I. Ryu, N. Sonoda, and D. P. Curran, *Chem. Rev.* **1996**, *96*, 177.

101. I. Ryu, in *Radicals in Organic Synthesis*, Vol. 2, P. Renaud and M. P. Sibi, Eds., Wiley-VCH, Weinheim, **2001**, pp. 22–43.

102. D. Nanni, in *Radicals in Organic Synthesis*, Vol. 2, P. Renaud and M. P. Sibi, Eds., Wiley-VCH, Weinheim, **2001**, pp. 44–61.

103. W. R. Bowman, C. F. Bridge, and P. Brookes, *Tetrahedron Lett.* **2000**, *41*, 8989.

104. H. Miyabe and T. Naito, *J. Synth. Org. Chem. Jpn.* **2001**, *59*, 33.

105. S. Kim, I. Y. Lee, J. Y. Yoon, and D. H. Oh, *J. Am. Chem. Soc.* **1996**, *118*, 5138.

106. A. L. J. Beckwith, and B. P. Hay, *J. Am. Chem. Soc.* **1989**, *111*, 2674.

107. S. L. Boyd and R. J. Boyd, *J, Phys, Chem. A* **2001**, *105*, 7096.

108. S. Z. Zard, *Angew. Chem., Int. Ed. Engl.* **1997**, *36*, 673.

109. C. Chatgilialoglu, A. Altieri, and H. Fischer, *J. Am. Chem. Soc.* **2002**, *124*, 12816.

110. J. Hartung, in *Radicals in Organic Synthesis*, Vol. 2, P. Renaud and M. P. Sibi, Eds., Wiley-VCH, Weinheim, **2001**, pp. 427–439.

111. J. Hartung and F. Gallou, *J. Org. Chem.* **1995**, *60*, 6706.

112. M. Newcomb, O. M. Musa, F. N. Martinez, and J. H. Horner, *J. Am. Chem. Soc.* **1997**, *119*, 4569.

113. T. A. Halgren, J. D. Roberts, J. H. Horner, F. N. Martinez, C. Tronche, and M. Newcomb, *J. Am. Chem. Soc.* **2000**, *122*, 2988.

114. K. U. Ingold, B. Maillard, and J. C. Walton, *J. Chem. Soc., Perkin Trans. 2* **1981**, 970.

115. F. N. Martinez, H. B. Schlegel, and M. Newcomb, *J. Org. Chem.* **1998**, *63*, 3618.

116. V. W. Bowry, J. Lusztyk, and K. U. Ingold, *J. Am. Chem. Soc.* **1991**, *113*, 5687.

117. M. H. Le Tadic-Biadatti, and M. Newcomb, *J. Chem. Soc., Perkin Trans. 2* **1996**, 1467.

118. S. Y. Choi and M. Newcomb, *Tetrahedron* **1995**, *51*, 657.

119. M. Newcomb, C. C. Johnson, M. B. Manek, and T. R. Varick, *J. Am. Chem. Soc.* **1992**, *114*, 10915.

120. F. N. Martinez, H. B. Schlegel, and M. Newcomb, *J. Org. Chem.* **1996**, *61*, 8547.

121. J. H. Horner, N. Tanaka, and M. Newcomb, *J. Am. Chem. Soc.* **1998**, *120*, 10379.

122. D. Colombani, *Prog. Polym. Sci.* **1999**, *24*, 425.

123. M. J. Davies and B. C. Gilbert, in *Advances in Detailed Reaction Mechanisms*, Vol. 1, 1st ed.; J. M. Coxon, Ed., JAI Press Inc., Greenwich, CT, **1991**, pp. 35–81.

124. E. Suarez and M. S. Rodriguez, in *Radicals in Organic Synthesis*, Vol. 2, M. P. Sibi, Ed., Wiley-VCH, Weinheim, **2001**, pp. 440–454.

125. M. Weber and H. Fischer, *J. Am. Chem. Soc.* **1999**, *121*, 7381.

126. M. Newcomb, P. Daublain, and J. H. Horner, *J. Org. Chem.* **2002**, *67*, 8669.

127. J. F. Bunnett, *Acc. Chem. Res.* **1978**, *11*, 413.

128. R. A. Rossi, A. B. Pierini, and A. B. Penenory, *Chem. Rev.* **2003**, *103*, 71.

129. A. Studer and M. Bossart, in *Radicals in Organic Synthesis*, Vol. 2, P. Renaud and M. P. Sibi, Eds., Wiley-VCH, Weinheim, **2001**, pp. 62–91.

130. C. Galli and J. F. Bunnett, *J. Am. Chem. Soc.* **1981**, *103*, 7140.

131. C. Amatore, M. A. Otturan, J. Pinson, J.-M. Saveant, and A. Thiebault, *J. Am. Chem. Soc.* **1985**, *107*, 3451.

132. G. Koltzenburg, G. Behrens, and D. Schulte-Frohlinde, *J. Am. Chem. Soc.* **1982**, *104*, 7311.

133. J. H. Horner, E. Taxil, and M. Newcomb, *J. Am. Chem. Soc.* **2002**, *124*, 5402.

134. E. Taxil, L. Bagnol, J. H. Horner, and M. Newcomb, *Org. Lett.* **2003**, *5*, 827.

135. A. L. J. Beckwith, D. Crich, P. J. Duggan, and Q. W. Yao, *Chem. Rev.* **1997**, *97*, 3273.

136. D. Crich, in *Radicals in Organic Synthesis*, Vol. 2, P. Renaud and M. P. Sibi, Eds., Wiley-VCH, Weinheim, **2001**, pp. 188–206.

137. D. Crich and K. Ranganathan, *J. Am. Chem. Soc.* **2002**, *124*, 12422.

138. P. Renaud, M. Gerster, and M. Ribezzo, *Chimia* **1994**, *48*, 366.

139. I. J. Rosenstein, in *Radicals in Organic Synthesis*, Vol. 1, P. Renaud and M. P. Sibi, Eds., Wiley-VCH, Weinheim, **2001**, pp. 50–71.

140. G. A. Russell, H. Tashtoush, and P. Ngoviwatchai, *J. Am. Chem. Soc.* **1984**, *106*, 4622.

141. S. Z. Zard, in *Radicals in Organic Synthesis*, Vol. 1, P. Renaud and M. P. Sibi, Eds., Wiley-VCH, Weinheim, **2001**, pp. 90–108.

142. D. H. R. Barton and J. C. Jaszberenyi, *Tetrahedron Lett.* **1989**, *30*, 2619.

143. W. Hartwig, *Tetrahedron* **1983**, *39*, 2609.

144. M. Newcomb, A. G. Glenn, and W. G. Williams, *J. Org. Chem.* **1989**, *54*, 2675.

145. D. A. Lindsay, J. Lusztyk, and K. U. Ingold, *J. Am. Chem. Soc.* **1984**, *106*, 7087.

146. E. J. Corey and S. G. Pyne, *Tetrahedron Lett.* **1983**, *24*, 2821.

147. S. Kim, K. S. Yoon, and Y. S. Kim, *Tetrahedron* **1997**, *53*, 73.

148. P. Dowd and W. Zhang, *Chem. Rev.* **1993**, *93*, 2091.

149. J. A. Howard and K. U. Ingold, *Can. J. Chem.* **1969**, *47*, 3797.

150. J. Saltiel and B. W. Atwater, *Adv. Photochem.* **1988**, *14*, 1.

151. B. Giese, in *Radicals in Organic Synthesis*, Vol. 1, P. Renaud and M. P. Sibi, Eds., Wiley-VCH, Weinheim, **2001**, pp. 381–399.

152. P. Renaud, in *Radicals in Organic Synthesis*, Vol. 1, P. Renaud and M. P. Sibi, Eds., Wiley-VCH, Weinheim, **2001**, pp. 400–415.

153. N. A. Porter, in *Radicals in Organic Synthesis*, Vol. 1, P. Renaud and M. P. Sibi, Eds., Wiley-VCH, Weinheim, **2001**, pp. 416–460.

154. M. P. Sibi and T. R. Rheault, in *Radicals in Organic Synthesis*, Vol. 1, P. Renaud and M. P. Sibi, Eds., Wiley-VCH, Weinheim, **2001**, pp. 461–478.

155. J. Stubbe and W. A. van der Donk, *Chem. Rev.* **1998**, *98*, 705.

CHAPTER 5

Non-Kekulé Molecules as Reactive Intermediates

JEROME A. BERSON

Department of Chemistry, Yale University, New Haven, CT

Reactive Intermediate Chemistry, edited by Robert A. Moss, Matthew S. Platz, and Maitland Jones, Jr.
ISBN 0-471-23324-2 Copyright © 2004 John Wiley & Sons, Inc.

1. INTRODUCTION

π-Conjugated non-Kekulé molecules,[1] do not conform to the standard rules of valence. For that reason, they have fascinated organic chemists as strange entities that lie on the borderline of existence. Attempts to synthesize such species can be traced back more than a century, and by now, scores of examples are known.

Workers in the field have been mindful of the potential for utility of these species in new reactions for organic synthesis and in new structures applicable to the design of molecular electronic devices. Such developments have in fact emerged, and progress in that area has been reviewed many times.[2-9] We must relinquish the opportunity to provide another review here, both because of lack of space and because the objectives and findings of that work are at some remove from our designated subject of "reactive intermediates." Rather, we focus our attention on the challenging problems that non-Kekulé molecules themselves present in experiment and theory. These include their synthesis, their structures and spin states, and their *chemical reactions*. Such investigations make possible significant tests of theory, which will be a recurrent theme. Even so limited, the subject is too large for complete treatment here. Accordingly, we concentrate on recent developments and on the background out of which they grew. Although many references to individual papers are given, the literature is so voluminous that when substantial reviews on a particular subject are readily available, the review may be cited here rather than the individual articles.

2. HISTORICAL BACKGROUND

A number of authors in the the field of non-Kekulé compounds have given summaries of one or another aspect of the historical record.[10-15] However, it may be helpful to relate briefly some of the main occurrences, events that in retrospect can be seen to mark the emerging (and sometimes temporally overlapping) eras in the field.

2.1. The Schlenk–Brauns Hydrocarbons

Earliest among these was the synthesis by Schlenk and Brauns[16,17] of the bis(tri-arylmethyls) **1** and **2**, the first non-Kekulé compounds.

1   **2**

They were prepared by adapting Gomberg's synthesis of triarylmethyls (e.g., tri-phenylmethyl),[18] that is, the reduction of triarylmethyl halides in inert atmosphere. Biradicals **1** and **2** persist under a carbon dioxide atmosphere in fluid solution at room temperature. Superficially, this might seem a not very significant extension of Gomberg's work, but as Schlenk and Brauns recognized, their biradicals differed in an important way from the triarylmethyls. The Gomberg monoradicals are *molecular fragments* and by definition cannot achieve a full Kekulé structure. The Schlenk–Brauns biradicals, on the other hand, formally have enough atoms to satisfy the rules of valence. Despite that capability, they exist with one missing bond, a fact that clearly signifies the appearance of a new kind of matter. This work was done in 1915, well before the advent of quantum chemistry, and at that time, under-standing and predicting the properties to be expected of these new species could at best be superficial.

2.2. Hund's Rule

The second phase began after quantum mechanical thinking had become estab-lished, even if mostly among physicists. Eugen Müller, in consultation with Friedrich Hund, was the first chemist to examine the paramagnetism of biradicals.[19,20] By magnetic susceptibility measurements with a Guoy balance, he showed that com-pound **1** was indeed paramagnetic, presumably because it exists as a triplet state. The latter finding was consistent with a molecular version of Hund's rule, which was a most gratifying result for the adherents of the new quantum mechanics.

A seminal paper of Longuet-Higgins in 1950[21] gave powerful impetus to the idea that the state of highest multiplicity should predominate in a π–conjugated non-Kekulé molecule. He showed by simple Hückel calculations that the four non-Kekulé molecules *m*-quinodimethane (*m*-xylylene, MQDM) **3**, trimethylene-methane (TMM) **4**, tetramethyleneethane (TME) **5**, and triangulene **6**, all had a pair of degenerate frontier orbitals occupied by only two electrons. Relying on Hund's rule, he predicted that all of these species should have triplet ground states. Hypothetical full-valence Kekulé isomers of these are shown as structures **7–10**.

The arrival of the third phase of non-Kekulé chemistry now awaited two neces-sary developments. First, the Guoy balance technique was of limited sensitivity and yielded no structural information about the source of the paramagnetism; thus, a most desirable development would be the appearance of a new and independent

Non-Kekulé

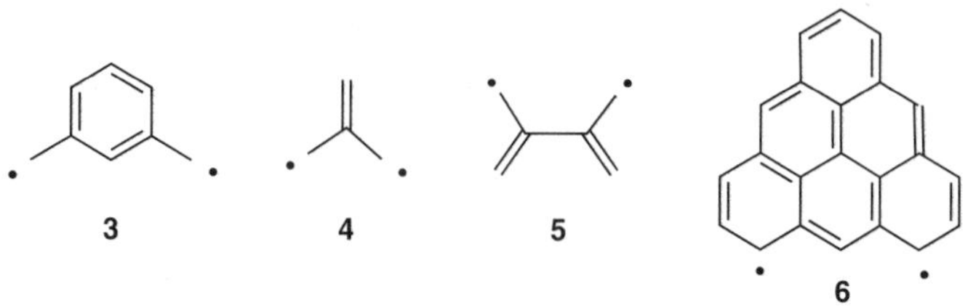

3	**4**	**5**	**6**

Full Valence

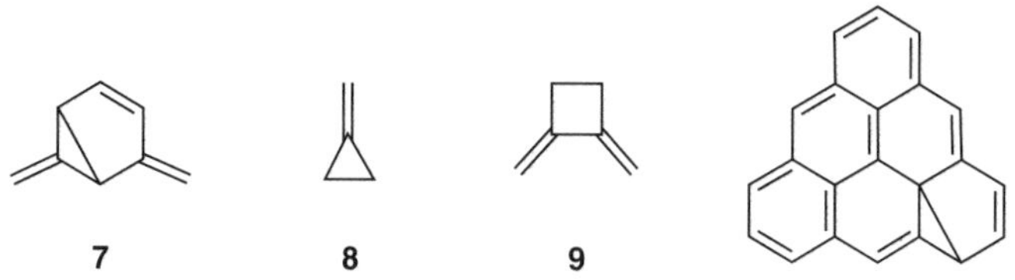

7	**8**	**9**	**10**

physical method. Second, Schlenk and Brauns had used sterically hindered and rather unreactive biradical derivatives of the triphenylmethyl group. The chemist who wished to reach general conclusions about non-Kekulé molecules had to venture beyond those secure boundaries into a realm of highly unstable species and would have to learn to prepare, keep, and study these molecules for extended periods, despite their instability.

2.3. Electron Spin Resonance Spectroscopy of Randomly Oriented Samples: Hund's Rule Vindicated?

A crucial methodological step forward was the discovery[22–24] that one could observe well-defined electron spin resonance (ESR) spectra of frozen solutions of triplet species in random orientation. By the early 1960s, spectra of the triplet states of a number of carbenes had been recorded. Thus, when Dowd[25] showed that photolysis of frozen matrices of the diazene (**11**) or the ketone (**12**) (Scheme 5.1) gave TMM (**4**), the spectroscopic tools for the characterization of this key non-Kekulé compound lay to hand. Trimethylenemethane was the first non-Kekulé molecule to be identified by ESR spectroscopy.

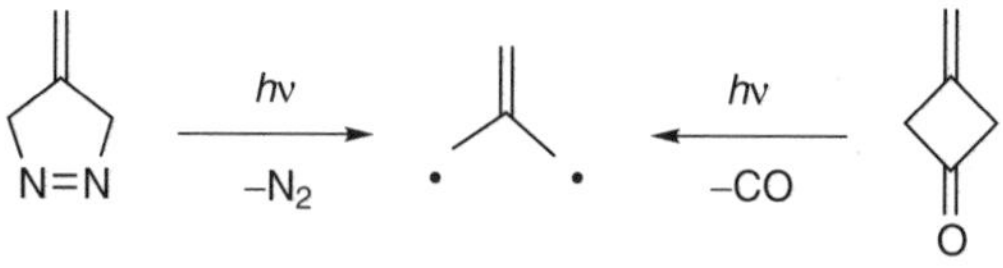

11	**4**	**12**

Scheme 5.1

Beginning in 1971 this third phase evolved into the examination of several other species, including the series of TMM derivatives based on 2-methylenecyclopentane-1,3-diyl (**14**) (Scheme 5.2) derived from the diazene (**13**).[26]

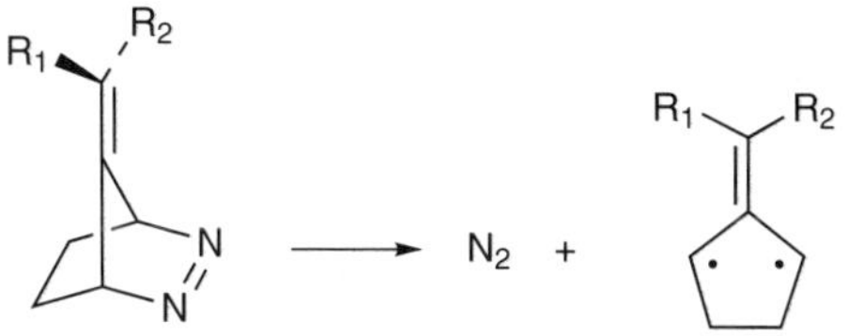

13   **14**

a: R_1, R_2 = H   d: R_1, R_2 = Ph
b: R_1, R_2 = Me   e: R_1 = Cl, R_2 = H
c: R_1 = Ph, R_2 = H   f: R_1 = OMe, R_2 = H

Scheme 5.2

The 1970s and 1980s saw the development of the chemistry of both the singlet and triplet states of **14** and its derivatives.[14,15,27] Modern theoretical methods provided important insights.[28] Also prepared and spectroscopically characterized during this period of rapid growth of the field were the non-Kekulé target MQDM (**3**),[29–32] which had been mentioned by Longuet-Higgins, and two of its close relatives, *m*-quinomethane (MQM, **15**),[33,34] and *m*-naphthoquinomethane (**16**).[34,35]

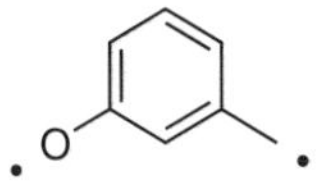

15 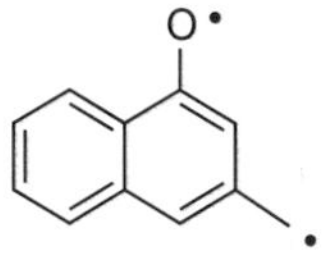   **16**

Two isomeric ring homologues of naphthoquinomethane also were synthesized during that phase, one (**17**) a triplet biradical in its ground state, and the other (**18**), a ground-state quintet tetraradical.[36–38]

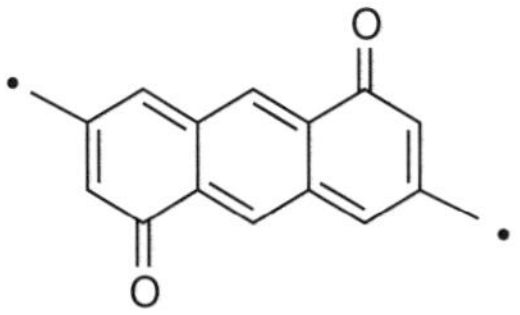

17 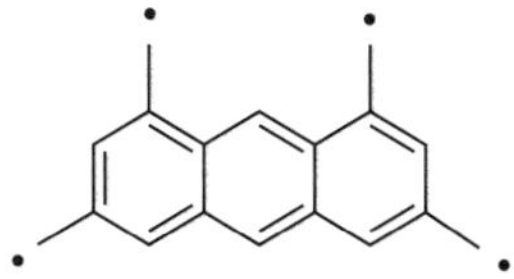   **18**

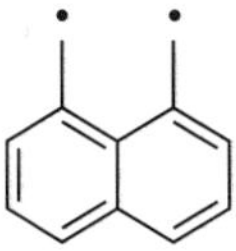

19 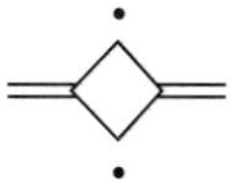   **20**

Another series of non-Kekulé compounds whose parent is 1,8-dimethylene-naphthalene (**19**)[39–46] dates from the same era. Extensive reviews[13,47] of that series are given elsewhere and will not be repeated here.

A non-Kekulé molecule conceptually formed by fusion of two TMM units and also predicted[48,49] to have a triplet ground state is 2,4-dimethylenecyclo-butane diyl (**20**), which ultimately was prepared by two independent syntheses.[50–52] Matrix ESR spectroscopy and gas-phase photodetachment photoelectron spectroscopy (PES)[53] (see Section 4.1.4) eventually agreed that the ground state is triplet.

Also dating from that period was the synthesis of another of the Longuet-Higgins molecules, tetramethyleneethane (**5**), which Dowd[54] reported in 1970 and assigned a triplet ground state in 1986[55] in view of its linear Curie plot. We discuss this molecule and its derivatives in Section 7.

That all these molecules apparently have triplet ground states might be taken as verification of Longuet-Higgins's predictions. However, we must keep in mind that the difficulty at this time of actually measuring the singlet–triplet separation forced the spin issue into binary form: Is the ground-state singlet or triplet? Even when the ground state had been assigned, one still did not know anything about how changes in structure affected the size of the gap.

2.4. The Dependence of Spin State Preference on Structure

This question, which was to initiate the fourth phase of our history, became urgent following the recognition that Longuet-Higgins's indiscriminate prediction of triplet ground states for all the non-Kekulé molecules might be too general. The first inkling of this insight had come decades earlier in a long-buried paper by Erich Hückel in 1936,[56,57] which was really a commentary on Eugen Müller's work in the same year on the Schlenk–Brauns hydrocarbons **1** and **2** (see above). The idea was almost forgotten in the literature until it was (quietly) resurrected by Baudet in 1971[58] and finally developed in 1977 by Borden and Davidson[59] and by Misurkin and Ovchinnikov[60,61] into a powerful new criterion for classifying non-Kekulé molecules. Major efforts in several laboratories were launched to test these new ideas, which now may be said to be firmly established.[62,63] We return later (Section 7) to these developments.

2.5. The Singlet–Triplet Gap Is Not the Whole Story

To this point, it would be fair to say that the dominant question was the comparison of experiment with theory by attempts to designate the ground state and if possible to determine the energy separation between it and the next higher state. To a considerable extent, this line of development was driven by the advances in theory itself, whose predictions were made with ever-increasing confidence, sophistication, and refinement. However, one should not conclude that this is the extent and sole purpose of research in the non-Kekulé domain. These molecules really are a new form of matter, and much progress has been made in the exploration of their physical and chemical properties, which differ significantly from those of the familiar monoradicals.

In the following pages of this chapter, brief introductions or literature references to the various experimental techniques are given. Closely related theoretical and computational work is described in Chapter 22 by Borden in this book.[64] The interplay of theory and experiment, as well as the mutually supporting roles of preparative "wet" chemistry and instrumental techniques, are emphasized.

The very nature of non-Kekulé species as reactive intermediates suggests that studies of them under conditions far from those used in conventional investigations of the synthesis and reactions of stable molecules are indispensable. These requirements frequently are met by immobilizing the species in crystalline hosts or randomly oriented matrices, as is described in Chapter 17 by Bally in this book.[65] Although some information available from crystal studies usually must be sacrificed in the random matrix technique, the latter is usually far more convenient, and most studies of non-Kekulé compounds in solids have used it.

2.6. Connecting the Spectroscopy of a Non-Kekulé Molecule to Its Structure

The generation of a non-Kekulé compound from an appropriate precursor may (and often does) generate other species as well. Consequently, spectroscopic investigations aimed at characterizing a non-Kekulé molecule may give misleading information if they detect the side product in addition to or instead of the target compound. Examples of this include an early attempt[66] to observe the ultraviolet–visible (UV–vis) absorption and emission spectra of the biradical 2-isopropylidenecyclopentane-1,3-diyl prepared from the corresponding diazene in a low-temperature matrix. Although some spectroscopic features were observed,[66] their interpretation was subsequently judged to be questionable, and the transitions were thought to be more probably associated with a side product.[40]

It is true that in some cases, the spectroscopic data on a reactive intermediate are so persuasive that the connection between structure and spectroscopic features is firm. However, in general this will not be the case, and additional spectroscopic or preparative criteria will have to be provided. So we are faced with the question: How can we connect the information obtained, for example, from observations in matrices or in solution-phase fast kinetic studies, to molecular structure? How do we know that the results of these experiments, using what we hopefully call "direct" methods, really pertain to the species we are trying to characterize? I attempt to deal with this issue in what follows. Since the methods used vary from one class of non-Kekulé species to another, specific classes are individually discussed, and special techniques are introduced as needed. Electron spin resonance spectroscopy has played such a pervasive role that it will be useful to give first a brief outline of that method.

3. ELECTRON SPIN (OR PARAMAGNETIC) RESONANCE SPECTROSCOPY IN MATRICES

Space limitations of this chapter preclude a thorough treatment of this topic. Readers who wish more information will benefit from the many reviews available elsewhere.[13,23,67–69] We will concentrate here on its practical applications to non-Kekulé molecules, its strengths, and its limitations.

The most detailed ESR spectroscopic information on paramagnetic species comes from studies in crystalline hosts, as exemplified in the ground-breaking work of Hutchison and Mangum.[70] A more conveniently applied technique involves matrix-immobilized, randomly oriented samples (also called powders). These give broader spectral lines than the crystalline preparations, and it is often more difficult to obtain hyperfine splittings from them. However, for many purposes, the powder spectra are extremely useful, as was soon demonstrated for paramagnetic metal complexes by Burns[24] and for the triplet states of organic species containing divalent carbon by the group at Bell Laboratories.[22]

It is important to note that even without hyperfine data, the powder spectra give valuable information about the carrier of the ESR spectrum. Both the molecular symmetry of the molecule and the effective distance between the unpaired electrons usually can be deduced from the spectra.

3.1. Zero-Field Splitting and the Appearance of ESR Spectra in Immobilizing Media

This topic has been extensively treated elsewhere.[13,23,67,68,71–73] For the present purpose, it may suffice to mention just a few of the conclusions that can be drawn from such spectra.

The transitions in the X-band ESR spectra of triplet species occur in two regions. The so-called $\Delta m_s = 1$ region represents transitions between energetically adjacent pairs of the three triplet sublevels. These are characterized by two so-called zero-field splitting parameters, D and E. The parameter D is inversely proportional to the cube of the average separation of the electron spins, and E is related to the molecular symmetry. The number of lines depends on the molecular symmetry. If all three magnetic axes of the molecular carrier of the spectrum are distinct, the spectrum in the $\Delta m_s = 1$ region will show six major resonances, plus any hyperfine lines that may be visible. If two of the principle axes are equivalent by symmetry, only four lines will be observed. In the latter case, the parameter E has the value of zero.[74]

Note here that in some instances, even when the molecule has lower symmetry, the value of E can be so small as to be indistinguishable from zero, especially with a randomly oriented sample. In that case again, only four of the expected six lines may be observed.[67] With this caution in mind, we can see that a non-zero E value may be interpreted confidently as indicative of a carrier with low symmetry, but the converse, an approximately zero value, could be due to true symmetry, or to an accidental equivalence of two axes. We return to this point in the discussion

(Section 5) of TME. Finally, in a few cases, the value of D happens to be just three times that of E, in which case only three lines will be observed.

A six- or four-line ESR spectrum that can be fitted to a triplet spin Hamiltonian is strong evidence that the species in the sample embodies two unpaired electron spins. Support for the presence of a triplet spin system often can be found in the weak $\Delta m_s = 2$ line,[75] which appears at one-half the field strength of the center of gravity of the $\Delta m_s = 1$ six-line pattern. This nominally "forbidden" $\Delta m_s = 2$ resonance results when the ESR spectrometer field and frequency produce a microwave quantum of energy just sufficient to jump the gap between the uppermost and lowermost triplet substates, that is, a transition over two quantum levels.

3.2. Biradicals or Radical Pairs?

However, what is not immediately obvious from this information alone is whether these spins are associated with a triplet *pair* of free radicals or with a true *molecular* triplet. For this purpose, the D value is indispensable, because from it, one can determine the average separation of the spins.

Gordy[68] and Gordy and Morehouse[76] derived an equation for calculating this average or effective distance based upon the assumption that the location of each spin dipole may be assigned approximately to a point.[77] The result may be stated as in Eq. 1, in which $2D$ in gauss is the separation between the two outermost of the six (or four) lines, and R is the effective interelectronic distance in angstrom units:

$$2D = 55,600/R^3 \qquad (1)$$

One of the useful applications of the Gordy–Morehouse equation is in just this problem of distinguishing true molecular triplets from radical pairs. Thus, many non-Kekulé triplets have $2D$ values between 450 and 600 G, corresponding to average interelectronic distances of $\sim$5 and 4.5 Å, respectively. Note that these distances would permit the two spins to reside within the boundaries of an ordinary molecule. Triplets with such separations may usually be assumed to be *molecular* triplets, rather than radical pairs. A radical pair with such a small separation of spins would have the two spin centers nearly in contact with each other. Distances of 4.5–5 Å would be too close to accommodate even a single molecule of the solvent in such frequently used matrices as Freons or perfluorocyclohexane. Ready mutual quenching of the pair then would be expected, thus precluding the observation of the spectrum. Moreover, one would expect that the interradical distance should vary within a given matrix, since the radical pairs are not constrained by covalent bonds to a fixed separation. Further, the effective distance should vary from one solvent matrix material to another. In fact, just this behavior has been observed[78] in matrix-immobilized preparations of radical pairs. These radical pairs show poorly resolved triplet spectra with broad lines, and the D values of such samples differ drastically with change of solvent. It is prudent, therefore, to make sure that the D value of a putative molecular triplet remains essentially constant in several different media.

3.3. Curie's Law and Its Application to Assignments of the Ground-State Multiplicity

In the context of ESR spectroscopy, the Curie law may be stated in its simplest form as $I = C/T$, where I is the intensity of an absorption line, T is absolute temperature, and C is a constant. A modified form of the law (Curie–Weiss law), $I = C/(T - \theta)$, sometimes is needed when the plot of I versus $1/T$ has a non-zero intercept. In both cases, the plot should be linear if the paramagnetic species responsible for the signal is not engaged in an equilibrium with other species of different multiplicity. The most common candidate for such other species is a singlet, with spin of zero.

Traditionally, it has been implicitly assumed that equilibrium between a triplet state and a lower lying singlet state would be rapidly established. However, instances of slow interconversion are now known and will be discussed in Section 7.2.1. If spin equilibrium with a more stable singlet is rapid, one would expect downward curvature of the plot at lower temperature. Bear in mind that if the singlet lies too far below the triplet (more than $\sim$2 kcal/mol), the amount of triplet present may be too small to detect by ESR. A linear plot has been the major criterion for a triplet ground state until quite recently, but as will become apparent, that may change soon. An accidental near degeneracy of singlet and triplet (less than $\sim$10-20 cal/mol separation) also will give rise to a linear plot.

For various reasons discussed elsewhere,[13] the Curie law plot is not informative at temperatures >77 K, so that measurements in the cryogenic region are imperative.

Finally, one uses low-temperature techniques for ESR measurements of non-Kekulé species not only because of the inherent high reactivity of the species, but also because unless the sample is immobilized or at least hampered in its motion by viscous solvent, rapid tumbling will cause the spectrum to suffer line shape distortions that impede the extraction of the zero-field splitting parameters.

4. SOME KEY EXAMPLES

4.1. Trimethylenemethane

4.1.1. ESR Spectrum. The TMM (**4**) derivatives had been proposed as reactive intermediates in the thermal rearrangements of methylenecyclopropanes,[14,79,80] but the first unequivocal spectroscopic characterization of any TMM, in fact, of any non-Kekulé compound, came from the ESR investigation by Dowd.[25] He synthesized TMM itself by irradiation of appropriate precursors **11** or **12** (see Scheme 5.1) at low temperatures in immobilizing media.

The zero-field values for TMM,[12,25] $D = 0.024$ cm^{-1} ($2D = 513$ G) and $E < 0.001$ cm^{-1}, were compatible with the expected triplet ground state and with an effective interelectron distance of $\sim$4.8 Å. Moreover, when the species was generated in a crystalline host, the lines were visibly further split into a seven-line hyperfine pattern with the binomial intensity relationship expected for six equivalent hydrogens. Later,[81] the spectrum was shown to follow the Curie law. All these data were compatible with a threefold symmetric structure and a triplet ground state for TMM.

The hyperfine splitting is certainly a significant observation, since it implies the presence of six *equivalent* hydrogens, as would be required by a planar structure with a threefold rotation axis perpendicular to the central carbon. Although the synthesis of the carrier of the ESR spectrum from two different precursors (Scheme 5.1) and the ESR data leave little room for doubt that the TMM structure is correct, one could quibble that the hyperfine result establishes threefold symmetry only permissively rather than decisively. It is conceivable that the seven-line pattern could result from an accidental equivalence of electron-nuclear coupling constants or from a rapid conformational averaging process in a TMM of lower symmetry. This possibility seems unlikely, in view of the unanimous agreement of several high-level quantum mechanical structure computations,[82] which point to a planar threefold symmetric configuration for triplet TMM.

Nevertheless, just that kind of accidental hyperfine equivalence may occur in the case of TME (**5**) (see below), which shows a nine-line pattern[54] suggestive of eight equivalent hydrogens. Neither planar nor twisted TME can have eight truly equivalent hydrogens, so the hyperfine pattern is due to some cause other than molecular symmetry.

Note that ESR spectroscopy gives no direct information whatever on the singlet state. In principle, from a nonlinear Curie plot, one can deduce the presence of the singlet and estimate the energy with which it is separated from the triplet. However, the Curie plot is a very insensitive method for detecting the presence of a singlet of slightly higher energy. Even when the Curie plot is linear, the singlet may be present but indetectable if its energy lies more than $\sim$0.05 kcal/mol above the triplet.[13] In some instances, which we discuss later, other spectroscopic tools may be applicable to this purpose.

4.1.2. The Chemistry of TMM: Ring Closure.

A few instances have been reported of chemical reactions in which the parent TMM molecule is suspected, with varying degrees of plausibility, to have been the reactive intermediate in chemical reactions. These examples have been extensively discussed in reviews.[12,14,27,83] The first example in which TMM is unequivocally involved is the reaction monitored by the disappearance of the triplet ESR signal in a matrix at low temperatures.[84,85] However, there has been some uncertainty about the actual chemical process responsible for this observation. Dowd and Chow[84,85] concluded that the actual chemical process is the ring closure to methylenecyclopropane (MCP). The primary evidence for this suggestion is the kinetic behavior of the reaction, which is first order. It is not easy to imagine another true unimolecular reaction of TMM. First-order kinetics also might be observed if the signal decay were simply the result of abstraction of a hydrogen atom from the matrix. This should have but apparently did not result in growth of the monoradical signal in the middle of the ESR spectrum. Dowd and Chow pointed out that the conversion of triplet TMM to singlet methylenecyclopropane required at some point a change of multiplicity. They wrote the mechanism ^{3}TMM → ^{1}TMM → ^{1}MCP for this process and concluded that the observed 7-kcal/mol barrier measured the energy gap between ^{3}TMM and ^{1}TMM.

This proposal leads to a still unresolved puzzle: Theoretical calculations[82] agree that the lowest singlet state of TMM, which has one of the methylene groups twisted out of the plane to a bisected conformation, lies some 0–7 kcal/mol below the planar singlet and 14–20 kcal/mol above the triplet. Thus, the computational results require that the activation energy for a ring closure to methylenecyclopropane by the Dowd–Chow mechanism be some 7–13 kcal/mol higher than that observed in the Dowd–Chow experiment. Clearly, if the calculations are properly reporting the situation, there is a substantial energy shortfall in the Dowd–Chow experiment for achieving the transformation via the pathway ^{3}TMM → ^{1}TMM → ^{1}MCP. It would appear that either the calculations overestimate the gap, or the experiment measures some other process. For further discussions of the issues, the reader is referred to reviews of the problem.[10,14,15,28] and to Section 4.1.4.

Additional spectroscopic information on the chemistry of the TMM triplet state now has been provided by the experiments of Maier et al.,[86] who have prepared TMM in substantial quantities by the irradiation of methylenecyclopropane in a xenon matrix at 10 K in the presence of codeposited halogen atoms. This experiment has permitted them to record the infrared (IR) spectrum of TMM, all the observable bands of which are in agreement with those calculated by ab initio methods. They also were able to supply direct evidence for a *photochemical* ring-closure reaction, ^{3}TMM → methylenecyclopropane, by irradiation of the biradical at 254 nm. The disappearance of the TMM IR absorptions is accompanied by growth of the methylenecyclopropane bands. Of course, this observation cannot be taken as a demonstration that the reaction reported by Dowd and Chow, namely, *thermal* cyclization of ^{3}TMM, actually occurs.

4.1.3. Bimolecular Trapping. Attempts to demonstrate bimolecular reactions of the parent unsubstituted TMM face a common difficulty: Even if adducts apparently formed by a transient-plus-reagent reaction are observed, how can one be sure that the bimolecular trapping step in the overall reaction actually involves the free transient in question and not the precursor or another transient? For example, the formation of a 35% yield[87] of 3-vinylmethylenecyclopentane (**21**, Scheme 5.3) in the photolysis of 3-methylenecyclobutanone **12** in liquid butadiene (**22**) might plausibly be ascribed to trapping of a transient TMM, but other pathways can be written and have not been excluded. In one such pathway, the sequence of trapping and decarbonylation is simply reversed: The Norrish I intermediate **23** might be trapped and the resulting adduct acyl biradical **24** might lose carbon monoxide to give the same product **21**.

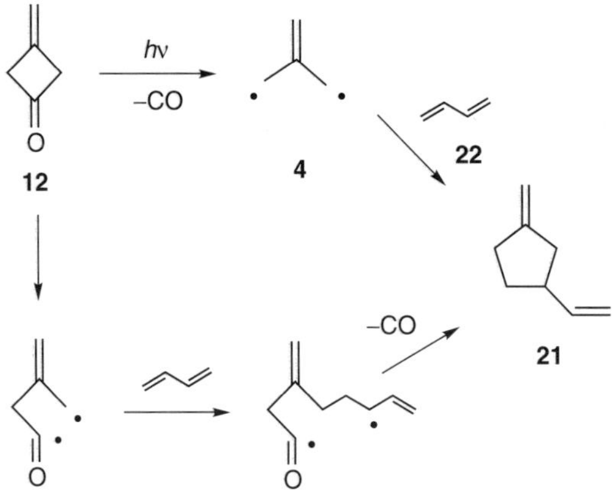

Scheme 5.3

4.1.4. Determination of the Singlet–Triplet Gap in TMM by Electron Photodetachment Photoelectron Spectroscopy: Principles of the Method.

The difficulty of determining the singlet–triplet gap in non-Kekulé compounds by ESR methods led a reviewer in 1988[13] to remark that "[t]he need for a more reliable, more broadly applicable method for the determination of [the singlet–triplet gap] should now be obvious. Perhaps no other single development would do more to advance the study of non-Kekulé molecules." In the past few years, just such a method has emerged from advances in the techniques of gas phase chemistry and physics of ions. These innovations have led to the development of the method of electron photodetachment spectroscopy (PES), notably in the laboratories of Lineberger, Ellison, and their colleagues at the Joint Institute for Laboratory Astrophysics (JILA) at the University of Colorado. When combined with chemical procedures for the generation of gas-phase negative ions, the technique has been applied to the problem of non-Kekulé molecules and other biradicals by the Colorado group and also by the late Robert Squires and Paul Wenthold and their colleagues at Purdue University. These efforts have now provided a major forward thrust in the direct measurement of the energy separations between spin multiplets.

In a simplified description, the experiment consists of three stages: (1) a radical anion formally related to the non-Kekulé neutral compound is generated by some suitable process; (2) the ions are mass selected and subjected to irradiation by a 351-nm laser to photodetach electrons; and (3) these electrons are allowed to pass through an energy analyzer that measures their kinetic energy. The last step generates a photoelectron spectrum, from which important information can be deduced about the vibrational states of the radical anion and about the binding energy with which the photoelectrons (PE) were held in the anion before they were detached. The amount of this energy will depend on the state of the neutral biradical produced in the photodetachment. If two spin states of the neutral differ in energy, as they will barring an accidental degeneracy, two separate band systems will appear in the PE spectrum. Conservation of energy requires that the sum of the energy of the reactants (the laser photon and the radical anion) must equal the sum of the energy of the products (the photoelectron produced and the particular state of the neutral produced, see Eqs. 2 and 3).

$$R^- + h\nu \rightarrow e^- + R \tag{2}$$

$$E_{h\nu} + E_{R^-} = E_{e^-} + E_R \tag{3}$$

Thus, the process that leads to the lower energy neutral must give a photoelectron of higher energy, whereas the process that leads to the higher energy neutral must give a photoelectron of lower energy. The separation between the band origins of the PE band corresponding to the two product neutral states directly gives the energy difference between them. Some authors report this directly as measured in kinetic energies of the electrons, and others report it as the "electron binding energy," which is simply the difference between the photon energy (3.51 eV) and the energy of the photoelectron. Clearly, if the PE data are reported as electron binding energies, the more stable neutral state will be represented by the lower photoelectron binding energy.

4.1.5. Application to TMM. A convincing example of the power of this method is the determination of the singlet–triplet separation in TMM. The TMM negative ion (**27**) is prepared by the gas-phase reaction shown in Scheme 5.4.[88] The reaction of 1,1-di[(trimethylsilyl)methyl]ethene (**25**) with fluoride ion in the gas phase removes one trimethylsilyl group and gives rise to the anion **26**, which in a subsequent step, is treated with molecular fluorine in the flowing afterglow source of a negative ion photoelectron spectrometer.

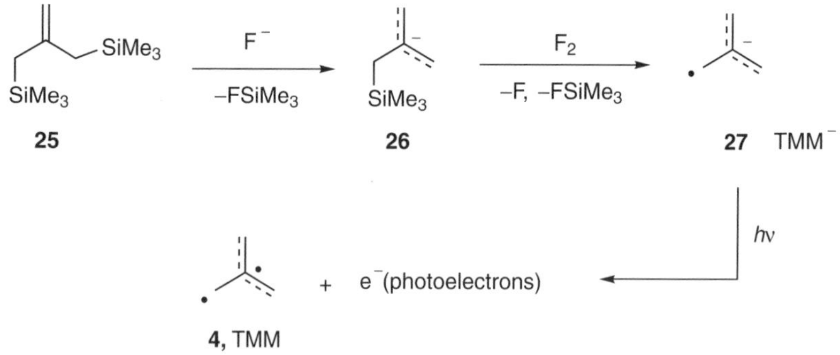

Scheme 5.4

This step generates the TMM radical anion **27**, which is selected for electron photodetachment of the mass 54 (TMM) anions from the beam.[82] Two PE band systems are observed in the PE spectrum in which the origin bands are separated by 0.699 eV (16.1 kcal/mol). Since computational results indicate the conformation of the radical anion to be planar, and since the photoelectron detachment is a Franck–Condon transition, the authors ascribe the observed separation to the energy difference between the planar forms of the singlet and triplet states of TMM. The energy separation between planar and the more stable bisected TMM neutral is calculated to be 0–3 kcal/mol, which suggests that the separation of the triplet from the lowest energy singlet is ~13–16 kcal/mol.

One can hardly argue with the conclusion that experiment now matches the theoretically calculated energy gap for the planar forms, ~17 kcal/mol. The picture is complicated somewhat by the fact that the gap between the triplet and the bisected singlet is not given directly by the experiment, and logically one could quibble that the extraction of this number has itself required a calculation. However, sustaining that argument requires that the calculations be entirely accurate for the planar forms but suddenly lurch into a large error for the gap between the triplet and the bisected singlet forms. Instead, it seems justified to move ahead with the authors' conclusion that this gap is 13–16 kcal/mol rather than the 7 kcal/mol proposed[84,85] earlier.

Note that the PE experiment does not identify by simple inspection which of the spin states is more stable, the singlet or the triplet. However, the assignment of the lower energy state to the triplet seems entirely reasonable, in view of Dowd's demonstration that the triplet is the ground state in matrices. Further confirmation of this may come from a complete vibrational analysis of the richly detailed PE spectrum.[82]

Further applications of these methods are discussed later in Section 5.1.

4.2. Ring-Constrained Derivatives of TMM

The study of these substances eventually was to evolve into a broad development of the chemistry of TMM derivatives. The original motivation was to expand earlier findings[89] on the stereospecific conrotatory thermal (300 °C) ring opening of the 2,3-dimethylbicyclo[2.1.0]pentanes to 2,5-heptadienes, plausibly via 3,4-dimethylcyclopentane-2,5-diyls. The planned follow-up experiment was to observe whether the insertion of a methylene group would generate a triplet TMM species (see Section 4.2.1), and thereby cause the ring opening to become nonstereospecific. This experiment was never completed, but observations in the early stages of it led to the study of several of such non-Kekulé compounds.

4.2.1. 2-Methylenecyclopentane-1,3-diyls 14: Synthesis and ESR Properties. The first member of the 2-methylenecyclopentane-1,3-diyl series to be synthesized was the 2-isopropylidene derivative **14b**,[26] (Scheme 5.2) followed by a number of others (**14a, 14c–f**), including the unsubstituted methylene compound **14a**.[27,90] Each of these substances can be generated from the corresponding diazene (**13**), either thermally or photochemically. This and many other related studies have been reviewed repeatedly,[14,15,27] and further studies on the ESR spectra of related

derivatives continue.[91] The most important conclusions may be briefly summarized (Scheme 5.5):

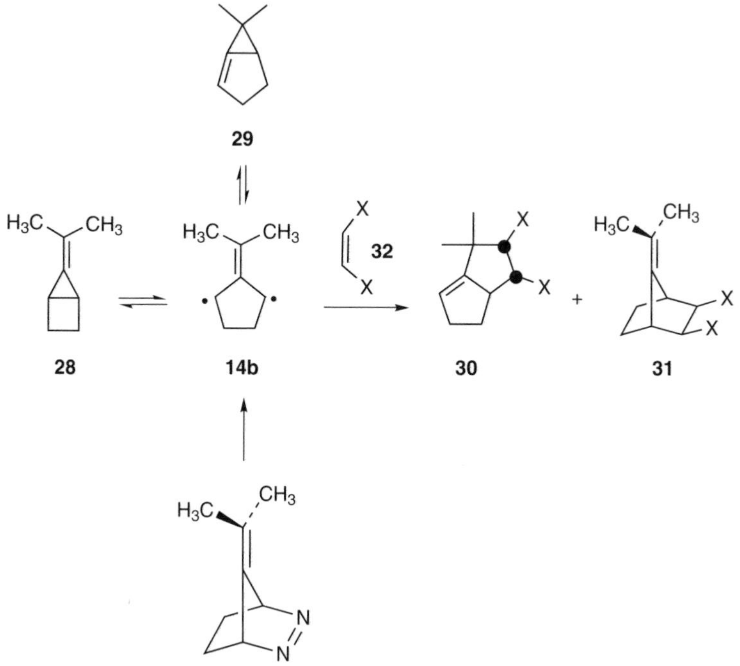

Scheme 5.5

1. Both singlet and triplet forms of the 2-alkylidenecyclopentane-1,3-diyls **14** undergo cycloadditions in high yield with alkenes **32** by characteristic stereochemical and regiochemical pathways. The singlet gives predominantly fused adducts **30** formed by concerted syn addition, whereas the triplet gives stereorandom and regiorandom (bridged **31** and fused **30**) adducts by a stepwise addition (Scheme 5.5). The regiochemistry in the singlet additions is consistently rationalized by a frontier orbital model.

2. The relative rates of reaction of the *singlet* TMM derivative [1]**14b** with a series of alkenes (**32**) parallel those of a conjugated diene with the same alkenes in Diels–Alder reactions. These relative rates also are well correlated by the frontier orbital model for a concerted reaction. The absolute rates of the biradical cycloadditions are many orders of magnitude greater than those of the model dienes. The relative rates of the alkenes in the cycloadditions of the *triplet* biradical [3]**14b**, on the other hand, follow the reactivity order of their addition reactions with monoradicals.

3. The bicyclic hydrocarbon full-valence isomer (**28**) of the 2-isopropylidene-cyclopentane-1,3-diyl (**15b**) can be prepared and kept at temperatures below

~210 K. It too gives the characteristic adducts obtained from the singlet, but the addition does not occur directly between the bicyclic hydrocarbon (**28**) and the alkene. Instead, the reaction occurs in two steps, first the reversible unimolecular ring opening of **28** to singlet biradical **14b**, followed by a bimolecular capture of the latter (Scheme 5.5). Another hydrocarbon isomer **29** can be prepared as a transient intermediate. Its thermal conversion to the biradical **14b** apparently occurs at even lower temperature.

4. In solution, the triplet biradical ^{3}**14b** dimerizes, and the dimeric products are formed with strong chemically induced nuclear polarization. The absolute rate of the dimerization at 146 K, as monitored in viscous solution by ESR spectroscopy, is just about that predicted by the spin-corrected encounter frequency under those conditions. The cycloaddition of the triplet with a typical alkene, acrylonitrile, also can be followed in this way.

5. Kinetic observations suggest that of the four species, the biradicals ^{1}**14b** and ^{3}**14b**, and the bicyclic hydrocarbons **28** and **29**, the triplet biradical ^{3}**14b** is the global energy minimum. Thus, the two related full-valency molecules **28** and **29** both have a *negative bond dissociation energy* (BDE).

6. Early experiments[92] used rather primitive methods to estimate an energy gap of 1–3 kcal/mol between the singlet and triplet form of the biradical **14b**. This gap was far out of line with the best calculations available at the time,[82] which predicted a gap of ~15 kcal/mol. Later, kinetic data, combined with some assumptions, permitted the estimation of the energy gap between the ground-state triplet and the singlet as at least 13 kcal/mol,[93] which is in good agreement with calculation and with the value later directly determined by the electron photodetachment PES value for TMM itself (see Section 4.1.4). Presumably, a similar measurement on the radical anion of 2-isopropylide-necyclopentane-1,3-diyl could confirm these conclusions. At present, such an experiment awaits the development of methods of gas-phase synthesis of the anion.

5. THE DEPENDENCE OF SPIN STATE ON MOLECULAR CONNECTIVITY: THE CASE OF TETRAMETHYLENEETHANE

As we have seen, Longuet-Higgins's 1950 paper[21] predicts that TME **5** should have a triplet ground state. This prediction seemed to find a confirmation in the experiments of Dowd et al.,[55] who observed the ESR spectrum, $D = 0.025$ cm^{-1}, $E = <0.001$ cm^{-1} upon irradiation of either of the matrix-immobilized precursors **33** or **34** (Scheme 5.6).

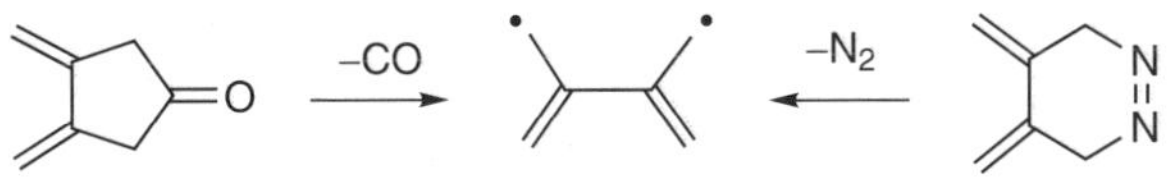

| 33 | 5 | 34 |

Scheme 5.6

However, it soon became clear that matters might not be so simple. There is a fundamental difference in the connectivity of the π centers in TME and that in TMM. Hückel[56,94] recognized a similar difference in the case of the two Schlenk–Brauns hydrocarbons **1** and **2**, and the theory was put on a modern basis and generalized by Borden and Davidson.[59]

Simply stated, the rule is that if the molecule can be mentally constructed (see Scheme 5.7) by joining two radical units at positions that have NBMO coefficients of (nominally) zero, as in the hypothetical formation of TME from two allyl radicals **35**, then the exchange interaction approaches zero.

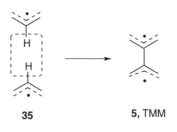

35   **5, TMM**

Scheme 5.7

Since it is the exchange interaction that determines to a large extent the energy separation of the multiplet states to first order, the physical basis for the triplet preference thus vanishes. In that case, other electron correlation effects tend to stabilize the singlet selectively.

The NBMOs of the singlet state of such a biradical are called *disjoint* in recognition of the fact that they can be considered to be confined to different sets of atoms. The net result is that in such molecules, the singlet–triplet gap will be small, and the singlet may become the ground state. Borden and Davidson[59] pointed out that TME (**5**) is the parent molecule of this non-Kekulé class, which we now call disjoint.

A second argument leading to the same conclusion comes from valence bond theory and can be expressed by a simple scheme involving the parity (starredness or unstarredness) of an alternant hydrocarbon or heteroatom substituted derivative thereof.[60,61] The conclusions are supported by other approaches involving parity arguments.[95] Moreover, they are extended to non-alternant systems in papers by other authors.[96–100] The generalized version of the rule may be stated in the following form: If the portion of the conjugated π system intervening between the unpaired spin centers has an even number of electrons, the molecule will be low spin, and if the number is odd, it will be high spin. The number of such electrons is calculated by assigning one electron to each intervening π center and performing the count on the resonance structure that minimizes the number of intervening π centers. Thus, the number is 2 in TME and 1 in TMM, corresponding to low and high spins, respectively.

Thus, a disjoint non-Kekulé compound like TME might well have a singlet ground state, since the singlet is expected to be stabilized by a favorable electron correlation effect. This conclusion seems to conflict with the ESR experiments on TME.

Since then, much further computational work has developed a more complex picture of the TME issue.[10,101] Currently, it is believed that both the singlet and triplet TME are twisted out of the planar configuration; the angle of twist for the triplet is ~50° (D_2 point group) and that for the singlet is 90° (D_{2d} point group). At the minimum energy triplet geometry, the triplet is predicted to be more stable by 1–1.5 kcal/mol. At a particular computational level (CI + DV2), the triplet minimum near 50° lies ~0.1 kcal/mol below the singlet at 90°, although the authors state that "the prediction of the relative stability of states lying within 2 kcal/mol of one another is a difficult proposition." Further computational effort has been devoted to this problem.[102,103]

5.1. The Singlet–Triplet Separation in Tetramethyleneethane in the Gas Phase

The TME radical anion can be generated in the gas phase by the reaction of O^- with 2,3-dimethylbutadiene, and the TME neutral biradical can be formed by electron photodetachment on the mass 80 ion by methods similar to those used in the case of TMM (see Section 4.1.4).[104] The spectrum obtained by monitoring the photoelectrons shows transitions of TME⁻ to *two* TME neutrals whose origin bands are separated by ~2 kcal/mol. The authors assign the lower energy species to the singlet and the higher energy one to the triplet on the basis of two lines of argument: (1) the intensities of the two band envelopes are in the ratio of ~3:1 in favor of the higher energy species, as is often approximately the case in PES leading to both singlet and triplet states, and (2) the lower energy species (singlet) has a much narrower Franck–Condon vibrational envelope than the higher energy one. This second result is consistent with the electronic structure computations, which predict a bisected geometry (D_{2d}) for both TME⁻ and TME singlet but a twisted (by ~50°) geometry for the triplet. Thus, a small geometry change is expected in the reaction TME⁻ → ¹TME and a large change for the reaction TME⁻ → ³TME.

These results of course are inconsistent with Dowd's finding of a triplet ground state from the matrix ESR spectrum. At present, the discrepancy is rationalized[104] with the hypothesis that the triplet is metastable in the matrix and that its conformational conversion to the more stable singlet is slow, thus allowing the observation of the ESR signal and the linear Curie plot. Other cases of what seem to be conformationally based triplet metastability in matrix ESR had been observed previously in 1993[105,106] and will be discussed later, together with further examples (see Section 7.2.1).

A six-line ESR spectrum has been observed[107] for the triplet state of the TME derivative 2,3-dimethylene-1,3-cyclohexadiene (**36**), which first had been generated in low-temperature matrices by Roth et al.[108,109]

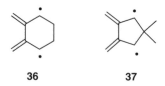

36   **37**

Molecular mechanics calculations[107] suggested that the molecule is planar, but later Hartree–Fock computations[110] concluded that the two allyl units are twisted with respect to each other by ~25°. Recently, Matsuda and Iwamura[111] used magnetometry (see Section 8) to show that the singlet and triplet states of 2,3-dimethylene-1,3-cyclohexadiene (**36**) are degenerate. This finding accounts for the observed linear Curie plot, which was previously interpreted[107] as indicative of a triplet ground state.

An example of a TME incorporated into a five-membered ring and, hence, presumably nearly if not actually planar is given by 5,5-dimethyl-2,3-dimethylene-1,3-cyclopentanediyl (**37**).[112] The 6-line ESR spectrum of this substance also shows a linear Curie plot and has been assigned to a triplet ground state.[112] At first glance this assignment would seem to be a refutation of the prediction of a singlet ground state for the parent TME, and the finding is so interpreted by the authors.[112]

The authors show that hydrocarbon precursors **38–40** of biradical **37** react with O$_2$ at elevated temperature to give the cyclic peroxides **41** (Scheme 5.8).

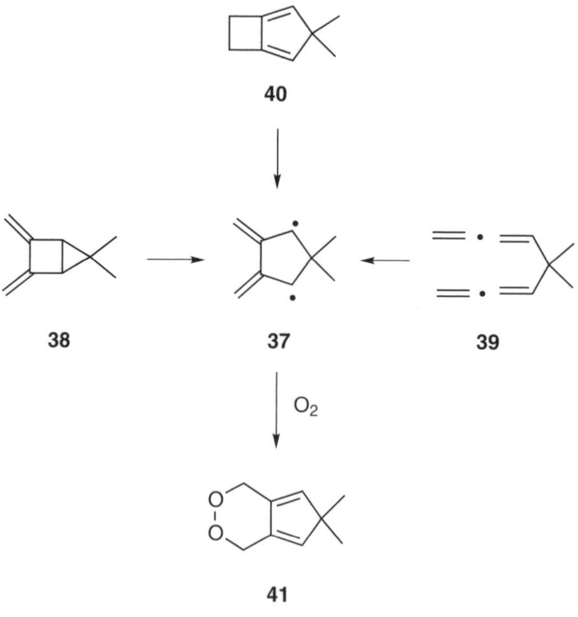

40

38 → **37** ← **39**

O$_2$

41

Scheme 5.8

On the assumptions that the triplet TMB biradical [3]**37** is the reactive intermediate and that its reaction with O_2 occurs at the encounter-controlled rate, the authors estimated that the triplet is more stable than the singlet by at least 4–5 kcal/mol,[112] or more if the diffusion-limited trapping rate assumed is actually lower.

Later single and double excitation configurational interaction (SD–CI) calculations by Nash et al.[113] show that the bridging CMe_2 group of **37**, which had been chosen to fasten together the ends of the TME unit, does not act as an electronically innocuous structural element but instead introduces a new factor that selectively stabilizes the triplet through hyperconjugation, making it the ground state by ~1 kcal/mol. This example illustrates the dangers of using model compounds to assess theoretical predictions when the point in question depends on small energy differences, a problem common to disjoint hydrocarbons as a class.

There remains a discrepancy between the experimental and computational values for the stability advantage of the triplet (4–5 vs. 1 kcal/mol, respectively). At this point, it is difficult to say whether this should be attributed to inaccuracy in one or in both.

We can summarize this section on TME and its hydrocarbon derivatives with the observation that it is not easy to test the major predictions of the theoretical model of disjoint non-Kekulé compounds. Whether or not the singlet is actually the ground state in any given case depends on subtle particularities of structure and conditions of measurement. Much of the contention of the last decade or more in this area focused on these difficulties, but many of those difficulties are suppressed in the case of tetramethylenebenzene (TMB).

6. TETRAMETHYLENEBENZENE

That this substance (**42**, Scheme 5.9), whose systematic name is 2,3,5,6-tetrakis-(methylene)-1,4-cyclohexanediyl, is a member of the disjoint series can be derived easily by application of the Ovchinnikov criterion,[60,61] which predicts a singlet ground state, as is shown in the scheme. Alternatively, one can consider the structure to be made up by the union of two pentadienyl units (**43**) at inactive sites. What

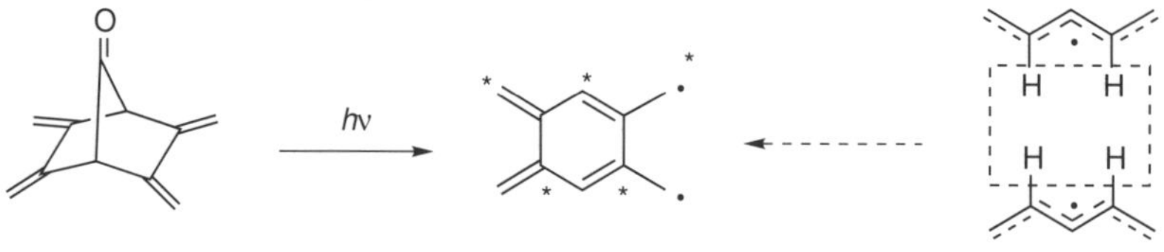

44   **42**   **43**

$(n^* - n)/2 = 0$
singlet

Scheme 5.9

cannot be deduced from this criterion alone is how great the separation of the singlet and triplet should be. However, a number of quantum mechanical calculations, both semiempirical[114,115] and ab initio[116,117] suggest the interesting possibility that the stabilization of the singlet in the 10-π-electron molecule TMB (singlet lower by 5–7 kcal/mol) should be even greater than that predicted, 1–2 kcal/mol, for the hypothetical planar disjoint 6-π-electron species TME. This prediction makes the TMB molecule an important test of the disjoint criterion, because the computed singlet–triplet (S-T) gap is large enough that even theoreticians would probably agree that a finding of a triplet ground state would clearly be a refutation of theory.

TMB (**42**) was first generated by Roth el al.[118] by photochemical decarbonylation of the ketone **44** in a low-temperature matrix. This preparation was intensely colored, with a main transition at ~490 nm and several subsidiary absorptions. Earlier π-CI quantum chemical computations[116] had predicted ultraviolet–visible (UV-vis) is transitions for the singlet and triplet states of TMB, and the bands observed by the Roth group were in better agreement with the predictions for the triplet. The preparation also showed a narrow ESR spectrum interpreted by the authors[118] as that of a triplet species with $D = 0.0042$ cm^{-1} and $E = 0.0009$ cm^{-1}, which gave a linear Curie plot. The authors assumed that the carriers of the UV–vis and ESR spectra were the same species, namely, triplet TMB. They concluded that TMB is a ground-state triplet, contrary to the disjoint theory and to the computational results described above.

However, a repetition of these experiments, supplemented by a variety of other evidence,[63,119,120] led to different conclusions. These studies showed that (1) the purple color and the ESR spectrum are associated with two *different* species; (2) the colored species is in fact TMB, but the ESR spectrum is associated not with the TMB triplet but rather with some still unidentified side product; (3) the major irradiation product of the ketone precursor **44** is singlet TMB, as was demonstrated by solid-state carbon-13 nuclear magnetic resonance ^{13}C NMR spectroscopy. This technique, developed by Zilm and co-workers,[121–124] is especially applicable to the distinction between a singlet and a paramagnetic species, since the latter, a triplet, for example, would give broadened lines in a completely different spectral region, if it can be observed at all, whereas a singlet should show a normal spectrum. The irradiation product from ^{13}CH$_2$-labeled **44** showed a ^{13}C NMR spectrum consisting of a single unbroadened resonance at 113 ppm, a normal position for a $=$CH$_2$ group. The signal intensity correlated well with the intensity of the 490-nm band in the UV–vis spectrum. Thus, the purple species is singlet TMB. No trace of the triplet spin state is formed in the photolysis or after prolonged storage at low temperature. These results are in accord with the theoretical prediction of a singlet ground state for TMB.

There remained the matter of the UV–vis spectra, which as we have seen, agreed better with the calculated[116] transitions of the triplet than with those of the singlet. This anomaly was resolved in a new calculation,[125] which went beyond the π-CI level used earlier[116] and incorporated correlation between the σ and π electrons of TMB.

In preparative reactions, alkenes such as fumaronitrile, 1,2-dichloroethylene, styrene, and diethyl fumarate were reported not to capture TMB (photochemically generated by continuous photolysis of ketone **44**).[126] However, O_2 was an efficient trapping agent. The authors[126] argued that since other singlet biradicals can be trapped with alkenes, the failure to trap TMB with these reagents pointed to the triplet biradical as the only reactive intermediate. These observations led them to conclude that the singlet–triplet equilibrium must lie strongly on the side of the triplet, in agreement with their interpretation of the ESR spectrum. However, the slow rate of the reaction of TMB singlet with alkenes[63] is now attributed to a phase mismatch between the biradical symmetric highest occupied molecular orbital (HOMO) and the alkene antisymmetric lowest unoccupied molecular orbital (LUMO) in the cycloaddition transition state, which prevents a facile cycloaddition.[63]

Roth et al.[126] derived a heat of formation for TMB by kinetically following the reaction of dicyclobutabenzene with O_2 in a gas-phase shock tube. They derived a heat of formation for TMB (**42**) on the assumption that the reaction passed through it as a reactive intermediate. The conclusions must be questioned, however, since a second important assumption of the analysis[126] was that the hypothetical biradical–dioxygen reaction occurred at the statistically modified encounter-limited rate, as the authors[126] believed to be appropriate for a *triplet* biradical. With TMB now having been identified as a *singlet*, this assumption becomes dubious. In fact, in solution, nanosecond time-resolved measurements[63] of the rate of disappearance of the flash photochemically generated TMB singlet show that the O_2 trapping rate is only about one-thousandth that of the encounter frequency. If, as seems likely, a similar factor exists in the gas-phase reaction, the reported[126] value of the heat of formation would require revision.

The O_2 trapping experiments also led to an estimate[126] that the triplet must lie at least 5 kcal/mol below the singlet. Again, this finding is inconsistent with other experimental[63] and theoretical[114–117] findings that all point to a singlet ground state.

Aside from the preceding argument based on the oxygenation kinetics,[126] no quantitative experimental estimates of the singlet–triplet separation in TMB have been reported. If a suitable precursor of the TMB radical anion in the gas phase can be made, photodetachment PES of TMB⁻ would become a likely way to solve this problem.

7. OTHER TESTS OF CONNECTIVITY THEORIES

7.1. *m*-Quinone Derivatives

An early attempt[36–38] to test the disjoint hypothesis compared the magnetic properties of two isomeric tricyclic *m*-quinonoid non-Kekulé molecules: **17**, formally a biradical with tetraradical resonance structures, and **18**, formally a tetraradical (Section 2.3). These molecules belong to the point groups C_{2h} and C_{2v}, respectively, and it will be mnemonically convenient to use those descriptors in what follows. The test derives from the recognition that the connectivities of the two molecules puts them into different classes. The C_{2v} molecule **18**, is a nondisjoint, unequal parity structure, which the Ovchinnikov rule predicts to have a ground-state spin of 2, that is, to be a quintet, whereas the C_{2h} molecule **17** is disjoint, and by that criterion should have $S = 0$, a ground-state singlet species.

Experimentally, the nondisjoint molecule **18**, as predicted, is clearly identifiable by ESR spectroscopy as a quintet. However, the disjoint isomer **17**, instead of being a singlet seems to be a triplet.[36,38] This surprising result is matched by an equally surprising intermediate neglect of differential overlap–configuration interaction (INDO–CI) semiempirical calculation[38] of a type that has been shown[114] to be quite reliable in reproducing high-level computations on other non-Kekulé systems. The INDO–CI calculations predict that the *triplet* of the disjoint isomer **17** should lie below the singlet by ~ 4 kcal/mol, in qualitative agreement with experiment.

Just which energetic factors conspire to produce this violation of the disjoint rule are not readily apparent. So far, the molecules **17** and **18** have proven too large for the application of the most advanced ab initio computational methods. One may hope that future technical advances in high-level computations may cast some light on this puzzle.

7.2. Heterocyclic Planar Tetramethyleneethane Derivatives

With the singlet and triplet states of TME now rather firmly established to be close in energy, the opportunity presents itself to tune the gap and thereby exert control over the magnetic and chemical properties of a biradical series. This modulation now has been studied in the series of heterocyclic TME derivatives (**46**) from the corresponding diazenes (**45**, Scheme 5.10). Reviews of this work are available.[127,128]

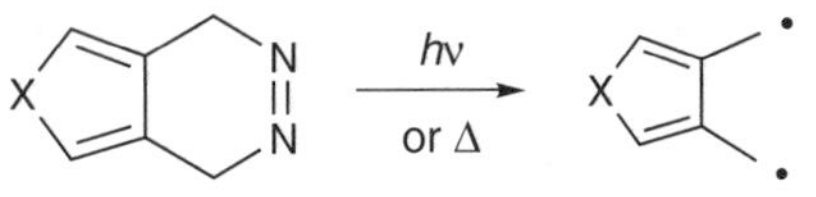

a. X = O	d. X = NMe
b. X = S	e. X = N t-BuCO
c. X = NH	f. X = SO$_2$C$_7$H$_7$

Scheme 5.10

The heteroatom p orbital in this series is symmetric with respect to the plane perpendicular to the heterocyclic ring and passing through it and the midpoint of the 3,4-bond. It perturbs the symmetric component of the two nominally degenerate π orbitals of the TME unit. The strength of this perturbation varies from one heteroatom to another, with the effect being strongest with the NH group, as is shown by a variety of semiempirical[106,114] and ab initio[129] calculations. These calculations suggest that the unsubstituted compound in all three heteroatom systems, O, S, and

NH (**46a–c**) should have a preference of a few kilocalories per mol for the singlet, but that substitution of a sufficiently strong electron-withdrawing group for the functional hydrogen in the NH compound can reduce the singlet–triplet gap to near zero.

Experiment shows that the furan and thiophene compounds **46a** and **b** and the pyrrole derivatives **46c–e** all have singlet ground states, as predicted.[127,128] The *N-p*-toluenesulfonyl derivative **46f**, however, is predicted to have a very small singlet–triplet gap, with the triplet more stable by $\sim$0.4–1.5 kcal/mol, depending on the method of calculation.[105,106,130] The calculations suggest that the tuning process should reach the crossover point from singlet to triplet in this region, and that consequently **46f** might have a detectable triplet state.

The outcome of this study confirms the predicted tuning effect. Irradiation of the diazene **45f** at 265 nm gives a well-defined triplet signal in the ESR spectrum that persists in the dark for weeks. Significantly, irradiation of **45f** *at 365 nm* gives a deep blue preparation that has no detectable ESR signal. Chemical evidence from trapping reactions[106,130] shows that both the triplet and singlet species have the *N*-tosyl-3,4-dimethylenepyrrole structure. The Curie plot of the triplet has a slight downward curvature at the lowest temperatures, which can be formally fitted to a gap $E_S - E_T$ of $\sim$ 0.02 kcal/mol. This curvature is not because of population of the blue singlet. Were that the case, the amount of triplet present in a blue singlet preparation would have been easily sufficient to detect in the ESR spectrum.[130] The predicted crossover to a populatable triplet thus is confirmed. Whether the triplet of **46f** is really the ground state becomes an oversimplified question in light of the unexpected persistence of both species, which are stable for weeks at 77 K without interconversion. This long-lived (or persistent) spin isomerism is further discussed in Section 7.2.1.

Another example of tuning the spin state by structural variation within a graded series has been provided by the Dougherty group.[131] Ab initio (6-31G* π-CISD) calculations on the model series of pyridinium cations **47–49** suggest that the ordering of the states in each compound is strongly dependent on the position of the pyridinium nitrogen. Although in the corresponding neutral pyridines, all the position isomers show triplet ground states, in the cationic systems, the ground state preference changes from definitely triplet in **47**, to near degeneracy in **48**, to slightly singlet in **49**. Thus, the 3,5- and 2,4-pyridinium units are predicted to be ferromagnetic couplers, but the 2,6-pyridinium unit should be an antiferromagnetic coupler. These simple analogues have not been studied so far, but the predictions have been tested in the series of pyridinium bis-TMM analogues (**50–52**). The ESR spectra are interpreted[131] as indicating quintet, quintet, and triplet ground states for **50**, **51**, and **52**, respectively, and the authors employ these results as support for the theoretical predictions.

7.2.1. Long-Lived (Persistent) Spin Isomerism.

The singlet of *N*-tosyl-3,4-dimethylenepyrrole (**46f**, Scheme 5.10) is a blue, ESR-silent species that is stable over many days in cold matrices. The triplet is also stable for days at low temperatures. This type of "long-lived spin isomerism" has been attributed[105,106,127,132] to a dependence of spin state on molecular conformation, probably determined by the torsional relationship of the pyrrole ring and the *p*-toluenesulfonyl substituent. Similar spin isomerism seems to appear in the cases of the adamantane bis(phenoxyl) derivative **53**,[133,134] the 1,3-phenylenebis(aminoxyl) biradical **54**,[135] and the 2-naphthyl(carbomethoxy)carbene **55**.[136]

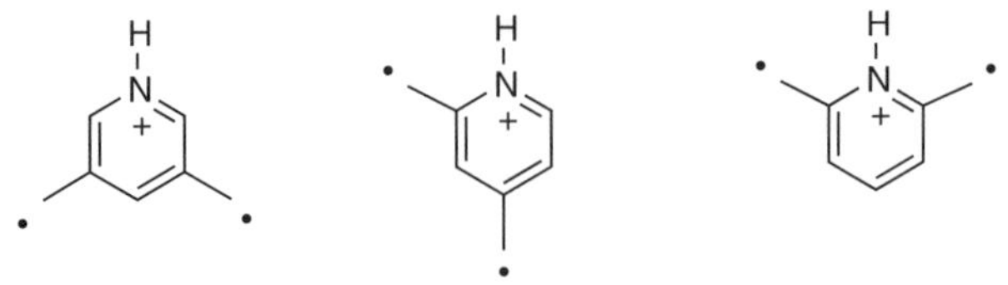

	47	**48**	**49**
$E_S - E_T$ (calc.) kcal/mol	6	0	−2

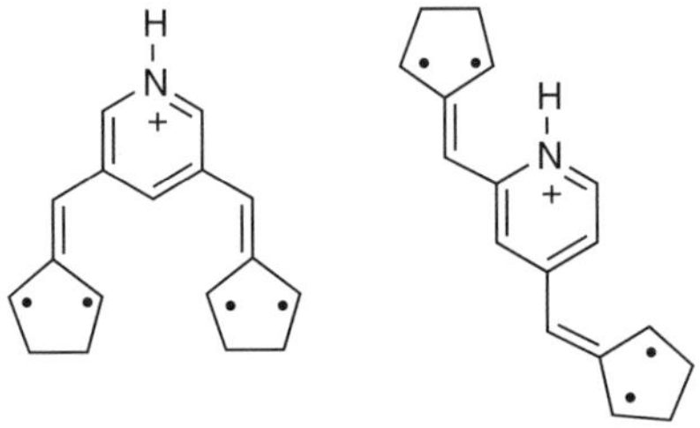

	50	**51**	**52**
Ground state obsd.	quintet	quintet	triplet

Although interconversions of the two spin states under matrix-immobilized spectroscopic conditions have not yet been observed in the cases of the pyrrole derivative **46f** and the bis(phenoxyl) **53**, reversible thermal isomerizations of the singlet and triplet states have been reported in the cases of the bis(aminoxyl) **54** and the carbene **55**.

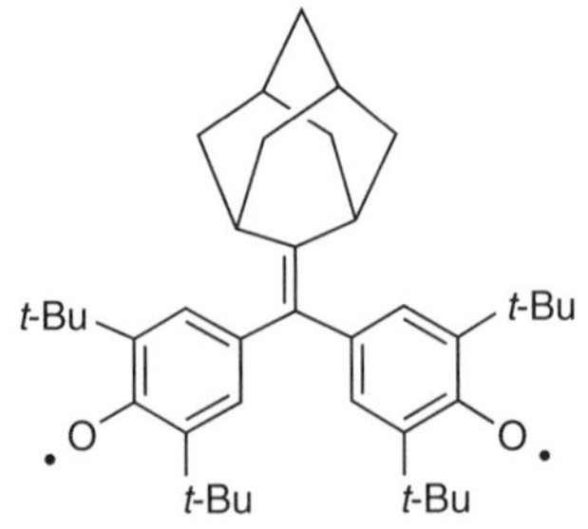

53	**54**	**55**

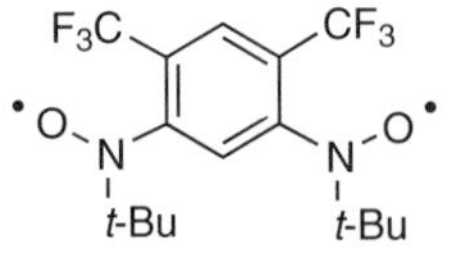

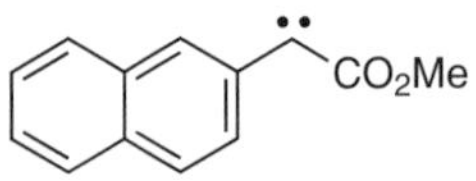

It should be mentioned that Cambi and Szego[137] discovered long-lived spin isomers in the 1930s in the field of transition metal complexes, and they have been extensively investigated since then.[138] In some cases, they are sufficiently stable for determination of their crystal structure by X-ray methods. The physical basis for the differences between spin isomers in these complexes seems not to be the result of rotational isomerism but instead to involve differences in the geometrical arrangement of the ligands around the metal atom. In several cases with hexavalent metal centers, these isomers correspond to local minima in the itinerary of distortions from octahedral symmetry. Stereochemical nomenclature is a notoriously contentious area, but at the risk of opprobrium from its cognoscenti, I would venture the suggestion that both the rotational isomerism proposed for the non-Kekulé systems and the distorted polyhedral isomerism of the complexes could be considered examples of conformational isomerism in which the change in conformation is accompanied by a change in spin state.

8. MEASUREMENT AND INTERPRETATION OF MAGNETIZATION AND MAGNETIC SUSCEPTIBILITY

Notable developments in technique[139–141] have made possible the elucidation of characteristics of paramagnetic materials that have not been readily accessible with ESR spectroscopy or gas-phase negative ion photodetachment PES. Foremost among these have been improvements in the design and construction of instruments for direct measurement of magnetization. The theories of magnetic properties are well established,[142–144] and many measurements of magnetization using Guoy or Faraday blances had been made, especially in the field of inorganic chemistry. The advent in recent years of superconducting quantum interference devices (SQUIDs) has improved the accuracy and convenience of such measurements. The direct magnetic methods can be applied to characterization of both small-molecule non-Kekulé compounds and oligomeric or polymeric paramagnetic substances. In favorable cases, it is possible to measure the separation between multiplet states and to recognize mixtures of them. We take a brief detour to introduce some of the key points here.

The alignment of paramagnetic molecules in an applied magnetic field is influenced by two opposing factors: the magnetic energy, $g\mu_B H_z$, which tends to align the spin moments with the field, and the energy of thermal motions, kT, which tend to randomize them. At sufficiently high field and low temperature, the magnetic susceptibility χ per mole of a paramagnetic sample is derived from the measured magnetization I in a field H by the relationship $\chi = I/H$. Several methods of handling the experimental data have been described, but for our purposes, it will be simplest to use the effective magnetic moment. It can be shown[142–144] that for those molecules in which spin-orbit coupling can be neglected, which will be the case for most of the systems in this chapter, the dependence of χ on the total spin S is usefully expressed by the effective magnetic moment μ_{eff} (Eq. 4), in which μ_B is the

Bohr magneton, N is the number of moles in the sample, and g is the (dimensionless) electron free-spin factor (g-factor), which has the value 2.0023.

$$\mu_{eff} = (3kT\chi/N)^{1/2} = g\mu_B[S(S+1)]^{1/2} \tag{4}$$

Experimentally, one thus expects that a triplet paramagnet ($S = 1$) should display $\mu_{eff} = 2.83\mu_B$, a quintet ($S = 2$) 4.90 μ_B, and so on. A mixture of two states of different spin should have a value intermediate between the two values. For two such states in equilibrium, the energy gap can be calculated from the value of χ and the Weiss temperature θ in the Curie–Weiss law (Eq. 5.)[139]

$$\chi = C/(T+\theta) \tag{5}$$

These methods have now been applied experimentally by many authors[2,3,8,131,139,145,146] to determine spin states and energy gaps in non-Kekulé molecules. They are nicely complementary to ESR spectroscopy. For example, although both techniques are in principle capable of determining $E_S - E_T$, ESR cannot be used if the gap is too large ($> \sim 0.5$ kcal/mol) or if the two states happen to be energetically degenerate, whereas these difficulties do not apply to direct magnetic measurements. On the other hand, ESR can report crucial structural information about the species, whereas in the direct magnetic measurements, molecules of the same spin multiplicity are indistinguishable. The ESR spectra become difficult to interpret when the total spin grows large, but direct magnetic measurements can afford values of S for multiple spins[3] or even for S values as large as 5000 in a high-spin polymer.[147]

9. WHERE THE DISJOINT AND PARITY-BASED PREDICTIONS DIFFER

The non-Kekulé compounds we have considered so far are all structurally related to TMM and TME, and the disjoint and parity methods for predicting qualitatively the ground-state spin give similar results. However, the two methods do not agree in the case of another type of non-Kekulé structure, of which the parent compound is the biradical 2,3,4-trimethylenepentane-1,5-diyl (**56**), commonly called penta-methylenepropane (PMP) (Scheme 5.11).

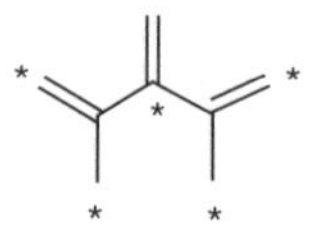

56
PMP

$(n^* - n)/2 = 1$

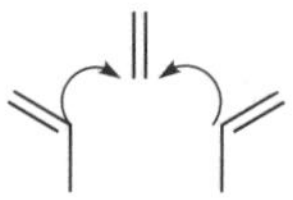

57

disjoint
connectivity

Scheme 5.11

The parity methods predict a high-spin ground state, but as Borden and Davidson[59] were the first to point out, PMP and its derivatives have disjoint connectivity **57** and, hence, should have a small multiplet gap with a likely singlet ground state.

So far, there have been no experimental tests of the subtle distinction between the parity and disjoint mnemonics on the parent PMP, which remains an unknown substance. However, a number of PMP derivatives have been synthesized, starting with the dicarbene **58**, a disjoint, equal-parity structure that was studied by Itoh as early as 1978[148,149] and shown by ESR spectroscopy to consist of both triplet and quintet states. These observations have been expanded in a series of variations, including some examples of disjoint-unequal parity cases, in the laboratories of Iwamura and co-workers[150–152] and of Lahti and co-workers.[153] These efforts now have produced important tests of the qualitative rules (see structures **58–62** and Scheme 5.12).

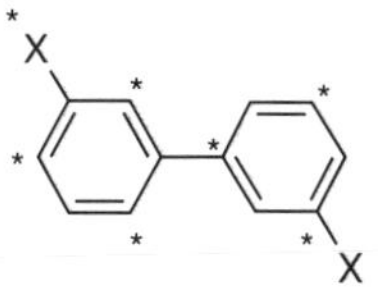

58: X = :CPh

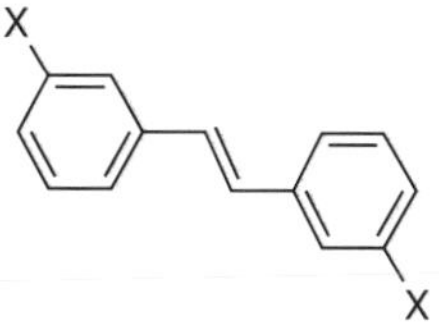

59a: X = :CPh
59b: X = N:
59c: X = N(*t*-Bu)O •

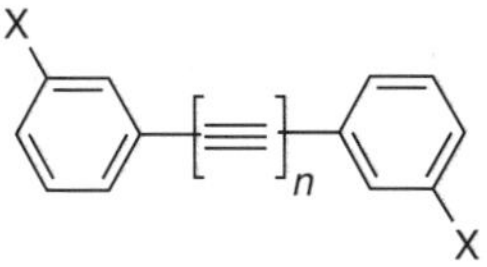

60a: X = N:, *n* = 1
60b: X = N:, *n* = 2

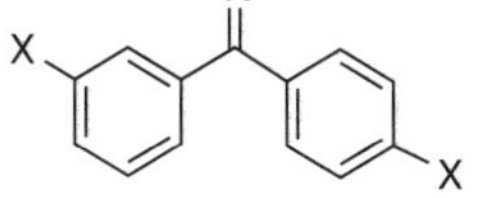

61a: X = N:, R = CH$_2$
61b: X = N(*t*-Bu)O, R = C(CH$_3$)$_2$

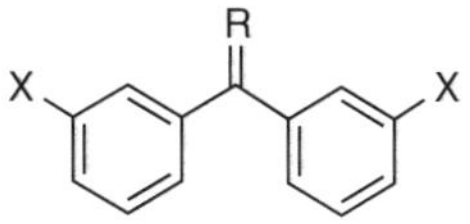

62a: X = N:, R = CH$_2$
62b: X = N(*t*-Bu)O, R = CH$_2$
62c: X = N:, R = O

Note that compounds (**58–61b**) are all equal-parity molecules and all disjoint, but compounds **62a–c** are of unequal parity but also disjoint by the Borden–Davidson criterion.

In each of the cases, ESR spectroscopy, or magnetic susceptibility measurements, or both, give evidence that more than one spin state is present and that one of them is a singlet. Although the energy gaps between the multiplets are very small (Scheme 5.12), the singlet is the ground state throughout the series. A

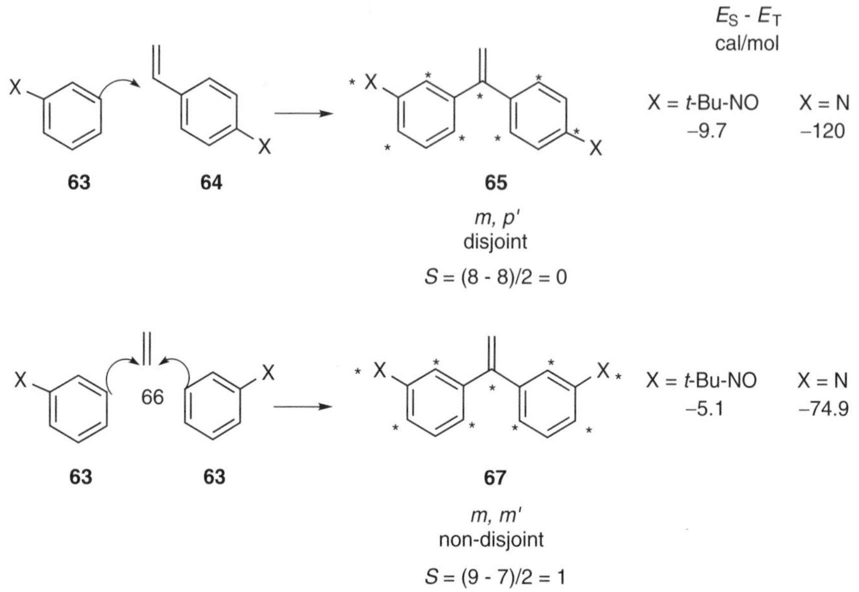

Scheme 5.12

number of factors might have been expected to perturb the test. For example, it seems likely that the aryl rings will be twisted out of coplanarity with the rest of the π system, which potentially could diminish electronic interactions determinative of the ground state. It is remarkable that even when the gap is as small as a few calories, the preference for the singlet persists.

10. CONCLUSION AND OUTLOOK

In principle, the domain of non-Kekulé molecules should extend over territory as broad as that of ordinary Kekulé molecules. Aside from the potentially huge applications in synthesis and in materials science, there are many intriguing theoretical and mechanistic questions to be explored.

One example that comes readily to mind is the mechanism of generation of high-spin non-Kekulé intermediates from spin-zero precursors, a transformation that requires a change in electronic multiplicity along the reaction path. Some of the questions that can be asked about such reactions include whether the high-spin intermediate is formed in a statistical mixture of spin substates or instead in predominantly one substate. Very little is known experimentally about such processes,[154] but the answers could provide much insight into the dynamics of reactions in which surface crossings must play a part.

A related area is the chemistry and physics of electronically excited non-Kekulé molecules. Many of the non-Kekulé molecules have absorption in the near-UV or even in the visible. The description of the excited states of these species offers a broad field of exploration for both experimentalists and theoreticians. That this will be a challenging problem already is suggested by the experience in the case of the excited states of TMB (see Section 6).

It may be fairly said that the field of non-Kekulé compounds has expanded from a few curious and esoteric molecules to an active domain of inquiry and a potential source of materials for practical application. Theory and experiment have interacted fruitfully in these developments, but some serious limitations remain. For example, for many years, experimentalists have operated under the imperative of structural simplification. The ideal is to construct molecules that embody the features of theoretical interest but are as free as possible of complicating impedimenta such as extra substituents or other structural elements left over from the synthesis itself and not readily removed.

There are several motivations for this minimalist program. One of them is a self-imposed if rather vaguely articulated wish to exclude perturbing features and to focus only on those aspects of chemical behavior seen as generalizable. This technique contains the implicit assumption that the "pure" or "unperturbed" test molecule really does meet those requirements, an assumption often not easily demonstrated.

Also, there is an esthetic element mixed into this motivation, which probably is derived from the tradition of stripped-down test molecules in physical organic chemistry, in which studies of the parent member of a series tend to be valued more highly than those of its derivatives. One thinks, for example, of the emphasis placed on studies of the unsubstituted molecules methylene, norbornyl cation, cyclobutadiene, tetrahedrane, benzyne, and so on. The higher valuation also is associated in some cases with the formidable difficulties experienced by experimentalists in the synthesis and observation of these species.

On the other hand, it is also true that in some cases, whole fields of investigation that were out of reach with the parent molecules opened up to scrutiny with derivatives. Examples given in this chapter include most of the cycloaddition chemistry of TMM and TME (Sections 4 and 5).

Nevertheless, the importance of the parent molecule remains undeniable, and the reason for it forms the second motivation for the minimalist approach: It is the traditional way to bring theory and experiment together for mutual comparison. For decades, chemists have been aware of the limitations on theoretical techniques imposed by computational capacity. Everyone dreamed of the day when the properties of any reasonably sized molecule could be calculated to any desired degree of accuracy. In fact, some computational enthusiasts announced about 25 years ago—prematurely, in my opinion—that the day already had arrived. Most chemists were skeptical then; they continued under the assumption that the most practical comparisons of theory and experiment would involve experiments on the simplest possible test molecules. Thus, those comparisons accepted the limitations that theory still operated under.

Today, however, one senses that there has been an exponential growth in computational power.[64] This change in circumstances raises the question of whether we should reexamine our attitudes. Is it possible that the time is at hand when theoreticians and experimentalists can share the burden of mutual adjustment more equally? Can experimentalists now—or at least soon—hope that the most powerful computational techniques will be applied to the many non-Kekulé test molecules and other species yet to be imagined that fall short of the classical minimalist ideal? As an experimentalist, I certainly hope so, and I venture the prediction that the coming postclassical era of physical organic research will be as rich in discoveries as the present classical one.

ACKNOWLEDGMENTS

It is a pleasure to acknowledge helpful correspondence and discussions with C. J. Cramer, G. B. Ellison, P. M. Lahti, M.S. Platz, R. R. Squires, and P. G. Wenthold.

SUGGESTED READING

Physical basis of Hund's rule: (a) J. A. Berson, in *The Chemistry of the Quinonoid Compounds*, Vol. II, S. Patai and Z. Rappoport, Eds., John Wiley & Sons, Inc., New York, 1988. (b) W. Kutzelnigg, "Friedrich Hund and Chemistry," *Angew. Chem. Intl. Ed. Engl.* **1996**, *35, 573*.

ESR spectroscopy: (a) J. E. Wertz and J. R. Bolton, in *Electron Spin Resonance: Elementary Theory and Practical Applications*; McGraw–Hill, New York, 1972. (b) J. A. Berson, in *The Chemistry of the Quinonoid Compounds*, Vol. II, S. Patai and Z. Rappoport, Eds., John Wiley & Sons, Inc., New York, 1988. (c) W. Gordy, in *Theory and Applications of Electron Spin Resonance*, Vol. 15, A. Weissberger series Ed., W. West, Ed., John Wiley & Sons, Inc., New York, 1980, p. 589. (d) E. Wasserman, W. A. Yager, and L. C. Snyder, "Electron spin resonance (E.S.R.) of the triplet states of randomly oriented molecules," *J. Chem. Phys.* **1964**, *41*, 1763.

REFERENCES

Added Notes

1. A.-T. Wu, et al.[155] have very recently observed triplet spectra from the irradiation of *N*-tosylpyrroles lacking the diazene group in various glasses at 4 K and have attributed these, without further structural evidence, to radical pairs resulting from the photocleavage of the N—Ts bond. The *D* values are essentially independent of the matrix or the temperature up to 77 K. In the case of the spectra from *N*-tosylpyrrole itself and the diazene **45f**, the *D* values correspond, when delocalization is taken into account, to actual separation between the partners of about 3.1–3.5 Å in the spin-dipolar approximation (with neglect of spin-orbit perturbations) (Section 3.2). The authors propose that the triplet spectra previously attributed[105,130] to *N*-tosyl-3,4-dimethylenepyrrole (**46f**) may be better interpreted as due to a radical pair resulting from N—Ts cleavage but leaving the diazene group of **45f** intact.

There are two groups of questions associated with this hypothesis. The first group concerns all of the spectra observed by Wu et al.[155] If the species responsible for the ESR signals observed are radical pairs, they behave quite differently from other radical pairs in frozen media (Section 3.2), which give unresolved ESR spectra and different spectral widths as a function of the matrix, resulting from site inequalities.

The separation of 3.1–3.5 Å is close to the sum (3.3 Å) of the van der Waals radii of S and N. One interpretative difficulty now becomes apparent: What keeps the caged radicals, which on this basis are essentially in physical contact, from recombining by formation of a covalent bond? Such close contacts of partners have been reported in cages in the crystalline state but apparently not in glassy media.

The second group concerns the properties of the species from the diazene **45f**, which in the interpretation of Wu et al. is thought to be a radical pair resulting from N—Tos cleavage only. However, the ESR spectrum obtained from photolysis of the model compound *N*-tosylpyrrole itself at 77 K is very weak or not observed. Under the same conditions, the diazene **45f** gives a signal at least 100 times as intense.[130] It is difficult to imagine why the quantum yield for radical-pair formation by N—Ts bond homolysis apparently should be increased to this extent by the presence of the remote nonreacting diazene function. Also, it is not obvious how the radical-pair hypothesis would account for the observation[130] that a species embodying the non-Kekulé biradical **46f** unit (shown not to be the blue singlet biradical) can be trapped chemically by thawing triplet-rich glasses from irradiation of diazene **45f**. Instead, the latter two observations are readily accommodated in the original interpretation by Bush et al.[105,130] that the ESR spectrum arises from a molecular triplet, **46f**, derived by photodeazetation of **45f**.

2. Useful applications of the Evans proton NMR shift method[156,157] have permitted the determination of effective magnetic moments μ_{eff} (see Section 8) and spin multiplicities for several high-spin radical–cation species.[158,159] This work makes available a simple procedure for such assignments when the spin carriers are stable in solution.

3. The closest approaches to triangulene **6** have been made by introduction of oxygen atoms as in the trianion biradical **68**, which has a triplet ground state. Several related derivatives have also been prepared.[160]

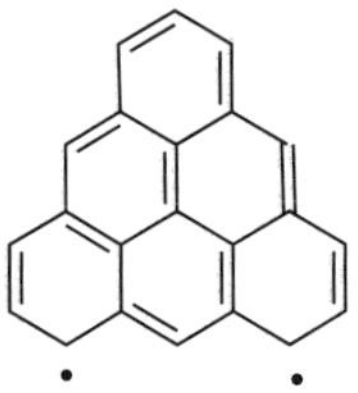

6 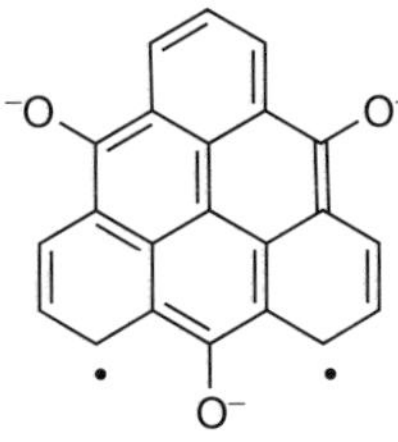   **68**

1. M. J. S. Dewar, *The Molecular Orbital Theory of Organic Chemistry* McGraw-Hill, New York, **1969**, p. 232.

2. P. M. Lahti, Ed., *Magnetic Properties of Organic Materials*, Marcel Dekker, New York, **1999**.

3. H. Iwamura, *Adv. Phys. Org. Chem.* **1990**, *26*, 179.

4. D. Gatteschi, O. Kahn, J. S. Miller, and F. Palacio, Eds., *Magnetic Molecular Materials*, Kluwer Academic, Dordrecht, The Netherlands, **1991**, p. 198E.

5. A. Rajca, *Chem. Rev.* **1994**, *94*, 871. A review.

6. S. J. Jacobs, D. A. Shultz, R. Jain, J. Novak, and D. A. Dougherty, *J. Am. Chem. Soc.* **1993**, *115*, 1744.

7. D. A. Dougherty, *Acc. Chem. Res.* **1991**, *24*, 88.

8. O. Kahn, *Molecular Magnetism*, VCH, Weinheim, Germany, **1993**.

9. J. S. Miller and A. J. Epstein, in *Molecule-Based Magnetic Materials*, Eds., M. M. Turnbull, T. Sugimoto, and L. K. Thompson), American Chemical Society, Washington, D.C., **1996**, 644, pp. 1ff.

10. C. J. Cramer, *J. Chem. Soc. Perkin 2* **1998**, 1007. This paper contains an appreciation of the important contributions of Paul Dowd to the study of trimethylenemethane and tetraamethyleneethane.

11. D. A. Hrovat, M. A. Murcko, P. M. Lahti, and W. T. Borden, *J Chem. Soc. Perkin 2* **1998**, 1037. This reference contains reminiscences of the collaboration of Dowd and Borden.

12. P. Dowd, *Acc. Chem. Res.* **1972**, *5*. A review.

13. J. A. Berson, in *The Chemistry of the Quinonoid Compounds*, Vol. II, Eds., S. Patai and Z. Rappoport), John Wiley & Sons, Inc., New York, **1988**. Chapter 10. This article gives a list of reviews on the EPR spectroscopy of high-spin molecules.

14. J. A. Berson, in *Rearrangements in Ground and Excited States*, P. d. Mayo, Ed., Academic Press, New York, **1980**, p. 311, and references reviewed therein.

15. J. A. Berson, in *Diradicals*, W. T. Borden, Ed., Wiley-Interscience, New York, **1982**, Chapter 4 and references reviewed therein.

16. W. Schlenk and M. Brauns, *Berchte* **1915**, *48*, 661.

17. W. Schlenk and M. Brauns, *Berchte* **1915**, *48*, 716.

18. M. Gomberg, *J. Am. Chem. Soc.* **1900**, *22*, 757.

19. E. Müller and I. Müller-Rodloff, *Justus Liebigs Ann. Chem.* **1935**, *517*, 134.

20. E. Müller and W. Bunge, *Ber. Dtsch. Chem. Ges.* **1936**, *69*, 2168.

21. H. C. Longuet-Higgins, *J. Chem. Phys.* **1950**, *18*, 265.

22. W. A. Yager, E. Wasserman, and R. M. R. Cramer, *J. Chem. Phys.* **1962**, *37*, 1148.

23. E. Wasserman and R. S. Hutton, *Acc. Chem. Res.* **1977**, *10*, 27.

24. G. Burns, *J. Appl. Phys.* **1961**, *32*, 2048.

25. P. Dowd, *J. Am. Chem. Soc.* **1966**, *88*, 2587.

26. J. A. Berson, R. J. Bushby, J. M. McBride, and M. Tremelling, *J. Am. Chem. Soc.* **1971**, *93*, 1544.

27. J. A. Berson, *Acc. Chem. Res.* **1978**, *11*, 446, and references reviewed therein.

28. W. T. Borden, in *Diradicals*, W. T. Borden, Ed., Wiley-Interscience, New York, NY **1982**, Chapter 1 and references cited therein.

29. E. Migirdicyan and J. Baudet, *J. Am. Chem. Soc.* **1975**, *97*, 7400.

30. B. B. Wright and M. S. Platz, *J. Am. Chem. Soc.* **1983**, *105*, 628.

31. J. L. Goodman and J. A. Berson, *J. Am. Chem. Soc.* **1984**, *106*, 1867.

32. J. L. Goodman and J. A. Berson, *J. Am. Chem. Soc.* **1985**, *107*, 5409.

33. M. Rule, A. R. Matlin, D. A. Dougherty, E. F. Hilinski, and J. A. Berson, *J. Am. Chem. Soc.* **1979**, *101*, 5098.

34. M. Rule, A. R. Matlin, D. E. Seeger, E. F. Hilinski, D. A. Dougherty, and J. A. Berson, *Tetrahedron* **1982**, *38*, 787.

35. J. L. Goodman, K. S. Peters, P. M. Lahti, and J. A. Berson, *J. Am. Chem. Soc.* **1985**, *107*, 276.

36. D. E. Seeger and J.A.Berson, *J. Am. Chem. Soc.* **1983**, *105*, 5144.

37. D. E. Seeger and J. A. Berson, *J. Am. Chem. Soc.* **1983**, *105*, 5146.

38. D. E. Seeger, P. M. Lahti, A. R. Rossi, and J. A. Berson, *J. Am. Chem. Soc.* **1986**, *108*, 1251.

39. J.-F. Muller, D. Muller, H. J. Dewey, and J. Michl, *J. Am. Chem. Soc.* **1978**, *100*, 1629.

40. M. Gisin, E. Rommel, J. Wirz, M. N. Burnett, and R. M. Pagni, *J. Am. Chem. Soc.* **1979**, *101*, 2216.

41. M. S. Platz, G. Carrol, F. Pierrot, J. Zayas, and S. Auster, *Tetrahedron* **1982**, *38*, 777.

42. M. Gisin and J. Wirz, *Helv. Chim. Acta* **1983**, *66*, 1556.

43. E. Hasler and E. Gassmann, J. Wirz, *Helv. Chim. Acta* **1985**, *68*, 777.

44. J. Ackermann, H. Angliker, E. Hasler, and J. Wirz, *Angew. Chem. Int. Ed. Engl.* **1982**, *21*, 618.

45. J. J. Fisher, J. H. Penn, D. Döhnert, and J. Michl, *J. Am. Chem. Soc.* **1986**, *108*, 1715.

46. M. N. Burnett, R. Boothe, E. Clark, M. Gisin, H. M. Hassaneen, R. M. Pagni, G. Persy, R. J. Smith, and J. Wirz, *J. Am. Chem. Soc.* **1988**, *110*, 2527.

47. J. Wirz, *Pure Appl. Chem.* **1984**, *56*, 1289.

48. E. Davidson, W. T. Borden, and J. Smith, *J. Am. Chem. Soc.* **1978**, *100*, 3299.

49. P. Du, D. A. Hrovat, and W. T. Borden, *J. Am. Chem. Soc.* **1989**, *111*, 3773.

50. G. Snyder and D. A. Dougherty, *J. Am. Chem. Soc.* **1986**, *108*, 299.

51. G. Snyder and D. A. Dougherty, *J. Am. Chem. Soc.* **1985**, *107*, 1774.

52. P. Dowd, and Y. H. Paik, *J. Am. Chem. Soc.* **1986**, *108*, 2788.

53. B. T. Hill, and R. R. Squires, *J. Chem. Soc. Perkin 2* **1998**, 1027.

54. P. Dowd, *J. Am. Chem. Soc.* **1970**, *92*, 1066.

55. P. Dowd, W. Chang, and Y. H. Paik, *J. Am. Chem. Soc.* **1986**, *108*, 7416.

56. E. Hückel, *Z. Phys. Chem. Abt. B* **1936**, *34*, 339.

57. J. A. Berson, *Chemical Creativity: Ideas from the Work of Woodward, Hückel, Meerwein, and Others*, Wiley-VCH, Weinheim, **1999**, pp. 55ff.

58. J. Baudet, *J. Chim. Phys. Phys.-Chim. Biol.* **1971**, *68*, 191.

59. W. T. Borden and E. R. Davidson, *J. Am. Chem. Soc.* **1977**, *99*, 4587. The origins of these arguments can be found in earlier discussions by Borden on the related case of the hypothetical square-planar cyclobutadiene: W. T. Borden, *J. Chem. Soc., Chem. Commun.* **1969**, 1968. W. T. Borden, *J. Am. Chem. Soc.* **1975**, *97*, 5968.

60. I. A. Misurkin and A. A. Ovchinnikov, *Russ. Chem. Rev. (Engl.Transl.)* **1977**, *46*, 967.

61. A. A. Ovchinnikov, *Theor. Chim. Acta* **1978**, *47*.

62. W. T. Borden, H. Iwamura, and J. A. Berson, *Acc. Chem. Res.* **1994**, *27*, 109.

63. J. H. Reynolds, J. A. Berson, K. K. Kumashiro, J. C. Duchamp, K. W. Zilm, J. C. Scaiano, A. B. Berinstain, A. Rubello, and P. Vogel, *J. Am. Chem. Soc.* **1993**, *115*, 8073.

64. W. T. Borden, in *Reactive Intermediate Chemistry*, R. A. Moss, M. Jones, and M. S. Platz, Eds., John Wiley & Sons, Inc., New York, **2004**, Chapter 22.

65. T. Bally, in *Reactive Intermediate Chemistry*, R. A. Moss, M. Jones, and M. S. Platz, Eds., John Wiley & Sons, Inc., New York, **2004**, Chapter 17.

66. N. J. Turro, M. J. Mirbach, N. Harrit, J. A. Berson, and M. S. Platz, *J. Am. Chem. Soc.* **1978**, *100*, 7653.

67. J. E. Wertz and J. R. Bolton, *Electron Spin Resonance: Elementary Theory and Practical Applications*, McGraw-Hill, New York, **1972**, Chapter 10.

68. W. Gordy, in *Theory and Applications of Electron Spin Resonance*, Vol. 15, A. Weissberger series Ed., W. West, Ed., John Wiley & Sons, Inc., New York, **1980**, p. 589.

69. E. Wasserman, L. C. Snyder, and W. A. Yager, *J. Chem. Phys.* **1964**, *41*, 1763.

70. C. A. Hutchison and B. W. Mangum, *J. Chem. Phys.* **1961**, *34*, 908.

71. W. Weltner, Jr., *Magnetic Atoms and Molecules*, Scientific and Academic ed., New York, **1983**.

72. A. Carrington and A. D. McLachlan, in *Introduction to Magnetic Resonance,* S. A. Rice, Ed., Harper's chemistry Series, Harper and Row, New York, **1967**.

73. S. P. McGlynn, T. Azumi, and M. Kinoshita, *Molecular Spectroscopy of the Triplet State,* Prentice-Hall, New York, **1969**.

74. E.Wasserman, L. C. Snyder, and W. A. Yager, *J. Chem. Phys.* **1964**, *41*, 1763.

75. M. S. deGroot and J. H. vanderWaals, *Mol. Phys.* **1963**, *6*, 545.

76. W. Gordy and R. Morehouse, *Phys. Rev.* **1966**, *151*, 207.

77. R. McWeeny, *J. Chem. Phys.* **1961**, *34*, 399.

78. R. L. Barcus, B. B. Wright, E. Leyva, and M. S. Platz, *J. Phys. Chem.* **1987**, *91*, 6677.

79. J. J. Gajewski, *Hydrocarbon Thermal Isomerizations,* Academic Press, New York, **1981**.

80. J. J. Gajewski, in *Mechanisms of Molecular Migrations*, B. Thyagarajan, Ed., John Wiley & Sons, Inc., New York, **1971**. 3, p. 11ff.

81. R. J. Baseman, D. W. Pratt, M. Chow, and P. Dowd, *J. Am. Chem. Soc.* **1976**, *98*, 5726.

82. P. G. Wenthold, J. Hu, R. R. Squires, and W. C. Lineberger, *J. Am. Chem. Soc.* **1996**, *118*, 475. This paper contains a list of references to computational results on TMM.

83. F. Weiss, *Q. Rev., Chem. Soc.* **1970**, *24*, 278.

84. P. Dowd and M. Chow, *J. Am. Chem. Soc.* **1977**, *99*, 2825, 6438.

85. P. Dowd and M. Chow, *Tetrahedron* **1982**, *38*, 799.

86. G. Maier, H. P. Reisenauer, K. Lanz, R. Tross, D. Jürgen, B. A. Hess, Jr., and L. J. Schaad, *Angew. Chem. Intl. Ed. Engl.* **1993**, *32*, 74.

87. P. Dowd, G. Sengupta, and K. Sachdev, *J. Am. Chem. Soc.* **1970**, *92*, 5726.

88. P. G. Wenthold, J. Hu, and R. R. Squires, *J. Am. Chem. Soc.* **1994**, *116*, 6961.

89. J. A. Berson, W. Bauer, and M. M. Campbell, *J. Am. Chem. Soc.* **1970**, *92*, 7515.

90. M. S. Platz, J. M. McBride, R. D. Little, J. J. Harrison, A. Shaw, S. E. Potter, and J. A. Berson, *J. Am. Chem. Soc.* **1976**, *98*, 5725.

91. M. Abe and W. Adam, *J. Chem. Soc. Perkin 2* **1998**, 1063. See references cited therein. See also Eq. 1 of p. 1063.

92. M. S. Platz and J. A. Berson, *J. Am Chem. Soc.* **1977**, *99*, 5178.

93. M. R. Mazur and J. A. Berson, *J. Am. Chem. Soc.* **1982**, *104*, 2217.

94. J. A. Berson, *Chemical Creativity: Ideas from the Work of Woodward, Hückel, Meerwein, and Others*, Wiley-VCH, Weinheim, **1999**, p. 55ff.

95. D. J. Klein, C. J. Nelin, S. Alexander, and F. A. matsen, *J. Chem. Phys.* **1982**, *77*, 3101.

96. N. Tyutyulkov, I. Kanev, O. Polansky, and J. Fabian, *Theor. Chim. Acta* **1977**, *46*, 191.

97. N. Tyutyulkov, S. Karabunarliev, and S. Ivanov, *Mol. Cryst. Liq. Cryst.* **1989**, *176*, 139. A review.

98. N. Tyutyulkov, F. Dietz, K. Müllen, M. Baumgarten, and S. Karabunarliev, *Theoret. Chim. Acta* **1995**, *86*, 353.

99. B. L. Prasad and T. P. Radhakrishnan, *J. Phys. Chem.* **1992**, *96*, 9232.

100. T. P. Radhakrishnan, *Chem. Phys. Lett.* **1991**, *181*, 455.

101. P. Nachtigall and K. D. Jordan, *J. Am. Chem. Soc.* **1993**, *115*, 270.

102. M. Filatov and S. Shaik, *J. Phys. Chem. A* **1999**, *103*, 8885.

103. J. Pittner, P. Nachtigall, and P. Carsky, *J. Phys. Chem. A* **2001**, *105*, 1354.

104. E. P. Clifford, P. G. Wenthold, W. C. Lineberger, G. B. Ellison, C. X. Wang, J. J. Grabowski, F. Vila, and K. D. Jordan, *J. Chem. Soc, Perkin 2* **1998**, 1015.

105. L. C. Bush, R. B. Heath, and J. A. Berson, *J. Am. Chem. Soc.* **1993**, *115*, 9830.

106. L. C. Bush, R. B. Heath, X. W. Feng, P. A. Wang, L. Maksimovic, A. I. H. Song, W. S. Chung, A. B. Berinstain, J. C. Scaiano, and J. A. Berson, *J. Am. Chem. Soc.* **1997**, *119*, 1406.

107. P. Dowd, W. Chang, and Y. H. Paik, *J. Am. Chem. Soc.* **1987**, *109*, 5284.

108. W. R. Roth, M. Biermann, G. Erker, K. Jelich, W. Gerhartz, and H. Görner, *Chem. Ber.* **1980**, *113*, 586.

109. W. R. Roth and G. Erker., *Angew. Chem. Int. Ed. Engl.* **1973**, *12*, 503.

110. Y. Choi, K. D. Jordan, Y. H. Paik, W. Chang, and P. Dowd, *J. Am. Chem. Soc.* **1988**, *110*, 7575.

111. K. Matsuda and H. Iwamura, *J. Chem. Soc. Perkin 2* **1998**, 1023.

112. W. R. Roth, U. Kowalczik, G. Maier, H. P. Reisenauer, R. Sustmann, and P. Müller, *Angew. Chem. Int. Ed. Engl.* **1987**, *26*, 1285.

113. J. J. Nash, P. Dowd, and K. D. Jordan, *J. Am. Chem. Soc.* **1992**, *114*, 10071.

114. P. M. Lahti, A. R. Rossi, and J. A. Berson, *J. Am. Chem. Soc.* **1985**, *107*, 2273.

115. P. M. Lahti, A. Ichimura, and J. A. Berson, *J. Org. Chem.* **1989**, *54*, 958.

116. P. Du, D. A. Hrovat, W. T. Borden, P. M. Lahti, A. R. Rossi, and J. A. Berson, *J. Am. Chem. Soc.* **1986**, *108*, 5072.

117. P. M. Lahti, A. R. Rossi, and J. A. Berson, *J. Am. Chem. Soc.* **1985**, *107*, 4362.

118. W. R. Roth, R. Langer, M. Bartmann, M. Stevermann, G. Maier, H. P. Reisenauer, R. Sustmann, and W. Müller, *Angew. Chem. Intl. Ed. Engl.* **1987**, *26*, 256.

119. J. H. Reynolds, J. A. Berson, J. C. Scaiano, and A. B. Berinstain, *J. Am. Chem. Soc.* **1992**, *114*, 5866.

120. J. H. Reynolds, J. A. Berson, K. K. Kumashiro, J. C. Duchamp, K. W. Zilm, A. Rubello, and P. Vogel, *J. Am. Chem. Soc.* **1992**, *114*, 763.

121. M. M. Greenberg, S. C. Blackstock, J. A. Berson, R. A. Duchamp, and K. W. Zilm, *J. Am. Chem. Soc.* **1991**, *113*, 2318.

122. K. W. Zilm, R. A. Merrill, G. G. Webb, M. M. Greenberg, and J. A. Berson, *J. Am. Chem. Soc.* **1987**, *109*, 1523.

123. K. W. Zilm, R. A. Merrill, M. M. Greenberg, and J. A. Berson, *J. Am. Chem. Soc.* **1987**, *109*, 1567.

124. M. M. Greenberg, S. C. Blackstock, J. A. Berson, R. A. Merrill, J. C. Duchamp, and K. W. Zilm, *J. Am. Chem. Soc.* **1991**, *113*, 2318.

125. D. A. Hrovat and W. T. Borden, *J. Am. Chem. Soc.* **1994**, *116*, 6327.

126. W. R. Roth, R. Langer, T. Ebbrecht, A. Beitet, and H.-W. Lennartz, *Chem. Ber.* **1991**, *124*, 2751.

127. J. A. Berson, *J. Mol. Struct. (Theochem)* **1998**, *424*, 21.

128. J. A. Berson, *Acc. Chem. Res.* **1997**, *30*, 238. A review.

129. P. Du, D. A. Hrovat, and W. T. Borden, *J. Am. Chem. Soc.* **1986**, *108*, 8086.

130. L. C. Bush, L. Maksimovic, X. W. Feng, P. A. Wang, H. S. M. Lu, and J. A. Berson, *J. Am. Chem. Soc.* **1997**, *119*, 1416.

131. A. P. West, Jr., S. K. Silverman, and D. A. Dougherty, *J. Am. Chem. Soc.* **1996**, *118*, 1452.

132. J. A. Berson, in *Magnetic Properties of Organic Materials*, P. M. Lahti, Ed., Marcel Dekker, New York, **1998**, Chapter 2.

133. D. A. Shultz, A. K. Boal, and G. T. Farmer, *J. Am. Chem. Soc.* **1997**, *119*, 3846.

134. D. A. Shultz, in *Magnetic Properties of Organic Materials*, P. M. Lahti, Ed., Marcel Dekker, New York, **1999**, p. 103.

135. A. Rajca, K. Lu, S. Rajca, and C. R. Ross, II, *Chem. Commun.* **1999**, 1249.

136. Z. Zhu, T. Bally, L. Stracener, and R. J. McMahon, *J. Am. Chem. Soc.* **1999**, *121*, 2863.

137. L. Cambi and L. Szego, *Ber. Dtsch. Chem. Ges.* **1931**, *64*, 2591. See also subsequent articles. I am indebted to R. Crabtree for calling this work to my attention.

138. P. Gütlich, A. Hauser, and H. Spiering, *Angew. Chem. Int. Ed. Engl.* **1994**, *33*, 2024.

139. H. Iwamura and N. Koga, *Acc. Chem. Res.* **1993**, *26*, 346.

140. C. J. O'Connor, in *Molecule-Based Magnetic Materials*, M. M. Turnbull, T. Sugimoto, and L. K. Thompson, Eds., American Chemical Society, Washington, DC, **1996**, 644 44ff.

141. P. A. Salyer and L. terHaar, in *Molecule-Based Magnetic Materials*, M. M. Turnbull, T. Sugimoto, and L. K. Thompson, Eds., American Chemical Society, Washington, DC **1996**, 644 p. 68ff.

142. R. L. Carlin, *Magnetochemistry* Springer, Heidelberg, Germany **1986**, pp. 5–18.

143. A. Earnshaw, *Introduction to Magnetochemistry* Academic Press, London, UK **1968**, pp. 1–10.

144. J. H. VanVleck, in *The Theory of Electric and Magnetic Susceptibilities,* R. H. Fowler, P. Kapitza, Eds., International Series of Monographs on Physics, Oxford University Press, Oxford, UK, **1932**.

145. E. Coronado, P. Delhaès, D. Gatteschi, and J. S. Miller, Eds., *Molecular Magnetism: From Molecular Assemblies to the Devices. NATO ASI Series,* Series E: Applied Sciences, Kluwer, Dordrecht, The Netherlands, **1996**, 321.

146. M. M. Turnbull, T. Sugimoto, and L. K. Thompson, Eds., *Molecule-Based Magnetic Materials,* ACS Symposium Series, American Chemical Society, Washington, DC, **1996**, 644.

147. A. Rajca, J. Wongsriratanakul, and S. Rajca, *Science* **2001**, *294*, 1503.

148. K. Itoh, *Pure Appl. Chem.* **1978**, *50*, 1251.

149. Y. Teki, T. Takui, M. Kitano, and K. Itoh, *Chem. Phys. Lett.* **1987**, *142*, 181.

150. T. Matsumoto, T. Ishida, N. Koga, and H. Iwamura, *J. Am. Chem. Soc.* **1992**, *114*, 9952.

151. T. Matsumoto, N. Koga, and H. Iwamura, *J. Am. Chem. Soc.* **1992**, *114*, 5448.

152. M. Murata and H. Iwamura, *J. Am. Chem. Soc.* **1991**, *113*, 5547.

153. C. Ling, M. Minato, P. M. Lahti, and H. vanWilligen, *J. Am. Chem. Soc.* **1992**, *114*, 9959.

154. W. P. Chisholm, H. L. Yu, R. Murugesan, S. I. Weissman, E. F. Hilinski, and J. A. Berson, *J. Am. Chem. Soc.* **1983**, *105*, 4419. This paper reports experiments of this kind in the formation of the TMM biradical **14** from the diazene **13** (see Scheme 5.2).

155. A.-T. Wu, Y.-J. Chen, and W.-S. Chung, *J. Am. Chem. Soc.* **2003**. Submitted for publication. I thank these authors for an advance copy of their paper.

156. D. F. Evans, *J. Chem. Soc.* **1959**, 2003.

157. D. H. Live and S. I. Chan, *Anal. Chem.* **1970**, *42*, 791.

158. K. R. Stickley, T. D. Selby, and S. C. Blackstock, *J. Org. Chem.* **1997**, *62*, 448.

159. T. D. Selby and S. C. Blackstock, *J. Am. Chem. Soc.* **1999**, *121*, 7152.

160. R. J. Bushby, in *Magnetic Properties of Organic Materials*, P. M. Lahti, Ed., Marcel Dekker, New York, **1999**, 179ff. A review.

CHAPTER 6

Organic Radical Ions

HEINZ D. ROTH

Department of Chemistry and Chemical Biology, Rutgers,
The State University of New Jersey, New Brunswick, NJ

1. HISTORICAL COMMENTS—ORIGINS OF RADICAL ION CHEMISTRY

Organic radical ions are recognized today as important intermediates in areas of chemistry ranging from intergalactic photochemistry, or a key role in photosynthesis,

Reactive Intermediate Chemistry, edited by Robert A. Moss, Matthew S. Platz, and Maitland Jones, Jr.
ISBN 0-471-23324-2   Copyright © 2004 John Wiley & Sons, Inc.

Figure 6.1. The Jovian moon Io; deep ultraviolet (UV) photolysis of its methane atmosphere proceeds with electron ejection, generating the molecular ion of methane (see color insert). NASA JPL Galileo program image from Voyager 1, http://www.jpl.nasa.gov/galileo/io/vgrio1.html

to natural occurrence as stable minerals. Some radical ions were observed as colored transients in the nineteenth century; their true nature was not recognized until the twentieth century, long after the discovery of trivalent carbon by Gomberg.[1] Even in the twenty-first century, the coverage of radical ions in elementary, or even advanced organic chemistry texts, leaves much to be desired. The following vignettes illustrate the range of radical ion chemistry.

The Jovian moon, Io, shows an orange hue (Fig. 6.1), which may be due to long-chain alkane radical cations. The atmosphere of Io consists mostly of methane; deep UV photolysis proceeds with electron ejection; thus, the molecular ion of methane was perhaps the earliest organic radical cation, generated by solar irradiation aeons ago.

On the planet Earth, the most important photoreaction occurs in green plants or in green or purple organisms. Their photochemical reaction centers contain a "special pair" of chlorins (cf. the purple bacterium *Rhodobacter sphaeroides*, Fig. 6.2).[2] Solar photons cause electron transfer and generate a radical ion pair. Within two picoseconds, the negative charge is transferred to a second chlorin, and from it to a quinone.[3-5]

A remarkable example of a persistent radical anion is the semiprecious stone, lapis lazuli, known and appreciated as a pigment since ancient times. The species imparting the blue hue is trisulfur radical anion, $S_3^{\bullet-}$, accompanied by variable fractions of $S_2^{\bullet-}$, which introduces a green tint; the sulfur radical anions are incarcerated

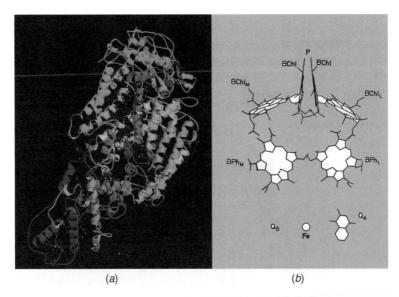

(a)   (b)

Figure 6.2. (a) Photosynthetic reaction center of *Rhodopseudomonas viridis* Reprinted from the Protein Data Bank, H. M. Berman, J. Westbrook, Z. Feng, G. Gilliland, T. N. Bhat, H. Weissig, I. N. Shindyalov, P. E. Bourne, *Nucleic Acids Res.* **2000**, 1, 235 (http://www.pdb.org/) PDB ID: IDXR, C. R. D. Lancaster, M. Bibikova, P. Sabatino, D. Oesterhelt, H. Michel, *J. Biol. Chem.* **2000**, 275, 39364.[2] (b) arrangement of the essential components in the purple bacterium *Rh. sphaeroides* (see color insert). [Adapted from Ref. 5.]

inside the small-pore zeolite sodalite, which immobilizes and protects these reactive species.[6,7]

As for radical ion chemistry in the laboratory, in 1835 Laurent treated benzil (in his view the "radical" benzoyl) with potassium—the resulting spontaneous ignition[8] did not encourage further experiments. Some 50 years later, Beckmann and Paul[9] reacted aromatic ketones and diketones with sodium. Because the colored solutions and the solid products were sensitive to air and moisture, they worked in a hydrogen atmosphere (Fig. 6.3). Twenty years later, Schlenk et al.[10,11] recognized the colored products as a new class of trivalent-carbon compounds. They proposed the term "metal ketyls"; the suffix "yl" was meant to indicate the "radical nature" of these substances. This view was finally confirmed by conductivity and magnetic susceptibility measurements in the 1930s[12,13] and electron spin resonance (ESR) since the 1950s.

In 1866, Berthelot[14] obtained a black dipotassium salt from naphthalene; the formula "$C_{20}H_8K_2$" reflects an incorrect atomic weight for carbon. Almost 50 years later Schlenk et al.[15] assigned the banded spectrum of a blue, transient species obtained from anthracene as a "monosodium addition product that contains trivalent carbon."

The first organic radical cations, derived from *p*-phenylenediamine, date back to Baeyer's Strassburg laboratory in 1875.[16] Wurster recognized these (Wurster's)

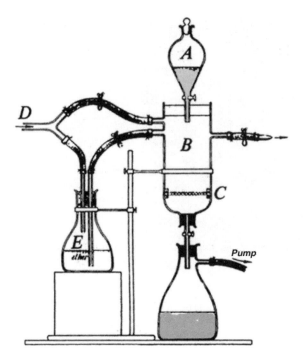

Figure 6.3. Apparatus for the reaction of alkali metals with carbonyl compounds and the separation of the solid adducts. [Adapted from Ref. 9.] *A*, separatory funnel in which the reaction is carried out; *B*, cylinder with inert atmosphere; *C*, perforated porcelain disc with filter paper; *D*, entry of inert gas (H_2 or CO_2) from Kipp apparatus; *E*, wash ether reservoir.

salts as oxidation products containing 1 equiv of an acid.[17,18] Willstätter and Piccard[19,20] formulated the colored salts as molecular aggregates with specific N$\cdots$N bonds, perhaps inspired by the molecular complexes, for example, between aromatic hydrocarbons and nitro compounds[21] or quinones,[22] which were emerging at the time.

Hantzsch recognized that radical cation "salts are not molecular complexes, but uniform, monomolecular chemical compounds with an unsaturated nitrogen or sulfur atom, whose unsaturated state would explain the intense color."[23] A decade later, Weitz confirmed that these species were monomolecular and contained an unpaired electron; he coined the term "Anionradikale" and "Kationradikale": "Both the salt and the cation have an odd number of electrons because of their radical-like composition." He recognized "... that the single positive charge is distributed between both halves of the cation ..." and considered this "strange charge distribution" to be "the cause of the deep color."[24]

In the 1930s, Michaelis compared radical cations with trivalent-carbon or divalent-nitrogen intermediates using potentiometric methods. He rationalized their unusual stability as follows: "The fact that such radicals are capable of existence at all, can be attributed to a particular symmetry of structure resulting in resonance;" a

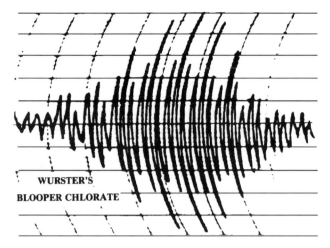

WURSTER'S
BLOOPER CHLORATE

Figure 6.4. Electron spin resonance spectrum of Wurster's blue ion. [Adapted from Ref. 28.]

number of "limiting states" contribute "a share to the resonating or mesomeric state;" ... "letters are atomic kernels, dashes are pairs of electrons, ... and the dot is a single electron."[25,26]

The advent of ESR opened several classes of moderately stable radical ions to scrutiny, namely, ketyls, semidiones, semiquinones, and those of aromatic systems.[27] Wurster's blue became an early target for an ESR study (Fig. 6.4).[28] Its tetramethyl derivative was chosen to study the degenerate electron transfer between an organic radical ion and its parent by nuclear magnetic resonance (NMR) line broadening.[29,30]

2. RADICAL ION GENERATION

A rich variety of reagents and methods have been applied to generate radical ions. As illustrated above, the first methods were chemical redox reactions. Radical anions have long been generated via reduction by alkali metals. Because of the high reduction potentials of these metals, the method is widely applicable, and the reductions are essentially irreversible.

Radical cations can be generated by many chemical oxidizing reagents, including Brønsted and Lewis acids, the halogens, peroxide anions or radical anions, metal ions or oxides, nitrosonium and dioxygenyl ions, stable aminium radical cations, semiconductor surfaces, and suitable zeolites. In principle, it is possible to choose a reagent with a one-electron redox potential sufficient for oxidation–reduction, and a two-electron potential insufficient for oxidation–reduction of the radical ion.

Certain zeolites, notably H-ZSM-5, spontaneously oxidize a range of substrates with oxidation potentials ≤ 1.65 V.[31,32] In the nearly cylindrical cavities of H-ZSM-5 rodlike molecules are protected against scavengers.[33-35] Zeolites also serve as supporting matrices for generating radical cations by ionizing radiation.

Among electrochemical techniques,[36,37] cyclic voltammetry (CV) utilizes a small stationary electrode, typically platinum, in an unstirred solution. The oxidation products are formed near the anode; the bulk of the electrolyte solution remains unchanged. The cyclic voltammogram, showing current as a function of applied potential, differentiates between one- and two-electron redox reactions. For reversible redox reactions, the peak potential reveals the half-wave potential; peak potentials of nonreversible redox reactions provide qualitative comparisons. Controlled-potential electrolysis or coulometry can generate radical ions for study by optical or ESR spectroscopy.

Generation of organic radical ions by radiolysis (Scheme 6.1) was pioneered by Hamill.[38-40] High-energy photons (X- or γ rays) eject electrons from appropriate matrices (R–X, e.g., freons), generating unstable radical cations (Eq. 1); scavenging by solute molecules (S) at low concentrations forms substrate radical cations (Eq. 2). The ejected electrons attach themselves to matrix molecules (Eq. 3), causing fragmentation to a halide ion and a free radical (Eq. 4); the halide ions may react with radical cations forming free radicals (Eq. 5).

$$R - X + \gamma \longrightarrow R - X^{+\bullet} + e^- \tag{1}$$

$$R - X^{+\bullet} + S \longrightarrow R - X + S^{+\bullet} \tag{2}$$

$$R - X + e^- \longrightarrow R - X^{-\bullet} \tag{3}$$

$$R - X^{-\bullet} \longrightarrow R^{\bullet} + X^- \tag{4}$$

$$S^{+\bullet} + X^- \longrightarrow S^{\bullet} - X \tag{5}$$

Scheme 6.1

Photoinduced electron transfer (PET; Scheme 6.2) is a mild and versatile method to generate radical ion pairs in solution,[41-43] exploiting the substantially enhanced oxidizing or reducing power of acceptors or donors upon photoexcitation. The excited state can be quenched by electron transfer (Eq. 7) before (aromatic hydrocarbons) or after intersystem crossing to the triplet state (ketones, quinones). The resulting radical ion pairs have limited lifetimes; they readily undergo intersystem crossing (Eq. 8), recombination of singlet pairs (Eq. 9), or separation by diffusion (Eq. 10), generating "free" radical ions.

$$A \xrightarrow{h\nu} {}^1A^* \longrightarrow {}^3A^* \tag{6}$$

$$^{1,3}A^* + D \longrightarrow {}^{1,3}\overline{A^{-\bullet}D^{+\bullet}} \tag{7}$$

$$^3\overline{A^{\bullet-}D^{+\bullet}} \underset{\longleftarrow}{\longrightarrow} {}^1\overline{A^{-\bullet}D^{+\bullet}} \tag{8}$$

$$^1\overline{A^{\bullet-}D^{+\bullet}} \longrightarrow A + D \tag{9}$$

$$^3\overline{A^{\bullet-}D^{+\bullet}} \longrightarrow {}^2A^{\bullet-} + {}^2D^{+\bullet} \tag{10}$$

Scheme 6.2

The change in free energy (ΔG) for electron-transfer (ET) reactions is given by an empirical relation (Eq. 11; E_T is the excited-state energy, E_{ox} and E_{red} are the one-electron redox potentials of donor and acceptor, respectively, and $e^2/\varepsilon a$ is a Coulombic term accounting for ion pairing).[44] The change in ΔG can be tuned (cf. Eq. 11) by variation of the solvent (polarity) and of the acceptor (reduction potential, excited-state energy).

$$\Delta G = -E_T - E_{red} + E_{ox} - e^2/\varepsilon a \qquad (11)$$

The redox potential of the acceptor excited state, $^*E_{red}$ (Eq. 12), defines its oxidative strength: PET is limited to substrates with oxidation potentials below $^*E_{red}$.

$$^*E_{red} = -E_T + E_{red} \qquad (12)$$

Competing reactions may introduce mechanistic ambiguities: Ketone or quinone triplet states abstract hydrogen atoms, forming neutral radicals. Also, many radical cations are proton donors and radical anions are comparably strong bases. Thus, geminate radical ion pairs may generate neutral radicals by proton transfer.

3. RADICAL ION DETECTION–OBSERVATION

It is useful to briefly review the methods of observing and studying radical ions, including their strengths and limitations. Mass spectrometry (MS) is the most generally applied technique for radical cations, less frequently for radical anions. Mass spectrometry identifies positive ions, usually generated by electron-impact ionization, by their trajectories through magnetic fields. Although it is a valuable analytical tool,[45,46] it provides little information on the structures of radical ions. In special instruments, ions of a particular m/z ratio can be selected and their ion molecule chemistry can be probed.

Photoelectron spectroscopy (PES) is also carried out in the gas phase: photons of known energy $(E_{h\nu})$, for example, the He(I)$_\alpha$ line (21.21 eV), ionize a substrate; the kinetic energy (E_{kin}) of the emitted electrons is measured and the vertical ionization potentials (I_v) derived (Eq. 13). The PES provides information on the energies of occupied molecular orbitals (MOs);[47] the highest occupied molecular orbital (HOMO) of the parent reveals the bond(s) likely to be weakened or broken upon ionization. The PES data reflect the geometries of the parent molecule and need not have any bearing on the *equilibrium* structure of the radical cation.

$$I_v = E_{h\nu} - E_{kin} \qquad (13)$$

Optical spectroscopy (OS) is naturally suited for studying (any type of) intermediate; absorption spectra characterize energy differences between occupied and unoccupied or singly occupied orbitals, including the HOMO–LUMO (lowest

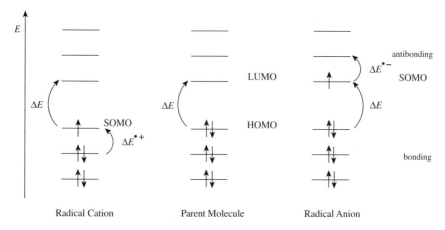

Figure 6.5. Schematic rationalization for lower energy (vis or near-IR) electronic transitions observed for open-shell species with singly occupied bonding (radical cations, left) or antibonding orbitals (radical anions, right).

unoccupied molecular orbital) gap. Emission spectra characterize the radiative decay of excited state species. Time-resolved (TR) spectra allow comparisons of the decay and rise of consecutive intermediates. The time resolution has evolved from millisecond[48,49] to nanosecond[50–53] and picosecond resolution[54–56] even femtosecond spectroscopy is now being practiced in many laboratories.[57] For further discussions on nanosecond, picosecond, and femtosecond spectroscopy, see Chapters 18 by J. C. Scaiano, 19 by E. Hilinski, and 20 by J. E. Baldwin in this volume.

The UV–visible (vis) spectra of many organic radical ions show significant bathochromic shifts relative to their precursors. The open-shell configurations of singly occupied bonding (radical cations) or antibonding orbitals (radical anions) introduce new electronic transitions of lower energies, in the visible or near-IR (cf. Fig. 6.5).[58]

Although exceedingly useful, the application of OS is not without problems. Optical spectra in condensed media consist of broad bands without identifying features; structural information can be derived by comparison with "known" species, or by high-resolution laser spectroscopy.[59] The TR–OS is often limited to wavelengths >300 nm, excluding some prototypes of the most interesting species.

Electron spin resonance can provide detailed information about free radical (ions) in condensed media. Transitions between the electron spin levels are stimulated by radiation at frequencies satisfying the resonance condition:

$$h\nu = g\,\mu\,H_0 \qquad (14)$$

(h is the Planck constant, g is a parameter characteristic for the radical under scrutiny, μ is the Bohr magneton, and H_0 is the applied field strength).

The (hyperfine) interaction between magnetic nuclei and unpaired electrons causes characteristic signal patterns; the spacing and relative intensities of the signals are indicative of the spin density distribution in the intermediate.[60,61] Steady-state ESR has provided detailed insights into structures of radicals and radical ions.

Time-resolved laser flash ESR spectroscopy[62–67] generates radicals with non-equilibrium spin populations and causes spectra with unusual signal directions and intensities. The signals may show absorption, emission, or both and be enhanced as much as 100-fold. Deviations from Boltzmann intensities, first noted in 1963,[68] are known as chemically induced dynamic electron polarization (CIDEP).[69–71] Because the splitting pattern of the intermediate remains unaffected, the CIDEP enhancement facilitates the detection of short-lived radicals. A related technique, fluorescence detected magnetic resonance (FDMR) offers improved time resolution and its sensitivity exceeds that of ESR. The FDMR experiment probes short-lived radical ion pairs, which form reaction products in electronically excited states that decay radiatively.[72]

Among NMR methods providing insight into radical ions,[73] chemically induced dynamic nuclear polarization (CIDNP) has proved especially useful; it results in enhanced transient signals, in absorption or emission; CIDNP effects were first reported in 1967;[74,75] their application was soon extended to radical ions.[76] The method lends itself to modest time resolution.[77,78]

The theory of CIDNP depends on the nuclear spin dependence of intersystem crossing in a radical (ion) pair, and the electron spin dependence of radical pair reaction rates. These principles cause a "sorting" of nuclear spin states into different products, resulting in characteristic nonequilibrium populations in the nuclear spin levels of geminate (in cage) reaction products, and complementary populations in free radical (escape) products. The effects are optimal for radical pairs with nanosecond lifetimes.

The quantitative theory[79–82] allows one to compute intensity ratios of CIDNP spectra from reaction and relaxation rates and characteristic parameters of the radical pair (initial spin multiplicity, μ), the individual radicals (electron g factors, hyperfine coupling constants, a); and the products (spin–spin coupling constants, J). Conversely, the patterns of signal directions and intensities for different nuclei of a reaction product reveal the hyperfine coupling constants of the corresponding nuclei in the radical cation. These results are often unambiguous because NMR chemical shifts clearly establish the identity of the coupled nuclei. Combined with PET as a method of radical ion generation, CIDNP has been the key to elucidating mechanistic details of important reactions, and provided insight into many short-lived radical cations with unusual structures, which had eluded other techniques. As all spectroscopic results, CIDNP results are not immune to misinterpretation.

The claim of *direct observation* is used occasionally as the ultimate panacea and to support claims of validity and authenticity. Direct observation means nothing more than that some property of an intermediate is measured during its lifetime, that is, the lifetime of the species exceeds the characteristic time scale of the method of observation. The simplest direct observable is absorption or emission of light, transitions to non- or antibonding states and radiative return to the ground state, which have little or no structural information. Problems inherent in relying on direct observation are evident in medieval accounts of unicorns or mermaids or post-Columbian accounts of Eldorado, all based on "direct" observation. Alas, they represent gross misinterpretations.

Frequently, the general nature or detailed structure of an intermediate is inferred from reaction products. Radical ions are invoked in reactions between electron donors and acceptors in polar solvents. Probing the fate of chirality, stereochemistry, or an isotopic label or substituent in the products of a chemical transformation exemplifies the classical approach to mechanism. Of course, this approach is not without shortcomings.

4. RADICAL ION STRUCTURES

Because radical ions are formed by removing an electron from the HOMO or adding an electron to the LUMO of a substrate, their singly occupied molecular orbitals (SOMOs) and their structures are determined to some extent by the nature of the HOMO or LUMO of the parent molecule. The relationship between parent and radical ion structure has been probed particularly for cations. Electron ejection in the gas phase (in PES spectroscopy) proceeds vertically, generating radical cations with the geometry of the parent. Some cations maintain structures closely related to the neutral parent even on longer time scales; others relax to lower-energy structures with changes in bond lengths and bond and dihedral angles. Particularly strained ring compounds may undergo major structural changes upon ionization. In the extreme case, a bond may break, giving rise to radical ions whose spin and charge are localized on two different atoms (or sections) of the molecule; such species are called distonic radical ions. Possible relationships between the geometries of radical cations and their precursors include the purely vertical relationship, minor changes (relaxation), or major changes in geometry (rearrangement, Fig. 6.6).

Depending on the HOMO or LUMO involved in the redox reaction, we differentiate between π, n, or σ donors and π or σ acceptors, respectively; in addition,

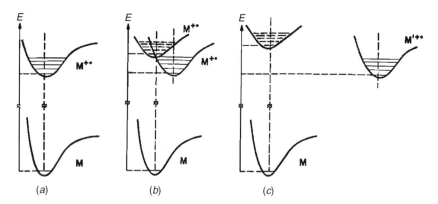

Figure 6.6. Schematic comparison between potential surfaces of radical cations and their precursors. (*a*) Vertical ionization without relaxation. (*b*) Vertical ionization followed by relaxation. (*c*) Vertical ionization followed by rearrangement.

some substrates contain a combination of functions. These families show different degrees of structural changes upon ET from or to the parent molecule. We will discuss selected radical ion structure types derived from the different families of donors–acceptors to illustrate their rich variety; we will emphasize the molecular features that determine these structures.

4.1. Radical Ions of π Donors

Among the radical ions of π donors, those of aromatic hydrocarbons were the first class to be investigated in detail, because they are reasonably stable and their spectra fall into a readily accessible range, allowing them to be characterized by ESR[83,84] and optical spectroscopy.[58] Their structures are closely related to their parents; this manifests itself in an interesting simple relationship between the PE spectra of ground-state parent molecules and the electronic spectra of their radical cations: The excitation energies, ΔE, of radical cations correspond approximately to differences in the ionization energies, ΔI, of the parent molecules (cf. Fig. 6.5).[85,86]

$$\Delta I_\mu = -\Delta E_\mu \qquad (15)$$

where μ are the indexes of MOs counting down from the HOMO and the index of the PE band, starting at the lowest ionization energy. The success of this simple relation is due to the fortuitous cancellation of errors inherent in the model.[86]

Of course, a close structural relationship between radical cations and parent molecules is not likely to hold generally, but it is a fair approximation for alternant hydrocarbons. Deviations have been noted: some stilbene radical cations have higher-lying excited states without precedent in the PE spectrum of the parent;[87,88] for radical cations of cross-conjugated systems (e.g., **1**) already the first excited state is without such precedent.[87,89] These states have been called "non-Koopmanns" states. Alkenes also feature major differences between parent and radical cation electronic structures.

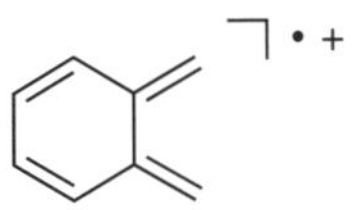

1·+

Of the radical ions derived from aromatic hydrocarbons, we mention the ions of benzene and tetracene. For benzene, both the positive and negative ions have been characterized by ESR spectroscopy. The radical anion shows seven evenly spaced lines ($a_H = 0.341$ mT),[90,91] suggesting that the spin density is distributed evenly over the six carbons (or that there is a fast equilibrium between structures corresponding to the two degenerate antibonding SOMOs; Fig. 6.7). Introducing a single D is sufficient to disturb the equilibrium ($a_H = 0.3983$ mT, 4H; $a_H = 0.3454$ mT,

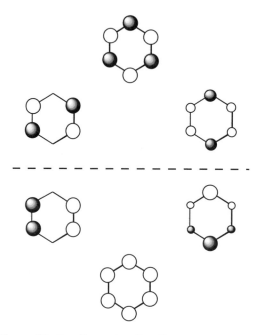

Figure 6.7. Bonding and antibonding MOs of benzene.

1H; $a_D = 0.056$ mT); D shows some preference for a position in a nodal plane.[92] This trend is more pronounced for a methyl group, which shows a significant preference for the nodal plane ($a_H = 0.5553$ mT, H-2,6; $a_H = 0.5687$ mT, H-3,5; $a_H = 0.0129$ mT, H-4; $a_H = 0.02$ mT, 3H).[93]

The benzene radical cation ($g = 2.00242$) also shows seven evenly spaced lines ($a_H = 0.444$ mT);[94] indicating that the spin density is distributed evenly over the six carbons or that a fast equilibrium between structures corresponding to two degenerate bonding SOMOs contribute equally; Fig. 6.7). The coupling constants for the cation are slightly larger than for the anion; apparently, the nuclei interact more efficiently with an unpaired electron in a bonding than in an antibonding SOMO. A methyl group affects the hyperfine coupling pattern differently ($a_H = 1.8$ mT, Me; $a_H = 0.21$ mT, H-2,6; $a_H = 1.18$ mT, H-4): for the cation the methyl group prefers a position of high spin density rather than in the nodal plane.[95]

The radical ions of tetracene have been studied by ESR and optical spectroscopy. The ESR data reveal the spin density distribution; again, the splittings in the anion ($a_H = 0.143$ mT, H-1,4,7,10; $a_H = 0.116$ mT, H-2,3,8,9; $a_H = 0.416$ mT, H-5,6,11,12)[96] are overall smaller than those of the cation ($a_H = 0.168$ mT, H-1,4,7,10; $a_H = 0.102$ mT, H-2,3,8,9; $a_D = 0.501$ mT, H-5,6,11,12),[97] indicating that the unpaired spin in the antibonding SOMO interacts less efficiently than in the bonding SOMO. The spin densities are higher in the center and fall off somewhat toward the outer rings. The optical spectra of the anion and the cation are remarkably similar (Fig. 6.8). Such similarity is to be expected, because HOMO and

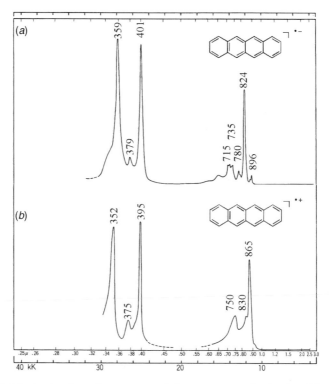

Figure 6.8. Electronic absorption spectra of the tetracene radical anion (*a*) and radical cation (*b*). [Adapted from Ref. 58.]

LUMO are "paired orbitals" according to a (pairing) theorem derived by McConnell and McLachlan (cf. Fig. 6.5).[98] Adding an electron to or removing an electron from an aromatic hydrocarbon occurs with minimal changes in structure.

As mentioned for the relationship between the PE spectrum of a parent molecule and the electronic spectrum of its radical cation, any close correspondence between the electronic spectra of anions and cations or their hyperfine coupling patterns holds only for alternant hydrocarbons. The anions and cations of nonalternant hydrocarbons (e.g., azulene) have significantly different hyperfine patterns. Azulene radical anion has major hyperfine splitting constants (hfcs) on carbons 6, and 4,8 ($a_H = 0.91$ mT, H-6; $a_H = 0.65$ mT, H-4,8; $a_H = 0.38$ mT, H-2);[99] in contrast, the radical cation has major hfcs on carbons 1 and 3 ($a_H = 1.065$ mT, H-1,3; $a_H = 0.152$ mT, H-2; $a_H = 0.415$ mT, H-5,7; $a_H = 0.112$ mT, H-6).[100]

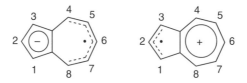

The divergent hyperfine patterns of the two ions can be ascribed to their 6π aromatic substructures, cyclopentadienide anion, and cycloheptatrienylium cation, respectively.

4.2. Radical Ions of *n* Donors

Radical cations of *n* donors are derived typically from substrates containing one or more N, O, or S atoms; they are substituted frequently with alkene or arene moieties. Among these systems, we mention only a few examples, including two radical ions derived from 1,4-diazabicyclo[2.2.2]octane (**2**) and the tricyclic tetraaza compound (**3**). For both ions, ESR as well as OS/PES data were measured. The bicyclic system ($a_N = 1.696$ mT, 2N; $a_H = 0.734$ mT, 12H)[101] shows perfect correspondence between the ΔEs of the radical ion and the ΔI's of its precursor.[102]

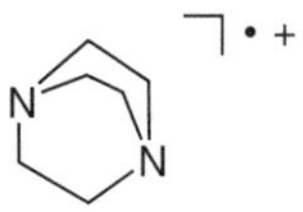

2 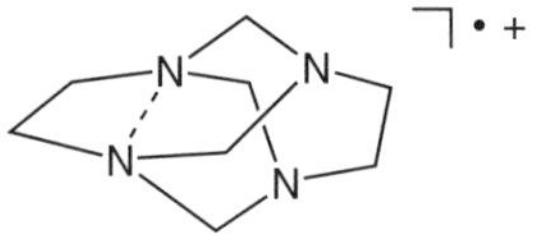 $^{\bullet\,+}$ **3** 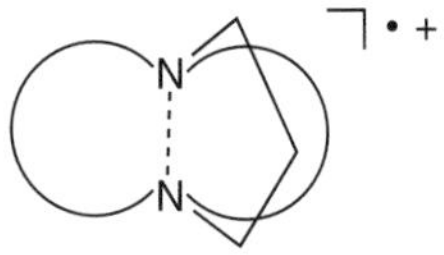 $^{\bullet\,+}$ **4** $^{\bullet\,+}$

In contrast, the radical cation of the tetracyclic system is significantly distorted: The parent system has D_{2d} symmetry and a b_2 HOMO, whereas the radical cation is distorted toward 2 equiv structures of C_{2v} symmetry (^2E), with a two-center three-electron N–N bond (**3**$^{\bullet\,+}$).[103,104] The ESR data ($a_N = 0.709$ mT, 4N; $a_H = 0.768$ mT, 8H, N–C–N; $a_H = 0.414$ mT, 8H, N–C–C)[101] support the rapid interconversion of the two structures. The structure of **3**$^{\bullet\,+}$ is one of many doubly or multiply bridged diaza compounds forming three-electron N–N bonds (e.g., **4**$^{\bullet\,+}$).[105] Many additional examples involving three-electron S–S or I–I bonds are also known.[106] Dioxetane radical cations (e.g., **5**$^{\bullet\,+}$), characterized by ESR spectroscopy as intermediates in oxygenations (cf., Section 5),[107] contain analogous three-electron O–O bonds.

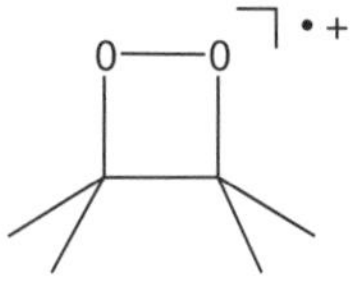

5 $^{\bullet\,+}$

Intermolecular σ-dimer cations are known for thioethers but dimeric amine cation species are not well characterized in solution. However, an example was detected in a zeolite;[108] the confinement in the narrow channels and the restricted diffusion favor interaction between the two entities.

4.3. Radical Ions of σ Donors

Alkanes are among the least reactive classes of compounds; they are poor electron acceptors (low electron affinities) as well as donors (high ionization potentials, viz., CH_4, 12.61 eV; $C_{10}H_{22}$, 9.65 eV).[109] The molecular anions of n-alkanes are especially unstable;[110] negative ion yields for simple alkanes are $\sim 10^4$ times lower than positive ion yields. Electron attachment results in small fragment ions (CH^-, CH_2^-, CH_3^-, C_2H^-).[111–113] n-Alkanes can be ionized by electron (MS) or He_α impact (PES) in the gas phase, but oxidation in solution is difficult. Alkane radical cations have become readily accessible only with the advent of matrix isolation techniques combined with ESR detection.

Of the small alkane radical cations $CH_4^{\bullet+}$ has potential significance as an interstellar species and may have played a role in the chemical evolution preceding the origins of life.[114] In the laboratory, however, $CH_4^{\bullet+}$ has long been elusive; it was finally generated by discharge ionization in a neon matrix.[115] An ESR quintet ($a_H = 5.48$ mT) suggested four equivalent protons, that is, a species with D_{2d} symmetry. However, $CH_2D_2^{\bullet+}$ showed a 1 : 2 : 1 triplet of 1 : 2 : 3 : 2 : 1 quintets ($a_H = 12.17$ mT; $a_D = 0.222$ mT, corresponding to an $a_H = 1.4$ mT); nonequivalent pairs support a species with C_{2v} symmetry. Therefore, C_{2v} symmetry was assigned to $CH_4^{\bullet+}$ and $CH_2D_2^{\bullet+}$; the averaged geometry of $CH_4^{\bullet+}$ was ascribed to dynamic Jahn–Teller (JT) distortion (Fig. 6.9).[115] The C_{2v} symmetry of $CH_4^{\bullet+}$ is the same as that of $BH_4^\bullet$,[116,117] with which it is isoelectronic.

Ab initio calculations of $CH_4^{\bullet+}$ at the UHF/6-31G* level showed two elongated (119.6 pm; $a_H = 13.7$ mT) and two shortened C—H bonds (109.4 pm; $a_H = -1.7$ mT);[118] ^2D nuclei preferentially occupy the shorter bonds (Fig. 6.9).[115] The calculated structure depends critically on the basis set and the method used to account for electron correlation. For example, the D_{2d} structure is the global minimum using UB3LYP/6-31G*.

The ESR spectrum of $C_2H_6^{\bullet+}$ at 4 K in SF_6 shows a 1 : 2 : 1 triplet; two strongly coupled protons ($a_H = 15.25$ mT) support spin densities, $\rho = 0.3$, in two H_{1s} orbitals (σ delocalization); the remaining protons are weakly coupled (≤ 1 mT).[119] The large positive hyperfine coupling constant is reproduced well by B3LYP calculations.[120] The structure of $C_2H_6^{\bullet+}$ resembles diborane, B_2H_6, rather than $B_2H_6^-$, with which it is isoelectronic (Fig. 6.9).[121] At 77 K, the spectrum changed to a septet ($a = 5.04$ mT, $\frac{1}{3}$ of a at 4 K), indicating rapid equilibration of three equivalent distorted forms (dynamic JT effect), with a very small activation energy, $E_a = 250$ cal/mol.[119]

Propane radical cation (C_3H_8) gave rise to three different structures in different matrices.[122] These results are ascribed to the fact that the energies of several high-lying orbitals lie close to each other, so that small perturbations due to the matrix may alter their relative energies. In SF_6, a 1 : 2 : 1 triplet ($A = 9.8$ mT) is observed; the in-plane ^1H nuclei at C1 and C3 are strongly coupled (σ delocalization).[123] Compared to the parent the rotational axes of the Me groups are bent toward the central carbon and the two in-plane C—H bonds are lengthened ($^4b_1$ symmetry).

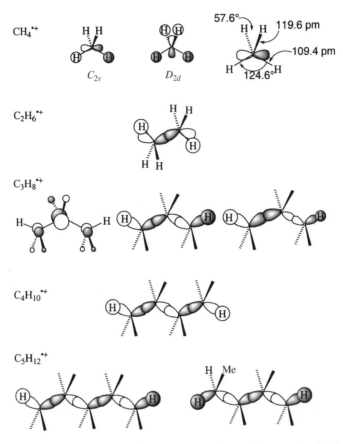

Figure 6.9. Top to bottom: Two possible Jahn–Teller distorted geometries for the methane radical cation and calculated geometry for its energy minimum;[115] schematic representation of the SOMO for ethane radical cation;[119] SOMOs of three different propane radical cations observed by ESR;[122] the SOMO of butane radical cation;[122] and SOMOs for two conformers of pentane radical cation.[122]

In Freon 113 two strong (10.5 mT) and four weaker hyperfine interactions (5.25 mT) were observed, assigned to a delocalized $C_3H_8^{•+}$ species with an antibonding pseudo-π orbital. The ^1H nuclei at C2 are strongly coupled, the protons in the C—C—C plane have negligible hfcs (Fig. 6.9).[122] A third structure type, obtained in C_3F_8, has one strongly coupled ($A = 8.4$ mT) and one moderately coupled proton ($A = 1.8$ mT); it was identified as a species with one lengthened C—C bond.[124,125]

In general, alkane radical cations are good examples of σ-delocalized species. Their SOMOs are σ orbitals spread over extended planar C—C systems and two in-plane ^1H nuclei at the terminal carbons (Fig. 6.9). The ESR spectra show $1 : 2 : 1$ triplets, due to strongly coupled terminal protons.[122,126–128] The splitting decreases from 10.5 mT for $C_3H_8^{•+}$ to 1.0 mT for $C_{10}H_{22}^{•+}$, because the unpaired electron spin is distributed over more and more carbon atoms. The hyperfine coupling constants of the inner protons and the out-of-plane protons of the terminal methyl groups are small.[123,129]

The actual ESR spectra are more complex because gauche conformers are also present.[122,130–134] In these species, the spin delocalization ceases at the gauche carbon, the unpaired electron is confined mostly to the longer fragment, and the in-plane ^1H nuclei are strongly coupled.[122,131] For example, *n*-pentane exists as an *s-trans,trans,trans-* and an *s-trans,trans,gauche*-conformer (Fig. 6.9). The extended radical cations of *n*-hexane ($A = 3.9$ mT) and *n*-octane ($A = 2.9$ mT) were observed exclusively inside pentasil zeolite (ZSM-5)[135] enforced by the geometry of the zeolite channels.[119]

4.4. Radical Ions of Strained Ring Compounds

Pronounced differences between radical cation structures and their parents must be expected for strained ring compounds. The HOMO or LUMO of these systems may be localized mainly in one bond, which may be weakened or actually break upon ionization. The oxidation potentials of strained ring compounds are lower than those of unstrained substrates because strain energy is released, resulting in noticeable changes in structure.

The potential energy surfaces of radical cations may differ from those of their neutral diamagnetic parents in three features: (1) reduced reaction barriers; (2) reduced or reversed free energy differences between isomers; and (3) energy minima on the radical cation surface may have geometries corresponding to transition structures on the parent potential surface. We will discuss a range of structures, focusing on the unusual ones.

Cyclopropane (D_{3h} symmetry) has a degenerate pair of in-plane e' orbitals (S, A). Vertical ionization leads to a doubly degenerate $^2E'$ state, and JT distortion results in two nondegenerate electronic states, $^2A_1$ and $^2B_2$ (C_{2v} symmetry),[136] corresponding to two different molecular structures. The $^2A_1$ state (orbital S singly occupied) corresponds to a structure with one lengthened C—C bond; it is lowest in energy for many cyclopropane radical cations (Fig. 6.10).

An ESR spectrum at 4.2 K shows two strongly coupled ^1H nuclei ($a_H = 2.04$ mT, β protons) and four less strongly coupled ones ($a_H = -1.17$ mT, α protons);[137] at 77 K, a single line was observed.[138,139] The low temperature spectrum supports the $^2A_1$ structure; at higher temperature, three equivalent $^2A_1$ structures are averaged by dynamic JT distortion. The $^2B_2$ structures are transition states; a_β (2.04 mT) and a_α (−1.17 mT) fortuitously average to $a_{avg} \sim 0$. A ring-opened structure, $7^{\bullet+}$, was observed in CF$_2$Cl-CFCl$_2$ matrices.[140]

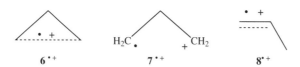

$6^{\bullet+}$ H$_2$C   CH$_2$ $7^{\bullet+}$ $8^{\bullet+}$

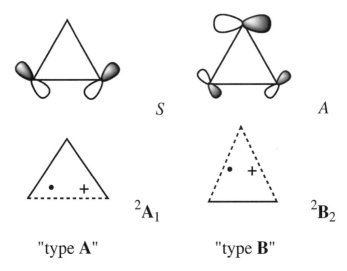

Figure 6.10. The degenerate pair of cyclopropane HOMOs, S and A (top) and schematic radical cation structures, type **A** and type **B**, of states $^2A_1$ and $^2B_2$, respectively, generated by removing an electron from one of these orbitals (bottom).

Ab initio calculations (UHF/6-31G*//MP2/6-31G*) on the $C_3H_6^{\bullet+}$ potential surface confirm the $^2A_1$ structure, but fail to support the existence of $7^{\bullet+}$.[141,142] The conversion of $6^{\bullet+}$ ($^2A_1$) to propene radical cation, $8^{\bullet+}$ ($\sim$10 kcal/mol below $6^{\bullet+}$), has a barrier of $\sim$30 kcal/mol (UMP2/6-31G*),[143] approximately one-half that measured for cyclopropane itself.[144] In contrast, $7^{\bullet+}$ rearranges to $8^{\bullet+}$ without a chemically significant barrier.[142] In the light of these results the observation of $7^{\bullet+}$ at cryogenic temperatures[140] must be ascribed to matrix forces (cf. the propane radical cations).

Because radical ions of structure type **A** are well established,[140,145,146] examples of an alternative structure are of special interest. Substitution at two carbons should stabilize the S orbital ($^2A_1$), whereas substitution at a single carbon should stabilize the A HOMO ($^2B_2$) by conjugation, homoconjugation, or hyperconjugation.[147]

According to ab initio calculations on methyl- and 1,1-dimethylcyclopropane,[148] hyperconjugation falls short of stabilizing type **B** structures; they are transition states, undergoing second-order JT-type distortions to unsymmetrical structures with one very long C—C bond. The hyperfine coupling constants ($a_2 = -1.43$ mT; $a_3 = 1.98$ mT; B3LYP/6-31G*//MP2/6-31G* calculations) support a structure of type **A**. However, conjugation with a vinyl[149] or phenyl group,[150,151] or homoconjugation in norcaradiene and derivatives[152] are sufficient to stabilize type **B** structures.

Phenylcyclopropane radical cation ($9^{\bullet+}$) has divergent hyperfine coupling constants for the secondary cyclopropane protons ($a_{\beta trans} = 0.78$ mT; $a_{\beta cis} = -0.07$ mT; CIDNP, B3LYP/6-31G* calculations), apparently because the cis protons are located in a nodal plane.[150] Similarly, vinylcyclopropane radical cation is stabilized by conjugation; it has two conformers, s-anti-$10^{\bullet+}$ and s-syn-$10^{\bullet+}$; both are type **B** structures, resembling a π complex between vinylmethylene and ethene.[149] Interestingly, the array, Cβ–Cα–C1, has been converted, in essence, to an allyl moiety.

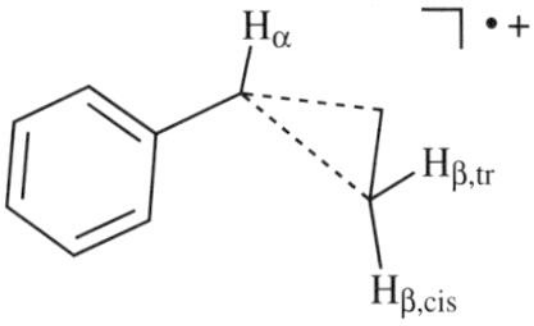

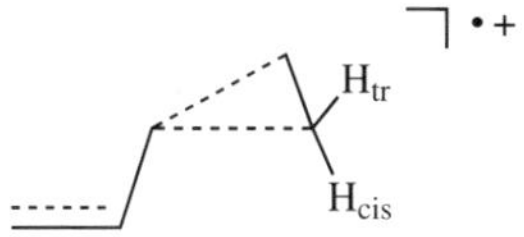

For radical cations of norcaradiene and derivatives,[152] the interaction of the cyclopropane in-plane e' orbitals with the butadiene frontier MO favors the type **B** structure. The assignments are based on ab-initio calculations, CIDNP results, and the ET photochemistry. The norcaradiene radical cation ($11a^{\bullet+}$) has a $^2A''$ electronic ground state (C_s symmetry). The C1–C6 bond is shortened on ionization (-3.4 pm) while the lateral bonds are lengthened ($+2.8$ pm). The delocalization of spin density to C7 ($\rho_7 = 0.246$; $\rho_{2,5} = 0.359$) and the hyperfine coupling constants of the cyclopropane moiety ($a_{1,6} = 1.36$ mT; $a_{7syn} = -0.057$ mT; $a_{7anti} = -0.063$ mT) support a type **B** structure.[152]

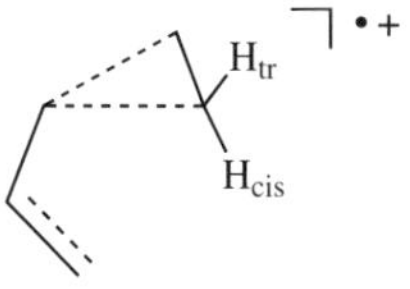

The assignment of an antisymmetrical cyclopropane SOMO to the radical cation ($12^{\bullet+}$) of benzobicyclo[4.1.0]hepta-2,4-diene rests on CIDNP effects, particularly on characteristic differences to those for cis-1,2-diphenylcyclopropane (Fig. 6.11).[145] The calculated carbon spin densities of $12^{\bullet+}$ on C2 ($\rho_2 = 0.355$), C5 ($\rho_5 = 0.153$), and C7 ($\rho_7 = 0.149$) and negative spin densities on the tertiary cyclopropane carbons ($\rho_1 = -0.009$, $\rho_6 = -0.007$)[152] are in qualitative agreement with the CIDNP effects.[146]

For tricyclo[4.3.1.0^{1,6}]deca-2,4-diene radical cation ($11b^{\bullet+}$, the spin density on the bridge carbon ($\rho_{10} = 0.203$, $\rho_{2,5} = 0.383$; $a_{10syn} = -0.54$ mT; $a_{10anti} = -0.48$ mT) support a type **B** structure (UHF/6-31G*; C_s symmetry imposed) and a noticeable contribution due to homoconjugation.[152]

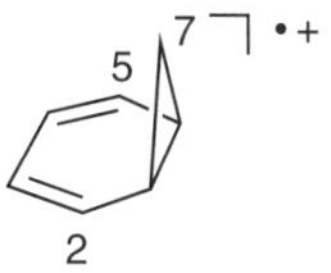

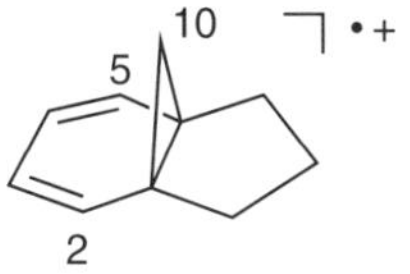

The case of cyclobutane radical cation presents another interesting structural problem. The parent has a puckered ring with D_{2d} symmetry; ET from one of its e orbitals leads to a JT unstable radical cation, which distorts to structures of D_{2d} and C_{2v} symmetry.[153] The ESR spectrum at 4 K ($a_H = 4.9$ mT, 2H; $a_H = 1.4$ mT,

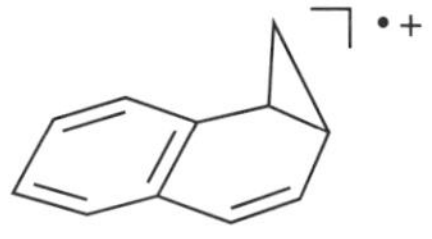

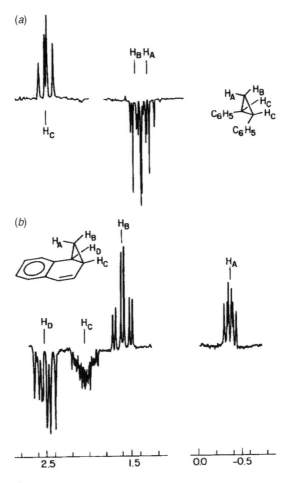

Figure 6.11. The ^{1}H CIDNP spectra (cyclopropane resonances only) observed during the photoreaction of chloranil with *cis*-1,2-diphenylcyclopropane (*a*) and benzonorcaradiene (*b*). The opposite signal directions observed for analogous protons provide evidence that the two radical cations belong to two different structure types.[145,146]

2 H; $a_H = 0.5$ mT, 4H) suggests an intermediate with C_{2v} symmetry (Fig. 6.12, type **D**).[154] At temperatures >77 K, a nearly isotropic nine-line spectrum ($a_H = 1.33$ mT) was observed, supporting a dynamic JT effect averaging all eight ^{1}H nuclei.

Four distorted minima were probed by calculations (Fig. 6.12)[155,156] rectangular (type **A**) and rhomboidal structures (type **B**), resulting from first-order JT distortion; and trapezoidal (type **C**) and irregular structures (type **D**) due to second-order JT distortion. A flexible rhombic structure emerged as most stable (QCISD-(T)/ 6-31G*//UMP2/6-31G*).[155] A recent assignment that the rhombic structure (4 equiv bonds, 157.3 pm) is a (very low-lying) transition structure between two parallelograms (2 pairs of bonds, 149.5 and 169.5 pm),[156] shows poor agreement between

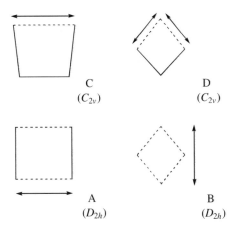

Figure 6.12. Possible structure types of cyclobutane radical cations.

the calculated splitting constants (a = 2.09, 0.29 mT) and the experimental ones (a = 4.9, 1.4 mT). This finding is unsettling, as calculated hyperfine couplings for many strained-ring systems typically agree well with experiment; a closer reproduction of experimental data by calculations appears desirable.

Cyclobutanes disubstituted in the 1,2-positions should favor structure-type C or a related distonic structure with one broken C–C bond. Calculations [QCISD-(T)/ 6-31G*//UMP2/6-31G*] suggest a trapezoidal structure for *trans*-1,2-dimethyl-cyclobutane radical cation.[155] This expectation is born out by experimental results such as the ET induced cis–trans-isomerization of 1,2-diaryloxycyclobutane (Ar = aryl), leading to **13**[.**+**], and likely involving the distonic radical cation (**14**[**•+**]) formed via a type C ion.[157]

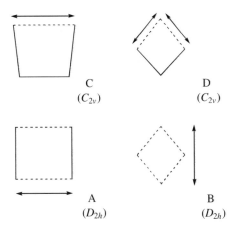

No discussion about strained-ring radical cations would be complete without the valence isomers quadricyclane (**15**[**•+**]) and norbornadiene, (**16**[**•+**]); **15** features two adjacent rigidly held cyclopropane rings, whereas **16** contains two ethene π systems well suited to probe through-space interactions.[158,159] Molecular orbital considerations suggest the antisymmetric combination of the ethene π orbitals (**16**) or cyclopropane Walsh orbitals (**15**) as respective HOMOs of the two parent molecules. The radical ions have different state symmetries and their SOMOs have different orbital symmetries.

The structures of the ions rest on CIDNP spectra delineating their hyperfine patterns,[160,161] ab initio calculations[162–164] ESR and ENDOR data[165] for **16**$^{\bullet+}$, and TR–ESR results for **15**$^{\bullet+}$.[166] Ab initio calculations (B3LYP/6-31G*//UMP2/6-31G*) reproduce positive and negative hyperfine coupling constants[164] satisfactorily. Each radical cation is related uniquely to the geometry of one of the precursors (Fig. 6.13).

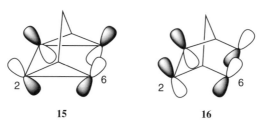

The unpaired spin density for either species resides on 4 equivalent carbons; the ^{1}H nuclei (H_α) at these centers have large negative hfcs due to π,σ spin polarization. The bridgehead protons (H_β) are quite different: a large positive hfc for **15**$^{\bullet+}$ is ascribed to hyperconjugation (π,σ spin delocalization); the very weak negative hfcs of **16**$^{\bullet+}$ is due to "residual" π,σ polarization; the hyperconjugative interaction is inefficient because the β protons lie in the nodal plane of the SOMO.[166] Large positive hfcs for the bridge protons (γ-H) of **16**$^{\bullet+}$ and a large negative hfc for the γ-H of **15**$^{\bullet+}$ are noted; an explanation would go beyond the scope of this chapter.

The unique bonding in bicyclobutane (**17**) and its unusual chemistry caused an early interest in its radical cation. The structure of **17**$^{\bullet+}$ rests on ESR/ENDOR (electron–nuclear double resonance) spectra ($a_{ax} = 7.7$ mT, 2H; $a_{eq} = 1.14$ mT, 4H).[167] The bridgehead carbons bear spin density and the transannular bond is lengthened (178.6 pm, MNDO; 174.3 pm, B3LYP/6-31G*//MP2/6-31G*) and the flap angle is increased (132°, MNDO; 133.6°, B3LYP/6-31G*//MP2/6-31G*).[167] The major hyperfine splitting was assigned to the axial protons (H_{ax}); the large difference between a_{ax} and a_{eq} supports a puckered geometry; no inversion occurs up to 160 K.

H   H

H_{eq} H_{eq}

H_{ax} H_{ax}

17$^{\bullet+}$ **18**$^{\bullet+}$

H_γ

H_β

H_α

Benzvalene (**18**) is a tricyclic benzene isomer containing a bicyclobutane ring system bridged by an ethylene moiety; its radical cation is accessible by PET or radiolysis. CIDNP indicated negative hfcs for the alkene protons (H_α), strong positive hfcs for the non-allylic bridgehead protons (H_γ), and negligible hfcs for the allylic bridgehead protons (H_β).[168] Accordingly, the spin and charge of **18**$^{\bullet+}$ are essentially localized in the alkene moiety, with efficient spin delocalization onto

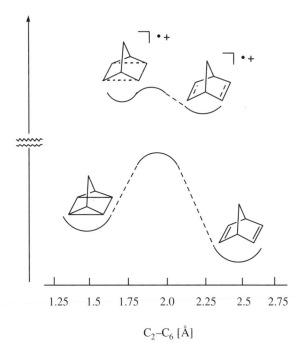

C_2–C_6 [Å]

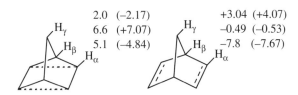

Figure 6.13. Schematic energy diagram for quadricyclane (**15**) and norbornadiene (**16**) and their radical cations. The respective minima on the two surfaces are in a unique relationship with characteristic changes in bond lengths and angles. Experimental[165,166] and calculated hyperfine coupling constants (in parentheses; G; B3LYP/6-31G*//MP2/6-31G*)[164] are shown below.

the non-allylic bridgehead ^{1}H nuclei. The assignment was confirmed by ESR/ ENDOR ($a = +2.79$ mT; H$_\alpha$; $a = -0.835$ mT; H$_\gamma$). The splitting of the allylic β protons ($a_\beta = -0.158$ mT; ENDOR)[169] reflects their position in the nodal plane of the π system.

These assignments are consistent with PES data[170] and supported by theoretical calculations;[164,170] in C_{2v} symmetry, **18**$^{\bullet+}$ has two low-lying radical cationic states, ^{2}**B**$_1$ and ^{2}**A**$_1$.[164] The ^{2}**B**$_1$ state is the ground state of **18**$^{\bullet+}$; the calculated hyperfine coupling constants (B3LYP/6-31G*//MP2/6-31G*) are compatible with chemically induced dynamic nuclear polarization (CIDNP) and ESR/ENDOR results. No spin density is found on the γ carbons; the transannular C—C bond is slightly shortened relative to the parent molecule.

The radical cation (**19**[•+]) of the strained bicyclo[2.1.0]pentane also has a puckered conformation, supported by one strongly coupled (flagpole) proton (a_{syn} = 4.49 mT).[171] Ab initio calculations indicate that the transannular bond retains some bonding and that the bridgehead carbons remain pyramidal.[164]

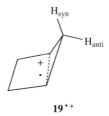

19[•+]

4.5. Radical Ions of 1,5-Hexadiene Systems

Radical cations derived from 1,5-hexadiene systems illustrate major differences between the potential surfaces of radical cations and neutral precursors. On the precursor potential surface, the states of intermediate geometry are saddle points (transition structures), but pronounced minima (Fig. 6.14) on the radical cation potential surface.

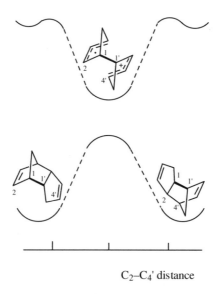

C_2–C_4' distance

Figure 6.14. Schematic cross-section through the potential surfaces of dicyclopentadiene (**20**) and its radical cation (**20**[•+]). The energy minimum on the radical cation surface corresponds to a saddle point (or a shallow minimum) on the potential surface of the precursor.

Dicyclopentadiene forms a radical cation (**20**$^{\bullet+}$) in which one of the bonds linking the monomer units is cleaved. The species contains two allyl moieties attached to a C_4 "spacer". Structure **20**$^{\bullet+}$ rests on an unmistakable CIDNP pattern[172–174] and is supported by an analysis of the electronic absorption spectrum.[175] The large energy gap in the OS of this ion ($\Delta E = 1.67$ eV) is incompatible with the photoelectron spectrum of the parent molecule ($\Delta E = 0.15$ eV), but it fits the ring-opened structure **20**$^{\bullet+}$.

Radical cations (**21**$^{\bullet+}$) derived from semibullvalene[176,177] or barbaralane[178] belong to a different structure type. The ESR spectrum of **21**$^{\bullet+}$ ($a = 3.62$ mT, 2H; $a = 0.77$ mT, 4H)[176] and strong CIDNP effects[177] support a structure in which two allylic moieties are held at a much closer "nonbonding" C–C distance ($2.2 - 2.3$ Å) than for **20**$^{\bullet+}$. At this distance, an interaction of the allyl moieties cannot be excluded.

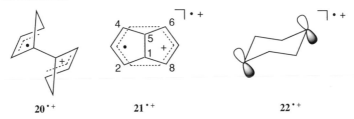

20$^{\bullet+}$	**21**$^{\bullet+}$	**22**$^{\bullet+}$

The third radical cation structure type is the cyclohexane-1,4-diyl radical cation (**22**$^{\bullet+}$) derived from 1,5-hexadiene. The free electron spin is shared between two carbons, which may explain the blue color of the species ("charge" resonance). Four axial β and two α protons are strongly coupled ($a = 1.19$ mT, 6H).[179,180]

4.6. Bifunctional or Distonic Radical Ions

At least one of the hexadiene radical cations (**20**$^{\bullet+}$) can be viewed as a species containing spin and charge in two separate (though equivalent) molecular fragments. This separation can be enforced in systems of lower symmetry; in this context, we mention two radical cations derived from 1,1-diaryl-2-methylenecyclopropane (**23**), and 6-methoxy-1,2,3,4,6-pentamethyl-5-methylenebicyclo [2.2.0]hex-2-ene (**25**).

The PET reaction of **23** results in the rapid equilibration of the *exo*-methylene and the secondary cyclopropane carbons. These findings were explained via a ring-opened trimethylenemethane radical cation (**24**$^{\bullet+}$). The rearrangement requires that, upon back ET, the diarylmethylene group couple with one of the allyl termini.[181] The CIDNP results indicate that the electron spin is localized in the allyl π system; the lack of polarization for the aromatic rings suggests that they are arranged orthogonal to the allyl group.[182]

23	**24**$^{\bullet+}$

Ar = aryl

6-Methoxy-1,2,3,4,6-pentamethyl-5-methylenebicyclo[2.2.0]hex-2-ene (**25**) gives rise to radical cation **26**[•+], containing a strained cyclobutenyl radical coupled to a methoxyallyl cation.[183] Resonance electron donation from the methoxy group stabilizes the charge in the allylic moiety. The significant reorganization upon oxidation is best illustrated by CIDNP effects of two pairs of methyl groups in the 1- and 4- and 2- and 3-positions, magnetically nearly indistinguishable in the parent molecule, but dramatically different in the radical cation.[183]

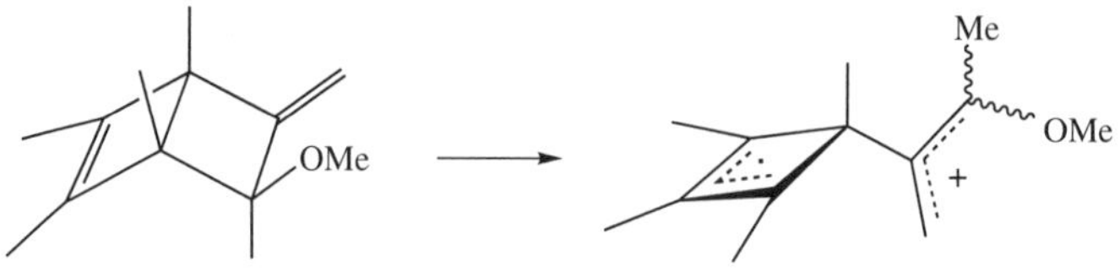

25   26·+

The bifunctional radical ions **24**[•+] and **26**[•+] are two examples of a family of species, commonly called "distonic," which enjoy significant current interest. Much of the research in distonic radical ions is being carried out in the gas phase, typically in tandem mass spectrometers with positive ions as the most frequent targets. However, distonic radical anions have also been studied, and some species are accessible in fluid solution[182,183] or in frozen matrices.

The term distonic was originally coined for species formed by a hydrogen shift in a vertical radical cation,[184,185] then redefined to designate radical cations derived formally by removing an electron from ylids, zwitterions, or diradicals.[186] It is important that the unpaired spin and the charge of these radical ions are localized in different functions. This restriction eliminates species, such as semiquinone and semidione radical anions, alkene radical cations, and a series of radical cations derived from strained ring systems, including the cyclohexanediyl radical cation **22**[•+] discussed above. This limitation favors species whose spin or charge are localized in a σ orbital. Alternatively, distonic radical ions can be designed with two functions in meta-disubstituted benzene systems that also will prevent mutual delocalization.

Not surprisingly, bifunctional species preceded the definition of the term distonic, both in the gas and condensed phases. For example, Hammond et al. studied the cage effect for radical pairs generated by decomposition of azo compounds. Among their targets was the doubly protonated amidine (**27**), whose decomposition yielded a pair of distonic radical cations (**28**[•+]).[187] Attachment of the spin-bearing carbon in the 2-position of the diazaallyl function ensures minimal delocalization of unpaired spin into the latter.

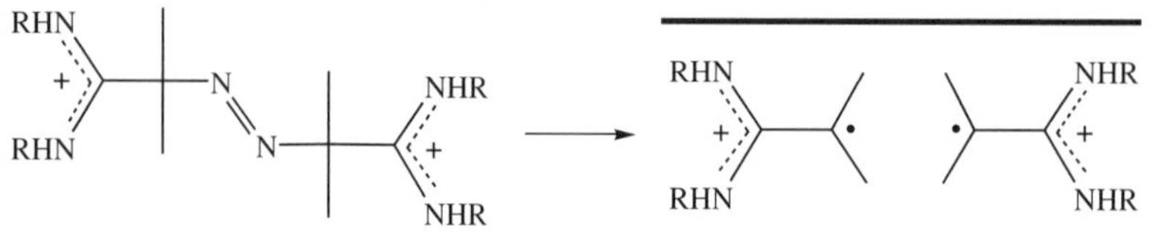

27   28·+

Diphenylfulvene is very readily converted to the distonic radical anion ($29^{\bullet -}$). The aromatic character of the 6-π-cyclopentadienide ring explains the stability of the distonic species.[188]

$29^{\bullet -}$

Several gas-phase reactions were known to involve bifunctional radical cations preceding the definition of the term distonic; we mention only the ring-opened oxirane radical cation, which will be discussed presently.[189]

Among simple hydrocarbon ions, the distonic trimethylene radical cation $\bullet CH_2CH_2CH_2^+$ ($7^{\bullet +}$) is accessible upon internal excitation of the cyclopropane molecular ion $6^{\bullet +}$,[190] or by loss of formaldehyde from the tetrahydrofuran (THF) molecular ion.[191] The gas-phase reactivity is clearly different from that of its isomers $6^{\bullet +}$ and $8^{\bullet +}$.[191]

The distonic radical cation ($14^{\bullet +}$) stabilized by delocalization of spin and charge into one aryl group each, was discussed above as a potential intermediate in the geometric isomerization of 1,2-diaryloxycyclobutane (**13**).[157]

Replacing a methylene group of cyclopropane by oxygen changes the system considerably. The PES data identify an oxygen n orbital (b_1), and a ring orbital (a_1), as the highest lying MOs.[192] The ET from b_1 would form species **30** ($^2B_1$), with spin and charge localized on oxygen whereas ET from a_1 would generate radical cation **31** ($^2A_1$), having spin and charge on the carbon atoms. Ring opening and rotation of one methylene group would form an asymmetric species ($32^{\bullet +}$) with spin and charge in two separate fragments; rotation of both methylene groups would generate a resonance stabilized oxallyl species ($33^{\bullet +}$), which has the positive charge on oxygen, where the SOMO has a node.

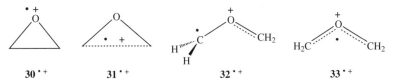

$30^{\bullet +}$ $31^{\bullet +}$ $32^{\bullet +}$ $33^{\bullet +}$

The radical ion observed in cryogenic matrices has a g-factor (2.0022–2.0024)[193] and β-hfcs ($a = 1.62$ mT) of a magnitude incompatible with an oxygen-centered radical ($30^{\bullet +}$), but is compatible with either the ring-closed $31^{\bullet +}$ or the oxallyl structure $33^{\bullet +}$.[193,194] A comparison between the electronic absorption[195] and ESR spectra[196] of the parent oxirane radical cation and those of tetramethyl-oxirane and 9,10-octalin oxide radical cations leave little doubt that the ions of simple oxiranes have ring-opened oxallyl structures (type **33**). Octalin oxide radical cation has a ring-closed structure of type **31**, which can be rationalized as a manifestation of Bredt's rule.[197] The subtle question remains whether the ring-opened radical cations have single-minimum or double-minimum potential surfaces.[198] The

ESR spectra in at least one matrix provide evidence for a "localized" ring-opened structure.[199]

In the gas phase, the distonic ion is formed by excitation of the cyclic ion,[189] by loss of formaldehyde from the 1,3-dioxolane molecular ion,[200] or from the ethylene carbonate molecular ion, by loss of CO_2 followed by reorganization. The ring-opened $C_2H_4O^{\bullet+}$ cation reacts with various neutral reagents;[201] details go beyond the scope of this chapter.

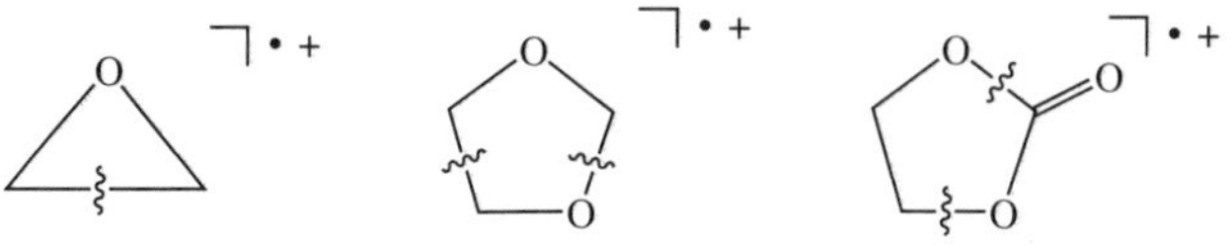

Oxetane gives rise to two radical cations depending on the medium and the reaction conditions. In frozen matrices, oxygen lone-pair ionization generates a π species (**34**[$^{\bullet+}$]) with the unpaired electron localized on the oxygen; **34**[$^{\bullet+}$] has a highly anisotropic and positively shifted g-factor ($g_{\parallel} = 2.0046$, $g_{\perp} = 2.0135$) and the four β protons are strongly coupled ($a_\beta = 6.6$ mT).[193,194] The ring-opened oxetane ion (**35**[$^{\bullet+}$]) can be generated in the gas phase by loss of formaldehyde from the 1,4-dioxane molecular ion.[202] Ion **35**[$^{\bullet+}$] reacts with a wide range of nucleophiles by transfer of $C_2H_4^{\bullet+}$; these reactions may be viewed as S_N2 substitutions with replacement of formaldehyde.[201]

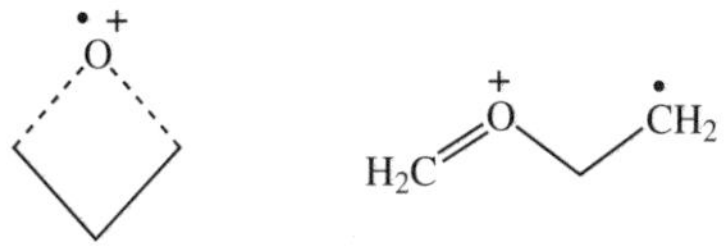

34[$^{\bullet+}$]   **35**[$^{\bullet+}$]

Distonic ions reacting primarily at the radical center provide the opportunity to study free radical reactions in the gas phase by MS. For example, 4-dehydroanilinium ions (**37**[$^{\bullet+}$], $m/z = 93$) were generated by multiple low-energy collisions of chloro-, bromo-, or iodoanilinium ions (**36**[$^+$], $m/z = 220$) with argon. This distonic species reacted with dimethyl sulfide by abstraction of thiomethyl (MeS•), that is, as a reactive radical with an inert charge site. However, **37**[$^{\bullet+}$] can be deprotonated by pyridine.[203,204]

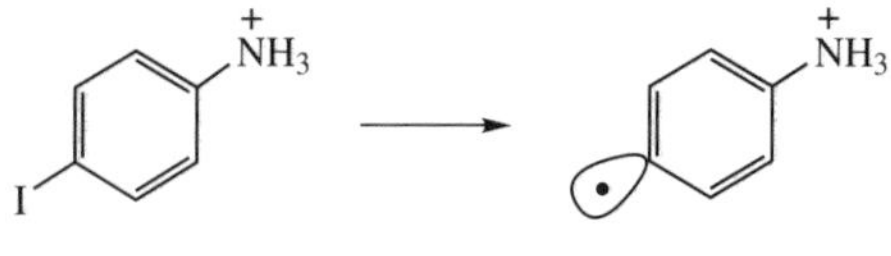

36[$^+$]   **37**[$^{\bullet+}$]

The distonic radical anions of o-, m-, and p-benzyne were crucial intermediates in an elegant determination of the S,T splitting of the corresponding benzynes. The ions are accessible by well-established (routine) gas-phase reactions: o-benzyne anion **38** was prepared from benzene and $O^{\bullet-}$ with loss of $H_2^{\bullet+}$, while the meta and para isomers (e.g., **39**) were prepared from 3- and 4-(bis-trimethylsilyl)benzene reacted, successively, with fluoride ion and molecular fluorine. Negative ion PES of these ions showed two series of transitions, characteristic for ionizations forming singlet and triplet benzynes. The differences in the ionization energies reveal the S,T splitting of the corresponding benzynes.[205,206]

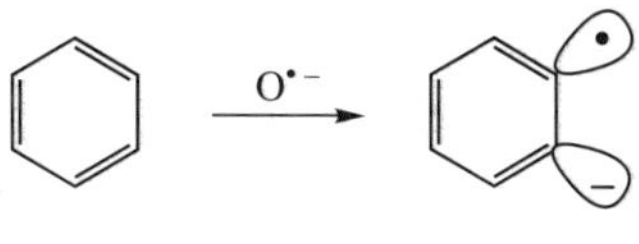

38 $^{\bullet-}$

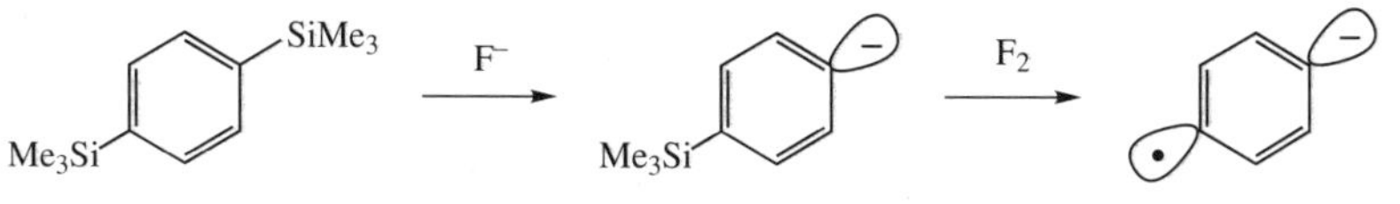

39 $^{\bullet-}$

Benzene derivatives such as *m*-methylanisole (**40**) can be converted to distonic carbene ions. Reaction of **40** with $O^{\bullet-}$ occurs with loss of $H_2^{\bullet+}$, generating the "conventional" carbene anion **41**$^{\bullet\bullet-}$; this anion reacts with molecular fluorine by dissociative ET, followed by nucleophilic attack of F^- on the methyl group, forming **42**$^{\bullet\bullet-}$. In contrast to phenylmethylene, **42**$^{\bullet\bullet-}$ has a singlet ground state; however, upon protonation it gives rise to the triplet state of *m*-hydroxyphenyl-methylene. This interesting reaction can be viewed as a spin-forbidden proton-transfer reaction.[207]

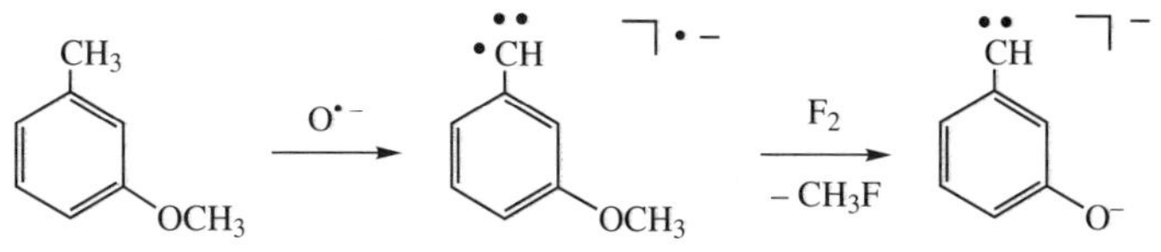

40   **41**$^{\bullet\bullet-}$ **42**$^{\bullet\bullet-}$

A distonic diradical cation (**44**$^{\bullet\bullet+}$) of *m*-benzyne was investigated as a model for benzyne reactivity. The species was prepared via ipso substitution of one bromine atom in the 1,3,5-tribromobenzene molecular ion (**43**$^{\bullet+}$) by 3-fluoropyridine and collision-activated or photoinduced dissociation of the remaining two bromine atoms.[208]

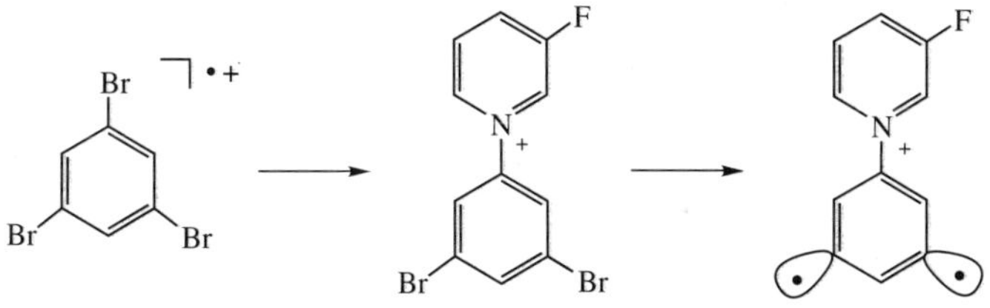

43 $^{\bullet+}$ **44**$^{\bullet\bullet+}$

The selection of distonic radical ions covered here is far from complete. Many additional distonic species have been characterized in frozen glasses, in solution, and in the gas phase. Alas, a more detailed coverage would go beyond the scope of this chapter.

5. RADICAL CATION REACTIONS: RELATIONSHIPS WITH OTHER INTERMEDIATES

Organic radical ions undergo a wide range of uni- and bimolecular reactions. While unimolecular reactions are limited to reorganization and bond cleavage reactions, bimolecular reactions cover a wide range of reaction types (Scheme 6.3). Because radical ions have an unpaired electron and a charge, they can undergo reactions typical of free radicals as well as ions. Radical ions react with: (a) alkenes and arenes (electron or hole transfer, cycloadditions, σ- or π-complex formation); (b) ionic, protic, or polar reagents (protonation, nucleo- or electrophilic capture); (c) free radicals (spin labeling); (d) radical ions of opposite charge (back electron transfer, proton, atom or group transfer, or coupling); and (e) radical ions of like charge (dimerization, disproportionation). These transformations convert radical

REACTIONS OF ORGANIC RADICAL IONS

I.   Unimolecular Reactions
    Electron, hydride, or alkyl Shift
    Bond reorganization
    Bond cleavage
II. Intra-Pair Reactions
    Back electron transfer
    Proton, atom, or group transfer
    Coupling
III. Bimolecular Reactions
    a. With neutral molecules
        Electron transfer
        Addition (to olefins), cycloaddition
        σ- or π- Complex formation
    b. With protic, ionic, or polar reagents
        Protonation/deprotonation
        Nucleo-/electrophilic capture
    c. With Radicals
        Spin labeling
        Coupling
        Oxygenation ($^3O_2$)
    d. Between Radical Ions of like Charge
        Dimerization
        Disproportionation

Scheme 6.3

ions into several classes of intermediates or products, which may retain the unpaired spin(s), the charge(s), spin and charge, or neither.

The competition between the various reactions depends on many factors, including the distance between the ions. Radical ion pairs generated by PET can be contact radical ion pairs (CRIP) or solvent separated radical ion pairs (SSRIP;

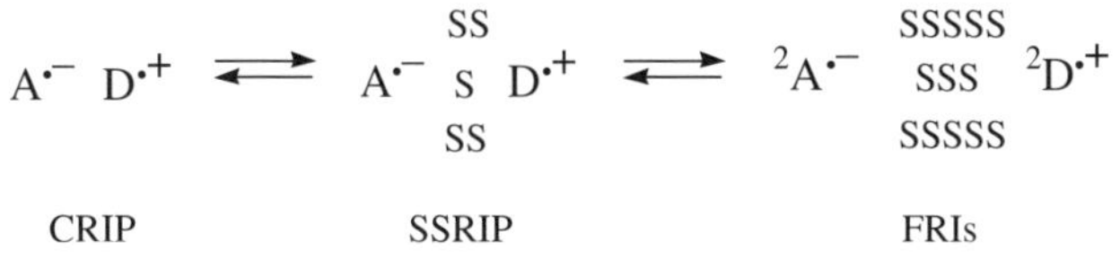

$$ A^{\cdot-} \; D^{\cdot+} \; \rightleftharpoons \; \overset{SS}{\underset{SS}{A^{\cdot-} \; S \; D^{\cdot+}}} \; \rightleftharpoons \; {}^{2}A^{\cdot-} \; \overset{SSSSS}{\underset{SSSSS}{SSS}} \; {}^{2}D^{\cdot+} $$

| CRIP | SSRIP | FRIs |

Scheme 6.4

Scheme 6.4, S denotes solvent molecules). Although they may diffuse apart generating "free" radical ions (FRI), reactions within the geminate pair can be quite fast. Unimolecular radical ion reactions compete with pair reactions, as do bimolecular reactions. The competition is aided by a spin multiplicity requirement governing electron return and coupling: in many cases, only singlet pairs can recombine or couple; pairs generated from triplet precursors must first undergo intersystem crossing and, therefore, have longer lifetimes. The rates of unimolecular reactions are determined by their (intrinsic) barriers; their efficiency depends on the rate of intersystem crossing and the efficiency of ion pair and bimolecular reactions; the latter can be "turned on or off" by the reagent concentration.

By various transformations, radical ions are related to several classes of intermediates or products (Scheme 6.5). The simplest relationship is that to their

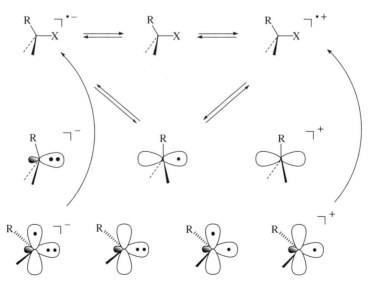

Scheme 6.5

parents; radical ions are readily interconverted with their neutral precursors by ET/ BET (back electron transfer). Similarly, BET converts distonic radical cations to biradicals or zwitterions. Some nondistonic radical cations generate (triplet) biradicals upon BET. Bond cleavage reactions, such as dissociation or fragmentation, can proceed in two directions: the charged (odd electron, OE) radical ion may generate a smaller radical ion by losing a neutral (even electron, EE) molecule, or cleave forming an OE free radical and an EE positive or negative ion. These reactions are well documented in the gas phase, but also in the condensed phase. For example, radical anions lose negative ions (Cl^-, CN^-), whereas radical cations lose positive ions (H^+, R^+). Bond cleavage reactions of cyclic radical ions generate bifunctional distonic species, in which spin and charge are localized (or delocalized) in separate segments of the molecule.

An intriguing relationship exists between carbene cations and anions, on the one hand, and the corresponding radical ions, on the other (Scheme 6.5). Although the reactivity of carbene ions is not explored in detail, one could envisage insertion and addition reactions similar to those of carbenes.

5.1. Unimolecular Radical Ion Reactions

Radical ions undergo various unimolecular reorganization reactions, of which we mention electron, hydride, or methyl shift, and several rearrangements. For example, radical ions of systems containing two donor or acceptor sites may undergo intramolecular ET, as observed for a series of radical anions of type A–Sp–biphenyl (**45**$^{•-}$) containing two acceptors linked by a 5a-androstan unit as spacer (Sp). The electron-transfer rates observed for the mono-anions of these systems, generated by radiolysis in cryogenic matrices, showed a striking deviation from the classical Brønsted relationship.[209,210]

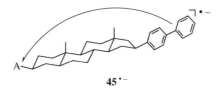

45$^{•-}$

Stereorigid radical cations may undergo stereospecific sigmatropic shifts; for example, the puckered ions, *anti-* and *syn-* 5-methyl-**19**$^{•+}$, undergo stereospecific hydride or methyl migration, respectively, forming the 1-methylcyclopentene ion (**46**$^{•+}$) and the 3-methyl isomer (**47**$^{•+}$).[171]

$$R_{syn} = H \quad \longleftarrow \quad \begin{matrix} R_{syn} \\ R_{anti} \end{matrix} \quad \longrightarrow \quad R_{syn} = Me$$

46$^{•+}$ 5-Methyl-**19**$^{•+}$ **47**$^{•+}$

Similarly, the radical cation (**48**[•+]) of sabinene generated the β-phellandrene radical cation (**49**[•+]).[211] This rearrangement occurs with high retention of optical purity, but only in the absence of nucleophiles (see below).

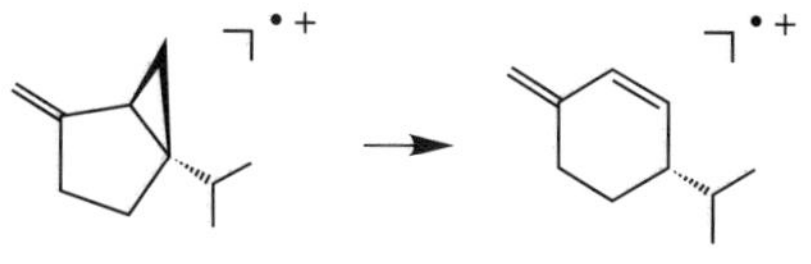

48[•+]   **49**[•+]

The 9,10-dicyanoanthracene sensitized irradiation of *cis*-stilbene results in nearly quantitative isomerization (>98%) to the trans isomer with quantum yields greater than unity.[212] Therefore, the isomerization was formulated as a free radical cation chain mechanism with two key features: (1) rearrangement of the *cis*-stilbene radical cation; and (2) electron transfer from the unreacted cis-olefin to the rearranged (trans-) radical cation.

Radiolysis of the diacetylene hexa-1,5-diyne (**50**) generates the hexa-1,2,4,5-tetraene radical cation (**51**[•+]) via a Cope rearrangement in the Freon matrix.[213]

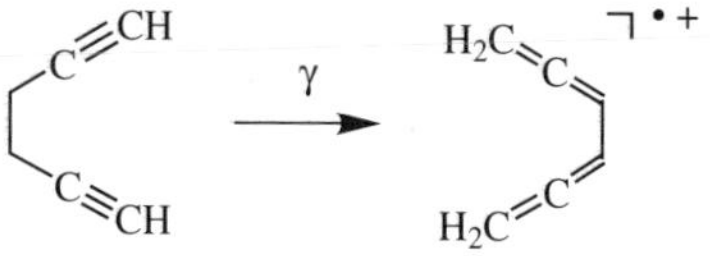

50   **51**[•+]

Intramolecular bond formations include (net) [2 + 2] cycloadditions; for example, diolefin **52**, containing two double bonds in close proximity, forms the cage structure **53**. This intramolecular bond formation is a notable reversal of the more general cycloreversion of cyclobutane type olefin dimers (e.g., **15**[•+] to **16**[•+]). The cycloaddition occurs only in polar solvents and has a quantum yield greater than unity.[214] In analogy to several cycloreversions[215,216] these results were interpreted in terms of a free radical cation chain mechanism.

52   **53**

The monocyclic 1,2,5,6-tetraphenylcycloocta-1,5-diene (**54**) undergoes a "cross"-cycloaddition,[217] forming a tricyclic product (**56**), most likely by 1,5-cyclization of **54**[•+] forming the bicyclic bifunctional radical cation **55**[•+] as an intermediate.

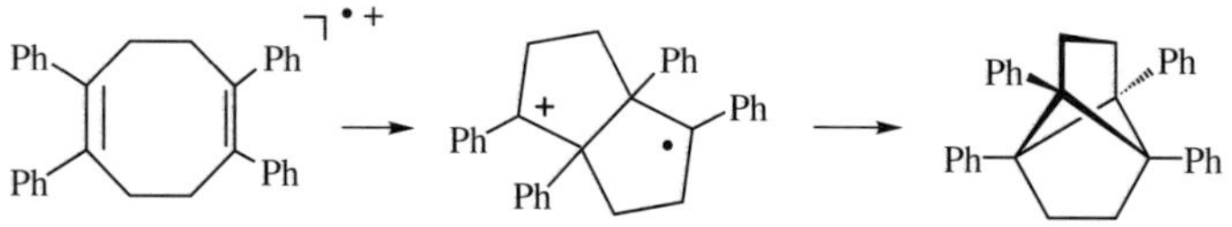

54[•+]   **55**[•+]   **56**

Radiolysis of deca-2,8-diyne (**57**) results in an interesting cycloaddition, forming a cyclobutadiene radical cation (**58**$^{\bullet+}$) at 77 K without requiring annealing at higher temperatures.[218]

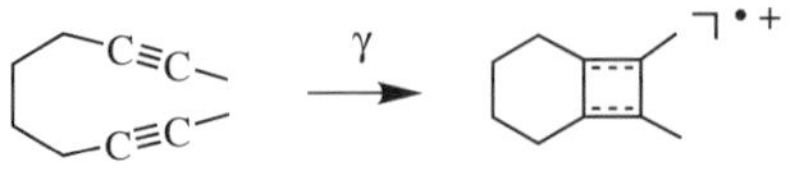

An ET induced rearrangement of tetra-*tert*-butyltetrahedrane generates the corresponding tetra-*tert*-butylcyclobutadiene radical cation (**60**$^{\bullet+}$) via **59**$^{\bullet+}$.[219]

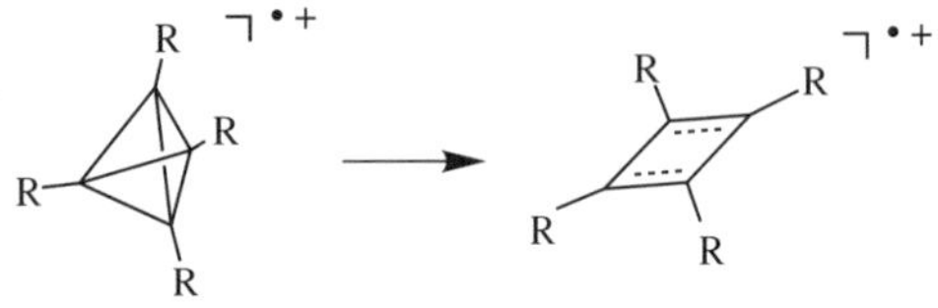

R = *t*-butyl

In addition to rearrangements, many radical ions undergo cleavage reactions. We have already encountered such a reaction: Electron attachment to halocarbon matrix molecules results in fragmentation to a halide ion and a free radical.

$$R-X + e^- \longrightarrow R-X^{\bullet-} \longrightarrow R^{\bullet} + X^-$$
$$R = haloalkyl$$

In the case of halocarbon matrices, the SOMO is of the σ* type and the cleavage is facilitated by the antibonding nature of the SOMO. In other ions, σ or σ* orbitals of scissile bonds interact with a π or π* orbital, causing them to be weakened. Accordingly, radical anions of benzyl halides may generate benzylic radicals with loss of a halide ion. Conversely, benzylsilane radical cations may form benzylic radicals with loss of a silyl cation (Fig. 6.15).

Concerning C–C bond cleavage reactions, the strained radical cation of quadricyclane (**15**$^{\bullet+}$) readily undergoes opening of the cyclobutane ring, forming the norbornadiene radical cation (**16**$^{\bullet+}$).[146] The ring opening of 1,2-diaryloxycyclobutane (**13**) forming the 1,4-bifunctional intermediate (**14**$^{\bullet+}$) was invoked to explain the electron transfer sensitized cis to trans-isomerization (see above).[155]

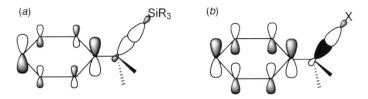

Figure 6.15. Schematic orbital diagrams explaining the weakening of a C–Si σ bond due to overlap with an adjacent π orbital (*a*) and the weakening of a C–X (halogen) bond due to overlap of the σ* orbital with an adjacent π* orbital (*b*).

Many distonic radical ions are involved in unimolecular dissociation reactions. For example, the McLafferty rearrangement of carbonyl compounds proceeds by C—C bond cleavage of distonic ions.[220] Alternatively, bond-cleavage reactions of cyclic molecular ions may generate distonic ions (see above).[200–202] For example, C—C bond cleavage of cyclic ketone molecular ions (e.g., **61**$^{\bullet+}$) generates **62**$^{\bullet+}$.[221]

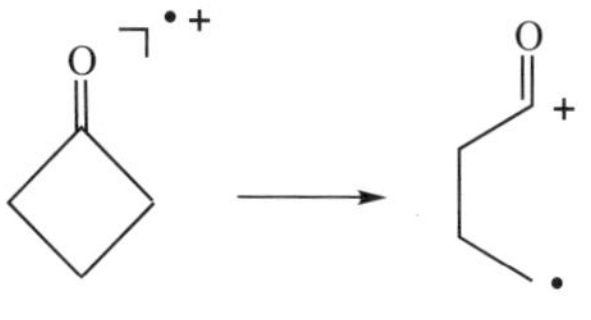

61$^{\bullet+}$ **62**$^{\bullet+}$

In analogy to the cleavage of quadricyclane radical ion, radical ions containing cyclobutane units frequently undergo cleavage of two C—C bonds. This finding is of interest in connection with the "photoreactivation" of DNA, damaged due to photochemical cyclobutane formation between adjacent thymine units. The DNA is restored by a photoreactivating enzyme, via ET to or from the pyrimidine dimer (**63**), that is, by cleavage of either the radical anion or cation.

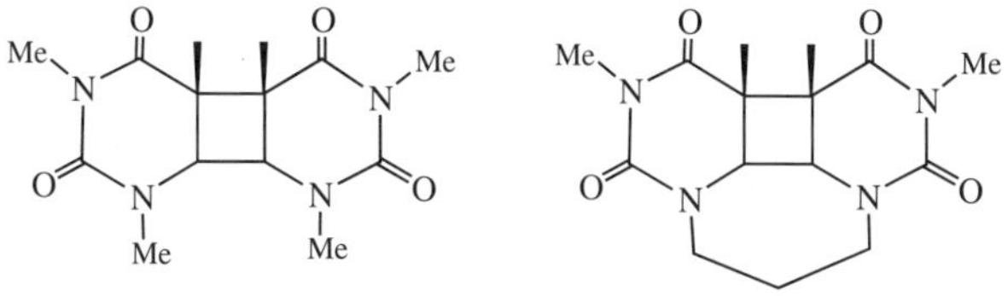

63   **64**

The mechanism of dimer cleavage has been probed in model systems,[222–225] including bifunctional ones in which a sensitizer (e.g., indole) is linked to the pyrimidine dimer.[226,227] Work on a linked dimer (**64**), suggested that a dimer radical cation is a discrete (short-lived) minimum.[225]

5.2. Intra-Pair Reactions

Intra-pair reactions include BET, proton, atom, or group transfer within the pair, or coupling (bond formation) of the geminate radical ions. Intra-pair reactions can be quite fast, particularly for singlet pairs and for CRIPs generated by PET from charge-transfer complexes (Scheme 6.4). In a majority of cases, only pairs of singlet spin multiplicity undergo BET; this process is an energy-wasting step that seriously limits the yield (and synthetic utility) of bimolecular radical ion reactions.

In some systems, triplet BET can occur, as deduced from time-resolved optical spectroscopy, magnetic field effects, CIDNP, or optoacoustic calorimetry.[228–234] Triplet BET is governed by energetic factors, which determine rates, and by the relative topologies of the potential surfaces of parent molecule, radical ions, and of accessible triplet or biradical states. Divergent topologies for different states may cause rearrangements.

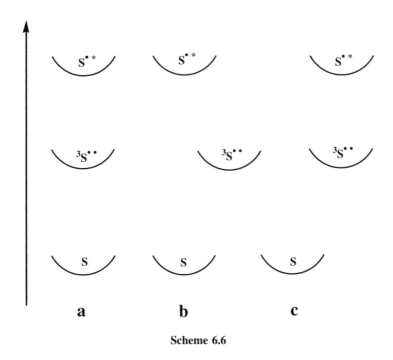

Scheme 6.6

The potential surfaces of the different states of a reagent may be related in three ways (Scheme 6.6): (a) the three surfaces have minima at closely related geometries; (b) the radical ion geometry is related to that of the parent, but the triplet state or biradical has a different geometry; (c) in systems giving rise to bifunctional–distonic radical ions (see above) triplet states or biradicals may have structures related to the radical ions.[235]

Typical aromatic donors and acceptors undergo only minor geometry changes upon oxidation or reduction or upon population of the triplet state; for these compounds, the reaction sequence ET followed by BET has no effect on the structure. If the triplet state or biradical belongs to a different structure type than radical ion and ground-state precursor, as is the case for *cis*- or *trans*-1,2-diphenylcyclopropane (**65**)[236,237] or norbornadiene (**16**)[238,239] BET may occur with cleavage[236,237] or formation[238,239] of one or more C–C bonds. In such cases, the sequence ET–BET may result in rearrangements.[235] For distonic radical ions (e.g., **24**$^{\bullet+}$) triplet BET will populate structurally related biradicals (e.g., **24**$^{\bullet\bullet}$), which may decay to rearranged products.[240,241]

H   Ph

cis-**65**   *trans*-**65**

The energetic requirements for triplet BET go beyond the trivial prerequisite that a triplet state of energy E_T, or a biradical of energy E_{BR}, exist below the ion-pair energy, ΔG^0_{SSRP}; intersystem crossing can be slow if the singlet–triplet gap is large. In fact, triplet recombination can become competitive upon either raising or lowering the ion-pair energy (using sensitizers with higher or lower excited state reduction potentials). For example, ET from **65** to triplet-chloranil generates an ion pair whose energy lies too low to access any triplet state;[236] using cyanoaromatics instead raises the pair energy and allows triplet recombination.[236] On the other hand, electron transfer from sabinene (**48**) to 1,4-dicyanobenzene generates an ion pair well above an available triplet state; however, triplet recombination is achieved by lowering the pair energy (using triphenylpyrylium ion as the electron acceptor).[242]

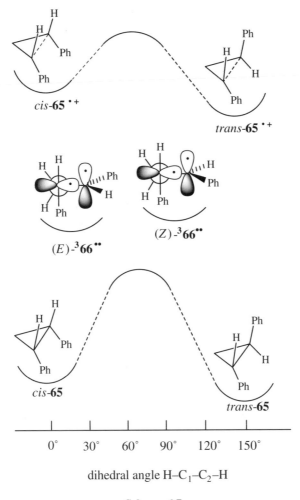

dihedral angle H–C$_1$–C$_2$–H

Scheme 6.7

We illustrate triplet BET resulting in isomerization with two examples: the reaction of *trans-* and *cis-***65** with either chloranil or 1,4-dicyanonaphthalene; and the reaction of 1,1-diaryl-2-methylenecyclopropane (**23**) with various sensitizers. The radical cations of *trans-* and *cis-***65**[•+] each are related uniquely to the geometry of their precursor, *trans-* and *cis-***65**. With chloranil as sensitizer, a relatively low-lying ion pair is formed; accordingly, BET occurs exclusively in singlet pairs, exclusively regenerating the reagent ground state. With 1,4-dicyanonaphthalene as sensitizer–acceptor, the identical radical cations *trans-* and *cis-***65**[•+], are formed; however, the pair energy is significantly higher. Therefore, BET in triplet pairs becomes feasible, populating a triplet species, (*E*)- or (*Z*)-[3]**66**[••], with a broken C—C bond. Upon intersystem crossing, the two rotamers regenerate both *cis-* and *trans-*[3]**65**. A schematic energy diagram is given in Scheme 6.7.

The second example of triplet BET in radical ion pairs involves the (ring-opened, distonic) trimethylenemethane radical cation (**24**[•+]) generated in the PET reaction of **23**.[181,182] The corresponding triplet biradical (**24**[••]) is sufficiently low in energy,[240] so that it is accessible by triplet BET.

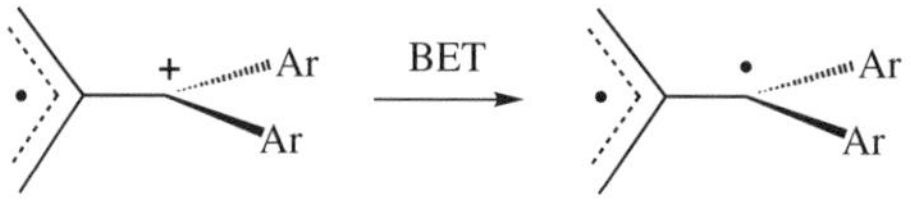

$$\textbf{24}^{\bullet\,+} \qquad\qquad \textbf{24}^{\bullet\bullet}$$

Ar = aryl

Radical ion pairs also react by proton, atom, or group transfer. We illustrate proton transfer in reactions of aromatic hydrocarbons with tertiary amines. These reactions cause reduction or reductive coupling. In the reduction of naphthalene, the initial ET is followed by H[+] transfer from cation to anion, forming **67**[•] paired with an aminoalkyl radical; the pair combines to generate **68**.[243,244]

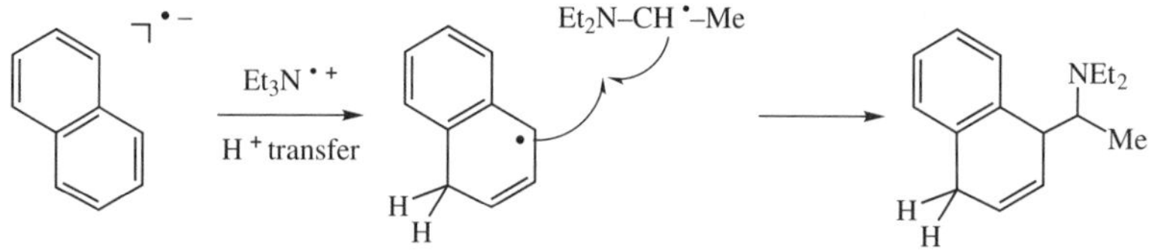

$$\textbf{67}^{\bullet} \qquad\qquad\qquad \textbf{68}$$

Similarly, the stilbene isomers (**69**) react with tertiary amines by ET followed by proton transfer and coupling, forming **70**.[245] During the irradiation of *cis*-stilbene in the presence of ethyldiisopropylamine, the *trans*-stilbene radical anion, *trans*-**69**[•−], was observed by Raman spectroscopy.[246] The ET mechanism is also supported by a pronounced dependence of the quantum yields on solvent polarity.

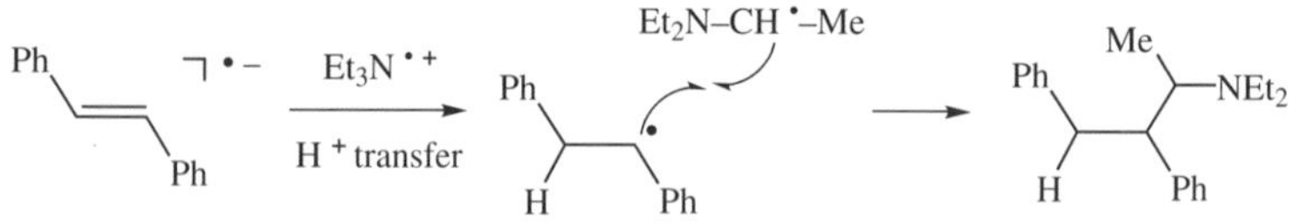

$$\textbf{69}^{\bullet-} \qquad\qquad\qquad\qquad \textbf{70}$$

The radical pair generated by proton transfer from tertiary amine radical cations to α,β-unsaturated ketone radical anions (e.g., **71**) couple in the β position, forming **72**.[247]

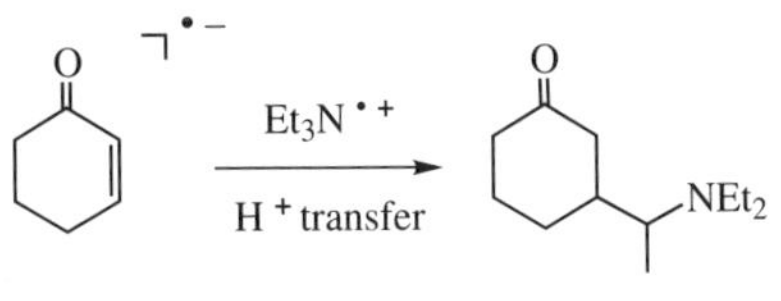

71$^{\bullet-}$ **72**

Alkene radical cations may transfer protons to cyanoaromatic radical anions, followed by coupling of the resulting radicals. For example, 1,4-dicyanobenzene and other cyano-aromatic acceptors form substitution products (e.g., **73**) with 2,3-dimethylbutene via coupling and loss of HCN.[248]

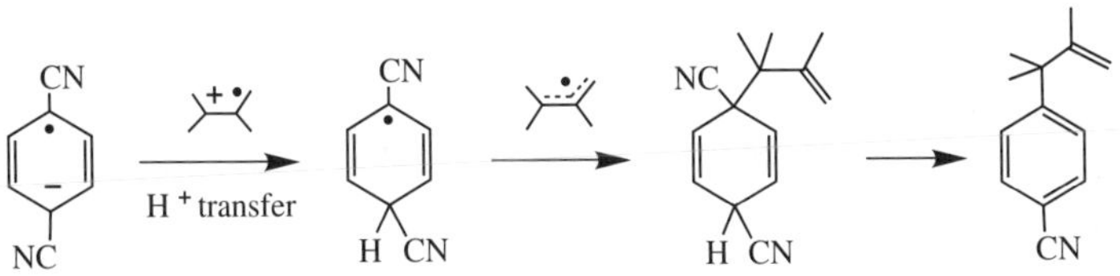

73

The third intra-pair reaction to be discussed involves bond formation between radical anion and cation without intervening H$^+$ transfer; both singlet and triplet radical ion pairs can couple. For example, the bifunctional radical cation **24**$^{\bullet+}$ generates two chloranil adducts, most likely via zwitterions (e.g., **74**$^{+-}$ and **75**$^{+-}$), initiated by forming a C—O bond. The CIDNP results indicate that **74** and **75** are formed from a singlet radical ion pair.[182] Adduct **75** is a minor product, as the major spin density of **24**$^{\bullet+}$ is located in the allyl function which, therefore, is expected to be the principal site of coupling.

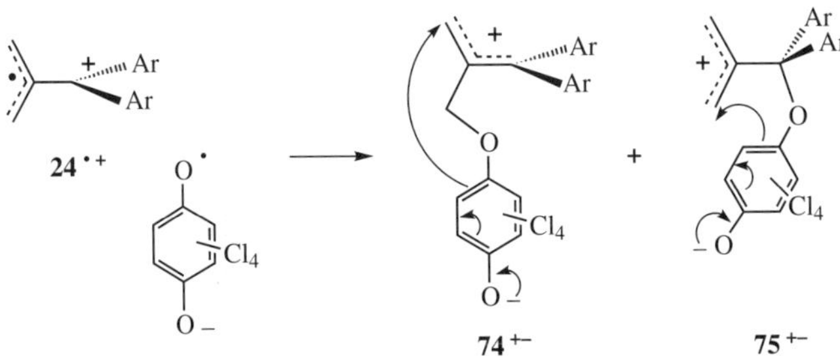

24$^{\bullet+}$ **74**$^{+-}$ **75**$^{+-}$

Ar = aryl

Chloranil forms two types of cycloadducts with 3,3-dimethylindene. In the early stages, oxetane (**76**) is formed via adduct **76**$^{\bullet\bullet}$, by addition of the carbonyl oxygen to the β carbon. However, oxetane (**76**) is only a minor product, because its radical cation is likely to fragment. Oxetane (**77**), in which the carbonyl oxygen is connected to the α carbon, is less favorable because one spin of **77**[··] is localized; **77** is built up, however, because **77**[·+] is less likely to undergo fragmentation. The CIDNP results indicate that **76**[··] is formed from a triplet pair, on a time scale comparable with BET in singlet pairs.[249]

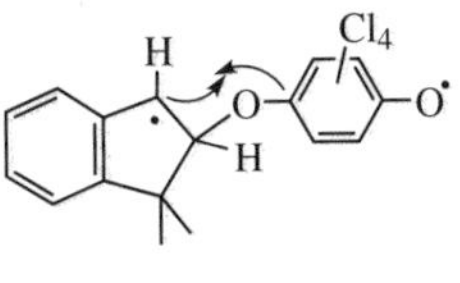

76[··] 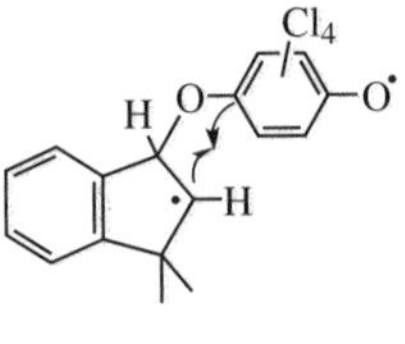   **77**[··]

When carried out in the presence of molecular oxygen, PET reactions between some donor–acceptor pairs yield oxygenated products. The radical anions of acceptors with appropriate reduction potentials reduce O_2 to superoxide ion, $O_2^{·-}$, which then couples to the cations.[250–252] In the PET reaction between *trans*-stilbene and 9-cyano-phenanthrene, the *trans*-stilbene radical cation was observed by optical spectroscopy and the 9-cyanophenanthrene radical anion by ESR; the anion spectrum decayed rapidly in the presence of oxygen.[251,253] Other details of the mechanism, for example, the spin multiplicity of the pair or the detailed mechanism of adduct formation, are not known.

The type of O_2 adduct depends on the donor structure. For example, tetraphenyloxirane forms an ozonide (**78**),[254] 1,4-bifunctional radical cations form dioxanes (**79**);[255] conjugated dienes form cyclic adducts (**80**);[256] and ergosteryl acetate (**81**) forms the 5a,8a-peroxide (**82**) at $-78\ °C$.[257]

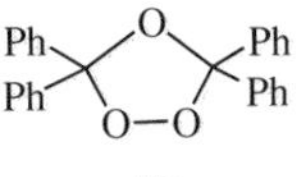

78 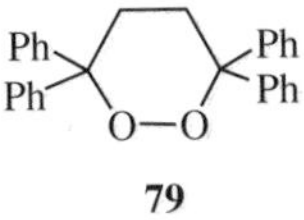   **79** 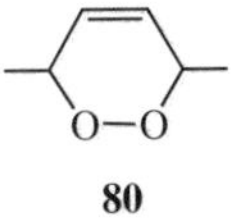   **80**

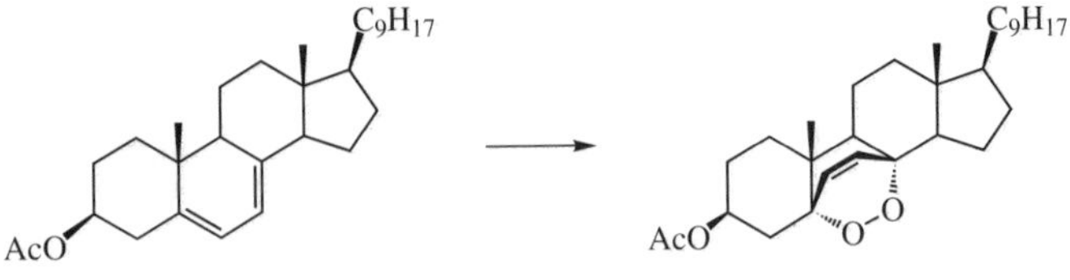

81   **82**

There are, however, additional mechanisms for the oxygenation of radical cations, which do not qualify as intra-pair reactions. For example, hindered, α-branched tetraalkylalkenes (e.g., biadamantylidene) form dioxetanes via a chain mechanism, initiated by chemical, electrochemical, or PET oxidation.[107,258,259] The hindered radical cation reacts with triplet oxygen, forming a bifunctional adduct that undergoes ring closure to a dioxetane radical cation of type $5^{\bullet+}$, with a three-electron bond (see above); ET from the alkene generates **5**; this step is efficient, because $5^{\bullet+}$ ($E_{ox} = 2.3$ V vs. SCE, -78 °C) is a better oxidant than the alkene radical cation ($E_{ox} = 1.6$ V vs. SCE, -78 °C).

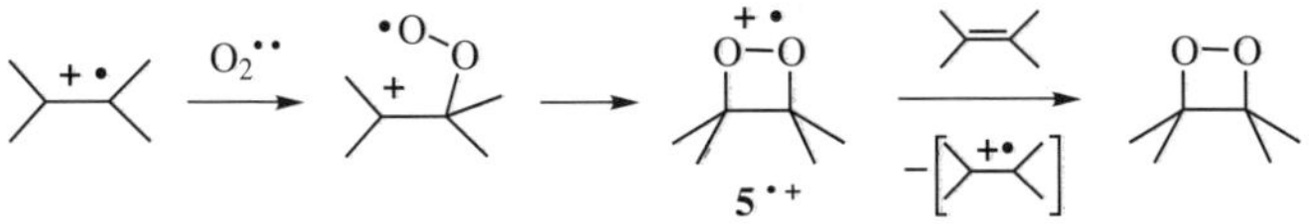

In some cases, the structures of oxygenation products have been crucial for assigning the structures of unusual radical cations. For example, the *endo*-peroxides (**83** and **85**) support the structures assigned to radical cations (**24**$^{\bullet+}$ and **84**$^{\bullet+}$) derived from 1,1-diaryl-2-methylenecyclopropane (**23**) and 2,5-diaryl-1,5-hexadiene, respectively.[241,255] Time-resolved spectroscopic data suggest that **83** is generated by coupling of triplet biradical (**24**$^{\bullet\bullet}$) with (triplet) molecular oxygen.[241]

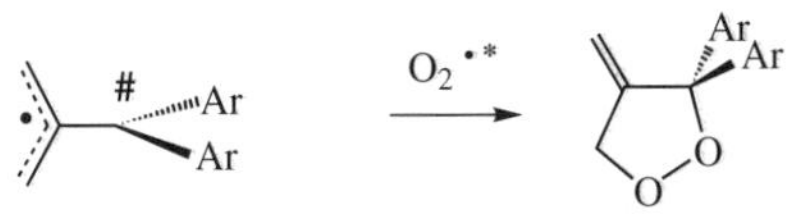

$24^{\bullet\,\#}; \# = +, \bullet \quad * = -, \bullet$ **83**

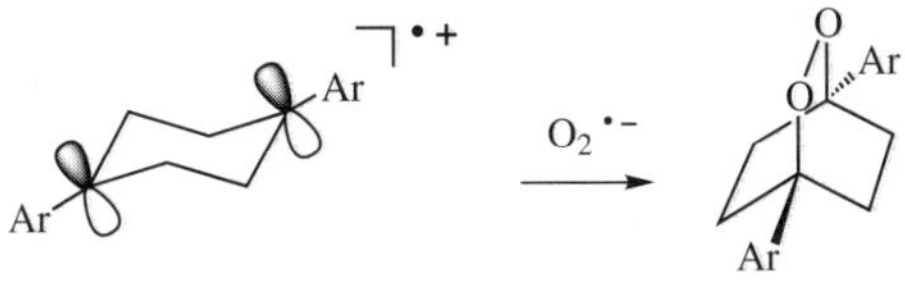

$84^{\bullet+}$ **85**

Ar = aryl

Radical ion pairs of triplet multiplicity can couple via charge recombination, equivalent to nucleophilic capture of the radical cation by the radical anion, yielding biradicals. In reactions of this type, radical cations with stereochemically distinct faces offer a choice of stereochemical approach; the isomers **15**$^{\bullet+}$ and **16**$^{\bullet+}$ provide illuminating examples. These species react with methanol in stereospecific fashion from the exo-face,[260,261] forming *exo*-methoxy-norbornyl and -nortricyclyl free radicals, which undergo rapid cyclopropylcarbinyl–butenyl interconversions. Likewise, benzoquinone radical anion approaches **15**$^{\bullet+}$ and **16**$^{\bullet+}$ stereoselectively from the exo-face.[262] The resulting biradicals (**86**$^{\bullet\bullet}$ and **87**$^{\bullet\bullet}$) undergo cyclopropylcarbinyl–butenyl interconversions, equilibrating **86**$^{\bullet\bullet}$ and **87**$^{\bullet\bullet}$ with **88**$^{\bullet\bullet}$. The "pre-oxetane" (**86**$^{\bullet\bullet}$) and the "pre-oxolane" (**88**$^{\bullet\bullet}$) form products **86** and **88**, respectively, after intersystem crossing. One of the primary biradicals (**87**$^{••}$) cannot cyclize, because the unpaired spins are mutually inaccessible.

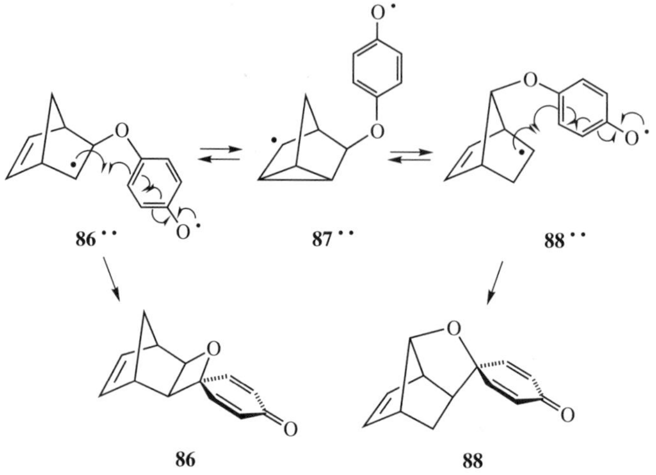

5.3. Bimolecular Radical Ion Reactions

The majority of radical ion reactions are bimolecular in nature, although some of these are merely variations of the unimolecular reactions discussed above, and many occur as pair reactions, albeit with a modified "partner." Radical ions may react with polar or nonpolar neutral molecules, with ions, with radicals, or with radical ions of like or opposite charge (Scheme 6.3). Alkene radical ions undergo a particularly rich variety of reactions, including additions and cycloadditions.

5.3.1. Reactions with Alkenes and Aromatics. Unsymmetrically substituted olefins form head-to-head dimers selectively. We mention the "classic" vinylcarbazole,[41] vinyl ethers,[157,263,264] indenes,[265–267] and *p*-methoxystyrene.[268] The regiochemistry of the addition is compatible with a stepwise mechanism proceeding via a singly linked 1,4-bifunctional radical cation, in which spin and charge are separated (see above). Several dimerizations have quantum yields greater than unity;[157,263,269] these results support radical cation chain processes.

1,1-Diphenylethylene radical cation forms a dimeric product (**90**) via a 1,4-bifunctional–distonic intermediate (**89**$^{•+}$) that undergoes an alternative (1,6-) cyclization with participation of one phenyl group.[270]

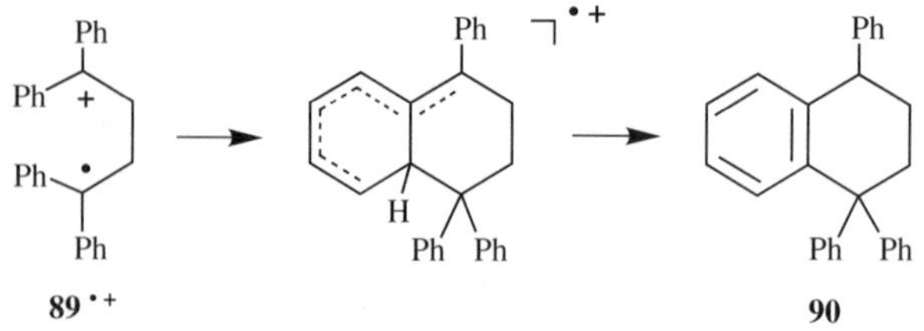

When generated in zeolites, alkene or arene radical cations react with the parent molecules to form π-dimer radical cations. For example, 2,3-dimethyl-1-butene[271] and benzene[272] formed **91•+** and **92•+**, respectively. The confinement and limited diffusion of the radical cation in the zeolite favor an interaction between a radical cation and a neutral parent in the same channel.

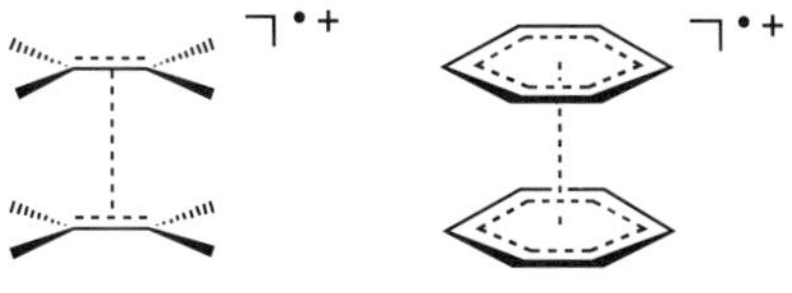

91 •+   **92 •+**

Similarly, fluorescence detected magnetic resonance effects observed during the pulse radiolysis of anthracene-d_{10} in the presence of 2,3-dimethyl-1-butene support the presence of 8 equivalent methyl groups. Because the splitting, $a_{di} = 0.82$ mT, was approximately one-half that of the monomer splitting, $a_{mon} = 1.71$ mT, the "sandwich" dimer **91•+** was invoked.[273]

However, ab initio calculations [QCISD-(T)/6-31G*//UMP2/6-31G*] on ethylene and its radical cation support an "anti"-π-complex, in which the two components are joined by one long bond (190 pm), rather than the "sandwich"-type π complex. The complex is connected to two transition states leading to a (rhombic) cyclobutane radical cation (see above) or, by 1,3-H-shift, to 1-butene radical cation.[274]

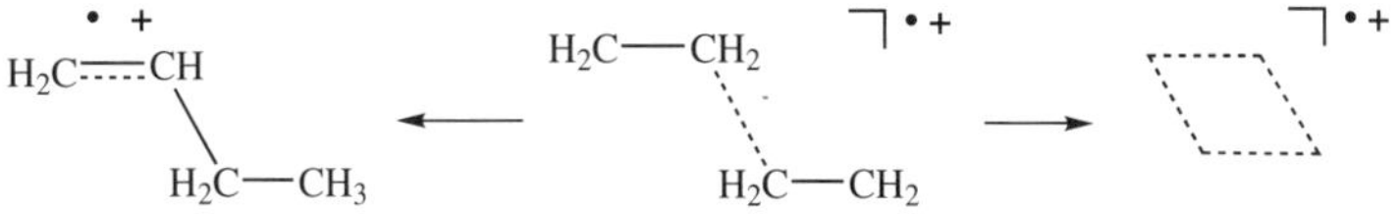

Irradiation of naphthalene in faujasite pores gave rise to loading-dependent nanosecond time-resolved diffuse reflectance spectra.[275] A structured band (680 nm) at low loadings was assigned to the monomer radical cation, whereas a broad band (590 nm) at higher loading was assigned to the dimer radical cation. The dimer structure is supported by a broad charge resonance band in the near-infrared (IR) ($\sim$1100 nm); it was assigned a "twisted" structure with partial overlap of the two π systems.

Some radical cations with spin and charge on heteroatoms form σ-dimer radical cations. For example, irradiation of acetophenone and aliphatic amines inside Na Y zeolite generated amine dimer radical cations (**93•+**; λ_{max} 480–580 nm).[108] Analogous σ-dimer radical cations are derived from thiols.[276]

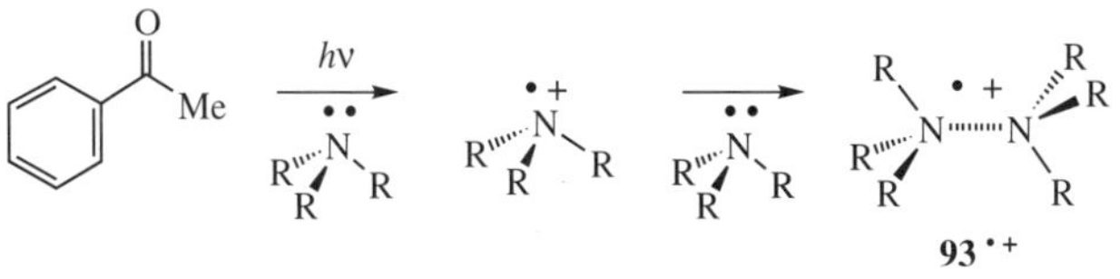

93 •+

Radical cations (**94**$^{\bullet+}$) of simple alkynes (e.g., butyne) generated by γ-irradiation in frozen matrices are stable at 77 K. Upon warming, changes in the ESR spectra support an interesting cycloaddition forming cyclobutadiene radical cations (**95**$^{\bullet+}$).[277]

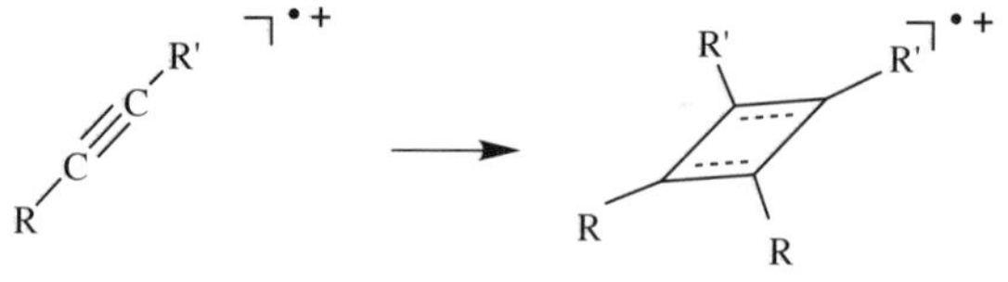

PET converts phenylacetylene to α-phenylnaphthalene; a 1,4-bifunctional dimer radical cation (**96**$^{\bullet+}$) is the key intermediate; 1,6-cyclization, followed by a hydrogen or hydride shift generates the final product.[278]

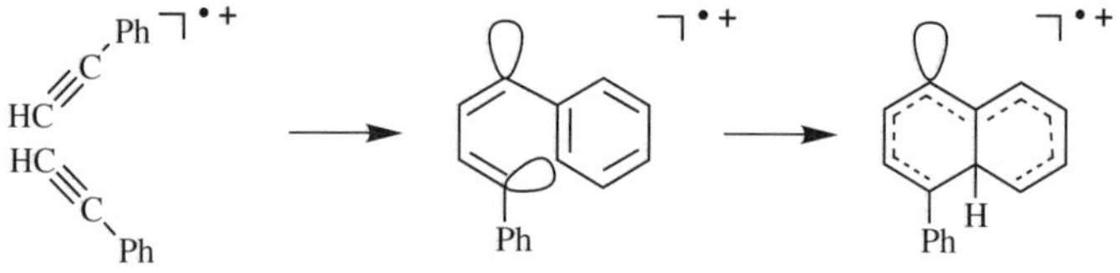

PET reactions of diolefins (viz., cyclohexadiene) generate [4 + 2] cycloadducts (e.g., *endo-* and *exo*-dicyclohexadiene).[279] This conversion occurs also upon radiolysis[280] or chemical oxidation [tris(*p*-bromophenyl)aminium salts; cation radical catalyzed Diels–Alder reaction]. The scope of this reaction and its synthetic utility have been delineated in detail; once again, a radical cation chain mechanism is indicated.[281]

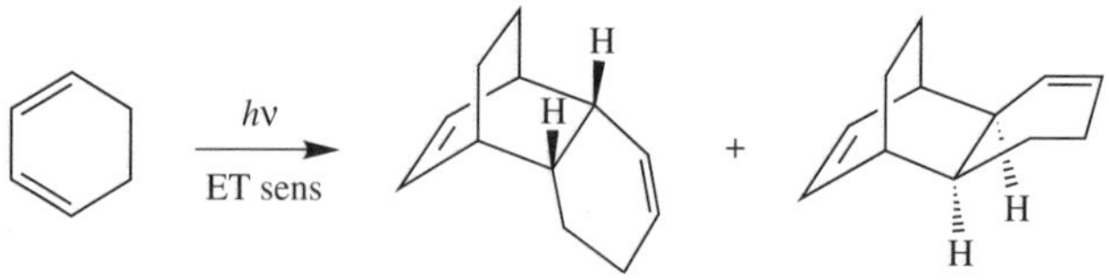

The mechanism of the cycloaddition appears to be concerted for various reagents; however, for several cases, radical cation cycloaddition–cycloreversions have a stepwise component. For example, CIDNP effects observed during the PET induced dimerization of spiro[2.4]heptadiene (**97**) identify a dimer radical cation with spin density only on two carbons of the dienophile fragment; this intermediate must be a doubly linked radical cation (**99**$^{\bullet+}$).[282,283] A pulsed laser experiment at high concentrations of **97** supports a second dimer radical cation; at high concentrations of **97**, the dimer cation is rapidly quenched by ET from the monomer. This rapid quenching allows the polarization of a second, singly linked dimer radical cation (**98**$^{\bullet+}$) to be observed.[282,283]

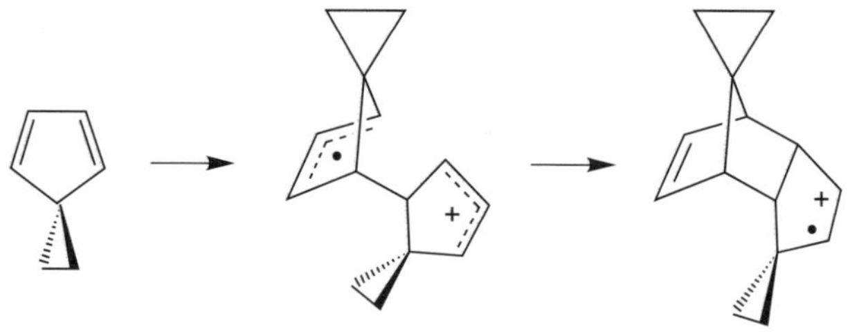

97   **98**$^{\bullet+}$ **99**$^{\bullet+}$

In some cases, a stepwise mechanism is indicated by randomization of the dienophile stereochemistry. For example, addition of *cis*-anethole radical cation (**100**$^{\bullet+}$) to cyclopentadiene produces comparable yields of four possible diastereoisomeric adducts (**102**) clearly supporting a distonic radical cation intermediate (**101**$^{\bullet+}$). Only products supporting the stepwise mechanim, that is, *trans,endo*- and *trans,exo*-**102**, are shown.[284]

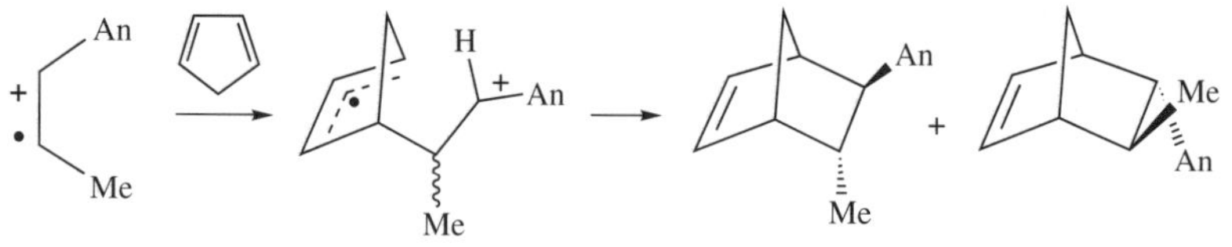

100$^{\bullet+}$ **101**$^{\bullet+}$ *trans,endo*-**102**   *trans,exo*-**102**

An = C$_6$H$_4$OMe

In the ET-catalyzed Diels–Alder reaction of indole with 1,3-cyclohexadiene a stepwise mechanism was derived on the basis of ^{13}C kinetic isotope effects. In light of B3LYP/6-31G* calculations, these effects support a stepwise mechanism, initiated by attack of the diene on the 3-position of indole.[285]

Of the many PET Diels–Alder reactions of potential synthetic utility we mention two reactions of vinylindole (**103**) catalyzed by 2,4,6-tris(4-methoxyphenyl)-pyrylium tetrafluoroborate. With cyclohexadiene, **103** reacts as a diene, giving rise to tetrahydrocarbazole (**104**);[286] with exocyclic dienes, **103** serves as dienophile generating a different tetrahydrocarbazole (**105**).[287] Molecular orbital calculations provide a rationale for the regio- and diastereoselectivities of these reactions.

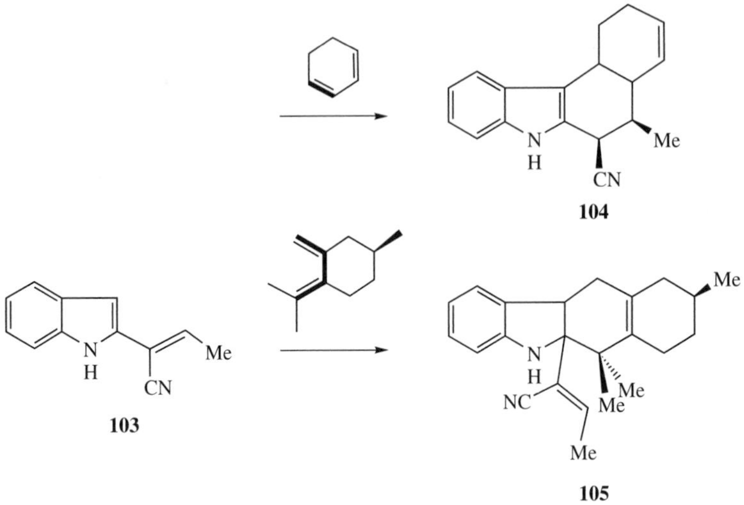

104

103

105

5.3.2. Reactions with Protic, Ionic, Polar Reagents.

The reactions of radical anions with proton donors include the reduction of arenes,[288] the well-known Birch reduction, as well as alkynes[289] by alkali metals in liquid ammonia. Both reactions have synthetic utility and belong to the few radical ion reactions included in elementary textbooks.

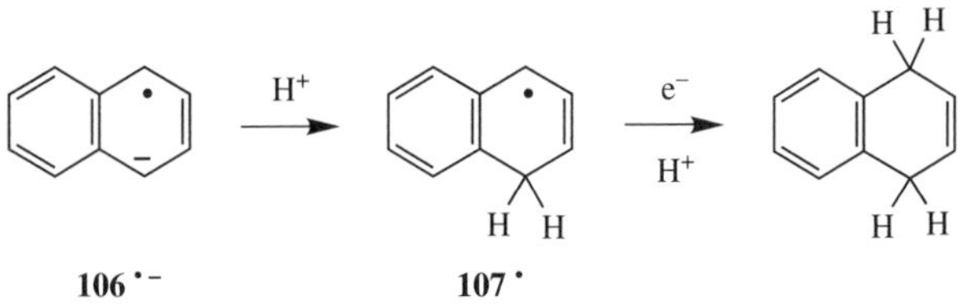

106•⁻   **107**•

Following the reduction of the substrates by a solvated electron, the solvent transfers a proton to the radical anions, **106**•⁻ and **108**•⁻; the resulting radicals, **107**• and **109**•, are then reduced again, and the anions, **107**⁻ and **109**⁻, are protonated once more. The regiochemistry of the protonation of **107**⁻ is kinetically controlled; the ready inversion of the alkenyl free radical **109**• is the key to the formation of the *trans*-alkene.

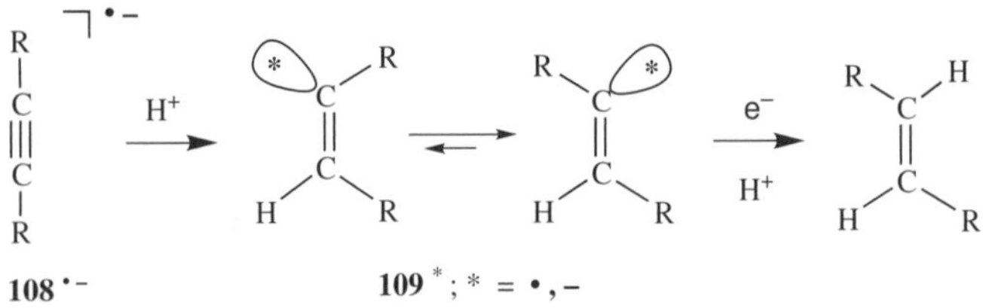

108•⁻   **109***; * = •, −

With ions or dipolar substrates, radical ions undergo nucleophilic or electrophilic capture. Nucleophilic capture is a general reaction for many alkene and strained-ring radical cations and may completely suppress (unimolecular) rearrangements or dimer formation. The regio- and stereochemistry of these additions are of major interest. The experimental evidence supports several guiding principles.

For example, the PET reaction of 1-phenylcyclohexene (**110**) in the presence of nucleophiles (KCN; acetonitrile / 2,2,2-trifluoroethanol)[270,290] proceeds by anti-Markovnikov addition of the nucleophile to the alkene.[270,290] A distonic oxonium radical cation, **111**$^{\bullet+}$, is a likely short-lived intermediate, which is rapidly deprotonated.

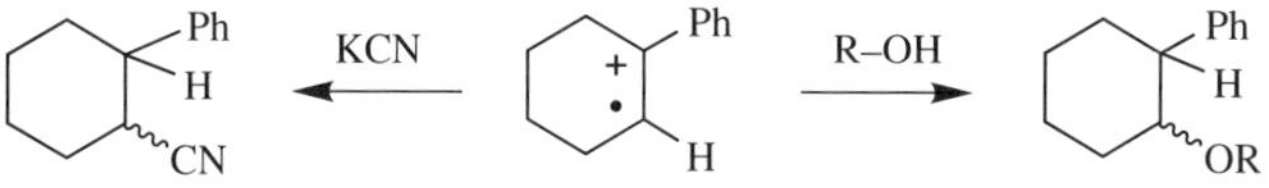

110$^{\bullet+}$

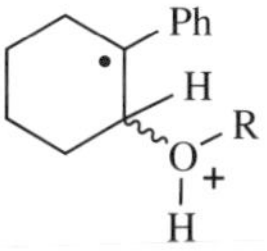

111$^{\bullet+}$

The nucleophilic capture of radical cations forms (free) radicals, one H atom shy of the adduct. The missing H may be introduced in one step, by hydrogen abstraction, or in two, involving successive reduction by the sensitizer radical anion and protonation. Both mechanisms have been observed, sometimes in competition with each other.

Substituted cyclopropane systems also undergo nucleophilic addition of suitable solvents (MeOH).[291] For example, the photoinduced ET reaction of 1,2-dimethyl-3-phenylcyclopropane (**112**, R = Me) with p-dicyanobenzene formed a ring-opened ether by anti-Markovnikov addition. The reaction occurs with essentially complete inversion of configuration at carbon, suggesting a nucleophilic cleavage of a "one-electron" cyclopropane bond, generating **113**$^{\bullet}$.[292] The retention of chirality confirms that the stereochemistry of the parent molecule is unperturbed in the radical cation **112**$^{\bullet+}$.

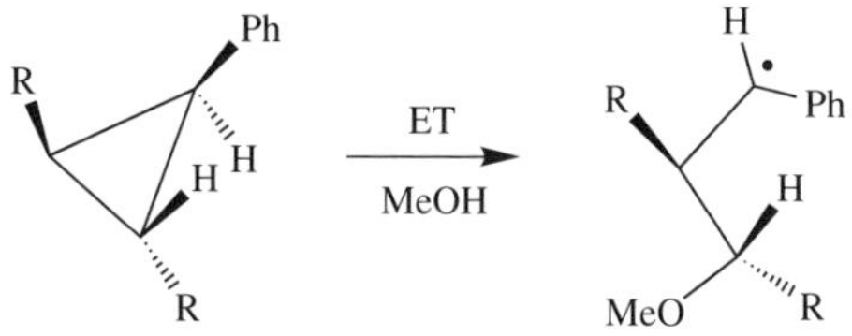

112   **113**$^{\bullet}$

Many of these reactions support a measure of "thermodynamic control" in nucleophilic capture: Conjugated radicals or products formed with release of ring strain are favored. For example, the addition of ethanol to radical cation **110**$^{\bullet+}$ is regiospecific, forming the more stable (benzylic) intermediate **111**$^{\bullet+}$; the capture of **112**$^{\bullet+}$ likewise forms a benzylic radical (**113**$^{\bullet}$). Radical cation **48**$^{\bullet+}$ generates a conjugated radical cation **49**$^{•+}$ via a sigmatropic shift, and an allyl radical **114**$^{•}$ upon nucleophilic attack;[211] both reactions form conjugated "products" with relief of ring strain. The high degree of regioselectivity in the nucleophilic capture of **48**$^{•+}$, **110**$^{•+}$, or **112**$^{•+}$ also reflects unfavorable energetics for the formation of alternative products.

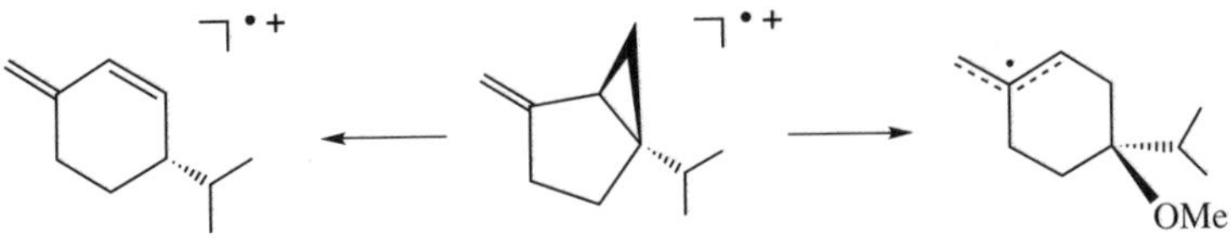

$$49^{•+} \quad\quad\quad 48^{•+} \quad\quad\quad 114^{•}$$

The nucleophilic capture of tricyclane radical cations **115**$^{•+}$ and **117**$^{•+}$ supports the role of "conventional" steric hindrance; **115**$^{•+}$ reacts at the 3° carbon ($\rightarrow$**116**$^{•}$),[293] whereas the chiral isomer **117**$^{•+}$ is captured by backside attack at the less hindered 3° carbon ($\rightarrow$**118**$^{•}$).[294] Both reactions are regio- and stereospecific and avoid attack at the neopentyl-type carbon (denoted by an asterisk).

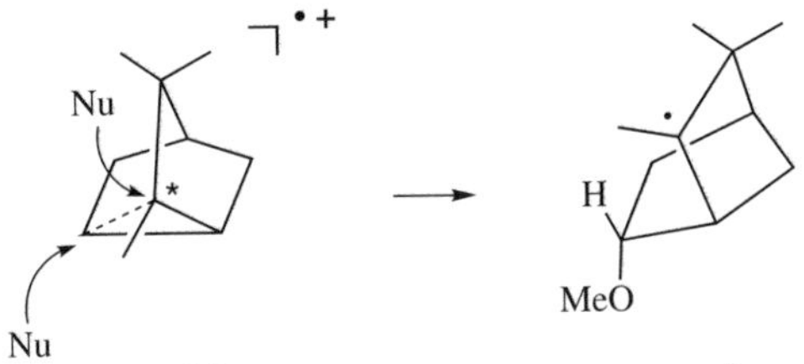

$$115^{•+} \quad\quad\quad 116^{•}$$

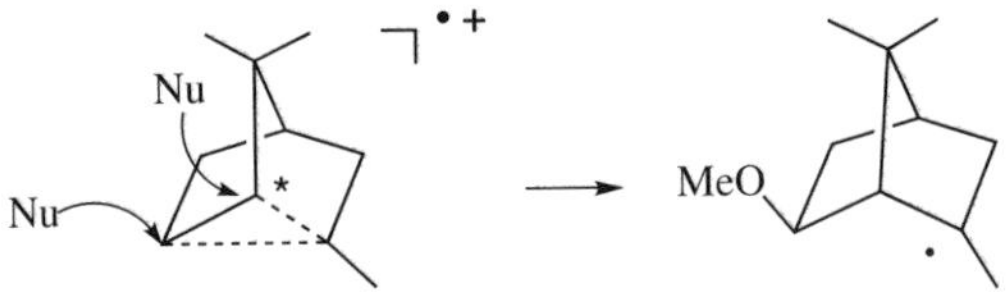

$$117^{•+} \quad\quad\quad 118^{•}$$

The radical cations **15**$^{•+}$ and **16**$^{•+}$ add methanol exclusively from the exo face forming *exo*-methoxynorbornyl and nortricyclyl free radicals, which undergo rapid cyclopropylcarbinyl–butenyl interconversions (**119**$^{•}$ $\rightleftharpoons$ **120**$^{•}$ $\rightleftharpoons$ **121**$^{•}$; see above).[260,261] The attack on **15**$^{•+}$ can be viewed as a backside attack of the nucleophile (Nu) on the weakened cyclopropane bond with inversion of configuration.

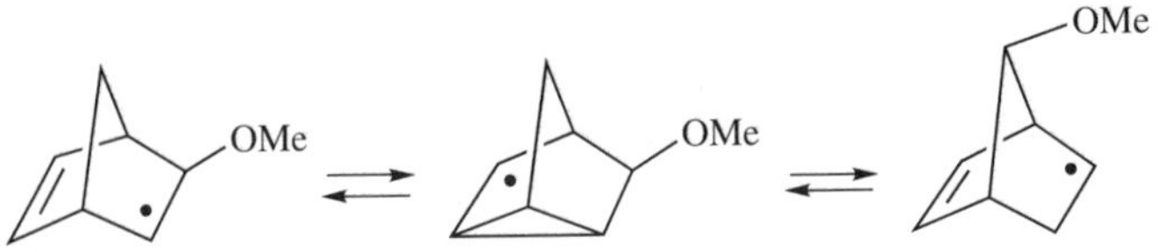

$$119^{•} \quad\quad\quad 120^{•} \quad\quad\quad 121^{•}$$

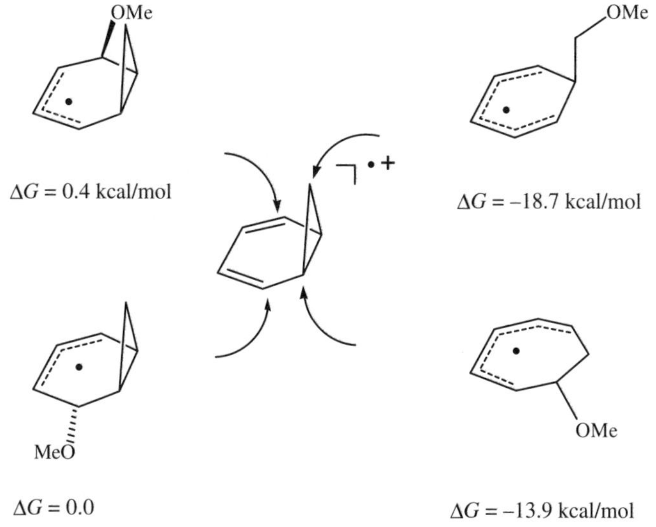

$\Delta G = 0.4$ kcal/mol

$\Delta G = -18.7$ kcal/mol

$\Delta G = 0.0$

$\Delta G = -13.9$ kcal/mol

Scheme 6.8

However, orbital factors may override thermodynamic control. For example, the regiochemistry of nucleophilic attack on the bridged norcaradiene radical cation **122**[•+] shows a significant deviation from thermodynamic control. Although attack on the cyclopropane ring should be favored by both release of ring strain and formation of delocalized free radicals (cf. Scheme 6.8), methanol attacks **122**[•+] selectively at C_2 (and C_5), generating **123**[•] and **124**[•]. There is little stereoselectivity: Products derived from **123**[•] and **124**[•] were formed in comparable yields.[152]

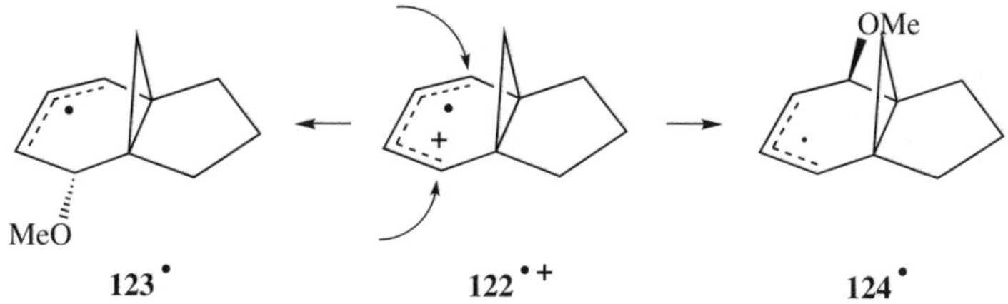

123[•]   **122**[•+]   **124**[•]

The bridged norcaradiene (**122**[•+]) appears particularly well suited to evaluate orbital and thermodynamic effects in radical cation reactions because these effects predict different reactivity patterns and because the orbital coefficients of SOMO and LUMO also differ significantly (Fig. 6.16). The two MOs potentially involved in the reaction have a substantially different distribution of orbital coefficients. The SOMOs of **122**[•+] have large orbital coefficients at C2,5 and C7 (C10), which are reflected in the hyperfine pattern of these species. In contrast, the principal orbital coefficients of the LUMOs are located at C2,5 and C3,4 and the orbital lobes at C7 (C10) offer no target for an attack by the nucleophile. Because the orbital coefficients of SOMO and LUMO differ, the norcaradiene system will elucidate whether

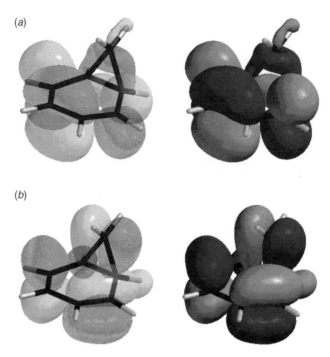

Figure 6.16. Spartan representation of SOMO (*b*) and LUMO (*a*) for norcaradiene radical cation (see color insert). [Adapted from Ref. 152.]

the regioselectivity of nucleophilic capture is governed by the SOMO, or by the nature and topology of the LUMO.

Orbital control was evaluated in several theoretical treatments. Curve crossing methodologies suggest that nucleophilic capture of radical cations requires "double excitation;"[295] the excitation energy can be small and the resulting barrier low.[296] Accordingly, nucleophiles attack the dibenzofuran radical cation at the site of the highest LUMO coefficient, which is also the site of highest spin density in the dibenzofuran triplet state.[297] Similarly, the stereochemical course of nucleophilic displacement of a σ bond (with inversion of configuration) was ascribed to involvement of the σ^* orbital of the weakened bond (the LUMO).[298,299]

The product-determining role of the LUMO can also explain the regioselective capture of other radical cations, including the nucleophilic attack on 1-aryl-2-alkylcyclopropanes (**112**$^{\bullet+}$). The SOMO and LUMO of disubstituted cyclopropane radical cations (e.g., 1,2-dimethylcyclopropane; Fig 6.17) suggest that the observed regioselectivity reflects electronic factors: capture at the unsubstituted cyclopropane carbon is unlikely, since neither SOMO nor LUMO have orbital coefficients at C3.[164]

Internal alcohol functions also may capture a radical cationic function in suitable substrates. The radical cation **125**$^{\bullet+}$ of a bicyclo[4.1.0]heptane system bearing a (3-hydroxybutyl) substituent and a *p*-tolylthio moiety in the 1- and 6-positions,

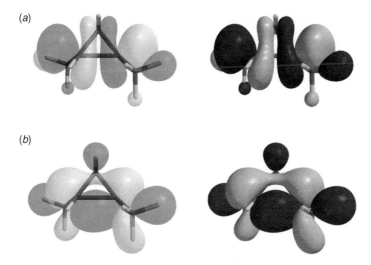

(a)

(b)

Figure 6.17. Spartan representation of SOMO (b) and LUMO (a) for 1,2-dimethylcyclopropane radical cation.[164]

respectively, generates the spiro compound **126** by regiospecific intramolecular nucleophilic capture.[300] This attack corresponds to a backside "substitution" of an intramolecular leaving group.

125$^{\bullet+}$ **126**

Tol = tohyl

The ET photochemistry of $(1R, 3S)$-$(+)$-*cis*-chrysanthemol (*cis*-**127**) proceeds via nucleophilic attack of the internal alcohol function on the vinyl group with simultaneous or rapid replacement of an isopropyl radical as an intramolecular leaving group, forming **128**$^{\bullet}$. This reaction is a mechanistic equivalent of an S_N2' reaction; the mode of attack underscores the major role of strain relief in governing nucleophilic capture in radical cations.[271]

127$^{\bullet+}$ **128**$^{\bullet}$

In contrast, homo-chrysanthemol (**129**) reacts exclusively by capture at the quaternary cyclopropane carbon, generating the five-membered cyclic ether (**130•**).[301] Apparently, the five-membered transition state leading to **130•** is significantly favored over the seven-membered one required for capture at the terminal carbon of the double bond of **129•+**.

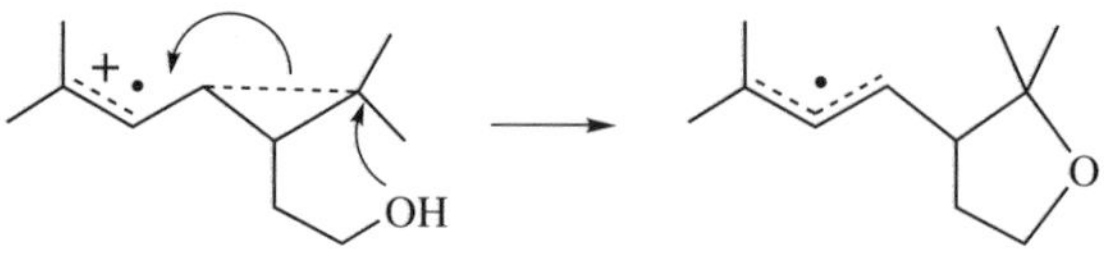

In addition to nucleophilic capture by alcohols, nonprotic nucleophiles also react with these intermediates. For example, the distonic dimer radical cation **96•+** can be trapped by acetonitrile; a hydride shift, followed by electron return, gave rise to the pyridine derivative **131**.[272] Similar acetonitrile adducts are formed in the electron-transfer photochemistry of terpenes such as α- and β-pinene[302] or sabinene.[303]

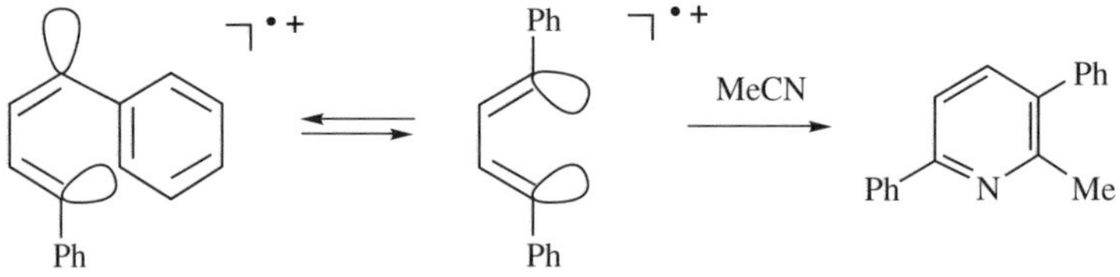

5.3.3. Reactions of Radical Anions With Radicals.

The coupling of arene or alkene radical anions with radicals is an important reaction, and one that has significant synthetic potential. For example, radicals formed by nucleophilic capture of radical cations couple with the acceptor radical anion, resulting in (net) aromatic substitution. Thus, the 1-methoxy-3-phenylpropyl radical (**113•**; R = H) couples with dicyanobenzene radical anion; loss of cyanide ion then generates the substitution product **132**.[291,304]

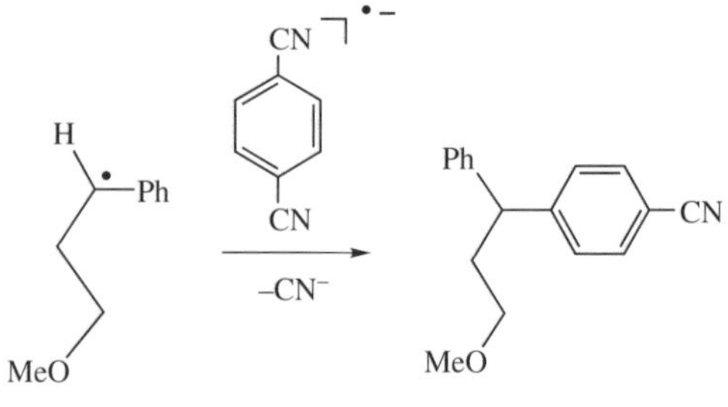

Coupling of alkyl radicals resulting from nucleophilic capture of alkenes with sensitizer radical anions in acetonitrile–methanol (3:1) was studied in detail.[305,306] The photochemical nucleophile olefin combination, aromatic substitution (photo-NOCAS) reaction, formulated below for 2,3-dimethylbutene–methanol–p-dicyano-benzene, has some synthetic utility. The final step, loss of cyanide ion, is not shown.

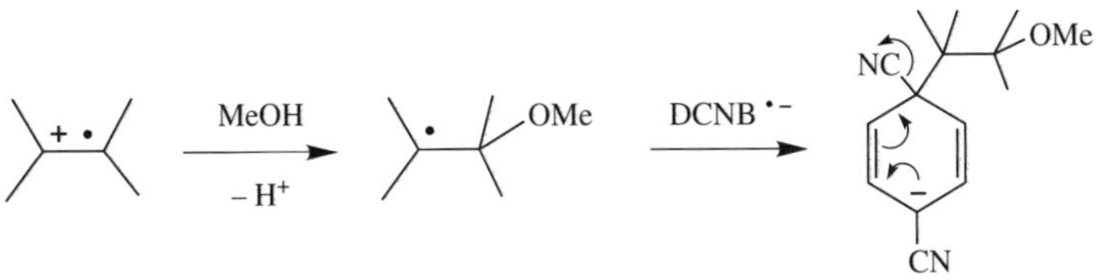

Similarly, radical **128°** couples with dicyanobenzene radical anion generating the substitution product **133** after loss of cyanide ion.

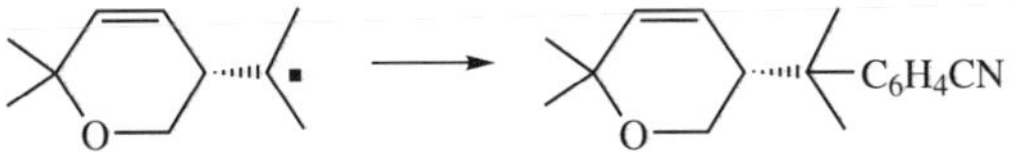

128°   **133**

Radicals of type **113°** also couple with (protonated) N-methylphthalimide anion.[307] The formation of ethers **132** and **134** and many related products shows that bimolecular nucleophilic capture is faster than coupling of the geminate radical ion pair.

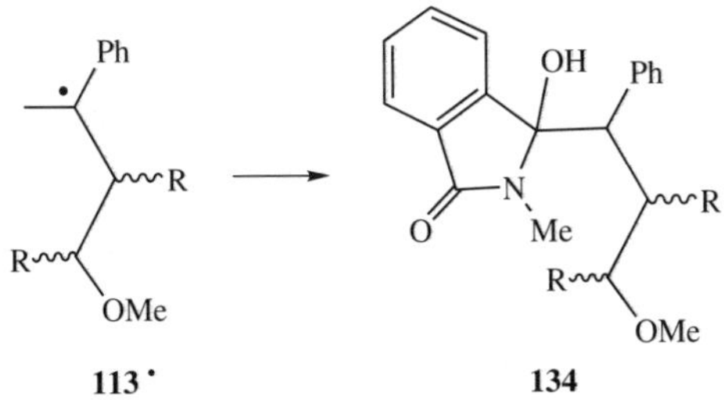

113°   **134**

In addition to nucleophilic capture of alkene or cyclopropane radical cations (see above) radicals may be generated by cleavage of C–X bonds, particularly C–Si bonds. Such cleavage is often assisted by a nucleophile. Because the radical is generated near the radical anion, to which it couples, the resulting C–C bond formation may be considered a reaction of a "modified" radical (ion) pair.

For example, allylic silanes react with photoexcited iminium salts by ET resulting in photoallylation.[308] Key steps in such reactions are the cleavage of the C–Si bond in a silyl-substituted radical cation, and the coupling of the resulting allyl radical to the aminoalkyl radical (e.g., **135°**) generating products such as **136**.

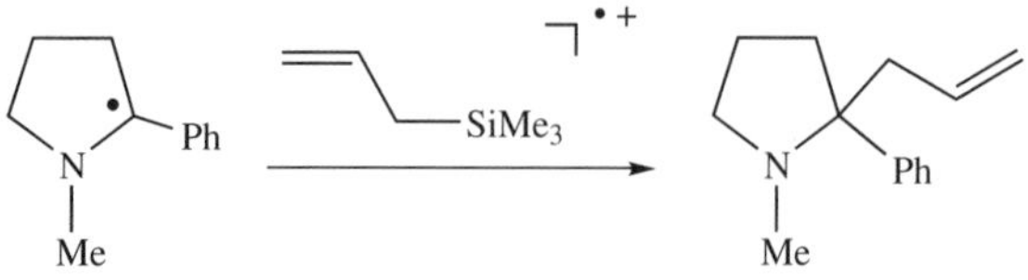

135°   **136**

ET induced allylation of dicyanoalkenes (e.g., **137**) is regioselective; the allyl radical couples to the carbon β to the cyano groups (→ **138**).[309] The efficiency of these reactions can be improved by using a cosensitizer, such as phenanthrene.[310]

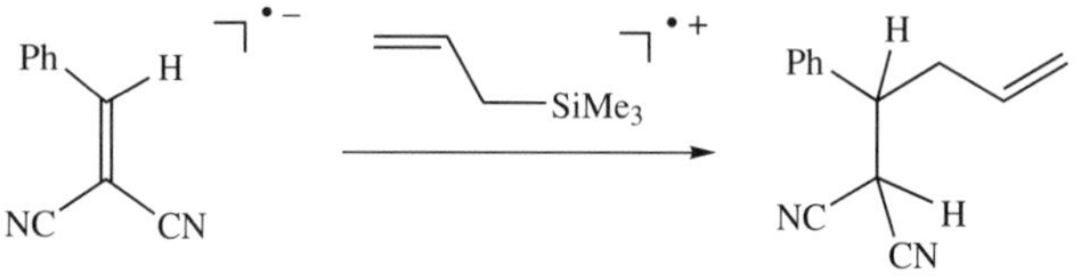

137°⁻   **138**

In addition to regiospecificity, appropriate substrates also show stereoselectivity. For example, the photoallylation of the rigid styrene (**139**) with phenanthrene as cosensitizer in acetonitrile, produced a 3 : 1 ratio of trans- to cis-allylated products.[311]

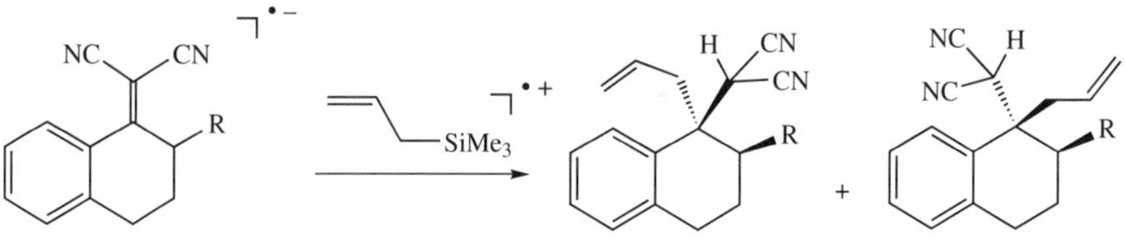

139   *trans*-**140**   *cis*-**140**

The role of the nucleophile was confirmed for a benzyldimethylsilane (**141**) bearing a nucleophile (OH, OMe) on an oligomethylene tether ($n = 3$–5). The radical cations of these silanes have very short lifetimes even in nonpolar solvents, and the resulting benzyl radicals couple with the radical anion.[312]

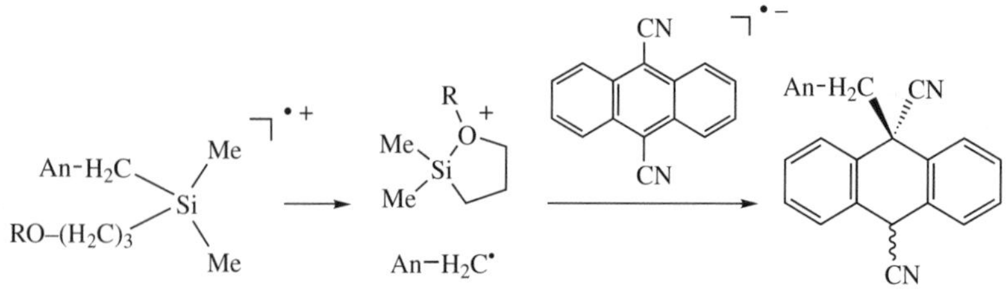

141°⁺

In analogy to the α-deprotonation of tertiary amine radical cations (see above), amines bearing an α-trimethylsilyl group may undergo heterolytic cleavage of the C–Si bond upon ET, particularly in the presence of nucleophiles. The resulting aminoalkyl radical may couple, for example, to the enone radical anion ($71^{\bullet-}$) generating **142**.[247]

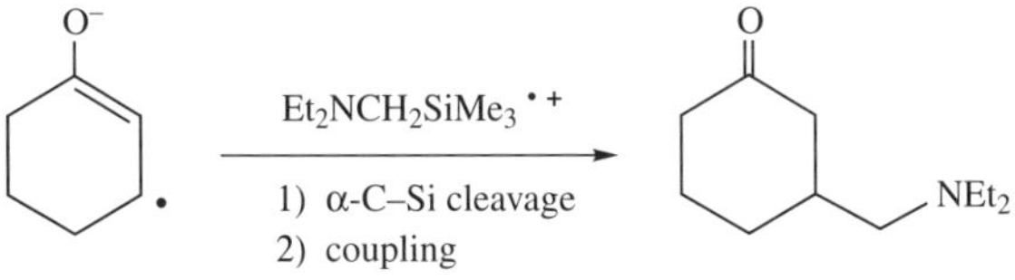

Interesting variations of these reactions are observed when the α-silylamine donor funtion is tethered to an enone. For example, the intramolecular ET reaction of **143** results in two divergent cyclizations. Methanol assists the cleavage of the C–Si bond; the resulting biradical anion **146**$^{\bullet\bullet-}$ couples; protonation and tautomerization then leads to **147**. In acetonitrile, on the other hand, transfer of an α proton to the enone radical anion function forms biradical **144**$^{\bullet\bullet}$; coupling and tautomerization then generates **145**.[247]

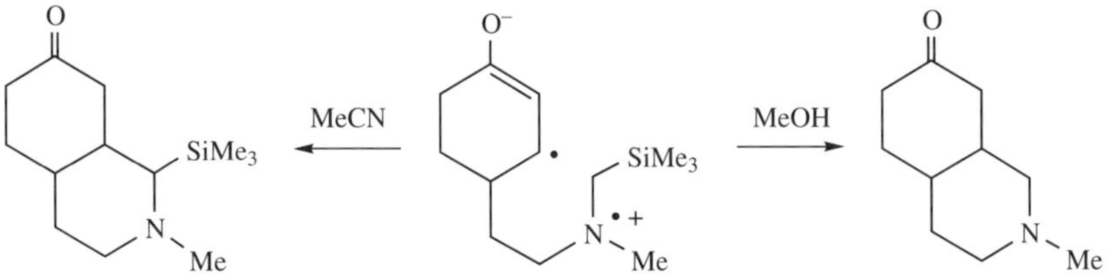

5.3.4. Reactions with Radical Ions of Like Charge.

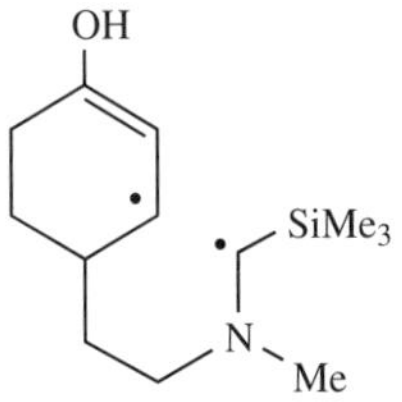

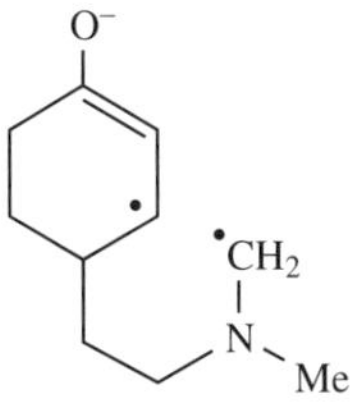

In the final section, we briefly mention reactions between radical ions of like charge. One of the long-standing problems of radical ion chemistry involves the actual structure of ketyls. Following extensive conductivity and magnetic susceptibility studies in the 1930s detailed ESR and optical studies have demonstrated the existence and interconversion of at least four distinct species: a free ketyl anion; a monomer ion pair; a paramagnetic dimer ion pair; and a diamagnetic dimer dianion. The latter is the logical precursor for the pinacols generated in these reactions. The system is further complicated by different degrees of solvation of the individual ions as well as the pairs. The equilibria between these species show pronounced solvent effects. For example, the characteristic colors of benzophenone or fluorenone ketyls are reduced in nonpolar solvents, a change ascribed to formation of the pinacolate dianions. However, the change is reversible once the nonpolar solvent is removed.[313] Details of these complex interactions exceed the scope of this chapter.

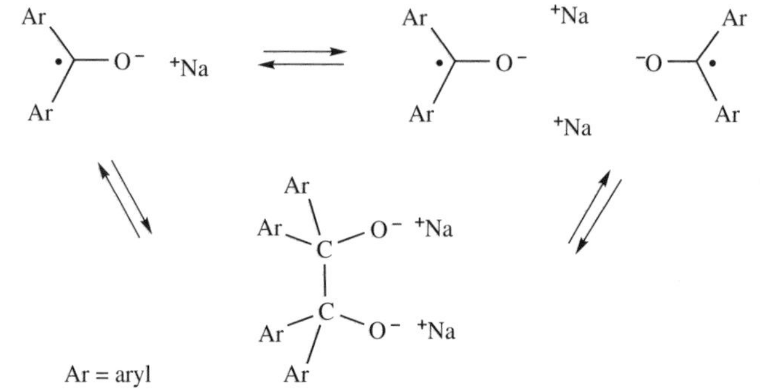

The acyloin condensation is closely related to the radical anion coupling forming pinacolate anions: two ester radical anions couple to form a dianion, which readily loses two alkoxide ions. The resulting diketone then is reduced by sodium, first to a semidione radical anion, then to the dianion.[314] Finally, aqueous work-up produces the acyloin. Acyloins are convenient precursors for the generation of semidione radical anions.[315]

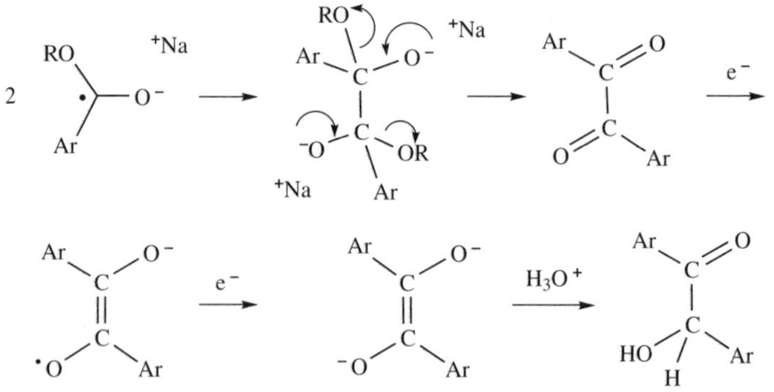

Pairs of radical ions of like charge also react by electron transfer (i.e., they disproportionate). One classic example involves reduction of tetraphenylethylene and subsequent ET between two tetraphenylethylene anions.[316] A more recent interesting example is that of cyclooctatetrene radical anion **148**[•−]. Alkali metals readily reduce the nonplanar cyclooctatetraene, generating a persistent planar radical anion

($a_H = -0.32$ mT, 8H);[317] **148**$^{\bullet-}$ disproportionates generating the aromatic 10-π-dianion **148**$^{2-}$ and the parent molecule.[318]

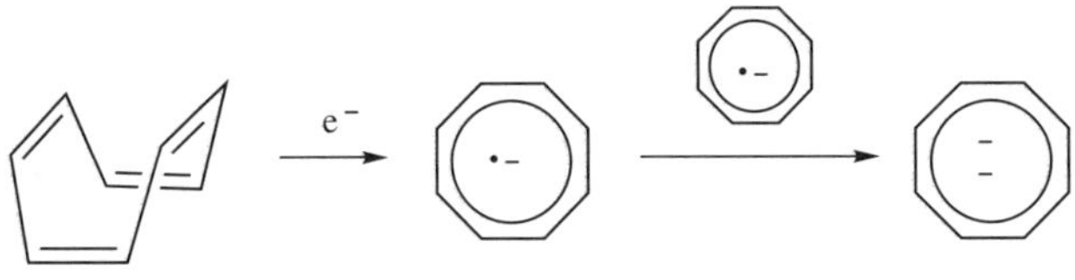

148   148 $^{\bullet-}$ 148 $^{--}$

In this section, too, we need to emphasize that space limitations preclude a more thorough treatment of these interesting species and their chemistry.

6. CONCLUSION AND OUTLOOK

The rich variety of radical ion reactions and the diversity of structure types portrayed in this chapter may lead the reader to the conclusion that this field has exhausted its growth potential, especially when one considers that the selection of topics and the depth of coverage offered here had to be limited. Nevertheless, radical ion chemistry remains in a phase of rapid development and can be expected to retain a high level of attention for the foreseeable future.

Dividing possible approaches to radical ion research (and chemical research in general) into three principal branches, structure, energetics, and kinetics, it is clear that the state of the art in these areas has not advanced to the same degree. Kinetics is perhaps the most matured due to the significant recent advances in time resolved spectroscopy. Concerning energetics, redox potentials, ionization potentials, and electron affinities provide the energetics of electron removal or attachment; in addition, some pair energies, and biradical and triplet energies are available from optoacoustic calorimetry and laser spectroscopy. As for structural features, hyperfine couplings and their signs, available from ESR or CIDNP studies, reveal a pattern of electron spin densities; no other properties can be *measured*, aside from the occasional crystallographic characterization of stable radical ion salts. Bond lengths and angles and a host of radical ion properties are, however, accessible by ab initio MO calculations.

After a period where the radical cation field owed significant advances to experimental studies, the development of density functional methodologies and ever increasing computing power has given theory a significant boost in recent years. Molecular orbital calculations have become an invaluable tool to evaluate possible structure types, and the interplay between experiment and theory has become a major force in radical ion chemistry. Particularly, the application of matrix isolation IR studies has received a significant boost from sophisticated state-of-the-art calculations. Ab initio techniques also should play a significant role in elucidating details of the interplay between electron transfer and structural reorganization. The extension of time resolved spectroscopy into as yet inaccessible spectral regions surely will yield major new contributions.

Continuing research in radical ion chemistry will yield many advances; the continuing development of new techniques and their application to previously studied systems may add some new facets to the reaction mechanisms. Of course, progress will also include the rediscovery of facts, structures, and mechanisms, derived or formulated over the past five decades, which can now be veiled in new terminology. In short, research in radical ion chemistry will flourish for some years to come.

SUGGESTED READING

H. D. Roth, "A History of Electron Transfer and Related Reactions," *Topics Curr. Chem.* **1990**, *156*, 1–19.

H. D. Roth, "Chemically induced dynamic nuclear polarization," in *Encyclopedia of Nuclear Magnetic Resonance*, Vol. 2, D. M. Grant, R. K. Harris, Eds., **1996**, John Wiley & Sons, Inc., New York, pp. 1337–1350.

N. L. Bauld, "Cation Radicals," in *Radicals, Ion Radicals, and Triplets*, Wiley-VCH, New York, **1997**, pp. 141–180.

N. L. Bauld, "Anion Radicals," in *Radicals, Ion Radicals, and Triplets*, Wiley-VCH, New York, **1997**, pp. 113–140.

Throughout this chapter, the author has cited *Accounts of Chemical Research* articles by leading researchers in the radical ion field, including those of Asmus,[276] Bauld and co-workers,[281] Courtneige and Davis,[277] Gerson,[167] Hoffmann,[158] Knight,[34] Ledwith,[41] Mangione and Arnold,[306] Mattes and Farid,[278] McLauchlan and Stevens,[62] Miyashi et al.,[255] Nelsen,[107] Ottolenghi,[230] Roth,[146] Shida et al.,[85] and Yoon and Mariano.[247] These articles are recommended to readers who may want a more detailed description of selected radical ion topics.

R. L. Smith, P. K. Chou, and H. I. Kenttämaa, "Structure and reactivity of selected distonic radical cations", in *Structure, Energetics and Dynamics of Organic Ions*, T. Baer, C. Y. Ng, and I. Pawis, Eds., John Wiley & Sons, Inc., New York, **1996**, p. 197.

REFERENCES

1. M. Gomberg, *J. Am. Chem. Soc.* **1900**, *22*, 757; *Ber. Dtsch. Chem. Ges.* **1900**, *33*, 3150.

2. J. Deisenhofer, O. Epp, K. Miki, R. Huber, and H. Michel, *J. Mol. Biol.* **1984**, *180*, 38; *Nature (London)* **1985**, *318*, 385.

3. N. W. Woodbury, M. Becker, D. Middendorf, and W. W. Parsons, *Biochemistry* **1985**, *24*, 7516.

4. J. L. Martin, J. Breton, A. J. Hoff, A. Migus, and A. Antonetti, *Proc. Natl. Acad. Sci. U.S.A.* **1986**, *83*, 957.

5. M. R. Wasielewski and D. M. Tiede, *FEBS Lett.* **1986**, *204*, 1986.

6. A. Goldbach, L. Iton, M. Grimsditch, and M. L. Saboungi, *J. Am. Chem. Soc.* **1996**, *118*, 2004.

7. D. Reinen and G.-G. Lindner, *Chem. Soc. Rev.* **1999**, *28*, 75.

8. A. Laurent, *Ann Chim.* **1835**, *59*, 367; *Justus Liebigs Ann. Chem.* **1836**, 17, 89.

9. E. Beckmann and T. Paul, *Justus Liebigs Ann. Chem.* **1891**, *266*, 1.

10. W. Schlenk and T. Weickel, *Ber. Dtsch. Chem. Ges.* **1911**, *44*, 1182.

11. W. Schlenk and A. Thal, *Ber. Dtsch. Chem. Ges.* **1913**, *46*, 2840.

12. S. Sugden, *Trans. Faraday Soc.* **1934**, *30*, 11.

13. R. N. Doescher and G. W. Wheland, *J. Am. Chem. Soc.* **1934**, *56*, 2011.

14. M. Berthelot, *Compt. Rend.* **1866**, *63*, 836.

15. W. Schlenk, J. Appenrodt, A. Michael, and A. Thal, *Ber. Dtsch. Chem. Ges.* **1914**, *47*, 473.

16. A. Baeyer, *Ber. Dtsch. Chem. Ges.* **1875**, *8*, 614.

17. C. Wurster, *Ber. Dtsch. Chem. Ges.* **1875**, *12*, 522.

18. C. Wurster and R. Sendtner, *Ber. Dtsch. Chem. Ges.* **1879**, *16*, 1803.

19. R. Willstätter and J. Piccard, *Ber. Dtsch. Chem. Ges.* **1908**, *41*, 1458.

20. J. Picccard, *Ber. Dtsch. Chem. Ges.* **1911**, *46*, 1843.

21. A. Werner, *Ber. Dtsch. Chem. Ges.* **1901**, *42*, 4324.

22. H. Haakh, *Ber. Dtsch. Chem. Ges.* **1901**, *42*, 4594.

23. A. Hantzsch, *Ber. Dtsch. Chem. Ges.* **1916**, *49*, 511.

24. E. Z. Weitz, *Elektrochem.* **1928**, *34*, 538

25. L. Michaelis, *Chem. Rev.* **1935**, *16*, 243.

26. L. Michaelis, M. P. Schubert, and S. Granick, *J. Am. Chem. Soc.* **1935**, *61*, 1981.

27. E. T. Kaiser and L. Kevan, Eds., *Radical Ions*, Interscience, New York, 1968.

28. S. I. Weissman, J. Townsend, D. E. Paul, and G. E. Pake, *J. Chem. Phys.* **1953**, *21*, 2227.

29. S. I. Weissman, *J. Chem. Phys.* **1954**, *22*, 1135.

30. C. R. Bruce, R. E. Norberg, and S. I. Weissman, *J. Chem. Phys.* **1954**, *24*, 473.

31. D. R. Corbin, D. F. Eaton, and V. Ramamurthy, *J. Am. Chem. Soc.* **1988**, *110*, 4848.

32. V. Ramamurthy, J. V. Caspar, and D. R. Corbin, *J. Am. Chem. Soc.* **1991**, *113*, 594.

33. T. Komatsu and A. Lund, *J. Phys. Chem.* **1972**, *76*, 1727.

34. L. B. Knight, *Acc. Chem. Res.* **1986**, *19*, 313.

35. X.-Z. Qin and A. D. Trifunac, *J. Phys. Chem.*, **1990**, *94*, 4751.

36. K. Yoshida, *Electrooxidation in Organic Chemistry*, John Wiley & Sons, Inc., New York, **1984**.

37. A. J. Bard and L. R. Faulkner, *Electrochemical Methods*, John Wiley & Sons, Inc., New York, **1980**.

38. P. S. Rao, J. R. Nash, J. P. Guarino, M. R. Ronayne, and W. H. Hamill, *J. Am. Chem. Soc.*, **1962**, *84*, 500.

39. J. P. Guarino, M. R. Ronayne, and W. H. Hamill, *J. Am. Chem. Soc.* **1962**, *84*, 4230.

40. J. P. Guarino and W. H. Hamill, *Radiation Res.* **1962**, *17*, 379.

41. A. Ledwith, *Acc. Chem. Res.* **1972**, *5*, 133.

42. S. L. Mattes and S. Farid, *Org. Photochem.* **1983**, *6*, 233.

43. J. Mattay, *Angew. Chem. Int. Ed. Engl.* **1987**, *26*, 825.

44. H. Knibbe, D. Rehm, and A. Weller, *Ber. Bunsenges. Phys. Chem.* **1969**, *73*, 839.

45. F. W. McLafferty and F. Tureček, *Interpretation of Mass Spectra*, 4th ed., University Science Books, Sausalito, CA, **1993**.

46. K. Levsen, *Fundamental Aspects of Organic Mass Spectrometry*, Verlag Chemie, New York, **1978**.

47. D. W. Turner, A. D. Baker, C. Baker, and C. R. Brundle, *Molecular Photoelectron Spectroscopy*, Wiley-Interscience, New York, **1970**.

48. R. G. W. Norrish and G. Porter, *Nature (London)* **1949**, *164*, 658.

49. G. Porter, *Proc. R. Soc.* **1950**, *200*, 284.

50. J. R. Novak and R. W. Windsor, *J. Chem. Phys.* **1967**, *47*, 3075.

51. G. Porter and M. R. Topp, *Nature (London)* **1968**, *220*, 1228.

52. A. Miller, *Z. Naturforsch. A* **1968**, *23*, 946.

53. P. M. Rentzepis, *Photochem. Photobiol.* **1968**, *8*, 579.

54. P. M. Rentzepis, *Chem. Phys. Lett.* **1968**, *2*, 117.

55. P. L. Dutton, K. J. Kaufmann, B. Chance, and P. M. Rentzepis, *FEBS Lett.* **1975**, *60*, 275.

56. M. G. Rockley, M. W. Windsor, R. J. Cogdell, and W. W. Parson, *Proc. Natl. Acad. Sci. U.S.A.* **1975**, *72*, 2251.

57. C. V. Shank, R. L. Fork, and R. T. Yen, *Springer Ser. Chem. Phys.* **1982**, *23*, 2.

58. T. Shida, *Electronic Absorption Spectra of Radical Ions*, Elsevier, Amsterdam, The Netherlands, **1988**.

59. V. E. Bondybey and T. A. Miller, *Molecular Ions, Spectroscopy, Structure, and Chemistry*, North Holland, Amsterdam, The Netherlands, **1983**.

60. A. Carrington and A. D. McLachlan, *Introduction to Magnetic Resonance*, Harper & Row, New York, **1967**.

61. J. E. Wertz and J. R. Bolton, *Electron Spin Resonance: Elementary Theory and Practical Applications*, McGraw-Hill, New York, 1972.

62. K. A. McLauchlan and D. G. Stevens, *Acc. Chem. Res.* **1987**, *21*, 54.

63. K. A. McLauchlan and D. G. Stevens, *Mol. Phys.* **1986**, *57*, 223.

64. S. Basu, K. A. McLauchlan, and G. R. Sealy, *J. Phys. E: Sci. Instrum.* **1983**, *16*, 767.

65. B. Smaller, J. R. Remko, and E. C. Avery, *J. Chem. Phys.* **1968**, *4*, 5147.

66. A. D. Trifunac, M. C. Thurnauer, and J. R. Norris, *Chem. Phys. Lett.* **1978**, *57*, 471.

67. A. D. Trifunac, J. R. Norris, and R. G. Lawler, *J. Chem. Phys.* **1979**, *71*, 4380.

68. R. W. Fessenden and R. H. Schuler, *J. Chem. Pys.* **1963**, *39*, 2147.

69. C. D. Buckley and K. A. McLauchlan, *Mol. Phys.* **1985**, *54*, 1.

70. K. A. McLauchlan, *Chem. Brit.* **1985**, *21*, 825.

71. F. J. Adrian, *Rev. Chem. Intermed.* **1986**, *7*, 173.

72. M. A. El Sayed, *MTP Intern. Rev. Science, Spectrosc.* **1972**, 119.

73. E. de Boer and H. van Willigen, *Progr. Nucl. Magn. Resonance Spectrosc.* **1967**, *2*, 111.

74. J. Bargon, H. Fischer, and U. Johnson, *Z. Naturforsch. A* **1967**, *22*, 1551.

75. H. R. Ward and R. G. Lawler, *J. Am. Chem. Soc.* **1967**, *89*, 5518.

76. H. D. Roth and A. A. Lamola, *J. Am. Chem. Soc.* **1974**, *96*, 6270.

77. G. L. Closs and R. J. Miller, *J. Am. Chem. Soc.* **1979**, *101*, 1639.

78. G. L. Closs and R. J. Miller, *J. Am. Chem. Soc.* **1981**, *103*, 3586.

79. G. L. Closs, *Adv. Magn. Reson.* **1974**, *7*, 157.

80. R. Kaptein, *Adv. Free Radical Chem.* **1975**, *5*, 319.

81. F. J. Adrian, *Rev. Chem. Intermed.* **1979**, *3*, 3.

82. J. H. Freed and J. B. Pedersen, *Adv. Magn. Reson.* **1976**, *8*, 2.

83. A. Berndt, M. T. Jones, M. Lehnig, L. Lunazzi, G. Placucci, H. B. Stegman, and K. Ulmschneider, "Organic Anion Radicals," in *Landolt Börnstein, Numerical Data and Functional Relationships in Science and Technology,* Vol. IX, Part d1, Springer-Verlag, Heidelberg, **1980**.

84. R. A. Forrester, K. Ishizu, G. Kothe, S. F. Nelson, H. Ohya-Nishiguchi, K. Watanabe, and W. Wilker, "Organic Cation Radicals and Polyradicals," in *Landolt Börnstein, Numerical Data and Functional Relationships in Science and Technology,* Vol. IX, Part d2, Springer-Verlag, Heidelberg, **1980**.

85. T. Shida, E. Haselbach, and T. Bally, *Acc. Chem. Res.* **1984**, *17*, 180.

86. T. Bally, in *Radical Ionic Systems*, A. Lund and M. Shiotani, Eds., Kluwer Academic, Dordrecht, The Netherlands, **1991**, p. 3.

87. P. Forster, R. Gschwind, E. Haselbach, U. Klemm, and J. Wirz, *Nouv. J. Chim.* **1980**, *4*, 365.

88. E. Haselbach, U. Klemm, R. Gschwind, T. Bally, L. Chassot, and S. Nitsche, *Helv. Chim. Acta* **1982**, *65*, 2464.

89. T. Bally, L. Neuhaus, S. Nitsche, E. Haselbach, J. Janssen, and W. Lüttke, *Helv. Chim. Acta* **1983**, *66*, 1288.

90. M. A. Komarynsky and S. I. Weissman, *J. Am. Chem. Soc.* **1975**, *97*, 1589.

91. G. V. Nelson and A. von Zelesky, *J. Am. Chem. Soc.* **1975**, *97*, 6279.

92. R. G. Lawler, J. R. Bolton, M. Karplus, and G. K. Fraenkel, *J. Chem. Phys.* **1967**, *47*, 2149.

93. F. Gerson, G. Moshuk, and M. Schwyzer, *Helv. Chim. Acta* **1971**, *54*, 1721.

94. M. K. Carter and G. Vincow, *J. Chem. Phys.* **1967**, *47*, 292.

95. T. Komatsu, A. Lund, and P. O. Kinell, *J. Phys. Chem.* **1972**, *76*, 367.

96. K. Möbius, *Z. Naturforsch. A* **1965**, *20*, 1093, 1102, 1117.

97. A. Reymond and G. K. Fraenkel, *J. Phys. Chem.* **1967**, *71*, 4570.

98. H. M. McConnel and R. E. Robertson, *J. Chem. Phys.* **1958**, *28*, 991; A. D. McLachlan, *Mol. Phys.* **1959**, *2*, 271.

99. F. Gerson, J. Heinzer, and E. Vogel, *Helv. Chim. Acta* **1971**, *54*, 1721.

100. R. M. Dessau and S. Shi, *J. Chem. Phys.* **1970**, *53*, 3169.

101. S. F. Nelsen, P. J. Hintz, J. M. Buschek, and G. R. Weisman, *J. Am. Chem. Soc.* **1975**, *97*, 4933.

102. T. Shida, Y. Nosaka, and T. Kato, *J. Phys. Chem.* **1978**, *82*, 695.

103. S. F. Nelsen and J. M. Buschek, *J. Am. Chem. Soc.* **1974**, *96*, 6424.

104. S. F. Nelsen, E. Haselbach, R. Gschwind, U. Klemm, and Z. Lanyiova, *J. Am. Chem. Soc.* **1978**, *100*, 4367.

105. S. F. Nelsen, R. W. Alder, R. B. Sessions, K.-D. Asmus, K.-O. Hiller, and M. Göbl, *J. Am. Chem. Soc.* **1980**, *102*, 1429.

106. M. Göbl, M. Bonifacic, and K.-D. Asmus, *J. Am. Chem. Soc.* **1984**, *106*, 5984.

107. S. F. Nelsen, *Acc. Chem. Res.* **1987**, *20*, 269.

108. J. C. Scaiano, S. Garcia, and H. Garcia, *Tetrahedron Lett.* **1997**, *38*, 5929.

109. S. G. Lias, *J. Phys. Chem. Ref. Data* **1988**.

110. K. D. Jordan and P. D. Burrow, *Chem. Rev.* **1987**, *87*, 557.

111. C. E. Melton, in *Mass Spectrometry of Organic Ions*, F. W. McLafferty, Ed., Academic Press, New York, **1963**, p. 163.

112. C. E. Melton, in *Principles of Mass Spectrometry and Negative Ions*, Marcel Dekker, New York, **1970**, p. 163.

113. B. K. Janousek and J. I. Brauman, in *Gas Phase Ion Chemistry*, Vol. II, M. T. Bowers, Ed., Academic Press, New York, **1979**, p. 53.

114. S. L. Miller and H. C. Urey, *Science* **1959**, *130*, 245.

115. L. B. Knight, Jr., J. Steadman, D. Feller, and E. R. Davidson, *J. Am. Chem. Soc.* **1984**, *106*, 3700.

116. M. C. R. Symons, T. Chen, and C. Glidewell, *J. Chem. Soc. Chem. Commun.* **1983**, 326.

117. T. A. Claxton, T. Chen, and M. C. R. Symons, *Faraday Disc. Chem. Soc.* **1984**, *78*, 1.

118. L. E. Eriksson, V. G. Malkin, O. L. Malkina, and D. R. Salahub, *Int. J. Quant. Chem*, **1994**, *52*, 879.

119. K. Toriyama, K. Nunome, and M. Iwasaki, *J. Am. Chem. Soc.* **1981**, *103*, 3591.

120. H. Zuilhof, J. P. Dinnocenzo, A. C. Reddy, and S. S. Shaik, *J. Phys. Chem.* **1996**, *100*, 15774.

121. T. A. Claxton, R. E. Overill, and M. C. R. Symons, *Mol. Phys.* **1974**, *27*, 701.

122. K. Toriyama, K. Nunome, and M. Iwasaki, *J. Chem. Phys.* **1982**, *77*, 5891.

123. K. Matsuura, K. Nunome, K. Toriyama, and M. Iwasaki, *J. Phys. Chem.* **1989**, *93*, 149.

124. K. Toriyama, *Chem. Phys. Lett.* **1991**, *177*, 39.

125. K. Toriyama, M. Okazaki, and K. Nunome, *J. Chem. Phys.* **1991**, *95*, 3955.

126. K. Toriyama, K. Nunome, and M. Iwasaki, *J. Am. Chem. Soc.* **1981**, *103*, 3591.

127. K. Toriyama, K. Nunome, and M. Iwasaki, *J. Phys. Chem.* **1981**, *85*, 2149.

128. J. T. Wang and F. Williams, *Chem. Phys. Lett.* **1981**, *82*, 177.

129. M. Iwasaki and K. Toriyama, *J. Am. Chem. Soc.* **1986**, *108*, 6441.

130. M. Tabata and A. Lund, *Rad. Phys. Chem.* **1984**, *23*, 545.

131. M. Lindgren, A. Lund, and G. Dolivo, *Chem. Phys.* **1985**, *99*, 103.

132. A. Lund, M. Lindgren, G. Dolivo, and M. Tabata, *Rad. Phys. Chem.* **1985**, *26*, 491.

133. G. Dolivo and A. Lund, *J. Phys. Chem.* **1985**, *89*, 3977.

134. G. Dolivo and A. Lund, *Z. Naturforsch. A* **1985**, *40*, 52–65.

135. K. Toriyama, K. Nunome, and M. Iwasaki, *J. Am. Chem. Soc.* **1987**, *109*, 4496.

136. R. Gleiter, *Top. Curr. Chem.* **1980**, *86*, 197.

137. M. Iwasaki, K. Toriyama, and K. Nunome, *J. Chem. Soc. Chem. Comm.* **1983**, 202.

138. K. Ohta, H. Nakatsuji, H. Kubodera, and T. Shida, *Chem. Phys.* **1983**, *76*, 271.

139. M. Iwasaki, K. Toriyama, and K. Nunome, *Faraday Disc. Chem. Soc.* **1984**, *78*, 19.

140. X. Z. Qin, L. D. Snow, and F. Williams, *J. Am. Chem. Soc.* **1984**, *106*, 7640.

141. D. D. M. Wayner, R. J. Boyd, and D. R. Arnold, *Can. J. Chem.* **1985**, *63*, 3283.

142. P. Du, D. A. Hrovat, and W. T. Borden, *J. Am. Chem. Soc.* **1988**, *110*, 3405.

143. A. Skancke, *J. Phys. Chem.* **1995**, *99*, 13886.

144. E. W. Schlag and B. S. Rabinovitch, *J. Am. Chem. Soc.* **1960**, *82*, 5996.

145. H. D. Roth and M. L. M. Schilling, *J. Am. Chem. Soc.* **1983**, *105*, 6805.

146. H. D. Roth, *Acc. Chem. Res.* **1987**, *20*, 343.

147. R. C. Haddon and H. D. Roth, *Croat. Chem. Acta* **1984**, *57*, 1165.

148. K. Krogh-Jespersen and H. D. Roth, *J. Am. Chem. Soc.* **1992**, *114*, 8388.

149. T. Herbertz and H. D. Roth, *J. Am. Chem. Soc.* **1998**, *120*, 11904.

150. T. Herbertz, P. S. Lakkaraju, H. D. Roth, G. Sluggett, and N. J. Turro, *J. Phys. Chem. A* **1999**, *103*, 11350.

151. J. P. Dinnocenzo, H. Zuilhof, D. R. Lieberman, T. R. Simpson, and M. W. McKechney, *J. Am. Chem. Soc.* **1997**, *119*, 994.

152. T. Herbertz, F. Blume, and H. D. Roth, *J. Am. Chem. Soc.* **1998**, *120*, 4591.

153. A. Almennigen, O. Bastiansen, and P. N. Skancke, *Acta Chim. Scand.* **1961**, *15*, 711.

154. K. Ushida, T. Shida, M. Iwasaki, K. Toriyama, and K. Numone, *J. Am. Chem. Soc.* **1983**, *105*, 5496.

155. P. Jungwirth, P. Carsky, and T. Bally, *J. Am. Chem. Soc.* **1993**, *115*, 5776.

156. O. Wiest, *J. Phys. Chem. A* **1999**, *103*, 7907.

157. T. R. Evans, R. W. Wake, and O. Jaenicke, in *The Exciplex*, M. Gordon and W. R. Ware, Eds., Academic Press, New York, **1975**, p. 345.

158. R. Hoffmann, *Acc. Chem. Res.* **1971**, *4*, 1.

159. E. Heilbronner and A. Schmelzer, *Helv. Chim. Acta* **1975**, *58*, 936.

160. H. D. Roth, M. L. M. Schilling, and G. Jones, *J. Am. Chem. Soc.* **1981**, *103*, 1246.

161. H. D. Roth and M. L. M. Schilling, *J. Am. Chem. Soc.* **1981**, *103*, 7210.

162. E. Haselbach, T. Bally, Z. Lanyiova, and P. Baertschi, *Helv. Chim. Acta* **1979**, *62*, 583.

163. K. Raghavachari, R. C. Haddon, and H. D. Roth, *J. Am. Chem. Soc.* **1983**, *105*, 3110.

164. T. Herbertz and H. D. Roth, **2001**, unpublished results.

165. F. Gerson and X.-Z. Qin, *Helv. Chim. Acta* **1989**, *72*, 383.

166. K. Ishiguro, I. V. Khudyakov, P. F. McGarry, N. J. Turro, and H. D. Roth, *J. Am. Chem. Soc.* **1994**, *116*, 6933.

167. F. Gerson, *Acc. Chem. Res.* **1994**, *27*, 63.

168. C. J. Abelt, H. D. Roth, and M. L. M. Schilling, *J. Am. Chem. Soc.* **1985**, *107*, 4148.

169. F. Gerson, A. Arnold, and U. Burger, *J. Am. Chem. Soc.* **1991**, *113*, 4359.

170. R. Gleiter, K. Gubernator, M. Eckert-Maksic, J. Spanget-Larsen, B. Bianco, G. Gandillion, and U. Burger, *Helv. Chim. Acta* **1981**, *64*, 1312.

171. F. Williams, Q. X. Guo, T. M. Kolb, and S. F. Nelsen, *J. Chem. Soc., Chem. Commun.* **1989**, 1835.

172. H. D. Roth and M. L. M. Schilling, *J. Am. Chem. Soc.* **1985**, *107*, 716.

173. H. D. Roth, M. L. M. Schilling, and C. J. Abelt, *Tetrahedron* **1986**, *42*, 6157.

174. H. D. Roth, M. L. M. Schilling, and C. J. Abelt, *J. Am. Chem. Soc.* **1986**, *108*, 6098.

175. T. Momose, T. Shida, and T. Kobayashi, *Tetrahedron* **1986**, *42*, 6337.

176. S. Dai, J. T. Wang, and F. Williams, *J. Am. Chem. Soc.* **1990**, *112*, 2835, 2837.

177. H. D. Roth, *Proc. IUPAC Symp. Photochem.* **1984**, *10*, 455.

178. H. D. Roth and C. J. Abelt, *J. Am. Chem. Soc.* **1986**, *108*, 2013.

179. Q. X. Guo, X. Z. Qin, J. T. Wang, and F. Williams, *J. Am. Chem. Soc.* **1988**, *110*, 1974.

180. F. Williams, Q. X. Guo, D. C. Bebout, and B. K. Carpenter, *J. Am. Chem. Soc.* **1989**, *111*, 4133.

181. Y. Takahashi, T. Mukai, and T. Miyashi, *J. Am. Chem. Soc.* **1983**, *105*, 6511.

182. T. Miyashi, Y. Takahashi, T. Mukai, H. D. Roth, and M. L. M. Schilling, *J. Am. Chem. Soc.* **1985**, *107*, 1079.

183. H. D. Roth, M. L. M. Schilling, and C. C. Wamser, *J. Am. Chem. Soc.* **1984**, *106*, 5023.

184. B. F. Yates, W. J. Bouma, and L. Radom, *J. Am. Chem. Soc.* **1984**, *106*, 5805.

185. W. J. Bouma, L. Radom, R. H. Nobes, and B. F. Yates, *Pure Appl. Chem.*, **1984**, *56*, 1831.

186. L. Radom, W. J. Bouma, and B. F. Yates, *Tetrahedron* **1986**, *22*, 6225.

187. G. S. Hammond and R. C. Neuman, *J. Am. Chem. Soc.* **1963**, *85*, 1501.

188. C. M. Camaggi, M. J. Perkins, and P. Ward, *J. Chem. Soc. B* **1971**, 2416.

189. W. J. Bouma, J. K. MacLeod, and L. Radom, *J. Chem. Soc., Chem. Commun.* **1978**, 724.

190. T. M. Sack, D. L. Miller, and M. L. Gross, *J. Am. Chem. Soc.* **1985**, *107*, 6795.

191. M. L. Gross, *J. Am. Chem. Soc.* **1972**, *94*, 3744.

192. H. Basch, M. B. Robin, N. A. Kuebler, C. Baker, and D. W. Turner, *J. Chem. Phys.* **1969**, *51*, 52.

193. L. D. Snow, J. T. Wang, and F. Williams, *Chem. Phys. Lett.* **1983**, *100*, 193.

194. M. L. M. Symons and B. W. Wren, *Tetrahedron Lett.* **1983**, *24*, 2315.

195. T. Bally, S. Nitsche, and E. Haselbach, *Helv. Chim. Acta* **1984**, *67*, 86.

196. F. Williams, S. Dai, L. D. Snow, X.-Z. Qin, T. Bally, S. Nitsche, E. Haselbach, S. Nelsen, and M. F. Teasley, *J. Am. Chem. Soc.* **1987**, *109*, 7526.

197. C. B. Quinn, J. R. Wiseman, and J. C. Calabrese, *J. Am. Chem. Soc.* **1973**, *95*, 6121.

198. D. Feller, E. R. Davidson, and W. T. Borden, *J. Am. Chem. Soc.* **1984**, *106*, 2513.

199. X.-Z. Qin, L. D. Snow, and F. Williams, *J. Phys. Chem.* **1985**, *89*, 3602.

200. R. H. Nobes, W. J. Bouma, J. K. MacLeod, and L. Radom, *Chem. Phys. Lett.* **1987**, *135*, 78.

201. K. M. Stirk, L. K. M. Kiminkinen, and H. I. Kenttämaa, *Chem. Rev.* **1992**, *92*, 1649.

202. D. Wittneben and H.-F. Grützmacher, *Int. J. Mass Spectrom. Ion Processes* **1990**, *100*, 545.

203. L. J. Chyall and H. I. Kenttämaa, *J. Am. Chem. Soc.* **1994**, *116*, 3135.

204. K. K. Thoen, R. L. Smith, J. J. Nousiainen, E. N. Nelson, and H. I. Kenttämaa, *J. Am. Chem. Soc.* **1996**, *118*, 8669.

205. P. G. Wenthold, J. A. Paulino, and R. R. Squires, *J. Am. Chem. Soc.* **1991**, *113*, 7414.

206. P. G. Wenthold, R. R. Squires, and W. C. Lineberger, *J. Am. Chem. Soc.* **1998**, *120*, 5279.

207. J. Hu, B. T. Hill, and R. R. Squires, *J. Am. Chem. Soc.* **1997**, *119*, 11699.

208. K. K. Thoen and H. I. Kenttämaa, *J. Am. Chem. Soc.* **1997**, *119*, 3832.

209. J. R. Miller, L. T. Calcaterra, and G. L. Closs, *J. Am. Chem. Soc.* **1984**, *106*, 3047.

210. G. L. Closs and J. R. Miller, *Science*, **1998**, *240*, 440.

211. H. Weng, Q. Sheik, and H. D. Roth, *J. Am. Chem. Soc.* **1995**, *117*, 10655.

212. F. D. Lewis, J. R. Petisce, J. D. Oxman, and M. J. Nepras, *J. Am. Chem. Soc.* **1985**, *107*, 203.

213. D. Sheng, R. S. Pappas, G.-F. Chen, Q. X. Guo, J. T. Wang, and F. Williams, *J. Am. Chem. Soc.* **1989**, *111*, 8759.

214. G. Jones, W. G. Becker, and S.-H. Chiang, *J. Photochem.* **1982**, *19*, 245.

215. T. R. Evans, R. W. Wake, and M. N. Sifain, *Tetrahedron Lett.* **1973**, 701.

216. K. Okada, K. Hisamitsu, T. Miyashi, and T. Mukai, *J. Chem. Soc., Chem. Commun.* **1982**, 974.

217. E. Hasegawa, T. Mukai, and K. Yanagi, unpublished results quoted by F. D. Lewis, in *Photoinduced Electron Transfer*, M. A. Fox and M. Chanon, Eds., Part C, 1, Elsevier, Amsterdam, The Netherlands, **1988**.

218. M. Shiotani, K. Ohta, Y. Nagata, and J. Sohma, *J. Am. Chem. Soc.* **1985**, *107*, 2562.

219. M. A. Fox, K. A. Campbell, S. Hünig, H. Berneth, G. Maier, K. A. Schneider, and K. D. Malsch, *J. Org. Chem.* **1982**, *47*, 3408.

220. S. Hammerum, *Mass Spectrom. Rev.* **1988**, *7*, 123.

221. K. M. Stirk and H. I. Kenttämaa, *J. Am. Chem. Soc.* **1990**, *112*, 5880.

222. H. D. Roth and A. A. Lamola, *J. Am. Chem. Soc.* **1972**, *94*, 1013.

223. A. A. Lamola, *Mol. Photochem.* **1972**, *4*, 107.

224. J. Kemmink, A. P. M. Eker, and R. Kaptein, *Photochem. Photobiol.* **1986**, *44*, 137.

225. T. Young, R. Nieman, and S. D. Rose, *Photochem. Photobiol.* **1990**, *52*, 661.

226. J. R. Van Camp, T. Young, R. F. Hartman, and S. D. Rose, *Photochem. Photobiol.* **1987**, *45*, 365.

227. S. T. Kim, R. F. Hartman, and S. D. Rose, *Photochem. Photobiol.* **1990**, *52*, 789.

228. A. Weller and K. Zachariasse, *J. Chem. Phys.* **1967**, *46*, 4984.

229. H. Grellmann, A. R. Watkins, and A. Weller, *J. Lumin.* **1970**, *1,2*, 678.

230. M. Ottolenghi, *Acc. Chem. Res.* **1973**, *6*, 153.

231. G. N. Taylor, cited in H. D. Roth, *Mol. Photochem.* **1973**, *5*, 91.

232. K. Schulten, H. Staerk, A. Weller, H. J. Werner, and B. Nickel, *Z. Phys. Chem. NF* **1976**, *101*, 371.

233. M. E. Michel-Beyerle, R. Haberkorn, W. Bube, E. Steffens, H. Schröder, N. J. Neusser, E. W. Schlag, and H. Seidlitz, *Chem. Phys.* **1976**, *17*, 139.

234. W. Bube, R. Haberkorn, and M. E. Michel-Beyerle, *J. Am. Chem. Soc.* **1978**, *100*, 5993.

235. H. D. Roth, *J. Photochem. Photobiol., C: Photochem Rev.* **2001**, *2*, 93.

236. H. D. Roth and M. L. M. Schilling, *J. Am. Chem. Soc.* **1980**, *102*, 7956.

237. S. B. Karki, J. P. Dinnocenzo, S. Farid, J. L. Goodman, I. Gould, and T. A. Zona, *J. Am. Chem. Soc.* **1997**, *119*, 431.

238. H. D. Roth, M. L. M. Schilling, and G. Jones, II, *J. Am. Chem. Soc.* **1981**, *103*, 1246.

239. A. Cuppoletti, J. P. Dinnocenzo, J. L. Goodman, and I. R. Gould, *J. Phys. Chem. A* **1999**, *103*, 11253.

240. H. Ikeda, T. Nakamura, T. Miyashi, J. L. Goodman, K. Akiyama, S. Tero-Kubota, A. Houmam, and D. D. M. Wayner, *J. Am. Chem. Soc.* **1998**, *120*, 5832.

241. H. Ikeda, T. Minegishi, H. Abe, A. Konno, J. L. Goodman, and T. Miyashi, *J. Am. Chem. Soc.* **1998**, *120*, 87.

242. M.-J. Climent, M. A. Miranda, and H. D. Roth, *Eur. J. Org. Chem.* **2000**, 1563.

243. M. Bellas, D. Bryce-Smith, and A. Gilbert, *J. Chem. Soc., Chem. Comm.* **1967**, 862.

244. D. Bryce-Smith, A. Gilbert, and C. Manning, *Angew. Chem. Int. Ed. Engl.* **1974**, *86*, 350.

245. F. D. Lewis and T.-I. Ho, *J. Am. Chem. Soc.* **1980**, *102*, 1751.

246. W. Hub, S. Schneider, F. Dörr, J. T. Simpson, J. D. Oxman, and F. D. Lewis, *J. Am. Chem. Soc.* **1982**, *104*, 2044.

247. U. C. Yoon and P. S. Mariano, *Acc. Chem. Res.* **1992**, *25*, 233.

248. D. R. Arnold, P. C. Wong, A. J. Maroulis, and T. S. Cameron, *Pure Appl. Chem.* **1980**, *52*, 2609.

249. P. M. Rentzepis, D. W. Steyert, H. D. Roth, and C. J. Abelt, *J. Phys. Chem.* **1985**, *89*, 3955.

250. J. Eriksen and C. S. Foote, *J. Am. Chem. Soc.* **1980**, *102*, 6083.

251. A. P. Schaap, K. A. Zaklika, B. Kaskar, and L. W.-M. Fung, *J. Am. Chem. Soc.* **1980**, *102*, 389.

252. L. T. Spada and C. S. Foote, *J. Am. Chem. Soc.* **1980**, *102*, 393.

253. S. L. Mattes and S. Farid, *J. Chem. Soc. Chem. Commun.* **1980**, 457.

254. A. P. Schaap, S. Siddiqui, G. Prasad, A. F. M. Magsudur-Rahman, and J. P. Oliver, *J. Am. Chem. Soc.* **1984**, *106*, 6087.

255. T. Miyashi, H. Ikeda, and Y. Takahashi, *Acc. Chem. Res.* **1999**, *32*, 815.

256. J. Eriksen, C. S. Foote, and T. L. Parker, *J. Am. Chem. Soc.* **1977**, *99*, 6455.

257. D. H. R. Barton, G. Leclerc, P. D. Magnus, and I. D. Menzies, *J. Chem. Soc. Chem. Commun.* **1972**, 447.

258. S. F. Nelsen and R. Akaba, *J. Am. Chem. Soc.* **1981**, *103*, 1096.

259. E. L. Clennan, W. Simmons, and C. W. Almgren, *J. Am. Chem. Soc.* **1981**, *103*, 2098.

260. P. G. Gassman and K. D. Olson, *Tetrahedron Lett.* **1983**, 19.

261. H. Weng and H. D. Roth, *J. Org. Chem.* **1995**, *60*, 4136.

262. M. Goez and I. Frisch, *J. Am. Chem. Soc.* **1995**, *117*, 10486.

263. S. Farid, S. E. Hartman, and T. R. Evans, in *The Exciplex*, M. Gordon and W. R. Ware, Eds., Academic Press, New York, **1975**, p. 327.

264. S. Kuwata, Y. Shigemitsu, and Y. Odaira, *J. Org. Chem.* **1973**, *38*, 3803.

265. R. A. Crellin and A. Ledwith, *Macromolecules* **1975**, *8*, 93.

266. M. Yasuda, C. Pac, and H. Sakurai, *Bull. Chem. Soc. Jpn.* **1980**, *53*, 502.

267. H. D. Roth and R. S. Hutton, *J. Phys. Org. Chem.* **1990**, *3*, 119.

268. M. Kajima, H. Sakuragi, and N. Tokumaru, *Tetrahedron Lett.* **1981**, 2889.

269. R. A. Crellin, M. C. Lambert, and A. Ledwith, *J. Chem. Soc., Chem. Commun.* **1970**, 682.

270. R. A. Neunteufel and D. R. Arnold, *J. Am. Chem. Soc.* **1973**, *95*, 4080.

271. T. Ichikawa, M. Yamaguchi, and H. Yoshida, *J. Phys. Chem.* **1987**, *91*, 6400.

272. P. L. Corio and S. Shih, *J. Phys. Chem.* **1971**, *75*, 3475.

273. M. F. Desrosiers and A. D. Trifunac, *J. Phys. Chem.* **1986**, *90*, 1560.

274. P. Jungwirth and T. Bally, *J. Am. Chem. Soc.* **1993**, *115*, 5783.

275. S. Hashimoto, *Chem. Phys. Lett.* **1996**, *262*, 292.

276. K.-D. Asmus, *Acc. Chem. Res.*, **1979**, *12*, 436.

277. J. L. Courtneidge and A. G. Davies, *Acc. Chem. Res.* **1987**, *20*, 90.

278. S. L. Mattes and S. Farid, *Acc. Chem. Res.* **1982**, *80*, 126.

279. J. Libman, *J. Chem. Soc., Chem. Comm.* **1976**, 361.

280. R. Schutte and G. R. Freeman, *J. Am. Chem. Soc.* **1969**, *91*, 3715.

281. N. Bauld, D. J. Bellville, B. Harirchian, K. T. Lorenz, R. A. Pabon, D. W. Reynolds, D. D. Wirth, H.-S. Chiou, and B. K. Marsh, *Acc. Chem. Res.* **1987**, *20*, 371.

282. H. D. Roth, M. L. M. Schilling, and C. J. Abelt, *J. Am. Chem. Soc.* **1986**, *108*, 6098.

283. H. D. Roth, M. L. M. Schilling, and C. J. Abelt, *Tetrahedron* **1986**, *42*, 6157.

284. N. Bauld and D. X. Gao, *J. Chem. Soc. Perkin Trans. 2*, **2000**, 931.

285. N. J. Saettel, O. Wiest, D. A. Singleton, and M. P. Meyer, *J. Am. Chem. Soc.* **2002**, *124*, 11552.

286. U. Haberl, E. Steckhan, S. Blechert, and O. Wiest, *Chem. Eur. J.* **1999**, *5*, 2859.

287. O. Wiest and E. Steckhan, *Angew. Chem.* **1993**, *105*, 932; *Angew. Chem., Int. Ed. Engl.*, **1993**, *32*, 901.

288. P. W. Rabideau and C. Marcinow, *Org. React.* **1992**, *23*, 1.

289. *Organic Electrochemistry*, H. Lund and M. M. Baizer, Eds., Marcel Dekker, New York, **1991**.

290. A. J. Maroulis, Y. Shigemitsu, and D. R. Arnold, *J. Am. Chem. Soc.* **1978**, *100*, 535.

291. V. R. Rao and S. S. Hixson, *J. Am. Chem. Soc.* **1979**, *101*, 6458.

292. J. P. Dinnocenzo, W. P. Todd, T. R. Simpson, and I. R. Gould, *J. Am. Chem. Soc.* **1990**, *112*, 2462.

293. D. R. Arnold and X. Du, *Can. J. Chem.* **1994**, *72*, 403.

294. M. Wlostowski and H. D. Roth, unpublished results.

295. A. Pross, *J. Am. Chem. Soc.* **1986**, *108*, 3537.

296. S. S. Shaik and A. Pross, *J. Am. Chem. Soc.* **1989**, *111*, 4306.

297. L. Eberson, M. P. Hartshorn, F. Radner, M. Merchan, and B. O. Roos, *Acta Chem. Scand.* **1993**, *47*, 176.

298. S. S. Shaik and J. P. Dinnocenzo, *J. Org. Chem.* **1990**, *55*, 3434.

299. S. S. Shaik, A. C. Reddy, A. Ioffe, J. P. Dinnocenzo, D. Danovich, and J. K. Cho, *J. Am. Chem. Soc.* **1995**, *117*, 3205.

300. Y. Takemoto, T. Ohra, H. Koike, S.-I. Furuse, C. Iwata, and H. Ohishi, *J. Org. Chem.* **1994**, *59*, 4727.

301. H. D. Roth and T. Herbertz, *J. Am. Chem. Soc.* **1993**, *115*, 9804.

302. D. R. Arnold and X. J. Du, *J. Am. Chem. Soc.* **1989**, *111*, 7666.

303. H. D. Roth, H. Weng, and V. Sethuraman, *J. Am. Chem. Soc.* **1994**, *116*, 7021.

304. K. Mizuno and J. Ogawa, *Chem. Lett.* **1981**, 741.

305. D. R. Arnold and M. S. Snow, *Can. J. Chem.* **1988**, *66*, 3012.

306. D. Mangione and D. R. Arnold, *Acc. Chem. Res.* **2002**, *35*, 297.

307. P. H. Mazzocchi, C. Somich, M. Edwards, T. Morgan, and H. L. Ammon, *J. Am. Chem. Soc.* **1986**, *108*, 6828.

308. K. Ohga and P. S. Mariano, *J. Am. Chem. Soc.* **1982**, *104*, 617.

309. K. Mizuno, M. Ikeda, and Y. Otsuji, *Chem. Lett.* **1988**, 1507.

310. K. Mizuno and Y. Otsuji, *Top. Curr. Chem.* **1994**, *169*, 301.

311. T. Nishiyama, K. Mizuno, Y. Otsuji, and H. Inoue, *Tetrahedron* **1995**, *51*, 6695.

312. W. P. Todd, J. P. Dinnocenzo, S. Farid, J. L. Goodman, and I. R. Gould, *Tetrahedron Lett.* **1993**, *34*, 2863.

313. N. Hirota, in *Radical Ions*, E. T. Kaiser and L. Kevan, Eds., John Wiley & Sons, Inc., New York, **1968**, p. 35.

314. J. March, *Advanced Organic Chemistry*, 4th edn., John Wiley & Sons, Inc., New York, **1992**, p. 1228.

315. G. A. Russel, in *Radical Ions*, E. T. Kaiser and L. Kevan, Eds., Interscience, New York, **1968**, p. 87.

316. J. F. Garst, in *Free Radicals*, Vol. I, J. K. Kochi, Ed., John Wiley & Sons, Inc., New York, **1973**, p. 518.

317. H. L. Strauss, T. J. Katz, and G. K. Fraenkel, *J. Am. Chem. Soc.* **1963**, *85*, 2360.

318. T. J. Katz and H. L. Strauss, *J. Chem. Phys.* **1960**, *32*, 1873.

CHAPTER 7

Singlet Carbenes

MAITLAND JONES, Jr.

Department of Chemistry, Princeton University, Princeton, NJ

ROBERT A. MOSS

Department of Chemistry and Chemical Biology, Rutgers, The State University
of New Jersey, New Brunswick, NJ

Reactive Intermediate Chemistry, edited by Robert A. Moss, Matthew S. Platz, and Maitland Jones, Jr.
ISBN 0-471-23324-2   Copyright © 2004 John Wiley & Sons, Inc.

1. INTRODUCTION

1.1. Prototypal Reactions: Addition and Insertion Reactions

Nearly 50 years ago Doering et al.[1a] and Dvoretzky et al.[1b] demonstrated the utterly unselective nature of the carbon–hydrogen insertion reactions of methylene, Hine[2] elucidated the intermediacy of dichlorocarbene in the hydrolysis of chloroform, and Doering and Hoffmann found that CCl_2 could be captured by alkenes to yield cyclopropanes. These seminal reports initiated a one-half century of vigorous research on the chemistry of carbenes, now described in a myriad of journal articles and reviews, as well as in a more manageable number of monographs. Among the latter are the classic "first generation" contributions of Hine[4] and Kirmse,[5] the two volume set edited by Moss and Jones,[6,7] the mammoth reference work edited by Regitz,[8] and more recent edited volumes by Platz,[9] Brinker,[10–12] and Bertrand.[13]

In *Reactive Intermediate Chemistry*, five chapters focus on various aspects of carbene chemistry: the present chapter (Chapter 7), devoted to singlet carbenes, Tomioka's chapter on triplet carbenes (Chapter 9 in this volume), Doyle's consideration of synthetic aspects of carbene (and nitrene) chemistry (Chapter 12 in this volume), Bertrand's essay on stable singlet carbenes (Chapter 8 in this volume), and Shevlin's discussion of the chemistry of atomic carbon (Chapter 10 in this volume). A major theme throughout is the impact of new methods, especially the direct observation of carbenes and high-level computation of their structures and energetics. Over the past 20 years, these techniques have provided increasingly detailed and quantitatively precise mechanisms for the reactions of carbenes and, indeed, reactive intermediates in general. In the specific area of singlet carbene chemistry, we largely focus our discussion on three prototypal carbene reactions: the intermolecular carbene–alkene addition reaction, and the intermolecular and intramolecular insertion reactions. In the discussion there will be some necessary "leakage" of triplet carbene chemistry in order to make comparisons.

1.2. Structure and Bonding

The simple formulation of a carbene as $:CR_2$ hides much of what has fascinated chemists about carbenes for the last five decades. The idea that a molecule as simple as $:CH_2$ can be interesting structurally is deliciously oxymoronic! What is there to say about a three-atom molecule such as methylene? The answer to that question is not only far more complex than it appears on the surface, but the developing of the experimental and theoretical answers has provided an instructive back and forth that showed many of the limitations of both experiment and theory. With the aid of hindsight, one can use the unraveling of the structure of methylene as an exemplar of why it is the cooperative application of theory and experiment that best advances our knowledge.

Let's start with those two nonbonding electrons. First of all, they can be either paired (singlet state) or unpaired (triplet state). For both singlet and triplet, we need first to know what orbitals contain those electrons. Of course, this first question has a geometric component—before we can explore the "where" question, we have to know what's possible and the answer to that question depends on geometry. Now the simplicity of the system helps—three points determine a plane and so the geometry question for methylene (and by extension for many other carbenes) quickly devolves to "linear or bent?" We will start with the simpler linear system, build a qualitative molecular orbital (MO) picture, and then bend those orbitals to get the angular system. Finally, we will have to see what theory says about the relative energies of these two forms for both the paired (singlet) and unpaired electron (triplet) cases.

The raw material consists of the atomic orbitals of carbon, $2s$, $2p_x$, $2p_y$ and $2p_z$, and the bonding and antibonding MOs of hydrogen, Ψ_B and Ψ_A. Considerations of energy and symmetry lead to the interactions shown in Figure 7.1. The interactions $\Psi_B \pm 2s$, and $\Psi_A \pm 2p_x$ give four of the six molecular orbitals of linear methylene.

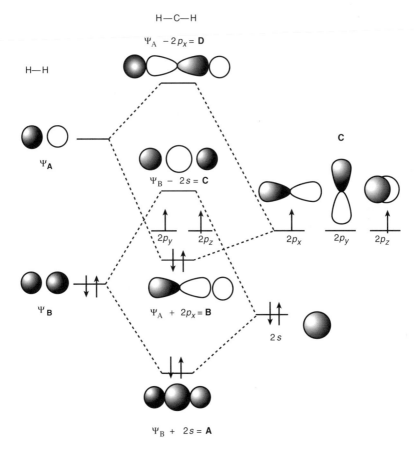

Figure 7.1

The remaining two are the unchanged $2p_y$ and $2p_z$ orbitals. There are six electrons to be placed in these orbitals, and for the linear structure, the lowest two, **A** and **B**, will be filled, and the $2p_y$, and $2p_z$ will have a single electron each. Hund's rule suggests that the triplet state with parallel spins will be lower in energy than the singlet state (Fig. 7.1). For an related, alternative valence-shell electron-pair repulsion (VESPR) theory treatment see Chapter 11 by Platz in this volume.

What happens to these orbitals as we deform the linear structure to construct the bent form (Fig. 7.2)? In two orbitals, **A** and **C**, a new bonding interaction develops

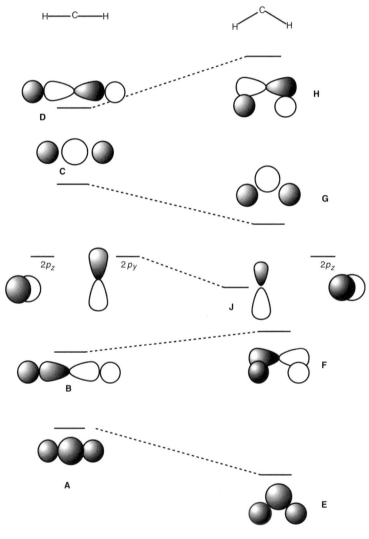

Figure 7.2

on bending to give **E** and **G**, and these orbitals will drop in energy as compared to the linear arrangement. In two others, **B** and **D**, a new antibonding interaction appears and these orbitals will rise in energy on bending to produce **F** and **H**. The $2p_z$ orbital is orthogonal to the plane determined by the three atoms and will not change in energy on bending. It is the last orbital, the $2p_y$, that is the most inter-esting, because it drops in energy on bending as it gains s character in **J**. Thus, the pattern of orbitals for a bent species is as shown in Figure 7.2.

Now we need to answer two questions: Where do the electrons go, and which spin state is lower for the bent carbene, singlet or triplet? There is a trade-off between orbital energy and electron–electron correlation (basically repulsion) between the possibilities shown in Figure 7.3. The lowest singlet (**1**) is surely the bent structure with the three lowest energy MOs filled, and the lowest triplet will have one electron in the third- and fourth-highest energy orbitals as in linear **2** or bent **3**. We now need to decide whether **2** or **3** is lower in energy, and, having done that, decide whether the "winner" is higher or lower than the lowest singlet (**1**). These are *not* easy questions to answer! Indeed, the chemical world argued for some decades over them. Finally, it became clear from increasingly sophisticated theoretical and experimental work that, at least for methylene the bent triplet (**3**) was the overall minimum, with a H—C—H angle of $\sim 137°$. The bent singlet state (**2**) lies some 9 kcal/mol higher, and the H—C—H angle has decreased to $\sim 102°$.[14] Nature has "chosen" to compromise, gaining some measure of the stabilization afforded by the lowering of orbital energy, while keeping the electrons farther apart than in a singlet state (Fig. 7.3). In addition, in the linear triplet the

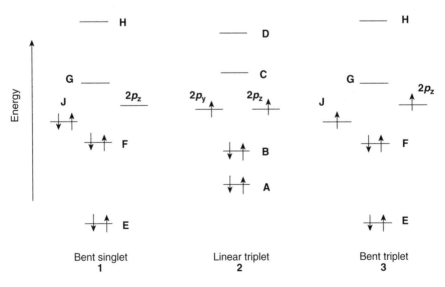

Figure 7.3

carbon–hydrogen bonds have more *s* character than in a bent arrangement, and are therefore stronger.

Of course, for carbenes more complex than methylene we will have to revisit these questions, and, as the competition between spin states is a close one, we can expect a good deal of variation in the answers for simple carbenes. For example, alkyl groups such as methyl can help stabilize the singlet through a hyperconjugative interaction with the empty $2p$ orbital, and sterically demanding groups, for example, phenyl, *tert*-butyl, or adamantyl, should favor the triplet state by demanding a wide R–C–R angle. Atoms with nonbonding electrons that can donate those electrons into the empty $2p$ orbital of a singlet are especially effective at producing ground-state singlets (Fig. 7.4). Difluorocarbene, :CF_2, is a classic example, but see Chapter 8 in this volume for a discussion of the extraordinarily stable aminocarbenes and phosphinocarbenes.

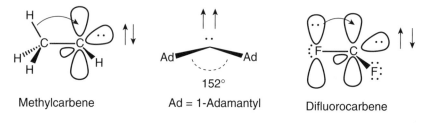

| Methylcarbene | Ad = 1-Adamantyl | Difluorocarbene |

Figure 7.4

Table 7.1 summarizes the ground states for a variety of structurally diverse carbene types and Table 7.2 gives a few specific values for the energy difference between singlets and triplets for some common carbenes. As you can see, in many cases the two reactive intermediates are very close in energy.

TABLE 7.1. Ground States for Typical Carbenes

Carbene	Ground State
:CH_2	Triplet
:CR_2	Singlet or triplet
H$\ddot{C}$COOR	Triplet
:$C(COOR)_2$	Triplet
:$C(C_6H_5)_2$	Triplet
:CAr_2	Triplet
H$\ddot{C}C_6H_5$	Triplet
:CX_2 (X=F, Cl, Br, I)	Singlet
H$\ddot{C}$X	Singlet
:$C(OR)_2$	Singlet
:$C(NR_2)_2$	Singlet

TABLE 7.2. Singlet–Triplet Gaps for a Few Typical Carbenes[a]

Carbene	Singlet–Triplet Gap	Method	References
:CH$_2$	9	Theory	14b,15
	9	Experiment	16,17,18
:CHCH$_3$	3–5	Theory	19
:C(CH$_3$)$_2$	−1.4	Theory	20a
	−1.6	Theory	20b
HC̈C$_6$H$_5$	2.5	Theory	21a
	2.3	Experiment	21b
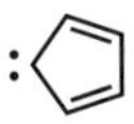 :⬠	4.3	Theory	22
:CF$_2$	−57	Experiment	23

[a] In kcal/mol; positive value indicates triplet ground state.

2. MORE DETAILED TREATMENTS OF STRUCTURE AND PROTOTYPAL REACTIONS

2.1. Addition Reactions

2.2.1. Philicity. A principal feature of the carbene–alkene addition reaction (Scheme 7.1) is the carbene's "philicity," that is the electronic character of its selectivity or response to the alkene's substituents.[24–26] Early work of Skell and Garner[27] and Doering and Henderson[28] showed that CBr$_2$ and CCl$_2$ preferentially

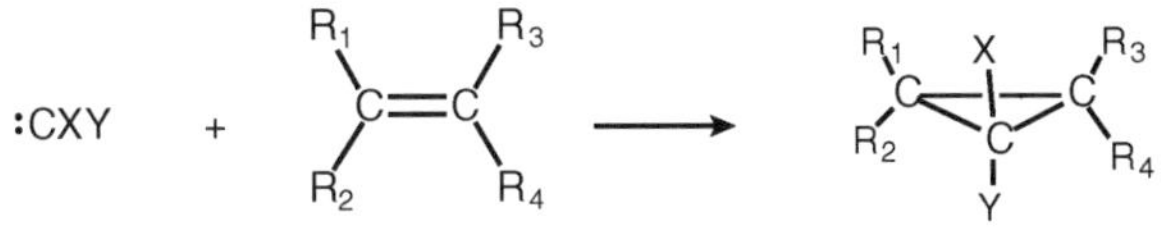

Scheme 7.1

reacted with more highly alkylated olefins (Fig. 7.5). Thus, these carbenes appeared to be *electrophiles* and transition states such as **4** were drawn[25] to represent π-electron donation from the alkene to the carbene (Fig. 7.5).

At that time, however, it was not possible to measure the absolute rates of carbene additions to alkenes, which were too rapid. Accordingly, relative reactivities

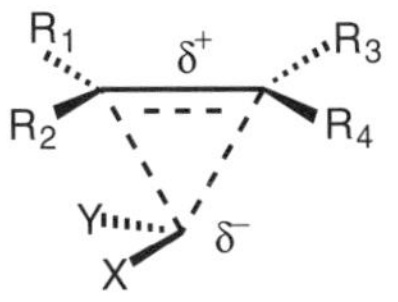

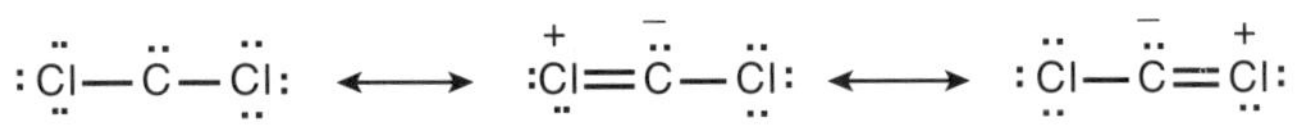

Figure 7.5

(approximately equal to relative rate constants) were determined. Alkenes a and b were allowed to compete for an insufficiency of carbene. From the mole ratio of the corresponding product cyclopropanes, corrected for the mole ratio of the initial alkenes, one obtained the relative reactivity k_a/k_b of the carbene toward the alkene pair.[29]

Relative reactivities for the electrophilic carbenes CH_3CCl and CCl_2 appear in Table 7.3,[26,30–33] in which they can be seen to react most rapidly with the most highly alkylated olefins, $(CH_3)_2C=C(CH_3)_2$, $(CH_3)_2C=CHCH_3$, less rapidly with disubstituted *trans*-butene (designated as the "standard" alkene, and assigned $k_{rel} = 1.00$), and least rapidly with such electron-poor olefins as methyl acrylate or acrylonitrile.

TABLE 7.3. Relative Reactivities of Carbenes Toward Alkenes[a]

Alkene	CH_3CCl[b]	CCl_2[c]	CH_3OCCl[d]	C_6H_5OCCl[e]	CH_3OCCH_3[f]
$(CH_3)_2C=C(CH_3)_2$	7.44	78.4	12.6	3.0	
$(CH_3)_2C=CHCH_3$	4.69				2.13
$(CH_3)_2C=CH_2$	1.92	4.89	5.43	7.3	2.18
trans-$CH_3CH=CHCH_3$	1.00	1.00	1.00	1.00	1.00
$CH_2=CHCOOCH_3$	0.078	0.060	29.7	3.7	362
$CH_2=CHCN$	0.074	0.047	54.6	5.5	686

[a] Data are at 25 °C except for CCl_2 (80 °C). All carbenes were generated from diazirines except CCl_2 which was produced by thermolysis of $C_6H_5HgCClBr$.
[b] References 30a,b.
[c] Reference 30b.
[d] Reference 31.
[e] Reference 32.
[f] Reference 33.

We also observe that CCl_2 shows greater kinetic discrimination than CH_3CCl over the set of substrate alkenes; the fastest/slowest relative reactivity ratios are 1668 for CCl_2 and 100 for CH_3CCl. This observation accords with simple considerations suggesting that CCl_2 is the more stabilized, and therefore less reactive carbene. As an approximately sp^2 hybridized singlet carbene with a vacant p orbital and a filled sp^2 orbital, CCl_2 is stabilized by the $Cl(3p) \rightarrow$ carbene$(2p)$ electron donation qualitatively represented by structure **5** (Fig. 7.5). Clearly, CCl_2 will be more stabilized than CH_3CCl, in which only one Cl atom is present.

A deeper understanding of carbenic philicity requires a more detailed representation of the addition reaction transition state than that afforded by structure **4**. Early MO calculations furnished structure **6** as representative of the transition state for addition of a singlet carbene to an alkene (Fig. 7.6).[34,35]

We may redraw **6** as **7a** and **7b**, in terms of frontier MOs.[25,26] Here we emphasize the highest occupied molecular orbital–lowest unoccupied molecular orbital (HOMO–LUMO) interactions that operate in the transition state: **7a** depicts the LUMO(carbene)/HOMO(alkene) or $p-\pi$ interaction; **7b** shows the HOMO (carbene)/LUMO(alkene) or $\sigma-\pi^*$ interaction. These formulations are especially

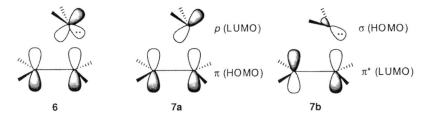

Figure 7.6

useful because they lend themselves to calculations of the relevant orbital energies, and transition state structures and energetics.[36,37]

Interaction **7a** features net electron donation from the alkene π orbital (HOMO) to the vacant carbene *p* orbital (LUMO), and tracks the electrophilic character of the carbene. Interaction **7b** represents electron donation from the filled carbene σ orbital (HOMO) to the vacant alkene π* orbital (LUMO) and reflects the carbene's nucleophilic character. Both interactions operate simultaneously in the addition transition state, but which one is dominant?

From the relative reactivities in Table 7.3, we conclude that **7a** is dominant for CH_3CCl and CCl_2, which exhibit electrophilic selectivity toward the alkenes. On the other hand, CH_3OCCH_3 displays strongly nucleophilic selectivity toward the electron-poor alkenes. Resonance donation from the methoxy oxygen atom to the carbene $2p$ orbital (**8**, Fig. 7.7) is sufficiently strong to render CH_3OCCH_3 a nucleophilic carbene for which σ → π* carbene to alkene electron donation is dominant in the transition state (viz., **7b**).

$$CH_3\ddot{\underset{..}{O}}-\ddot{C}-CH_3 \quad \longleftrightarrow \quad CH_3\overset{+}{\underset{..}{O}}=\overset{..}{C}-CH_3$$

8

Figure 7.7

If CCl_2 and CH_3OCCH_3 represent electrophilic and nucleophilic carbenes, respectively, in which the transition states predominantly resemble **7a** or **7b**, can we identify carbenes for which *both* p–π and σ–π* electronic interactions are comparably important? Such carbenes would react as electrophiles with electron-rich alkenes, but as nucleophiles with electron-poor alkenes; that is, they would be *ambiphiles*.[25] Indeed, Table 7.3 shows that CH_3OCCl[31] and C_6H_5OCCl[32] behave in just this manner. Their relative reactivities are high toward *both* electron-rich and electron-poor alkenes, but lower toward alkenes that are electronically "moderate" (*trans*-butene or 1-hexene[26]).

Our qualitative discussion has limitations. Why, for example, is CH_3OCCH_3 seemingly more nucleophilic than CH_3OCCl (Table 7.3)? If resonance donation by CH_3O (as in **8**) makes interaction **7b** dominant in the addition reactions of CH_3OCCH_3, why should additional resonance donation by Cl (as in **5**) not make

CH_3OCCl even more nucleophilic? Of course, we have not yet considered inductive effects; Cl is significantly more withdrawing inductively than OCH_3, and a more complete treatment must consider this effect. Indeed, empirical correlations of carbene relative reactivities have been developed based on linear free energy treatments that make use of the resonance (σ_{R+}) and inductive (σ_I) constants of a carbene's substituents (X and Y in CXY).[24-26,38] These correlations can be used to predict whether a given carbene should be electrophilic, ambiphilic, or nucleophilic.[25]

A more precise approach makes use of computational methods. We calculate the carbenic orbital energies[36] corresponding to the interactions expressed in **7a** and **7b**, and evaluate the differential orbital energies for each interaction, Eqs. 1 and 2.[26,39]

$$\Delta\varepsilon_E = \varepsilon_{CXY}^{LU} - \varepsilon_{C=C}^{HO} = p - \pi \tag{1}$$

$$\Delta\varepsilon_N = \varepsilon_{C=C}^{LU} - \varepsilon_{CXY}^{HO} = \pi - \sigma \tag{2}$$

where LU = LUMO and HO = HOMO

We insert the appropriate orbital energies into Eqs. 1 and 2, in which $\Delta\varepsilon_E$ and $\Delta\varepsilon_N$ represent the differential orbital energies corresponding to the electrophilic (**7a**) and nucleophilic (**7b**) interactions, respectively. Numerical results for CCl_2, CH_3OCCl, CH_3OCCH_3, and $(CH_3O)_2C$ appear in Table 7.4.[24,26]

According to frontier MO theory,[37] the stabilization energy of a cycloaddition transition state depends *inversely* on the differential energies of the interacting "frontier" MOs. If we neglect orbital overlap, a smaller $\Delta\varepsilon$ implies greater stabilization, a lower activation energy, and a higher rate constant. Inspection of $\Delta\varepsilon$ for CCl_2 in Table 7.4 shows that $\Delta\varepsilon_E < \Delta\varepsilon_N$ for each alkene substrate. Accordingly, transition state interaction **7a** should dominate the addition reactions of CCl_2, which should therefore exhibit electrophilic selectivity over the alkene set of Table 7.4. The experimental relative reactivities shown in Table 7.3 agree with this prediction.

TABLE 7.4. Differential Orbital Energies (eV) for Carbene–Alkene Additions[a]

Alkene	CCl_2[b]		CH_3OCCl[b]		CH_3OCCH_3[b]		$(CH_3O)_2C$[b]	
	$\Delta\varepsilon_E$	$\Delta\varepsilon_N$	$\Delta\varepsilon_E$	$\Delta\varepsilon_N$	$\Delta\varepsilon_E$	$\Delta\varepsilon_N$	$\Delta\varepsilon_E$	$\Delta\varepsilon_N$
$(CH_3)_2C=C(CH_3)_2$	8.58	13.71	10.73	13.09	12.31	11.68	12.36	13.08
$(CH_3)_2C=CHCH_3$	8.99	13.68	11.14	13.06	12.72	11.65	13.02	12.86
$(CH_3)_2C=CH_2$	9.55	13.63	11.70	13.01	13.28	11.60	13.33	13.00
trans-$CH_3CH=CHCH_3$	9.43	13.54	11.58	12.92	13.16	11.51	13.21	12.91
$CH_2=CHCOOCH_3$	11.03	12.24	13.18	11.62	14.76	10.21	14.81	11.61
$CH_2=CHCN$	11.23	11.65	13.38	11.03	14.96	9.62	15.01	11.02

[a] See Eqs. 1 and 2. Orbital energies can be found in Ref. 25, 33, and 36.
[b] The subscript E in ε_E = electrophilic. The subscript N in ε_N = nucleophilic.

In contrast, the $\Delta\varepsilon$ values for CH_3OCCH_3 in Table 7.4 reveal that $\Delta\varepsilon_N < \Delta\varepsilon_E$, so that interaction **7b** should dominate additions of this carbene. It should be nucleophilic, and Table 7.3 demonstrates that CH_3OCCH_3 does indeed exhibit pronounced reactivity toward electron-poor alkenes.

Most strikingly, the $\Delta\varepsilon$ values for CH_3OCCl in Table 7.4 "cross-over." For the electron-rich alkenes, $\Delta\varepsilon_E < \Delta\varepsilon_N$, whereas $\Delta\varepsilon_N < \Delta\varepsilon_E$ for the electron-poor alkenes. Here is a clear prediction that CH_3OCCl should behave as an ambiphile, reacting rapidly with both electron-rich and electron-poor alkenes, but reacting more slowly with alkenes of modest electronic properties. Again, the relative reactivities of Table 7.3 are consonant with these ideas.

The computed values of $\Delta\varepsilon_E$ and $\Delta\varepsilon_N$ also predict that dimethoxycarbene should be a nucleophilic carbene. Experimentally, dimethoxycarbene does not add at all to electron-rich alkenes (preferring to dimerize instead), but does add readily to electron-poor methyl acrylate and acrylonitrile.[40] Many other nucleophilic reactions of $(CH_3O)_2C$ and related dialkoxycarbenes have been investigated and reviewed by Warkentin.[41]

In summary, carbenic electrophilicity is favored by a low-lying HOMO or σ orbital, which makes electron donation by the carbene unfavorable, and by a low-lying, accessible LUMO or p acceptor orbital. Conversely, carbenic nucleophilicity is favored by a high-lying HOMO for efficient electron donation, and a high-lying, inaccessible LUMO.[24,26] Other quantitative treatments of carbenic philicity appear in the work of Houk and co-workers,[36] Schoeller and Brinker,[42] Platz and co-workers,[43] Zollinger,[44] Fukui and co-workers,[45] and Garcia-Garibay and co-worker.[46] Their work is reviewed in Ref. 24, and is in general agreement with the conclusions presented here.

Sander applied DFT (B3LYP) theory to carbenic philicity, computing the electron affinities (EA) and ionization potentials (IP) of the carbenes.[47,48] The EA tracks the carbene's electrophilicity (its ability to accept electron density), whereas the IP represents the carbene's nucleophilicity (its ability to donate electron density). This approach parallels the differential orbital energy treatment. Both EA and IP can be calculated for any carbene, so Sander was able to analyze the reactivity of "super" electrophilic carbenes such as difluorovinylidene (**9**)[49] which is sufficiently electrophilic to insert into the C—H bond of methane. It even reacts with the H—H bond of dihydrogen at temperatures as low as 40 K, Scheme 7.2)!

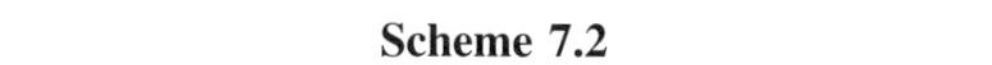

$$F_2C=C: \; + \; H—H \; \longrightarrow \; F_2C=CH_2$$
$$\mathbf{9}$$

Scheme 7.2

At the opposite end of the philicity spectrum, nucleophilic carbenes have proven useful in synthesis. Warkentin[41] pioneered the thermolysis of oxadiazolines as precursors for $(CH_3O)_2C$ and related dioxacarbenes (Scheme 7.3). Dimethoxycarbene generated from an oxadiazoline undergoes a variety of intermolecular reactions.[41] One example is the ring enlargement of strained cyclic ketones, for example, cyclobutanone.[50] In this reaction, the nucleophilic carbene initiates the ring expansion by

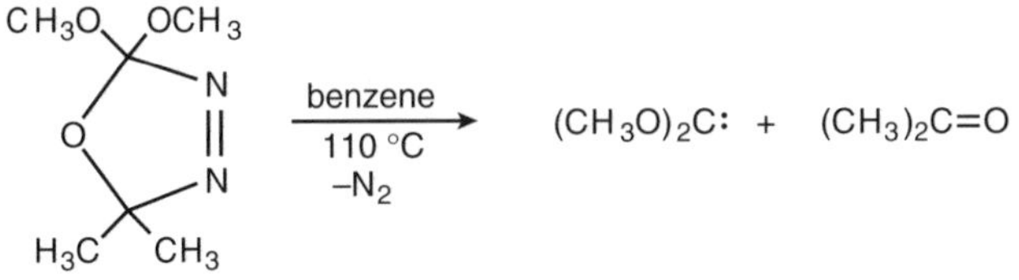

Scheme 7.3

attack on the carbonyl group of cyclobutanone. A zwitterionic intermediate, **10**, leads to the ring-expanded, masked diketone (**11**, Scheme 7.4). When either of two substrate carbons can be involved in the expansion, selectivity favors migration of the more substituted carbon.

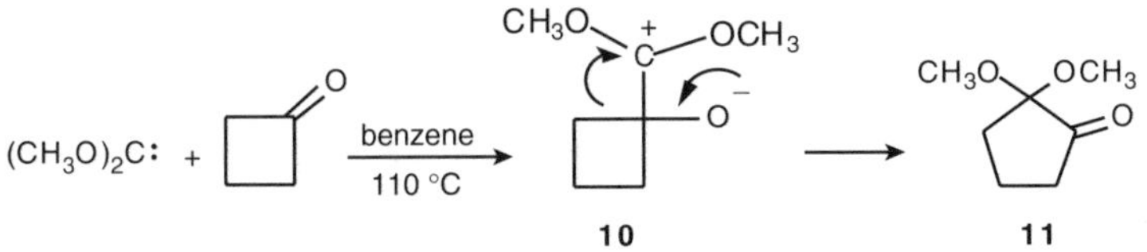

Scheme 7.4

Similarly, Rigby and coworkers[51] used the oxadiazoline method to generate dithiacarbenes, which initiate nucleophilic attacks on isocyanates (Scheme 7.5). In this case, nucleophilic attack on the isocyanate "carbonyl" leads to ring closure via intermediate **12**. The proximate product then undergoes an apparent N–H insertion reaction with a second carbene to yield the final product. The philicity of the N–H insertion is unclear; it could be initiated by "electrophilic" attack of the ambiphilic $(C_3H_7S)_2C$ on the substrates's nitrogen lone pair, or it might originate in N–H proton abstraction by a "nucleophilic" carbene acting as a base. Related reactions of $(CH_3O)_2C$ with isocyanates are known.[52]

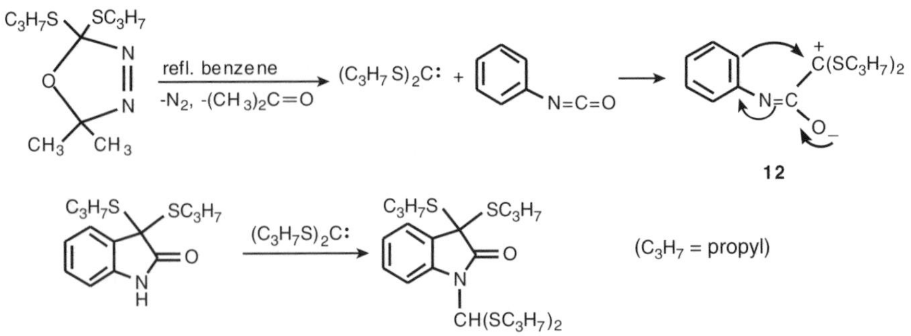

Scheme 7.5

Finally, addition reactions of the isolable phosphasilylcarbenes (**13**) to such electron-poor substrates as methyl acrylate, $C_4F_9CH=CH_2$, and styrene afford cyclopropanes. The additions of **13a** to (*E*)- or (*Z*)-β-deuteriostyrene are stereospecific, and the competitive additions of **13b** to ring-substituted styrenes exhibit nucleophilic selectivity, consistent with singlet, nucleophilic carbene addition (Fig. 7.8).[53]

$$(R_2N)_2P-\ddot{C}-Si(CH_3)_3$$

13a, R = *i*-C$_3$H$_7$; **b**, R = cyclohexyl

Figure 7.8

2.1.2. Rates and Activation Parameters.

The first condensed-phase absolute rate measurement for a carbene–alkene addition was reported by Closs and Rabinow in 1976: flash lamp photolysis of diphenyldiazomethane generated (triplet) diphenylcarbene, which added to butadiene (in benzene) with $k = 6.5 \times 10^5\,M^{-1}s^{-1}$ at 25 °C.[54] However, most singlet carbene additions to olefins are faster, and their addition reactions are too rapid for the time resolution afforded by a flash lamp.

The development of laser flash photolysis (LFP), with flash times of 10–20 ns, made possible the measurement of many absolute rate constants for singlet carbene additions. The initial reports focused on chlorophenylcarbene (C_6H_5CCl)[55] and singlet and triplet fluorenylidene.[56] Consider C_6H_5CCl: LFP at 351 nm of chlorophenyldiazirine (**14**) in isooctane containing a given concentration of an alkene such as $(CH_3)_2C=C(CH_3)_2$, affords a transient ultraviolet (UV) signal for C_6H_5CCl at ~310 nm that decays as the carbene adds to the alkene. In the absence of an alkene, the carbene can decay by dimerization, by reaction with the diazirine precursor, or by reaction with adventitious water. When sufficient alkene trap is present, however, these pathways are suppressed. From the time dependence of the carbene signal's decay, we obtain a pseudo-first order rate constant (k_{obs}) for carbene–alkene addition. Variation of the alkene concentration and repetition of the LFP experiment yields additional values of k_{obs}, and correlation of k_{obs} with [alkene] gives a straight line, the slope of which is the second-order rate constant for the addition of C_6H_5CCl to $(CH_3)_2C=C(CH_3)_2$, in this case, $k = 2.8 \times 10^8\,M^{-1}s^{-1}$ Scheme 7.6).[57]

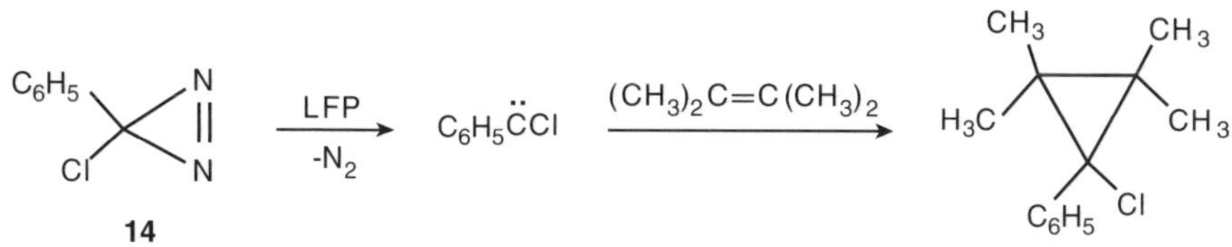

Scheme 7.6

This "direct" method works quite well as long as the carbene has an accessible and sufficiently strong UV signature. Many absolute rate constants for carbene additions were measured in this way.[24,26,57] Data for several of these species appear in Table 7.5.[57–62]

TABLE 7.5. Absolute Rate Constants for Carbene–Alkene Additions[a]

Alkene	C_6H_5CF[b]	C_6H_5CCl[b]	CH_3OCCl[c]	$C_6H_5COCH_3$[d]	CH_3OCOCH_3[e]
$(CH_3)_2C{=}C(CH_3)_2$	1.6×10^8	2.8×10^8	4.2×10^3		
$(CH_3)_2C{=}CHCH_3$	5.3×10^7	1.3×10^8			4.7×10^3
$(CH_3)_2C{=}CH_2$			1.8×10^3	4.0×10^4	4.8×10^3
$trans\text{-}CH_3CH{=}CHC_2H_5$	2.4×10^6	5.5×10^6	3.3×10^{2f}	3.8×10^{3f}	2.2×10^{3f}
$n\text{-}C_4H_9CH{=}CH_2$	9.3×10^5	2.2×10^6		2.8×10^3	
$CH_2{=}CHCOOCH_3$	1.4×10^6	5.1×10^6	9.8×10^3	6.6×10^5	7.9×10^5
$CH_2{=}CHCN$	2.3×10^6	7.0×10^6	1.8×10^4	1.7×10^6	1.5×10^6
$CH_2{=}CClCN$	1.2×10^8	2.1×10^8	5.6×10^5	3.4×10^7	4.9×10^7

[a] Rate constants are in $M^{-1}\,s^{-1}$ and were measured at 23–25 °C in alkene–hydrocarbon solvents.
[b] Data from Ref. 57–60.
[c] Data from Ref. 61.
[d] Data from Ref. 59, 61, and 62.
[e] Data from Ref. 33 and 61.
[f] Data are for *trans*-butene.

The data reveal a very large kinetic span for carbene–alkene addition reactions with k ranging from $> 10^8\,M^{-1}\,s^{-1}$ for additions of C_6H_5CCl or C_6H_5CF to $(CH_3)_2C{=}C(CH_3)_2$ to $3.3 \times 10^2\,M^{-1}s^{-1}$ for the addition of CH_3OCCl to *trans*-pentene. This million-fold variation testifies to the great modulating power of carbenic substituents on carbenic reactivity.

Both C_6H_5CCl and C_6H_5CF behave as ambiphiles toward the alkene set of Table 7.5. They react rapidly with both electron-rich $(CH_3)_2C{=}C(CH_3)_2$ and electron-poor alkenes $CH_2{=}CClCN$; their slowest reactions are with the mono-alkylethylene, 1-hexene. Although ambiphilicity is an intrinsic property of any singlet carbene,[25,26] the *absolute* reactivity of, for example, C_6H_5CCl is much greater than that of the classic ambiphile CH_3OCCl.[31] The former reacts ~67,000 times more rapidly with $(CH_3)_2C{=}C(CH_3)_2$, 375 times more rapidly with $CH_2{=}CClCN$, and ~17,000 times more rapidly with *trans*-pentene. The CH_3O for C_6H_5 "swap" that formally converts C_6H_5CCl into CH_3OCCl is responsible for very significant resonance stabilization of the carbene (see **8**), and a concomitant decrease in reactivity.

Even the F for Cl change transmuting C_6H_5CCl into C_6H_5CF is accompanied by a diminution in reactivity; C_6H_5CF reacts about one-half as rapidly as C_6H_5CCl with the alkenes of Table 7.5. This lowering of reactivity can be attributed to greater stabilization of C_6H_5CX when X=F, as opposed to X=Cl. Fluorine is a better resonance electron donor to the vacant carbene $2p$ orbital than chlorine, and this factor (rather than the greater electronegativity of fluorine) appears to be dominant.[63]

What about CCl_2, the paradigmatic electrophilic carbene?[2,3,28] It can be generated by photoextrusion from its phenanthrene adduct (**15**, Scheme 7.7).[64] Although CCl_2 lacks a strong UV signal to follow directly, one can use Jackson and Platz's method[65a] in which CCl_2 addition to pyridine (affording a UV active ylide) competes with CCl_2 addition to an alkene substrate. The desired second-order rate constant for the CCl_2–alkene addition can be extracted from the dependence of the apparent LFP rate constant for ylide formation as a function of alkene concentration (at a constant concentration of pyridine). For a detailed exposition of the kinetic principles underlying this methodology, the reader is referred to Chapter 18 in this volume.

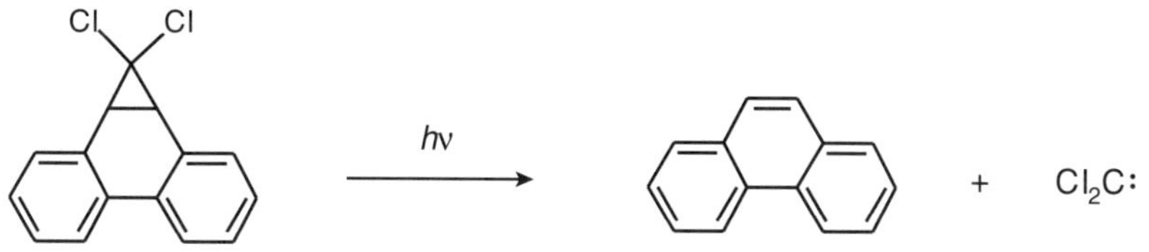

15

Scheme 7.7

Rate constants $(M^{-1}s^{-1})$ were thus measured for dichlorocarbene additions to $(CH_3)_2C=C(CH_3)_2$ (3.8×10^9), $(CH_3)_2C=CHCH_3$ (2.2×10^9), *trans*-pentene (6.3×10^7), cyclohexene (3.5×10^7), and 1-hexene (1.1×10^7). Dichlorocarbene is again seen to be an electrophile, as we concluded from the relative reactivity data of Table 7.3, and the differential orbital energies of Table 7.4. Suprisingly, although CCl_2 is somewhat more selective than C_6H_5CCl [e.g., k_{rel} for $(CH_3)_2C=C(CH_3)_2/$ n-BuCH=CH_2 is 345 for CCl_2 vs. 127 for C_6H_5CCl], CCl_2 is also marginally more reactive than C_6H_5CCl. The k_{obs} values for CCl_2–alkene additions exceed those of C_6H_5CCl in each case by about a factor of 10.

On the other hand, the CH_3O for Cl exchange that converts CCl_2 into CH_3OCCl is accompanied by about a 10^5-fold decrease in reactivity (Table 7.5). The $O(2p)$–$C(2p)$ resonance electron donation of CH_3O is much more effective than the analogous $Cl(3p)$–$C(2p)$ interaction. Another way to approach this reactivity comparison is to note that CCl_2 is a very reactive electrophile because of its low-lying, accessible LUMO orbital $(0.3\ eV)^{36}$ whereas the LUMO of CH_3OCCl $(2.5\ eV)^{36}$ is raised by resonance donation from CH_3O and is much less accessible. Indeed, $k = 3.3 \times 10^2\ M^{-1}\ s^{-1}$ for the addition of CH_3OCCl to *trans*-butene is currently the lowest known bimolecular rate constant for a condensed phase carbene–alkene addition.[61]

Similar analyses can be made for the addition reactions of $C_6H_5COCH_3$ and CH_3COCH_3 (Table 7.5). Both carbenes are ambiphiles/nucleophiles that are more reactive than CH_3OCCl. Thus, the Cl for CH_3 exchange that transforms CH_3OCCH_3 into CH_3OCCl stabilizes the carbene and decreases its reactivity, especially toward the electron-poor alkenes of Table 7.5, where it can be seen that CH_3OCCl is ~80 times less reactive than CH_3COCH_3. A similar attenuation of reactivity accompanies the Cl for C_6H_5 swap that converts $C_6H_5COCH_3$ into CH_3OCCl.

Moreover, we see that the nucleophilic properties of CH_3COCH_3 are pronounced (Tables 7.3 and 7.4): the combination of a relatively high-lying, filled HOMO donor orbital (-9.4 eV)[33] and a high-lying, poorly accessible vacant p orbital (LUMO, $4.0 \text{ eV})$[33] makes CH_3COCH_3 a good nucleophile and a poor electrophile. The computed differential orbital energies for additions of CH_3COCH_3 to the alkenes of Table 7.4 are dominated by the nucleophilic $\Delta\varepsilon_N$ term.

The CH_3O for CH_3 substituent swap that transforms CH_3COCH_3 into $(CH_3O)_2C$ results in additional stabilization of the carbene and lower reactivity. Thus, $(CH_3O)_2C$, with $\varepsilon_{LU} = 4.3 \text{ eV}$,[40] no longer adds to electron-rich alkenes such as $(CH_3)_2C=CHCH_3$, dimerizing instead, whereas its nucleophilic additions to electron-poor olefins, such as $CH_2=CClCN$ and $CH_2=CHCN$, are $\sim$100–1000 times slower than those of CH_3COCH_3.[33,40]

If the resonance stabilization of singlet carbenes is increased even more, for example, by replacing the CH_3O substituents of $(CH_3O)_2C$ by amino groups [as in $(R_2N)_2C$] or phosphorus substituents (**13**), stable, isolable, nucleophilic carbenes can be obtained. These species are considered in Chapter 8 in this volume.

Given our ability to measure absolute rate constants for carbene additions, variable temperature studies readily afford activation parameters.[57,63] Initial studies of C_6H_5CX additions gave very low E_a values, $\sim$1 kcal/mol for reactions with 1-hexene and *trans*-pentene.[66] Most surprisingly, the E_a values for C_6H_5CCl additions to $(CH_3)_2C=C(CH_3)_2$ and $(CH_3)_2C=CHCH_3$ were *negative* (-1.7 and -0.8 kcal/mol, respectively). The reaction rates increased as temperature decreased. The preexponential (A) factors were low $(2–6 \times 10^7 \, M^{-1} \, s^{-1})$, indicative of an unfavorable entropy of activation.

The most convincing explanation for the negative activation energies was offered by Houk et al.,[67] who computed ΔH, ΔS, and ΔG for additions of CBr_2, CCl_2, and CF_2 to $(CH_3)_2C=C(CH_3)_2$ and $(CH_3)_2C=CH_2$. With CBr_2, a very reactive carbene, and the highly reactive $(CH_3)_2C=CHCH_3$, the addition reaction was so exothermic that ΔH continually decreased all along the reaction coordinate. Therefore, $\Delta H^{\ddagger}$ and E_a were negative. However, the loss of the translational, vibrational, and rotational entropy required in the transition state for addition led to an unfavorable and dominant entropy of activation $(\Delta S^{\ddagger} < 0)$ which, in turn, created a free energy barrier to addition $(\Delta G^{\ddagger} > 0)$. With the more stable and less reactive carbene, CF_2, E_a, and $\Delta H^{\ddagger}$ were positive, adding to the $\Delta G^{\ddagger}$ barrier.[67] The key role of entropy in carbenic additions, highlighted by Houk, was earlier noted by Skell and Cholod,[35b] and by Giese.[68]

A wide-ranging experimental study of carbene–alkene addition activation parameters examined a series of substituted arylhalocarbenes (**16**) in reactions with $(CH_3)_2C=C(CH_3)_2$ and 1-hexene (Fig. 7.9).[69]

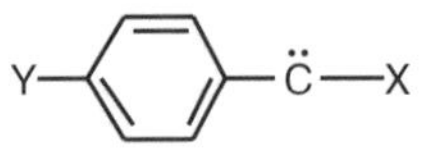

$$Y-\!\!\left\langle \underset{}{\bigcirc} \right\rangle\!\!-\ddot{C}-X$$

16

Figure 7.9

"Extreme" cases were reactions of the least stabilized, most reactive carbene (Y = CF_3, X = Br) with the more reactive alkene $(CH_3)_2C=C(CH_3)_2$, and the most stabilized, least reactive carbene (Y = CH_3O, X = F) with the less reactive alkene (1-hexene). The rate constants, as measured by LFP, were 1.7×10^9 and $5.0 \times 10^4\,M^{-1}\,s^{-1}$, respectively, spanning an interval of 34,000. In agreement with Houk's ideas,[67] the reactions were entropy dominated ($\Delta S^{\ddagger} \sim -22$ to -29 e.u.). The $\Delta G^{\ddagger}$ barriers were 5.0 kcal/mol for the faster reaction and 11 kcal/mol for the slower reaction, mainly because of entropic contributions; the $\Delta H^{\ddagger}$ components were only -1.6 and $+2.5$ kcal/mol, respectively.[69] Despite the dominance of entropy in these reactive carbene addition reactions, a kind of "*de facto* enthalpic control*" operates. The entropies of activation are all very similar, so that in any comparison of the reactivities of alkene pairs (i.e., k_{rel}), the rate constant ratios reflect differences in $\Delta\Delta H^{\ddagger}$, which ultimately appear in $\Delta\Delta G^{\ddagger}$. Thus, carbenic *philicity*, which is the pattern created by carbenic *reactivity*, behaves in accord with our qualitative ideas about structure–reactivity relations, as modulated by substituent effects in both the carbene and alkene partners of the addition reactions.[24]

Finally, volumes of activation were measured for the additions of C_6H_5CCl to $(CH_3)_2C=C(CH_3)_2$ and *trans*-pentene in both methylcyclohexane and acetonitrile.[70] The measured absolute rate constants increased with increasing pressure; $\Delta V^{\ddagger}$ ranged from -10 to -18 cm^3/mol and were independent of solvent. These results were consistent with an early, and not very polar transition state for the addition reaction.

2.1.3. Symmetry of the Transition State.
The asymmetric, "non-least-motion" transition state (**6**) for carbene–alkene cycloaddition has been discussed in qualitative[1,71] and quantitative[34,36,67,72] terms for many years. As shown in Fig. 7.6, the transition state features carbene *p*–alkene π^* orbital interactions.[24–26,36,42–46] Hoffmann first presented a theoretical analysis supporting a "π-approach" transition state (**6**), which also rationalizes the electrophilic character of CH_2 or CCl_2 addition reactions.[34] He also showed that the symmetrical "least-motion" approach (**17**) was forbidden by orbital symmetry.[34] Subsequent computational studies supported Hoffmann's conclusions.[36,67,72] Note that although the trajectory of the addition reaction begins with the geometry represented by **6**, the carbene must later pivot toward geometry **17** in order to complete the cyclopropanation (Fig. 7.10).

Keating et al.[73] used natural abundance isotope effects to define the "direction" of the asymmetry for the addition of CCl_2 to 1-pentene, and presented complementary

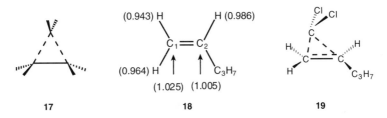

Figure 7.10

B3LYP/6-31G* computations. Thus, 3 mol of 1-pentene was cyclopropanated to 89–93% conversion with CCl_2 (generated by the phase-transfer-catalyzed reaction of $CHCl_3$ and NaOH). The *unreacted* 1-pentene was isolated and brominated, and the resulting dibromide was analyzed by ^{13}C and ^{2}H nuclear magentic resonance (NMR) spectroscopy. The C and H isotopic abundances were compared to those in dibromide derived from a sample of the initial 1-pentene, leading to the isotope effects shown for a typical experiment in structure **18** (Fig. 7.10).

Of course, the transition state for addition to an unsymmetrical alkene must be unsymmetrical in some fashion. Thus, the $^{12}C/^{13}C$ primary kinetic isotope effects at C1 and C2 are unequal. However, the larger isotope effect at C1 implies more advanced binding at this carbon atom in the transition state (**19**). Note too the inverse α–secondary (k_H/k_D) isotope effects at the hydrogens bonded to C1 and C2. These carbons rehybridize from sp^2 toward sp^3 as the addition proceeds, so that $k_H/k_D < 0$ is expected. The greater effect at *H*-C1 versus *H*-C2 is also consistent with greater carbene binding to the alkene's terminal C in the transition state.[73]

The experimental results agree with electrophilic addition of CCl_2 to 1-pentene. Stronger bonding at C1 in the transition state will impose most of the partial positive charge at C2, the more substituted carbon, where it can be better dispersed by inductive and hyperconjugative electron release from the alkyl substituent. The B3LYP/6-31G* (and 6-311$^+$G*) computations for the addition of CCl_2 to propene support the experimental results. The optimum computed transition state geometry is shown in **20**[74] for a carbene–alkene separation of 2.4 Å (Fig. 7.11). The computed $\Delta G^{\ddagger}$ for addition via **20** is ~10.1 kcal/mol, preferable to alternative geometry **21**,[74] in which $\Delta G^{\ddagger}$ ~12.6 kcal/mol at a separation of 2.3 Å.[73]

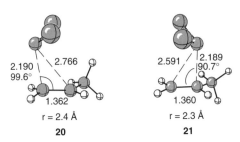

Figure 7.11. B3LYP/6-31G* geometries for CCl_2 plus propene mode A (structure **20**) and mode B (structure **21**) addition at constrained values of *r*. Bond lengths are given in angstroms.

The energetically "preferred" direction of the asynchronous CCl_2 addition transition state with propene (**20**) places the carbene's σ orbital closer to C1, and the substituents on the carbene directed toward C2.[73,75] The asymmetry of bonding to the carbene is evident in **20**, and computed carbon and hydrogen isotope effects for **20** (and for an analogous structure with a 2.5-Å carbene–alkene separation) agree very well with the experimental isotope effects.[73]

In Section 2.1, we have not considered the vast literature devoted to the addition reactions of *carbenoids*.[76] Typically, these compounds are organometallic molecules such as $(C_6H_5)_2CBrLi$, $C_6H_5CHBrLi$, $LiCCl_3$, and ICH_2ZnI (Simmons–Smith reagent) that react with olefins to give cyclopropanes just as do the corresponding free carbenes, $(C_6H_5)_2C$, C_6H_5CH, CCl_2, and CH_2, respectively. Carbenoids usually add electrophilically to alkenes. An extensive review of carbenoid philicity has recently been published by Boche and Lohrenz.[77]

2.1.4. Stepwise versus Concerted Addition.

One of the factors that makes carbene chemistry fascinating is that regardless of which species is first formed, reactivity may issue from either the singlet or triplet state. In this chapter (and in Chapter 9 in this volume), we will have to consider the spin state of the reacting species in essentially all reactions. It is not enough to assume that the first-formed intermediate is the reacting species. Indeed, the singlet generally has an enormous advantage over the triplet. Nearly all molecules are ground-state singlets—all electrons are paired. Thus, they can be formed directly from singlets but not from triplets. Reactions of singlet carbenes are often exceptionally exothermic, and the formation of two new bonds in a single step from a singlet implies transition states that are very reactant-like (Hammond postulate), and thus with very low barriers. Nearly every time, when a singlet is close enough in energy to be formed from a triplet, it will be the intermediate that does the chemistry and forms the products. We will point out a few exceptions in this chapter, and be sure to see Chapter 9 in this volume as well.

As noted above, singlet reactions in general, and addition reactions in particular, must benefit greatly from the exothermicity of the formation of two sigma bonds. That idea must have been behind the early assumption, made by both the Skell[27,35c] and Doering groups,[78] that the observation of a stereospecific addition would be diagnostic for the singlet state. Of course, that assumption not only posits that the addition of a singlet carbene would be one step, but that the necessarily stepwise addition of the triplet would *not* be stereospecific. In turn, those ideas assume that in the intermediate diradical formed in the first step of a triplet addition, rotation about a single bond would be faster than spin inversion (intersystem crossing—"isc") and closure Scheme 7.8).

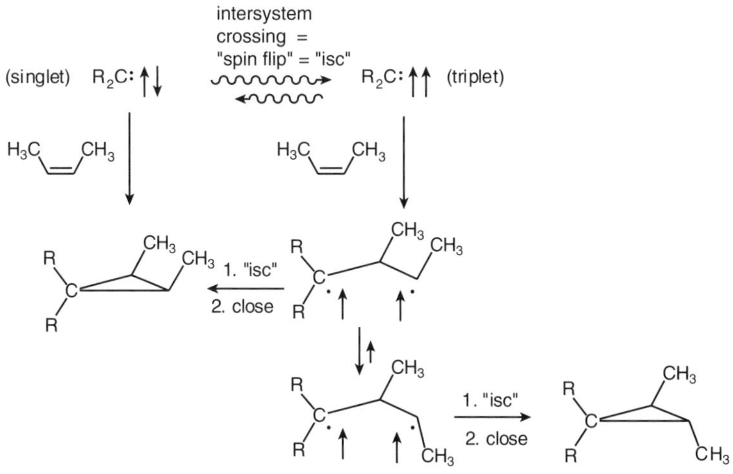

Scheme 7.8

Surprisingly, the critical experiment has been done infrequently over the last one-half of a century! The requirements for an experiment that truly speaks to the issue at hand are that one be able to see the results of addition of *both* spin states of a *single* carbene, and these requirements rarely have been met. For example, the direct irradiation of methyl diazomalonate leads to the stereospecific addition expected of a singlet carbene, whereas the photosensitized decomposition of the diazo compound leads to formation of the triplet carbene and loss of the stereochemical relationship originally present in the reacting alkene. Rotational equilibration in the intermediate seems to be complete, as it makes no significant difference whether cis or trans alkene is used as starting material (Scheme 7.9).[79]

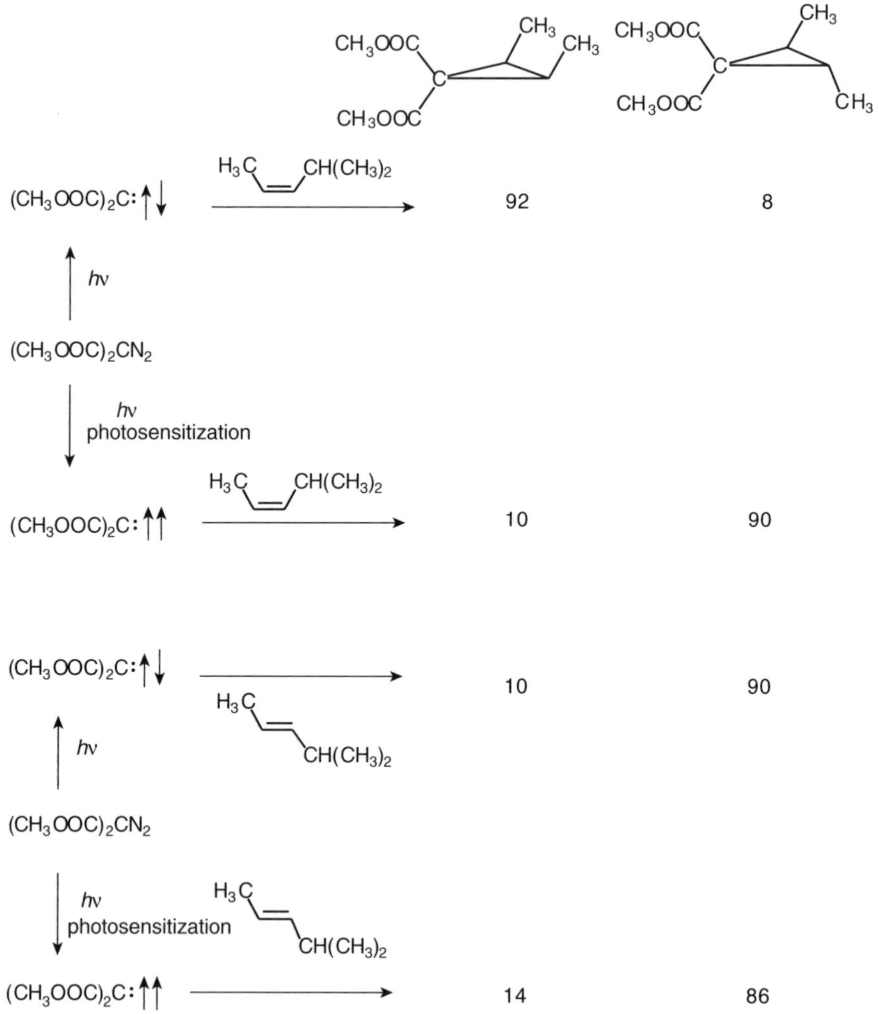

Scheme 7.9

The critcal experiment has also been attempted many times for the parent carbene, methylene. The difficulties of working in the gas phase, in which product-
scrambling hot molecule rearrangements are all-too common, and of finding a
good source of the triplet carbene in solution meant that a truly convincing experiment waited until 1987. Turro et al.[80] decomposed diazirine through direct and
photosensitized irradiation to refine a much earlier experiment of Hammond
et al.[81] that used diazomethane as methylene source. The data shown in
Scheme 7.10 are consistent with what the "Skell hypothesis" would dictate: Reactions of the singlet state are stereospecific, whereas those of the lower energy triplet
scramble the stereochemical relationships present in the reactant alkene. In this
case, rotation appears not to be significantly faster than intersystem crossing and
closure.

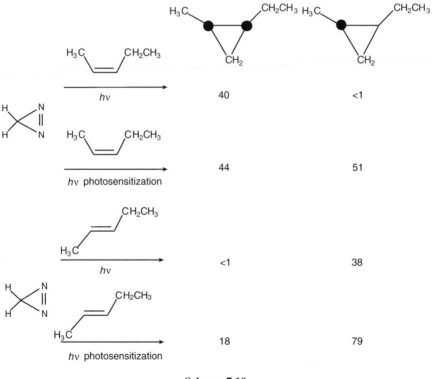

Scheme 7.10

It is also possible to induce intersystem crossing from an initially formed singlet
state to a lower energy triplet by forcing the singlet state to suffer collisions with an
inert medium (no easy task for a species as reactive as methylene!).[80,82]

Despite the somewhat thin underpinnings, the observation of stereospecific or
nonstereospecific addition remains the most widely used experimental test of singlet or triplet reactivity.

Can one find counterexamples—addition reactions—or any reactions—in which singlets react in a stepwise fashion? In principle, the answer to this question must be yes—we need only find a reaction for which an intermediate, either diradical or dipolar, is strongly stabilized, whereas the transition state for the concerted reaction is destabilized. As the stepwise reaction becomes easier and the concerted reaction becomes more difficult, eventually the activation energy for the stepwise path will fall below that for the concerted reaction. How would one go about searching for such reactions? Brinker has suggested that the addition of dihalocarbenes to 1,2-diarylcyclopropenes[83a] (and 1,2-diarylcyclobutenes[83b]) involves a stepwise addition. The transition state for cyclopropanation is clearly destabilized by the strain inherent in the incipient bicyclo[1.1.0] system, and the halogens and aryl groups can stabilize a dipolar intermediate, as in **22**. The reaction reveals itself not through stereochemistry, but by the formation of a ring-opened product **23** (Scheme 7.11).

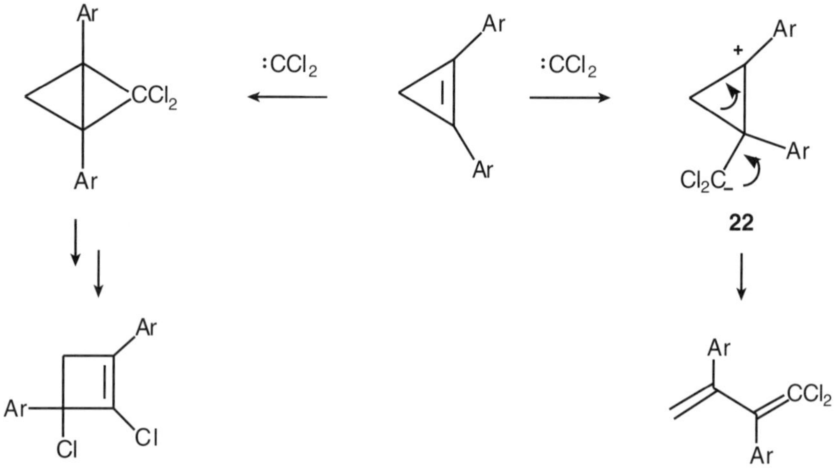

Scheme 7.11

The relevant products are the dienes (**23**) which are the major products. The cyclobutenes are very likely formed from an intermediate bicyclo[1.1.0]butane. Three pathways leading to **23** are considered A, B, and C (Scheme 7.12).

Path C is surely unlikely for several reasons, chief among them is that **23** is formed even at −45 °C, at which temperature the bicyclobutane would surely be stable. The real problem is to rule out the non-carbene mechanism, path A. Brinker cleverly does this by examining the dienes formed from **24** and **25** (Scheme 7.13). The products are nicely rationalized through intermediates **22a** and **22b** in which the direction of the initial addition is determined by the stabilizing or

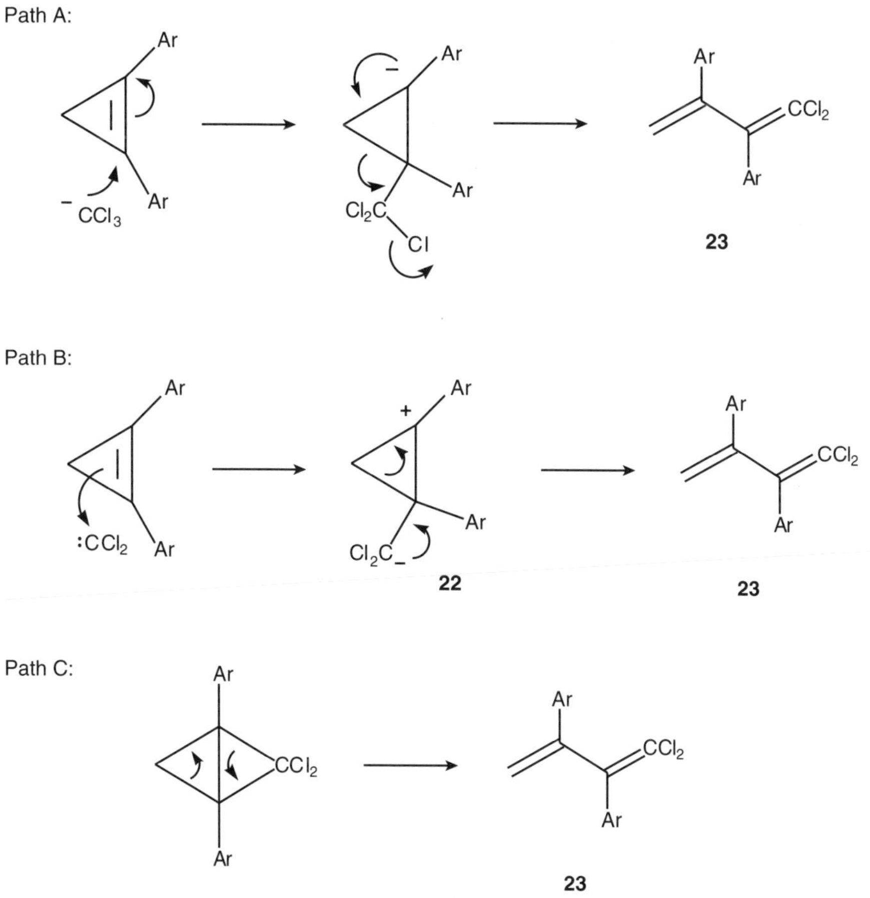

Path A:

Path B:

Path C:

Scheme 7.12

destabilizing effect of the substituent. Methoxy will stabilize an adjacent positive charge, but trifluoromethyl will destabilize it (Scheme 7.13). Were the reaction to take place through an anionic addition (Path A), the opposite results would be expected.

Does this experiment provide an airtight case for stepwise addition of dichloro-carbene? Not quite, as Brinker specifically recognizes in Ref. 83. The regioselec-tivity observed could come from polar transition states, and does not quite *require* an intermediate (Scheme 7.14). The case is strong, however.

Such a reaction, in which two bonds are broken in a single step, has precedent in the two-bond pluck mechanism proposed for the reactions of bicyclobutanes (and quadricyclanes) with carbenes (Scheme 7.15).[84a] In this reaction the transition state

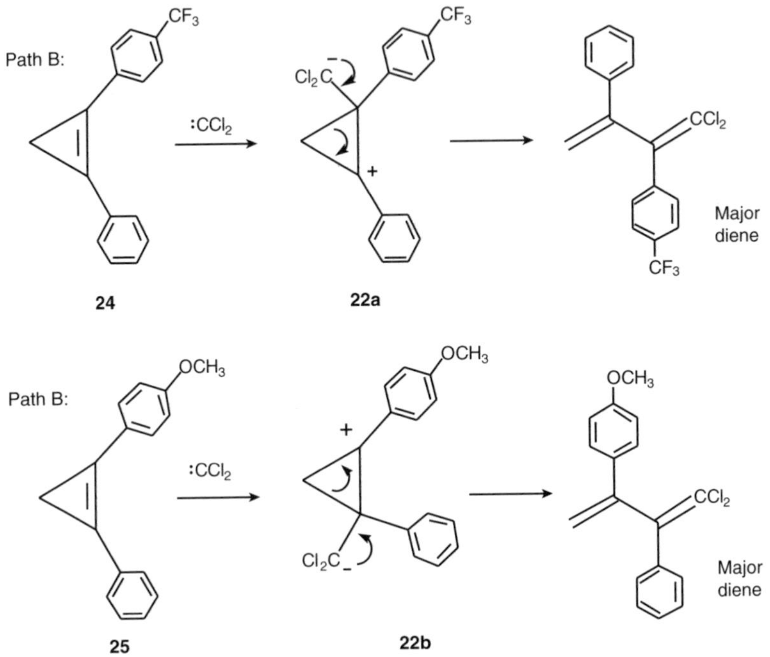

Scheme 7.13

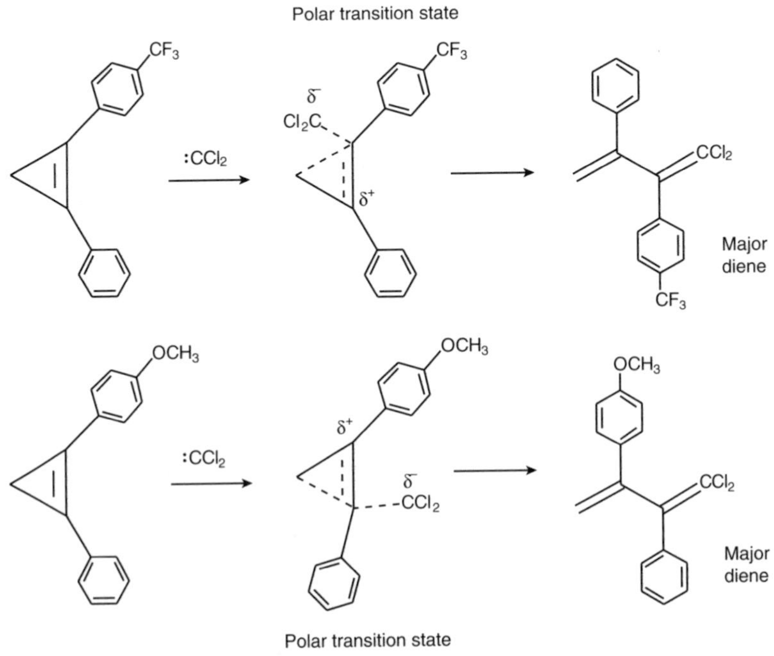

Scheme 7.14

is necessarily unsymmetrical, and potentially polar. See Ref. 84b for a counter-proposal.

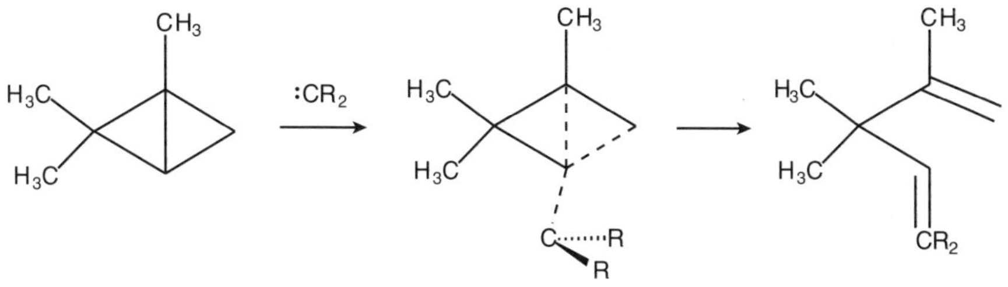

Proposed transition state

Scheme 7.15

What about the general question, Are there other stepwise reactions of singlets in which it is likely that the transition state for a two-step reaction falls below that for a concerted process? The answer here, is "not many," but there are some. For example, vinyl-substituted cyclopropylcarbenes give a variety of products that strongly suggests the intermediacy of a diradical. The principle remains the same—the energetic advantage of the usual concerted reactions must be overcome by special stabilization—delocalization by the vinyl group in this case—of an intermediate.

Cyclopropylcarbenes (see Section 2.4 for more on this reaction, and a caution) ring expand to give cyclobutenes in what is likely a concerted reaction. When a vinyl group is attached to the cyclopropane ring in the proper way, new products appear that are best rationalized by diradical intermediate **26** (Scheme 7.16).[85]

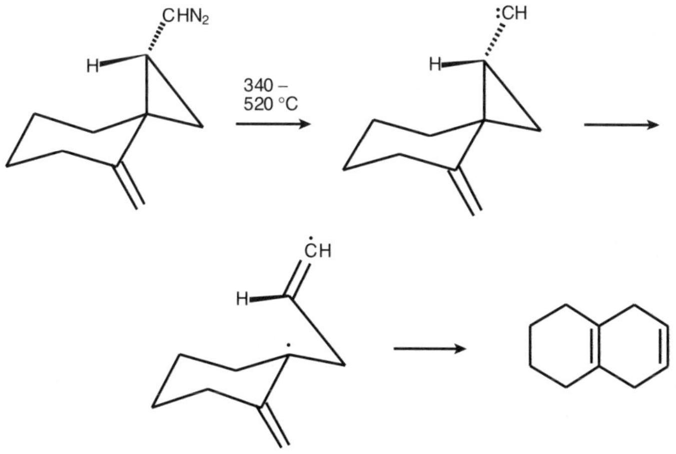

26

Scheme 7.16

2.2. Carbon–Hydrogen Insertion Reactions

2.2.1. Intermolecular Insertions. Singlet carbenes undergo insertion reactions with X–H bonds such as O–H (alcohols), N–H (amines), Si–H (silanes), and so on. The reactions with alcohols can be extremely fast.[86] Here, however, we focus on the C–H insertion reactions of singlet carbenes,[1] in which carbon–carbon bonds are created.[87]

In 1956, Doering et al. reported that methylene (CH_2) inserted into the C–H bonds of pentane, 2,3-dimethylbutane, and cyclohexene with no discrimination (other than statistical) between chemically different sites; CH_2 was "classed as the most indiscriminate reagent known in organic chemistry."[1,87] Doering[88] and Kirmse[89] also demonstrated that the C–H insertion reactions of CH_2 in solution were direct, single barrier concerted processes with transition states that could be represented as **27** (Fig. 7.12). In particular, they did not proceed via initial H abstraction to give radical pair intermediates that subsequently recombined. (Triplet carbene C–H "insertions," however, do follow abstraction–recombination, radical pair mechanisms, as demonstrated in classic experiments of Closs and Closs[90] and Roth[91] (see Chapter 9 in this volume).

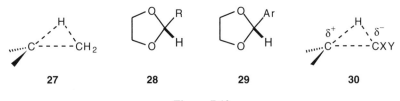

Figure 7.12

Given the high reactivity and lack of selectivity of singlet CH_2 toward C–H bonds, let us focus on more stabilized and selective carbenes such as CCl_2 and C_6H_5CCl. Insertions of singlet CCl_2 can be efficient. For example, CCl_2 generated by the phase-transfer-catalyzed reaction of NaOH and $CHCl_3$ in benzene inserted into the bridgehead C–H of adamantane in 91% yield (based on consumed adamantane).[92] Similarly, CCl_2 inserted into the indicated C–H bond (bold) of ketals **28**, in which R included a large variety of alkyl groups.[93] Note that the reactive C–H is both tertiary and activated by two α-oxygen atoms. With the related substrates (**29**) a Hammett study of CCl_2 relative reactivities as a function of aryl substituents gave $\rho = -0.63$ (vs. σ^+), indicating that CCl_2 insertion is electrophilic and imposes δ^+ on the substrate carbon, analogous to CCl_2 addition to alkenes.[94]

A qualitative representation of the transition state for singlet carbene insertion might be rendered as **30**, in which the appropriate partial charges are indicated. This formulation was offered by Seyferth et al.[95] to account for their observations of selective CCl_2 insertions into the tertiary C–H bonds of various simple hydrocarbons,[96] as well as the α-CH bonds of ethers. Indeed, in a Hammett study of CCl_2 insertions into substituted cumenes **31** (with CCl_2 generated by extrusion from $C_6H_5HgCCl_2Br$ at 80 °C), Seyferth and Cheng[97] found $\rho = -1.9$ (vs. σ) or -0.89 (vs. σ^+), consistent with transition state **30** (Fig. 7.13).

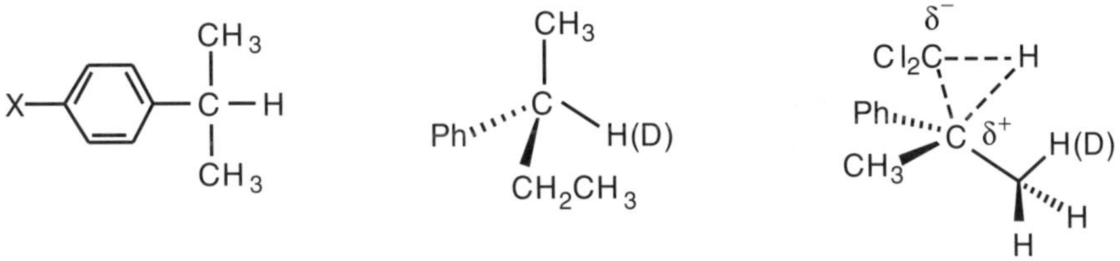

Figure 7.13

Moreover, CCl_2 insertion into the benzylic tertiary C—H of (S)-2-butylbenzene (**32**) occurred with retention, as expected for *direct* C—H insertion.[97] Finally, the C—D analogue of **32** gave k_H/k_D for CCl_2 insertion as 2.5. This relatively small primary kinetic isotope effect, together with the low Hammett ρ value, suggest an "early" transition state, in which relatively little charge separation has developed.

Similar conclusions attend the insertions of CCl_2 (from the thermolysis of $Cl_3CCOONa$ at 120 °C) into α-deuteriocumene and cumene β-d_6, in which the primary $k_H/k_D = 2.6$,[98] similar to Seyferth's finding with **32**, and the β-secondary kinetic isotope effect is 1.20–1.25 for six deuteriums.[98] Here, hyperconjugation at the β-CH (CD) bonds is thought to stabilize the partial cationic charge at the reaction center in transition state **33**.

Polar transition states **30** or **33** appear to be rather general for singlet carbene insertions. For example, Hammett analysis of the Rh carbenoid-mediated benzylic insertion of the arylcarbomethoxycarbene **34** into p-substituated ethylbenzene derivatives gave $\rho = -1.27$ (vs. σ^+).[99] Dichlorocarbene insertions into the Si—H bonds of silacumenes **35** are similarly characterized by $\rho = -0.63$ (vs. σ).[100] In all cases, $\rho < 0$ implies the substrate carbon(δ^+)/carbene carbon(δ^-) polarization depicted in **30** and reflective of electrophilic carbene attack on the C—H or X—H bond. The relatively low magnitude of ρ suggests an early transition state in which charge development at the substrate carbon is small. Note that ρ is smaller for CCl_2 insertions into the Si—H bond of **35** than the C—H bond of **31**; Si does not delocalize the developing positive charge in the transition state into the adjacent aryl ring as well as does C (Fig. 7.14).[97]

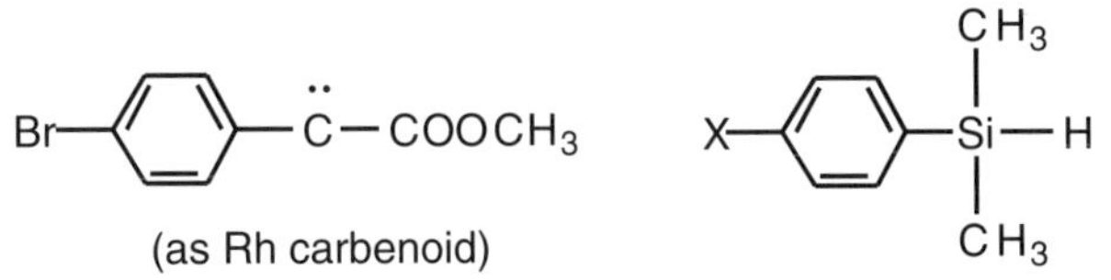

(as Rh carbenoid)

34   **35**   **36**

Figure 7.14

Not all singlet carbene insertions are concerted. For example, Roth showed that the attack of singlet CH_2 on the C—Cl bond of CCl_4 proceeded via initial chlorine abstraction to give a radical pair **36**, which could either collapse to Cl_3CCH_2Cl (minor pathway) or diffuse to free radicals that followed other pathways. Chemically induced dynamic nuclear polarization (CIDNP) demonstrated that the Cl_3CCH_2Cl "insertion" product was largely formed by this two-step, singlet radical pair reaction.[101] Analogous *singlet* abstraction–recombination mechanisms do not appear to occur with C—H bonds.

More recently, Jones and co-workers[102] found evidence for C—H bond activation by appropriately situated carbon–carbon bonds; that is, C—C hyperconjugative stabilization of δ^+ at the C insertion center. For example, the tertiary C—H bonds in substrates **37–39** were particularly reactive toward CCl_2; the interacting C—C bonds presumed responsible for the C—H reactivity are indicated by boldface (Fig. 7.15).

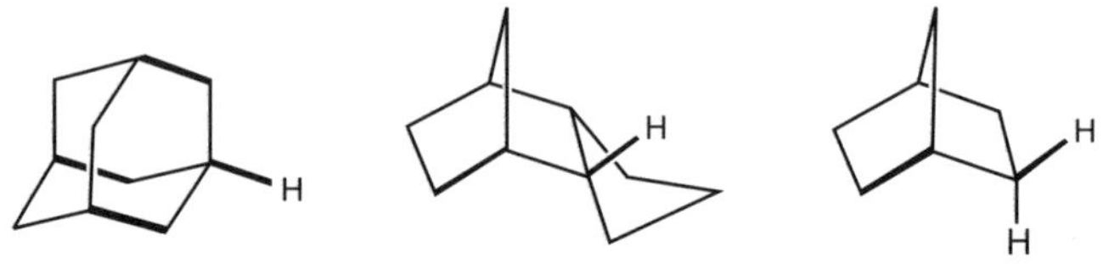

Figure 7.15

Moreover, the equatorial C—H bond of **40** was six times more reative toward CCl_2 than the axial C—H bond of diastereomer **41**. This observation is consistent with C—C hyperconjugative stabilization that can operate in **40**, but not in **42**.[102] Note that no CCl_2 insertion occurred in the tertiary C—H bonds α to the *tert*-butyl groups of **40** or **41** This lack of reactivity is presumably the result of adverse steric factors as well as lack of hyperconjugation (Fig. 7.16).

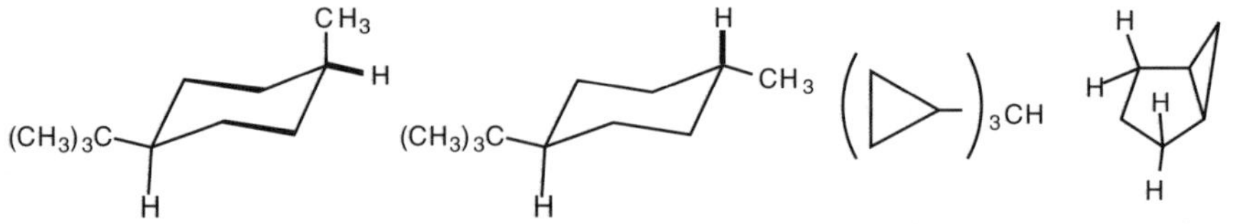

Figure 7.16

Dichlorocarbene insertions into C—H bonds α to cyclopropyl groups were also studied.[102,103] As with C—C hyperconjugation, the orientation of the target C—H bond relative to the activating unit is crucial. Thus, the indicated C—H bond of tricyclopropylmethane (**42**) in which the cyclopropyl groups enjoy significant rotational freedom, is five times less reactive toward CCl_2 than is the indicated C—H bond of substrate **38**.[102] On the other hand, the fixed cyclopropyl unit of **43** activated the α hydrogens toward CCl_2 insertion.[103] Dichlorocarbene attacked the endo or cis C—H bonds 4.3 times faster than the exo or trans C—H bonds. This kinetic preference was attributed to better alignment of the endo C—H bond with the neighboring electron-rich cyclopropyl C—C bond (and the Walsh orbitals thereof), and better stabilization of the partial positive charge that develops in the transition state for endo C—H insertion.[103]

The LFP method has been used to determine absolute rate constants for intermolecular carbene insertion reactions. Given the more convenient diazirine precursor (**14**) available for C_6H_5CCl, as opposed to the phenanthrene precursor (**15**) of CCl_2, as well as the directly observable character of C_6H_5CCl, LFP studies have focused on C_6H_5CCl.[104–107]

Rate constants were determined for C_6H_5CCl insertions into Si—H, N—H, and C—H bonds.[104,105] The C—H substrates included cumene (**31**, X = H, $k = 1.7 \times 10^5\,M^{-1}\,s^{-1}$), ethylbenzene ($8.2 \times 10^4\,M^{-1}\,s^{-1}$), and toluene ($7.5 \times 10^3\,M^{-1}\,s^{-1}$).[104] These C—H insertions are several orders of magnitude slower than the alkene additions of C_6H_5CCl summarized in Table 7.5. Other interesting substrates include *cis,cis*-1,3,5-trimethylcyclohexane (**44**), adamantane (**37**), and cyclohexane (**45**). On a per-H basis, the rate constants for C_6H_5CCl insertion were 1.0×10^5, 1.3×10^5, and $0.06 \times 10^5\,M^{-1}\,s^{-1}$, respectively (Fig. 7.17).[106] The tertiary C—H bonds of **44** and **37** are slightly less reactive than the tertiary and benzylic C—H of cumene, but they are 15–20 times more reactive than the secondary C—H bonds of cyclohexane. These observations agree with the charge distributions depicted in transition states **30** and **33**.

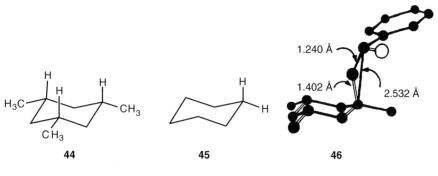

Figure 7.17

Activation parameters determined for C_6H_5CCl insertions into **44** and **37** (in benzene solution) were $E_a = 2.9$ and 3.2 kcal/mol, with $\Delta S^{\ddagger} = -25.6$ and -24.2 e.u., respectively.[106] These activation energies are slightly greater than those observed in additions of C_6H_5CCl to alkenes,[66] but the very negative entropies of activation are comparable to those observed in C_6H_5CCl additions,[57] and reflect the entropy decrease that occurs when two reactant molecules combine to form a single product molecule.[57,67,69]

A kinetic isotope effect was measured for the insertion of C_6H_5CCl into the *sec*-C—H (C—D) bonds of cyclohexane and cyclohexane-d_{12}; $k_H/k_D \sim 3.8$ at 23 °C.[106]

This value is slightly larger than that for insertion of CCl_2 into the benzylic, tertiary C–H (C–D) position of 2-butylbenzene (2.5 at 80 °C)[97] or cumene (2.6 at 120 °C).[98] All three reports, however, are consistent with appreciable C–H or C–D bond breaking in the transition state for insertion.

Calculations at the B3LYP/6-31G* level applied to the C_6H_5CCl-**45** insertion afforded transition state **46**, in which the target C–H bond has lengthened from 1.10 to 1.40 Å, and bonding to the carbenic carbon is well established at a separation of 1.24 Å.[106] The resemblance of **46** to the qualitative constructs **30** and **33** is clear, but the computed activation parameters ($E_a = 7.1$ kcal/mol, $\Delta S^{\ddagger} = -39.8$ e.u.) are not in very good agreement with the measured values (see above).

Finally, a study of C_6H_5CCl insertion into the tertiary C–H bond of substituted adamantanes **47** (Fig. 7.18, X = H, OCH_3, $COOCH_3$, Cl, CN) gave a good Taft-type correlation between k_{ins} and σ_I, the inductive substituent constant of X.[107] The σ value (-1.5) is again consistent with polar transition state **30**, and indicates that significant electronic effects connect the C–X dipole and the insertion site. These effects may be propagated through space, through the intervening σ bonds, or by a combination of modalities.[107] Computational studies at the B3LYP/6-31G* level provide an insertion transition state consistent with **30**, **33**, and **46**. In particular, δ^+ on the adamantyl carbon was computed at $+0.18$–0.20, and the negative charge on the carbenic carbon was -0.18 to -0.19 in the transition state for this electrophilic insertion reaction.

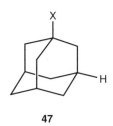

47

Figure 7.18

2.2.2. Intramolecular Insertions: Rearrangements.

These are signature reactions of singlet alkylcarbenes. Here, the intrinsic instability and high reactivity of the carbenes can be harnessed to the formation of highly strained olefins or multicyclic products, often granting access to ring systems not easily obtained by other means. Intramolecular insertions and rearrangements are therefore of great synthetic and mechanistic interest and have been frequently reviewed.[108–115] Here we will briefly consider the kinetics and activation parameters of several representative intramolecular insertion reactions of carbenes, the role of "bystander" substituents, and the possible intervention of tunneling.

Alkylcarbenes generally lack useful UV signals for LFP studies, but they can be indirectly visualized by the pyridine ylide method in which their intramolecular reactions compete kinetically with capture of the carbene by added pyridine.

Reaction with pyridine leads to the formation of a UV-active pyridinium ylide.[65] Rate constants for the alkylcarbene reaction(s) can be extracted from the intercept of the linear correlation of k_{obs} for ylide formation versus the pyridine concentration.[65]

Consider first the 1,2-H shift that converts chloromethylcarbene (**48**) into vinyl chloride (Scheme 7.17). The LFP experiments show that the H shift occurs with $k = 1.2 - 3.0 \times 10^6 \, s^{-1}$ in isooctane, cyclohexane, or dichloroethane at 21–25 °C.[116] The rearrangement is fast, but not ultrafast; carbene (**48**) has a lifetime of ~ 0.3–1 μs under these conditions. Variable temperature kinetics studies give $E_a = 4.9$ kcal/mol (isooctane) or 6.8 kcal/mol (dichloroethane), with unfavorable entropies of activation of -16 or -11.5 e.u., respectively.[116b]

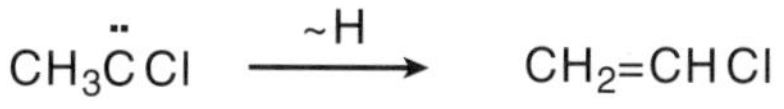

$$CH_3\ddot{C}Cl \xrightarrow{\sim H} CH_2=CHCl$$

48

Scheme 7.17

This seemingly simple rearrangement has been subjected to extensive theoretical analysis.[117–119] Qualitatively, a three-centered transition state (Fig. 7.19, TS **49**) can be drawn for this reaction. The similarity to intermolecular carbene insertion TS **30** is apparent. The 1,2-H shift can be regarded as an intramolecular carbene insertion into an adjacent C–H bond.

An important question is the charge distribution in **49**, and the change in charge

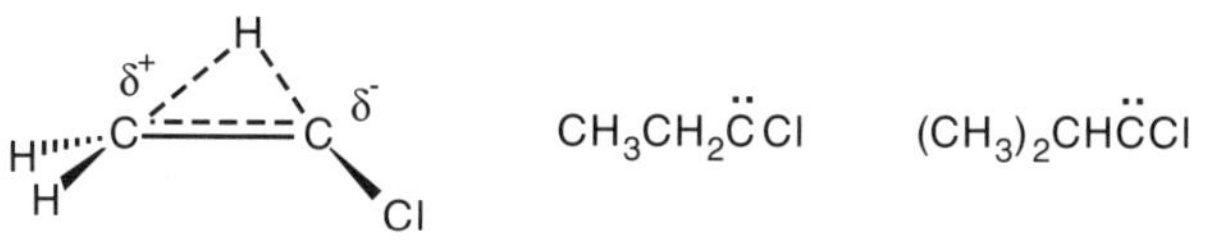

Figure 7.19

on the three key atoms as carbene **48** goes to TS **49**. Carbene 1,2-H shifts are usually called "hydride" shifts, although some high-level computations indicate that the migrating H may actually become more positive in TS **49** than it is in carbene **48** (e.g., by +0.09 in heptane[119]). Negative charge does develop on the carbenic center (-0.21 in **49**) as electron density shifts away from the methyl carbon (+0.14 unit) and the migrant H (+0.09).[118,119] The polar transition state is stabilized in polar solvents, which increase the rate of rearrangement.

Because the methyl carbon of **48** becomes more positive in TS **49**, replacement of its hydrogens by electron-releasing substituents will mitigate the charge increase and "drive" the H shift. Indeed, successive replacements of the hydrogens of **48** by methyl groups (affording carbenes **50** and **51**, Fig. 7.19) increase $k(1,2$-H) to $1.3 \times 10^8 \, s^{-1}$ (isooctane, 25 °C)[120] and $>10^8 \, s^{-1}$ (isooctane, -90 °C).[121] These rate-enhancing methyl substituent effects have been termed "bystander assistance."[122]

The generally accepted order of 1,2-shift tendencies or "migratory aptitudes" for alkylcarbenes is $(R) = H > C_6H_5 > CH_3$.[18,122,123] However, taking bystander assistance by substituents G_1 and G_2 into account, Nickon suggested that the intrinsic migratory aptitude order is $C_6H_5 > H > CH_3$.[122] This conclusion has been supported computationally.[118] The point is that bystander assistance by C_6H_5 (Scheme 7.18, **52**, $G_1 = C_6H_5$; $R = H$) stabilizes positive charge at the migration origin and accelerates the 1,2-H shift, whereas there is no comparable stabilization and drive for C_6H_5 shift when $G_1 = H$ and $R = C_6H_5$. Nickon has provided numerical "bystander assistance factors" for various groups (G) in 1,2-H shift rearrangements.[122]

52

Scheme 7.18

Bystander substituents directly influence the migrating group at the migration origin. We can also identify effects of "spectator substituents" (X in **52**) at the migration terminus. Spectator groups "tune" the electronic environment at the divalent carbon through resonance and inductive effects and modulate its electrophilicity, reactivity, and lifetime. Spectator substituents alter the rates of 1,2-rearrangements but do not change the order of migratory aptitudes or bystander assistance effects.[114] Computational studies show that the activation energies for 1,2-H shifts in CH_3CX are linearly related to the electron donating power ($\sigma°R$) of spectator X;[123] E_a increases as X becomes a better electron donor.

Computational studies of the CH_3CCl to vinyl chloride rearrangement (Scheme 7.17) provide an activation energy that can be compared to those measured by LFP experiments.[118,119] The gas phase computed E_a is $\sim$11.5 kcal/mol,[118,119] which is reduced to $\sim$9.3 kcal/mol in (simulated) heptane.[119] The experimental value in isooctane is 4.9 kcal/mol.[116b] Some of the 4.4 kcal/mol difference between the computed and observed E_a can be narrowed if quantum mechanical tunneling (QMT) is included in the calculations: The migrant H atom can "tunnel" through the activation barrier as well as climb over it.[116b,117,119] The QMT considerations reduce the computed E_a for 1,2-H shift to 6.9 kcal/mol in heptane[119] in reasonable agreement with the observed value.

Charge development in TS **49** makes the transition state more polar than CH_3CCl itself and subject to stabilization by a polar solvent. Indeed, the computed E_a for the rearrangement of CH_3CCl to vinyl chloride in dichloroethane (including tunneling) is 6.0 kcal/mol, 0.9 kcal/mol less than in heptane,[119] so, we expect the 1,2-shift to be faster in the more polar solvent. Such effects are rather general.[124]

Another well-studied 1,2-H shift is the rearrangement of benzylchlorocarbene (**53**) to (Z) and (E)-β-chlorostyrene (Scheme 7.19).[110,114,125,126] In the 0–31 °C temperature range, $k_{1,2\text{-H}} \sim 6 \times 10^7\,\text{s}^{-1}$ with $E_a = 4.5$–4.8 kcal/mol in isooctane.[127] At lower temperature (to -80 °C) Arrhenius studies led to curved correlations of ln $k_{1,2\text{-H}}$ versus $1/T$ that were initially attributed to QMT.[128] However, the intervention of competing reactions of **53** with solvent isooctane and with its precursor diazirine biased the kinetic results at lower temperatures and led to curvature in the Arrhenius plot.[126] In tetrachloroethane, in which the 1,2-H shift of **53** is relatively clean even at lower temperatures, a linear correlation of ln $k_{1,2\text{-H}}$ and $1/T$ was found from 3.2 to -71.4 °C, affording $E_a = 3.2$ kcal/mol and log $A = 10.0\,\text{s}^{-1}$,[126] in good agreement with values determined in CHCl$_3$.[128] Although tunneling does not seem to be important in the 1,2-H shift of **53** in solution near ambient temperature,[126] tunneling is dominant when the rearrangement occurs in Ar or Xe matrices at 10–30 K, because the "normal" activated process is vanishingly slow.[129]

Scheme 7.19

The 1,2-H shift in **53** is of interest in connection with the possible intervention of kinetically significant carbene–alkene complexes and 1,2-H shifts that may occur in excited states of the nitrogenous carbene precursor as well as in the carbenes.[110,114,115] These questions will be further discussed in Section 2.4.

An important example of a 1,2-C migration (C—C insertion) is the ring expansion of chlorocyclopropylcarbene (**54**) to 1-chlorocyclobutene (Scheme 7.20).[127,130–132] The 1,2-C shift takes precedence over 1,2-H shift. Chloromethylenecyclopropane (**55**), the putative product of a 1,2-H shift, is not formed. Interaction of the electron-rich "bent" cyclopropane C—C bond(s) with the vacant p orbital of carbene **54** leads to a more favorable rearrangement pathway than the alternative σ$_{\text{C-H}/p}$ interaction leading to **55**.

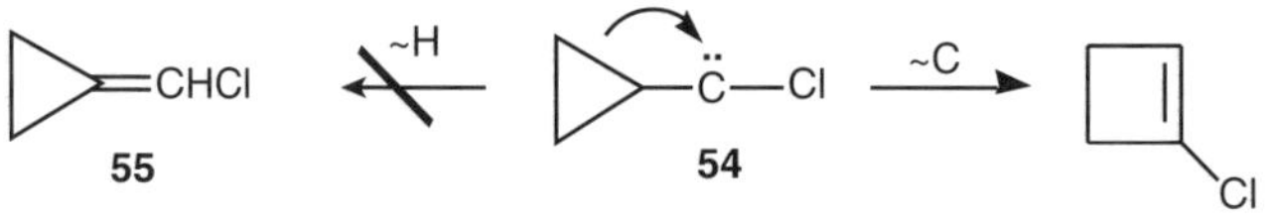

Scheme 7.20

In hydrocarbon solvents, $k_{1,2\text{-C}}$ ranges from $\sim 4 \times 10^5$–$1.5 \times 10^6\,\text{s}^{-1}$, according to several measurements.[130,131] The E_a has been variously reported as ~ 3.0 kcal/mol[131] or 7.4 kcal/mol.[130] The former value, measured by four different approaches, is probably the more reliable one. For comparison, the analogous rearrangement of cyclopropylfluorocarbene to 1-fluorocyclobutene occurs with

$k_{1,2-C} = 1.4 \times 10^5 \text{ s}^{-1}$; $E_a = 4.2$ kcal/mol (pentane, 23 °C).[132] Note the spectator substituent effect; F is a better electron donor than Cl, and therefore slows the 1,2-C shift and raises E_a, relative to **54**.

The 1,2-C shift in **54** exhibits an unfavorable $\Delta S^{\ddagger}$ (−23 e.u.) that was attributed to dynamic effects.[131] However, a very detailed theoretical study of this rearrangement is not in good agreement with the experimental results.[119] The computed E_a (8.2 kcal/mol in simulated isooctane) is ∼5 kcal/mol higher than the experimental value, and the computed $\Delta S^{\ddagger}$ is close to zero. With cyclopropylfluorocarbene, there are even larger disparities between the observed and computed activation parameters. The experimental measurements for cyclopropylfluorocarbene[132] were reproduced[119] with no significant change, so that reason(s) for the divergence between the measured and computed activation parameters remain unknown. An extraordinarily thorough discussion of the problem appears in Ref. 119.

We conclude this section with a brief consideration of dimethylcarbene [CH_3CCH_3 (**56**)], the simplest dialkylcarbene, in which a 1,2-H shift produces propene. Although methylene (CH_2), and methylcarbene (CH_3CH) are ground-state triplets, dimethylcarbene is computed to be a ground-state singlet: calculated singlet–triplet differential energies (ΔE_{S-T}, kcal/mol) are CH_2 (9), CH_3CH (3–5), and **56** (−1.4).[14–18,22,23,133] As mentioned at the beginning of this chapter, the electron-releasing methyl substituents stabilize a singlet carbene center (with its vacant, electronegative p orbital), relative to a triplet carbene center (with singly occupied sp^2 and p orbitals).

In hydrocarbon solvents, the 1,2-H shift in **56** occurs with $k_{1,2-H} = 0.5 − 2.4 \times 10^8 \text{ s}^{-1}$.[134] The E_a is computed to be 7.3 kcal/mol. This 1,2-H shift is considerably faster than that of CH_3CCl (considered above), for which $k_{1,2-H} = 1.2 − 3.0 \times 10^6 \text{ s}^{-1}$. Therefore, there is a strong spectator effect of the Cl substituent, which stabilizes CH_3CCl by resonance electron donation into the carbene's p orbital and thus slows the 1,2-H shift. The spectator CH_3 substituent of **56** is a poorer electron donor than Cl and is less effective in stabilizing CH_3CCH_3, so that its 1,2-H shift is more rapid. However, the spectator methyl group of **56** strongly affects the 1,2-H shift relative to that of CH_3CH. The computed E_a values for CH_3CH and CH_3CCH_3 are 1.9 and 7.3 kcal/mol, respectively.[133]

The lifetime of **56** in hydrocarbon solvents at 25 °C is only several nanoseconds, so that its *intermolecular* chemistry should be difficult to observe. Indeed, intermolecular C−H insertion reactions of **56** are inefficient, although the carbene can be captured (competitively with 1,2-H shift) by insertions into O−H or N−H bonds or by addition to isobutene.[135]

The lifetime of **56** can be extended by deuteration: CD_3CCD_3 lives 3.2 times longer in pentane than does CH_3CCH_3. This longevity is the result of a substantial primary kinetic isotope effect where $k_{1,2-H} > k_{1,2-D}$.[134b] Tunneling is important here: the H shift occurs in part by tunneling, a pathway not as available to the D shift. On the other hand, polar solvents increase $k_{1,2-H}$ and decrease the lifetime of **56**, in accord with the increase in charge stabilization in the 1,2-H shift transition state.[134b]

2.3. Singlet–Triplet Equilibration

As mentioned earlier, when the triplet and singlet states lie close together in energy, as they often do, one can expect equilibration and, possibly, reactions of both spin states. Except in special circumstances, the exothermicity of singlet reactions (remember that two new bonds can be formed from a singlet) gives the singlet a great advantage over the triplet. An impressive array of spectroscopic and theoretical tools has been brought to bear on the case of 2-naphthyl(carbomethoxy)carbene (**57**). A combination of LFP, time-resolved infrared (IR), and UV/vis and electron spin resonance (ESR) spectroscopies, carried out in five different laboratories in Japan, Switzerland, and the United States has clarified the picture in great detail.[136a] Photolysis of the diazo compound leads initially to the singlet state, which rapidly decays to the planar, ground-state triplet. Triplet carbene (**57T**) is mainly formed in the two geometries shown, reflecting the preference of the starting diazo compound for the (*E,E*) and (*Z,E*) forms. Photolysis of **57T** at 515 nm leads to the perpendicular singlet state (**57S**). In the perpendicular geometry, the filled $\sim sp^2$ orbital of the divalent carbon can overlap with the π system of the carbonyl, and the destabilizing interaction between the empty $2p$ orbital of the carbene and this π system is minimized. At 12 K, or on photolysis at 450 nm, **57S** reverts to **57T** (Scheme 7.21).

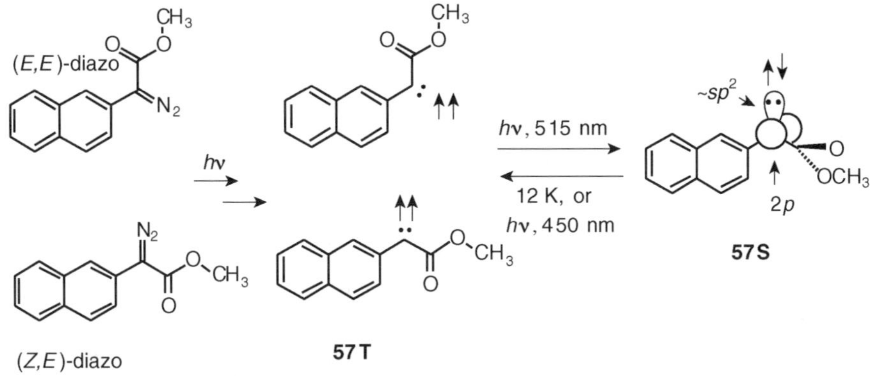

Scheme 7.21

The infrared studies allow an estimation of the (solvent dependent) energy gap between the lower triplet **57T** and the higher singlet **57S** as only 0.2 ± 0.1 kcal/mol. Computations find a larger gap (4.5 kcal/mol), but computations do not reflect the differential stabilization of the singlet (with its cation-like empty orbital) state by solvent.

In this case (but not all others, see Section 2.4), it is the singlet carbene that is responsible for intramolecular carbon–hydrogen insertion to give **58** and Wolff rearrangement to give **59** (Scheme 7.22). Note that direct formation of **59** from the favored (*E,E*) and (*Z,E*) forms of the diazo compound is unlikely because the migrating methoxy group would have to displace the departing nitrogen from

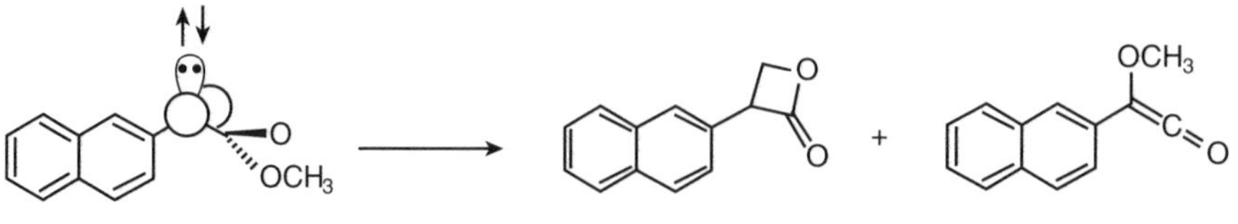

58 (C—H insertion)   **59** (C—C insertion = Wolff rearrangement)

57S

Scheme 7.22

the front side. This kind of phenomenon was first noticed many years ago by Kaplan and Meloy.[136b]

Although the equilibration of singlets and triplets has been many times inferred from classical studies of reaction products, and taken as a given by the community, this tour de force work for the first time allows the direct observation of the many interrelated reactive intermediates involved when a carbene precursor is irradiated.

2.4. Carbene Mimics

Throughout this chapter, we have almost always ignored the role of the carbene precursor. Carbenes are generally made from diazo compounds, or from a variety of surrogate diazo compounds including diazirines, tosylhydrazone salts, and aziridyl imines, all of which probably decompose through nonisolable diazo compounds. Not surprisingly, it turns out that diazo compounds have a rich chemistry of their own, especially when irradiated. Moreover, that chemistry often closely resembles the reactions of carbenes. Much of intramolecular "carbene chemistry" is, in fact, diazo compound chemistry.

As early as 1964 Frey observed that the ratio of 1,1-dimethylcyclopropane and 2-methyl-2-butene, the products from intramolecular reactions of *tert*-butyl diazomethane, was strongly dependent on the method used to decompose the diazo compound (Table 7.6).[137] The response of the community of carbene chemists was largely to ignore Frey's data, and to continue to treat diazo compounds as uncomplicated sources of carbenes. So they *may* be in intermolecular reactions, but they are far from benign in intramolecular chemistry, as we shall now see.

Modarelli and Platz and his co-workers[138] used the pyridine ylide technique[65] to determine that much of the intramolecular chemistry attributed to simple alkylcarbenes such as methylcarbene and dimethylcarbene is, in fact, owing to reactions of excited nitrogenous precursor molecules. As we have seen, many carbenes, even those with lifetimes so short that they cannot be visualized directly by LFP experiments, can be trapped by pyridine to form ylides, which absorb strongly in the conveniently accessible visible region of the electromagnetic spectrum. As the concentration of pyridine is increased, the signal coming from the ylide saturates. Under such conditions every carbene formed from the starting material is converted into an ylide. If products of "carbene" reactions persist under such conditions, they

TABLE 7.6. Products from "*tert*-Butylcarbene"

Source	Conditions	% 1,1-Dimethyl-cyclopropane	% 2-Methyl-2-butene	Reference
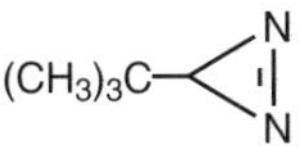 $(CH_3)_3C$—(diazirine)	*hv*	51	49	137b
	hv	50	50	137c
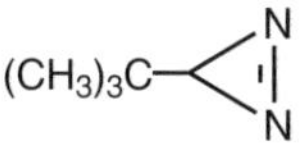 $(CH_3)_3C$—(diazirine)	160 °C	92	8	137b
$(CH_3)_3C$—CHN_2	130 °C	89	11	137c
60	*hv*, 25 °C	90	10	141a
60	*hv*, −78 °C	100	0	141a
61	C atoms, −78 °C	100	0	141b
60	*hv*, 25 °C, decane	90	10	150
60	*hv*, 25 °C, THF [a]	59	41	150

[a] Tetrahydrofuran = THF.

must in reality owe their formation to some other intermediate, presumably the precursor in some form. In this fashion, it has been shown that photolysis of dimethyldiazirine yields an excited state that forms trappable dimethylcarbene in only low yield. Most of the propene that is formed as the major product comes from direct reaction of the excited state.[134b,138] This phenomenon is general: In the photochemical decompositions of nitrogenous precursors to carbenes, the chemistry resulting from many alkylcarbenes has been shown to be accompanied by significant amounts of precursor chemistry (Scheme 7.23). Other, ring-opened excited states may also intervene in the reaction sequence of Scheme 7.23. Occasionally, even thermal reactions involve precursor chemistry.[139]

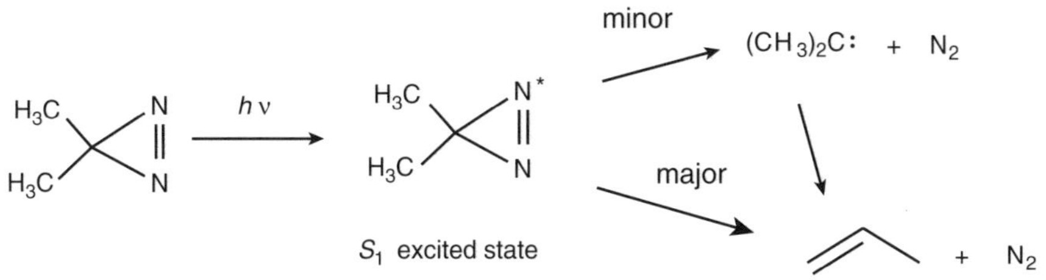

Scheme 7.23

A completely different approach was used to probe the reactivity of *tert*-butylcarbene, one of Frey's original examples.[137] Table 7.6 shows the varying products of thermal and photochemical decomposition of the diazo compound. It would appear that carbon–hydrogen insertion and carbon–carbon insertion are about equally facile in the carbene presumed to be formed in photolytic reactions. Even in 1964, this observation should have seemed strange (as it clearly did to

Frey), as there was no known example of an intermolecular carbon–carbon insertion.[1] The ensuing decades have not revealed such a reaction, despite some hard searching.[140] Why should the intramolecular version of carbon–carbon insertion compete so favorably with carbon–hydrogen insertion? It doesn't. Two alternative sources of *tert*-butylcarbene, photochemical decomposition of **60**[141a] and deoxygenation of pivaldehyde (**61**)[141b] have shown that the real carbene chemistry is that found in the thermal decomposition of the diazo compound (Scheme 7.24; cf. with Table 7.6).[141] Photochemical decomposition, as Frey suspected in 1964, is complicated by a substantial direct rearrangement to 2-methyl-2-butene in the excited diazo compound, mimicking a carbon–carbon insertion of the carbene.

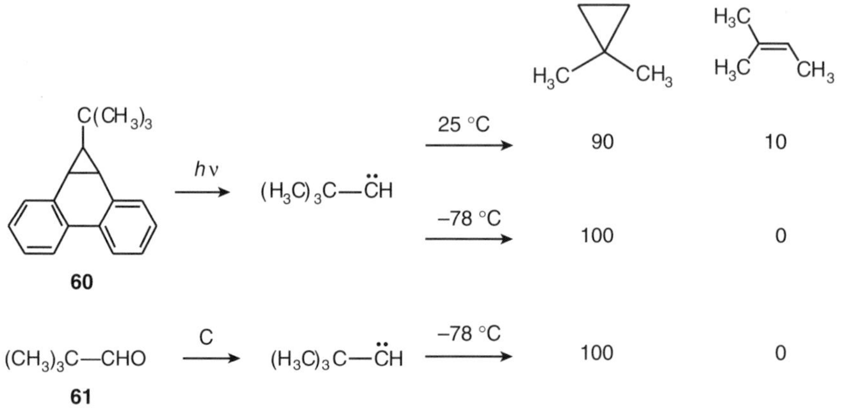

Scheme 7.24

A similar story attends the chemistry of cyclopropylcarbenes. In the 1960s it was shown that cyclopropyl methyl diazomethane gives almost exclusively the product of apparent carbon–carbon insertion.[142] Surely we are now suspicious, and rightly so, as neither photolysis of **62** nor deoxygenation of cyclopropyl methyl ketone reproduces these results. Instead, carbon–hydrogen insertion dominates (Scheme 7.25). Once again, a direct reaction of photoexcited diazo compound appears to be the dominant source of the product of ring expansion (carbon–carbon insertion).

Huang and Platz[143] used a variation of the pyridine ylide technique[65] to determine that ring expansion was a property of the parent diazo compound, (*tert*-butyl)cyclopropyl diazomethane (**63**) not (*tert*-butyl)cyclopropylcarbene itself. They used increasing amounts of 2,3-dimethyl-2-butene to scavenge the carbene formed on photolysis of the cyclopropyldiazirine. The yield of ring expanded product, 3-*tert*-butylcyclobutene, was *not* diminished significantly with increasing alkene concentration, thus showing that ring expansion, again a putative carbon–carbon insertion, was not the result of a carbene reaction. If the cyclobutene were a carbene product, its yield would fall off as carbene reacted with the alkene. Once again, direct ring expansion of a nitrogenous precursor is implicated. The take home

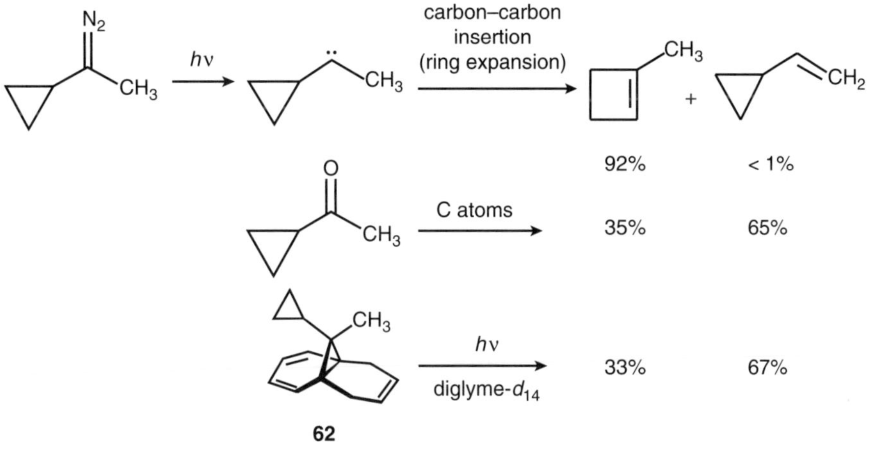

Scheme 7.25

lesson here is that one must always consider whether precursor chemistry is mas-querading as carbene chemistry, especially when carbon–carbon insertion appar-ently leads to the product (Scheme 7.26).

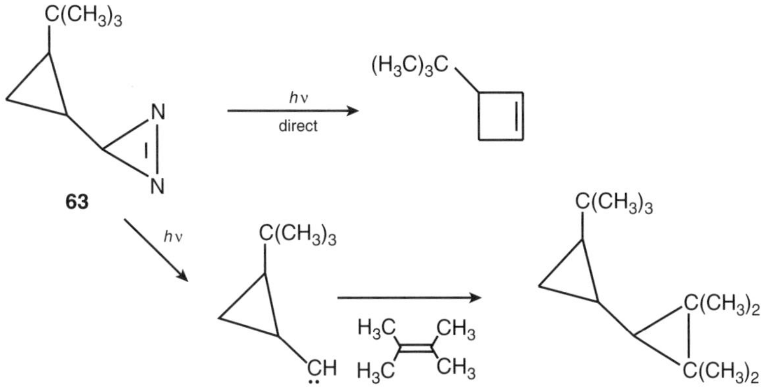

Scheme 7.26

A carbene mimic was also at the center of a rather long-standing controversy on the role of carbene–alkene complexes as potential intermediates in carbene reac-tions. The arguments began when it was found that photolysis of benzylchlorodiaz-irine led to (Z)- and (E)-β-chlorostyrenes through *two* intermediates. There was no simple linear relationship between the yields of the two styrenes and the concentra-tion of an added trapping agent, 2-methyl-2-butene. Instead, the plots were curved and this observation requires two intermediates. It was originally suggested that the carbene and a complex between carbene **53** and the alkene were the two intermedi-ates (Scheme 7.27).[125]

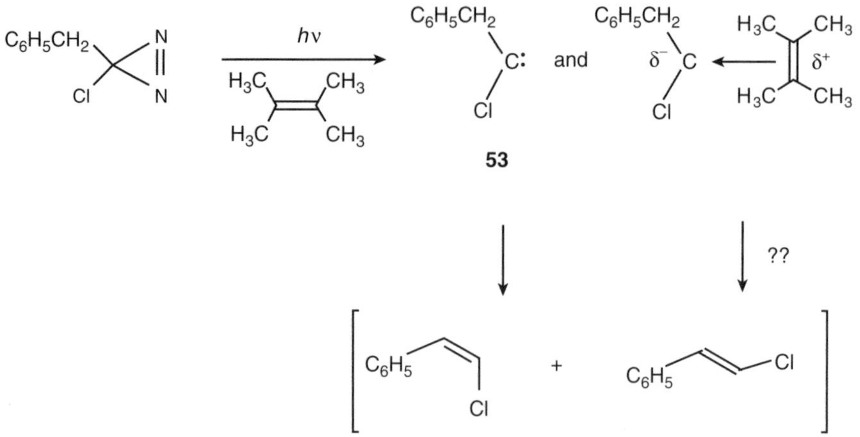

Scheme 7.27

Complexes were also considered for other carbenes[144] and even suggested for reactions in alkane solution(!)[128] A counterproposal, backed by the inability of theory to find support for such stable complexes,[67b] held that the second source was not the carbene–alkene complex but instead was the diazo compound, formed from isomerization of excited diazirine.[145] Other LFP studies reinforced the need for two intermediates, but could not finally resolve the question of carbene complexes versus diazo compound.[125,127a,128] However, the question is now settled in this case, and, by extension, in others as well. No curved plots appear when the benzylchlorocarbene is produced from **64**, a non-nitrogenous source (Scheme 7.28).[146] The second source of the styrenes is not a carbene–alkene complex; this notion is dead, at least for benzylchlorocarbene.

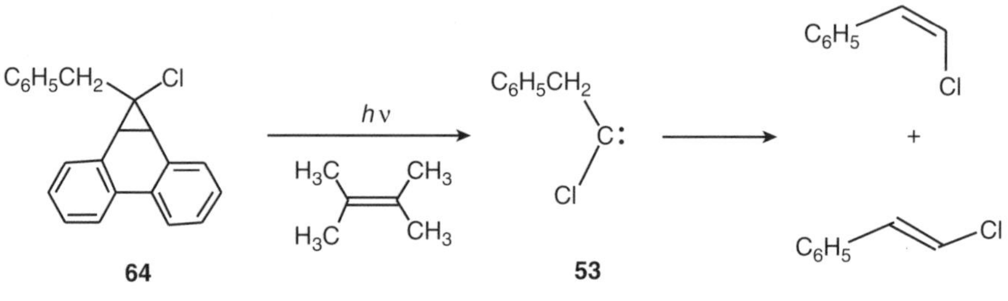

Scheme 7.28

All this is not to say that dialkylcarbenes are incapable of the reactions formerly attributed to them exclusively. They often—usually—are able to do the reactions, even in cases in which diazo compound chemistry pre-empts their doing so. For example, homocubylidenes (**65**) are not the first-formed intermediates from the diazohomocubane precursor (**66**). The bridgehead alkenes, homocubenes (**67**) are. However, it was possible to use a complex kinetic analysis involving both the pyridine ylide technique[65] and the alternative hydrocarbon precursor **68** to show that in the parent system the two reactive intermediates **65** and **67** are in equilibrium once one of them is formed. Indeed, the equilibrium constant is ~1 (Scheme 7.29).[147–149] Carbene **65** can do carbon–carbon insertions.[147–149]

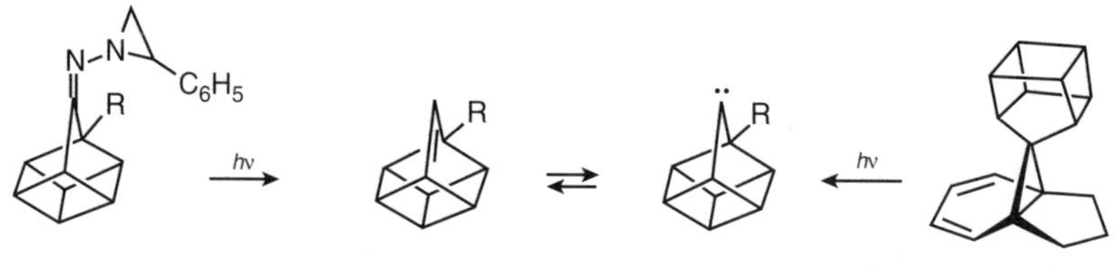

66, R = C$_6$H$_5$, H)   **67**   **65**   **68** (R = H)

Scheme 7.29

2.4.1. Solvent Effects.
A variation on this theme involves solvent-mediated carbene reactions. Once again, it is intramolecular chemistry that is most strongly affected. In the direct reactions of the nitrogen-containing precursors, the nitrogen functions as a leaving group. When a carbene reacts with a nucleophile such as pyridine to form an ylide,[65] a new potential leaving group is produced. Nucleophiles produce ylides or more weakly bonded ylide-like complexes, that have a substantial impact on the observed chemistry. For example, generation of *tert*-butylcarbene in an alkane gives the usual 90:10 ratio of 1,1-dimethylcyclopropane and 2-methyl-2-butene (Table 7.6). By contrast, Ruck found that the same reaction in THF produces a ratio of 59:41. Other nucleophilic solvents, including benzene, produced similar changes in ratio—the percentage of alkene always increased.[150] Presumably, the "excess" 2-methyl-2-butene comes from a concerted rearrangement as shown in **69** in Scheme 7.30.

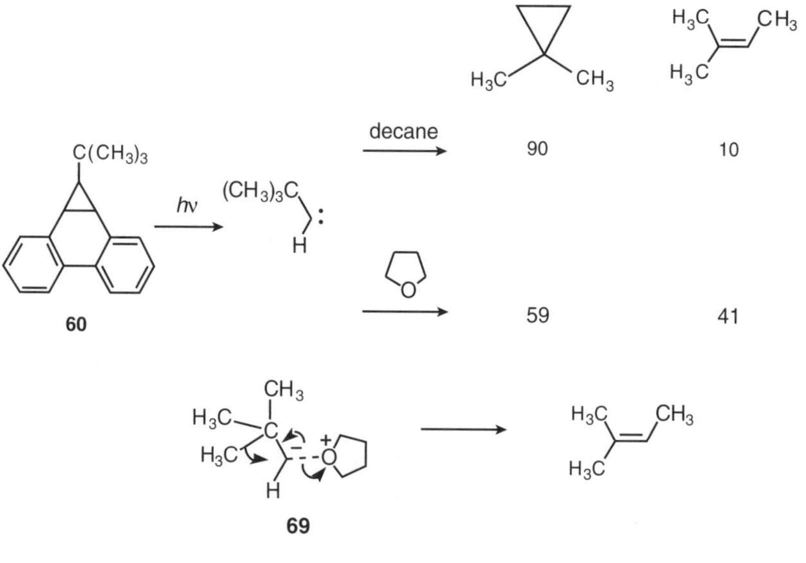

Scheme 7.30

Note that despite the death of the carbene–alkene complex in the study of benzylchlorocarbene (**53**) (see above), benzene is able to modulate the intramolecular reactivity of *tert*-butylcarbene.[150] Some sort of complex must be involved here. Benzene complexes with carbenes have been proposed before. Kahn and Goodman[151] found a transient species on photolysis of diazomethane in benzene, and attributed it to a complex. Moss et al.[152] found that benzene modulated the ratio of intramolecular rearrangement to intermolecular addition for three different carbenes (**53**), chloropropylcarbene, and chlorocyclopropylcarbene, and proposed that a carbene–benzene complex **70** favored the intramolecular rearrangement (Scheme 7.31). Their proposal was bolstered by ab initio calculations that found such stable complexes for CCl_2 and CH_3CCl.

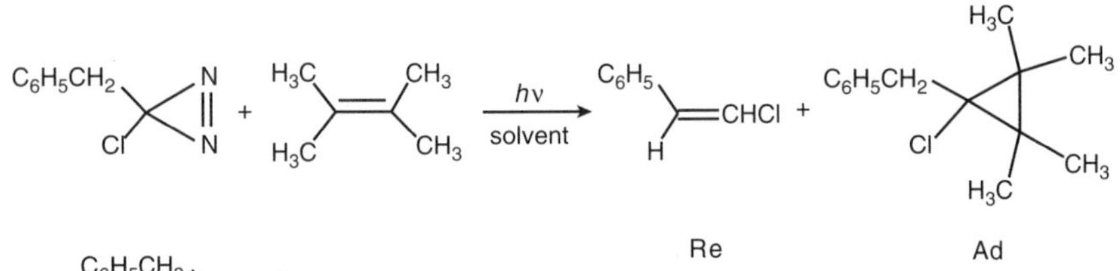

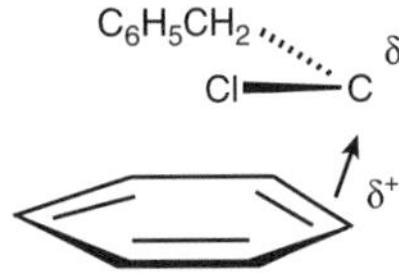

70

solvent	Re/Ad
isooctane	1.20
benzene	2.54
anisole	3.70

Scheme 7.31

There are suggestions that carbene mimics are at work in intermolecular chemistry as well. The data are fragmentary and rare, but cannot be dismissed. For example, it was suggested long ago that excited diazomalonic ester apparently can undergo intermolecular insertions.[153] There are some anomalies in this work—for example, allylic carbon–hydrogen bonds are apparently less reactive than unconjugated carbon–hydrogen bonds—but recent LFP work has supported the idea that intermolecular chemistry is possible for excited states of diazo compounds.[154] If carbon–hydrogen insertion is possible, then it would seem that cyclopropanation, so far assumed to be a carbene reaction, might be similarly afflicted by reactions of carbene mimics. This is surely an area for further work, but there is an ancient suggestion that such is the case.[155] Someone should make dicarbomethoxycarbene from a hydrocarbon source and compare its reactivity with that of the intermediate formed from the diazo compound.

2.5. The Phenylcarbene Rearrangement: The Chemistry of an Incarcerated Carbene

In solution, phenylcarbene behaves normally; it undergoes routine intermolecular cycloadditions and insertions, for example.[5,6] In the gas phase, in which intermolecular reactions are difficult or impossible, phenylcarbene produces a set of dimers that implies rearrangement to a seven-membered ring, either **71** or **72** in Scheme 7.32).[156] The determination of just what that seven-membered ring species is took a lot of work, and has culminated in the beginning of a new aspect of reactive intermediate chemistry—the separation of highly reactive species by incarceration in molecular prisons that protect other molecules from them and vice versa.

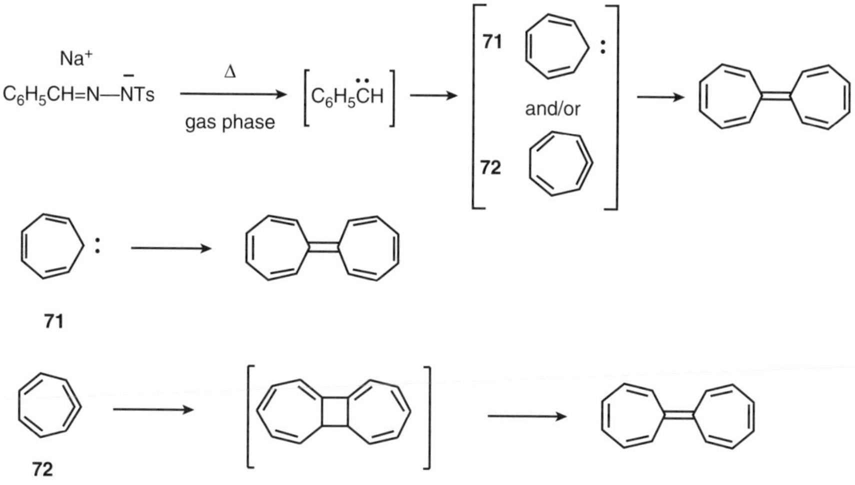

Scheme 7.32

The beginnings of the phenylcarbene rearrangement actually go back to the ancient German literature, which describes some quite amazing reactions of what we now recognize to be carbenes. For example, in 1913 Staudinger and Endle reported that generation of diphenylcarbene by what we would now call Flash Vacuum Pyrolysis led to fluorene.[157] Their proposed mechanism, a double abstraction process, though incorrect, was certainly reasonable at the time. In 1970, Myers et al.[158] showed that the real mechanism involved what has come to be known as the phenylcarbene rearrangement (Scheme 7.33).

Staudinger and Endel, Ref. 157

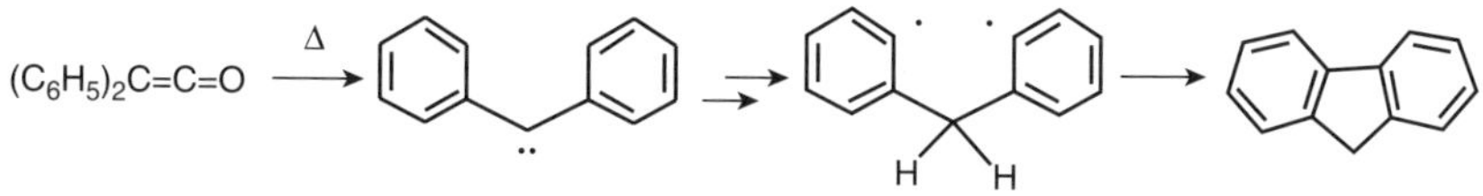

Myers, et al., Ref. 158

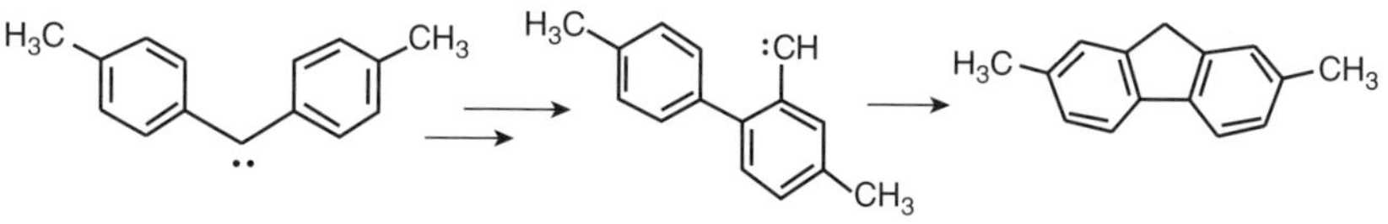

Scheme 7.33

As early as 1964, Vander Stouw and Shechter discovered that in the gas-phase *p*-tolylcarbene led to benzocyclobutene and styrene (Scheme 7.34). This work languished for some unknown reason in Vander Stouw's dissertation and was not published in the primary literature until several years later.[159]

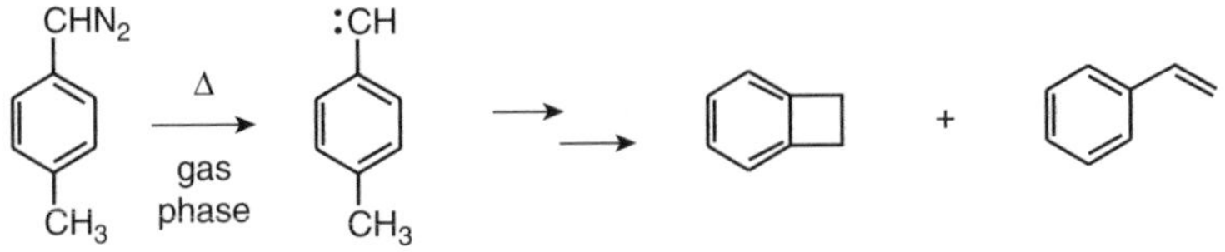

Scheme 7.34

Prompted by W. M. Jones' work on the formation of heptafulvene from phenylcarbene in the gas phase,[156] which implied a seven-membered ring intermediate, and by the implications of the possible reversibility of the process, Baron et al.[160] rediscovered Vander Stouw's cryptic work and showed that all three possible tolylcarbenes gave benzocyclobutene and styrene, albeit in different ratio from the ortho isomer than from the meta or para species.[161] Baron et al.[160] proposed a mechanism in which a methylcycloheptatrienylidene (**CH₃-71**) was the active seven-membered species, and which additionally featured the intermediacy of methylbicyclo[4.1.0]-heptatrienes **73** and **73′** (Scheme 7.35).

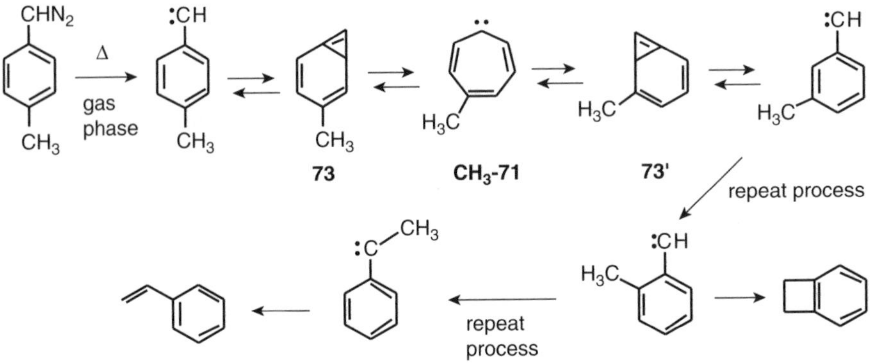

Scheme 7.35

Over the years it became apparent, especially from matrix-isolation studies, that the seven-membered ring species was better formulated as cycloheptatetraene (**72**), a cyclic allene, rather than as the carbene (**71**). The question of the intermediacy of **73** remains unresolved: Theory supports Baron's early suggestion, but the molecule itself has never been detected in a simple system.

The formation of benzocyclobutene and styrene in different ratios from the three isomeric tolylcarbenes is not easily explained by the Baron mechanism,[160] and led to more than one clever (but ultimately wrong) alternative proposal.[161] The question is best resolved by assuming that in the ortho system hydrogen atom transfer in the diazo compound leads to "extra" benzocyclobutene

(Scheme 7.36).[162] Some "carbene" product in the ortho isomer (the excess benzo-cyclobutene) actually comes not from the carbene, but from the carbene precursor (see Section 2.4 for much more on this general subject).

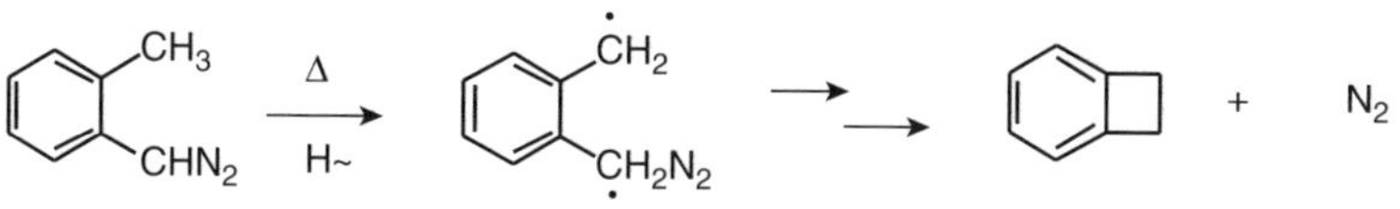

Scheme 7.36

Ultimately, the seven-membered intermediate was identified as cyclohepta-1,2,4,6-tetraene (**72**) or put more properly, this intermediate was indentified when phenyldiazomethane was photolyzed in an argon matrix.[163] But the cycloheptatet-raene (**72**) can be detected, not only at 10 K in argon, but even at +100 °C! The trick is to sequester the carbene precursor, in this case phenyldiazirine, inside a molecular cage.[164] Reaction of the carbenes with the inside of the cage is discour-aged by deuterating the cage, thus raising the barrier to insertion. If photolysis of the diazirine is carried out at low temperature, and reaction with the inside of the cage is thwarted by deuteration, the reactive carbene ring expands to give the twisted allene, cycloheptatetraene (**72**). The NMR studies in a chiral cage show that the barrier to interconversion of the two allenes, over a transition state that must be the planar cycloheptatrienylidene (**71**), is >19.6 kcal/mol (Scheme 7.37).[164,165]

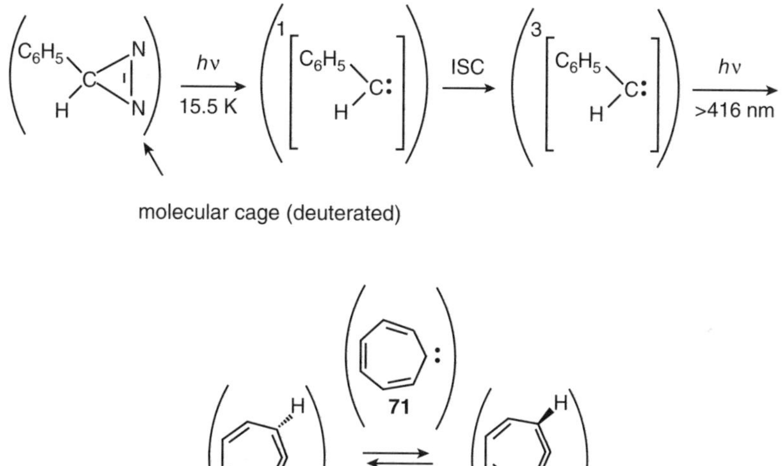

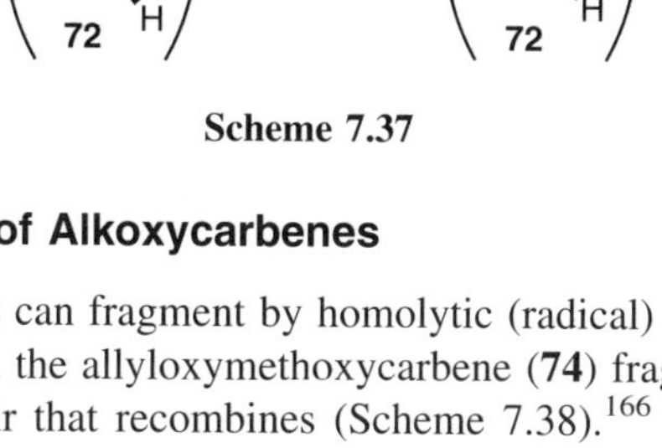

Scheme 7.37

2.6. Fragmentation of Alkoxycarbenes

Acyclic alkoxycarbenes can fragment by homolytic (radical) or heterolytic (ionic) pathways. For example, the allyloxymethoxycarbene (**74**) fragments in benzene at 110 °C to a radical pair that recombines (Scheme 7.38).[166] The radical pair can

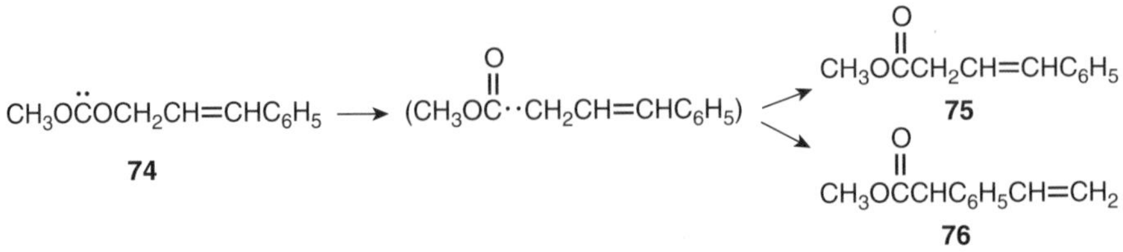

$$CH_3O\overset{\cdot\cdot}{C}OCH_2CH=CHC_6H_5 \longrightarrow (CH_3O\overset{O}{\overset{\|}{C}}\cdot\cdot CH_2CH=CHC_6H_5)$$

74

$$CH_3O\overset{O}{\overset{\|}{C}}CH_2CH=CHC_6H_5$$

75

$$CH_3O\overset{O}{\overset{\|}{C}}CHC_6H_5CH=CH_2$$

76

Scheme 7.38

diffuse apart and the individual radicals can be trapped. Alternatively, recombination produces the allylic isomers **75** and **76** in a 2:1 ratio. When the initial carbene is **77**, the allylic isomer of **74**, fragmentation affords **75** and **76** in a 1:2 ratio. Thus, the radical pair intermediates generated by fragmentation of **74** and **77** retain some regiochemical "memory" of their origin; recombination is competitive with motional scrambling of the radicals (Fig. 7.20).

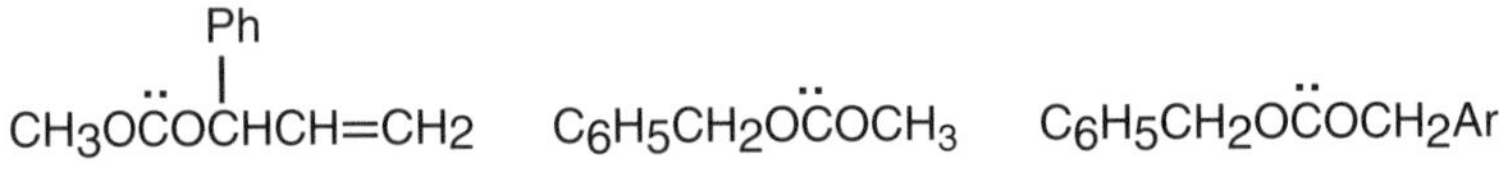

$$CH_3O\overset{\cdot\cdot}{C}O\overset{\overset{\text{Ph}}{|}}{C}HCH=CH_2 \qquad C_6H_5CH_2O\overset{\cdot\cdot}{C}OCH_3 \qquad C_6H_5CH_2O\overset{\cdot\cdot}{C}OCH_2Ar$$

77   **78**   **79**

Figure 7.20

Related homolytic fragmentations were reported for benzyloxymethoxycarbene (**78**)[167a] and aryloxybenzyloxycarbenes (**79**, Fig. 7.20).[167b] The former fragments at 100 °C to methoxycarbonyl and benzyl radicals, which can be trapped with tetramethylpyridinenitroxyl (TEMPO).[167a] Carbene **79** has two possible fragmentations: to $C_6H_5CH_2$ and $ArCH_2OC=O$ radicals or to $C_6H_5CH_2OC=O$ and $ArCH_2$ radicals. Electron-withdrawing substituents on Ar direct the fragmentation toward $ArCH_2$ ($\rho = 0.7$ vs. σ^+), indicating that benzylic radicals are stabilized by electron withdrawal.[167b]

The fragmentation of bis(benzyloxycarbene) (**79**, Ar = C_6H_5) is apolar and homolytic,[167] but the fragmentation of benzyloxychlorocarbene (**80**) appears to be polar and heterolytic.[168,169] Loss of CO produces benzyl chloride in a process likely to involve ion pair **81**, at least in polar solvents such as CH_3CN (Fig. 7.21).[169] When fragmentation occurs in a nucleophilic solvent such as methanol, solvent capture competes with ion pair collapse yielding both $C_6H_5CH_2OCH_3$ (57%) and $C_6H_5CH_2Cl$ (43%).[168] Note that fragmentation of **80** is faster than trapping of this (ambiphilic) carbene by methanol. The LFP rate constant for the fragmentation is $\sim 4 \times 10^5 \, s^{-1}$ in CH_3CN,[170] and the computed activation energy in simulated CH_3OH is only ~ 1.5 kcal/mol.[171]

$$C_6H_5CH_2O\overset{\cdot\cdot}{C}Cl \qquad\qquad [C_6H_5CH_2^+ \; OC \; Cl^-]$$

80   **81**

Figure 7.21

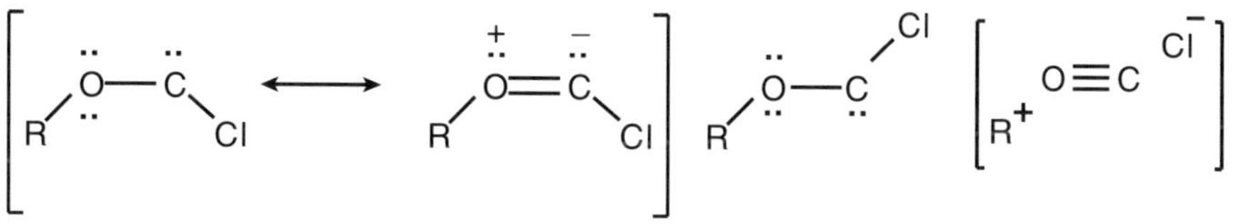

$$\left[\begin{array}{c} \underset{R}{\ddots}O-\underset{\ddots}{C}\diagdown_{Cl} \end{array} \longleftrightarrow \begin{array}{c} \overset{+}{\underset{R}{\ddots}}O=\underset{\ddots}{\overset{-}{C}}\diagdown_{Cl} \end{array} \right] \qquad \underset{R}{\ddots}O-C\diagup^{Cl} \qquad \left[\begin{array}{c} O\equiv C \quad \overset{-}{Cl} \\ \underset{R}{\overset{+}{}} \end{array} \right]$$

    82   **82'**   **83**   **84**

Figure 7.22

A complication is that alkoxychlorocarbenes can exist in two conformations, cis (**82**) and trans (**83**), which interconvert only slowly because of the partial O—C double-bond character shown in **82'** (Fig. 7.22).

Fragmentation and collapse of **82** to RCl will be less encumbered by an intervening molecule of CO than the fragmentation of **83**, which must give rise to ion pair **84**, in which CO "insulates" R^+ from Cl^-. It may be that such CO separated ion pairs are scavenged by CH_3OH to give $ROCH_3$, whereas **82** efficiently collapses to RCl through very short-lived ion pairs, or even concertedly in pentane or vacuum. In pentane, for example, $C_6H_5CH_2OCCl$ readily affords $C_6H_5CH_2Cl$ even though ion pairs are now highly unlikely because of a lack of solvation. Indeed, computations indicate that fragmentation of cis-**80** in pentane directly gives $C_6H_5CH_2Cl + CO$.[173]

Fragmentation of alkoxychlorocarbenes provides a new entry to carbocation–ion pair chemistry that complements the classical solvolytic method. Because of the high rate constants for fragmentation $(10^4–10^6 \, s^{-1})$[170] and the low activation energies $(<10 \, kcal/mol)$[171] the ion pairs are generated in polar solvents with "memory" of their precursor carbenes. Thus, subtle differences in product distributions can be observed in the 2-norbornyl cation–chloride ion pairs generated by the fragmentations of exo- or endo-2-norbornyloxychlorocarbenes.[174]

The 1-norbornyl cation (**85**) is highly strained and unstable because of its inability to assume planarity at the cationic carbon atom. Accordingly, the solvolysis (70 °C, HOAc) of 1-norbornyl triflate proceeds very slowly via **85** $(k = 6.5 \times 10^{-8} \, s^{-1})$ with $E_a = 28.2 \, kcal/mol$.[175] In contrast, fragmentation of carbene **86** via [**85** OC Cl^-] occurs with $k = 3.3 \times 10^4 \, s^{-1}$ in dichloroethane at 25 °C, and $E_a = 9.0 \, kcal/mol$ (Fig. 7.23).[176] Thus, there is a rate acceleration of 5×10^{11} for carbene fragmentation relative to triflate solvolysis.[176]

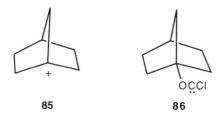

       85   **86**

Figure 7.23

3. CONCLUSION AND OUTLOOK

In Conan Doyle's "The Hound of the Baskervilles," Dr. Mortimer, describing the death of Sir Charles Baskerville, informs Sherlock Holmes and Dr. Watson that footprints of the perpetrator were found beside the dead man's body: "Mr. Holmes, they were the footprints of a gigantic hound."

Over the past one-half of a century, the chemistry of carbenes in particular (and reactive intermediates in general) has developed in parallel to the plot of Doyle's tale. At first, only the "footprints" of the culprit were visible: the structure of reaction products, the relative rates of their formation, the fates of isotopic labels, and other, often ingenious, but still indirect indicators were used to construct an ident-a-kit portrait of the various reactive intermediates.

However, the advent of very fast spectroscopic techniques, such as nanosecond and picosecond LFP, now makes it possible to observe the "hound" itself, while the ever-increasing power of computational methods permits remarkably accurate calculations of the structures and energies associated with carbenes and other reactive intermediates, and often of the potential energy surfaces on which their reactions occur.

To what can we look forward in the future? The impact of spectroscopy and computation will strengthen. Reactions of excited-state singlet and triplet carbenes may be explored by very fast (ps) or even ultrafast (fs) spectroscopy. Solvent modulation of carbenic reactivity will be probed on a molecular level, and perhaps turned to synthetic advantage. The related problem of carbene–arene π complexes will be further elucidated. Dynamics will be increasingly applied to compute the reaction channels available to highly energetic carbenes reacting on relatively flat energy surfaces. Discrepancies between computed and experimental activation energies in such "simple" but fundamental carbenic reactions as the 1,2-carbon migration will be resolved. More precise experimental and computational descriptions will become available for excited-state carbene precursors (carbene mimics), and their reactivity will be better differentiated from that of the carbenes themselves. Continued studies of carbenes constrained in molecular capsules will provide NMR and other spectra of metastable carbenes, while the boundary between these species and stable, isolable singlet carbenes will blur.

Dr. Watson's description of the final confrontation with the Hound of the Baskervilles is vivid and memorable: "A hound it was, an enormous coal-black hound, but not such a hound as mortal eyes have ever seen. Fire burst from its open mouth, its eyes glowed with a smoldering glare, its muzzle and hackles and dewlap were outlined in flickering flame. Never in the delirious dream of a disordered brain could anything more savage, more appalling, more hellish be conceived than that dark form and savage face which broke upon us out of the wall of fog."

Yet, in the end, the hound proved mortal, not supernatural; its unearthly "flickering flame" the result of a "cunning preparation" of phosphorus. (For other cunning preparations of phosphorus, see the discussion of phosphinidenes in Chapter 11 in this volume.) We too have peeled back much of the mystery and cleared away most of the fog that once surrounded the structures, electronic states, and reactions of carbenic intermediates. In doing so, chemists' inductive and deductive powers have rivaled those of the master detective (himself a notable chemist). Our adventures will surely continue.

ACKNOWLEDGMENT

It is a pleasure to acknowledge the critical and helpful reading of this manuscript by Matthew S. Platz. Needless to say, any errors, and the inevitable infelicities that remain are our responsibility alone.

SUGGESTED READING

U. H. Brinker, Ed., *Advances in Carbene Chemistry*, Vol. 2, JAI Press, Stamford, CT, **1998**.

U. H. Brinker, Ed., *Advances in Carbene Chemistry*, Vol. 3, Elsevier, Amsterdam, The Netherlands, **2001**.

G. Bertrand, Ed., *Carbene Chemistry: From Fleeting Intermediates to Powerful Reagents*, FontisMedia, Lausanne, Dekker, New York, **2002**.

R. A. Moss, "Carbenic Selectivity in Cyclopropanation Reactions," *Acc. Chem. Res.* **1980**, *13*, 58.

R. A. Moss, "Carbenic Selectivity Revisited," *Acc. Chem. Res.* **1989**, *22*, 15.

F. Mendez and M. A. Garcia-Garibay, "A Hard–Soft Acid–Base and DFT Analysis of Singlet-Triplet Gaps and the Addition of Singlet Carbenes to Alkenes," *J. Org. Chem.* **1999**, *64*, 7061.

J. E. Jackson and M. S. Platz, in *Advances in Carbene Chemistry*, U. H. Brinker, Ed., "Laser Flash Photolysis Studies of Ylide-Forming Reactions of Carbenes," Vol. 1., JAI Press, Greenwich CT, **1994**, pp. 89ff.

K. N. Houk, N. G. Randan, and J. Mareda, "Theoretical Studies of Halocarbene Cycloaddition Selectivities. A New Interpretation of Negative Activation Energies and Entropy Control of Selectivity," *Tetrahedron*, **1985**, *41*, 1555.

T. V. Albu, B. J. Lynch, D. G. Truhlar, A. C. Goren, D. A. Hrovat, W. T. Borden, and R. A. Moss, "Dynamics of 1,2-Hydrogen Migration in Carbenes and Ring Expansion in Cyclopropylcarbenes," *J. Phys. Chem. A* **2002**, *106*, 5323.

A. Nickon, "New Perspectives on Carbene Rearrangements: Migratory Aptitudes, Bystander Assistance, and Geminal Efficiency," *Acc. Chem. Res.* **1993**, *26*, 84.

REFERENCES

1. (a) W. von E. Doering, R. G. Buttery, R. G. Laughlin, and N. Chaudhuri, *J. Am. Chem Soc.* **1956**, *78*, 3224. (b) D. B. Richardson, M. C. Simmons, and I. Dvoretzky, *J. Am. Chem. Soc.* **1960**, *82*, 5001.

2. J. Hine, *J. Am. Chem. Soc.* **1950**, *72*, 2438.

3. W. von E. Doering and A. K. Hoffmann, *J. Am. Chem. Soc.* **1954**, *76*, 6162.

4. J. Hine, *Divalent Carbon.* Ronald Press, New York, 1964.

5. (a) W. Kirmse, *Carbene Chemistry,* Academic Press, New York, 1964. (b) W. Kirmse, *Carbene Chemistry,* 2nd ed., Academic Press, New York, 1971.

6. M. Jones, Jr. and R. A. Moss, Eds., *Carbenes,* Vol. 1, Wiley-Interscience, New York, **1973**.

7. R. A. Moss and M. Jones, Jr., Eds., *Carbenes,* Vol. 2, Wiley-Interscience, New York, **1975**.

8. M. Regitz, Ed., *Carbene (Carbenoide) Methoden der Organische Chemie (Houben-Weyl),* Vol. E19b, Thieme, Stuttgart, **1989**.

9. M. S. Platz, Ed., *Kinetics and Spectroscopy of Carbenes and Biradicals,* Plenum, New York, **1990**.

10. U. H. Brinker, Ed., *Advances in Carbene Chemistry,* Vol. 1, JAI Press, Greenwich, CT, **1994**.

11. U. H. Brinker, Ed., *Advances in Carbene Chemistry,* Vol. 2, JAI Press, Stamford, CT, **1998**.

12. U. H. Brinker, Ed., *Advances in Carbene Chemistry,* Vol. 3, Elsevier, Amsterdam, The Netherlands, **2001**.

13. G. Bertrand, Ed., *Carbene Chemistry: From Fleeting Intermediates to Powerful Reagents,* FontisMedia, Lausanne, Dekker, New York, **2002**.

14. (a) R. Hoffmann, G. D. Zeiss, and G. W. Van Dine, *J. Am. Chem. Soc.* **1968**, *90*, 1485. (b) I. Shavitt, *Tetrahedron,* **1985**, *41*, 1531 and references cited therein.

15. C.-H. Hu, *Chem. Phys. Lett.* **1999**, *309*, 81.

16. A. R. W. McKellar, P. R. Bunker, T. J. Sears, K. M. Evenson, and R. J. Saykally, *J. Chem. Phys.* **1983**, *79*, 5251.

17. D. G. Leopold, K. K. Murray, and W. C. Lineberger, *J. Chem. Phys.* **1984**, *81*, 1048.

18. D. G. Leopold, A. E. S. Miller, and W. C. Lineberger, *J. Chem. Phys.* **1985**, *83*, 4849.

19. M. M. Gallo and H. F. Schaefer, *J. Phys. Chem.* **1992**, *96*, 1515; S. Khodabandeh and E. A. Carter, *J. Phys. Chem.* **1993**, *97*, 4360.

20. (a) J. Matzinger and J. P. C. Fülscher, *J. Phys. Chem.* **1995**, *99*, 10747. (b) C. A. Richards, Jr., S.-J. Kim, Y. Yamaguchi, and H. F., Schaefer, III, *J. Am. Chem. Soc.* **1995**, *117*, 10104.

21. (a) J. Matzinger, T. Bally, E. V. Patterson, and R. J. McMahon, *J. Am. Chem. Soc.* **1996**, *118*, 1535. (b) A. Admasu, A. D. Gudmundsdottir, and M. S. Platz, *J. Phys. Chem. A* **1997**, *101*, 3832.

22. G. Maier and H. P. Reisenauer, Ref. 12, pp. 119–120, and references cited therein.

23. W. Sander, Ref. 13, p. 8; K. K. Murray, D. G. Leopold, T. M. Miller, and W. C. Lineberger, *J. Chem. Phys.* **1988**, *89*, 5442.

24. For a recent review of carbenic philicity, see R. A. Moss, in Ref. 13, pp. 57ff.

25. R. A. Moss, *Acc. Chem. Res.* **1980**, *13*, 58.

26. R. A. Moss, *Acc. Chem. Res.* **1989**, *22*, 15.

27. P. S. Skell and A. Y. Garner, *J. Am. Chem. Soc.* **1956**, *78*, 5430.

28. W. Doering and W. A. Henderson, Jr., *J. Am. Chem. Soc.* **1958**, *80*, 5274.

29. For a large collection of such data, see R. A. Moss, in Ref. 6, pp. 153ff.

30. (a) R. A. Moss and A. Mamantov, *J. Am. Chem. Soc.* **1970**, *92*, 6951. (b) R. A. Moss and R. C. Munjal, *Tetrahedron Lett.* **1979**, *20*, 4721.

31. R. A. Moss, M. Fedorynski, and W.-C. Shieh, *J. Am. Chem. Soc.* **1979**, *101*, 4736.

32. R. A. Moss, L. A. Perez, J. Wlostowska, W. Guo, and K. Krogh-Jespersen, *J. Org. Chem.* **1982**, *47*, 4177.

33. R. S. Sheridan, R. A. Moss, B. K. Wilk, S. Shen, M. Wlostowski, M. A. Kesselmayer, R. Subramanian, G. Kmiecik-Lawrynowicz, and K. Krogh-Jespersen, *J. Am. Chem. Soc.* **1988**, *110*, 7563.

34. R. Hoffmann, *J. Am. Chem. Soc.* **1968**, *90*, 1475.

35. (a) R. Hoffmann, D. M. Hayes, and P. S. Skell, *J. Phys. Chem.* **1972**, *76*, 664. See also Ref. 27, 28 and (b) P. S. Skell and M. S. Cholod, *J. Am. Chem. Soc.* **1969**, *91*, 7131. (c) P. S. Skell and R. C. Woodworth, *J. Am. Chem. Soc.* **1956**, *78*, 4496.

36. N. G. Rondan, K. N. Houk, and R. A. Moss, *J. Am. Chem. Soc.* **1980**, *102*, 1770.

37. Further references to the frontier MO treatement of carbene (and related) addition reactions can be found in Ref. 26, note 39.

38. R. A. Moss, C. B. Mallon, and C.-T. Ho, *J. Am. Chem. Soc.* **1977**, *99*, 4105.

39. Spectroscopically determined olefinic LUMO (π^*) and HOMO (π) orbital energies are used; (Ref. 25, 26).

40. R. A. Moss, M. Wlostowski, S. Shen, K. Krogh-Jespersen, and A. Matro, *J. Am. Chem. Soc.* **1988**, *110*, 4443.

41. J. Warkentin, in Ref. 11, pp. 245ff.

42. W. W. Schoeller and U. H. Brinker, *Z. Naturforsch.* **1980**, *35b*, 475. W. W. Schoeller, *Tetrahedron Lett.* **1980**, *21*, 1505, 1509. W. W. Schoeller, N. Aktekin, and H. Friege, *Angew. Chem. Int. Ed. Engl.* **1982**, *21*, 932.

43. N. Soundararajan, M. S. Platz, J. E. Jackson, M. P. Doyle, S.-M. Oon, M. T. H. Liu, and S. M. Agnew, *J. Am. Chem. Soc.* **1988**, *110*, 7143.

44. H. Zollinger, *J. Org. Chem.* **1990**, *55*, 3846.

45. H. Fujimoto, S. Ohwaki, J. Endo, and K. Fukui, *Gazz. Chim. Ital.* **1990**, *120*, 229; H. Fujimoto, *Acc. Chem. Res.* **1987**, *20*, 448.

46. F. Mendez and M. A. Garcia-Garibay, *J. Org. Chem.* **1999**, *64*, 7061.

47. W. Sander, C. Kötting, and R. Hübert, *J. Phys. Org. Chem.* **2000**, *13*, 561.

48. W. Sander, in Ref. 13, pp. 1ff.

49. W. Sander and C. Kötting, *Chem. Eur. J.* **1999**, *5*, 24.

50. P. C. Venneri and J. Warkentin, *Can. J. Chem.* **2000**, *78*, 1194.

51. J. H. Rigby and M. D. Danca, *Tetrahedron Lett.* **1999**, *40*, 6891; J. H. Rigby and W. Dong, *Org. Lett.* **2000**, *2*, 1673; J. H. Rigby and S. Laurent, *J. Org. Chem.* **1999**, *64*, 1766.

52. R. W. Hoffmann and M. Reiffen, *Chem. Ber.* **1977**, *110*, 49; R. W. Hoffmann and M. Reiffen, *Chem. Ber.* **1976**, *109*, 2565; R. W. Hoffmann, K. Steinbach, and B. Dittrich, *Chem. Ber.* **1973**, *106*, 2174.

53. S. Goumri-Magnet, T. Kato, H. Gornitzka, A. Baceiredo, and G. Bertrand, *J. Am. Chem. Soc.* **2000**, *122*, 4464.

54. G. L. Closs and B. E. Rabinow, *J. Am. Chem. Soc.* **1976**, *98*, 8190.

55. N. J. Turro, J. A. Butcher, Jr., R. A. Moss, W. Guo, R. C. Munjal, and M. Fedorynski, *J. Am. Chem. Soc.* **1980**, *102*, 7576.

56. J. J. Zupanic and G. B. Schuster, *J. Am. Chem. Soc.* **1980**, *102*, 5958.

57. See R. A. Moss and N. J. Turro, in Ref. 9, pp. 213ff.

58. D. P. Cox, I. R. Gould, N. P. Hacker, R. A. Moss, and N. J. Turro, *Tetrahedron Lett.* **1983**, *24*, 5313.

59. R. A. Moss, H. Fan, L. M. Hadel, S. Shen, J. Wlostowska, M. Wlostowski, and K. Krogh-Jespersen, *Tetrahedron Lett.* **1987**, *28*, 4779.

60. R. A. Moss, W. Lawrynowicz, L. M. Hadel, N. P. Hacker, N. J. Turro, I. R. Gould, and Y. Cha, *Tetrahedron Lett.* **1986**, *21*, 4125.

61. R. A. Moss, C.-S. Ge, J. Wlostowska, E. G. Jang, E. A. Jefferson, and H. Fan, *Tetrahedron Lett.* **1995**, *36*, 3083.

62. R. A. Moss, S. Shen, L. M. Hadel, G. Kmiecik-Lawrynowicz, J. Wlostowska, and K. Krogh-Jespersen, *J. Am. Chem. Soc.* **1987**, *109*, 4341.

63. For further discussions of C_6H_5CX reactivity, see I. R. Gould, N. J. Turro, J. Butcher, Jr., C. Doubleday, Jr., N. P. Hacker, G. F. Lehr, R. A. Moss, D. P. Cox, W. Guo, R. C. Munjal, L. A. Perez, and M. Fedorynski, *Tetrahedron* **1985**, *41*, 1587.

64. J. E. Chateauneuf, R. P. Johnson, and M. M. Kirchhoff, *J. Am. Chem. Soc.* **1990**, *112*, 3217.

65. J. E. Jackson, N. Soundararajan, M. S. Platz, and M. T. H. Liu, *J. Am. Chem. Soc.* **1988**, *110*, 5595. See J. E. Jackson and M. S. Platz, in Ref. 10, pp. 89ff; M. S. Platz, in Ref. 13, pp. 27ff.

66. N. J. Turro, G. F. Lehr, J. A. Butcher, Jr., R. A. Moss, and W. Guo, *J. Am. Chem. Soc.* **1982**, *104*, 1754.

67. (a) K. N. Houk, N. G. Rondan, and J. Mareda, *J. Am. Chem. Soc.* **1984**, *106*, 4293; (b) *Tetrahedron* **1985**, *41*, 1555.

68. B. Giese and C. Neumann, *Tetrahedron Lett.* **1982**, *23*, 3357. B. Giese, W.-B. Lee, and C. Neumann, *Angew. Chem. Int. Ed.* **1982**, *21*, 310.

69. R. A. Moss, W. Lawrynowicz, N. J. Turro, I. R. Gould, and Y. Cha, *J. Am. Chem. Soc.* **1986**, *108*, 7028.

70. N. J. Turro, M. Okamoto, I. R. Gould, R. A. Moss, W. Lawrynowicz, and L. M. Hadel, *J. Am. Chem. Soc.* **1987**, *109*, 4973.

71. W. R. Moore, W. R. Moser, and J. E. LaPrade, *J. Org. Chem.* **1963**, *28*, 2200.

72. B. Zurawski and W. Kutzelnigg, *J. Am. Chem. Soc.* **1978**, *100*, 2654; J. F. Blake, S. G. Wierschke, and W. L. Jorgensen, *J. Am. Chem. Soc.* **1989**, *111*, 1919; A. E. Keating, M. A. Garcia-Garibay, and K. N. Houk, *J. Am. Chem. Soc.* **1997**, *119*, 10805.

73. A. E. Keating, S. R. Merrigan, D. A. Singleton, and K. N. Houk, *J. Am. Chem. Soc.* **1999**, *121*, 3933.

74. Structures **20** and **21** reproduced from Ref. 73 with permission from A. E. Keating, S. R. Merrigan, and K. N. Houk, *J. Am. Chem. Soc.*, **1999**, *121*, 3933. Copyright © 1999 the American Chemical Society.

75. See Ref. 35 (Hoffmann, Hayes, and Skell), in which such geometries are also analyzed.

76. G. L. Closs and L. E. Closs, *Angew. Chem.* **1962**, *74*, 431; G. L. Closs and R. A. Moss, *J. Am. Chem. Soc.* **1964**, *86*, 4042.

77. G. Boche and J. C. W. Lohrenz, *Chem. Rev.* **2001**, *101*, 697.

78. W. von E. Doering and P. LaFlamme, *J. Am. Chem. Soc.* **1956**, *78*, 5447.

79. M. Jones, Jr., W. Ando, M. E. Hendrick, A. Kulczycki, Jr., P. M. Howley, K. R. Hummel, and D. S. Malament, *J. Am. Chem. Soc.* **1972**, *94*, 7469.

80. N. J. Turro, Y. Cha, and I. R. Gould, *J. Am. Chem. Soc.* **1987**, *109*, 2101.

81. K. R. Kopecky, G. S. Hammond, and P. A. Leermakers, *J. Am. Chem. Soc.* **1962**, *84*, 1015.

82. M. Jones, Jr. and K. R. Rettig, *J. Am. Chem. Soc.* **1965**, *87*, 4013; M., Jr., Jones and K. R. Rettig, *J. Am. Chem. Soc.* **1965**, *87*, 4015.

83. (a) J. Weber and U. H. Brinker, *Angew. Chem. Int. Ed. Engl.* **1997**, *36*, 1623. (b) R. A. Wagner and U. H. Brinker, *Chem. Lett.* **2000**, *3*, 246. See also N. C. Yang and T. A. Marolewski, *J. Am. Chem. Soc.* **1968**, *90*, 5644, who first proposed this mechanism.

84. (a) J. E. Jackson, G. B. Mock, M. L. Tetef, G.-x. Zheng, and M. Jones, Jr., *Tetrahedron* **1985**, *41*, 1453; (b) J. Weber, L. Xu, and U. H. Brinker, *Tetrahedron Lett.* **1992**, *33*, 4537.

85. J. M. Cummins, T. A. Porter, and M. Jones, Jr., *J. Am. Chem. Soc.* **1998**, *120*, 6473; J. M. Cummins, I. Pelczer, and M. Jones, Jr., *J. Am. Chem. Soc.* **1999**, *121*, 7595.

86. Singlet carbene O–H reactions have been very thoroughly reviewed by Kirmse: W. Kirmse, in Ref. 10, pp.1ff.; W. Kirmse, in Ref. 12, pp. 1ff.

87. Excellent reviews of early research on C–H insertion reactions of carbenes may be found in Refs. 5 and 6.

88. W. von E. Doering and H. Prinzbach, *Tetrahedron* **1959**, *6*, 24.

89. W. Kirmse and M. Buschoff, *Chem. Ber.* **1969**, *102*, 1098.

90. G. L. Closs and L. E. Closs, *J. Am. Chem. Soc.* **1969**, *91*, 4549; G. L. Closs and A. D. Trifunac, *J. Am. Chem.* Soc. **1969**, *91*, 4554.

91. H. D. Roth, *J. Am. Chem. Soc.* **1972**, *94*, 1761.

92. I. Tabushi, Z.-i. Yoshida, and N. Takahashi, *J. Am. Chem. Soc.* **1970**, *92*, 6670.

93. K. Steinbeck, *Chem. Ber.* **1979**, *112*, 2402; K. Steinbeck and J. Klein, *J. Chem. Res. (S)* **1980**, 94.

94. K. Steinbeck and J. Klein, *J. Chem. Res. (S)* **1978**, 396.

95. D. Seyferth, V. M. Mai, and M. E. Gordon, *J. Org. Chem.* **1970**, *35*, 1993.

96. D. Seyferth, J. M. Burlitch, K. Yamamoto, S. S. Washburne, and C. J. Attridge, *J. Org. Chem.* **1970**, *35*, 1989.

97. D. Seyferth and Y. M. Cheng, *J. Am. Chem. Soc.* **1973**, *95*, 6763.

98. R. A. Pascal, Jr. and S. Mischke, *J. Org. Chem.* **1991**, *56*, 6954.

99. H. M. L. Davies, Q. Jin, P. Ren, and A. Yu. Kovalevsky, *J. Org. Chem.* **2002**, *67*, 4165.

100. D. Seyferth, R. Damrauer, J. Y.-P. Mui, and T. F. Jula, *J. Am. Chem. Soc.* **1968**, *90*, 2944.

101. H. D. Roth, *J. Am. Chem. Soc.* **1971**, *93*, 1527.

102. I. R. Likhotvorik, K. Yuan, D. W. Brown, P. A. Krasutsky, N. Smyth, and M. Jones, Jr., *Tetrahedron Lett.* **1992**, *33*, 911.

103. L. Xu, W. B. Smith, and U. H. Brinker, *J. Am. Chem. Soc.* **1992**, *114*, 783. This reference contains an excellent bibliography of dihalocarbene C–H insertion reactions.

104. M. P. Doyle, J. Taunton, S.-M. Oon, M. T. H. Liu, N. Soundararajan, M. S. Platz, and J. E. Jackson, *Tetrahedron Lett.* **1988**, *29*, 5863.

105. J. E. Jackson, N. Soundararajan, M. S. Platz, M. P. Doyle, and M. T. H. Liu, *Tetrahedron Lett.* **1989**, *30*, 1335.

106. R. A. Moss and S. Yan, *Tetrahedron Lett.* **1998**, *39*, 9381.

107. R. A. Moss and S. Yan, *Org. Lett.* **1999**, *1*, 819.

108. See Ref. 5b, pp. 236 *ff.*

109. W. J. Baron, M. R. De Camp, M. E. Hendrick, M. Jones, Jr., R. H. Levin, and M. B. Sohn, in Ref. 6, pp. 19 *ff.*

110. See R. A. Moss, in Ref. 10, pp. 59 *ff.*

111. See R. Bonneau and M. T. H. Liu, in Ref. 11, pp. 1 *ff.*

112. See M. Jones, Jr., in Ref. 11, pp. 77 *ff.*

113. See M. S. Platz, in Ref. 11, pp. 133 *ff.*

114. See D. C. Merrer and R. A. Moss, in Ref. 12, pp. 53 *ff.* The Appendix to this reference contains extensive tables of rates and activation parameters for the intramolecular insertions of many alkylcarbenes.

115. See G. Szeimies, in Ref. 12, pp. 269 *ff.*

116. (a) R. Bonneau and M. T. H. Liu, *J. Am. Chem. Soc.* **1989**, *111*, 5973. (b) E. J. Dix, M. S. Herman, and J. L. Goodman, *J. Am. Chem. Soc.* **1993**, *115*, 10424; J. A. LaVilla and J. L. Goodman, *J. Am. Chem. Soc.* **1989**, *111*, 6877.

117. J. W. Storer and K. N. Houk, *J. Am. Chem. Soc.* **1993**, *115*, 10426.

118. A. E. Keating, M. A. Garcia-Garibay, and K. N. Houk, *J. Phys. Chem. A* **1998**, *102*, 8467.

119. T. V. Albu, B. J. Lynch, D. G. Truhlar, A. C. Goren, D. A. Hrovat, W. T. Borden, and R. A. Moss, *J. Phys. Chem. A* **2002**, *106*, 5323.

120. S. Celebi, S. Levya, D. A. Modarelli, and M. S. Platz, *J. Am. Chem. Soc.* **1993**, *115*, 8613; R. Bonneau, M. T. H. Liu, and M. T. Rayez, *J. Am. Chem. Soc.* **1989**, *111*, 5893.

121. M. T. H. Liu and R. Bonneau, *J. Am. Chem. Soc.* **1996**, *118*, 8098.

122. A. Nickon, *Acc. Chem. Res.* **1993**, *26*, 84.

123. J. D. Evenseck and K. N. Houk, *J. Phys. Chem.* **1990**, *94*, 5518.

124. M. H. Sugiyama, S. Celebi, and M. S. Platz, *J. Am. Chem. Soc.* **1992**, *114*, 966.

125. M. T. H. Liu, *Acc. Chem. Res.* **1994**, *27*, 287.

126. D. C. Merrer, R. A. Moss, M. T. H. Liu, J. T. Banks, and K. U. Ingold, *J. Org. Chem.* **1998**, *63*, 3010.

127. (a) M. T. H. Liu and R. Bonneau, *J. Am. Chem. Soc.* **1990**, *112*, 3915; (b) R. Bonneau, M. T. H. Liu, and M. T. Rayez, *J. Am. Chem. Soc.* **1989**, *111*, 5973.

128. M. T. H. Liu, R. Bonneau, S. Wierlacher, and W. Sander, *J. Photochem. Photobiol., A: Chem.* **1994**, *84*, 133.

129. S. Wierlacher, W. Sander, and M. T. H. Liu, *J. Am. Chem. Soc.* **1993**, *115*, 8943.

130. M. T. H. Liu and R. Bonneau, *J. Phys. Chem.* **1989**, *93*, 7298.

131. R. A. Moss, G. J. Ho, S. Shen, and K. Krogh-Jespersen, *J. Am. Chem. Soc.* **1990**, *112*, 1638. See also G.-J. Ho, K. Krogh-Jespersen, R. A. Moss, S. Shen, R. S. Sheridan, and R. Subramanian, *J. Am. Chem. Soc.* **1989**, *111*, 6875.

132. R. A. Moss, G.-J. Ho, and W. Liu, *J. Am. Chem. Soc.* **1992**, *114*, 959.

133. B. T. Hill, Z. Zhu, A. Boeder, C. M. Hadad, and M. S. Platz, *J. Phys. Chem. A* **2002**, *106*, 4970.

134. (a) J. P. Pezacki, P. Couture, J. A. Dunn, J. Warkentin, P. D. Wood, J. Lusztyk, F. Ford, and M. S. Platz, *J. Org. Chem.* **1999**, *64*, 4456; (b) D. A. Modarelli, S. Morgan, and M. S. Platz, *J. Am. Chem. Soc.* **1992**, *114*, 7034.

135. I. R. Likhotvorik, E. Tippmann, and M. S. Platz, *Tetrahedron Lett.* **2001**, *42*, 3049.

136. (a) Z. Zhu, T. Bally, L. L. Stracener, and R. J. McMahon, *J. Am. Chem. Soc.* **1999**, *121*, 2863; Y. Wang, T. Yuzawa, H. Hamaguchi, and J. P. Toscano, *J. Am. Chem. Soc.* **1999**, *121*, 2875; J.-L. Wang, I. Likhotvorik, and M. S. Platz, *J. Am. Chem. Soc.* **1999**, *121*, 2883; (b) F. Kaplan and G. K. Meloy, *J. Am. Chem. Soc.* **1966**, *88*, 950.

137. (a) H. M. Frey, *Pure Appl. Chem.* **1964**, *9*, 527. (b) A. M. Mansoor and I. D. R. Stevens, *Tetrahedron Lett.* **1966**, 1733; (c) K.-T. Chang and H. Shechter, *J. Am. Chem. Soc.* **1979**, *101*, 5082.

138. M. S. Platz, in Ref. 11, p. 133ff.

139. J. M. Fox, J. E. Gillen Sacheri, K. G. L. Jones, M. Jones, Jr., Jones, P. B. Shevlin, B. Armstrong, and R. Sztyrbicka, *Tetrahedron Lett.* **1992**, *33*, 5021.

140. G. Wu, M. Jones, Jr., W. von E. Doering, and L. H. Knox, *Tetrahedron*, **1997**, *53*, 9913.

141. (a) H. Glick, I. R. Likhotvorik, and M. Jones, Jr., *Tetrahedron Lett.* **1995**, *36*, 5715; (b) B. M. Armstrong, M. L. McKee, and P. B. Shevlin, *J. Am. Chem. Soc.* **1995**, *117*, 3689.

142. W. Kirmse, B. G. von Bulow, and H. Schepp, *Justus Liebigs Ann. Chem.* **1966**, *41*, 691.

143. H. Huang and M. S. Platz, *J. Am. Chem. Soc.* **1998**, *120*, 5990.

144. R. A. Moss and N. J. Turro, Ref. 9, p. 213ff.

145. M. S. Platz, Ref. 11, p. 164.

146. M. Nigam, M. S. Platz, B. Showalter, J. Toscano, R. Johnson, S. C. Abbot, and M. M. Kirchhoff, *J. Am. Chem. Soc.* **1998**, *120*, 8055.

147. (a) P. E. Eaton and K. L. Hoffman, *J. Am. Chem. Soc.* **1987**, *109*, 5285; (b) P. E. Eaton and R. B. Appell, *J. Am. Chem. Soc.* **1990**, *112*, 4055.

148. W. R. White, M. S. Platz, N. Chen, and M. Jones, Jr., *J. Am. Chem. Soc.* **1991**, *113*, 4981; D. A. Hrovat and W. T. Borden, *J. Am. Chem. Soc.* **1996**, *118*, 1535.

149. See also R. T. Ruck and M. Jones, Jr., *Tetrahedron Lett.* **1998**, *39*, 4433.

150. R. T. Ruck and M. Jones, Jr., *Tetrahedron Lett.* **1998**, *39*, 2277.

151. M. I. Kahn and J. Goodman, *J. Am. Chem. Soc.* **1995**, *117*, 6635.

152. R. A. Moss, S. Yan, and K. Krogh-Jespersen, *J. Am. Chem. Soc.* **1998**, *120*, 1088; K. Krogh-Jespersen, S. Yan, and R. A. Moss, *J. Am. Chem. Soc.* **1999**, *121*, 6269.

153. D. S. Wulfman, B. Poling, and R. S. McDaniel, Jr., *Tetrahedron Lett.* **1975**, 4519.

154. J. P. Toscano, Ref. 11, p. 231ff.

155. See Reference 79.

156. R. C. Joines, A. B. Turner, and W. M. Jones, *J. Am. Chem. Soc.* **1969**, *91*, 7754.

157. H. Staudinger and R. Endle, *Chem. Ber.* **1913**, *46*, 1437; F. O. Rice and J. D. Michaelsen, *J. Phys. Chem.* **1962**, *66*, 1535.

158. J. A. Myers, R. C. Joines, and W. M. Jones, *J. Am. Chem. Soc.* **1970**, *92*, 4740.

159. G. G. Vander Stouw, A. R. Kraska, and H. Shechter, *J. Am. Chem. Soc.* **1972**, *94*, 1655.

160. W. J. Baron, M. Jones, Jr., and P. P. Gaspar, *J. Am. Chem. Soc.* **1970**, *92*, 4739.

161. For a nice review of the early work, and provocative mechanistic suggestions, see: P. P. Gaspar, J.-P. Hsu, S. Chari, and M. Jones, Jr., *Tetrahedron*, **1985**, *41*, 1479.

162. O. L. Chapman, J. W. Johnson, R. J. McMahon, and P. R. West, *J. Am. Chem. Soc.* **1988**, *110*, 501.

163. P. R. West, O. L. Chapman, and J.-P. LeRoux, *J. Am. Chem. Soc.* **1982**, *104*, 1779; R. J. McMahon, C. J. Abelt, O. L. Chapman, J. W. Johnson, C. L. Kreil, J.-P. LeRoux, M. Mooring, and P. R. West, *J. Am. Chem. Soc.* **1987**, *109*, 2459; O. L. Chapman and C. J. Abelt, *J. Org. Chem.* **1987**, *52*, 121.

164. R. Warmuth, *Eur. J. Org. Chem.* **2001**, 423.

165. R. Warmuth and M. A. Marvel, *Chem. Eur. J.* **2001**, *7*, 1209.

166. P. C. Venneri and J. Warkentin, *J. Am. Chem. Soc.* **1998**, *120*, 11182.

167. (a) N. Merkley, M. El-Saidi, and J. Warkentin, *Can. J. Chem.* **2000**, *78*, 356. (b) N. Merkley and J. Warkentin, *Can. J. Chem.* **2000**, *78*, 942.

168. R. A. Moss, B. K. Wilk, and L. M. Hadel, *Tetrahedron Lett.* **1987**, *28*, 1969.

169. R. A. Moss, *Acc. Chem. Res.* **1999**, *32*, 969.

170. R. A. Moss, L. A. Johnson, S. Yan, J. P. Toscano, and B. M. Showalter, *J. Am. Chem. Soc.* **2000**, *122*, 11256.

171. S. Yan, R. R. Sauers, and R. A. Moss, *Org. Lett.* **1999**, *1*, 1603.

172. M. A. Kesselmayer and R. S. Sheridan, *J. Am. Chem. Soc.* **1986**, *108*, 99, 844.

173. R. A. Moss, Y. Ma, F. Zheng, R. R. Sauers, T. Bally, A. Maltsev, J. P. Toscano, and B. M. Showalter, *J. Phys. Chem. A*, **2002**, *106*, 12280.

174. R. A. Moss, F. Zheng, R. R. Sauers, and J. P. Toscano, *J. Am. Chem. Soc.* **2001**, *123*, 8109.

175. R. C. Bingham and P. v. R. Schleyer, *J. Am. Chem. Soc.* **1971**, *93*, 3189.

176. R. A. Moss, F. Zheng, J.-M. Fedé, Y. Ma, R. R. Sauers, J. P. Toscano, and B. M. Showalter, *J. Am. Chem. Soc.* **2002**, *124*, 5258.

CHAPTER 8

Stable Singlet Carbenes

GUY BERTRAND

UCR-CNRS Joint Research Chemistry Laboratory, (UMR 2282)
Department of Chemistry, University of California, Riverside, CA

1. INTRODUCTION

As early as 1835, attempts to prepare the parent carbene (CH_2) by dehydration of methanol had been reported.[1] It is interesting to note that at that time the

Reactive Intermediate Chemistry, edited by Robert A. Moss, Matthew S. Platz, and Maitland Jones, Jr.
ISBN 0-471-23324-2   Copyright © 2004 John Wiley & Sons, Inc.

tetravalency of carbon was not established, and therefore the existence of stable carbenes was considered to be quite reasonable. At the very beginning of the twentieth century, Staudinger and Kupfer demonstrated that carbenes, generated from diazo compounds or ketenes, were highly reactive species.[2] It quickly became clear that their six valence electron shell, which defied the octet rule, was responsible for their fugacity. As a consequence, the quest for stable carbenes became an unreasonable target, and indeed remained so for quite some time! In the 1950s, Breslow[3] and Wanzlick,[4] realized that the stability of a carbene could be dramatically enhanced by the presence of amino substituents, but they were not able to isolate a "monomeric" carbene. It was only in 1988 that our group reported the synthesis of a stable carbene, namely, a (phosphino)(silyl)carbene.[5]

Since this discovery, a few persistent triplet carbenes[6] have been prepared and other types of stable singlet carbenes have been isolated. Triplet carbenes are discussed in Chapter 9 by H. Tomioka in this volume, and therefore this chapter will be focused on singlet carbenes.

2. SINGLET VERSUS TRIPLET GROUND STATE?

The divalent carbon atom of carbenes has only six electrons in its valence shell. Four are used to make σ bonds with the two substituents, and therefore two non-bonding electrons remain. To understand, the difference between the singlet and the triplet state, let us consider a prototypal carbene. The carbon atom can be either linear or bent, each geometry is describable by a specific hybridization. The linear geometry implies an sp hybridized carbene center with two nonbonding degenerate orbitals (p_x and p_y). Bending the molecule breaks this degeneracy and the carbon atom adopts an sp^2-type hybridization: the p_y orbital remains almost unchanged (it is usually called p_π), while the orbital that starts as a pure p_x orbital is stabilized because it acquires some s character (it is therefore called σ) (Fig. 8.1). The carbene ground-state multiplicity is related to the relative energy of the σ and p_π orbitals. The singlet ground state is favored by a large σ-p_π separation; Hoffmann assumed

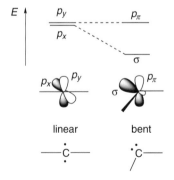

Figure 8.1. Relationship between the carbene bond angle and the nature of the frontier orbitals.

that a value of at least 2 eV is necessary to impose a singlet ground state, whereas a value <1.5 eV leads to a triplet ground state.[7]

Given these conditions, the influence of the substituents on the carbene ground-state multiplicity can be easily analyzed in terms of steric and electronic effects.

The electronic stabilization of the triplet state is at a maximum when the carbene frontier orbitals are degenerate, so that a linear geometry will favor the triplet state. A way to force a linear geometry is to increase the steric bulk of carbene substituents.[8] Dimethylcarbene has a bent singlet ground state (111°),[9] while the di(*tert*-butyl)-[10] and diadamantyl-[11] carbenes are triplets. In contrast, cyclopropenylidene[12] and even cyclopentylidene[13] have singlet ground states due, at least in part, to angular constraint.

Does that mean that all carbenes featuring a broad carbene bond angle will have a triplet ground state? The answer is clearly no. Indeed, steric effects dictate the ground-state spin multiplicity as far as the electronic effects are negligible, which is rarely the case.[14] The influence of the substituents' electronegativity on the carbene multiplicity was recognized relatively early on,[15,16] and reexamined more recently.[17] It is now well established that σ-electron-withdrawing substituents favor the singlet versus the triplet state. In particular, Harrison et al.[15a,c] showed that the ground state goes from triplet to singlet when the substituents are changed from electropositive lithium (triplet favored by 23 kcal/mol) to hydrogen (triplet favored by 11 kcal/mol) to electronegative fluorine (singlet favored by 45 kcal/mol), although mesomeric effects certainly also play a role for the latter element. Indeed, inductive effects play a minor role on the ground-state spin multiplicity compared to mesomeric effects, which involve of the interaction of the carbon orbitals (σ, p_π) and appropriate p or π orbitals of one or two carbene substituents. Substituents interacting with the carbene center can be classified into two types, namely, D (for π-electron-donating groups such as −F, −Cl, −Br, −I, −NR$_2$, −PR$_2$, −OR, −SR, etc.), and W (for π-electron-withdrawing groups such as −COR, −CN, −BR$_2$, −SiR$_3$, −PR$_3^+$, etc.). Importantly, D-C-D, most D-C-W, and some W-C-W carbenes are predicted to have singlet ground states; however, their electronic structure and geometry can vary a lot, as shown in Figure 8.2.

(D,D)-Carbenes are predicted[16,17] to be bent and the donation of the D substituent lone pairs results in a polarized four-electron three-center π system. The C−D

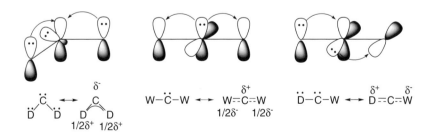

Figure 8.2. Electronic effects of the substituents for D-C-D, W-C-W, and D-C-W carbenes (D=π-electron-donating substituent; W=π-electron-withdrawing substituent).

bonds acquire some multiple bond character, which implies that (D,D)-carbenes are best described by the superposition of two ylidic structures with a negative charge at the carbene center. The most representative carbenes of this type are the transient dimethoxy-[18] and dihalocarbenes,[19] and the stable diaminocarbenes that will be described in detail in this chapter.

Most of the (W,W)-carbenes are predicted to be linear[16,17] and this substitution pattern results in a polarized two-electron three-center π system. Here also, the C−W bonds have some multiple bond character; these (W,W)-carbenes are best described by the superposition of two ylidic structures featuring a positive charge at the carbene carbon atom. The most studied carbenes of this type are the transient dicarbomethoxycarbenes[20] and the "masked" diborylcarbenes.[21] Since no carbenes of the latter type have yet been isolated, they are not included in this chapter. Lastly, the quasilinear (D,W)-carbenes combine both types of electronic interaction. The D substituent lone pair interacts with the p_y orbital, while the W substituent vacant orbital interacts with the p_x orbital. These two interactions result in a polarized allene-type system with DC and CW multiple bonds. Good examples of this type of carbene are given by the transient halogenocarboethoxycarbenes[22] and by the stable (phosphino)(silyl)- and (phosphino)(phosphonio)carbenes (see below).

3. HOW THE FIRST TWO FAMILIES OF STABLE CARBENES WERE DISCOVERED

3.1. From the Curtius Bis(carbene)–Acetylene Analogy to (Phosphino)(silyl)carbenes

The first carbene ever isolated was **Ia**, which was prepared using the most classical route to transient carbenes, namely, the decomposition of diazo compounds. The [bis(diisopropylamino)phosphino](trimethylsilyl)diazomethane precursor (**1a**) was obtained by treatment of the lithium salt of trimethylsilyldiazomethane with 1 equiv of bis(diisopropylamino)chlorophosphine.[23] Dinitrogen elimination occurs by photolysis (300 nm) or thermolysis (250 °C under vacuum)[5] affording carbene **Ia** as a red oily material in 80% yield (Scheme 8.1). Carbene **Ia** is stable for weeks at room temperature and can even be purified by flash distillation under vacuum (10^{-2} Torr) at 75–80 °C.

We have to confess that when we first carried out the decomposition of the diazo derivative **1a**, we did not hope that the carbene **Ia** would be stable. In fact, based on

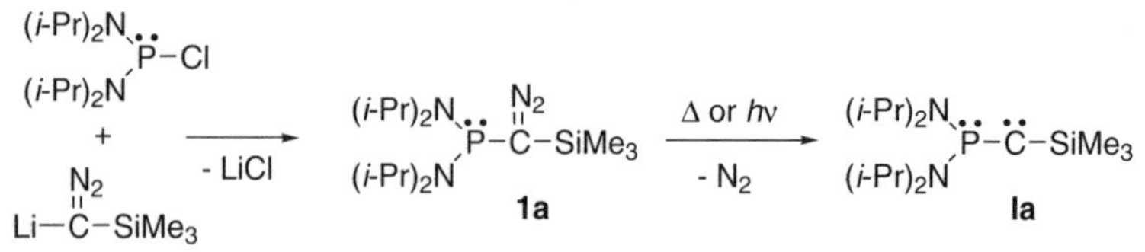

Scheme 8.1

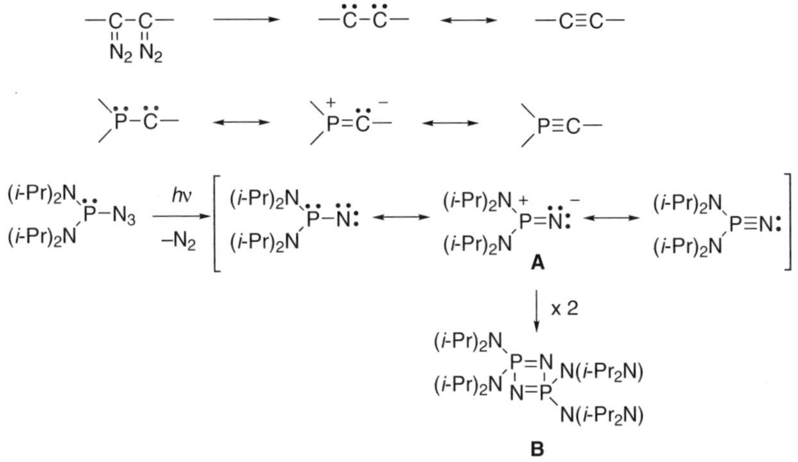

Scheme 8.2

a very well-known reaction described at the end of the nineteenth century by Curtius[24] and on our previous work concerning phosphinonitrenes (**A**),[25] we were trying to demonstrate that a phosphinocarbene would behave as a phosphaacetylene (Scheme 8.2). Indeed, Curtius had shown that α,α'-bis(diazo) derivatives spontaneously lost two molecules of N_2 giving rise to the corresponding alkynes. As diazo compounds are precursors of carbenes, this result clearly showed that an α,α'-bis (carbene) is an alkyne. Like carbenes, tricoordinated-trivalent phosphorus atoms possess a lone pair of electron and, to some extent, an accessible vacant orbital of the σ^* type. Therefore, it was reasonable to believe that the phosphorus–carbon bond of α-phosphinocarbenes would feature some multiple-bond character. This hypothesis was reinforced by our own work concerning phosphinonitrenes. Indeed, we had shown that the decomposition of phosphine azides led to transient phosphinonitrenes (**A**) featuring a strong multiple-bond character, as shown by the obtention of the $2+2$ dimer, namely, a cyclodiphosphazene (**B**, Scheme 8.2).[25]

When we carried out the photolysis of **1a** in the presence of trimethylchlorosilane and dimethylamine, as trapping agent, clean reactions occurred giving the corresponding adducts formally resulting from a 1,2-addition across the polarized PC-multiple bond.[23] We only recognized later[5a] that these adducts more likely resulted from a carbene insertion into the A—X bonds of the trapping agent, followed by 1,2-shifts (Scheme 8.3).

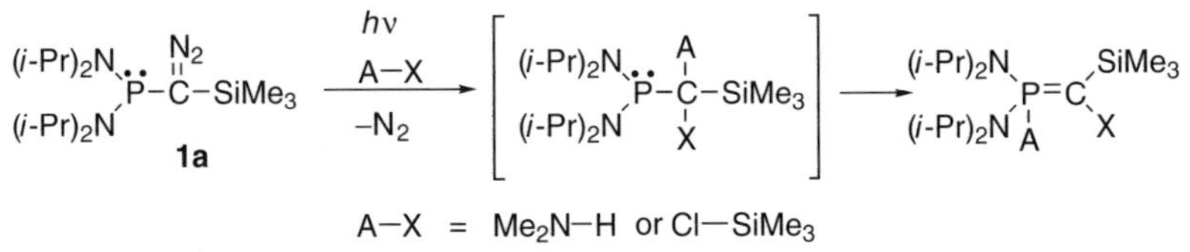

A—X = Me$_2$N—H or Cl—SiMe$_3$

Scheme 8.3

We were so focused on the PC multiple-bond character of **Ia**, that in a footnote of the preliminary communication,[23] we wrote: *"According to the ^{31}P NMR, besides numerous products, a major species ($\delta^{31}P = -41$ ppm) was formed, in the absence of trapping agents or in the presence of dimethylbutadiene, cyclohexene or dimethylsulfur. However, because of the extreme instability of this compound (maybe the four-membered ring dimer), we have not been able to characterize it."* We needed 3 years to first realize that the major species was indeed **Ia**,[5a] and another year to find out that it was the first stable carbene. Indeed, in 1989 we published a paper entitled: *"[Bis(diisopropylamino)phosphino]trimethylsilylcarbene: A stable nucleophilic carbene."*[5b]

3.2. From Wanzlick Equilibrium to Diaminocarbenes

The search for stable diaminocarbenes dates back to the early 1960s and is associated with the name of Wanzlick.[4,26] At that time the preparation of the 1,3-diphenyl imidazolidin-2-ylidenes (**IIIa**) was first examined (Scheme 8.4). Precursors of **IIIa** included the dimeric and electron-rich olefin {**IIIa**}$_2$ and the chloroform adduct **3a** of the desired carbene. By means of cross-coupling experiments, however, it was shown that {**IIIa**}$_2$ was not in equilibrium with the two carbene units.[27] On the other

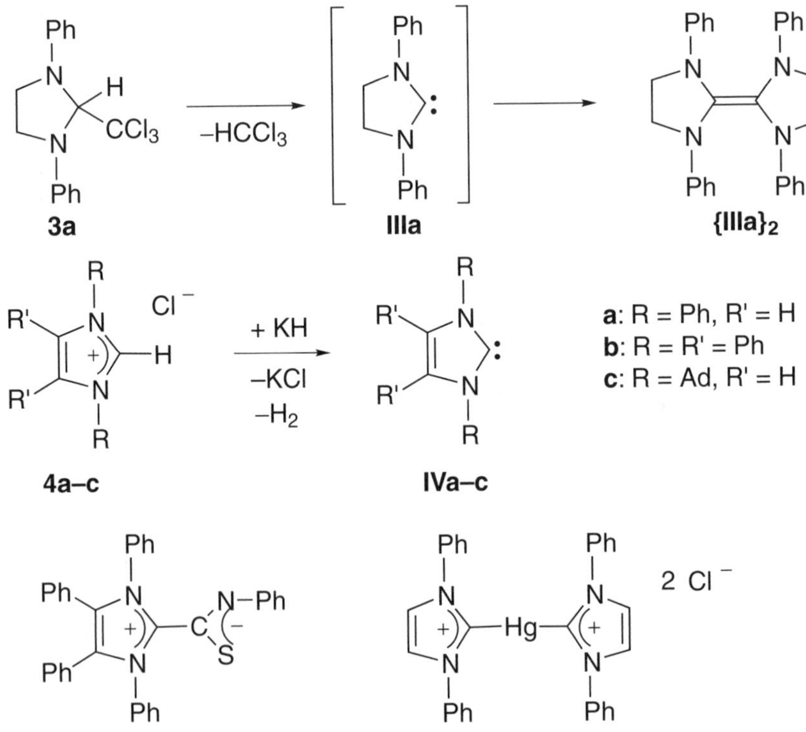

Scheme 8.4

hand, it is certain that the C=C double bond of {**IIIa**}$_2$ is easily cleaved with electrophiles with liberation of **IIIa**.[28] In any case, the carbene-stabilizing influence of the two amino groups was clearly recognized and interestingly Wanzlick also mentioned that an even larger stabilizing effect should result from the incorporation of a double bond in the ring, owing to "the participation of the aromatic resonance structures."[4] Wanzlick et al.[29] did demonstrate in 1970 that imidazolium salts such as **4a,b** could be deprotonated by potassium *tert*-butoxide to afford the corresponding imidazol-2-ylidenes (**IVa,b**), which were trapped but not isolated. With phenylisothiocyanate and mercury acetate as trapping agents the corresponding zwitterion[29a] and carbene–mercury complex,[29b] respectively, were isolated (Scheme 8.4).

Some 30 years later, Wanzlick's work was reinvestigated by Arduengo et al.,[30] who were able to prepare and to isolate the imidazol-2-ylidene **IVc** (R = adamantyl) in near quantitative yield (Scheme 8.4). The deprotonation of the 1,3-di-1-adamantylimidazolium chloride (**4c**) was carried out with sodium or potassium hydride, in the presence of catalytic amounts of either *t*-BuOK or the dimethyl sulfoxide (DMSO) anion. Carbene **IVc** is thermally stable in the solid state (colorless crystals, mp 240–241 °C).

Very interestingly, in 1998, Arduengo published a paper[31] entitled "1,3,4,5-Tetraphenylimidazol-2-ylidene: the realization of Wanzlick's dream." A modification of the procedure published by Wanzlick makes it possible to isolate the carbene **IVb**. In the conclusion, it is mentioned that "the inconvenient physical properties of the carbene, possible problems with respect to purity of the starting material, and the then widely accepted idea that imidazol-2-ylidenes are too labile to be isolated in pure form probably all contributed to the fact that Wanzlick et al. did not actually isolate **IVb**." To be fair to Wanzlick, we have to realize that at that time the research facilities were quite different, otherwise this great scientist might have been the first to isolate a singlet carbene.

4. SYNTHESIS AND STRUCTURAL DATA FOR STABLE SINGLET CARBENES

This section will be divided into three parts following the chronological order of the discovery of the different types of stable carbenes. The first two parts deal with carbenes featuring two heteroatom substituents, first of type D-C-W, then of type D-C-D. Then, we will discuss stable carbenes, which have been discovered in the last 2 years in our group, and that feature only one electronically active heteroatom substituent.

4.1. Carbenes with a π-Electron-Donating and a π-Electron-Withdrawing Heteroatom Substituent (D-C-W)

Two families of carbenes which fall into this category are the (phosphino)(silyl)carbenes **I** and the (phosphino)(phosphonio)carbenes **II**.

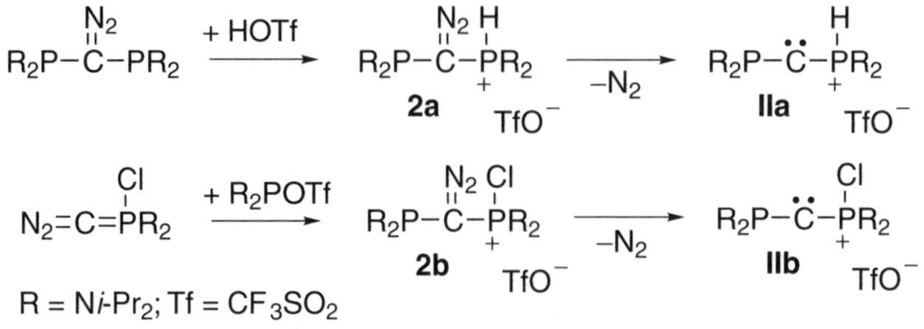

$$R_2P-\overset{\overset{N_2}{\|}}{C}-PR_2 \xrightarrow{\ +\ HOTf\ } R_2P-\overset{\overset{N_2}{\|}}{\underset{\underset{TfO^-}{+}}{C}}-\overset{H}{\underset{}{PR_2}} \xrightarrow[-N_2]{} R_2P-\overset{\cdot\cdot}{\underset{\underset{TfO^-}{+}}{C}}-\overset{H}{\underset{}{PR_2}}$$

$$\text{2a} \qquad\qquad\qquad \text{IIa}$$

$$N_2{=}C{=}\overset{Cl}{\underset{}{PR_2}} \xrightarrow{\ +\ R_2POTf\ } R_2P-\overset{\overset{N_2}{\|}}{\underset{\underset{TfO^-}{+}}{C}}-\overset{Cl}{\underset{}{PR_2}} \xrightarrow[-N_2]{} R_2P-\overset{\cdot\cdot}{\underset{\underset{TfO^-}{+}}{C}}-\overset{Cl}{\underset{}{PR_2}}$$

$$\text{2b} \qquad\qquad\qquad \text{IIb}$$

$$R = Ni\text{-}Pr_2;\ Tf = CF_3SO_2$$

Scheme 8.5

Since the discovery of the stable carbene **Ia**, many other (phosphino)(silyl) carbenes have been prepared, all of them using diazo precursors,[32] but only a few of them are stable at room temperature. The silyl group at the carbene center can be replaced by an isoelectronic and isovalent phosphonio substituent without dramatic modification. Indeed, the stable (phosphino)(phosphonio)carbenes **IIa**[33a] and **IIb**[33b] were synthesized from the corresponding diazo precursors **2a** and **2b** in 76 and 85% yields, respectively (Scheme 8.5).

The (phosphino)(silyl)carbenes (**I**) are all characterized by high-field nuclear magnetic resonance NMR chemical shifts for phosphorus (-24 to -50 ppm) and silicon (-3 to -21 ppm), and low-field chemical shifts for carbon (78–143 ppm) with large coupling constants to phosphorus (147–203 Hz). Classical shielding arguments indicate an electron-rich phosphorus atom, or equally, an increase in coordination number. The silicon atom seems also to be electron rich, whereas the carbon has a chemical shift in the range expected for a multiply bonded species. The values of the coupling constants are difficult to rationalize, but classical interpretation of the NMR data indicates that the P–C bond of (phosphino)(silyl) carbenes has some multiple bond character. The (silyl)- and (phosphonio)(phosphino)carbenes **I** and **II** are very similar, as deduced from the similarity of their NMR spectroscopic features.

The exact nature of the bonding system in phosphinocarbenes was clarified to some extent by the X-ray analyses performed on the (phosphino)(phosphonio) carbene (**IIa**),[33a] and more recently on the (phosphino)(silyl)carbene **Ib**.[34]

The molecular structure of **Ib** (Fig. 8.3) shows that the ring and the PCSi fragment are coplanar (maximum deviation from the best plane: 0.03 Å), the PC bond length [1.532(3) Å] is short, and the PCSi framework is bent [152.6(3)°]. The presence of a strongly polarized $P^{\delta+}C^{\delta-}$ fragment is suggested by the short $SiC_{carbene}$ [1.795 Å compared with 1.86–1.88 Å for Si–Me] and PN bond distances [1.664(2) Å]. These data indicate an interaction of the phosphorus lone pair with the formally vacant orbital of the carbene, and might suggest that the carbene lone pair interacts with the σ^* orbitals of both the silyl and the phosphino groups (negative hyperconjugation). Therefore, at that stage, form **A** could be ruled out, but all the others were reasonable.

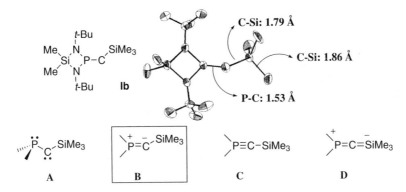

Figure 8.3. Oak ridge thermal ellipsoid plot (ORTEP) view of carbene **Ib** and possible representations for (phosphino)(silyl)carbenes.

In order to have more of an insight into the electronic nature of (phosphino)(silyl)carbenes, an electron localization function (ELF) analysis[34] of the model compound $[(H_2N)_2PCSiH_3]$ (**Ic**) was carried out. We found a bent structure for the energy minimum of **Ic**, however, the linear structure was only 1 kcal/mol higher in energy. For the ground-state bent form [Fig. 8.4(a)], the lone pair on the carbon atom is directed away from both the phosphorus and silicon, indicating that neither the triple bond (form **C**) nor the cumulene (form **D**) structure is the best formulation for (phosphino)(silyl)carbenes (**I**). Since the PC double bond is clearly evident, **Ic** has to be regarded as the phosphavinylylide (form **B**). Importantly, the isosurfaces representing the PC double bond are bent toward phosphorus, indicating that the phosphorus is reluctant to delocalize its lone pair into the formally vacant orbital of the carbene center. This finding will explain the observed reactivity of (phosphino)(silyl)carbenes (**I**) (see below). Interestingly, the linear form of **Ic** [Fig. 4(b)] also features the typical pattern for a PC double bond. The stretched shape of the isosurface is an indication of SiC double-bond character, and since this isosurface is perpendicular to that attributed to the PC double bond, the linear form of **Ic** is best described by the cumulenic structure (**D**).

In summary, the phosphino group clearly acts as π-donor substituent, while to some extent the silyl acts as π attractor due to the aptitude of silicon for hypervalency: (phosphino)(silyl)carbenes are push–pull carbenes. Importantly, when

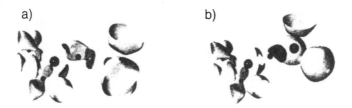

Figure 8.4. ELF plots of $(H_2N)_2PCSiMe_3$ (**Ic**): isosurfaces 0.85: (*a*) bent form of **Ic**; (*b*) linear form of **Ic**.

compared to nitrogen, phosphorus is much more reluctant to achieve a planar configuration with sp^2 hybridization.[35,36] The ensuing smaller stabilizing effect of phosphorus compared to nitrogen on the carbene center (see Section 4.3) is illustrated by the small singlet–triplet gap predicted for the (phosphino)(silyl)carbenes (5.6–13.9 kcal/mol)[37] compared to that calculated for acyclic as well as cyclic diaminocarbenes (58.5–84.5 kcal/mol).[38] This result means that the commitment of the lone pair to donation into the vacant orbital on the divalent carbon atom is less definitive for phosphorus than for nitrogen, and thus the phosphinocarbenes retain more of a divalent-carbon behavior; this is well illustrated by their reactivity (see Section 5.2).

4.2. Carbenes with Two π-Electron-Donating Heteroatom Substituents (D-C-D)

Apart from imidazol-2-ylidenes (**IV**), eight other types of carbenes are included in this category: imidazolidin-2-ylidenes (**III**),[39] tetrahydropyrimid-2-ylidene (**V**),[40] benzimidazol-2-ylidene (**VI**),[41] 1,2,4-triazol-5-ylidene (**VII**),[42] 1,3-thiazol-2-ylidenes (**VIII**),[43] as well as acyclic diamino- **IX**,[44] aminooxy- **X**,[45] and aminothiocarbenes (**XI**)[45] (Fig. 8.5).

Different synthetic routes have been used to prepare these carbenes (Scheme 8.6). The most common procedure is the deprotonation of the conjugate acid. In early experiments, sodium or potassium hydride, in the presence of catalytic amounts of either *t*-BuOK or the DMSO anion were used.[30] Then, Herrmann et al.[46] showed that the deprotonation occurs much more quickly in liquid ammonia as solvent (homogeneous phase), and many carbenes of type **IV** have been prepared following this procedure. In 1993, Kuhn and Kratz[47] developed a new and versatile approach to the alkyl-substituted N-heterocyclic carbenes **IV**. This original synthetic strategy relied on the reduction of imidazol-2(3*H*)-thiones with potassium in boiling tetrahydrofuran (THF). Lastly, Enders et al.[42] reported in 1995 that the 1,2,4-triazol-5-ylidene (**VIIa**) could be obtained in quantitative yield from the corresponding 5-methoxy-1,3,4-triphenyl-4,5-dihydro-1*H*-1,2,4-triazole by thermal elimination (80 °C) of methanol in vacuo (0.1 mbar).

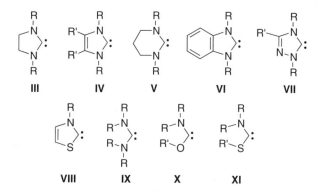

Figure 8.5. The different types of stable D-C-D carbenes.

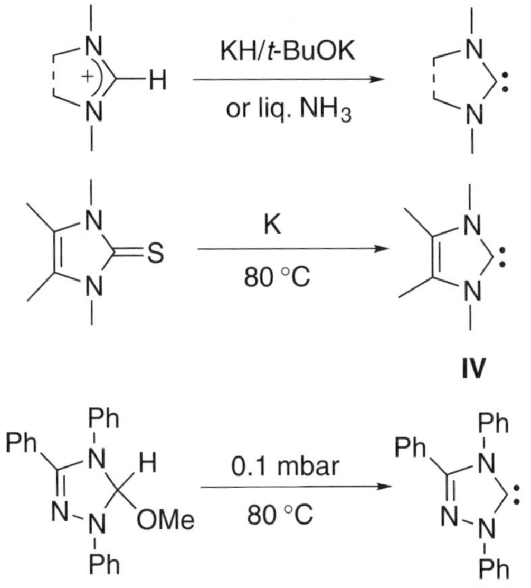

Scheme 8.6

Dixon and Arduengo[48] explained the extraordinary stability of the carbene **IVc** as mainly resulting from the inductive effect of the neighboring nitrogen atoms. The nitrogen lone pairs and the C=C double bond were supposed to ensure sufficient kinetic stability because of their high electron density, which was supported by a variety of experimental techniques.[49] A subsequent study by Cioslowski[50] even came to the conclusion that the π-donation of the nitrogen lone pairs played only a minor role. However, in 1996, Apeloig[51a] and Frenking[51b] and their co-workers independently investigated the importance of the aromaticity in carbenes (**IV**). According to structural, thermodynamic, and magnetic criteria, as well as the π populations and ionization potentials, cyclic electron delocalization does indeed occur in the imidazol-2-ylidenes (**IV**). This finding has been confirmed by inner-shell electron energy loss spectroscopy.[52] Although this aromatic character is less pronounced than in the imidazolium salt precursors **4**, it affords an additional stabilization of ∼ 25 kcal/mol.[51a]

Note that the other type of aromatic carbene isolated by Enders et al.,[42] namely, the 1,2,4-triazol-5-ylidene (**VIIa**) is stable enough to be prepared by thermal elimination at 80 °C, and it became the first carbene to be commercially available.

However, aromaticity is not the major stabilizing factor for carbenes of types **IV** and **VII**. More important is the interaction of the carbene center with the π-donating σ-attracting amino substituents. This is the reason that in recent years other diaminocarbenes were isolated, including acyclic diaminocarbenes of type **IX**.[44]

With an amino group present, other π-electron-donor groups can be used as the second substituent, as shown by the preparation of 1,3-thiazol-2-ylidenes (**VIII**),[43] acyclic aminooxy- **X**,[45] and aminothio-carbenes **XI**.[45] However, the superior π-donor ability and therefore the superior stabilizing effect of amino versus alkoxy groups (and thio groups) has been evidenced experimentally. Indeed, the bis-(dimethylamino)carbene $Me_2N-C-NMe_2$ can be observed by NMR spectroscopy at room temperature,[53] while the dimethoxycarbene MeO—C—OMe has only been characterized in matrices at low temperature (lifetime in solution at room temperature: 2 ms).[18]

The carbene carbon of **III–XI** resonates at rather low field in ^{13}C NMR spectra ($\delta = 205–300$ ppm) compared to the corresponding carbon atom of the cationic precursors ($\delta = 135–180$ ppm).

The solid-state structures of derivatives of type **III**, **IV**, and **VI–X** have been elucidated by single-crystal X-ray diffraction studies. The bond angles observed at the carbene centers (100–110°) are in good agreement with those expected for singlet carbenes of this type. The larger value observed in the acyclic bis(diisopropylamino)carbene of type **IX** (121.0°) probably results from severe steric effects. The nitrogen atoms of the amino group are always in a planar environment, and the N—$C_{carbene}$ bond lengths are rather short (1.32–1.37 Å). It is noteworthy that similar structural data are observed for their iminium salt precursors, the N—C bond lengths being only a little bit shorter (1.28–1.33 Å). These data as a whole indicate that the N—$C_{carbene}$ bonds have some multiple-bond character, which results from the donation of the nitrogen lone pairs to the carbene vacant orbital. This finding is confirmed by the large barriers to rotation about the N—$C_{carbene}$ bond determined for **IX** and **X** (13 and at least 21 kcal/mol, respectively) by variable temperature solution NMR experiments.[44,45,54] Therefore, diaminocarbenes are best described by resonance forms **B** and **C**, which may be summarized by structure **D** (Fig. 8.6). For aminothio and aminooxycarbenes the S—$C_{carbene}$ and O—$C_{carbene}$ bonds have less π character, and therefore, the best representation for these monoaminocarbenes is provided by resonance form **B**, in which oxygen and sulfur act as σ-electron-withdrawing moieties.

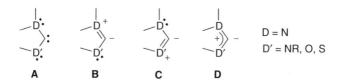

$$D = N$$
$$D' = NR, O, S$$

A   **B**   **C**   **D**

Figure 8.6. Resonance structures for D-C-D′ carbenes.

4.3. Carbenes with One Electronically Active Heteroatom Substituent

The isolation of stable singlet diheteroatom-substituted carbenes represents a spectacular synthetic achievement. However, their reactivity (see Section 5) is strongly influenced by the interaction of the two heteroatom substituents with the carbene center and therefore is somewhat different from that of their transient cousins. Thus it was tempting to try preparing carbenes with only one heteroatom substituent.

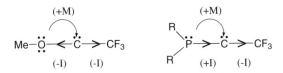

Figure 8.7. Mesomeric and inductive effects of substituents on (methoxy)(trifluoromethyl)-carbene and (phosphino)(trifluoromethyl)carbene.

In the beginning of the 1990s Moss et al.[55a] and Dailey and co-worker,[55b] independently studied the (methoxy)(trifluoromethyl)carbene ($MeO-C-CF_3$). This push–pull carbene was, however, highly unstable and electronically indiscriminate in its reaction with alkenes. This behavior was attributed to the pull inductive effect ($-I$) of the CF_3 group, which predominates over the push resonance effect ($+M$) of the MeO group. Recognizing that the methoxy group also has a pull inductive effect ($-I$), we chose to introduce a phosphino group in its place.[56] This substituent features both resonance ($+M$) and inductive push ($+I$) effects and, in addition, provides considerably greater steric bulk (Fig. 8.7).

Photolysis (300 nm) of [bis(dicyclohexylamino)phosphino](trifluoromethyl) diazomethane (**12a**) at $-60\,°C$ in pentane does not afford carbene **XIIa** but its dimer {**XIIa**}$_2$, which precipitates from the solution as orange crystals (Scheme 8.7). However, based on the well-known Platz' method for observing "invisible" carbenes through ylide formation,[57] we carried out the irradiation of **12a** in THF or diethylether (300 nm, $-60\,°C$) and cleanly generated the desired carbene **XIIa**, which is stable for days in solution at $-30\,°C$. The chemical behavior of **XIIa** exactly matches that of its transient congeners (see Section 5). Even subtle effects observed with transient carbenes can be reproduced with **XIIa**. Just as an example, Jones,[58a] Moss,[58b,c] Goodman,[58d] and their co-workers recently showed that transient singlet carbenes (such as chlorocarbenes) interact weakly

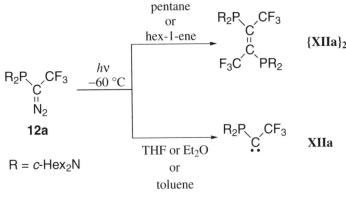

Scheme 8.7

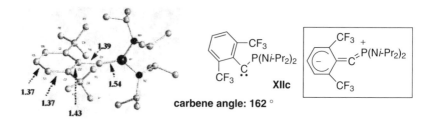

Figure 8.8. Molecular structure of push–pull carbene **XIIc**.

with aromatics, but not with simple olefins. For example, this interaction led to an extension of the benzylchlorocarbene lifetime from 23 ns (isooctane) to 285 ns (benzene).[58c] Similarly, we found that photolysis of **12a** in hex-1-ene at −60 °C led to the carbene dimer {**XIIa**}$_2$, while in toluene we obtained the carbene **XIIa**, which is stable up to −30 °C (Scheme 8.7).

Calculations[59] carried out on the simplified model compound **XIIb** (R = NH$_2$), predict the molecule to be significantly bent (P—C—C bond angle: 126.4°), the phosphorus planar, and the P—C bond length short (1.584 Å) as expected for this type of singlet carbene. The singlet–triplet energy gap was predicted to be small (12 kcal/mol), but in favor of the singlet state, and the energy barrier for the dimerization negligible, in perfect agreement with the experimental results.

We then investigated the possibility of replacing the σ-attracting CF$_3$ group (-I) by the bulky 2,6-bis(trifluoromethyl)phenyl group, which is both a σ and a π attractor (−I, −M). For all solvents used, photolysis of the diazo precursor at −10 °C afforded the corresponding carbene **XIIc**, which is stable for weeks at room temperature both in solution and in the solid state (melting point 68–70 °C).[56] The molecular structure of **XIIc** (Fig. 8.8) shows that the phosphorus atom is in a planar environment and the P1—C1 bond length [1.544(3) Å] is short, as expected because of the donation of the phosphorus lone pair into the carbene vacant orbital. The aromatic ring is perpendicular to the C$_{carbene}$PNN plane allowing the delocalization of the carbene lone pair into the ring, which is apparent from the very large angle at the carbene center (162°). Clearly, the aryl group acts as an electron-withdrawing group and helps in the stabilization of the carbene center. In other words, carbene **XIIc** has to be considered a push–pull carbene.

The isolation of **XIIc** was of importance since it demonstrated for the first time that monoheteroatom-substituted carbenes could be stable, indefinitely, at room temperature. The next step was to investigate whether a single electronically active substituent is sufficient to isolate a carbene.

Since the carbene **XIIc** is stable thanks to the presence of two substituents of opposite electronic properties, which preserve the electroneutrality of the carbene center, it was tempting to use a single substituent, which on its own would be both an electron donor and an electron acceptor; the amino group was the obvious choice.

The iminium salt **13a**, bearing a bulky *tert*-butyl group at nitrogen and a 2,4,6-tri-*tert*-butylphenyl group at carbon was deprotonated with potassium hydride at

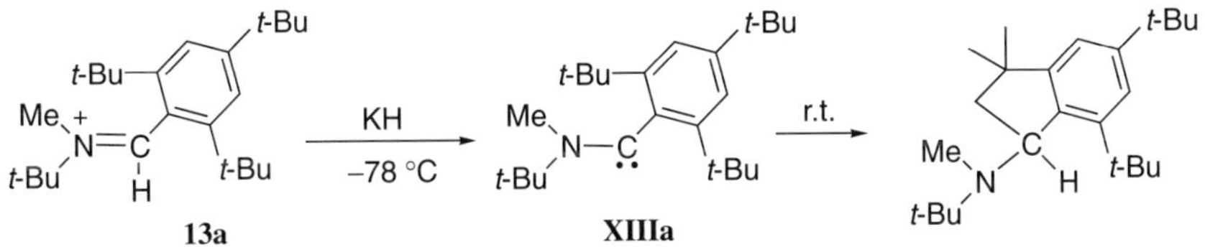

Scheme 8.8

−78 °C in THF, which cleanly led to the formation of (amino)(aryl)carbene **XIIIa** (Scheme 8.8).[60] A ^{13}C NMR signal at δ = 314.2 ppm leaves no doubt of the formation of **XIIIa**. Carbene **XIIIa** is stable for days in solution at −50 °C but undergoes a C—H insertion reaction at room temperature within a few hours, giving rise to the 4,6-di-*tert*-butyl-1,1-dimethyl-3-(methyl-*tert*-butylamino)indane as the major product. It is interesting to note that this reaction, typical of transient singlet and triplet carbenes,[61] has never been observed for diamino carbenes. This striking difference demonstrates the less perturbed character of carbene **XIIIa**.

Since C—F bonds are inert toward insertion of any type of carbene,[55b,62] we then replaced the ortho-*tert*-butyl groups of the aryl ring by trifluoromethyl substituents. Following the same procedure as described above, the (amino)(aryl)carbene (**XIIIb**) was prepared and isolated at room temperature in almost quantitative yield (m.p. 16 °C). The ^{13}C NMR resonance for the carbene carbon of **XIIIb** (303 ppm) is very similar to that of **XIIIa** (314 ppm) suggesting an analogous electronic structure. It is interesting to compare the molecular structure of **XIIIb** (Fig. 8.9) with those of the push–push diaminocarbenes (**IX**) and push–pull (phosphino)(aryl)carbene (**XIIc**) (Fig. 8.8). For both aminocarbenes, the nitrogen atom is in a planar environment but the N—$C_{carbene}$ bond length of **XIIIb** (1.28 Å) is even shorter than that observed for acyclic diaminocarbenes (**IX**) (1.37 Å), which indicates a stronger donation of the nitrogen lone pair into the vacant carbene orbital. In marked contrast with the cumulenic system **XIIc**, the $C_{carbene}$–C_{aro} bond distance is long (**XIIIb**: 1.45 Å; **XIIc**: 1.39 Å) and the carbene bond angle rather acute (**XIIIb**: 121°; **XIIc**: 162°). These data clearly indicate that, in contrast to what is observed for **XIIc**, the potentially π-acceptor 2,6-bis(trifluoromethyl)phenyl group does not strongly interact with the carbene lone pair of **XIIIb** and is therefore a *spectator*.

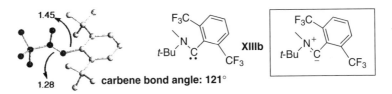

Figure 8.9. Molecular structure of carbene **XIIIb**.

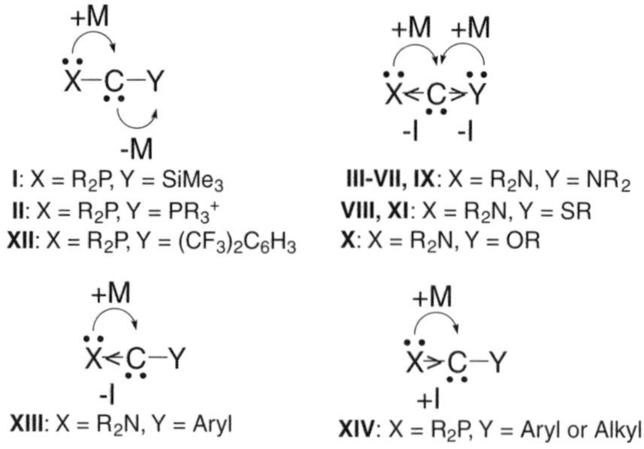

+M

$$\overset{\cdot\cdot}{X}-\overset{\cdot\cdot}{\underset{\cdot\cdot}{C}}-Y$$

−M

I: X = R₂P, Y = SiMe₃
II: X = R₂P, Y = PR₃⁺
XII: X = R₂P, Y = (CF₃)₂C₆H₃

+M +M

$$\overset{\cdot\cdot}{X}<\overset{}{C}>\overset{\cdot\cdot}{Y}$$

−I   −I

III–VII, IX: X = R₂N, Y = NR₂
VIII, XI: X = R₂N, Y = SR
X: X = R₂N, Y = OR

+M

$$\overset{\cdot\cdot}{X}<\overset{}{\underset{\cdot\cdot}{C}}-Y$$

−I

XIII: X = R₂N, Y = Aryl

+M

$$\overset{\cdot\cdot}{X}>\overset{}{\underset{}{C}}-Y$$

+I

XIV: X = R₂P, Y = Aryl or Alkyl

Figure 8.10. Different modes of stabilization of carbenes.

All of the stable singlet carbenes discussed so far follow, to some extent, Pauling's predictions[16b] that the substituents should preserve the electroneutrality of the carbene center (Fig. 8.10), which is obvious for push–pull carbenes **I**, **II**, and **XII**, in which the carbene bears both a π-donating and a π-withdrawing substituent. Diaminocarbenes, as well as aminooxy- and aminothio-carbenes **III–XI** have two π-donor substituents, and are sometimes referred to as push–push carbenes, but the amino group also acts as a strong σ-electron-withdrawing substituent. Lastly, carbene **XIII** features a spectator substituent and an amino group, which is again both a π donor and σ attractor and raises the question: Is electron donation sufficient to stabilize a carbene, as shown for **XIV** (Fig. 8.10)?

We first investigated the synthesis of a (phosphino)carbene featuring an electron-rich aryl group.[63] The (phosphino)(mesityl)carbene (**XIVa**) was generated by photolysis at −50 °C of a THF solution of the corresponding diazo compound. The ^{31}P and ^{13}C NMR chemical shifts of **XIVa** (^{31}P δ −23.5; ^{13}C δ 151.1) compare well with those observed for neutral push–pull (phosphino)carbenes **I** and **XIIa** (^{31}Pδ −40 to −20 ppm; ^{31}C δ120 to 150 ppm). The only noticeable difference is the magnitude of the coupling constant between the phosphorus and the carbene center, which is significantly smaller for **XIVa** ($^{1}J_{P–C} = 65$ Hz than for **I** ($^{1}J_{P–C} = 150$–200 Hz) and **XII** ($^{1}J_{P–C} = 271$ Hz). Although these data are difficult to rationalize, they are likely to indicate some peculiarity in the electronic structure of **XIVa**. Pale yellow crystals of **XIVa** (mp 148 °C) suitable for an X-ray diffraction study were obtained by cooling a saturated pentane solution to −80 °C (Fig. 8.11).

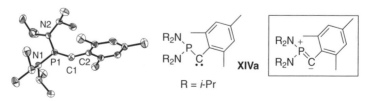

$$R = i\text{-Pr}$$

Figure 8.11. Molecular structure of carbene **XIVa**.

As expected because of the donation of the phosphorus lone pair into the carbene vacant orbital, the phosphorus atom is in a planar environment, and the P1C1 bond length is short [1.564(3) Å]. The P1C1C2 angle [148.7(2)°] is significantly smaller than in the push–pull system **XIIa** [162.1(3)°]; the C1C2 bond length [1.438(3) Å] is in the range typical for $C(sp^2)$–$C(sp^2)$ single bonds, and is longer than that observed for **XIIa** [1.390(4) Å]. These results strongly suggest the absence of delocalization of the carbene lone pair into the aromatic ring. In other words, the carbene center interacts only with the phosphino group, while the mesityl substituent remains merely an electronic spectator.

Direct observation of singlet (alkyl)carbenes usually requires matrix isolation conditions.[64] Using the π-donor and σ-attractor methoxy substituent, Moss and co-workers[65] could characterize the (methoxy)(methyl)carbene (MeOCMe) by ultraviolet (UV) and infrared (IR) spectroscopies, but only in a nitrogen matrix (at 10 K) or in solution thanks to a nanosecond time-resolved LFP technique ($t_{1/2} < 2\,\mu s$ at 20 °C). The remarkable stability of carbene **XIVa** both in the solid state and in solution (no degradation observed after several weeks at room temperature), prompted us to investigate the preparation of (phosphino)(alkyl)carbenes.

Photolysis (254 nm, −80 °C) of a toluene solution of the (phosphino)(*tert*-butyl)diazomethane afforded the (phosphino)(*tert*-butyl)carbene (**XIVb**, Scheme 8.9).[63] This alkylcarbene could indeed be characterized spectroscopically by multinuclear NMR spectroscopy at low temperature (−50 °C). The ^{31}P NMR chemical shift (δ −36.4) is in the expected range, while the ^{13}C NMR chemical shift (δ 186.3) is the highest reported for a (phosphino)carbene. Although **XIVb** rearranges within few minutes above −10 °C to afford the 1,2-amino-migration product as the major product, it is still observable by ^{31}P NMR up to −20 °C. We thus decided to reduce the steric hindrance of the spectator alkyl group, and hence studied the photolysis of the (phosphino)(methyl)diazomethane (254 nm, −85 °C, toluene solution). The corresponding (phosphino)(methyl)carbene (**XIVc**) was characterized spectroscopically at −85 °C. The ^{31}P and ^{13}C NMR data for **XIVc** (^{31}P δ −17.8; ^{13}C δ 164.8, $^1J_{P-C} = 44$ Hz) are similar to those for **XIVb**. Despite the poor steric protection and the potential to undergo 1,2-shifts,[66] the

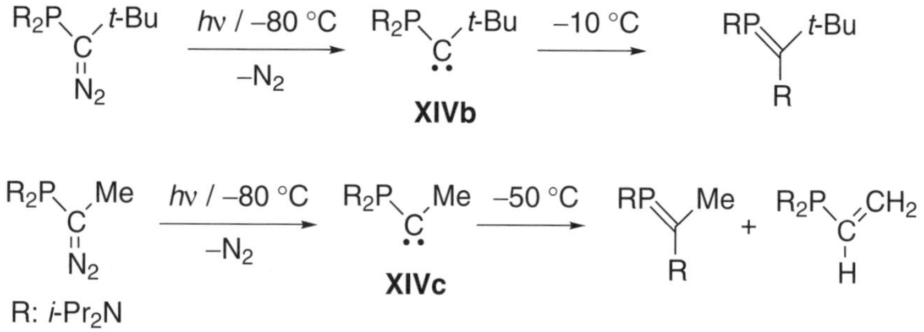

Scheme 8.9

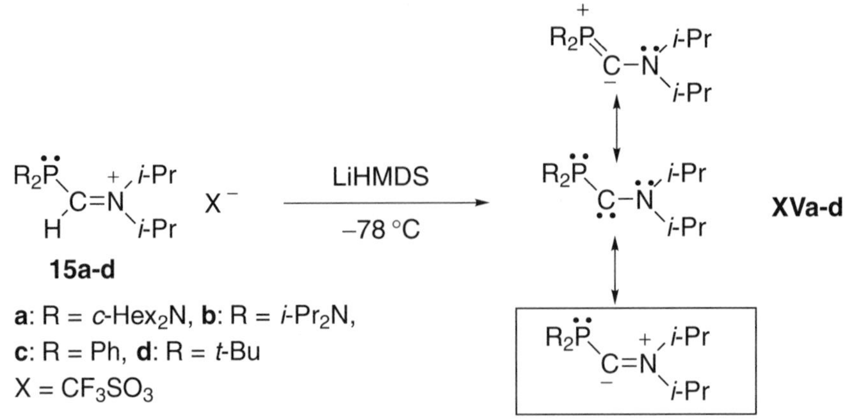

Scheme 8.10

methylcarbene (**XIVc**) could be observed by ^{31}P NMR up to $-50\,^\circ$C ($t_{1/2} \sim 10$ min at $-50\,^\circ$C), where it quickly isomerizes (Scheme 8.9).

As we have seen, all known stable singlet carbenes feature either an amino or a phosphino substituent. The preparation of carbenes **XIV** demonstrates that the presence of a phosphino substituent allows for the spectroscopic characterization of singlet carbenes under standard laboratory conditions, when the steric bulk of the spectator substituent is decreased even to the size of a methyl group. Therefore, it was of interest to find out if a phosphino group was more efficient for stabilizing a carbene center than an amino group, or vice versa. The latter situation would imply that a broad range of stable mono- aminocarbenes would be easily available. We approached this problem by synthesizing the (phosphino)(amino)carbenes (**XV**).[67]

Carbenes **XVa–d** were cleanly generated at $-78\,^\circ$C by deprotonation of the corresponding phosphinoiminium salts **15a–d** with the lithium salt of hexamethyldisilylazane (Scheme 8.10), and were characterized by multinuclear NMR spectroscopy at $-30\,^\circ$C. The main feature of the NMR spectra of **XVa–d** is the very low field values of the ^{13}C chemical shifts of the carbene carbon atoms (δ: 320–348 ppm, J_{PC}: 22–101 Hz). These signals are even at lower field than those observed for the other known aminocarbenes (210–300 ppm) and in a totally different range from those for phosphinocarbenes (70–180 ppm). In all cases, ^{1}H NMR spectra showed the presence of two different isopropyl groups on the nitrogen atom bound to the carbene center, which indicates the absence of rotation about the C—N bond. All these NMR data strongly suggest that only the amino substituent interacts with the carbene center, the phosphino group merely remaining a spectator substituent.

In the case of **XVa** (R = c-Hex$_2$N), orange crystals suitable for an X-ray diffraction study were obtained by cooling a saturated ether solution to $-30\,^\circ$C (Fig. 8.12). The pyramidalization of the phosphorus atom (sum of bond angles $= 304.5^\circ$) and the long P1C1 bond length (1.856 Å), which is in the range for PC single bonds, demonstrate that the phosphino group is indeed a spectator substituent. As

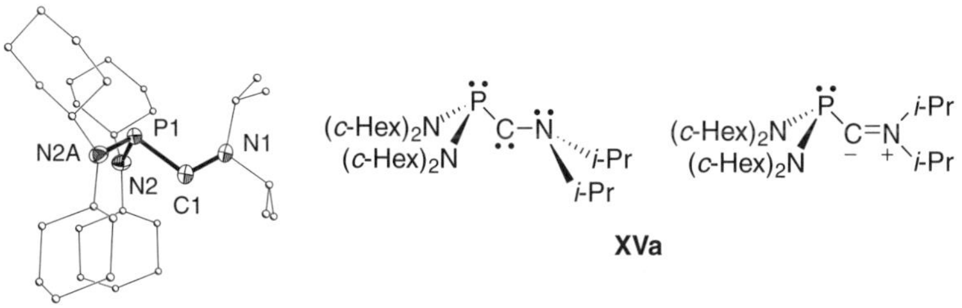

Figure 8.12. Molecular structure of carbene **XVa**.

expected, the nitrogen atom is in a planar environment (sum of bond angles = 359.6°), and the N1—C1 bond length (1.296 Å) is short. Finally, the carbene bond angle is small (116.5°) as expected for aminocarbenes and in contrast with (phosphino)carbenes.

The stability of carbenes **XVa–d** is dependent on the nature of the phosphorus substituents. All the carbenes are stable for days at $T < -20$ °C; the most stable is **XVd** ($R = t$-Bu), which can be stored in solution for a few days at 20 °C.

From these data, it can be stated that an amino group is much more efficient for stabilizing a carbene center than a phosphino group. *These results also resolve a controversy: If an aminocarbene is considered a carbene despite the interaction between the amino group and the carbene center, a phosphinocarbene should also be considered a carbene.*

5. REACTIVITY OF STABLE SINGLET CARBENES

Whereas triplet carbenes exhibit radical-like reactivity, singlet carbenes are expected to show nucleophilic as well as electrophilic behavior because of the lone pair and vacant orbital. This section will be focused on the most typical reactions involving stable singlet carbenes: dimerizations, cyclopropanations, and formation of ylides and reverse ylides with Lewis bases and acids, respectively. Since, as recently noted by Herrmann:[68] "a revolutionary turning point in organometallic catalysis is emerging," a part of this section will devoted to the role of stable singlet carbenes as ligands for transition metal catalysts.

5.1. Dimerization

First, it should be repeated that despite early[69a] and recent claims,[69b] no examples of thermal dissociation of alkenes into carbenes (the reverse reaction of the carbene dimerization) are known.[70] However, in their recent paper, Lemal and co-worker[70] pointed out that electrophiles can catalyze the dissociation of tetraaminoalkenes.

In the Carter and Goddard formulation,[71] the strength of the C=C double bond resulting from the dimerization of singlet carbenes should correspond to that of a canonical C=C double bond (usually that of ethene) minus twice the singlet–triplet

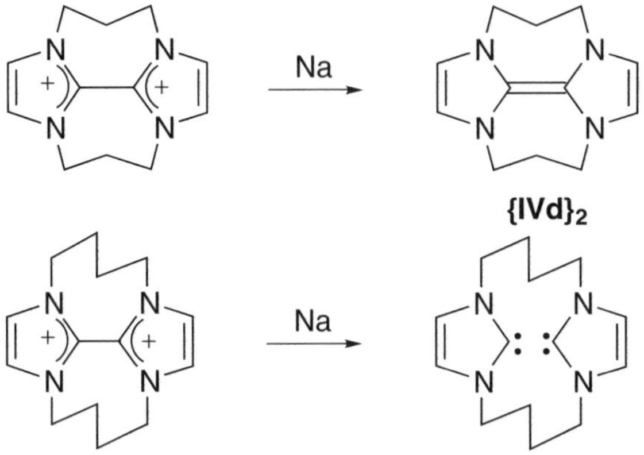

Scheme 8.11

energy gap for the carbene. For example, the singlet–triplet splitting in the parent imidazol-2-ylidene (**IV**) has been calculated to be ~ 85 kcal/mol;[72] accordingly, one expects the CC bond strength in the dimer to be approximately only $[172 - (2 \times 85)] = 2$ kcal/mol. A convincing experimental proof of the weakness of the C=C bond in tetraamino alkenes comes from the very elegant work by Taton and Chen.[73] The tightly constrained {**IVd**}$_2$ exists as the tetraazafulvalene, whereas the homologue with longer tethers dissociates into a biscarbene (**IVe**, Scheme 8.11). Perhaps equally remarkable is the fact that the central C=C bond in {**IVd**}$_2$ is of normal length (1.337 Å), even though the strength of this bond can be only a few kilocalories per mole.

Similarly to imidazol-2-ylidene (**IV**), the calculated value for the energy of dimerization of Enders-type carbene **VII** is only 9.5 kcal/mol.[74] These remarkably small values, at least partially due to the loss of aromaticity in the carbene dimers {**IV**}$_2$ and {**VII**}$_2$, highlight the difficulty of dimerization of such carbenes. In contrast, in the case of the parent acyclic diaminocarbene **IX**, Heinemann and Thiel[72] found a dimerization energy of ~ 45 kcal/mol. This poses another question What is the value of the energy barrier for the dimerization? Nowadays, the dimerization of singlet carbenes is believed to follow a nonleast motion pathway that involves the attack of the occupied in-plane σ lone pair of one singlet carbene center on the out-of-plane vacant p_π orbital of a second carbene (Fig. 8.13).[71,75] Calculations

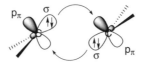

Figure 8.13. Schematic representation of the mechanism for the dimerization of singlet carbenes.

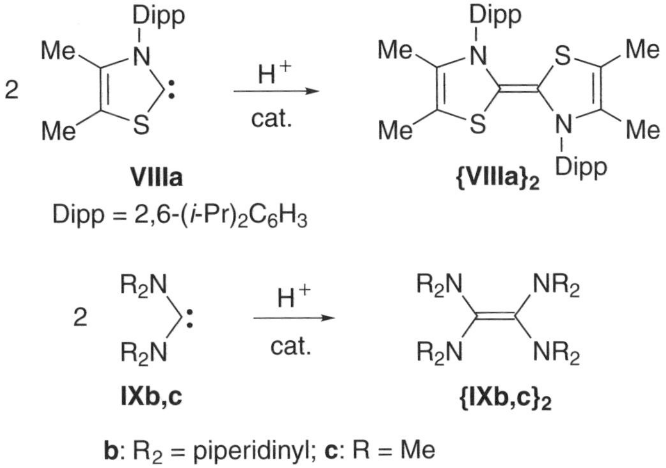

Scheme 8.12

regarding the dimerization path indicate a significant barrier of $\sim$ 19.4 kcal/mol for the parent 1,2,4-triazol-5-ylidene (**VII**),[74] and Alder estimated the ΔG^* for the dimerization of the bis(N-piperidinyl)carbene (**IXb**) and bis(dimethylamino)carbene (**IXc**) to be >25 kcal/mol.[53,76] These large values are not surprising because the dimerization reaction involves the carbene vacant orbital that is very high in energy due to the donation of the nitrogen lone pairs.

However, besides the work by Wanzlick, several examples of aminocarbene dimerizations have been reported.[43,53,77] Noteworthy is the isolation of both the thiazol-2-ylidene (**VIIIa**) and of its dimer {**VIIIa**}$_2$[43] and the spectroscopic characterization by Alder and Blake of the bis(N-piperidinyl)carbene (**IXb**) and of the tetrakis(N-piperidinyl)ethene ({**IXb**}$_2$)[53] (Scheme 8.12).

Interestingly, Arduengo observed that **VIIIa** was stable with respect to dimerization in the absence of a Brønsted or Lewis acid catalyst.[43] Similarly, in the absence of an acid catalyst, dimerization of **IXb** is extremely slow and is first order in carbene.[53] Therefore, the observed formal dimerization of **VIIIa** and **IXb,c** does not involve the coupling of two carbenes, but the nucleophilic attack of one carbene upon its conjugate acid, followed by proton elimination, as already suggested by Chen and Jordan[78] (Scheme 8.13). It is important to keep in mind that even N,N-dialkylimidazolium ions have pK_a values of $\sim$ 24 in DMSO,[79] and based on the calculated proton acidity, Alder estimated the pK_a values for acyclic diaminocarbenes to be from 2 to 6 pK_a units higher than for imidazol-2-ylidenes.[54]

Another important issue has recently been raised by Alder. He proved that diaminocarbenes, including aromatic carbenes, coordinate alkali metal ions (coming from the base), which is important with regard to the rate of dimerization.[40] The metal ions might act as a Lewis acid catalyst for dimerization, as observed for protons, but alternatively, strong complexation might stabilize the carbene center and prevent its dimerization. This question is still open to debate.

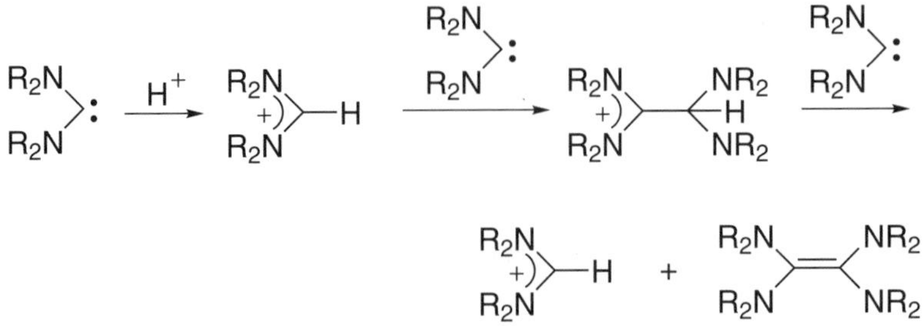

Scheme 8.13

Only one example of dimerization of a phosphinocarbene has been reported. As already briefly mentioned, carbene **XIIa** is stable for days in solution at $-30\,°C$ in THF or toluene solution, but in pentane or upon evaporation of the donor solvent, even at $-50\,°C$, a dimerization generating exclusively the alkene {**XIIa**}$_2$ is observed.[56] Interestingly, upon warming the THF solution of **XIIa** to $-20\,°C$, a clean rearrangement occurs affording a cumulene, with no trace of the carbene dimer {**XIIa**}$_2$ (Scheme 8.14). These observations are in perfect agreement with previous work on transient carbenes. It has been shown that the extent of 1,2-migration processes increases, relative to intermolecular reaction (like the dimerization), as the solvent is changed from an alkane to a donor solvent.[80]

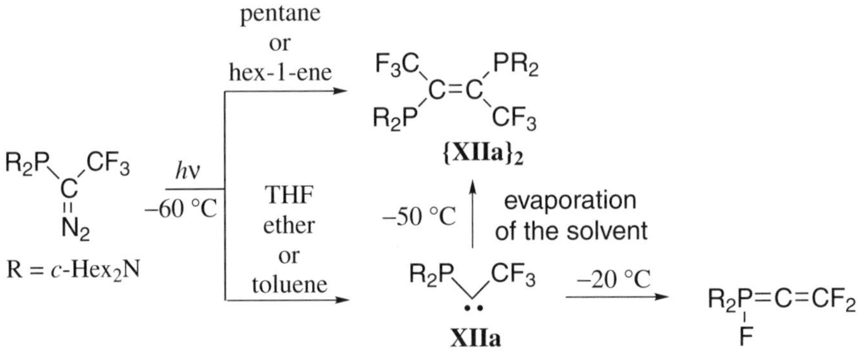

Scheme 8.14

5.2. Cyclopropanation

In a recent paper,[58c] Moss wrote: "The addition of a carbene to an alkene with the formation of a cyclopropane is perhaps the most fundamental of cycloaddition reactions, as well as a basic component of the synthetic armamentarium." Indeed, cyclopropanation reactions involving transient carbenes[81] or even transition metal carbene complexes have been widely studied.[82] Both singlet and triplet transient carbenes undergo cyclopropanation reactions, although with a totally different

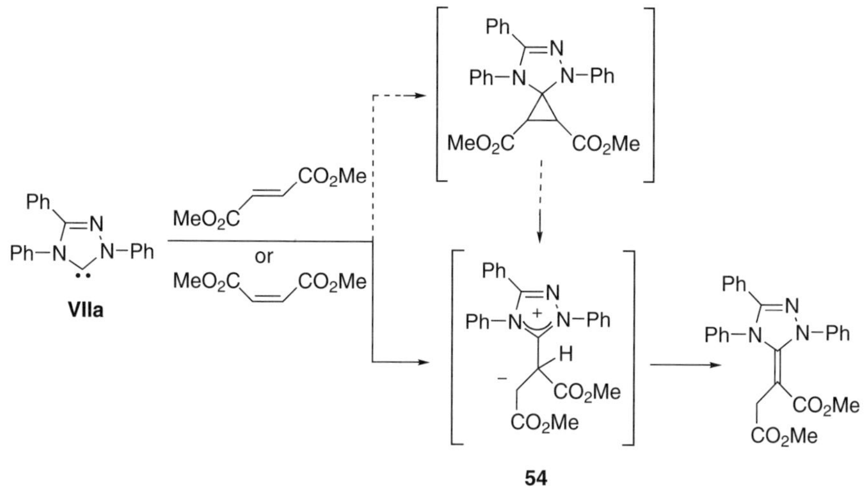

Scheme 8.15

mechanism, which is apparent from the stereochemistry of the reaction: with singlet carbenes the stereochemistry about the original carbon–carbon double bond is maintained, while with triplet carbenes the stereochemical information is lost.[83]

1,2,4-Triazol-5-ylidene (**VIIa**) reacts with diethyl fumarate and also diethyl maleate, not giving the corresponding cyclopropane, but the methylenetriazoline derivative (Scheme 8.15).[42] According to Enders et al.,[42] a [1 + 2]-cycloaddition first occurs, leading to the transient cyclopropane. Then, a ring opening would lead to a zwitterionic derivative, which would undergo a 1,2-H shift [AM1 calculations predict a strongly negative reaction enthalpy ($\Delta H = -18$ kcal/mol) for the rearrangement of the cyclopropane to the isolated product].[42] However, it is quite clear that a mechanism directly leading to the zwitterionic species also explains the experimental results.

In marked contrast, we used the cyclopropanation reaction of (phosphino)(silyl)-carbene (**Ia**) with methyl acrylate to demonstrate the carbene nature of our compound.[5b] More recently, we studied in detail the stereochemistry of this type of reaction.[84] The (phosphino)(silyl)carbene (**Ia**) reacted efficiently with 3,3,4,4,5,5,6,6,6-nonafluorohex-1-ene and (Z)- and (E)-2-deuteriostyrene giving the corresponding cyclopropanes in good yields. The stereochemical outcome was such that all monosubstituted alkenes gave exclusively the syn isomer (with respect to the phosphino group), and the addition of disubstituted alkenes was totally stereospecific (Scheme 8.16).

The stereospecificity observed with (Z)- and (E)-2-deuteriostyrene presents convincing evidence for the concerted nature of the cyclopropanation reaction, and therefore the genuine singlet carbene nature of stable (phosphino)(silyl)carbenes (**I**). The diastereoselectivity is at first glance surprising. Indeed, it is clear that steric factors cannot govern the observed selectivity since the bis(amino)phosphino group

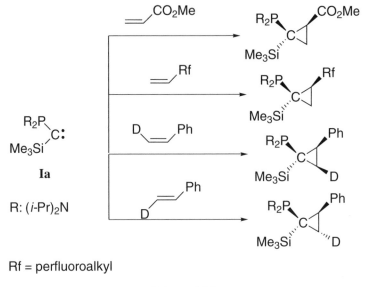

Rf = perfluoroalkyl

Scheme 8.16

is at least as sterically demanding as the trimethylsilyl group. Since the syn-selectivity observed with the carbene **Ia** is not due to steric factors, the obvious other possibility is to invoke orbital controlled reactions. Compared **Ia** is a nucleophilic carbene, so that the more dominant orbital interaction, in the transition state, is between the highest occupied molecular orbital ($HOMO_{carbene}$) and the lowest unoccupied molecular orbital ($LUMO_{alkene}$), but this set of orbitals does not explain the observed syn-selectivity. However, in the course of computational and experimental work on the addition of carbenes to alkenes, the significance of a second pair of orbital interactions in the addition geometry has been pointed out.[85] Thus, we believe that the secondary orbital interaction $LUMO_{carbene} - HOMO_{alkene}$ explains the observed selectivity. Due to donation of the phosphorus lone pair, the $LUMO_{carbene}$ has some $\pi^*(PC)$ character and shows significant bonding overlap between the phosphorus center and the alkene substituent, as indicated schematically in Figure 8.14. In other words, like the endo selectivity in the Diels–Alder

HOMO$_{carbene}$ - LUMO$_{olefin}$ LUMO$_{carbene}$ - HOMO$_{olefin}$

Figure 8.14. The HOMO–LUMO interactions between a phosphinocarbene (**I**) and acrylate.

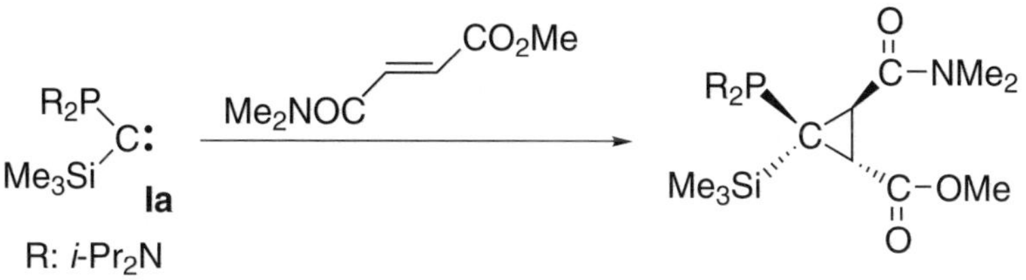

Scheme 8.17

reaction, the high stereoselectivity observed could be rationalized on the basis of favorable "secondary orbital interactions."

The effectiveness of the secondary orbital interactions on the selectivity of the cyclopropanation reactions was demonstrated by the formation of a single cyclopropane isomer in the reaction of **Ia** with (E)–Me$_2$NCOCH=CHCO$_2$Me (Scheme 8.17). Simple Hückel calculations show that the highest coefficient of the HOMO of the alkene is located at the amido group, and thus despite the smaller steric demand of the ester group, the amido substituent lies *syn* with respect to the phosphino group.

Reactions related to cyclopropanation can also be carried out with (phosphino)(silyl)carbenes (**I**). For example, benzaldehyde reacts with **Ia** at 0 °C leading to the corresponding epoxide, again as only one diastereomer.[5b] Even more striking are the reactions with benzonitrile and *tert*-butylphosphaalkyne that lead initially to azirine[86] and phosphirene.[87] Both three-membered heterocycles subsequently undergo ring expansion reactions affording azaphosphete and diphosphete, respectively (Scheme 8.18). This reaction is a new route for the synthesis of heterocyclobutadienes,[88] and this demontrates the usefulness of (phosphino)(silyl)carbenes (**I**) for the synthesis of novel species.

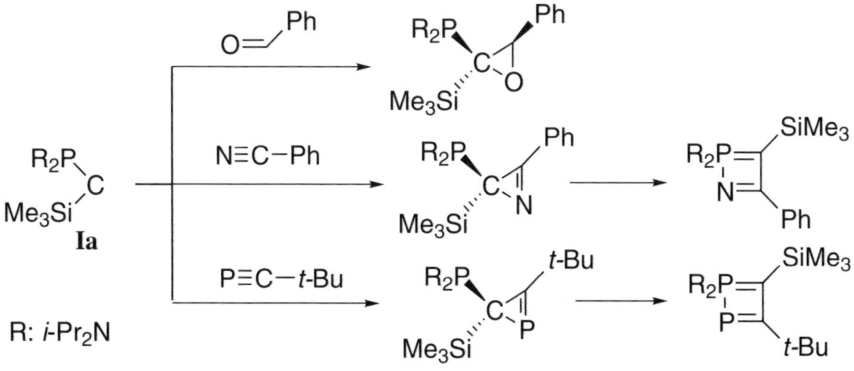

Scheme 8.18

5.3. Reactions with Lewis Bases and Acids

Because of the presence of a lone pair and a vacant orbital, singlet carbenes are supposed to be able to react with both Lewis bases and acids. Transient electrophilic carbenes are known to react with Lewis bases to give *normal ylides* (Scheme 8.19). For example, carbene–pyridine adducts have been spectroscopically characterized and used as a proof for the formation of carbenes,[57] and the reaction of transient dihalogenocarbenes with phosphines is even a preparative method for C-dihalogeno phosphorus ylides.[89] Little is known about the reactivity of transient carbenes with Lewis acids.

In marked contrast, because of the nucleophilicity of the stable amino- and phosphino-carbenes, numerous examples of carbene-Lewis acid adducts (*reverse ylides*) (Scheme 8.19) have been reported, whereas a very few examples of normal ylides are known.

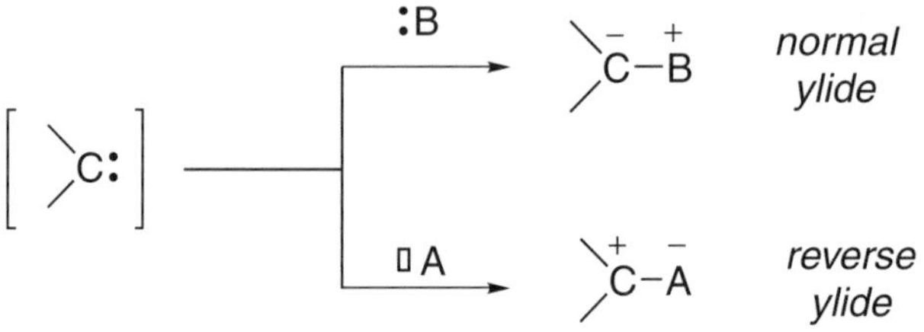

Scheme 8.19

We have recently shown that instantaneous and quantitative formation of the corresponding phosphorus ylides occurred when one equivalent of trimethylphosphine was added at low temperature to carbenes **Ia**[90] and **XIIa**[56] (Scheme 8.20). The formation of phosphorus ylides in these reactions is of particular significance since it is certainly the most striking evidence for the presence of an accessible vacant orbital at the carbene center of (phosphino)carbenes, as expected for singlet carbenes.

Because of the strong interaction of the nitrogen lone pair with the carbene vacant orbital, stable diamino carbenes do not interact with Lewis bases, such as phosphines, to give ylides. Instead, carbenes **IV** depolymerize cyclopolyphosphines $[(PPh)_5$ and $(PCF_3)_4]$ and cyclopolyarsines $[(AsPh)_6$ and $(AsC_6F_5)_4]$ to produce adducts of the type "carbene-ER",[91] while with chlorodiphenylphosphine the corresponding phosphino-imidazolium salt as been isolated.[92] It is also interesting to

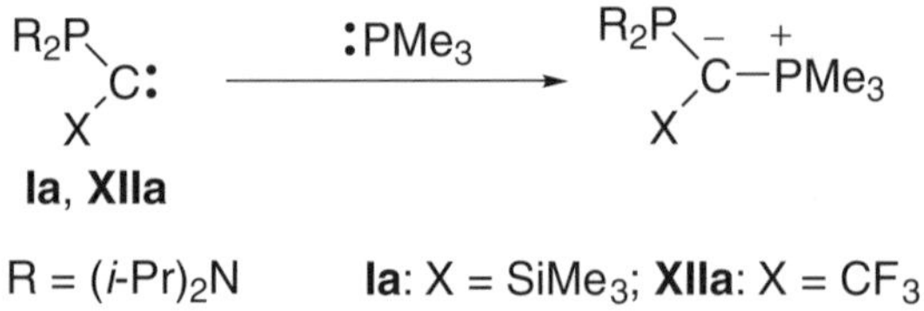

Scheme 8.20

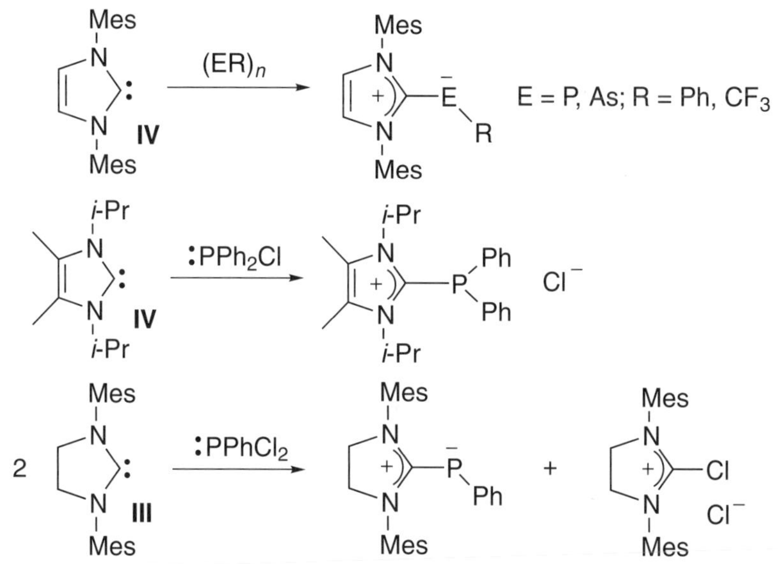

Mes = 2,4,6-trimethylphenyl

Scheme 8.21

note that imidazolidin-2-ylidene (**III**) reacts with dichlorophenylphosphine affording the carbene-phosphinidene adduct along with the 2-chloro-imidazolidinium chloride[91b] (Scheme 8.21).

The previous reactions demonstrate the high nucleophilicity of diaminocarbenes, and thus it is not surprising that a variety of reverse ylides can be prepared. Some selected examples are given hereafter.

The first isolated adduct between an aminocarbene and a Lewis acid was obtained by reacting an imidazol-2-ylidene (**IV**) with iodopentafluorobenzene (Scheme 8.22).[93] In contrast to classical halonium methylides (obtained from transient electrophilic carbenes and halogen centers), which feature a characteristically small C—X—C angle, this adduct has an almost linear C—I—C framework (178.9°). This finding is consistent with a 10e-I-2c (hypervalent) bonding at the iodine center, and a *reverse ylide* nature.

With iodine, the imidazol-2-ylidenes (**IV**) form stable adducts (Scheme 8.22),[94] in which the carbene clearly acts as a basic σ donor, just like a tertiary phosphine. Interestingly, the molecular structure of this adduct may be considered as an isolated transition state that models the nucleophilic attack of the carbene on the iodine molecule.

Halogen abstraction from hypervalent sulfur halides has also been reported for the imidazol-2-ylidenes (**IV**) (Scheme 8.22). This reaction gives a nice example of the synthetic utility of *N*-heterocyclic carbenes. Indeed, this adduct is the first structurally characterized derivative featuring the chlorosulfite ion (SO_2Cl^-).[95]

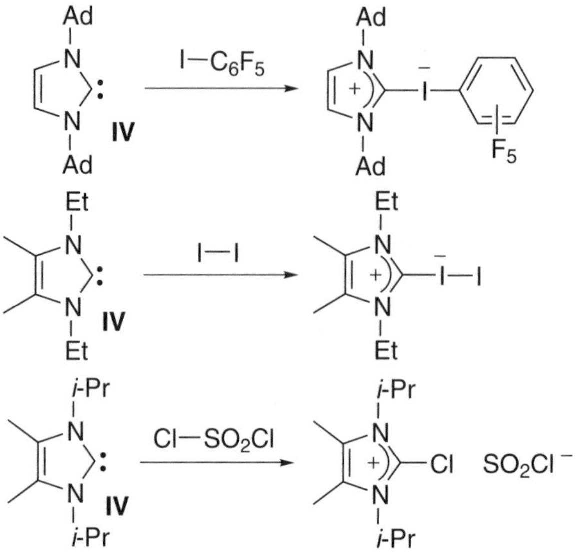

Scheme 8.22

The first group 13 (IIIA) element–carbene complex to be reported was an imidazol-2-ylidene-alane complex (Scheme 8.23).[96] Based on NMR data, it was suggested that the imidazole fragment has an electronic structure that is intermediate between those of the free carbene and imidazolium ion. The use of an imidazol-2-

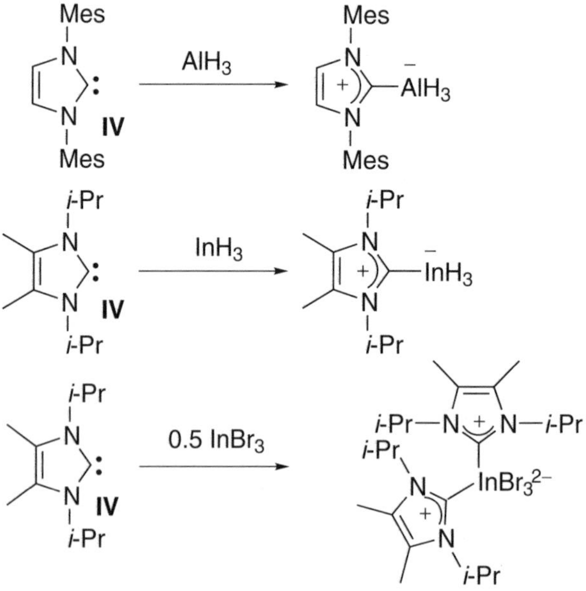

Scheme 8.23

ylidene (**IV**) has also allowed the isolation of the first structurally authenticated indium trihydride complex.[97] The stability of these alane and indane complexes is especially remarkable considering the fact that they contain a potential hydride donor adjacent to a potentially electrophilic center. This unusual stability clearly demonstrates that the vacant orbital at the carbon is very high in energy and there-fore, not available, which explains, for example, why no cyclopropanation occurs with diaminocarbenes and especially imidazol-2-ylidenes (**IV**). Also, in the indium series, Jones and co-workers[98] obtained biscarbene complexes, which highlights the propensity of indium to achieve higher coordination numbers than those of alumi-num and gallium for which only 1:1 adducts are known.

In marked contrast, the stability or even the existence of (phosphino)(silyl) carbene-group 13 (IIIA) complexes is strongly related to the element but also the substituents of the Lewis acid. So far, only the aluminum, gallium, and indium trichloride adducts have been isolated (Scheme 8.24).[99] When trimethyl-alumi-nium, -gallium, and -indium were used, instead of the corresponding trichloride derivatives, phosphorus ylides were obtained in good yields; no intermediates were spectroscopically detected.[99,100] Although numerous so-called stabilized phosphorus ylides, in which the negative charge is delocalized into an organic or organometallic framework have been studied, these ylides were the first examples of phosphorus ylides C-substituted by a group 13 (IIIA) element to be studied by X-ray diffraction. These compounds are of interest since they can also be considered as group 13 (IIIA) element–carbon double bonded compounds C substituted by a phosphonio group. For example, the gallium–carbon bond length (1.93 Å) is remarkably short.

The lower stability of (phosphino)(silyl)carbene–group 13 (IIIA) element Lewis acid adducts compared to that of the corresponding imidazol-2-ylidene adducts is easily rationalized by considering the inferior ability of phosphorus to stabilize a positive charge in the α-position. In other words, once complexed, the carbene

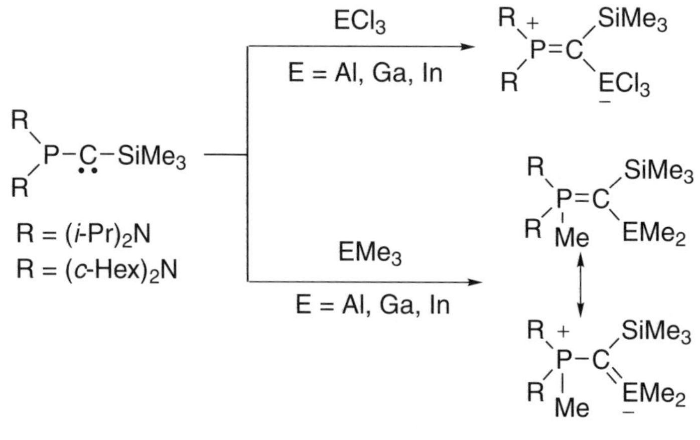

Scheme 8.24

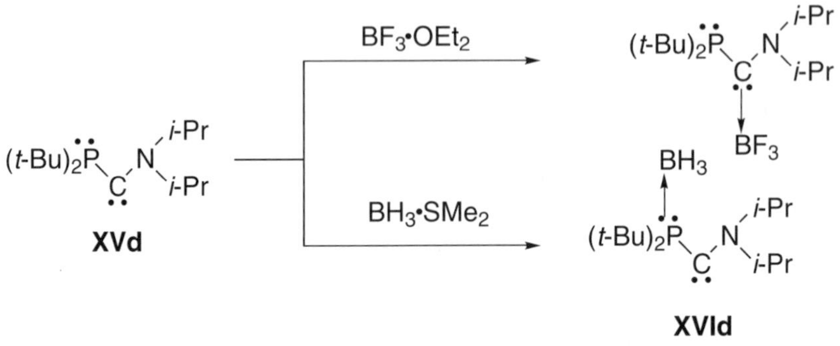

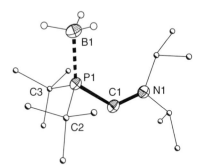

Scheme 8.25

center of phosphinocarbenes still has an accessible vacant orbital, which is not the case for the aminocarbenes and their adducts.

Interestingly, highly regioselective reactions were also observed by addition of Lewis acids to the (amino)(phosphino)carbene (**XVd**). Indeed with one equivalent of $BF_3 \cdot OEt_2$ the quantitative formation of the carbene–BF_3 complex was observed, whereas the softer Lewis acid BH_3 interacts selectively with the phosphorus lone pair to afford a new stable carbene **XVId** (Scheme 8.25).[67]

The structure of this new carbene (**XVId**), which is perfectly stable at room temperature both in solution and in the solid state (mp 100–102 °C), was confirmed by an X-ray diffraction study (Fig. 8.15). Here is a nice example of preparation of a stable carbene from a stable carbene!

Figure 8.15. Molecular structure of carbene **XVId**.

5.4. Stable Carbene–Transition Metal Complexes: Applications in Catalysis

Recently, spectacular achievements have been reported in the catalysis arena using stable cyclic diaminocarbenes **III**, **IV**, and **VII**, now called N-heterocyclic carbenes (NHC), as new types of ligands for transition metal catalysis.[68,101–103] However, it is important to remember that NHC–transition complexes have been known since 1968[104] and their organometallic chemistry investigated by Lappert in the 1960s.[105]

As recently written by Herrmann,[68] "NHC are not just phosphine mimics, there is increasing experimental evidence that NHC–metal catalysts surpass their phosphine-based counterparts in both activity and scope of application." Additionally, NHC are more strongly bound to the metal (thus avoiding the necessity for the use of excess ligand), the catalysts are less sensitive to air and moisture, and have proved remarkably resistant to oxidation.[106] The efficiency of NHC in catalysis is largely due to their strong σ-donor property, which can be superior to that of the best phosphine donor ligands.[107] Note also that the first transition complexes of acyclic diaminocarbenes have been recently prepared and show promising catalytic properties.[108]

In marked contrast, direct complexation of (phosphino)(silyl)carbenes **I** has not yet been reported,[109] and only a few phosphinocarbene complexes are known.[110] The reluctance of carbenes **I** to act as ligands has recently been rationalized theoretically.[111] However, in 2002, the first complexes featuring the stable (aryl) (phosphino)carbene **XIIc**[112] and (amino)(phosphino)carbenes **XVd** have been prepared.[113]

5.4.1. Electronic Structures. So far, carbene complexes have been divided into two types according to the nature of the formal metal–carbon double bond (Fig. 8.16). The metal–carbon bond of Fischer-type carbene complexes is a donor–acceptor bond and formally results from the superposition of carbene to metal σ-donation and metal to carbene π-back donation (Fig. 8.16). In contrast, the metal–carbene bond of Schrock-type complexes is essentially covalent and it formally results from the interaction of a triplet carbene with a triplet metal fragment (Fig. 8.16). In relation to these different bonding situations, Fischer-complexes are generally formed with a low-valent metal fragment and a carbene bearing at least a π-donor group, whereas Schrock complexes are usually formed with metals in a high oxidation state and carbene ligands bearing alkyl substituents.

Due to the presence of two π-donor substituents at the carbene center, the NHC complexes may be classified, at a first glance, as Fischer-type compounds. However, in contrast to usual Fischer-type complexes, NHC bind to transition metals only through σ donation, π-back-bonding being negligible. Photoelectron

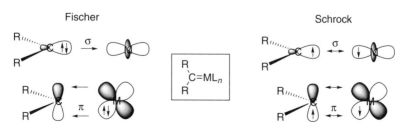

Figure 8.16. Fischer versus Schrock carbenes.

spectroscopy (PES) coupled with density functional calculations have demonstrated that even for group 10 (VIII) metals, bonding occurs very predominantly through σ donation from the carbene lone pair.[114] These peculiar binding properties are easily understandable since the energy of the vacant p_π orbital at the carbene center is considerably increased by the strong N $\rightarrow$ C π donation. The ratio of σ donation to π-back-donation for Fe(CO)$_4$ bonded heteroatom-substituted carbenes increases in the order :C(OR)R< :C(NR$_2$)R < :C(NR$_2$)$_2$ ≈ imidazolidin-2-ylidenes ≈ imidazol-2-ylidenes. The π-acceptor ability of NHC lies between that of nitriles and pyridine.

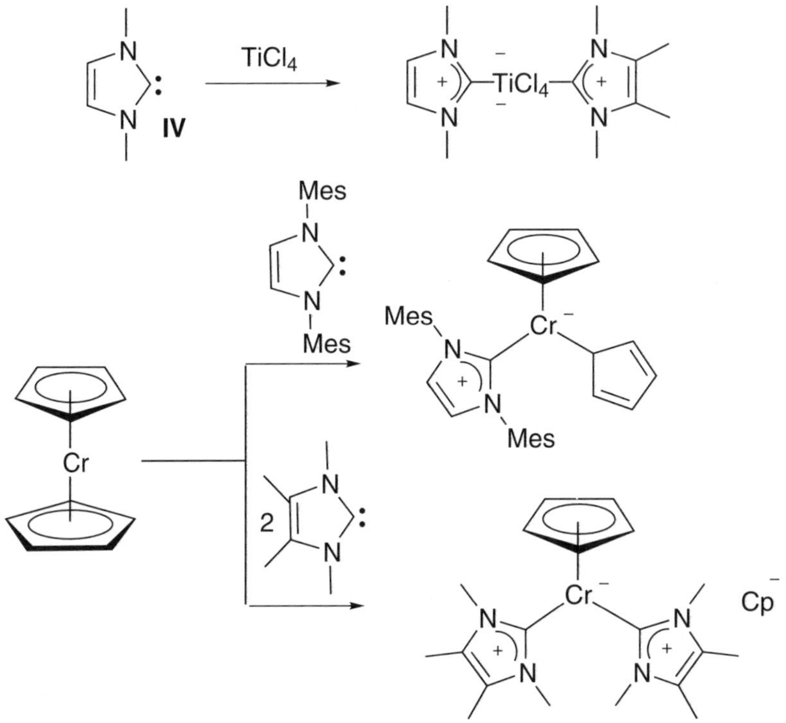

Scheme 8.26

Given these statements, it is not surprising that NHC complexes of almost all the transition metals have been prepared. In particular, metals incapable of π-back-donation such as titanium were only involved in Schrock-carbene complexes until the stable Fischer-type complexes were prepared from TiCl$_4$ and imidazol-2-ylidenes (**IV**).[115] The electronic properties of these NHC are also well illustrated in metallocene chemistry: (a) 14-electron chromium(II) complexes have been isolated,[116] (b) the displacement of a Cp ligand of chromocene and nickellocene can be achieved by imidazol-2-ylidenes (**IV**), giving bis(carbene) complexes[116b] (Scheme 8.26).

The X-ray diffraction studies performed on many complexes also reveal the pure σ-donor character of the *N*-heterocyclic carbenes. Indeed, elongated single M—C bonds are generally observed. At the same time, the internal ring angle at the carbon atom is slightly larger in coordinated than in free carbenes, although not as large as that in the related imidazolium salts. Similarly, the C—N bond distances lie between those of the free carbenes and imidazolium salts. These data as a whole suggest that the three- and five-center π delocalization in the imidazolidin-2-ylidenes (**III**) and imidazol-2-ylidenes (**IV**), respectively, is increased by coordination of the carbene center.

Among several hundreds of transition metal complexes featuring NHC ligands, a very few complexes were recently prepared by Herrmann and co-workers[108] using the acyclic bis(diisopropylamino)carbene (**IX**). Interestingly enough, they found that (a) even though the free carbene **IX** is more sensitive than the NH carbenes, the stability toward air and moisture of the corresponding metal complexes is similar to that of complexes of NHcarbene ligands; (b) the carbene ligand **IX** induces even higher electron density at the metal center than the saturated NHC. To date, **IX** is the most basic known carbene ligand. It is also worth mentioning that in addition to the classical η^1-coordination, acyclic diaminocarbenes **IX** can side-on coordinate to transition metals (Scheme 8.27).

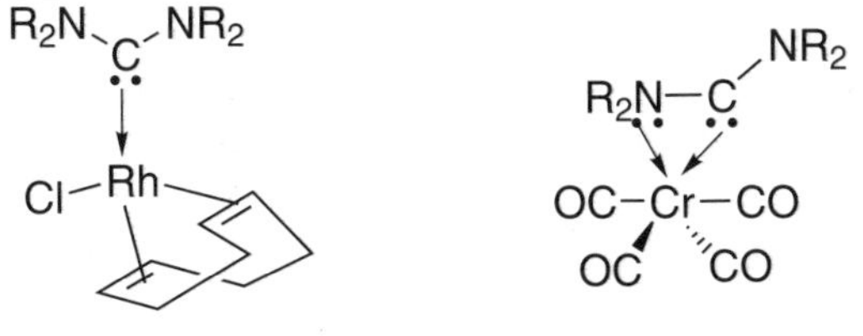

Scheme 8.27

Lastly, rhodium complexes featuring the [2,6-bis(trifluoromethyl)phenyl](phosphino)carbene (**XIIc**)[112] and the (amino)(phosphino)carbene (**XVd**)[113] have been prepared (Scheme 8.28).

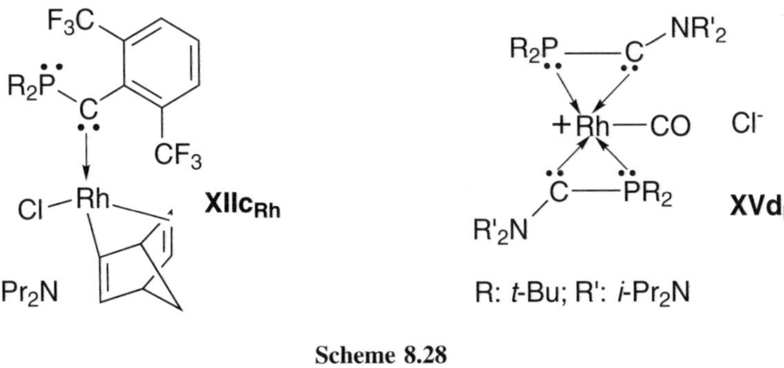

Scheme 8.28

As predicted by Schoeller et al.,[111] coordination of **XIIc** induces a considerable contraction of the bond angle about the carbene center (from 162° in the free carbene **XIIc** to 119° in **XIIc**$_{Rh}$). The carbene-rhodium bond distance [2.096(7) Å] is

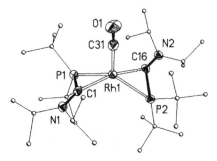

Figure 8.17. Molecular structure of the (amino)(phosphino)carbene rhodium complex XVd_{RH}.

in the range typical for C—Rh single bonds, and is even slightly longer than that observed for related NHC-rhodium complexes (2.00–2.04 Å). These data suggest that $XIIc_{Rh}$ is best regarded as a Fischer-type carbene complex; the carbene–metal interaction consists almost exclusively of donation of the carbene lone pair into an empty metal-based orbital. Back-donation from the metal to the carbene center is negligible compared to that from the phosphorus lone pair. Note that in the case of $XIIc_{Rh}$, the inherent chelation of the norbornadiene (nbd) ligand prevents the formation of an η^2-complex. Such a coordination mode has been found in the (amino)(phosphino)carbene–rhodium complex XVd_{Rh}, the structure of which was unambiguously established by an X-ray diffraction study (Fig. 8.17).

5.4.2. Catalytic Properties. In recent years, NHC ligands have led to numerous breakthroughs in different highly useful reactions such as the Heck,[106a,117] Suzuki,[118] Sonogashira,[119] Kumada[120] and Stille couplings,[121] aryl amination,[106c,122a] and amide α-arylation,[122b] and hydrosilylation.[123]

A good illustration is given by the ruthenium-catalyzed olefin metathesis methodologies,[103,124] one of the most powerful routes for the construction of carbon–carbon bonds. Although the potential of ruthenium catalysts was described in the 1960s,[125] the first breakthroughs occurred with the isolation by Grubbs of well-defined catalysts bearing phosphine ligands.[126] It has been found that the replacement of one phosphine by an imidazol-2-ylidene **IV**,[127] or even better by the saturated NHC analogue **III**[128] (Fig. 8.18) afforded catalysts that displayed performance that was previously possible only with the most active early metal systems,[129] but with an incomparable functional group tolerance. In other words,

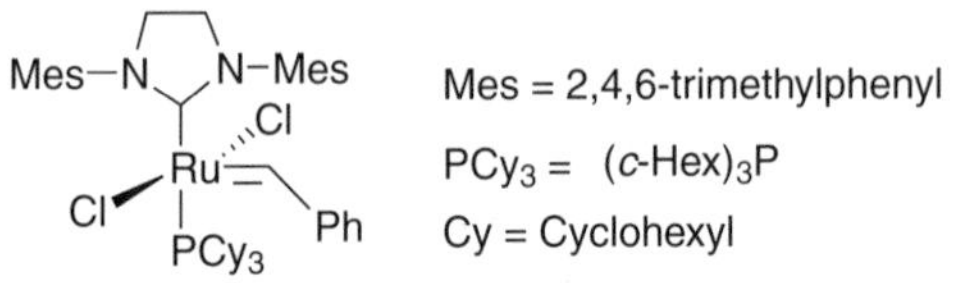

Mes—N N—Mes   Mes = 2,4,6-trimethylphenyl
        Cl   PCy₃ = (c-Hex)₃P
   Ru   Cy = Cyclohexyl
  Cl   Ph
    PCy₃

Figure 8.18. The new generation of Grubbs' catalyst.

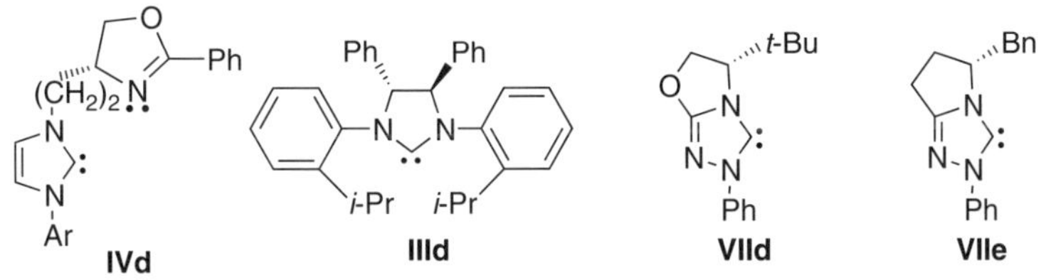

Figure 8.19. Some examples of chiral NHC ligands.

thanks to NHC ligands, this new generation of ruthenium catalysts combines the best characteristics of early and late metal-derived systems in a single species.

Apart from a few examples, stereoselective synthesis has to date not been thoroughly investigated with NHC-based catalysts,[130] although some recent reports are quite encouraging in this respect. An iridium(I) complex of **IVd** (Fig. 8.19) gave enantioselectivities of up to 98% in asymmetric hydrogenations of (E)-aryl alkenes.[131] These enantioselectivities are especially notable because asymmetric hydrogenations of this kind are difficult and there are only a few practical systems for achieving this transformation. Moreover, it was reported that competing mechanistic pathways that diminish the enantioselectivity of the process, were less prevalent than for analogous phosphine complexes. Using a chiral ruthenium catalyst featuring **IIId**, high enantiomeric excesses (up to 90%) were observed in the desymmetrization of achiral trienes.[132] Note also, that recent papers by Enders and Kalfass[133a] and Rovis and co-workers[133b] describe free triazolinylidenes **VIId** and **VIIe** as the most efficient catalysts to date for the asymmetric benzoin condensation and intramolecular Stetter reaction, respectively, with enantiomeric excesses >90%.

As mentioned above, complexes bearing the saturated NHC ligand **III** perform much better as olefin metathesis catalysts than those bearing the unsaturated analogue **IV**. However, recent calorimetric studies of reaction enthalpies between reactions involving saturated **III** and unsaturated ligands **IV** show that **III** is indeed a better donor, but that the difference in their relative bond dissociation energies (BDE) is very small, 1 kcal/mol.[134] This implies that even a small difference in the donor capabilities of the ligands will effect significant changes in a given catalytic system. The situation is even more puzzling, because Herrmann and co-workers[135] found that alkyl substituents at the nitrogen of NHC produces catalysts that are dramatically less active than their aryl counterparts, which are of course poorer σ donors; steric hindrance also plays an important role. After an extensive comparative study concerning metathesis catalysts featuring ligand **III** or **IV**, Fürstner et al. concluded that "no single catalyst outperforms all others in all possible applications."[136]

This short analysis supports: (a) the statement by Herrmann that "from the work in numerous academic laboratories and in industry, a revolutionary turning point in organometalic catalysis is emerging" thanks to NHC;[68] (b) the conclusion by Grubbs that "the exact mechanism of these powerful new catalysts remains

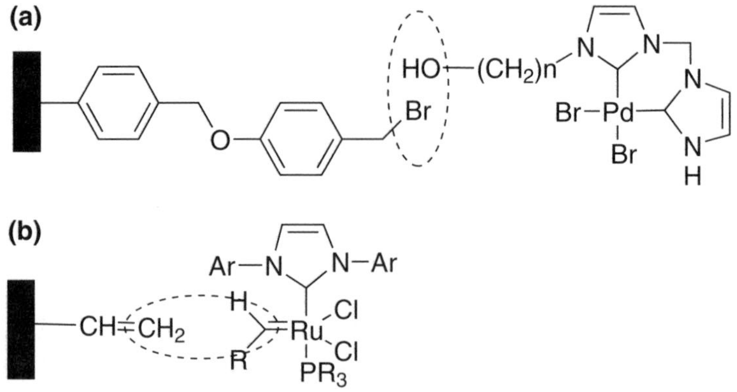

(a)

(b)

Scheme 8.29

unclear"... and that "these developments illustrate the extent to which it is possible to tune a metal center by modifying the ligand environment".[103]

Maximizing recovery of the catalyst is a key issue, especially because residual metal may cause decomposition of the product over time and increase toxicity of the final material. The NHC are particularly suitable ligands for immobilization, since as already mentioned they are strongly bound to the metals, do not undergo dissociation during the catalytic process, and are remarkably air and water stable. The first polymer-supported NHC catalysts were reported in 2000.[137] The synthetic strategies can be divided into two types. Either the transition metal catalyst is first synthesized and then grafted to the support (Scheme 8.29), or alternatively a carbene precursor is first grafted to the support and only then is the carbene generated and coordinated to the metal (Scheme 8.30).

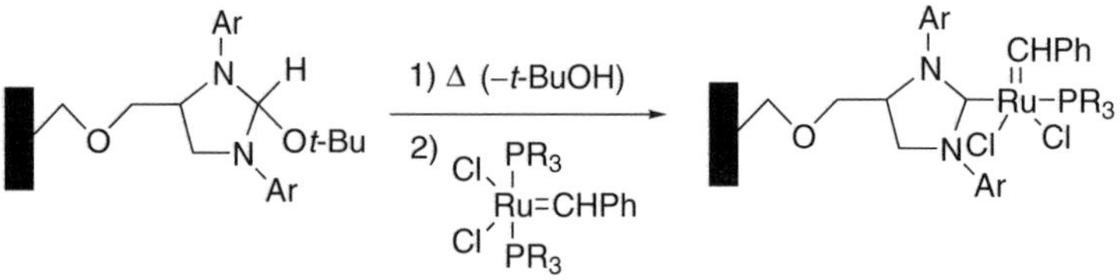

Scheme 8.30

Using the first approach, Herrmann and co-workers[137a] obtained goods results in the Heck coupling reactions, using palladium complexes grafted on the Wang resin via one of the amino substituents of the NHC. As expected, hardly any catalyst leaching was observed during the catalytic reactions (Scheme 8.29a).

Hoveyda and co-workers[137b] immobilized an olefin metathesis catalyst on monolithic sol–gel and claimed that the catalytic material is easily recyclable. Barrett and co-workers[137c] prepared a "recyclable boomerang polymer supported catalyst for olefin methathesis" by grafting the preformed catalyst to a polystyrene resin. Very elegantly, the "classical" carbene that initiates the metathesis is itself anchored to the vinylated polystyrene resin by a metathetical step (Scheme 8.29b). Note that in this case the NHC is not directly bound to the support and thus the precatalyst becomes soluble during the course of the reaction and must be recaptured by the polymer.

The alternative strategy for heterogenization has been pursued by Blechert and co-workers,[137d] for a polymer-supported olefin metathesis catalyst. A polymer-anchored carbene precursor was prepared by coupling an alkoxide to a cross-linked polystyrene Merrifield-type resin. Subsequently, the desired polymer-bound carbene complex was formed by thermolytically induced elimination of *tert*-butanol while heating the precursor resin in the presence of the desired transition metal fragment (Scheme 8.30).

Similarly, Buchmeister and co-workers[137e] functionalized a monolithic carrier by an imidazolium carbene precursor and, following generation of the carbene by deprotonation, prepared the active ruthenium catalyst.

6. CONCLUSION AND OUTLOOK

From this chapter, it is clear that the common statement that carbenes only occur as transient reactive intermediates is no longer valid.

Due the nature of the substituents, all the stable singlet carbenes exihibit some carbon-heteroatom multiple-bond character and for some time their carbene nature has been a subject of controversy. One has to keep in mind that apart from dialkylcarbenes, all the transient singlet carbenes present similar electronic interactions. As early as 1956, Skell and Garner[138] drew the transient dibromocarbene in its ylide form based on *"the overlap of the vacant p-orbital of carbon with the filled p orbitals of the bromine atoms"* (Scheme 8.31).

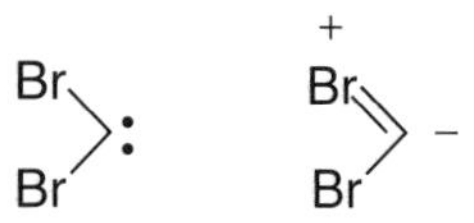

Scheme 8.31

Up to 2000, the number and variety of stable carbenes have been limited by the perceived necessity for two strongly interacting substituents. The preparation of stable or persistent (aryl)- or (alkyl)-(phosphino)carbenes as well as (aryl)(amino)-carbenes demonstrates that a single electron-active substituent allows the spectroscopic characterization of singlet carbenes under standard laboratory conditions. It has been shown that an amino substituent is more efficient than a phosphino substituent to stabilize a carbene center and that the steric bulk of the spectator substituent can be decreased even to the size of a methyl group in the phosphino series, so that a broad range of "observable" carbenes will be readily available.

In the last couple of years, it has been demonstrated that NHC strongly bind transition metal centers, leading to extremely active and robust catalysts, which often outperform their phosphine-based analogues. However, the structure of NHC can only be modified through variation in the nitrogen and carbon backbone substituents. Consequently, the steric demands of these ligands can easily be varied using the *N*-substituents, but their electronic character can only be modified slightly. Indeed, the only significant variation possible is a choice between using either a saturated or an unsaturated backbone.

The availability of stable carbenes featuring a broad range of substituents directly adjacent to the carbene center will allow for great variation of their electronic properties to be achieved and therefore catalytic activities of the resulting carbene complexes should be readily tuned.

Major breakthroughs can also be expected in asymmetric catalysis, because enantiomerically pure carbene complexes, with the source of asymmetry close to the metal center, will be accessible.

Stable carbenes are not laboratory curiosities any longer, they are powerful tools for a variety of chemists.

SUGGESTED READING

G. Bertrand, Ed., *Carbene Chemistry: From Fleeting Intermediates to Powerful Reagents*, Marcel Dekker, New York, **2002**.

W. A. Herrmann, "*N*-Heterocyclic Carbenes: A New Concept in Organometallic Catalysis," *Angew. Chem. Int. Ed. Engl.* **2002**, *41*, 1290.

A. J. Arduengo, III and T. Bannenberg, "Nucleophilic Carbenes and their Applications in Modern Complex Catalysis," *The Strem Chemiker*, **2002**, *XVIV*, 2.

L. Jafarpour and S. P. Nolan, "Transition-Metal Systems Bearing a Nucleophilic Carbene Ancillary Ligand: From Thermochemistry to Catalysis," *Adv. Organomet. Chem.* **2001**, *46*, 181.

D. Enders and H. Gielen, "Synthesis of Chiral Triazolinylidene and Imidazolinylidene Transition Metal Complexes and First Application in Asymmetric Catalysis," *J. Organomet. Chem.* **2001**, *617*, 70.

T. M. Trnka and R. H. Grubbs, "The Development of $L_2X_2Ru=CHR$ Olefin Metathesis catalysts: An Organometallic Success Story," *Acc. Chem. Res.* **2001**, *34*, 18.

D. Bourissou, O. Guerret, F. P. Gabbaï, and G. Bertrand, "Stable Carbenes," *Chem. Rev.* **2000**, *100*, 39.

REFERENCES

1. J. B. Dumas and E. Peligot, *Ann. Chim. Phys.* **1835**, *58*, 5.

2. H. Staudinger and O. Kupfer, *Ber. Dtsch. Chem. Ges.* **1912**, *45*, 501.

3. R. Breslow, *J. Am. Chem. Soc.* **1958**, *80*, 3719.

4. H.-W. Wanzlick, *Angew. Chem.* **1962**, *74*, 129.

5. (a) A. Igau, H. Grutzmacher, A. Baceiredo, and G. Bertrand, *J. Am. Chem. Soc.* **1988**, *110*, 6463. (b) A. Igau, A. Baceiredo, G. Trinquier, and G. Bertrand, *Angew. Chem. Int. Ed. Engl.* **1989**, *28*, 621.

6. (a) H. Tomioka, *Acc. Chem. Res.* **1997**, *30*, 315. (b) H. Tomioka, in *Advances in Carbene Chemistry*, Vol. 2, U. H. Brinker, Ed., JAI Press, Stamford, **1998**, p. 175. (c) H. Tomioka, E. Iwamoto, H. Takura, and K. Hirai, *Nature (London)* **2001**, *412*, 626. (d) H. Tomioka, T. Watanabe, M. Hattori, N. Nomura, and K. Hirai, *J. Am. Chem. Soc.* **2002**, *124*, 474.

7. R. Gleiter and R. Hoffman, *J. Am. Chem. Soc.* **1968**, *90*, 1475.

8. B. C. Gilbert, D. Griller, and A. S. Nazran, *J. Org. Chem.* **1985**, *50*, 4738.

9. G. B. Schuster, *Adv. Phys. Org. Chem.* **1986**, *22*, 311.

10. D. R. Myers, V. P. Senthilnathan, M. S. Platz, and, J. Jones, Jr., *J. Am. Chem. Soc.* **1986**, *108*, 4232.

11. J. E. Gano, R. H. Wettach, M. S. Platz, and V. P. Senthilnathan, *J. Am. Chem. Soc.* **1982**, *104*, 2326.

12. (a) H. P. Reisenauer, G. Maier, A. Reimann, and R. W. Hoffmann, *Angew. Chem. Int. Ed. Engl.* **1984**, *23*, 641. (b) T. J. Lee, A. Bunge, and H. F. Schaefer, III, *J. Am. Chem. Soc.* **1985**, *107*, 137.

13. G. Xu, T. M. Chang, J. Zhou, M. L. McKee, and P. B. Shevlin, *J. Am. Chem. Soc.* **1999**, *121*, 7150.

14. (a) R. Hoffman, G. D. Zeiss, and G. W. Van Dine, *J. Am. Chem. Soc.* **1968**, *90*, 1485. (b) N. C. Baird and K. F. Taylor, *J. Am. Chem. Soc.* **1978**, *100*, 1333.

15. (a) J. F. Harrison, *J. Am. Chem. Soc.* **1971**, *93*, 4112. (b) C. W. Bauschlicher, Jr., H. F. Schaefer, III, and P. S. Bagus, *J. Am. Chem. Soc.* **1977**, *99*, 7106. (c) J. F. Harrison, C. R. Liedtke, and J. F. Liebman, *J. Am. Chem. Soc.* **1979**, *101*, 7162. (d) D. Feller, W. T. Borden, and E. R. Davidson, *Chem. Phys. Lett.* **1980**, *71*, 22.

16. (a) W. W. Schoeller, *Chem. Commun.* **1980**, 124. (b) L. Pauling, *Chem. Commun.* **1980**, 688.

17. K. I. Irikura, W. A. Goddard, III, and J. L. Beauchamp, *J. Am. Chem. Soc.* **1992**, *114*, 48.

18. (a) R. A. Moss, M. Wlostowski, S. Shen, K. Krogh-Jespersen, and A. Matro, *J. Am. Chem. Soc.* **1988**, *110*, 4443. (b) X. M. Du, H. Fan, J. L. Goodman, M. A. Kesselmayer, K. Krogh-Jespersen, J. A. La Villa, R. A. Moss, S. Shen, and R. S. Sheridan, *J. Am. Chem. Soc.* **1990**, *112*, 1920.

19. (a) R. A. Mitsch, *J. Am. Chem. Soc.* **1965**, *87*, 758. (b) R. A. Moss and C. B. Mallon, *J. Am. Chem. Soc.* **1975**, *97*, 344. (c) S. Koda, *Chem. Phys. Lett.* **1978**, *55*, 353.

20. (a) J. L. Wang, J. P. Toscano, M. S. Platz, V. Nikolaev, and V. Popik, *J. Am. Chem. Soc.* **1995**, *117*, 5477. (b) P. Visser, R. Zuhse, M. W. Wong, and C. Wentrup, *J. Am. Chem. Soc.* **1996**, *118*, 12598.

21. (a) A. Berndt, *Angew. Chem. Int. Ed. Engl.* **1993**, *32*, 985. (b) A. Berndt, D. Steiner, D. Schweikart, C. Balzereit, M. Menzel, J. H. Winkler, S. Mehle, M. Unverzagt, T. Happel, P. v. R. Schleyer, G. Subramanian, and M. Hofmann, in *Advances in Boron Chemistry*, W. Siebert, Ed., The Royal Society of Chemistry, Cambridge; **1997**, p. 61. (c) M. Menzel, H. J. Winckler, T. Ablelom, D. Steiner, S. Fau, G. Frenking, W. Massa, and A. Berndt, *Angew. Chem. Int. Ed. Engl.* **1995**, *34*, 1340.

22. R. A. Moss, C. B. Mallon, and C. T. Ho, *J. Am. Chem. Soc.* **1977**, *99*, 4105.

23. A. Baceiredo, G. Bertrand, and G. Sicard, *J. Am. Chem. Soc.* **1985**, *107*, 4781

24. Th. Curtius, *Ber. Dtsch. Chem. Ges.* **1889**, *22*, 2161.

25. (a) G. Sicard, A. Baceiredo, G. Bertrand, and J. P. Majoral, *Angew. Chem. Int. Ed. Engl.* **1984**, *23*, 459. (b) A. Baceiredo, G. Bertrand, J. P. Majoral, G. Sicard, J. Jaud, and J. Galy, *J. Am. Chem. Soc.* **1984**, *106*, 6088.

26. For reviews: (a) R. W. Hoffmann, *Angew. Chem. Int. Ed. Engl.* **1968**, *7*, 754. (b) J. Hocker and R. Merten, *Angew. Chem. Int. Ed. Engl.* **1972**, *11*, 964.

27. For a review: N. Wiberg, *Angew. Chem. Int. Ed. Engl.* **1968**, *7*, 766.

28. D. M. Lemal, R. A. Lovald, and K. J. Kawano, *J. Am. Chem. Soc.* **1964**, *86*, 2518.

29. (a) H.-W. Wanzlick and H.-J. Schonherr, *Liebigs Ann. Chem.* **1970**, *731*, 176. (b) H.-J. Schonherr and H.-W. Wanzlick, *Chem. Ber.* **1970**, *103*, 1037.

30. A. J. Arduengo, III, R. L. Harlow, and M. Kline, *J. Am. Chem. Soc.* **1991**, *113*, 36.

31. A. J. Arduengo, III, J. R. Goerlich, R. Krafczyk, and W. J. Marshall, *Angew. Chem. Int. Ed. Engl.* **1998**, *37*, 1963.

32. For reviews: (a) G. Bertrand, in "*Carbene Chemistry: From Fleeting Intermediates to Powerful Reagents*," G., Bertrand Ed., Marcel Dekker, New York, **2002**, p. 177. (b) D. Bourissou, O. Guerret, F. P. Gabbaï, and G. Bertrand, *Chem. Rev.* **2000**, *100*, 39. (c) D. Bourissou and G. Bertrand, *Advances in Organomet. Chem.* **1999**, *44*, 175. (d) R. Réau and G. Bertrand, in *Methoden der Organischen Chemie (Houben-Weyl)*, M. Regitz, Ed., Georg Thieme Verlag: Stuttgart, **1996**, *E17a*, p. 794. (e) G. Bertrand and R. Reed, *Coord. Chem. Rev.* **1994**, *137*, 323.

33. (a) M. Soleilhavoup, A. Baceiredo, O. Treutler, R. Ahlrichs, M. Nieger, and G. Bertrand, *J. Am. Chem. Soc.* **1992**, *114*, 10959. (b) P. Dyer, A. Baceiredo, and G. Bertrand, *Inorg. Chem.* **1996**, *35*, 46.

34. T. Kato, H. Gornitzka, A. Baceiredo, A. Savin, and G. Bertrand, *J. Am. Chem. Soc.* **2000**, *122*, 998.

35. (a) J. Kapp, C. Schade, A. M. El-Nahasa, and P. v. R. Schleyer, *Angew. Chem. Int. Ed. Engl.* **1996**, *35*, 2236. (b) C. Schade and P. v. R. Schleyer, *Chem. Commun.* **1987**, 1399.

36. S. Goumri, Y. Leriche, H. Gornitzka, A. Baceiredo, and G. Bertrand, *Eur. J. Inorg. Chem.* **1998**, 1539.

37. (a) M. T. Nguyen, M. A. McGinn, and A. F. Hegarty, *Inorg. Chem.* **1986**, *25*, 2185. (b) M. R. Hoffmann and K. Kuhler, *J. Chem. Phys.* **1991**, *94*, 8029. (c) D. A. Dixon, K. B. Dobbs, A. J. Arduengo, III, and G. Bertrand, *J. Am. Chem. Soc.* **1991**, *113*, 8782. (d) L. Nyulaszi, D. Szieberth, J. Reffy, and T. Veszpremi, *J. Mol. Struct. (THEOCHEM)* **1998**, *453*, 91.

38. C. Heinemann, T. Müller, Y. Apeloig, and H. Schwarz, *J. Am. Chem. Soc.* **1996**, *118*, 2023. C. Boehme and G. Frenking, *J. Am. Chem. Soc.* **1996**, *118*, 2039.

39. A. J., Arduengo, III, J. Goerlich, and W. Marshall, *J. Am. Chem. Soc.* **1995**, *117*, 11027.

40. R. W. Alder, M. E. Blake, C. Bortolotti, S. Bufali, C. P. Butts, E. Linehan, J. M. Oliva, A. G. Orpen, and M. J. Quayle, *Chem. Commun.* **1999**, 241.

41. F. E. Hahn, L. Wittenbecher, R. Boese, and D. Bläser, *Chem. Eur. J.* **1999**, *5*, 1931.

42. D. Enders, K. Breuer, G. Raabe, J. Runsink, J. H. Teles, J. P. Melder, K. Ebel, and S. Brode, *Angew. Chem. Int. Ed. Engl.* **1995**, *34*, 1021.

43. A. J. Arduengo, III, J. R. Goerlich, and W. J. Marshall, *Liebigs Ann.* **1997**, 365.

44. R. W. Alder, P. R. Allen, M. Murray, and G. Orpen, *Angew. Chem. Int. Ed. Engl.* **1996**, *35*, 1121.

45. R. W. Alder, C. P. Butts, and A. G. Orpen, *J. Am. Chem. Soc.* **1998**, *120*, 11526.

46. (a) W. A. Herrmann, M. Elison, J. Fischer, C. Köcher, and G. R. J. Artus, *Chem. Eur. J.* **1996**, *2*, 772. (b) W. A. Herrmann, C. Köcher, L. J. Goossen, and G. R. J. Artus, *Chem. Eur. J.* **1996**, *2*, 1627.

47. N. Kuhn and T. Kratz, *Synthesis* **1993**, 561.

48. D. A. Dixon and A. J. Arduengo, III, *J. Phys. Chem.* **1991**, *95*, 4180.

49. (a) A. J. Arduengo, III, D. A. Dixon, K. K. Kumashiro, C. Lee, W. P. Power, and K. W. Zlim, *J. Am. Chem. Soc.* **1994**, *116*, 6361. (b) A. J. Arduengo, III, H. V. R. Dias, D. A. Dixon, R. L. Harlow, W. T. Klooster, and T. F. Koetzle, *J. Am. Chem. Soc.* **1994**, *116*, 6812. (c) A. J. Arduengo, III, H. Bock, H. Chen, M. Denk, D. A. Dixon, J. C. Green, W. A. Herrmann, N. L. Jones, M. Wagner, and R. West, *J. Am. Chem. Soc.* **1994**, *116*, 6641.

50. J. Cioslowski, *Int. J. Quantum Chem., Quant. Chem. Symp.* **1993**, *27*, 309.

51. (a) C. Heinemann, T. Müller, Y. Apeloig, and H. Schwarz, *J. Am. Chem. Soc.* **1996**, *118*, 2023. (b) C. Boehme and G. Frenking, *J. Am. Chem. Soc.* **1996**, *118*, 2039.

52. J. F. Lehmann, S. G. Urquhart, L. E. Ennis, A. P. Hitchcock, K. Hatano, S. Gupta, and M. K. Denk, *Organometallics* **1999**, *18*, 1862.

53. R. W. Alder and M. E. Blake, *Chem. Commun.* **1997**, 1513.

54. R. W. Alder, M. E. Blake, and J. M. Oliva, *J. Phys. Chem. A* **1999**, *103*, 11200.

55. (a) R. A. Moss, T. Zdrojewski, and G. Ho, *Chem. Commun.* **1991**, 946. (b) D. L. S. Brahms and W. P. Dailey, *Chem. Rev.* **1996**, *96*, 1585.

56. C. Buron, H. Gornitzka, V. Romanenko, and G. Bertrand, *Science* **2000**, *288*, 834.

57. (a) J. E. Jackson and M. S. Platz, *Adv. Carbene Chem.* **1994**, *1*, 89. (b) A. Admasu, A. D. Gudmundsdottir, M. S. Platz, D. S. Watt, S. Swiatkowski, and P. J. Crocker, *J. Chem. Soc., Perkin Trans. 2* **1998**, 1093.

58. (a) R. T. Ruck and M. Jones, Jr., *Tetrahedron Lett.* **1998**, *39*, 2277. (b) K. Krogh-Jespersen, S. Yan, and R. A. Moss, *J. Am. Chem. Soc.* **1999**, *121*, 6269. (c) R. A. Moss, S. Yan, and K. Krogh-Jespersen, *J. Am. Chem. Soc.* **1998**, *120*, 1088. (d) M. I. Khan and J. L. Goodman, *J. Am. Chem. Soc.* **1995**, *117*, 6635.

59. W. W. Schoeller, *Eur. J. Org. Chem.* **2000**, *2*, 369.

60. S. Sole, H. Gornitzka, W. W. Schoeller, D. Bourissou, and G. Bertrand, *Science* **2001**, *292*, 1901.

61. K. Hirai, K. Komatsu, and H. Tomioka, *Chem. Lett.* **1994**, 503.

62. H. Tomioka and K. Taketsuji, *Chem. Commun.* **1997**, 1745.

63. E. Despagnet, H. Gornitzka, A. B. Rozhenko, W. W. Schoeller, D. Bourissou, and G. Bertrand, *Angew. Chem. Int. Ed. Engl.* **2002**, *41*, 2835.

64. U. H. Brinker and M. G. Rosenberg, in *Advances in Carbene Chemistry*, Vol. 2, U. H. Brinker, Ed., JAI Press, Stamford, **1998**, p. 29.

65. R. S. Sheridan, R. A. Moss, B. K. Wilk, S. Shen, M. Wlostowski, M. A. Kesselmayer, R. Subramanian, G. Kmiecik-Lawrynowicz, and K. Krogh-Jespersen, *J. Am. Chem. Soc.* **1988**, *110*, 7563.

66. R. Bonneau and M. T. H. Liu, in *Advances in Carbene Chemistry*, Vol. 2, U. H. Brinker, Ed., JAI Press, Stamford, **1998**, p. 1.

67. N. Merceron, K. Miqueu, A. Baceiredo, and G. Bertrand. *J. Am. Chem. Soc.* **2002**, *124*, 6806.

68. W. A. Herrmann, *Angew. Chem. Int. Ed. Engl.* **2002**, *41*, 1290.

69. (a) H.-W. Wanzlick and H.-J. Kleiner, *Angew. Chem.* **1961**, *73*, 493. (b) M. K. Denk, K. Hatano, and M. Ma, *Tetrahedron Lett.* **1999**, *40*, 2057.

70. Y. F. Liu and D. M. Lemal, *Tetrahedron Lett.* **2000**, *41*, 599.

71. E. A. Carter and W. A. Goddard, III, *J. Phys. Chem.* **1986**, *90*, 998.

72. C. Heinemann and W. Thiel, *Chem. Phys. Lett.* **1994**, *217*, 11.

73. T. A. Taton and P. Chen, *Angew. Chem. Int. Ed. Engl.* **1996**, *35*, 1011.

74. G. Raabe, K. Breuer, D. Enders, and J. H. Teles, *Z. Naturforsch.* **1996**, *51a*, 95.

75. (a) G. Trinquier and J. P. Malrieu, *J. Am. Chem. Soc.* **1987**, *109*, 5303. (b) J. P. Malrieu and G. Trinquier, *J. Am. Chem. Soc.* **1989**, *111*, 5916. (c) H. Jacobsen and T. Ziegler, *J. Am. Chem. Soc.* **1994**, *116*, 3667.

76. R. Alder, in *"Carbene Chemistry: From Fleeting Intermediates to Powerful Reagents,"* G. Bertrand, Ed., Marcel Dekker, New York, **2002**, p. 153.

77. (a) M. K. Denk, A. Thadani, K. Hatano, and A. J. Lough, *Angew. Chem. Int. Ed. Engl.* **1997**, *36*, 2607. (b) Z. Shi, V. Goulle, and R. P. Thummel, *Tetrahedron Lett.* **1996**, *37*, 2357.

78. Y.-T. Chen and F. Jordan, *J. Org. Chem.* **1991**, *56*, 5029.

79. (a) R. W. Alder, P. R. Allen, and S. J. Williams, *Chem. Commun.* **1995**, 1267. (b) Y.-J. Kim and A. Streitwieser, *J. Am. Chem. Soc.* **2002**, *124*, 5757.

80. For a particularly clear example of this effect, the reader is referred to Ref. 58a.

81. (a) R. A. Moss, *Acc. Chem. Res.* **1989**, *22*, 15. (b) *Carbocyclic Three-Membered Ring Compounds*, Vol. E17a. *Methoden der Organische Chemie (Houben-Weyl)*, A. de Meijere, Ed., Thieme Verlag, Stuttgart, **1996**.

82. (a) M. Brookhart and W. B. Studabaker, *Chem. Rev.* **1987**, *87*, 411. (b) D. F. Harvey and D. M. Sigano, *Chem. Rev.* **1996**, *96*, 271. (c) H. W. Frühauf, *Chem. Rev.* **1997**, *97*, 523.

83. P. S. Skell, *Tetrahedron* **1985**, *41*, 1427.

84. S. Goumri-Magnet, T. Kato, H. Gornitzka, A. Baceiredo, and G. Bertrand, *J. Am. Chem. Soc.* **2000**, *122*, 4464.

85. (a) A. E. Keating, S. R. Merrigan, D. A. Singleton, and K. N. Houk, *J. Am. Chem. Soc.* **1999**, *121*, 3933. (b) N. G. Rondan, K. N. Houk, and R. A. Moss, *J. Am. Chem. Soc.* **1980**, *102*, 1770.

86. G. Alcaraz, U. Wecker, A. Baceiredo, F. Dahan, and G. Bertrand, *Angew. Chem. Int. Ed. Engl.* **1995**, *34*, 1246. V. Piquet, A. Baceiredo, H. Gornitzka, F. Dahan, and G. Bertrand, *Chem. Eur. J.* **1997**, *3*, 1757.

87. M. Sanchez, R. Réau, C. J. Marsden, M. Regitz, and G. Bertrand, *Chem. Eur. J.* **1999**, *5*, 274. R. Armbrust, M. Sanchez, R. Réau, U. Bergstrasser, M. Regitz, and G. Bertrand, *J. Am. Chem. Soc.* **1995**, *117*, 10785.

88. G. Bertrand, *Angew. Chem. Int. Ed. Engl.* **1998**, *37*, 270.

89. (a) H. J. Bestmann and R. Zimmermann, in *Methoden der Organischen Chemie (Houben-Weyl)*, M. Regitz, Ed., Georg Thieme Verlag, Stuttgart, **1982**, Vol. E1, p. 616. (b) A. W. Johnson, W. C. Kaska, K. A. O. Starzewski, and D. A. Dixon, in *Ylides and Imines of Phosphorus*, John Wiley & Sons, Inc., New York, **1993**, p. 115.

90. S. Goumri-Magnet, O. Polischuck, H. Gornitzka, C. J. Marsden, A. Baceiredo, and G. Bertrand, *Angew. Chem. Int. Ed. Engl.* **1999**, *38*, 3727.

91. (a) A. J. Arduengo, III, H. V. R. Dias, and J. C. Calabrese, *Chem. Lett.* **1997**, 143. (b) A. J. Arduengo, III, J. C. Calabrese, A. H. Cowley, H. V. R. Dias, J. R. Goerlich, W. J. Marshall, and B. Riegel, *Inorg. Chem.* **1997**, *36*, 2151.

92. N. Kuhn, J. Fahl, D. Blaser, and R. Boese, *Z. Anorg. Allg. Chem.* **1999**, *625*, 729.

93. A. J., Arduengo, III, M. Kline, J. C. Calabrese, and F. Davidson, *J. Am. Chem. Soc.* **1991**, *113*, 9704.

94. N. Kuhn, T. Kratz, and G. Henkel, *Chem. Commun.* **1993**, 1778.

95. N. Kuhn, H. Bohnen, D. Bläser, R. Boese, and A. H. Maulitz, *Chem. Commun.* **1994**, 2283.

96. A. J. Arduengo, III, H. V. R. Dias, J. C. Calabrese, and F. Davidson, *J. Am. Chem. Soc.* **1992**, *114*, 9724.

97. D. E. Hibbs, M. B. Hursthouse, C. Jones, and N. A. Smithies, *Chem. Commun.* **1998**, 869.

98. S. J. Black, D. E. Hibbs, M. B. Hursthouse, C. Jones, K. M. A. Malik, and N. A. Smithies, *J. Chem. Soc., Dalton Trans.* **1997**, 4313.

99. A. Cowley, F. Gabbaï, C. Carrano, L. Mokry, M. Bond, and G. Bertrand, *Angew. Chem. Int. Ed. Engl.* **1994**, *33*, 578.

100. K. H. v. Locquenghien, A. Baceiredo, R. Boese, and G. Bertrand, *J. Am. Chem. Soc.* **1991**, *113*, 5062.

101. W. A. Herrmann and C. Köcher, *Angew. Chem. Int. Ed. Engl.* **1997**, *36*, 2163.

102. L. Jafarpour and S. P. Nolan, *Adv. Organomet. Chem.* **2001**, *46*, 181.

103. T. M. Trnka and R. H. Grubbs, *Acc. Chem. Res.* **2001**, 34, 18.

104. (a) K. Öfele, *J. Organomet. Chem.* **1968**, *12*, P42. (b) K. Öfele and C. G. Kreiter, *Chem. Ber.* **1972**, *105*, 529.

105. D. J. Cardin, B. Cetinkaya, and M. F. Lappert, *Chem. Rev.* **1972**, *72*, 545.

106. (a) E. Peris, J. A. Loch, J. Mata, and R. H. Crabtree, *Chem. Commun.* **2001**, 201. (b) M. Albrecht, J. R. Miecznikowski, A. Samuel, J. W. Faller, and R. H. Crabtree, *Organometallics* **2002**, *21*, 3596. (c) M. S. Viciu, R. M. Kissling, E. D. Stevens, and S. P. Nolan, *Org. Lett.* **2002**, *4*, 2229.

107. I. Huang, H.-J. Schanz, E. D. Stevens, and S. P. Nolan, *Organometallics* **1999**, *18*, 2370.

108. (a) K. Denk, P. Sirsch, and W. A. Herrmann, *J. Organomet. Chem.* **2002**, *649*, 219. (b) M. Tafipolsky, M. Scherer, K. Ofele, G. Artus, B. Pedersen, W. A. Herrmann, and S. G. McGrady, *J. Am. Chem. Soc.* **2002**, *124*, 5865.

109. For attempted coordination of an anionic di(phosphino)carbene, see: A. Fuchs, D. Gudat, M. Nieger, O. Schmidt, M. Sebastian, L. Nyulaszi, and E. Niecke, *Chem. Eur. J.* **2002**, *8*, 2188.

110. (a) E. O. Fischer and R. Restmeier, *Z. Naturforsch.* **1983**, *38b*, 582. (b) F. R. Kreissl, T. Lehotkay, C. Ogric, and E. Herdtweck, *Organometallics* **1997**, *16*, 1875. (c) S. Dovesi, E. Solari, R. Scopelliti, and C. Floriani, *Angew. Chem. Int. Ed. Engl.* **1999**, *38*, 2388.

111. (a) W. W. Schoeller, D. Eisner, S. Grigoleit, A. J. B. Rozhenko, and A. Alijah, *J. Am. Chem. Soc.* **2000**, *122*, 10115. (b) W. W. Schoeller, A. J. B. Rozhenko, and A. Alijah, *J. Organomet. Chem.* **2001**, *617*, 435.

112. E. Despagnet, K. Miqueu, H. Gornitzka, P. W. Dyer, D. Bourissou, and G. Bertrand, *J. Am. Chem. Soc.* **2002**, *124*, 11834.

113. N. Merceron, K. Miqueu, A. Baceiredo, and G. Bertrand, unpublished results.

114. J. C. Green, R. G. Scurr, P. L. Arnold, and F. G. N. Cloke, *Chem. Commun.* **1997**, 1963.

115. W. A. Herrmann, K. Öfele, M. Elison, F. E. Kühn, and P. W. Roesky, *J. Organomet. Chem.* **1994**, *480*, C7. N. Kuhn, T. Kratz, D. Bläser, and R. Boese, *Inorg. Chim. Acta* **1995**, *238*, 179.

116. (a) M. H. Voges, C. Romming, and M. Tilset, *Organometallics* **1999**, *18*, 529. (b) C. D. Abernethy, J. A. C. Clyburne, A. H. Cowley, and R. A. Jones, *J. Am. Chem. Soc.* **1999**, *121*, 2329.

117. (a) W. A. Herrmann, M. Elison, J. Fischer, C. Kocher, and G. R. J. Artus, *Angew. Chem. Int. Ed. Engl.* **1995**, *34*, 2371. (b) C. L. Yang, H. M. Lee, and S. P. Nolan, *Org. Lett.* **2001**, *3*, 1511. (c) M. V. Bakler, B. W. Skelton, A. H. White, and C. C. Williams, *J. Chem. Soc., Dalton Trans.* **2001**, 111.

118. (a) A. Furstner and A. Leitner, *Synthesis* **2001**, *2*, 290. (b) C. W. K. Gstottmayr, V. P. W. Bohm, E. Herdweck, M. Grosche, and W. A. Herrmann, *Angew. Chem. Int. Ed. Engl.* **2002**, *41*, 1363.

119. (a) S. Caddick, F. G. N. Cloke, G. K. B. Clentsmith, P. B. Hitchcock, D. McKerrecher, L. R. Titcomb, and M. R. V. Williams, *J. Organomet. Chem.* **2001**, *617*, 635. (b) C. L. Yang and S. P. Nolan, *Organometallics,* **2002**, *21*, 1020.

120. (a) W. P. Bohm, C. W. K. Gstottmayr, T. Weskamp, and W. A. Herrmann, *Angew. Chem. Int. Ed. Engl.* **2001**, *40*, 3387. (b) W. P. Bohm, T. Weskamp, C. W. K. Gstottmayr, and W. A. Herrmann, *Angew. Chem. Int. Ed. Engl.* **2000**, *39*, 1602.

121. G. A. Grasa and S. P. Nolan, *Org. Lett.* **2001**, *3*, 119.

122. (a) S. R. Stauffer, S. W. Lee, J. P. Stambuli, S. I. Hauck, and J. F. Hartwig, *Org. Lett.* **2000**, *2*, 1423. (b) S. Lee and J. F. Hartwig, *J. Org. Chem.* **2001**, *66*, 3402.

123. I. E. Marko, S. Sterin, O. Buisine, R. Mignani, P. Branland, B. Tinant, and J. P. Declercq, *Science* **2002**, *298*, 204.

124. A. Fürstner, *Angew. Chem. Int. Ed. Engl.* **2000**, *39*, 3013.

125. F. W. Michelotti and W. P. Keaveney, *J. Polym. Sci.* **1965**, *A3*, 895.

126. S. T. Nguyen, L. K. Johnson, R. H. Grubbs, and J. W. Ziller, *J. Am. Chem. Soc.* **1992**, *114*, 3974.

127. (a) M. Scholl, T. M. Trnka, J. P. Morgan, and R. H. Grubbs, *Tetrahedron Lett.* **1999**, *40*, 2247. (b) T. Weskamp, F. J. Kohl, W. Hieringer, D. Gleich, and W. A. Herrmann, *Angew. Chem. Int. Ed. Engl.* **1999**, *38*, 2416. (c) L. Ackermann, A. Fürstner, T. Weskamp, F. J. Kohl, and W. A. Herrmann, *Tetrahedron Lett.* **1999**, *40*, 4787. (d) J. Huang, E. D. Stevens, S. P. Nolan, and J. L. Petersen, *J. Am. Chem. Soc.* **1999**, *121*, 2674.

128. M. Scholl, S. Ding, C. W. Lee, and R. H. Grubbs, *Org. Lett.* **1999**, *1*, 953.

129. R. R. Schrock, *Tetrahedron* **1999**, *55*, 8141.

130. D. Enders and H. Gielen, *J. Organomet. Chem.* **2001**, *617*, 70.

131. M. T. Powell, D. R. Hou, M. C. Perry, X. H. Cui, and K. Burgess, *J. Am. Chem. Soc.* **2001**, *123*, 8878.

132. T. J. Seiders, D. W. Ward, and R. H. Grubbs, *Org. Lett.* **2001**, *3*, 3225.

133. (a) D. Enders and U. Kalfass, *Angew. Chem. Int. Ed. Engl.* **2002**, *41*, 1743. (b) M. S. Kerr, J. Read de Alaniz, and T. Rovis, *J. Am. Chem. Soc.* **2002**, *124*, 10298.

134. A. C. Hillier, W. J. Sommer, B. S. Yong, J. L. Petersen, L. Cavallo, and S. P. Nolan, *Organometallics*, **2003**, in press.

135. T. Weskamp, F. J. Kohl, W. Hieringer, D. Gleich, and W. A. Herrmann, *Angew. Chem. Int. Ed. Engl.* **1999**, *38*, 2416.

136. A. Fürstner, L. Ackermann, B. Gabor, R. Goddard, C. W. Lehmann, R. Mynnot, F. Stelzer, and O. R. Thiel, *Chem. Eur. J.* **2001**, *7*, 3236.

137. (a) J. Schwartz, V. P. W. Bohm, M. G. Gardiner, M. Grosche, W. A. Herrmann, W. Hieringer, and G. Raudaschl-Sieber, *Chem. Eur. J.* **2000**, 1773. (b) J. S. Kingsbury, S. B. Garber, J. M. Giftos, B. L. Gray, M. M. Okamoto, R. A. Farrer, J. T. Fourkas, and A. H. Hoveyda, *Angew. Chem. Int. Ed. Engl.* **2001**, *40*, 4251. (c) M. Ahmed, T. Arnauld, A. G. M. Barrett, D. C. Braddock, and P. A. Procopiou, *Synlett* **2000**, 1007. (d) S. C. Schurer, S. Gessler, N. Buschmann, and S. Blechert, *Angew. Chem. Int. Ed. Engl.* **2000**, *39*, 3898. (e) M. Mayr, B. Mayr, and M. R. Buchmeiser, *Angew. Chem. Int. Ed. Engl.* **2001**, *40*, 3839.

138. (a) P. S. Skell and A. Y. Garner, *J. Am. Chem. Soc.* **1956**, *78*, 3409 and 5430. (b) P. S. Skell and R. C. Woodworth, *J. Am. Chem. Soc.* **1956**, *78*, 4496 and 6427.

CHAPTER 9

Triplet Carbenes

HIDEO TOMIOKA

Chemistry Department for Materials, Faculty of Engineering, Mie University,
Tsu, Mie 514-8507, Japan

Reactive Intermediate Chemistry, edited by Robert A. Moss, Matthew S. Platz, and Maitland Jones, Jr.
ISBN 0-471-23324-2   Copyright © 2004 John Wiley & Sons, Inc.

1. RELATIONSHIP BETWEEN STRUCTURE AND THE SINGLET–TRIPLET ENERGY GAP

Carbenes have two free electrons, and therefore can have two electronic states depending on the direction of spin of the electrons. There is a singlet state that has antiparallel spins and a triplet state that has parallel spins. Since the two electronic states have different electronic configurations, each state exhibits different reactivity and is affected differently by substituents. Generally speaking, the singlet state of the carbene undergoes concerted reactions with high efficiency leading to many useful products and, hence, has played an important role in synthetic organic chemistry. The triplet state, on the other hand, is a less reactive and selective reagent and, hence, is a less attractive intermediate from the standpoint of organic synthesis. Nonetheless, it has attracted much attention in materials science as a unit from which to construct organic ferromagnetic compounds.

In the reactions involving reactive intermediates, the most stable form usually plays a crucial role in controling the reaction pathway. However, in the reactions of carbenes, it is not always ground-state multiplicity that is involved. This idea is especially true in the reactions of a triplet ground-state carbenes with a small singlet–triplet (S–T) energy gap. The primary factor that influences the reaction is the energy gap between singlet and triplet states (ΔE_{ST}). Although we will review the reactions of triplet carbenes as studied by modern methods in this chapter, it is essential to know how ΔE_{ST} of the carbene in question changes with its structure.[1] So, we first overview the relationship between structure and ΔE_{ST}, and then we will see how characteristic reactions of triplet state carbene will be changed depending on ΔE_{ST} and reaction conditions. Finally, we will see how and to what extent those highly reactive species can be stabilized.

1.1. Overview

Here we will first overview the structural differences between singlet and triplet carbenes, and then we will give a more quantitative analysis based on theoretical studies. The four lowest energy configurations of carbenes have electronic configurations described as $\sigma^1 p^1$, σ^2, or p^2. The electron spins in the $\sigma^1 p^1$ configuration may be paired, a singlet, or parallel to form a triplet, while the σ^2 and p^2 configurations must be an electron-paired singlet. Thus, the triplet state has the $\sigma^1 p^1$ configuration, while σ^2 is generally thought to be the lowest energy configuration for the singlet (Fig. 9.1).

In a singlet state carbene, the electron–electron Coulombic repulsion would be severe, since two electrons are constrained to the same small molecular orbital (MO). On the other hand, the triplet configuration is stabilized by relief of the Coulombic repulsion and "exchange repulsion." However, the separation of electrons into different MOs does not come without a cost. Thus, the magnitude of the energy difference between the triplet and singlet states (the S–T splitting, ΔE_S) is roughly equal to the electron–electron repulsion minus the energy required to promote an electron from the σ- to the p-nonbonding orbital. In other words, as the energy separation between σ and p states increases, the promotion energy becomes large enough to overcome the repulsion energy, while if the spacing is small, the species will still have a triplet ground state. A small difference between the energies of S_0 and T_1 may easily be overturned by the effects of substituents on the carbene center. The factors that influence the spacing can be analyzed in terms of electronic and steric effects.

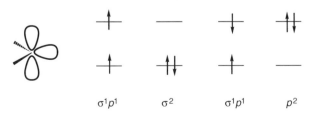

Figure 9.1. Electronic states of carbene.

Throughout this chapter, T–S splittings and ΔE_{ST}'s imply energy differences in the sense that $E_S - E_T$. Thus, positive values indicate that the triplet is lower in energy than the singlet.

1.1.1. Electronic Effects.

Because of more favorable overlap, the interaction of the carbon $2p$ orbital with substituent p or π orbitals is expected to dominate. The σ (sp^n) orbital that lies in the nodal plane of the substituent p or π orbital will only interact with substituent σ orbitals, and then only weakly. Thus its energy is mostly unperturbed by the substituent. A simple way to analyze the effect of the perturbation by substitution is to superimpose the orbitals of a prototypal carbene on those of the π system of the substituent.

Substituents interacting with a π system can be classified into three classes, namely, X: (π-electron donors such as $-NR_2$, $-OR$, $-SR$, $-F$, $-Cl$, $-Br$, and $-I$), Z (π-electron acceptors such as $-COR$, $-SOR$, $-SO_2R$, $-NO$, and $-NO_2$), and C (conjugating groups such as alkenes, alkynes, or aryl).[2]

As shown in Figure 9.2(a), an X: substituent, which has a p orbital, or other suitable doubly occupied orbital that will interact with the π bond, raises the $2p$ orbital of the carbene, thereby increasing the separation of the $2p$ and sp^n (σ) orbitals. The ground state of an X:-substituted carbene becomes a singlet, and many carbenes in this class are known. The most familiar of these are the halocarbenes.

The Z and C substituents having a p or π^* orbital and evenly spaced π and π^* orbitals, respectively, either lower the $2p$–sp^n gap or leave it about the same, as shown in Figure 9.2(b and c). In either case, the ground state for these carbenes is expected to be T_1 although the magnitude of ΔE_{ST} may vary. It has been demonstrated by electron paramagnetic resonance (EPR) studies that most aryl- and diarylcarbenes have triplet ground states. From a valence bond viewpoint, it can be said that electron–donor groups stabilize the electrophilic singlet carbene

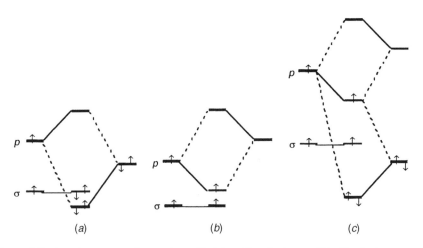

Figure 9.2. A carbene center interacting with (a) an X: substituent, (b) a Z substituent, and (c) a C substituent.

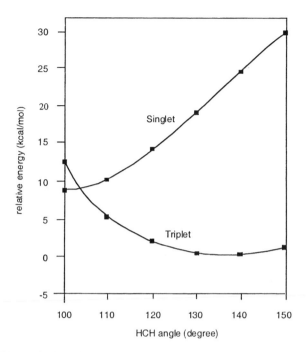

Figure 9.3. Change in the relative energy of singlet and triplet methylene with respect to <HCH at Becke 3LYP/TZ2P.

more than they do the radical-like triplet, while electron-withdrawing groups destabilize the singlet and lead to a greater ΔE_{ST}.

1.1.2. Steric Effects. The magnitude of ΔE_{ST} is expected to be sensitive to the carbene carbon bond angle. A linear carbene has two degenerate p orbitals, an arrangement that is calculated to provide the maximum value of ΔE_{ST}. Bending the carbene removes the orbital degeneracy and reduces ΔE_{ST}. As the carbene–carbon bond angle is contracted, the σ orbital gains s character and consequently moves even lower in energy. The smaller the bond angle, the more energy it takes to promote an electron from the σ to the p orbital, and the smaller ΔE_{ST} becomes.

This effect is shown more quantitatively by calculations for methylene (Fig. 9.3).[3,4] The calculations predict that the energy of singlet methylene will drop below that of the triplet state for carbenes with bond angles less than $\sim 100°$. On the other hand, theory also suggests that opening of the central angle strongly destabilizes the singlet state, but requires very little additional energy for the triplet, thus making ΔE_{ST} larger.

1.2. Theoretical Predictions

1.2.1. Electronic Effects. Theoretical calculations (B3LYP/6-311 + G**// B3LYP/6-31G*) of a series of substituted phenylcarbenes (PCs, **1**) give more

TABLE 9.1. ΔE_{ST} for Substituted Phenylcarbenes (PCs, 1)^a

| 1 | Substituents | ΔE_{ST} (kcal/mol) | |
		para	meta
a	NH$_2$	0.7	5.2
b	OH	1.7	5.3
c	Me	4.1	5.2
d	F	4.1	6.6
e	H	5.4	5.4
f	Cl	5.3	6.4
g	CF$_3$	8.1	6.9
h	CO$_2$Me	8.2	5.7
i	CN	9.0	7.2
j	CHO	9.3	6.1
k	NO$_2$	10.3	7.8

^{a} B3LYP/6-311 + G**//B3LYP/6-31G*. Single-point energy using the B3LYP/6-31G*-optimized geometry.

quantitative pictures (Table 9.1).[5] For example, p-aminophenylcarbene (p-**1a**) is predicted to have ΔE_{ST} of 0.7 kcal/mol, which is 4.7 kcal/mol smaller than that of parent PC (**1e**). On the other hand, p-nitrophenylcarbene (p-**1k**) is predicted to have a ΔE_{ST} of 10.3 kcal/mol, 4.9 kcal/mol larger than that of **1e**. A good correlation for ΔE_{ST} is observed with σ_p^+ and the resultant slope (ρ) is 5.0. When the mesomeric effect of π donation is removed, as in the meta-substituted PCs, this effect is greatly attenuated. Similar correlation with σ_m is linear with a smaller ρ value of 3.3. The spin density at the carbene center does not depend on substitution, and therefore does not appear to contribute to the observed differential ΔE_{ST} in these carbenes. Moreover, the charge on both the singlet and triplet carbenes gives a good correlation to σ_p^+. The slope for this correlation is 0.03 for the singlet and 0.02 for the triplet, indicating that the singlet has slightly more charge separation than the triplet state does. Thus, the origin of this effect is preferential interaction of aryl substituents with the singlet rather than the triplet species. This difference is expected, as the singlet is conjugated to the π system via an empty p orbital, whereas the triplet presents only a singly occupied p orbital to the π system. In accord with this expectation, on phenyl substitution of methylene to form phenyl-carbene, the carbene center gains 0.07 e$^-$ in the singlet, while in the triplet, the C loses 0.02 e$^-$. Thus the phenyl stabilizes the singlet with respect to the triplet(ΔE_{ST} decreases from 10 to 5 kcal/mol on going from CH$_2$ to PhCH).

As expected, the angle about the carbene center in the singlet states (106.2–106.6°) is smaller than that in the triplet (134.1–134.8°).

1.2.2. Hyperconjugation Effects.
Singlet carbenes are isoelectronic with carbocations, and the same effects that stabilize carbocations will also stabilize singlet

carbenes. Triplet carbenes have a singly occupied p orbital, as is the case for radicals. While both carbocations and radicals are stabilized by hyperconjugation, the magnitude of stabilization is much smaller for the radicals. Theoretical calculations exactly predict the difference in the magnitude of the effect for dialkylcarbenes.[4] Methylene (**2**) is known to have a triplet ground state, with a singlet lying higher in energy by 10 kcal/mol. Compared to methylene, both the singlet and the triplet are stabilized by alkyl substituent(s), but the magnitude of the effect for the singlets is about twice as large as that for the triplets. Thus, ΔE_{ST} usually decreases as the alkyl groups are introduced on methylene; the S–T gap decreases from 2 to −2 kcal/mol in going from MeCH (**3**) to MeCMe (**4**) (B3LYP/TZ2P). The hyperconjugation between the empty p orbital and the alkyl substituent(s) is most pronounced for singlet t-BuCH. The *tert*-butyl group stabilizes singlet t-BuCH (**5**) by 17.8 kcal/mol relative to methylene (B3LYP/TZ2P). The C1(carbene)–C2–C3 bond angle, which is a measure of the amount of hyperconjugation, has decreased to 82° from 109.5° for the ideal tetrahedral carbon.

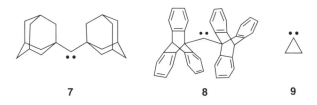

1.2.3. Steric Effects. However, when two bulky groups are introduced, the steric repulsion between the substituents widens the CCC bond angle and influences the S–T splitting, increasing the p character of the doubly occupied sp^2 orbital and destabilizing the singlet. For example, ΔE_{ST} increases from 2 to 5 kcal/mol on going from t-BuCH (**5**) to t-BuCt-Bu (**6**) (B3LYP/TZ2P). The second *tert*-butyl group stabilizes the triplet state by 12.5 kcal/mol.[4,6] This change is interpreted in terms of the diminished magnitude of the stabilizing effect of the second *tert*-butyl group in the singlet and the large stabilization of triplet by the second *tert*-butyl group. The ΔE_{ST} value is calculated to be 5.19 kcal/mol, and bond angles at the singlet (θ_S) and triplet states (θ_T) are predicted to be 125.1 and 133.9°, respectively.[4]

TABLE 9.2. ΔE_{ST} for Smaller and Medium-Size Cyclic Carbenes[9]

Carbenes		ΔE_{ST} (kcal/mol)
10	n	
	0	-25
	1	-6
	2	~ 0
	3	4
11	n	
	0	-7
	1	-5
	2	-2

Similarly, di(adamanthyl)carbene (**7**)[7] and di(triptycyl)carbene (**8**)[8] are predicted to have triplet ground states. The S–T gap and θ_S/θ_T calculated for **7** (B3LYP/6-31G*) and **8** (B3LYP/6-31G*) are 9.3 kcal/mol and 125°/149°, and 14.0 kcal/mol and 129.3°/153.3°, respectively.

The effect of the bond angle on ΔE_{ST} is easily understood by comparing the predicted values for three typical carbenes. Thus, in cyclopropylidene (**9**) in which the angle is constrained by a ring to $<70°$, the singlet state is greatly stabilized. While alkyl groups stabilize the singlet state by hyperconjugation, two bulky alkyl groups stabilize the triplet state more than the singlet. The S–T gap and θ_S/θ_T calculated for **9**, **4** and **6** are -12.9 kcal/mol and 59.7°/69.1°, -0.2 kcal/mol and 113.5°/133.9°, and 5.16 kcal/mol and 125.1°/141.9° (B3LYP/TZ2P), respectively.[4]

The effect of the bond angle on the S–T gap of the carbene is more systematically investigated by calculating a series of conjugated (**10**) and unconjugated cycloalkenyldenes (**11**) (G2(MP2, SVP), Table 9.2).[9] The results indicate that, as the size of the ring decreases, the singlet state becomes stabilized with respect to the triplet. For the conjugated system (**10**), the break-even point occurs with the six-membered ring (**10**, $n = 2$) in which the triplet and singlet states are close in energy. However, in nonconjugated systems (**11**), the singlet state is still the ground state even in the seven-membered ring (**11**, $n = 2$).

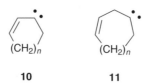

10   **11**

This difference is explained in terms of the difference in the stabilization of singlet and triplet states by vinyl and alkyl substitution. The enthalpies (kcal/mol) calculated for the following isodesmic reactions for the singlet (S) and triplet (T) states of vinyl- and ethylcarbenes (Eqs. 1–3) suggest that a vinyl group stabilizes the singlet somewhat more than the triplet compared to a hydrogen substituent, but that an ethyl group stabilizes the singlet significantly more than the triplet, presumably because of hyperconjugation. The last equation indicates that a vinyl group stabilizes the triplet more than the singlet, as compared to an alkyl substitution.

$$
\overset{\bullet\bullet}{\underset{}{CH}} \text{—} + CH_4 \longrightarrow \overset{\bullet\bullet}{CH_2} + \quad \text{//} \qquad S:18.2\,/\,T:17.6 \tag{1}
$$

$$
\overset{\bullet\bullet}{\underset{}{CH}} \text{—} + CH_4 \longrightarrow \overset{\bullet\bullet}{CH_2} + \quad S:16.5\,/\,T:5.3 \tag{2}
$$

$$
\overset{\bullet\bullet}{\underset{}{CH}} \text{—} + \quad \longrightarrow \quad \overset{\bullet\bullet}{\underset{}{CH}} \text{—} + \quad \text{//} \qquad S:1.7\,/\,T:12.3 \tag{3}
$$

2. GENERATION AND REACTIONS OF TRIPLET CARBENES

Most carbenes are relatively easily generated by photolysis of nitrogeneous precursors such as diazo compounds[10] or diazirines.[11] The reaction is clean, as nitrogen is the only byproduct and it is also very efficient as nitrogen evolution is a highly exothermic reaction. Therefore carbenes can be easily generated even under very inert conditions, such as in a noble gas matrix at very low temperatures.

Although carbenes have two electronic states of different stability, it is not always the ground-state multiplicity that is involved in the reaction. What are then factors that control the reaction pattern of carbenes?

Take the reaction of carbenes generated by photolysis of diazo compounds (Scheme 9.1), for example. Direct irradiation of a diazo compound (**12**) is believed to generate the carbene initially in singlet state (^{1}**13**) via the singlet excited state of the precursor. Triplet sensization, on the other hand, is presumed to give the triplet

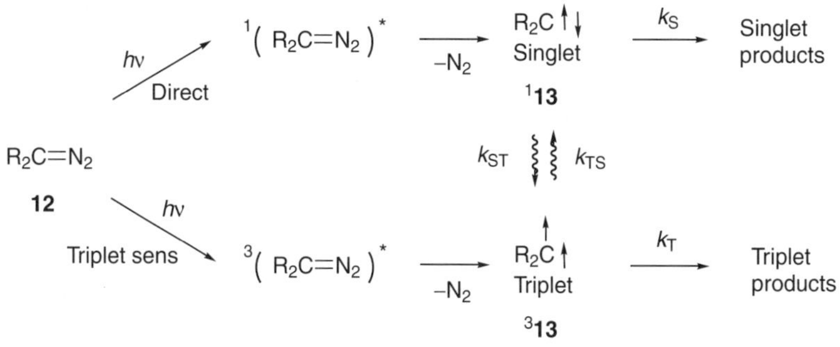

Scheme 9.1

carbene (^{3}**13**) directly via the triplet diazo compound without first forming its singlet state. However, when the energy difference between the states is small, thermal reactivation and reaction from the upper state can compete with the reaction of the ground state. Thus, even though a carbene is formed in the triplet ground state, it may react predominantly from the singlet state, which has similar but higher energy.

Generally speaking, the rate of reaction in the singlet state (k_S) is larger than that of the triplet (k_T). The rate of singlet to triplet (k_{ST}) intersystem crossing and the reverse rate (k_{TS}) are related to ΔE_{ST}. Thus, $k_{ST} > k_{TS}$ if ΔE_{ST} is large, but $k_{ST} \cong k_{TS}$ when ΔE_{ST} is less than ~ 3 kcal/mol. In the latter case, singlet–triplet equilibration is usually assumed. The multiplicity that is involved in the reaction can be summarized as follows.

1. Carbenes with a singlet ground state far below the triplet react in the singlet state even if generated by triplet-sensitized photolysis.
2. Carbenes with a triplet ground state well separated from the singlet ($\Delta E_{ST} > 5$ kcal/mol) react in a multiplicity determined by the method of generation.
3. Carbenes with a triplet ground state with a small energy separation ($\Delta E_{ST} < \sim 3$ kcal/mol) react in the singlet state regardless of the method of generation.

Of course, the rate constants (k_S and k_T) are dependent on the substrates and, hence, the above criteria should be taken only as a general guide. For example, if one chooses a quencher that efficiently reacts with the triplet state, such as O$_2$, carbenes with triplet ground states react efficiently with the quencher to give products such as the corresponding ketones, regardless of ΔE_{ST} values (see Section 6).

3. DIRECT OBSERVATION OF TRIPLET CARBENES

Carbenes are highly reactive species and, hence, their direct observation requires considerable effort. One way to observe them is to use the matrix isolation technique.[12] In this technique, carbenes can be generated by irradiation of an appropriate precursor within a glass or more ordered inert gas matrix at very low temperature. The low temperature of the experiments stops or slows reactions of the carbene with the matrix materials. Also, the rigidity of the medium prevents diffusion and dimerization of the carbene. A second way to observe those highly reactive species is to conduct an experiment on a very short time scale.[13] Such experiments rely on the rapid photochemical generation of the carbenes with a short pulse of light and the detection of the carbene with a probe beam. These pump–probe experiments can be performed on time scales ranging from pico- to milliseconds. The low temperature and short time scale measurements support each other in the identification of the detected intermediates.

3.1. Spectroscopic Characterization of Triplet Carbenes in Matrices at Low Temperature

Matrix isolation coupled with some conventional spectroscopic method [usually electron paramagnetic resonance (EPR), ultraviolet (UV), and infrared (IR)][12] is used to detect and characterize the species of interest. There are several advantages in studying the chemical behavior of reactive species under matrix-isolation conditions. Thus, the inert material of the matrix essentially shuts down any intermolecular reactions giving the opportunity to examine the intrinsic reactivity of the reactive intermediate. By annealing the matrix, it is possible to see if low-barrier isomerizations are available to the investigated species. Doping the matrix with some reagent and/or using less inert material for the matrix (e.g., oxygen or methane) gives insight into several intermolecular chemical pathways that may be available. Finally, molecules in an inert matrix can be essentially considered as isolated, making the comparison of experimental and computational data more straightforward.

3.1.1. EPR Spectroscopy. The structure of triplet carbenes is unequivocally characterized by EPR zero-field splitting (ZFS) parameters, which are analyzed by D and E values. In a simple model, the ZFS parameters D and E for a triplet depend on the distance between electrons with parallel spins as given by equation (4):

$$D \propto \left\langle \frac{r^2 - 3z^2}{r^5} \right\rangle \approx \left\langle \frac{1}{r^5} \right\rangle, \qquad E \propto \left\langle \frac{y^2 - x^2}{r^5} \right\rangle \tag{4}$$

where r is the distance between the two spins, x, y, and z are the components of r along the x, y, and z axis, respectively. The parameter D measures the magnetic dipole interaction along the z axis and is related to the average $1/r^3$ as shown above. A high value of D implies a large spin–spin interaction and proximity of the two spins. The parameter E, on the other hand, is a measure of the difference between similar magnetic dipole interactions along the x and y axes. A molecule with three different axes should have a finite E, whereas this quantity vanishes for linear molecules with degenerate p orbitals.[14–16]

More plainly, the more the two electrons are delocalized in carbenes with a conjugated π system, the smaller the value of the D will be. On the other hand, increasing the bond angle at the carbene center leads to a higher p-orbital contribution and a smaller value for E. Although the values D and E depend on the electronic distribution, it has been shown that there is a good correlation between the E/D ratio and the bond angle at the divalent carbon atom.

However, be aware that the ratio of E/D is not always a reliable guide to the structure of triplet carbenes. For example, diphenylcarbene and fluorenylidene have very different geometries but almost exactly the same D and E values.[15a] If one really wants to learn about the geometry, one must label the carbene carbon with ^{13}C and measure the hyperfine coupling constants.

The temperature dependence of the intensity of the EPR signals for a triplet carbene can provide some information about the ordering and energetics of the lowest singlet and triplet spin states. No EPR signal from the triplet carbene will be observed for a carbene with a singlet ground state far below the triplet. When the energy difference between the singlet and triplet spin state is small, then an EPR spectrum can be observed regardless of the ground-state multiplicity. However, the intensity of magnetization (I) should deviate from the behavior predicted by the Curies law[17] according to Eq. 5.

$$I = \frac{C}{T}[3 + \exp(-\Delta E_{ST}/RT)]^{-1} \tag{5}$$

This law means that the intensity becomes smaller as the temperature is raised. Deviations from linearity could indicate a temperature-dependent equilibrium between a triplet and a singlet. Conversely, a linear Curie plot is taken as evidence for a triplet ground state far below the singlet, but such observations require that the carbene be stable to a certain extent at elevated temperature.

3.1.1.1. Triplet Carbene Structure and ZFS Parameters. The ZFS parameters are thus shown to provide information on the molecular and electronic structures of triplet carbenes. We will see next how the parameters change systematically by examining a series of triplet carbenes.

a. EFFECTS OF CARBENIC SUBSTITUENTS. Some D and E values for typical triplet carbenes are collected in Table 9.3. The ZFS parameters for methylene,[18] the parent compound of all carbenes, clearly indicate that it has a bent structure. The bond angle is estimated to be 136°,[18a] which is in good agreement with most theoretical calculations.[1]

Introduction of aryl groups on methylene results in a significant decrease in D values; thus D values decrease from 0.69 to 0.515 on going from methylene (**2**) to phenylcarbene (**1e**).[19] The D values decrease further as the aromatic ring is changed from phenyl (**1e**) to naphthyl (**12**) to anthryl (**13**).[20] These trends are interpreted in terms of increase in spin delocalization into the aromatic rings. It is interesting to note here that there are only small changes in E/D values among those monoaryl-carbenes, indicating that the central bond angle of the carbenes are not affected significantly by a change in those aromatic rings.

b. EFFECTS OF REMOTE SUBSTITUENTS. Effects of para substituents on the EPR spectrum of triplet diphenylcarbenes (^{3}DPCs, ^{3}**14**) have been investigated (Table 9.4, Scheme 9.2).[21] Two trends become obvious when the D values are compared for di-para substituted DPCs. First, substitution generally causes a decrease in D over that in the parent molecule. This effect is obviously due to extended π delocalization of spin density. To estimate the relative abilities of substituents to delocalize the spin, sigma-dot substituent constants ($\sigma\cdot$) have been proposed.[22] Among the various approaches to the definition of a $\sigma\cdot$ scale, Arnold's $\sigma_\alpha\cdot$ scale

TABLE 9.3. The ZFS Parameters[a] for Some Typical Triplet Carbenes

| Carbenes | D | E | E/D | Reference |
	(cm^{-1})			
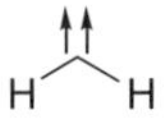 (3)	0.69	0.003	0.004[b]	18
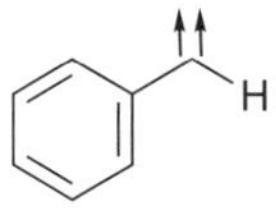 (1e)	0.5150	0.0251	0.04873[c]	19
(α-12)	(Z) 0.4347	0.0208	0.0424	20
	(E) 0.4555	0.0202	0.0433	
(β-12)	(Z) 0.4926	0.0209	0.0424	20
	(E) 0.4711	0.0243	0.0516	
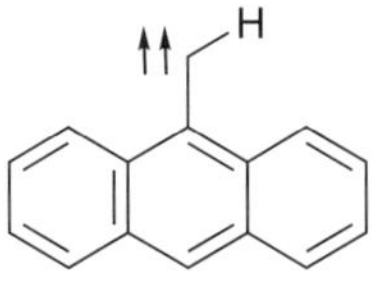 (13)	0.3008	0.0132	0.0439	19

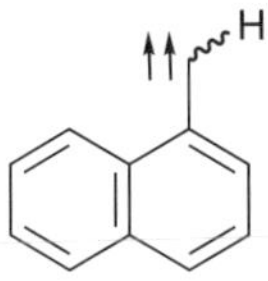

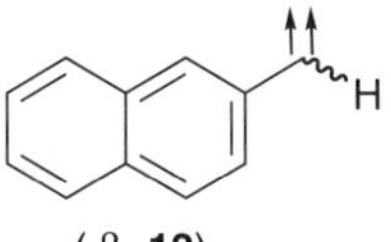

[a]Measured in benzophenone at 77 K unless otherwise noted.
[b]Measured in xenon at 4.2 K.
[c]Measured in p-dichlorobenzene.

is the most suitable for the analysis of the substituent effect on the D values of DPCs, since the scale is a nonkinetic measure of radical stabilizing effects based on hyperfine coupling constants in the benzylic radical.[22] Actually, the D values correlate reasonably well with the $\sigma_\alpha\cdot$ scale.[23]

Second, the decrease in D is largest when ^{3}DPC is substituted with one para electron-withdrawing group (e.g., NO$_2$) and one para' electron-donating group (e.g., NMe$_2$). The decrease in D in these unsymmetrically disubstituted ^{3}DPCs is always larger than that predicted taking the sum of the effects in the monosubstituted derivatives. These observations are explained in terms of merostabilization, a

TABLE 9.4. The ZFS Parameters[a] for Para, Para′-Disubstituted Diphenylcarbenes (14)

Carbenes	Substituents		D	E
14	X	Y	(cm^{-1})	
a	H	H	0.4088	0.0170
b	H	OMe	0.4043	0.0191
c	OMe	OMe	0.4022	0.0189
d	H	CN	0.3906	0.0193
e	CN	CN	0.3879	0.0178
f	H	NMe_2	0.3876	0.0168^b
g	NMe_2	NMe_2	0.3748	0.0180^c
h	H	NO_2	0.3778	0.0173^d
i	NO_2	NO_2	0.3773	0.0177^b
j	CN	NMe_2	0.3518	0.0163^d
k	NO_2	NMe_2	0.3351	0.0164^b

[a]Measured in 4:1 methylcyclohexane: isopentane matrix at 77 K unless otherwise noted.
[b]Measured in tetrahydrofuran (THF) matrix.
[c]Measured in 4:1 methylcyclohexane: THF matrix.
[d]Solid salt was used.

term first suggested by Katritzky and co-workers[24] to describe increased delocalization in radicals for which reasonable charge-separated resonance structures can be drawn (Scheme 9.2). These charge-separated resonance structures contribute only in unsymmetrically disubstituted DPCs containing strong electron-withdrawing and donating groups.

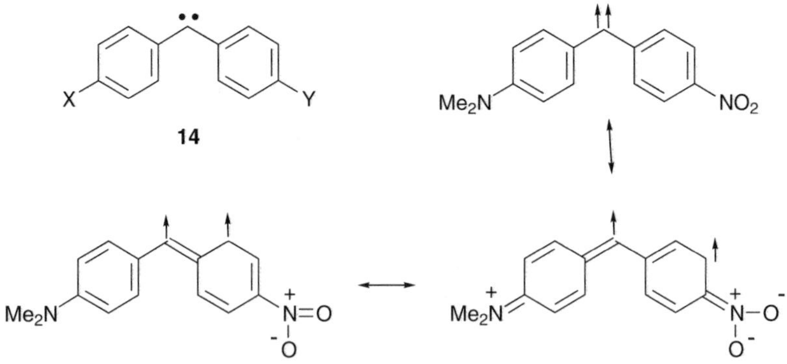

Scheme 9.2

3.1.1.2. Geometrical Isomerism. Triplet carbenes in which the divalent carbon atom is substituted with an sp^2 hybridized carbon atom may exist in two rotomeric forms that are stable at very low temperatures. In favorable cases, EPR spectra of these carbenes exhibit two sets of triplet signals with significantly different ZFS parameters.

The magnitude of the ZFS parameter D is largely determined by the spin–spin dipolar interaction of the two electrons at the divalent carbon atom. Accordingly, the fraction of the π spin density located at the carbenic center can be estimated from the D value of the carbene. In spite of the predominance of this one-center interaction, the spin density at atoms several bonds removed from the divalent carbon atom can also have a significant effect on the ZFS parameters.

This effect was first observed for the pairs of 1- and 2-naphthylcarbenes (**12**).[20] Since then, reports of geometric isomerism in triplet carbenes have appeared with increasing frequency; two sets of triplet signals having similar but nonidentical ZFS parameters are observed.[25] The spectra are assigned to the two conformations of the carbene in which the σ orbital at the divalent carbon and the aromatic moiety are coplanar. When the distribution of the spin in the π orbital is unsymmetric, the dipole spin–spin interaction of the p electron with the electron localized in the σ orbital is different for the two conformations. Consequently, the ZFS parameters will be different and in cases in which the differences are sufficiently large, it is possible to observe the spectra of the two isomers.

The assignment of the ZFS parameters to a specific conformer is made possible by a point spin model,[26] with individual contributions, D_i, due to π-spin densities, ρ_i, at the individual carbon atoms, C_i, namely,

$$D_i \propto \rho_i \frac{r_i^2 - 3z_i^2}{r_i^5} \approx \frac{1}{r^3} \tag{6}$$

where r_i is the distance between the divalent carbon C_{di} and a carbon, C_i, bearing π-spin density, and z_i is the z coordinate of C_i. As mentioned above, carbenes will have noticeable secondary contributions to $|D/hc|$ due to the nearest carbons with π electron spin densities. Take 2-naphthylcarbene (β-**12**) for instance (Scheme 9.3). The p electron density at the 1-position is twice that at the 3-position ($\rho_3 \approx 1/2\rho_1$). For the (E) isomer, C1 lies close to the z axis so that r is approximately parallel to z. Hence, $z_1 \propto r_1$, and there is a noticeable negative contribution to $|D/hc|$. A similar inspection of the (Z) isomer shows that C3 lies close to the z axis ($z_3 = r_3$), whereas C1 is far from the z axis ($z_1 > r_1$). Therefore, the negative contribution to $|D/hc|$ is smaller. Based on these considerations, the (E)-conformer must have the smaller $|D/hc|$ value (see Table 9.3).

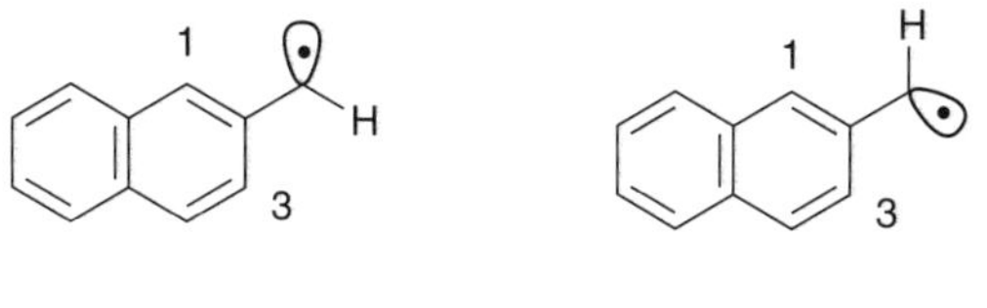

(*E*) - **12**   (*Z*) - **12**

Scheme 9.3

3.1.1.3. Geometrical Changes upon Annealing. Geometrical changes upon annealing the matrix are observed especially for sterically hindered carbenes. For example, the E/D value obtained for di(1-naphthyl)carbene (**15**) in methyltetrahydrofuran (MTHF) at 15 K is 0.0109/0.3157. A new set of triplet signals with a smaller E/D value (0.0051/0.2609) ascribable to the linear configuration of ^{3}**15** is observed at the expense of the original peaks when the matrix is annealed to ~80 K. This observation is interpreted in terms of steric strain in triplet carbenes.[27] Thus, when the carbene is formed in a rigid matrix at low temperature, it should have the bent geometry presumably dictated by that of the precursor. Even if the thermodynamically most stable geometry of the carbene is different from that imposed at birth, the rigidity of the matrix prevents it from assuming its minimum energy geometry. However, when the matrix is softened on annealing, the carbene relaxes to a structure that is closer to linear, as evidenced by the substantial reducing in E. The small reductions in D are also consistent with this picture, as they indicate that the unpaired electrons are more efficiently delocalized in the relaxed geometry. Since E values for ^{3}DPC (**14**) are essentially insensitive to the matrix, suggesting that the carbene has achieved its relaxed geometry regardless of the environment, the observations indicate that the carbene undergoes expansion of the central C—C—C angle to gain relief from steric compression in soft matrices or upon an annealing of harder matrices.

Similar dependence of the ZFS parameters on the rigidity of matrices is observed for many other sterically congested triplet carbenes and hence can be considered as an indication of such steric strain.[15c]

3.1.2. Ultraviolet/Visible Spectroscopy.[12,15a,b]

Unlike the EPR spectrum, the optical spectrum reveals virtually no structural information. Thus, irradiation of a carbene precursor at low temperature creates new absorption features, but it is difficult to assign these bands to a particular structure with certainty. Comparison of the optical and EPR spectrospcopic results obtained under identical conditions can corroborate the simultaneous existence of both the carbene and the absorbing species. The data play an important role in the interpretation of the transient spectroscopic results obtained in time-resolved ultraviolet/visible (UV–vis) spectroscopy. These short time scale experiments help to confirm the spectral assignments made on the basis of the low-temperature measurement.

Spectroscopic studies of DPCs (**14a**) have been most extensively done in an Ar matrix as well as in organic matrices. The wealth of the data may be due to the fact that a precursor, diphenyldiazomethane, is available that can be prepared and handled in a relatively simple way. Most of DPCs are stable in an organic matrix at liquid nitrogen temperature.

Generally speaking, the spectra of DPCs consist of two features, intense absorption bands in the UV region and weak, broad, and sometimes structured bands in the visible region. There is a qualitative resemblance between the spectra of DPCs and those of diphenylmethyl radicals. As a rule of thumb, DPCs are usually blueshifted by 20–30 nm with respect to the corresponding radicals. For example, DPC has absorptions with maxima near 300 and 465 nm and the diphenylmethyl radical exhibits maxima at 336 and 515 nm. If it is assumed that the p_z electron is not involved in the transitions, then the π system of DPCs can be considered as that of an odd-alternant hydrocarbon radical having 13 π electrons. In systems of this type, a strong absorption band due to $\pi-\pi^*$ transition is expected.

As expected, the absorption bands due to polynuclear aromatic carbenes are red-shifted as π conjugation is developed. For example, triplet di(1-naphthyl)carbene (**15**)[28] and di(9-anthryl)carbene (**16**)[29] show a rather strong and structured UV–vis absorption bands at 300–450 and 350–500 nm, respectively.

15   **16**

3.1.3. Fluorescence.

Triplet DPCs (**14**) show strong emission in a low-temperature matrix.[15a,b] For example, for DPC (**14a**), emission at $\lambda_{max} = 482$ nm ($\tau = 123$ ns) with a quantum yield of 0.33 is observed in a methylcyclohexane glass at 77 K. The emission is assigned to fluorescence arising from a radiative transition from the first excited triplet level to the triplet ground level ($T_1 - T_0$). Substituted DPCs also show fluorescence, but the subsitution leads to a red-shift in the emission regardless of whether they have electron releasing or withdrawing properties.

Again, there is a resemblance between the emission of DPCs and those of the diphenylmethyl radicals. The emission maximum of DPCs is blue-shifted by 20–50 nm with respect to that of the radicals. For example, diphenylmethyl radical exhibits its emission at 535 nm. Another characteristic difference between these two species is that the lifetime of ^{3}DPC* is significantly shorter than that of the first excited state of the diphenylmethyl radical.[30,31]

3.1.4. IR Spectroscopy.

The IR spectra of triplet carbenes can be obtained by using matrix isolation techniques.[32] Fortunately, a matrix IR spectrum consists of a series of sharp bands showing no rotational structure. This relative simplicity is the result of the rigidity of the matrix and the minimization of intermolecular interactions. Thus, it is often possible to discriminate individual species in mixtures with great ease.

Like UV–vis spectra, IR spectra do not give concrete evidence for the multiplicity of carbene in question and, hence, comparison with EPR spectra is highly desired. However, IR spectroscopic data encode a lot of structural information and can be analyzed with the help of computational methods, thus aiding in the identification of the observed species. Since geometries of the singlet and triplet carbenes are usually different, their IR spectra are expected to be different. One can often assign a multiplicity of the carbene if one can calculate theoretical

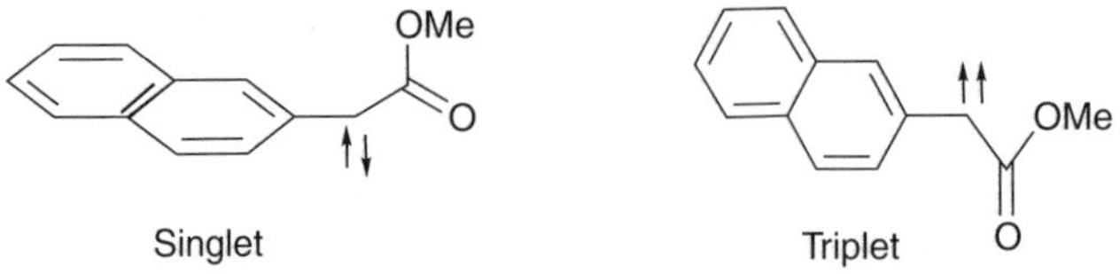

Scheme 9.4

IR bands for the both states. For example, singlet and triplet states of 2-naphthyl-(methoxycarbonyl)carbenes (NMCs, **17**) show very different IR spectra (Scheme 9.4)[33] (1590, 1625, 1640 cm^{-1} for ^{1}NMC and 1660 cm^{-1} for ^{3}NMC). Both states are observable in this case. In the singlet state, the carbomethoxy group assumes a conformation perpendicular to the naphthylcarbene plane to avoid destabilization of the empty carbenic 2p atomic orbital by the electron-withdrawing carbonyl group, while in the triplet, the methoxycarbonyl group is in the same plane to delocalize an unpaired electron. For this reason, there is a barrier between the two states and hence both of them are observable under these conditions.

The other advantage of matrix IR spectroscopy is that one can follow the subsequent reaction (either thermally or photochemically) of triplet carbenes relatively precisely, again with the aid of computational methods.[32,34,35] This complementary approach will reinforce the assignment of triplet species. For example, triplet carbenes react with oxygen even at very low temperature to give oxidation products such as carbonyl oxides (see Section 6), which are easily characterized by comparing the IR spectra with those calculated.

Matrix IR data, like UV–vis, play an important role in the interpretation of the transient spectroscopic results obtained in the time-resolved IR spectroscopy described below (Sections 4.2 and 4.3).

3.2. Time-Resolved Spectroscopy in Solution at Room Temperature

A second way to observe those highly reactive species appeared later than the one developed for low temperatures. The new method provided an important opportunity absent from the low-temperature experiments, that is, the ability to measure their rates of reaction under normal conditions. In the late 1960s, Moritani et al.[36] detected transient absorption bands of triplet dibenzocyclopentadienylidene (**18**) and later in the 1970s, Closs and Rabinow[37] observed transient UV–vis absorption bands of ^{3}DPCs (**14**). Both groups used conventional flash photolysis. However, with this technique, the measurements were restricted to microsecond time resolution. With the advent of the laser flash photolysis (LFP) technique, nanosecond time resolution became widely available, and made the measurement of the absolute rate constants for the reactions involving carbenes and related species very easy.[38] The methods provide much useful kinetic information including reactivities and energetics of carbenes, as we will see in this chapter.

Time-resolved UV–vis spectroscopy is the most common technique and has been widely used, but other time-resolved spectroscopies such as IR and EPR have become available as well.

3.2.1. Time-Resolved UV–Vis (TRUV–Vis) Spectroscopy.

In LFP, a solution of a photolabile precursor such as a diazo compound or diazirine is photolyzed with a pulse of a laser to give the transient of interest. Once formed, the transient is monitored by an optical detection system consisting of a xenon lamp, monochromator, and photomultiplier tube. The output of the photomultiplier is fed to a transient digitizer and the signal is finally fed to a computer for storage and kinetic analysis. Reaction kinetics are investigated at a fixed wavelength by monitoring changes in the optical absorption of a transient with time. The spectrum of a transient is reconstructed from measurements of these time profiles at a series of wavelengths.

Assignment of the transient absorption spectrum typically is done by reference to the low-temperature spectra described above. It is also important to analyze the chemical reactions of the intermediates. For example, triplet carbenes are known to react with O_2 very efficiently to give ketones. So, if the detected intermediate can be observed to react with O_2, there is additional evidence for assignment as a triplet carbene (see Section 6.5).

Rate measurements are straightforward if the carbene can be monitored directly. As a rule, the decay of carbene absorption is pseudo-first order due to rearrangement and/or reaction with the solvent. In the presence of a quencher, the decay is accelerated and the rate constant k_q is obtained from a plot of k_{obs} versus the concentration of the quencher (Eq. 7). Carbenes that contain a UV chromophore (e.g., DPC) are readily observed and their decay kinetics during the reaction can be readily followed by LFP.

$$k_{obs} = k_o + k_q[Q] \tag{7}$$

However, carbenes such as alkylcarbenes that contain no chromophore are generally transparent in the most useful UV region. Those spectroscopically invisible carbenes can be monitored by the ylide method. Here, the carbenes react with a nucleophile Y: competitively with all other routes to decay to form a strongly absorbing and long-lived ylide (Scheme 9.5). In the presence of an additional

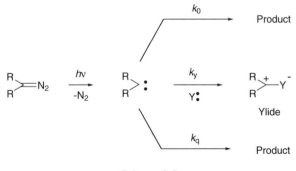

Scheme 9.5

quencher (q), the observed pseudo-first-order rate constant for ylide formation is given by Eq. 8. A plot of k_{obs} versus the concentration of the quencher at the constant concentration of the nucleophile will provide k_q. If the growth rate of ylide absorption is too fast to be monitored, relative rates can still be obtained by a Stern–Volmer approach (Eq. 9). In this case, the yield of ylide, measured as the change in optical density (ΔOD), decreases in the presence of a carbene quencher.

$$k_{obs} = k_o + k_y[Y:] + k_q[Q] \tag{8}$$

$$\frac{\Delta OD_0}{\Delta OD} = 1 + \frac{k_q[Q]}{k_y[Y:]} = 1 + k_q\tau[Q] \tag{9}$$

By plotting $1/\Delta OD$ as a function of [Q], the ratio of $k_q/k_y[Y:]$ can be derived. This ratio corresoponds to the $k_q\tau$ term of the Stern–Volmer equation. Here τ is the lifetime of the carbene in the absence of quenchers.

When monitoring the transient due to triplet carbenes is difficult because of the inherent weak nature of the bands and/or severe overlapping with the absorption bands of the parent diazo compounds, it is more convenient to follow the dynamics of the triplet carbene by measuring the rate of the products formed by reaction of triplet carbenes with quenchers such as radicals (Section 5.3) and carbonyl oxides (Section 6.5). In this case, note that the observed rate constant (k_{obs}) of a triplet carbene reaction is the sum of the decay rate constants of the triplet. These may include decay via an associated but invisible singlet with which the triplet is in rapid equilibrium. Thus in general,

$$k_{obs} = k_T + k_S K \tag{10}$$

where k_T and k_S are singlet and triplet rate constants and K is the equilibrium constant: the ratio of the singlet equilibrium population to that of triplet. Unfortunately, k_{obs} cannot be dissected further by LFP and this process must be accomplished by product analysis.

3.2.2. Time-Resolved IR Spectroscopy.

More recently, time-resolved IR (TRIR) experiments have been used to characterize species with lifetimes of micro- and even nanoseconds.[39] Since IR spectra provide structural information in more detail than UV, this technique will be more powerful than TRUV–vis if one can find a carbene that can be detected and studied by this technique. To date, however, only one carbene has been studied by using TRIR. The matrix IR study shows that the planar triplet and twisted singlet states of 2-naphthyl(methoxycartbonyl) carbenes (NMC, **17**) show distinctly different IR bands (see Section 3.1.4). Both [1]NMC and [3]NMC are detected by TRIR in solution and their kinetics have been studied. Such experiments provide clear cut data for the reaction kinetics as well as energetics of both states (see Sections 4.2 and 4.3).[40a,b]

4. EXPERIMENTAL ESTIMATION OF S–T GAPS

One of the main issues in carbene chemistry is to measure the S–T gap, especially in large systems such as DPCs. We have seen that high-level theory can predict the S–T gap of these large systems fairly precisely. However, these computations mostly estimate values in the gas phase. It has been shown that the S–T gap is sensitive to the solvent polarity (see Section 4.3). Therefore, accurate measurements of carbene singlet–triplet energy gaps are both fundamentally and mechanistically very important. As the absolute rate constants for the reactions involving carbenes are available, attempts to estimate the gap become a more tempting issue. In this section, we will see how those values are determined.

4.1. Preequilibrium Mechanism

The quenching of a triplet carbene reaction with methanol is frequently used as the standard means of probing the singlet–triplet gap. It is widely believed that singlet carbenes insert readily into the O—H bonds of methanol, while the triplet states undergo hydrogen abstraction from the C—H bonds.[41] The behavior of diarylcarbenes appears to violate this rule.

The LFP of diphenyldiazomethane (DDM) in a variety of solvents produces triplet diphenylcarbene (^{3}DPC, ^{3}**14a**), whose transient absorption is readily monitored. The optical absorption spectrum of ^{3}DPC is quenched by methanol and yields the product of O—H insertion, suggesting that ^{3}DPC is quenched by the O—H bond of methanol. The quenching rate constant (k_T) is determined to be $6.8 \times 10^6 \ M^{-1}s^{-1}$ in benzene.[37]

The results are interpreted in terms of a preequilibrium mechanism[42] in which ^{3}DPC equilibrates with ^{1}DPC before it can interact with O—H bond to give benzhydryl methyl ether. According to this mechanism, the observed rate constant for the reaction of ^{3}DPC with methanol is

$$k_{obs} = \frac{k_{TS}k_S}{k_{ST} + k_S[\text{MeOH}]} \tag{11}$$

At relatively low concentration of methanol, with $k_{ST} > k_S$ [MeOH], k_{obs} reduces to

$$k_{obs} = \frac{k_{TS}}{k_{ST}}k_S = k_S K \qquad K = \frac{k_{TS}}{k_{ST}} = \frac{[\text{S}]}{[\text{T}]} \tag{12}$$

If one assumes that the rate constant of ^{1}DPC with methanol is at the diffusion controlled rate ($k_S = 5 \times 10^9 \ M^{-1}s^{-1}$), measurement of k_{obs} immediately yields the lower limit of K to be 0.002. Since $\Delta G_{ST} = -RT \ln K$, the maximum free energy separation between ^{1}DPC and ^{3}DPC can be calculated to be 3.9 kcal/mol. If one assumes that the only entropic difference between the spin states is due to multiplicity ($R \ln 3$), then ΔH_{ST} is estimated to be 3 kcal/mol for DPC in benzene.

This approach is further elaborated by measuring unknown rate constants, k_{ST}, k_{TS}, and k_S.[43] Since only the triplet ground state of DPC could be observed spectroscopically, as the singlet is invisible, estimations of ΔG_{ST} are obtained with a combination of pico- and nanosecond TRUV–vis measurements, chemical quenching, and triplet sensitized experiments. Thus, picosecond laser induced fluorescence is used to measure the rate of formation of ^{3}DPC upon photolysis of DDM and, hence, the singlet–triplet interconversion rate constant $[k_{ST} = (3.22 \pm 0.23) \times 10^9 \text{ s}^{-1}$ in acetonitrile]. Methanol and isoprene are then used as selective ^{1}DPC and ^{3}DPC traps, respectively, to determine other rate constants. The rate constant of ^{3}DPC with isoprene is obtained by nanosecond LFP by monitoring the decay of the transient signal due to ^{3}DPC as a function of [isoprene]. Finally, the rate constants k_S and k_{TS} are estimated by competitive quenching experiments obtained by photolysis of two different carbene precursors in the presence of both isoprene and methanol. Kinetic and thermodynamic parameters obtained from the competitive quenching experiments are in good agreement under the various quenching conditions. The values of ΔH_{ST} are sensitive to solvent polarity and are much larger in isooctane (3.8–4.1 kcal/mol) than in acetonitrile (2.2–2.8 kcal/mol) (see Section 4.3).

This method is criticized later (see Section 4.2), but has been employed to estimate ΔG_{ST} for several diarylcarbenes systematically (Table 9.5). For example, widening of the carbene angle is predicted to enhance ΔG_{ST}. Opening of the angle is achieved by sterically crowding the ortho position of diphenylcarbenes.[44] The quenching rate constant by methanol is shown to decrease as more methyl groups are introduced at the ortho position of diphenylcarbenes (**19a–c**). The ΔG_{ST} parameter is estimated to increase on going from DPC to (2,4,6-trimethylphenyl)-(2-methylphenyl)carbene (**19b**). Methanol (1.2 M) fails to quench the transient absorption of triplet dimesitylcarbene (**19c**), indicating a triplet quenching rate constant of $<2 \times 10^3 \ M^{-1}\text{s}^{-1}$. The quenching rate constant of the singlet state by methanol is estimated by a Stern–Volmer study of the yield of triplet signal as a function of methanol concentration, to be $1.4 \times 10^9 \ M^{-1}\text{s}^{-1}$ (this value is to be compared with the rate constant of ^{1}**14a** of $1.3 \times 10^{10} \ M^{-1}\text{s}^{-1}$ under the same conditions). Apparently, steric hindrance slows down the singlet reaction, but not enough to account for the observed lack of triplet carbene reactivity with methanol. These results indicate that there is a large ΔH_{ST} and that k_{TS} is comparatively small. From the forward rate constant of intersystem crossing ($k_{ST} = 1.05 \times 10^{10} \ M^{-1}\text{s}^{-1}$), the lower limit of the S–T free energy can be estimated to be 8 kcal/mol for dimesitylcarbene (**19c**).

Conversely, decreasing the bond angle by bringing the rings closer should destabilize the triplet. This aim is achieved by producing a series of cyclophane diarylcarbenes in which phenyl rings are linked by an alkyl chain of 9–12 methylene units (**20**).[45] The bond angle is contracted by shortening the tether in these cyclophane diarylcarbenes. As one may expect, k_{obs} jumps by a factor of 320 upon decreasing the bridge length from 12 to 9. Assuming that k_S is diffusion controlled, K is thought to be the sole determining factor for the variation in the observed rates with chain length.

TABLE 9.5. Rate Constants for Reactions of Triplet Carbenes with Methanol and ΔG_{ST}

Carbenes				Solvent[a]	$k_q K$ $(M^{-1} s^{-1})$	ΔG_{ST} (kcal/mol)	Reference
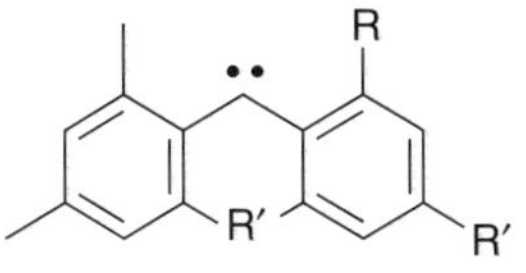 **19**		R	R'				
	a : H	H		i-C$_8$H$_{18}$	8.2×10^4	6.5	44
	b : Me	H		i-C$_8$H$_{18}$	2.1×10^4	7.3	
	c : Me	Me		c-C$_5$H$_{10}$	$< 2.0 \times 10^3$	~8	
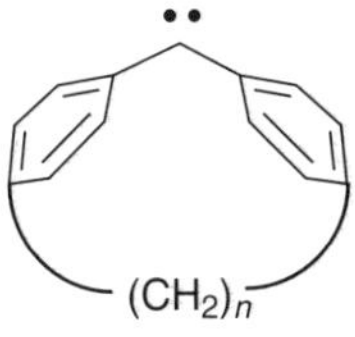 **20**		n					
		12		THF	2.0×10^8	1.9	45
		11			2.1×10^8	1.8	
		10			6.8×10^8	1.2	
		9			2.0×10^9	0.5	
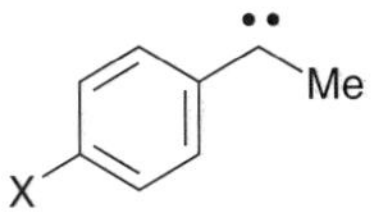 **21**		X					
		H		n-C$_7$H$_{16}$	2.00×10^7	2.0	46
		Br			6.61×10^6	3.0	
		OMe			1.01×10^8	1.4	
		CF$_3$			2.56×10^6	3.5	
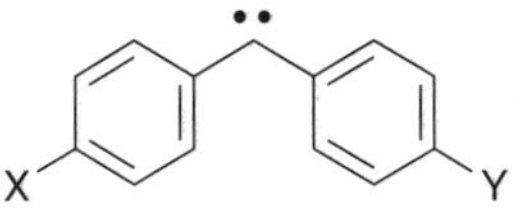 **14**		X	Y				
	a	H	H	c-C$_6$H$_{12}$	1.2×10^7	3.6	47
	l	Ph	H		7.9×10^6	3.8	
	m	CO$_2$Me	H		1.3×10^6	4.9	
	d	CN	H		8.7×10^5	5.1	
	n	Me	H		3.6×10^7		
	o	Me	Me		1.0×10^8		

[a]Isooctane = i-C$_8$H$_{18}$, c-C$_5$H$_{10}$ = cyclopentane, n-C$_7$H$_{16}$ = n-heptane, c-C$_6$H$_{12}$ = cyclohexane.

Theory predicts that electron-donating substituents in the para position of phenyl-carbene should stabilize the singlet, thereby reducing the magnitude of ΔG_{ST}, while electron-withdrawing substituents should enhance it. The prediction has been tested by using the pyridine probe method. Thus methylphenylcarbenes (**21**) having a series of para substituents (Br, OMe, and CF$_3$) are generated in the presence of pyridine and the apparent bimolecular rate constant of reaction of the carbene with pyridine (k_{py}^{app}) is measured.[46] If only the singlet carbene reacts with pyridine, the rate constant can be equated to the equilibrium constant K as

$$k_{py}^{app} = k_{py} K \qquad (13)$$

If k_{py} for ^{1}**21** is assumed to be 1×10^9 $M^{-1}s^{-1}$, the equilibrium constant (K) and free energy gap (ΔG_{ST}) can be deduced as shown in Table 9.5.

Similar attempts to estimate ΔG_{ST} have been carried out for p-substituted DPCs (**14**) by measuring the rate constant with methanol.[47] The data are in line with the predictions. Modest retarding rate effects are found with π acceptor substituents in the 4-position of DPCs. Such groups help to delocalize an unpaired electron of the triplet carbene, while destabilizing the singlet state. Conversely, electron-donating substituents accelerate the rate of triplet carbene reaction with methanol (Table 9.5).

Effects of electronic factors on ΔG_{ST} have been examined more systematically for a series of cyclic aromatic carbenes incorporated into a presumably planar ring of five or six atoms. The carbene bond angle in those examples is not expected to change much and the diversity of chemical behavior must then be primarily associated with electronic changes. The parameter ΔG_{ST} of each carbene has been estimated from the rate constants for reactions of the electronic states and are summarized in Table 9.6.[48]

TABLE 9.6. Equilibrium Constant (K_{eq}) and ΔG_{ST} Estimated for Cyclic Aromatic Carbenes

Carbenes[a]	Solvent[b]	$K_{eq}{}^{c}$	ΔG_{ST} (kcal/mol)	Reference
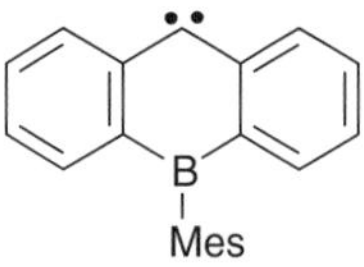 **22** (BA)	PhH	≥ 4000	≥ 5.2	49
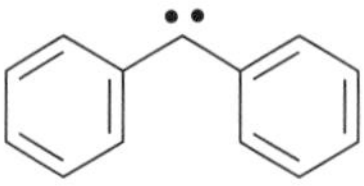 **14a** (DPC)	c-C$_6$H$_{12}$	200	3.2	51
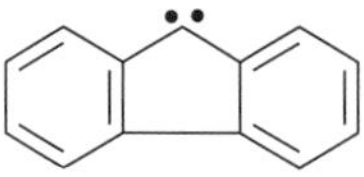 **23** (FL)	MeCN	20	1.9	52
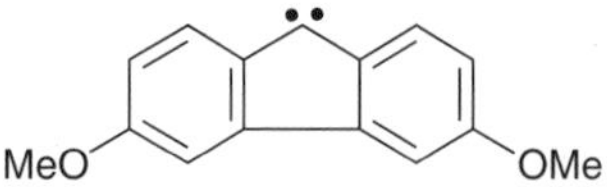 **23a** (DMFL)	PhH	~ 0.03	-2	53
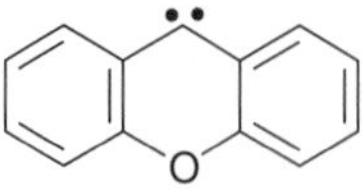 **24** (XA)	n-C$_5$H$_{12}$	≤ 0.0002	≤ -5	50

[a]Mesylate = Mes, FL = fluorenylidene, DMFL = dimethoxyfluorenylidene.
[b]Cyclohexane = c-C$_6$H$_{12}$, n-C$_5$H$_{12}$ = n-Pentane.
[c]$K_{eq} = k_{ST}/K_{TS}$.

For example, the reaction of mesityl(bora)anthrylidene (BA, **22**) with 2-propanol gives only a 17% yield of an O–H insertion product. The remainder of the products arise from radical-recombination processes attributable to the triplet carbene. No ether is formed when the diazo precursor is photolyzed in the presence of triplet sensitizer. The absolute rate constant of the reaction of ^{3}BA with 2-propanol obtained from LFP is $3.4 \times 10^6 \ M^{-1}s^{-1}$, which is much smaller than that of DPC. These observations mean that ^{1}BA is not easily accessed from ^{3}BA, probably because of a large ΔG_{ST}. Although k_{ST} and k_{TS} could not be directly obtained, an estimate is possible based on the available data. The fact that sensitized photolysis give no ether product indicates that $k_{TS} < k_T$ [2-propanol]. The latter term is obtained by LFP and thus $k_{TS} < 2 \times 10^6 \ s^{-1}$. From a Stern–Volmer plot of ether yield versus 2-propanol concentration, k_{ST}/k_S is obtained from the slope. Coupling this result with a rise time of <100 ps for ^{3}BA limits k_{ST} to at most $1 \times 10^{10} \ s^{-1}$ and yields a rate constant of $k_S = 6.6 \times 10^8 \ M^{-1}s^{-1}$ for reaction of ^{1}BA with 2-propanol. This gives $K > 4,000$, or $\Delta G_{ST} > 5.2$ kcal/mol.[49]

On the other hand, the reaction of xanthylidene (XA, **24**) with *tert*-butyl alcohol gave an ether almost exclusively. An LFP study showed that the reaction proceeded at a nearly diffusion-controlled rate ($3.4 \times 10^9 \ M^{-1}s^{-1}$). In contrast to the behavior of BA, triplet sensitized formation of XA in the presence of alcohol gives exactly the same results as does the direct irradiation. More interestingly, XA does not react measurably with O_2. The lifetime of ^{1}XA is the same in O_2 saturated cyclohexane as it is in solutions that have been deoxygenated. If ^{3}XA were in rapid equilibrium with ^{1}XA, then O_2 should shorten the apparent lifetime of ^{1}XA by reacting with the triplet. These observations, coupled with the fact that XA generated in matrix at low temperature shows no EPR signals, suggest that XA has a singlet ground state with the triplet significantly higher in energy. An estimate of ΔG_{ST} for XA can be obtained based on its reactivity with O_2. With the knowledge that triplet carbenes react irreversibly with O_2 with a rate constant close to the diffusion limit (k_{diff}), K_{eq} can be expressed as

$$K_{eq} = \frac{k_{diff}[O_2]}{k_{obs}^{O_2} - k_{obs}^{N_2}} \tag{14}$$

where $k_{obs}^{O_2}$ is the rate constant for the reaction of ^{1}XA in oxygen-saturated solution, and $k_{obs}^{N_2}$ is the rate constant in the absence of O_2. Since no significant difference in these rate constants can be detected, a limit, set in part by the experimental uncertainty, indicates $K_{eq} > 1 \times 10^4$. This value corresponds to $\Delta G_{ST} < -5$ kcal/mol.[50]

A similar experimental determination of ΔG_{ST} has been carried out for a series of cyclic diarylcarbenes, and results are summarized in Table 9.6.[49–53] The observations clearly support the theoretical prediction that electron-donating and electron-withdrawing groups have an opposite influence on the magnitude of ΔG_{ST}. For example, for BA(**22**), which is at one extreme among those carbenes studied, the aromatic LUMO is significantly lowered by the presence of the vacant aromatic orbital of the boron, and thus mixed with the carbene p orbital. This interaction lowers the energy of this nonbonding orbital, thus resulting in an increase in ΔG_{ST}. An analogous explanation is applied for analysis of XA (**24**), another extreme carbene, where the occupied aromatic orbital is raised by the electron-donating ability of oxygen and is mixed with the carbene p orbital. In this case, the splitting of the orbitals in the mixed state is sufficiently increased beyond that of a prototype carbene to make ΔG_{ST} negative. Other aromatic carbenes (**14**, **23**, and **23a**) fall into intermediate positions in this range.

4.2. Surface-Crossing Mechanism

For triplet ground-state carbenes with relatively small ΔG_{ST}, chemistry often arises from the higher lying, but more reactive singlet carbene. As mentioned above, this behavior is explained usually in terms of a preequilibrium mechanism, in which the equilibrium between singlet and triplet carbenes is rapid relative to reaction from either multiplicity. This mechanism predicts that the observed barrier (E_a) for a singlet carbene reaction is given by the actual activation barrier of the reaction ($\Delta H^{\ddagger} \sim 1\text{--}3$ kcal/mol) plus the energy required to populate the singlet from the lower energy triplet carbene ($\Delta H_{ST} = 4.2$ and 2.5 kcal/mol in isooctane and aceto-nitrile, respectively). However, the activation energy for the reaction of ^{3}DPC with methanol is measured to be 3.61 and 1.66 kcal/mol in isooctane and acetonitrile, respectively, which are smaller than that predicted by the preequilibrium mechanism.[54]

There are a several explanations for this discrepancy. One resolution would be if the entropic difference between ^{1}DPC and ^{3}DPC is larger than $R\ln3$. The polar nature of singlet carbene could result in a more ordered solvent. Any increase in ΔG_{ST} would reduce ΔH_{ST}. Moreover, k_{ST} was not measured under the actual conditions used to determine ΔG_{ST}.

However, Griller et al. proposed that ^{3}DPC can react directly with methanol to produce the ether, without intervention of ^{1}DPC.[54] This behavior is involves a sur-face crossing mechanism in which triplet carbene can react directly with alcohols, with surface crossing occurring after the carbene has begun to interact with the O—H bond. In the surface crossing mechanism, the triplet carbene surface crosses the singlet carbene $\rightarrow$ product surface at a point below the energy of the singlet carbene, leading to an observed E_a that is lower than the sum of $\Delta H^{\ddagger}$ and ΔH_{ST}.

Therefore it is highly desirable to observe and monitor simultaneously the trans-ient absorption bands due to both singlet and triplet states. This goal is realized in the TRIR studies of 2-naphthyl(methoxycartbonyl)carbenes (**17**), in which both ^{1}**17** and ^{3}**17** are detected in solution and their kinetics studied. Thus, the rate of decay of both a triplet carbene signal (1650 cm^{-1}) and a singlet carbene signal (1584 cm^{-1}) is monitored as a function of methanol concentration. Note that in this case, quenching rate constants observed at 1650 cm^{-1} (^{3}**17**) and 1584 cm^{-1} (^{1}**17**) for the reagent corresponds to that of each state. The rate constants are essentially the same within experimental error.[40a] This behavior is the first direct evidence to support a preequilibrium mechanism. The rate constants observed for the reaction with 2,3-dimethyl-2-butene are also the same, indicating that alkene also reacts with an already equilibrated singlet/triplet mixture of carbene **17**.

Previous studies with the TRUV method to estimate ΔG_{ST} have employed a combination of product studies and kinetic measurements and have been based on assumptions concerning the spin-selective reaction of carbenes. However, the TRIR study with **17** allows a direct experimental estimate of a carbene singlet–triplet gap in solution for the first time. In this case, since the relative intensities of singlet and triplet carbene are directly related to the concentration of each state, an equilibrium constant and free energy difference can be directly derived. The well-separated signals observed at 1650 (^{3}**17**) and 1584 cm^{-1} (^{1}**17**) are used to estimate the ratio of ^{3}**17**/^{1}**17**, which is determined to be 2.1 at 21 °C in Freon-113 solution. This value leads to an equilibrium constant of 1.4 ± 0.2 at 21 °C and a free energy difference of 0.2 ± 0.1 kcal/mol, with the triplet lower in energy.

4.3. Effect of Solvent on the S–T Energy Gap

This value is significantly lower than that calculated by DFT theory ($\Delta H_{ST} = 4.52$ kcal/mol). It has been shown that singlet DPC is stabilized relative to triplet DPC in polar solvent, presumably as a result of the zwitterionic nature of singlet carbenes.[55] The singlet state is polar and will be stabilized in polar solvent whereas the less polar triplet will not experience such stabilization. Thus, it is expected that the singlet–triplet energy gap will decrease as the solvent polarity increases. A possible explanation for the difference between the calculated (gas phase) and experimental (Freon-113 solution) value of ΔH_{ST} is that the singlet is preferentially stabilized in solution. This idea has been examined by generating the same carbene in four solvents (hexane, Freon-113, methylene chloride, and acetonitrile) of different polarity.[40b] The ratio of ^{3}**17**/^{1}**17** decreases as the solvent polarity increases, and ΔG_{ST} decreases from 0.3 to -0.3 kcal/mol on going from hexane to acetonitrile (Table 9.7). There is a good correlation between ΔG_{ST} and the solvent polarity

TABLE 9.7. Experimental^a and Theoretical^b Thermochemical Parameters for 2-Naphthyl(methoxycarbonyl)carbene (17)[40b]

Solvent	ΔG_{ST} (kcal/mol)	ΔH_{ST} (kcal/mol)	ΔS_{ST} (cal/mol · K)
n-C$_6$H$_{14}$	0.3 ± 0.09^a (0.4 ± 0.1) $[0.9]^b$	0.8 ± 0.3^a $[1.4]^b$	1.5 ± 1.3^a $[1.7]^b$
Freon-113	0.2 ± 0.1^a $[2.2]^b$	0.7 ± 0.8^a $[1.3]^b$	1.7 ± 3.0^a $[-2.9]^b$
CH$_2$Cl$_2$	0.1 ± 0.1^a (-0.06 ± 0.1) $[-1.1]^b$	0.2 ± 0.6^a (-0.2 ± 0.7) $[0.5]^b$	-0.3 ± 2.4^a (-0.5 ± 2.7) $[5.2]^b$
MeCN	$(-0.2 \pm 0.08)^a$ $[-0.7]^b$	$(-0.4 \pm 0.6)^a$ $[-0.6]^b$	$(-1.5 \pm 2.5)^a$ $[0.3]^b$

^{a}Data obtained in the 1680–1540-cm^{-1} spectral region. The data in parentheses are those obtained in the 1260–1140 cm^{-1} spectral region.
^{b}The values shown in [ ] calculated at B3LYP/6-311 + G**//B3LYP/6-31G* level using the PCM solvation model and corrected to 298 K.

parameter $E_T(30)$, indicating that the specific carbene–solvent complexes do not play a significant role in determining carbene stability for the solvent used.

Furthermore, variable-temperature experiments allow investigation of the enthalpy (ΔH_{ST}) and entropy (ΔS_{ST}) difference between ^{3}**17** and ^{1}**17**. The results indicate that ^{3}**17** is favored enthalpically in hexane ($\Delta H_{ST} = 0.8 \pm 0.3$ kcal/mol), but ^{1}**17** is favored in acetonitrile ($\Delta H_{ST} = -0.4 \pm 0.6$ kcal/mol). The data are in good agreement with calculations using the PCM (the polarizable continuum model) solvation model at the B3LYP/6-311 + G**//B3LYP/6-31* level ($\Delta H_{ST} = 1.4$ and -0.6 kcal/mol in hexane and acetonitrile, respectively). These values indicate that both solvent and geometrical effects on the entropic difference between singlet and triplet carbene should be considered when assessing energetic difference between carbene spin states.

The rate of formation of ^{3}DPC upon photolysis of diphenyldiazomethane (**30**), and hence the singlet–triplet interconversion rate constant (k_{ST}) is measured also by picosecond laser-induced fluorescence in several solvents of different polarity.[56] The intersystem crossing rate constant depends strongly on the solvent polarity, the rate being larger in the less polar solvents. Again a linear correlation exists between the logarithm of the rate and the empirical solvent polarity parameters, $E_T(30)$. Since intersystem crossing depends on the energy gap, this correlation is interpreted as indicating that as the solvent polarity increases, ΔG_{ST} decreases and k_{ST} increases.

Such solvent stabilization effects on the reactivity of singlet carbenes in equilibrium with their triplet ground states have also been observed experimentally in other arylcarbene derivatives.[57]

5. HYDROGEN ATOM ABSTRACTION

5.1. Product Studies

Triplet carbenes have a singly occupied p orbital, as is the case for radicals, and hence react like those radicals. Hydrogen atom transfer reactions are fundamental reaction pathways of triplet carbenes. The reaction of a triplet carbene with a hydrocarbon is quite analogous to the free radical hydrogen atom transfer process (Scheme 9.6).

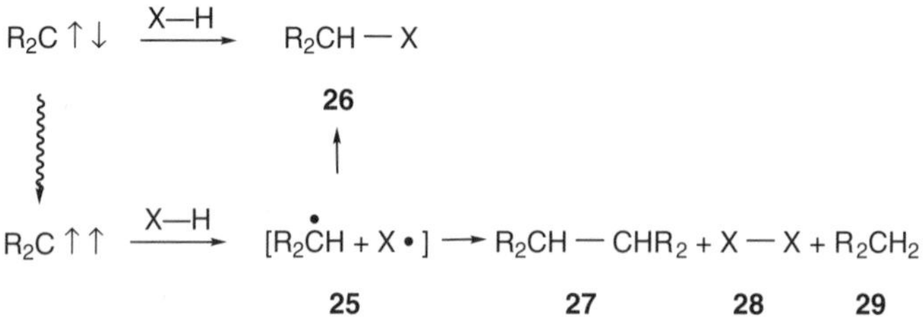

Scheme 9.6

When generated in a hydrogen-donating solvent such as a hydrocarbon, the triplet abstracts a hydrogen atom. This reaction results in generation of radical pairs (**25**) with triplet multiplicity. In fluid solution the members of the radical pairs generated by hydrogen abstraction by the triplet carbenes from the solvent usually diffuse apart to form a product mixture consisting of a C—H insertion product (**26**), radical recombination dimers (**27**, **28**), and double-hydrogen abstraction products (**29**). Therefore, the "C—H insertion" product formed from triplet states is usually accompanied by a large amount of "escaped products" (**27–29**).

For example, in the photolysis of (**30**) in toluene solution, the product of insertion of DPC into the benzylic C—H bonds, 1,1,2-triphenylmethane (**31**), was accompanied by substantial amounts of 1,1,2,2-tetraphenylethane (**32**) and bibenzyl (**33**).[58a] When solvents such as cyclohexane are used, tetraphenylethane (**32**) is formed as the major product, indicating that direct C—H insertion in the singlet state is not the main process in most diarylcarbenes (Scheme 9.7).[59,60] In contrast, 9-cyclohexylfluorene (**37**) is produced by photolysis of diazofluorene (**36**) in cyclohexane as a main product (65%) along with a small amount of escaped products (**38** and **39**). One can estimate in this case that at most 14% of **37** arises from free radical processes. Similarly, direct or sensitized photolysis of diazomalonate in 2,3-dimethylbutane gives C—H insertion products, but in the triplet-sensitized

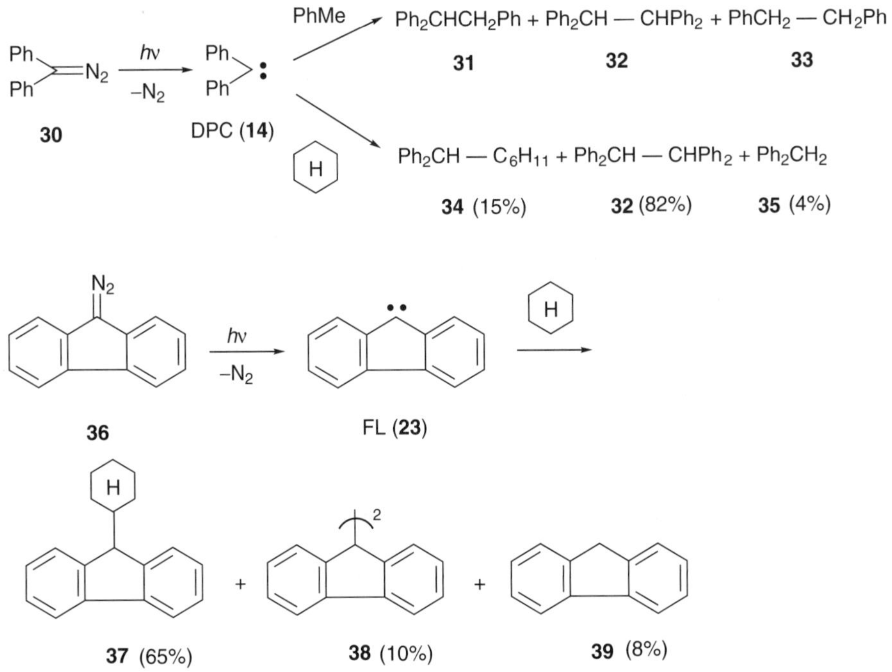

Scheme 9.7

photolysis, radical dimer and double-hydrogen abstraction product rather than the insertion product were the major compounds formed.[61]

A simple and straightforward way to distinguish between a direct insertion process and one going through free radicals is by a cross-over experiment. Cyclohexane and cyclohexane-d_{12} are often used as the probe for crossover, and hence the reactive multiplicity of the carbenes in question (Scheme 9.8).

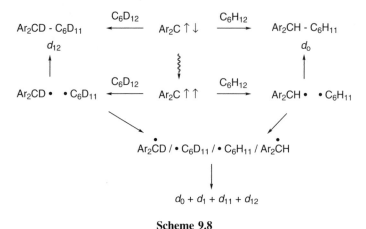

Scheme 9.8

Thus, if a mixture of these hydrocarbons reacts with a singlet state, then the addition product will consist entirely of undeuterated (d_0) and deuterated(d_{12}) compounds. If some of the reaction occurs by the hydrogen abstraction–recombination route, then some crossed products (d_1 and/or d_{11} compounds) will be formed. By using this technique, it has been shown that, while some of the DPC-cyclohexane adduct (**34**) is formed by combination of radical pairs, there is no evidence of crossover product present in the FL–cyclohexane adduct (**37**).[59] The results are interpreted by assuming rapid spin state equilibration relative to the reaction of either spin state with solvent. The larger amount of singlet chemistry of FL relative to DPC can then be explained if the S–T gap is smaller in FL than in DPC.

Other differences between singlet (concerted) insertion and triplet (abstraction–recombination) carbene "insertion" are seen in selectivity, stereochemistry, and the kinetic deuterium isotope effect. The triplet states are more selective in C—H insertion than the singlets. For example, the triplet shows higher tertiary to primary selectivity than the singlet in the insertion reaction with 2,3-dimethylbutane. Singlet carbene is shown to insert into C—H bond with retention of configuration, while racemization is expected for triplet insertion reaction from the abstraction–recombination mechanism. For example, the ratios of diastereomeric insertion product in the reaction of phenylcarbene with *rac*- and *meso*-2,3-dimethylbutanes are 98.5:1.5 and 3.5:96.5, respectively.[62]

It is generally observed that the isotope effect for abstraction of hydrogen by a triplet carbene is $\sim$2–9, whereas insertion into a carbon–hydrogen bond by a singlet carbene proceeds with an isotope effect of $\sim$1–2.[49,53,63]

It is generally accepted that a singlet carbene undergoes a C—H insertion (in a concerted manner) with a triangular transition state, while a triplet insertion favors a linear transition state in its abstraction of H. For example, in the reaction of 2-(*o*-methylenephenyl)bicyclo[2.2.1]heptane (**40**), insertion occurs predominantly into the *exo*-C—H bond, giving the strained product (**41**, Scheme 9.9). This result indicates that the carbene (singlet) can approach the *exo*-C—H bond from the side, but not from the endo side.[64]

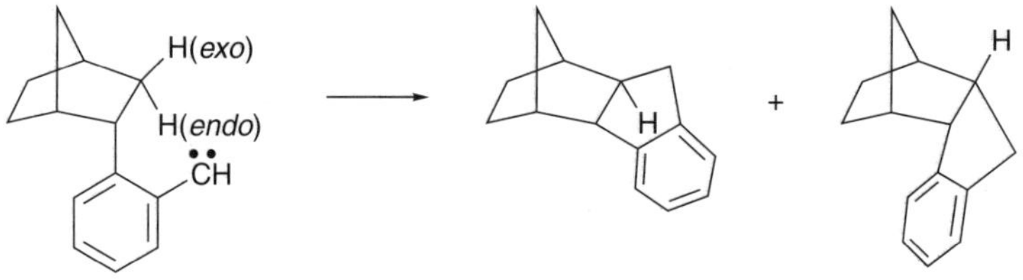

40   **41 (50%)**   **42 (8%)**

Scheme 9.9

The difference in the TS between the singlet and triplet insertion becomes crucial in intramolecular processes, in which approach to a C—H bond is restricted by ring strain. Photolysis of *o*-(*n*-butyl)phenyldiazomethane gives five- and six-membered ring compounds as the major products, both of which are thought to originate from the singlet carbene.[64] However, a similar reaction of optically active [2-(1-methylpropyl)oxy]phenylcarbene (**43**) affords a five-membered product (**44**) almost exclusively, but with the significant loss of enantiomeric purity (~30% ee) (Scheme 9.10). This result is explained in terms of hydrogen abstraction by

43   **44 (~30% ee)**

Scheme 9.10

triplet followed by recombination of the radical pairs, rather than a concerted insertion from the singlet. Presumably, the concerted insertion reaction of the singlet involves the *p* orbital as well as the σ orbital of the carbene. For the interaction of both orbitals with δ-C—H bonds, rotation must occur about the bond connecting the divalent carbon to the benzene ring, with concomitant loss of benzylic stabilization and deformation of bond angles. In contrast, transfer of the δ hydrogen to the half-filled, in-plane σ orbital of the triplet carbene can proceed by way of a favorable six-membered transition state in which the benzylic resonance is not disturbed.[62]

5.2. Chemically Induced Dynamic Nuclear Polarization

Chemically induced dynamic nuclear polarization (CIDNP)[65] is a very powerful tool for establishing the existence of radical pair intermediates and their spin. CIDNP has reinforced the view that singlet carbene undergoes direct insertion into C—H bonds and that the triplet abstracts hydrogen.

When radical pair reactions or reactions involving diradicals are carried out in the NMR probe, CIDNP produces strong emission and abnormally intense absorption signals in nuclear magnetic resonance (NMR) spectra. These abnormal signal intensities are associated with deviation from the Boltzmann distribution of the nuclear spin states. The spins of the nuclei attached to radical pairs or diradicals can promote or inhibit the probability of bond formation upon encounter because of their effect on the probability that these species will be in the singlet electronic spin state. The products of the radical pair or diradical reactions preserve the nuclear spins of their precursors long enough for the spin state population to be measured and recorded.

The radical pair theory of CIDNP depends on the assumption that only singlet radical pairs couple to give stable products. If a triplet pair is formed first, some will cross to singlets and couple before the pair correlation is lost by diffusion. If a singlet pair is formed initially, some will combine directly, some will be separated by diffusion, and some will cross to triplets, thereby increasing their prospects for diffusive separation. The electronic magnetic moments of the triplets will couple with the nuclear magnetic moments of nearby protons, producing complex magnetic quantum states. Those having the same symmetry as the singlet radical pairs will undergo the most rapid singlet–triplet transitions. Bond formation to produce a stable product leaves a nonequilibrium distribution of nuclear spin states in both the reaction product and the remaining radical pairs that may dissociate and react in other ways. As a result, the sense of nuclear polarization in products of combination of geminate radical pairs will depend on whether the pair was born as a singlet or a triplet.

The signs of net polarization ($\Gamma ne +$ for absorption and $-$ for emission) are given by $\Gamma ne = \mu \varepsilon \Delta g A i$ according to the first-order treatment of Kaptein, where μ, ε, and $A i$ labels indicate the multiplicity of precursor ($-$ for singlet precursor and $+$ for triplet precursor), the type of the product ($+$ for geminate products and $-$ for scavenging products) and the sign of hyperfine coupling of the nuclei i under consideration, respectively. Based on the treatment, one can predict signs of polarization in NMR of the product derived from either the singlet or triplet radical pair (net absorption/A or emission/E for the net effect and EA:emission at the lower field and absorption in the higher field or AE for the multiple effect).[66]

Observation of emission and absorption in the NMR benzylic proton signal from an irradiated solution of diphenyldiazomethane in toluene was the first example showing the importance of the CIDNP technique.[58] Triplet diphenylcarbene generated by photolysis of diphenyldiazomethane in toluene abstracts a hydrogen atom to generate the triplet radical pair, which either recombines to give **31**, or diffuses apart, ultimately to produce dimers of each fragment (**32**, **33**, Scheme 9.7,

Eq. 15). On the other hand, thermolysis of azo compound (**45**) is known to generate the same radical pair in the singlet state and gives the same products in a different ratio (Eq. 16). The ^{1}H NMR spectrum of the coupling product **31** shows that the sense (AE) of polarization observed for the carbene reaction is opposite to that observed (EA) for the thermolysis of azo compound. This result implies that the coupling product **31** in the photolysis is formed by recombination of benzyl and diphenylmethyl generated from a precursor believed to be triplet DPC.

$$Ph_2C \uparrow \uparrow + PhCH_3 \longrightarrow [\overset{\bullet}{Ph_2CH} + \overset{\bullet}{PhCH_2}]^3 \longrightarrow Ph_2CH - CH_2Ph \tag{15}$$

$$\textbf{31}$$

$$Ph_2CH - N = N - CH_2Ph \overset{\Delta}{\longrightarrow} [Ph_2CH \bullet \bullet \bullet CH_2Ph]^1 \longrightarrow Ph_2CH - CH_2Ph \tag{16}$$

$$\textbf{45} \qquad\qquad\qquad\qquad\qquad\qquad \textbf{31}$$

Similar CIDNP experiments with triplet methylene generated by triplet sensitized photolysis in toluene gives a CIDNP signal consistent with the formation of ethylbenzene by in-cage recombination of methyl and benzyl radicals formed from a precursor believed to be triplet methylene (AE polarized ethylbenzene). The CIDNP experiments with the singlet methylene, on the other hand, show no polarized spectrum. The failure to detect a CIDNP signal is interpreted as indicating that no radicals are formed with a lifetime long enough to permit nuclear-spin-dependent intersystem crossing to compete with diffusion out of the cage. This result in turn supports the occurrence of direct insertion by singlet methylene.[67]

The CIDNP experiments have been carried out to demonstrate an intervention of triplet radical pairs in many hydrogen abstraction reactions involving triplet carbenes. The following examples in reactions with halo compounds and ethers are of particular interest.

When a methylene precursor is photolyzed in $CDCl_3$, either by direct or benzophenone triplet-sensitized irradiation, different polarized NMR signals are observed in the two cases. Chlorine atom abstraction by singlet methylene is deduced from the proton signal of Cl_2DC-CH_2Cl (**47**) observed during direct irradiation (Eq. 17). Hydrogen abstraction by triplet methylene is indicated by the enhanced absorption observed for the hydrogens of Cl_3C-CDH_2 (**48**) in the sensitized reaction (Eq. 18).

$$CH_2 \uparrow \downarrow + CDCl_3 \longrightarrow (\overset{-}{CH_2} - \overset{+}{ClCDCl_2}) \longrightarrow [\bullet CH_2Cl \bullet CDCl_2]^1$$

$$\textbf{46}$$

$$\longrightarrow ClH_2C - CDCl_2 \tag{17}$$

$$\textbf{47}$$

$$CH_2 \uparrow \uparrow + CDCl_3 \longrightarrow [\bullet CH_2D \bullet CCl_3]^3 \longrightarrow H_2DC - CCl_3 \tag{18}$$

$$\textbf{48}$$

No CIDNP signals from **47** are observed in the experiment in which polarized **48** is produced. This result indicates that while triplet methylene abstracts hydrogen, singlet carbene abstracts a chlorine atom.[68] The reason why singlet abstracts chlorine atom is explained in terms of formation of an intermediate chloronium ylide (**46**). A similar singlet radical pair formation from singlet carbene is also shown in the reaction with ethers. Thermolysis of diazoacetate in benzyl ethyl ether gives C—O insertion product (**50**). Strongly polarized signals due to the C—O insertion product are observed in ^{1}H and ^{13}C NMR spectra and are explained in terms of a singlet radical pair. The formation of oxonium ylide (**49**) followed by homolysis of C—O bond is proposed (Eq. 19). Again, no CIDNP is observed for the product derived from the same ylide by a nonradical mechanism, such as the Hoffmann type β-elimination of olefin.[69]

$$MeO_2CCH \uparrow\downarrow + PhCH_2OCH_2CH_3 \longrightarrow MeO_2C\overset{-}{C}H-\overset{+}{O}-CH_2Ph$$

$$\underset{\textbf{49}}{} \quad CH_2CH_3$$

$$\longrightarrow [\ MeO_2C\overset{\bullet}{C}HOCH_2CH_3 \bullet CH_2Ph\ ]^1 \longrightarrow MeO_2CCH-CH_2Ph \qquad (19)$$

$$\underset{\textbf{50}}{} \quad OCH_2CH_3$$

On the other hand, in the reaction of methylene with CCl_4, a considerable fraction of Cl_3CCH_2Cl ^{1}H NMR signals is found to be polarized both in direct and triplet-sensitized photolysis. It is deduced that both singlet and triplet methylene appear capable of abstracting chlorine atom from CCl_4 (Eqs. 20 and 21).[70]

$$CH_2 \uparrow\downarrow + CCl_4 \rightarrow [ClH_2C\bullet + \bullet CCl_3]^1 \rightarrow ClH_2C-CCl_3 \qquad (20)$$

$$CH_2 \uparrow\uparrow + CCl_4 \rightarrow [ClH_2C\bullet + \bullet CCl_3]^3 \rightarrow ClH_2C-CCl_3 \qquad (21)$$

5.3. Laser Flash Photolysis Studies

The LFP techniques make it possible to observe directly the hydrogen atom abstraction process by triplet carbenes from appropriate H donors. The techniques also give useful kinetic information.

For example, LFP of diphenyldiazomethane in cyclohexane shows a strong absorption at 314 nm attributable to triplet diphenylcarbene (^{3}DPC). As the 314-nm carbene signal decays, a new species showing a strong absorption band with λ_{max} 334 nm ($\tau = \sim 1\mu s$) is formed. This band is attributable to the diphenylmethyl radical. The decay of triplet DPC at 314 nm kinetically correlates with the growth of the diphenylmethyl radical, indicating that the triplet DPC decays by abstracting a hydrogen atom from the solvent to generate the radical. The growth of the diphenylmethyl radical derived from DPC gives absolute rate constants for reaction of the carbene with cyclohexane ($k = 5.7 \times 10^5$ s^{-1}).[71]

The observed rate constant (k_{obs}) of a triplet carbene reaction is the sum of all decay rate constants of the triplet. These may include decay via an associated but invisible singlet with which the triplet is in rapid equilibrium (see Eq. 10). Since the product studies indicate that radical-derived products are formed almost exclusively in this case, the growth rates reflect essentially pure triplet processes.

The monitoring techniques are used to obtain the absolute rate constants of DPC with other substrates (Q) known to react with triplet carbenes. The experiments are carried out at several reagent concentrations and the experimental pseudo-first-order rate constant, k_{obs}, is plotted against the substrate concentration. It can be shown that k_{obs} is expressed by Eq. 22

$$k_{obs} = k_0 + k[RH] + k_q[Q] \tag{22}$$

where k_q is the rate constant of DPC with the substrate and k_o includes all pseudo-first-order reactions DPC may undergo in any given solvent in the absence of substrates. A plot of the observed pseudo-first-order rate constant of the formation of the radical against [substrate] is linear, and the slope of this plot yields the absolute rate constant for the reaction of ^{3}DPC with the substrate (Table 9.8).[72] Kinetic isotope effects are determined to be 2.3 and 6.5 for cyclohexane and toluene, respectively.

The results reveal the nature of triplet DPC reaction in a quantitative manner. Thus, triplet DPC reacts with cyclohexane and THF at least 100 times as rapidly as does methyl or benzhydryl radical. Triplet DPC is ∼30 times less reactive with THF than is the phenyl radical. Thus, DPC has a reactivity toward hydrogen abstraction intermediate between those of methyl and phenyl radicals.

The activation parameters for hydrogen abstraction by DPC from diethyl ether and toluene have been measured ($E_a = 3.9$ kcal/mol, $A = 10^{7.8}$ M^{-1} s^{-1} and $E_a = 3.6$ kcal/mol, $A = 10^{7.2}$ M^{-1} s^{-1}, respectively). The preexponential factors for reaction of triplet benzophenone with cyclohexane ($10^{8.66}$ M^{-1} s^{-1}),[73] *tert*-butoxy radical with THF ($10^{8.7}$ M^{-1} s^{-1}),[74] *tert*-butyl radical with tri-*n*-butyltin hydride ($10^{8.43}$ M^{-1} s^{-1}),[75] and phenyl with tri-*n*-butyltin hydride (10^{10} M^{-1} s^{-1})[76] are 1 to 3 orders of magnitude larger than those observed for reaction with DPC (**14a**) and DBC (**18**). These values are considerably smaller than expected on the basis of the corresponding data available in free radical hydrogen atom transfer

TABLE 9.8. Rate Constants (at 300 K) and Arrhenius Parameters for the Reaction of DPC (14a) with Various Hydrogen Donor Substrates

Substrate[a]	Solvent[a]	k (M^{-1} s^{-1})	E_a (kcal/mol)	$\log A$ (M^{-1} s^{-1})	Reference
c-C$_6$H$_{12}$	c-C$_6$H$_{12}$	5.7×10^{5b}	2.5 ± 0.4	7.5 ± 0.3^a	71
THF	PhH	$(1.4 \pm 0.2) \times 10^5$			71
c-C$_6$H$_{10}$	c-C$_6$H$_{12}$	$(2.8 \pm 0.1) \times 10^5$			71
1,4-CHD	c-C$_6$H$_{12}$	$(1.0 \pm 0.1) \times 10^7$			71
PhCH$_3$	PhCl	2.89×10^4	3.6 ± 1.7	7.2 ± 1.3	72
Et$_2$O	PhCl	7.07×10^4	3.9 ± 0.6	7.8 ± 0.5	72

[a]Cyclohexane = c-C$_6$H$_{12}$, c-C$_6$H$_{10}$ = cyclohexene, 1,4-CHD = 1,4-Cyclohexadiene.
[b]In reciprocal seconds.

TABLE 9.9a. Rate Constants for the Reaction of Triplet Carbenes with Hydrogen Donor Substrates

Carbenes			Substrate[a]	k^b or (τ^c)	Reference
14	X	X	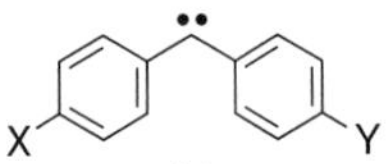 $c\text{-}C_6H_{12}$		47
	a H	H		(1.5)	
	p H	Br		(2.1)	
	q Br	Br		(1.6)	
	r H	Cl		(2.0)	
	s Cl	Cl		(2.1)	
	o Me	Me	(2.0)	(1.2)	
	t Me	CN		(3.8)	
	d H	CN		(5.4)	
	m H	CO_2Me		(5.0)	
	l H	Ph		(7.8)	
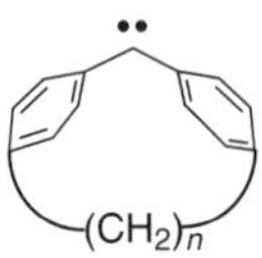 **(20)**	n		1,4-CHD		45
	12			7.3×10^6	
	11			7.0×10^6	
	10			1.2×10^7	
	9			2.3×10^7	

[a] Cyclohexane:$c\text{-}C_6H_{12}$, 1,4-CHD:1,4-cyclohexadiene.
[b] In $M^{-1}\,s^{-1}$. c In s^{-1}.

processes[77] and much lower than the calculated value (7 kcal/mol) for the presumably much more reactive triplet state of methylene. The activation energy (14.3 kcal/mol) for hydrogen atom abstraction by methyl radical from methane[78] is about twice that calculated for the triplet carbene reaction. A contribution of quantum mechanical tunneling (QMT) to the atom transfer is suggested and this view is supported by measuring the temperature dependence of the kinetic isotope effects (see Section 5.4).

The rate constants observed for hydrogen atom abstraction by several other carbenes are summarized in Table 9.9.[79–84] The effect of substituents on hydrogen

TABLE 9.9b. Rate Constants for the Reaction of Triplet Carbenes with Hydrogen Donor Substrates

Carbenes	Substrate[a]	$K(M^{-1}\,s^{-1})$	K_H/K_D	Reference
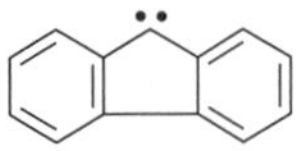 **(23)**	$c\text{-}C_6H_{12}$	$(7.7 \pm 3) \times 10^{7b}$	$1.2 \sim 2.1$	52
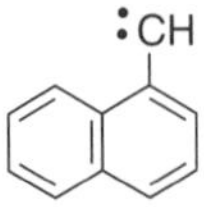 **(α-12)**	$c\text{-}C_6H_{12}$	8.8×10^5	1.3	82

TABLE 9.9b (*Continued*)

Carbenes	Substrate[a]	$K(M^{-1}\,s^{-1})$	K_H/K_D	Reference
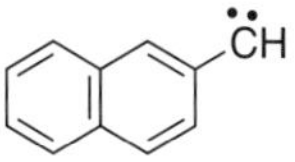 (β-**12**)	c-C_6H_{12}	1.66×10^6	2.6	83
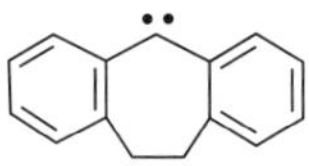 (**18**)	Et_2O	1.3×10^5		72,79
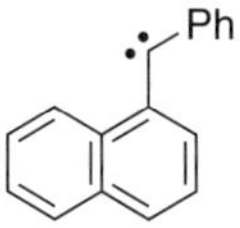 (α-**51**)	1,4-CHD	$(1.1 \pm 0.9) \times 10^6$		52
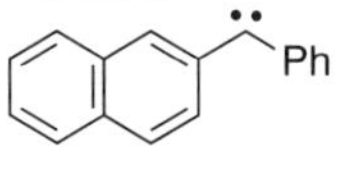 (β-**51**)	1,4-CHD	$(4.6 \pm 0.3) \times 10^6$		80
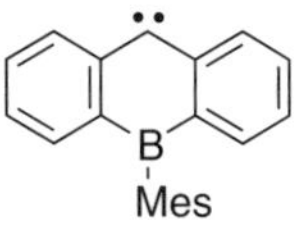 (**22**)	c-C_6H_{12}	$(8.1 \pm 0.8) \times 10^5$	7	49
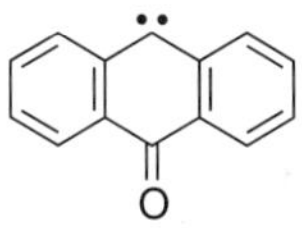 (**52**)	c-C_6H_{12}	$(7.2 \pm 0.5) \times 10^6$	6.5	48,81
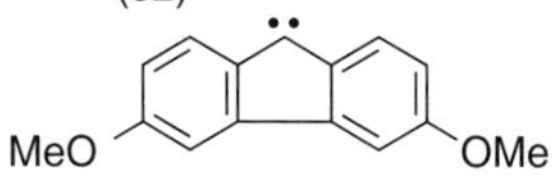 (**23a**)	c-C_6H_{12}	$(7.5 \pm 1.2) \times 10^6$	1.7	53
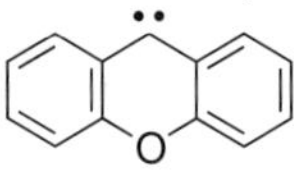 (**24**)	c-C_6H_{12}	$< 1.7 \times 10^2$		50
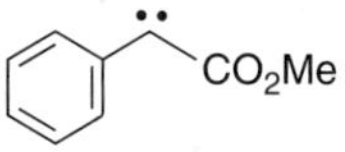 (**53**)	n-C_6H_{12}	2.2×10^7		84

[a]Cyclohexane:c-C_6H_{12}, 1,4-CHD:1,4-Cyclohexadiene.
[b]In $M^{-1}\,s^{-1}$. c In s^{-1}.

atom abstraction by DPC (**14**) from cyclohexane is examined.[47] The results show carbene substitution has a minimal effect on the growth rate of the radical, which is in sharp contrast with that observed for the reaction with methanol, where the rate changes by a factor of $\sim$100 among the substituents studied (see Section 4.1). Only for the cases in which considerable delocalization into substituents is possible are the τ values longer. This difference may represent enhanced stability of the triplet because of delocalization. These observations are in line with the radical nature of the triplet.

Behavior of FL (**23**) toward the C—H bond is in contrast with that of DPC (**14a**). Picosecond LFP of diazofluorene (**36**) in cyclohexane generates a transient absorption band at 470 nm ascribable to FL. As time proceeds, a band at 497 nm appears, while the 470-nm absorption is still present. On a nanosecond time scale, the two bands at 470 and 497 nm are observed and assigned to 9-fluorenyl radical (FLH $\cdot$) ($\tau = 1.4$ ns), indicating that **23** rapidly abstracts a hydrogen atom ($k = 7.7 \times 10^7$ M^{-1} s^{-1}) from cyclohexane to give FLH. Although the growth of the radical can be observed, product analysis indicates that triplet abstraction of hydrogen is a minor pathway in this case (see Section 5.1). Examination of the relative yields of 9-fluorenyl and 9-fluorenyl-9-d in neat cyclohexane and cyclohexane-d_{12} by measuring the absorbance change at 497 nm 100 ns after laser irradiation of **36**, leads to a ratio of yields of $\sim$2. This value is too small to attribute to ^{3}FL and is attributed to ^{1}FL. The small S–T gap ($\Delta G_{ST} = 1.9$ kcal/mol) and high reactivity of FL allow singlet insertion reactions to predominate.[52]

Reaction of dimethoxyFL (**23a**), which is known to have a singlet ground state ($\Delta G_{ST} = -2$ kcal/mol), in cyclohexane supports this idea. These LFP studies in cyclohexane and cyclohexane-d_{12} indicate that a similar small kinetic isotope effect of 1.7 is observed for the decay of the transient absorption bands associated with **23a** and the corresponding radical.[53]

Effects of the magnitude of ΔG_{ST} on the hydrogen atom abstraction reactivity are seen more clearly in the reaction of cyclophane DPCs (**20**) with 1,4-cyclohexadiene. Thus, k_{obs} increases by a factor of 7 as the number of methylene units is decreased and ΔG_{ST} decreases from 3 to 0.5 kcal/mol. This observation is interpreted in terms of some participation of the singlet for those carbenes with sufficient small energy gap.[45]

So, triplet carbenes are classified into two cases in terms of the behavior toward C—H bonds depending on the magnitude of ΔG_{ST}. One is exemplified by DPC and its derivatives in which there are triplet ground states with a rather large ΔG_{ST}. Other carbenes in these groups are dibenzocycloheptadienylidene (**18**, $\Delta G_{ST} = \sim 5$ kcal/mol), BA (**22**, $\Delta G_{ST} > 5.2$ kcal/mol), and anthronylidene (**52**, $\Delta G_{ST} = 5.8$ kcal/mol).

Fluorenylidene and its substituted derivatives are typical examples of the other class. Other carbenes showing similar behavior are monoarylcarbenes such as mononaphthylcarbenes (**12**) and phenyl(methoxycarbonyl)carbene (**53**, $\Delta G_{ST} = \sim 0.2$ kcal/mol).

These observations suggested that the reactivity differences toward C—H bonds are primarily due to changes in the S–T gap (and thus K) and only to a lesser extent to a variation in triplet reactivity. However, this rough guide is not applicable for carbenes that are perturbed by conjugated π systems. For example, the reaction of XA (**24**, $\Delta G_{ST} = < -5$ kcal/mol) with cyclohexane is sluggish and the C—H insertion product is obtained only at elevated temperatures. This reaction is thermally activated and does not compete with other bimolecular reaction of **24** at room temperature. The absence of cross-product in the isotope tracer experiments (Scheme 9.8) shows that the insertion product is not formed by an encounter of free cyclohexyl radical with xanthenyl radical, indicating that the interaction of **24** with hydrocarbons appears to be primarily the result of a singlet rather than a triplet carbene. The reduced reactivity of **24** toward C—H bonds is explained in terms of the reduced electrophilicity of this carbene.[50]

Conversely, triplet anthronylidene (**52**) abstracts a H atom from cyclohexane with a rate comparable to that of FL, while ΔG_{ST} of anthronylidene is much larger than that of FL.[48,81] This finding is probably explained in terms of increased electrophilicity of triplet anthronylidene as a result of a π delocalization of one of the unpaired electrons. Both IR and EPR spectroscopic studies of oxocyclohexadienylidene suggest an extensive delocalization of one unpaired electron. Moreover, halogenated oxocyclohexadienylidene derivatives are shown to undergo insertion into H_2 and CH_4 even at very low temperature (see Section 5.4.2).

5.4. Hydrogen Atom Tunneling[85]

5.4.1. Product Studies.   Alkenes are known as diagnostic reagents for spin state of reacting carbenes. Thus, the reaction of a singlet carbene with an olefin usually results in the formation of a cyclopropane through stereospecific addition to the C—C double bond, while a triplet carbene gives rise to a nonstereospecific addition product (see Section 7).

The experiments carried out by Moss and his associates >30 years ago, which triggered the research in this field and can be regarded as landmarks, reveal that reactions of arylcarbenes with solidified alkenes at 77 K are completely different from those expected based on well-established fluid solution-phase chemistry.

Photolysis of monophenyldiazomethane (**54**), for example, in *cis*-but-2-ene (**55**) solution at 0 °C results in the formation of *cis*-1,2-dimethyl-3-phenylcyclopropanes (**56**) as syn and anti-mixtures along with small amounts of the trans isomer of the cyclopropane (**57**), and the C—H insertion products, 3-benzylbut-1-ene (**58**) and 5-phenylpent-2-enes (**59**). Product distributions are changed dramatically when the irradiation is carried out in solidified butene at 77 K. Here the formal C—H insertion products, that is, **58–59** are increased at the expense of the cyclopropane (**56**). The starting alkene configuration is retained (Scheme 9.11).[86] Similar dramatic changes in the product distributions in going from liquid to solid are observed in the reaction of other arylcarbenes with other alkenes, although the extent of the change is somewhat sensitive to the structure of carbenes as well as alkenes.[87]

What is the cause of these dramatic changes? The formation of the cyclopropane (**56**) is reasonably explained in terms of stereospecific addition of singlet monophenylcarbene (**55**) generated by photolysis of the diazomethane (**54**), while the

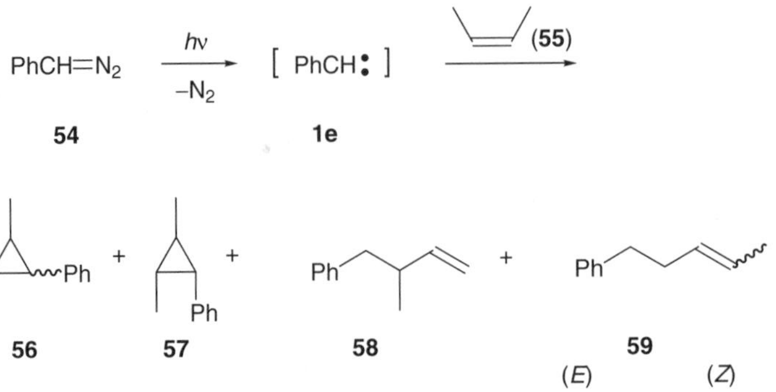

$$PhCH=N_2 \xrightarrow[-N_2]{hv} [PhCH\colon] \xrightarrow{\hspace{2cm}} (55)$$

54   **1e**

	56	**57**	**58**	**59**	
				(E)	(Z)
0 °C	92.7%	2.1%	1.4%	0.4%	3.5%
−196 °C	43.9	3.2	18.5	4.9	29.0

Scheme 9.11

formation of the pentene [(Z)-**59**] can be interpreted as indicating that the singlet **1e** undergoes insertion into the C—H bonds of the methyl group of the butene. Alternatively, triplet states generated by the intersystem crossing of the singlet might abstract the allylic hydrogen from the butene, followed by recombination of the resulting radical pairs. The formation of **58** and (E)-**59**, on the other hand, is clearly understood in terms of the hydrogen abstraction–recombination (a–r) mechanism occuring in the triplet carbene. The appearance of these "radical" products in significant amounts suggests that the triplet states are responsible for the formation of the pentene (**59**).

This assignment is unambiguously supported by the labeling experiments. Thus, DPC (**14**) reacts with $^{13}CH_2=CMe_2$ at low temperatures to produce, in addition to the cyclopropane (**60**), 1,1-dimethyl-4,4-diphenylbut-1-ene (**61**) in which $^{13}C_3/^{13}C_1$ label distributions are found to be 50:50 and 28:72 at −77 and −196 °C, respectively (Scheme 9.12). The equal distribution of ^{13}C between C1 and C3 at −77 °C

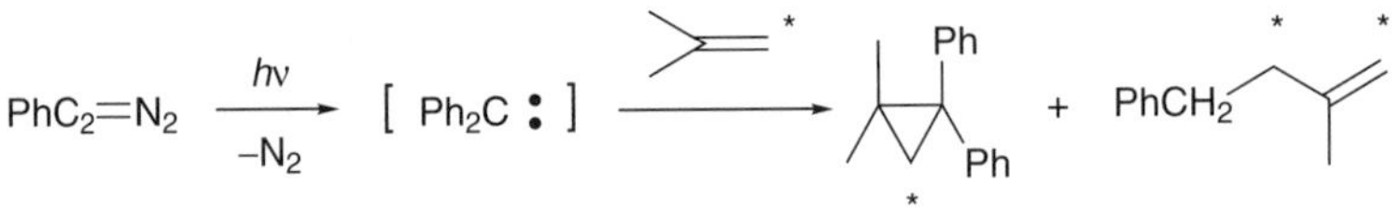

$$PhC_2=N_2 \xrightarrow[-N_2]{hv} [Ph_2C\colon]$$

14   **60**   **61**

Scheme 9.12

establishes an abstraction–recombination mechanism with complete equilibration of the radical pair. The results also eliminate the mechanism involving triplet carbene addition, followed by hydrogen migration for the formation of the butene (**61**), since such a mechanism would require an excess of ^{13}C at C3.[87b]

A series of experiments using alkene matrices clearly suggests that in a rigid matrix at low temperatures, triplet states of arylcarbenes undergo abstraction of allyl hydrogen from the matrix alkenes to form arylmethyl-allyl radical pairs. These then undergo recombination, after equilibration within the matrix, to give final products (Scheme 9.13).

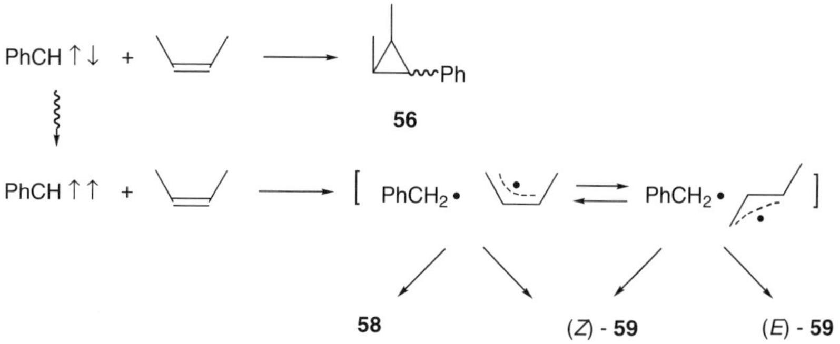

Scheme 9.13

Alcohols are also used as diagnostic reagents for the spin state of reacting carbenes. Thus, singlet carbenes are trapped very efficiently by the O—H bond to produce ethers, whereas the triplets undergo abstraction of hydrogen atom from the C—H bonds, not from the OH bond, leading to product mixtures resulting from radical pairs. Eventually, double hydrogen abstraction products and radical dimers are formed (see Scheme 9.6).

The reaction patterns of arylcarbenes with solidified alcohol at 77 K are also completely different from those observed in alcohol solution. For example, generation of phenylcarbene (**1e**) in methanol matrices at 77 K results in the formation of alcohol (**63**) at the expense of benzyl methyl ether (**62**), which is the exclusive product in the reaction in alcoholic solution at ambient temperatures (Scheme 9.14).[88] A similar dramatic increase in the CH "insertion" products is observed in the reaction involving other carbenes with alcohols.[89]

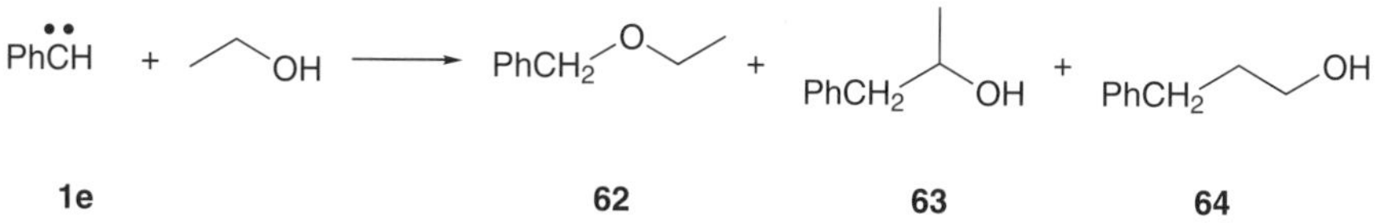

	62	**63**	**64**
0 °C	83.5%	~ 0%	0%
−196 °C	20.8	64.2	4.8

Scheme 9.14

By analogy with the mechanism proposed for the reaction with alkenes, C—H insertion product formation can be explained in terms of a H abstraction–recombination process of triplet arylcarbenes. The observations that ground-state singlet carbenes, for example, chlorophenylcarbene (**67**), produce only O—H insertion product even at 77 K again suggests that the ground-state triplet intervenes in the formation of C—H insertion products.

5.4.2. Kinetics of Hydrogen Atom Transfer.

Reactions in alcohols in alkene matrices clearly indicate that, as the temperature is lowered, the reactions of the singlet come to be suppressed and the singlet undergoes intersystem crossing to generate the triplet ground state. The triplet then abstracts hydrogen from the matrix molecules to produce ultimately C—H insertion products. In a fluid solution at room temperatures, most carbenes react in the singlet state regardless of the ground state multiplicity.

It is not easy to explain why the triplet reactions that are energetically much less favored than those of the singlets become dominant at low temperature. Based on E_a and log A measured for triplet carbene abstraction (see Section 5.3), one can estimate the rate constant at 77 K to be $<10^{-5}$ M^{-1} s^{-1}, suggesting that triplet carbene reactions in matrices at 77 K should not occur. Obviously, reactions of carbenes within matrices are controlled by factors that are not operating in solution phase, as one might expect from dramatic changes in reaction conditions.

An explanation for this conundrum is provided by kinetic experiments using EPR. The EPR spectra of simple triplet carbenes such as DPC (**14**) and FL (**23**) are produced by brief photolysis of the corresponding diazo compounds in a series of organic matrices at low temperatures. The rate of signal disappearance is measured immediately after photolysis is discontinued. The decay of the triplet carbenes is attributed to hydrogen-atom abstraction from the matrix based mainly on the observation that the rate of carbene decay roughly follows the expected order of hydrogen-atom-donating ability of the matrix. It is also decreased upon perdeuteration of the matrix. Decay is not observed in perfluorinated matrices, which have no abstractable hydrogens. The decay curves of the triplet carbenes are not single exponential functions because of a multiple-site problem, as is observed in the dynamic behavior in polycrystalline solids. An absolute rate constant for a specific site is determined based on the assumption that the first 20% of the carbene decay corresponds to the decay of the carbene in a single site, and both the Arrhenius activation energy and preexponential factors are determined. The data listed in Table 9.10 suggest that the parameters are much smaller than expected

TABLE 9.10. Pseudo-First-Order Rate Constants and Arrhenius Parameters for the Decay of Triplet Ph$_2$C: (14a) in Organic Glasses

Solvent[a]	Temp (°C)	$10^3 k$ (s$^{-1/2}$)	log A (s$^{-1/2}$)	E_a (cal/mol)
MeOH—MTHF	−184	51	−1.464	470
PrOH—MTHF	−184	41	−0.328	920
MTHF	−184	61	0.690	1291
PhMe	−196	35.6	0.85	1219
C$_6$D$_5$CD$_3$	−196	5.2	1.14	1773
c-C$_6$H$_{12}$	−137	32		

[a]Methyltetrahydrofuran = MTHF, c-C$_6$H$_{12}$ = cyclohexane.

for classical hydrogen atom transfer. Thus, QMT is proposed to be the mechanism for the matrix hydrogen atom transfer. The observation that the carbenes are completely stable in CCl_4 matrices at 77–100 K suggests that the light hydrogen atom can tunnel, but the more massive Cl atom cannot.[90]

The members of the radical pairs generated by the hydrogen atom abstraction of the triplet carbenes usually diffuse apart in fluid solution. Product mixtures consisting of radical dimers, double hydrogen abstraction products and so on are formed (Scheme 9.6). In a rigid matrix, however, the members of the pairs are not able to diffuse apart (owing to the limited diffusibility within the matrix) and therefore recombine with high efficiency to give the C—H "insertion" products upon thawing the matrix.

A hydrogen atom transfer reaction in which the reaction mechanism changes from a completely classical process in a soft warm glass to a completely quantum mechanical tunneling process in a cold hard glass has been more evidently demonstrated by using a technique with greater time resolution than a conventional EPR method: LFP. Thus, LFP of diazofluorene in a number of glasses generates singlet FL(**23**), which undergoes intersystem crossing to triplet **23**.[91]

In contrast to the EPR studies, exponential decays of the triplet **23** are observed. The Arrhenius treatment of the data obtained in hydrogen donating glasses shows that there are two regions; a steep region at high temperatures and a flat, temperature-insensitive region with very low-activation parameters. Upon extrapolation of the high-temperature data to the low-temperature regime, one finds that the observed rate is hundreds of times faster than predicted. No such break in the Arrhenius plot is observed for the decays of **23** in perhalogenated solvents, which have no abstractable hydrogens.[92] Thus, these broken Arrehnius plots observed in hydrogen donating solvents are best explicable in terms of a change in the reaction mechanism.

5.4.3. Why Can the Triplet Find a Tunneling Pathway?

The exact reason why triplets can find a tunneling pathway, whereas singlet carbenes cannot is not clear. However, recent investigations on the low-temperature hydrogenation of carbenes provide some clues concerning this issue. It has been shown that a carbene with a triplet ground state, for example, phenyltrifluoromethylcarbene (**65**) undergoes hydrogenation when generated at 10 K in 2% H_2/Ar matrix, followed by warming to 30 K in the dark to give the corresponding reduction product (**66**, Scheme 9.15). On the other hand, a carbene with singlet ground state, for example, phenyl(chloro)carbene (**67**) does not react at all upon annealing in an H_2-doped Ar matrix even at 35 K (Scheme 9.15). Moreover, **65** is completely unreactive with D_2 under comparable conditions, suggesting the presence of a large kinetic deuterium isotope effect. The results are interpreted as indicating that the triplet reacts through hydrogen-tunneling abstraction followed by recombination of the resultant radical pair to give the reduction product, whereas the singlet, requiring concerted addition, does not undergo tunneling reaction under cryogenic conditions.[93]

Energetics of carbenes and hydrogen reactions have been calculated at the B3LYP/6-31G** level of theory. They indicate that all the H_2 additions are

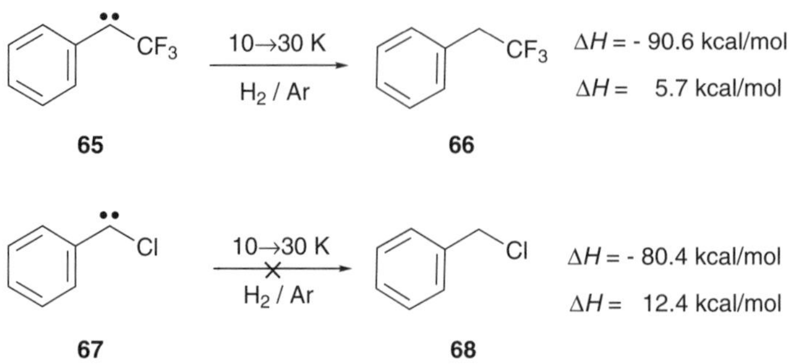

Scheme 9.15

be very exothermic ($-80{\sim}-90$ kcal/mol), including those of chlorophenylcarbene (**67**). Thus overall exothermicity does not appear to be the reason for the lack of reactivity of the singlet carbene. Concerted addition of H_2 to **67** is calculated to have a barrier of 12.4 kcal/mol, while hydrogen abstractions from H_2 by triplet carbenes (**65**) are predicted to have significantly lower energy barriers (5.7 kcal/mol). Based on these calculations, it is suggested that the direct insertion of singlet **67** via quantum mechanical tunneling is less facile than stepwise reaction of the triplet carbenes **65** with H_2, either because of a higher classical barrier or because of the lower probability of QMT involving two hydrogens simultaneously. However, much more reactive singlet carbenes, such as difluorovinylidene ($F_2C{=}C{:}$) appear to surmount the restrictions on concerted additions even at very low temperatures.[94]

On the other hand, it has been shown that even at very low temperature highly electrophilic triplet carbenes undergo thermal H_2 additions in H_2–Ar matrix as well as C—H insertion into CH_4. For example, triplet tetrafluoro-4-oxocyclohexa-2,5-dienylidene (**69a**) generated in an H_2-doped Ar matrix reacts with H_2 even at 35 K to give tetrafluoro-4-oxocyclohexa-2,5-dienone (**70a**). It also reacts with CH_4 at 40 K to give a formal C—H insertion product (**71a**). Interestingly, no deuterium isotope effects are observed in this case; the same reaction takes place at qualitatively the same rate in D_2 and CD_4. The reactivity decreases on going from tetrachloro (**69b**) to tetrahydro derivatives (**69c**). The tetrahydro derivative does not react at all under the same conditions (Scheme 9.16). DFT calculations (B3LYP/6-31G(d)) predict that the reaction of carbene (**69**) with H_2 to give cyclohexadienenone (**70**) is highly exothermic (91.4 and 73.8 kcal/mol for the fluoro and chloro derivatives, respectively) but do not reveal a transition state for the reaction. These calculations suggest a thermal reaction with an extremely small or absent barrier. This unusual reaction is explained in terms of the philicity of triplet carbene.[95] It has been proposed that the electron affinities (EA) can be used as a measure of the carbene philicity. The EA values were calculated [B3LYP/6-311++G(d,p)] for fluoro (**69a**), chloro (**69b**), and hydro (**69c**) derivatives are 3.32, 3.08, and 2.05 eV, respectively.[96] The EAs for phenyl chlorocarbene (singlet, ^{1}**67**) and diphenylcarbenes (^{3}**14**) are 1.52 and 1.48 eV, respectively. These are well below the electrophilicity of the hydro derivative (**69c**).

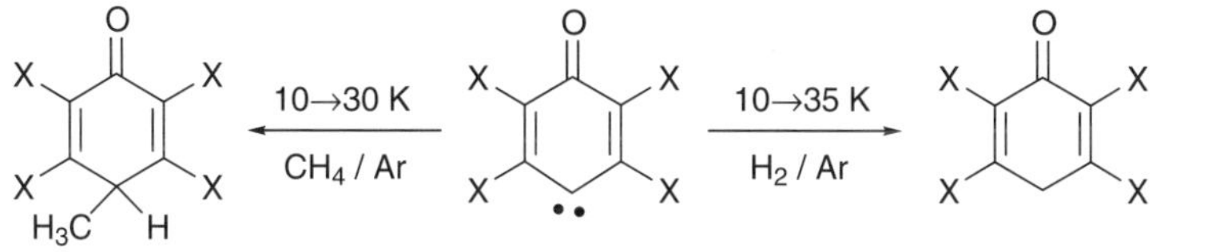

71 69 70

ΔH = −78.8 kcal/mol | a : X = F | ΔH = −91.4 kcal/mol
ΔH = −58.8 kcal/mol | b : X = Cl | ΔH = −73.8 kcal/mol
c : X = H

Scheme 9.16

5.4.4. Tunneling in Intramolecular Reactions. Carbenes undergo intramolecular hydrogen insertions efficiently, when the structure allows. For example, 1,2-hydrogen migration forming an alkene is one of the most frequently documented rearrangement reactions of carbenes. It has been demonstrated that QMT plays an important role in some of intramolecular hydrogen migration processes.

2-Methylphenylcarbene (**72**) gives rise to *o*-quinodimethane (**73**). This reaction is explained in terms of an intramolecular 1,4-hydrogen shift (Scheme 9.17). This

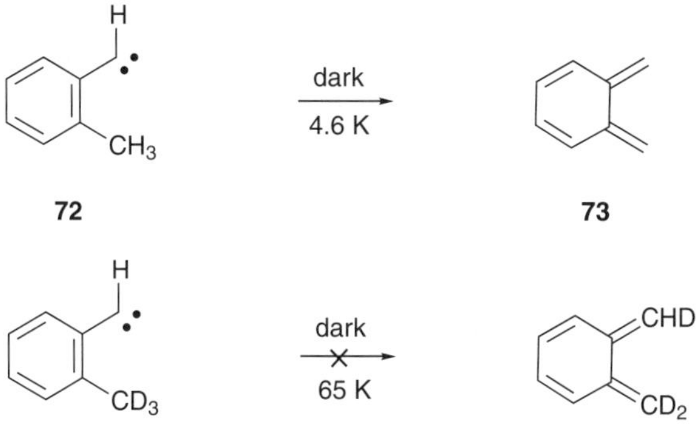

72 73 72 - d_3 73 - d_3

Scheme 9.17

migration reaction was investigated in great detail by using matrix isolation spectroscopic techniques. Thus, ^{3}**72** was generated in an argon matrix at low temperature and characterized by IR and UV–vis spectroscopy. Triplet carbene (**72**) is found to decay thermally to give singlet *o*-quinodimethane (**73**) even at temperatures as low as 4.6 K. The kinetics of carbene disappearance follow the standard $(\text{time})^{1/2}$ dependence, because of the multiple reaction sites in the matrix. The rate was monitored as a function of temperature from 4.6 to 30.0 K. A small temperature dependence of the rate and the nonlinear Arrhenius plot were noted. Moreover, 2(trideuteriomethyl)phenylcarbene (**72**-d_3) was found to be thermally stable even at 19 K in an Ar matrix. In a Xe matrix, it was stable even at 59 K for

1150 min. These observations implicate a QMT mechanism for alkene formation.[97] A similar QMT mechanism is proposed in a thermal 1,4-H shift of mesitylcarbene at 11 K.[98]

In contrast, 1-phenylethylidene (**74**), which can undergo 1,2-hydrogen migration to form styrene (**75**), was found to be thermally stable in argon or xenon matrix at 10 K. The carbene decay to give styrene only when warming to 65 K in xenon matrix (Scheme 9.18). From the disappearance rate constant for **74**, a energy barrier

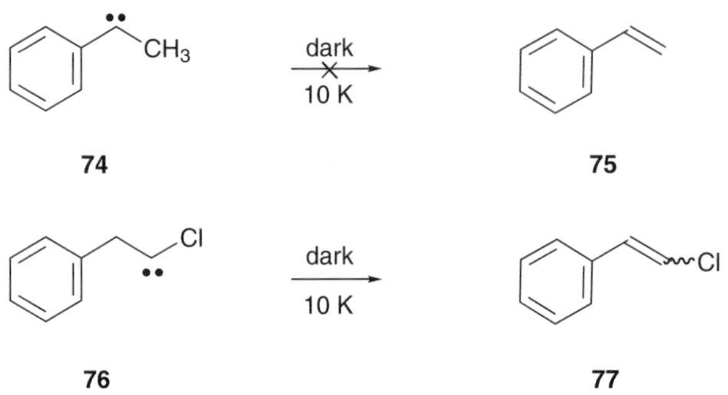

Scheme 9.18

of 4.7 kcal/mol (at 65 K) was estimated. Thus, the thermal rearrangement of triplet 1-phenylethylidene to styrene likely occurs upon thermal population of the S_1 state of **74** at 65 K.[97]

A 1,2-hydrogen shift in benzylchlorocarbene (**76**), which has a singlet ground state, forms β-chlorostyrene (**77**). This reaction has been found to involve a QMT. Thus, the carbene was found to decay thermally at temperature as low as 10 K. In contrast, deuterated carbene (**76**-d_2) was stable at higher temperatures (up to 42 K) (Scheme 9.18). Arrhenius plots of the decay kinetics were obtained by following the reaction between 10 and 34.5 K in an argon matrix. Again the small temperature dependence of the rate at lower temperature and the curved Arrhenius plot at higher temperature zone were noted. From extrapolation of the data at room temperature, a rate of 10^{-2} s^{-1} at 80 K is expected. This value indicates that tunneling is important at low temperature.[98]

The difference in thermal stability of singlet benzylchlorocarbene (**76**) compared with that of triplet methylphenylcarbene (**74**) was noted, and was interpreted as being caused by the different spin states. In solution, a minimum barrier of 4.3 kcal/mol was assessed for the triplet carbene, which is in the range of the value which had been measured for **76** in solution. Tunneling at low temperatures was excluded for the triplet species since it is stable up to 30 K in argon. Intersystem crossing is not necessary in the reaction of the singlet species. Therefore, spin control may play an important role in the triplet molecule, resulting in a minimum activation barrier in the order of magnitude of the singlet–triplet gap.[99]

Singlet carbenes are also known to undergo intramolecular C—H insertion when structurally favored. For example, *tert*-butylchlorocarbene (**78**) gives rise to

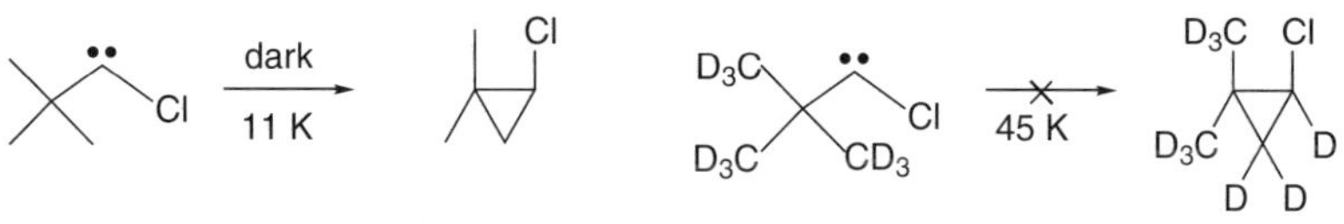

78   79   78 - d_9 79 - d_9

Scheme 9.19

1,1-dimethyl-2-chlorocyclopropane (**79**) as a result of insertion of carbene into the C—H bond of methyl group (Scheme 9.19). When **78** was generated in a nitrogen matrix at low temperature, it was detected by IR and UV–vis spectroscopy. However, the carbene decay on standing at 11 K in the dark produces the cyclopropane. The insertion rate was insensitive to temperature; warming the matrix from 11 to 30 K caused no abrupt increase in the disappearance of the carbene. The deuterated carbene (**78-d_9**) was found to be thermally stable as the carbene survived up to 45 K. Thus, there is a minimal temperature dependence of the reaction at low temperatures and an unusual isotopic sensitivity of the kinetics all of this evidence suggests that the 1,3-CH insertion reaction of the carbene **78** involves QMT. The formation of C—C bond during the reaction also suggests that heavy atom tunneling might be involved.[100]

5.4.5. Tunneling Reactions at Elevated Temperatures.

Thus far, we have seen QMT reactions of carbenes at very low temperature. Many examples of QMT in reactions at elevated temperatures have also been reported. In fact, high-temperature QMT was demonstrated long before the low-temperature examples were discovered.

At cryogenic temperatures, the tunneling reaction proceeds mainly through the zero-point level and is sometimes referred to as deep tunneling. Low-temperature QMT is easily recognizable because deep tunneling is many orders of magnitude faster than the classical process, which is negligible at very low temperature. At elevated temperatures tunneling can occur through vibrational levels close to the top of the barrier. Consequently, the rates of a classical process and of a QMT process at ambient temperature may be comparable. A contribution of QMT to the overall rate of reaction at elevated temperatures can be revealed by measurement of kinetic isotope effects as a function of temperature. Differential activation energies greater than the difference in carbon–hydrogen and carbon–deuterium bond zero point energies [$Ea(D)-Ea(H) > 1.2$ kcal/mol], and unusual ratio of preexponential factors ($A_H/A_D < 0.7$) signal a contribution of QMT to the overall rate of a reaction at elevated temperatures.[101]

In order to determine whether QMT may contribute to the overall reaction of diarylcarbenes with hydrogen atom donors in solution at ambient temperature, kinetic isotope effects for the benzylic hydrogen atom abstractions of the triplet states of several diarylcarbenes with toluene–toluene-d_8 in fluid solution were determined over the temperature ranges of −75 to 135 °C. The results are very much dependent on the structure of the carbene (Table 9.11).[102] The differential

TABLE 9.11. Differential Activation Parameters of Triplet Carbene H(D) Atom Transfer

Carbenes	k_H/k_D	T (range) (°C)	$E_a(D) - E_a(H)$ (kcal/mol)	A_H/A_D
1-Naphthyl(phenyl)carbene (α-51)	8.38	−50 + 150	1.77	0.42
2-Chlorodiphenylcarbene (14u)	8.00	−50 + 120	1.69	0.43
2-Trifluoromethyldiphenylcarbene (14v)	6.75	−25 + 120	1.88	0.31
Diphenylcarbene (14a)	7.00	−75 + 135	1.37	0.62
Dibenzocycloheptadienylidene (18)	5.26	−75 + 50	1.46	0.48
4-Biphenylphenylcarbene (14l)	5.58	−75 + 50	1.20	0.75
Fluorenylidene (23)	4.08	−75 + 25	0.80 ± 0.20	1.05 ± 0.10
Anthronylidene (52)	6.09	−75 + 25	0.89 ± 0.20	1.35 ± 0.10

kinetic isotope effects observed for 1-naphthylphenylcarbene (α-51), 2-chlorodiphenylcarbene (14u), and 2-trifluoromethyldiphenylcarbene (14v) were much larger than predicted by complete loss of all zero-point energy in the transition state. This observation indicates that there is a contribution of QMT to the H(D) atom transfer process. For diphenylcarbene (14a), dibenzocycloheptadienylidene (18), and 4-biphenylphenylcarbene (14l), the differential kinetic isotope effects were barely consistent with a completely classical atom-transfer reaction. For fluorenylidene (23) and anthronylidene (52), the data were completely consistent with a purely classical atom-transfer process. The lifetime of the carbenes undergoing QMT are probably longer than those of the other carbenes because of steric hindrance about the carbene carbon. The substituent may also widen the bond angle at the carbene carbon. These effects will decrease the amount of s character in the singly occupied orbitals of the carbene and lower the reactivity still further. Thus, the trends are interpreted as indicating that the relatively slow rate of the classical atom-transfer reaction in the hindered carbenes allows a QMT pathway to contribute more prominently to the overall hydrogen atom-transfer rate.[102]

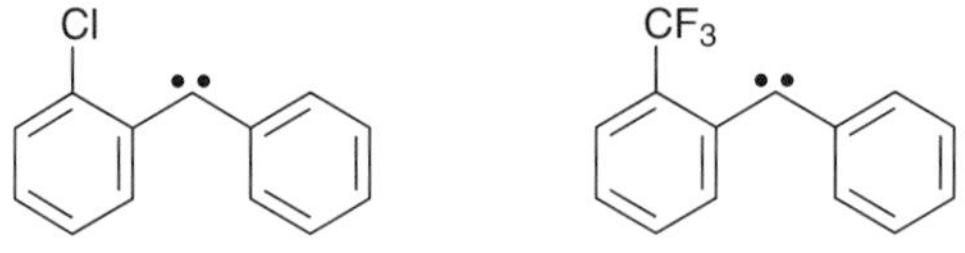

14u   14v

A contribution of QMT in solution at room temperature is also observed in 1,2-hydrogen rearrangement of a singlet carbene.[103,104]

6. REACTIONS WITH OXYGEN

Molecular oxygen is an important participant in reactions of triplet carbene because of its triplet ground state and its ubiquity as an impurity in reaction systems. Reaction of triplet carbenes with molecular oxygen is very efficient because the reaction is spin allowed process. So, the reaction with oxygen is sometimes used to judge the multiplicities of carbenes.

6.1. Product Studies

The photooxidation of diphenyldiazomethane yielded up to 73% benzophenone and no other products containing oxygen could be detected (Scheme 9.20). It was

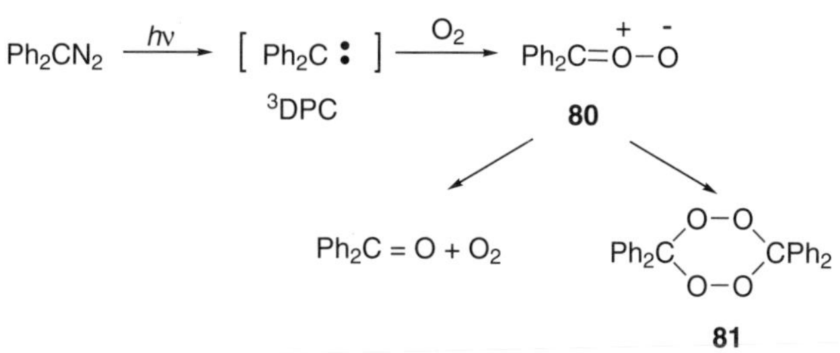

Scheme 9.20

postulated that the primary adduct arising from photooxidation is benzophenone oxide (**80**, Criegee intermediate), which transfers its terminal O atom to another molecule of carbene or diazomethane.[60] When the photooxidation was carried out in solid chlorobenzene at -78 °C, a tetroxane (**81**), a dimer of the oxide was isolated in 15% yield (Scheme 9.20). It was shown that the elimination of O_2 from two molecules of the oxide does not proceed to a significant extent.[105] More recent results show that if no trapping reagents are present, the extrusion of O_2 from two separate oxides becomes the main reaction pathway for the oxide. The formation of the tetroxane does not occur in solution at room temperature.[106] Other oxidation products arising from the oxidation are esters.[107]

6.2. Matrix Isolation Studies

Carbonyl oxides have been detected by matrix isolation and time-resolved laser spectroscopy. In a matrix, carbonyl oxides are easily generated upon annealing of O_2-doped inert gas matrix containing triplet carbenes from 10 to 30–40 K (Scheme 9.21). The carbonyl oxides (**80**) are characterized by a broad and strong absorption in the 380–460-nm range. In most cases, the matrix turns intense yellow to red. The observed absorbance is assigned to a π–π* transition in accord with a CNDO/S computation that predicts n–π* transitions in the region 600–800 nm. The latter absorption is very weak and has not been directly measured experimentally, but its presence is indicated by the fact that photochemical conversion of the carbonyl oxide can be initiated by irradiation of this range (see below).[108]

Infrared spectroscopy has also been used to characterize carbonyl oxides in matrix isolation. The carbonyl oxides are identified by their intense O—O stretching

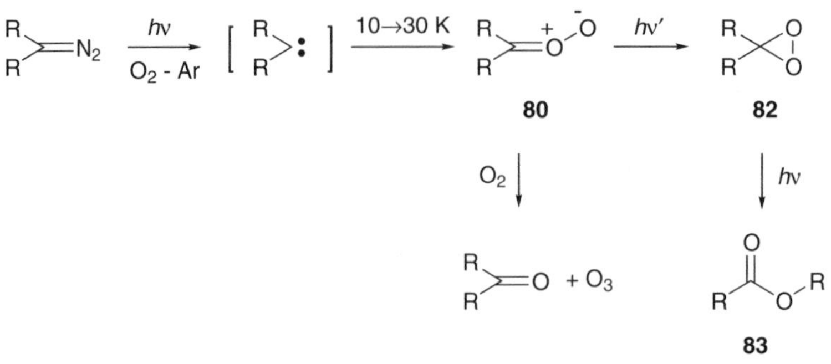

Scheme 9.21

vibrations ~900–1000 cm^{-1}. Isotope labeling with ^{18}O in one and both oxygens has been used to assign vibrational bands as well as to confirm the structure. For example, the IR bands observed for cyclopentadienone oxide at 1014, 947, and 940 cm^{-1} exhibit a large shift (-28, -22, and -15 cm^{-1}, respectively) on double labeling with ^{18}O and are therefore assigned to vibrations with a considerable O—O stretching component. There are two $^{16}O^{18}O$ isotopomers for this species, as evidenced by the fact that the band at 1014 cm^{-1} is split into two (1001 and 995 cm^{-1}), clearly indicating that the oxygen atoms are not equivalent. These observations eliminate a dioxirane structure (**82**), which should show only one $^{16}O^{18}O$ isotopomer.[109]

These bands are in good accord with the calculated spectra for the oxides. Theoretical studies of the oxides aimed at the determination of electronic structures. They tried to determine whether these species are diradicals or not (Eq. 23).

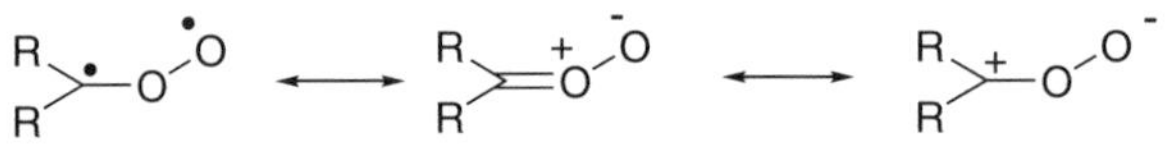

$$(23)$$

It has been shown that different methods may ascribe different bond lengths to the O—O and C—O bonds and that the medium and substituents affect the electronic behaviors of carbonyl oxides.[110] For example, recent computational studies (B3LYP/6-31 + G (d, p)) of carbonyl oxides, *syn*- and *anti*-methyl carbonyl oxides and dimethylcarbonyl oxides in gas and solution reveal that dipolar character increases with the number of methyl groups, and the ionic configuration is stabilized in a polar medium. These effects result in a weakened O—O bond and an increased double-bond character in the CO bond.[111]

Carbonyl oxides are extremely photolabile even under matrix conditions and irradiation with red light (600 nm) rapidly produces dioxiranes (**82**).[109] The dioxiranes are stable under these conditions but at 400-nm irradiation are converted into esters (**83**) or lactones. Ketones have been observed as byproducts in the carbene–O$_2$ reactions in frozen matrices. Since the reaction of triplet carbene with O$_2$ is very exothermic ($\sim$47 kcal/mol for the formation of H$_2$C—O—O), carbonyl oxides are generated in vibrationally excited states. Their stabilization occurs either through relaxation to the thermal ground state or by extrusion of an O atom and subsequent formation of a carbonyl compound. The ratio of carbonyl oxide to ketone is strongly dependent on the substituents at the carbene center. Especially in the case of hydrogen atom substituents, the extrusion of oxygen atoms predominates. Some of the oxygen atoms react with O$_2$ to give O$_3$, especially in matrices with high O$_2$ concentration (Scheme 9.21).

The conversion of triplet carbenes and O$_2$ to the corresponding oxides and the following reactions are easily followed spectroscopically, and therefore are used to characterize the reaction of triplet carbenes.

Singlet carbenes generated in an O$_2$-doped inert gas matrix also react with O$_2$ to give the corresponding oxides (**80b**, Scheme 9.22). For example, chlorophenyl-carbene (**67**) reacts with O$_2$ to give the oxides. However, the reaction is much

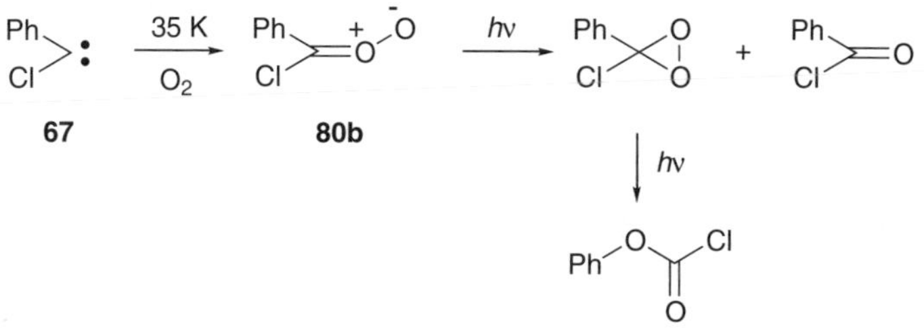

Scheme 9.22

less efficient than similar reactions of triplet carbene. Under conditions where the triplet carbenes are completely converted into carbonyl oxide, **67** does not react perceptibly.[112] The origin of this spin-forbidden process is not clear. A complex formation between singlet carbene and O$_2$ might facilitate intersystem crossing before addition. From the fact that the more electron-deficient p-nitrophenylcarbene reacts markedly faster than the parent chlorocarbenephenyl both in matrix and in solution (see below), it is suggested that the reaction is dominated by the philicity of the carbene. The electrophilic attack of the carbene on the O$_2$ molecule is the rate-determining step. On the other hand, decrease in ΔE_{ST} by an electron-withdrawing nitro group may facilitate the reaction from the upper lying triplet state. It is interesting to note that apparent spin-forbidden reactions of matrix-isolated triplet carbenes with (singlet) CO$_2$, CO, and N$_2$ have been observed.[109,113] Further work to clarify these points is required.

6.3. Dimesityl ketone Oxide

Carbonyl oxides are usually highly unstable and can be detected only by matrix isolation spectroscopy at low temperature or by LFP techniques. However, they can be stabilized through steric protection. Dimesityl ketone oxide (**80c**, $\lambda_{max} =$ 398 nm) generated by the reaction of triplet dimesitylcarbene (**19c**) with O$_2$ at

77 K is found to be stable even at −79 °C for many hours in the dark and is characterized by ^{13}C and ^{1}H NMR spectroscopy (in CFCl$_3$/(CF$_2$Br)$_2$).[114] A ^{13}C resonance at δ 211.1 ppm is in good agreement with the chemical shift calculated (using the IGLO method) for the parent carbonyl O-oxide H$_2$C—O—O (δ 227 ppm)[115] and is assigned to the carbonyl carbon of dimesitylketone oxide (**80c**). The number of methyl groups observed in the ^{1}H NMR spectrum indicates that the C—O—O moiety is configurationally stable at −70 °C. Exposure to daylight or irradiation with λ > 455 nm rapidly causes complete bleaching of the yellow solution of the oxide. After the solution is warmed to room temperature, the ketone, ester (**83c**), and dioxirane (**82c**) are isolated. Dimesityldioxirane (**82c**) is also stable and can be obtained as a colorless crystalline material, completely stable at −20 °C. Irradiation of the dioxirane with λ > 420 nm light results in the formation of the ester (**83c**), which is not formed from the oxide either thermally or photochemically (Scheme 9.23). These results confirm the photochemistry observed in solid argon at low temperature.

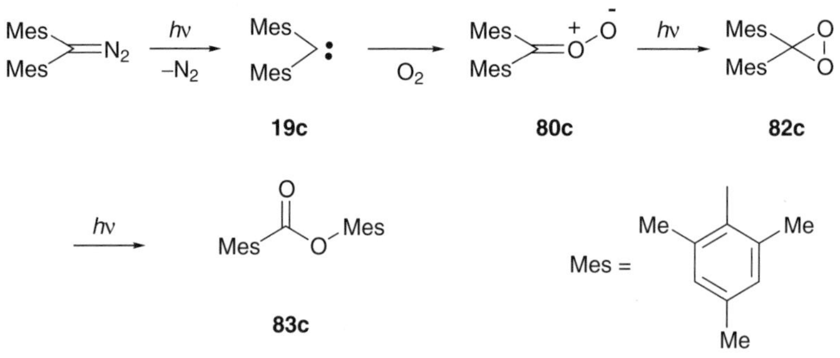

Scheme 9.23

6.4. Emission

The emission of a strong chemiluminescence was observed during warm-up from 77 K of organic glasses containing triplet DPC and a traces of O$_2$.[116] The chemiluminescence was identified as the phosphorescence of benzophenone. Since chemiluminescence decay was faster than consumption of the carbene, it was proposed that oxygen transfer from carbonyl oxides to triplet carbenes results in the formation of the ketones in electronically excited states. In O$_2$-doped Ar matrices, chemiluminescence starts at temperatures as low as 15 K. Chemiluminescence decay was faster than the oxidation of the carbenes, but IR spectroscopy excluded carbonyl oxides as oxygen-transfer reagents. A direct reaction of the oxygen atom, produced by decomposition of carbonyl oxides, with triplet carbenes, producing a C=O bond in a very exothermic reaction, was proposed as the chemiluminescence step (Scheme 9.24).[117]

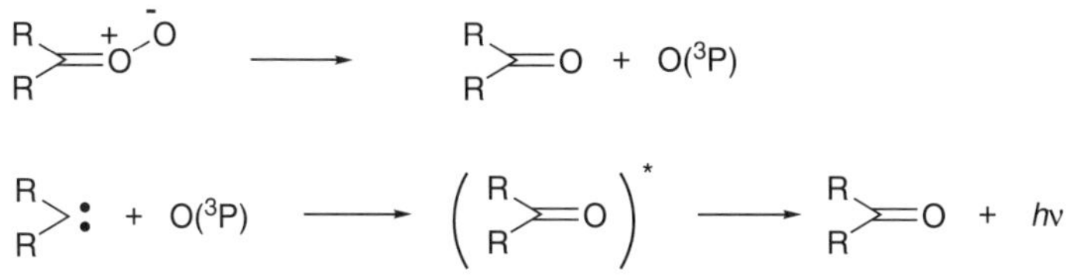

Scheme 9.24

6.5. Laser Flash Photolysis Studies

Carbonyl oxides are also easily detected by LFP with fast TRUV–vis spectroscopy. For example, LFP of α-diazophenylacetate in deaerated Freon-113 generates a transient absorption band at $\lambda < 270$ nm ascribable to triplet methoxycarbonylphenylcarbene (**53**), which shows a pseudo-first-order decay with lifetime of 460 ns. When LFP is carried out in aerated solvents, a new transient band appears at 410 nm at the expense of the transient band due to triplet carbene. The decay rate of triplet carbene increases as the concentration of oxygen increases. This correspondence indicates that the triplet carbene is trapped by oxygen to form the carbonyl oxides (**53-O$_2$**). This result confirms that the transient absorption quenched by oxygen

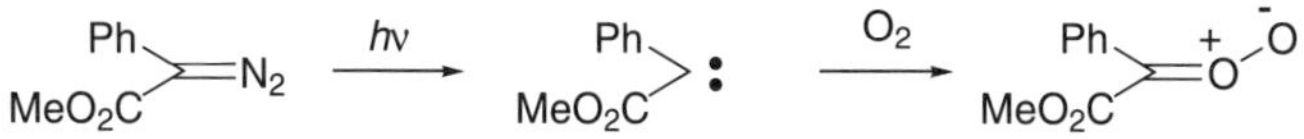

Scheme 9.25

belongs to the triplet carbene (Scheme 9.25). The apparent built-up rate constant, k_{obs}, of the carbonyl oxide is essentially identical to that of the carbene, and k_{obs} is expressed as given in Eq. 24,[84]

$$k_{obs} = k_0 + k_{O_2}[O_2] \tag{24}$$

where k_0 represents the rate of decay of triplet carbene in the absence of oxygen and k_{O_2} is the quenching rate constant of triplet carbene with oxygen. The plot of the observed pseudo-first-order rate constant of the formation of the carbonyl oxide is linear. From the slope of this plot, the rate constant for the quenching of triplet carbene by oxygen is determined to be $8.6 \times 10^8\ M^{-1}\,s^{-1}$.

Absolute rate constants for the representative triplet carbene–O$_2$ reaction in solution at room temperature are listed in Table 9.12.[118–123] The second-order rate constants are close to $10^9\ M^{-1}s^{-1}$, near the diffusion-controlled rate. Thus, the reaction with O$_2$ is taken as evidence for the presence of the triplet state of the carbene.

TABLE 9.12. Bimolecular Rate Constants for the Reaction of Carbenes (R–C–R′) with Oxygen

Carbenes	R	R′	Solvents[a]	$k_{O_2}\,(M^{-1}\,s^{-1})$	Reference
14a	Ph	Ph	MeCN	$(5.0 \pm 0.1) \times 10^9$	118
53	Ph	CO$_2$Me	Freon-113	8.6×10^8	84
23	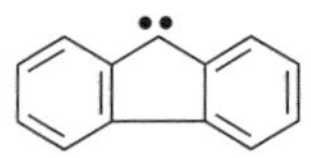		Freon-113	$(2.02 \pm 0.08) \times 10^8$	118
19c	Mes	Mes	MeOH	$(1.46 \pm 0.04) \times 10^8$	118
			PhH	$(0.83 \pm 0.08) \times 10^8$	
α-12	1-NC	H	c-C$_6$H$_{12}$	$(2.1 \pm 0.3) \times 10^9$	119
β-51	2-NC	Ph	c-C$_6$H$_{12}$	$(2.6 \pm 0.2) \times 10^{10}$	120
α-51	1-NC	Ph	c-C$_6$H$_{12}$	$(5.4 \pm 0.4) \times 10^9$	120
52	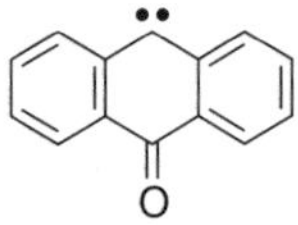		PhH	$(9.8 \pm 1.2) \times 10^8$	81
52a	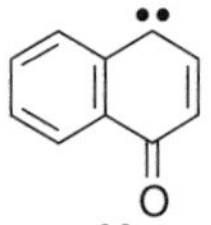		Freon-113	$(7.3 \pm 0.6) \times 10^9$	121
69c	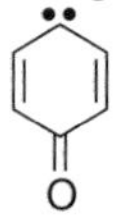		Freon-113	$(4.4 \pm 0.8) \times 10^{10}$	121
69d	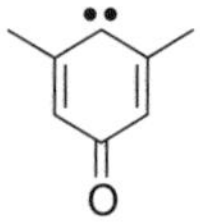		Freon-113	$(2.7 \pm 0.4) \times 10^7$	121
7	Ad	Ad	c-C$_6$H$_{12}$	$\sim 3 \times 10^6$	122
24	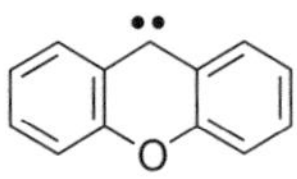		n-C$_5$H$_{12}$	~ 0	50
67b	4-NO$_2$C$_6$H$_4$	Cl	i-C$_8$H$_{18}$	$\sim 3 \times 10^6$	123

[a]Cyclohexane: c-C$_6$H$_{12}$, n-C$_5$H$_{12}$: n-pentane, i-C$_8$H$_{18}$: isooctane.

It is interesting to note here that the value of the rate constant for DPC (**14a**), 1-NC (**α-12**), methoxycarbonylphenylcarbene (**53**), and FL (**23**) with O$_2$ are 5×10^9, $(3.5 \pm 0.7) \times 10^9$, 8.6×10^8, and $(1.4 \pm 0.2) \times 10^9\ M^{-1}\,s^{-1}$, respectively. The difference in the rate constants is not as large as that observed for the hydrogen atom abstraction rate constants for those carbenes (Table 9.9) and do not reflect the difference in the magnitude of ΔG_{ST}. The reason is probably because the rate constant of triplet carbenes is very fast and because the singlet states do not interact with triplet oxygen because of the spin restriction.

Singlet carbenes usually do not react with O_2. For example, LFP of 3-chloro-3-phenyldiazirine has shown that the reactions of chlorophenylcarbene (**67**) with numerous substrates are insensitive to the presence of oxygen, indicating that the rate constant for chlorophenylcarbene with O_2 must be $< 10^4 M^{-1}s^{-1}$. However, singlet chloro p-nitrophenylcarbene (**67b**) is readily scavenged by O_2, as evidenced by the decrease in lifetime in the presence of increasing O_2 concentration and the appearance of a new broad transient absorption in the 350–500-nm range ($\lambda_{max} = 400$ nm). A plot of the reciprocal lifetimes for the carbene decay versus oxygen concentration yields the quenching rate constant, $k_{O_2} = 2.24 \times 10^7 M^{-1} s^{-1}$. This value is two orders of magnitude smaller than that for the reaction of triplet carbenes with O_2. Continuous irradiation of the diazirine with O_2 in isooctane gave several oxidation products including p-$NO_2C_6H_4CHO$ and p-$NO_2C_6H_4COCl$. However, when the photolysis is carried out in the presence of (Z)-4-methyl-2-pentene at 25 °C under O_2, all additions take place stereo-specifically with no discernible effect by O_2, again suggesting that the reaction of singlet carbene with O_2 is very inefficient.[123] The reason why only chloro-p-nitrophenylcarbene reacts with O_2, while unsubstituted **67** does not, is perplexing.

Carbonyl oxide formation can also be used to monitor carbenes that are transparent in the most useful UV region. For example, most alkylcarbenes are spectroscopically invisible. The ground-state singlet alkylcarbenes are generally monitored by trapping with pyridine to form ylides that show an intense absorption band ~400 nm. However, this technique cannot be used to monitor the ground-state triplet dialkylcarbenes bearing rather bulky groups. For example, LFP of diadamantyldiazomethane in degassed cyclohexane does not produce a UV–vis active transient intermediate. However, LFP in aerated solution produces a transient spectrum exhibiting a maximum at 307 nm with a rise time of 200 ns. The transient band is attributable to carbonyl oxide (**7-O_2**). The absorption maximum of the oxide is shifted by roughly 100 nm relative to benzophenone oxide and other aryl carbonyl oxides because of the lack of conjugation with an aromatic ring. Since diadamantylcarbene (**7**) has been shown to react with methanol in its singlet state, it is expected that the yield of **7-O_2** will be reduced while its rate of formation will be increased upon addition of methanol. Stern–Volmer analysis of the methanol quenching data and the slope of the observed rate constant of **7-O_2** formation versus [MeOH] gives a rate constant of $(2.0 \pm 0.4) \times 10^7 M^{-1}s^{-1}$ and $(3.16 \pm 0.81) \times 10^7 M^{-1}s^{-1}$, respectively. Thus, diadamantylcarbene (**7**) and DPC (**14**) react with the singlet carbene quencher methanol with about the same rate constant, probably at spin equilibrium.[7b]

Compared to the time scale of their formation, the carbonyl oxides are quite long lived (10^{-5}–10^{-3} s), and so their subsequent reactions can be monitored kinetically. For most of the carbonyl oxides, the decay is best fit to a second-order rate law, indicating a bimolecular decomposition pathway. For benzophenone oxide, the ketone is the major product at room temperature, and no dimer can be detected. A bimolecular process involving O_2 extrusion from two molecules of the oxides is suggested under these conditions.

6.6. Reaction with Tetramethylpiperidine *N*-Oxide

The reaction of triplet carbenes with a persistent nitroxide such as 2,2,6,6-tetra-methylpiperidine *N*-oxide (TEMPO, **84**) to form benzophenone would be spin allowed and >100-kcal/mol exothermic (Scheme 9.26). The reaction has a few parallels in free radical chemistry, such as the reaction of *tert*-butoxyl with carbon monoxide (to yield CO_2) or with phosphorus (III) substrates to yield P(V) products.[124]

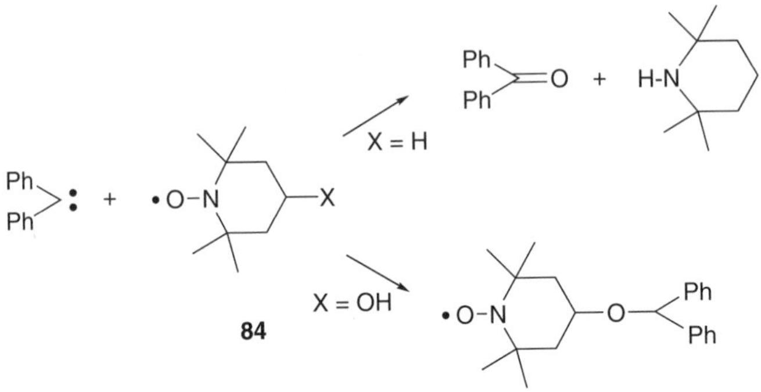

Scheme 9.26

Photolysis of diphenyldiazomethane in MeCN solution in the presence of suffi-cient **84** yields benzophenone in >90% yield and tetramethylpiperidine in equi-molar amounts. A similar photolysis with 4-hydroxy-TEMPO (**84**, X = OH) gives benzophenone and ether (**85**) in ~16:1 ratio, indicating that attack at the nitroxide center predominated even though it is more hindered than the OH group. In the case of cyclohexanol, which is used as a reference substrate for 4-hydroxy-TEMPO, the reaction with DPC gives >90% yield of the expected ether.[125]

Triplet DPC is readily detectable at 315 nm and is quenched by TEMPOs. The bimolecular rate constant (k_q) for carbene scavenging is determined by plotting the pseudo-first-order rate constant for carbene decay as a function of substrate concen-tration. The values of k_q (2–3×10^8 $M^{-1} s^{-1}$) are practically the same for the reac-tion of all TEMPOs employed and are approximately one-tenth of that for carbene scavenging by O_2. One expects one-third of the encounters to have the appropriate spin configuration for the reaction. Thus the process is, while fast, well below the diffusion limit.

The reaction with 4-hydroxy-TEMPO leading to the ether is particularly inter-esting as the reaction could be spin allowed (i.e., doublet + triplet → doublet) if sufficient interaction between the O—H and nitroxide centers takes place. However, the EPR parameters for 4-hydroxy-TEMPO suggest that the interaction between the two sites is small.[126] The magnitude of the relaxation of spin conservation rules seems unclear, but the kinetic results show virtually no effect. The rate constant for insertion at the O—H bond is ~2×10^7 $M^{-1} s^{-1}$, which is essentially the same in the case of cyclohexanol. This result may suggest that the dominant factor controlling differences in reactivity between singlet and triplet carbenes appears to be their orbital occupancy, rather than the overall multiplicity.[125]

7. ADDITION TO DOUBLE BONDS

A singlet carbene adds in concerted fashion to an alkene to form a cyclopropane stereospecifically, while the reaction of a triplet is stepwise and nonstereospecific (Skell–Woodworth rule). This difference is because the intermediate triplet 1,3-diradical undergoes bond rotation before it intersystem crosses to the singlet. The intervention of a diradical intermediate can be proven by using an alkene bearing a substituent susceptible to radical rearrangement. For example, the reaction of a triplet carbene with dicyclopropylethylene (**86**) results in formation of a product (**88**) resulting from cyclopropane ring opening of the intermediate diradical (**87′**) (Scheme 9.27).[127]

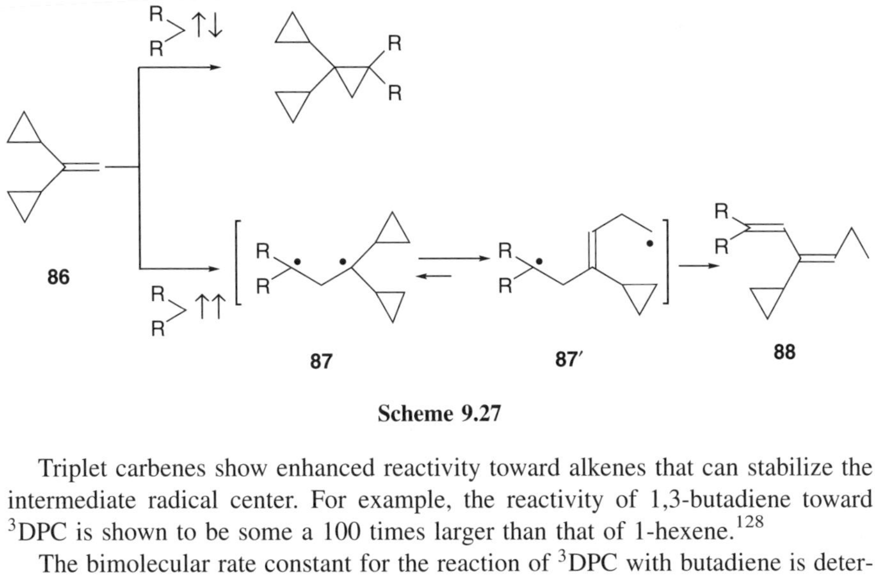

Scheme 9.27

Triplet carbenes show enhanced reactivity toward alkenes that can stabilize the intermediate radical center. For example, the reactivity of 1,3-butadiene toward ^{3}DPC is shown to be some a 100 times larger than that of 1-hexene.[128]

The bimolecular rate constant for the reaction of ^{3}DPC with butadiene is determined to be $6.5 \times 10^5\ M^{-1}\,s^{-1}$. Isoprene can be employed as a selective trap for triplet carbenes. Styrene is also shown to be an efficient trap for triplet carbene. (E)-β-Deutero-α-methylstyrene (**89**) is a very convenient reagent to diagnose the multiplicity of the reacting carbene because it reacts with both singlet and triplet carbenes with different stereochemical outcomes. The stereochemistry of the adduct cyclopropane (**90**) can be easily judged by ^{1}H NMR (Scheme 9.28). For example, BA (**22**) reacts with styrene with total loss of stereochemistry, while in the reaction with dimethoxy FL (**23a**), the expected cyclopropane is obtained with complete retention of stereochemistry. The rate constants for the additions are $(1.2 \pm 0.2) \times 10^7$ and

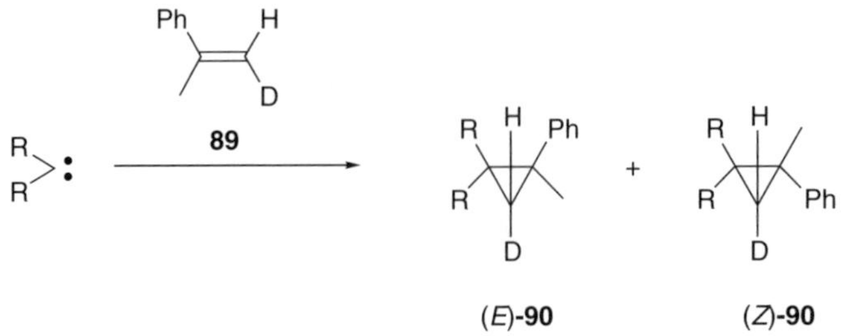
Ph
H
D
89
R
R
R
H
Ph
R
D
+
R
H
R
Ph
D
(E)-90
(Z)-90

Scheme 9.28

$(1.3 \pm 0.1) \times 10^8\ M^{-1}\,s^{-1}$, respectively, indicating that the addition of triplet is slightly less efficient than that of the singlet. Since the reactivity difference between the singlet and triplet is not large, retention of configuration depends on S–T gap of the reacting carbenes. Retention of configuration, with FL (**23**) is 49%, while that of 2,3-benzofluorenylidene (BFL, **23b**) is 80%.[129] The parameter ΔG_{ST} for FL and BFL is <1.9 and <1.0 kcal/mol, respectively.

An alkene bearing a heteroatom also shows multiplicity selectivity. Moderately electrophilic singlet carbenes interact with the lone-pair electrons on the heteroatom to generate an ylide by accepting an electron pair in the empty π orbital, and this interaction is more efficient than that with π electron of the double bond. Thus,

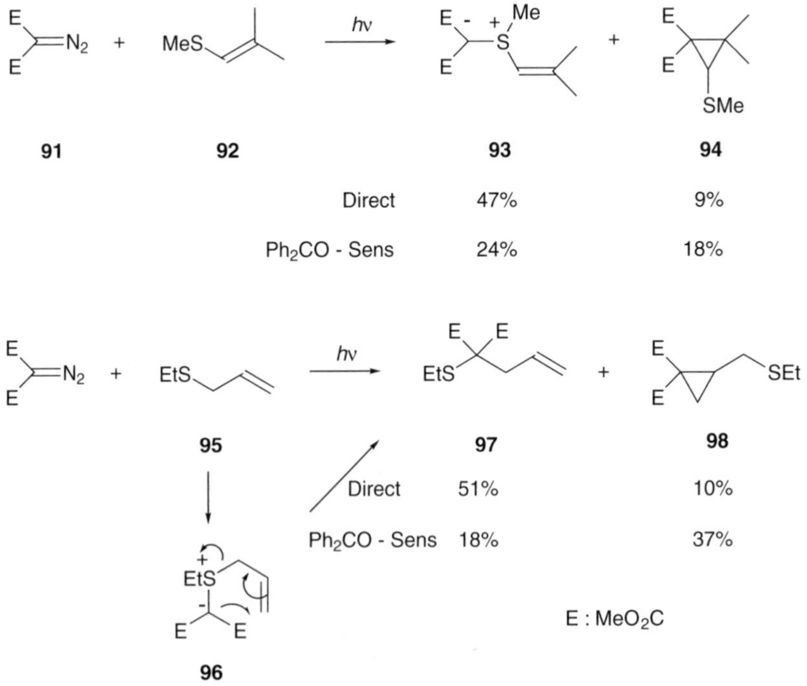
E
=N2
E
+
MeS
hν
E
-
+
Me
S
E
+
E
E
SMe
91
92
93
94
Direct
47%
9%
Ph2CO - Sens
24%
18%
E
=N2
E
+
EtS
hν
E
E
EtS
+
E
E
SEt
95
97
98
Direct
51%
10%
Ph2CO - Sens
18%
37%
+
EtS
-
E
E
96
E : MeO2C

Scheme 9.29

the reaction with alkenes usually results in the formation of the product by way of the ylidic intermediates. On the other hand, for the triplet state with a half-filled orbital, the most reactive center in the alkene is still the olefinic π orbitals. For example, direct photolysis of diazomalonate (**91**) with vinyl sufide (**92**) affords the sulfur ylide (**93**) as a main product along with a small amount of cyclopropane (**94**). However, cyclopropane formation is increased when the photolysis is carried out in the presence of a triplet sensitizer. Similar photolysis with allyl sulfide (**95**) produces the formal S–C insertion product (**97**) as a main product, obviously formed as a result of Claisen-type rearrangement of the sulfur ylide (**96**) along with a cyclopropane (**98**). Again, triplet-sensitized photolysis results in a marked increase in the yield of the cyclopropane (Scheme 9.29).[130]

The transition state of singlet carbene cycloaddition to alkenes involves an electrophilic approach of the vacant p orbital to the π bond of alkenes. By contrast, the first step of the triplet addition process may involve the in-plane σ orbital of the carbene. As in the case of C–H insertion (see Section 5.1), the difference in the transition structure between the singlet and triplet cycloaddition becomes important in the intramolecular process, especially when approach to a double bond is restricted by ring strain. Direct photolysis of (E)-2-(2-butenyl)phenyldiazomethane (**99**) in the presence of methanol gives 1-ethenyl-1,1a,6,6a-tetrahydrocycloprop [a]indene [**100**, 29%, (E/Z) = 10:1] and 1-(2-butenyl)-2-(methoxymethyl)benzene (**101**, 67%). Triplet-sensitized photolysis results in a marked increase in the indene (52%, E/Z = 1.3:1) at the expense of the ether formation (4%) (Scheme 9.30). On the other hand, direct photolysis of phenyldiazomethane in an equimolar mixture of

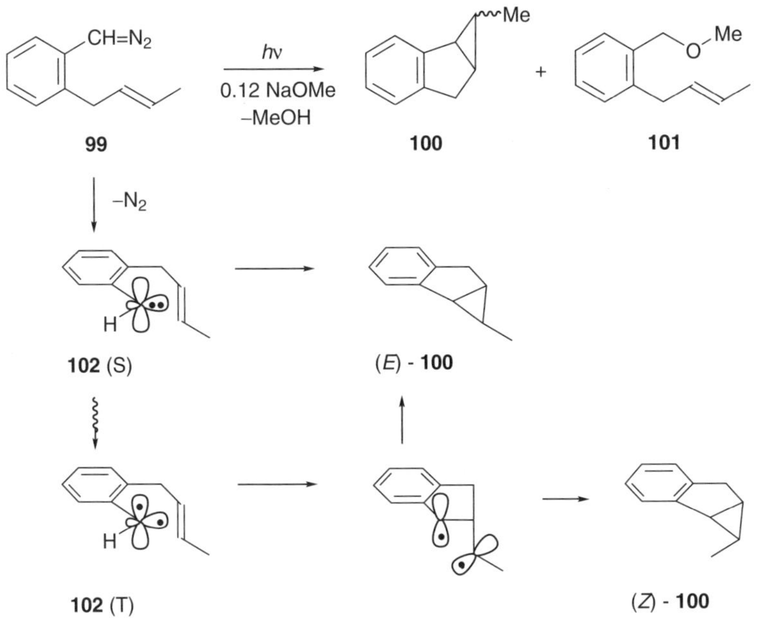

Scheme 9.30

methanol and (E)-2-butene gives methyl benzyl ether (30%) and 2,3-dimethyl-1-phenylcyclopropane (70%, $(E/Z) = 68.9/0.3$). The triplet-sensitized photolysis results in ether (27%) along with the cyclopropane [73%, $(E/Z) = 71.5/1.6$], indicating a very slight change in the product distribution.[131]

The results are explained as indicating that the addition of triplet arylcarbenes to intramolecular double bonds is accelerated by factor of 300–800 relative to intermolecular addition. The intramolecular addition reactions of singlet arylcarbenes exhibit much smaller rate enhancements. The most stable planar conformer of singlet (**102**) cannot interact with the π bond of an allyl group attached to the ortho position. Rotation about the bond connecting the divalent carbon to the ring must occur in order for an electrophilic approach to take place. This rotation will result in the loss of benzylic stabilization. In marked contrast, the first step of the triplet addition can take place with no rotation of the divalent carbon.[131]

8. EXCITED STATES OF TRIPLET CARBENE[132]

8.1. Product Studies

Some triplet DPCs are stable in organic glass matrices at 77 K as long as the matrix conditions are retained. Thus, they can be photoexcited upon continued irradiation. Therefore an electronically excited state of the triplet can be generated. Actually, rather strong triplet–triplet fluorescence is observed for DPC in a hydrocarbon matrix at 77 K. In these studies, it was noted that the intensity of the emission decreased rapidly at prolonged photolysis time with concomitant appearance of solvent and diphenylmethyl radicals. Thus, it was proposed that the products from prolonged exposure times result from reaction of photoexcited DPC. More recently, it has been shown that in the photolysis of diphenyldiazomethane (**30**) in ethanol at 77 K, the ratio of the C—H to O—H insertion markedly increased with increasing

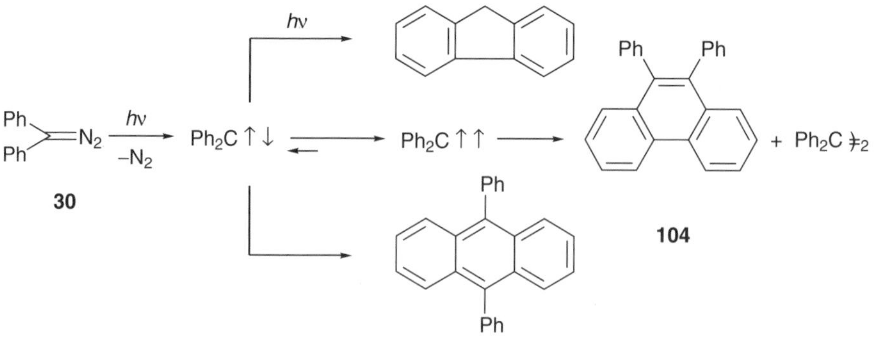

Scheme 9.31

light intensity and photolysis time. The results are interpreted as indicating that an excited triplet DPC reacts with the matrix by hydrogen atom abstraction to give ultimately the C—H insertion product (see Section 5.4.1).[133]

In addition to the rather trivial differences mentioned above, laser irradiation can also lead to products as a result of reexcitaion of the carbenes. Thus, excitation of **30** in isooctane with a pulse of the 249-nm line from a KrF excimer laser results in the formation of 9,10-diphenylanthrancene (**103**), 9,10-diphenylphenanthrene (**104**), and fluorene, in addition to tetraphenylethylene (Scheme 9.31). Conventional lamp irradiation of **30** results in the formation of benzophenone azine as a major product. None of the products mentioned above are detected. Moreover, the yield of both **103** and fluorene increased markedly with increased laser power. While the details of the mechanism of this reaction are not certain yet, it is clear from the dependence on laser power that some of these products arise from carbene photochemistry.[134]

8.2. Spectroscopic Studies

More direct evidence for the intervention of excited states of triplet carbenes in reactions in solution is obtained by spectroscopic studies. Thus, picosecond lasers make it possible to study the quenching of carbene fluorescence by various substrates in solution at room temperature. Diphenylcarbene is generated upon laser photolysis of **30** and a second UV laser pulse is time delayed by 8 ns and is used to excite the carbene, thereby producing the excited triplet DPC (Scheme 9.32). The fluorescence of ^{3}DPC* is then monitored with a streak camera. The fluorescence

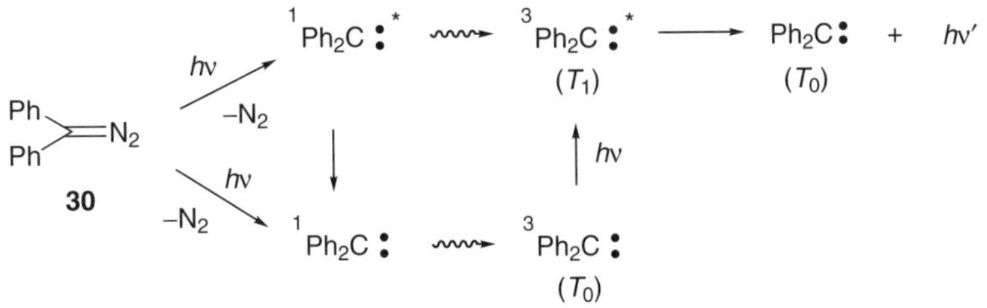

Scheme 9.32

lifetime of ^{3}DPC* in acetonitrile with no quencher is found to be 3.8 ns. The decay rate is markedly increased upon addition of methanol and isoprene.[135] A plot of the decay constant against methanol concentration is used to obtain the bimolecular rate constant. From the slope of the lines, values of $(3.1 \pm 0.4) \times 10^8 \ M^{-1} s^{-1}$ for methanol and $(2.1 \pm 0.3) \times 10^9 \ M^{-1} s^{-1}$ for isoprene are obtained. These values are to be compared to those known for the reaction of the lowest singlet and ground triplet states of DPC. The observed rapid diffusion-controlled reaction of methanol with DPC ($\sim 2 \times 10^{10} \ M^{-1} s^{-1}$) is thought to occur via the low-lying singlet state. The reaction of excited ^{3}PDC with methanol is thus approximately two orders of magnitude slower than that of ^{1}DPC. On the other hand, there is roughly a factor of 10^4 difference in the rate constants between the ground ^{3}DPC (3.5×10^5 M^{-1} s^{-1}) and the excited ^{3}PDC* $(3.1 \pm 0.4) \times 10^8$ M^{-1} s^{-1} in the reactivity of isoprene.[135]

Fluorescence of ^{3}PDC* is also quenched by amines. The ordering of reactivity is tertiary > secondary > primary, which follows inversely the ionization potential (Table 9.13). The results are explained as indicating that ^{3}PDC* undergoes photo-reduction by amines, thereby forming triplet charge-transfer intermediates as the primary step in quenching. Therefore, the mechanism of the ^{3}PDC* reaction is not the same as the proposed mode of reaction of ^{1}PDC, which involves direct formation of an ylide intermediate by electrophilic attack on the lone-pair electrons of the amine (Table 9.13).[136]

Fluorescence quenching studies can establish the rate constant at which a certain substrate interacts with the excited carbene, but they cannot provide any independent mechanistic information. Absorption studies are somewhat more informative in that the primary product of reaction can sometimes be detected directly. In the reaction of di-p-tolylcarbene with CCl$_4$, the radical, (MeC$_6$H$_4$)$_2$CCl, obviously formed as a result of abstraction of Cl atom from the substrate, is detected. Its formation can be monitored to give a rate constant of 1.1×10^9 M^{-1} s^{-1} for the excited state, which should be compared with a rate of $\sim 2 \times 10^6$ M^{-1} s^{-1} for ground-state triplet DPC.[137]

The rather short lifetime (a few nanoseconds) of the triplet excited carbene makes extensive studies of intermolecular processes difficult. However, the excited-state lifetime (60 ns) of triplet dimesitylcarbene (**19c**) is exceptionally large, probably because of decreased efficiency of both intermolecular and intramolecular deactivation pathways. Intermolecular rate constants for the reaction with CCl$_4$, O$_2$ and 1,4-cyclohexadiene have been determined.[138]

TABLE 9.13. Bimolecular Rate Constants for the Reaction of Excited Triplet Diarylcarbenes with Substrates[a]

Ar in Ar–C–Ar	Substrate	Solvents	k $(M^{-1}$ s$^{-1})$	Reference
Ph	MeOH	MeCN	$(3.1 \pm 0.4) \times 10^8$	135b
	MeOD		$(2.02 \pm 0.08) \times 10^8$	
	i-PrOH		$(1.46 \pm 0.04) \times 10^8$	
	t-BuOH		$(0.83 \pm 0.08) \times 10^8$	
	Isoprene		$(2.1 \pm 0.3) \times 10^9$	135a
	DABCO		$(2.6 \pm 0.2) \times 10^{10}$	136
	NEt$_3$		$(5.4 \pm 0.4) \times 10^9$	
	HNEt$_2$		$(9.8 \pm 1.2) \times 10^8$	
	t-BuNH$_2$		$(2.7 \pm 0.4) \times 10^7$	
4-CH$_3$C$_6$H$_4$	CCl$_4$	i-C$_8$H$_{18}$	1.1×10^9	137
Mes	CCl$_4$	PhH	$(7.3 \pm 0.6) \times 10^9$	138
	O$_2$		$(4.4 \pm 0.8) \times 10^{10}$	
	1,4-CHD		$\sim 3 \times 10^6$	

[a] 1,4-Diaza[2.2.2]bicyclooctane = DABCO, 1,4-CHD = 1,4-cyclohexadiene, i-C$_8$H$_{18}$ = isooctane.

The quenching rate constant by O_2 is $(4.4 \pm 0.8) \times 10^{10}$ M^{-1} s^{-1}. For comparison, the ground-state carbene reacts with a rate constant of 1.9×10^8 M^{-1} s^{-1}. The increase in rate of over two orders of magnitude between ground and excited carbenes is much larger than the enhancement observed for free radicals. The high rate constant for the excited carbene is interesting because it suggests that the process cannot have any significant spin restrictions. Thus, 5/9 of the encounters will be quintet events, which must lead to the ground state of both oxygen and the carbene. One third of the encounters are triplet encounters, which requires one of the reactants to convert into its singlet state. Whether the reaction leads to the carbene singlet or singlet oxygen has not been determined yet.[139]

Carbene fluorescence in solution is usually red shifted by $\sim$25–30 nm with respect to the band position observed in matrix at 77 K. This shift is attributed to emission from nonequilibrated conformations at low temperature. In matrices, the carbene is produced in a locked conformation similar to that for the precursor diazo compound but, in solution, it approaches the thermodynamically favored configuration. This difference has been demonstrated by variable temperature EPR studies of sterically congested carbenes (see Section 3.1.1.3). So, in solution, the equilibrium conformation is reached rapidly and only fluorescence from the relaxed state is observed. In support of this suggestion, the shift for dimesitylcarbene is smaller than for other carbenes, indicating that shifts are smaller when the carbene structure is such that it restricts conformational change.

8.3. Geometry of Excited Triplet State Carbenes

Conventional EPR techniques have been successfully used to measure the D and E values of matrix-isolated carbenes in the ground triplet state because the steady-state concentration of triplet species is sufficiently high in the system. The technique cannot be used, however, for excited species having triplet lifetimes of the order of 10–100 ns, since their steady-state concentration is too low. The D parameters are estimated from the external magnetic field effect on the T–T fluorescence decay in a hydrocarbon matrix at low temperature. The method is based on the effect of the Zeeman mixing on the radiative and nonradiative decay rates of the T_1–T_0 transition in the presence of a weak field. The D values are estimated by fitting the decay curve with that calculated for different D values. The D (T_1) values estimated for nonplanar DPC (c_1 symmetry) is 0.20 cm^{-1}.[140,141]

Direct detection of ^{3}DPC* is made by time-resolved EPR spectroscopy. In this method, ^{3}DPC is first generated by photolysis of **30** in a hydrocarbon matrix at 16 K and is excited by a 465-nm laser, which corresponds to a T–T absorption of the T_0 state of DPC. The transient triplet spectrum of the species having a decay rate of 160 ns at 16 K is assigned to the EPR spectrum of ^{3}DPC*. The ZFS parameters are determined by computer simulation to be $D = 0.201$ m^{-1} and $E = 0.0085$ cm^{-1}.[142] The D values observed by different methods are essentially identical.

Examination of the data indicates that for all the carbenes studied, the D (T_1) values are significantly smaller than the D (T_0) values. Since D is proportional to

$1/r^3$, r being the average separation between the two unpaired electrons, the observed decrease of D indicates that the average separation of r is significantly larger in the T_1 state than in the T_0 state. The electronic structure of the excited T_1 state involves the promotion of the unpaired electrons from the π orbital to a π^* orbital. Calculations performed on DPC (**14a**) and 2-naphthylphenylcarbene (β-**51**) indicate that this π^* orbital has a minor contribution from the carbenic $2p\pi$ atomic orbital. Thus $\pi - > \pi{:}^*$ excitation results in a delocalization of the unpaired onto the rings and to a decrease of the corresponding spin density on the carbenic center, while the spin density originating from the σ unpaired electrons is retained. As a consequence, the average separation between the two unpaired electrons is increased in the T_1 state relative to T_0 ground state, leading to a D (T_1) values significantly smaller than the $D(T_0)$.

8.4. Reactivity Differences between Triplet and Excited Triplet Carbenes

It is very interesting to examine the origin of the big difference in the reactivity between the ground and excited triplet carbene. A simple model for ^{3}DPC, ^{1}DPC, and ^{3}DPC* is given by the orbital filling scheme involving the highest occupied nonbonding molecular orbitals (σ, p) and the antibonding (π^*) orbital of the carbene. The ZFS parameters support this idea. It can be seen by this representation that ^{1}DPC and ^{3}DPC* will be similar to each other insofar as they both possess an empty low-lying orbital. The ground ^{3}DPC, on the other hand, has no empty low-lying orbitals available to it (Scheme 9.33). This simple scheme offer an explanation for the fact that the ^{3}PDC* would undergo certain reactions that ^{3}PDC would not, and ^{3}PDC* would appear to resemble ^{1}PDC in certain of its reactions.[136]

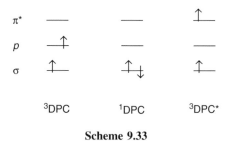

Scheme 9.33

It was shown that the patterns of the relative rates of reaction of ^{1}DPC and ^{3}DPC* with alcohols are essentially identical and followed the relative acidity of the alcohols (MeOH > i-PrOH > t-BuOH) and showed a kinetic deuterium isotope effect on reaction with the OH bond (Table 9.13). These results indicate that ^{3}DPC* attacks the O—H bond rather than the C—H bond of the alcohol. If the C—H bonds of the alcohol were attacked by ^{3}DPC*, as in the ground triplet state reaction, then one would expect 2-propanol would react faster than methanol. Lack of any discernable quenching of ^{3}DPC* by diethyl ether and THF indicates that ^{3}DPC* does not appear to react with alcohol by initial attack on the heteroatom to generate an ylide type species, and that simple C—H abstraction α to the oxygen is not important. It is proposed therefore that ^{3}DPC* reacts with alcohols in a manner analogous to the ^{1}DPC reaction with alcohol. It has been demonstrated that ^{1}DPC reacts with methanol by accepting a proton on the unpaired electrons in the σ orbital to generate a carbocation intermediate. This mechanism cannot be opperating in the reaction of ^{3}DPC*. Perhaps the greater energy of 58 kcal/mol available for ^{3}DPC* account for the reaction.[136]

On the other hand, the formation of a dimeric product in the laser photolysis of **30** in the absence of quenchers can be qualitatively interpreted by taking into account the spin distribution in ^{3}DPC*. Thus, the reaction of ^{3}DPC* with ^{3}DPC according to Scheme 9.34 will lead to the observed products. Such reactions would be favored in the presence of an intense excimer laser, which creates a high concentration of ^{3}DPC*.[140]

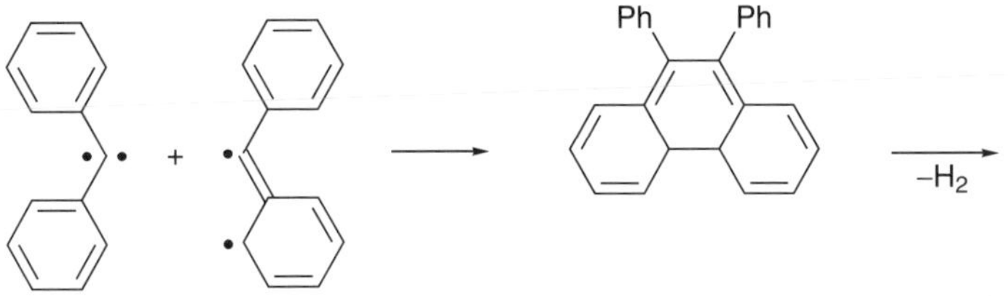

Scheme 9.34

9. PERSISTENT TRIPLET CARBENES

Now that we know the nature of triplet carbenes[143] in great detail, it will be very challenging to attempt to stabilize this highly reactive species so that they can be isolated under normal conditions. This goal is especially intriguing because triplet carbenes are importent from a practical viewpoint; Because of its unusual high spin state, triplet carbenes are an attractive spin source for organomagnetic materials.

Reactive species can be stabilized either thermodynamically or kinetically. Thermodynamic stabilization is usually done electronically by perturbation of the conjugated π system. As we have seen, electronic effects usually play an important role in stabilizing the singlet state. The stabilized singlet state becomes less reactive. They can be even isolated under ambient conditions in some cases.[144] Kinetic stabilization, on the other hand, is usually achieved by retarding the decay processes of the species in question. Sterically bulky substituents are introduced around the reactive center in order to prevent it from reacting with external reagents. Triplet states are also stabilized by electronic effects and their reactions through the upper lying singlet state may be suppressed. In this light, kinetic stabilization using steric protecting groups should be more effective for generating persistent triplet carbenes. Moreover, the introduction of sterically bulky groups around the carbenic center must expand the carbene bond angle, which results in the thermodynamic stabilization of the triplet state relative to the singlet state.

This method was shown to be very useful by Zimmerman and Paskovich in 1964 in their attempts to prepare a hindered divalent species completely unreactive toward external reagents.[145] They generated dimesitylcarbene (**19c** = **106b**) and bis(2,4,6-trichlorophenyl)carbene (**106d**).[145] Although those carbenes were not stable enough to be isolated, they exhibited unusual chemical properties. Thus, in solution at room temperature, these carbenes did not react with the parent diazo compound to give azine but dimerized instead to give tetrakis(aryl)ethylene (**107**) in 70–80% yield. Dimesitylcarbene decayed by intramolecular attack at a *o*-methyl group to form benzocyclobutene (**110**), a reaction that is not observed for 2-methyl-diphenylcarbenes under similar conditions (Scheme 9.35).

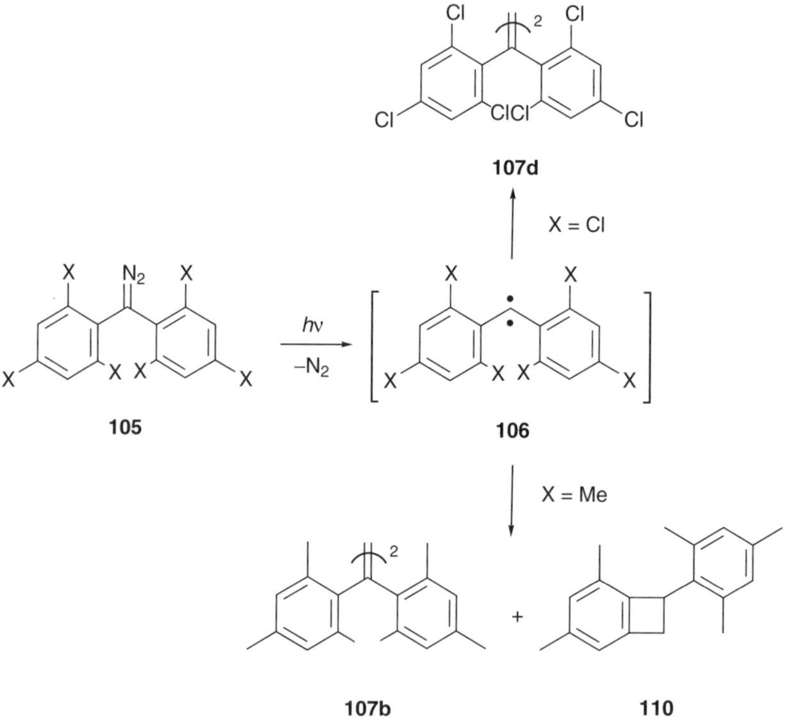

Scheme 9.35

The formation of olefinic dimerization products as the main product is rare in the decomposition of diazo compounds, whereas formation of ketazine is virtually omnipresent. The authors explained these data by assuming that the hindered diarylcarbenes do not have accessible singlet counterparts, because the singlet would require a smaller carbene angle and incur severe aryl–aryl repulsion. As a result of severe steric hindrance and consequent resistance to external attack by solvent, the hindered triplet diarylcarbene concentration builds up to the point at which dimerization occurs.

These considerations clearly suggest that kinetic stabilization is a far better way to stabilize the triplet states of carbenes than thermodynamic stabilization. It is also important to note that thermodynamic stabilization usually results in the perturbation of electronic integrity of the reactive center, as has been seen in the case of phosphinocarbene (**111**)[144a] and imidazol-2-ylidene (**112**).[144b] On the other hand, kinetic stabilization affects the original electronic nature only slightly. According to the definition, a carbene is a molecule in which the two unpaired electrons can be localized on one carbon center. Therefore, if the unpaired electrons are excessively delocalized, one may wonder whether one should still call the species a carbene.[146] In this regard, a carbene stabilized by steric protection can be regarded as a "real" stable carbene.

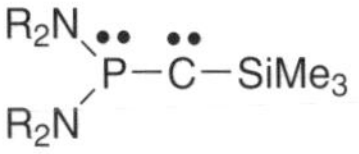

111 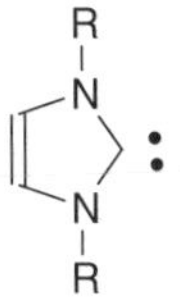   **112**

Triplet diphenylcarbene is chosen as a prototypical triplet system as it has been the most extensively studied of diarylcarbenes and abundant basic data are available. Alkyl, halo and fluoroalkyl groups have been used as kinetic protectors for ^{3}DPCs (Scheme 9.36).

9.1. Triplet Diphenylcarbenes Protected by Alkyl Groups

The *tert*-butyl group is known as one of the most effective kinetic protectors and many reactive molecules have been stabilized and even isolated by using this group. For example, the divalent center of silylenes and germylenes, the heavy atom analogues of carbenes, have been shown to be blocked by *tert*-butyl groups.

Photolysis of [2,4,6-tris(*tert*-butylphenyl]phenyldiazomethane (**105a**) results in the formation of 3-phenylindane (**113**) almost exclusively. This product is most probably produced from the photolytically generated carbene (**106a**) through insertion into the C—H bonds of the *o-tert*-butyl groups (Scheme 9.36). The LFP of **105a** in degassed benzene in the presence of benzophenone as a triplet energy sensitizer produced transient absorption bands (340 nm) ascribable to the triplet carbene (^{3}**106a**), which followed a first-order decay($k_i = 7.97 \times 10^3$ s^{-1} at 20 °C). This observation means that ^{3}**106a** decays mainly by abstracting hydrogen intramolecularly from the methyl group of an *o-tert*-butyl group to form the indane (**113**). The lifetime was estimated to be 125 µs, which is only 60 times longer than that observed for "parent" DPC ($\tau = 2$ µs) under the identical conditions.[147]

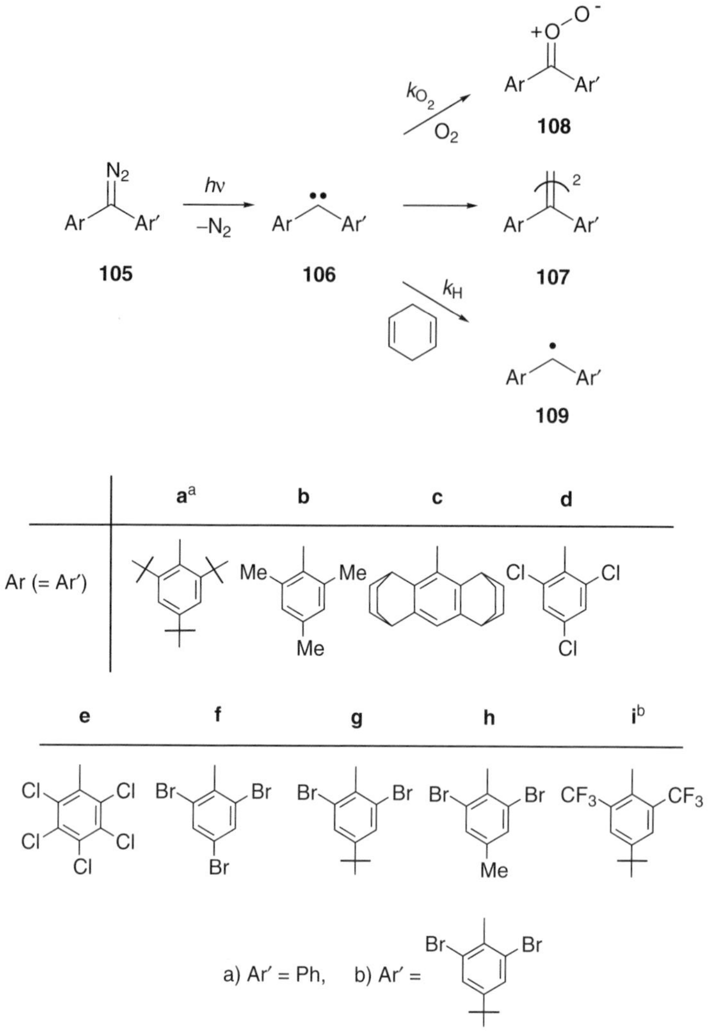

Scheme 9.36

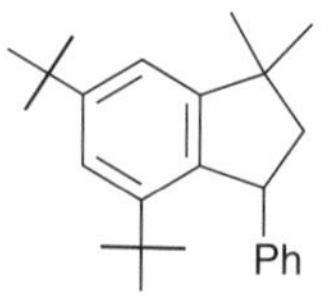

113 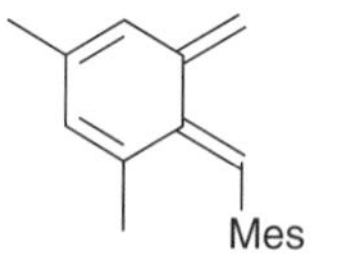   114

The methyl group is the smallest alkyl group and is generally not considered as an effective kinetic protector. However, Zimmerman and Paskovich[145] showed that decay pathways of dimesitylcarbene are suppressed both inter- and intramolecularly by nearby methyl groups (Scheme 9.35).

The LFP of the precursor diazomethane (**105b**) in degassed benzene produced a transient band (330 nm) easily ascribable to the triplet carbene (^{3}**106b**). This band disappeared more slowly than those of *tert*-butylated **106a**. The decay kinetics of the transients indicate that the absorption at 330-nm decays within 1 s to generate a new species with a maximum ~370 nm, which is too long lived to be monitored by the LFP system. The decay of the initial bands is kinetically correlated with the growth of the secondary species. Product analysis of the spent solution showed the presence of **107b** and **110** (Scheme 9.35). The initially formed transient with a maximum at 330 nm is assigned to the carbene and the second to *o*-quinodimethane (**114**). The latter species is probably formed as a result of the 1,4-H shift from the *o*-methyl to the carbene center of ^{3}**106b** and undergoes cyclization leading to benzocyclobutene (**110**, Scheme 9.35). The decay rate (k_d) of the carbene was determined to be 1.1×10 s^{-1}, while the growth rate (k_i) was 1.5 s^{-1}. From the decay curve, a half-life ($t_{1/2}$) of ^{3}**106b** was estimated to be ~160 ms, while the lifetime (τ) based on k_i was 666 ms.[148,149] This lifetime is roughly more than three orders of magnitude greater than that of DPCs protected by *tert*-butyl groups.

Those studies suggest that acyclic alkyl groups encounter a limitation when used as protecting groups for triplet carbenes because, as they are brought closer to the carbene center in order to shield it from the external regents, they are easily trapped by the reactive center that they are expected to protect. In this light, it is crucial to develop a protecting group that is sterically congesting but unreactive toward triplet carbenes.

Bicycloalkyl groups are regarded as an attractive protector because bridgehead C—H bonds are relatively unsusceptible to hydrogen abstraction,[150] and yet the bridging chains will act as protectors. Thus, bis[octahydro-1,4,:5,8-di(ethano)-anthryl]carbene (**106c**)[151] (Scheme 9.36) showed a significantly smaller E/D value (0.0265) than that observed for "parent" DPC ($E/D = 0.0464$) or **106b**, an open-chain "counterpart" of **106c** (Table 9.14). This observation suggests that the central angle of **106c** is significantly expanded because of the steric influence of the bicyclohexyl groups. The LFP studies have shown that triplet **106c** exhibited a transient band at ~320–330 nm and a weak broad band at ~480 nm. The transient absorption bands decayed in second-order kinetics ($2k/\varepsilon l = 8.4$ s^{-1}), but no new absorption was formed (Table 9.14). The approximate half-life of ^{3}**39** is estimated to be 1.5 s, which is to be compared with that of ^{3}**106b** ($t_{1/2} = 0.16$ s). Thus, the formation of all-hydrocarbon triplet carbenes having a half-life over a second under normal conditions was realized for the first time.

Bicyclohexyl groups act as an ideal kinetic protector of triplet carbene not only by quenching the intramolecular hydrogen-donating process but also by inhibiting dimerization of the carbene center.

TABLE 9.14. Zero-Field Splitting Parameters[a] and Kinetic Parameters[b] for Persistent Triplet Diarylcarbenes (106)

Carbenes (106)	D	E	$t_{1/2}(\tau)^c$	k_{O_2}	k_{CHD}	Reference
	(cm^{-1})		(s)	(M^{-1} s^{-1})		
a	0.338	∼0	0.0001	1.2×10^8		147
b	0.355	0.0116	0.16	2.0×10^8	4.6×10^2	148
c	0.400	0.0106	(1.5)	2.3×10^8	3.6×10^2	151
d	0.364	0.0115	0.018		3.5×10^3	154
e	0.402	0.0139	0.028		6.2×10^3	154
f	0.396	0.0295	1	1.1×10^7	7.4×10^2	157
g	0.397	0.0311	16	2.1×10^7	5.3×10^2	157
h	0.346	0.0275	0.2	2.3×10^7	3.2×10^2	157
i	0.381	0.023	940	8.6×10^5	1.0×10	164

[a] In MTHF at 77 K.
[b] In benzene at room temperature.
[c] Indicated by half-life ($t_{1/2}$) or lifetime (τ) depending on the mode of decay.

Note here that the reactivity (k_{O_2} or k_{CHD}) of **106c** toward typical triplet quenchers (O$_2$ or CHD) is somewhat higher than that of **106b** (Table 9.14). Comparison of the optimized geometries of **106c** and **106b** obtained by the AM1/UHF method reveals that there is slightly more space around the carbene center in **106c** than there is in **106b**. This observation suggests that **106c** is subject to attack of a small particle such as hydrogen more easily than **106b**.

9.2. Triplet Diphenylcarbenes Protected by Halogens

Alkyl groups are attractive kinetic protectors for triplet carbenes. However, they are potentially reactive toward triplet carbenes and, hence, will not be able to shield the reactive center completely. In this respect, we need to explore protecting groups that are almost completely unreactive toward triplet carbenes. Halogens are generally reactive toward the singlet state of carbene but are not reactive with the triplet.

The enormous effect of perchlorophenyl groups on the stability of arylmethyl radicals has been well documented by a series of reports by Ballester.[152] Thus, perchlorotriphenylmethyl has been shown to have a half-life on the order of 100 years in solution at room temperature in contact with air. It has therefore been termed an inert free radical. Even perchlorodiphenyl(chloro)methyl has been shown to be stable.[153] The perchlorophenyl group may be expected to exert a similar stabilizing effect on triplet DPCs.

The stability of bis(2,4,6-trichlorophenyl)carbene[145,154] generated by Zimmerman in 1964, was measured by using LFP. The transient absorption due to ^{3}**106d**, however, was found to disappear disappointingly quickly (within ∼100 ms). The decay was found to be second order, in accordance with the product analysis data, showing that dimerization to form an ethylene is the main decay pathway for ³**106d** under these conditions. From the decay curve, the dimerization rate constant (k_d) was determined to be $1.2 \times 10^7 \ M^{-1} \ s^{-1}$.[154] The half-life was determined to be 18 ms, which is four orders of magnitude larger than that observed for parent DPC but is about one-tenth that of ³**106d** (Scheme 9.35, Table 9.14).

Perchlorinated DPC ³**106e**[154] also gave a carbene dimer when generated in degassed benzene and decayed in second-order kinetics. The dimerization rate constant (k_d) was found to be $2.5 \times 10^6 \ M^{-1} \ s^{-1}$, which is approximately one order of magnitude smaller than that determined for ³**106e**. This observation suggests that the dimerization rate is sharply decreased as four additional chlorine groups are introduced at the meta positions, obviously because of the buttressing effect. The half-life is estimated to be 28 ms.

On the other hand, k_{CHD} increases only slightly as more chlorine atoms are introduced. This result is somewhat surprising in the light of the marked decrease in the dimerization rate caused by the four meta buttressing chloro substituents. It can be understood to indicate that electrophilicity of the carbene center is increased as more chlorine groups are introduced on the phenyl rings (Table 9.14). Thus, the attack of a bulky substrate (e.g., chlorinated DPC) on the carbenic center must be severely restricted, while the rate of abstraction of a very small hydrogen atom from a very efficient H donor is still controlled by the electrophilicity of the carbenic center.

Perchlorophenyl groups, which have an enormous effect on the stability of arylmethyl radicals, have only a slight effect on the stability of DPCs. This fact suggests how unstable triplet DPCs are and how much more difficult they are to stabilize than arylmethyl radicals.

Bromine groups appear to be more promising as a protecting group toward a triplet carbene center because their van der Waals radius is similar to that of methyl (Br, 1.85 Å; Me 1.75 Å). More importantly, the C—Br bond length (1.85 Å) is longer than C—C (Me, 1.50 Å).[155] This observation suggests that the o-bromo groups can overhang the reactive site more effectively. In addition, the value of ξ_1, which gives a measure of the strength of the spin-orbit interaction, is increased rather dramatically in going from chlorine (586 cm⁻¹) to bromine (2460 cm⁻¹).[156] Therefore intersystem crossing from the nascent singlet carbene to the triplet should be accelerated by introducing a bromine atom.

Bis(2,4,6-tribromophenyl)carbene (**106f**)[157] was easily generated by photolysis of the precursor diazomethane (**105f**) and was characterized by EPR spectroscopy (Scheme 9.14). The triplet carbene generated in a degassed solution at room temperature decayed very slowly, persisting for at least 30 s. The decay was found to be second order $(2k/\varepsilon l = 8.9 \ s^{-1})$. The parameter $t_{1/2}$, is estimated to be 1 s (Table 9.14).

The reaction patterns observed for ³**106f** were analogous to those observed with chlorinated and methylated DPCs. However, the structure and yield of the dimer were notably different from those observed for other persistent DPCs (e.g., **106b** and **106c,d**), which produced tetraarylethylenes in fairly good yield ($\sim$80%). Generation of ³**106f** in a degassed benzene solution resulted in the formation of a tarry matter, from which a relatively small amount (20%) of dimeric product was isolated. The dimer was identified as a phenanthrene derivative (**115**) rather than a simple carbenic dimer, tetraarylethylenes (**107f**). A possibility is that triplet carbenes undergo dimerization to form tetraarylethylenes (**107f**), which subsequently cyclize to lead to phenanthrene. The fact that the carbene dimer is formed in a complicated form in surprisingly small amount suggests that the carbene center is effectively blocked by four o-Br groups.

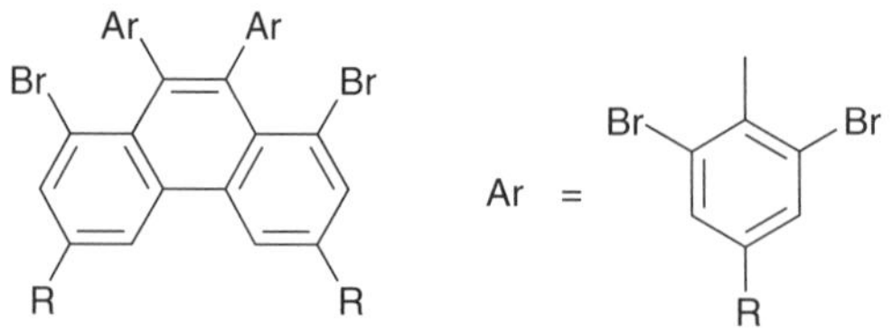

115

The observations clearly indicated that bromine groups shield the carbene center better than methyl groups, as expected. One may wonder whether through-space interaction between the triplet carbene center and the o-bromine groups may play a role in stabilizing the triplet. The EPR data clearly indicate, however, that there is no such interaction at least in a matrix at low temperature.

Anomalous effects of para-substituents on the stabilities of triplet brominated DPC were observed in this case (Scheme 9.36, the Table 9.14).[157b,c] The LFP studies show that bis(2,6-dibromophenyl)carbene ^{3}**106g**, which has *tert*-butyl groups at para positions, decayed very slowly: much slower than ^{3}**106f**. It took >200 s before all the signals due to ^{3}**106g** disappeared completely. The decay was found to be second order ($2k/\varepsilon l = 0.36$ s^{-1}), and $t_{1/2}$ was estimated to be 16 s, suggesting that ^{3}**106g** is some 20 times longer lived than ^{3}**106f**. In marked contrast, bis(2,6-dibromophenyl)carbene ^{3}**106h**, which has methyl groups at the para positions, decayed rapidly within 1 s. The decay was found to be first order in this case, and the lifetime is estimated to be 0.21 s ($t_{1/2} = 0.22$ s).

What is the origin for this anomalous effects of para substituents? The inspection of ZFS parameters clearly indicates that there are no significant changes in E/D values as the para substituents are varied (Table 9.14). These observations indicate that the para substituents do not have much effect on the geometries of brominated DPCs. The reactivities of brominated DPCs toward typical triplet quenchers, that is, oxygen and 1,4-cyclohexadiene, are also not significantly affected by the para substituents (Table 9.14).

These observations clearly support the naïve idea that the para substituents exert almost no effect at remote carbenic center of brominated DPCs, at least in terms of steric congestion. Furthermore, electronic effects do not seem important enough to change the reactivity of brominated DPCs so significantly. For example there is rather small distribution in the Hammett and spin-delocalizing substituent constants.

Benzene is recognized as a very unreactive solvent, especially for triplet carbenes. Therefore, the most reactive reactants under these conditions must be the triplet carbenes themselves. However, the products obtained from the photolysis in benzene consist of a highly complex mixture containing small amounts of carbene dimers. It is then possible that the simple dimerization of brominated DPCs must suffer from severe steric repulsion and, therefore, the carbene is forced to react at other positions. The most probable reactive sites are the aromatic rings, where spin can be delocalized.

Trityl radicals are known to undergo either methyl para or para–para coupling, depending on the substituent patterns.[158] Thus, it is likely that the brominated DPCs also undergo similar coupling. The coupling reactions of trityl radicals are not suppressed by "reactive" substituents. For example, (p-bromophenyl)diphenylmethyl undergoes methyl-para coupling.[159] On the other hand, tri(p-tolyl)methyl undergoes rapid disproportionation to yield tri(p-tolyl)methane and a quinoid compound that rapidly polymerizes.[160] However, coupling at the para positions is retarded by the presence of a *tert*-butyl group at this position.[161]

9.3. Triplet Diphenylcarbenes Protected by Trifluoromethyl Groups

The trifluoromethyl group has been regarded as an ideal kinetic protector of carbenes because it is much bigger than methyl and bromine groups and, more importantly, C—F bonds are known to be almost the only type of bond unreactive toward carbenes.[162] For example, trifluoromethylated DPCs generated in the gas phase at high temperature ($350\ ^\circ C/10^{-4}$ Torr) gave trifluoromethylated fluorene derivatives as a result of carbene–carbene rearrangement. The CF_3 group remained intact even at this high temperature.[163]

However it is almost impossibly difficult to prepare a precursor diphenyldiazomethane bearing four ortho CF_3 groups. The precursor we were able to prepare is diphenyldiazomethane (**105i**), which has two ortho bromine groups in addition to two CF_3 groups (Scheme 9.36).

Triplet carbene (**106i**)[164] was easily characterized by EPR spectra recorded after irradiating the diazomethane precursor in MTHF glass at 77 K. The triplet signals were found to be persistent even at a significantly higher temperature. For example, a significant decay of the triplet signals began only at 210 K in MTHF. In more viscous solvents, ^{3}**106i** was able to survive at much higher temperature. In triacetin, for example, no appreciable changes were observed even at 230 K. A measurable decay of the signals was observed only at 273 K (0 °C), at which the "first-order" half-life ($t_{1/2}$) was ~10 min (Table 9.14).

To estimate the lifetime of triplet carbene **106i** under our standard condition, that is, in degassed benzene at 20 °C, LFP of **105i** was carried out. However, the lifetime of ^{3}**106i** was too long to be monitored by the LFP system, and a conventional UV–vis spectroscopic method was more conveniently employed in this case. The transient signals due to ^{3}**106i** did not disappear completely even after 1 h under these conditions. The decay was analyzed as a sum of two exponential decays

$(k_1 = 1.9 \times 10^{-3} \text{ s}^{-1}; \tau_1 = 9.0 \text{ min and } k_2 = 1.0 \times 10^{-5} \text{ s}^{-1}; \tau_2 = 1602 \text{ min}).$[165] Since the half-life estimated by dynamic ESR studies is in the order of 10 min, we assume that the first-decaying component is assigned to the triplet carbene, while the slower decaying one is assigned to a secondary product formed during the decay pathway of the triplet carbene. Thus, a triplet carbene having a lifetime over several minutes was realized for the first time.[164]

The rate constants of ^{3}**106i** with typical triplet quenchers, that is, oxygen and 1,4-cyclohexadiene, were estimated and compared with those observed for the longest-lived triplet carbene known thus far. It is clear that reactivities decrease by ~2 orders of magnitude as two *o*-bromine groups in ^{3}**106g** are replaced with two CF$_3$ groups.

9.4. Triplet Polynuclear Aromatic Carbenes

Introduction of aryl groups on methylene results in a significant decrease in D values. For example, D decreases from 0.69 to 0.515 on going from methylene to phenylcarbene.[13] The D values decrease further as the aromatic ring is changed from phenyl to 1-naphthyl to 9-anthryl (Table 9.3). Similar trends are seen in diarylcarbene systems. The D values of DPC, di-1-naphthylcarbene (**15**) and di-9-anthrylcarbene (**16a**) are 0.4055, 0.3157, and 0.113 cm^{-1},[13] respectively. These data clearly suggest that the unpaired electrons are more extensively delocalized as the number of the aromatic rings is increased, and hence the triplet states are thermodynamically stabilized.

Among the many triplet diarylcarbenes known, triplet di-9-anthrylcarbene (**16a**)[166] shows the smallest D (0.113 cm^{-1}) and E (0.0011 cm^{-1}) values ever reported. This observation means that the carbene has an almost linear and perpendicular geometry with extensive delocalization of the unpaired electrons into the anthryl portions of the molecule.[167] The extensive delocalization is expected to stabilize this carbene thermodynamically, while the perpendicular geometry of the anthryl group stabilizes the carbene center kinetically, through shielding with the four *peri*-hydrogens. That is, the electronic factors and the molecular structure of the molecule seem ideal for the formation of a stable triplet carbene.

In spite of those highly favorable structural factors, ^{3}**16a** is very ephemeral.[168] Its lifetime in degassed benzene is 0.5 μs,[169] which is shorter even than that of parent triplet DPC. Product analysis studies have shown that ^{3}**16a** forms a trimer of dianthrylcarbene (**116**) as the main product (50–60%) (Scheme 9.37).[170] The trimer is the one formed as a result of a threefold coupling at position 10 of the carbene. This observation suggests that delocalization of the unpaired electrons in ^{3}**16a** leads to their leaking out from the carbene center to position 10, where sufficient spin density builds up for the trimerization to take place. At the same time, the lack of formation of olefin-type dimers through coupling of two units of ^{3}**16a** at their carbene centers indicates that the carbene center itself is indeed well shielded and stabilized.

The above observations indicate that the stability of ^{3}**16a** would increase if the trimerization reaction were somehow suppressed. Thus, an attempt was made to generate a dianthrylcarbene having a substituent at position 10.

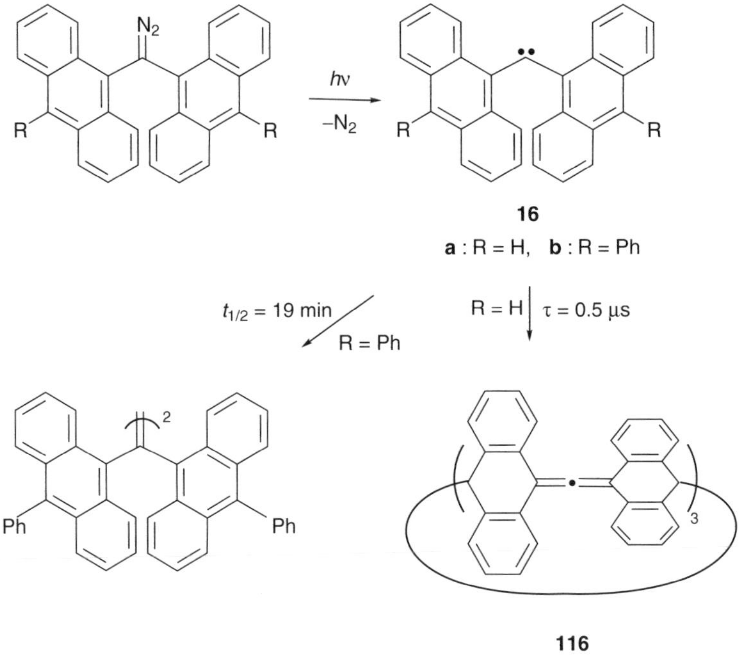

Scheme 9.37

The triplet carbene (**16b**) generated by photolysis of bis[9-(10-phenyl)anthryl]-diazomethane in MTHF glass at 77 K gave EPR signals very similar to those observed for dianthrylcarbene (**16a**) (Scheme 9.37). The ZFS parameters ($D = 0.105$ cm^{-1}, $E = 4.4 \times 10^{-4}$ cm^{-1}) derived from the signals are essentially the same as those obtained for ^{3}**16a**. These observations indicate that the phenyl groups are not in the same plane as the anthryl rings owing to the repulsion between *ortho*- and *peri*-hydrogens. The EPR signals of **16a** and **16b** differ only in their thermal stability. When the MTHF glass containing **16a** was warmed gradually, the signals due to **16a** started to disappear at ∼90 K, whereas no significant decay of the signals of ^{3}**16b** was observed up to 240 K. The signal of **16b** started to decay only at ∼270 K (∼0 °C), but was still visible when heating to 300 K (∼27 °C).

To probe the stability of ^{3}**16b** under normal conditions, the precursor diazomethane was photolyzed in degassed benzene at room temperature. The bands due to ^{3}**16b** decayed cleanly, showing isosbestic points, persisting for more than 3h before disappearing completely. The decay curve was analyzed in terms of second-order kinetics ($2k/\varepsilon l = 5.2 \times 10^{-4}$ s^{-1}). The half-life for ^{3}**16b** was estimated to be 19 min. This carbene is the longest-lived triplet thus far generated in our research group.[29]

9.5. Toward Persistent High-Spin Polycarbenes

Triplet carbenes are regarded as one of the most effective sources of high spin. The magnitude of the exchange coupling between the neighboring centers is large, and the photolytic production of polycarbenes is possible even in solid solution at cryogenic temperature under helium. Therefore, *in situ* magnetic measurements are easy if poly(diazo) precursors are available.[171,172] Actually, Iwamura and co-workers prepared a "starburst" type nonadiazo compound, 1,3,5-tris[[3′,5′-bis[phenyl(1,1-diazanediyl)-methyl]phenyl] (1,1-diazanediyl)benzene (**117**), and have demonstrated that the nine diazo groups are photolyzed therefrom at low temperature to give a pentadecet ground state (**118**) (Scheme 9.38).[173] However, those systems

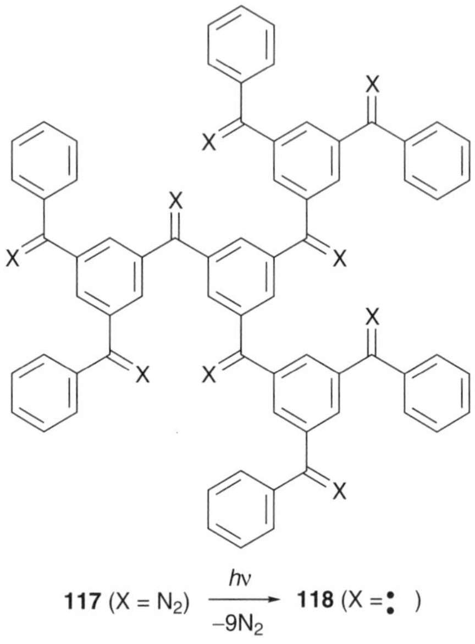

117 (X = N$_2$) $\xrightarrow[\text{–9N}_2]{h\nu}$ **118** (X = ••)

Scheme 9.38

lack the stability necessary for characterization under ambient conditions and have an inherent drawback for further extension to usable magnetic materials. Moreover, the synthesis of the precursor diazo compounds becomes more difficult and laborious as the number of carbene units increases.

In order to overcome these difficulties for realization of usable macroscopic spins, we have to find a way to stabilize triplet carbene centers and then to connect them in ferromagnetic fashion. As we have fairly stable triplet carbenes, the next step should be to explore a way to connect them while retaining a robust π-spin polarization. The topology-mediated molecular design of neutral organic high-spin systems with π conjugation has been well established in the field of purely organic magnetics.[172] In order to use polybrominated triplet DPCs as persistent spin units, and to construct as high spins as possible, we designed 1,3,5-tris-(ethynyl)phenyl as a two-dimensional linker.

The reaction scheme to realize the coupling necessary to generate the persistent carbene precursor **122** is illustrated in Scheme 9.39.[174] Diazo groups are generally labile, and, hence, these groups are usually introduced at the last step of the

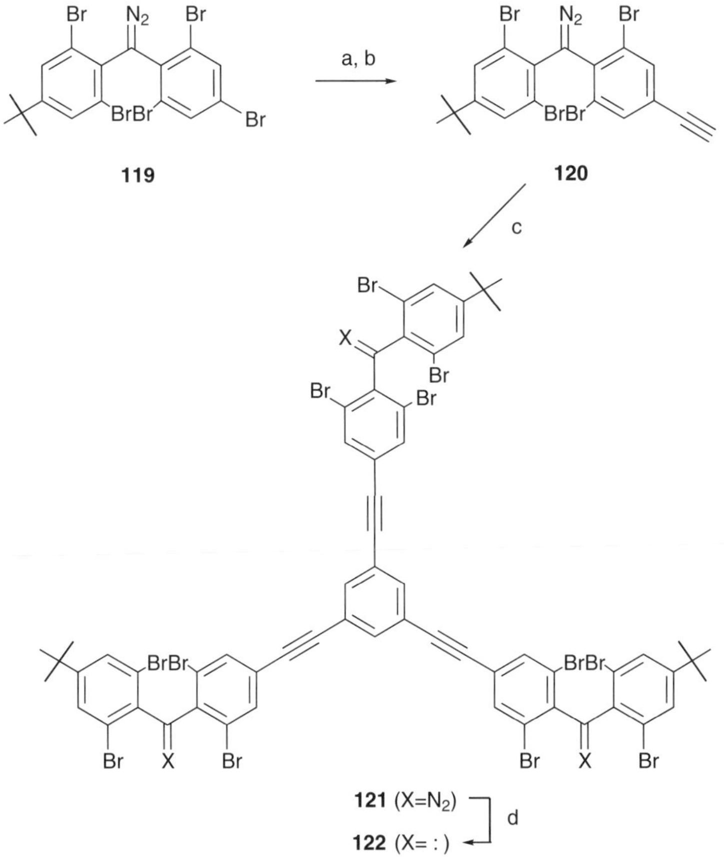

(a) Me$_3$SiC≡CH/(Ph$_3$P)$_2$PdCl$_2$/CuI; (b) KOH/MeOH; (c) 1,3,5-I$_3$C$_6$H$_3$/(Ph$_3$P)$_2$PdCl$_2$/CuI; (d) hν (λ>300 nm).

Scheme 9.39

synthesis when preparing not only mono(diazo) but also poly(diazo)compounds.[174] Application of this strategy to these highly sterically congested diazo compounds, prepared mostly by base treatment of *N*-nitrosocarbamates, is almost impossibly difficult. Fortunately, however, the polybromodiphenyldiazomethanes **119** and **120** were found to be stable enough to survive in the presence of Pd(0) and CuI under Sonogashira coupling reaction conditions.[175] The four bromine groups at the ortho positions effectively protect the carbene center from the external reagents and are also able to protect the diazo carbon effectively.

Moreover, each of the four ortho carbons is also protected from the approach of external reagents by these bromine groups. This stabilization enabled us to introduce connecting groups at the para position of the monomer diazo compound (e.g., **119**) to give the functionalized monomer **120**, and then to connect it to an appropriate linker satisfying the topological requirement for intramolecular ferromagnetic spin alignment. In practice, we were able to introduce an ethynyl functional group very selectively at the para position in a reasonable yield. Deprotection of the trimethylsilyl group proceeded almost quantitatively to give **120**. Subsequent coupling with 1,3,5-triiodobenzene took place smoothly to produce tris(diazo) compound **121**, in which three carbene precursor units were properly introduced so as to generate tris(carbene) (**122**), a molecule connected in a ferromagnetic fashion.[173]

Photolysis of **121** in MTHF glass at 5 K gave a fine-structure EPR spectrum that was different from that observed on the photolysis of the corresponding monodi-azomethane. The salient features of organic high-spin species apparently failed to appear in the observed spectrum. Canonical peaks dominate in the $g \sim 2$ range, and the key peaks appearing in the wings away from the $g \sim 2$ region are only vague. It is obvious that conventional continuous wave EPR (CW EPR) spectroscopy fails to give evidence of the spin identification of the tris(carbene).

In order to identify the spin multiplicity of the tris(carbene), field-swept two-dimensional electron spin transient nutation (2D-ESTN) spectroscopy was used.[176] This technique is based on pulsed fourier transform (FT) EPR spectro-scopic methods and is capable of elaborating straightforward information on elec-tronic and environmental structures of high-spin species even in amorphous materials, information that conventional CW EPR cannot provide. The nutation spectra unequivocally demonstrated that the observed fine structure spectrum is due to a septet spin state.

A nutation peak arising from any doublet species of byproducts was not detected, implicating a remarkable chemical stability of the tris(carbene) **122**. Furthermore, the marked thermal stability of **122** manifested itself during the annealing process. A significant change in the EPR signal shape was observed when the matrix temperature was raised >140 K and maintained for 15 min. New fine-structure signals were formed irreversibly with temperature. They were also shown to originate in a septet ground state by 2D-ESTN spectroscopy. This result is in sharp contrast with that observed for septet-state benzene-1,3,5-tris(phenylmethylene), which has been shown to be persistent only up to 85 K[177] in rigid glasses.

These studies show that the diazo compounds prepared to generate persistent triplet carbenes are also very persistent and therefore survive during the further chemical procedures to functionalize these precursor molecules. The building blocks thus prepared are very useful as they can be introduced very easily on the properly designed p-toplogical linkers, ad libitum. Moreover, the undesirable bimo-lecular side reactions that generate chemical defects among the topologically con-trolled high-spin ($S = 1$) centers for extended spin alignment are rigorously avoided. These findings open a new door to the design and preparation of new persistent high-spin molecules for more practical organic ferromagnetics.

10. CONCLUSION AND OUTLOOK

Much of the interest in carbene chemistry derives from the complexity induced by the presence of two electronic states, singlet and triplet. In this respect, it is crucial to know the energy gap between the two states with great accuracy in order to understand and control apparently complex reactions of carbenes.

We have seen that high-level computational studies are able to predict ΔE_{ST} of even fairly large systems in the gas phase quite precisely. They are also able to analyze the structural factors, influencing the relative stability (conjugation with π system, hyperconjugation, carbene bond angle, and so on). On the other hand, in the light of the experimental suggestions that the singlet state is stabilized relative to the triplet state in polar solvents, it is highly desirable to estimate ΔE_{ST} in solution with an equal accuracy.

Recent development of the time-resolved spectroscopic techniques has enabled us to study the dynamic properties of these highly reactive species in a quantitative manner, and provides an opportunity to estimate ΔE_{ST} experimentally. Although we are still not able to obtain all four key rate constants (k_{ST}, k_{TS}, k_S, and k_T) of a carbene to determine its ΔG_{ST}, a rough experimental evaluation of ΔE_{ST} has been successfully made for several arylcarbenes with rather small S–T gaps based on the assumption of preequilibrium mechanism. The data, along with the theoretical estimates, have helped us understand the nature of carbene and its chemistry in a quantitative way. However, the critical assumption can be criticized, and a new mechanism, surface crossing, is proposed. A response to this criticism can be offered if the kinetic data can be measured under the conditions under which the key reaction is carried out. At present, only in one case has the direct observation of both states been made simultaneously by using TRIR techniques. It will not be long before those key rate constants become available for more carbenes with the aid of spectroscopic methods with faster time resolution.

The other issue involved in triplet carbene chemistry is the involvement of a higher lying singlet state especially when the final products from the both states are the same. This situation is seen especially in the reaction with C–H bonds. Although CIDNP provides unequivocal evidence for the spin state of precursor carbenes, the interpretation of reactivity obtained by time-resolved spectroscopy is always hampered by the intervention of a higher lying singlet state, especially when the S–T gap is small. Moreover, philicity of not only singlet but also triplet state must also to be taken into account.

The role of spin in the reaction is an especially interesting and important issue in carbene chemistry. Apparent violation of the spin selection rule, such as the reaction of ground state singlet carbene with triplet molecular oxygen in matrices as well as in solution, and the reaction of triplet ground-state carbenes with (singlet) CO, CO_2, and N_2 in matrices, are challenging issues for the near future.

Finally, a persistent triplet carbene under normal conditions is still an on-going target, although a fairly stable one with a half-life over several minutes is available. Although the recent synthesis of stable singlet counterparts at room temperature greatly encourages the attempt, the inherent nature of triplet state, such as high reactivity toward oxygen makes the attempt far more difficult. However, it is obviously not beyond our wildest dream. The realization of a stable triplet carbene and its use as a spin source to construct organic magnetic materials will be another example to demonstrate that the research in physical organic chemistry eventually results in production of a idea for the practical world.

SUGGESTED READING

W. T. Borden, Ed.; *Diradicals*, John Wiley & Sons, Inc., New York, **1982**.

C. Wentrup, *Reactive Intermediates*, John Wiley & Sons, Inc., New York, **1984**.

G. B. Schuster, "Structure and Reactivity of Carbenes Having Aryl Substituents," *Adv. Phys. Org. Chem.* **1986**, *22*, 311.

M. S. Platz and V. M. Maloney, "Laser Flash Photolysis Studies of Triplet Carbenes," in *Kinetics and Spectroscopy of Carbenes and Biradicals*, M. S. Platz, Ed., Plenum, New York, **1990**, pp. 239–352.

W. Kirmse, "Carbenes and O-H Bond," in *Advances in Carbene Chemistry*, Vol. 1, U. Brinker, Ed., JAI Press, Greenwich, CT, **1994**, pp. 1–57.

W. Kirmse, "Carbene Protonation," in *Advances in Carbene Chemistry*, Vol. 3, U. Brinker, Ed., JAI Press, Greenwich, CT, **2001**, pp. 1–51.

M. S. Platz, "Atom-Transfer Reactions of Aromatic Carbenes," *Acc. Chem. Res.* **1988**, *21*, 236.

H. Iwamura, "High-Spin Organic Molecules and Spin Alignment in Organic Molecular Assemblies," *Adv. Phys. Org. Chem.* **1990**, *26*, 179.

REFERENCES

1. For reviews of general reactions of carbenes, see (a) W. Kirmse, *Carbene Chemistry*, 2nd ed. Academic Press, New York, **1971**. (b) R. A. Moss and M. Jones, Jr., Ed., *Carbenes*. Vol. 1 and 2, John Wiley & Sons, Inc., New York, **1973** and **1975**. (c) M. Regitz, Ed., *Carbene(oide), Carbine*, Thieme, Stuttgart, **1989**. (d) C. Wentrup, *Reactive Intermediates*, John Wiley & Sons, Inc., New York, **1984**, pp. 162–264.

2. I. Fleming, *Frontier Orbitals and Organic Chemical Reactions*, John Wiley & Sons, Inc., New York, **1976**. See also A. Rauk, *Orbital Interaction Theory of Organic Chemistry*, John Wiley & Sons, Inc., New York, **1994**, pp. 126–141.

3. J. F. Harrison, *J. Am. Chem. Soc.* **1971**, *93*, 4112.

4. H. M. Sulzbach, E. Bolton, D. Lenoir, P. v, R. Schleyer, and H. F., III. Schaefer, *J. Am. Chem. Soc.* **1996**, *118*, 9908.

5. C. M. Geise and C. M. Hadad, *J. Org. Chem.* **2000**, *65*, 8348.

6. J. E. Gano, R. H. Wettach, M. S. Platz, and V. P. Senthilnathan, *J. Am. Chem. Soc.* **1982**, *104*, 2326.

7. (a) D. R. Myers, V. P. Senthilnathan, M. S. Platz, and M. Jones, Jr., *J. Am. Chem. Soc.* **1986**, *108*, 4232. (b) S. Morgan, M. S. Platz, and M. Jones, Jr., *J. Org. Chem.* **1991**, *56*, 1351.

8. E. Iiba, K. Hirai, H. Tomioka, and Y. Yoshioka, *J. Am. Chem. Soc.* **2002**, *124*, 14308.

9. A. Nicolaides, T. Matsushita, and H. Tomioka, *J. Org. Chem.* **1999**, *64*, 3299.

10. M. Regitz and G. Maas, *Diazo Compounds. Properties and Synthesis*, Academic Press, Orlando, FL, **1986**.

11. M. T. H. Liu, Ed., *Chemistry of Diazirines*, Vol. I and II, CRC Press, Boca Raton, FL, **1986**.

12. W. Sander, G. Bucher, and S. Wierlacher, *Chem. Rev.* **1993**, *93*, 1583.

13. M. S. Platz and V. M. Maloney, in *Kinetics and Spectroscopy of Carbenes and Biradicals*, M. S. Platz, Ed., Plenum, New York, **1990**, pp. 239–352.

14. E. Wasserman, L. C. Snyder, and W. A. Yager, *J. Chem. Phys.* **1964**, *41*, 1763.

15. See for reviews: (a) A. M. Trozzolo, *Acc. Chem. Res.* **1968**, *1*, 329. (b) A. M. Trozzolo and E. Wasserman, in *Carbenes*, M. Jones, Jr. and R. A. Moss, Ed., John Wiley & Sons, Inc., New York, 1975, Vol. 2, pp. 185–206. (c) H. Tomioka, in *Advances in Strain and Interesting Organic Molecules*, Vol. 8, B. Halton, Ed., JAI Press, Greenwich, CT, **2000**, pp. 83–112.

16. For more detail description of EPR, see, (a) A. Carrington and A. D. McLachlan, *Introduction to Magnetic Resonance*, Harper International, New York, **1967**. (b) J. E. Wertz and J. R. Bolton, *Electron Spin Resonance*, McGraw-Hill, New York, **1972**.

17. C. P. Poole, Jr., *Electron Spin Resonance*, John Wiley & Sons, Inc., New York, **1967**.

18. (a) R. A. Bernheim, H. W. Bernard, S. P. Wang, L. S. Wood, and P. S. Skell, *J. Chem. Phys.* **1970**, *53*, 1280. (b) E. Wasserman, W. A. Yager, and V. Kuck, *Chem. Phys. Lett.* **1970**, *7*, 409.

19. E. Wasserman, A. M. Trozzolo, and W. A. Yager, *J. Chem. Phys.* **1964**, *40*, 2408.

20. A. M. Trozzolo, E. Wasserman, and W. A. Yager, *J. Am. Chem. Soc.* **1965**, *87*, 129.

21. (a) R. W. R. Humphreys and D. R. Arnold, *Can. J. Chem.* **1979**, *57*, 2652. (b) D. R. Arnold and W. R. Humphreys, *J. Chem. Soc., Chem. Commun.* **1978**, 181.

22. (a) J. M. Dust and D. R. Arnold, *J. Am. Chem. Soc.* **1983**, *105*, 1221 and 6531. (b) D. D. M. Wayner and D. R. Arnold, *Can. J. Chem.* **1984**, *62*, 1164. (c) D. D. M. Wayner and D. R. Arnold, *Can. J. Chem.* **1985**, *63*, 2378.

23. Y. Hu, K. Hirai, and H. Tomioka, *J. Phys. Chem. A* **1999**, *103*, 9280; Y. Hu, Y. Ishikawa, K. Hirai, and H. Tomioka, *Bull. Chem. Soc. Jpn.* **2001**, *74*, 2207.

24. R. W. Baldock, P. Hudson, and A. R. Katritzky, *J. Chem. Soc., Perkin Trans. 1* **1974**, 1422. (b) A. R. Katritzky, *J. Chem. Soc., Perkin Trans. 1* **1974**, 1427.

25. (a) R. S. Hutton, M. L. Manion, H. D. Roth, and E. Wasserman, *J. Am. Chem. Soc.* **1974**, *96*, 4680. (b) R. S. Hutton and H. D. Roth, *J. Am. Chem. Soc.* **1978**, *100*, 4324. (c) H. D. Roth and R. S. Hutton, *Tetrahedron* **1982**, *41*, 1564.

26. R. S. Hutton and H. D. Roth, *J. Am. Chem. Soc.* **1982**, *104*, 7395.

27. H. Tukada, T. Sugawara, S. Maruta, and H. Iwamura, *Tetrahedron Lett.* **1986**, *27*, 235.

28. T. Koshiyama, K. Hirai, and H. Tomioka, *J. Phys. Chem. A* **2002**, *106*, 879.

29. H. Tomioka, E. Iwamoto, H. Itakura, and K. Hirai, *Nature (London)* **2001**, *412*, 626.

30. Y. Ono and W. R. Ware, *J. Phys. Chem.* **1983**, *87*, 4426.

31. D. Weir and J. C. Scaiano, *Chem. Phys. Lett.* **1986**, *128*, 156. J. C. Scaiano and A. D. Weir, *Can. J. Chem.* **1988**, *66*, 491.

32. See for reviews: (a) I. R. Dunkin, *Chem. Soc. Rev.* **1980**, *9*, 1. (b) R. S. Sheridan, *Organic Photochemistry*, Vol. 8, A. Padwa, Ed., Dekker, New York, **1987**, pp. 159–248.

33. Z. Zhu, T. Bally, L. L. Stracener, and R. J. McMahon, *J. Am. Chem. Soc.* **1999**, *121*, 2863.

34. H. Tomioka, *Bull. Chem. Soc. Jpn.* **1998**, *71*, 1501.

35. (a) W. Sander and H. F. Bettinger, in *Advances in Carbene Chemistry*, Vol. 3, U. Brinker, Ed., JAI Press, Greenwich, CT, **2001**, pp. 160–203. (b) W. Sander and A. Kirschfeld, in *Advances in Strain and Organic Chemistry*, Vol. 4, B. Halton, Ed., JAI Press, Grennwich, CT, **1995**, pp. 1–80. (c) W. Sander, in *Carbene Chemistry*, G. Bertrand, Ed., Fontis Media

S. A., Lausanne, **2002**, pp. 1–25. (d) G. Maier and H. P. Reisenauer, in *Advances in Carbene Chemistry*, Vol. 3, U. Brinder, Ed., JAI Press, Greenwich, CT, **2001**, pp. 116–157.

36. (a) I. Moritani, S. Murahashi, H. Asitaka, K. Kimura, and H. Tsubomura, *J. Am. Chem. Soc.* **1968**, *90*, 5918. (b) I. Moritani, S. Murahashi, N. Nishino, and H. Tsubomura, *Tetrahedron Lett.* **1966**, 373.

37. G. L. Closs and B. E. Rabinow, *J. Am. Chem. Soc.* **1976**, *98*, 8190.

38. See for reviews (a) R. A. Moss and N. J. Turro, in *Kinetic and Spectroscopy of Carbenes and Biradicals*, M. S. Platz, Ed., Plenum, New York, **1990**, pp. 213–238. (b) R. A. Moss, *Acc. Chem. Res.* **1989**, *22*, 15. (c) D. Griller, A. S. Nazran, and J. C. Scaiano, *Tetrahedron* **1985**, *41*, 1525. (d) D. Griller, A. S. Nazran, and J. C. Scaiano, *Acc. Chem. Res.* **1984**, *17*, 283. (e) R. A. Moss, in *Advanced in Carbene Chemistry*, U. Brinker, Ed., JAI Press: Grennwich, CT, **1994**, Vol. 1, pp. 60–88. (f) J. E. Jackson and M. S. Platz, in *Advances in Carbene Chemistry*, Vol. 1, U. Brinker, Ed., JAI Press: Greenwich, CT, **1994**, pp. 90–160. (g) M. S. Platz, in *Advances in Carbene Chemistry*, Vol. 2, U. Brinker, Ed., JAI Press, Greenwich, CT, **1998**, pp. 134–174. (h) J. P. Toscano, in *Advances in Carbene Chemistry*, Vol. 2, U. Brinker, Ed., JAI Press, Greenwich, CT, **1998**, pp. 215–244. (i) D. C. Merrer and R. A. Moss, in *Advances in Carbene Chemistry*, Vol. 3, U. Brinker, Ed., JAI Press, Greenwich, CT, **1998**, pp. 54–114. (j) M. S. Platz, in *Carbene Chemistry*, G. Bertrand, Ed., Fontis Media S. A., Lansanne, **2002**, pp. 27–56.

39. (a) K. Iwata and H. Hamaguchi, *Appl. Spectrosc.* **1990**, *44*, 1431. (b) T. Yuzawa, C. Kato, M. W. George, and H. Hamaguchi, *Appl. Spectrosc.* **1994**, *48*, 684.

40. (a) Y. Wang, T. Yuzawa, H. Hamaguchi, and J. P. Toscano, *J. Am. Chem. Soc.* **1999**, *121*, 2875. (b) Y. Wang, C. M. Hadad, and J. P. Toscano, *J. Am. Chem. Soc.* **2002**, *124*, 1761. (c) C. M. Geise, Y. Wang, O. Mykhaylova, B. T. Frink, J. P. Toscano, and C. M. Hadad, *J. Org. Chem.* **2002**, *67*, 3079. (d) J.-L. Wang, I. Likhotvorik, and M. S. Platz, *J. Am. Chem. Soc.* **1999**, *121*, 2882. (e) G. C. Hess, B. Kohler, I. Likhotvorik, J. Penon, and M. S. Platz, *J. Am. Chem. Soc.* **2000**, *122*, 8087.

41. See for reviews: (a) W. Kirmse, in *Advances in Carbene Chemistry*, Vol. 1, U. Brinker, Ed., JAI Press, Greenwich, CT, **1994**, pp. 1–57. (b) W. Kirmse, In *Advances in Carbene Chemistry*, Vol. 3, U. Brinker, Ed., JAI Press: Greenwich, CT, **2001**, pp. 1–51.

42. D. Bethell, J. Hayes, and A. R. Newall, *J. Chem. Soc., Perkin Trans. 2* **1974**, 1307.

43. K. B. Eisental, N. J. Turro, E. V. Sitzmann, I. R. Gould, G. Hefferon, J. Lagan, and Y. Cha, *Tetrahedron* **1985**, *41*, 1543.

44. B. C. Gilbert, D. Griller, and A. S. Nazran, *J. Org. Chem.* **1985**, *50*, 4738.

45. R. Alt, I. R. Gould, H. A. Staab, and N. J. Turro, *J. Am. Chem. Soc.* **1986**, *108*, 6911.

46. M. H. Sugiyama, S. Celebi, and M. S. Platz, *J. Am. Chem. Soc.* **1992**, *114*, 966.

47. L. M. Hadel, V. M. Maloney, M. S. Platz, W. G. McGimpsey, and J. C. Scaiano, *J. Phys. Chem.* **1986**, *90*, 2488.

48. For review, G. B. Schuster, *Adv. Phys. Org. Chem.* **1986**, *22*, 311.

49. S. C. Lapin, B.-E. Brauer, and G. W. Schuster, *J. Am. Chem. Soc.* **1984**, *106*, 2092.

50. S. C. Lapin and G. B. Schuster, *J. Am. Chem. Soc.* **1985**, *107*, 4243.

51. E. V. Sitzman and K. B. Eisenthal, in *Applications of Picosecond Spectroscopy to Chemistry*: Reidel Publishing, Dordrecht, The Netherlands, **1983**, pp. 41–63.

52. (a) P. B. Grasse, B.-E. Brauer, J. J. Zupancic, K. J. Kaufman, and G. B. Schuster, *J. Am. Chem. Soc.* **1983**, *105*, 6833. (b) D. Griller, L. M. Hadel, A. S. Nazran, M. S. Platz, P. C. Wong, T. G. Savino, and J. C. Scaiano, *J. Am. Chem. Soc.* **1984**, *106*, 2227.

53. C. Chuang, S. C. Lapin, A. K. Schrock, and G. B. Schuster, *J. Am. Chem. Soc.* **1985**, *107*, 4238.

54. (a) D. Griller, A. S. Nazran, and J. C. Scaiano, *J. Am. Chem. Soc.* **1984**, *106*, 198. (b) D. Griller, A. S. Nazran, and J. C. Scaiano, *Tetrahedron* **1985**, *41*, 1525.

55. L. Salem and C. Rowland, *Angew. Chem., Int. Ed. Engl.* **1972**, *11*, 92.

56. (a) J. G. Langan, E. V. Sitzmann, and K. B. Eisenthal, *Chem. Phys. Lett.* **1984**, *110*, 521. (b) E. V. Sitzmann, J. G. Langan, and K. B. Eisenthal, *J. Am. Chem. Soc.* **1984**, *106*, 1868.

57. (a) A. S. Admasu and M. S. Platz, *J. Phys. Org. Chem.* **1992**, *5*, 123. (b) M. A. Garcia-Garibay, *J. Am. Chem. Soc.* **1993**, *115*, 7011. (c) M. A. Garcia-Garibay, C. Theroff, S. H. Shin, and J. Jernelius, *Tetrahedron Lett.* **1993**, *34*, 8415. (d) K. R. Motschiedler, J. P. Toscano, and M. A. Garcia-Garibay, *Tetrahedron Lett.* **1997**, *38*, 949.

58. (a) G. L. Closs and L. E. Closs, *J. Am. Chem. Soc.* **1969**, *91*, 4549. (b) G. L. Closs and A. D. Trifunac, *J. Am. Chem. Soc.* **1969**, *91*, 4554.

59. T. G. Saoino, V. P. Senthilnathan, and M. S. Platz, *Tetrahedron* **1986**, *42*, 2167.

60. W. Kirmse, L. Horner, and H. Hoffmann, *Liebigs Ann. Chem.* **1958**, *614*, 19.

61. M. Jones, Jr., W. Ando, M. E. Hendrick, A. Kulczycki, Jr., P. M. Howley, K. M. Hummel, and D. S. Malament, *J. Am. Chem. Soc.* **1972**, *94*, 7469.

62. (a) W. Kirmse and I. S. Özkir, *J. Am. Chem. Soc.* **1992**, *114*, 7590. (b) W. Kirmse, I. S. Özkir, and D. Schnitzler, *J. Am. Chem. Soc.* **1993**, *115*, 792.

63. D. Bethell, A. R. Newall, and D. Whittaker, *J. Chem. Soc. B.* **1971**, 23.

64. C. D. Gutsche, G. L. Bachman, W. Udell, and S. Bäuerlein, *J. Am. Chem. Soc.* **1971**, *93*, 5172.

65. See for reviews, (a) H. Fischer, *Fortschr. Chem. Forsch.* **1971**, *24*, 1. (b) H. R. Ward, *Acc. Chem. Res.* **1972**, *5*, 18. (c) R. G. Lawler, *Acc. Chem. Res.* **1972**, *5*, 25.

66. (a) R. Kaptein, *J. Am. Chem. Soc.* **1972**, *94*, 6251. (b) R. Kaptein, *J. Chem. Soc., Chem. Commun.* **1971**, 732.

67. H. D. Roth, *J. Am. Chem. Soc.* **1972**, *94*, 1761.

68. H. D. Roth, *J. Am. Chem. Soc.* **1971**, *93*, 4935.

69. H. Iwamura, Y. Imahashi, K. Kushida, K. Aoki, and S. Satoh, *Bull. Chem. Soc. Jpn.* **1976**, *49*, 1690.

70. H. D. Roth, *J. Am. Chem. Soc.* **1971**, *93*, 1527.

71. L. M. Hadel, M. S. Platz, and J. C. Scaiano, *J. Am. Chem. Soc.* **1984**, *106*, 283.

72. R. L. Barcus, M. S. Platz, and J. C. Scaiano, *J. Phys. Chem.* **1987**, *91*, 695.

73. M. V. Encinas and J. C. Scaiano, *J. Am. Chem. Soc.* **1981**, *103*, 6393.

74. J. C. Scaiano and V. Malatesta, *J. Org. Chem.* **1982**, *47*, 1455.

75. C. Chargilialoglu, K. U. Ingold, and J. C. Scaiano, *J. Am. Chem. Soc.* **1981**, *103*, 7737.

76. L. J. Johnston, J. Lusztyk, D. D. M. Wayner, A. N. Abeywickreyma, A. L. Beckwith, J. C. Scaiano, and K. U. Ingold, *J. Am. Chem. Soc.* **1985**, *107*, 4595.

77. C. N. Bauschlicher, C. F. Bender, and H. F. III., Schaeffer, *J. Am. Chem. Soc.* **1976**, *98*, 3072.

78. J. M. Tedder, *Tetrahedron* **1982**, *38*, 3072.

79. L. H. Hadel, M. S. Platz, B. B. Wright, and J. C. Scaiano, *Chem. Phys. Lett.* **1984**, *105*, 539.

80. Y. Fujiwara, M. Sasaki, Y. Tanimoto, and M. Itoh, *Chem. Phys. Lett.* **1988**, *146*, 133.

81. K. W. Field and G. B. Schuster, *J. Org. Chem.* **1988**, *53*, 4000.

82. G. W. Griller and K. A. Horn, *J. Am. Chem. Soc.* **1987**, *109*, 4919.

83. K. A. Horn and J. E. Chateauneuf, *Tetrahedron* **1985**, *41*, 1465.

84. Y. Fujiwara, Y. Tanimoto, M. Itoh, K. Hirai, and H. Tomioka, *J. Am. Chem. Soc.* **1987**, *109*, 1942.

85. See for reviews (a) H. Tomioka, *Res. Chem. Intermed.* **1994**, *20*, 605. (b) B. B. Wright, *Tetrahedron* **1985**, *41*, 1517. (c) M. S. Platz, in *Kinetic and Spectroscopy of Carbenes and Biradicals*, M. S. Platz, Ed., Plenum, New York, **1990**, pp. 143–211. (d) M. S. Platz, *Acc. Chem. Res.* **1988**, *21*, 236.

86. R. A. Moss and U.-H. Dolling, *J. Am. Chem. Soc.* **1971**, *93*, 954.

87. (a) R. A. Moss and M. A. Joyce, *J. Am. Chem. Soc.* **1977**, *99*, 1263 and 7399. (b) R. A. Moss and J. K. Huselton, *J. Am. Chem. Soc.* **1978**, *100*, 1314. (c) R. A. Moss and M. A. Joyce, *J. Am. Chem. Soc.* **1978**, *100*, 4475. (d) T. G. Savino, K. Kanakarajan, and M. S. Platz, *J. Org. Chem.* **1986**, *51*, 1305.

88. H. Tomioka and Y. Izawa, *J. Am. Chem. Soc.* **1977**, *99*, 6128.

89. (a) H. Tomioka, S. Suzuki, and Y. Izawa, *Chem. Lett.* **1982**, 843. (b) M. S. Platz, V. P. Senthilnathan, B. B. Wright, and C. W. McCurdy, Jr., *J. Am. Chem. Soc.* **1982**, *104*, 6494. (c) B. B. Wright, V. P. Senthilnathan, M. S. Platz, and C. W. McCurdy, Jr., *Tetrahedron Lett.* **1984**, *23*, 833. (d) B. B. Wright and M. S. Platz, *J. Am. Chem. Soc.* **1984**, *106*, 4175. (e) H. Tomioka, H. Okuno, and Y. Izawa, *J. Chem. Soc., Perkin Trans. 2* **1980**, 1634. (f) H. Tomioka, T. Inagaki, and Y. Izawa, *J. Chem. Soc., Chem. Commun.* 1976, 1023. (g) H. Tomioka, T. Inagaki, S. Nakamura, and Y. Izawa, *J. Chem. Soc., Perkin Trans. 1* **1979**, 130.

90. (a) V. P. Senthilnathan and M. S. Platz, *J. Am. Chem. Soc.* **1980**, *102*, 7637. (b) V. P. Senthilnathan and M. S. Platz, *J. Am. Chem. Soc.* **1981**, *103*, 5503. (c) C.-T. Lin and P. P. Gaspar, *Tetrahedron Lett.* **1980**, *21*, 3553.

91. J. Ruzicka, E. Leyva, and M. S. Platz, *J. Am. Chem. Soc.* **1992**, *114*, 897.

92. B. B. Wright, K. Kanakarajan, and M. S. Platz, *J. Phys. Chem.* **1985**, *89*, 3574.

93. P. Zuev and R. S. Sheridan, *J. Am. Chem. Soc.* **2001**, *123*, 12343.

94. (a) W. Sander and C. Kötting, *Chem. Eur. J.* **1999**, *5*, 24. (b) C. Kötting and W. Sander, *J. Am. Chem. Soc.* **1999**, *121*, 8891.

95. (a) W. Sander, R. Hübert, E. Kraka, J. Gräfenstein, and D. Cremer, *Chem. Eur. J.* **2000**, *6*, 4567. (b) H. H. Wenk, R. Hübert, and W. Sander, *J. Org. Chem.* **2001**, *66*, 7994.

96. W. Sander, C. Kötting, and R. Hübert, *J. Phys. Org. Chem.* **2000**, *13*, 561.

97. R. J. McMahon and O. L. Chapman, *J. Am. Chem. Soc.* **1987**, *109*, 683.

98. A. Admasu, M. S. Platz, A. Marcinek, J. Michalak, A. D. Gudmundsdottir, and J. Gebicki, *J. Phys. Org. Chem.* **1997**, *10*, 207.

99. S. Wierlacher, W. Sander, and M. T. H. Liu, *J. Am. Chem. Soc.* **1993**, *115*, 8943.

100. (a) P. S. Zuev and R. S. Sheridan, *J. Am. Chem. Soc.* **1994**, *116*, 4123. (b) P. S. Zuev, R. S. Sheridan, T. V. Albu, D. G. Truhlar, D. A. Hrovat, W. T. Borden, *Science* **2003**, *299*, 867.

101. J. Bigeleisen, *J. Phys. Chem.* **1952**, *56*, 823.

102. M. W. Schaffer, E. Leyva, N. Soundararajan, E. Chang, D. H. S. Chang, V. Capuano, and M. S. Platz, *J. Phys. Chem.* **1991**, *95*, 7273.

103. E. J. Dix, M. S. Herman, and J. L. Goodman, *J. Am. Chem. Soc.* **1993**, *115*, 10424.

104. J. W. Storer and K. N. Houk, *J. Am. Chem. Soc.* **1993**, *115*, 10426.

105. P. D. Bartlet and T. G. Traylor, *J. Am. Chem. Soc.* **1962**, *84*, 3408.

106. M. Girard and D. Griller, *J. Phys. Chem.* **1986**, *90*, 6801.

107. K. Ishiguro, Y. Hirano, and Y. Sawaki, *J. Org. Chem.* **1988**, *53*, 5397.

108. See for reviews; (a) W. Sander, *Angew. Chem., Int. Ed. Engl.* **1990**, *29*, 344. (b) W. H. Bunnelle, *Chem. Rev.* **1991**, *91*, 335.

109. (a) G. A. Bell and I. R. Dunkin, *J. Chem. Soc., Chem. Commun.* 1983, 1213. (b) I. R. Dunkin and C. J. Shields, *J. Chem. Soc., Chem. Commun.* **1986**, 154. (c) I. R. Dunkin and G. A. Bell, *Tetrahedron* **1985**, *41*, 339.

110. R. D. Bach, J. L. Andres, A. L. Owensby, H. B. Schlegel, and J. J. W. McDouall, *J. Am. Chem. Soc.* **1992**, *114*, 7207.

111. C. Selcuki and V. Aviyente, *Chem. Phys. Lett.* **1998**, *288*, 669.

112. G. A. Ganzer, R. S. Sheridan, and M. T. H. Liu, *J. Am. Chem. Soc.* **1986**, *108*, 1517.

113. O. L. Chapman and R. S. Sheridan, *J. Am. Chem. Soc.* **1979**, *101*, 3690.

114. (a) W. Sander, A. Kirschfeld, W. Kappert, S. Muthusamy, and M. Kiselewsky, *J. Am. Chem. Soc.* **1996**, *118*, 6508. (b) W. Sander, K. Schroeder, S. Muthusamy, A. Kirschfeld, W. Kappert, R. Boese, E. Kraka, C. Sosa, and D. Cremer, *J. Am. Chem. Soc.* **1997**, *119*, 7173.

115. D. Cremer and M. Schindler, *Chem. Phys. Lett.* **1987**, *133*, 293.

116. (a) A. M. Trozzolo, R. W. Murray, and E. Wasserman, *J. Am. Chem. Soc.* **1962**, *84*, 4990. (b) E. Wasserman, L. Barash, and W. A. Yager, *J. Am. Chem. Soc.* **1965**, *87*, 4974. (c) N. J. Turro, J. A. J. Butcher, and G. J. Hefferon, *Photochem. Photobiol.* **1981**, *34*, 517.

117. (a) W. Sander, *J. Org. Chem.* **1988**, *53*, 121. (b) W. Sander, *J. Org. Chem.* **1989**, *54*, 333. (c) W. Sander, *Spectrochim. Acta* **1987**, *43A*, 637.

118. J. C. Scaiano, W. G. McGimpsey, and H. L. Casal, *J. Org. Chem.* **1989**, *54*, 1612.

119. R. L. Barcus, L. M. Hadel, L. J. Johnston, M. S. Platz, T. G. Savino, and J. C. Scaiano, *J. Am. Chem. Soc.* **1986**, *108*, 3928.

120. Y. Fujiwara, M. Sasaki, Y. Tanimoto, and M. Itoh, *Chem. Phys. Lett.* **1988**, *146*, 133.

121. B. R. Arnold, J. C. Scaiano, G. F. Bucher, and W. W. Sander, *J. Org. Chem.* **1992**, *57*, 6469.

122. S. Morgan, M. S. Platz, M. Jones, Jr., and D. R. Myers, *J. Org. Chem.* **1991**, *56*, 1351.

123. M. T. H. Liu, R. Bonneau, and C. W. Jefford, *J. Chem. Soc., Chem. Commun.* **1990**, 1482.

124. (a) E. A. Lissi, J. C. Scaiano, and A.-E. Villa, *Chem. Commun.* **1971**, 457. (b) B. P. Roberts, *Adv. Free Radical Chem.* **1983**, *6*, 225–289. (c) W. G. Bentrude, in "*Free Radicals*," Vol. 2, J. K. Kochi, Ed., John Wiley & Sons, Inc., New York, **1973**, pp. 620–652.

125. H. L. Casal, N. H. Werstiuk, and J. C. Scaiano, *J. Org. Chem.* **1984**, *49*, 5214.

126. A. R. Forrester, in *Landott-Börnstein, Magnetic Properties of Free Radicals*, Vol. 9, Springer-Verlag, Berlin, **1979**, Chapter 6.

127. N. Shimizu and S. Nishida, *J. Am. Chem. Soc.* **1974**, *96*, 6451.

128. R. M. Etter, H. S. Skovronek, and P. S. Skell, *J. Am. Chem. Soc.* **1959**, *81*, 1008.

129. P. B. Grasse, J. J. Zupancic, S. C. Lapin, M. P. Hendrich, and G. B. Schuster, *J. Org. Chem.* **1985**, *50*, 2352.

130. (a) W. Ando, H. Higuchi, and T. Migita, *J. Org. Chem.* **1977**, *42*, 3365. (b) W. Ando, K. Nakayama, K. Ichibori, and T. Migita, *J. Am. Chem. Soc.* **1969**, *91*, 5164. (c) W. Ando, S. Kondo, K. Nakayama, K. Ichibori, H. Kohda, H. Yamamoto, I. Imai, S. Nakaido, and T. Migita, *J. Am. Chem. Soc.* **1972**, *94*, 3870.

131. (a) W. Kirmse and G. Homberger, *J. Am. Chem. Soc.* **1991**, *113*, 3925. (b) G. Homberegr, A. E. Dorigo, W. Kirmse, and K. N. Houk, *J. Am. Chem. Soc.* **1989**, *111*, 475.

132. See for review; J. C. Scaiano, in *Kinetics and Spectroscopy of Carbenes and Biradicals*; M. S. Platz, Ed., Plenum, New York, **1990**, pp. 353–368.

133. E. Leyva, R. L. Barcus, and M. S. Platz, *J. Am. Chem. Soc.* **1986**, *108*, 7786.

134. N. J. Turro, M. Aikawa, J. A. Butcher, Jr., and G. W. Griffin, *J. Am. Chem. Soc.* **1980**, *102*, 5127.

135. (a) Y. Wang, E. V. Sitzmann, F. Novak, C. Dupuy, and K. B. Eisenthal, *J. Am. Chem. Soc.* **1982**, *104*, 3238. (b) E. V. Sitzmann, Y. Wang, and K. B. Eisenthal, *J. Phys, Chem.* **1983**, *87*, 2283.

136. (a) E. V. Sitzmann, J. Langan, and K. B. Eisenthal, *Chem. Phys. Lett.* **1983**, *102*, 446. (b) K. A. Horn and B. D. Allison, *Chem. Phys. Lett.* **1985**, *116*, 114.

137. L. J. Johnston and J. C. Scaiano, *Chem. Phys. Lett.* **1985**, *116*, 109.

138. J. C. Scaiano and D. Weir, *Chem. Phys. Lett.* **1987**, *141*, 503.

139. J. C. Scaiano, M. Tanner, and D. Weir, *J. Am. Chem. Soc.* **1985**, *107*, 4396.

140. See for review; E. Migirdicyan, B. Kozankiewicz, and M. S. Platz, in *Advances in Carbene Chemistry*, Vol. 2, U. Brinker, Ed., JAI Press, Greenwich, CT, **1998**, pp. 97–132.

141. A. Després, V. Lejeune, E. Migirdicyan, A. Admasu, M. S. Platz, G. Berthier, O. Parisel, J. P. Flament, I. Baraldi, and F. Momicchioli, *J. Phys. Chem.* **1993**, *97*, 13358.

142. K. Akiyama, S. Tero-Kubota, and J. Higuchi, *J. Am. Chem. Soc.* **1998**, *120*, 8269.

143. See for reviews (a) H. Tomioka, *Acc. Chem. Res.* **1997**, *30*, 1315. (b) H. Tomioka, in *Advances in Carbene Chemistry*, Vol. 2, U. Brinker, Ed., JAI Press, Greenwich, CT, **1988**, pp. 175–214.

144. See for reviews (a) D. Bourissou, O. Guerret, F. P. Gabbai, and G. Bertrand, *Chem. Rev.* **2000**, *100*, 39. (b) A. J. III., Arduengo, *Acc. Chem. Res.* **1999**, *32*, 913.

145. H. E. Zimmerman and D. H. Paskovich, *J. Am. Chem. Soc.* **1964**, *86*, 2149.

146. (a) M. Regitz, *Angew. Chem., Int. Ed. Engl.* **1991**, *30*, 674. (b) R. Dagani, *Chem. Eng. News* **1991**, *Jan. 28*, 19; **1994**, *May 2*, 20. (c) C. Heinemann, T. Müller, Y. Apeloig, and H. Schwartz, *J. Am. Chem. Soc.* **1996**, *118*, 2023. (d) C. Beohme and G. Frenking, *J. Am. Chem. Soc.* **1996**, *118*, 2039. (e) W. Kirmse, *Angen. Chem. Int. Ed.* **2003**, *42*, 2117.

147. K. Hirai, K. Komatsu, and H. Tomioka, *Chem. Lett.* **1994**, 503.

148. H. Tomioka, H. Okada, T. Watanabe, K. Banno, K. Komatsu, and K. Hirai, *J. Am. Chem. Soc.* **1997**, *119*, 1582.

149. H. Tomioka, H. Okada, T. Watanabe, and K. Hirai, *Angew. Chem., Int. Ed. Engl.* **1994**, *33*, 873.

150. V. R. Koch and G. J. Gleicher, *J. Am. Chem. Soc.* **1971**, *93*, 1657.

151. (a) H. Tomioka, H. Mizuno, H. Itakura, and K. Hirai, *J. Chem. Soc., Chem. Commun.* **1997**, 2261. (b) H. Itakura, H. Mizuno, K. Hirai, and H. Tomioka, *J. Org. Chem.* **2000**, *65*, 8797.

152. M. Ballester, *Acc. Chem. Res.* **1985**, *18*, 380; *Adv. Phys. Org. Chem.* **1989**, *25*, 307, 321.

153. M. Ballester, J. Riera, J. Castafier, C. Badfa, and J. M. Monsó, *J. Am. Chem. Soc.* **1971**, *93*, 2115.

154. (a) H. Tomioka, K. Hirai, and C. Fujii, *Acta Chem. Scand.* **1993**, *46*, 680. (b) H. Tomioka, K. Hirai, and T. Nakayama, *J. Am. Chem. Soc.* **1993**, *115*, 1285.

155. R. S. Rowland and R. Taylor, *J. Phys. Chem.* **1996**, *100*, 7384.

156. S. C. Murov, I. Carmichael, and G. L. Hug, *Handbook of Photochemistry*, Marcel Dekker, New York, 1993.

157. (a) H. Tomioka, T. Watanabe, K. Hirai, K. Furukawa, T. Takui, and K. Itoh, *J. Am. Chem. Soc.* **1995**, *117*, 6376. (b) H. Tomioka, M. Hattori, and K. Hirai, *J. Am. Chem. Soc.* **1996**, *118*, 8723. (c) H. Tomioka, T. Watanabe, M. Hattori, N. Nomura, and K. Hirai, *J. Am. Chem. Soc.* **2002**, *124*, 474.

158. Reviews of triarylmethyls: (a) V. D. Sholle and E. G. Rozantsev, *Russ. Chem. Rev.* **1973**, *42*, 1011; (b) J. M. McBride, *Tetrahedron* **1974**, *30*, 2009.

159. S. T. Bowden and T. F. Watkins, *J. Chem. Soc.* **1940**, 1249.

160. (a) C. S. Marvel, W. H. Rieger, and M. B. Mueller, *J. Am. Chem. Soc.* **1939**, *61*, 2769. (b) C. S. Marvel, M. B. Mueller, C. M. Himel, and J. F. Kaplan, *J. Am. Chem. Soc.* **1939**, *61*, 2771.

161. A. Rajca and S. Utamupanya, *J. Org. Chem.* **1992**, *113*, 2552.

162. (a) J. A. Kerr, A. W. Kirk, B. V. O'Grady, and A. F. Trotman-Dickenson, *Chem. Commun.* **1967**, 365. (b) G. O. Pritchard, J. T. Bryant, and R. L. Thommarson, *J. Phys. Chem.* **1965**, *69*, 2804. (c) J. A. Kerr, B. V. O'Grady, and A.-F. Trotman-Dickenson, *J. Chem. Soc. A*, **1961**, 1621.

163. H. Tomioka and H. Taketsuji, *J. Chem. Soc., Chem. Commun.* **1997**, 1745.

164. K. Hirai and H. Tomioka, *J. Am. Chem. Soc.* **1999**, *121*, 10213.

165. The decay curve was analyzed in terms of second-order kinetics ($2k/\varepsilon l = 1.7 \times 10^{-3}$ s^{-1}) in our original paper (see Ref. 164) but we found that the curve was best analyzed as a sum of two exponential decays.

166. E. Wasserman, V. J. Kuck, W. A. Yager, R. S. Hutton, F. D. Greene, V. P. Abegg, and N. M. Weinshenker, *J. Am. Chem. Soc.* **1971**, *93*, 6355.

167. (a) H. Tomioka, J. Nakijima, H. Mizuno, E. Iiba, and K. Hirai, *Can. J. Chem.* **1999**, *77*, 1066. (b) H. Itakura, and H. Tomioka, *Org. Lett.* **2000**, *2*, 2995.

168. D. J. Astles, M. Girard, D. Griller, R. J. Kolt, and D. D. M. Wayner, *J. Org. Chem.* **1988**, *53*, 6053.

169. K. Hirai, Y. Nozaki, and H. Tomioka, to be published.

170. Y. Takahashi, M. Tomura, K. Yoshida, S. Murata, and H. Tomioka, *Angew. Chem., Int. Ed. Engl.* **2000**, *39*, 3478.

171. (a) K. Itoh, *Chem. Phys. Lett.* **1967**, *1*, 235. (b) E. Wasserman, R. W. Murray, W. A. Yager, A. M. Trozzolo, and G. Smolinski, *J. Am. Chem. Soc.* **1967**, *89*, 5076.

172. See for review (a) A. Rajca, *Chem. Rev.* **1994**, *94*, 871. (b) H. Iwamura, *Adv. Phys. Org. Chem.* **1990**, *26*, 179.

173. K. Matsuda, N. Nakamura, K. Inoue, N. Koga, and H. Iwamura, *Bull. Chem. Soc. Jpn.* **1996**, *69*, 1483.

174. H. Tomioka, M. Hattori, K. Hirai, K. Sato, D. Shiomi, T. Takui, and K. Itoh, *J. Am. Chem. Soc.* **1998**, *120*, 1106.

175. K. Sonogashira, in *Comprehensive Organic Synthesis*, Vol. 3, B. M. Trast and I. Fleming, Eds., Pergamon Press, Oxford, **1991**, pp. 521–549.

176. T. Takui, K. Sato, D. Shiomi, K. Itoh, T. Kaneko, E. Tsuchida, and H. Nishide, in *Magnetism; A Supramolecular Function*, O. Kahn, Ed., Kluwer Academic Publishers, Dordrecht, The Netherlands, 1996, pp. 249–280.

177. K. Matsuda, K. Takahashi, K. Inoue, N. Koga, and H. Iwamura, *J. Am. Chem. Soc.* **1995**, *117*, 5550.

CHAPTER 10

Atomic Carbon

PHILIP B. SHEVLIN

Department of Chemistry, Auburn University, Auburn University, AL

Reactive Intermediate Chemistry, edited by Robert A. Moss, Matthew S. Platz, and Maitland Jones, Jr.
ISBN 0-471-23324-2   Copyright © 2004 John Wiley & Sons, Inc.

1. HISTORICAL BACKGROUND AND SCOPE

Many of the intermediates in organic reactions derive their high reactivity from the fact that they involve low-valent forms of carbon that exhibit a thermodynamic drive to form a tetravalent carbon. In this connection, zero valent atomic carbon can be considered the ultimate in low valent carbon centered reactive intermediates. Much of the attraction of this fascinating species results from the fact that, in its drive to form four bonds, it often traverses other reactive species such as carbines, carbenes, and free radicals.

Many of the early studies of atomic carbon involved atoms generated in nuclear reactions with the pioneering work of Wolf[1] and Wolfgang[2,3] dominating in these investigations. This "hot atom chemistry" formed the basis of the important use of ^{11}C in nuclear medicine. In 1966, Skell et al.[4] developed the use of the carbon arc to generate and study the reactions of carbon atoms and published an impressive array of papers documenting new C atom reactions. Since that time, a variety of methods of producing and investigating C atom reactions has been reported. Although the chemistry of atomic carbon has been the subject of several reviews,[1–6] the last of these appeared in 1980 and it is clearly time to review the progress that has been made in this important field since that time. While this chapter will concentrate on reactions that have been reported since 1980, earlier studies will be mentioned in order to compare reactivities of C atoms generated by various methods. Although atomic carbon has been identified as an extraterrestrial species[7] and numerous studies of its rate of reaction with simple molecules under extraterrestrial conditions have been reported, this aspect of C atom chemistry will not be included. This chapter will concentrate on C atom reactions in which products and intermediates have been identified and will not consider kinetic studies in detail.

1.1. Energies and Spin States in Carbon Atom Reactions

The heats of formation (in kcal/mol) of several representative neutral carbon centered reactive intermediates are listed in Eq. 1. It is not surprising that the energies of these species are inversely proportional to the number of bonds to carbon, rendering C atoms both difficult to generate and highly reactive. The energetics are further complicated by the fact that the ground state of atomic carbon is a triplet $[C(^3P)\Delta H_f^\circ = 171$ kcal/mol] with two metastable singlet excited states $[C(^1D)$ $\Delta H_f^\circ = 201$ kcal/mol] and $[C(^1S)\Delta H_f^\circ = 233$ kcal/mol]. Since many of the C atom reactions that have been reported involve the ^1D state, this species brings an additional 30 kcal of energy to its reactions. Thus, C atom reactions are usually highly exothermic and generate products with a great deal of excess energy. This exothermicity is especially interesting when the initial product of a C atom reaction is another reactive intermediate. For example, when a carbon inserts into a C—H bond to generate a carbene, the carbene is formed with >100 kcal/mol of excess energy. Since carbenes are themselves high-energy species, it is unusual to generate them with so much excess energy and unique reactions are possible. Such considerations will be discussed in detail below.

$$
\begin{array}{ccccc}
:C: & H-\ddot{C}: & H{\displaystyle\mathop{C}^{..}}H & H{\displaystyle\mathop{C}^{|}_{..}}H & H{\displaystyle\mathop{C}^{H}}H \\
171\ \text{kcal} & 141\ \text{kcal} & 92\ \text{kcal} & 35\ \text{kcal} & -18\ \text{kcal}
\end{array} \tag{1}
$$

2. METHODS OF GENERATING ATOMIC CARBON

2.1. Nucleogenic Carbon and Nuclear Medicine

As mentioned above, the first studies of C atom reactions involved the generation of this species by nuclear reactions. Reactions that have been used include $^{14}N(n,p)$ ^{14}C,[8] $^{12}C(n,2n)$ ^{11}C,[9] $^{12}C(\gamma,n)$ ^{11}C,[10] $^{14}N(p,\alpha)$ ^{11}C,[11] $^{12}C(p,pn)$ ^{11}C,[12] $^{10}B(d,n)$ ^{11}C,[13] and $^{11}B(p,n)$ ^{11}C.[13] Since yields in these nuclear reactions are low, only highly sensitive radiochromatographic techniques can be used to identify the products of the C atom reactions. The technique takes advantage of the fact that these reactions generate either ^{14}C or ^{11}C, both of which are radioactive. The fact that only products that retain the reacting C atom can be identified is a drawback to this method from a mechanistic standpoint. However, an impressive number of substrates has been investigated and products are generally similar to those from C atoms generated by other methods. The fact that reactions of these nucleogenic C atoms often generate products containing ^{11}C has found considerable practical advantage in nuclear medicine.[14] Since ^{11}C is a positron emitter, it can be counted in the body with noninvasive positron emission tomography (PET) techniques. The use of ^{11}C in nuclear medicine generally involves the production of a simple molecule such as ^{11}CO, $^{11}CO_2$, or $^{11}CN^-$. These labeled species are then incorporated into an organic molecule of biological interest that is injected into the subject and imaged by PET. These sophisticated studies are often made more challenging by the fact that ^{11}C has a 20.4-min half-life and syntheses of relevant biomolecules must be rapid. Although a detailed discussion of the use of ^{11}C in nuclear medicine is beyond the scope of this chapter, a particularly clever example reported by Lifton and Welch[15] will be mentioned. These workers generate $^{11}CO_2$, which they expose to a green leaf in the presence of light. Photosynthesis converts the $^{11}CO_2$ into ^{11}C labeled glucose, which is separated, injected into the patient, and imaged by PET to monitor brain metabolism.

2.2. Graphite Vaporization Methods

2.2.1. The Carbon Arc. The use of the carbon arc to investigate the reactions of atomic carbon was first reported by Skell and Harris in 1965[16] and was reviewed by Skell et al.[4] in 1973. This technique was employed extensively by the Skell group and has more recently become an important method of investigating C atom reactions in our research group. Figure 10.1 shows a diagram of a typical carbon arc apparatus for generating and studying the reactions of C atoms. The reaction chamber is evacuated to 1 m Torr, and a carbon arc is struck between two graphite rods attached to water-cooled brass electrodes. Atomic carbon is vaporized by the arc and deposited on the cooled walls of the reaction vessel along with substrate introduced through an inlet tube. Although various low-temperature baths have been used to cool the reactor walls, most reactions have been run at liquid nitrogen temperatures (77 K). At the conclusion of the reaction, the reactor is warmed and the products are removed and analyzed by the traditional instrumental methods of organic chemistry [gas chromatography (GC), GC–mass spectrometry (GC/MS), nuclear magnetic resonance (NMR), infrared (IR), etc.). An advantage of this method is the fact that it is relatively easy to produce carbon vapor enriched in ^{14}C[17] and ^{13}C.[18] The use of ^{13}C atoms allows an NMR determination of the position of the attacking carbon in products facilitating mechanistic evaluations.[18]

While the carbon arc method yields products in amounts that are easily characterized, there is a number of caveats of which one must be aware. Since the carbon arc operates at extremely high temperatures ($>2000\ °C$) and emits copious amounts of light, there is the very real possibility of pyrolysis and/or photolysis of both substrate and products. These problems may be minimized by carrying out control experiments in which pyrolysis and photolysis products are identified and excluded. Maximum yields in carbon arc reactions are obtained when carbon and substrate are cocondensed. However, this technique can result in pyrolysis of substrate, which can be avoided by alternately depositing substrate and carbon on the cold reactor walls. Often both methods are employed in order to identify pyrolysis products. Since the carbon arc results in removal of macroscopic pieces of graphite from the rods, it is impossible to measure product yields based on actual carbon evaporated.

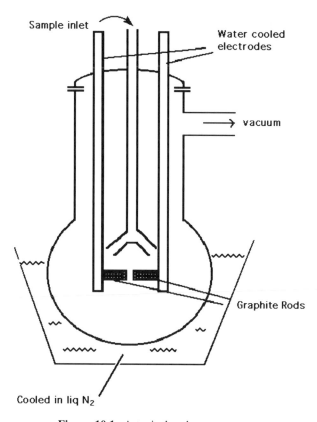

Figure 10.1. A typical carbon arc reactor.

Another complication in carbon arc studies is the fact that molecular carbon species, C_2 and C_3, are also produced when the arc is struck under high vacuum.[19] This "problem" can be a positive aspect of these reactions as it allows studies of the reactions of the molecular carbon species. When the arc is struck in the presence of an inert gas, the situation is more complex with fullerenes resulting.[20] We have not observed fullerene formation from a carbon arc under high vacuum.[21]

2.2.2. Resistive Heating of Graphite. Another method of generating atomic carbon is the simple resistive heating of graphite or a carbon fiber.[22,23] These reactions have also been carried out by cocondensing carbon with substrate. However, this method leads only to ground state $C(^3P)$ atoms,[23,24] which are relatively unreactive toward organic compounds under these conditions. In contrast, the carbon arc generates both ground-state triplet and excited singlet C atoms that react with organics in many interesting ways.

2.2.3. Laser Evaporation Methods.
The reactions of atomic carbon have also been investigated by evaporation from graphite with a laser.[25] While this method has received little attention in studies of C atom reactivity, it is important in mass spectroscopic investigations of carbon clusters[26] and in molecular beam studies of the reactivity of C(^3P).[27]

2.3. Atomic Carbon from Chemical Precursors

While evaporative methods of generating C atoms utilize bulk carbon and a great deal of thermal or photochemical energy, another approach is to use the exothermicity of the decomposition of a suitable precursor to produce C atoms. These methods yield what are called chemically generated carbon atoms and will be outlined below.

2.3.1. Carbon Suboxide Photolysis.
In principle, carbon suboxide (**1**) can be used as a precursor to atomic carbon and two molecules of carbon monoxide as shown in Eq. 2. However, this reaction is endothermic by 141 kcal/mol[28] and can only be realized in the vacuum ultraviolet (UV) at wavelengths that destroy most organic substrates. However, photolysis of **1** at 1470 Å produces C atoms in a low-temperature matrix.[29] The short wavelength flash photolysis of **1** coupled with atomic absorption has been used to measure the rate constants for various spin states of carbon with simple substrates.[30]

$$O=C=C=C=O \xrightarrow{h\nu} \; :C: \; + \; 2CO \qquad (2)$$

1

Photolysis of **1** at longer wavelengths is an excellent way of producing :C=C=O, which acts as a C atom donor to alkenes (Eq. 3).[31]

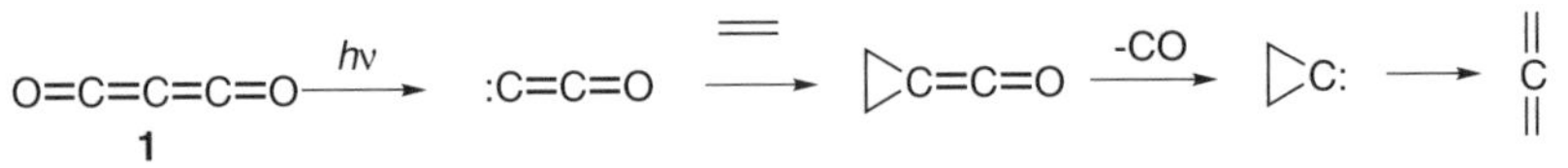

$$O=C=C=C=O \xrightarrow{h\nu} \; :C=C=O \longrightarrow \; \triangleright\!C=C=O \xrightarrow{-CO} \; \triangleright\!C: \longrightarrow \; \overset{||}{\underset{||}{C}} \qquad (3)$$

1

2.3.2. Cyanogen Azide Photolysis on Low-Temperature Matrices.
The photolysis of cyanogen azide (**2**), proceeding through cyanonitrene (**3**), has been reported as a method of studying C atom reactions on low-temperature matrices (Eq. 4).[32] Since decay of carbon to its ground state is rapid under matrix isolation conditions, the method can only be used to monitor the reactions of C(^3P).

$$N_3-C\equiv N \xrightarrow{h\nu} N_2 + N-C\equiv N \xrightarrow{h\nu} C=N=N \xrightarrow{h\nu} N_2 + C \qquad (4)$$

2   **3**

2.3.3. Diazo Compounds as Carbon Atom Precursors.

An attractive strategy for the generation of a C atom employs a diazo compound as a chemical precursor. Diazo compounds are well known to generate carbenes with the loss of the stable N_2 molecule. The trick is to design a diazo compound in which cleavage of the remaining two bonds in the initial carbene is energetically feasible. Two such diazo compounds have been designed and used in this way. The first is the quadricyclic diazo compound **4**, which has been proposed to decompose to carbene **5**, and thence to a C atom with the loss of benzene (Eq. 5).[33] Since **4** is rather tedious to synthesize and its thermolysis also produces benzene, which is reactive toward C, this interesting compound has received only limited attention as a C atom precursor.

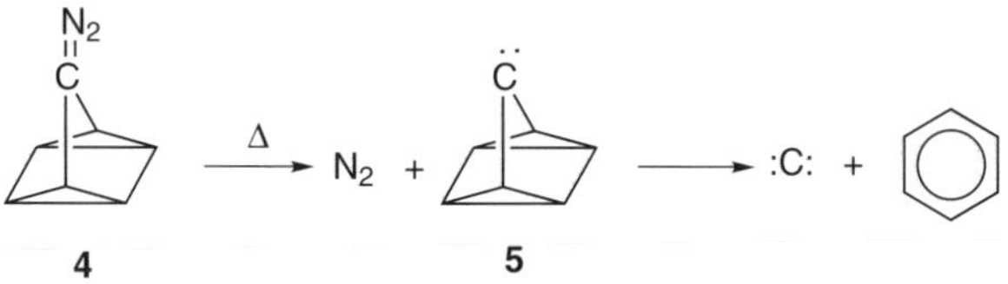

$$(5)$$

A more extensively investigated precursor to chemically generated C atoms is diazotetrazole (**6**), which is easily prepared from readily available 5-aminotetrazole (**7**).[34] In this method, **7** is converted into the corresponding diazonium chloride **8**, which is coated on the walls of a flask and pyrolysed in the presence of a gaseous substrate (Eq. 6). This technique has the drawback that **8** is extremely explosive and only small quantities can be prepared at a time.[34c] However, the synthesis of **7** with a labeled carbon is quite simple, allowing convenient evaluation of the fate of the reacting carbon.[34b]

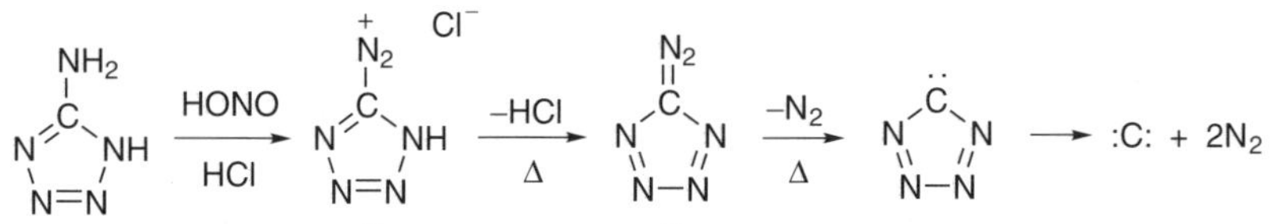

$$(6)$$

A potential problem in the use of diazo compounds as C atom precursors is the fact that intermediates in these reactions may act as C donors with the free atoms not involved. Indeed, the timing of the reactions in Eq. 6 is unknown and some of these intermediates may be bypassed in the thermolysis of **8**. However, a comparison of the reactions of carbon from **8** with those of nucleogenic and arc generated carbon reveals quite similar products from many different substrates and provides circumstantial evidence for free C atoms in the decomposition of **8**.

2.4. Production of Carbon Atoms in Molecular Beams

Recently, several interesting papers describing the reactions of atomic carbon with organic substrates in molecular beams have appeared.[27] In these studies, a beam of C atoms is generated by laser evaporation of graphite and crossed with a beam of substrate molecules. For the most part, these molecular beam studies have involved the generation of $C(^3P)$ and its reaction with simple organics. While triplet carbon condensed on a low-temperature surface tends to react with itself rather than with organic substrates, such is not the case when a beam of $C(^3P)$ interacts with organics and a rich chemistry has been observed.

2.5. Miscellaneous Methods of Generating Carbon Atoms

The reaction of CBr_4 with potassium is reported to generate free C atoms and the rate constants for reaction with methane, ethylene, and benzene have been reported.[35] The reaction of nitrogen atoms with CN radicals has also been used as a C atom source.[36] Carbon atoms have also been produced by passing organics through a microwave discharge.[37]

3. REACTIONS OF ATOMIC CARBON

3.1. Overview of Carbon Atom Reactivity

As mentioned in the introduction, a hallmark of C atom chemistry is the large amount of energy that these species bring to their reactions. These energetics often result in products with enough energy to undergo further reactions and one must always be alert to this possibility. The initial product of a C atom reaction is generally another neutral organic reactive intermediate that often has a great deal more energy than the same species formed by conventional chemistry. This fact must always be considered when examining the chemistry of these intermediates.

Since many of the reactions of C atoms are extremely exothermic, it may be that they proceed without an enthalpic barrier. Thus, selectivities observed in C atom reactions may result from free energy barriers in which entropy considerations are the major factor. In discussions of C atom reactions, we shall see that carbenes are often intermediates. Two ways in which carbenes can be produced in C atom reactions are C–H insertion (Eq. 7) and deoxygenation of carbonyl compounds (Eq. 8). In several cases, the same carbene has been generated by both methods. When this comparison has been made, the reactions will be discussed together even though they represent different aspects of C atom reactivity.

$$R-H \quad + \; C \quad \longrightarrow \quad R\overset{\cdot\cdot}{\underset{}{C}}{}_{H} \tag{7}$$

$$R\overset{O}{\underset{}{\overset{\|}{C}}}{}_{R'} \quad + \; C \quad \longrightarrow \quad CO \; + \; R\overset{\cdot\cdot}{\underset{}{C}}{}_{R'} \tag{8}$$

R, R' = organic and/or H

3.2. Reactions of Carbon Atoms with Inorganic Substrates

Reaction of $C(^1D)$ with H_2 gives $CH + H^{38}$ and $C(^3P)$ is thought to produce $^3CH_2.^{29,39}$ Triplet carbon reacts rapidly with oxygen to give carbon monoxide and $O(^3P).^{10,40}$ This extremely facile reaction of $C(^3P)$ can be used to deduce the spin state of a reacting carbon.[39a] For example, carbon may be allowed to react with a substrate, the products identified, and the reaction repeated in the presence of oxygen. If the product yields are not reduced in the presence of oxygen, these products are taken to result from the reaction of singlet C atoms. These experiments have led to the conclusion that the majority of products from the reactions of nucleogenic, arc generated, and chemically generated carbon are from the singlet state. The formation of CO in these systems indicates that $C(^3P)$ is formed but is rather unreactive toward singlet substrates. Carbon atoms react with N_2 to give $CN + N,^{10}$ with CO to yield $C_2O,^{10,41}$ and with NO to give $CN + O.^{10}$ Insertion of C into $B-B,^{42,43}$ $B-Cl,^{42}$ $B-H,^{44}$ $B-C,^{42}$ $Si-H,^{45}$ and $S-Cl^{45}$ bonds has been reported. The reaction of C atoms in a low-temperature matrix with $HCl,^{46}$ $HF,^{47}$ $Cl_2,^{48}$ and $F_2,^{49}$ is reported to give the corresponding halo and dihalocarbenes.

3.2.1. Formation of Amino Acid Precursors in the Reaction of Carbon Atoms with Ammonia.

When arc generated carbon is cocondensed with ammonia at 77 K, the volatile products are hydrogen cyanide and methylamine. Hydrolysis of the nonvolatile products generates a number of simple amino acids (Eq. 9).[50] The mechanism of this reaction involves initial C—H insertion to give aminocarbene (**9**), which either rearranges to methylenimine (**10**) or loses H_2 to give HNC.[51] Rearrangement of HNC produces HCN. Addition of HCN to **10** gives aminoacetonitrile (**11**), which may be hydrolyzed to glycine (**12**) (Eq. 10). The methylamine results from the addition of methylene (formed by H abstraction) to ammonia or H abstraction by **9**. A series of reactions with methylamine analogous to those in Eq. 10 leads to alanine (**13**) and N-methylglycine (**14**). The aspartic acid (**15**) is most likely formed by a second addition of C to **10** to give a ketenimine that subsequently adds ammonia and two molecules of HCN. When water is cocondensed with $C + NH_3$, serine is formed in addition to the products in Eq. 9.

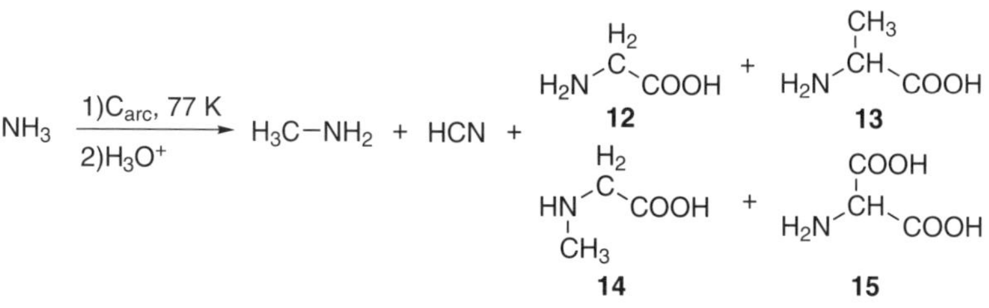

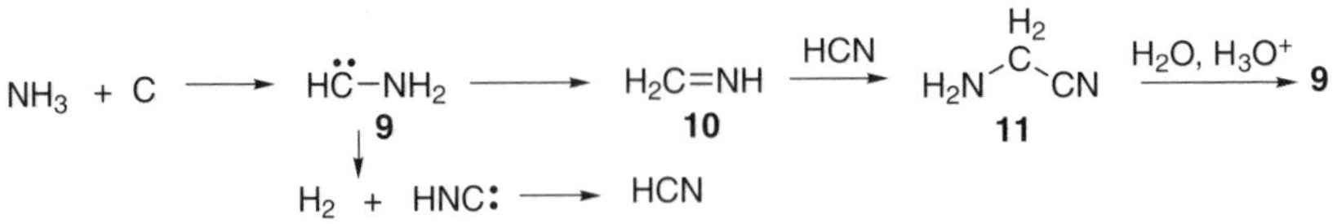

$$NH_3 + C \longrightarrow H\ddot{C}-NH_2 \xrightarrow{} H_2C=NH \xrightarrow{HCN} H_2N\overset{\overset{\displaystyle H_2}{C}}{\diagdown}CN \xrightarrow{H_2O,\ H_3O^+} 9$$
$$\underset{9}{\phantom{H\ddot{C}-NH_2}} \qquad \underset{10}{} \qquad \underset{11}{}$$
$$H_2 + HNC\!: \longrightarrow HCN$$

$$\tag{10}$$

3.2.2. Formation of Carbohydrates in the Reaction of Carbon Atoms with Water.

When carbon is cocondensed with water at 77 K, formaldehyde and aldoses up to C_5 are isolated.[52] The mechanism of this reaction involves an initial C–H insertion to give hydroxycarbene **16**, which migrates a hydrogen producing formaldehyde (Eq. 11).[53] An interesting addition of **16** to the carbonyl group in formaldehyde generates glyceraldehyde **17**.[54] Subsequent additions of **16** to product hydroxyaldehydes builds up the aldoses whose identity up to C_5 has been confirmed by GC–MS (Eq. 12).[52] The use of D_2O and excess formaldehyde as substrates generates aldoses with D on all carbons except the terminal methylene group supporting the mechanism in Eq. 12.

$$H_2O + C \longrightarrow \underset{16}{H\ddot{C}-OH} \longrightarrow H_2C=O \tag{11}$$

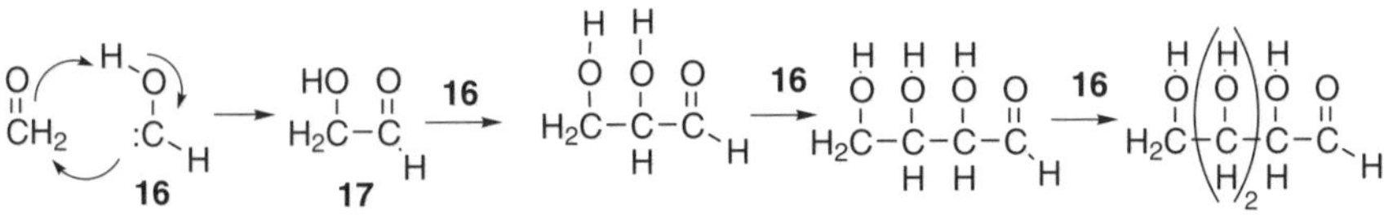

$$\tag{12}$$

A cycloaddition analogous to that in Eq. 12 also occurs with aminocarbene **9**. Thus, addition of formaldehyde to the cocondensate of $C + NH_3$ generates serine **18** by the mechanism in Eq. 13.[55] This reaction is the key step in the formation of serine when water is added to the $C + NH_3$ reaction (with $C + H_2O$ producing the $CH_2=O$). Addition of acetaldehyde to the $C + NH_3$ cocondensate results in threonine and allothreonine (Eq. 14).[55]

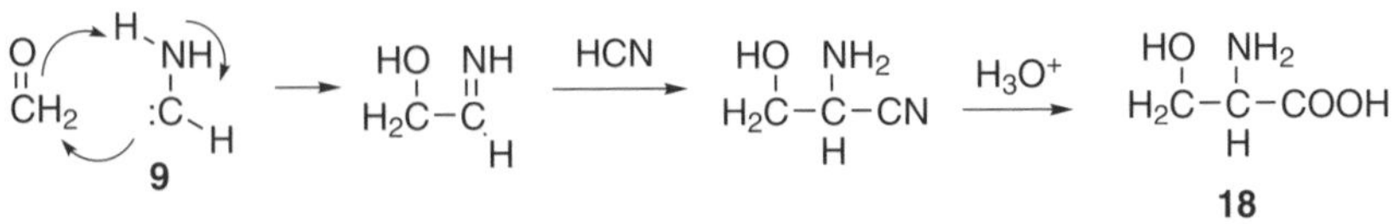

$$\tag{13}$$

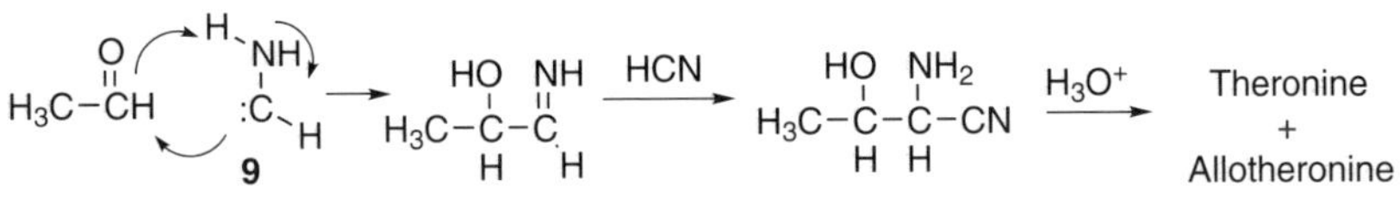

$$(14)$$

The fact that simple biochemicals are generated in these C atom reactions provides a potential pathway to these molecules under extraterrestrial conditions. It is easy to imagine interstellar C atoms condensing on a cold surface with known interstellar molecules such as NH_3, H_2O, HCN, and $CH_2=O$.

3.3. Reaction of Carbon Atoms with Simple C–H Bonds

Like many singlet carbenes, nucleogenic, arc generated and chemically generated C atoms react with aliphatic C–H bonds by insertion. In the simplest case, reaction of chemically generated C atoms with methane yields ethylene and acetylene.[56] When a mixture of CH_4 and CD_4 is used, product analysis indicates that the acetylene results from H abstraction followed by dimerization of the CH, while the ethylene results from C–H insertion followed by H migration in the carbene (Eq. 15). It seems probable that CH is formed in all reactions of carbon with hydrocarbons as acetylene is invariably produced in these reactions.

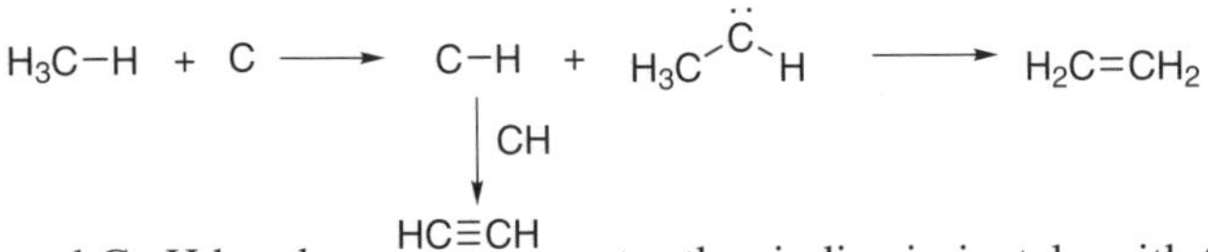

$$(15)$$

Unactivated C–H bonds appear to react rather indiscriminately with C atoms to give the corresponding carbene. For example, reaction of chemically generated carbon with propane gives products that are consistent with initial insertion into both the 1° and 2° C–H bonds to give the corresponding carbenes (Eq. 16).[34c] The products in Eq. 16, although in slightly different ratios, are also formed in the reaction of arc generated[57] and nucleogenic carbon[58] with propane.

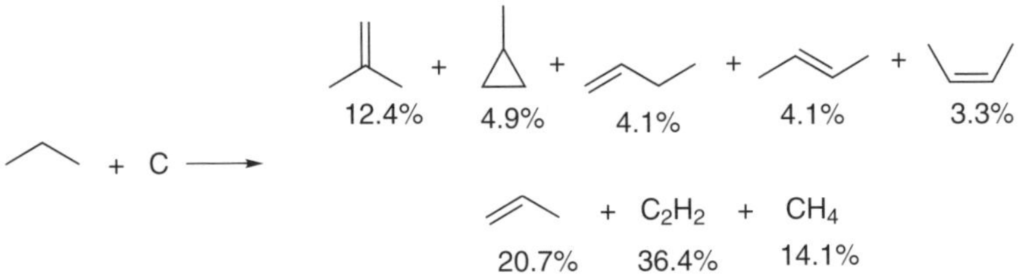

$$(16)$$

3.4. Reaction of Carbon Atoms with Alkenes

In analogy with carbenes, atomic carbon reacts with alkenes by double-bond addition (DBA) to give cyclopropylidenes such as **19**, which undergo the known ring

opening to cumulenes. In the simplest case, reaction of nucleogenic,[59] arc generated,[60] and chemically produced C atoms[34a] with ethylene gives allene and propyne. The ratios of allene to propyne from the ^{11}C reaction (3.7, gas phase; 2.6,
condensed phase), from the carbon arc (2.6 condensed phase) and from tetrazole
decomposition (1.6, gas phase) are in reasonable agreement. Carbon-11 reacts
with ethylene to give allene that is predominately center labeled and propyne
labeled on both the 1- and 2-carbons.[61] These results were rationalized by assuming
competing double-bond addition to form allene, some of which rearranged to
propyne, and C—H insertion to give vinylcarbene **20**, which rearranges to propyne
and allene (Eq. 17). The situation is complicated by the fact that **20** is expected to
rearrange to cyclopropene and this product is not reported. It may be that the cyclopropene rearranges to allene and methylacetylene under the energetic reaction
conditions.

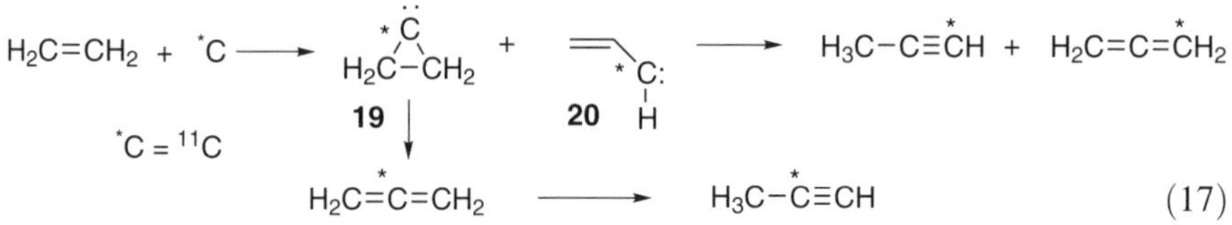

$$H_2C=CH_2 + \overset{\centerdot}{C} \longrightarrow \underset{\underset{\mathbf{19}}{H_2\overset{\centerdot}{C}-CH_2}}{\overset{\overset{\centerdot\centerdot}{\overset{*}{C}}}{}} + \overset{}{\underset{\underset{\mathbf{20}\;\;H}{}}{\overset{*}{C:}}} \longrightarrow H_3C-C\equiv\overset{*}{C}H + H_2C=C=\overset{*}{C}H_2$$

$$\overset{\centerdot}{C} = {}^{11}C$$

$$\underset{\mathbf{19}}{\big\downarrow}$$

$$H_2\overset{*}{C}=C=CH_2 \longrightarrow H_3C-\overset{*}{C}\equiv CH \tag{17}$$

Reaction of atomic carbon with alkenes generally involves both DBA and vinyl
C—H insertion. An interesting example is the reaction of C atoms with styrene in
which the major products are phenylallene (**21**) and indene (**22**).[62] The synthesis of
a number of specifically deuterated styrenes and the measurement of the deuterium
isotope effects on the **21**/**22** ratio led to the conclusion that **21** was formed by DBA
followed by ring expansion and by C—H(D) insertion into X_a and followed by rearrangement of the resultant *trans*-vinylcarbene (**23**). The indene was formed by
C—H(D) insertion into X_b followed by cyclization of the resultant *cis*-vinylcarbene
(**24**) (Eq. 18). An examination of the product ratios and their label distributions
when ^{13}C atoms are used leads to the conclusion that the ratio of C=C addition
to C—H insertion is 0.72:1 in this case.

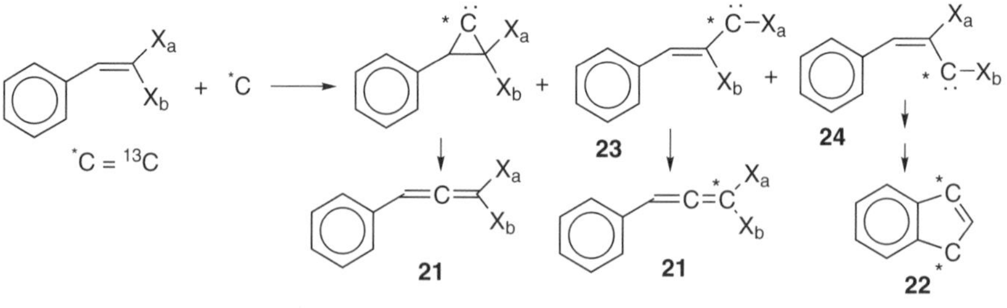

21:**22** = 6.9 (X_a, X_b = H); = 9.9 (X_a = H, X_b = D); = 3.6 (X_a = D, X_b = H)

$$\tag{18}$$

In order to confirm these mechanistic conclusions, carbenes **23** and **24** have been generated by the C atom deoxygenation of (*E*)- and (*Z*)-3-phenylpropenal (**25** and **26**). Deoxygenation of **25** gives **21** and **22** in a 8.5:1 ratio, while deoxygenation of **26** yields **21** and **22** in a 0.13:1 ratio (Eqs. 19 and 20). These results indicate that carbenes **23** and **24** undergo intramolecular reaction faster than they interconvert.

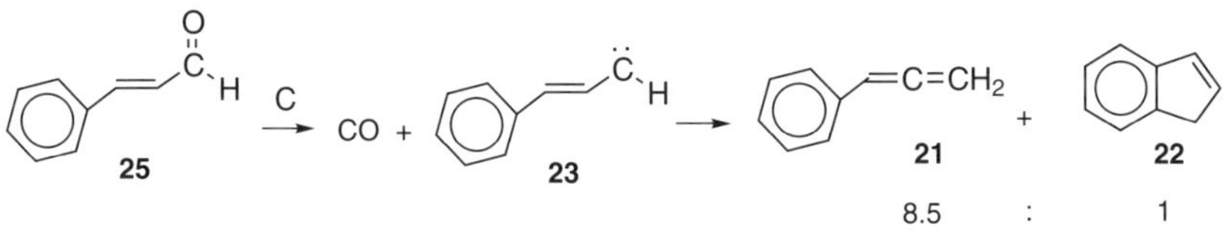

$$(19)$$

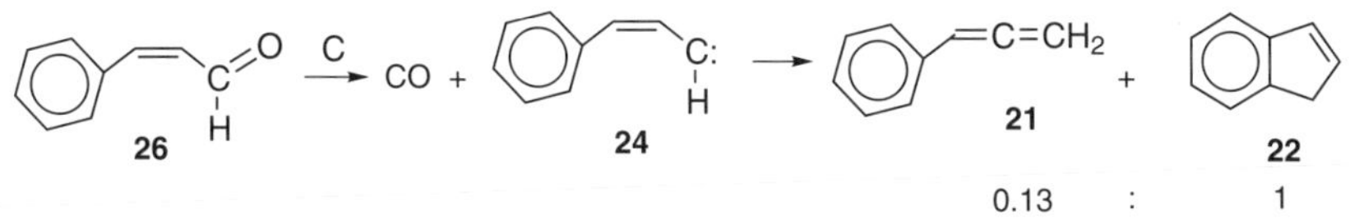

$$(20)$$

In a related reaction, arc generated ^{13}C atoms react with **22** to generate naphthalene labeled in the 2-position. Double labeling experiments and calculations implicate an initial C—H insertion to produce 2-indenylcarbene **27** followed by ring expansion and H migration (Eq. 21). Independent production of **27** by C atom deoxygenation of the corresponding aldehyde also generates naphthalene (Eq. 21).[63]

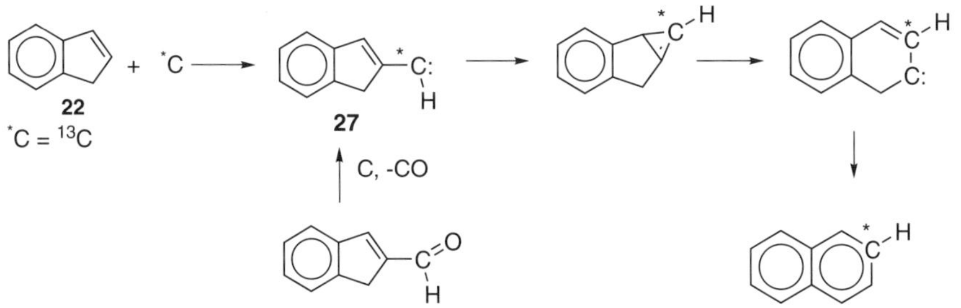

$$(21)$$

3.4.1. Reaction of Carbon Atoms with Cycloalkenes to Give Cyclic Cumulenes.

Since carbon is known to react with acyclic double bonds to generate cumulenes via a cyclopropylidene, a straightforward extension takes advantage of the high energy of atomic carbon as a potential route to cyclic cumulenes. Thus, reaction of arc generated carbon with cyclopentene in an attempt to prepare 1,2-cyclohexadiene (**28**) by DBA via an intermediate cyclopropylidene results in ethylene, vinylacetylene (**29**), and benzene as the only isolable products.[64] Attempts to trap **28** were unsuccessful. The ethylene and **29** were postulated to arise from the known retro Diels–Alder reaction of an intermediate **28**.[65] However, the use of ^{13}C atoms revealed that the label in **29** was distributed on both alkyne carbons and the vinyl CH. Hence, an additional pathway to **28** involving an initial C–H insertion to give vinylcarbene (**30**), which then ring expands, was postulated. Since the label is unequally distributed between C1 and C3 in **29**, it seem likely that a direct cleavage of **30** to ethylene and **29** competes with ring expansion (Eq. 22). In this case, the ratio of DBA to C–H insertion was 0.87:1.[64]

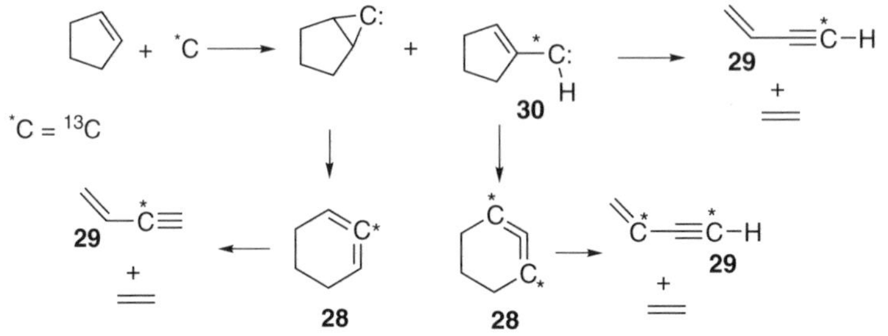

$$(22)$$

The competing ring expansion and cleavage in carbene **30** was confirmed by generating the deuterium labeled carbene **30a** by the C atom deoxygenation of aldehyde **31** (Eq. 23).[64] While the origin of the benzene in these reactions has not been established, several pathways to energetic cyclohexadiene followed by loss of hydrogen may be envisioned.

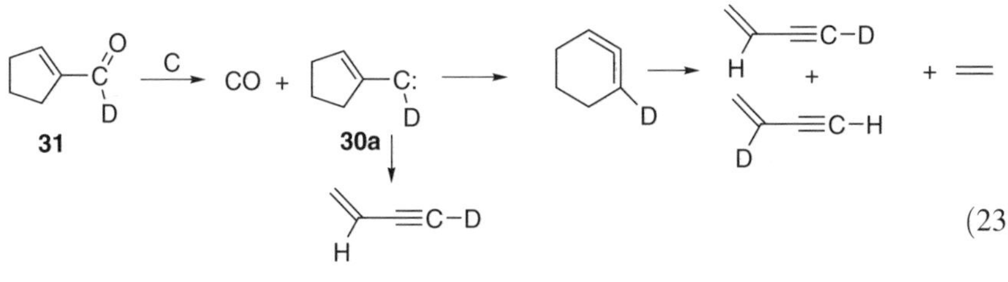

$$(23)$$

Carbon atoms react with norbornadiene **32** in a manner analogous to their reaction with cyclopentene.[66] Thus, DBA and vinyl C–H insertion both generate bicyclo[3.2.1]octa-2,3,5-triene (**33**), which undergoes a known[67] [3.3] sigmatropic rearrangement to *endo*-6-ethynylbicyclo[3.1.0]hex-3-ene (**34**). When ^{13}C atoms are employed, the label distribution indicates that the three pathways in Eq. 24 are involved. The DBA gives tricyclopropylidene (**35**), which opens to **33** (path a). The C–H insertion gives vinylcarbene (**36**), which either cleaves directly to **34** (path b) or ring expands to **33** (path c). In this case, the ratio of DBA to C–H insertion by C atoms is 1.08:1. In neither **28** nor **33** is cumulene formation from the carbene expected to be reversible. These reactions are calculated to be endothermic by 27 and 30.5 kcal/mol, respectively.

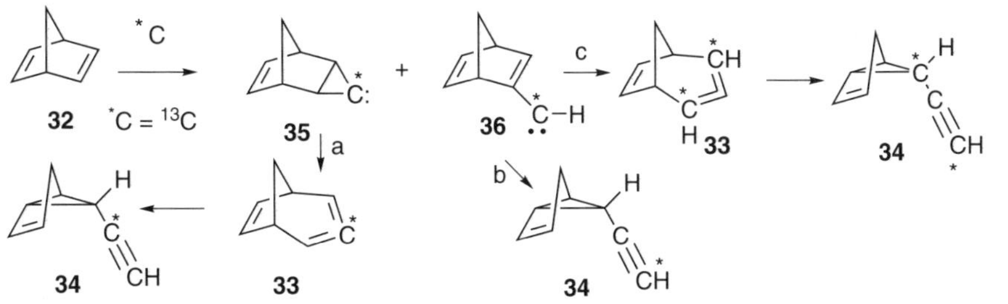

$$(24)$$

The reaction of arc generated ^{13}C atoms with cyclooctatetraene (**37**) produces indene (**22**) in which the label is on C_9.[68] An initial DBA followed by ring opening to cyclononapentaene (**38**) is proposed. In analogy with known reactions of **38**,[69] electrocyclic closure followed by H migration leads to **22**-$^{13}C_9$ (Eq. 25).

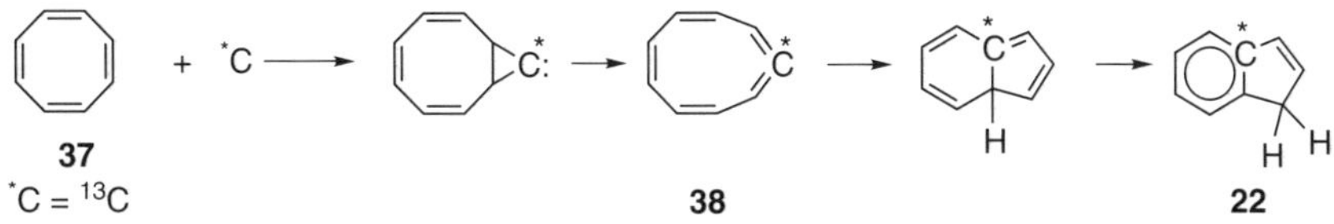

$$(25)$$

3.5. Halomethylidene Formation in the Reaction of Carbon Atoms with Carbon–Halogen Bonds

In testing the limits of C atom reactivity, the extremely strong C–F bond provides an interesting case. Although C atoms, like carbenes, fail to insert into C–F bonds, unlike carbenes, they abstract fluorine in a practical method for generating and investigating the reactions of fluoromethylidene, CF.[70,71] When an alkene is added to the C + CF_4 reaction mixture, the CF may be trapped by DBA to give a fluorocyclopropyl radical (Eq. 26). Like singlet carbenes, CF adds to double bonds in a stereospecific manner to give cyclopropyl radicals in which the original stereochemistry of the alkene is preserved.[71] When the intermediate fluorocyclopropyl radical has sufficient strain energy, electrocyclic ring opening can occur.[72]

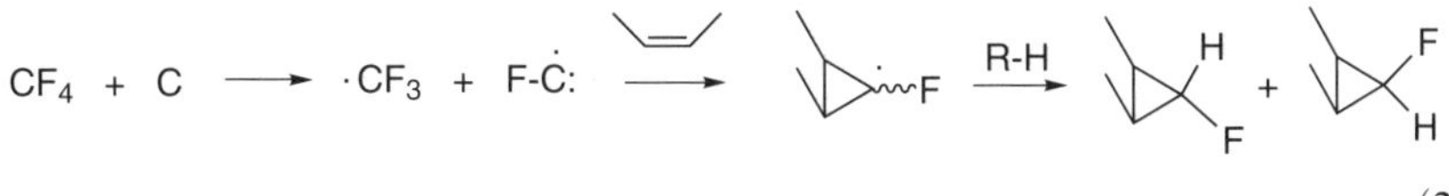

$$(26)$$

Reaction of CF with benzene generates the 7-fluoronorcaradien-7-ly radical (**39**), which abstracts hydrogen (from added isobutane) and opens to 7-fluorocyclohepta-triene (**40**). Cycloheptatriene (**10**) is trapped as tropylium fluoroborate (**41**) by the addition of BF_3 (Eq. 27).[73] An additional product of CF + benzene is fluorobenzene (**42**), in which labeling studies demonstrate that the attacking carbon contains the fluorine in **42**. The interesting transfer of CH in Eq. 28 is proposed to account for the formation of **42**.[73,74]

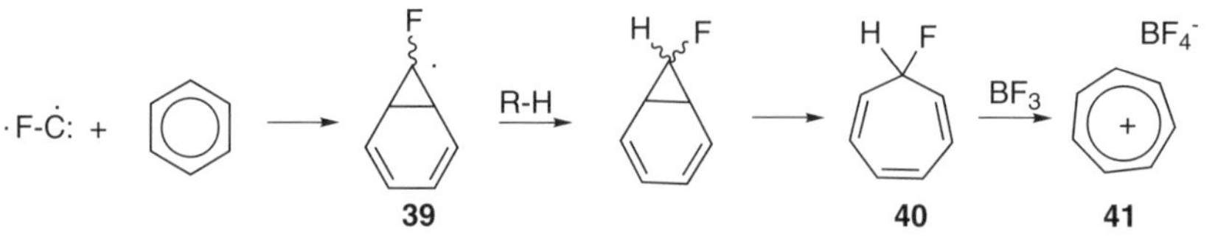

$$(27)$$

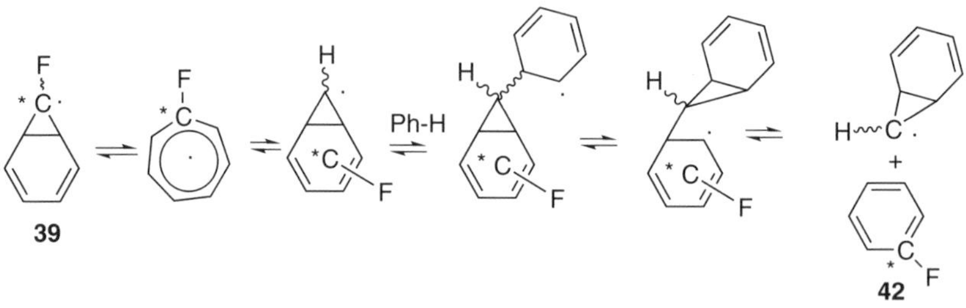

$$(28)$$

Like carbenes, C atoms insert into the weaker C—Cl and C—Br bonds. For exam-ple, reaction of C with CCl_4 may be rationalized by assuming an initial C—Cl inser-tion to give the chlorocarbene **43** (Eq. 29).[16] However, C also abstracts Cl from CCl_4 to give CCl, which has been trapped by cyclohexene to give 7-chloronor-caran-7-yl radicals (**44**). These radicals abstract both H and Cl from substrate (Eq. 30).[75] Control experiments demonstrate that CCl_2 is not involved in this reac-tion. Reaction of C with chlorofluorocarbons results in abstraction of both F and Cl with the latter predominating as would be expected from a consideration of the relative bond strengths. An example is shown in Eq. 31.[75] Abstraction of halogen by C can also be used to generate CBr and CI. Thus, reaction of C with $CBrF_3$ and CIF_3 in the presence of cyclohexene results in trapping products of the bromo- and iodonorcaranyl radicals, respectively.[76]

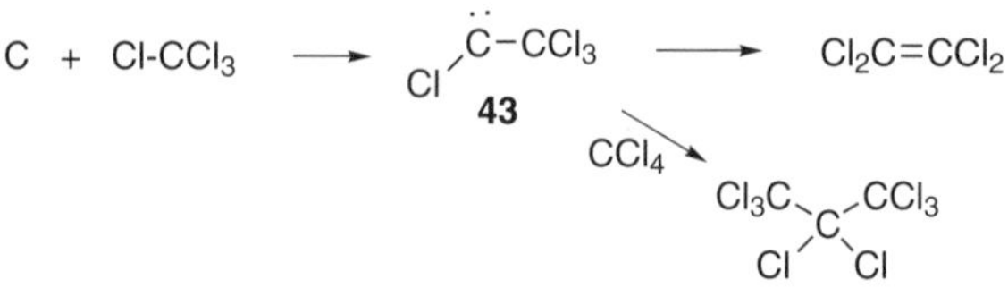

$$(29)$$

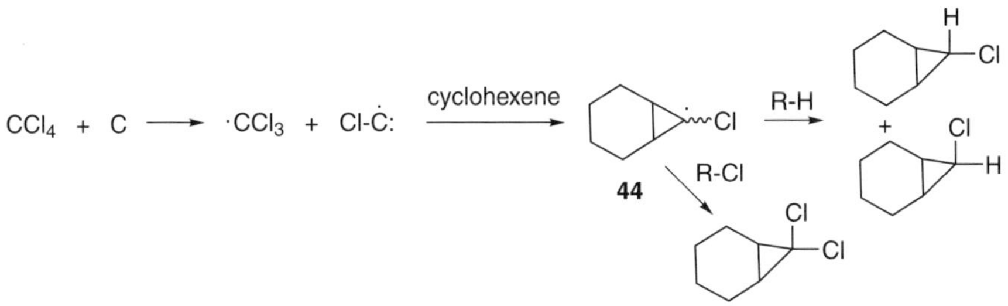

$$(30)$$

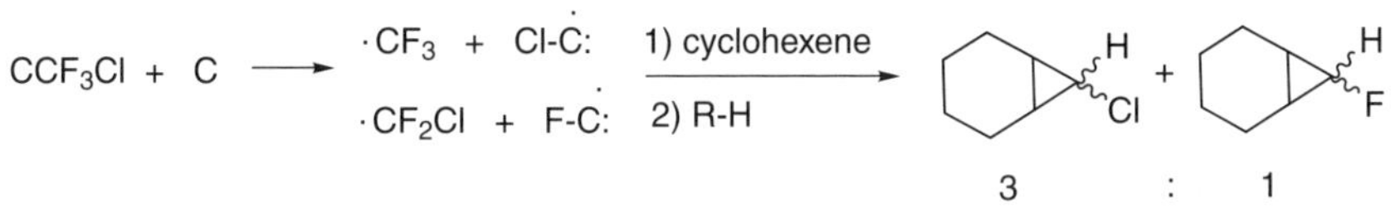

$$(31)$$

Although this method of generating halomethylidenes has received relatively little attention, at present it is the only way to examine the chemistry of these interesting reactive intermediates. If the chemistry of monovalent carbon intermediates is to be elaborated, this appears to be the best way to produce these species.

3.6. Reaction of Carbon with Aromatic and Heteroaromatic Compounds

3.6.1. Reaction of Carbon with Benzene and Substituted Benzenes. The reaction of atomic carbon with benzene has a long history.[18,77] Products that have been identified in various studies are benzocyclopropene (**45**), toluene (**46**), cycloheptatriene (**47**), diphenylmethane (**48**), heptafulvalene (**49**), and phenylcycloheptatriene (**50**) (Eq. 32). Of these products, only **45** results exclusively from carbon + benzene. However, **46–50** can be rationalized by assuming an initial C–H insertion to give phenylcarbene **51** and/or DBA to generate norcaradienylidene **52**, which opens to cycloheptatetraene **53**.

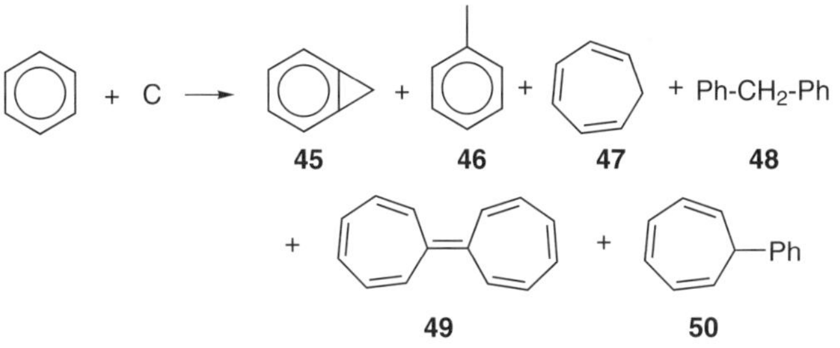

$$(32)$$

Although the situation is complicated by the well-known phenylcarbene rearrangement that interconverts **51** and **53**,[78] the use of labeled carbon allows an evaluation of initial C atom reactivity. Thus, initial C—H insertion places the label on a carbon bearing a hydrogen while DBA labels the quaternary carbon (Eq. 33).

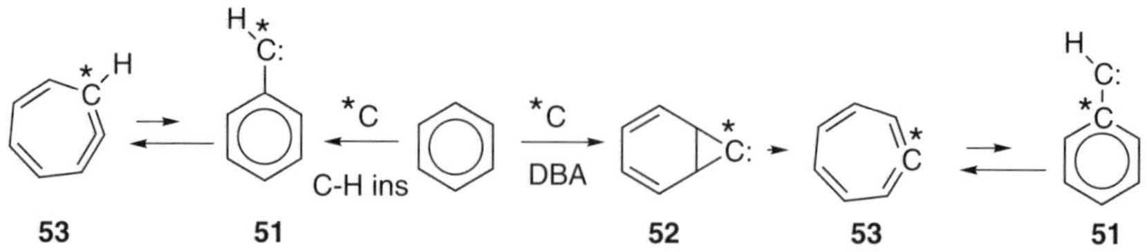

$$(33)$$

In the reaction of nucleogenic ^{11}C atoms with toluene, the methyl group traps an initially formed o-tolylcarbene (**54**) as benzocyclobutene (**55**).[79] A partial degradation of **55** and an examination of the label distribution indicated that 43% of **55** arose from **54** formed by an initial C—H insertion (Eq. 34). The remainder of the label in **55** was in the ring, indicating the initial formation of the m- and p-tolylcarbenes and/or a methylcycloheptatetraene.

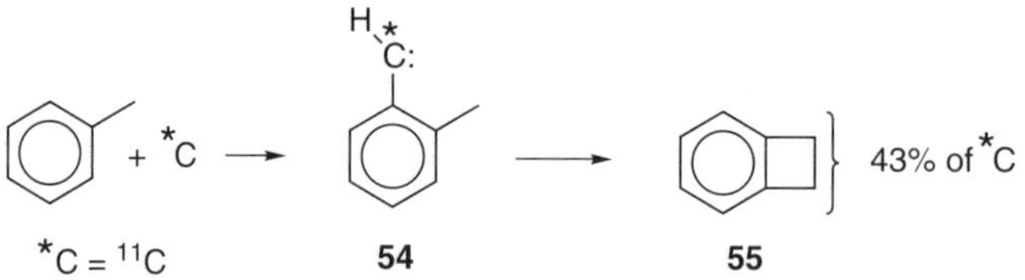

$$(34)$$

In a similar study, arc generated ^{13}C atoms were allowed to react with tert-butylbenzene in order to use the tert-butyl group as an intramolecular trap for the carbene.[77i] This reaction gave dimethylindane-3-^{13}C (**56**) and 3,3-dimethyl-3-phenylpropene-1-^{13}C (**57**). The label distribution in **56** indicates that it arises from an initial insertion into an ortho C—H bond to give o-tert-butylphenylcarbene (**58**), while **57** results from insertion into a methyl group C—H (Eq. 35). The fact that

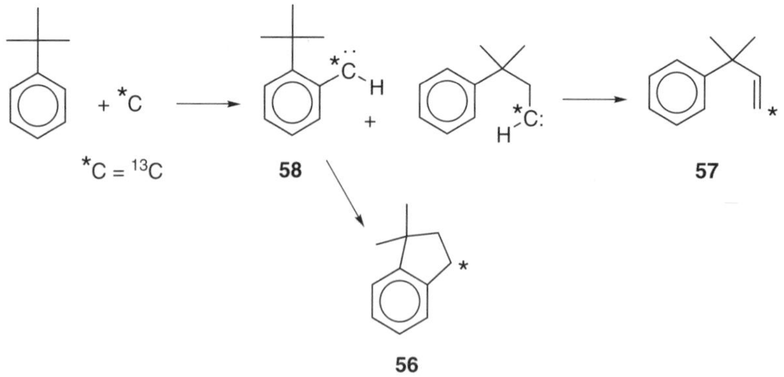

$$(35)$$

CH groups in the aromatic ring of **56** are unlabeled indicates that neither the *m*- nor *p*-*tert*-butylphenylcarbene **59** and **60**, if formed, rearranges to **58**. Addition of HBF_4 to the cold reactor bottom following the $^{13}C + tert$-butylbenzene reaction results in the formation of *tert*-butyltropylium fluoroborate (**61**) labeled in the 3- and 4-positions in a 1:2 ratio, and no excess ^{13}C in the 2-position (Eq. 36). These results are consistent with formation of carbenes **58–60** by C—H insertion followed by intramolecular trapping of **58**, and rearrangement of **59** and **60** to *tert*-butyl-cycloheptatetraenes (**62**), which are trapped by HBF_4. Calculations indicate that the first step of the phenylcarbene rearrangement, ring expansion of **51**, must surmount a barrier of ~13 kcal/mol to give **53**.[80] However, rearrangement back to **51** (in this case interconverting **58–60**) must cross a barrier of ~30 kcal/mol, and it may be that initial ring expansion of **59** and **60** leads to cycloheptatetraenes **62** which, lacking the energy to rearrange further, are trapped by HBF_4. A similar result is observed when **58** and **60** are generated by deoxygenation of the corresponding aldehydes. Thus, **58** is trapped intramolecularly while **60** ring expands to a *tert*-butylcycloheptatetraene that is trapped as the tropylium salt.

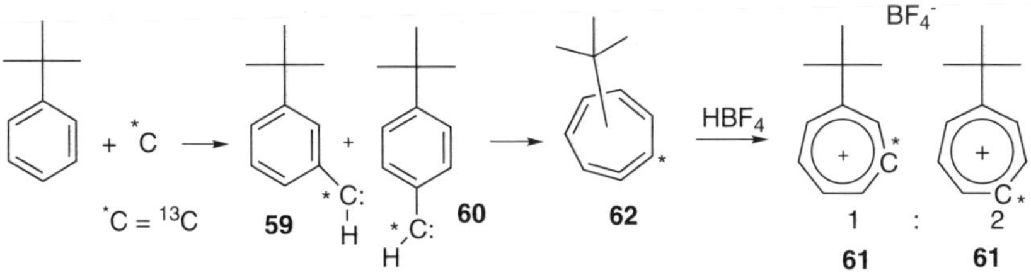

$$(36)$$

An intermediate cycloheptatetraene can also be trapped in the addition of carbon to benzene itself. When ^{13}C atoms react with benzene-d_6 and HBF_4 is added, tropylium fluoroborate (**41**), containing deuterium on the labeled carbon, is observed. This result is consistent with an initial C—D insertion by carbon to give **51**-d_6, which ring expands to **53**-d_6, and is trapped by HBF_4 (Eq. 37).[77i]

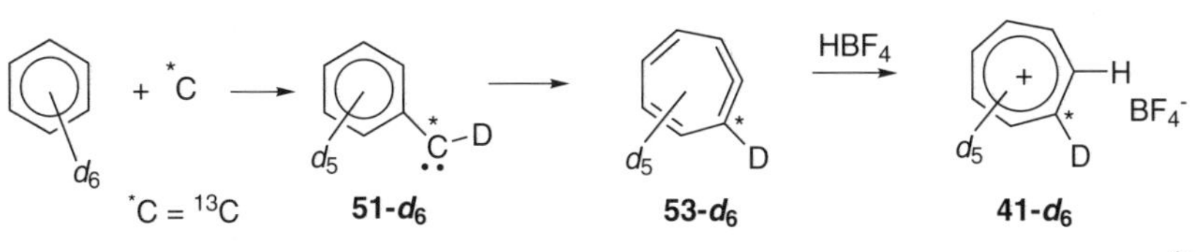

$$(37)$$

Several other substituted benzenes have been reacted with carbon and the substituted phenylcarbene or cycloheptatetraene have been trapped in an intramolecular reaction. Thus, phenol reacts with C atoms to give tropone (**63**) and traces of the *o*-, *m*-, and *p*-cresols.[81] In this case, an initial *o*-, *m*-, or *p*-hydroxyphenylcarbene (**64**) is postulated to ring expand to a hydroxycycloheptatetraene that traps itself by intramolecular proton transfer (Eq. 38). A competing reaction of **64** is intersystem crossing (isc) to the triplet that abstracts hydrogen to produce the cresols. Carbenes **64**, produced by C atom deoxygenation of the corresponding aldehydes behave in a similar manner. It has also been reported that generation of carbenes **64** from the pyrolysis of the corresponding tetrazoles at 530–700 °C leads to **63**.[82]

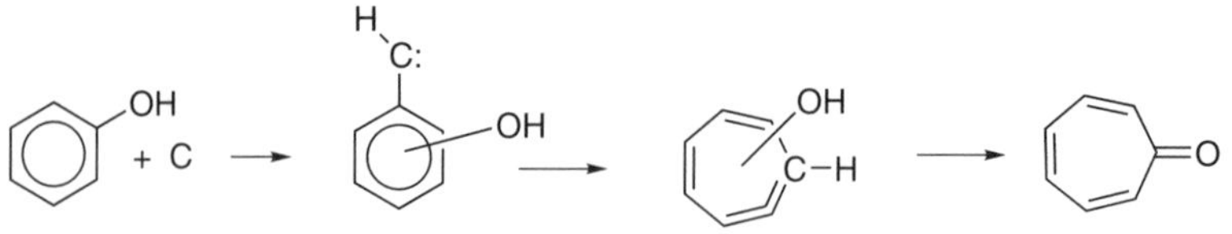

$$(38)$$

Reaction of arc generated C atoms with anisole gives phenyl vinyl ether, benzodihydrofuran (**65**), from trapping of the *o*-methoxyphenylcarbene (**66a**), the *o*-, *m*-, and *p*-methylanisoles, and methoxytropylium ion from trapping of the methoxycycloheptatetraenes (Eq. 39).[83] When carbenes **66a–c** are generated by C atom deoxygenation of the corresponding anisaldehydes, **65** again results along with products of ring expansion, which are trapped by the addition of HBF$_4$. It is interesting that calculations (B3LYP/6-311+G**//B3LYP/6-31G*) predict that the lowest energy reaction of **66a** is intramolecular trapping to give **65** ($\Delta H^\ddagger = 7.0$ kcal/mol) with ring expansion ($\Delta H^\ddagger = 15.4$ kcal/mol) not expected to be competitive.[83] However, ring-expansion products, which may result from the high energy of **66a** from the C atom reactions, predominate in this system.

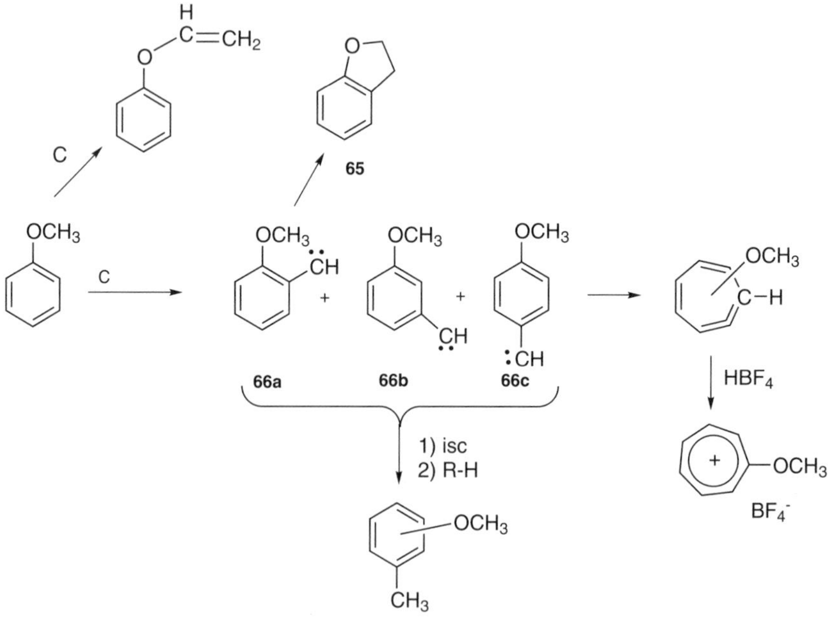

$$(39)$$

Cocondensation of arc generated carbon with naphthalene (**67**) at 77 K generates cyclobuta[*de*]naphthalene (**68**) and the 1- and 2-methylnaphthalenes (**69a,b**).[84] The intermediacy of the 1- and 2-naphthylcarbenes (**70a,b**) is postulated. Intersystem crossing produces triplet **70a,b** that abstract hydrogen, while intramolecular trapping of singlet **70a** gives **68** in a known reaction (Eq. 40).[85] Generation of **70a** and **70b** independently by the deoxygenation of the corresponding naphthaldehydes results in the formation of **68** and **69a,b**. The fact that **68** is formed when 2-naphthaldehyde is deoxygenated indicates that ^{1}**70b** rearranges to ^{1}**70a** under the energetic reaction conditions. This rearrangement has been shown to occur at high temperatures[85] and has been studied computationally.[86]

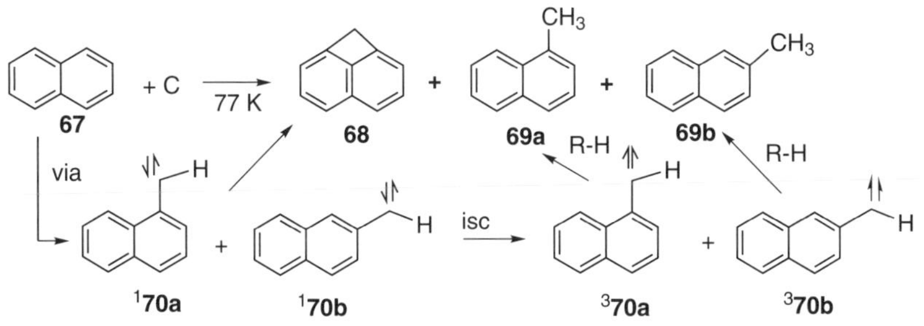

$$(40)$$

Cocondensation of C with a 1:1 mixture of **67**-d_0 and **67**-d_8 generates **68**-d_0 and **68**-d_8 in a 0.12 ratio as a result of a large k_H/k_D for isc in **70a,b**. This unusually large k_H/k_D is postulated to result from the generation of the singlet state of **70a,b** with excess energy and it is the isotope effect upon the degradation of the energetic carbene to the triplet state that is observed. The large isotope effect on isc leads to more triplet products from **70a,b**-d_0 than from **70a,b**-d_8. Thus, when **70a,b** are trapped with the 2-butenes, 4–18% of the carbene adducts are formed nonstereospecifically from C + **67**-d_0, while addition of the carbenes from C + **67**-d_8 to the 2-butenes produces only stereospecific adducts (Eq. 41). A similar result is observed when triplet **70a,b** are trapped with oxygen (Eq. 42).[84]

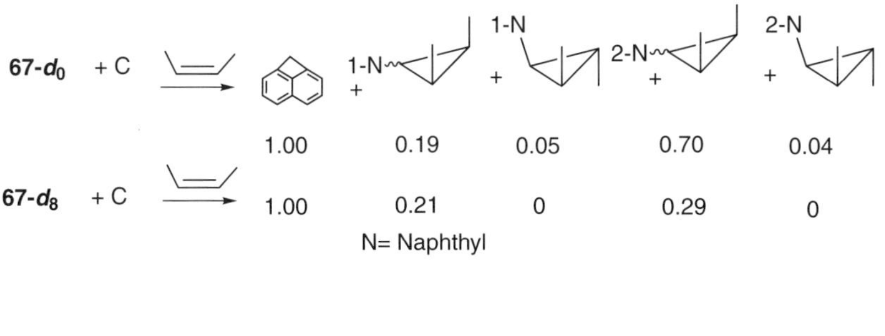

$$(41)$$

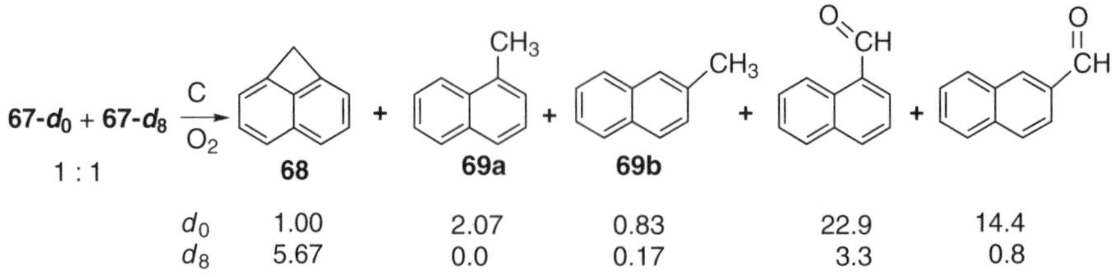

67-d0 + 67-d8
1 : 1
C
O2
CH3
CH3
O
CH
O
CH
68
69a
69b
d0 1.00 2.07 0.83 22.9 14.4
d8 5.67 0.0 0.17 3.3 0.8

(42)

3.6.2. Reaction of Atomic Carbon with Pyrrole. Cocondensation of arc generated C with pyrrole **71** gives pyridine **72**.[18] This clean reaction is useful to insure that the system is functioning properly when setting up a new carbon arc reactor. Double labeling experiments with ^{13}C atoms and **71**-1-d_1 produce **72** with both the ^{13}C and the D on C_3 (Eq. 43).[18] While it is tempting to propose that this reaction proceeds via 1-aza-2,3,5-cyclohexatriene (**73**), which undergoes a 1,5 hydrogen shift (path a Eq. 43), calculations (MP2/6-31G*) show that the intermediate is planar and is best represented as the aromatic dehydropyridinium species **74**, which adds D in an intermolecuiar acid–base reaction (path b Eq. 43). The intramolecular acid–base reaction in Eq. 43 was confirmed by the addition of MeOD to the C + **71**-d_0 reaction.[18]

71, X=H, D
+ *C
C:
a
b
73
72
74
X–N
+
72

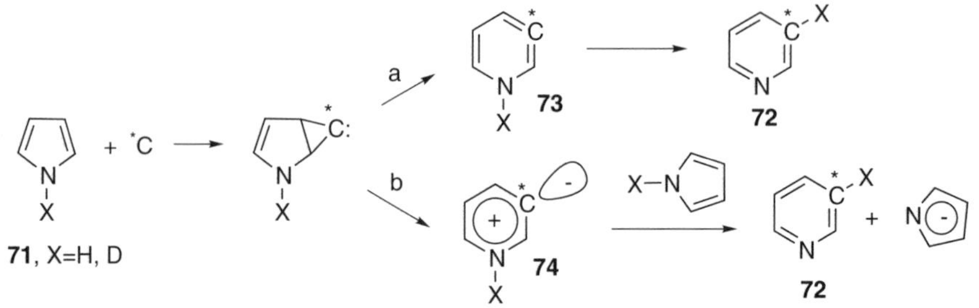

(43)

The dehydropyridinium species **74** represents a new class of reactive intermediate that is conveniently accessed through C atom chemistry. When *N*-methylpyrrole (**75**) is used as a substrate in this reaction, the lack of an acidic hydrogen permits trapping a portion of the nucleophilic dehydropyridinium by CO_2 (Eq. 44).[87]

+ C
CO2
COO-
CH3
75

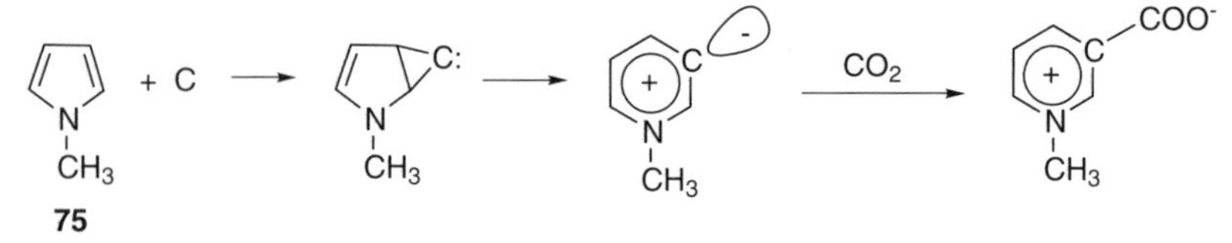

(44)

3.6.3. *Reaction of Carbon Atoms with Thiophene.*

The reaction of atomic carbon with thiophene (**76**) is more complex than that with **71**.[88] The use of ^{13}C atoms indicates that an important pathway involves addition of C to a double bond to give thiabicyclopropylidene (**77**), which ring opens to thiacyclohexa-2,3,5-triene (**78**), now calculated to be an energy minimum.[88,89] Addition of H(D)Cl during the reaction traps **78** as the 3 labeled (^{13}C and D) thiopyrylium ion (**79**). When acid is absent, **78** rearranges to the 2-thienylcarbene (**80**), labeled predominately on the quaternary carbon that is trapped by thiophene to give **81**. An additional product of acid trapping is the 2 labeled thiapyrylium ion **82** that was postulated to result from protonation of thiacyclohexa-3,5-dien-2-ylidene (**83**). Calculations ([QCISD(T)/6-311+G(3df,2p)]//B3LYP/6-32(d)) indicate that **78** rearranges to **80** via ring opening to thial (**84**), which subsequently closes to **80**.[89] In a reaction analogous to the phenylcarbene rearrangement, **80** rearranges to **83** via a bicyclopropene intermediate (Eq. 45).

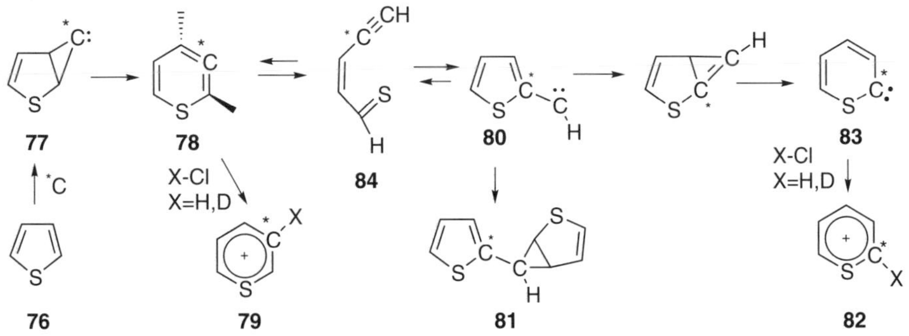

$$(45)$$

Calculations identify an additional low-energy pathway in the reaction of carbon with **76** in which 1,4-addition leads to ylid **85** followed by rearrangement to thio-ketone **86** (Eq. 46).[89] This pathway has not been examined experimentally.

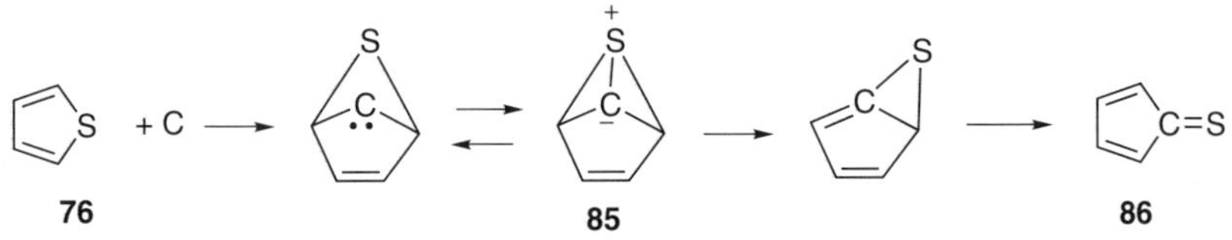

$$(46)$$

3.6.4. *Reaction of Carbon Atoms with Furan.*

Attempts to trap oxacyclo-hexatriene in the reaction of arc generated C atoms with furan (**87**) in analogy with the reaction of carbon with **76** led to polymerization of **87**, precluding identification of other products. However, reaction of chemically generated ^{13}C atoms with **87** gave 2-penten-3-nyal (**88**) labeled on the 4 carbon.[90] Aldehyde **88** presumably arises from ring opening of an intermediate oxacyclohexatriene formed by DBA of C to **87** (Eq. 47).

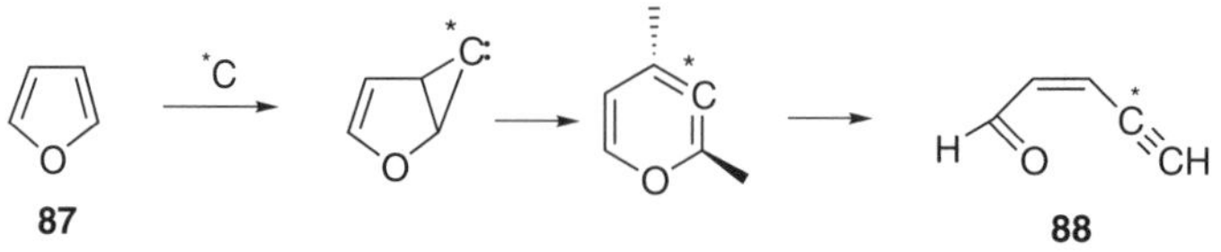

(47)

87

88

3.6.5. C–H Insertion versus Double-Bond Addition in the Reaction of Carbon with Aromatics.
The factors that determine if C atoms react with an aromatic by C–H insertion or DBA are not yet understood. Benzene and substituted benzenes are postulated to react by C–H insertion although detailed labeling studies have only been carried out in benzene, toluene, and *tert*-butylbenzene. No C–H insertion is observed in the reaction of carbon with **71**, while C–H insertion is a minor pathway when carbon reacts with **76** and **87**.

3.7. Deoxygenation by Atomic Carbon

3.7.1. Reaction of Carbon Atoms with Alcohols and Ethers.
The electrophilicity of atomic carbon and the exothermicity of carbon monoxide formation in its reactions facilitates attack on, and removal of oxygen by C atoms. Deoxygenation of ethers, alcohols, and carbonyl compounds has been reported. This process is generally a highly exothermic reaction, which is likely to generate products with excess energy.

The reaction of nucleogenic and arc generated C atoms with alcohols gives products of deoxygenation, O–H, and C–H insertion. For example, cocondensation of arc generated C with methanol gives the products in Eq. 48.[91] Experiments with specifically deuterated methanols confirm that the acetaldehyde is a product of a C–H insertion while the dimethoxymethane results from two consecutive O–H insertions. Deoxygenation produces CO and an unidentified fragment (presumably methyl radicals).

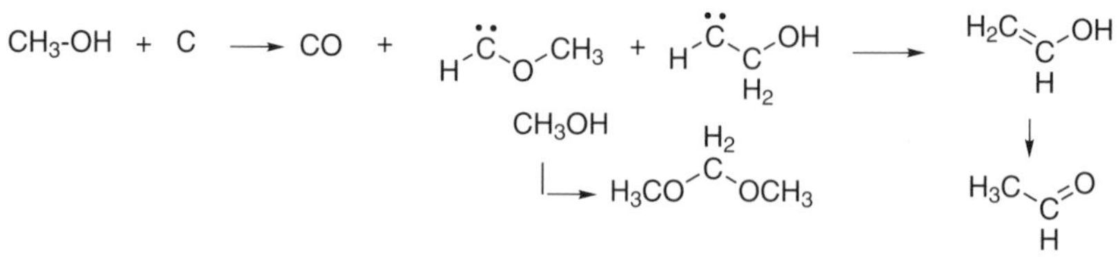

(48)

The deoxygenation of ethers by C atoms was first studied by McKay and Wolfgang[92] who reported a high yield of ^{11}CO in the reaction of ^{11}C with oxirane. Deoxygenation of oxiranes was subsequently reported for both arc and chemically generated carbon. An investigation of the stereochemistry of the deoxygenation of *cis*-2,3-dimethyloxirane (**89**) revealed that arc generated carbon gave a $(Z)/(E)$ 2-butene ratio of 1.5:1,[93] while chemically generated carbon produced the butenes in a 5.6:1 ratio (Eq. 49).[94] It may be that some of the nonstereospecificity in the carbon arc reaction results from photolysis of the products in the arc. Deoxygenation of the 2,4-dimethyloxetanes has been shown to be nonstereospecific.[95]

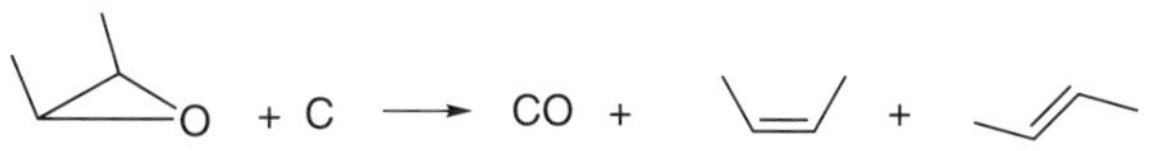

$$\text{89} + \text{C} \longrightarrow \text{CO} + \text{} + \text{} \tag{49}$$

$$
\begin{array}{ccll}
1.5 & : & 1 & \text{from C arc} \\
5.6 & : & 1 & \text{from tetrazole}
\end{array}
$$

The deoxygenation of tetrahydrofuran (THF, **90**), which yields ethylene and carbon monoxide,[96,97] is an interesting case. While this and other deoxygenations might be expected to proceed through an yild intermediate and a biradical as shown in Eq. 50, calculations (MP2/6-31G*) indicate that neither ylid **91** nor biradical **92** is an intermediate in this reaction.[97] These calculations reveal a concerted removal of oxygen that proceeds to carbon monoxide and two molecules of ethylene without barrier. Experimental evidence that **91** is not an intermediate is provided by the fact that reaction of carbon with a mixture of **90** and **90**-d_8 generates ethylene and ethylene-d_8 in a 2.7:1 ratio.[97] This secondary isotope effect of 1.13 (per D) would not be expected if **91** (or **92**) were an intermediate.

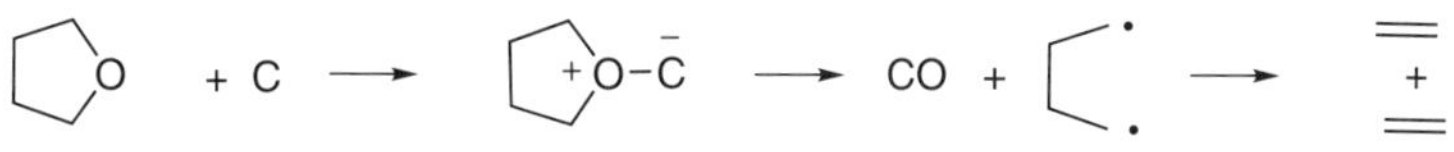

$$\text{90} + \text{C} \longrightarrow \text{91} \longrightarrow \text{CO} + \text{92} \longrightarrow \tag{50}$$

Deoxygenation of acyclic ethers gives CO and a radical pair. Thus, reaction of arc generated C with diethyl ether produces CO and a pair of ethyl radicals, which disproportionate to ethane and ethylene and dimerize to from butane.[98] However, deoxygenation is not always the exclusive pathway followed in C atom reactions with ethers. We have seen that furan, with its reduced electron density on oxygen reacts exclusively at the double bond. The reaction of 2,5-dihydrofuran (**93**) with arc generated C produces CO, H$_2$, C$_2$ H$_2$, and 1,3-butadiene and was postulated to proceed through energetic 2-buten-1,4-diyl (**94**), which cleaves to H$_2$ and C$_2$H$_2$ and decays to butadiene (Eq. 51).[96] However, reaction of chemically generated C with

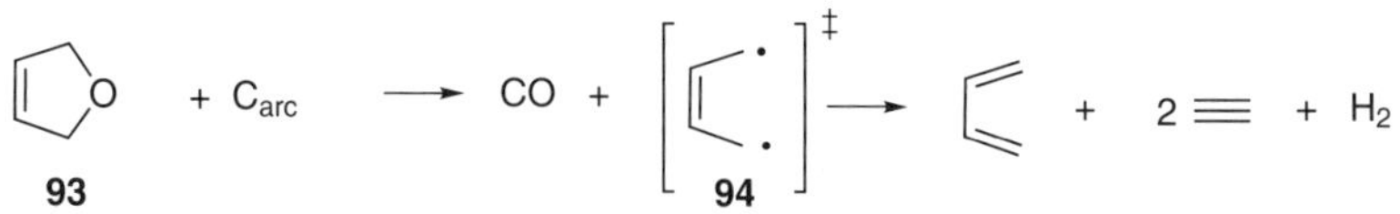

$$\text{93} + \text{C}_{arc} \longrightarrow \text{CO} + \left[\text{94} \right]^{\ddagger} \longrightarrow + \ 2 \equiv + \ \text{H}_2 \tag{51}$$

93 gives CO, butadiene, C$_2$H$_2$, and formaldehyde in a reaction postulated to involve deoxygenation and a competing C—H insertion followed by carbene cleavage

(Eq. 52).[99] It seems likely that the reaction of arc generated C with **93** actually follows this same pathway with the H_2 and some CO resulting from photolysis of formaldehyde in the carbon arc.

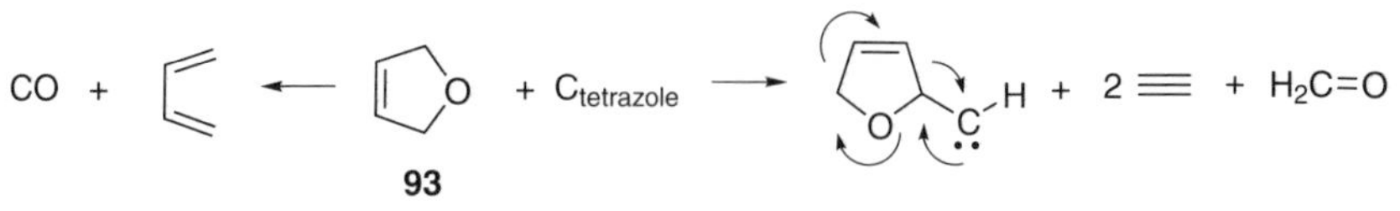

$$CO + \quad \longleftarrow \quad + C_{tetrazole} \quad \longrightarrow \quad H + 2 \equiv + H_2C=O$$

93

$$(52)$$

3.7.2. The Formation of Carbenes by Carbon Atom Deoxygenation of Carbonyl Compounds.

Skell and Plonka[100] were the first to report that deoxygenation of carbonyl compounds by arc-generated carbon is an efficient method of preparing carbenes. For example, 2-butanone is deoxygenated to 2-butanylidene, which yields products similar to those observed when this carbene is generated by other methods (Eq. 53).[100a] Figure 10.2 shows structures of and literature references to many of the carbenes that have been generated by this method. Since carbenes are not produced by thermal C atoms and yields are unaffected by added oxygen, the deoxygenating C is thought to be a singlet, which therefore produces singlet carbenes. This method of carbene generation can be useful in producing carbenes which are difficult to generate by conventional methods. Examples include the interesting cyclobutenylidene **95** (Eq. 54).[101]

$$(53)$$

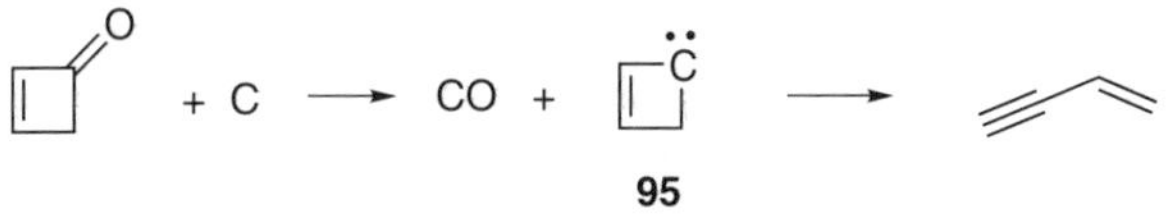

95

$$(54)$$

Although carbenes are an extensively studied class of energetic intermediates, investigations of their chemistry are often complicated by the fact that one must take care to insure that it is the reactions of the free carbene rather than those of a precursor that are being studied.[102] This difficulty seems to be particularly severe in the case of nitrogenous carbene precursors (diazo compounds and diazirines) in which the precursor, either in its ground or excited state, also yields carbene-type products. For this reason, it is often deemed prudent to avoid the use of nitrogenous precursors. In this connection, carbonyl deoxygenations along with other alternatives such as the cheleotropic extrusion of carbenes from cyclopropanes[103] have proved useful. However, as pointed out earlier, C atom deoxygenations are highly exothermic and are expected to produce carbenes with a great deal of excess energy.

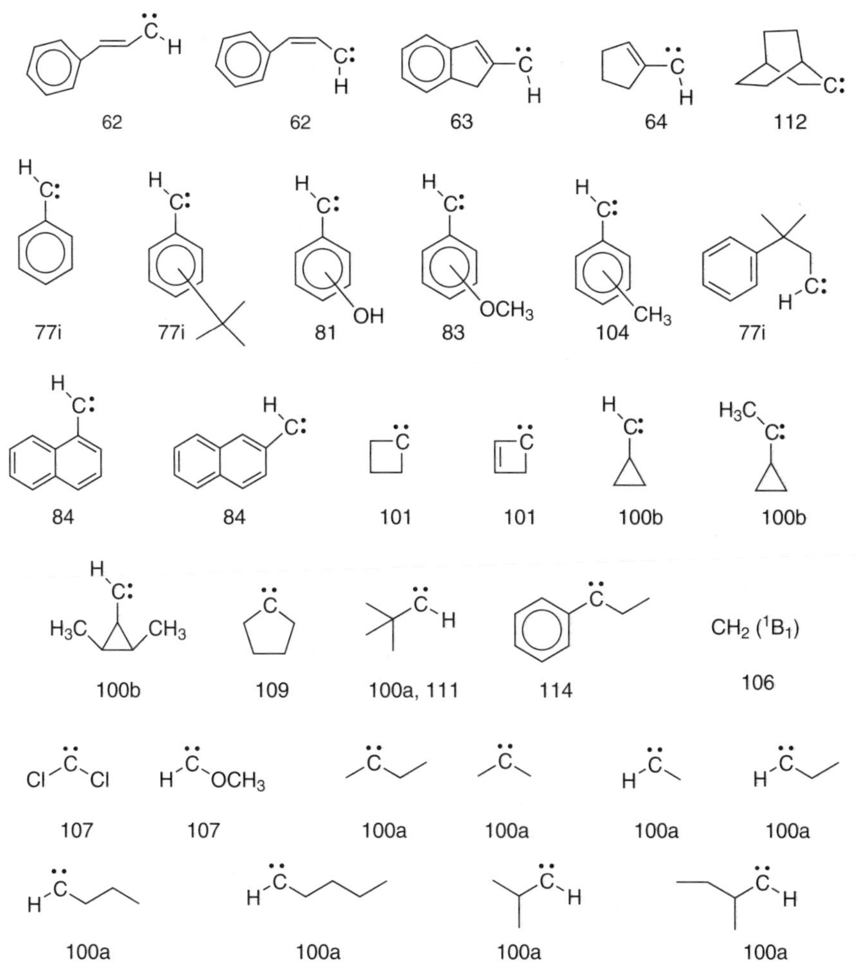

Figure 10.2. Carbenes that have been produced by the deoxygenation of carbonyl compounds. Numbers below the structures refer to the literature references in the text.

While this can lead to interesting new carbene reactions, it may also complicate comparisons with the same carbene generated by conventional methods.

A dramatic example of the production of energetic carbenes by C atom deoxygenations is provided by the deoxygenation of the tolualdehydes **96a–c** to the corresponding tolylcarbenes **97a–c** at 77 K (Eq. 55).[104] When the *m-* and *p-*tolylcarbenes **97b** and **97c** are generated from the diazo compounds and pyrolysed above 650 °C, they rearrange to the *o*-tolylcarbene (**97a**), which is trapped as benzocyclobutene or rearranges further to 1-phenylethylidene, and thence to styrene.[105]

The fact that this same rearrangement occurs in **97a–c** produced by deoxygenation at 77 K is strong evidence for energetic carbenes in this system.

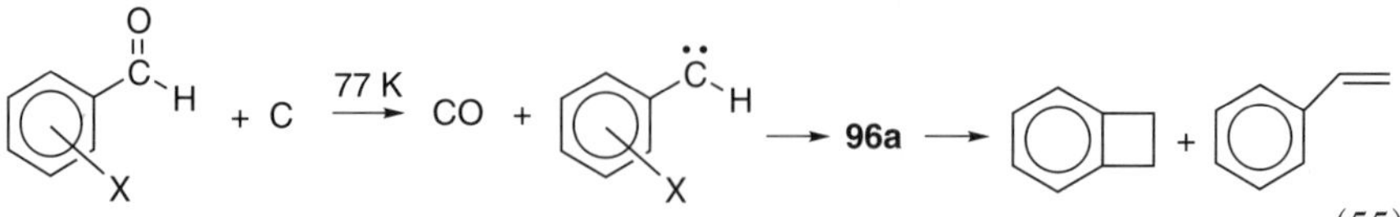

96a, X = *ortho* CH$_3$
96b, X = *meta* CH$_3$
96c, X = *para* CH$_3$

97a, X = *ortho* CH$_3$
97b, X = *meta* CH$_3$
97c, X = *para* CH$_3$

$$(55)$$

Deoxygenation of formaldehyde,[106] methylformate,[107] and phosgene[107] leads to methylene, methoxycarbene, and dichlorocarbene respectively, which can be trapped (Eq. 56). The methylene produced from reaction of chemically generated C atoms with formaldehyde adds nonstereospecifically to the 2-butenes in the gas phase (Eq. 57).[106] Since yields are not effected by added O$_2$ and the reaction becomes stereospecific upon addition of 200 Torr N$_2$, triplet methylene is not involved. The fact that increasing the pressure of the 2-butene does not alter the stereochemistry leads to the conclusion that nonstereospecificity does not result from isomerization of energetic dimethylcyclopropane products. Instead, the reaction is postulated to produce CH$_2$ in its excited ($^1B_1$) singlet state, a species in which nonstereospecific addition to alkenes has been predicted.[108] The addition of the inert N$_2$ degrades the excited singlet to its ground ($^1A_1$) state, which then adds stereospecifically. Since increasing the pressure of the 2-butene to 200 Torr only increases reactant concentration, nonstereospecific addition prevails. Since deoxygenation of formaldehyde to CH$_2$ ($^1A_1$) is exothermic by 98 kcal/mol and CH$_2$ ($^1B_1$) lies 21.1 kcal above CH$_2$ ($^1A_1$), production of excited singlet CH$_2$ is energetically feasible in this system.

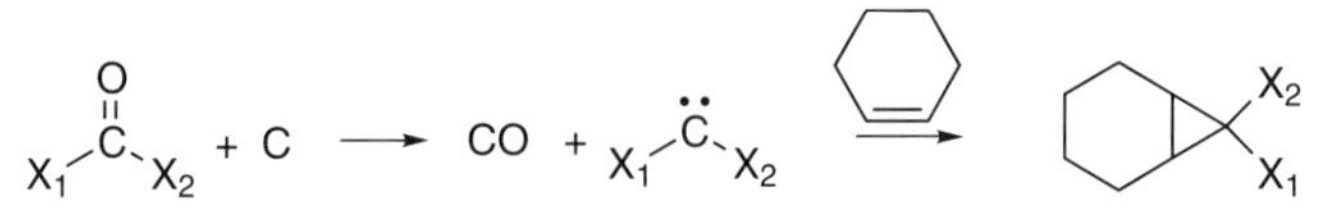

$X_1 = X_2 = H$;
$X_1 = X_2 = Cl$;
$X_1 = H, X_2 = OCH_3$;

$$(56)$$

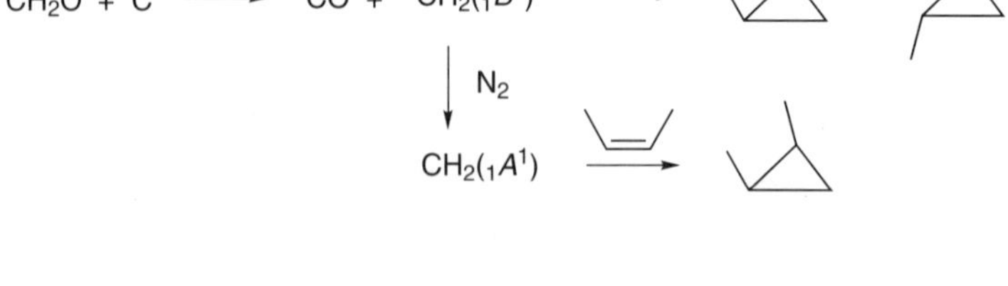

$$(57)$$

When cyclopentanone (**98a**) is deoxygenated to cyclopentanylidene (**99a**) with C atoms produced both in an arc and by tetrazole decomposition, an unusual carbene cleavage to ethylene and allene occurs along with rearrangement to cyclopentene (Eq. 58).[109] Since this cleavage is not observed when **99** is generated by tosylhydrazone decomposition, it has been attributed to energetic **99**. Calculations (B3LYP/6-311+G(d)+ZPC) indicate that the cleavage proceeds via the open shell singlet **99**, which cleaves to biradical **100**. The lack of concert in the cleavage reaction is demonstrated by the fact that *cis*-3,4-dideuterocyclopentane (**98b**) is deoxygenated to *cis*-3,4-dideuterocyclopentylidene (**99b**), which cleaves nonstereo-specifically to *cis*- and *trans*-1,2-dideuteroethylene.[109]

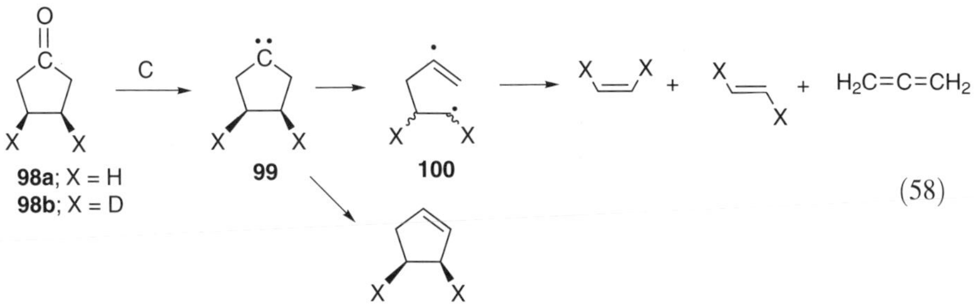

98a; X = H
98b; X = D

99   100

$$\text{(58)}$$

Carbene generation by C atom deoxygenation has been useful in answering questions concerning the intermediacy of free carbenes in certain systems. For example, *tert*-butylcarbene (**101**) from several precursors gives 1,1-dimethylcyclo-propane (**102**) by C—H insertion, and 2-methyl-2-butene by C—C insertion.[100a,110] However, calculations ([QCISD(T)/6-31+G(2d,p)]// MP2/6-31G(d)) indicate that C—C insertion should not be competitive with C—H insertion in this carbene.[111] In agreement with this prediction, deoxygenation of 2,2-dimethylpropanal (**103**) by C atoms, at 77, 158, and 195 K, yields only the C—H insertion product (Eq. 59).[111] A similar result is observed when free carbene **101** is generated by cheleotropic extrusion from a tricyclopropane.[103]

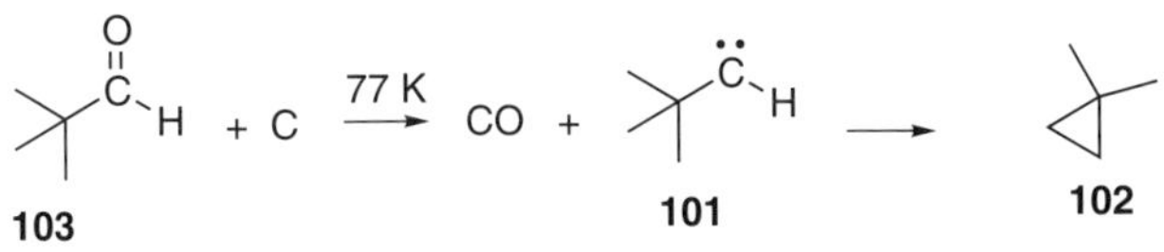

103   101   102

$$\text{(59)}$$

Other free carbenes that have been generated in this way and their reactions compared with those from other precursors include bicyclo[2.2.2]octylidene,[112] cyclopropylmethylcarbene,[113] and 1-phenylpropylidine (**104**).[114] In the case of **104**, the same ratio of (*E*) to (*Z*)-1-phenylpropene was produced when the carbene was generated by deoxygenation of the corresponding ketone and by the phenylcar-bene rearrangement (Eq. 60).[114] Thus, the method represents a viable alternative to more conventional routes to carbenes with the caveat that the high energies involved in these deoxygenations may bring about unexpected reactions.

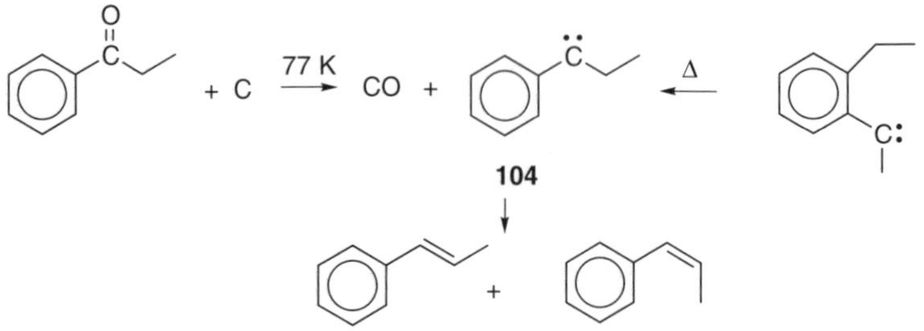

$$(60)$$

3.7.3. Phenylnitrene by the Deoxygenation of Nitrosobenzene.

The cocondensation of carbon with nitrosobenzene (**105**) yields aniline (**106**), azobenzene (**107**), azoxybenzene (**108**), and phenylisocyanate (**109**) (Eq. 61).[115] In analogy with carbonyl deoxygenations to carbenes, this reaction is postulated to give singlet phenylnitrene ^{1}**110**, which decays to ^{3}**110**. Hydrogen abstraction by ^{3}**110** gives **106**, dimerization yields **107**, and reaction of **105** with ^{3}**110** produces **108**. It seems likely that **109** arises from addition to the N—O double bond in a reaction competitive with deoxygenation. Attempts to trap azacycloheptatetraene (**111**), the product of ring expansion of ^{1}**110** are successful when the C + **105** reaction is carried out at 195 K but not at 77 K (Eq. 61).

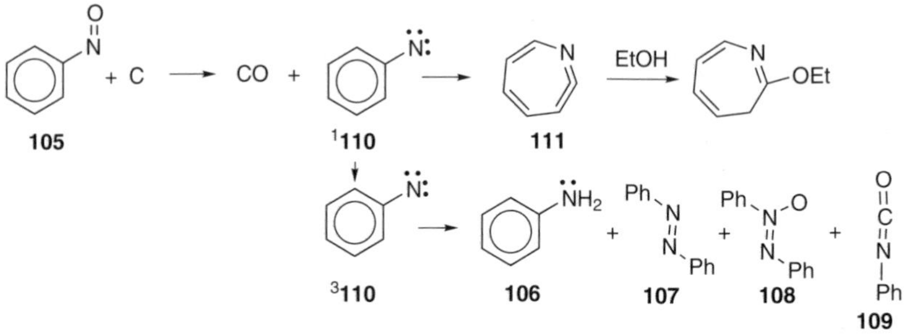

$$(61)$$

This temperature effect reflects the known fact that ring expansion of ^{1}**110** to **111** is only competitive with isc to ^{3}**110** at higher temperatures.[116] The reaction of C atoms with pyridine is also a route to singlet and triplet **110** presumably via the pyryidylcarbenes (**112**) (Eq. 62).[115]

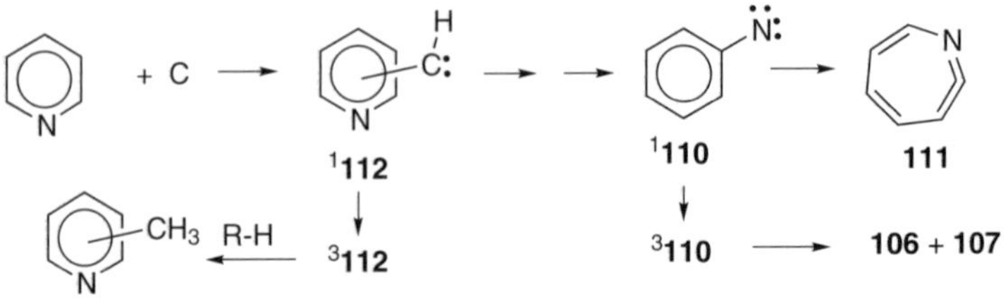

$$(62)$$

3.8. Other Reactions of Carbon Atoms with Lone Pairs

Since carbon monosulfide is a higher energy species than carbon monoxide, desulfurizations by atomic carbon are expected to be less energetic. This prediction has been born out in a series of experiments by Skell et al.[96] For example, desulfurization of tetrahydrothiophene leads to ethylene and cyclobutane in a 10:1 ratio. The fact that cyclobutane is generated in the desulfurization of tetrahydrothiophene but not in the deoxygenation of THF, has been attributed to the lower exothermicity of the desulfurizations. The generation of carbenes, presumably with less energy than those from deoxygenation, by desulfurization of thiocarbonyl compounds has not yet been attempted.

Reaction of arc generated carbon with aziridines generates alkenes in a nonstereospecific manner along with concurrent production of HCN.[117] The reaction of arc generated carbon with trimethylamine appears to proceed by attack on the nitrogen lone pair to give ylid **113**, which has limited stability at 77 K.[118] Addition of oxygen to a 77 K matrix of carbon + trimethylamine results in the formation of carbon monoxide, indicating that **113** functions as a triplet C atom donor (Eq. 63).

$$-\overset{|}{\underset{|}{N}}: \;+\; C \;\xrightarrow{77\ K}\; -\overset{|}{\underset{|}{N}}{}^{+}-\bar{C} \;\xrightarrow[77\ K]{O_2}\; CO \;+\; -\overset{|}{\underset{|}{N}}: \tag{63}$$

113

4. MOLECULAR BEAM STUDIES OF CARBON ATOM REACTIONS

There have been several recent reports of the interaction of C atoms with substrates in molecular beams.[27] In these experiments, a beam of C atoms is generated by laser evaporation of graphite, the energy of the beam is selected, and the beam is crossed with a substrate beam of known energy. Products are identified mass spectrometrically and their energies measured. Under the single collision conditions of these investigations, initial products cannot dissipate their energy and a C—H bond cleavage invariably occurs (Eq. 64). The observed energy release in the reaction is then correlated with high quality calculations to deduce the structure of the product.

$$C \;+\; R\text{-}H \;\longrightarrow\; \left[\, C\text{-}R\text{-}H \,\right]^{\ddagger} \;\longrightarrow\; \text{"C-R"} \;+\; H\cdot \tag{64}$$

Substrates that have been studied using these beam techniques include acetylene,[119] ethylene,[120] propene,[121] propyne,[122] allene,[123] propargyl radicals,[124] 2-butyne,[125] 1,2-butadiene,[126] H_2S,[127] and benzene.[27,128] The reaction of $C(^3P)$ with benzene is of interest as initial reactivity is different from that observed in other carbon + benzene reactions discussed earlier. In this case, initial addition gives complex **114** between triplet carbon and benzene. Ring opening of **114** to triplet **53** is followed by loss of H to give the 1,2-dehydrocycloheptatrienyl radical **115** (Eq. 65).

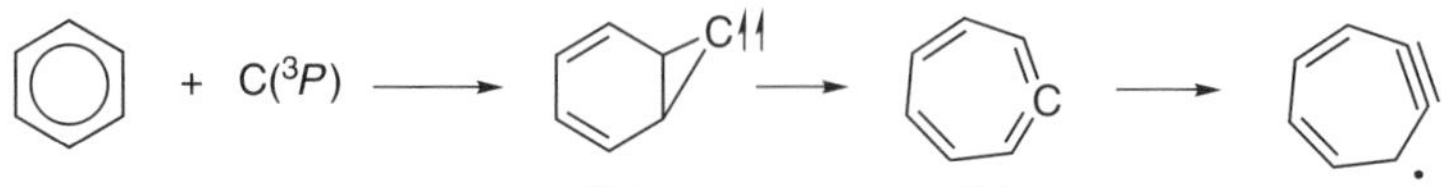

$$ \text{(65)} $$

In molecular beam studies of C atom reactions, it is also possible to select higher energy portions of the beam that contain $C(^1D)$. This selection has been accomplished in a study of the reaction of singlet carbon with acetylene that proceeds with loss of hydrogen to give **116** (Eq. 66).[119a] In contrast, $C(^3P)$ reacts with acetylene by two channels producing both **116** and **117** (Eq. 67).[119b]

$$ \text{(66)} $$

$$ \text{(67)} $$

Since molecular beam investigations of C atom reactivity differ from other C atom studies with regard to dissipation of reaction exothermicity, it is difficult to compare them with other methods. A hallmark of carbon molecular beam studies is the fact that addition of C is followed by loss of H. This reaction path is unimportant in other reactions of C atoms.

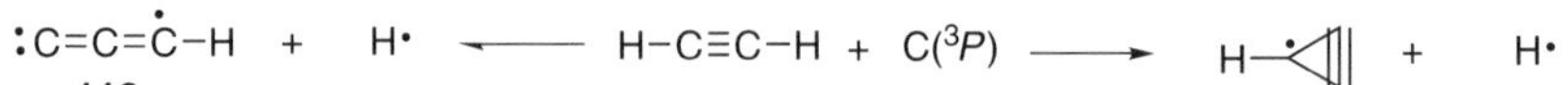

5. CONCLUSION AND OUTLOOK

It is clear that atomic carbon exhibits a rich and exotic chemistry. Indeed, as one of the most energetic species in organic chemistry, such behavior is anticipated. Only a small portion of this chemistry has been explored, leaving a vast area of C atom chemistry awaiting further research. It is probable that the ideal photochemical precursor to C atoms has not yet been developed. Such a molecule would produce C upon gas-phase photolysis at convenient wavelengths.

The continuing interest in carbene chemistry and the fact that C atoms serve as convenient carbene precursors makes this research area attractive. Although carbonyl compounds are conveniently deoxygenated to carbenes, the analogous deoxygenation of acid chlorides and amides to generate chloro- and aminocarbenes has not been explored. The fact that many carbenes produced by this method will have excess energy and may be in electronically excited states should provide the basis for many exciting experiments. The chemistry of energetic nitrenes from carbon + nitroso compounds should be investigated. The deoxygenation of ethers should lead to many radical pairs that can be compared with the same radical pair generated in thermolysis reactions. Deoxygenation of metal carbonyls seems a viable route to metal carbide complexes. The chemistry of carbenes produced by C atom desulfurizations should be investigated. These are just a few of the many possible future directions in C atom chemistry. Future researchers should be encouraged by the fact that is difficult to imagine a molecule that will not react with atomic carbon.

SUGGESTED READING

For a discussion of the formation of biomolecules in carbon atom reactions see: P. B. Shevlin, D. W. McPherson, and P. Melius, "The Reaction of Atomic Carbon with Ammonia. The Mechanism of Formation of Amino Acid Precursors," *J. Am. Chem. Soc.* **1983**, 105, 488 and G. Flanagan, S. N. Ahmed and P. B. Shevlin, "The Formation of Carbohydrates in the Reaction of Atomic Carbon with Water," *J. Am. Chem. Soc.* **1992**, 114, 3892.

For a discussion of the possible intermediacy of electronically excited carbenes in carbon atom reactions see: G. Xu, T.-M. Chang, J. Zhou, M. L. McKee, and P. B. Shevlin, "An Unusual Cleavage of an Energetic Carbene," *J. Am. Chem. Soc.* **1999**, 121, 7150.

The reaction of carbon atoms with aromatic rings is discussed by B. M. Armstrong, F. Zheng, and P. B. Shevlin, "The Mode of Attack of Atomic Carbon on Benzene Rings," *J. Am. Chem. Soc.*, **1998**, 120, 6007. This reactivity may be contrasted with that observed when a beam of $C(^3P)$ reacts with benzene as described by I. Hahndorf, Y. T. Lee, R. I. Kaiser, L. Vereecken, J. Peeters, H. F. Bettinger, P. R. Schreiner, P. von R. Schleyer, W. D. Allen, and H. F. Schaefer, III. "A combined crossed-beam, ab initio, and Rice–Ramsperger–Kassel–Marcus investigation of the reaction of carbon atoms $C(^3Pj)$ with benzene, $C_6H_6(X\ ^1A_{1g})$ and d_6-benzene, $C_6D_6(X\ ^1A_{1g})$," *J. Chem. Phys.* **2002**, *116*, 3248.

For a combined experimental and computational approach to carbon atom chemistry see C. M. Geise, C. M. Hadad, F. Zheng, and P. B. Shevlin, "An Experimental and Computational Evaluation of the Energetics of the Isomeric Methoxyphenylcarbenes Generated in Carbon Atom Reactions," *J. Amer. Chem. Soc.* **2002**, *124*, 355.

REFERENCES

1. A. P. Wolf, *Adv. Phys. Org. Chem.* **1964**, 2, 201.
2. R. Wolfgang, *Prog. React. Kinet.* **1965**, 3, 97.
3. R. Wolfgang, *Adv. High. Temp. Chem.* **1971**, 4, 43.
4. P. S. Skell, J. Havel, and M. J. McGlinchey, *Acc. Chem. Res.* **1973**, 6, 97.
5. C. MacKay, in *Carbenes*, Vol. II, R. A. Moss, M. Jones, Jr., Eds., Wiley-Interscience, New York, **1975**, pp. 1–42.
6. P. B. Shevlin, in *Reactive Intermediates*, Vol. I, R. A. Abramovitch, Ed., Plenum Press, New York, **1980**, pp. 1–36.
7. F. P. Israel and F. Baas, *Astron. Astrophys.* **2002**, *383*, 82.

8. A. P. Wolf and R. C. Anderson, *J. Am. chem. Soc.* **1955**, *77*, 1608.

9. B. Suryanarayana and A. P. Wolf, *J. Phys. Chem.* **1958**, *62*, 1369.

10. J. Dubrin, C. MacKay, M. L. Pandow, and R. Wolfgang, *J. Inorg. Nucl. Chem.* **1964**, *25*, 2113.

11. H. J. Ache and A. P. Wolf, *Radiochim. Acta* **1966**, *6*, 33.

12. G. Stocklin, H. Stangl, D. R. Christman, J. B. Cumming, and A. P. Wolf, *J. Phys. Chem.* **1963**, *67*, 1735.

13. A. P. Wolf and C. S. Redvanly, *Int. J. Appl. Radiat. Isotopes* **1977**, *28*, 29.

14. R. A. Ferrieri and A. P. Wolf, *Radiochim. Acta* **1983**, *34*, 69.

15. J. F. Lifton and M. J. Welch, *Radiation Res.* **1971**, *45*, 35.

16. P. S. Skell and R. F. Harris, *J. Am. Chem. Soc.* **1965**, *87*, 5807.

17. P. S. Skell and R. F. Harris, *J. Am. Chem. Soc.* **1966**, *88*, 5933.

18. C. J. Emanuel and P. B. Shevlin, *J. Am. Chem. Soc.* **1994**, *116*, 5991.

19. P. S. Skell, L. Jackman, S. Ahmed, M. L. McKee, and P. B. Shevlin, *J. Am. Chem. Soc.* **1989**, *111*, 4422.

20. W. Krätschmer, L. B. Lamb, K. Fostiropoulos, and D. R. Huffman, *Nature (London)* **1990**, *347*, 354.

21. T-M. Chang, A. Naim, S. N. Ahmed, G. Goodloe, and P. B. Shevlin, *J. Am. Chem. Soc.* **1992**, *114*, 7603.

22. J. Drowart, R. P. Burno, G. De Mario, and M. G. Inghram, *J. Chem. Phys.* **1959**, *31*, 1131.

23. P. S. Skell and P. W. Owen, *J. Am. Chem. Soc.* **1971**, *94*, 1578.

24. W. A. Chupka and M. G. Inghram, *J. Chem. Phys.* **1953**, *21*, 1313.

25. J. F. Verdieck and A. W. Mau, *H. Chem. Commun.* **1969**, 226.

26. H. W. Kroto, J. R. Heath, S. C. O'Brien, R. F. Curl, and R. E. Smalley, *Nature (London)* **1985**, *318*, 162.

27. I. Hahndorf, Y. T. Lee, R. I. Kaiser, L. Vereecken, J. Peeters, H. F. Bettinger, P. R. Schreiner, P. von R. Schleyer, W. D. Allen, and H. F. Schaefer, III, *J. Chem. Phys.* **2002**, *116*, 3248 and references cited therein.

28. M. W. Chase, Jr., *J. Phys. Chem. Ref. Data, Monograph 9* **1998**, 1.

29. N. G. Moll and W. E. Thompson, *J. Chem. Phys.* **1966**, *44*, 2684.

30. D. Husain and A. X. Ioannou, *J. Photochem. Photobiolo., A* **1999**, *129*, 1–7 and references cited therein.

31. (a) K. Bayes, *J. Am. Chem. Soc.* **1961**, *83*, 3712. (b) D. G. Williamson and K. D. Bayes, *J. Am. Chem. Soc.* **1968**, *90*, 1957.

32. D. E. Milligan and M. E. Jacox, *J. Chem. Phys.* **1966**, 44, 2684.

33. P. B. Shevlin and A. P. Wolf, *Tetrahedron Lett.* **1970**, 3987.

34. (a) P. B. Shevlin, *J. Am. Chem. Soc.* **1972**, 94, 1397. (b) S. L. Kammula and P. B. Shevlin, *J. Am. Chem. Soc.* **1974**, *96*, 7830. (c) S. L. Kammula, and P. B. Shevlin, *J. Am. Chem. Soc.* **1977**, *99*, 2167.

35. (a) A. Bergeat and J-C. Loison, *Phys. Chem. Chem. Phys.* **2001**, *3*, 2038. (b) N. Galland, F. Caralp, M-T. Rayez, Y. Hannachi, J-C. Loison, G. Dorthe, and A. Bergeat, *J. Phys. Chem. A* **2001**, 105, 9893.

36. D. Kley, N. Washida, K. H. Becker, and W. Groth, *Chem. Phys. Lett.* **1972**, *15*, 45.

37. (a) F. F. Martinotti, M. J. Welch, and A. P. Wolf, *Chem. Commun.* **1968**, 115. (b) A. P. Wolf, E. Y. Y. Lam, and P. P. Gaspar, *J. Phys. Chem.* **1971**, 75, 445.

38. (a) D. C. Scott, J. De Juan, D. C. Robie, D. Schwartz-Lavi, and H. Reisler, *J. Phys. Chem.* **1992**, *96*, 2509. (b) G. M. Jursich and J. R. Wiesenfeld, *Chem. Phys. Lett.* **1984**, *110*, 14.

39. (a) C. MacKay, J. Nicholas, and R. Wolfgang, *J. Am. Chem. Soc.* **1967**, *89*, 5758. (b) T. L. Rose, *J. Phys. Chem.* **1972**, *76*, 1389.

40. D. Husain and A. N. Young, *J. Chem. Soc. Faraday Trans. II* **1975**, *71*, 525.

41. S. L. Kammula and P. B. Shevlin, *J. Am. Chem. Soc.,* **1973**, *95*, 4441.

42. J. E. Dobson, P. M. Tucker, R, Schaeffer, and F. G. A. Stone, *J. Chem. Soc. A* **1969**, 1882.

43. S. R. Prince and R. Schaeffer, *Chem Commuun.* **1968**, 451.

44. Q. Ye, J. Li, M. Jones, Jr., H. Wu, M. L. McKee, and P. B. Shevlin, *Tetrahedron Lett.* **2002**, *43*, 735.

45. J. Binenboym and R. Schaeffer, *Inorg. Chem.* **1970**, *9*, 1578.

46. M. E. Jacox and D. E. Milligan, *J. Chem. Phys.* **1968**, *48*, 2265.

47. M. E. Jacox and D. E. Milligan, *J. Chem. Phys.* **1967**, *47*, 703.

48. M. E. Jacox and D. E. Milligan, *J. Chem. Phys.* **1969**, *50*, 3252.

49. M. E. Jacox and D. E. Milligan, *J. Chem. Phys.* **1967**, *47*, 1626.

50. (a) P. B. Shevlin, D. W. McPherson, and P. Melius, *J. Am. Chem. Soc.* **1981**, *103*, 7006. (b) P. B. Shevlin, D. W. McPherson, and P. Melius, *J. Am. Chem. Soc.* **1983**, *105*, 488.

51. D. W. McPherson, M. L. McKee, and P. B. Shevlin, *J. Am. Chem. Soc.* **1983**, *105*, 6493.

52. G. Flanagan, S. N. Ahmed, and P. B. Shevlin, *J. Am. Chem. Soc.* **1992**, *114*, 3892.

53. S. N. Ahmed, M. L. McKee, and P. B. Shevlin, *J. Am. Chem. Soc.* **1983**, *105*, 3942.

54. S. N. Ahmed, M. L. McKee, and P. B. Shevlin, *J. Am. Chem. Soc.* **1985**, *107*, 1320.

55. D. W. McPherson, K. Rahman, I. Martinez, and P. B. Shevlin, *Origins of Life* **1987**, *17*, 275.

56. G. H. Jeong, K. J. Klabunde, O-G. Pan, G. C. Paul, and P. B. Shevlin, *J. Am. Chem. Soc.* **1989**, *111*, 8784.

57. P. S. Skell and R. R. Engel, *J. Am. Chem. Soc.* **1966**, *88*, 4883.

58. G. Stocklin and A. P. Wolf, *J. Am. Chem. Soc.* **1963**, *85*, 229.

59. C. MacKay, P. Polak, H. E. Rosenberg, and R. Wolfgang, *J. Am. Chem. Soc.* **1962**, *84*, 308.

60. P. S. Skell, J. E. Villuame, J. H. Plonka, F. A. Fagone, *J. Am. Chem. Soc.* **1971**, *93*, 2699.

61. M. Marshall, C. MacKay, and R. Wolfgang, *J. Am. Chem. Soc.* **1964**, *86*, 4741.

62. P. B. Shevlin and H. Wu, *Book of Abstracts*, 219th ACS National Meeting, San Francisco, CA, March 26–30, **2000**, ORGN-455.

63. Ting Chen, Ph. D. Thesis, Auburn University, December, **2001**.

64. Guopin Xu, Ph. D. Thesis, Auburn University, December, **1999**.

65. (a) A. Runge and W. Sander, *Tetrahedron Lett.* **1986**, *27*, 5835. (b) A. Z. Bradley and R. P. Johnson, *J. Am. Chem. Soc.* **1997**, *119*, 9917.

66. (a) Haitao Wu, Ph. D. Thesis, Auburn University, December, **2001**. (b) Jin M. S. Zeng, Thesis, Auburn University, June, **1999**.

67. (a) M. Balci, and W. M. Jones, *J. Am. Chem. Soc.* **1981**, *103*, 2874. (b) R. G. Bergman and V. Rajadhyaksha, *J. Am. Chem. Soc.* **1970**, *92*, 2163. (c) P. K. Freeman and K. E. Swenson, *J. Org. Chem.* **1982**, *47*, 2033.

68. W. Pan and P. B. Shevlin, *J. Am. Chem. Soc.* **1996**, *118*, 10004.

69. E. E. Waali and N. T. Allison, *J. Org. Chem.* **1979**, *44*, 3266.

70. (a) D. Blaxell, C. MacKay, and R. Wolfgang, *J. Am. Chem. Soc.* **1969**, *92*, 50. (b) R. D. Finn, H. J. Ache, and A. P. Wolf, *J. Phys. Chem.* **1970**, *74*, 3194.

71. M. Rahman, M. L. McKee, and P. B. Shevlin, *J. Am. Chem. Soc.* **1986**, *108*, 6296.

72. C. J. LaFrancois and P. B. Shevlin, *Tetrahedron* **1997**, *53*, 10071.

73. R. Sztyrbicka, M. M. Rahman, M. E. D'Aunoy, and P. B. Shevlin, *J. Am. Chem. Soc.* **1990**, *112*, 6712.

74. This reaction finds analogy in the work of M. Pomerantz and A. S. Ross, *J. Am. Chem. Soc.* **1975**, *97*, 5850 and references cited therein.

75. C. J. LaFrancois and P. B. Shevlin, *J. Am. Chem. Soc.* **1994**, *116*, 9405.

76. (a) Christopher J. LaFrancois, Ph. D. Thesis, Auburn University, June, **1996**.

77. (a) A. P. Wolf, C. S. Redvanly, and R. C. Anderson, *Nature (London)* **1955**, *176*, 831. (b) A. G. Schrodt and W. F. Libby, *J. Am. Chem. Soc.* **1956**, *78*, 1267. (c) B. Suryanarayana and A. P. Wolf, *J. Phys. Chem.* **1958**, *62*, 1369. (d) J. L. Sprung, S. Winstein, and W. F. Libby, *J. Am. Chem. Soc.* **1965**, *87*, 1812. (e) T. Rose, C. MacKay, and R. Wolfgang, *J. Am. Chem. Soc.* **1967**, *89*, 1529. (f) R. L. Williams and A. F. Voigt, *J. Phys. Chem.* **1969**, *73*, 2538. (g) R. M. Lemmon, *Acc. Chem. Res.* **1973**, *6*, 65 and references cited therein. (h) K. A. Biesiada, C. T. Koch, and P. B. Shevlin, *J. Am. Chem. Soc.* **1980**, *102*, 2098. (i) B. M. Armstrong, F. Zheng, and P. B. Shevlin, *J. Am. Chem. Soc.* **1998**, *120*, 6007.

78. For reviews with many references see: (a) W. M. Jones, in *Rearrangements in Ground and Excited States*, Vol. 1., P. de Mayo, Ed., Academic, New York, **1980**. (b) C. Wentrup, in *Methoden der Organischen Chemie (Houben-Weyl)*, Vol. E19b, M. Regitz, Ed., G. Thieme, Stuttgart, **1989**, p. 824. (c) P. P. Gaspar, J.-P. Hsu, S. Chari, and M. Jones, Jr., *Tetrahedron* **1985**, *41*, 1479. (d) M. S. Platz, *Acc. Chem. Res.* **1995**, *28*, 487.

79. P. P. Gaspar, D. M. Berowitz, D. R. Strongin, D. L. Svoboda, M. B. Tuchler, R. A. Ferrieri, and A. P. Wolf, *J. Phys. Chem.* **1986**, *90*, 4691.

80. (a) S. Matzinger, T. Bally, E. V. Patterson, and R. J. McMahon, *J. Am. Chem. Soc.* **1996**, *118*, 1535. (b) M. W. Wong and C. Wentrup, *J. Org. Chem.* **1996**, *61*, 7022. (c) P. R. Schreiner, W. L. Karney, P. v. R. Schleyer, W. T. Borden, T. P. Hamilton, and H. F. Schaefer, III, *J. Org. Chem.* **1996**, *61*, 7030. (d) C. J. Cramer, F. J. Dulles, and D. E. Falvey, *J. Am. Chem. Soc.* **1994**, *116*, 9787.

81. F. Sevin, I. Sökmen, B. Düz, and P. B. Shevlin, *Tetrahedron Lett.* **2003**, *44*, 3405.

82. (a) M. Jones, Jr. and A. H. Golden, *J. Org. Chem.* **1996**, *61*, 4460. (b) A. Kumar, R. Narayanan, and H. Shechter, *J. Org. Chem.* **1996**, *61*, 4462.

83. C. M. Geise, C. M. Hadad, F. Zheng, and P. B. Shevlin, *J. Am. Chem. Soc.* **2002**, *124*, 355.

84. F. Zheng, M. L. McKee, and P. B. Shevlin, *J. Am. Chem. Soc.* **1999**, *121*, 11237.

85. (a) J. Becker and C. Wentrup, *Chem. Commun.* **1980**, 190. (b) T. A. Engler and H. Shechter, *Tetrahedron Lett.* **1982**, 23, 2715.

86. Y. Xie, P. R. Schreiner, P. v. R. Schleyer, and H. F. Schaefer, *J. Am. Chem. Soc.* **1997**, *119*, 1370.

87. W. Pan and P. B. Shevlin, *J. Am. Chem. Soc.* **1997**, *119*, 5091.

88. W. Pan, M. Balci, and P. B. Shevlin, *J. Am. Chem. Soc.* **1997**, *119*, 5035.

89. M. L. McKee and P. B. Shevlin, *J. Am. Chem. Soc.* **2001**, *123*, 9418.

90. S. F. Dyer and P. B. Shevlin, *J. Am. Chem. Soc.* **1979**, *101*, 1303.

91. P. S. Skell and R. F. Harris, *J. Am. Chem. Soc.* **1969**, *91*, 4440.

92. C. MacKay and R. Wolfgang, *Radiochim. Acta.*, **1962**, *1*, 42.

93. J. H. Plonka and P. S. Skell, *Chem. Commun.*, **1970**, 1108.

94. R. H. Parker and P. B. Shevlin, *Tetrahedron Lett.* **1975**, 2167.

95. P. S. Skell, K. J. Klabunde, and J. H. Plonka, *Chem. Commun.* **1970**, 1109.

96. P. S. Skell, K. J. Klabunde, J. H. Plonka, J. S. Roberts, and D. L. Williams-Smith, *J. Am. Chem. Soc.*, **1973**, *95*, 1547.

97. M. L. McKee, G. C. Paul, and P. B. Shevlin, *J. Am. Chem. Soc.* **1990**, *112*, 3374.

98. P. S. Skell, J. H. Plonka, and R. R. Engel, *J. Am. Chem. Soc.*, **1967**, *89*, 1748.

99. P. C. Martino and P. B. Shevlin, *Tetrahedron Lett.* **1982**, 2547.

100. (a) P. S. Skell and J. H. Plonka, *J. Am. Chem. Soc.*, **1970**, *92*, 836. (b) P. S. Skell and J. H. Plonka, *Tetrahedron* **1972**, *28*, 3571.

101. S. F. Dyer, S. L. Kammula, and P. B. Shevlin, *J. Am. Chem. Soc.* **1977**, *99*, 8104.

102. (a) R. A. Moss and W. Liu, *Chem. Commun.* **1993**, 1597. (b) D. A. Modarelli, S. Morgan, and M. S. Platz, *J. Am. Chem. Soc.* **1992**, *114*, 7034. (c) N. Chen, M. Jones, Jr., W. R. White, and M. S. Platz, *J. Am. Chem. Soc.* **1991**, *113*, 4981. (d) R. A. Seburg and R. J. McMahon, *J. Am. Chem. Soc.* **1992**, *114*, 7183. (e) H. Tomioka, H. Kitagawa, and Y. Izawa, *J. Org. Chem.* **1979**, *44*, 3072. (f) N. Yamamoto, F. Bernardi, A. Bottoni, M. Olivucci, M. A. Robb, and S. Wilsey, *J. Am. Chem. Soc.* **1994**, *116*, 2064. (g) M. T. H. Liu, *Acc. Chem. Res.* **1994**, *27*, 287. (h) S. Çelebi, S. Leyva, D. Modarelli, and M. S. Platz, *J. Am. Chem. Soc.* **1993**, *115*, 8613.

103. H. C. Glick, I. R. Likhotvorik, and M. Jones, Jr., *Tetrahedron Lett.* **1995**, *36*, 5715.

104. M. Rahman and P. B. Shevlin, *Tetrahedron Lett.* **1985**, *26*, 2559.

105. W. J. Baron, M. Jones, Jr., and P. P. Gaspar, *J. Am. Chem. Soc.* **1970**, *92*, 4739.

106. S. N. Ahmed and P. B. Shevlin, *J. Am. Chem. Soc.* **1983**, *105*, 6488.

107. P. S. Skell and J. H. Plonka, *J. Am. Chem. Soc.*, **1970**, *92*, 2160.

108. R. Hoffmann, *J. Am. Chem. Soc.* **1968**, *90*, 1475.

109. G. Xu, T.-M. Chang, J. Zhou, M. L. McKee, and P. B. Shevlin, *J. Am. Chem. Soc.* **1999**, *121*, 7150.

110. (a) H. M. Frey and I. D. R. Stevens, *J. Chem. Soc.* **1965**, 3101. (b) A. M. Mansoor and I. D. R. Stevens, *Tetrahedron Lett.* **1966**, 1733. (c) M. Fukushima, M. Jones, Jr., and U. H. Brinker, *Tetrahedron. Lett.* **1982**, *23*, 3211. (d) M. J. Goldstein and W. R. Dolbier, Jr., *J. Am. Chem. Soc.* **1965**, *87*, 2293. (e) K.-T. Chang and H. Shechter, *J. Am. Chem. Soc.* **1979**, *101*, 5082.

111. B. M. Armstrong, M. L. McKee, and P. B. Shevlin, *J. Am. Chem. Soc.* **1995**, *117*, 3685.

112. Q. Ye, M. Jones, Jr., T. Chen, and P. B. Shevlin, *Tetrahedron Lett.* **2001**, *42*, 6979.

113. D. M. Thamattoor, M. Jones, Jr., W. Pan, and P. B. Shevlin, *Tetrahedron Lett.* **1996**, *37*, 8333.

114. J. M. Fox, J. E. G. Scacheri, K. G. H. Jones, M. Jones, Jr., P. B. Shevlin, B. Armstrong, and R. Sztyrbicka, *Tetrahedron Lett.* **1992**, *33*, 5021.

115. B. M. Armstrong and P. B. Shevlin, *J. Am. Chem. Soc.* **1994**, *116*, 4071.

116. M. S. Platz, in *Kinetics and Spectroscopy of Carbenes and Biradicals*, M. S. Platz and V. M. Maloney, Eds., Plenum Press, New York **1990**, pp. 306 ff.

117. J. F. Villaume and P. S. Skell, *J. Am. Chem. Soc.* **1972**, *94*, 3455.

118. D. W. McPherson, M. L. McKee, and P. B. Shevlin, *J. Am. Chem. Soc.* **1984**, *106*, 2712.

119. (a) R. I. Kaiser, C. Ochsenfeld, M. Head-Gordon, Y. T. Lee, and A. G. Suits, *J. Chem. Phys.* **1997**, *106*, 1729. (b) R. I. Kaiser, A. M. Mebel, and Y. T. Lee, *J. Chem. Phys.* **2001**, *114*, 231. (c) L. Cartechini, A. Bergeat, G. Capozza, P. Casavecchia, G. G. Volpi, W. D. Geppert, C. Naulin, and M. Costes, *J. Chem. Phys.* **2002**, *116*, 5603.

120. R. I. Kaiser, Y. T. Lee, and A. G. Suits, *J. Chem. Phys.* **1996**, *105*, 8705.

121. R. I. Kaiser, D. Stranges, H. M. Bevsek, Y. T. Lee, and A. G. Suits, *J. Chem. Phys.* **1997**, *106*, 4945.

122. R. I. Kaiser, D. Stranges, Y. T. Lee, and A. G. Suits, *J. Chem. Phys.* **1996**, *105*, 8721.

123. R. I. Kaiser, A. M. Mebel, A. H. H. Chang, S. H. Lin, and Y. T. Lee, *J. Chem. Phys.* **1999**, *110*, 10330.

124. (a) T. L. Nguyen, A. M. Mebel, S. H. Lin, and R. I. Kaiser, *J. Phys. Chem. A* **2001**, *105*, 11549. (b) R. I. Kaiser, W. Sun, A. G. Suits, and Y. T. Lee, *J. Chem. Phys.* **1997**, *107*, 8713.

125. L. C. L. Huang, H. Y. Lee, A. M. Mebel, S. H. Lin, Y. T. Lee, and R. I. Kaiser, *J. Chem. Phys.* **2000**, *113*, 9637.

126. N. Balucani, H. Y. Lee, A. M. Mebel, Y. T. Lee, and R. I. Kaiser, *J. Chem. Phys.* **2001**, *115*, 5107.

127. (a) R. I. Kaiser, C. Ochsenfeld, M. Head-Gordon, and Y. T. Lee, *J. Chem. Phys.* **1999**, *110*, 2391. (b) R. I. Kaiser, W. Sun, and A. G. Suits, *J. Chem. Phys.* **1997**, *106*, 5288.

128. (a) H. F. Bettinger, P. v. R. Schleyer, H. F. Schaefer, III, P. R. Schreiner, R. I. Kaiser, and Y. T. Lee, *J. Chem. Phys.* **2000**, *113*, 4520. (b) R. I. Kaiser, I. Hahndorf, L. C. L. Huang, Y. T. Lee, H. F. Bettinger, P. v. R. Schleyer, H. F. Schaefer, III, and P. R. Schreiner, *J. Chem. Phys.* **1999**, *110*, 6091.

CHAPTER 11

Nitrenes

MATTHEW S. PLATZ

Department of Chemistry, The Ohio State University, Columbus, OH

Reactive Intermediate Chemistry, edited by Robert A. Moss, Matthew S. Platz, and Maitland Jones, Jr.
ISBN 0-471-23324-2   Copyright © 2004 John Wiley & Sons, Inc.

1. INTRODUCTION

Compounds containing neutral, monovalent nitrogen atoms are known as nitrenes.[1] The parent structure, NH, is also called imidogen. Because most stable compounds of neutral nitrogen have a valence of 3, it is no surprise that nitrenes typically are very short lived, reactive intermediates. A short history of nitrenes has been presented by Lwowski[2] who points out that they were first proposed by Tiemann in 1891[3] as transient intermediates in the Lossen rearrangement.

Nitrenes are commonly generated by decomposition of organic and inorganic azides, although other types of precursors are known.[1] As azides are bound to many elements, many types of nitrenes are known or can be imagined. This chapter will be limited to only those nitrenes commonly encountered in organic chemistry.

The term "nitrene" is reminiscent of "carbene." In fact, the two types of intermediates have more than a superficial resemblance. Carbenes and nitrenes have two bonds fewer than most stable compounds of carbon and nitrogen, respectively. As we will see, many of the properties of nitrenes are best appreciated on comparison with carbenes.

Nitrenes are involved or imagined in many useful transformations in classical synthetic organic chemistry.[4] Physical organic chemists seek to understand the role of nitrenes in these reactions and how their structures control their reactivity. Biochemists sometime append azide groups to the natural ligands of biological macromolecules.[5] Photolysis of the azide–biomolecule complex often leads to covalent attachment of the ligand to the biomolecule. This technique, invented by Singh et al.[6] with carbenes, and later by Knowles and Bayley[7] with nitrenes, has come to be known as photoaffinity labeling. Materials chemists also seek to attach probes to the surfaces of polymers using reactions of nitrenes.[8] Industrial scientists use azide photochemistry and nitrenes in lithography and as photoresists.[9] Thus, the chemistry of nitrenes attracts the interest of a diverse group of scientists and to understand their properties requires the multidisciplinary contributions of organic chemists, spectroscopists, kineticists, and theoreticians. The goal of this chapter is to interweave these contributions rather than to present the historical development of the field. Space does not permit a comprehensive review and coverage reflects the biases of the author.

1.1. Imidogen (NH)

The nitrene nitrogen atom of imidogen is *sp* hybridized. Both the NH bond and the sigma lone pair of electrons use nitrogen orbitals that are rich in $2s$ character.

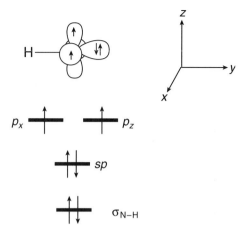

In triplet imidogen, there are two electrons with parallel spins, singly occupying pure p_x and p_z orbitals. This state is designated as $^3\Sigma^-$ by spectroscopists.

Because of the high symmetry of the NH molecule, the singlet state of imidogen cannot be classified as either open or closed shell. The "closed-shell" and "open-shell" singlets have exactly the same energy and form the two components of a doubly degenerate $^1\Delta$ state.[10]

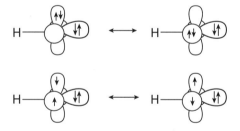

The triplet state of NH is 36 kcal/mol lower in energy than the singlet state.[11] The triplet state is favored because, on the average, electrons with parallel spin spend less time in proximity with each other than electrons with antiparallel spin. Consequently, the Coulombic electron–electron repulsion in the triplet state is less severe than in the singlet state (cf. Borden, Chapter 22 in this volume.) As we will see in vinyl- and phenylnitrene, delocalization of an unpaired electron by conjugation dramatically stabilizes the singlet relative to the triplet states of nitrenes.

The singlet–triplet splitting of NH was determined experimentally by spectroscopy of neutral NH[11] and by negative ion photoelectron spectroscopy (PES) of the NH anion.[12] In the latter experiment, the anion NH$^-$ is prepared in the gas phase and exposed to monochromatic ultraviolet (UV)-laser light. This photolysis leads to ejection of photoelectrons whose kinetic energies (E_K) are analyzed. As the energy of the incident photons ($h\nu$) is known, the binding energies (E_B) of the detached photoelectrons can be deduced and plotted in the form of a spectrum.

$$h\nu = E_K - E_B$$

The imidogen anion (NH$^-$) can ionize to form either the singlet or triplet nitrene, thus, the difference in the kinetic energies of the photoelectrons leading to ^{1}NH and ^{3}NH is just the singlet–triplet splitting of NH.

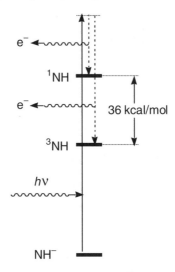

Analogous negative ion PES experiments reveal that the singlet–triplet splitting of methylene (CH$_2$) is only 9.05 kcal/mol.[13] This difference is easily understood on the basis of a simple molecular orbital (MO) diagram.

It is useful to imagine CH$_2$ with a bond angle of 120°, that is, sp^2 hybridization. The aufbau procedure directs the last two electrons into the lower energy sp^2 orbital to form a closed-shell electron configuration. According to valence-shell electron-pair repulsion (VSEPR) theory, lone-pair–bond-pair repulsion is stronger than bond-pair–bond-pair repulsion, and this idea explains why the HCH bond angle of singlet methylene is reduced to 105°.[14]

In triplet methylene, one nonbonding electron resides in the in-plane sp^2 orbital and the second is in the out-of-plane p_π orbital. Using the same VSEPR arguments, bond-pair–bond-pair repulsion is now greater than the repulsion between a bonding pair of electrons and the singly occupied sp^2 orbital. Thus the HCH bond angle of triplet methylene expands to 135°.[15]

Triplet methylene is 9.05 kcal/mol lower in energy than singlet methylene even though both nonbonding electrons of the singlet state reside in the lower energy $sp^{\sim 2}$ orbital in the singlet state. The reason for this is that the two nonbonding electrons experience more severe Coulombic repulsion when they both reside in the same sp^2 orbital than when they are placed in two separate orbitals. If the difference in Coulombic repulsion outweighs the sp^2/p_π orbital energy gap (which is the case in CH_2), then the triplet becomes the ground state.

In NH, the "last" two electrons will reside in two degenerate pure p orbitals. The singlet state of NH still suffers from increased electron–electron repulsion, relative to the triplet, but unlike in $^1CH_2$, does not benefit from placing two electrons into a lower energy orbital rich in $2s$ character. Thus, the singlet–triplet (ST) splitting of NH is much larger than that of CH_2. This rule is general. In many cases, the ST splittings of carbenes are so small that the two states rapidly and reversibly interconvert. Relaxation of a singlet to the corresponding triplet nitrene, however, is always irreversible (when the triplet is the ground state) because of the large ST gap.

Simple MO pictures also explain the differences in the bond dissociation energies (BDE) of methane and ammonia, and explain different triplet carbene and nitrene reactivities toward hydrogen atom donors.[10,16]

The first BDE of ammonia is slightly larger (3.5 kcal/mol) than that of methane, but the second BDE of methane is substantially larger (13.3 kcal/mol) than that of ammonia.

$$\Delta H \text{ (kcal/mol)}$$

CH_4	$\longrightarrow$	$CH_3\cdot$ + H $\cdot$		104.7
CH_3	$\longrightarrow$	$\cdot CH_2$ + H $\cdot$		110.3
NH_3	$\longrightarrow$	$NH_2\cdot$ + H $\cdot$		108.2
NH_2	$\longrightarrow$	$\cdot NH$ + H $\cdot$		97.0

In the first bond dissociation of CH_4 and NH_3 the hybridization of the central atom is changing from sp^3 to sp^2 in CH_3 and approximately to sp^2 in NH_2. When $\cdot CH_3$ fragments to $^3CH_2 + \cdot H$, there is little change in hybridization. However, when $\cdot NH_2$ fragments to $NH + \cdot H$, there is substantial rehybridization. The rehybridization of NH to sp stabilizes ^3NH by lowering the energy of the nonbonding electron pair. This requirement makes the second BDE of ammonia less endothermic. The result is that triplet methylene reacts exothermically with methane to yield two methyl radicals, whereas the reaction of ^3NH with methane to give $\cdot NH_2 + \cdot CH_3$ is endothermic. Therefore triplet nitrenes are much less reactive toward hydrogen atom donors than triplet carbenes.

$$\Delta H \text{ (kcal/mol)}$$

$^3CH_2$ + CH_4	$\longrightarrow$	2 $CH_3\bullet$		-5.6
^3NH + CH_4	$\longrightarrow$	$\bullet NH_2$ + $CH_3\bullet$		+7.7

Imidogen can be generated in the gas phase by photolysis or by pyrolysis of hydrazoic acid (HN_3) although other precursors find occasional use.[17] Exposure of HN_3 to either 248 or 266-nm light generates ^1NH almost exclusively in its lowest singlet state with a quantum yield near unity.

Singlet NH inserts into the CH bonds of hydrocarbons, much like singlet methylene (see Chapter 7 in this volume). Triplet NH abstracts hydrogen atoms from hydrocarbons to form aminyl ($NH_2^{\bullet}$) radicals and alkyl radicals in the same manner as triplet methylene, in spite of the fact that the reactions of $^3CH_2$ are *exothermic*, whereas some reactions of ^3NH are *endothermic*, depending on the alkane![19] Absolute rate constants for many of these processes have been measured in the gas phase. However, the gas-phase chemistry of methylene is much more developed than that of imidogen.[17-19]

Hydrazoic acid is a reagent in the well-known Schmidt reactions in solution, but imidogen is probably not involved in these transformations.[20] Systematic mechanistic studies of the photochemistry of HN_3 in solution and the chemistry of NH with hydrocarbons, particularly cis- and trans- alkenes as per the detailed studies of CH_2,[19] have not been performed.

1.1.1. Gas-Phase Spectroscopy.

1.1.1. Gas-Phase Spectroscopy. The 336-nm absorption band of triplet imidogen was first observed by Eder[21] in 1892 and has been the subject of numerous subsequent studies. The singlet state of NH absorbs at 324 nm. The CASSCF (6,5)/ CASPT2 level of theory predicts transitions at 323 and 293 nm for singlet and triplet imidogen, respectively.[22] In each spin state, an electron is promoted from the "*sp*" hybrid lone pair to a singly occupied $2p$ orbital as shown below for ^1NH.

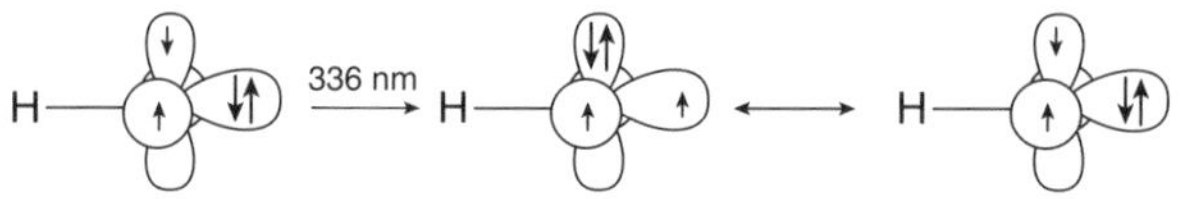

The singlet and triplet absorptions of imidogen are very similar because the same orbitals are involved in excitation of each spin state. As the transitions are localized on the nitrogen atom, these bands are characteristic of nitrenes in general and will appear in the same spectral region in alkyl- and arylnitrenes. These bands are diagnostic of triplet alkyl- and arylnitrenes. Because this wavelength is in a convenient range for laser flash photolysis (LEP) studies, certain alkylnitrenes and many arylnitrenes have now been detected in solution.

Singlet and triplet imidogen have not yet been detected in solution by LFP methods. Future effort should be expended in this direction and it seems safe to predict that such work will be reported in the near future.

Although singlet and triplet imidogen have similar absorption spectra, singlet and triplet methylene do not.[15] In fact, most carbenes have rather poor chromophores for UV–vis detection by LFP and must be "visualized" by trapping with pyridine to form ylides.[23] The exceptions are arylcarbenes, which have $\pi \rightarrow \pi^*$ transitions localized on the aromatic π system.[24]

1.1.2. Matrix Isolation Spectroscopy. Gaseous hydrazoic acid and xenon (1/200) were condensed on a CsI disk cooled to 28–35 K. Exposure of the mixture to 254-nm radiation led to the consumption of HN_3 and the formation of new vibrational bands at 3131.8, 3120.6, and 3109.0 cm^{-1} assigned to triplet NH. The use of xenon as the matrix host is crucial to the success of the experiment. The heavy atom host accelerated intersystem crossing in either the excited state of HN_3 or ^1NH, which led to good yields of ^3NH.[25]

Matrix isolated ^3NH reacts with NO, an excellent nitrene trap, to form *trans*-HNNO, which was characterized by infrared (IR) spectroscopy.[25] Triplet imidogen also reacts with oxygen in matrices.

The electron paramagnetic resonance (EPR) spectrum of triplet imidogen immobilized in a cryogenic matrix has not yet been observed. One negative attempt was reported in argon, krypton, and xenon matrices in 1967.[26] It would seem to be worthwhile for a new attempt in a xenon matrix.

2. ALKYLNITRENES

Negative ion PES of methylnitrene anion demonstrates that triplet methylnitrene is lower in energy than the singlet by 31.2 kcal/mol.[27] These results are in excellent agreement with theoretical predictions.[28] Thus, a methyl group preferentially stabilizes the singlet state, relative to the triplet state by ∼4 kcal/mol, compared to hydrogen.

Singlet methylnitrene is converted into methyleneimine in a very exothermic process.[28] Analysis of the vibrational structure of the PES spectrum indicates that the species detected belongs to a resonance rather than a true minimum on the singlet MeN potential surface.[27]

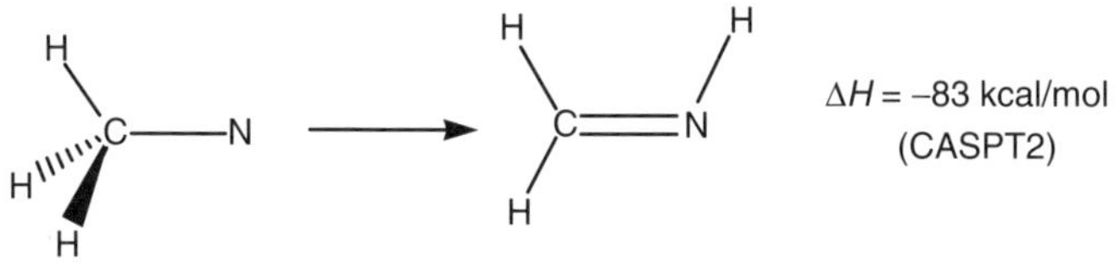

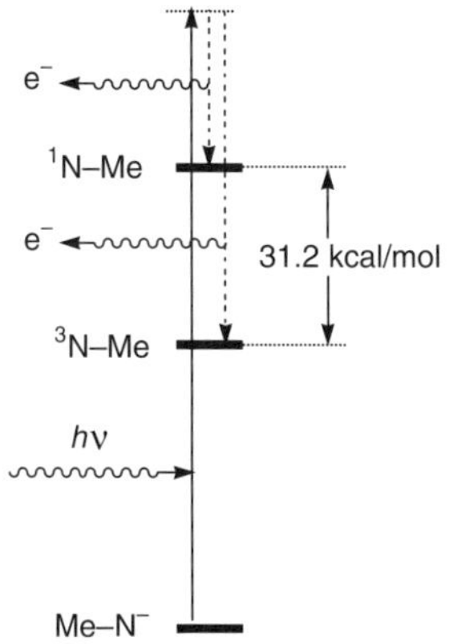

The key question, then, is whether singlet methylnitrene is a minimum on the potential energy surface and is a true reactive intermediate with a finite lifetime. Most early calculations found singlet methylnitrene to be protected from rearrangement by sizeable barriers.[29] The most recent work, using the CAS(10,8)-MP2 method, predicts that the barrier to rearrangement of singlet methylnitrene is only 1.4 kcal/mol, up from 0.5 kcal/mol at the (12/11) CASSCF/cc-pVTZ level. (CASPT2 predicts a barrier of 2.8–3.8 kcal/mol depending on the basis set.[27]) So it seems clear that, if singlet methylnitrene does exist as a true reactive intermediate, it will not live long enough to undergo bimolecular reactions and will not be detectable in solution by nanosecond spectroscopy. In fact, singlet methylnitrene has so far eluded detection by femtosecond spectroscopy.[30]

These conclusions were anticipated by product studies. Alkyl azides are readily available and their thermal and photochemical decomposition reactions have been studied.[31,32] In general, light and heat induced decomposition of methyl azide does not produce a MeN species that can be intercepted in respectable yields with a bimolecular trap.[31,32] For example, attempts to trap MeN with cyclohexane as solvent produced only a 0.4% yield of adduct.[33] Photolysis of CH_3N_3 or CD_3N_3 at cryogenic temperatures fails to produce an IR spectrum attributable to triplet methylnitrene. The IR spectrum of $CH_2=NH$ (or $CD_2=ND$) is observed instead.[34]

The intramolecular chemistry of alkylnitrenes resembles that of alkylcarbenes in general.[18,19,35]

The photochemistry of *tert*-butylazide in a nitrogen matrix at 12 K was followed by IR spectroscopy. Only one product, trimethylimine, was observed.[36] Thus, the removal of alpha hydrogen from the alkyl azide does not suppress 1,2-migration reactions.

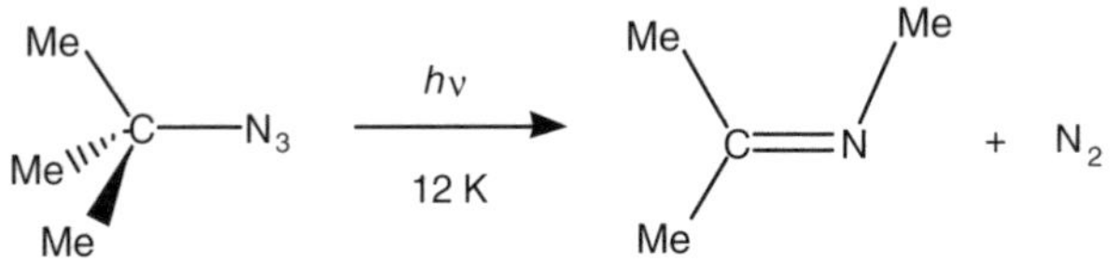

In fact, rearrangements of singlet alkylnitrenes generated in frozen matrices cannot even be suppressed by developing strain in imine products such as **1–3**, although some triplet nitrene is detected by EPR and UV–vis spectroscopy upon low-temperature photolysis of 1-azidonorbornane.[37]

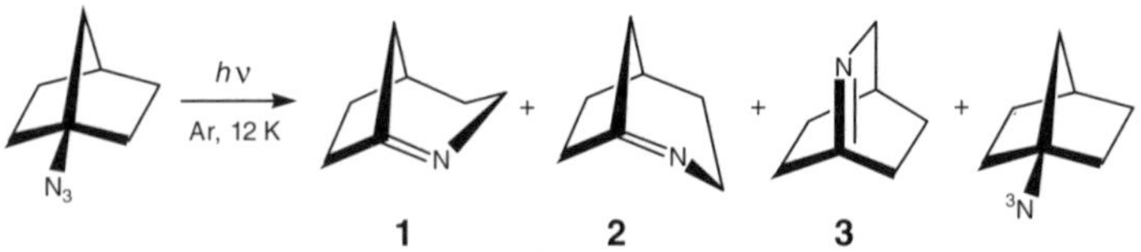

Photolysis of methyl azide immobilized in a rigid cryogenic matrix does not produce methylnitrene in its triplet ground state, as revealed by EPR spectroscopy.[38,39] This observation is in good agreement with Milligan's study of this system using matrix IR spectroscopy.[34] However, triplet sensitized photolysis of methyl azide in a frozen matrix does produce the persistent EPR spectrum of triplet methylnitrene.[38,39] This experiment again demonstrates the low barrier to isomerization of singlet methylnitrene, which is bypassed upon triplet sensitization.

The solution-phase photochemistry of nine alkylazides was studied by Kyba and Abramovitch.[40] Photolysis of tertiary alkyl azides leads cleanly to 1,2 alkyl migration.

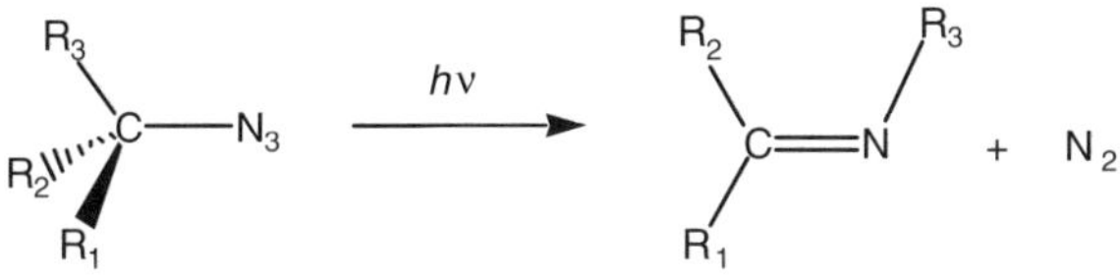

Photolysis of alkyl azides bearing pendant aryl groups does not lead to intramolecular trapping of a nitrene by the aryl group. However, intramolecular capture of the putative nitrene is observed upon pyrolysis of the azide precursor. These observations convinced Kyba and Abramovitch[40] that 1,2-migration is concerted with loss of nitrogen from the excited state of the azide, but that a free singlet nitrene is formed upon thermal decomposition. The chemistry of acyl azides will be shown later to exhibit the opposite pattern.

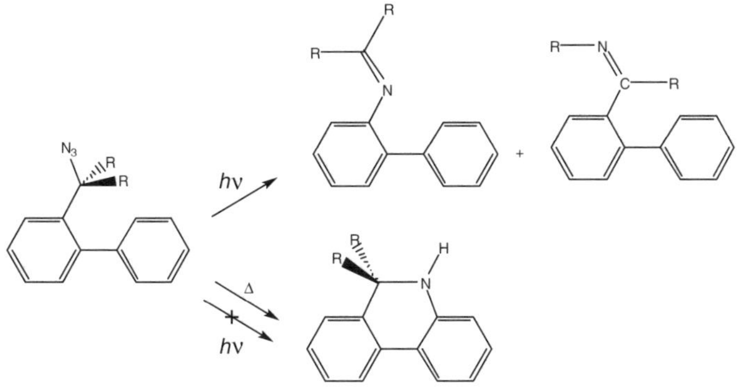

Triplet sensitized LFP studies of simple alkyl azides and systematic product analyses in solution under these conditions have not been reported, although the triplet energies of alkyl azides are 75–80 kcal/mol[41] and the quantum yields for azide disappearance with appropriate triplet sensitizers approach unity.

Recently, however, examples of intramolecular triplet sensitization have been described. Alkyl aryl ketones with pendant azido groups in the alkyl moiety were irradiated into the benzoyl chromophore. The excited alkylphenyl ketones undergo intersystem crossing to their triplet states within picoseconds and they can then transfer their triplet energy to the nearby alkyl azido group. This process produces the triplet alkylnitrene that can be detected by transient UV–vis and matrix spectroscopy.[42]

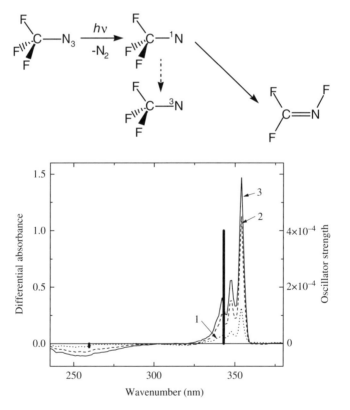

Photolysis (254 nm) of trifluoromethyl azide in argon or alkane at 4–10 K without sensitization produces trifluoromethyl nitrene in its ground triplet state, which has been characterized by both UV–vis (Fig. 11.1) and EPR spectroscopy.[43]

Figure 11.1. Differential electronic absorption spectra produced by irradiation of CF_3N_3 in an Ar matrix at 12 K with 254-nm light of 5 min (1), 25 min (2) and 45 min, (3) duration. The calculated absorption bands (TD-B3LYP/6-31*) of triplet CF_3N (positive) and CF_3N_3 (negative) are depicted as solid vertical lines. [Reproduced with permission from N. P. Gritsan, I. Likhotvorik, Z. Zhendog, and M. S. Platz, *J. Phys. Chem. A* **2001**, *105*, 3039.]

It is presumed that the barrier-to-1,2-fluorine migration in the singlet nitrene substantially exceeds the analogous barrier-to-hydrogen shift. This difference allows a portion of the nascent singlet nitrene to relax to the lower energy triplet in the matrix.

Triplet methylnitrene has been produced in the gas phase in a corona discharge. This process has allowed its highly resolved absorption spectrum to be recorded and analyzed. [44]

Polyfluorination does seem to suppress rearrangements of alkyl azides and to extend the lifetimes of the corresponding singlet nitrenes. Photolysis of **5** in cyclohexane produces insertion adduct **6**.[45]

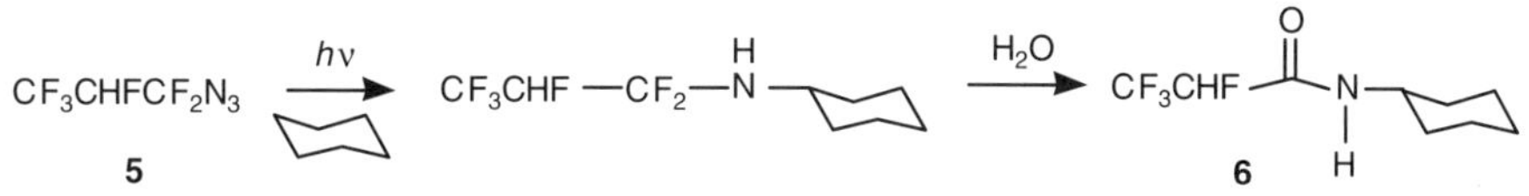

3. CARBONYLNITRENES AND RELATED ACYLNITRENES

Treatment of amides with bromine in alkaline medium promotes the Hofmann rearrangement, which may or may not involve a free nitrene intermediate.[46] The oxidation of primary amides with lead tetraacetate and the resulting Lossen rearrangement also produces isocyanates, with the possible intervention of an acylnitrene.[3,47]

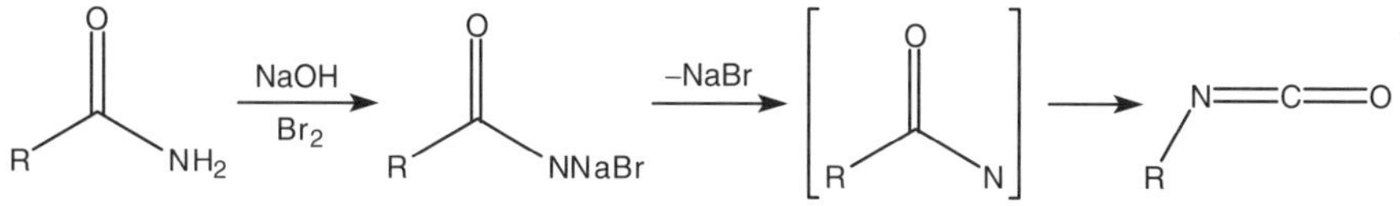

Pivaloylazide (**7**) has received considerable study by Lwowski and Tisue.[48] Pyrolysis of this compound in neat cyclohexane, cyclohexene, or 2-methylbutane leads to *tert*-butylisocyanate in nearly quantitative yields.

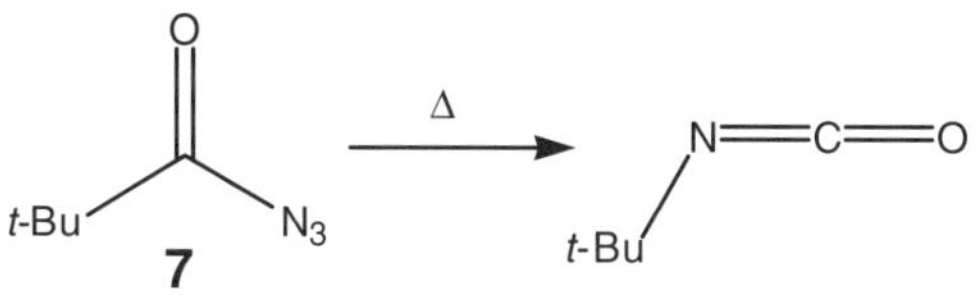

From this result, one can argue that the Curtius rearrangement is a concerted process.

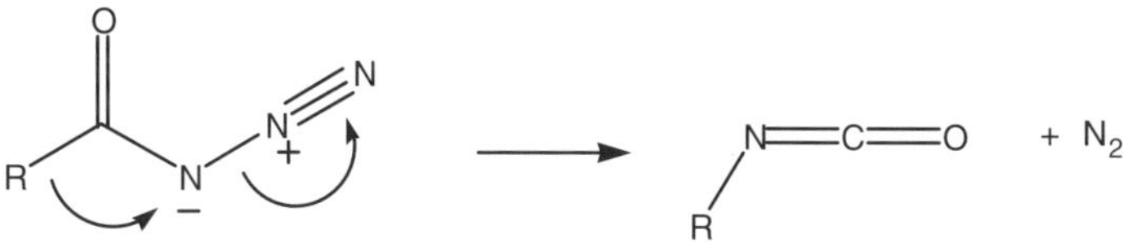

Alternatively, one could conclude the reaction is stepwise, but that the rearrangement of the acylnitrene is much faster than bimolecular chemistry.

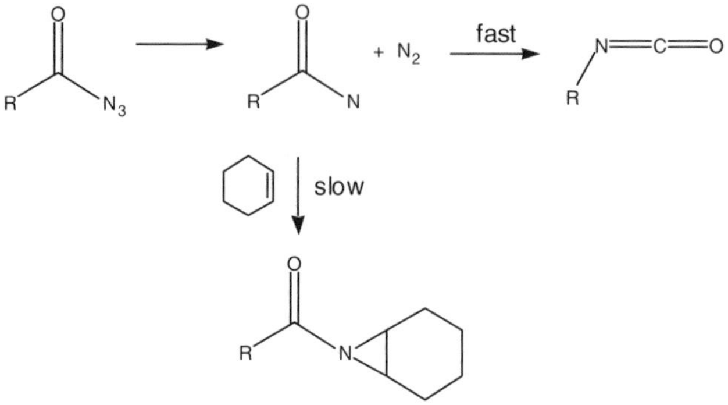

However, Lwowski and co-workers[49] demonstrated that *photolysis* of **7** in cyclohexane, cyclohexene- or 2-methylbutane *does* lead to the formation of adducts. Therefore, the acylnitrene **8** is a trappable intermediate. The thermal Curtius rearrangement does *not* involve free nitrenes and it must be a concerted process.

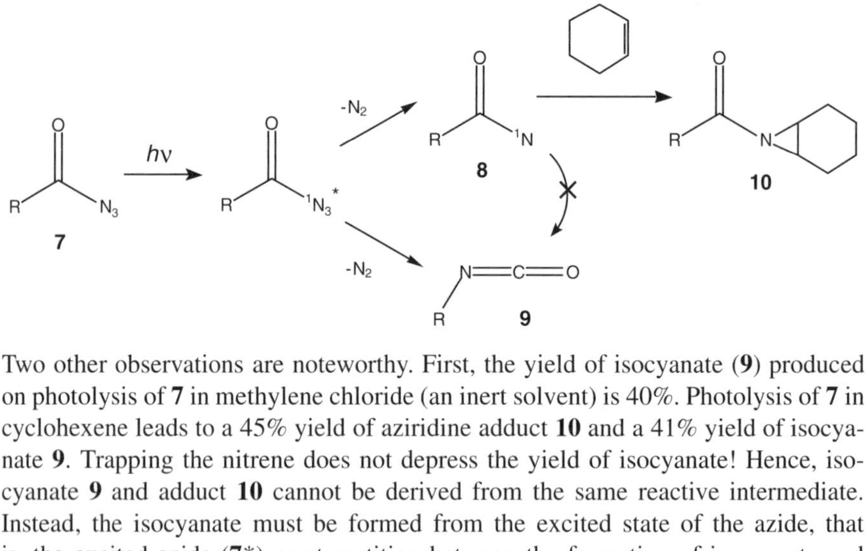

Two other observations are noteworthy. First, the yield of isocyanate (**9**) produced on photolysis of **7** in methylene chloride (an inert solvent) is 40%. Photolysis of **7** in cyclohexene leads to a 45% yield of aziridine adduct **10** and a 41% yield of isocyanate **9**. Trapping the nitrene does not depress the yield of isocyanate! Hence, isocyanate **9** and adduct **10** cannot be derived from the same reactive intermediate. Instead, the isocyanate must be formed from the excited state of the azide, that is, the excited azide (**7***) must partition between the formation of isocyanate and nitrene.

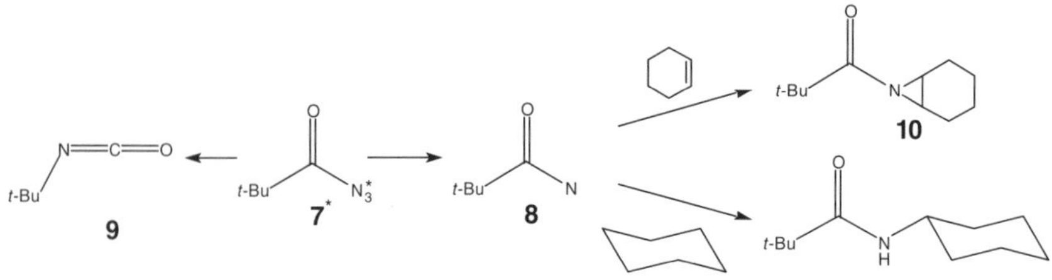

Similar conclusions concerning the photochemistry of diazo carbonyl compounds and diazirines have been reached using similar experiments and logic.[50]

The second interesting point to note is that photolysis of 2-naphthoylazide (**11**, where 2-Np = 2-naphthoyl) in the presence of *cis*-4-methyl-2-pentene produces azirine **12** *stereospecifically*.[51]

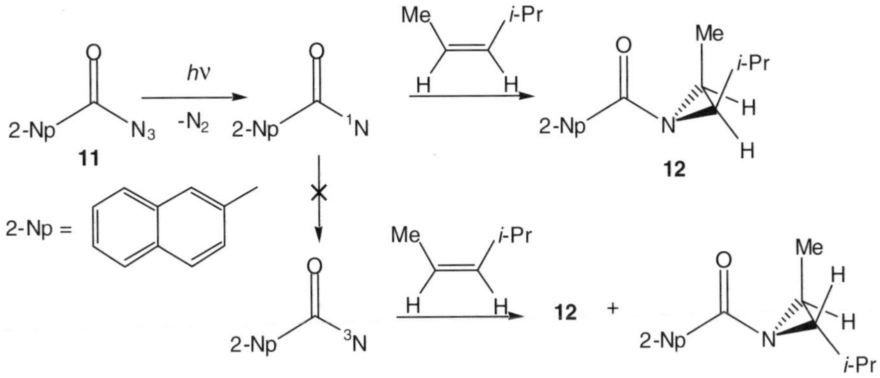

This experiment indicates that only the singlet acylnitrene is trapped because it is either the ground state of the nitrene or it relaxes very slowly to a triplet state. However, if the alkene is diluted to a concentration of only 0.01 *M*, the aziridination reaction remains stereospecific. Extending the nitrene lifetime by diluting the concentration of the trapping agent does not lead to relaxation of a putative excited singlet nitrene to its putative lower energy triplet state.[51]

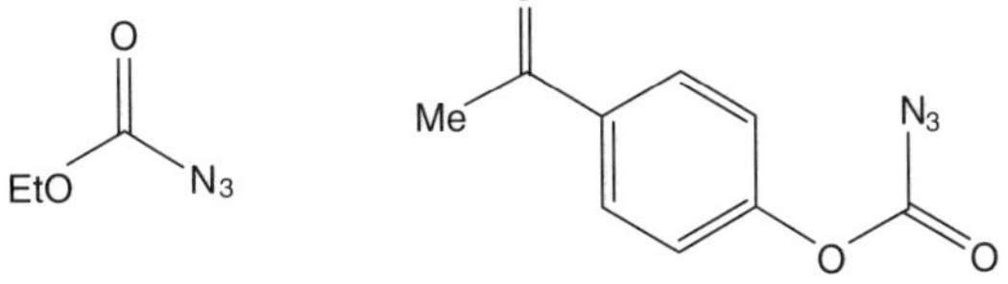

The observation of a triplet ESR spectrum produced by photolysis of RC(O)—N₃ in a frozen matrix would prove that RC(O)—N has a triplet ground state. Although photolysis of azide esters **13**[39] and **14**[52] in matrices does produce triplet nitrene EPR signals, photolysis of RC(O)N₃ under the same conditions, does not.[51] Thus, one is left again to wonder whether the singlet is the ground state of the nitrene or simply does not find a competitive pathway for intersystem crossing to a lower energy triplet state. These issues were clarified by triplet sensitization experiments by Autrey and Schuster.[51]

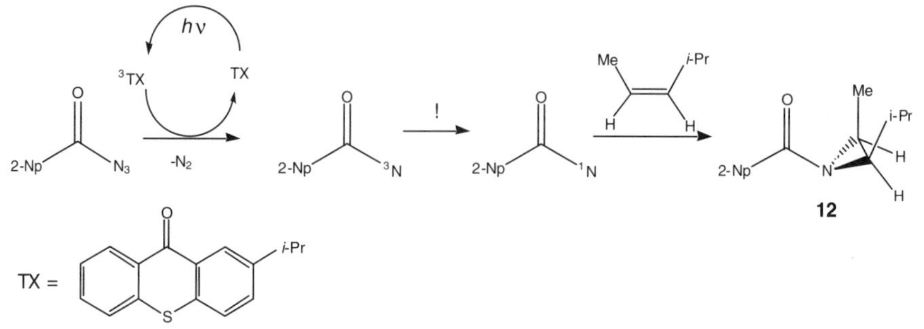

A mixture of 2-naphthoylazide and sensitizer 2-isopropylthioxanthone ($E_T =$ 65 kcal/mol) was irradiated ($\lambda > 385$ nm), conditions under which the sensitizer absorbed >95% of the light. Laser flash photolysis experiments demonstrated that the triplet sensitizer is quenched by 2-naphthoylazide at a diffusion controlled rate.

Direct photolysis of 2-naphthoylazide (**11**) in cyclohexane produces the corresponding isocyanate (**15**) in 50% yield along with C—H insertion adduct **16**.

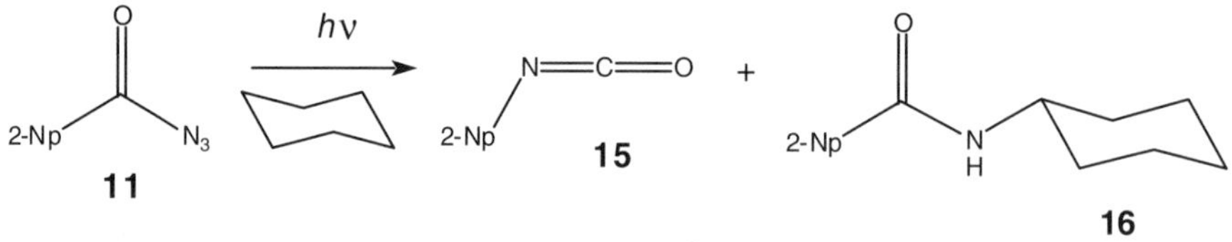

In the presence of 0.2 *M cis*-4-methyl-2-pentene, the yield of isocyanate **15** is unchanged and a 46% yield of aziridine **12** is observed. Upon sensitized photolysis, the yield of isocyanate drops to 5%, demonstrating that triplet sensitization has been achieved and that the isocyanate is a product of reaction of the singlet excited state of the azide precursor. However, triplet sensitized photolysis of **11** in the presence of *cis*-4-methyl-2-pentene gives aziridine **12** with complete retention of stereospecificity! When the triplet acyl nitrene is generated intentionally the products of singlet nitrene reactions are obtained! Thus, singlet 2-naphthoylnitrene must be lower in energy than the corresponding triplet nitrene and it must rearrange rather slowly to isocyanate![52] Direct photolysis of **11** in the presence of an alkene (e.g., cyclohexene) can be summarized as shown.

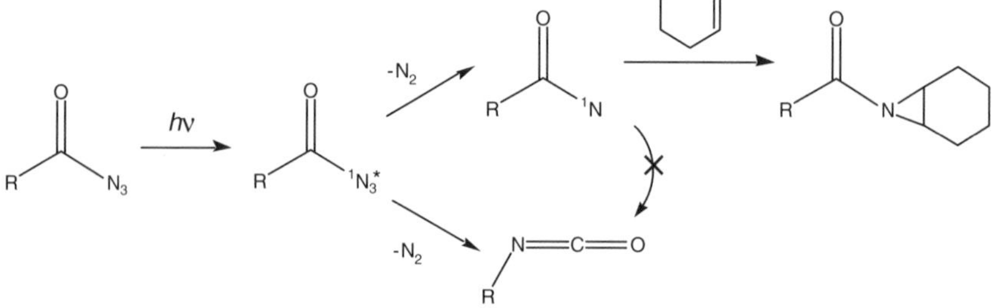

In subsequent work, Sigman et al.,[52] studied aroyl azides **14** and **17–18**.

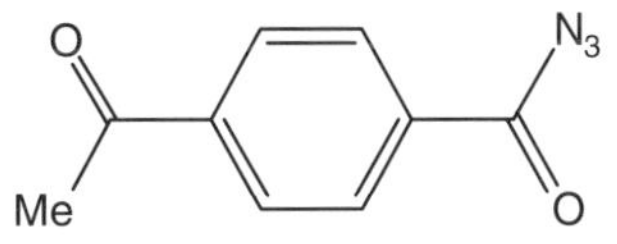

17 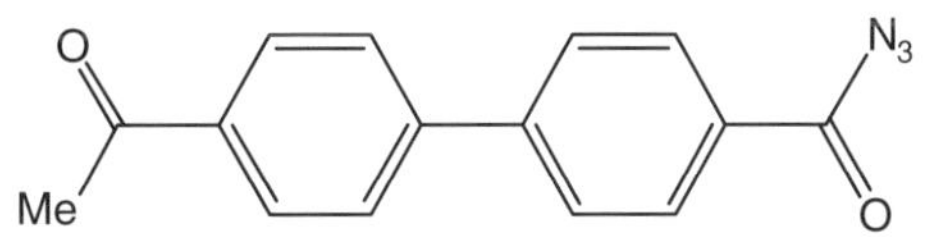   **18**

Each of these azides contains an internal sensitizing group. Upon LFP the triplet states of all three azides are observed. However, in the case of **17** and **18**, only singlet nitrene derived products are formed. Photolysis of **17** and **18** at 8 K in fluorolube fails to produce persistent triplet EPR spectra. It seems clear that these aroyl nitrenes also have singlet ground states. As mentioned previously, photolysis of **14** does produce a triplet nitrene EPR spectrum.

Recent high-level calculations support the conclusion that benzoyl nitrene has a singlet ground state.[53,54] According to CCSD(T)/cc-pVTZ calculations on HCNO the distance between the oxygen and nitrogen atoms and the C—N bond length are unusually small, whereas the C—O bond length is much longer than in a carbonyl group in the singlet state (in contrast to the triplet state where all these values are normal). The minimum energy geometry of ^{1}HCON is intermediate between that of an acylnitrene and an oxazirine. Presumably, it is this bonding interaction that lowers the energy of the closed-shell singlet nitrene–oxazirine and this hybrid species becomes more stable than the corresponding triplet nitrene. Singlet benzoylnitrene–oxazirine has recently been observed as a persistent species in argon by IR spectroscopy. Its vibrational spectrum is in good agreement with theory.[54]

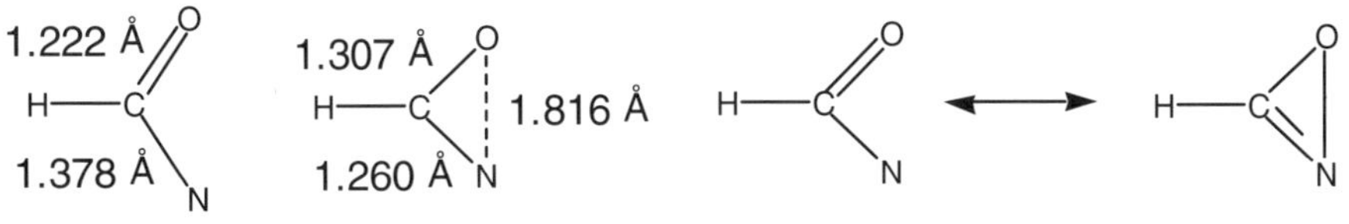

^{3}HCON ^{1}HCON

3.1. Nitrene Esters

Carboethoxynitrene (**19**) can be generated in solution by base-catalyzed α-elimination of **20** or upon photolysis of azide **13**. The nitrene undergoes addition to double bonds and formal CH insertion reactions.[55]

The reaction of carboethoxynitrene with *cis*- and *trans*-4-methyl-2-pentene was studied as a function of alkene concentration. At large alkene concentrations, aziridination is stereospecific. Upon dilution of the alkene, the stereospecificity is lost.[56] These results are completely analogous to studies of carbenes in which a stereospecific singlet intermediate is produced initially, and subsequently relaxes to a less selective, lower energy triplet intermediate.[57]

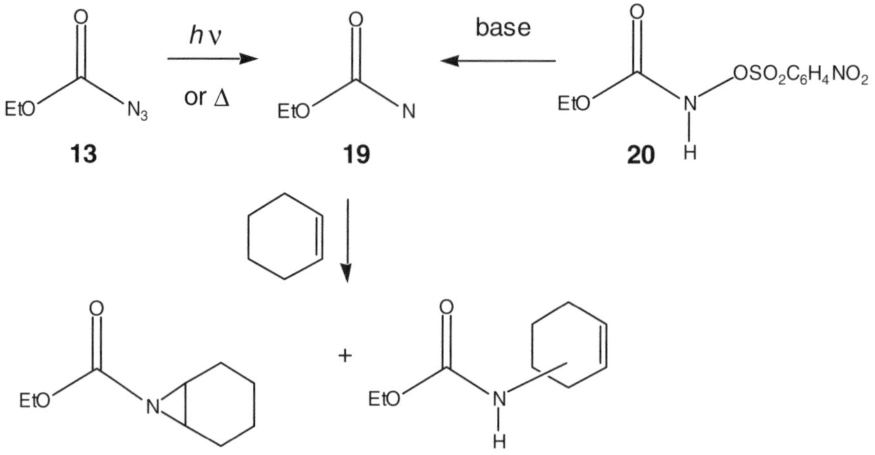

The results with carboethoxynitrene are quite different from those obtained with pivaloyl azide **7** and the aroylnitrenes. Furthermore, photolysis of azidoesters **13**[39] and **14**[52] in contrast to pivalolyl and aroylazides at cryogenic temperatures produces persistent nitrene EPR signals. Thus, it is clear that the nitrene esters have triplet ground states.

Why does the nitrene ester have a triplet ground state in contrast to the acetyl and benzoyl analogues? This author speculates that the N—O bonding interaction in the singlet nitrene ester is weaker than that in the singlet acetylnitrene because such an interaction destroys resonance within the ester group.

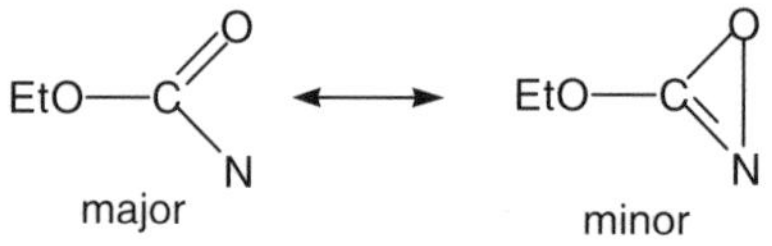

Nitrene esters have not yet been detected by matrix IR spectroscopy. IR spectroscopy is less sensitive than EPR spectroscopy and IR observes all the photoproducts, not just the paramagnetic triplet species of interest.

Buron and Platz[58] recently studied the photochemistry of **13** in solution by LFP. The triplet state of **19** absorbs at ~400 nm in 1,1,2-trifluorotrichloroethane with a lifetime of 1–2 µs. The triplet is formed within 10 ns of the laser pulse. Relaxation of the singlet to the triplet state of **19** is fast relative to the related process in arylnitrenes and is comparable to a carbenic process.[59] As we will see later when we discuss intersystem crossing rates of singlet arylnitrenes, this difference is most likely due to the closed-shell electronic configuration of the singlet state of **19**.

Nanosecond time resolved infrared (TRIR) spectroscopy has recently become available to physical organic chemists.[60] This spectroscopy is an attractive tool for studying carbonyl nitrenes. Such work is in progress in several laboratories[58] and it seems safe to predict that this technique will be greatly contributing to our understanding of carbonylnitrenes in solution in the near future.

3.2. Sulfonylnitrenes

We will follow the lead of Lwowski[61] and classify sulfonyl- and phosphorylnitrenes as a type of carbonylnitrenes.

Pyrolysis or photolysis of methanesulfonyl azide in 2-methyl butane (RH) produces formal C—H insertion adducts attributed to reactions of the thermally generated nitrene. The selectivity of this nitrene for 3°/2°/1° C—H bonds was 6/2.3/1.0.[62]

Sulfonylnitrenes, like acylnitrenes, insert into the OH bonds of alcohols.

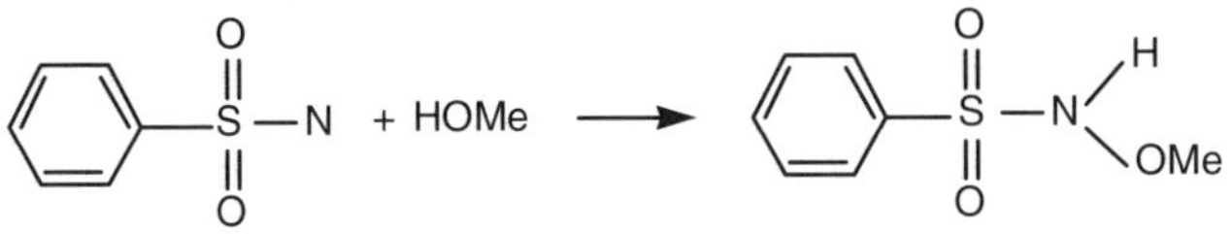

However, product studies indicate that the chemistry of sulfonyl azides and nitrenes is complicated by rearrangement to form intermediates, such as **21**.[61]

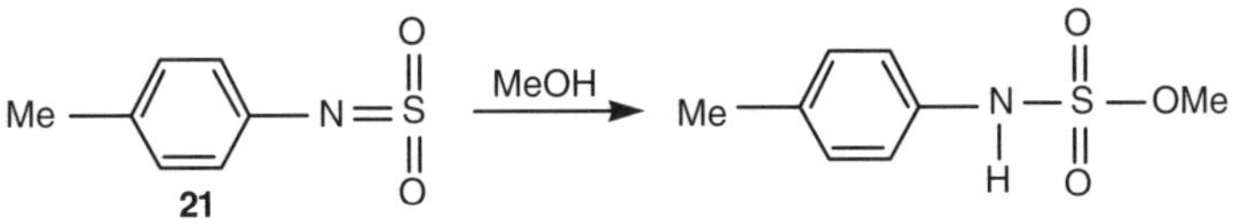

Fragmentation of arylsulfonyl nitrenes to form an arylnitrene and SO_2 is also known.[61] Presumably, this process involves a prior rearrangement to a species analogous to **21**.

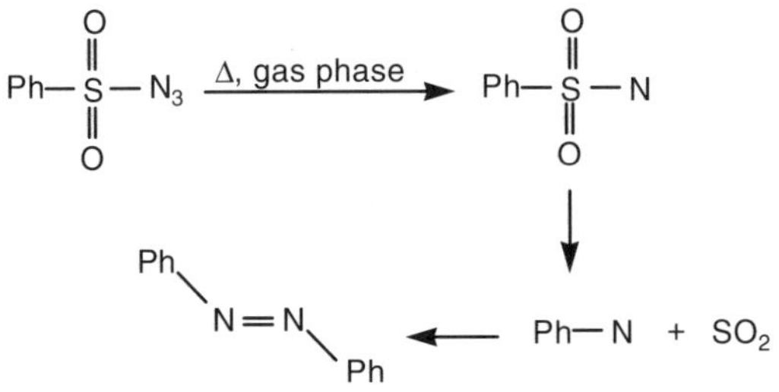

The dimerization of phenylnitrene to form azobenzene is a characteristic reaction of the triplet state of an arylnitrene. Thus, some intermediate formed along the reaction coordinate must undergo intersystem crossing to the triplet manifold. Photolysis (254 nm) of tosylazide[63,64] at 77 K in fluorolube or EPA (2:5:5 ethanol: 2-methylbutane/diethyl ether) glass produces a persistent EPR spectrum assigned to the triplet state of p-toluenesulfonylnitrene $|D/hc| = 1.471$ cm^{-1}. However, upon extended photolysis of the sulfonylnitrenes, the EPR spectrum of tolylnitrene $|D/hc| = 0.9761$ cm^{-1} is detected.

Photolysis (254 nm) of tosylazide in EPA produces a species with a sharp absorption maximum at 313 nm that also absorbs broadly between 250 and 600 nm. The carrier of the spectrum was associated with triplet tosylnitrene on the basis of the EPR work.[64]

Laser flash photolysis (266 nm) of tosylazide in cyclohexane produces a transient absorption (325 nm strong, 450 nm weak) with a lifetime on the order of microseconds.[64] The lifetime of the transient is shortened to 328 ns in aerated cyclohexane. The same transient was also observed in ethanol and methanol with lifetimes of 8 and 9 μs, respectively. Oxygen again shortened the transient lifetime. Maloney and co-workers[64] attributed the transient signals to tosylnitrene in its triplet ground state.

3.3. Phosphorylnitrenes

Photolysis of azidophosphodiesters **22** releases a nitrene that also inserts efficiently into the CH bonds of alkanes. When R′H = 2,3-dimethylbutane is employed the yield of CH insertion adducts is almost 60%, with the remainder the amide **24**.[65]

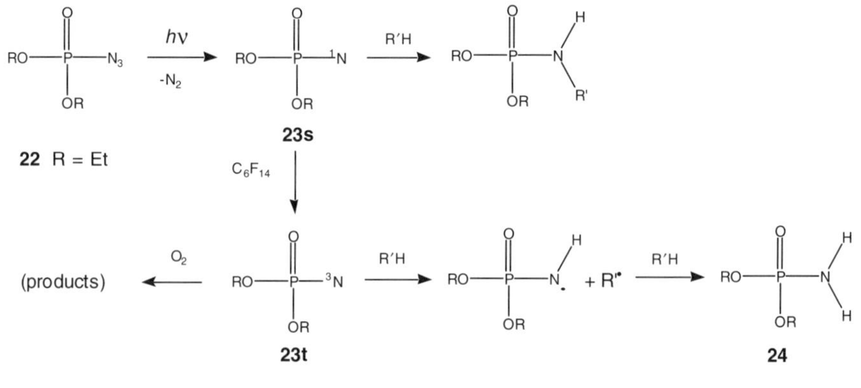

When 2,3-dimethylbutane (DMB) is used, both 3° and 1° formal C—H insertion reactions are possible. In neat DMB, the 3°/1° ratio obtained with the azidodiethylester precursor **22** is nearly statistical, indicating that a very unselective, highly reactive, and short-lived species has been trapped. The ratio of adducts does not change upon dilution with perfluorohexane, although the yield of adducts drops and there is a small increase in the yield of amide **24**. The presence of oxygen has no effect on the yield of CH insertion adducts with DMB, but depresses the yield of amide **24** from 40.1 to 6.3%. Maslak[65b] attributed the nitrene–alkane insertion adducts to C—H insertion reactions of the singlet nitrene **23s**, and considered amide **24** to be a diagnostic product derived from triplet nitrene **23t** reactions. The data are consistent with initial formation of a short-lived, highly reactive singlet nitrene that inserts rapidly into primary and tertiary C—H bonds. Maslak found that a plot of the ratio of singlet-to-triplet nitrene derived products versus the concentration of RH in perfluorohexane was not linear. Increasing the concentration of singlet nitrene trap cannot completely suppress the formation of triplet nitrene derived products. This observation led Maslak to conclude that there is some inter-system crossing in the singlet excited state of the azide. This viewpoint was strongly supported by the observation that 1–3% isoprene, an excellent quencher of triplet excited states, inhibits the formation of phosphoramide.[65]

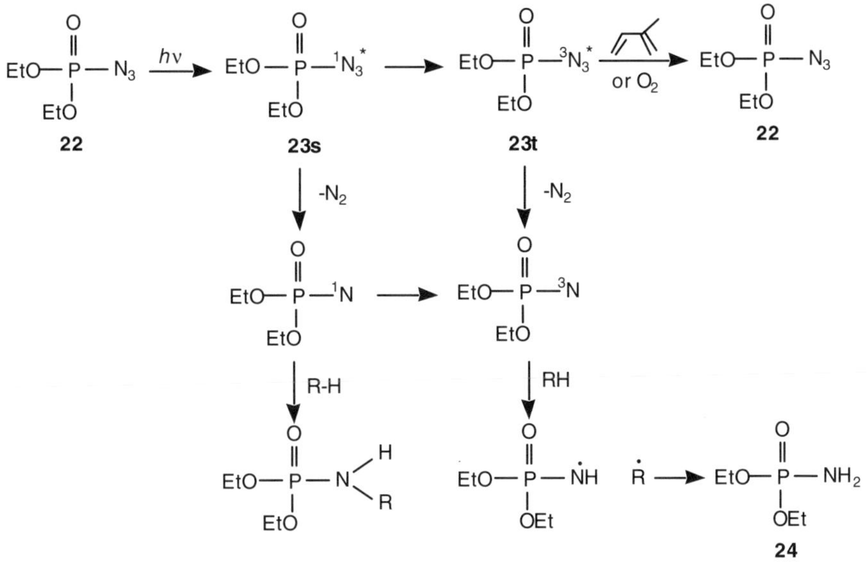

Assuming that singlet nitrene reacts with alkanes at near diffusion controlled rates allowed deduction of a rate constant of singlet-to-triplet nitrene intersystem cross-ing (ISC) of 2–8×10^8 s^{-1}. This ISC rate is slower than in carbenes, but signifi-cantly faster than with arylnitrenes, which are discussed in a subsequent section.

Maloney and co-workers[66] studied diphenoxyphosphorylnitrene by direct obser-vational techniques. Irradiation (75 s, 254 nm) of **25** in EPA glass at 77 K produced persistent triplet EPR signals with zero-field parameters $|D/hc| = 1.5148$ cm^{-1} and $|E/hc| = 0.00739$ cm^{-1}, which are rather similar to those of triplet carboethoxy-nitrene.[39]

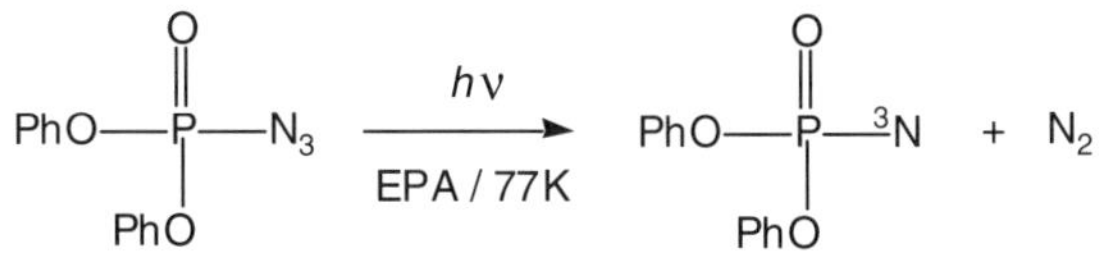

Photolysis of **25** in ether–isopentane–ethanol (EPA) at 77 K produced a species that absorbs at 336 nm. The carrier of this spectrum is persistent for at least 1 h at 77 K, but disappears upon annealing. Based on the EPR spectrum and the similarity of the UV spectrum to those of triplet alkylnitrenes, the species absorbing at 336 nm was attributed to the triplet phosphorylnitrene.[66]

Laser flash photolysis (266 nm) of **25** in cyclohexane produced a weak transient absorption at 345 nm. The same transient was observed in ethanol, but the signal intensity was stronger. The transient lifetimes in ethanol-d_6 were 3.8 and 4.1 µs, respectively.

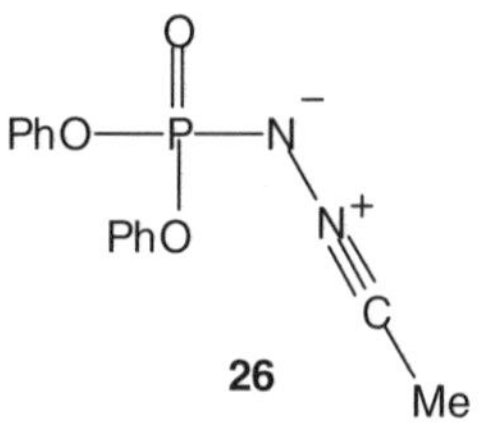

PhO—P—N
‖
O
PhO

26   Me

The presence of oxygen had no effect, outside of experimental error, on the transient lifetime in ethanol. On the basis of the solution and glassy matrix work the transient absorbing at 345 nm was assigned to the triplet nitrene. LFP of **25** in acetonitrile produced a longer lived transient ($\tau > 150$ µs) also absorbing near 345 nm, which may be the result of formulation of a nitrene–solvent ylide.

4. VINYLNITRENES

A simple picture of the nitrogen-centered orbitals of triplet vinylnitrene is shown below in which p_z is used to form the sp hybrid orbital on the nitrene nitrogen, p_π conjugates with the π system of the double bond, and p_σ is orthogonal to that π system. Consequently, the open- and closed-shell configurations of singlet vinylnitrene are no longer degenerate because conjugation slightly raises the energy of p_π relative to p_σ.

Parasuk and Cramer's calculations indicate that triplet vinylnitrene is 15 kcal/mol more stable than the lowest energy singlet state.[67] The lowest energy singlet state is open shell and resembles a 1,3-biradical.[67,68]

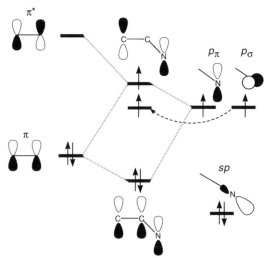

Parasuk and Cramer[67] predict that the closed-shell singlet state is 40 kcal/mol higher in energy than the triplet ground state. These predictions have not yet been confirmed by negative ion PES, but it is clear that the vinyl group reduces the singlet triplet separation (by 20 kcal/mol relative to that of imidogen or methyl-nitrene, as a result of the delocalization of one of the unpaired electrons and the corresponding preferential reduction of electron–electron repulsion in the singlet state.

Theory indicates that the electronic structure of singlet vinylnitrene differs significantly from that of singlet vinylcarbene and other carbenes, in that it has an open-shell biradical-like structure. The orbitals at the termini of the biradical are disposed at 90° to one another, therefore the p_π, p_σ splitting is so small that the nitrene prefers the open-shell configuration to minimize electron–electron repulsion between the electrons of opposite spin. The closed-shell, singlet carbene-like configuration p_y^2 is 25 kcal/mol higher in energy than the open-shell configuration.[67]

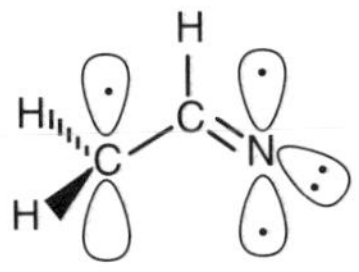

The above MO picture explains why bimolecular chemistry of parent singlet vinylnitrene is unknown.[69] The barrier to cyclization of singlet vinylnitrene is very low. The situation with vinylnitrene is analogous to that in methylnitrene and it is not clear if either of these singlet nitrenes are true reactive intermediates with finite lifetimes. The (4,4) CASSCF 6-31G* calculations predict that singlet vinylnitrene is not even a minimum. Rather it is the transition state for the interchange of the enantiotopic hydrogens of 2*H*-azirine. The open-shell biradical structure of vinylnitrene makes bimolecular chemistry less favorable[68] because the concerted bimolecular insertion and cycloaddition processes available to closed-shell singlet carbenes, are not available to the open-shell nitrene singlet.

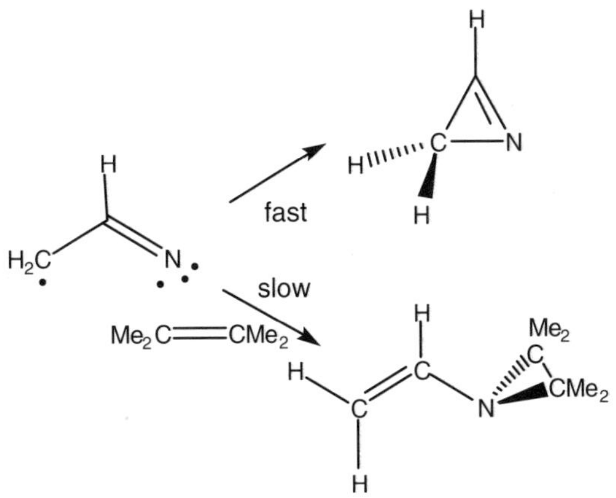

Unsurprisingly, singlet vinylnitrene has never been observed by time resolved spectroscopy and triplet vinylnitrene has yet to be observed by either time resolved or matrix spectroscopy.

5. PHENYLNITRENE AND SIMPLE DERIVATIVES

The chemistry of phenylcarbene and related aryl- and diarylcarbenes has been studied extensively since the 1950s. Analysis of arylcarbene product mixtures revealed the broad mechanistic picture.[19,70] The singlet and triplet states of phenylcarbene are in rapid equilibrium. The higher energy singlet carbene reacts rapidly and stereospecifically with alkenes to form cyclopropanes. The triplet ground state reacts less rapidly and without stereospecificity—but it abstracts hydrogen atoms from hydrocarbons to form radical pairs and reacts very rapidly with oxygen, as do free radicals.[19,70]

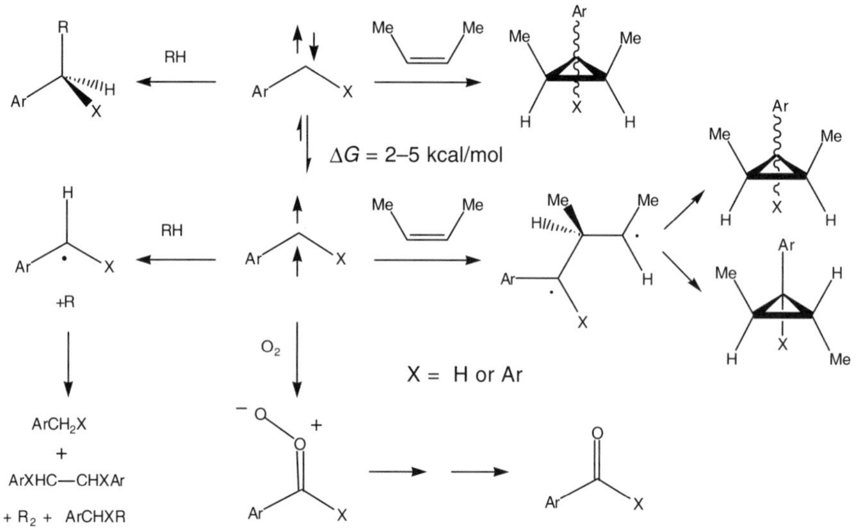

Matrix and time-resolved spectroscopy confirmed the pattern revealed previously by chemical analysis and placed it on a quantitative basis. Singlet and triplet aryl-carbenes [X = H, aryl (Ar)] generally have triplet ground states 0–5 kcal/mol below the singlet state.[70,71] Intersystem crossing from singlet-to-triplet carbene takes place in hundreds of picoseconds.[59] In phenylcarbene itself, the singlet–triplet splitting is 2–4 kcal/mol in solution[70,71] and spin state interconversion in both directions is faster than bimolecular processes.[19,70] Thus, most of the chemistry proceeds through the singlet state while the lower energy, less reactive triplet serves as a reservoir for the more reactive singlet carbene.[19,70]

Understanding the chemistry of arylnitrenes proceeded more slowly than that of arylcarbenes because product studies were less informative. Unlike carbenes, much of our current understanding of arylnitrenes has come from direct observational methods.

Photolysis or pyrolysis of most aryl azides in hydrocarbons leads to polymeric tars instead of diagnostic insertion products and aziridines.[72]

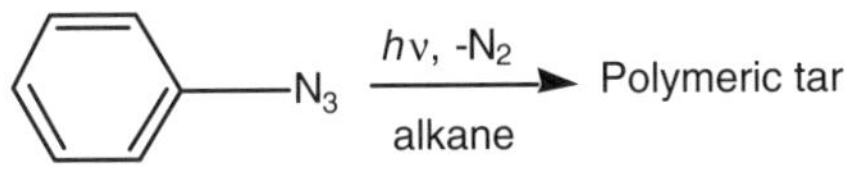

The strongest chemical evidence for the existence of singlet arylnitrenes comes not from the parent system, but from intermolecular capture of ortho-substituted aryl azides,[73] and intermolecular capture of certain highly electron-deficient arylnitrenes.[74,75]

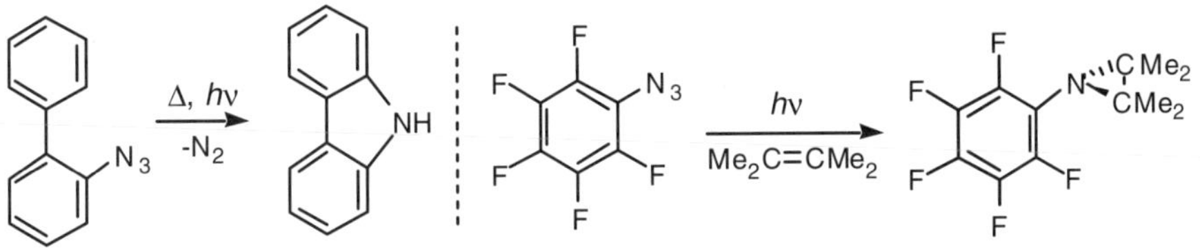

Phenylazide was first synthesized by Greiss in 1864.[76] In 1912, Wolf studied the pyrolysis of phenyl azide in aniline.[77] The product of this reaction, azepine (**27**), was identified by Huisgen and co-worker in 1958.[78] Eight years later, Doering and Odum[79] demonstrated that azepine (**27**) is formed upon photolysis of phenylazide in diethylamine[79] and in 1977 Carroll et al.[80] discovered the formation of **28** upon photolysis of phenyl azide in the presence of ethanethiol.

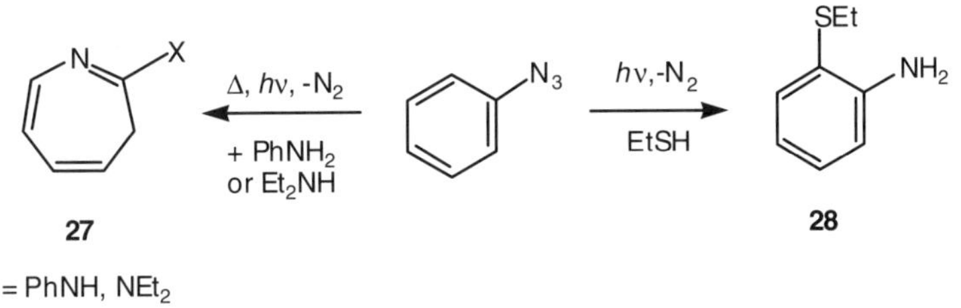

X = PhNH, NEt$_2$

This finding led to general acceptance of the view that either azirine (**29**) and/or ketenimine (**30**) are the trappable reactive intermediates produced upon photolysis of phenyl azide in solution.[72]

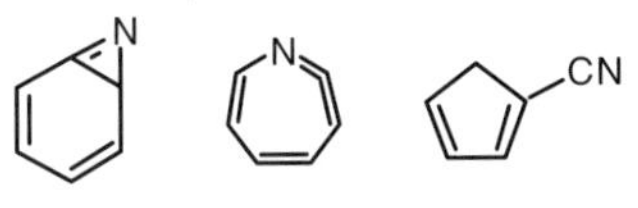

Wentrup and co-worker[81] studied the gas-phase pyrolysis of various phenylnitrene precursors and discovered several interesting stable products, including cyano-cyclopentadiene **31** formed in a ring contraction process.

Reiser et al.[82] examined the photolysis of aryl azides in a low-temperature glass. Excitation of the precursor led to the formation of new UV–vis bands that disappeared upon annealing the glass. These transitions were attributed to the ground state of the triplet arylnitrenes. Shortly thereafter, Wasserman[39] demonstrated that photolysis of phenyl azide immobilized in an organic glass cooled to 77 K produced the EPR spectrum of triplet phenylnitrene (^{3}PN) in good agreement with the conclusions of the low-temperature UV–vis spectroscopic studies.

However, in 1978, Chapman and LeRoux[83] discovered that photolysis of phenyl azide, matrix isolated in argon at 10 K, produces a persistent species with a strong vibrational band at 1880 ± 10 cm^{-1}. The carrier of this species was most reasonably assigned to ketenimine **30** rather than benzazirine **29** or triplet phenylnitrene. This result implies that it is the ketenimine **30** and not benzazirine **29** that is trapped with amines to form the 3H-azepines (**27**) that had been isolated earlier. It does, however, raise the question as to why two groups observed triplet phenylnitrene by low temperature spectroscopy while a third observed ketenimine **30**.

To add to the confusion, various groups reported that gas-phase photolysis of phenyl azide produced the absorption and emission spectra of triplet phenylnitrene.[84,85] These observations were reconciled by the work of Leyva et al.[86] who discovered that the photochemistry of phenyl azide in the presence of diethylamine was very sensitive to temperature. Above 200 K, azepine **30** is formed, but <160 K, azobenzene, the product of triplet nitrene dimerization, is produced. The ketenimine can react with itself or with phenyl azide to produce a polymer,[87] which can be converted into an electrically conducting material.[75] Gritsan and Pritchina[88] pointed out that at high-dilution ketenimine **30** can interconvert with singlet phenylnitrene which eventually relaxes to the lower energy triplet that subsequently dimerizes to form azobenzene.

Levya et al.[86] also demonstrated that triplet phenylnitrene is very light sensitive. Thus, triplet phenlynitrene subsequently and inadvertently absorbs light employed to decompose phenyl azide at low temperature and then the triplet nitrene rearranges to ketenimine **30**. This idea explains the formation and detection of ketenimine **30** in argon by Chapman and LeRoux.[83] Later Sheridan and Hayes[89] found that if phenyl azide is decomposed by 340 instead of 254-nm light, then triplet phenylnitrene is the major persistent primary photoproduct formed in argon at 10 K, which allowed this group to assign its IR spectrum.

What of the gas-phase results?[84,85] Cullin et al. demonstrated that triplet phenylnitrene is *not* the UV–vis active species detected in the gas phase. That species is actually the cyanocyclopentadienyl radical![90] The UV–photolysis of phenyl azide produces singlet phenylnitrene with excess vibrational energy, which can not be shed by collisions with solvent molecules. Thus, the hot nitrene explores the PhN surface and eventually finds the global minimum, cyanocyclopentadiene, which can shed its excess energy by losing a hydrogen atom to form the cyano-cyclopentadienyl radical. These results are in excellent agreement with Wentrup's gas-phase pyrolysis studies.[81] To date, phenylnitrene has not been detected in the gas phase; neither has phenylcarbene.

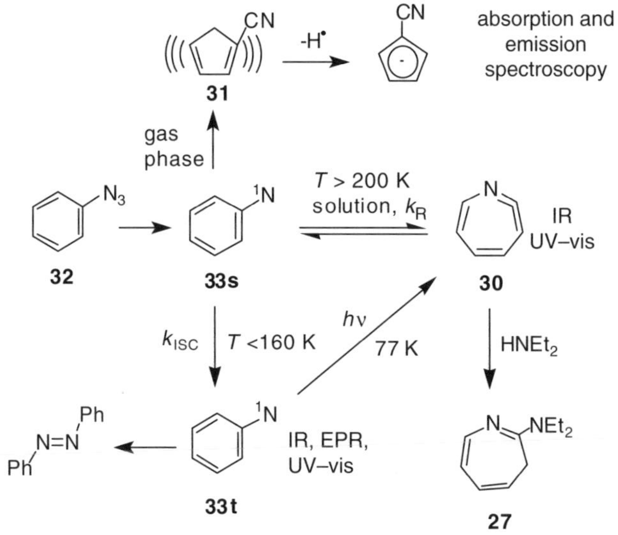

Photolysis of phenyl azide (**32**) produces singlet phenylnitrene (**33s**), but what happens next depends on temperature and phase. In the gas-phase **33s** isomerizes to cyanocyclopentadiene **31**, in solution at ambient temperature, it isomerizes to ketenimine **30** and in cryogenic matrices, singlet phenylnitrene isomerizes to triplet phenylnitrene (**33t**).

5.1. Computational Chemistry

In 1992, the Borden and co-workers[91] and Schaefer and co-workers[92] predicted that the singlet–triplet splitting of singlet phenylnitrene is 18.5 kcal/mol in the gas phase, a result that was confirmed by the negative ion photodetachment studies of Ellison and co-workers.[93] Thus, the phenyl group dramatically lowers the singlet–triplet splitting of the nitrene relative to imidogen ($\Delta E_{ST} = 36$ kcal/mol).[11,12] The situation is reminiscent of vinylnitrene, which has a calculated ST splitting of only 15 kcal/mol.[67] The lowest energy singlet states of both vinyl- and phenylnitrene have open-shell biradical-like configurations.

Although ΔE_{ST} of phenylnitrene is small in comparison to that of imidogen, it is still very large compared to phenylcarbene.[70,71] Thus, while singlet and triplet phenylcarbene interconvert rapidly,[19,70] singlet to triplet intersystem crossing of phenylnitrene in solution is irreversible.[73,86,88] The differences in ΔE_{ST} of singlet and triplet nitrenes have been discussed above with parent methylene and imidogen, and the same arguments apply to their aryl derivatives.

Singlet phenylcarbene is a closed-shell species, but singlet phenylnitrene has an open-shell electron configuration. This observation explains why singlet phenylcarbene reacts more readily with electron pairs than does singlet phenylnitrene. The empty orbital of phenylcarbene can coordinate with pairs of electrons and react rapidly with C—H bonds. This type of interaction and facile insertion is not possible with singlet phenylnitrene, which has an open-shell singlet state. Basically, phenylcarbene is a strong Lewis acid, whereas singlet phenylnitrene is not.

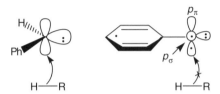

The UV–vis spectrum of triplet phenylnitrene, obtained by brief photolysis of phenyl azide in glassy ether–pentane–alcohol is shown in Figure 11.2.

Pople–Parr–Pariser (PPP) calculations in 1986 indicated that the visible band of triplet phenylnitrene at 500 nm is the result of promotion of an electron from the

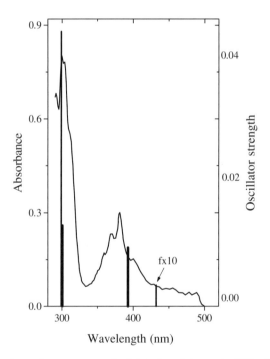

Figure 11.2. The absorption spectrum of triplet phenylnitrene in EPA glass at 77 K. The computed positions and oscillator strengths (f, right-hand axes) of the absorption bands are depicted as solid vertical lines. For very small oscillator strength, the value multiplied by 10 is presented ($f \times 10$). [Reproduced with permission from N. P. Gritsan, Z. Zhu, C. M. Hadad, and M. S. Platz, *J. Am. Chem. Soc.* **1999**, *121*, 1202. Copyright © 1999 American Chemical Society.]

TABLE 11.1. Maxima (in nm) of the Most Intense Absorption Bands in the Electronic Absorption Spectra of Substituted Singlet and Triplet Phenylnitrenes (near UV and vis)

Substituent	Singlet Nitrene	Triplet Nitrene[a]
4-F	365	a
4-Cl	360	a
4-Br	361	a
2,4,6-triBr	395	326, 340
4-I	328	a
4-Me	365	315
4-CF$_3$	~320	a
4-COMe	334	a
4-Ph	345	320
4-(4′-azidophenyl)	380	a
2-Me	350	a
2,6-diMe	350	297, 310
2,4,6-triMe	348, 366	319
2-F	342	294, 315
2,6-diF	331, 342	313
2,3,4,5,6-pentaF	330	315
2-CN	382	328
2,6-diCN	385, 405	341

[a] Spectrum was not detected—a.

singly occupied p_σ-AO into the π^* system.[86] The transition from the sp lone pair on nitrogen to a π^* orbital is obscured by a $\pi \rightarrow \pi^*$ transition. On the other hand, the $sp \rightarrow \pi^*$ and $\pi \rightarrow \pi^*$ transitions are very well separated in perfluorophenyl nitrene. The $sp \rightarrow p_\sigma$ transition at 308 nm of triplet phenylnitrene, which is very characteristic of triplet nitrenes has been observed in NH (336 nm), MeN (317 nm), and CF$_3$N (342, 347.5, 354 nm). More advanced calculations performed by Gritsan et al.[94] are in excellent agreement with the experimental results for phenylnitrene. Table 11.1 lists the absorption maxima of several triplet arylnitrenes.

In 1997, Karney and Borden[68] predicted that the rearrangement of **35s** to ketenimine **30** proceeds in two steps, via the azirine intermediate **29**. The initial reaction should be thought of as simply the cyclization of a 1,3 biradical as discussed for singlet vinylnitrene. Singlet phenylcarbene is not a biradical, and therefore has a larger barrier to cyclization than does singlet phenylnitrene. Because the bimolecular reactions of singlet phenylcarbene are very fast in solution, its thermal cyclization to a benzocyclopropene (see the discussion on the interconversion of pyridylcarbene and phenylnitrene, Section 5.9) occurs only in the gas phase at high temperature.[95]

Since the ring opening of **29** to the didehydroazepine **30** is very rapid, the first step of the unimolecular rearrangement of singlet phenylnitrene **33s** is rate determining. The calculated potential surface is shown in Figure 11.3. Thus, it is not surprising that **29** has so far evaded spectroscopic detection although the IR

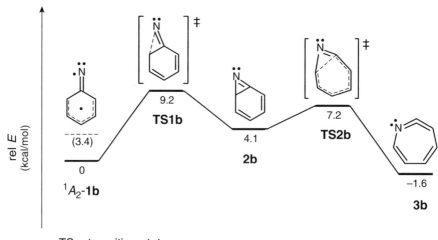

TS = transition state

Figure 11.3. Energetics of the ring expansion of singlet phenylnitrene calculated at the CASPT2 (8,8)/6-311G(2d,p)//CASSCF(8,8)/6-31G* level. [Reproduced with permission from W. L. Karney and W. T. Borden, *J. Am. Chem. Soc.* **1997**, *119*, 1378. Copyright © 1997 American Chemical Society.]

spectrum of the perfluorinated analog of **29** in a cryogenic matrix has been reported.[96] Benzazirine has, however, been intercepted with ethanethiol, in solution.[80]

According to calculations, there is essentially no barrier to cyclization of singlet vinylnitrene,[68] but at the same level of theory, the corresponding barrier in singlet phenylnitrene is predicted to be 9 kcal/mol. Thus, the latter species should be more easily detected in solution. Taking into account that this level of theory overestimates the barrier to cyclization of singlet vinylnitrene by 3.4 kcal/mol, one can extrapolate that the best prediction of the barrier to cyclization of singlet phenylnitrene is close to 6 kcal/mol.

Singlet phenylnitrene is longer lived than singlet vinylnitrene because the aromaticity of the benzene ring of singlet phenylnitrene is lost upon cyclization. For the same reason, the phenylnitrene rearrangement is much less exothermic than the cyclization of vinylnitrene to azirine. The product of rearrangement of vinylnitrene is azirine, an isolable species, benzazirine is not. The latter species rapidly ring opens to form a didehydroazepine.

5.2. Laser Flash Photolysis Studies

Sundberg and co-workers[97] studied the photochemistry of phenyl azide by conventional flash photolysis in 1974. They detected the transient UV absorption of ketenimine **30** and measured its absolute rate constant of reaction with diethylamine to form the 1H azepine, which subsequently rearranges to Doering and Odum's 3H-azepine (**27**).[79]

Schrock and Schuster[98] confirmed and extended this work using LFP techniques in 1984. Schuster and co-workers[99] later employed flash photolysis with IR detection to confirm that ketenimine **30** is the reactive intermediate present on the microsecond time scale.

In 1997, two groups simultaneously reported that LFP of phenyl azide **32** or compounds **34** and **35** produces a previously undetected transient with $\lambda_{max} = 350$ nm and a lifetime of ≈ 1 ns at ambient temperature.[100,101] The transient decays at the same rate that cyclic ketenimine **30** is formed, implying that the newly detected transient is singlet phenylnitrene **33s**.

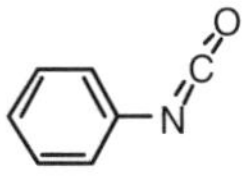

34 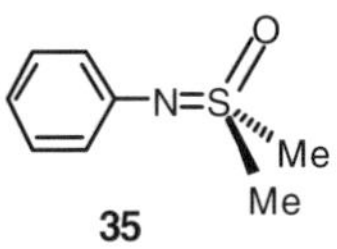   **35**

Laser flash photolysis (266 nm) of phenyl azide in pentane at 233 K produces a transient absorption spectrum with two sharp bands with maxima at 335 and 352 nm (Fig. 11.4). Spectrum 1 was measured, point by point, 2 ns after the laser pulse. In later work, the spectrum of **33s** was reinvestigated and an additional very weak, long wavelength absorption band at 540 nm was observed (Spectrum 2).[94] The transient spectrum of Figure 11.4 was assigned to singlet phenylnitrene in its lowest open-shell electronic configuration ($^1A_2$).

The electronic absorption spectrum of **33s** in the $^1A_2$ state was calculated at the CASPT2 level and is in good agreement with the transient spectrum (Fig. 11.5). The two lowest excited singlet states of **33s** are both of A_1 symmetry and electronic transitions into these states are predicted to occur at 1610 and 765 nm. However, they are forbidden by electric dipole selection rules and thus neither of these transitions has been detected. In the absorption spectrum of **33s**, the only intense absorption band is located ~ 350 nm, ~ 40 nm from the intense 308-nm band in **33t**. This band has a long tail out to 450 nm and displays some fine structure that may be associated with the vibrations of the phenyl ring in **33s**. The strongest absorption band in **33s**, predicted at 368 nm by CASPT2, is due to the transition to the 2^1B_1 exited state. The main configuration involved in this transition is similar to that of 2^3B_1 state, that is, it arises by promotion of an electron from the sp lone-pair orbital on nitrogen to the singly occupied nitrogen p_σ orbital.

The electronic absorption spectra of **33s** and **33t** are very similar (Figs. 2 and 4), but all of the calculated and experimental bands of **33s** exhibit a red shift compared to those of **33t**. This result is very reasonable because both these species have very similar open-shell electronic configurations ($^3A_2$ and $^1A_2$). The absorption maxima of several singlet arylnitrenes are given in Table 11.1.

5.3. Dynamics of Singlet Phenylnitrene

The decay of **33s** was monitored in pentane at 350 nm over a temperature range of 150–270 K, which allowed direct measurement of k_{ISC} and determination of

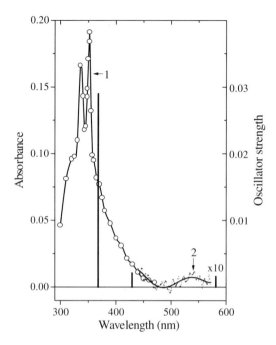

Figure 11.4. Transient spectrum of singlet phenylnitrene produced upon LFP of phenyl azide. Spectrum 1 was recorded 2 ns after the laser pulse (266 nm, 35 ps) at 233 K. Long-wavelength band (2) was recorded with an optical multichanal analyzer at 150 K (with a 100-ns window immediately after the laser pulse, 249 nm, 12 ns). The computed positions and oscillator strengths (f, right-hand axes) of the absorption bands are depicted as solid vertical lines. For very small oscillator strength, the value multiplied by 10 is presented ($f \times 10$). [Reproduced with permission from N. P. Gritsan, Z. Zhu, C. M. Hadad, and M. S. Platz, *J. Am. Chem. Soc.* **1999**, *121*, 1202. Copyright © 1999 American Chemical Society.]

accurate barriers to cyclization. The lifetime of **33s** at 298 K in CH_2Cl_2 was found to be ~0.6 ns.[94]

The formation of the products [cyclic ketemimine (**30**) and/or triplet nitrene (**33t**)] was monitored at 380 nm. The decay of singlet phenylnitrene and the growth of the products are exponential and can be analyzed to yield an observed rate constant k_{OBS}. An Arrhenius treatment of these data (open circles), is presented in Figure 11.5. The magnitude of k_{OBS} decreases with decreasing temperature until ~170 K, whereupon it reaches a value of $\sim 3.2 \times 10^6$ s^{-1}. Below this temperature, k_{OBS} remains constant.[94] The breakpoint in the Arrhenius plot is ~180–200 K, which is in exactly the same temperature range in which the solution phase chemistry changes from the trapping of ketenimine **30** with diethylamine to the dimerization of **33t**.[86] Thus, the low-temperature data of Figure 11.5 were associated with k_{ISC}, the rate constant for intersystem crossing of singlet to triplet phenylnitrene, and the high temperature data with k_R, the rate constant for rearrangement of **33t**.

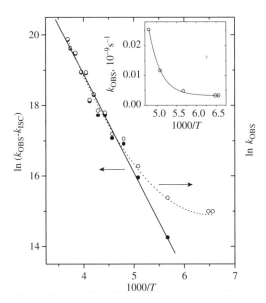

Figure 11.5. Arrhenius treatment of the k_{OBS} data (open circles) and k_R data (filled circles) for singlet phenylnitrene deduced upon assuming that k_{ISC} is independent of temperature. Insert: temperature dependence of k_{OBS} data. [Reproduced with permission from N. P. Gritsan, Z. Zhu, C. M. Hadad, and M. S. Platz, *J. Am. Chem. Soc.* **1999**, *121*, 1202. Copyright © 1999 American Chemical Society.]

The temperature-independent rate constant observed at low temperature (3.2×10^6 s^{-1}) is the rate constant for ISC to the triplet ground state (k_{ISC}). The rate constant k_{OBS} is equal to $k_R + k_{ISC}$, where k_R is the absolute rate constant for rearrangement for singlet phenylnitrene to benzazirine (**29**). As $k_{OBS} = k_R + k_{ISC}$, it is possible to deduce values of k_R as a function of temperature and to obtain its associated Arrhenius parameters. Indeed, an Arrhenius plot of $k_R = k_{OBS} - k_{ISC}$ is linear (Fig. 11.5, solid circles) with an activation energy for rearrangement of $E_a = 5.6 \pm 0.3$ kcal/mol and preexponential factor $A = 10^{13.1 \pm 0.3}$ s^{-1}.[94]

The calculated barrier for the cyclization of **33t** to **29** is 6 kcal/mol after taking into consideration that the CASPT2 method overestimates the energy difference between open- and closed-shell states by ~3 kcal/mol.[68] The correspondingly adjusted predicted 6 kcal/mol barrier is in nearly exact agreement with the experimental value (5.6 ± 0.3 kcal/mol).[94]

5.4. Intersystem Crossing Rates (k_{ISC})

Singlet phenylnitrene undergoes ISC three orders of magnitude more slowly than arylcarbenes.[59,94] There are at least three reasons why arylcarbenes do ISC much faster than singlet phenylnitrene. The rate of a radiationless transition increases as the energy separation between the two states goes to zero.[102] The calculated gas-phase singlet–triplet splitting of phenylcarbene (PC) is 2–4 kcal/mol;[71] whereas in phenylnitrene it is 18 kcal/mol.[91–93] Carbenes are divalent and have a bending mode with which to couple singlet and triplet surfaces, a vibration that is, of course, lacking in monovalent nitrenes. However, the most important factor is probably the electronic structure of the respective singlet intermediates. Singlet phenylcarbene has a closed-shell electronic structure with one filled and one empty nonbonding orbital. In such an "ionic" singlet, spin–orbit coupling (SOC) is a particularly effective mechanism of intersystem crossing.[102–106] Singlet phenylnitrene, on the other hand, is an open-shell covalent singlet. The SOC is forbidden in this case and is ineffective in promoting ISC.

5.5. para-Substituted Derivatives of Phenylnitrene

Values of k_{ISC} for singlet para-substituted phenylnitrenes are given in Table 11.2.[107] The ISC rate constant for singlet p-bromo phenylnitrene is about seven times larger than that of parent **33s** and the p-fluoro and p-chloro analogues. This difference is easily attributable to a small heavy atom effect. The heavy atom effect of iodine is even larger than that of bromine, as expected, and raises the ISC rate by more than a factor of 20, relative to parent **33s**.

A very large acceleration in k_{ISC} is observed with p-methoxy and dimethylamino substituents (Table 11.2). This observation is consistent with the solution-phase photochemistry of p-methoxy and p-dimethylaminophenyl azides.[107–111] The Me, CF_3, acetyl, fluoro, and chloro substituents are not sufficiently strong π donors or acceptors to significantly influence the size of k_{ISC} (Table 11.2), but the strong π

TABLE 11.2. Kinetic Parameters of Para-Substituted Singlet Arylnitrenes in Pentane[a]

Para-X	τ_{295K} (ns)	k_{ISC} ($\times 10^6$ s^{-1})	E_a (kcal/mol)	Log A (s^{-1})
H	~1	3.2 ± 0.3	5.6 ± 0.3	13.1 ± 0.3
Me	~1	5.0 ± 0.4	5.8 ± 0.4	13.5 ± 0.2
CF_3	1.5	4.6 ± 0.8	5.6 ± 0.5	12.9 ± 0.5
C(O)Me	5.0	8 ± 3	5.3 ± 0.3	12.5 ± 0.3
F	~0.3	3.5 ± 1.4	5.3 ± 0.3	13.2 ± 0.3
Cl	~1	3.9 ± 1.5	6.1 ± 0.3	13.3 ± 0.3
Br	~3	17 ± 4	4.0 ± 0.2	11.4 ± 0.2
I	[b]	72 ± 10	[b]	[b]
OMe	<1	>500	[b]	[b]
CN	8 ± 4	6 ± 2	7.2 ± 0.8	13.5 ± 0.6
Ph	15 ± 2	12 ± 1	6.8 ± 0.3	12.7 ± 0.3
NMe[c]	0.12	8300 ± 200	[b]	[b]
NO$_2$[d]	<20	>50	[b]	[b]

[a]See Ref. 95.
[b]Not measured.
[c]In toluene.
[d]In benzene.

donating p-methoxy and p-dimethylamino groups have a huge influence on k_{ISC}.[107,111] The increased ionic character in these open-shell singlet biradicals, as shown in a resonance structure below, can increase spin–orbit coupling and the rate constant of intersystem crossing.

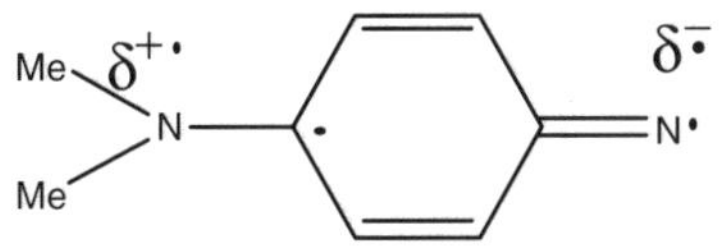

The electron-withdrawing substituents (CF_3, COMe, CN, and NO_2) have a smaller, but measurable influence on k_{ISC}. It is interesting to note that both donating and withdrawing substituents accelerate the rate of ISC.

5.6. ortho- and meta-Substituted Derivatives of Phenylnitrene

Intersystem crossing rate constants of ortho- and meta-substituted singlet phenylnitrenes are presented in Table 11.3.[107–111] Mono- and di-o-fluorine substituents have no influence on ISC rate constants.[109] No effect with *meta, meta*-difluoro substitution is observed either. Pentafluoro substitution has no effect on k_{ISC} in pentane although a modest acceleration is observed in the more polar solvent methylene chloride.[107–111]

TABLE 11.3. Intersystem Crossing Rate Constants of Ortho and Meta Substituted Phenylnitrenes[a]

Substituent	Solvent	k_{ISC} (10^6 s^{-1})
2-Methyl-	Pentane	10 ± 1
2,6-Dimethyl-	Pentane	15 ± 3
2,6-Dimethyl-	$CF_2ClCFCl_2$	30 ± 8
2,4,6-Trimethyl-	Pentane	20 ± 1
	$CF_2ClCFCl_2$	29 ± 3
2-Fluoro-	Pentane	3.3 ± 0.5
3,5-Difluoro-	Pentane	3.1 ± 1.5
2,6-Difluoro-	Hexane	2.4 ± 0.3
2,6-Difluoro-	CCl_4	2.7 ± 0.3
2,3,4,5,6,-Pentafluoro-	Pentane	3.3 ± 1.5
	CH_2Cl_2	10.5 ± 0.5
2-Cyano-	Pentane	2.8 ± 0.3
2,6-Dicyano-	CH_2Cl_2	4.5 ± 0.5
	Pentane	6.2 ± 0.8
	THF[b]	5.9 ± 1.5
2-Pyrimidyl-	CH_2Cl_2	800 ± 200

[a]See Refs. 96–98.
[b]Tetrahydrofuran = THF.

An o-cyano group has little influence on k_{ISC}, but two o-cyano groups slightly accelerate intersystem crossing.[107–110] Singlet arylnitrenes with electron-withdrawing groups in the para-position have little influence on the rate constant of ISC.

An o-methyl group accelerates intersystem crossing relative to singlet p-tolyl nitrene. Two o-methyl groups are more effective than one at accelerating intersystem crossing. Singlet 2,4,6-trimethylphenylnitrene undergoes intersystem crossing about as readily as 2,6-dimethylphenylnitrene. These results are consistent with the general trend that electron donating groups (methyl, methoxy, dimethylamino) accelerate intersystem crossing. These groups increase the zwitterionic character of the singlet nitrene relative to the parent system, which facilitates a spin–orbit coupling mechanism of intersystem crossing.

5.7. Cyclization to Azirines

The cyclic ketenimine **30** is the major trappable reactive intermediate present in solution when phenyl azide (at moderate concentrations) is decomposed photolytically at 298 K. The rate of decay of singlet phenylnitrene **33s** is equal to the rate of formation of the cyclic ketenimine. The first step, cyclization to benzazirine (**29**) is rate determining, and is followed by fast electrocyclic ring opening to cyclic ketenimine **30**.

For most aryl azides, the rate constants of singlet nitrene decay and product formation (triplet nitrene and/or ketenimine) are the same. Thus, in all these phenylnitrenes cyclization to substituted benzazirines is the rate-limiting step of the process of isomerization to ketenimine, as is the case for the parent phenylnitrene. The only known exception, o-fluorophenylnitrene, will be discussed in the next section.

The activation parameters for cyclization of para-substituted singlet phenylnitrenes are presented in Table 11.2.[107–110] It is readily seen from that table that polar substituents such as para Me, CF_3, halogen, and acetyl have little influence on k_R. This noneffect is not very surprising given that theory predicts that singlet phenylnitrene has an open-shell electronic structure.[91,92,106] Cyclization of singlet phenylnitrene only requires that the nitrogen atom bends out of the molecular plane, so that the singly occupied σ nonbonding molecular orbital (NBMO) can interact with the singly occupied π NBMO.[68,106] Azirine formation may therefore simply be regarded as the cyclization of a quinoidal 1,3-biradical, which originally has two orthogonal, antiparallel spins. Thus, polar effects are not anticipated.

Two para-substituents, phenyl[112] and cyano depress k_R and retard the rate of cyclization significantly (Table 11.2).[110] p-Phenyl and p-cyano are both radical stabilizing substituents. These conjugative substituents reduce the spin density on the carbon ortho to the nitrene nitrogen. The reduced spin density at carbons ortho to the nitrogen lowers the rate at which the 1,3-biradical cyclizes. The effect with p-cyano and p-biphenyl singlet phenylnitrene is quite dramatic. The lifetimes of these singlet nitrenes at ambient temperature are 8 and 15 ns, respectively, and the activation barriers to cyclization are 7.2 and 6.8 kcal/mol, respectively.

Sundberg, et al.[113] demonstrated that simple alkyl substituents direct cyclization away from the substituent.

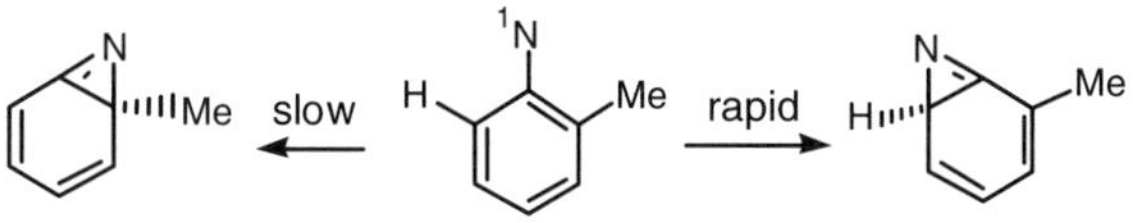

A single o- or p-methyl substituent has no influence on the rate of cyclization of the singlet tolylnitrene to the favored azirine.[108] The methyl group has no bystander effect on benzazirine formation. Cyclization of 2,6-dimethylphenyl or 2,4,6-tri-methylphenylnitrenes necessarily proceeds toward a carbon bearing a substituent. A steric effect raises the barrier to cyclization by 1.5–2.0 kcal/mol, in excellent agreement with the predictions of Karney and Borden.[114] The steric effect extends the lifetime of 2,6-dimethylphenylnitrene at ambient temperature to 13 ns in Freon-113 and of 2,4,6-trimethylphenylnitrene to 8 ns, in the same solvent (Table 11.4).[108]

A cyano group is a smaller substituent than methyl, thus cyclization toward and away from a cyano-substituted carbon should be more evenly balanced. Consistent with this hypothesis, Lamara et al.[115] found that singlet o-cyanophenylnitrene (**36s**) undergoes ring expansion to afford not only **37**, the product formed by cyclization away from the cyano substituent, but also **38**, the product formed by cyclization toward the cyano group. Similar results have been found in the ring expansion of singlet o-acetylphenylnitrene.[116]

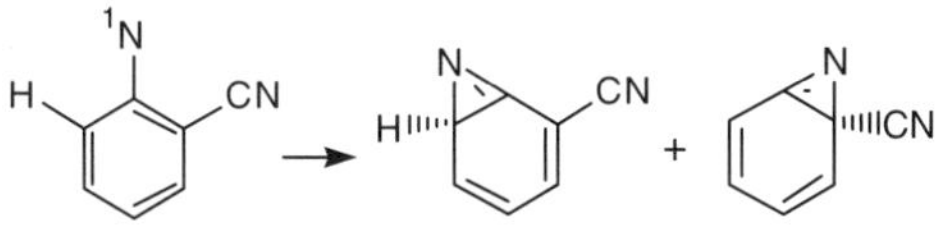

TABLE 11.4. Summary of Kinetic Results for Singlet Methyl-Substituted Phenylnitrenes[a]

Substituent	$\tau(295)$ (ns)	$\log A$ (s^{-1})	E_a (kcal/mol)	Solvent
4-Methyl	$\sim 1^b$	13.2 ± 0.2	5.8 ± 0.4	C_5H_{12}
2-Methyl	$\sim 1^b$	12.8 ± 0.3	5.3 ± 0.4	C_5H_{12}
2,6-Dimethyl	12 ± 1	13.0 ± 0.3	7.0 ± 0.3	C_6H_{14}
2,6-Dimethyl	13 ± 1	12.9 ± 0.3	7.5 ± 0.5	$CF_2ClCFCl_2$
2,4,6-Trimethyl	8 ± 1	13.4 ± 0.4	7.3 ± 0.4	$CF_2ClCFCl_2$

[a]See Ref. 96.
[b]Lifetime estimated by extrapolation of the data to 295 K.

TABLE 11.5. Kinetic Parameters of Fluorosubstituted Singlet Phenylnitrenes[a]

Substituent	τ_{298} (ns)	$\log A$ (s^{-1})	E_a (cal/mol)	Solvent
H	~1	13.1 ± 0.3	5600 ± 300	C_5H_{12}
2-Fluoro-	8 ± 1	13.0 ± 0.3	6700 ± 300	C_5H_{12}
	10 ± 2			CH_2Cl_2
	10 ± 2			$CF_2ClCFCl_2$
4-Fluoro-	~0.3	13.2 ± 0.3	5300 ± 300	C_5H_{12}
3,5-Difluoro-	~3	12.8 ± 0.3	5500 ± 300	C_5H_{12}
2,6-Difluoro-	240 ± 20	11.5 ± 0.5	7300 ± 700	C_6H_{14}
	260 ± 20	12.0 ± 1.2	8000 ± 1500	CCl_4
2,3,4,5,6-Pentafluoro-	56 ± 4	12.8 ± 0.6	7800 ± 600	C_5H_{12}
	32 ± 3	13.8 ± 0.3	8800 ± 400	CH_2Cl_2
Perfluoro-4-biphenyl	260 ± 10	13.2 ± 0.2	9400 ± 400	CH_2Cl_2
	220 ± 10	12.5 ± 0.4	8900 ± 300	MeCN
4-CONHC$_3$H$_8$-2,3,5,6-Tetrafluoro	210 ± 20	12.0 ± 0.2	7500 ± 300	MeCN

[a]See Ref. 97.

5.8. Fluoro-Substituted Singlet Phenylnitrenes

Abramovitch et al.[74] and Banks et al.[75] discovered that unlike most arylnitrenes, polyfluorinated arylnitrenes have a bountiful bimolecular chemistry. Perfluorophenylnitrene reacts with diethylamine to form a hydrazine, with tetramethylethylene to form an aziridine and forms robust adducts with benzene and even cyclohexane.[117] Polyfluorinated arylnitrenes are useful reagents in synthetic organic chemistry,[117] in photoaffinity labeling,[5] and for the covalent modification of polymer surfaces.[8]

Laser flash photolysis of a series of fluorinated aryl azides produces the transient spectra of the corresponding singlet nitrenes.[109] With the exception of singlet o-fluorophenylnitrene (**39s**), the rate of decay of the singlet nitrene was equal to the rate of formation of the reaction products, for example, didehydroazepines and triplet nitrenes. Values of k_{ISC} and the Arrhenius parameters for azirine formation are summarized in Table 11.5.

Placement of fluorine substituents at both ortho positions raises the barrier to cyclization by ~3 kcal/mol, relative to the unsubstituted system. The work of Leyva and Sagredo[118] demonstrated, in fact, that cyclization of the singlet nitrene **39s** proceeds away from the fluorine substituent. The steric argument predicts that a single o-fluorine substituent will have little influence on the rate of conversion of **39s** to **40**, since cyclization occurs at the unsubstituted ortho-carbon.[114]

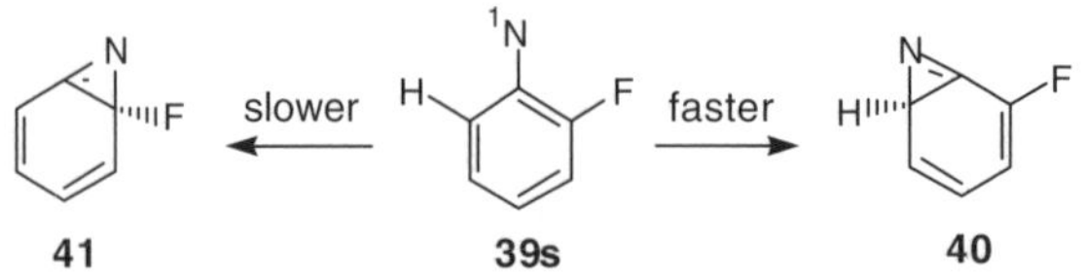

41   slower   39s   faster   40

Figure 6.1. The Jovian moon Io; deep ultraviolet (UV) photolysis of its methane atmosphere proceeds with electron ejection, generating the molecular ion of methane. NASA JPL Galileo program image from Voyager 1, http://www.jpl.nasa.gov/galileo/io/vgrio1.html

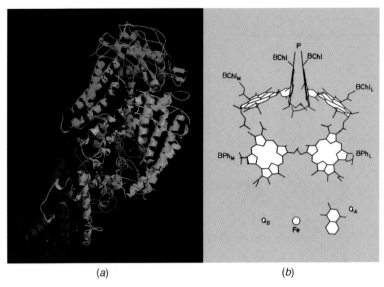

Figure 6.2. (*a*) Photosynthetic reaction center of *Rhodopseudomonas viridis* Reprinted from the Protein Data Bank, H. M. Berman, J. Westbrook, Z. Feng, G. Gilliland, T. N. Bhat, H. Weissig, I. N. Shindyalov, P. E. Bourne, *Nucleic Acids Res.* **2000**, 1, 235 (http://www.pdb.org/) PDB ID: IDXR, C. R. D. Lancaster, M. Bibikova, P. Sabatino, D. Oesterhelt, H. Michel, *J. Biol. Chem.* **2000**, 275, 39364.[2] (*b*) arrangement of the essential components in the purple bacterium *Rh. sphaeroides*. [Adapted from Ref. 5.]

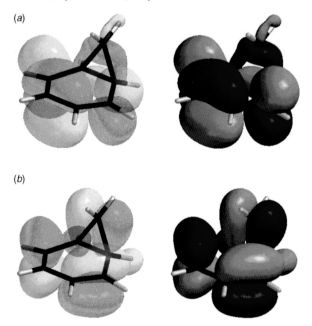

Figure 6.16. Spartan representation of SOMO (*b*) and LUMO (*a*) for norcaradiene radical cation. [Adapted from Ref. 152.]

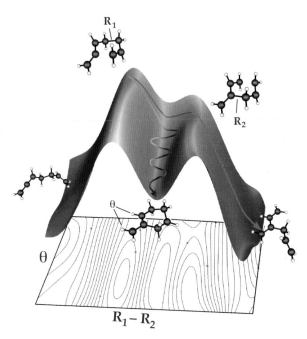

Figure 21.2. Schematic PE surface for the rearrangement of 1,2,6-heptatriene. The geometrical coordinates, θ and R_1–R_2 are defined as follows: θ is the dihedral angle between the H8–C1–H9 and C4–C3–H10 planes. R_1 is the C4–C5 distance and R_2 is the C2–C7 distance. See Scheme 21.3 for atom numbering.

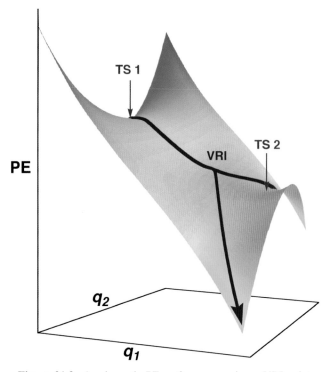

Figure 21.3. A schematic PE surface possessing a VRI point.

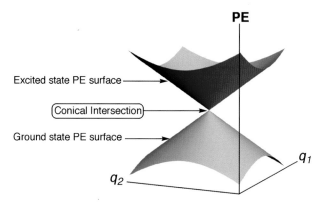

Figure 21.5. A schematic representation of a conical intersection between two electronic states of a molecule. Coordinates q_1 and q_2 are the nonadiabatic coupling vector and the gradient difference vector, along which the degeneracy between the states is lifted.

However, the barrier to this process is larger (outside of experimental error) than that in the parent system (Table 11.5). In fact, the lifetime of singlet *o*-fluorophenylnitrene (**39s**) at 298 K is 8–10 times longer than that of the parent **33s** and 20–30 times longer than that of *p*-fluorophenylnitrene. Unlike a methyl group, a single *o*-fluorine atom exerts a small, but significant bystander effect on remote cyclization that can not be simply steric in origin.

In order to understand this substituent effect, the atomic charges for the different centers were calculated using CASSCF(8,8)/6-31G* wave functions and the natural population analysis (NPA) method of Reed, Weinstock et al. (1985).[119a] Fluorine substitution makes the adjacent carbon very positively charged (+0.48 e). In the transformation of *o*-fluorophenylnitrene **39s** to **40** or **41**, there is an increase in positive charge at the (ipso) carbon bearing the nitrogen. The increased activation barrier to cyclization for *o*-fluorophenylnitrene (**39s**) relative to **33s** or *p*-fluorophenylnitrene is therefore due to a greater amount of positive–positive charge interaction between the ortho and ipso carbons in the transition state.

In the case of insertion toward fluorine (**41**), there is an even greater amount of positive–positive charge repulsion between the ortho and ipso carbons than in the transition state and this effect is responsible, in part, for a higher activation barrier for insertion toward F to form **41** than away from fluorine to form **40**. Therefore, the origin of the pronounced influence of ortho,ortho-difluoro substitution on the lifetime of singlet arylnitrene and the increased activation energy of its cyclization is the result of combination of the steric effect and the extraordinary electronegativity of fluorine atom.

As mentioned previously, unique kinetic results were obtained upon LFP of *o*-fluorophenyl azide,[109] in that the singlet nitrene decays faster than the ketenimine is formed. This finding requires the presence of an intermediate, presumably benzazirine **40**, between the singlet nitrene and ketenimine **42**. The data could be interpreted by assuming that azirine **40** reverts easily to singlet nitrene according to the scheme below.[109] The equilibrium constant is equal to the ratio of [**40**]/[**39s**] and was deduced to be ~0.5 with ΔG ~350 cal/mol. Younger and Bell have also reported a system in which a benzazirine and ketenimine interconvert.[120]

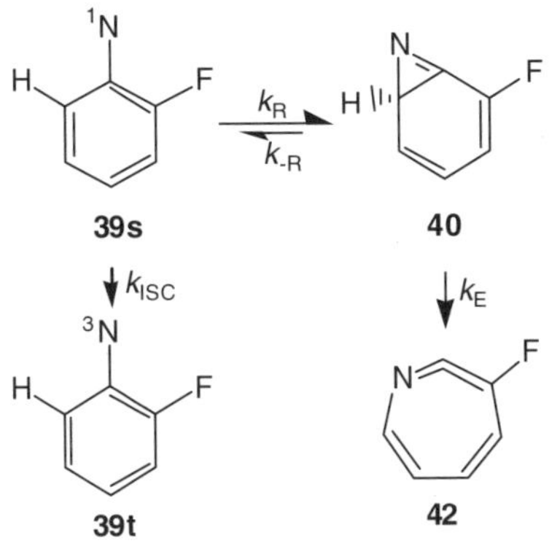

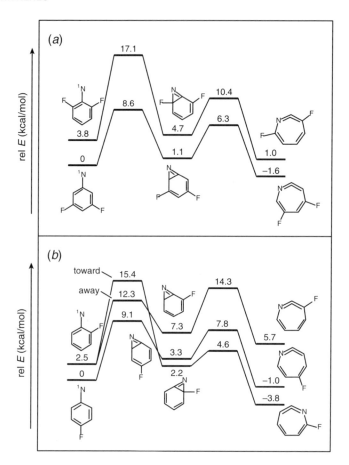

Figure 11.6. Relative energies (in kcal/mol) of species involved in the ring expansions of singlet fluoro-substituted phenylnitrenes calculated at the CASPT2/cc-pVDZ//CASSCF(8,8)/ 6-31G* level. (*a*) Difluorinated phenylnitrenes. (*b*) Monofluorinated phenylnitrenes. [Reproduced with permission from N. P. Gritsan, A. D. Gudmundsdottir, D. Tigelaar, Z. Zhu, W. L. Karney, C. M. Hadad, and M. S. Platz, *J. Am. Chem. Soc.* **2001**, *123*, 1951. Copyright © 2001 American Chemical Society.]

The rate constant of the ring-opening reaction k_E was measured and the Arrhenius parameters were found to be $A = 10^{13.5 \pm 0.4} \, M^{-1} \, s^{-1}$ and $E_a = 9.00 \pm 0.5$ kcal/mol. The proposed mechanism is supported by modern theoretical methods, which reveal that the reverse of **40** to **39s** is exothermic and proceeds over a very small barrier (Fig. 11.6).

5.9. Interconversions of Pyridylcarbene and Phenylnitrene

Gas-phase pyrolysis of phenyldiazomethane releases phenylcarbene, which undergoes the "phenylcarbene rearrangement."[95] Phenylcarbene cyclizes to form a cyclopropene, which subsequently undergoes a cyclohexadiene–hexatriene electrocyclic ring opening to form a symmetric cycloheptatetraene. This species can reform the original cyclopropene or the cyclopropene in which the label has apparently moved into the ring before subsequently forming stable products. The label can eventually migrate into the meta and para positions by additional "phenylcarbene" rearrangements.

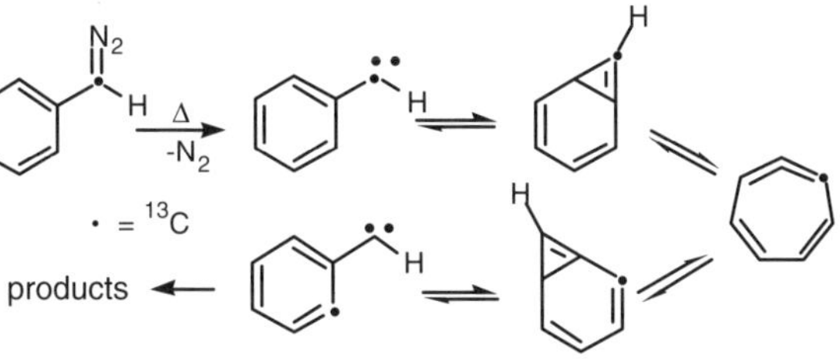

Similarly the pyridylcarbenes rearrange to form phenylnitrenes.[81]

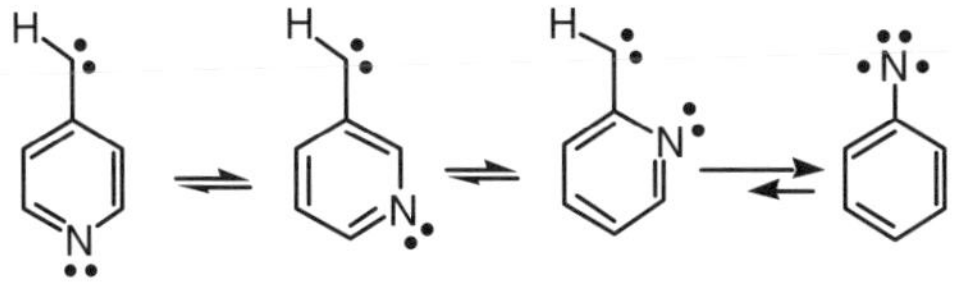

As illustrated by the methyl derivative, the pyridylcarbene gives high yields of aryl nitrene derived products indicating that phenylnitrene is substantially lower in energy than the isomeric pyridylcarbenes, as first suggested by Wentrup and coworkers.[81]

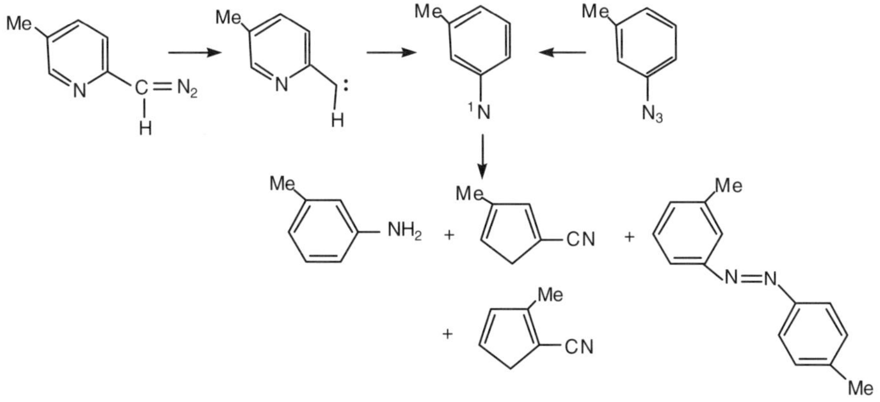

Indeed, CASPT2/6-31G* calculations of Kemnitz et al.[121] find that triplet phenyl nitrene is 26.4 kcal/mol lower in energy than the isomeric triplet *m*-pyridylcarbene.

As the singlet–triplet splitting of phenylnitrene is 18 kcal/mol, one can deduce that singlet phenylnitrene is 11.4–13.4 kcal/mol more stable than singlet pyridylcarbene.[121] However, the corresponding radicals, anilines, and substituted pyridines have very similar energies by CASPT2/6-31G*. Hence, the large difference in enthalpy between phenylnitrene and pyridylcarbene is not the result of intrinsic differences between benzenes and pyridines and their ability to delocalize an unpaired electron, but must reflect a basic difference in stability between carbenes and nitrenes.[121]

This finding is further illustrated in the following isodesmic reactions:

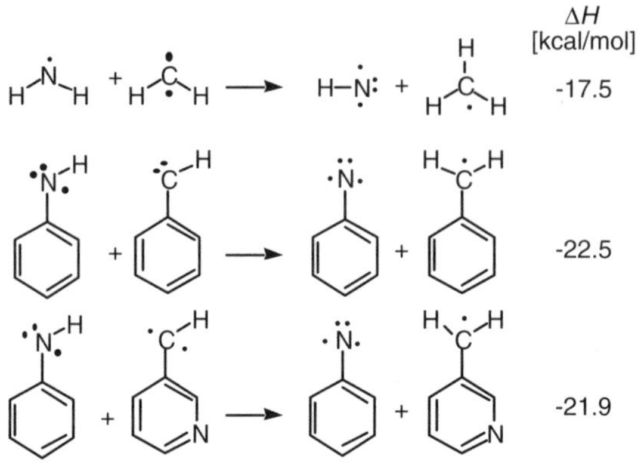

Phenylnitrene is intrinsically more stable than the isomeric pyridylcarbene. Karney and Borden explained that the reason why the N—H BDEs in RNH• radicals are smaller than the C—H BDEs of RCH•₂ radicals, that is, rehybridization, is also responsible for the fact that nitrenes are thermodynamically more stable than carbenes.[121] The lone pair of electrons of a nitrene reside in a low-energy sp hybrid orbital. This effect dramatically stabilizes nitrenes relative to carbenes in which the nonbonding electrons reside in either pure p or pseudo sp^2 orbitals.

6. POLYCYCLIC ARYLNITRENES

6.1. Naphthylnitrenes

Extrapolation of the chemistry of phenylnitrene leads to qualitatively accurate predictions of the properties of the naphthylnitrenes. The naphthylnitrenes have triplet ground states that have been characterized by low-temperature EPR and UV–vis spectroscopy.[39,122] The singlet–triplet splitting has not yet been measured, but Tsao's CASPT2(12,12) + ZPE calculations predict energy separations of 13.9 and 16.6 kcal/mol for 1- and 2-naphthylnitrene, respectively.[123]

One can predict that singlet 1-naphthylnitrene will cyclize more readily than phenylnitrene because the resonance energy per aromatic ring is lower in naphthalene than benzene, but by the same token, 1-naphthylnitrene should cyclize more slowly than vinylnitrene.

In principle, singlet 1-naphthylnitrene can cyclize to two distinct azirines **43** and **44**, but **43** is clearly much lower in energy than isomer **44**, which lacks aromaticity in either ring. One can therefore predict that **43** will be formed preferentially.

Ring opening of azirine **43**, to ketenimine **45** is most likely endothermic, in dramatic contrast to the PhN system, because ring opening of azirine **43** converts a species with one aromatic ring into a ketenimine that has no aromaticity. In the PhN system, a nonaromatic azirine opens to a nonaromatic ketenimine (on the other hand, ring opening of azirine (**44**) will be exothermic, because aromaticity is restored in one of the rings).

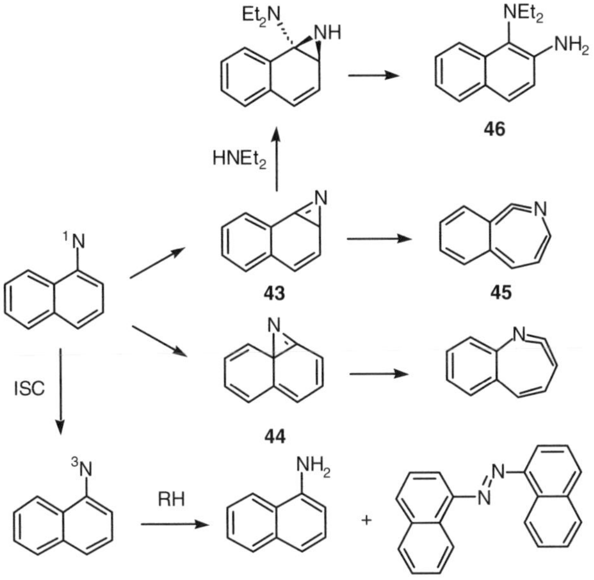

Experimental observations support these views. Photolysis of 1-naphthylazide in the presence of diethylamine and tetramethylethylenediamine (TMEDA) yields azirine, but no ketenimine-derived adducts at ambient temperature.[124] In the presence of diethylamine but in the absence of TMEDA, good yields of 1-aminonaphthalene and 1,1′-azo-naphthalene, products attributable to the triplet nitrene are observed. Good yields of **46** are also achieved when the photolysis of 1-naphthylazide and diethylamine is performed at −60 °C in the absence of TMEDA.[125] Presumably, lowering the temperature extends the lifetime of azirine **43** by reducing its rate of reversion to singlet 1-naphthylnitrene more than it retards the rate of its reaction with diethylamine.

Laser flash photolysis of 1-naphthylazide produces a transient with IR bands at 1728 cm^{-1}. The carrier of this signal is attributed to azirine **43**.[123] The TRIR spectrum of **43** is in excellent agreement with the earlier observation of **43** as a persistent species (1730 cm^{-1}) in argon by Dunkin and Thomson.[126] In that study it was not possible to ascertain whether a single azirine or a mixture of azirines was formed. Recently, however, Maltsev et al. have reported a mono complete matrix IR study.[127] This group demonstrated that only azirine **43** is formed and its vibrational spectrum is in excellent agreement with the predictions of theory. It is also clear that photolysis of the matrix-isolated azirine leads to formation of a ketenimine ($\bar{v} = 1910$–1930 cm^{-1}).[127]

Tsao has used TRIR spectroscopy to determine that the lifetime of azirine in **43** at ambient temperature is 2.6 μs,[123] which is in excellent agreement with the work of Shrock and Schuster,[128] who studied the same system years earlier by LFP with UV–vis detection.

No transient absorption >350 nm is detected upon LFP of 1-naphthylazide.[128] A band with absorption maxima at 370 nm is formed with a time constant of 2.8 μs after the laser pulse. The carrier of the 370-nm absorption reacts over >100 μs to form azonaphthalene. The carrier of the 370-nm absorption is identified as triplet 1-naphthylnitrene that has previously been characterized as a persistent species at 77 K by UV–vis[122] ($\lambda_{max} = 367$ nm) and EPR spectroscopy.[39] Azirine **43**, detected by TRIR spectroscopy must not absorb significantly >350 nm, a fact that was established later by the matrix isolation studies of Wentrup's and Bally's groups.[127] Assuming a rapidly equilibrating mixture of azirine and nitrene, and given that $k_{isc} = 1 \times 10^7$ s^{-1} (determined by Tsao by LFP at 77 K and assumed to have the same value at 298 K),[123] then $K = [\text{singlet nitrene}]/[\text{azirine } \mathbf{43}] = 0.038$ at 298 K.

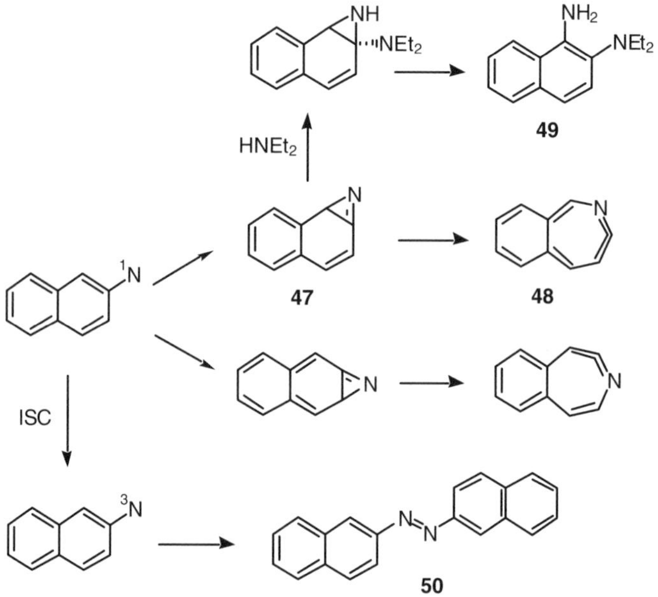

The 2-naphthylnitrene story is similar. Two azirines (**47** and **48**) can be formed, but azirine **47** is preferred and this azirine will form ketenimine **48** only reluctantly.

The azirine **47** has been detected as a persistent species by matrix isolation spectroscopy ($v = 1708, 1723, 1736$ cm^{-1}).[126] The same species has also been detected in solution by LFP with IR detection ($\tau = 150$ μs).[123] Azirine **47** can be intercepted with diethylamine to form **49** in 94% yield when the amine concentration is 1.45 *M*.[128] This yield is a much greater than can be achieved with 1-naphthylazide at ambient temperature. Adduct **49** is not formed upon triplet sensitized photolysis of 2-naphthylazide and diethylamine, instead azonaphthalene **50** is formed.[128] This product is also formed in 45% yield upon direct photolysis in the absence of amine.

In the absence of trapping reagents, azirine **47** reverts to singlet-2-naphthylnitrene, which ultimately relaxes to the triplet. Azirine **47** has a lifetime >150 μs, which indicates that the [nitrene]/[azirine] equilibrium constant is very much smaller in 2-naphthyl than in 1-naphthylnitrene. In fact, upon LFP of 2-naphthylazide, no transient signals absorbing >350 nm are observed immediately after the laser pulse. Unlike in the 1-naphthyl case, the transient spectrum of 2-naphthylnitrene does not grow in on the microsecond time scale, but the growth of azo compound is observed on the millisecond time scale.[128]

Shrock and Schuster also studied the pyrenylnitrenes and obtained results very similar to those described above for the naphthyl systems. Other polycyclic nitrenes can be imagined, but have received far less comprehensive study.[128]

6.2. Biphenylnitrenes

The LFP of *p*-biphenyl azide produces singlet-*p*-biphenylnitrene.[112] The phenyl group has little influence on the electronic spectra of either singlet or triplet *p*-biphenylnitrene (Table 11.1), but it extends the singlet nitrene lifetime from 1 to 17 ns at ambient temperature by "diluting" the spin density ortho to the nitrene nitrogen.

In 1951, Smith and Brown discovered that the decomposition of *o*-biphenyl azide produces carbazole in excellent yield.[73a,b] Swenton et al.[129] demonstrated that photochemical formation of carbazole was a singlet nitrene process. In contrast, triplet sensitized photolysis of *o*-azidobiphenyl produces the corresponding azo compound. Direct photolysis in the presence of diethylamine produces an azepine in addition to carbazole, and the latter product is formed even in neat amine.[130]

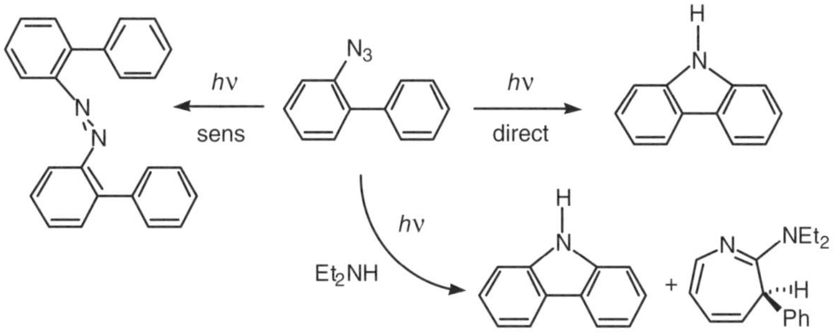

Triplet *o*-biphenylnitrene has been observed as a persistent species at 77 K by UV–vis[131] and EPR spectroscopy.[39] Tsao has detected singlet *o*-biphenylnitrene by LFP at 77 K ($\lambda_{max} = 410$ nm, $\tau = 59$ ns), which decays cleanly to the triplet nitrene at this temperature.[123]

The photochemistry of *o*-azidobiphenyl has been studied by conventional[132] and by LFP methods with both UV–vis and IR detection of the intermediates.[133]

Thereby, it was found that the formation of carbazole proceeds on more than a single time scale. The following mechanistic picture accounts for many observations.[133]

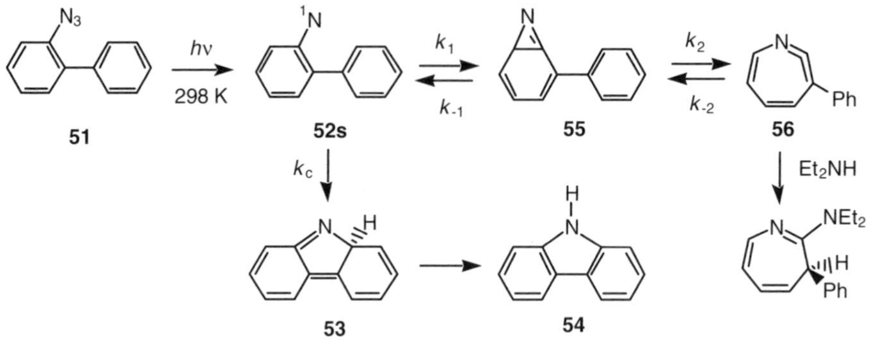

Singlet *o*-biphenylnitrene **52s** cleanly relaxes to the lower energy triplet state of **52t** at 77 K, but at ambient temperature it likely partitions to form a mixture of azirine **55** and isocarbazole **53** within a few nanoseconds of a laser pulse. Some azirine **55** expands to ketenimine **56**, which can be trapped with diethylamine.

In the absence of amine, the ketenimine–azirine singlet nitrene species can equilibrate and, eventually, the singlet nitrene can cyclize to form carbazole. Berry and co-workers[132] independently monitored the growth of carbazole ($\lambda_{max} = 289.4$ nm) by this process. In cyclohexane, some carbazole was formed this way with an observed rate constant of 2.2×10^3 s^{-1} at 300 K over a barrier of 11.5 kcal/mol. Tsao and co-workers[133] recently used TRIR spectroscopy to show that ketenimine decay equals the rate of carbazole formation.

Some of the initially formal singlet nitrene **52s** can also cyclize to isocarbazole **53**. Isocarbazole **53** absorbs broadly in the visible ($\lambda_{max} \approx 430$ nm) and decays to carbazole by a symmetry allowed 1,5-hydrogen shift with a time constant of 70 ns.[133]

7. POLYNITRENES

Interest in the preparation of high-spin organic compounds has led to the matrix isolation of polynitrenes, such as **57–61**.[134] These species have been studied primarily by EPR spectroscopy, but increasing use is being made of matrix IR and UV–vis spectroscopy. Density functional theoretical calculations have been used to assign the vibrational spectra that have been observed. Polynitrenes are under active study by material scientists interested in the development of organic magnets.

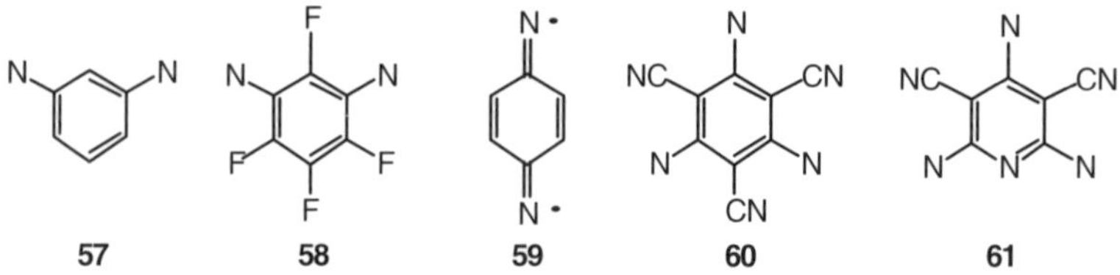

Mixed-carbene–nitrene compounds have also been prepared by Tomioka and co-workers (X = H, halogen).[135]

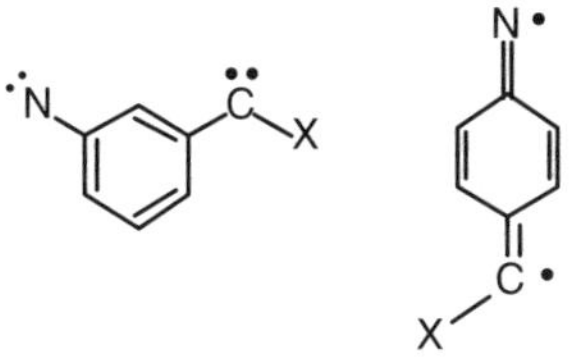

8. AMINO- AND OXONITRENES

In 1984, Sylwester and Dervan found that photolysis of carbamoyl azide **62** in a glass of 2-methyltetrahydrofuran (2MTHF) at 80 K led to the formation of a violet species showing a highly structured absorption band between 500 and 700 nm (vibrational progression of 1300 cm^{-1}). The violet photoproduct is neither trans ($\lambda_{max} = 386$ nm) nor *cis*-diimide ($\lambda_{max} = 260$ nm), and was assigned to 1,1-diazene **63**.[136] Ab initio calculations predict H_2NN to have a singlet ground state with a short ($r = 1.25$ Å) N–N bond, indicating that this species is best regarded as a 1,1-diazene rather than an aminonitrene. The generalized valence bond method indicates that the lowest triplet state is pyramidal and has less N–N double-bond character ($r = 1.37$ Å) and that $H_2N=N$ has an excited singlet state S_1 at 2.2 eV or 566 nm.[137]

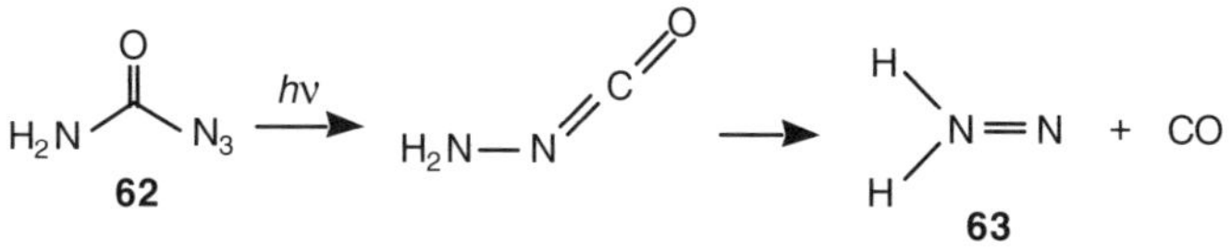

Sylwester and Dervan's assignment was supported by the observation that the violet photoproduct decomposes to N_2 and H_2 upon warming the glass or upon further irradiation of the glass. Furthermore, photolysis of **62** in argon at 10 K leads to the disappearance of the IR spectrum of the azide and the appearance of new bands at 2865, 2808, 2141, 1863, 1574, and 1003 cm^{-1}. The 2141-cm^{-1} band is due to the formation of carbon monoxide. The band at 1574 is assigned to the $^{14}N=^{14}N$ stretch of 1,1-diazene. Photolysis of ^{15}N labeled **62** led to a new vibration at 1548 cm^{-1} attributed to the $^{14}N=^{15}N$ stretch in good agreement with Hooke's law.

Davis and Goddard[137] calculated that the heat of formation (298 K) of singlet H_2NN is 90.1 kcal/mol. This value is only 14 kcal/mol below that of molecular nitrogen and two hydrogen atoms, and is 54 kcal/mol above $\Delta H_f(298$ K) for *trans*-diimide determined experimentally by Willis et al.[138] The relative energies of 1,1- and *trans*-1,2-diazene have been calculated by several methods; these studies have been reviewed by Parsons and Dykstra.[139] It was found using the self-consistent electron pairs method with the TZ + 2p basis set spanning a space up to 176,000 configurations that 1,1-diazene **63** is 24.5 kcal/mol higher in energy than *trans*-1,2-diazene, but that **63** is protected from decay to *trans*-diimide by a barrier of 58.1 kcal/mol. Kemper and Buck[140] also studied this rearrangement and concluded that bimolecular hydrogen exchange is much easier than a unimolecular 1,2-hydrogen shift. These authors found many similarities between diimide chemistry and formaldehyde photochemistry. The chemistry of aminonitrenes has been reviewed by Lemal[141] and by Ioffe and Kuznetsov.[142]

8.1. Stable Aminonitrenes

The Dervan group discovered that addition of *tert*-butyl hypochlorite to a solution of 1-amino-2,2,6,6-tetramethylpiperidine and triethylamine in ether at −78 °C produces triethylammonium chloride as an insoluble white precipitate and an intense purple solution. The purple coloration, which is stable for hours at −78 °C, fades in minutes at 0 °C and was assigned to the 1,1-diazene **64**.[143,144]

64   **65**

Examination of the optical spectrum of the filtered purple solution gave a structured absorption band with maxima at 514 and 543 nm. This position is remarkably close (566 nm) to the $n-\pi^*$ electronic transition predicted by Davis and Goddard[137] for the parent system $H_2N=N$. As expected for an $n-\pi^*$ transition, the position of the absorption maximum is solvent dependent. In dichloromethane solution, λ_{max} is 541 nm, in 2-propanol it is 526 nm. The blue shift of 15 nm is completely consistent with the $n-\pi^*$ absorptions of isoelectronic carbonyl compounds.

The identity of **64** was established by IR spectroscopy. At −78 °C, the IR spectrum has a strong band at 1595 cm^{-1} that disappears on warming to 25 °C. The stretching frequency is similar to that of trans-azo compounds (1576 cm^{-1}), which can, however, only be observed by Raman spectroscopy. Upon ^{15}N labeling of the amino group in the piperidine precursor, the band of the photoproduct shifts to 1569 cm^{-1}, in agreement with a Hooke's law calculation for **64**.

Schultz, Dervan, and co-workers[145] subsequently reported the synthesis and characterization of the five-membered ring analogue **65**. As before, the oxidation of the appropriate hydrazine at −78 °C gave a clear, colored solution of the aminonitrene. This 1,1-diazene absorbs at 497 nm (in CH_2Cl_2) and 487 nm (in 2-propanol). The IR spectrum shows a strong absorption at 1638 cm^{-1} that disappears on warming to 25 °C. The ^{15}N isotopomer has a vibration at 1612 cm^{-1}, a shift of 26 cm^{-1}.

Both aminonitrenes **64** and **65** were also studied by low-temperature proton nuclear magnetic resonance (^1H NMR) spectroscopy.[145] The spectrum of the chromatographed six-membered ring diazene **64** at −78 °C showed signals due to a tetrazene in addition to absorptions at δ 1.15 and 2.15. Warming the samples resulted in the disappearance of the diazene bands with the concomitant increase of the resonances due to the tetrazene. The same experiment with the five-membered ring compound **65** gave resonances at δ 1.05 and 2.32 at −78 °C. That one can observe an NMR spectrum is convincing evidence that the singlet is the ground state. The NMR of a triplet nitrene most likely can not be observed because of very rapid spin relaxation.

8.2. Oxonitrenes

Oxygen substituted nitrenes have received much less study than aminonitrenes, but there have been some noteworthy recent developments. Toscano and co-workers[146] studied the photochemistry of diazenium diolates. The quantum yield of photodegradation was 0.10.

The presence of oxygen significantly altered the distribution of products formed. A major new product was the nitrate ester. Under argon, the yield of benzaldoxime (R = phenyl) was 66%, but in the presence of oxygen, the yield of this product goes to zero. The data indicated the following mechanism involving benzyloxynitrene:

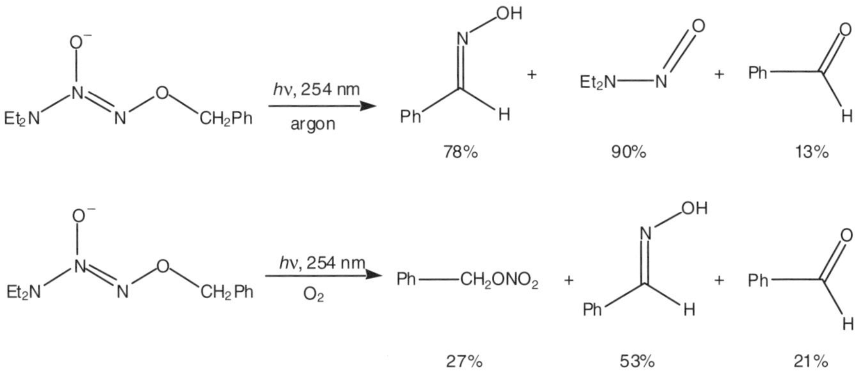

The putative benzyloxynitrene can be intercepted with tetramethylethylene to form the expected aziridine. Time-resolved IR spectroscopy was unable to detect the O-nitrene, but detected the presence of $PhCH_2N{=}O$ formed with a time constant of 250 ns after the laser pulse. Thus, the lifetime of benzyloxynitrene is also 250 ns. The TRIR spectroscopic studies indicated that benzyloxynitrene reacts with oxygen a rate constant of 10^{9-10} $M^{-1}s^{-1}$. This value strongly suggests that the O-nitrene, in contrast to the N-nitrenes has a triplet ground state.

9. PHOSPHINIDINES

The ground state of PH has triplet multiplicity, as does imidogen, but the singlet–triplet splitting of the phosphorus compound is only 22 kcal/mol,[147] down 14 kcal/mol from the nitrogen analogue.[11,12] The valence electrons of PH reside in $3n$ levels, which are more diffuse than the $2n$ level orbitals in NH. Thus, the electron–electron repulsion in singlet–PH is reduced, and this lowers the energy of singlet–PH relative to triplet–PH.

The open- and closed-shell configurations of singlet–NH and PH are components of a doubly degenerate state.[10] As discussed in Sections 7.4 and 7.5, a vinyl group lowers the symmetry and removes the degeneracy[67] and the open- and closed-shell singlet states of vinylnitrene lie 15 and 40 kcal/mol, respectively, above the ground triplet state.[67] The open-shell singlet resembles a 1,3-biradical. The delocalization of one of the unpaired electrons reduces the electron–electron repulsion and stabilizes this configuration. The closed-shell configuration of singlet vinylnitrene is zwitterionic and does not enjoy decreased electron–electron repulsion.

In vinylphosphinidine, the open- and closed-shell singlets lie 17 and 23 kcal/mol above the ground triplet state. The energies of the two singlet states of vinylphosphinidine are comparable because electron–electron repulsion in the $3p$ orbitals is reduced and because the overlap of $2p$ and $3p$ orbitals of carbon and phosphorus is poor and the resulting carbon–phosphorus double bond is relatively weak.[67]

Mathey[148] discovered that flash vacuum pyrolysis of vinylphosphirane **66** at 700 °C produced phosphapropyne **68**, presumably by rearrangement of vinylphosphinidine **67**.

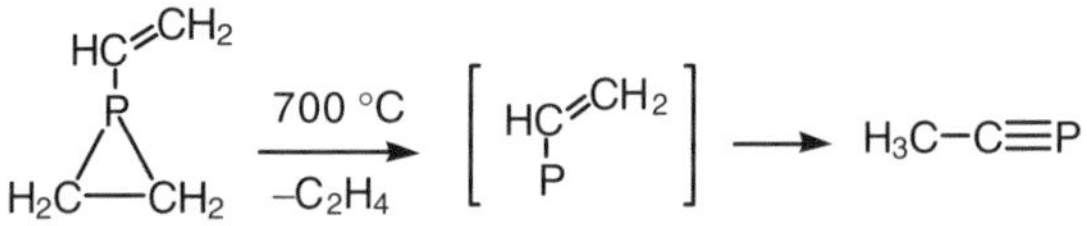

This rearrangement is unusual because hydrogen usually migrates toward electron-deficient centers as calculated for vinylcarbene.

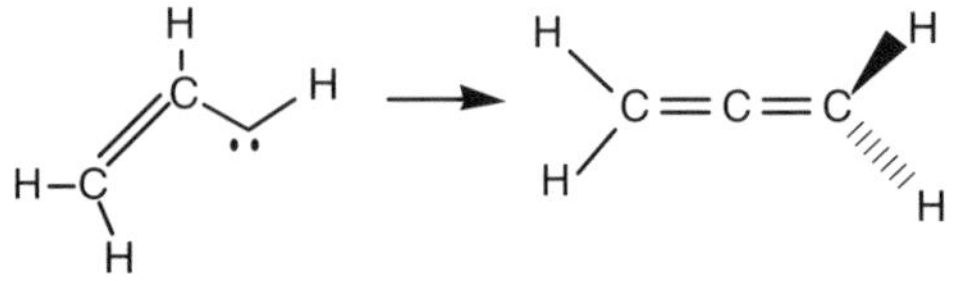

This observation led Berger et al.[149] to investigate the C_2H_3P singlet surface by computational methods. Cyclization of vinylphosphinidine to a three-membered ring species and isomerization to a phosphaallene are the kinetic products of rearrangement. The global minimum on the surface is the phosphapropyne. The barrier to phosphapropyne formation is 6.9 kcal/mol greater than the barrier to isomerization of the vinylphosphinidine to the phosphapropyne, which is $\sim$20 kcal/mol.

In phenylnitrene, the open-shell singlet lies below the closed-shell singlet, but the opposite trend is observed in singlet phenylphosphinidine. In phenylphosphinidine, the closed-shell singlet configuration has been found by two groups to be 4 kcal/mol lower in energy than the open-shell singlet state.[150] It is predicted that the electronic structure of phenylphosphinidine is more reminiscent of singlet phenylcarbene than is singlet phenylnitrene! However, the most recent calculations of Galbraith et al.[151] place the open-shell singlet of phenylphosphinidine slightly below the closed-shell singlet.

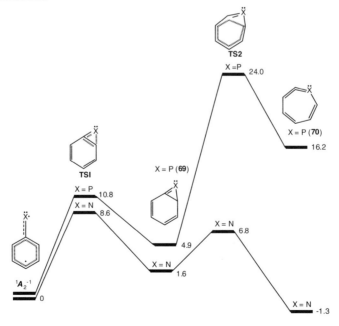

The latter group also studied the rearrangements of open-shell singlet phenyl- and vinylphosphinidine at a common theoretical level – (8/8)CASPT2. The surfaces for X = P and N are quite different. Ring closure of X = P is slightly more endothermic and less rapid than that of phenylnitrene. The major difference is that whereas ring opening of benzazirine to azacycloheptatetraene is both fast and exothermic, the analogous process for X = P is very endothermic and very slow. Theory predicts that singlet phosphinidine may reversibly cyclize to **69**. This species should be trappable and detectable, but it will not expand to the seven-membered ring species **70**. The difference in these systems can be attributed to the difference in the strengths of the σ bonds formed to nitrogen and phosphorus in these reactions.

There is chemical evidence for the formation of phenylphosphinidine in solution.[152]

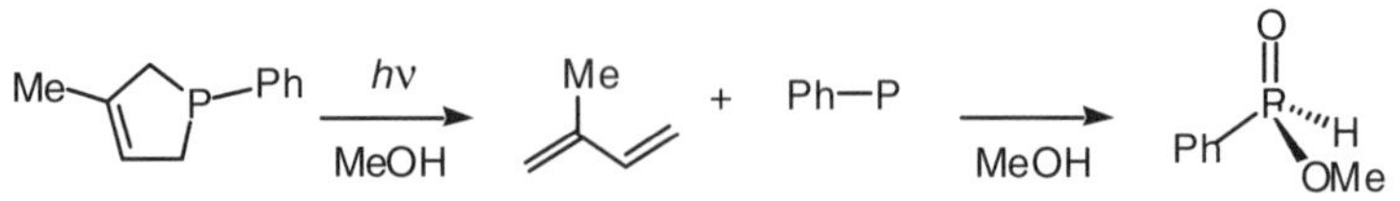

Hammond et al.[153] reported data that indicate that phenylphosphinidine is relatively long lived and likely has a triplet ground state.

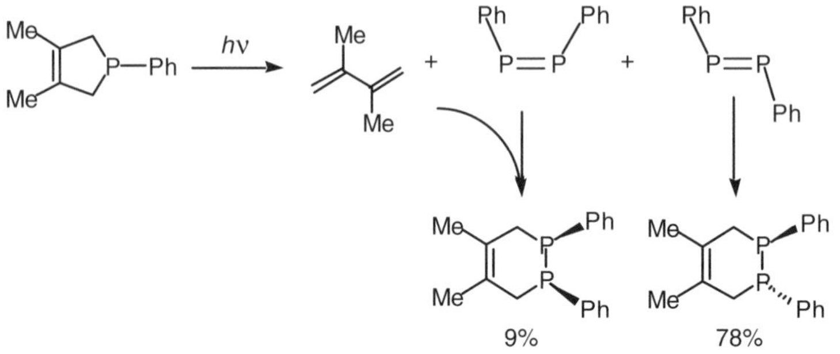

Phospacyclopropanes appear to be excellent precursors but Mathey[148] has pointed out that one can write reasonable mechanisms to the formation of these products that do not involve the intermediacy of arylphosphinidines.

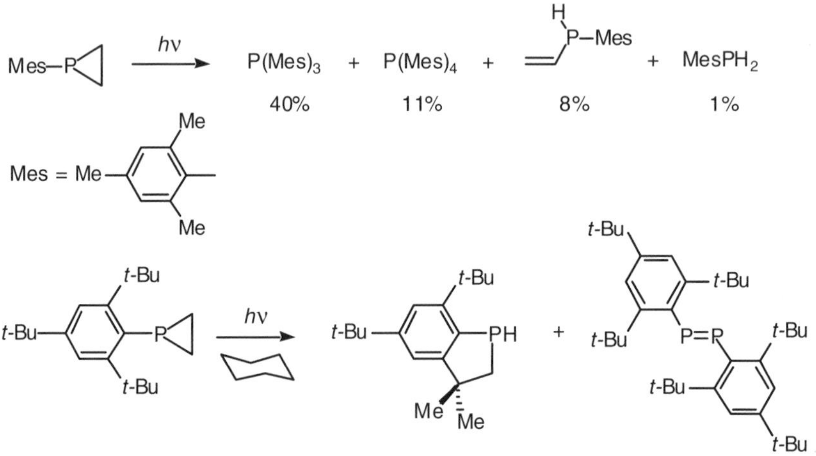

Spectacular proof of the existence of arylphosphinidines has been provided by Gaspar and co-workers[154] and Weissman et al.[155] A frozen solution of **71** in methylcyclohexane glass was exposed to 254-nm radiation. The glass turned yellow upon photolysis. Thawing the glass led to the formation of phosphine **72** in the absence of trap, and to the formation of **73** in the presence of 3-hexyne. When the frozen, previously irradiated sample was placed in an EPR spectrometer a triplet signal at

11,492 G was observed with $|D/hc| = 3.521$ cm^{-1}. This value is much larger than the $|D/hc|$ value of phenylnitrene of ≈ 1 cm^{-1}. This difference is most likely the result of a heavy atom effect of phosphorus, relative to nitrogen, on the second-order spin–orbit contribution.[155]

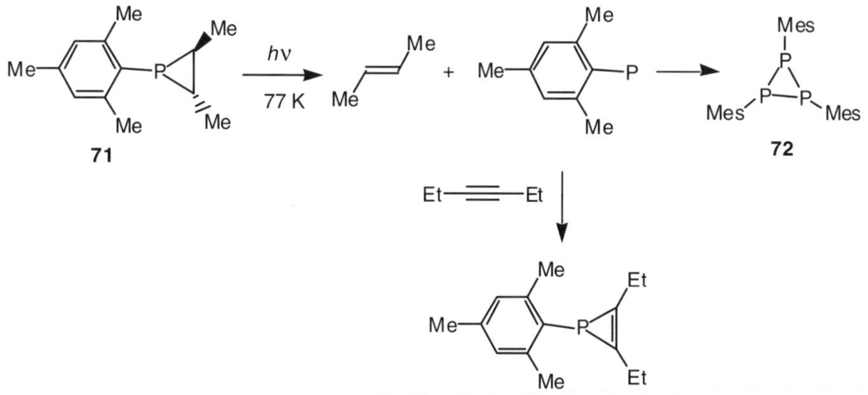

Recently, Protasiewicz and co-workers[156] reported a new precursor of arylphosphinidines.

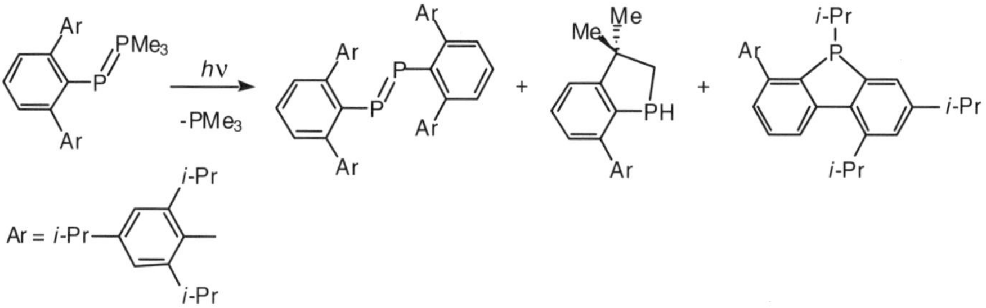

10. CONCLUSION AND OUTLOOK

In recent years, LFP studies of nitrenes by nanosecond time-resolved UV–vis spectroscopy, working closely with theoretical calculations, have provided a comprehensive understanding of the chemistry, kinetics, and spectroscopy of nitrenes. The nuances of arylnitrenes are now very well understood and acylnitrenes are becoming well understood. It can be safely predicted that in the coming years time-resolved IR spectroscopy of acylnitrenes, benzazirines and ketenimines, and matrix spectroscopy of these species will, in concert with theory, provide a deeper appreciation of the properties of these intermediates. There is no doubt that these studies will assist synthetic organic chemists, biochemists performing photoaffinity labeling studies, and materials scientists seeking to functionalize polymer surfaces.

Important unresolved questions remain to be answered. For example, are singlet alkyl- and vinylnitrenes true intermediates with finite lifetimes or do they correspond to nonstationary points on the potential energy surface? Photolysis and pyrolysis of azides occasionally leads to very different distributions of stable products. Is the inability to trap singlet nitrenes upon photolysis of certain alkyl azides (in contrast to their pyrolysis) due to a conical intersection in the azide excited state that leads directly to product?[157,158a] Is the failure to trap an acylnitrene upon pyrolysis of pivaloyl azide (in contrast to its photolysis) due to a dynamical effect, as discussed in Chapter 21 in this volume?[158b] Azides absorb strongly in the UV and typically very weakly in the visible region. Exposure of an azide to UV light provides this precursor with ~ 100 kcal/mol of energy, far more energy than is necessary to break the $N=N$ bond of ~ 35 kcal/mol.[31,32] This difference leads one to wonder whether electronically or vibrationally excited nitrenes are produced upon UV photolysis and whether some chemistry long attributed to relaxed singlet nitrene species may actually originate from excited states. Hence, it seems safe to predict that future research will make good use of femtosecond spectroscopy. Thus, more work remains to be done and the next 10 years of research in nitrene chemistry should be just as exciting as the recent past.

ACKNOWLEDGMENTS

The author is pleased to acknowledge the outstanding contributions of graduate and postdoctoral students too numerous to mention and senior collaborators such as Nina Gritsan, Christpher Hadad, Thomas Bally, Jakob Wirz, and Wes Borden. Thanks go to Professor Bally, Professor Borden, Professor Cramer, Professor Smith, and Professor Sundberg, and to Eric Tippmann and Meng-Lin Tsao for a critical reading of the manuscript. The generous support of research in the Platz laboratory by the National Science Foundation, the National Institutes of Health, and the Soros Foundation are gratefully acknowledged.

SUGGESTED READING

W. Lwowski, Ed., *Nitrenes*, John Wiley & Sons, Inc., New York, **1970**.

E. F. V. Scriven, Ed., *Azides and Nitrenes*, Academic Press, New York, **1984**.

C. Wentrup, *Reactive Molecules*, John Wiley & Sons, Inc., New York, **1984**.

G. B. Schuster and M. S. Platz, *Adv. In Photochem.* **1992**, *17*, 69.

N. P. Gritsan and M. S. Platz, *Adv. Phys. Org. Chem.* **2001**, *36*, 255.

REFERENCES

1. W. Lwowski, Ed., *Nitrenes*, John Wiley & Sons, Inc., New York, **1920**.

2. W. Lwowski, *Nitrenes*, W. Lwowski, Ed., John Wiley & Sons, Inc., New York, **1920**, p. 1.

3. X. Tiemann, *Bereche*, **1891**, *24*, 4162.

4. M. B. Smith and J. March, *Advanced Organic Chemistry*, fifth ed., John Wiley & Sons, Inc., New York, **2001**.

5. (a) H. Bayley, *Photogenerated Reagents in Biochemistry and Molecular Biology*, Elsevier, Amsterdam, the Netherlands, **1983**. (b) S. A. Fleming, *Tetrahedron* **1995**, *51*, 12479.

6. A. Singh, E. R. Thornton, and F. H. Westheimer, *J. Biol. Chem.* **1962**, *237*, 3006.

7. H. Bayley and J. R. Knowles, *Methods Enzymol.* **1977**, *46*, 69.

8. (a) J. F. W. Keana and S. X. Cai, *J. Fluorine Chem.* **1989**, *43*, 151. (b) J. F. W. Keana and S. X. Cai, *J. Org. Chem.* **1990**, *55*, 2034. (c) S. X. Cai, and J. F. W. Keana, *Bioconjugate Chem.* **1991**, *2*, 28. (d) S. X. Cai, D. R. Glenn, and J. F. W. Keana, *J. Org. Chem.* **1992**, *57*, 1299.

9. D. Breslow, in *Azides and Nitrenes* E. F. V. Scriven, Ed., Academic Press, New York, **1984**, p. 491.

10. W. L. Karney and W. T. Borden, *Advanced Carbene Chemistry* Vol. 3, Elsevier, New York, **2001** p. 205.

11. P. W. Fairchild, G. P. Smith, D. R. Crosly, and J. B. Jeffries, *Chem. Phys. Lett.* **1984**, *107*, 181.

12. P. C. Engelking and W. C. Lineberger, *J. Chem. Phys.* **1976**, *65*, 4323.

13. D. G. Leopold, K. K. Murray, A. E. S. Miller, and W. C. Lineberger, *J. Chem. Phys.* **1985**, *83*, 4849.

14. R. J. Gillespie, *J. Am. Chem. Soc.* **1960**, *82*, 5978.

15. I. Shavitt, *Tetrahedron* **1985**, *41*, 1531.

16. M. S. Platz, *Acc. of Chem. Res.* **1995**, *28*, 487.

17. W. Hack, in *N-Centered Radicals* Z. B. Alfassi, Ed., John Wiley & Sons, Inc., New York, **1998**, p. 413.

18. W. Kirmse, *Carbene Chemistry*, Academic Press, New York, **1971**.

19. W. J. Baron, M. R. DeCamp, M. E. Hendrick, M., Jr., Jones, R. H. Levin, and M. B. Sohn, in *Carbenes*, Vol. I, M., Jr., Jones and R. A. Moss, Eds., John Wiley & Sons, Inc., New York, **1973**, p. 1.

20. X. Banthorpe, *Chemistry of the Azide Group*, S. Patai, Ed., John Wiley & Sons, Inc., New York, **1971**, p. 405.

21. J. M. Eder, *Montash. Chem.* **1892**, *12*, 86.

22. D. R. Yarkony, H. F. Schaefer, III, and S. Rothernberg, *J. Am. Chem. Soc.* **1974**, *96*, 5974.

23. J. E. Jackson and M. S. Platz, in *Advanced Carbene Chemistry 1*, Vol. 1, U. Brinker, Ed., **1994**, JAI Press, Inc., Greenwich, CN, p. 89.

24. R. A. Moss and N. J. Turro, *Kinetics and Spectroscopy of Carbenes and Biradicals*, M. S. Platz, Ed., Plenum, New York, **1990**, p. 213.

25. S. L. Laursen, J. E. Grace, Jr., R. L. Dekock, and S. A. Spronk, *J. Am. Chem. Soc.* **1998**, *120*, 12583.

26. P. H. H. Fischer, S. W. Charles, and C. A. McDowell, *Can. J. Chem. Phys.* **1967**, *46*, 2162.

27. M. J. Travers, D. C. Cowles, E. P. Clifford, G. B. Ellison, and P. C. Engelking, *J. Phys. Chem.* **1999**, *111*, 5349.

28. C. R. Kemnitz, G. B. Ellison, W. L. Karney, and W. T. Borden, *J. Am. Chem. Soc.* **2000**, *122*, 1098.

29. (a) D. R. Yarkony, H. F. Schaefer, III, and S. Rothenberg, *J. Am. Chem. Soc.* **1974**, *96*, 5974. (b) J. Demuyanck, D. J. Fox, Y. Yamaguchi, and H. F. Schaefer, III, *J. Am. Chem. Soc.* **1980**, *102*, 6204. (c) J. A. Pople, K. Raghavachari, M. J. Frisch, J. S. Binckley, and P. v. R. Schleyer, *J. Am. Chem. Soc.* **1983**, *105*, 6389. (d) M. T. Nguyen, *Chem Phys. Lett.* **1985**, *117*, 290. (e) C. Richards, Jr., C. Meredith, S.-J. Kim, G. E. Quelch, and H. F. Schaefer, III, *J. Chem. Phys.* **1994**, *100*, 481. (f) M. T. Nguyen, D. Sengupta, and T.-K. Ha, *J. Phys. Chem.* **1996**, *100*, 6499.

30. J. H. Glowina, J. Misewich, and P. P. Serokin, in *Supercontinuum Laser Sources*, R. R. Alfono, Ed., Springer Verlag, New York, **1989**, p. 337.

31. E. P. Kyba, in *Azides and Nitrenes*, E. F. V. Scriven, Ed., Academic Press, New York, **1984**, p. 2.

32. F. D. Lewis and W. H. Saunders, Jr., in *Nitrenes* W. Lwowski, Ed., John Wiley & Sons, Inc., New York, **1970**, p. 47.

33. W. Pritzkow and D. Timm, *J. Prakt. Chem.* **1966**, *32*, 178.

34. D. W. Milligan, *J. Chem. Phys.* **1961**, *35*, 149.

35. W. Pritzkow and D. Timm, *J. Prakt. Chem.* **1966**, *32*, 178.

36. I. R. Dunkin and P. C. P. Thomson, *Tetrahedron Lett.* **1980**, *21*, 3813.

37. J. G. Radziszewski, J. W. Downing, C. Wentrup, P. Kaszynski, M. Jawdosiuk, P. Kovacic, and J. Michl, *J. Am. Chem. Soc.* **1985**, *107*, 2799.

38. L. Barash, E. Wasserman, and W. A. Jager, *J. Chem. Phys.* **1967**, *89*, 3931. For a correction, see R. F. Ferrante, *J. Chem. Phys.*, **1987**, *86*, 25.

39. E. Wasserman, *Prog. Phys. Org. Chem.*, **1971**, *8*, 319.

40. E. P. Kyba and R. A. Abramovitch, *J. Am. Chem. Soc.* **1980**, *102*, 735.

41. F. D. Lewis and W. H. Saunders, *J. Am. Chem. Soc.* **1968**, *90*, 3828; 7031; 7033.

42. S. M. Mandel, J. A. K. Bauer, and A. Gudmundsdottir, *Org. Lett.* **2001**, *3*, 523.

43. N. P. Gritsan, I. Likhotvorik, Z. Zhu, and M. S. Platz, *J. Phys. Chem. A* **2001**, *105*, 3039.

44. P. G. Carrick and P. C. Engelking, *J. Chem. Phys.* **1984**, *81*, 1661.

45. R. E. Banks, D. Berry, M. J. McGlinchey, and G. J. Moore, *J. Chem. Soc. (C).* **1970**, 1017.

46. E. S. Wallis, *Org. Reactions* **1946**, *3*, 267.

47. E. Bauer, *Angew. Chem. Int. Ed., Engl.* **1974**, *13*, 376.

48. W. Lwowski and G. T. Tisue, *J. Am. Chem. Soc.* **1965**, *82*, 4022.

49. (a) G. T. Tisue, S. Linke, and W. Lwowski, *J. Am. Chem. Soc.* **1967**, *89*, 6303. (b) S. Linke, G. T. Tisue, and W. Lwowski, *J. Am. Chem. Soc.* **1967**, *89*, 6308.

50. H. Huang and M. S. Platz, *J. Am. Chem. Soc.* **1998**, *120*, 5990.

51. T. Autrey and G. B. Schuster, *J. Am. Chem. Soc.* **1987**, *109*, 5814.

52. M. E. Sigman, T. Autrey, and G. B. Schuster, *J. Am. Chem. Soc.* **1988**, *110*, 4297.

53. N. P. Gritsan and E. A. Pritchina, *Mendeleev Comm.* **2001**, 94.

54. E. A. Pritchina, N. P. Gritsan, A. Maltsev, T. Bally, T. Autrey, Y. Liu, Y. Wang, and J. P. Toscano, *Phys. Chem. Chem. Phys.* **2003**, *5*, 1010.

55. (a) W. Lwowski and T. W. Mattingly, Jr., *J. Am. Chem. Soc.* **1965**, *87*, 1947. (b) W. Lwowski and F. P. Woerner, *J. Am. Chem. Soc.* **1965**, *87*, 5491.

56. J. S. McConaghy and W. Lwowski, *J. Am. Chem. Soc.* **1965**, *87*, 5490.

57. M. Jones, Jr., K. R. Rettig, *J. Am. Chem. Soc.* **1965**, *87*, 4013, 4015.

58. C. Buron and M. S. Platz, *Org. Lett.*, **2003**, *5*, 3383.

59. (a) E. V. Sitzmann, J. Langan, and K. B. Eisenthal, *J. Am. Chem. Soc.* **1984**, *106*, 1868. (b) P. B. Grasse, B.-E. Brauer, J. J. Zupancic, K. J. Kaufmann, and G. B. Schuster, *J. Am. Chem. Soc.* **1983**, *105*, 6833.

60. J. P. Toscano, *Adv. Photochem.* **2001**, *26*, 41.

61. W. Lwowski, in *Azides and Nitrenes*, E. F. V. Scriven, Ed., Academic Press, New York, **1984**, p. 205.

62. (a) T. Shingaki, M. Inagaki, N. Torimoto, and M. Takebayashi, *Chem. Lett.* **1972**, 1181. (b) D. S. Breslow, E. I. Edwards, E. C. Linsay, and H. Omura, *J. Am. Chem. Soc.* **1976**, *98*, 4268.

63. (a) G. Smolinsky, E. Wasserman, and W. A. Yager, *J. Am. Chem. Soc.* **1962**, 84, 3320. (b) R. M. Moriarty, M. Rahman, and G. King, *J. Am. Chem. Soc.* **1966**, 88, 842.

64. J.-C. Garay, V. Maloney, M. Marlow, and P. Small, *J. Phys. Chem.* **1996**, *100*, 5788.

65. (a) R. Breslow, A. Feiring, and F. Herman, *J. Am. Chem. Soc.* **1974**, *96*, 5937. (b) P. Maslak, *J. Am. Chem. Soc.* **1989**, *111*, 820.

66. M. Houser, S. Kelley, V. Maloney, M. Marlow, K. Steininger, and H. Zhou, *J. Phys. Chem.* **1995**, *99*, 7946.

67. V. Parasuk and C. J. Cramer, *Chem. Phys. Lett.* **1996**, *260*, 7.

68. W. L. Karney and W. T. Borden, *J. Am. Chem. Soc.* **1997**, *119*, 1378.

69. A. Hassner, in *Azides and Nitrenes*, E. F. V. Scriven, Ed., Academic Press, New York, **1984**, p. 205.

70. (a) T. A. Baer and C. D. Gutsche, *J. Am. Chem. Soc.* **1971**, *93*, 5180. (b) R. A. Moss and U.-H. Dolling, *J. Am. Chem. Soc.* **1971**, *93*, 954. (c) A. Admasu, A. D. Gudmundsdottir, and M. S. Platz, *J. Phys. Chem. A* **1997**, *101*, 3832.

71. (a) S. Matzinger, T. Bally, E. V. Patterson, and R. J. McMahon, *J. Am. Chem. Soc.* **1996**, *118*, 1535. (b) M. W. Wong and C. Wentrup, *J. Org. Chem.* **1996**, *61*, 7022. (c) P. R. Schreiner, W. L. Karney, P. v. R. Schleyer, W. T. Borden, T. P. Hamilton, and H. F. Schaefer, III, *J. Org. Chem.* **1996**, *61*, 7030.

72. P. A. S. Smith, in *Azides and Nitrenes*, E. F. V. Scriven, Ed., Academic Press, New York, **1984**, p. 8.

73. (a) P. A. S. Smith and B. B. Brown, *J. Am. Chem. Soc.* **1951**, *73*, 2438. (b) P. A. S. Smith and B. B. Brown, *J. Am. Chem. Soc.* **1951**, *73*, 2435. (c) P. A. S. Smith and J. H. Hall, *J. Am. Chem. Soc.* **1962**, *84*, 1632.

74. (a) R. A. Abramovitch, S. R. Challand, and E. F. V. Scriven, *J. Am. Chem. Soc.* **1972**, *94*, 1374. (b) R. A. Abramovitch, S. R. Challand, and E. F. V. Scriven, *J. Org. Chem.* **1975**, *40*, 1541.

75. (a) R. E. Banks and G. R. Sparkes, *J. Chem. Soc., Perkin Trans. 1* **1972**, 1964. (b) R. E. Banks and N. D. Venayak, *J. Chem. Soc., Chem. Commun.* **1980**, 900. (c) R. E. Banks and A. Prakash, *Tetrahedron Lett.* **1973**, 99. (d) R. E. Banks and A. Prakash, *J. Chem. Soc., Perkin Trans. 1* **1974**, 1365. (e) R. E. Banks and I. M. Madany, *J. Fluorine Chem.* **1985**, *30*, 211.

76. P. Greiss, *Proc. R. Soc. (London)* **1863–1864**, *13*, 375.

77. L. Wolf, *Ann.* **1912**, *394*, 59.

78. (a) R. Huisgen, D. Vossius, and M. Appl, *Chem. Ber.* **1958**, *91*, 1. (b) R. Huisgen and M. Appl, *Chem. Ber.* **1958**, *91*, 12.

79. W. Doering and R. A. Odum, *Tetrahedron* **1966**, *22*, 81.

80. S. E. Carroll, B. Nay, E. F. V. Scriven, H. Suschitzky, and D. R. Thomas, *Tetrahedron Lett.* **1977**, 3175.

81. (a) W. D. Crow and C. Wentrup, *Tetrahedron Lett.* **1968**, 6149. (b) C. Wentrup, *J. Chem. Soc., Chem. Commun.* **1969**, 1386 (c) M. Kuzaj, H. Lüerssen, and C. Wentrup, *Angew. Chem. Int. Ed., Engl.* **1986**, *25*, 480. (d) C. Wentrup, *Topics Curr. Chem.* **1976**, *62*, 173.

82. (a) A. Reiser, G. Bowes, and R. J. Horne, *Trans. Faraday Soc.* **1966**, *62*, 3162. (b) A. Reiser and U. Frazer, *Nature (London)* **1965**, *208*, 682.

83. O. L. Chapman and J. P. LeRoux, *J. Am. Chem. Soc.* **1978**, *100*, 282.

84. K. Ozawa, T. Ishida, K. Fuke, and K. Kaya, *Chem. Phys, Lett.* **1988**, *150*, 249.

85. G. Porter and B. Ward, *Proc. R. Soc., London, A* **1968**, *303*, 139.

86. E. Leyva, M. S. Platz, G. Persy, and J. Wirz, *J. Am. Chem. Soc.* **1986**, *108*, 3783.

87. E. W. Meijer, S. Nijhuis, and F. C. B. M. van Vroonhoven, *J. Am. Chem. Soc.* **1988**, *110*, 7209.

88. (a) N. P. Gritsan and E. A. Pritchina, *J. Inf. Rec. Mat.* **1989**, *17*, 391. (b) N. P. Gritsan and E. A. Pritchina, *Russ. Chem. Rev.* **1992**, *61*, 500.

89. J. S. Hayes and R. S. Sheridan, *J. Am. Chem. Soc.* **1990**, *112*, 5879.

90. (a) D. W. Cullin, N. Soundarajan, M. S. Platz, and T. A. Miller, *J. Phys. Chem.* **1990**, *94*, 8890. (b) D. W. Cullin, N. Soundarajan, M. S. Platz, and T. A. Miller, *J. Phys. Chem.* **1990**, *94*, 3387.

91. D. A. Hrovat, E. E. Waali, and W. T. Borden, *J. Am. Chem. Soc.* **1992**, *114*, 8698. See also O. Castell, V. M. García, C. Bo, and R. Caballolo, *J. Comput. Chem.* **1996**, *17*, 42.

92. S.-J. Kim, T. P. Hamilton, and H. F. Schaefer, *J. Am. Chem. Soc.* **1992**, *114*, 5349.

93. (a) M. J. Travers, D. C. Cowles, E. P. Clifford, and G. B. Ellison, *J. Am. Chem. Soc.* **1992**, *114*, 8699. (b) R. N. McDonald and S. J. Davidson, *J. Am. Chem. Soc.* **1993**, *115*, 10857.

94. N. P. Gritsan, Z. Zhu, C. M. Hadad, and M. S. Platz, *J. Am. Chem. Soc.* **1999**, *121*, 1202.

95. (a) G. G. Vander Stouw, Ph.D. Thesis, The Ohio State University, 1964. (b) R. C. Joines, A. B. Turner, and W. M. Jones, *J. Am. Chem. Soc.* **1969**, *91*, 7754. (c) P. P. Gaspar, J. P. Hsu, S. Chari, and M. Jones, Jr., *Tetrahedron* **1985**, *41*, 1479. (d) P. C. Schissel, M. E. Kent, D. J. McAdoo, and E. Hedaya, *J. Am. Chem. Soc.* **1970**, *92*, 2147.

96. J. Morawietz and W. Sander, *J. Org. Chem.* **1996**, *61*, 4351.

97. B. A. DeGraff, D. W. Gillespie, and R. J. Sundberg, *J. Am. Chem. Soc.* **1973**, *95*, 7491.

98. A. K. Schrock and G. B. Schuster, *J. Am. Chem. Soc.* **1984**, *106*, 5228.

99. (a) Y.-Z. Li, J. P. Kirby, M. W. George, M. Poliakoff, and G. B. Schuster, *J. Am. Chem. Soc.* **1988**, *110*, 8092. (b) C. J. Shields, D. R. Chrisope, G. B. Schuster, A. J. Dixon, M. Poliakoff, and J. J. Turner, *J. Am. Chem. Soc.* **1987**, *109*, 4723.

100. N. P. Gritsan, T. Yuzawa, and M. S. Platz, *J. Am. Chem. Soc.* **1997**, *119*, 5059.

101. R. Born, C. Burda, P. Senn, and J. Wirz, *J. Am. Chem. Soc.* **1997**, *119*, 5061.

102. J. Michl, and V. Bonačić-Koutecky, *Electronic Aspects of Organic Photochemistry*, John Wiley & Sons, Inc., New York, **1990**.

103. L. Salem and C. Rowland, *Angew. Chem. Int. Ed., Engl.* **1972**, *11*, 92.

104. J. Michl, *J. Am. Chem. Soc.* **1996**, *118*, 3568.

105. (a) H. E. Zimmerman and A. G. Kutateladze, *J. Am. Chem. Soc.* **1996**, *118*, 249. (b) H. E. Zimmerman and A. G. Kutateladze, *J. Org. Chem.* **1995**, *60*, 6008.

106. T. W. J. Johnson, M. B. Sullivan, and C. J. Cramer, *Int. J. Quant. Chem.* **2001**, *85*, 492.

107. N. P. Gritsan, D. Tigelaar, and M. S. Platz, *J. Phys. Chem. A* **1999**, *103*, 4465.

108. N. P. Gritsan, A. D. Gudmundsdottir, D. Tigelaar, and M. S. Platz, *J. Phys. Chem. A* **1999**, *103*, 3458.

109. (a) N. P. Gritsan, A. D. Gudmundsdottir, D. Tigelaar, Z. Zhu, W. L. Karney, C. M. Hadad, and M. S. Platz, *J. Am. Chem. Soc.* **2001**, *123(9)*, 1951.

110. N. P. Gritsan, I. Likhotvorik, M.-L. Tsao, N. Celebi, M. S. Platz, W. L. Karney, C. R. Kemnitz, and W. T. Borden, *J. Am. Chem. Soc.* **2001**, *123*, 1425.

111. T. Kobayashi, H. Ohtani, K. Suzuki, and T. Yamaoka, *J. Phys. Chem.* **1985**, *89*, 776.

112. M.-L. Tsao, N. P. Gritsan, T. James, and M. S. Platz, unpublished research at The Ohio State University.

113. R. J. Sundberg, S. R. Suter, and M. Brenner, *J. Am. Chem. Soc.* **1972**, *94*, 513.

114. W. L. Karney and W. T. Borden, *J. Am. Chem. Soc.* **1997**, *119*, 3347.

115. K. Lamara, A. D. Redhouse, R. K. Smalley, and J. R. Thompson, *Tetrahedron* **1994**, *50*, 5515.

116. M. A. Berwick, *J. Am. Chem. Soc.* **1971**, *93*, 5780.

117. (a) R. Poe, J. Grayzar, M. J. T. Young, E. Leveya, K. Schnapp, and M. S. Platz, *J. Am. Chem. Soc.* **1991**, *113*, 3209. (b) R. Poe, K. Schnapp, M. J. T. Young, J. Grayzar, and M. S. Platz, *J. Am. Chem. Soc.* **1992**, *114*, 5054. (c) A. Marcinek and M. S. Platz, *J. Phys. Chem.* **1993**, *97*, 12674. (d) A. Marcinek, M. S. Platz, S. Y. Chan, R. Floresca, K. Rajagopalan, M. Golinski, and D. Watt, *J. Phys. Chem.* **1994**, *98*, 412. N. P. Gritsan, H. B. Zhai, T. Yuzawa, D. Karweik, J. Brooke, and M. S. Platz, *J. Phys. Chem. A* **1997**, *101*, 2833.

118. E. Leyva and R. Sagredo, *Tetrahedron* **1998**, *54*, 7367.

119. (a) A. E. Reed, R. B. Weinstock, and F. A. Weinhold, *J. Chem. Phys.* **1985**, *83*, 735. (b) A. E. Reed, F. A. Weinhold, and L. A. Curtiss, *Chem. Rev.* **1988**, *88*, 899.

120. C. G. Younger and R. P. Bell, *J. Chem. Soc., Chem. Commun.* **1992**, 1359.

121. C. R. Kemnitz, W. L. Karney, and W. T. Borden, *J. Am. Chem. Soc.* **1998**, *120*, 3499.

122. (a) A. Reiser, G. Bowes, and R.-J. Horne, *Trans. Faraday Soc.* **1966**, *62*, 3162. (b) A. Reiser and R. Morley, *Trans. Far. Soc.* **1968**, *64*, 1806.

123. Tsao, M.-L., Ph.D. Thesis, The Ohio State University (2003).

124. (a) A. Reiser, F. W. Willets, G. C. Terry, V. Williams, and R. Marley, *Trans. Faraday Soc.* **1968**, *64*, 3265. (b) P. A. S. Smith and J. H. Boyer, *J. Am. Chem. Soc.* **1951**, *73*, 2626. (c) S. E. Hilton, E. F. V. Scriven, and H. Suschitzky, *J. Chem. Soc., Chem. Commun.* **1974**, 853. (d) B. Iddon, H. Suschitzky, and D. S. Taylor, *J. Chem. Soc. Perkins Trans. 1* **1974**, 579. (e) S. E. Carroll, B. Nay, E. F. V. Scriven, and H. Suschitzky, *Synthesis* **1975**, 710. (f) R. N. Carde and G. Jones, *J. Chem. Soc. Perkin Trans. 1.* **1975**, 519. (g) B. Iddon, M. W. Pickering, H. Suschitzky, and D. S. Taylor, *J. Chem. Soc. Perkins Trans. 1* **1975**, 1686. (h) J. Rigaudy, C. Igier, and J. Barcelo, *Tetrahedron Lett.* **1979**, 1837. (i) B. Nay, E. F. V. Scriven, H. Suschitzky, and Q. U. Khan, *Synthesis* **1979**, 757. (j) G. F. Bettinetti, E. Fasani, G. Minoli, and S. Pietra, *Gazz. Chim. Ital.* **1979**, *109*, 175. (k) P. T. Gallagher, B. Iddon, and H. Suschitzky, *J. Chem. Soc. Perkin Trans. 1* **1980**, 2362.

125. E. Leyva and M. S. Platz, *Tetrahedron Lett.* **1987**, *28*, 11.

126. I. R. Dunkin and P. C. P. Thomson, *J. Chem. Soc., Chem. Commun.* **1980**, 499.

127. A. Maltsev, T. Bally, M.-L. Tsao, M. S. Platz, A. Kutin, M. Vosswinkel, and C. Wentrup, *J. Am. Chem. Soc.*, submitted.

128. A. K. Schrock and G. B. Schuster, *J. Am. Chem. Soc.* **1984**, *106*, 5234.

129. J. Swenton, T. Ikeler, and B. Williams, *J. Am. Chem. Soc.* **1970**, *92*, 3103.

130. R. J. Sundberg, M. Brenner, S. R. Suter, and B. P. Das, *Tetrahedron Lett.* **1970**, *31*, 2715. (b) R. J. Sundberg and R. W. Heintzelman, *J. Org. Chem.*, **1974**, *39*, 2546. (c) R. J. Sundberg, D. W. Gillespie, and B. A. DeGraff, *J. Am. Chem. Soc.* **1975**, 97, 6193.

131. (a) A. Reiser, H. Wanger, and G. Bowes, *Tetrahedron Lett.* **1966**, *23*, 629. (b) A. Reiser, F. W. Willets, G. C. Terry, V. Williams, and R. Marley, *Trans. Faraday Soc.*, **1968**, *4*, 3265 (c) A. Reiser, G. Bowes, and R. J. Horne, *Trans. Faraday Soc.* **1966**, *62*, 3162. (d) V. A. Smirnov and S. B. Brichkin, *Chem. Phys. Lett.* **1982**, *87*, 548.

132. (a) P. A. Lehman and R. S. Berry, *J. Am. Chem. Soc.* **1973**, *95*, 8614. (b) R. J. Sundberg, D. W. Gillespie, and B. A. DeGraff, *J. Am. Chem. Soc.* **1975**, *97*, 6193.

133. M.-L. Tsao, N. Gritsav, T. R. James, M. S. Platz, D. A. Hrovat, and W. T. Borden, *J. Am. Chem. Soc.* **2003**, *125*, 9343.

134. (a) K. Itoh, *Chem. Phys. Lett..* **1967**, *1*, 235. (b) P. M. Lahte, M. Minato, and C. Ling, *Cryst. Liq. Cryst. Sci. Tech. Series A* **1995**, *271*, 147. (c) S. V. Chapyshev, A. Kuhn, M. W. Wong, and C. Wentrup, *J. Am. Chem. Soc.* **2000**, *122*, 1572. (d) S. V. Chapyshev, R. Walton, J. A. Sanborn, and P. M. Laht, *J. Am. Chem. Soc.* **2000**, *122*, 1580. (e) K. Haider, N. Soundararajan, M. Shaffer, and M. S. Platz, *Tetrahedron Lett.* **1989**, *30*, 1225.

135. T. Enyo, A. Nicolaides, and H. Tomioka, *J. Org. Chem.* **2002**, *67*, 5578.

136. A. P. Sylwester and P. B. Dervan, *J. Am. Chem. Soc.* **1984**, *106*, 4648.

137. J. H. Davis and W. A. Goddard, III, *J. Am. Chem. Soc.* **1977**, *99*, 7111.

138. C. Willis, F. P. Lussing, and R. A. Back, *Can. J. Phys.* **1976**, *54*, 1.

139. C. A. Parsons and C. E. Dykstra, *J. Chem. Phys.* **1979**, *71*, 3025.

140. M. J. H. Kemper and H. M. Buck, *Can. J. Chem.* **1981**, *59*, 3044.

141. D. M. Lemal, in *Nitrenes*, W. Lwowski, Ed., Wiley-Interscience, New York, **1970**.

142. B. V. Ioffi and M. Kuznetsov, *Russ. Chem. Rev. (Engl. Trans.)* **1972**, *41*, 131.

143. W. D. Hinsberg III and P. B. Dervan, *J. Am. Chem. Soc.* **1978**, *100*, 1608.

144. W. D. Hinsberg III and P. B. Dervan, *J. Am. Chem. Soc.* **1979**, *101*, 6142.

145. (a) P. G. Schultz and P. B. Dervan, *J. Am. Chem. Soc.* **1980**, *102*, 878. (b) W. D. Hinsberg III, P. G. Schultz, and P. B. Dervan, *J. Am. Chem. Soc.* **1982**, *104*, 766. (c) P. B. Dervan, M. E. Squillacote, P. M. Lahti, A. P. Sylwester, and J. D. Roberts, *J. Am. Chem. Soc.* **1980**, *103*, 1120. (d) P. G. Schultz and P. B. Dervan, *J. Am. Chem. Soc.* **1981**, *103*, 1563.

146. A. Srinivasan, N. Kebede, J. E. Saavedra, A. V. Nikolaitchik, D. A. Brady, E. Yourd, K. M. Davies, L. K. Keefer, and J. P. Toscano, *J. Am. Chem. Soc.* **2001**, *123*, 5465.

147. K. P. Huber and G. Herzberg, *Molecular Spectra and Molecular Structure* Vol. 4, Van Nostrand-Reinhold, New York, **1961**, p. 534.

148. F. Mathey, in *Multiple Bonds and Low Coordination in Phosphorus Chemistry*, M. Regitz and O. J. Scherer, Eds., Thieme, Stuttgart, German, **1990**, p. 34.

149. (a) D. J. Berger, P. P. Gaspar, P. LeFloch, F. Mathey, and R. Grev, *Organometallics* **1996**, *15*, 4904. (b) D. J. Berger, P. P. Gaspar, R. S. Grev, and F. Mathey, *Organometallics* **1994**, *13*, 640.

150. (a) T. P. Hamilton, A. G. Willis, and S. D. Williams, *Chem. Phys. Lett.* **1995**, *246*, 59. (b) M. T. Nguyan, A. Vankeer, L. A. Eriksson, and L. G. Vanquickenborne, *Chem. Phys. Lett.* **1996**, *254*, 307.

151. J. M. Galbraith, P. P. Gaspar, and W. T. Borden, *J. Am. Chem. Soc.* **2002**, *124*, 11669.

152. H. Tomioka, S. Nakamura, T. Ohi, and Y. Izawa, *Bull. Chem. Soc. Jpn.* **1976**, *49*, 3707.

153. P. J. Hammond, G. Scott, and C. D. Hall, *J. Chem. Soc. Perkin Trans. 2* **1982**, 205.

154. X. Li, D. Lei, M. Y. Chiang, and P. P. Gaspar, *J. Am. Chem. Soc.* **1992**, *114*, 8256.

155. X. Li, S. I. Weissman, T.-S. Lin, and P. P. Gaspar, *J. Am. Chem. Soc.* **1994**, *116*, 7899.

156. S. Shah, M. C. Simpson, R. C. Smith, and J. D. Protasiewicz, *J. Am. Chem. Soc.* **2001**, *123*, 6925.

157. J. F. Arenas, J. C. Otero, A. Sánchez-Gálvez, J. Soto, and P. Viruela, *J. Chem. Phys.* **1999**, *111*, 551.

158. (a) J. F. Arenas, J. I. Marcos, J. C. Otero, A. Sánchez-Gálvez, and J. Soto, *J. Chem. Phys.* **1999**, *111*, 551. (b) B. K. Carpenter, in *Reactive Intermediate Chemistry*, R. A. Moss, M. S. Platz, and M. Jones, Jr., Eds., John Wiley & Sons, Inc., New York, **2004**, Chapter 21.

CHAPTER 12

Synthetic Carbene and Nitrene Chemistry

MICHAEL P. DOYLE

College of Science, Department of Chemistry, University of Arizona, Tucson, AZ

1. INTRODUCTION

Carbenes and nitrenes were already well known and characterized by the early 1960s,[1] but they were widely regarded as curious highly reactive species rather than potentially valuable synthetic intermediates.[2] The problems with their uses

Reactive Intermediate Chemistry, edited by Robert A. Moss, Matthew S. Platz, and Maitland Jones, Jr.
ISBN 0-471-23324-2   Copyright © 2004 John Wiley & Sons, Inc.

were reactivity and, especially, selectivity. The unique transformations for which they were well known—addition to carbon–carbon multiple bonds and insertion into carbon–hydrogen bonds—showed little regioselectivity or stereoselectivity, and whether or not the reaction proceeded through a singlet or triplet intermediate was also uncertain. In the same time period, however, two independent developments transformed these mechanistic curiosities into valued units for organic synthesis.

Working with diazo compounds, known since the early 1900s to undergo loss of dinitrogen when treated with copper or copper salts, Yates described in 1952[3] the possibility that transition metals could form an intermediate that combined units of the diazo compound and the metal (Eq. 1, L = ligand) and acted like a carbene in addition and insertion reactions. Somewhat later, but independently, E. O. Fischer isolated and characterized stable metal carbenes that could also undergo cyclopropanation reactions.[4] They were derived from transition metals on the left side of the

$$N_2C\begin{smallmatrix}R\\R'\end{smallmatrix} + ML_n \xrightarrow{-N_2} (L_nM)(CRR')$$

$$(CO)_5M=\begin{smallmatrix}R\\OMe\end{smallmatrix}$$

Metal carbenoid   (1)

M = Cr, Mo, W
"Fischer carbene"

periodic table, contained an alkoxy substituent, could be applied to organic synthesis as stoichiometric reagents, and were chemically and physically well defined.[5] Thus the search for synthetically viable ways to utilize carbenes was being pulled in two directions. Catalytic methods using diazo compounds, and allied metal carbenoid processes such as the Simmons–Smith reaction [alkene + $CH_2I_2(Zn-Cu)$ → cyclopropane], looked to new developments in catalysis, especially with catalysts that were soluble in the reaction medium and, eventually, those that possessed chiral ligands.[6] With metal carbenes in the new emerging field of organometallic chemistry, emphasis was placed on the preparation of stable analogues and their chemistries.[7] From these investigations arose the understandings and applications of electrophilic and nucleophilic metal carbenes (**1a–c**).

$$L_n\bar{M}-\overset{+}{C}RR' \longleftrightarrow L_nM=CRR' \longleftrightarrow L_n\overset{+}{M}-\bar{C}RR'$$

1a   **1b**   **1c**

Electrophilic   Nucleophilic
metal carbene   metal carbene

The philicity of carbenes and control of selectivity in their reactions became, in retrospect, an enabling concept. But how does one control philicity through the metal and its bound ligands? The metal itself contributes to the philicity of the carbene by its kind and degree of d-orbital back-bonding.[8] The ligands attenuate this effect by their electronic and steric influences. And having an open coordination site on the metal because of ligand dissociation is important to their reactions (Scheme 12.1). Herein was a discussion that lasted more than a decade.[9] Metal carbenes were implicated in both cyclopropanation (Eq. 2)[10] and metathesis reactions

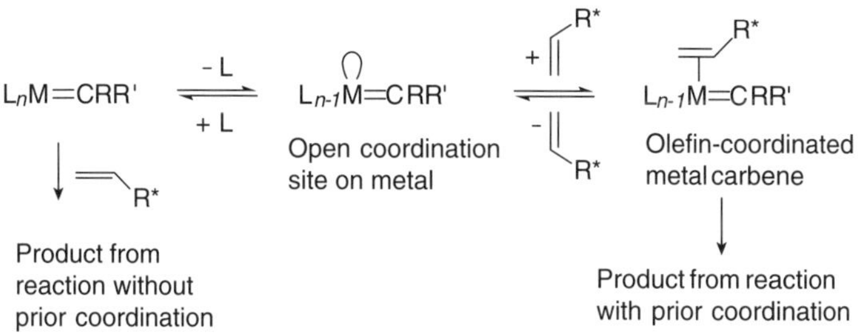

Scheme 12.1

(Eq. 3).[11] Could they both arise through the same mechanism but differ in the philicity of the metal carbene, or did one proceed through a pathway that was different from the other? The answers came reluctantly over an extended period, but they paved the way to the synthetic uses that we enjoy today.

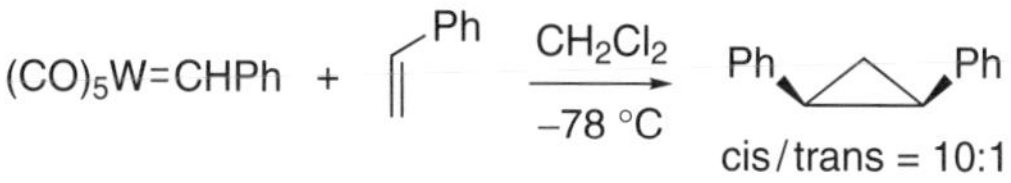

$$(CO)_5W=CHPh \ + \ \overset{Ph}{\underset{}{\|}} \quad \xrightarrow[-78\ °C]{CH_2Cl_2} \quad Ph \bigtriangleup Ph \qquad (2)$$
$$cis/trans = 10:1$$

$$\overset{Me}{\underset{}{\|}} \ + \ \overset{Me}{\underset{}{\|}} \quad \xrightarrow[\substack{\text{"Phillips} \\ \text{triolefin} \\ \text{process"}}]{catalyst} \quad \| \ + \ MeCH=CHMe \qquad (3)$$

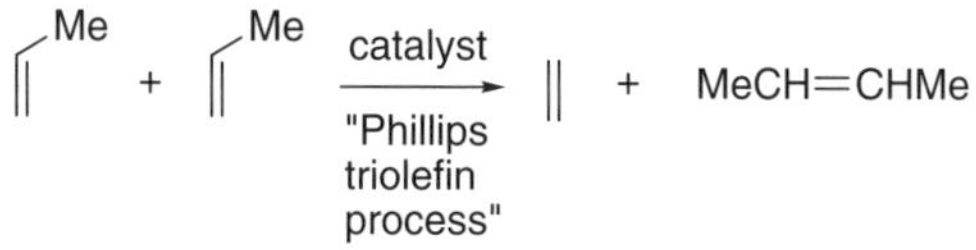

2. ELECTROPHILIC AND NUCLEOPHILIC METAL CARBENES

Synthetic uses came from two directions, neither based on fundamentals but both moving the field from uncertainty to the effective control of parameters. Electrophilic metal carbenes were recognized from their ability to undergo addition to electron-rich alkenes such as vinyl ethers or styrene (Eq. 4),[6] but not electron-poor

$$N_2CHCOOEt \ + \ \overset{R}{=\!\!\diagup} \quad \xrightarrow{ML_n} \quad \overset{\bigtriangleup}{R \quad COOEt} \qquad (4)$$

rate: R = OEt > Ph > Me(CH$_2$)$_3$ >> COOMe

alkenes that included α,β-unsaturated carbonyl compounds. Nucleophilic metal carbenes, eventually understood to form and be produced from metallocyclobutane intermediates, exhibited an orientation expected of a nucleophilic carbene (Scheme 12.2). Those catalysts that form electrophilic metal carbenes have lower oxidation states [e.g., Cu(I), Rh(II), W(II)], whereas those that undergo metathesis have higher oxidation states (e.g., **4–7** with **6** and **7** possessing a nitrogen-stabilizing carbene ligand). Ligand dissociation from Ru by the phosphine initiates a metathesis

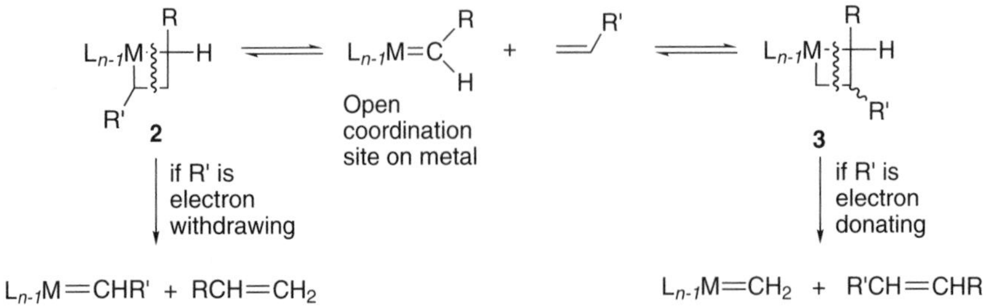

Scheme 12.2

reaction, but there is no evidence that ligand dissociation is an essential step in cyclopropanation reactions.

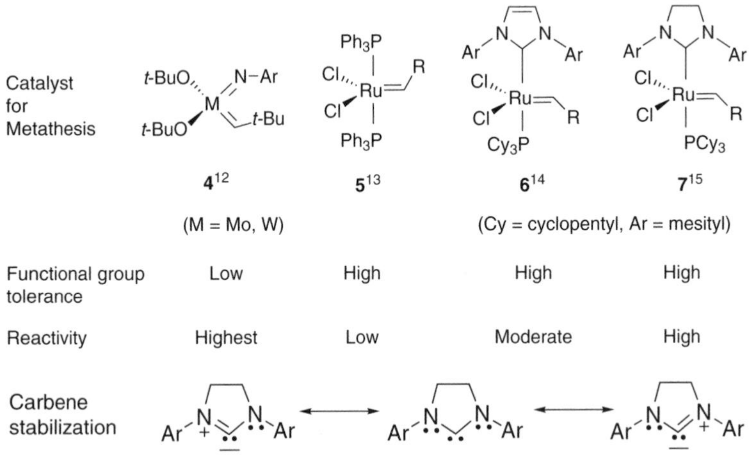

Understanding the structure and dynamics of metal carbenes related to Fischer carbenes provides insights into their reactivities and selectivities. For example, their stability decreases when electron-donating substituents are replaced by hydrogen or, especially, electron-withdrawing substituents:[6]

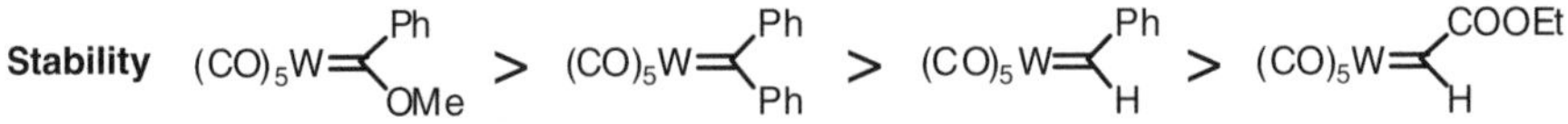

Furthermore, the opening of a coordination site for olefin association is a necessary condition for metathesis, but not for cyclopropanation (Eq. 5).[16]

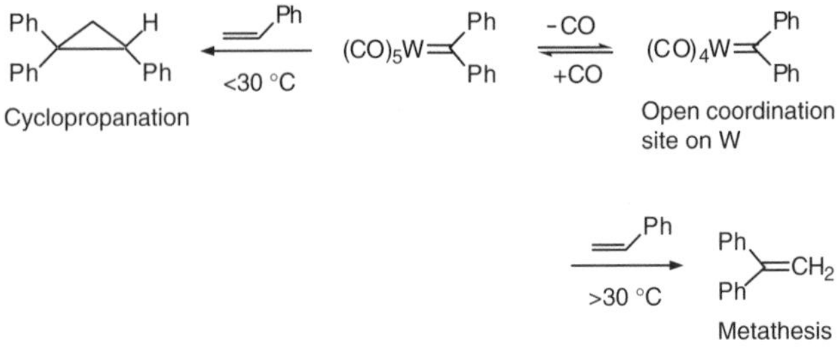

$$\tag{5}$$

2.1. Catalysis of Diazo Decomposition

By 1960, there was recognition that copper salts could cause the loss of dinitrogen from diazocarbonyl compounds with addition of the resulting carbene intermediate to a carbon–carbon double bond to form a cyclopropane product. That this reaction, first reported by G. Stork in 1961 (Eq. 6),[17] could occur in an intramolecular fashion and thus avoid the formation of isomers, ushered in the first significant synthetic

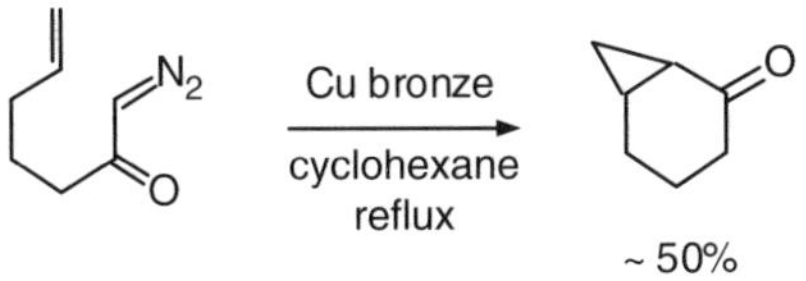

$$(6)$$

applications beyond pyrethroid syntheses. Extensions of this methodology led to the preparation of a large number of natural products (e.g., **8–10**),[18] but neither

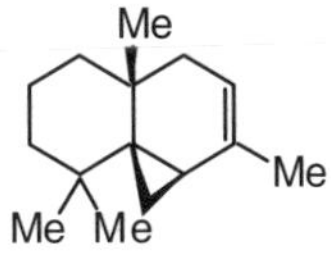

8	9	10
Thujopsene[19]	Aristolone[20]	Presqualene alcohol[21]

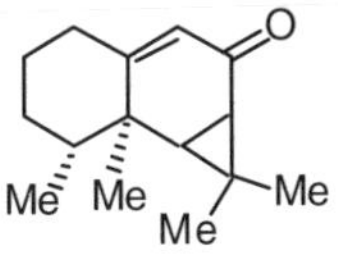

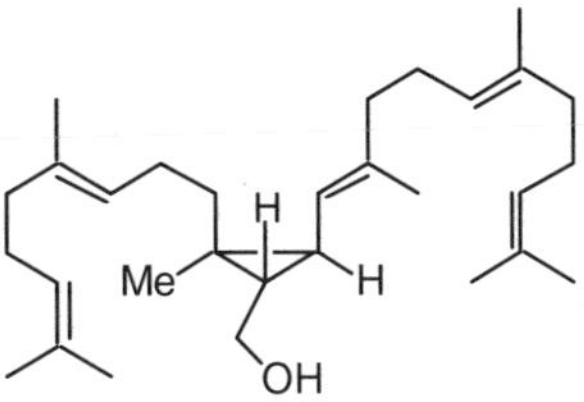

the mechanism of this transformation, nor methods to control reaction selectivity, were well understood. It was during this decade that new catalysts were developed, eventually resulting in those now recognized to be the most reactive and selective for cyclopropanation (**11–16**).[6]

Achiral Catalysts for Diazo Decomposition	Cu(acac)₂ Low reactivity stable	CuOTf High reactivity easily oxidized	Rh₂(OAc)₄ Moderate reactivity stable	PdCl₂ High reactivity stable

Chiral Catalysts for Diazo Decomposition		
11[22]	**12**[23]	**13**[24]
Moderate reactivity easily oxidized	Moderate reactivity stable	Low reactivity stable

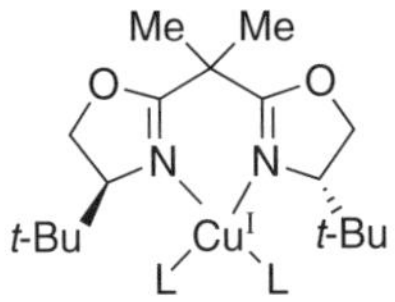

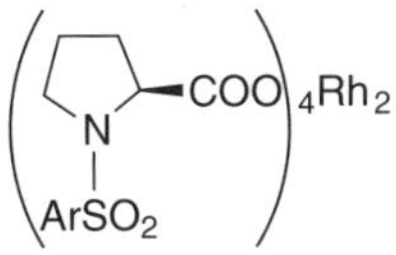

Ar = aromatic

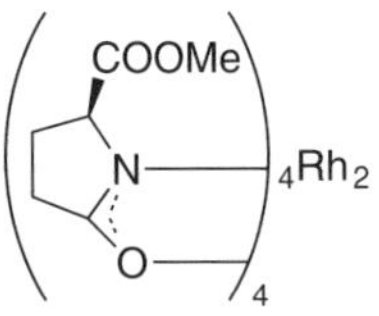

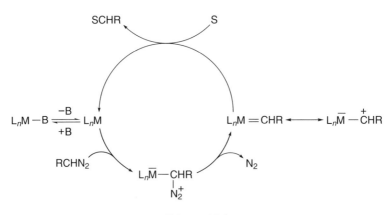

14[25]	**15**[26]	**16**[27]
M = Ru(NO)Cl	Moderate reactivity	Moderate reactivity
Low reactivity	stable	activated with
requires photoinitiation		*N*-methylimidazole

The mechanism for diazo decomposition is now widely understood.[6] The ligated metal, with an open coordination site and acting as a Lewis acid, undergoes electrophilic addition to the diazo compound. Loss of dinitrogen then forms the intermediate metal carbene that is able to transfer the carbene from the metal to a substrate and thereby regenerate the catalytically active ligated metal (Scheme 12.3). It is in the carbene transfer step that selectivity is achieved. The rate-limiting step is either electrophilic addition or loss of dinitrogen.

Scheme 12.3

The transfer of the carbene may be to any of a variety of substrates and occurs in an intermolecular or intramolecular fashion. Cyclopropanation is perhaps the best known catalytic transformation (Eq. 7),[28] but carbon–hydrogen insertion (Eq. 8),[29] ylide formation and rearrangement or cycloaddition (Eq. 9),[26] and addition to multiple bonds other than C=C (Eq. 10)[30] are also well established.[6,31–33]

Cyclopropanation

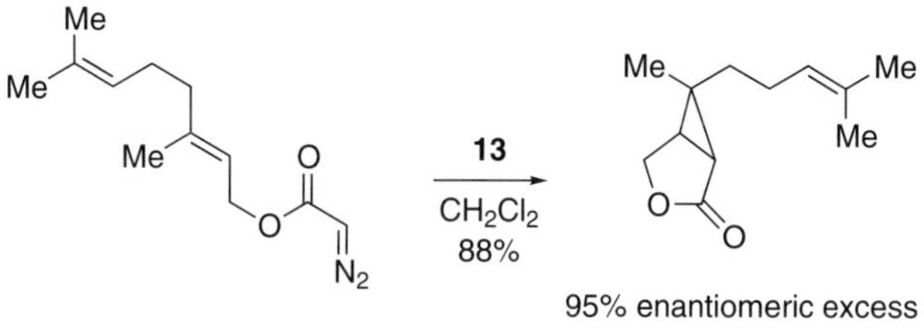

$$\xrightarrow[\substack{CH_2Cl_2 \\ 88\%}]{\textbf{13}}$$

(7)

95% enantiomeric excess (ee)

C—H Insertion

$$\xrightarrow[\substack{-50\,°C,\ 67\%}]{\textbf{12}\ \text{in hexane}}$$

(**12**: Ar = p-C$_{12}$H$_{25}$C$_6$H$_4$)

97% ee
74 : 26 diastereomer ratio (dr)

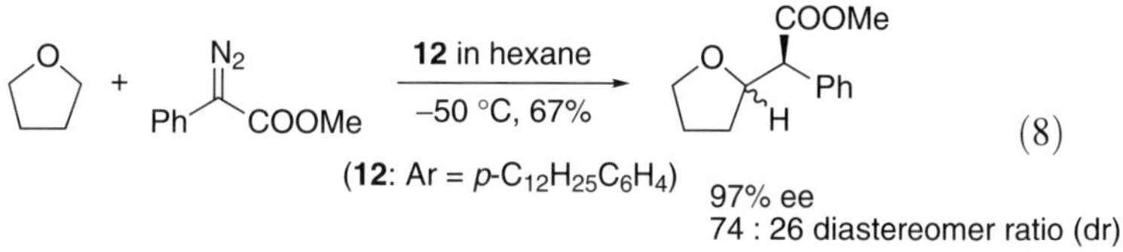

(8)

Carbonyl ylide
formation/cycloaddition

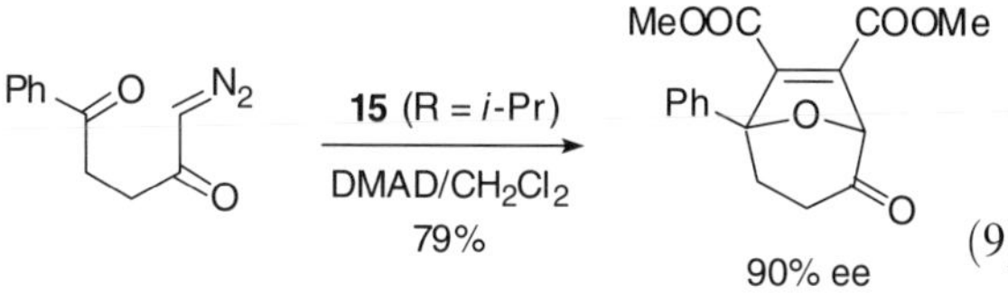

$$\xrightarrow[\substack{DMAD/CH_2Cl_2 \\ 79\%}]{\textbf{15}\ (R = i\text{-Pr})}$$

90% ee

(9)

DMAD = dimethylacetylenedicarboxylate

Addition

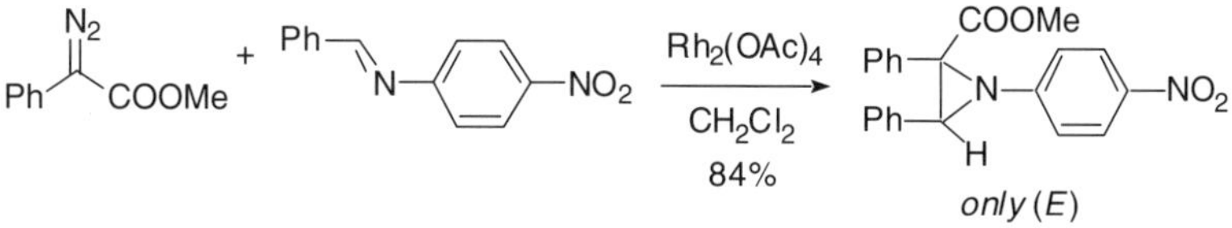

$$\xrightarrow[\substack{CH_2Cl_2 \\ 84\%}]{Rh_2(OAc)_4}$$

only (E)

(10)

Diazocarbonyl compounds are optimum for these transformations, and they may be readily prepared by a variety of methods. The use of iodonium ylides (**17**) has also been developed,[34] but they exhibit no obvious advantage for selectivity in carbene-transfer reactions. Enantioselection is much higher with diazoacetates than with diazoacetoacetates (**18**).

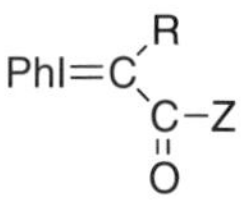

17

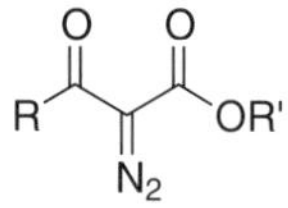

18

2.2. Stable Transition Metal Carbene Complexes

In contrast to transition metal carbene complexes generated catalytically, those of the early transition metals are generally stable (Fischer-type carbenes) and undergo

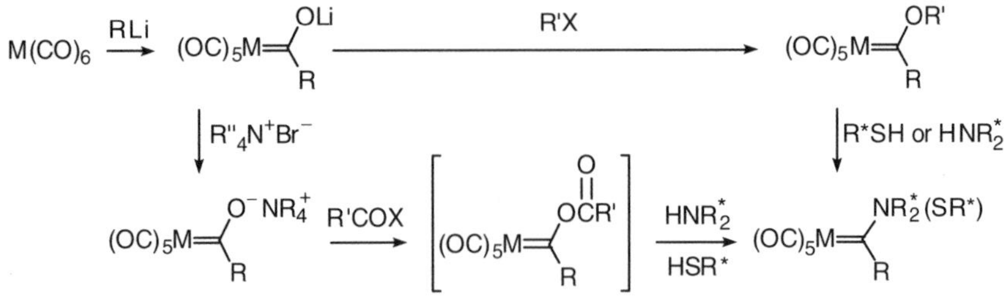

Scheme 12.4

stoichiometric reactions. They are typically formed by addition of a metal alkyl to the bound carbonyl of $M(CO)_6$, where M is a transition metal in the chromium triad (Scheme 12.4).[7,35,36] The difference between these metal carbenes and those generated catalytically via Scheme 12.3 is the presence of an electron-donating substituent (OR, NR_2, SR) on the carbene rather than an electron-withdrawing substituent that stabilizes the reactant diazo compound (COOR, NO_2).

Reactions are generally initiated by the loss of CO from the metal carbene. Benzannulation (Eq. 11)[37] is one of the best known transformations, as is cyclopro-

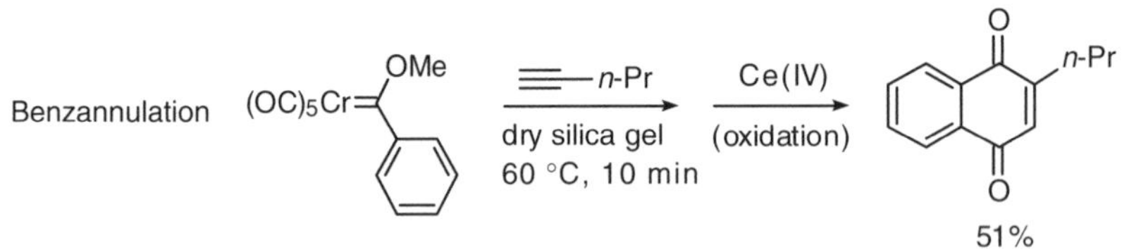

panation, but transformations resulting in the formation of four-, five-, and seven-membered rings can also be achieved (e.g., Eqs. 12,[38] 13[39]). These reactions are

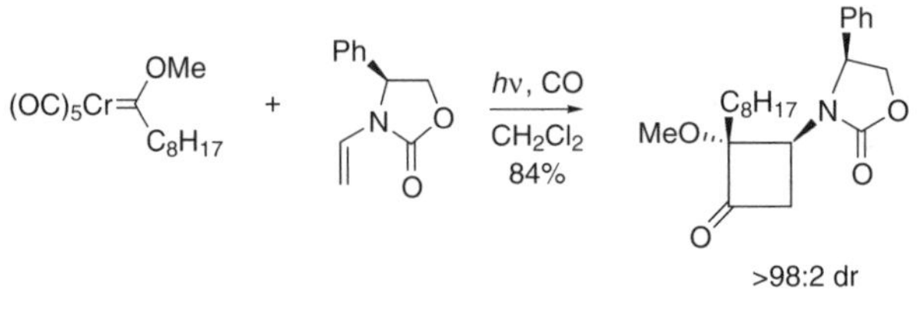

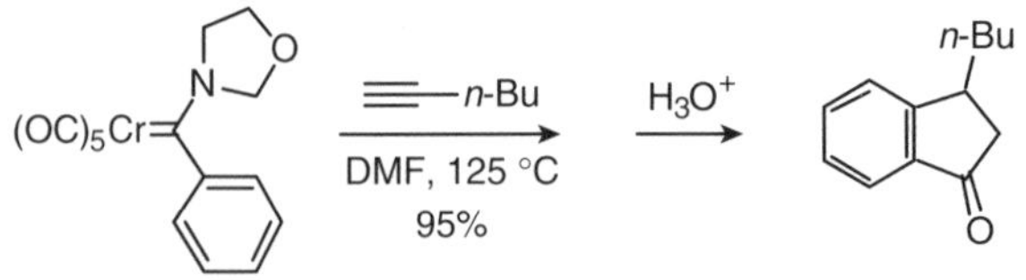

DMF = dimethylformamide

initiated by the loss of CO from the metal to open a coordination site for complexation with an alkyne or alkene (Scheme 12.5). The key steps, however, arise from

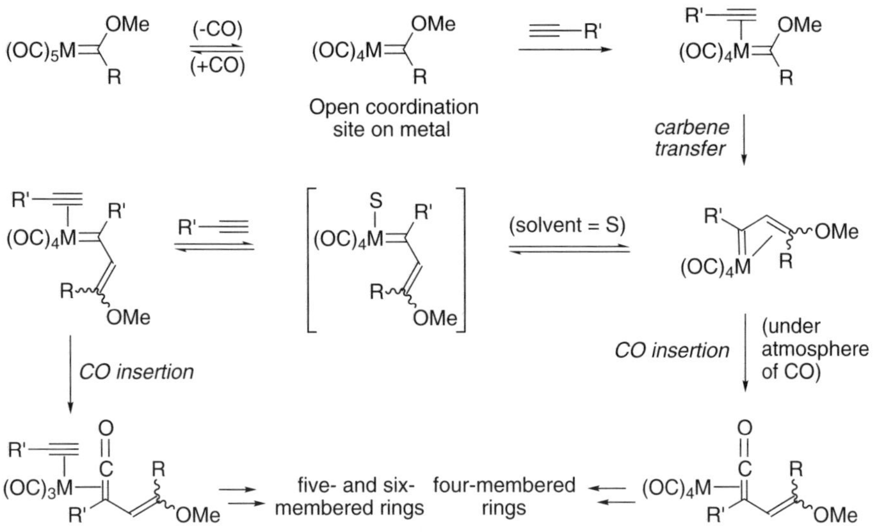

Scheme 12.5

insertion of the carbene into CO to form the intermediate ketene from which products are derived. Control of CO insertion (pressure and temperature) determines the eventual fate of the reaction.

2.3. Metathesis

One of the most useful reactions in organic synthesis, metathesis, had its origins with double-bond scrambling reactions in industrial chemistry and in the search for polymerization catalysts that could transform cyclic alkenes to polyalkenes (Eq. 14).[40,41] In the polymerization reactions, known as "ring-opening metathesis polymerization," the propagating species is a metalloalkene (metal carbene), and a metallocyclobutane is the presumed reaction intermediate. Here the catalytically active metal species remains attached to the growing end of the polymer chain, and the polymer is referred to as a "living polymer," whose characteristics include narrow molecular weight distribution and a facility for block copolymer formation.[12,42,43] The catalysts that are desirable are those for which cyclopropanation and β-hydride elimination (e.g., $L_nM=CHCH_2R \rightarrow L_nM + H_2C=CHR$) are minimized.

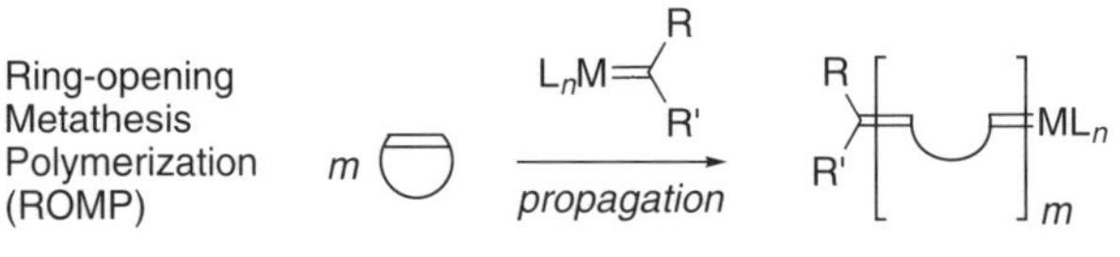

$$(14)$$

Classical metathesis such as that for the "Phillips triolefin process" (Eq. 3) or the "Shell higher olefins process (SHOP)" (e.g., Eq. 15)[44] set the stage for the next major advance in this field—ring-closing metathesis (Eq. 16).[11c,d,45] Here the low reactivity of unstrained olefins toward ring-opening metathesis is a major

SHOP $\qquad$ MeCH=CHMe + $C_{10}H_{21}CH=CHC_{10}H_{21}$ $\underset{}{\overset{\text{catalyst}}{\rightleftharpoons}}$ $2MeCH=CHC_{10}H_{21}$

$$(15)$$

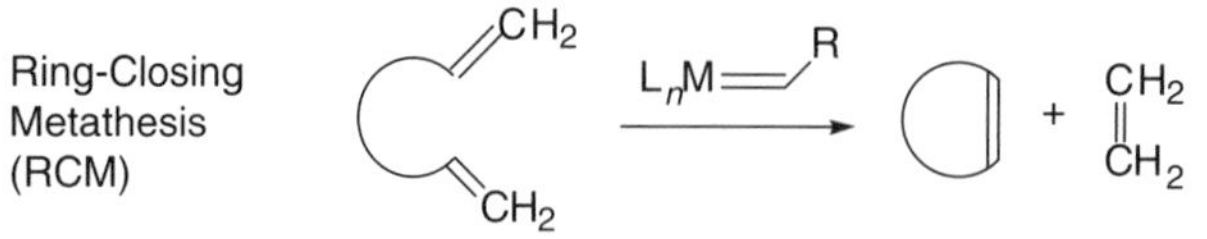

Ring-Closing Metathesis (RCM)

$$(16)$$

consideration in the synthetic viability of the process. The formation of five-, six-, or seven-membered rings, and even macrocyclic compounds, can be made to occur with the appropriate catalysts (Eqs. 17,[46] 18[47]).

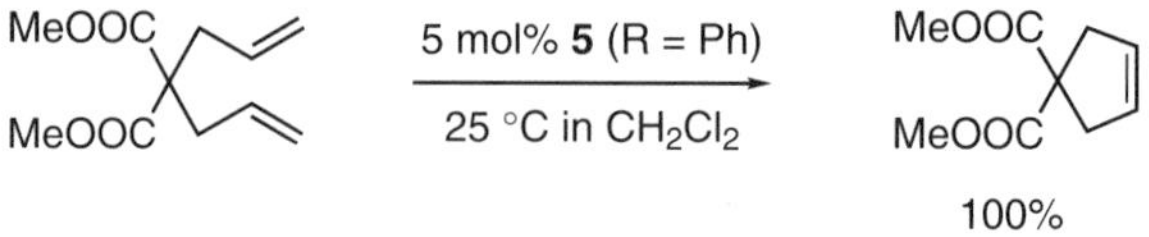

$$(17)$$

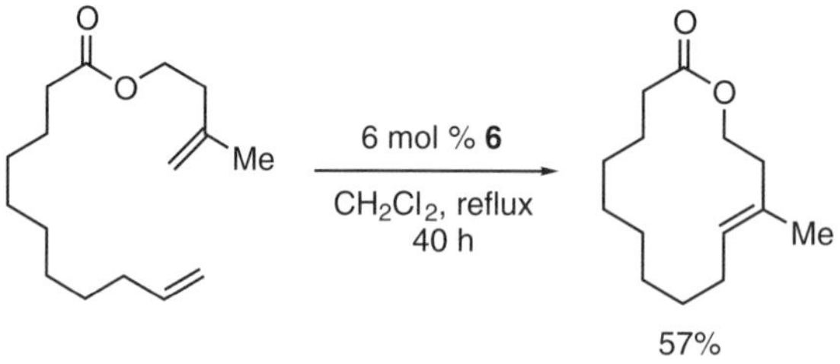

$$(18)$$

As indicated by the examples, ruthenium catalysts **5–7** can be used for compounds containing a broad range of functional groups, especially those containing oxygen, and they are also tolerant to water. The greater challenge is the use of these catalysts for "cross-metathesis" for which SHOP (Eq. 15) is one example, and Eq. 19[48] represents a recent success.

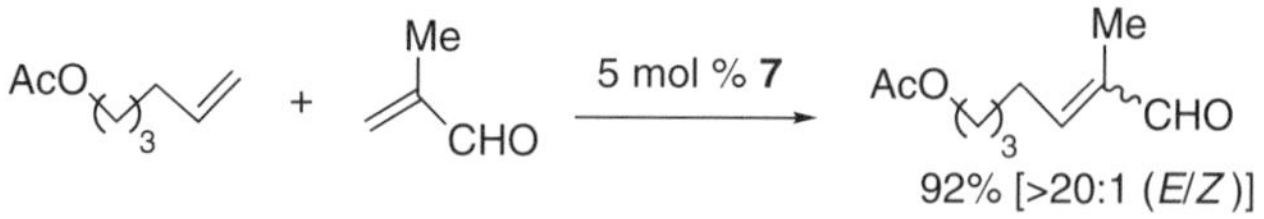

$$(19)$$

2.4. Other Alkylidenes

There is a variety of reagents that undergo carbene-related transformations that do not fit into the categories of nucleophilic and electrophilic metal carbenes described earlier. Those that are the most versatile for organic synthesis, like the Tebbe reagent (**19**, Cp $= C_5H_5$)[49] or "Zn=CH$_2$" in the Simmons–Smith reaction,[50] occur as adducts. Their reactions are typical of nucleophilic or electrophilic metal carbenes even though these intermediates may not ever be involved in the mechanism of the transformation.

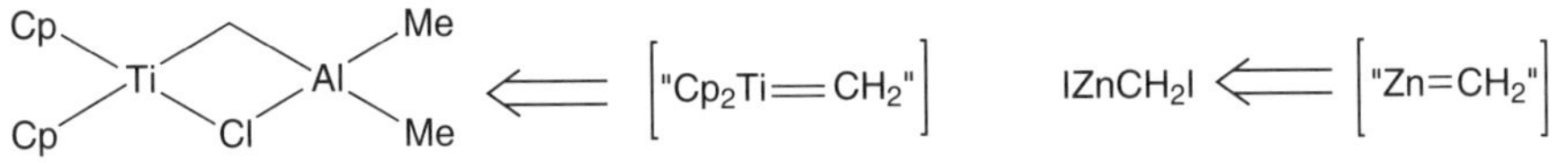

19

Tebbe reagent

Reactive intermediate
in Simmons–Smith reaction

The Tebbe reagent, first reported in 1978,[51] has been used extensively for stoichiometric carbonyl oxygen replacement by methylene (e.g., Scheme 12.6 and Eq. 20).[52] Aldehydes, ketones, esters, and even selected amides, undergo this transformation. And, since the Tebbe reagent is a source of the nucleophilic Cp$_2$Ti=CH$_2$, ring-closing metathesis is the outcome when a second equivalent is used on the diene intermediate. Understanding the Tebbe reagent did, in fact, lead to a basic understanding of homogenous metathesis catalysts.[53]

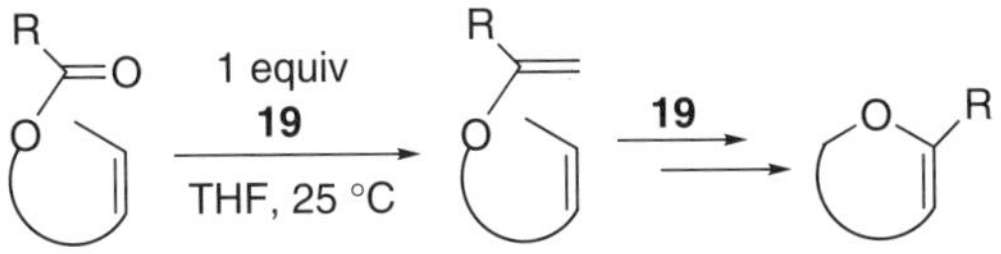

THF = tetrahydrofuran

Scheme 12.6

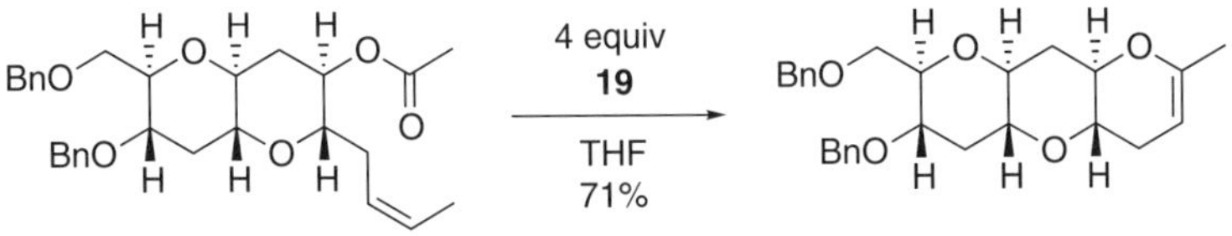

(20)

Bn = benzyl

The Simmons–Smith reaction[54] and its variants[55] are widely used for the stereospecific synthesis of cyclopropane compounds. The methodology involves the use of copper-treated zinc metal (the zinc–copper couple) with diiodomethane to add methylene to a carbon–carbon double bond. Alternative use of diazomethane in catalytic reactions does not offer the same synthetic advantages and is usually avoided because of safety considerations.[6] As significant as is the Simmons–Smith reaction for cyclopropane formation, its employment for organic synthesis was markedly advanced by the discovery that allylic and homoallylic hydroxyl groups accelerate and exert stereochemical control over cyclopropanation of alkenes[56] (e.g, Eq. 21),[57] and this acceleration has been explained by a transition state model

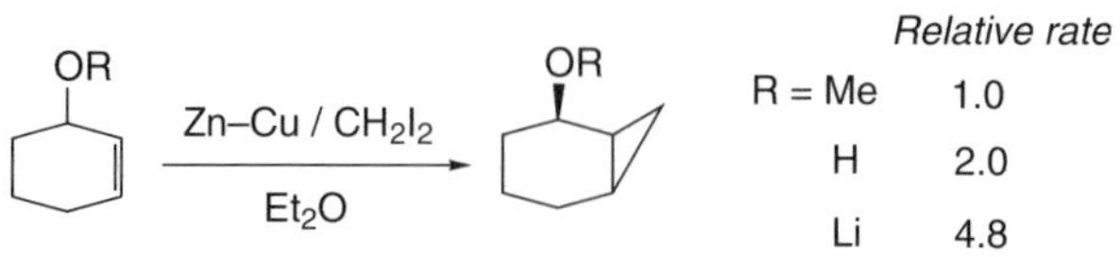

Relative rate

R = Me	1.0
H	2.0
Li	4.8

$$(21)^{57}$$

(**20** and **21**) in which a Lewis acid (ZnX_2) is associated to oxygen.[50b,58] This latter understanding has allowed the redesign of the Simmons–Smith reaction for

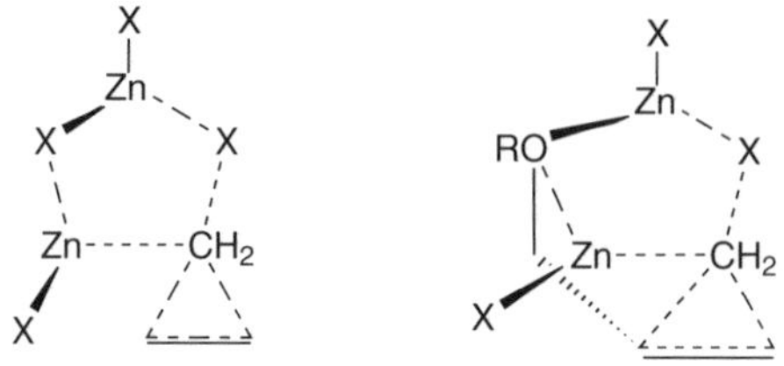

20   **21**

optimum synthetic advantage (e.g., Eq. 22,[59] 23[60]). As noted, reagent modification provides significant control of reaction pathway and selectivity. Enantiocontrol is possible with the use of chiral catalysts such as **22**.[60]

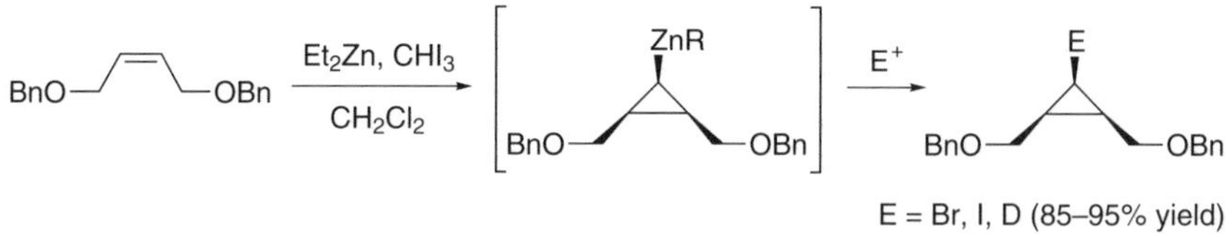

E = Br, I, D (85–95% yield)

$$(22)$$

TADDOL-Ti(O*i*-Pr)₂ (**22**)

Zn(CH₂I)₂

85%

92% ee

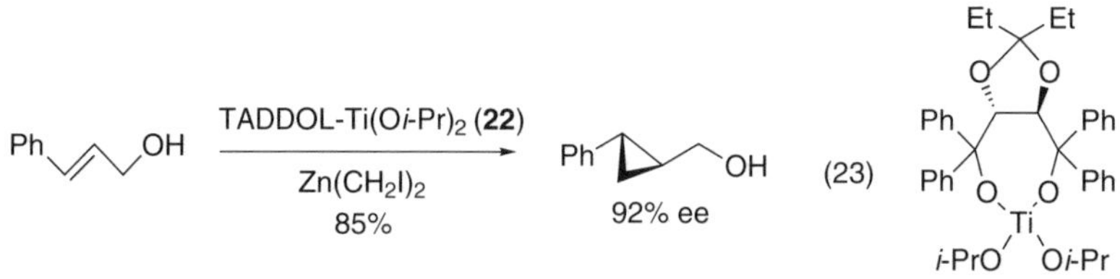

(23)

TADDOL-Ti(O*i*-Pr)₂

22

3. SYNTHETIC ADVANTAGES OF METAL CARBENES

3.1. Applications of Catalytic Methods with Diazo Compounds

The nature of the diazo compound and the identity of the catalyst are important variables for the success of these transformations.[6] Catalytic methods are not successful with simple diazo compounds such as CH_2N_2 or aliphatic diazo compounds because of a variety of competing reactions. However, diazocarbonyl compounds, which are easy to prepare and handle,[61] offer an abundance of uses for organic synthesis. Their reactivity for diazo decomposition is dependent on the number and nature of the groups that stabilize the diazo unit (Scheme 12.7). Ethyl diazoacetate, which is available commercially, has been investigated to the greatest extent, but phenyl- and vinyldiazoacetates have provided a wealth of pathways not available with the simpler diazo compounds.

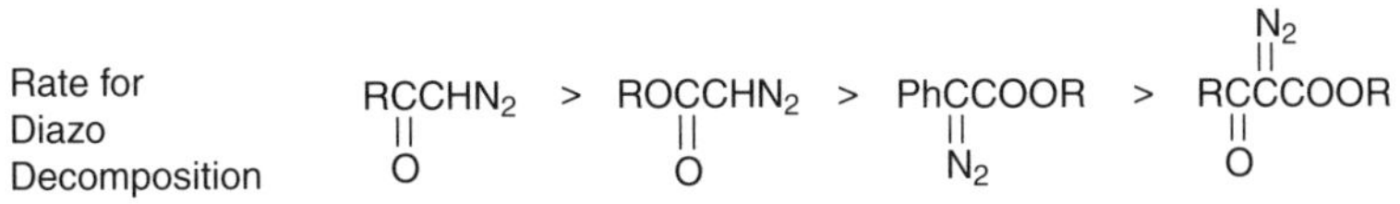

Rate for Diazo Decomposition

$$\underset{O}{RCCHN_2} \; > \; \underset{O}{ROCCHN_2} \; > \; \underset{N_2}{PhCCOOR} \; > \; \overset{N_2}{\underset{O}{RCCCOOR}}$$

Scheme 12.7

3.1.1. Cyclopropanation and Cyclopropenation.

At this time, intermolecular cyclopropanation with alkyl diazoacetates is best accomplished with cobalt catalysts **16**[27] or **23**[62] to achieve high enantiocontrol with high selectivity for the formation of the trans-disubstituted cyclopropane isomer (Eq. 24). To form the

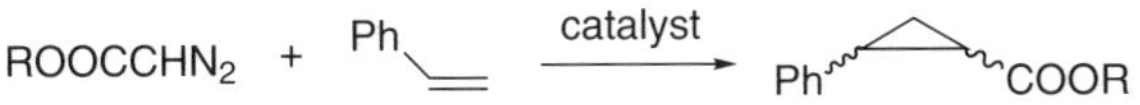

$$ROOCCHN_2 \; + \; Ph\diagdown\!\!= \; \xrightarrow{\text{catalyst}} \; Ph\cdots\!\!\!\triangle\!\!\!\cdots COOR \tag{24}$$

R	Catalyst	Yield (%)	trans/cis	%ee
t-Bu	**23**	80	96:4	93(*t*)
Et	**16** (Ar = Me$_3$C$_6$H$_2$)	99	91:9	96(*t*)
Et	**11**	77	73:27	99(*t*)
t-Bu	**14** (M = Co)	89	2:98	98(*c*)

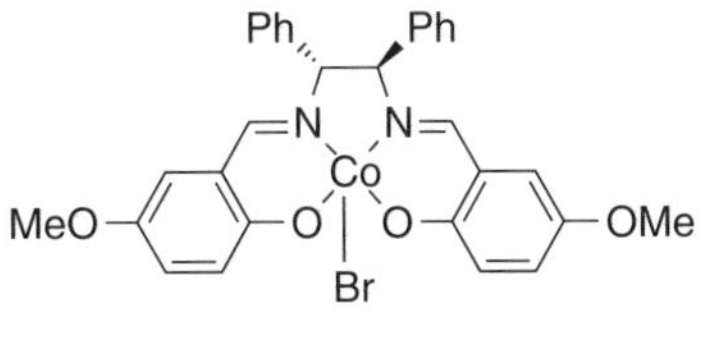

23

cis-disubstituted cyclopropane isomer, the use of **14** (M = Co) gives optimum results for diastereocontrol and enantiocontrol.[63] Enantioselectivity for intramolecular cyclopropanation of allyl diazoacetates is best achieved with chiral dirhodium(II) carboxamidate catalysts (≥94% ee, e.g., Eq. 7),[6,28] and chiral dirhodium(II) carboxylates give the highest enantioselectivities for intermolecular cyclopropanation reactions of phenyldiazoacetates (e.g., Eq. 25).[23,31] The lesson here is that different chiral catalysts exhibit different selectivities for different applications. The selectivities are often based on the barriers to the approach of the

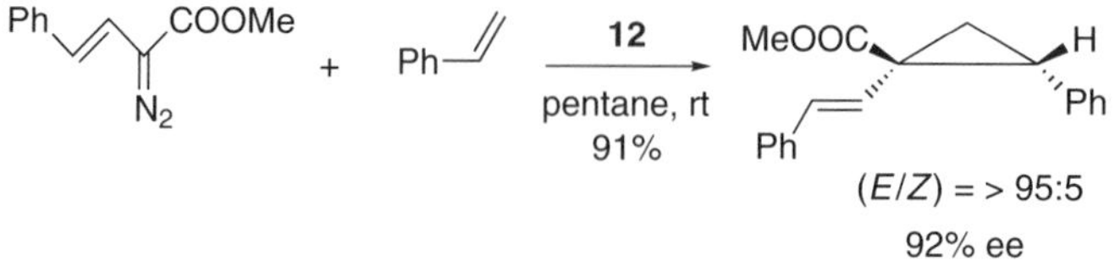

$$(25)$$

$(E/Z) = > 95:5$

92% ee

carbon–carbon double bond to the metal carbene center (e.g., **24** and **25**, where L is a linker), and the outcome is becoming predictable.[64] For example, copper catalysts

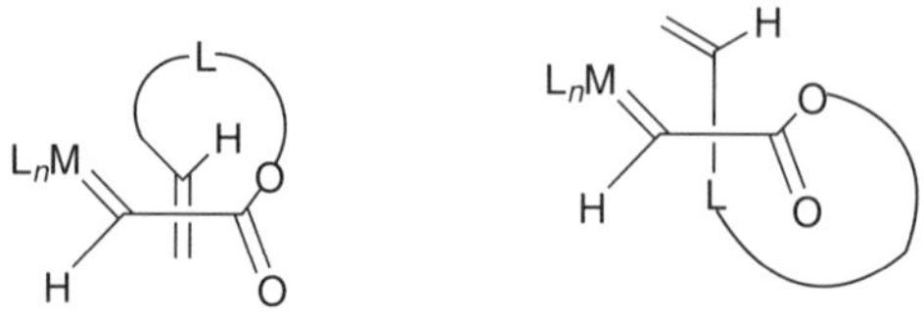

24   **25**

prefer **24**, whereas rhodium catalysts prefer approach **25**. Applications in total synthesis include the synthesis of the antidepressant sertraline (**26**),[65] the cyclopropane–NMDA receptor antagonist milnacipran (**27**),[66] and sirenin (**28**).[67]

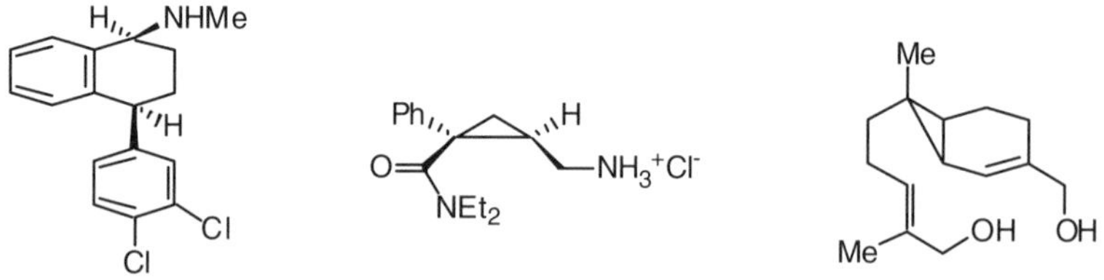

Sertraline (**26**)   (–)-milnacipram (**27**)   Sirenin (**28**)

The addition to a carbon–carbon triple bond results in the formation of cyclopropene products, and with diazoacetates the catalyst of choice for intermolecular addition is the dirhodium(II) carboxamidate **13** (e.g., Eq. 26).[68] The reactions are general, except for phenylacetylene whose cyclopropene product undergoes [2 + 2]-cycloaddition, and selectivities are high. However, high selectivities have not been reported for reactions with allenes.

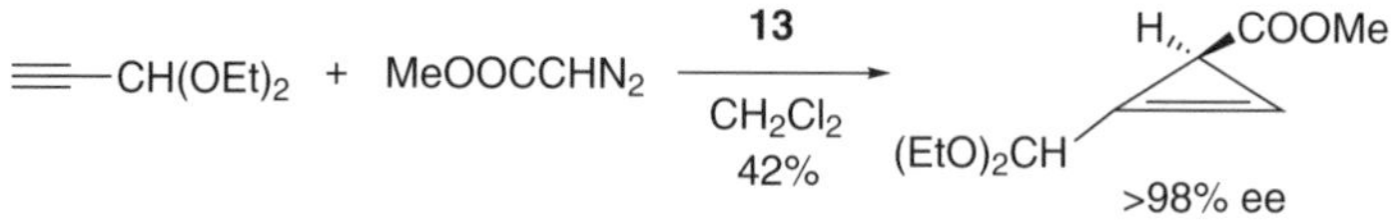

$$(26)$$

>98% ee

Intramolecular cyclopropanation reactions are not limited to the formation of five- to seven-membered rings, as once believed. They occur with high stereocontrol and yield for reactions that produce large rings.[69] Intramolecular cyclopropenation is also a facile process with ring sizes of 10 or higher. High levels of enantiocontrol can be achieved in these reactions with catalysts appropriate to the transformation (e.g., Eqs. 27[70], 28[71]). Note that the choice of catalyst (Eq. 28) can influence chemoselectivity as well as stereoselectivity. As the length of the chain increases, selectivity approaches outcomes that can be predicted from inter-molecular reactions.

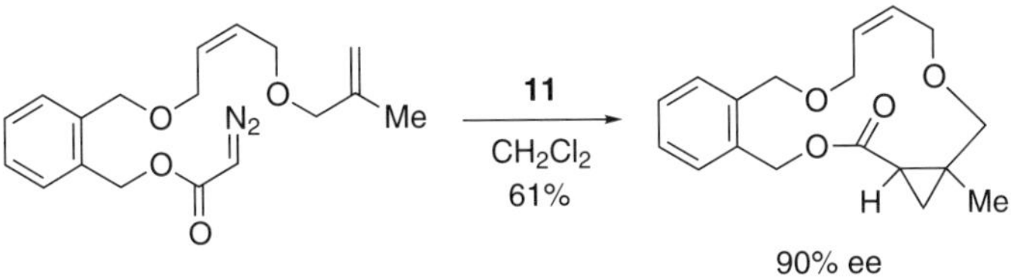

Beyond these systems, challenges in stereocontrol remain for both inter- and intramolecular cyclopropanation reactions with diazoketones, diazoketoesters (**18**), diazomalonates, and diazomethane.[6,31–33,72] Although some progress has been made in intramolecular reactions of diazoketones, with selected examples having high % ee values,[73] enantiocontrol is generally low to moderate for these systems.

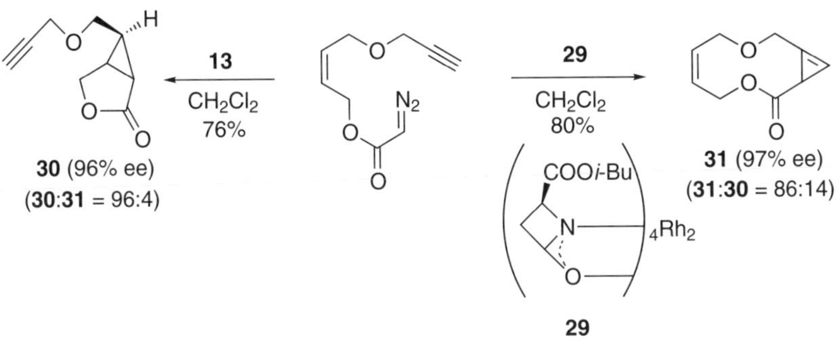

3.1.2. Insertion.
One of the unique advantages of dirhodium(II) catalysts in synthesis is their ability to effect carbon–hydrogen insertion reactions.[6,32,33,74] For intramolecular reactions insertion into the gamma position is virtually exclusive, and only when this position is blocked or deactivated does insertion occur into the beta or delta C—H position (e.g., Eqs. 29,[75] 30,[76] and 31[77]). The electrophilic character of these insertion reactions is suggested by the C—H bond reactivity in competitive experiments ($3° > 2° \gg 1°$)[78] and by the enhancement due to

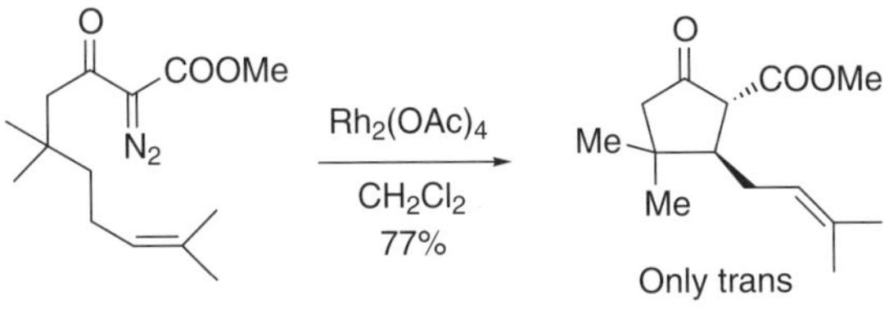

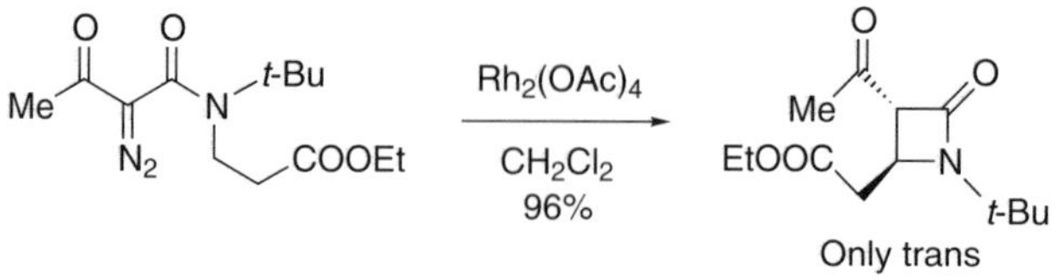

(30)

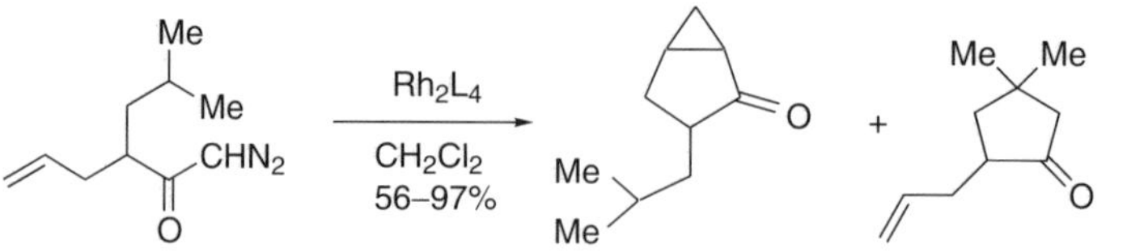

Only trans

(31)

pfb = CF$_3$CF$_2$CF$_2$COO	Rh$_2$(pfb)$_4$	0	100
	Rh$_2$(OAc)$_4$	44	55
cap 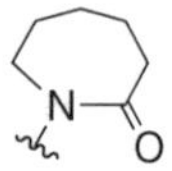 =	Rh$_2$(cap)$_4$	100	0

heteroatoms such as oxygen[79] or nitrogen. A recent theoretical treatment[80] confirmed the mechanistic proposal (Scheme 12.8) that C—C and C—H bond formation with the carbene carbon proceeds as the ligated metal dissociates.[81] As indicated by the influence of ligands on selectivity in Eq. 31, one transformation may be turned on and the other turned off with the proper selection of catalyst.

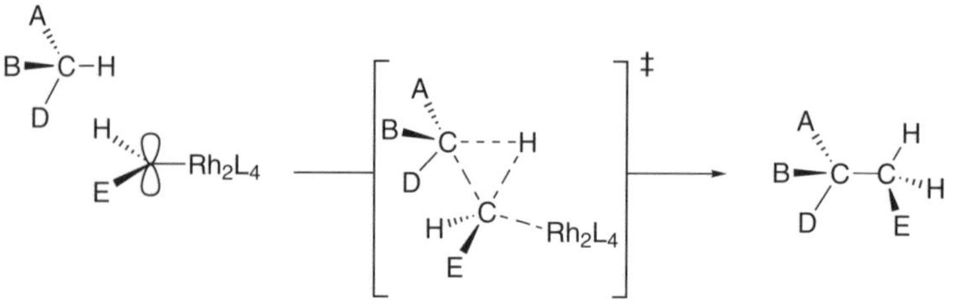

Scheme 12.8

Enantiocontrol in C—H insertion reactions is highly effective for diazoacetates and diazoacetamides,[32] and although various chiral catalysts have been used, **15**[82] and **29**[83] have proven to be the most selective. Examples of biologically active compounds that have been prepared in >90% ee by this methodology include the lignan lactones (e.g., enterolactone **30**),[84] the sugar-based 2-deoxyxylolactone **31**,[85] and the γ-aminobutyric acid (GABA) receptor agonist (*R*)-baclofen **32**.[86]

Until recently, intermolecular C—H insertion reactions were more a curiosity than a synthetically productive undertaking. Davies and Antoulinakis[87] discovered in the late 1990s that aryl- and vinyldiazoacetates undergo intermolecular insertion with a wide variety of hydrocarbons in high yield. With Rh$_2$(*S*-DOSP)$_4$ (**12**, Ar = C$_{12}$H$_{25}$C$_6$H$_4$) moderate-to-high enantioselectivities have been achieved, but diastereoselection is often low to moderate (e.g., Eq. 32[88], 33[89]). Note that a

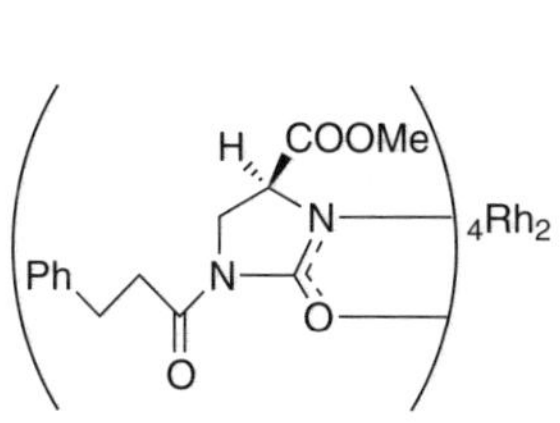

Rh$_2$(4S-MPPIM)$_4$ **(29)**

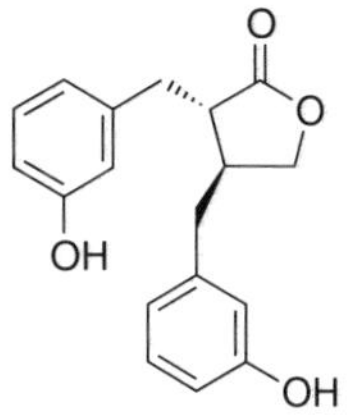

(−)-Enterolactone **(30)**

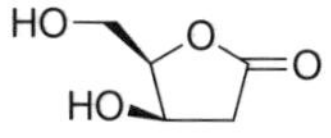

2-deoxyxylolactone **(31)**

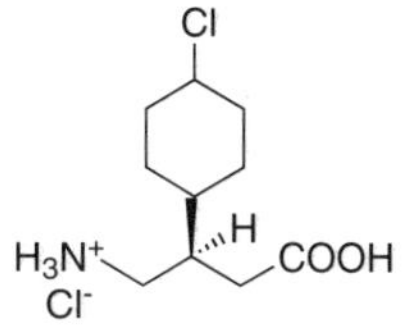

(R)-(−)Baclofen **(32)**

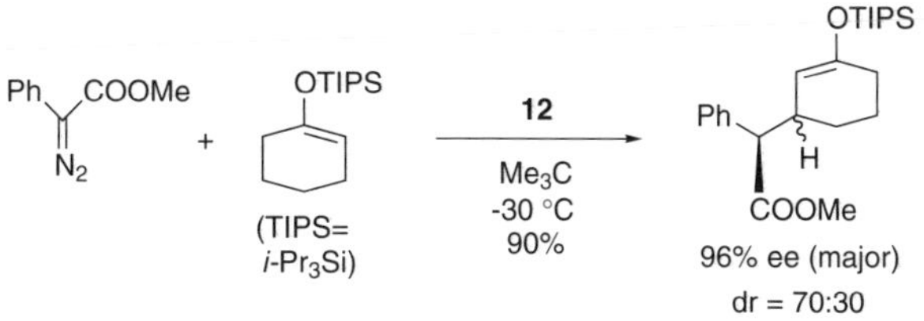

$$(32)$$

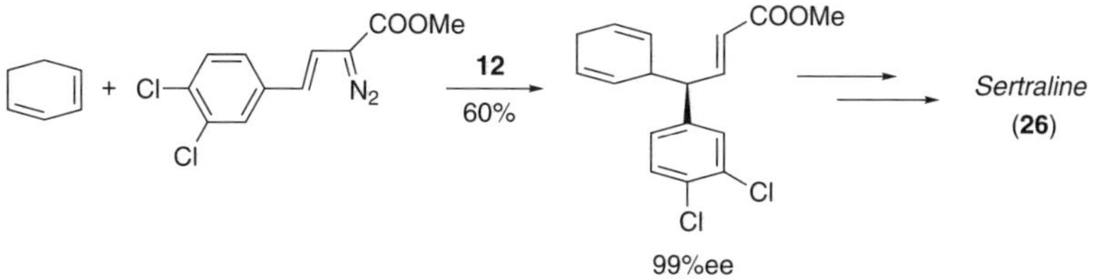

$$(33)$$

most unusual rearrangement must occur in the course of Eq. 33. The relative rates of insertion using methyl phenyldiazoacetate with catalysis by **12** (Scheme 12.9) suggest significant charge separation in the transition state;[88] for comparison, addition to styrene had a relative rate of 24,000.

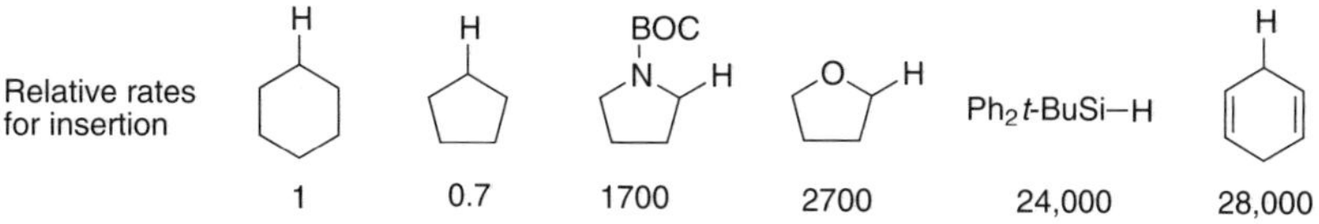

Relative rates for insertion

| 1 | 0.7 | 1700 | 2700 | 24,000 | 28,000 |

BOC = benzyloxycarbonyl

Scheme 12.9

3.1.3. Ylide Formation and Reactions. As electrophiles, metal carbene intermediates in catalytic reactions will capture Lewis bases to form ylide intermediates that are themselves subject to a variety of transformations (e.g., Eqs. 10, 34, 35).[90–92] Alternative base-promoted methodologies do not have the generality afforded by these catalytic methods. In the [2,3]-sigmatropic rearrangement depicted in Eq. 34[91] the transition metal catalysis is obviously associated with the ylide (**34**) in the product forming step, and this was a surprise when first reported. However, this is not the case with all oxygen-centered ylides, and rarely with nitrogen or sulfur ylides. The trapping of carbonyl ylides (e.g., **35**) by dipolarophiles such as DMAD (MeOOCC≡CCOOMe) provides versatility to the overall transformation, and high enantioselectivity has been achieved in one case (Eq. 9).[26] Although catalytic entry into ylides has been widely investigated and is known to be highly versatile for synthesis,[90] stereoselectivity in ylide transformations remains a significant challenge.

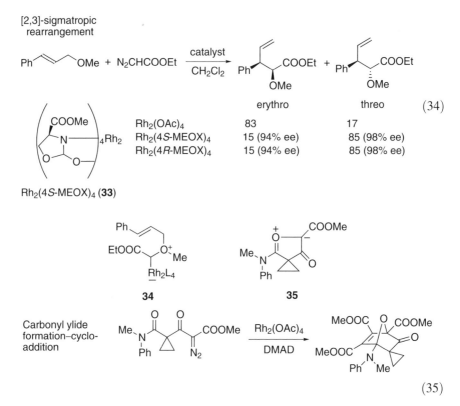

[2,3]-sigmatropic rearrangement

	erythro	threo
$Rh_2(OAc)_4$	83	17
$Rh_2(4S\text{-MEOX})_4$	15 (94% ee)	85 (98% ee)
$Rh_2(4R\text{-MEOX})_4$	15 (94% ee)	85 (98% ee)

(34)

$Rh_2(4S\text{-MEOX})_4$ (**33**)

34

35

Carbonyl ylide formation–cycloaddition

(35)

3.1.4. Other Transformations. Transition metal catalysts also promote reactions of diazocarbonyl compounds that are significantly different from standard addition, insertion, and ylide transformations (e.g., Eqs. 36–39).[6,93–96] These demonstrate the enormous versatility of diazo compounds as metal carbene precursors in organic synthesis. In most cases, the substrate to catalyst ratio is 100, but ratios up to 10,000 have been reported, so catalyst cost is not a major factor in potential pharmaceutical uses.

Aromatic Substitution

(36)

(36)[93]

Aromatic Cycloaddition

(37)[94]

N—H Insertion

(38)[95]

Cycloaddition

(39)[96]

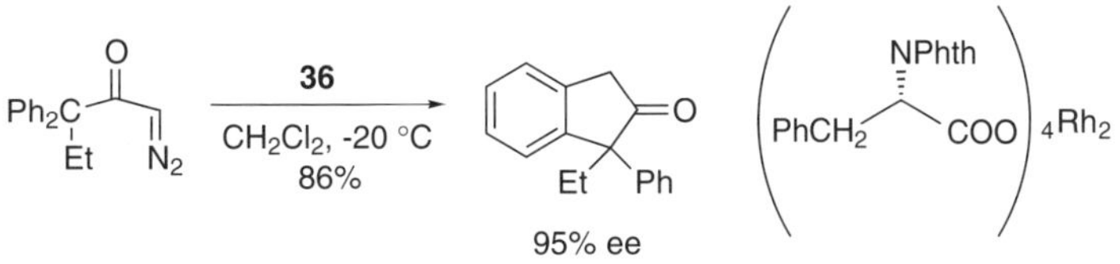

3.2. Synthetic Versatility of Fischer Carbene Complexes

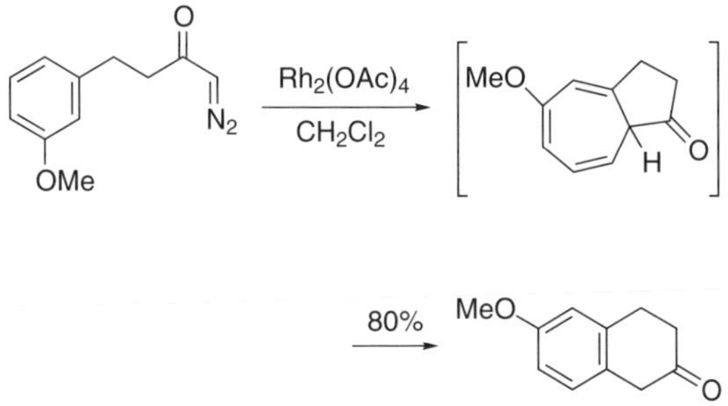

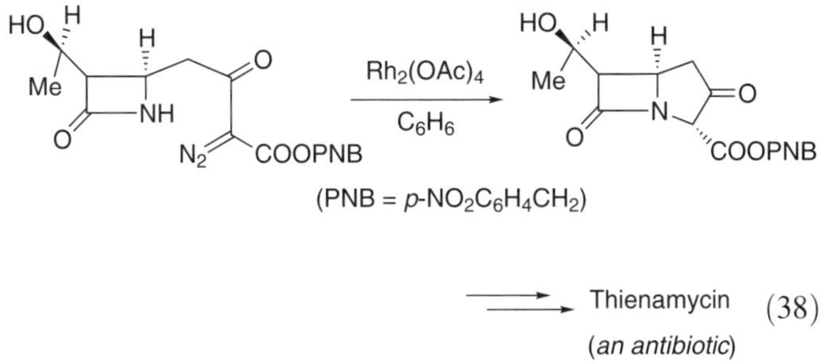

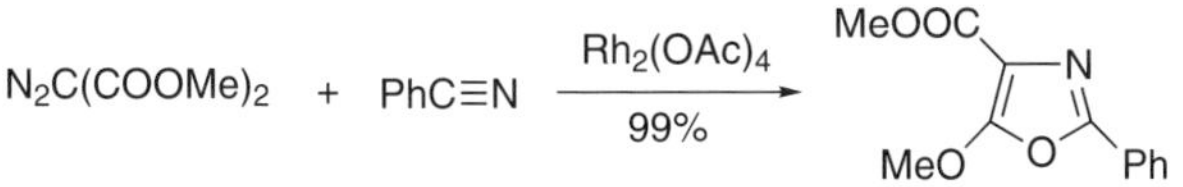

Stoichiometric metal carbene reagents undergo reactions that typify those of their catalytic counterparts,[7,9,97] but it is often the appendages and carbon monoxide ligands of the metal that provide the synthetic versatility of these reagents. Both cyclopropanation (Eq. 40)[98] and carbon–hydrogen insertion (Eq. 41)[99] are well

Cyclopropanation

(40)

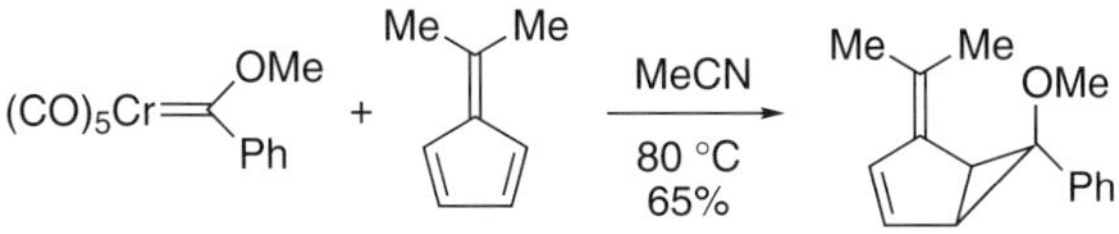

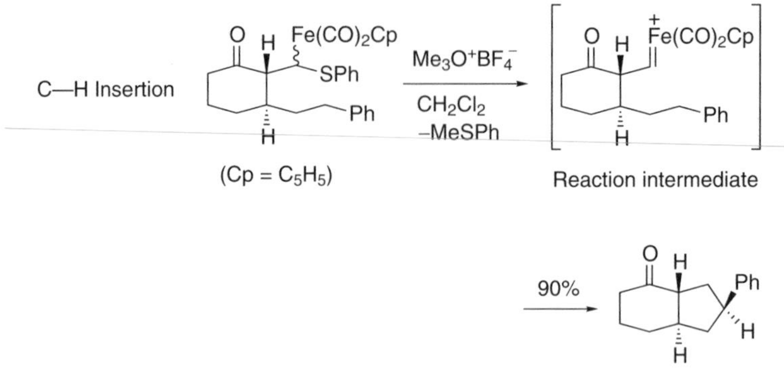

C—H Insertion

(Cp = C$_5$H$_5$)   Reaction intermediate   (41)

known but, as expected from relative metal carbene stabilities, cyclopropanation is sluggish with Fischer carbenes, and they do not undergo C—H insertion. As indicated by Eq. 41, the nonheteroatom stabilized metal carbenes are sufficiently reactive to undergo C—H insertion.

The major direction taken with Fischer carbenes, however, has been annulation reactions (e.g., Eqs. 11–13) rather than cyclopropanation and insertion. Here, the dissociation of carbon monoxide initiates the sequence of events that lead to product (e.g., Eq. 42).[100] Alternatively, an unsaturated unit conjugated with the carbene

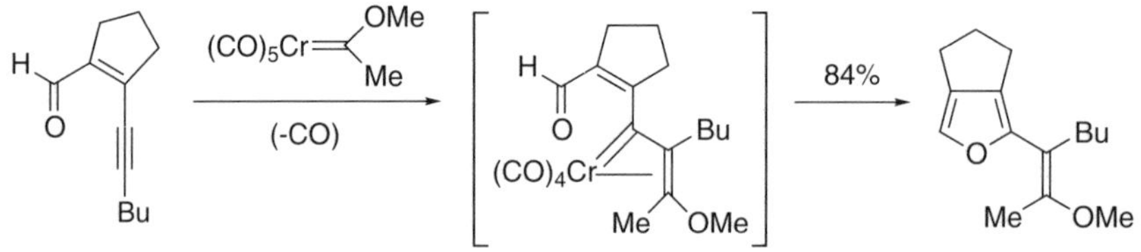

(42)

controls the reaction (e.g., Eq. 43).[101] This is a mature area for synthesis and includes macrocyclization[102] and [3+2] cycloaddition.[103]

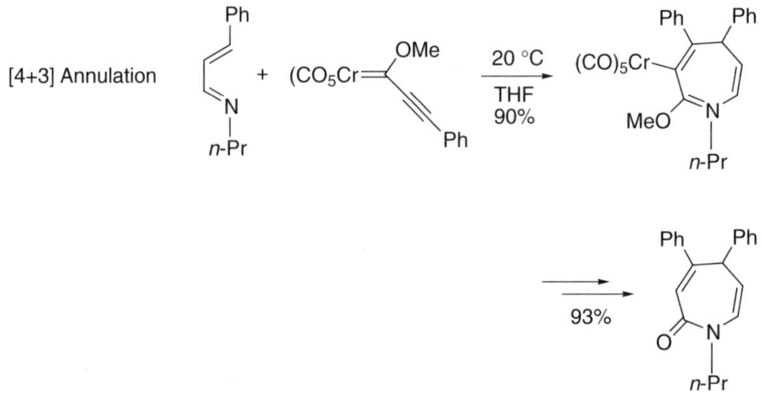

[4+3] Annulation

(43)

3.3. Synthetic Advantages of Ring-Closing Metathesis

Functional group tolerance and its suitability for ring formation of any practical size beyond four has made ring-closing metathesis (RCM) one of the most valuable synthetic methodologies now available. Examples include RCM directed toward the synthesis of natural products (**37–39**) as well as specific strategies for ring construction based on RCM.[104] Among the challenges that remain are stereoselectivity, and here use of the N-heterocyclic carbene-ligated ruthenium (**7**) shows considerable

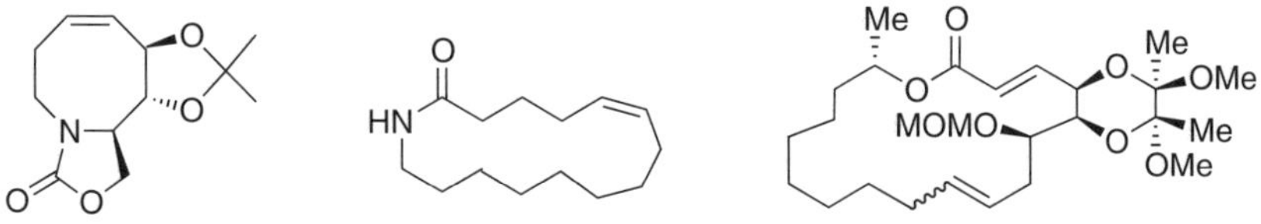

37 ⟶ (+)-aristraline[105]   **38** ⟶ motuporamines[106]   **39** ⟶ (+)-aspicilin[107]

promise (Eq. 44).[108] Modest enantioselection has been achieved with use of **41** (Eq. 45),[109] but much higher selectivities (up to 98% ee for **40**)[110] have been

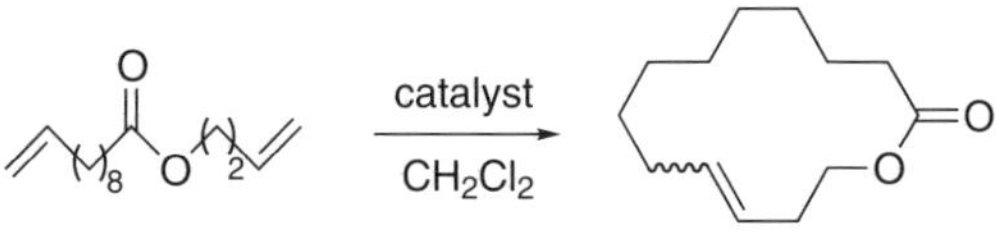

7	40 min	(*E/Z*) = 11.5
6	5 h	(*E/Z*) = 4.5

(44)

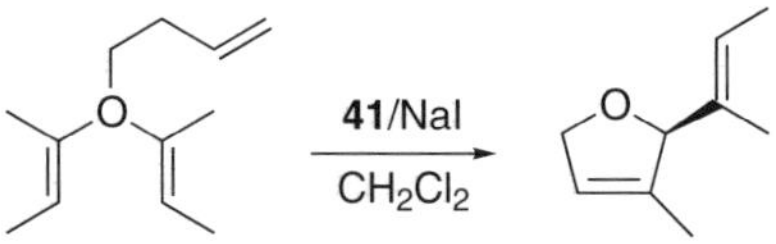

40: 85% ee

(45)

obtained with **42** and its biphenyl analogues (e.g., Eq. 46).[111] Catalytic asymmetric ring-opening metathesis coupled with subsequent cross-metathesis has also been achieved in a clever transformation (Eq. 47)[112] that takes advantage of the susceptibility of norbornene to ring opening.

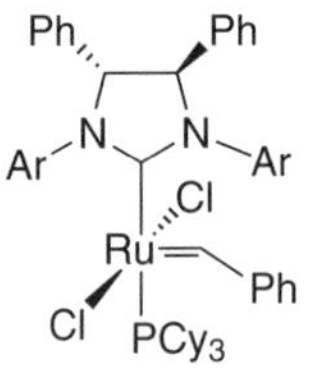

41

(Ar = *o*-MeC$_6$H$_4$)

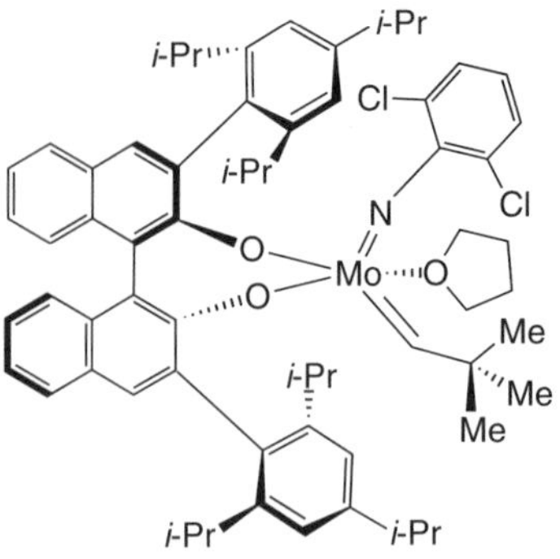

42

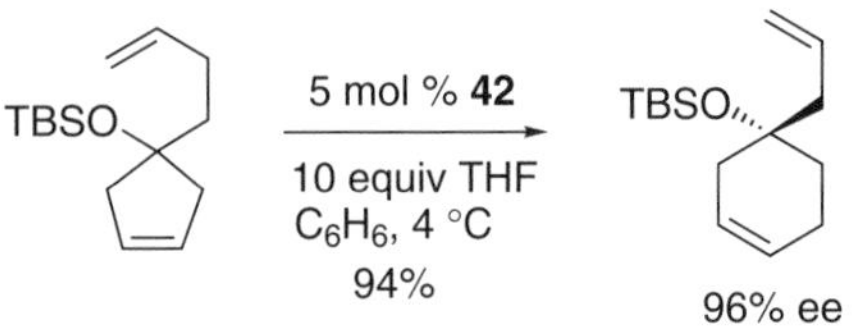

$$\text{(46)}$$

96% ee

TBS = *tert*-butyldimethysilyl

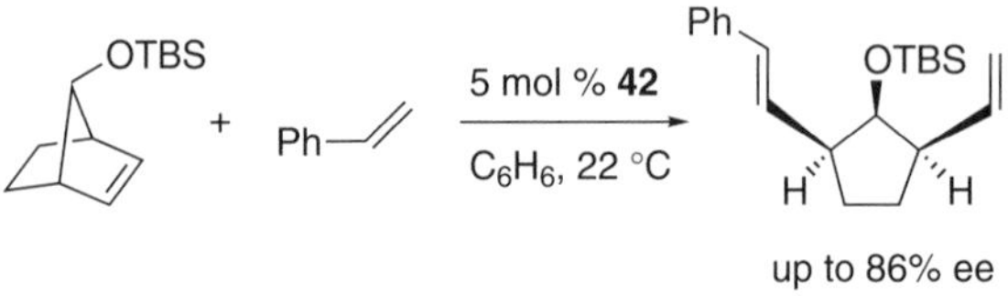

$$\text{(47)}$$

up to 86% ee

New applications continue to demonstrate the enormous versatility of RCM for organic synthesis. Examples include triple ring closing (Eq. 48)[113] and alkyne metathesis,[114] an example being that of cross-metathesis that provides an efficient synthetic strategy for prostaglandin E$_2$ (Eq. 49).[115] Amines and alcohols deactivate metathesis catalysts, but their protection as ethers, esters, and amides allows them to be incorporated into the designated transformation.

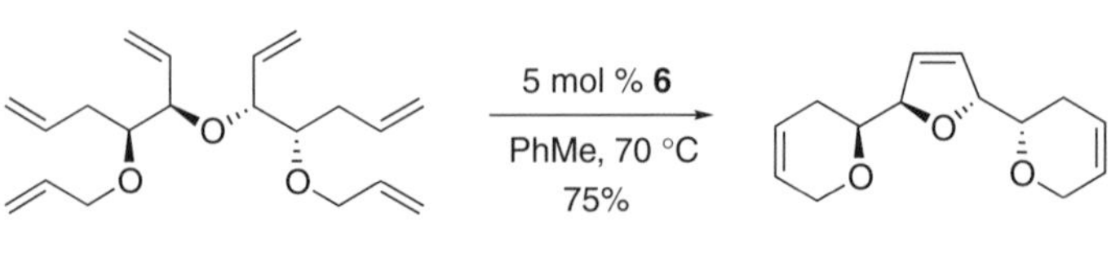

$$\text{(48)}$$

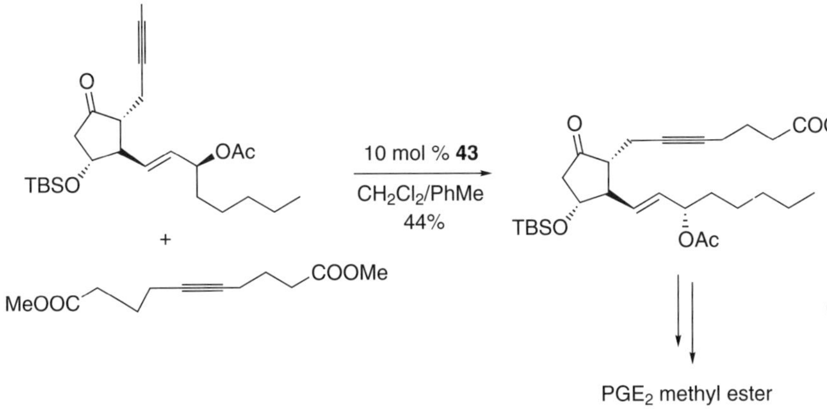

$$(49)$$

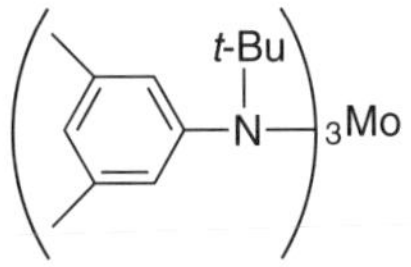

4. METAL NITRENES IN ORGANIC SYNTHESIS

If metal carbene chemistry can be said to be mature, metal nitrene chemistry is in its infancy. Although the first report of a catalytic process used benzenesulfonyl azide,[116] high temperatures were required, and no one has yet provided a synthetically viable method to use azides as sources of nitrenes. Instead, iminophenyliodinanes (**44**), formed from the corresponding sulfonamide by oxidation with diacetoxyiodobenzene, PhI(OAc)$_2$,[117] and chloramine-T or bromamine-T (**45**) are the standard precursors for nitrenes.

Catalytic methods are suitable for nitrene transfer,[118] and many of those found to be effective for carbene transfer are also effective for these reactions. However, 5- to 10-times more catalyst is commonly required to take these reactions to completion, and catalysts that are sluggish in metal carbene reactions are unreactive in nitrene transfer reactions. An exception is the copper(II) complex of a 1,4,7-triazacyclononane for which aziridination of styrene occurred in high yield, even with 0.5 mol% of catalyst.[119] Both addition and insertion reactions have been developed.

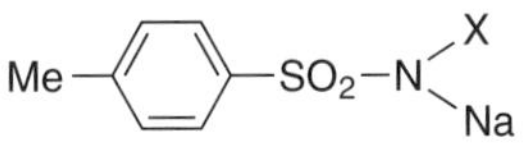

However, comparable transformations to those of stable Fischer carbenes are not available.

Copper catalysts were the first to be employed, and with the bis(oxazoline)-ligated catalyst **46** enantioselectivites up to 97% have been achieved (e.g., Eq. 50).[120] In contrast to metal carbene reactions using diazo compounds, α,β-unsaturated substrates undergo reaction with the nitrene intermediate. In one study, the

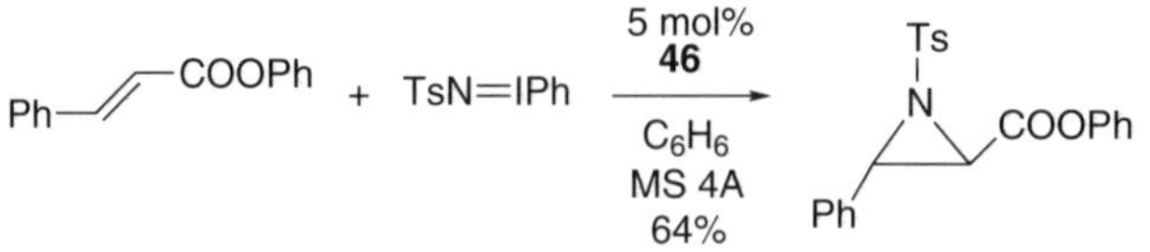

$$\text{Ph}\diagup\diagup\text{COOPh} + \text{TsN=IPh} \xrightarrow[\substack{\text{C}_6\text{H}_6 \\ \text{MS 4A} \\ 64\%}]{\substack{5 \text{ mol}\% \\ \mathbf{46}}} \text{Ph}\diagdown\underset{\text{N}}{\overset{\text{Ts}}{\triangle}}\text{COOPh} \qquad (50)$$

induction of enantioselectivity is proposed to occur by ion pairing of the cationic copper catalyst (**46**) in a chiral pocket formed from the chiral borate formed with 1,1'-bi-2-naphthol (BINOL).[121] Chiral bis(salicylidene)ethylenediamine(salen) complexes (**47**) with CuOTf also gave high selectivities for aziridination.[122] With

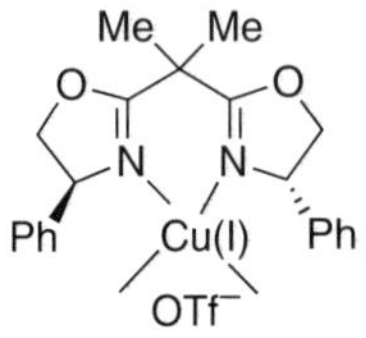

46

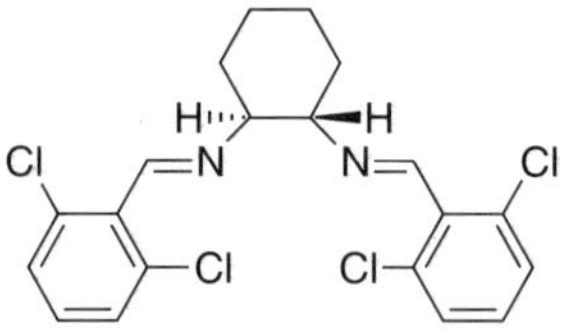

47

dirhodium(II) catalysts, the highest selectivities (50–75% ee) were achieved with rhodium(II) bis(naptholphosphate) **48** and the more reactive NsN=IPh (Ns = p-NO$_2$C$_6$H$_4$SO$_2$).[123]

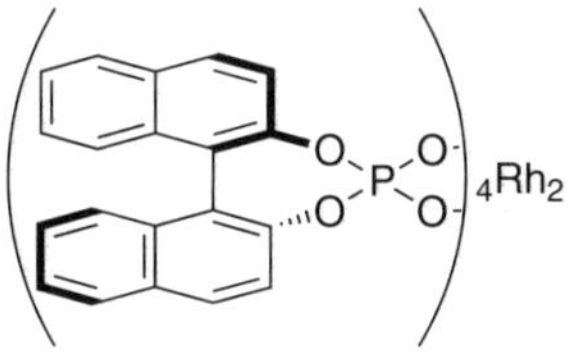

48

Thus far, enantioselective intramolecular aziridination via metal nitrene intermediates has not been successful. Bromamine-T has recently been shown to be a viable source of nitrene for addition to alkenes in copper halide catalyzed reactions,[124] and so has iodosylbenzene (PhI=O)[125] that forms **44** *in situ*. Conceptually, aziridination does not necessarily fall between cyclopropanation and epoxidation, as some have suggested. Instead, metal nitrene chemistry has unique problems and potential advantages associated with the electron pair at nitrogen that are yet to be fully overcome.

The mechanism of the copper-catalyzed aziridination of alkenes using **44** as the nitrene source has been described with the steps shown in Scheme 12.10 contributing

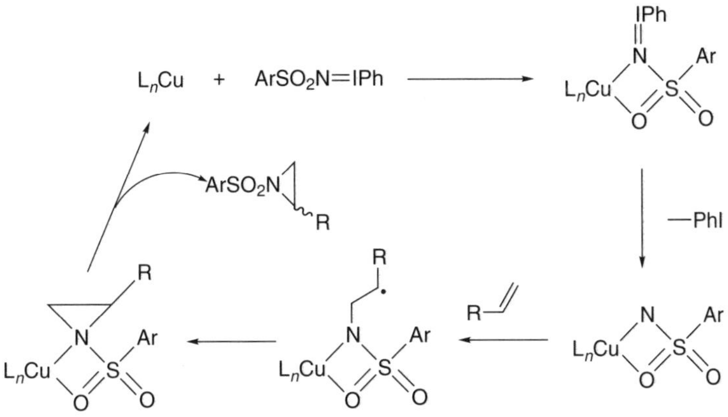

Scheme 12.10

to the overall pathway.[126] This study explained why both Cu(I) and Cu(II) salts yield the same active species in a catalytic cycle that involves Cu(I)/Cu(III) and radical intermediates, as first postulated by Jacobsen and co-workers.[122a,b] This pathway explains why metal nitrene access via azides is so difficult to achieve.

Like carbene insertions into carbon–hydrogen bonds, metal nitrene insertions occur in both intermolecular and intramolecular reactions.[118,127,128] For intermolecular reactions, a manganese(III) *meso*-tetrakis(pentafluorophenyl)porphyrin complex gives high product yields and turnovers up to 2600;[128] amidations could be effected directly with amides using $PhI(OAc)_2$ (Eq. 51). The most exciting development in intramolecular C–H reactions thus far has been the oxidative cyclization of sulfamate esters (e.g., Eq. 52),[129] as well as carbamates (to oxazolidin-2-ones),[130] and one can expect further developments that are of synthetic advantage.[131] Still the challenge of imposing enantiocontrol on these reactions is daunting.

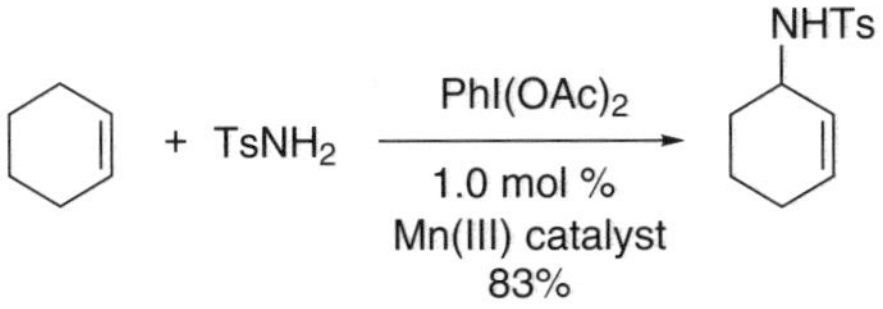

$$(51)$$

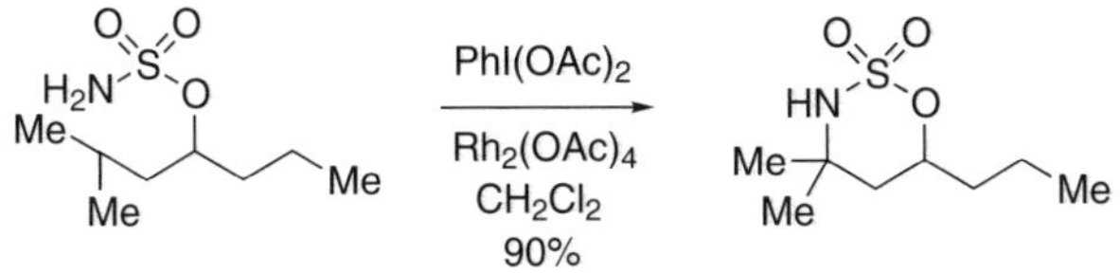

$$(52)$$

Developing analogues to carbene metathesis with nitrene chemistry is just now being explored, but few examples offer significant promise. Transfer of a nitrene to a metal from a diazene has been reported,[132] and reactions of imido complexes with aldehydes and imines are known,[133] but a catalytic approach is not yet evident.

5. CONCLUSION AND OUTLOOK

In contrast to considerations of 50 years ago, today carbene and nitrene chemistries are integral to synthetic design and applications. Always a unique methodology for the synthesis of cyclopropane and cyclopropene compounds, applications of carbene chemistry have been extended with notable success to insertion reactions, aromatic cycloaddition and substitution, and ylide generation and reactions. And metathesis is in the lexicon of everyone planning the synthesis of an organic compound. Intramolecular reactions now extend to ring sizes well beyond 20, and insertion reactions can be effectively and selectively implemented even for intermolecular processes.

Key to these applications has been the control of selectivity–chemoselectivity, regioselectivity, diastereoselectivity, and, especially, enantioselectivity. Here the design of chiral catalysts, begun already in the 1960s, has allowed access to one product when multiple products would, in the past, have been expected. Both electronic and steric control are important, and different metal ions with their associated chiral ligands can have unexpected effects.

Catalytic processes using diazo chemistry and stoichiometric methods with Fischer carbenes are complimentary for the introduction of a substituted carbene into a molecule. For methylene addition, however, there is no viable alternative to the modified Simmons–Smith reaction. Ring–closing metathesis, and its ROMP counterpart, have matured so fast that even now they rank among the most useful synthetic methodologies in organic chemistry.

So what is left to be accomplished? During the current decade one can expect further asymmetric applications and catalyst designs for metathesis reactions, a maturing of chiral catalyst development for cyclopropanation and insertion with increasing synthetic applications, and decreased reliance on traditional Fischer carbenes in synthesis. Major changes remain for ylide applications, especially those that can be enantioselective, in catalytic carbene chemistry, and advances in nitrene chemistry that are comparable to those achieved over the years in carbene chemistry are in their infancy.

Major challenges remain in catalyst development. There will be continuing efforts to increase turnover numbers and rates and to perform these reactions under environmentally friendly conditions. Transition metals, especially Cu, Rh, Ru, Co, Mo, and Zn are effective today; will other metals with attendant ligands be found whose electronic and steric properties are superior to those currently optimized?

Carbene delivery in catalytic reactions remains a challenge. Although diazocarbonyl compounds are relatively safe, and numerous commercial processes have used and continue to employ these materials, methods for diazo transfer using azides are of concern, and cost-effective alternatives are not evident. Also elusive are structures that could deliver stabilized carbenes, not unlike those of Fischer carbenes, in catalytic processes.

SUGGESTED READING

H. M. L. Davies and E. G. Antoulinakis, "Intermolecular Metal-Catalyzed Carbenoid Cyclopropanations," *Org. React. (N. Y.)* **2001**, *57*, 1.

M. P. Doyle and D. C. Forbes, "Recent Advances in Asymmetric Catalytic Metal Carbene Transformations," *Chem. Rev.* **1998**, *98*, 911.

A. Fürstner, "Olefin Metathesis and Beyond," *Angew. Chem. Int. Ed. Engl.* **2000**, *39*, 3012.

D. M. Hodgson, F. Y. T. M. Pierand, and P. A. Stupple, "Catalytic Enantioselective Rearrangements and Cycloadditions Involving Ylides from Diazo Compounds," *Chem. Soc. Rev.* **2001**, *30*, 50.

H. Lebel, J.-F. Marcoux, C. Molinaro, and A. B. Charette, "Stereoselective Cyclopropanation Reactions," *Chem. Rev.* **2003**, *103*, 977.

P, Müller, "Transition Metal-Catalyzed Nitrene Transfer: Aziridination and Insertion," in *Advances in Catalytic Processes*, Vol. 2, M. P. Doyle, Ed., JAI Press, Greenwich, CT, pp. 113 *f*.

T. M. Trnka and R. H. Grubbs, "The Development of $L_2X_2Ru=CHR$ Olefin Metathesis Catalysts: An Organometallic Success Story," *Acc. Chem. Res.* **2001**, *34*, 18.

W. D. Wulff, "Transition Metal Carbene Complexes: Alkyne and Vinyl Ketene Chemistry," in *Comprehensive Organometallic Chemistry II*, Vol. 12, L. S. Hegedus, Ed., Pergamon, Tarrytown, NY, 1995, Chapter 5.3, pp. 469 *f*.

REFERENCES

1. J. Hine, *Physical Organic Chemistry*, 2nd ed., McGraw-Hill, New York, **1962**.

2. J. March, *Advanced Organic Chemistry: Reactions, Mechanisms and Structures*; McGraw-Hill, New York, **1968**.

3. P. Yates, *J. Am. Chem. Soc.* **1952**, *74*, 5376.

4. E. O. Fischer and K. H. Dötz, *Chem. Ber.* **1970**, *103*, 1273.

5. K. H. Dötz, *Angew. Chem., Int. Ed. Engl.* **1984**, *23*, 587. K. H. Dötz, *Transition Metal Carbene Complexes*; VCH Publishing, New York, **1983**.

6. M. P. Doyle, M. A. McKervey, and T. Ye, *Modern Catalytic Methods for Organic Synthesis with Diazo Compounds*, John Wiley & Sons, Inc., New York, **1998**.

7. W. D. Wulff, in *Advances in Metal-Organic Chemistry*, Vol. 1, L. Liebeskind, Ed., JAI Press, Greenwich, CT, **1989**.

8. (a) R. H. Crabtree, *The Organic Chemistry of the Transition Metals,* 3rd ed., John Wiley & Sons, Inc., New York, **2000**. (b) J. P. Collman, L. S. Hegedus, J. R. Norton, and R. G. Finke, *Principles and Applications of Organotransition Metal Chemistry*, 2nd ed., University Science Books, Herndon, VA, **1987**.

9. M. Brookhart and W. B. Studabaker, *Chem. Rev.* **1987**, *87*, 411.

10. C. P. Casey, S. W. Polichnowski, A. J. Shustermann, and C. R. Jones, *J. Am. Chem. Soc.* **1979**, *101*, 7282.

11. (a) R. R. Schrock and G. W. Parshall, *Chem. Rev.* **1976**, *76*, 243. (b) P. J. Davidson, M. F. Lappert, and R. Pearce, *Chem. Rev.* **1976**, *76*, 219. (c) T. M. Trnka and R. H. Grubbs, *Acc. Chem. Res.* **2001**, *34*, 18. (d) H. Eleuterio, *CHEMTECH* **1991**, 92. (e) F. Z. Dorwald and

R. H. Grubbs, *Metal Carbenes in Organic Synthesis*, VCH Publishing, New York, **1999**.

12. R. R. Shrock, *Acc. Chem. Res.* **1990**, *23*, 158.

13. S. T. Nguyen, L. K. Johnson, R. H. Grubbs, and J. W. Ziller, *J. Am. Chem. Soc.* **1992**, *114*, 3974.

14. (a) M. Scholl, T. M. Trnka, J. P. Morgan, and R. H. Grubbs, *Tetrahedron Lett.* **1999**, *40*, 2247. (b) J. Huang, E. D. Stevens, S. P. Nolan, and J. L. Peterson, *J. Am. Chem. Soc.* **1999**, *121*, 2674.

15. (a) M. Scholl, S. Ding, C. W. Lee, and R. H. Grubbs, *Org. Lett.* **1999**, *1*, 953. (b) L. Jafarpour, A. C. Hillier, and S. P. Nolan, *Organometal.* **2002**, *21*, 442.

16. C. P. Casey and M. C. Cesa, *Organometal.* **1982**, *1*, 87.

17. G. Stork and J. Ficini, *J. Am. Chem. Soc.* **1961**, *83*, 4678.

18. S. D. Burke and P. A. Grieco, *Org. React. (N. Y.)*, **1979**, *26*, 361.

19. K. Mori, M. Ohki, A. Kobayashi, and M. Matsui, *Tetrahedron*, **1970**, *26*, 2815.

20. F. Medina and A. Majarrez, *Tetrahedron* **1964**, *20*, 1807.

21. D. H. Rogers, E. C. Yi, and C. D. Poulter, *J. Org. Chem.* **1995**, *60*, 941.

22. D. A. Evans, K. A. Woerpel, M. M. Hinman, and M. M. Faul, *J. Am. Chem. Soc.* **1991**, *113*, 726.

23. (a) H. M. L. Davies, P. R. Bruzinski, D. H. Lake, N. Kong, and M. J. Fall, *J. Am. Chem. Soc.* **1996**, *118*, 6897. (b) M. A. McKervey, and T. Ye, *J. Chem. Soc., Chem. Commun.* **1992**, 823.

24. (a) M. P. Doyle, R. J. Pieters, S. F. Martin, R. E. Austin, C. J. Oalmann, and P. J. Müller, *J. Am. Chem. Soc.* **1991**, *113*, 1423. (b) M. P. Doyle, W. R. Winchester, M. N. Protopopova, A. P. Kazula, and L. J. Westrum, *Org. Syn.* **1996**, *73*, 13. (c) For applications with polymer supported catalyst: M. P. Doyle, D. J. Timmons, J. S. Tumonis, H.-M. Gau, and E. C. Blossey *Organometal.* **2002**, *21*, 1747.

25. T. Uchida, R. Irie, and T. Katsuki, *Tetrahedron* **2000**, *56*, 3501.

26. S. Kitagaki, M. Anada, O. Kataoka, K. Matsuno, C. Umeda, N. Watanabe, and S.-i. Hashimoto, *J. Am. Chem. Soc.* **1999**, *121*, 1417.

27. T. Ikeno, M. Sato, H. Sekino, A. Nishizaka, and T. Yamada, *Bull. Chem. Soc. Jpn.* **2001**, *74*, 2139.

28. M. P. Doyle, R. E. Austin, A. S. Bailey, M. P. Dwyer, A. B. Dyatkin, A. V. Kalinin, M. M. Y. Kwan, S. Liras, C. J. Oalmann, R. J. Pieters, M. N. Protopopova, C. E. Raab, G. H. P. Roos, Q.-L. Zhou, and S. F. Martin, *J. Am. Chem. Soc.* **1995**, *117*, 5763.

29. H. M. L. Davies, T. Hansen, and M. R. Churchill, *J. Am. Chem. Soc.* **2000**, *122*, 3063.

30. M. P. Doyle, W. Hu, and D. J. Timmons, *Org. Lett.* **2001**, *3*, 933.

31. H. M. L. Davies and E. G. Antoulinakis, *Org. React. (N. Y.)* **2001**, *57*, 1.

32. M. P. Doyle and D. C. Forbes, *Chem. Rev.* **1998**, *98*, 911.

33. M. P. Doyle, in *Catalytic Asymmetric Synthesis*, 2nd ed., I. Ojima, Ed., Wiley-VCH, New York, **2000**.

34. P. Müller, and C. Boléa, *Helv. Chim. Acta* **2001**, *84*, 1093.

35. W. D. Wulff, in *Comprehensive Organometallic Chemistry II*, Vol. 12, L. S. Hegedus, Ed., Pergamon, Tarrytown, N.Y., **1995**, Chapter 5.3.

36. L. S. Hegedus, in *Comprehensive Organometallic Chemistry II*, Vol. 12, L. S. Hegedus, Ed., Pergamon, Tarrytown, N.Y., **1995**, Chapter 5.4.

37. J. P. A. Harrity, W. J. Kerr, and D. Middlemiss, *Tetrahedron* **1993**, *49*, 5565.

38. B. B. Snider, *Chem. Rev.* **1998**, *88*, 793.

39. A. Yamashita, *Tetrahedron Lett.* **1986**, *27*, 5915.

40. G. W. Parshall and S. D. Ittel, *Homogeneous Catalysis*, 2nd ed., John Wiley & Sons, Inc., New York, **1992**.

41. J. S. Moore, in *Comprehensive Organometallic Chemistry II*, Vol. 12, L. S. Hegedus, Ed., Pergamon, Tarrytown, N.Y., **1995**, Chapter 12.2.

42. R. H. Grubbs and W. Tumas, *Science* **1989**, *243*, 907.

43. (a) M. R. Buchmeiser, *Chem Rev.* **2000**, *100*, 1565. (b) U. H. F. Bunz, *Acc. Chem. Res.* **2001**, *34*, 998.

44. K. J. Ivin and J. C. Mol, *Olefin Metathesis and Metathesis Polymerization*; Academic Press, London, **1997**.

45. (a) M. Schuster and S. Blechert, *Angew. Chem., Int. Ed. Engl.* **1997**, *36*, 2036. (b) R. H. Grubbs and S. Chang, *Tetrahedron* **1998**, *54*, 4413. (c) A. Fürstner, *Angew. Chem., Int. Ed. Engl.* **2000**, *39*, 3012.

46. (a) P. Schwab, M. P. France, J. W. Ziller, and R. H. Grubbs, *Angew Chem., Int. Ed. Engl.* **1995**, *34*, 2039. (b) For applications with a polymer-supported catalyst: L. Jafarpour and S. P. Nolan, *Org. Lett.* **2000**, *2*, 4075.

47. A. Fürstner, O. R. Thiel, and L. Ackerman, *Org. Lett.* **2001**, *3*, 449.

48. A. K. Chatterjee, J. P. Morgan, M. Scholl, and R. H. Grubbs, *J. Am. Chem. Soc.* **2000**, *122*, 3783.

49. J. R. Stille, in *Comprehensive Organometallic Chemistry II*, Vol.12, L. S. Hegedus, Ed., Pergamon, Tarrytown, N.Y., **1995**, Chapter 5.5.

50. (a) W.-H. Fang, D. L. Phillips, D. Wang, and Y.-L. Li, *J. Org. Chem.* **2002**, *67*, 154. (b) E. Nakamura, A. Hirai, and M. Nakamura, *J. Am. Chem. Soc.* **1998**, *120*, 5844.

51. F. N. Tebbe, G. W. Parhsall, and G. S. Reddy, *J. Am. Chem. Soc.* **1978**, *100*, 3611.

52. K. C. Nicolaou, M. D. D. Postema, and C. F. Claiborne, *J. Am. Chem. Soc.* **1996**, *118*, 1565.

53. D. A. Straus and R. H. Grubbs, *Organometal.* **1982**, *1*, 1658.

54. A. B. Charette, in *Organozinc Reagents: A Practical Approach*, P. Knochel, and P. Jones, Eds., Oxford University Press, Oxford, **1999**.

55. S. E. Denmark, J. P. Edwards, and S. R. Wilson, *J. Am. Chem. Soc.* **1992**, *114*, 2592.

56. A. H. Hoveyda, D. A. Evans, and G. C. Fu, *Chem. Rev.* **1993**, *93*, 1307.

57. D. Cheng, T. Kreethadumrongdat, and T. Cohen, *Org. Lett.* **2001**, *3*, 2121. The first report on the directing ability of hydroxyl groups was S. Winstein, J. Sonnenberg, and L. deVries, *J. Am. Chem. Soc.* **1959**, *81*, 6523.

58. A. B. Charette, A. Beauchemin, and S. Francoeur, *J. Am. Chem. Soc.* **2001**, *123*, 8139.

59. A. B. Charette, A. Gagnon, and J.-F. Fournier, *J. Am. Chem. Soc.* **2002**, *124*, 386.

60. A. B. Charette, C. Molinaro, and C. Brochu, *J. Am. Chem. Soc.* **2001**, *123*, 12168.

61. M. Regitz and G. Maas, *Diazo Compounds: Properties and Syntheses*, Academic Press, New York, **1986**.

62. T. Fukuda and T. Katsuki, *Tetrahedron* **1997**, *53*, 7201.

63. T. Niimi, T. Uchida, R. Irie, and T. Katsuki, *Tetrahedron Lett.* **2000**, *41*, 3647.

64. (a) M. P. Doyle and W. Hu, *J. Org. Chem.* **2000**, *65*, 8839. (b) M. P. Doyle, W. Hu, B. Chapman, A. B. Marnett, C. S. Peterson, J. P. Vitale, and S. A. Stanley, *J. Am. Chem. Soc.* **2000**, *122*, 5718.

65. E. J. Corey and T. G. Grant, *Tetrahedron Lett.* **1994**, *35*, 5373.

66. M. P. Doyle and W. Hu, *Adv. Synth. Catal.* **2001**, *343*, 299.

67. T. G. Grant, M. C. Noe, and E. J. Corey, *Tetrahedron Lett.* **1995**, *36*, 8745.

68. (a) M. P. Doyle, M. N. Protopopova, P. Müller, D. Ene, and E. A. Shapiro, *J. Am. Chem. Soc.* **1994**, *116*, 8492. (b) P. Müller and H. Imogaï, *Tetrahedron: Asymmetry* **1998**, *9*, 4419.

69. M. P. Doyle and W. Hu, *Synlett* **2001**, 1364.

70. M. P. Doyle, C. S. Peterson, M. N. Protopopova, A. B. Marnett, D. L. Parker, Jr., D. G. Ene, and V. Lynch, *J. Am. Chem. Soc.* **1997**, *119*, 8826.

71. M. P. Doyle, D. G. Ene, C. S. Peterson, and V. Lynch, *Angew. Chem. Int. Ed. Engl.* **1999**, *38*, 700.

72. M. P. Doyle and M. N. Protopopova, *Tetrahedron* **1998**, *54*, 7919.

73. (a) R. Tokunoh, B. Fähndrich, and A. Pfaltz, *Synlett* **1995**, 491. (b) M. Barberis, J. Perez-Prieto, S.-E. Stiriba, and P. Lahuerta, *Org. Lett.* **2001**, *3*, 3317, 4325.

74. D. F. Taber, in *Houben-Weyl: Methods of Organic Chemistry*, Vol. E2la, G. Helmchen, Ed., Georg Thieme Verlag, Stuttgart, Germany, **1995**, Chapter 1.2.

75. D. F. Taber and E. H. Petty, *J. Org. Chem.* **1982**, *47*, 4808.

76. M. P. Doyle, M. S. Shanklin, S.-M. Oon, H.-Q. Pho, F. R. van der Heide, and W. R. Veal, *J. Org. Chem.* **1988**, *53*, 3384.

77. A. Padwa, D. J. Austin, A. T. Price, M. A. Semonis, M. P. Doyle, M. N. Protopopova, W. R. Winchester, and A. Tran, *J. Am. Chem. Soc.* **1993**, *115*, 8669.

78. D. F. Taber and R. E. Ruckle, Jr., *J. Am. Chem. Soc.* **1986**, *108*, 7686.

79. P. Wong and J. Adams, *J. Am. Chem. Soc.* **1994**, *116*, 3296.

80. E. Nakamura, N. Yoshikai, and M. Yamanaka, *J. Am. Chem. Soc.* **2002**, *124*, 7181.

81. M. P. Doyle, L. J. Westram, W. N. E. Wolthuis, M. M. See, W. P. Boone, V. Bagheri, and M. M. Pearson, *J. Am. Chem. Soc.* **1993**, *115*, 958.

82. H. Saito, H. Oishi, S. Kitagaki, S. Nakamura, M. Anada, and S. Hashimoto, *Org. Lett.* **2002**, *4*, 3887.

83. M. P. Doyle, Q.-L. Zhou, C. E. Raab, G. H. P. Roos, S. H. Simonsen, and V. Lynch, *Inorg. Chem.* **1996**, *35*, 6064.

84. J. W. Bode, M. P. Doyle, M. N. Protopopova, and Q.-L. Zhou, *J. Org. Chem.* **1996**, *61*, 9146.

85. M. P. Doyle, J. S. Tedrow, A. B. Dyatkin, C. J. Spaans, and D. G. Ene, *J. Org. Chem.* **1999**, *64*, 8907.

86. M. P. Doyle and W. Hu. *Chirality* **2002**, *14*, 169.

87. H. M. L. Davies and E. G. Antoulinakis, *J. Organometal. Chem.* **2001**, *617*, 47.

88. H. M. L. Davies and P. Ren, *J. Am. Chem. Soc.* **2001**, *123*, 2070.

89. H. M. L. Davies, D. G. Stafford, and T. Hansen, *Org. Lett.* **1999**, *1*, 233.

90. (a) A. Padwa and S. F. Hornbuckle, *Chem. Rev.* **1991**, *91*, 263. (b) D. M. Hodgson, F. Y. T. M. Pierand, and P. A. Stupple, *Chem. Soc. Rev.* **2001**, *30*, 50.

91. M. P. Doyle, D. C. Forbes, M. M. Vasbinder, and C. S. Peterson, *J. Am. Chem. Soc.* **1998**, *120*, 7653.

92. A. Padwa, J. P. Snyder, E. A. Curtis, S. M. Sheehan, K. J. Worsencroft, and C. O. Kappe, *J. Am. Chem. Soc.* **2000**, *122*, 8155.

93. N. Watanabe, Y. Ohtake, S.-i. Hashimoto, M. Shiro, and S. Ikegami, *Tetrahedron Lett.* **1995**, *36*, 1491.

94. M. Kennedy, M. A. McKervey, A. R. Maguire, S. M. Tuladhar, and M. F. Twohig, *J. Chem. Soc., Perkin Trans 1* **1990**, 1047.

95. T. N. Salzmann, R. W. Ratcliffe, B. G. Christensen, and F. A. Bouffard, *J. Am. Chem. Soc.* **1980**, *102*, 6161.

96. R. Connell, F. Scavo, P. Helquist, and B. Akermark, *Tetrahedron Lett.* **1986**, *27*, 5559.

97. (a) K. Ruck-Braum, M. Mikulas, and P. Amrhein, *Synthesis* **1999**, 727. (b) W. Petz, *Iron-Carbene Complexes*, Springer-Verlag: Berlin, 1993.

98. J. Barluenga, S. Martinez, A. L. Suárez-Sobrino, and M. Tomás, *J. Am. Chem. Soc.* **2002**, *124*, 5948.

99. S. Ishii, S. Zhao, G. Mehta, C. J. Knors, and P. Helquist, *J. Org. Chem.* **2001**, *66*, 3449.

100. J. W. Herndon and H. Wang, *J. Org. Chem.* **1998**, *63*, 4564.

101. J. Barluenga, M. Tomás, E. Rubio, J. A. López-Pelegrin, S. Garcia-Granda, and P. Pertierra, *J. Am. Chem. Soc.* **1996**, *118*, 695.

102. H. Wang and W. D. Wulff, *J. Am. Chem. Soc.* **1998**, *120*, 10573.

103. H. Kagoshima, T. Okamura, and T. Akiyama, *J. Am. Chem. Soc.* **1998**, *121*, 4516.

104. J. D. White, P. Hrnciar, and F. T. Yokochi, *J. Am. Chem. Soc.* **1998**, *120*, 7359.

105. W. P. D. Goldring and L. Weiler, *Org. Lett.* **1999**, *1*, 1471.

106. D. J. Dixon, A. C. Foster, and S. V. Ley, *Org. Lett.* **2000**, *2*, 123.

107. (a) C. W. Lee, T.-L. Choi, and R. H. Grubbs, *J. Am. Chem. Soc.* **2002**, *124*, 3224. (b) H. Bieräugel, T. P. Jansen, H. E. Shoemaker, H. Hiemstra, J. H. van Maarseveen, *Org. Lett.* **2002**, *4*, 2673. (c) T.-L. Chio, C. W. Lee, A. K. Chatterjee, and R. H. Grubbs, *J. Am. Chem. Soc.* **2001**, *123*, 10417.

108. C. W. Lee and R. H. Grubbs, *Org. Lett.* **2000**, *2*, 2145.

109. T. J. Seiders, D. W. Ward, and R. H. Grubbs, *Org. Lett.* **2001**, *3*, 3225.

110. R. R. Schrock, J. Y. Jamieson, S. J. Dolman, S. A. Miller, P. J. Bonitatebus, Jr., and A. H. Hoveyda, *Organometal.* **2002**, *21*, 409.

111. X. Teng, D. R. Cefalo, R. R. Schrock, and A. H. Hoveyda, *J. Am. Chem. Soc.* **2002**, *124*, 10779.

112. D. S. La, E. S. Sattely, J. G. Ford, R. R. Schrock, and A. H. Hoveyda, *J. Am. Chem. Soc.* **2001**, *123*, 7767.

113. M. P. Heck, C. Baylon, S. P. Molan, and C. Mioskowski, *Org. Lett.* **2001**, *3*, 1989.

114. A. Fürstner, *Angew. Chem. Int. Ed. Engl.* **2000**, *39*, 3012.

115. A. Fürstner and C. Mathes, *Org. Lett.* **2001**, *3*, 221.

116. H. Kwart and A. A. Kahn, *J. Am. Chem. Soc.* **1967**, *89*, 1951.

117. Y. Yamada, T. Yamamoto, and M. Okawara, *Chem. Lett.* **1975**, 361.

118. P. Müller, in *Advances in Catalytic Processes*, Vol. 2, M. P. Doyle, Ed., JAI Press, Greenwich, CT, **1997**.

119. J. A. Halfen, J. K. Hallman, J. A. Schultz, and J. P. Emerson, *Organometal.* **1999**, *18*, 5435.

120. D. A. Evans, M. M. Faul, M. T. Bilodeau, B. A. Anderson, and D. M. Barnes, *J. Am. Chem. Soc.* **1993**, *115*, 5328.

121. D. B. Llewellyn, D. Adamson, and B. A. Arndtsen, *Org. Lett.* **2000**, *2*, 4165.

122. (a) K. Li, K. R. Conser, and E. N. Jacobsen, *J. Am. Chem. Soc.* **1995**, *117*, 5889. (b) Z. Li, R. W. Quan, and E. N. Jacobsen, *J. Am. Chem. Soc.* **1995**, *117*, 5889. (c) For biaryl Schiff bases: K. M. Gillespie, C. J. Sanders, P. O'Shaugnessy, I. Westmoreland, C. P. Thickitt, and P. Scott, *J. Org. Chem.* **2002**, *67*, 3450.

123. (a) P. Müller, C. Baud, Y. Jacquier, M. Moran, and I. Nägeli, *J. Phys. Org. Chem.* **1996**, *9*, 341. (b) P. Müller, C. Baud, and I. Nägeli, *J. Phys. Org. Chem.* **1998**, *11*, 597.

124. B. M. Chandra, R. Vyas, and A. V. Bedekar, *J. Org. Chem.* **2001**, *66*, 30.

125. P. Dauban, L. Saniere, A. Tarradu, and R. H. Dodd, *J. Am. Chem. Soc.* **2001**, *2*, 4165.

126. P. Brandt, M. J. Södergren, P. G. Andersson, and P. Norrby, *J. Am. Chem. Soc.* **2000**, *122*, 8013. See also: M. M. Díaz-Requejo, P. J. Pérez, M. Brookhart, and J. L. Templeton, *Organometal.* **1997**, *16*, 4399.

127. (a) P. Müller, C. Baud, and Y. Jacquier, *Can. J. Chem.* **1998**, *76*, 738. (b) S.-M. Au, J.-S. Huang, W.-Y. Yu, W.-H. Fung, and C.-M. Che, *J. Am. Chem. Soc.* **1999**, *121*, 9120.

128. X.-Q. Yu, J.-S. Huang, X.-G. Zhou, and C.-M. Che, *Org. Lett.* **2000**, *2*, 2233.

129. C. G. Espino, P. M. Wehn, J. Chow, and J. Du Bois, *J. Am Chem. Soc.* **2001**, *123*, 6935.

130. C. G. Espino, and J. Du Bois, *Angew. Chem. Int. Ed. Engl.* **2001**, *40*, 598.

131. A. Padwa and T. Stengel, *Org. Lett.* **2002**, *4*, 2137.

132. M. A. Aubart and R. G. Bergman, *Organometal.* **1999**, *18*, 811.

133. W.-D. Wang and J. H. Espenson, *Organometal.* **1999**, *18*, 5170.

CHAPTER 13

Nitrenium Ions

DANIEL E. FALVEY

Department of Chemistry and Biochemistry, University of Maryland, College Park, MD

Reactive Intermediate Chemistry, edited by Robert A. Moss, Matthew S. Platz, and Maitland Jones, Jr.
ISBN 0-471-23324-2   Copyright © 2004 John Wiley & Sons, Inc.

1. INTRODUCTION

1.1. Definitions of Nitrenium Ions and Various Subclasses

A nitrenium ion is a molecular species characterized by a nitrogen atom that bears a formal positive charge and two covalent bonds. Thus the simplest example of this class of reactive intermediates is NH_2^+, which is also known as the imidogen ion. The higher homologues occur when one or both of the hydrogen atoms is substituted with larger groups, including alkyl, aryl, halides, and so on. Nitrenium ions are similar to nitrenes in that both classes possess a nitrogen atom with only six valence electrons (Fig. 13.1). However, they are distinct in that nitrenium ions are dicoordinate and cationic, whereas nitrenes are monocoordinate and neutral. Nitrenium ions are similar to carbenium ions in that they are cationic, but of course, have nitrogen, rather than carbon as the atom formally bearing the positive charge. Nitrenium ions are related to carbenes in that both species have a dicoordinate central atom, which in turn has two nonbonding orbitals and two nonbonding electrons. Indeed, like carbenes, nitrenium ions can exist in either ground-state singlet or triplet states, depending on the R and R$'$ groups, Structures of several specific nitrenium ions are provided in Figure 13.2, including the parent system **1**, methylnitrenium ion **2**, phenylnitrenium ion **3**, 4-methoxyphenylnitrenium ion **4**, *N*-acetyl-*N*-(2-fluorenyl)nitrenium ion **5**, 4-(4$'$-*N*,*N*-dimethylaminophenyl)phenylnitrenium ion **6**, and 1,3-dimethyltriazolium ion **7**.

This simple definition of nitrenium ion may seem reasonably straightforward. However, a detailed consideration reveals that there is some ambiguity in any definition. In fact, most nitrenium ions exhibit significant delocalization of the positive charge from nitrogen onto the ligands. This is especially true when the ligand is a π donor such as a benzene ring. However, calculations on singlet alkylnitrenium ions show that there is substantial hyperconjugation from the vicinal sigma bonds. Consider the series of ions depicted in Figure 13.2. While it is clear that NH_2^+, by anyone's definition, can safely be regarded as a nitrenium ion, the case becomes increasingly unclear as one moves from left to right in the figure. Phenylnitrenium ion **3** is, of course, a very short lived and highly reactive intermediate. However, density functional theory (DFT) calculations [which have been validated by time-resolved infrared (IR) measurements on some substituted analogues] show that this species experiences considerable charge delocalization into the phenyl ring as well as bond order alternation that would be expected for an imino-cyclohexadienyl cation.[1] The arylnitrenium ion **6** can be regarded as an *N*-methylated derivative

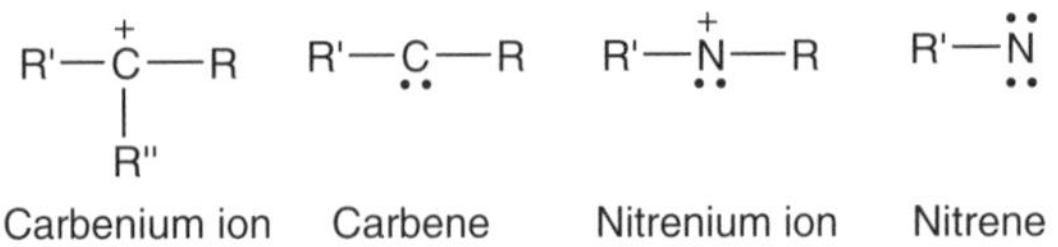

Figure 13.1. Nitrenium ions and related species.

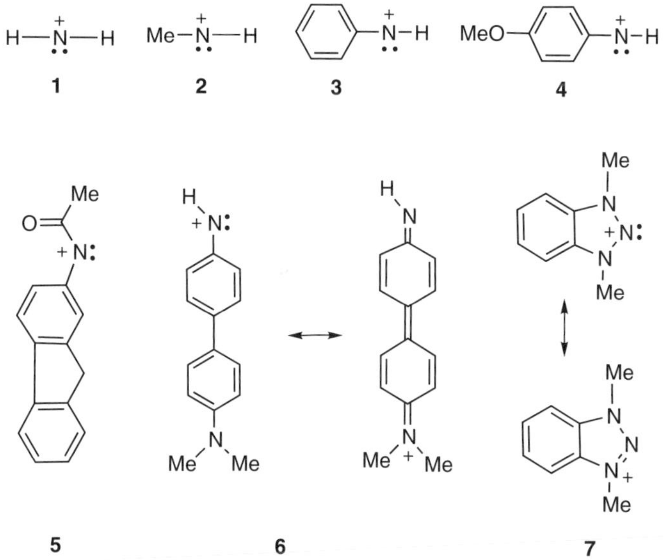

Figure 13.2. Examples of nitrenium ions.

of benzidine diimine. Yet Dicks et al.[2] found that it reacts in the same way as **3–5**, albeit with highly attenuated rate constants. The triazolium ions (e.g., **7**) have been known for decades to be stable isolable salts.[3] Yet Boche and co-workers[4,5] argued that these species are nitrenium ions by virtue of the resonance form that assigns a positive charge to the central nitrogen. In fact, Boche's study of these species was undoubtedly inspired by Wanzlik and Schikora[6] and the work of Arduengo et al.[7] on stable carbenes possessing analogous structures. However, many would argue that these ions are not nitrenium ions. They are readily isolable and are no more reactive toward nucleophiles than are pyridinium ions or any other quaternary nitrogenous heterocycle. Furthermore, they are ground-state singlets with very high energy triplet states.

This raises the question: When is a nitrenium ion not a nitrenium ion? Or to state it another way: When is the degree of charge delocalization so great that the species in question is no longer suitably considered a nitrenium ion. In principle, one could propose either a structure- or reactivity-based definition of nitrenium ion. As seen in the discussion above any definition based on qualitative valence bond structures faces the problem of judging which canonical structure is "better." One might do better with ab initio quantum chemistry calculations. However, the parameters that can recovered from such calculations (bond angles, charge distributions, orbital energies, etc.) are likely to vary in a more or less continuous fashion throughout the series. Thus a quantitative structural definition would have to rely on an arbitrary, and ultimately subjective definition of charge distribution or other structural features.

Likewise any reactivity-based definition would require some arbitrary choice of reaction rate. Which reaction would be used? What minimal rate would be required? What conditions of temperature, solvent, or acidity would be chosen for the definition? As a practical matter, such a definition could not be applied to any unknown species for the simple reason that reaction rates are, at this point in time, notoriously difficult to predict from first principles.

For these reasons, a permissive definition of nitrenium ions is chosen. A nitrenium is any species that can, through valence bond representations, be depicted as having a dicoordinate positively charged nitrogen. Such a definition has the advantage of being readily applied to any species for which a valence bond representation can be depicted. On the other hand, such a definition does encompass a number of species with rather different properties.

Nitrenium ions can be divided into several subclasses on the basis of their ligands. Of these, the most widely studied group is the arylnitrenium ions, which are characterized by having an aromatic ring attached to the nitrogen. Likewise alkyl-, halo-, and heteroaryl nitrenium ions are those species having, respectively, alkyl groups, halogens, and heteroaromatic rings attached to the nitrogen.

There is an additional subclass of nitrenium ions where the positively charged nitrogen atom is doubly bound to one ligand (**8**, Fig. 13.3). These can be regarded as nitrenium analogues of vinylidene carbenes. By analogy, such species are proposed to be named vinylidene nitrenium ions. Very little is known about such species. Vinylidene nitrenium ions would be formed in the Beckman rearrangement were it to occur in a stepwise fashion. The α-cyanocarbeniun ions could also be considered members of this family on the basis of resonance structure **9**, shown in Figure 13.3. In fact, such species have been characterized by nuclear magnetic resonance spectroscopy (NMR) in superacid media.

Because singlet nitrenium ions bear a formal nonbonding electron pair, they can be regarded as Lewis bases. In principle, it should be possible to derive a dicationic species through protonation (**10**, Fig. 13.4) of a nitrenium ion, or by removal of two

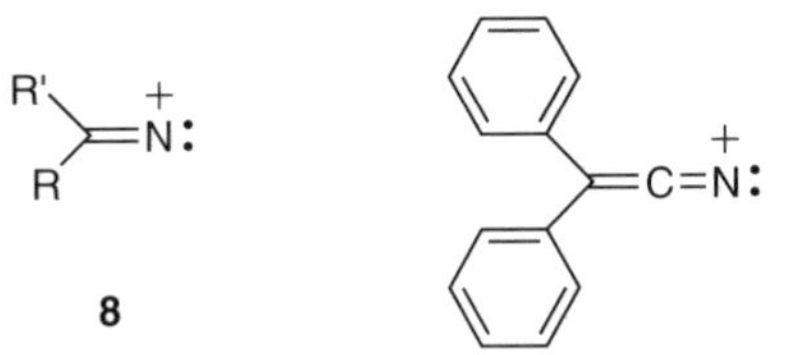

Figure 13.3. Vinylidene nitrenium ions.

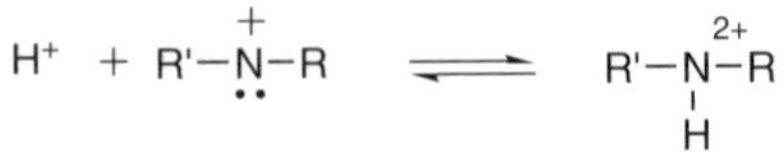

Figure 13.4. Nitrenium dications.

electrons from an amine. The author proposes that such species be called nitrenium dications. While nitrenium dications may appear to be nearly inaccessible on the basis of electrostatic considerations, it will be shown below that the more delocalized arylnitrenium ions can be protonated to form dications under relatively mild conditions.

1.2. Scope of This Chapter and Previous Reviews

Nitrenium ions (or "imidonium ions" in the contemporaneous nomenclature) were described in a 1964 review of nitrene chemistry by Abramovitch and Davis.[8] A later review by Lansbury[9] in 1970 focused primarily on vinylidine nitrenium ions. Gassmann's[10] 1970 review was particularly influential in that it described the application of detailed mechanistic methods to the question of the formation of nitrenium ions as discrete intermediates. McClelland[11] reviewed kinetic and lifetime properties of nitrenium ions, with a particular emphasis on those studied by laser flash photolysis (LFP). The role of singlet and triplet states in the reactions of nitrenium ions was reviewed in 1999.[12] Photochemical routes to nitrenium ions were discussed in a 2000 review.[13] Finally, a noteworthy review of arylnitrenium ion chemistry by Novak and Rajagopal[14] has recently appeared.

The purpose of this chapter is to describe the current knowledge of nitrenium ion properties and chemical reactions. As such, it will focus on experiments where the role of the nitrenium ion as a discrete intermediate has been well established, or at least widely presumed to occur. A critical survey of chemical processes that might involve these intermediates is well beyond the intended scope of this chapter. In addition, this chapter will emphasize nitrenium ion study subsequent to 1984. This is a logical starting point for two reasons. First, Abramovitch and Jeyaraman's[15] review covers the field up to that point. Although there have been several reviews written subsequently, a later starting point would lead to a discussion lacking in context and continuity. Second, in the period following 1983, the existence of discrete nitrenium ions in certain reactions became less controversial and investigators began to delve into the question of the properties of nitrenium ions as true intermediates. Of course, this transition did not occur instantaneously. Indeed it will be seen in the following discussion that the formation of true nitrenium ions in many reactions is far from a settled question.

1.3. Relevance of Nitrenium Ions

As a class, nitrenium ions are rather poorly characterized relative to similar reactive intermediates such as carbenes and carbenium ions. This situation alone is sufficient to motivate many fundamental studies into their structures and behavior, There are also several practical considerations that motivate their study. The following is intended as a brief overview of these latter areas.

Nitrenium ions, particularly the arylnitrenium ions, have been proposed as intermediates in deoxyribonucleic acid (DNA) damaging reactions that can ultimately convert a normal cell into a cancer cell. Carcinogenesis is a complex phenomenon,

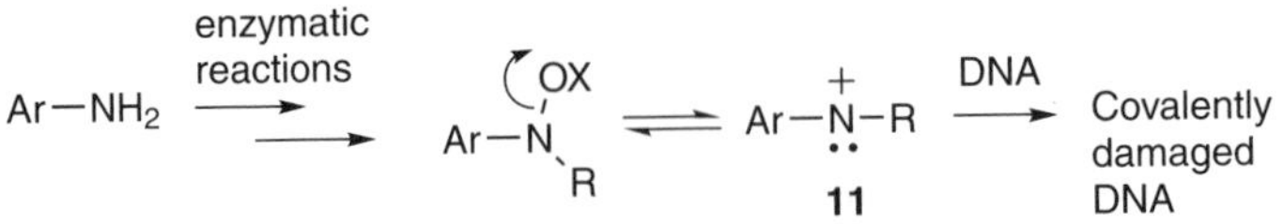

Figure 13.5. Nitrenium ions in DNA damaging reactions.

involving a number of processes beyond the initial DNA damaging step. The reader is referred elsewhere for a more comprehensive discussion of chemical carcinogenesis, including issues such as DNA repair, and specific genetic targets and cellular protection mechanisms.[16–20]

Certain aromatic amines were first recognized, on the basis of epidemiological data, to be correlated with bladder cancer, an otherwise rare condition. It is now known that the amines themselves are not the genotoxic agents. Instead it appears that these amines are converted by several tissue-specific enzymatic processes into arylnitrenium ions (**11**).[18] It is these species that are thought to attack DNA at guanine bases, producing covalent adducts. A simplified mechanism is provided in Figure 13.5. It should be pointed out that the role of nitrenium ions in this process is somewhat controversial and alternative mechanisms involving cation radicals have also been advanced.

Both arylnitrenium and alkoxynitrenium ion intermediates have been incorporated into strategies for the chemical synthesis of complex molecules. There are several diverse reaction pathways employed in these reactions. Thus it is difficult to summarize this chemistry with a single reaction scheme. In general, though, the nitrenium ions serve as nitrogen-based electrophiles and form bonds either at the nitrogen or at a mesomeric positive center with various nucleophiles.[21–26] The most promising (i.e., regiospecific and high yielding) reactions explicitly designed to use nitrenium ion intermediates, involve the nitrenium ion center covalently linked to a π nucleophile, such as an aromatic ring (**12**, Fig. 13.6).[24–26]

Polyaniline is a repeating unit of benzene rings each joined by an N–H group (**13**, Fig. 13.7). This polymer and its derivatives are of interest because they are electrically conductive when doped with oxidants. This material is prepared by oxidation of aniline electrochemically,[27–30] enzymatically,[31,32] or with simple chemical reagents.[33–35] Polyaniline can be formally regarded as a polymer of

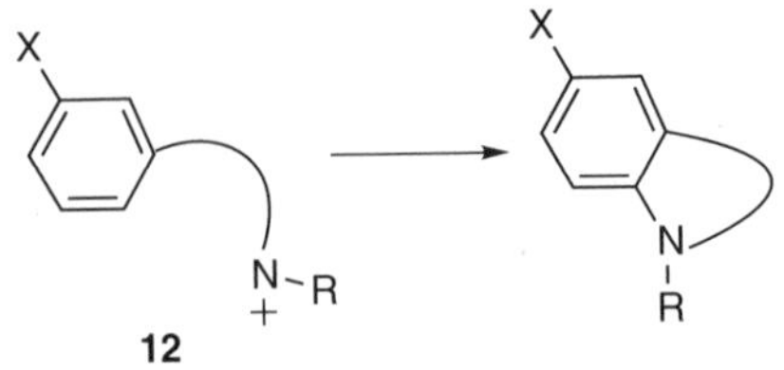

Figure 13.6. Nitrenium ions in synthesis.

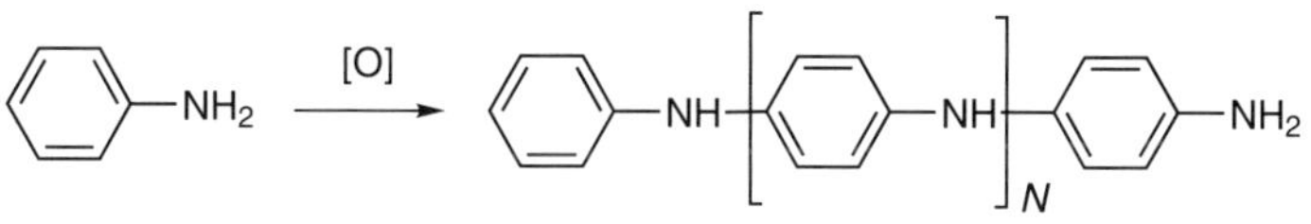

13

Figure 13.7. Polyaniline formation.

phenylnitrenium ion. As such it occurred to early workers that the polymer forms when oxidation of aniline gives phenylnitrenium ion, which in turn adds to an aniline unit to provide phenylbenzidine.[27,30] The chain could elongate through further oxidation to give a substituted nitrenium ion that would incorporate another aniline, and so on. Others have argued that the mechanism can be explained by radical coupling processes.[29,33,34]

1.4. Nitrenium Ion History: Highlights of Pre-1984 Studies

Prior to the 1980s, most studies of nitrenium ions dealt with the question of their discrete existence. In essence, any transformation that goes from one stable molecule through a nitrenium ion intermediate, could also be formulated to occur in a concerted fashion (Fig. 13.8). This creates something of a "chicken-and-egg" problem. In the absence of any unambiguous data on nitrenium ion behavior, it was impossible to argue that observation of a given reactivity pattern was indicative of nitrenium ion formation. Likewise, one could not identify characteristic nitrenium ion reactions, because there were no conditions universally agreed to form nitrenium ions.

It is now apparent that nitrenium ions occur in a variety of contexts including the so-called Bamberger rearrangement [36–38] (i.e., acid-catalyzed isomerization of N-hydroxyaniline **14** to 4-aminophenol **15**, Fig. 13.9), and the aforementioned catabolism of arylamines. The earlier workers, however, for the most part did not explicitly consider the possibility of such a species. Heller et al.[39] were the first to present kinetic evidence for such an intermediate. Interestingly, they chose to represent it as the iminocyclohexadienyl cation (**16**), "With the customary ellipsis of allowing a single valency structure to represent mesomeric systems."

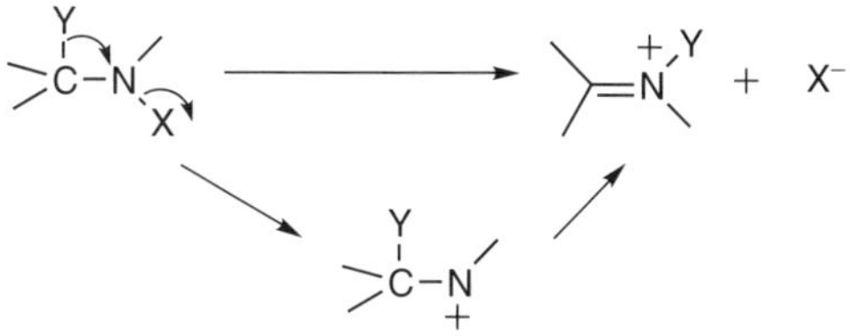

Figure 13.8. Nitrenium ion formation versus concerted rearrangement.

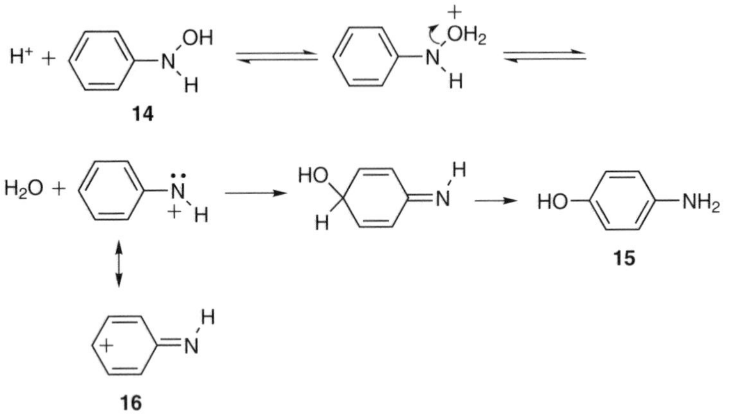

Figure 13.9. Bamberger rearrangement.

Gassman et al.[40,41] provided further evidence for the existence of the arylnitrenium ion as a discrete intermediate by means of kinetic studies (Fig. 13.10). They found that the solvolysis rates of a series of substituted N-chloroaniline derivatives (**17**) depend dramatically on the electron-donating and -accepting characteristics of the substituent.[42–44] Electron-donating groups provided a substantial increase in the rate of the reaction. A fit to the Hammett equation gave a ρ^+ value of -6.35. This value was considered to be consistent with the formation of a fully charged, cationic intermediate.

McEwan and co-worker[45] examined the acid-catalyzed decomposition of unsymmetrical benzhydryl azides **18** and some related species (Fig. 13.11). The aryl migration step showed very little discrimination between aryl rings with electron-donating and those with electron-withdrawing substituents. This low selectivity was judged to be more consistent with formation of a discrete nitrenium ion intermediate (**19**). These workers reasoned that a concerted migration would exhibit higher selectivity toward donor-substituted arenes, because in that mechanism the electrons from the arene would participate in the reaction.

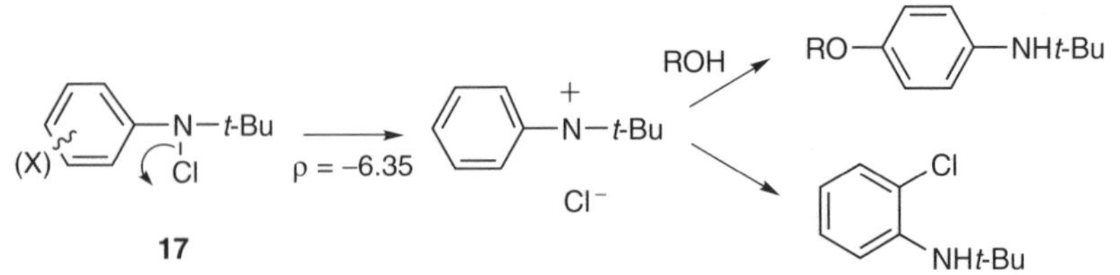

Figure 13.10. Gassman's linear free energy relationship (LFER) experiment on N-chloroamines.

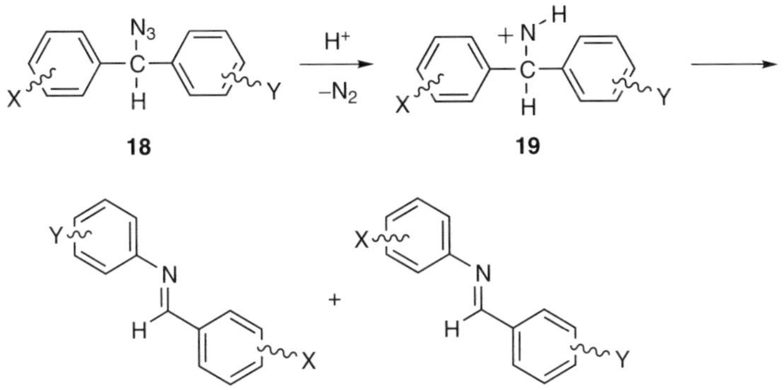

Figure 13.11. Nitrenium ions in the rearrangment and elimination of benzhydryl azides.

Other evidence advanced in favor of alkyl nitrenium ions came from Ag⁺ promoted isomerization and solvolysis of various cyclic and bicyclic *N*-chloroamines (e.g., **20**).[40,41,46] A representative example is shown in Figure 13.12. It was argued that the silver ion abstracted the chloride from the substrate creating a singlet nitrenium ion (**21**). Products were derived from 1,2-alkyl shifts to the apparent nitrenium nitrogen and subsequent trapping by either the solvent (**22**) or the chloride ion. While these products are consistent with formation of a free nitrenium ion, it was also acknowledged that they could be formed in a concerted fashion, whereby alkyl migration occurred concertedly with elimination of the chloride ion.

The formation of the parent amine **23** in these solvolysis reactions was considered to be the most definitive evidence for formation of a discrete nitrenium ion. Gassman and Cryberg[41] postulated the following. (a) Initial Cl—N bond heterolysis would occur adiabatically, generating the singlet nitrenium ion **21**. (b) The triplet

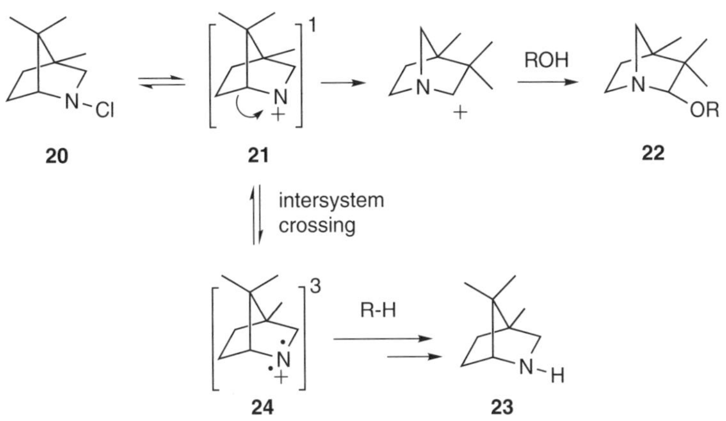

Figure 13.12. Evidence for a discrete nitrenium ion intermediate.

state **24** was either the ground state, or else it was very close in energy to the singlet. (c) The trapping of the singlet nitrenium ion (accompanied by the alkyl migration) would compete with relaxation to the triplet state. Given these three assumptions, it was reasoned that collision of the nitrenium ion with heavy atom molecules would accelerate the singlet to triplet interconversion. This so-called heavy atom effect is, a well established phenomenon in photochemistry[47] and photophysics,[48] but at the time it was unknown in nitrenium ion chemistry.

To test for the heavy atom effect, Gassman and Cryberg[41] compared solvolysis reactions carried out with and without bromoform added as the heavy atom agent. Enhanced yields of the parent amine **23** were detected in the presence of both bromoform and 1,4-dibromobenzene, suggesting the operation of a heavy atom effect, and thus the formation of a discrete nitrenium ion. Later workers[49–51] questioned these findings. The proposed formation of the parent amine could also be explained by a competing process in which the Cl—N bond of a protonated or metalated form of the amine fragments in a homolytic fashion. Regardless of the interpretation of these experiments, Gassman should be credited with firmly advancing the idea that singlet and triplet nitrenium ions could exist and that their chemical reactions would differ. As we shall see below, this idea has withstood the test of time, even if the original experiments that inspired it remain controversial.

Nitrenium ions can be viewed as products from two-electron oxidation of amines (Fig. 13.13) followed by loss of a proton. Thus they need to be considered as intermediates in the oxidation of amines. In two early studies, diarylnitrenium ions were shown to have formed in the oxidation of diarylamines. Svanholm and Parker[52] carried out cyclic voltammetry on N,N-di-(2,4-methoxyphenyl)amine (**25**) in acetonitrile with alumina added to suppress any adventitious nucleophiles. The voltammogram revealed two sequential oxidation processes: (1) formation of the cation radical **26**, and (2) either the nitrenium ion **27** or its conjugate acid. In strongly acidic solution the latter was sufficiently stable that its absorption spectrum could be recorded.

Serve[53] carried out similar experiments with bis(2,4-dimethoxyphenyl)amine and bis(2,4,6-trimethoxyphenyl)amine. In the presence of 2,5-lutidine, a nonnucleophilic base, he detected a two-electron oxidation wave in cyclic voltammetry

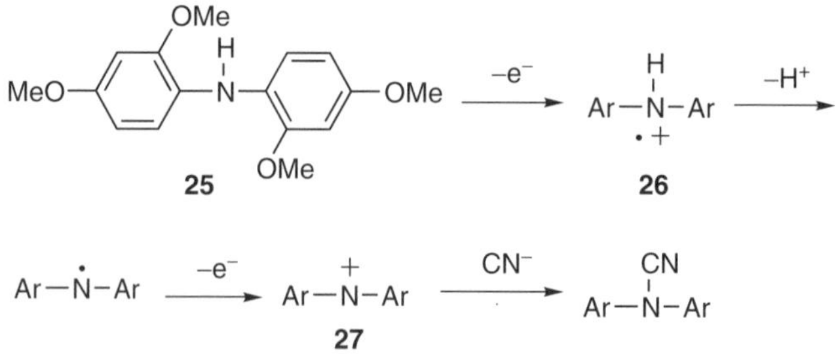

Figure 13.13. Electrochemical oxidation of hindered amines gives nitrenium ions.

(CV) experiments. The product of this oxidation was identified as the corresponding nitrenium ion on the basis of its visible absorption band. The latter was shown to be distinct from the nitrenium dication band that was generated under acidic conditions. Also, the cyanide adduct (Fig. 13.13) was isolated and characterized by spectroscopy (^{1}H NMR).

2. THEORETICAL TREATMENTS OF NITRENIUM IONS

The inherent instability of nitrenium ions makes them attractive candidates for study using various quantum-based computational methods. By using such methods, it is possible to predict a variety of molecular properties with useful accuracy at a reasonable cost. These properties include geometries, charge distributions, heats of formation, and relative energies of the singlet and triplet states, which are rather difficult to determine experimentally for short-lived species such as nitrenium ions. Generally, experimental measurements most readily provide stable product distributions, and in some cases, reaction rate constants and electronic spectra. Although it is also possible to theoretically determine transition state energies and geometries, and thus predict decay pathways, such calculations are presently either too demanding of computational resources or else so approximate as to be, at best, a qualitative guide. Thus computational studies are generally able to compliment experimental measurements. On the other hand, the lack of a common data set makes it difficult to assess the accuracy of the various computational approaches. A future challenge remains in finding experiments that either provide computationally accessible data and/or in developing computational procedures that can predict experimentally accessible data.

2.1. Parent, Alkyl-, and Halonitrenium Ions

In general, nitrenium ions have two low-energy electronic states of interest, the triplet state (designated triplet σp in Fig. 13.14), in which the nonbonding electrons have parallel spins and occupy separate orbitals, and the singlet state (designated singlet σ^2), in which the two nonbonding electrons are paired in the same orbital. The remaining singlets, σp and p^2, are generally much higher in energy and are thus

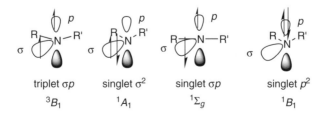

	triplet σp	singlet σ^2	singlet σp	singlet p^2
	$^3B_1$	$^1A_1$	$^1\Sigma_g$	$^1B_1$

Figure 13.14. Electronic configurations of simple nitrenium ions.

not particularly relevant to chemical behavior. However, one report has attributed certain solution reactions of an arylnitrenium ion to the p^2 state (called a "π nitrenium ion" in that report).[54]

Due to its simplicity, the parent nitrenium ion NH_2^+ has been subjected to the most detailed theoretical treatments.[55-63] Currently, there is generally good agreement between calculations and high-resolution gas-phase spectra of this simple species.[59,64,65] The lowest energy state for NH_2^+ is the $^3B_1$ state.[62] This state has a quasilinear geometry, which means that its lowest energy configuration is bent ($\sim 152°$), but that the linear transition state is lower in energy than the zero-point vibrational energy of system. In other words, the ion could be described as bent, but floppy. The N—H bond lengths are computed to be 1.041 Å, making them longer than the typical ammonia bond length of 1.020 Å. As might be expected on the basis of a simple valence-shell electron-pair repulsion (VSEPR) model, the $^1A_1$ state of NH_2^+ has a bent geometry with a bond angle of 107–108° and a pronounced barrier to inversion. For this state, the computed bond lengths are 1.055 Å. The excited singlet state ($^1B_1$) corresponds to double occupation of an out-of-plane molecular orbital (MO) allowing the bond angle to open to 156.4° with a bond distance of 1.041 Å. Finally, the open-shell singlet ($^1\Sigma_g$), which has the same orbital occupation as the triplet, gives a nearly linear geometry of 175° with a bond distance of 1.043 Å.

Alklyation has two effects.[66] First, hyperconjugation from the adjacent C—H bonds substantially reduces the singlet–triplet energy gap. As seen in Table 13.1, there is a 16 kcal/mol decrease in the gap when going from NH_2^+ to $MeNH^+$. The additional methyl group in the dimethyl derivative has a much less pronounced effect on the singlet–triplet gap, in this case diminishing it by a further 4.6 kcal/mol relative to the monomethyl system. The reduced effect of the second methyl group has been attributed to a partially compensating steric effect–the $^1A_1$ state has a 119.7° central bond angle compared with 112.1° in the monomethyl and 108° in

TABLE 13.1. Singlet–Triplet Energy Gaps and Geometric Parameters for Simple Nitrenium Ions

R^1	R^2	ΔE_{st}	RNR' Angle (°) Triplet/Singlet	Method	Reference
H	H	29.16	152.82/108.38	CASSCF-MRCI	62
Me	H	13.2	150.4/112.1	CCSD(T)	66
Me	Me	8.6	143.3/119.7	CCSD(T)	66
F	H	−0.29	125.7/105.1	B3LYP/6-311G	58
Cl	H	5.02	133.7/108.7	B3LYP/6-311G	58
CN	H	27.5	180.0/120.4	B3LYP/6-311G	58
F	F	−57.3	124.8/107.6	MP4/6-311G	68
Cl	Cl	−19.8	137.0/117.3	MP4/6-311G	68
Azi^a	Azi^a	10.7	81.2/63.4	BLPY/B2//MCSCF/B1	70

^{a} Aziridinium ion, **28** in Figure 13.16.

the parent. Also consistent with increased steric strain in the dimethylnitrenium ion singlet state is the increased N—C bond length of 1.396 Å compared with 1.364 Å in the monomethyl. In contrast the triplet N—C bond lengths are much less affected by additional substitution, increasing by only 0.005 Å in the dimethyl relative to the monomethyl nitrenium ion.

Interestingly, a recent gas-phase photoelectron spectroscopy (PES) study of dimethylnitrenium ion by Wang et al.[67] would seem to be in conflict with the earlier calculations. These workers measured ionization of the dimethylamine radical to give various electronic states of the nitrenium ion. The first ionizing transition, which they attribute to formation of the nitrenium ion's $^1A_1$ state, occurs with a potential of 9.01 V. The transition to the $^3B_1$ occurs at 9.65 V. If these assignments, which generally agree with DFT calculations run by the same group, are correct, then dimethylnitrenium ion is a ground-state singlet, with a rather substantial energy gap (-17 kcal/mol). Of course, this study reports only the vertical gap, whereas the gap reported by Cramer and co-workers[66] was adiabatic.

It is important to note that for methyl nitrenium ion the singlet state is not predicted at higher levels of theory to show a potential energy minimum. Rather this species is a transition structure that eliminates H_2 without a barrier (Fig. 13.15). It seems likely that the difficulties encountered in attempting to study alkyl nitrenium ions might be traced to their propensity to rearrange or eliminate.

Gonzales et al.[58] calculated energies and singlet–triplet energy gaps for some halogen and cyano substituted nitrenium ions. The effect of these substituents is to stabilize the singlet in preference to the triplet. This is attributed to the π-donation and σ-withdrawing effects of these groups. The cyano group has the more modest effect, reducing the gap by $\sim$1.6 kcal/mol. The halogens, fluorine, and chlorine have more pronounced effects reducing the gap by $\sim$29 and 24 kcal/mol, respectively. Depending on the theoretical method used the monohalogenated systems are either ground-state singlets or triplets, but in either case the energy difference between the states is small. The dihalo species are clearly ground-state singlets.[68] Interestingly, the triplet state of the CN derivative gives a linear geometry and is probably best represented as a protonated 1,3-diradical.

Gonzales and co-workers[69] extended their work on simple nitrenium ions, examining the reactions of several small, monosubstituted nitrenium ions (XNH^+, where X = H, F, Cl, CN, and Me) with water. These computational studies (QCISD(T)/6–311++G**) treated only the singlet states of the nitrenium ion. According to these calculations, each singlet nitrenium ion adds to water to form an O-protonated hydroxylamine intermediate ($R_2N-OH_2^+$). No activation barrier was detected for

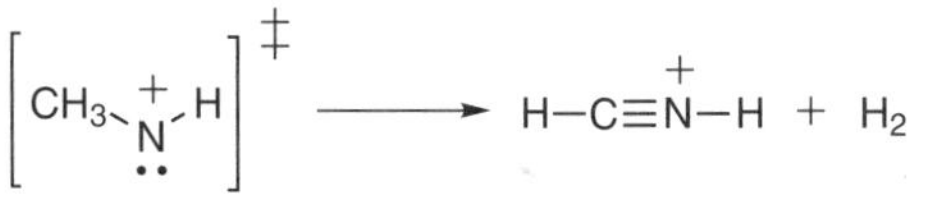

$$\left[CH_3 \underset{\overset{\cdot\cdot}{N}}{\overset{+}{\diagdown}} H \right]^{\ddagger} \longrightarrow H-C{\equiv}\overset{+}{N}-H \; + \; H_2$$

2

Figure 13.15. Singlet CH_3NH^+ is predicted to dissociate without a barrier.

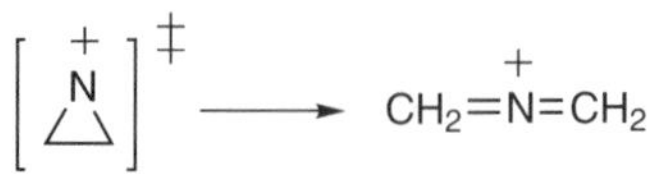

Figure 13.16. Singlet aziridenium ion is predicted to be a transition structure.

this complexation, which is followed by a slower unimolecular 1,2-proton shift to form the more stable N-protonated hydroxylamine (R_2HN^+-OH). The transition states for the proton-transfer steps were located for each case and found to be 22 to 33 kcal/mol above the intermediate. The effect of aqueous solvation was also modeled using an isodensity polarizable continuum model. A qualitatively similar mechanism was predicted.

Cramer and Worthington[70] examined the aziridinium ion (**28**, Fig. 13.16) and, despite its small bond angle, discovered it to be a ground-state triplet with a singlet–triplet energy gap of 10.7 kcal/mol. As with the dialkyl systems, the singlet was found to be a transition structure, which spontaneously underwent ring opening. While the acute bond angle might be expected to favor the singlet state, it is also the case that there is less effective overlap between the filled C—H σ orbitals and the p-type nonbonding orbital (hyperconjugation). This latter effect tends to destabilize the singlet state allowing this species to maintain a triplet ground state.

2.2. Aryl- and Heteroarylnitrenium Ions

The earliest computational studies on phenylnitrenium ion by the groups of Ford et al.,[71] Glover and Scott,[72] and Cramer and co-workers,[1,73] all concluded that the ground state of this species was singlet and planar, and that there was considerable charge delocalization from the nitrogen into the phenyl ring (Fig. 13.17). These qualitative results have withstood the test of time. Subsequent post-Hartree Fock (HF) and DFT (density functional theory) treatments of these species have refined the energies and geometries somewhat. The best recent value for ΔE_{st} is -18.8 kcal/mol, with a central angle of $111°$ for the singlet.[74] A DFT study by Cramer et al.[75,76] showed that the triplet state has a nonplanar geometry, with the N—H bond perpendicular to the phenyl ring. In fact, this species is described as a "protonated triplet nitrene."

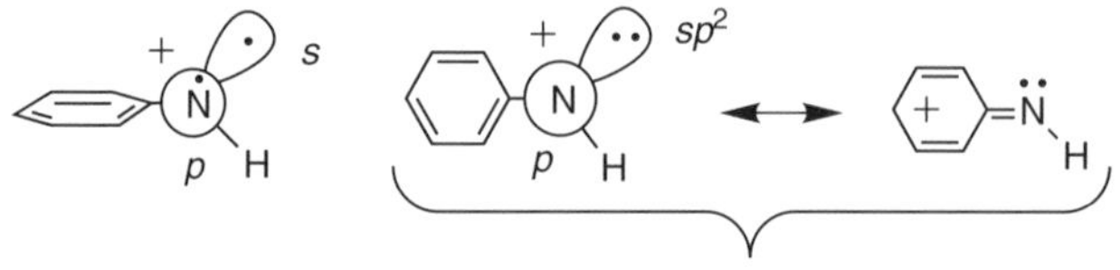

Triplet   Singlet (planar)

Figure 13.17. The DFT calculations show singlet phenylnitrenium ion is planar and the triplet is perpendicular.

Generally speaking, substitution of an aromatic ring on the nitrenium ion center acts to stabilize the singlet state. This follows because the filled π orbitals on the aromatic ring raise the energy of the unfilled p orbital on the nitrenium ion center relative to the in-plane sp^2 orbital. Thus it follows that additional π-donating groups on the aromatic ring are likely to further stabilize the singlet. Likewise, π acceptors are likely to counteract the effect of phenyl substitution and destabilize the singlet relative to the triplet.

These effects were examined in a DFT study by Sullivan et al.,[74] who computed ΔE_{st} for a series of 4-substituted phenylnitrenium ions. The effects mostly followed the predicted trend. Interestingly, however, the nitro group stabilizes the triplet far less than the carbonyl groups (formyl, acetyl, carbomethoxy). The differences being 2.0 kcal/mol for the nitro compared with the unsubstituted system, and 6.6 kcal/mol for the formyl. The halogens, fluorine and chlorine, slightly stabilized the singlet, whereas alkoxy and amino groups provided substantial stabilization for the singlet. The results are given in Table 13.2. Another interesting result from the same study was an assessment of the effects of solvation on the stabilities of the singlet and triplet states. In most cases, it was found that the singlet state was preferentially stabilized relative to the triplet by ~1–4 kcal/mol.

DFT-based calculations were carried out on a series of heteroarylnitrenium ions with nitrogens at various positions in the aromatic ring. Some results from this study are shown in Table 13.3.[77] Several interesting results emerge. First, in contrast to the phenyl systems, the heteroaromatic nitrenium ions are slightly nonplanar in the singlet state and planar as triplets. Second, ring nitrogens stabilize the triplet

TABLE 13.2. Singlet–Triplet Energy Gaps in 4-Substituted Phenylnitrenium Ions ($X-C_6H_4NH^+$)[a]

X	ΔE_{st}, (kcal/mol)
NH_2	−27.8
NMe_2	−27.7
OMe	−26.7
OH	−25.4
F	−22.2
Me	−21.9
Cl	−21.3
H	−18.8
CF_3	−17.9
CN	−17.5
NO_2	−16.8
CO_2Me	−16.5
CO_2H	−16.2
COMe	−13.5
CHO	−12.2

[a] See Ref. 74.

TABLE 13.3. Singlet–Triplet Energy Gaps in Heteroarylnitrenium Ions $(ArNH^+)^a$

Ar	ΔE_{st} (kcal/mol)
2-Pyridyl	−14.5
3-Pyridyl	−6.0
4-Pyridiyl	−3.6
4-Pyrimidyl	5.4
2-Pyrimidyl	7.0
5-Pyrimidyl	0.8
2-(1,3,5)-Triazinyl	10.9

^{a} See Ref. 77.

state relative to the singlet. In fact, the triazinyl and all of the pyrimidyl nitrenium ions are predicted to be ground-state triplets. The pyridyls also show smaller gaps than the phenyl system, but are predicted to be ground-state singlets. As with the arylnitrenium ions, solvation is predicted to stabilize the singlet by 0.5–5 kcal/mol relative to the triplet state. Finally, the heteroaromatic systems with ortho ring nitrogens (**29**) rearrange in a stepwise fashion, from the singlet state, forming N-heteroarylnitriles (**30**, Fig. 13.18). These reactions occur in the singlet state. The predicted barriers range from 3–25 kcal/mol, with the triazenyl systems being the most facile.

More recent computational efforts have focused on mapping out reaction mechanisms, and to the extent possible, predicting product distributions in reactions. Novak and Lin[78] determined the relative hydroxylation energies for a series of arylnitrenium ions and then attempted to relate these to experimental azide/water selectivity ratios (see Section 5). The former quantity was determined using a series of isodesmic reactions, wherein the ΔE for hydroxide transfer from the adduct phenylnitrenium ion (**31**, Fig. 13.19) was used as a gauge for the relative hydrolytic stability of these nitrenium ions. The calculations (RHF/6-31G*) showed a good correlation between the hydroxylation energies and the product distributions observed when the same species were generated in aqueous solution in the presence of azide ion. Presumably, the rate of water addition correlates with the energy of hydroxide addition, and the rate constant for azide addition is constant.

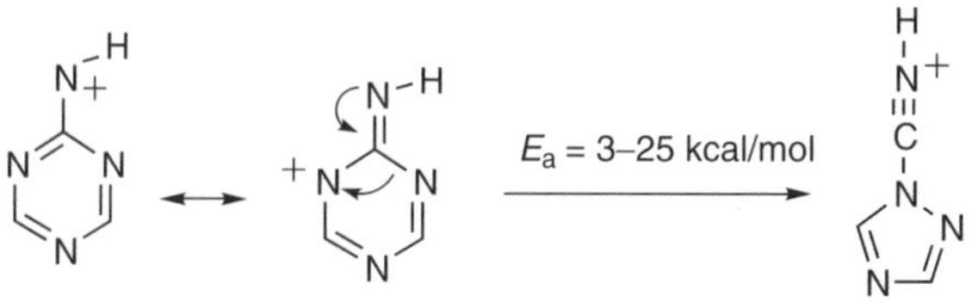

Figure 13.18. Rearrangement predicted by DFT for 2-(1,3,5)-triazolylnitrenium ion.

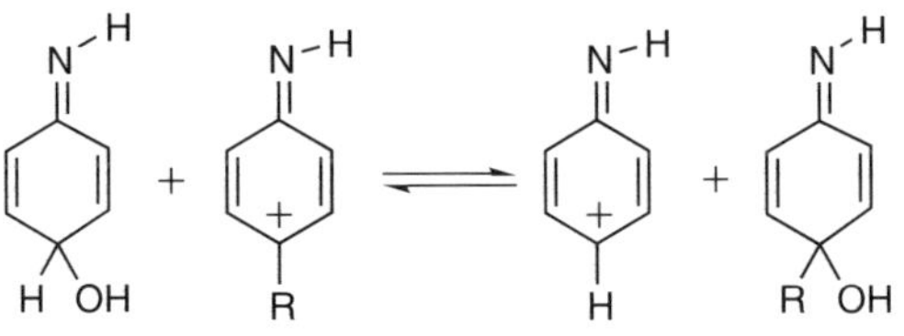

Figure 13.19. Isodesmic reactions used to assess relative stability of arylnitrenium ions toward addition by water.

More recently, Sullivan and Cramer[77] used DFT methods to model the addition of water to phenylnitrenium ion and N-acetylphenylnitenium ion (**32**, Fig. 13.20). This process was found to be endoergonic in the gas phase, presumably because the resulting oxonium ion **33** has a more concentrated positive charge than does the more delocalized singlet **32**. However, when the effect of solvation was modeled, using a dielectric continuum, the reaction was predicted to be exoergonic and barrierless. The authors point out that a small barrier might be present due to specific solvent–solvent and solvent–solute interactions, which are not considered in their solvation model.

Interestingly, it was also found that N-acetyl-N-arylnitrenium ions cyclize to form protonated benzoxazole derivatives (**34**).[77] This reaction was found to have barriers of 4–11 kcal/mol in the gas phase. Although this finding has yet to be verified experimentally, it is significant in view of the fact that these nitrenium ions are frequently used in mechanistic studies of DNA damage.

One general, and as yet unsolved problem, is understanding and predicting the regiochemistry of arylnitrenium ion addition to DNA bases. This reaction is complicated because phenylnitrenium ion has four potential sites of addition (Fig. 13.21): the nitrenium center as well as the ortho and para ring carbons. Likewise, DNA has numerous sites for electrophilic addition. In guanine bases alone, there is a possibility for addition to N7, O6, N2, or C8.

Ford and Thompson[79] used semiempirical methods to evaluate the relative stabilities of various nitrenium ion guanine adducts where the exocyclic 2-amino group of guanine is coupled to nitrogen and various ring positions of phenyl or

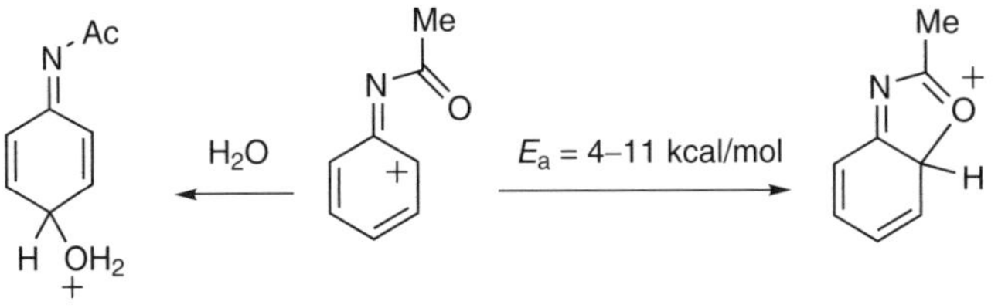

Figure 13.20. Reactions predicted by DFT for N-acetyl-N-arylnitrenium ions.

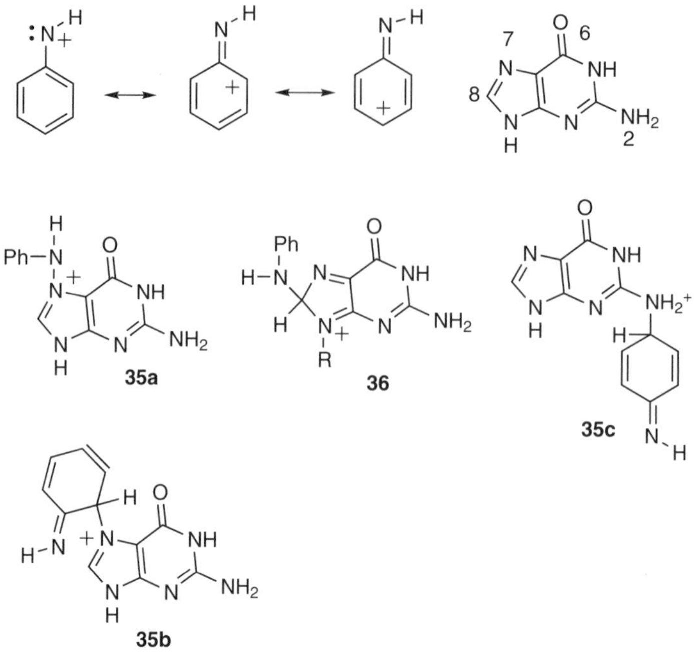

Figure 13.21. Reactions of phenylnitrenium ion with guanine.

other aromatic nitrenium ions. In the case of the 2-naphthylnitrenium ion, where there are experimental data available, the most stable adduct by AM1 calculations corresponds to the most prevalently formed isomer. The authors extend the AM1 analysis to other arlynitrenium ions where the experimental product distributions are unknown.

Parks et al.[80] used DFT methods to calculate the relative energies for 22 isomeric adducts of phenylnitrenium ion with guanine. The four most stable adducts are shown in Fig. 13.21. These are in order, the N—N7 adduct **35a**, the N—C8 adduct **36**, the o-N7 adduct **35b**, and the p-N7 adduct **35c**. This study did not treat the transitions states for the various addition reactions, but it was noted that the lowest energy adducts corresponded to the most prevalent adducts in biochemical systems. The implication is that either the various pathways are under thermodynamic control or that the transitions state energies track the corresponding adduct energies. Transitions states for two possible pathways yielding the observed N—C8 adduct were compared and revealed that a stepwise shift from **35b** to the N—C8 **36** was more feasible than a migration of **35a** to the same product.

In order to increase the experimental benchmarks available for comparison with theoretical calculation, McIlroy et al.,[81] determined the singlet triplet energy splitting in a diaminosubstituted nitrenium ion, 1,3-dimethylbenzotriazolium ion **37** (Fig. 13.22). This species can be viewed a nitrenium ion by virtue of mesomer **37a**, which places a positive charge on the central nitrogen. Because this species

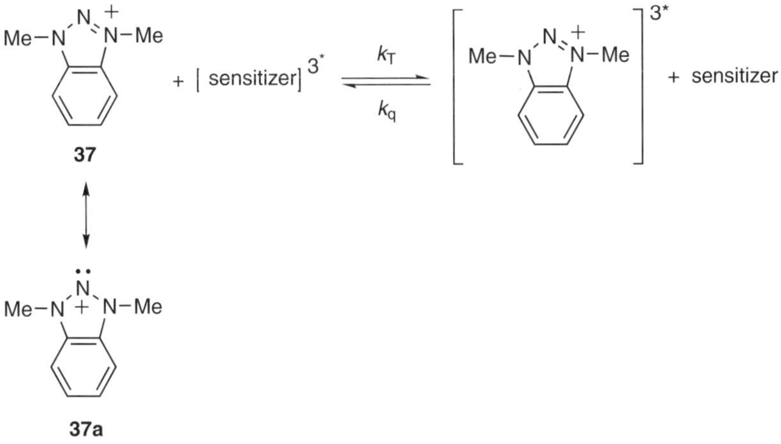

Figure 13.22. Singlet–triplet gap in stable nitrenium ion probed by energy transfer experiments.

is stable, and likely has a very large singlet–triplet gap, it is possible to measure ΔE_{st} using conventional photophysical techniques. This was done using a series of triplet energy-transfer experiments, wherein the rate constants for triplet energy transfer to the nitrenium ion were measured using a series of triplet sensitizers of varying triplet energy. The result of these studies provided a ΔE_{st} of 64 ± 4 kcal/ mol in excellent agreement with DFT calculations.

3. METHODS OF NITRENIUM ION GENERATION

3.1. Thermal Methods

It is perhaps obvious that reliable experimental information regarding any reactive intermediate requires a method for generating the species in question. Moreover, it is necessary that such methods provide the intermediate cleanly, with minimal side reactions. Likewise, it is desirable that such methods be generally applicable, providing routes to as many examples as possible and under a wide variety of conditions. In other words, methods that generate only very specific structures, and/or generate them under a very narrow range of conditions are less useful. Until relatively recently, the study of nitrenium ions has been inhibited by the paucity of well-characterized methods for their generation.

The first route to nitrenium ions is the acid-catalyzed dehydroxylation of N-hydroxylamines (Fig. 13.23).[39,82–84] The chief advantage to this route is that N-hydroxylamines are stable and relatively easy to prepare, but the hydroxylamine route has three deficiencies for systematic studies of the nitrenium ion intermediates. First, the reaction is limited to polar and acidic media. Second, it is only known for arylnitrenium ions. Finally, the most serious limitation, is the complexity

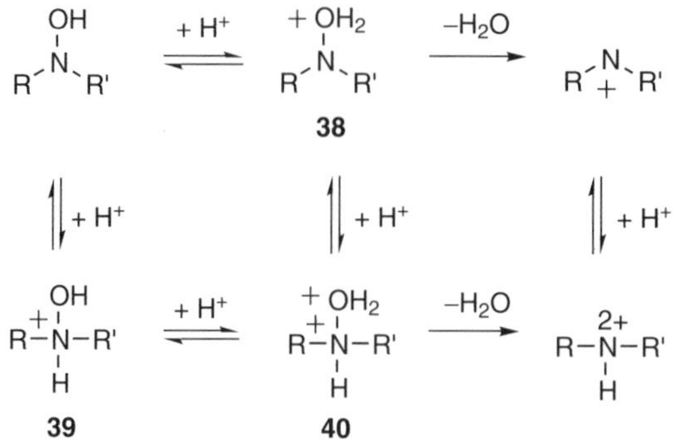

Figure 13.23. Nitrenium ion formation by heterolysis of *N*-hydroxylamines in strong acid.

of the decomposition mechanism. For simple hydroxylamines there are two basic sites: the nitrogen and the oxygen. Of these, the nitrogen is generally considered to be the more basic site. Thus any mechanistic study needs to account for *O*-protonated **38**, *N*-protonated **39**, and diprotonated precursor **40**, as well as the formation of the nitrenium ion and its dication.[85]

Nitrenium ions have also been generated through the decomposition of azides under acidic conditions (e.g., trifluoroacetic acid–arene solvent mixtures).[86,87] There are two potential pathways for the formation of the nitrenium ion from the precursors (Fig. 13.24). The first involves initial dissociation of the azide **41** to give a singlet nitrene **42**, followed by proton transfer to the latter to yield the primary nitrenium ion **43**. The second involves acid-induced decompostion of the azide, whereby preprotonation of the azide (**44**) forms the primary nitrenium ion in a direct manner. As with the hydroxylamine route, this method is limited to acidic or protic media.

It is also possible to generate similar nitrenium ions from arylazides and Lewis acids such as AlCl₃. Takeuchi et al.[88] found that when this decomposition was carried out in an aromatic solvent (e.g., benzene) good yields of *N,N*-diarylamines could be achieved. This was particularly true when the arylazides had electron withdrawing groups, which contrasts with the same arylnitrenium ions generated under protic conditions. Such metallonitrenium ions, or "nitreniumoids" have seen little study.

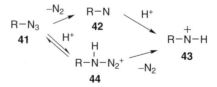

Figure 13.24. Nitrenium ion formation by decomposition of azides in acidic solution.

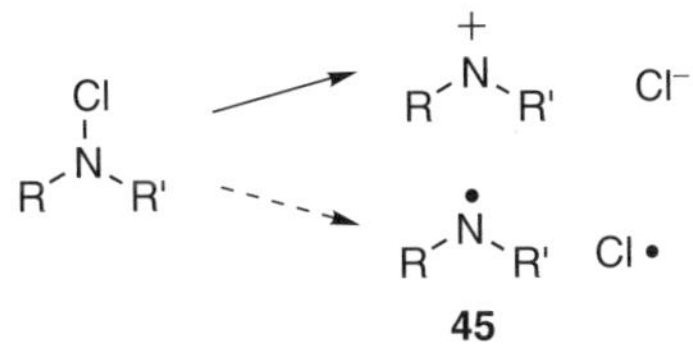

Figure 13.25. Nitrenium ions through heterolysis of *N*-chloroamines.

Gassman pioneered the use of *N*-chloroamines for the formation of both alkyl and arylnitrenium ions.[10,40–44,46,89] Several examples are given in Figs. 13.10 and 13.12. In polar media, these precursors heterolyze to give chloride and the corresponding nitrenium ion (Fig. 13.25). The use of an anionic leaving group eliminates the requirement for strong acid. One disadvantage is that chloride is a very reactive nucleophile and tends to trap many nitrenium ions at the diffusion limit. In fact, when arylnitrenium ions are generated by this route, isomers result from internal return of the chloride ion.[44,90] This problem can be avoided, to some extent, through the use of Ag(I) ions that form stable salts with chloride. Another potential pitfall is the possibility of competing N—Cl homolysis, which generates aminyl radicals **45**, rather than the desired nitrenium ion.[91] This competing process complicates the elucidation of nitrenium ion behavior. It is probably for this reason that this route has been used only rarely in the last two decades.

Most recent investigations aimed specifically at the study of nitrenium ions have employed the heterolytic cleavage of various esters of hydroxylamines **46** (Fig. 13.26).[50,92–102] This method has the advantage of not requiring the introduction of acids to promote the reaction. Also, because the N—O bond is stronger than the N—Cl, it is less likely to result in homolysis. With sterically unhindered esters (such as acetylhydroxylamines), complications can arise from acyl-transfer reactions, which generate the corresponding *N*-hydroxylamine **47**.[103–105] For this reason, sterically hindered esters, such as pivaloyl **46** (Fig. 13.26, where R″ = *t*-Bu), are often employed. As with the chlorides, internal return can also be a major decay pathway, particularly in less polar media.

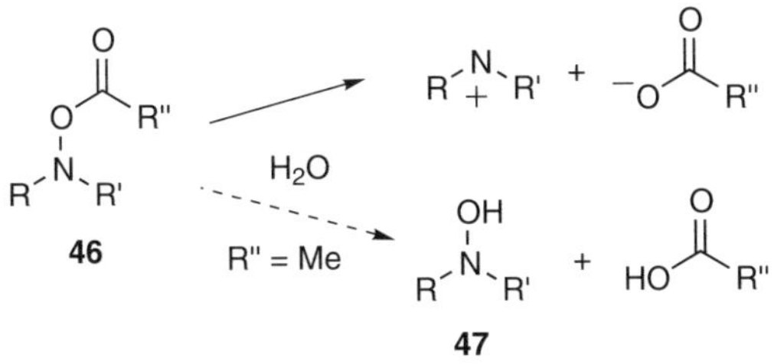

Figure 13.26. Nitrenium ions through heterolysis of *N*-hydroxylamine esters.

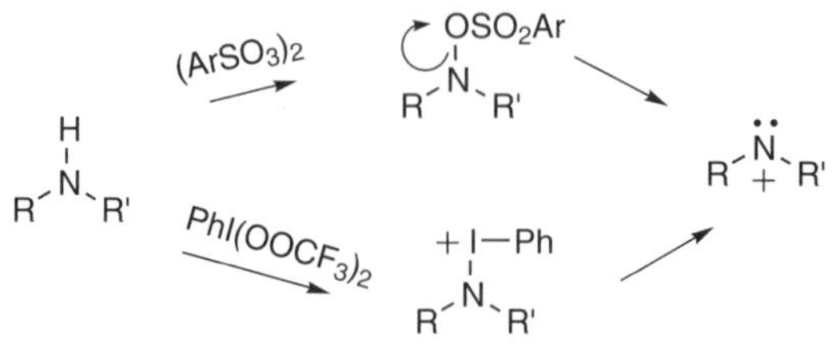

Figure 13.27. Nitrenium ions from chemical oxidants.

Various chemical oxidants have been applied directly to amines in order to generate nitrenium ion products (Fig. 13.27). For example Hoffmann and Christophe[51] treated O-alkylhydroxylamines with m-(trifluoromethylphenyl)sulfonyl peroxide and generated stable products that they reasoned originated from the formation of a alkoxynitrenium ion. The latter was inferred to be formed from fragmentation of the corresponding sulfonate ester, although this precursor was not detected under the reaction conditions. Wardrop and Basak,[25] Wardrop and Zhang,[26] as well as Kikugawa and Kawase,[106,107] used a similar method wherein a N-alkoxyamide is treated with bis(trifluoroacetoxy)iodobenzene. Again, neither the nitrenium ion nor its presumed ester precursor were isolated, but their formation was inferred from the final stable products. This general approach has proven to be very useful in natural product synthesis.

Katrizky et al.[108,109] employed N-aminopyridinium ions[110,111] (**48**, Fig. 13.28) as thermal precursors of nitrenium ions. This route to nitrenium ions is attractive in that there is no net generation of charge. That is, a positively charged reactant gives a positively charged product (**49**) and a neutral byproduct (**50**). Thus it could be useful for generating these intermediates in nonpolar media. Unfortunately, these salts are generally less readily available than azides, hydroxylamines, or hydroxylamine esters. Also, their formation generally seems to require elevated temperatures and/or long reaction times. In some cases, concerted reactions, rather than the formation of free nitrenium ions, have been inferred.

3.2. Photochemical Methods

There has been considerable interest in photochemical methods for producing nitrenium ions. Photochemical routes hold several significant advantages over

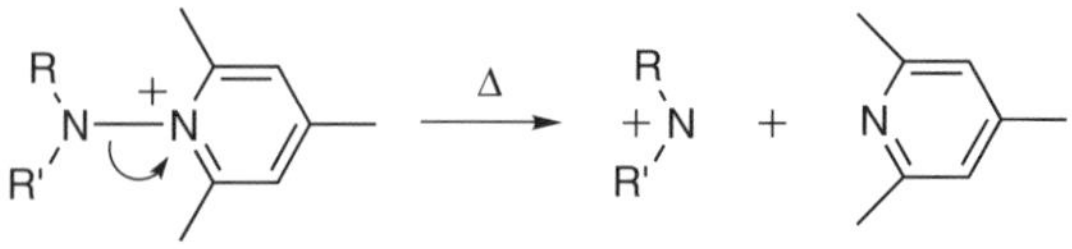

Figure 13.28. Nitrenium ions from the thermolysis of N-aminopyridium ions.

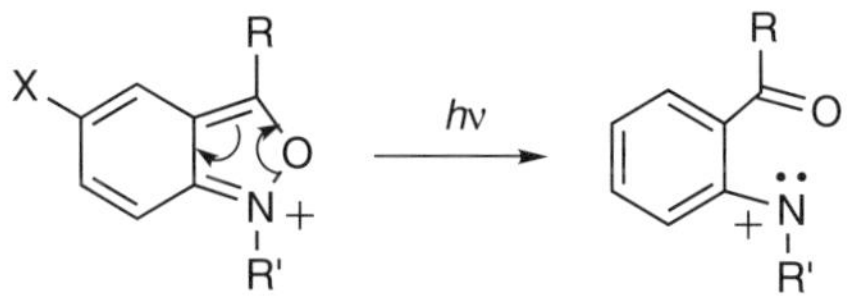

Figure 13.29. Arylnitrenium ions from the photolysis of anthranilium ions.

conventional routes. First, photochemistry is required if fast kinetic methods such as LFP are to be applied to the study of these intermediates. Second, as described below, photochemical routes are capable of generating nonequilibrium distributions of the singlet and triplet states. In this way, it is possible to kinetically trap the higher energy state and study its reactivity.

The earliest example of a photochemical reaction explicitly intended to produce a nitrenium ion is apparently Haley's study of the photochemistry of N-alkylanthranilium ions (also called N-alkyl-2,1-benzoisoxazolium ions (**51**), Fig. 13.29).[112] It was shown that photolysis of these heterocycles effects an electrocyclic rearrangement, cleaving the N—O bond, and producing the nitrenium ion isomer **52**. Subsequent studies by Hansen, Giovaninni, and their co-workers[113–116] showed that photolysis of anthranils under highly acidic conditions (H_2SO_4 or HBr solvent) led to formation of the primary nitrenium ions. Under these conditions, the nitrogen in the precursor is protonated (**51**, $R' = H$) and ring opening leads directly to the primary nitrenium ion. More recent studies have shown that triplet state arylnitrenium ions can be generated through triplet energy transfer.[117–119]

Although the anthranilium route proved useful in a variety of applications, it is capable of producing only a limited range of nitrenium ion structures. The nitrenium ions are limited to those where the central nitrogen carries an aromatic ring along with either a proton or an alkyl group. The aromatic ring, in turn, inevitably bears an ortho carbonyl substituent. A second limitation is that the process is reversible. Several experiments have shown that both singlet[117] and triplet[119] nitrenium ions formed in this manner recyclize to regenerate the anthranilium ion.

McClelland et al.[120,121] developed an azide-based photochemical route to primary nitrenium ions (Fig. 13.30). It had been previously demonstrated that singlet carbenes can be protonated under moderately acidic conditions.[122–124] It was then demonstrated that the analogous protonation of singlet nitrenes could occur. In McClelland's route, the latter is produced through photolysis of an arylazide. Once the singlet nitrene is formed it is consumed by three competing processes:[125–127] (1) intersystem crossing producing the triplet nitrene, (2) ring expansion producing either a benzazirine or a cyclic ketenimine, and (3) proton transfer producing the corresponding primary nitrenium ion. McClelland has examined the pH dependence of this reaction and found that 2-fluorenylnitrene, for example, is protonated in water with pH < 10.[121] In other cases more acidic conditions are required.[128]

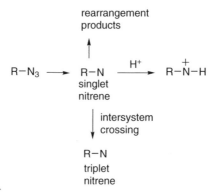

Figure 13.30. Primary nitrenium ions through azide photolysis–nitrene protonation.

Thus nitrenium ion formation is favored by a spin-induced barrier to the deprotonation (Fig. 13.31). Obviously, the feasibility of this route requires that the protonation rate constant exceeds the sum of the ring-expansion and intersystem crossing rate constants. This assumption is the case for arylazide-derived nitrenium ions. For these species, the corresponding nitrenes are generally ground-state triplets, whereas the corresponding arylnitenium ions are ground state singlets. Protonation of triplet nitrenes has yet to be demonstrated.

McClelland's route has proven extremely useful.[2,129–137] Aryazides are relatively easy to prepare, and azide photolysis produces N_2 as the only byproduct. The disadvantages to this approach are (a) only primary, singlet-state nitrenium ions are accessible in this way; (b) the nitrenium ions can only be generated in polar and protic media (usually acidic aqueous solutions), and (c) the azido group does not significantly shift the absorption band of the aromatic ring to which it is attached. Consequently, it is often necessary to use low wavelength ultraviolet (UV) light to generate the nitrenium ion, which can create problems if various

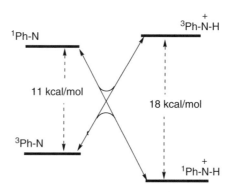

Figure 13.31. Protonation of singlet nitrenes to give singlet nitrenium ions.

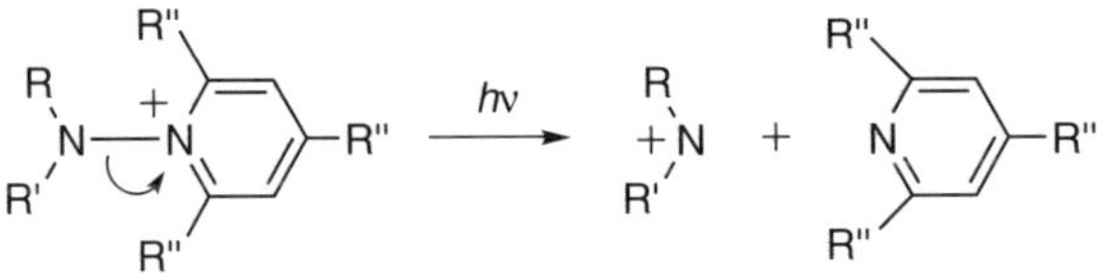

53 54

R'' = Me or Ph.

Figure 13.32. Photolysis of *N*-aminopyridinium ions.

trapping agents (such as electron-rich aromatic compounds) have overlapping absorption bands. However many biochemically relevant nitrenium ions are in fact singlet, aromatic substituted species and the important DNA damaging reactions occur in aqueous media.

Photolysis of *N*-aminopyridinium ions (**53**) also can be used to create nitrenium ions.[110,111,138–147] In this reaction, photolysis and N–N bond heterolysis generates a neutral pyridine fragment (**54**) and the corresponding nitrenium ion (Fig. 13.32). This route is, in principle, capable of producing any primary or secondary nitrenium ion depending on the substitution in the precursor. Salts **53** typically exhibit long wavelength charge-transfer absorption bands in the near-UV or low wavelength visible region of the spectrum. For example, 1-(*N*,*N*-diphenylamino)pyridinium ion has an absorption maximum at 380 nm.[145] High wavelength absorption bands allow for the selective photolysis of the precursor in the presence of traps, such as arenes, which might otherwise compete for the light.

The principal disadvantage to the aminopyridinium ion route is the accessibility of the precursors. None are available commercially, and most require multistep syntheses giving relatively low yields. Another potential pitfall is the formation of the pyridine byproduct (**54**). Pyridines can function as nucleophiles, attacking the nitrenium ion and creating complex mixtures. Finally, pyridinium ions are electron deficient and can serve as good ground-state electron acceptors. Many of the stable products generated from nitrenium ion reactions are amines and are relatively easy to oxidize. Thus, a potential problem is secondary reaction, whereby primary photoproducts are oxidized by the precursor.

Davidse et al.[148] described the photolysis of *N*-chloroamine and hydroxylamine esters (**55**) to form nitrenium ions by heterolysis (Fig. 13.33). As noted above, these reactants also yield nitrenium ions through thermal chemistry, albeit more slowly.

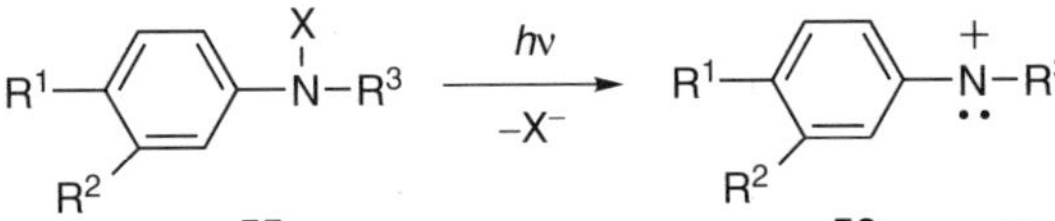

55 56

X = OSO_2^-, Cl

R^1 = Aryl, R^2 = H or CH_2-aryl, R^3 = Ac

Figure 13.33. Arylnitrenium ions through photoheterolysis.

Photolysis in polar media (e.g., H_2O–MeCN mixtures) results in the same products observed from thermal generation. In addition, however, the parent amine, which is not observed in the thermal reactions, is formed photochemically. This finding suggests that there may be a competition between heterolysis and homolysis in the photochemical reaction. It has also been suggested that the amine might result from formation of the triplet nitrenium ion. In any case this competing process along with the instability of the precursors has limited interest in this photochemical route.

3.3. Electrochemical Methods

Nitrenium ions can be generated from the electrochemical oxidation of amines (Figs. 13.34 and 13.35). Rieker et al.[149–152] used a similar procedure to generate 4'-dimethylamino-4-(2,6-di-*tert*-butyl)biphenylylnitrenium ion **58** from precursor **57**. Both this nitrenium ion and its corresponding dication **59** are sufficiently stable to be characterized by NMR and UV–vis spectroscopy.

The only nitrenium ions that have been directly characterized by electrochemical generation are those that are kinetically stabilized by resonance and steric hindrance. The main limitation to the electrochemical method appears to be competing polymerization processes (Fig. 13.7). It is known that electrolysis of aniline and its simple derivatives provides poly(aniline).[27,29,153] The polymerization mechanism is controversial, but the more recent experiments point to the coupling of amine cation radicals. Others have suggested that the elongation step occurs via addition of a nitrenium ion to an unconverted amine. In any case, the use of the electrochemical method for detailed study of nitrenium ions obviously requires that the competing polymerization process be suppressed.

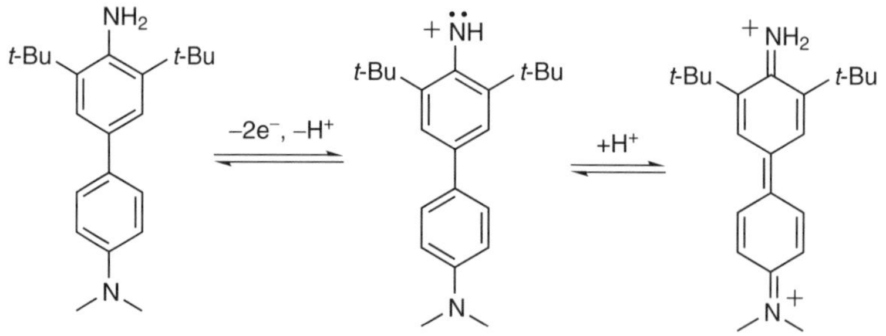

Figure 13.34. Nitrenium ions from electrochemical oxidation of benzidine derivatives.

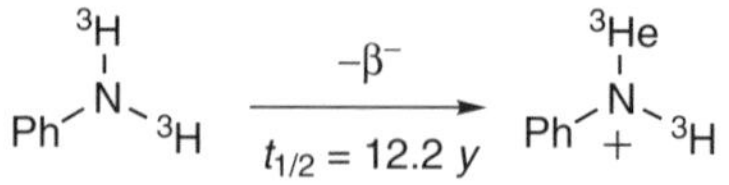

Figure 13.35. Nitrenium ions through β-decay of tritiated amines.

Nefedov et al.[154] reported the generation and study of nitrenium ions by a novel method: radiochemical decay (Fig. 13.35). Tritium is known to decay by β-particle emission, whereby a neutron is converted into a proton, and has the effect of converting the tritium into ^{3}He. Thus a tritium-substituted amine (**60**), upon radiolytic decay, is converted into a nitrenium ion–He complex (**61**). It is assumed that the helium atom is weakly bound and dissociates. These workers used this method to generate phenylnitrenium ion in benzene and toluene. One interesting property of this route to nitrenium ions is that it is possible to generate a truly "free" nitrenium ion. The β-particle byproduct is ejected with sufficient energy that it can be presumed not to interact with the nitrenium ion. On the other hand, this nuclear decay has a half-life of 12.2 years. Thus, high conversion of the starting material over practical time scales is not possible. For example, the study of phenylnitrenium ion in benzene required a reaction time of several months.

4. REACTIONS OF NITRENIUM IONS

There have been numerous specific reactions attributed to nitrenium ions over the years. Space constraints prohibit an exhaustive catalogue, so that the reactions described below should be understood to be typical examples that highlight important trends, rather than a comprehensive listing. In particular, we emphasize reactions where the involvement of the nitrenium ion has been well established or at least widely acknowledged. As noted above, there are numerous reactions of electrophilic nitrogen species in which the involvement of a nitrenium ion is possible but where the data are insufficient to permit any firm conclusion. Interested readers are encouraged to consult the Additional Reading section for further examples.

4.1. Singlet-State Rearrangement and Elimination Reactions

By analogy to carbenium ions, it can be expected that alkylnitrenium ions will experience 1,2-alkyl and hydride shifts. These would presumably be orbital symmetry allowed reactions that convert the nitrenium ion into a more stable iminium ion. This idea is supported by several theoretical studies suggesting that simple

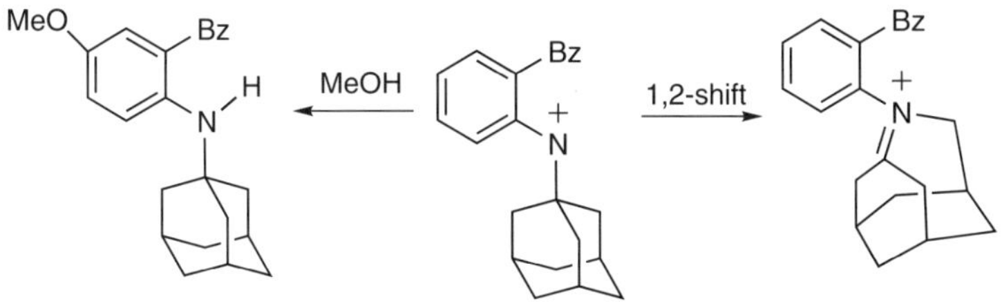

Bz = benzoyl

Figure 13.36. 1,2-Alkyl shift of an alkylarylnitrenium ion.

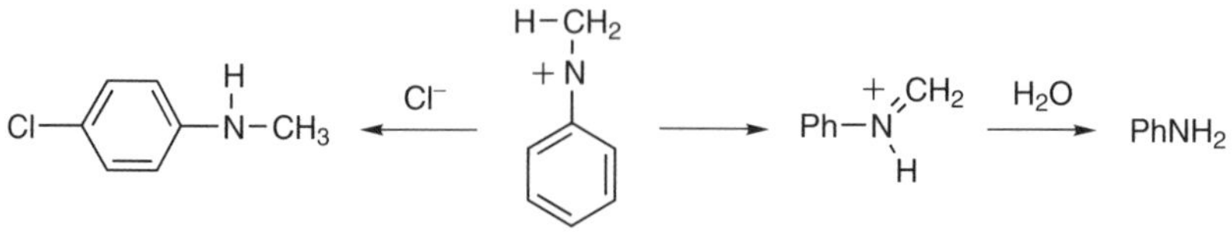

Figure 13.37. A 1,2-hydride shift in a *N*-methyl-*N*-arylnitrenium ion.

singlet alkylnitrenium ions should rearrange or eliminate, in some cases with no detectable barrier.[66,73,155]

The observation of apparent 1,2-alkyl and hydride shift products was the basis of much discussion of alkylnitrenium ion chemistry several decades ago. In fact, several reactions (e.g., Fig. 13.12) reported by Gassman and others, were ascribed to nitrenium ions. Unfortunately, it is in practice very difficult to determine whether such products resulted from reaction of a free nitrenium ion or whether they occurred concertedly with the dissociation of the leaving group.

The situation is different for alkylarylnitrenium ions. Haley first demonstrated that a *N*-1-adamantyl-*N*-(2-benzoylphenyl) nitrenium ion **62** (Fig. 13.36) could be generated by photolysis of an anthranilium ion.[112] In this case, rearrangement of the adamantyl ring, to give products from hydrolysis of **63**, competes with addition reactions, giving **64**. Similar experiments carried out on *N*-*tert*-butyl-*N*-arylnitrenium ions also showed that the 1,2-shift of a methyl group competes with additions of nucleophiles to the aryl ring.[117,156,157] In the latter cases, the formation of the alkylarylnitrenium ion was verified by LFP.

Hydride shifts are more difficult to ascertain. This process gives a *N*-protonated iminium ion, which is the conjugate acid of an imine. The latter could also result when proton transfer occurs from the alkylnitrenium ion to a base. Chiapperino and Falvey[143] examined the reactions of *N*-methyl-*N*-phenylnitrenium ion **65** (Fig. 13.37) in the presence and absence of chloride. It was found the yield of aniline (resulting from hydrolysis of the product iminium ion) was unaffected by added base. This finding ruled out a deprotonation process and led to the conclusion that a 1,2-hydride shift had occurred. Cramer et al.[76] modeled this process using ab initio methods.

Diarylnitrenium ions have not been studied in detail, but work on Ph_2N^+ (**66**, Fig. 13.38) shows that, in the absence of nucleophiles, the primary decay pathway

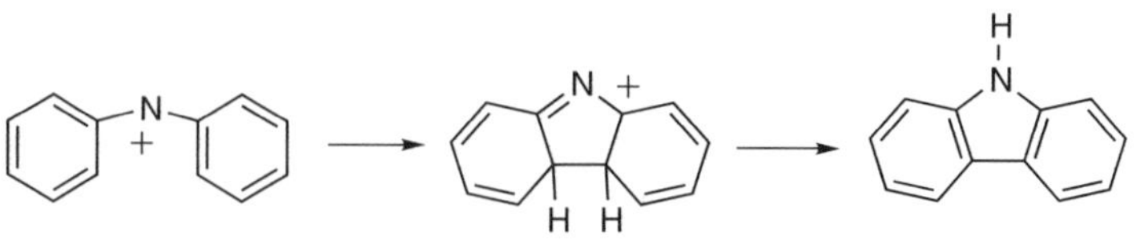

Figure 13.38. Cyclization of a diarylnitrenium ion.

is a cyclization reaction to form carbazole **68**.[144,145] This apparently proceeds via a Nazarov-like cyclization to give the carbozole tautomer **67**. The latter undergoes a hydrogen shift and deprotonation to give the observed product. Typically, the isolated yields of carbazole are low because it reacts with nitrenium ion at the diffusion limited rate giving a poorly defined set of oxidized oligomeric products. Maximal yields of carbazole are obtained only when the reaction is carried out under highly dilute conditions.[144]

4.2. Singlet-State Reactions with *n* Nucleophiles

The only well-characterized addition of an *n* nucleophile to a nonaromatic nitrenium ion is the reaction of singlet NH_2^+ with water or methanol to provide, respectively, NH_2OH and NH_2OMe (Fig. 13.39).[158]

In many studies, singlet arylnitrenium ions **69** have been generated in aqueous solutions. The primary decay route under such conditions is addition of water (or alcohols) to the para (**70**) and ortho positions (**71**) of the aromatic ring (Fig. 13.40).[39,44,78,89,159–161] Addition of water or hydroxide to the nitrenium center has not been observed. This reaction pathway is very general and most *n*-type nucleophiles (N_3^-, halides, alcohols, amines) show analogous behavior. Mixtures of ortho and para adducts are seen with the more reactive nucleophiles which exhibit less selectivity, whereas less reactive nucleophiles show a preference for para addition.

There are some minor discrepancies in the experimental literature regarding the regiochemistry of this addition. Some workers report a firm preference for para addition, even in situations where the para position is substituted.[160] Other studies of similar, but not identical nitrenium ions, show ortho products as the major adducts.[162] It is likely that many of these differences can be traced to the reversibility of the addition process. For example, the initial adduct formed from water addition to arylnitreniums is an oxonium ion (**72** or **73**), a species that can also

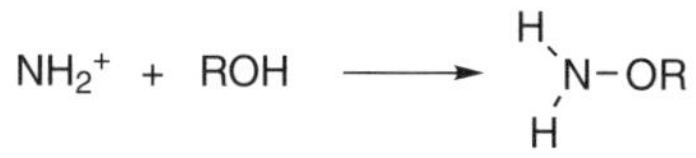

$$NH_2^+ \ + \ ROH \ \longrightarrow \ \overset{H}{\underset{H}{\diagdown}}N-OR$$

Figure 13.39. Addition of water and alcohols to the singlet parent nitrenium ion.

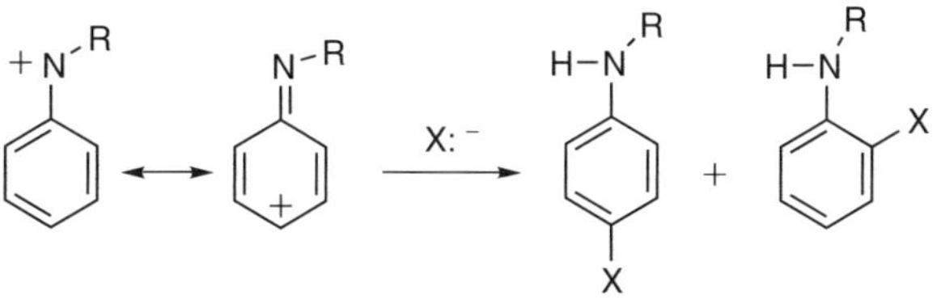

Figure 13.40. Addition of nucleophiles to the ring carbons of arylnitrenium ions.

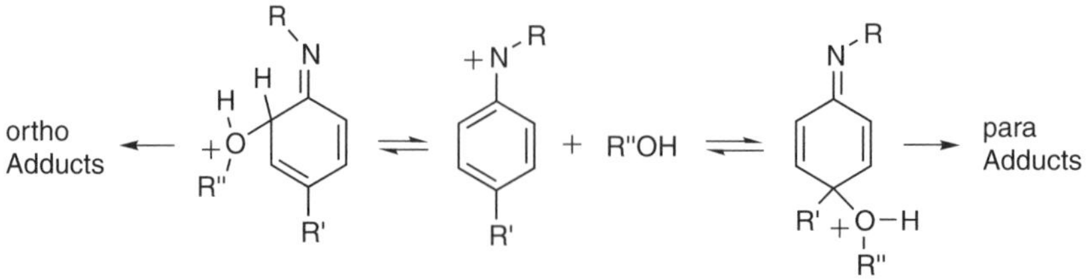

Figure 13.41. Addition of alcohols and water to arylnitrenium ions.

serve as a good leaving group (Fig. 13.41). Thus in some cases, especially under conditions where the concentration of base is low, the isolated adducts might reflect a thermodymic rather than a kinetic preference.

These addition reactions require formation of an imino-cyclohexadiene intermediate (Fig. 13.41). In cases where the ipso substituent is a proton, tautomerization to form the substituted aniline derivative is fast, and such intermediates have not been isolated. On the other hand, in situations where the nucleophile adds to a substituted ring position, the intermediate can undergo further secondary reactions. For example Novak et al.[160] showed that the 4-biphenylylnitrenium ion reacts with water forming the imine cyclohexadiene intermediate **74**, which in turn experiences an acid-catalyzed phenyl shift reaction to **76** via **75** (Fig. 13.42).

In cases where the substituent is a halide, the initial adduct is particularly reactive. For example, trapping of N-(4-chlorophenyl)-N-methylnitrenium ion **77** with methanol gives a variety of products (**79–81**).[163] These result from various homolytic and heterolytic fragmentations of the intermediate **78**. (Fig. 13.43).

With Br$^-$[164] and other reducing nucleophiles such as thiols,[165] there is a competing inner-sphere reduction pathway that competes with adduct formation (Fig. 13.44). The initial step, as outlined in Figure 13.43, involves addition of the nucleophile to the ring to give **82**. However, instead of rearranging, an additional Br$^-$ combines with the original Br to form Br$_2$ and the conjugate base of the parent amine **83**. Thus the net effect of this process is a reduction of the nitrenium ion.

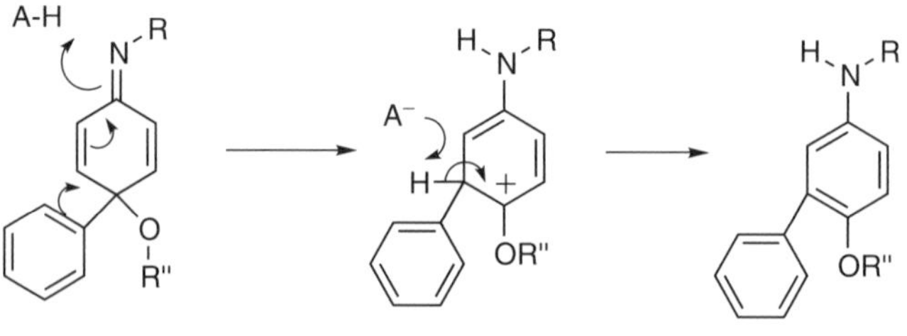

Figure 13.42. Addition followed by a 1,2-phenyl shift.

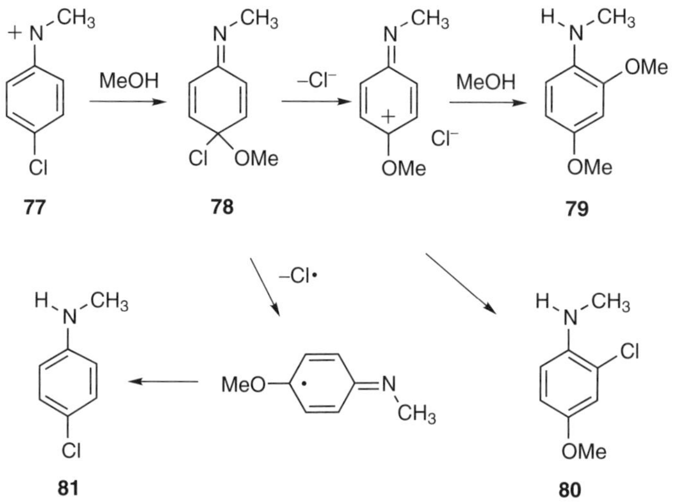

Figure 13.43. Addition and secondary reactions of *N*-methyl-*N*-(4-chlorophenyl)nitrenium ion.

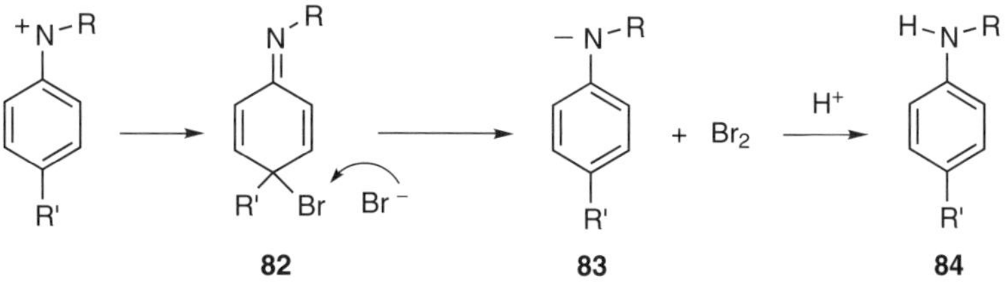

Figure 13.44. Bromide reduction of arylnitrenium ions.

In some cases, the initial adduct can combine with an additional nucleophile to give a diadduct. This process is particularly likely when the driving force for aromatization of the monoadduct is weak or absent. For example, Novak has shown that heteroarylnitrenium ion **85** is trapped at first by water and then by an additional nucleophile to give the diadduct **87** (Fig. 13.45).[166] Likewise, Bose et al.[167] reported the dihydroxylation of the nitrenium ion **88** derived from stilbene

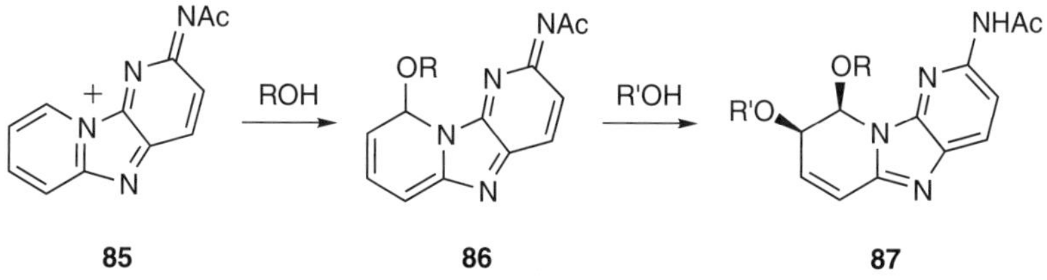

Figure 13.45. Addition of two nucleophiles to a heteroarylnitrenium ion.

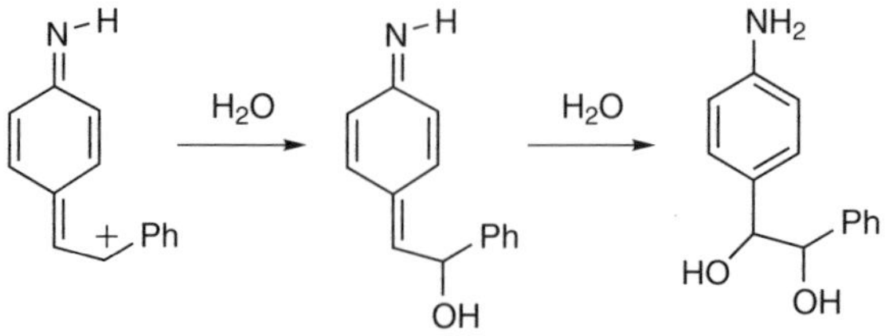

Figure 13.46. Sequential addition of water to a 4-stilbenylnitrenium ion.

(Fig. 13.46). The initial addition to the β-alkenyl carbon gives a imine quinone methide **89**. The latter rearomatizes upon addition of another water molecule to give dihydroxyl adduct **90**.

4.3. Singlet-State Reactions with π Nucleophiles

The addition of nitrenium ions to π nucleophiles (arenes and alkenes) has attracted attention as a potentially useful synthetic transformation. Takeuchi et al.[140–142] and Srivastava et al.[158] showed that the parent nitrenium ion is capable of adding to simple aromatics such as benzene, toluene and anisole (**91**) to give aniline derivatives (**92a–c**, Fig. 13.47). Unfortunately, these reactions give a distribution of regioisomers as well as products derived from competing hydride and/or hydrogen atom transfer.

Arylnitrenium ions are likewise capable of adding to π nucleophiles. With substituted aromatics (e.g., toluene) there exists the possibility of three reactive sites on the nitrenium ion (the nitrogen, ortho- and para-ring positions), along with up to three possible sites on the arene (ortho, para, and meta in the case of a monosubstituted trap). Thus in a typical case there is the possibility of nine distinct regioisomers. Obviously, any synthetic utility of such chemistry relies on the ability of the reagents to react in a selective manner.

An example of this is illustrated by a recent study of Takeuchi et al.,[83,87] summarized in Figure 13.48. The phenylnitrenium ion was generated, in the presence of anisole, from N-hydroxylaniline using triflouroacetic acid and polyphosphoric acid.

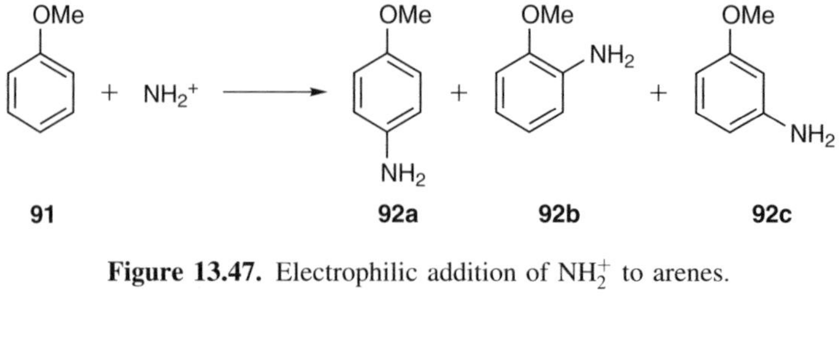

Figure 13.47. Electrophilic addition of NH_2^+ to arenes.

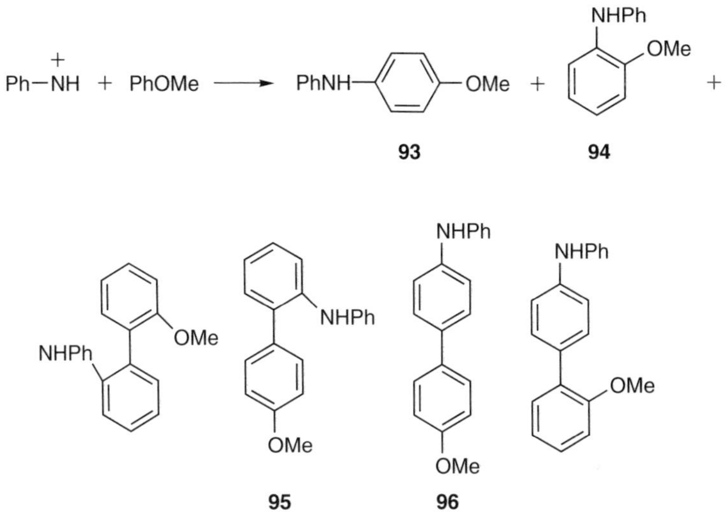

Figure 13.48. Addition of phenylnitrenium ion to anisole.

In general, six of the nine potential isomers are found in significant yields. There is a slight preference for coupling between the nitrenium ion's nitrogen and the para position on anisole to give **93**, followed by **94**, then **95** and **96**.

In a recent study by McIlroy and Falvey[168] on *N,N*-diphenylnitrenium ion [generated by photolysis of the corresponding *N*-amino(2,4,6-trimethylpyridinium) ion], it was shown that Ph_2N^+ is trapped by 1,3,5-trimethoxybenzene (**97**) to give adducts dominated by coupling to the 4-position on the nitrenium ion **99**, along with nearly equal amounts of the *N* (**98**) and ortho adducts (**100**) (see Fig. 13.49). Apparently the additional aryl ring suppresses reaction at N and, to some extent, at the ortho positions through steric hindrance. Obviously, further development of this synthetic strategy will require more selective reactions.

Chiapperino et al.[162] examined the behavior of *N*-methyl-*N*-4-biphenylylnitrenium ion **101** (R = Me) with a series of arenes. In most cases, the reactions gave predominantly ortho adduct along with significant amounts of the reduction

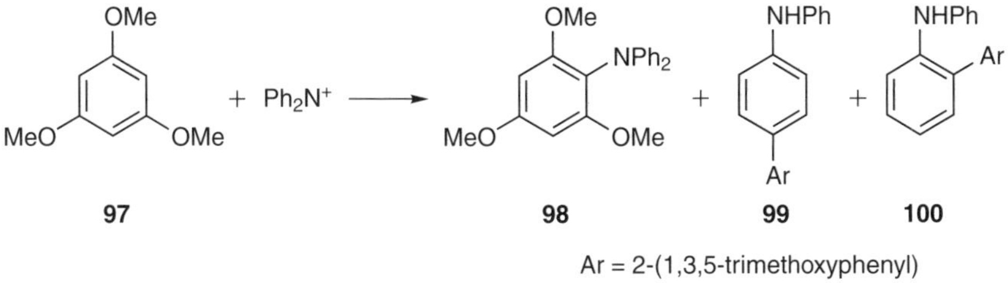

Ar = 2-(1,3,5-trimethoxyphenyl)

Figure 13.49. Addition of *N,N*-diphenylnitrenium ion to 1,3,5-trimethoxybenzene.

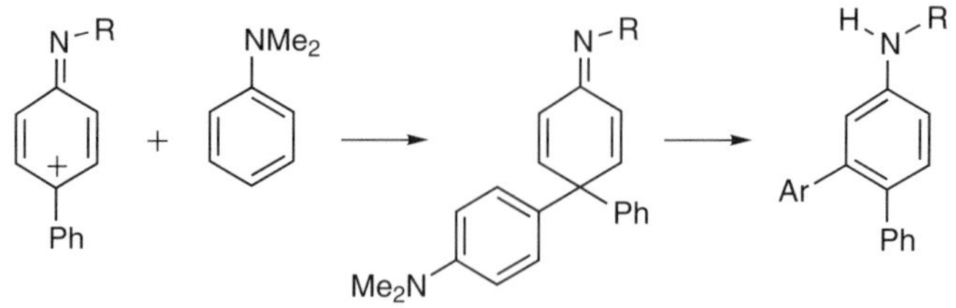

Figure 13.50. Addition–migration of *N,N*-dimethylaniline to 4-biphenylylnitrenium ion.

product, *N*-methyl-*N*-4-biphenylyamine. In some cases, N adducts were detected, but they were minor products, if they were formed at all. In this study, no adducts derived from para addition were found. It was argued that the initial adduct resulting from para addition suffered a relatively slow homolytic dissociation to give the *N*-methyl-*N*-(4-biphenylyl)aminyl radical. The latter abstracted an additional hydrogen atom to provide the reduction product.

Novak et al.[169] examined the addition of *N,N*-dimethylaniline to 4-biphenylynitrenium ion (**101**, R = H) and found significant yields of meta adduct **103**. This product was attributed to an initial addition coupling the para position of the nitrenium ion to the para position of the aniline to give **102**, followed by an acid-catalyzed 1,2-migration of the trap to the meta position (Fig. 13.50). In this case, the solvent was water. In the previous case, the solvent was acetonitrile. Presumably, the difference in the fate of the initial para adduct is governed by the solvent. The 1,2-shift requires a general base, which is abundant in the buffered aqueous solutions used by Novak et al.[169] The homolytic dissociation is more probable in aprotic media such as MeCN.

More promising from a synthetic point of view are the reactions of nitrenium ions that are covalently linked to aromatic substrates. The constraints provided by the linker generally reduce or eliminate the problem of forming multiple regioisomers. Wardrop et al.[25,26] applied the cyclization of acylalkoxynitrenium ions (e.g., **104**) and report good yields of *spiro*-*N*-alkoxypyrolidone derivatives (e.g., **105**), which are intermediates in the synthesis of several natural products, such as TAN1251A (Fig. 13.51).

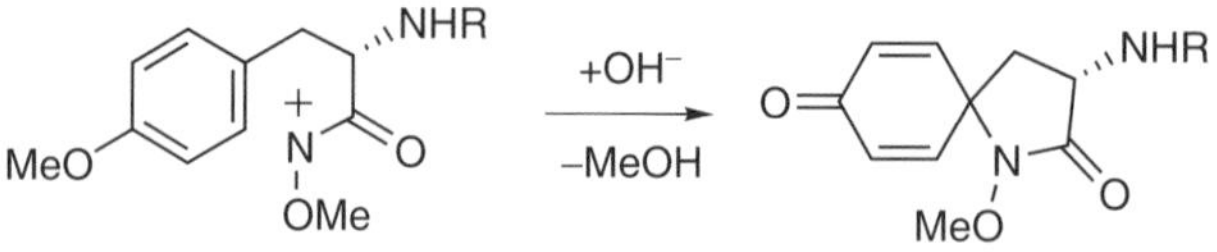

Figure 13.51. Cyclization of a *N*-acyl-*N*-alkoxynitrenium ion in the synthesis of TAN1251A.

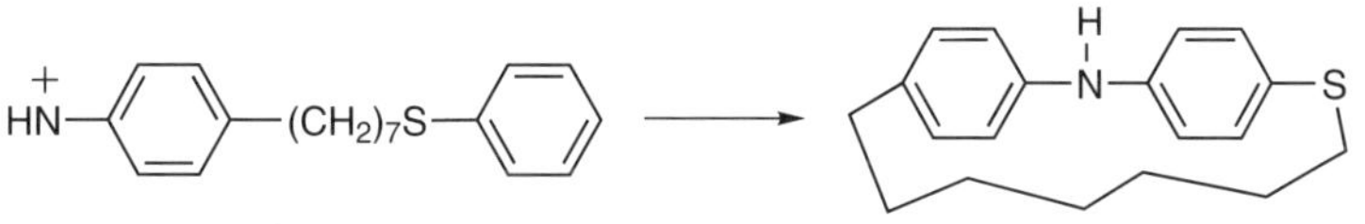

Figure 13.52. Macrocyclization through nitrenium ion/arene coupling.

Analogous methods have been applied to the synthesis of macrocycles. Abramovitch et al.[24] obtained fair yields of a macrocyclic diarylamine (**107**) through N to para cyclization of a arylnitrenium ion (**106**) onto a thiobenzene group (Fig 13.52). In a related study, Abramovitch and Ye[170] attempted the addition of an arylnitrene derivative to a tethered alkoxybenzene ring **108**. The reaction actually produced the phenanthrene derivative **109**. This apparent result is from the coupling of the meta position of the arylnitrenium ion to the 4-position of the proximal arylether. It is argued that the addition proceeds by an initial para to para cyclization to a five membered ring (**110**). Acylation by trifluoroacetic anhydride induces a 1,2-shift to give **111**. The latter is deprotonated to give the observed product (Fig. 13.53).

There are several examples of arylnitrenium ion additions to alkenes (Fig. 13.54). For example, Dalidowicz and Swenton[171] trapped N-acetyl-N-(4-methoxyphenyl)-nitrenium ion **112** with 3,4-dimethoxy-1-propenylbenzene and isolated products resulting from addition of the alkene to the ortho position of the nitrenium ion, giving cation **113**, followed by its cyclization onto N (yielding **114**) or the acetyl carbonyl group (yielding **115**).

Similar results were obtained when diphenylnitrenium ion was trapped with various silylenol ethers and silyl ketene acetals (e.g., **116**).[172] In these experiments, a distribution of N-(**117**), p- (**118**), and o- (**119**) adducts were generated (Fig. 13.55). The ortho adducts underwent a cyclization reaction, producing an indolone derivative.

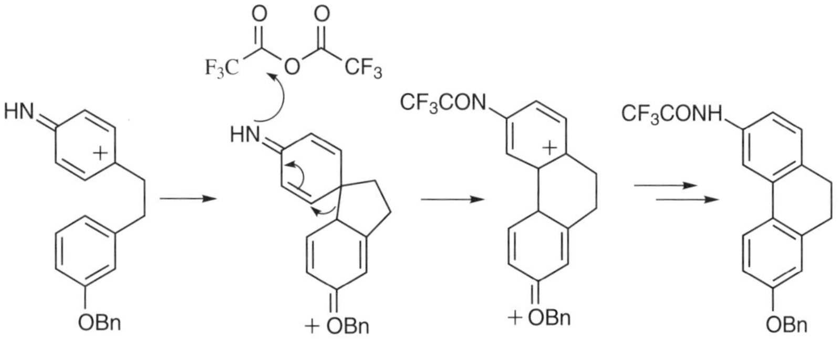

Figure 13.53. Cyclization–rearrangement of a arylnitrenium ion.

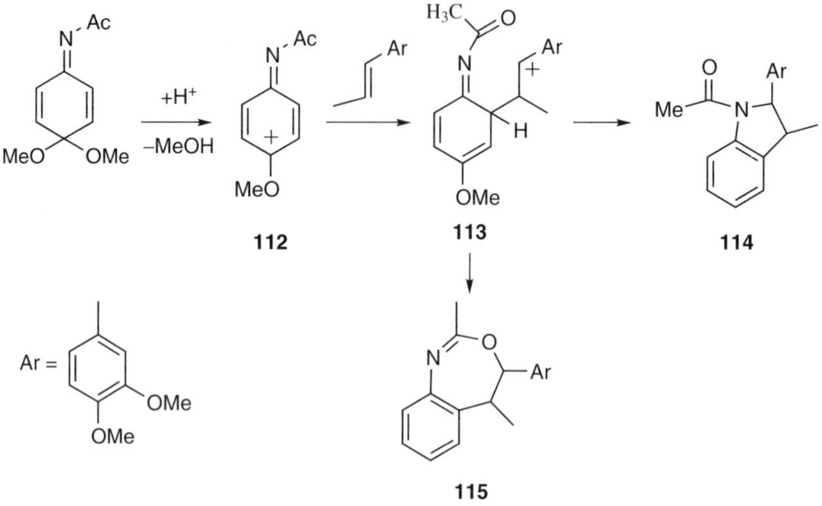

Figure 13.54. Addition of an arylnitrenium ion to an alkene.

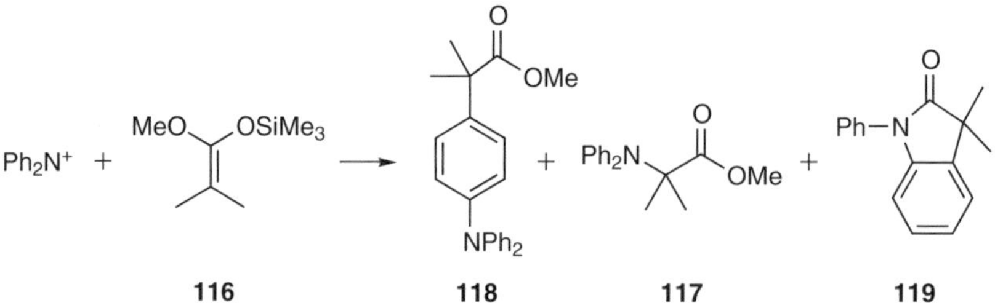

Figure 13.55. Addition of *N,N*-diphenylnitrenium ion to alkenes.

4.4. Singlet-State Reactions with Hydride Donors

This process has not been studied in detail. It has been shown that diphenylnitrenium ion reacts with various hydrocarbons and metal hydrides to give diphenyl amine.[144] An analysis of the rate constants for these processes showed that the reaction was most likely a hydride transfer, rather than a hydrogen atom transfer (Fig. 13.56). Novak and Kazerani[173] found a similar process in their study of the decay reaction of heteroarylnitrenium ions.

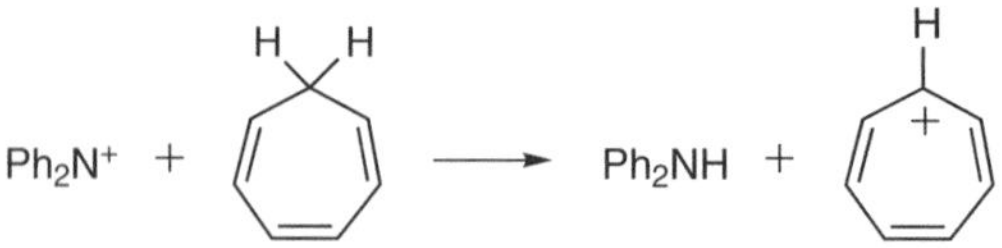

Figure 13.56. Hydride transfer.

4.5. Triplet-State Hydrogen Atom Transfer Reactions

Relatively little is known about the chemical reactions of triplet state nitrenium ions. In fact, the only reaction that is unambiguously assigned to this spin state is hydrogen atom transfer. The triplet state has two unpaired electrons. Intuitively, one might suspect that it would behave in a way similar to a free radical. To the extent that unambiguous data are available, this does seem to be the case.

N-*tert*-Butyl-N-(2-acetylphenyl)nitrenium ion was generated by photolysis of the corresponding anthranilium ion. This photolysis first creates a singlet state of the *precursor*. The latter partitions between ring opening, providing the singlet-state nitrenium ion, and intersystem crossing, providing the excited triplet state of the precursor. The latter undergoes a ring opening creating the triplet nitrenium ion (Fig. 13.57).[117,156,157]

These reactions give three types of stable products: (1) adducts from the addition of nucleophiles (alcohols, water, halides) to the benzene ring, (**120**); (2) an iminium ion (**121**), apparently resulting from the 1,2-shift of methyl group, and (3) the parent amine (**122**). It was argued that **120** and **121** are specific products of the singlet state, whereas **122** is derived from the triplet state. These assignments are supported by several additional experiments. For example, it was shown that the yield of the triplet product was enhanced when the photolysis was carried out using triplet sensitization, but it was suppressed when a triplet quencher (a molecule that deactivates the excited triplet state of the precursor by an energy-transfer mechanism) was employed.[117] Several other anthranilium ion generated nitreniuim ions were investigated in this manner and they all showed qualitatively similar behavior.[118,156,157,174,175]

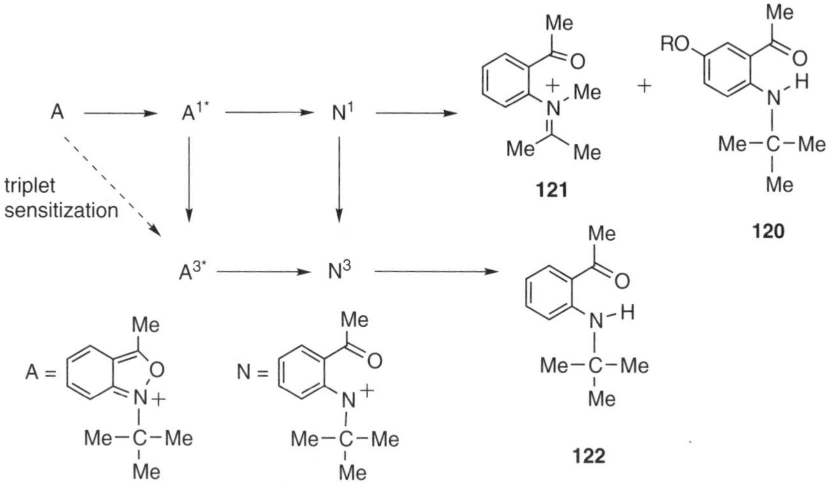

Figure 13.57. Spin-specific chemistry of arylnitrenium ions.

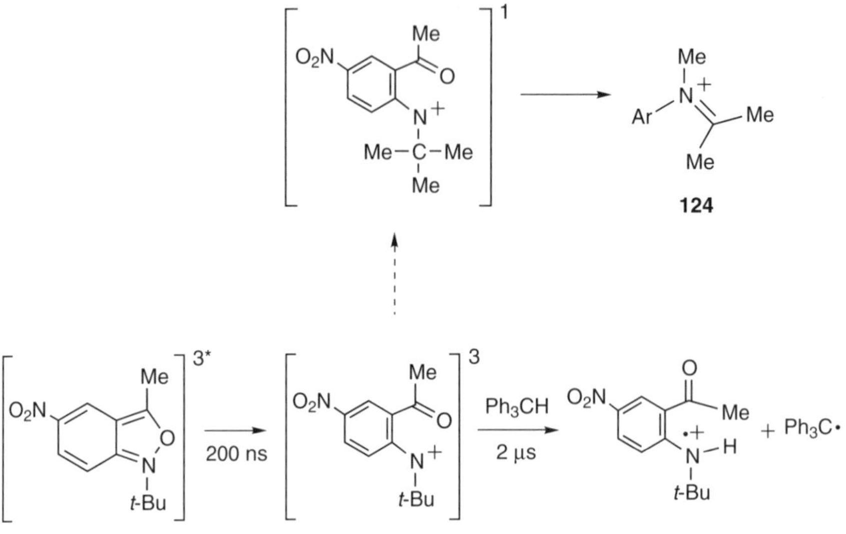

Figure 13.58. Evidence for a triplet ground-state arylnitrenium ion.

The 4-nitro derivative in the anthranilium ion series showed qualitatively differ-ent behavior (Fig. 13.58).[119] For one thing, it proved impossible to trap it using any external nucleophile. On the other hand, significant yields of the iminium ion **124** and the parent amine were observed. Only the parent amine was observed when the decomposition was carried out through triplet energy transfer. Laser flash photoly-sis experiments were not successful at detecting the nitrenium ion, but instead showed that the excited triplet state of the precursor **125** was formed. Addition of a hydrogen atom donor, Ph$_3$CH, resulted in formation and detection of the corresponding radical, Ph$_2$C•. The latter's growth did not correlate with the disap-pearance of triplet state **125**. Instead, there was a ~2-μs delay (extrapolated to [Ph$_3$CH] = 0 M) between the disappearance of **125** and formation of Ph$_3$C•. The delay time corresponds to the lifetime of the triplet nitrenium ion. That no singlet products (i.e., **124**) were formed in the triplet sensitization experiment led to the conclusion that either the nitrenium ion **123** was a ground-state triplet or else that the singlet was ground state but that intersystem crossing was anomalously slow in this system.

The parent nitrenium ion (NH$_2^+$) is firmly established as a ground-state triplet: both extensive ab initio calculations as well as PES experiments all agree that the singlet–triplet energy gap is 30 kcal/mol. There have been several investigations on its behavior in solution. Takeuchi et al.[141] showed that this species could be gener-ated by photolysis of 1-amino-(2,4,6-triphenylpyridinium) ion. These photolyses were carried out in the presence of various aromatic compounds. It was found that the triplet state abstracted hydrogen atoms from traps such as toluene

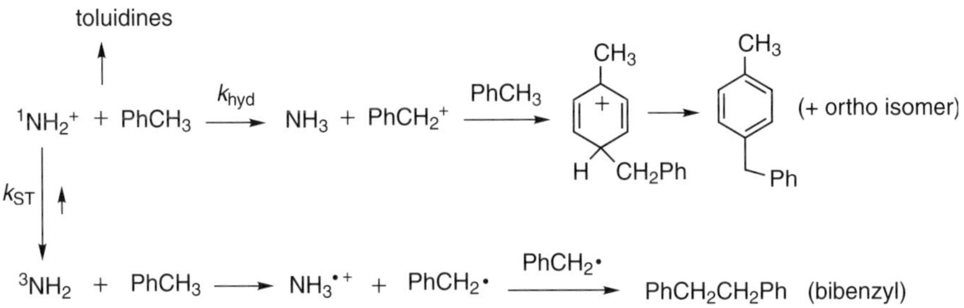

Figure 13.59. Hydrogen atom transfer to triplet nitrenium ion.

(Fig. 13.59). Srivastava et al.[158] carried out similar experiments and demonstrated that the singlet was formed initially upon photolysis. This species either inserts into the OH bond of water or it suffers intersystem crossing to the lower energy triplet. As in the former study, the only reaction found for the triplet was hydrogen atom transfer.

4.6. Intersystem Crossing

The interconversion of singlet states and triplet states is known as intersystem crossing. To date, a direct observation of this process has not been reported other than for the decay of the excited triplet state of the highly stabilized nitrenium ion **7** (Fig. 13.2). In several cases, intersystem crossing has been inferred from studies of product distributions. For example, the photolytic generation of NH_2^+ creates the singlet state which abstracts a hydride from toluene to give the benzyl cation (Fig. 13.59). The latter adds to an additional molecule of toluene to give ortho and para benzyltoluenes. In contrast, the triplet abstracts a hydrogen atom from toluene. The resulting benzyl radicals dimerize in a head-to-head fashion giving bibenzyl. It was found that the ratio of bibenzyl to the benzyltoluenes increased with the addition of an unreactive diluent to the toluene trap. An analysis of these data yields a ratio of the intersystem crossing rate constant, k_{ST}, to the rate constant for hydride transfer from the singlet, k_{hyd}, of $k_{ST}/k_{hyd} = 31\,M$.[158] Assuming a diffusion limited rate constant for hydride transfer, $k_{hyd} = 1\text{--}10 \times 10^9\,M^{-1}\,s^{-1}$, it is possible to estimate k_{ST} as between $3 \times 10^{11}\,s^{-1}$ and $3 \times 10^{12}\,s^{-1}$. Direct ultrafast measurements of these values would be desirable.

5. SPECTROSCOPIC AND KINETIC STUDIES OF NITRENIUM IONS

5.1. Brief Summary of Methods

Many studies of carbenium ions have made use of superacid media for their generation and characterization. Attempts to apply this method to the study of nitrenium ions have been largely unsuccessful. The reason is that singlet nitrenium

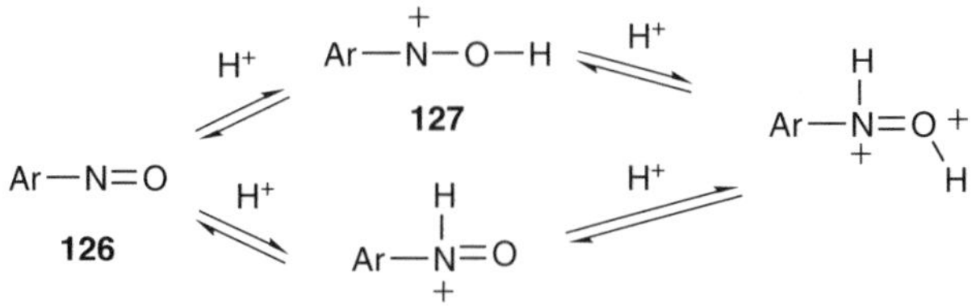

Figure 13.60. Attempt to protonate nitrosocompound gives dication.

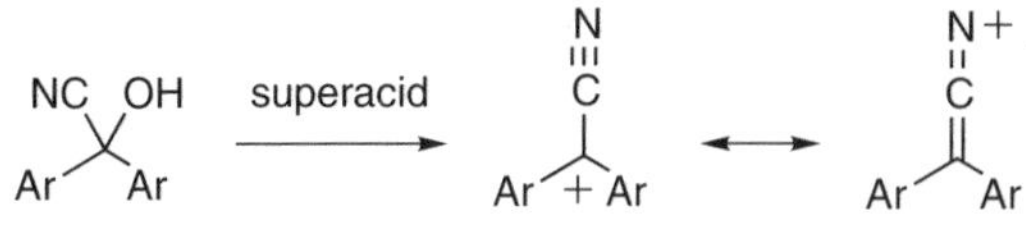

Figure 13.61. Vinylidene nitrenium ion stabilized in superacid media.

ions can serve as weak bases due to their nonbonding electron pairs. For example, attempts to generate N-hydroxyl-N-phenylnitrenium ion **127** through the protonation of N-nitrosobenzene **126** resulted in the formation of the dication (Fig. 13.60).[176] Olah and co-workers[177–179] did succeed in creating alkylidenenitrenium ions. This result was accomplished by treating cyanohydrins **128** with superacid, creating α-cyanocarbenium ion **129a**, which can be regarded as a nitrenium ion on the basis of resonance structure **129b** (Fig. 13.61).

A major concern in nitrenium ion research has been to characterize the lifetimes and reaction rates of these species in solution. Two techniques have been the mainstays of this research. The oldest and most generally applicable method to characterize arylnitrenium ion lifetimes in solution is to measure the relative yields of the stable products. The ratio of rate constants can then be inferred from the product ratios.

One specific embodiment of this approach has been to use the azide (az) clock method (Fig. 13.62).[180–182] Azide ion (N_3^-) is a very strong nucleophile (Nu) and is thus assumed to react with most arylnitrenium ions at the diffusion—limited rate,

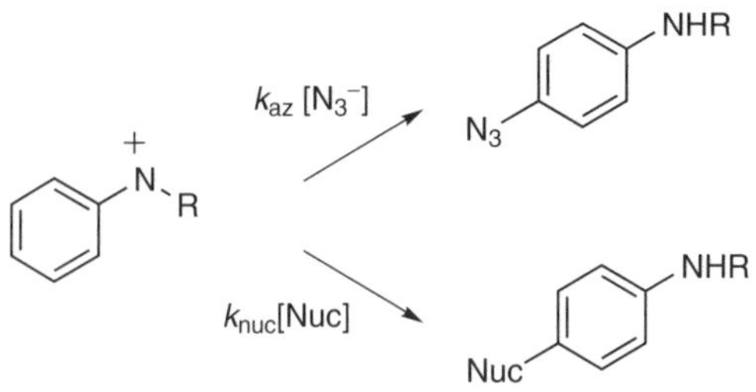

Figure 13.62. Azide clock method for determining arylnitrenium ion reaction rate constants.

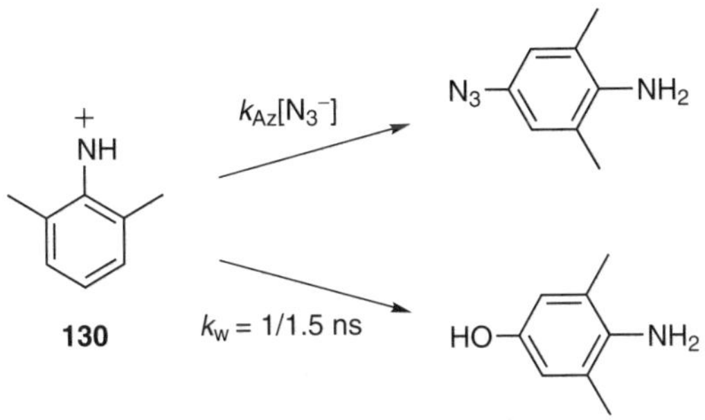

Figure 13.63. Lifetime of a nitrenium ion in aqueous solvent.

$5 \times 10^9 \, M^{-1} \, s^{-1}$. Direct measurements of azide trapping reactions have generally validated this assumption (see Chapter 2). Except for highly stabilized nitrenium ions, k_{az}, is seen to fall in a narrow range of $4–10 \times 10^9 \, M^{-1} \, s^{-1}$.[11,131,183] The yield of azide adducts relative to the competing products can be used to estimate the rate constants for formation of the latter.

In view of the role of arylnitrenium ions in carcinogenic DNA damage, there has been much interest in their lifetimes in aqueous solution. Arylnitrenium ions that are rapidly consumed through hydrolysis are unlikely to have lifetimes sufficient for them to diffuse through the cell, penetrate the nucleus and damage the DNA. Fishbein and McClelland[84,184,185] trapped the 2,6-xylylnitrenium ion (**130**) generated in the Bamberger rearrangement (acid-catalyzed decomposition of the corresponding hydroxylamine) with N_3^-. On the basis of product ratios and the assumption that the reaction of azide was diffusion limited, it was concluded that the lifetime of this nitrenium ion in aqueous solution was ~1.5 ns. Such a short lifetime implies that this nitrenium ion (Fig. 13.63) would probably not be able to efficiently attack DNA in an aqueous cellular environment.

Several representative selectivity values $S(= \log[k_{az}/k_w])$ derived in this manner are listed in Table 13.4. In the main, para-substituted aromatic rings and alkenyl

TABLE 13.4. Relative Selectivities ($S = \log k_{az}/k_2$) Derived from Azide Trapping of $R^1R^2N^{+a}$

R^1	R^2	S
4-Biphenylyl	Ac	2.97
4-Biphenylyl	H	3.45
2-Fluorenyl	Ac	4.76
2-Fluorenyl	H	5.07
4-Ethoxyphenyl	Ac	2.73
4-Ethoxyphenyl	H	3.68
4(4′-Methoxybiphenylyl)	H	6.59
4(4′-Fluorobiphenylyl)	H	3.85

[a] See Ref. 14.

groups show the most profound effect on the selectivity. This effect is seen in the series of fluorenyl and 4-biphenylyl nitrenium ions that have S mostly in excess of 3.5, ranging as high as 6.59. Alkoxy groups, which are generally regarded as stronger donors, show more modest effects on the selectivity, falling in the range of 2.7–3.7. Alkyl groups are even less effective in conferring aqueous stability, showing S values <1.0. N-Alkyl and N-acetyl groups have only modest effects on S usually increasing or decreasing it by <1 relative to H.

The azide clock approach has the advantage of being universally applicable to any arylnitrenium ion that can be generated in aqueous solution (azide salts are typically insoluble or sparingly soluble in nonaqueous media). This corresponds to the conditions regarded as most relevant to their toxicological behavior. The method can also be adapted to the study of trapping rate constants for reagents other than water. The major disadvantage of this approach is that it only can give accurate rate constants in cases where the assumption of diffusion limited trapping is valid and preassociation of the azide with the precursor does not occur,

Laser flash photolysis methods have also been applied to the study of nitrenium ion trapping rates and lifetimes. This method relies on short laser pulses to create a high transient concentration of the nitrenium ion, and fast detection technology to characterize its spectrum and lifetime The most frequently used detection method is fast UV–vis spectroscopy. This method has the advantage of high sensitivity, but provides very little specific information about the structure of the species being detected. More recently, time-resolved infrared (TRIR) and Raman spectroscopies have been used in conjunction with flash photolysis methods. These provide very detailed structural information, but suffer from lower detection sensitivity.

5.2. Ultraviolet–Visible Spectra of Nitrenium Ions

The first nitrenium ion to be examined using any flash photolysis method was the 4-dimethylaminophenylnitrenium ion (**131**).[186] However, the workers who carried out these experiments apparently did not consider this species to be a nitrenium ion. It was generated by pulsed Xe lamp photolysis from the corresponding azide (**132**, Fig. 13.64), and was detected through its absorption at 325 nm. As the resonance structure **131** implies, this nitrenium ion (or quinonediimine) is especially stable. In fact, its lifetime exceeds 100 ms at neutral pH, and it decays through hydrolysis of the C=N bond to give iminoquinone (**133**).

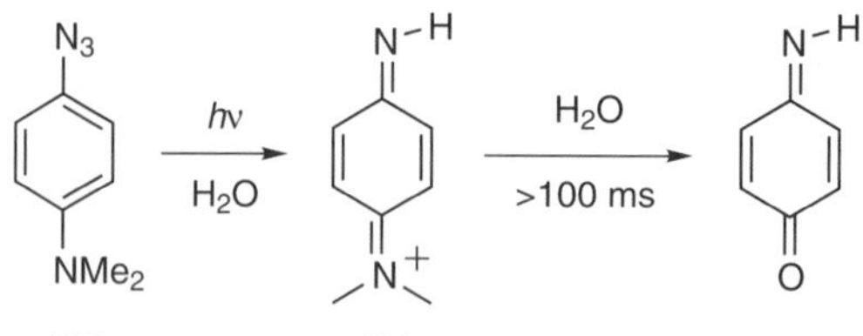

Figure 13.64. First arylnitrenium ion detected by direct spectroscopic methods.

TABLE 13.5. *N-tert*-Butyl-*N*-aryl Nitrenium Ions Generated through LFP of Anthranilium Ion Salts[a]

X^b	λ_{max}	τ in MeCN (ns)	k_{trap} (Trap)^c
Cl	385	0.14	3.2×10^6
Br	395	0.13	1.3×10^6
CH_3	460	0.40	3.1×10^7
Ph	470	30	6.2×10^4
OMe	500	600	2.1×10^5

[a] See Refs. 157 and 175.
[b] Aryl substituent para to the nitrenium ion.
[c] The trap is water.

The study of more typically reactive nitrenium ions had to await the application of faster methods. Nearly two decades later, nanosecond LFP methods were applied to the ring opening of anthranilium ions (Fig. 13.29).[174] A series of alkylarylnitrenium ions were characterized in this manner and the results are compiled in Table 13.5.[117,157,175] These species in general had much shorter lifetimes ($<50\,\mu s$) than the quinone diimine. The 4-halonitrenium ions of this class show absorption maxima at wavelengths <400 nm, along with a weak tail that extends into the visible. For the 4-methyl derivative, only the weak tail can be detected. The UV maximum apparently overlaps with the absorption band of the starting material. Substitution by a phenyl ring or an alkoxy group shifts the absorption maximum into the visible region of the spectrum. Interestingly, it is the 4-phenyl substituent that has the most profound effect on the stabilization of the nitrenium ion toward attack by H_2O.

The next development in direct detection of nitrenium ions came from McClelland et al.[120] who applied the azide method to LFP measurements. This permitted the direct detection of those arylnitrenium ions implicated in carcinogenic DNA damage. McClelland's approach proved to be particularly useful in the study of 4-aryl[2,129,131] and 4-alkoxy substituted phenylnitrenium ions.[130] Apparently, the corresponding singlet nitrenes are sufficiently long lived to allow for protonation in aqueous solution. Several arylnitrenium ions studied by this route are described in Table 13.6.

TABLE 13.6. Arylnitrenium Ions (Ar–N$^+$–H) Generated from Arylazide Photolysis in Protic Media

Ar	λ_{max}	τ (μs)	Reference
4-Biphenylyl	465	$0.35/H_2O$	183
2-Fluorenyl	460	$75/H_2O$	183
$4'$-MeO-4-Biphenylyl	500	$633/H_2O$	183
2-(1-Methylimidazolyl)	235	$100,000/H_2O$	183
4-Stilbenyl	520	$0.16/H_2O$	167
4-MeO-phenyl	300	$0.8/H_2O$	183
2,4,6-Tribromophenyl	420, 600	$0.2/MeCN/H_2O$	125
4-hydroxymethyl-2,3,5,6-tetrafluorophenyl	375	$>1000/MeCN$	128

TABLE 13.7. Absorption Maxima (nm), Lifetimes, and Trapping Rate Constants $(M^{-1}\,s^{-1})$ of Nitrenium Ions Characterized by Laser Flash Photolysis of N-Aminopyridinium Salts[a]

R^1	R^2	λ_{max}	$\tau(\mu s)$	k_{trap} (Trap)
Ph	Ph	435, 690	1.6 (MeCN)	6.1×10^5 (H_2O)
4-Tolyl	4-Tolyl	425,680	280 (MeCN)	1.9×10^4 (H_2O)
4-Biphenylyl	Me	460	24 (MeCN)	9.3×10^4 (H_2O)
4-Chlorophenyl	Me	340	0.95 (MeCN)	8.9×10^6 (H_2O)
4-MeO-phenyl	Me	320	100 (MeCN)	5.5×10^4 (H_2O)
4-Tolyl	Me	330	7.4 (MeCN)	2.3×10^7 (H_2O)
1-Naphthyl	Me	500	0.8 (MeCN)	1.7×10^8 (H_2O)

[a] See Refs. 145, 146, and 163.

The aminopyridinium route has been employed in flash photolysis studies of aryl as well as diarylnitrenium ions. Several examples of nitrenium ion species, along with their absorption maxima and some trapping rate constants are given in Table 13.7. To the extent the data are comparable, there is good agreement with the behavior of nitrenium ions generated by the azide route. For example, the 4-biphenylyl systems from the azide protonation and N-aminopyridinium routes both give absorption maxima at 460 nm and live for several microseconds in water. Likewise, the 4-methoxyphenyl systems show maxima at 300 nm (from azide) and 320 nm (from aminopyridinium ion). The discrepancy in this case can be attributed to the N-methyl substituent, present in the aminopyridinium route, but absent in the azide experiment.

5.3. Infrared and Raman Spectra of Nitrenium Ions

A promising recent development in the study of nitrenium ions has been the introduction of time-resolved vibrational spectroscopy for their characterization. These methods are based on pulsed laser photolysis. However, they employ either time resolved IR (TRIR) or time-resolved resonance Raman (TRRR) spectroscopy as the mode of detection. While these detection techniques are inherently less sensitive than UV–vis absorption, they provide more detailed and readily interpretable spectral information. In fact, it is possible to directly calculate these spectra using relatively fast and inexpensive DFT and MP2 methods. Thus, spectra derived from experiment can be used to validate (or falsify) various computational treatments of nitrenium ion structures and reactivity. In contrast, UV–vis spectra do not lend themselves to detailed structural analysis and, moreover, calculating these spectra from first principles is still expensive and highly approximate.

The first study of this nature involved the application of TRIR to diphenylnitrenium ion (Fig. 13.65).[187] This species gave strong distinct transient IR bands at 1568, 1440, and 1392 cm^{-1}. It was further demonstrated that the decay of these signals followed the same kinetics found by UV–vis detected LFP for the same

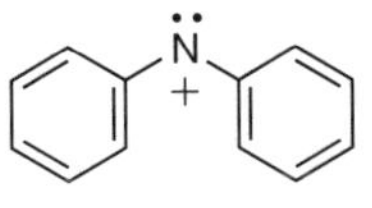

$$\nu = 1568, 1440, 1392 \text{ cm}^-$$

Figure 13.65. Infrared bands for diphenylnitrenium ion.

species. The 1392-cm^{-1} band shifted by 16 cm^{-1} to lower frequencies when the central nitrogen was enriched with ^{15}N. On this basis it was assigned to an asymmetric CNC vibration, analogous to the 1320-cm^{-1} band observed for the parent amine. The remaining bands were unperturbed by ^{15}N substitution and thus assigned to C—C stretches (Fig. 13.65).

Several 4-substituted phenylnitrenium ions **134/135** were also examined by TRIR and the results were compared with IR spectra generated from DFT calculations (Fig. 13.66).[163] Each of these aryntrenium ions shows prominent bands in the region 1580–1628 cm^{-1}. These signals were assigned to symmetric aromatic C=C stretches. The frequencies of these bands correlate with the expected degree of charge delocalization. For example, the 4-methoxy derivative was expected to show the most charge delocalization and exhibit the highest degree of quinoidal character. This leads to an increase in bond alternation in the aromatic ring (i.e., resonance form **135** is favored) and thus the highest frequency for the C=C. In contrast, the 4-chloro substituent renders the ring less susceptible to delocalization, and exhibits a C=C frequency closer to that of an unperturbed aromatic ring (1604 cm^{-1}). The 4-phenyl deriviative shows two bands: a more quinoidal band at 1612 cm^{-1}, attributed to the more perturbed proximal ring, and a more aromatic band at 1584 cm^{-1} attributed to the less perturbed distal ring.

The DFT calculations (BPW91/cc-pVDZ) on the same singlet nitrenium ions give excellent agreement with these measurements. For calculated and experimental C=C stretches, deviation between experiment and calculation is generally <5 cm^{-1}. On the other hand, similar calculations carried out for the triplet states showed significantly greater deviation from the measured values.

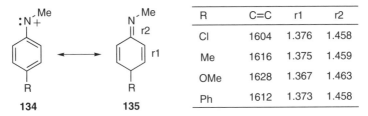

R	C=C	r1	r2
Cl	1604	1.376	1.458
Me	1616	1.375	1.459
OMe	1628	1.367	1.463
Ph	1612	1.373	1.458

Figure 13.66. Bond lengths (Å) and aromatic C=C stretching frequencies (cm^{-1}) for arylnitrenium ions.

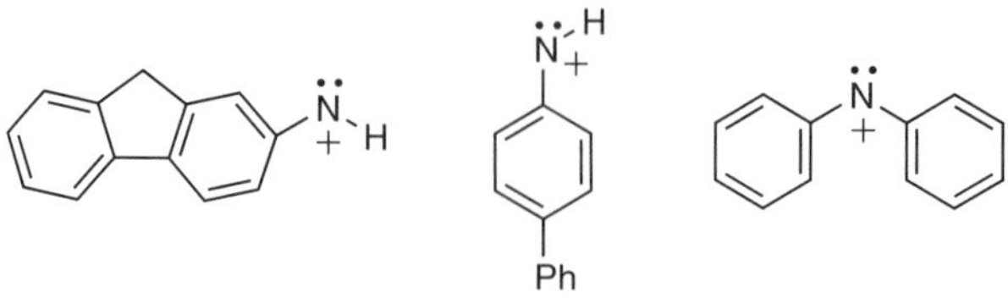

136	137	66
1637	1625	1604
1607	1607	1525
1567	1496	1478
1486	1422	1455

Figure 13.67. Selected Raman bands (in cm^{-1}) for arylnitrenium ions.

Phillips and co-workers[188] introduced the technique of TRRR to these studies. They reported detailed TRRR studies and DFT calculations on 2-fluorenynitrenium ion (**136**), the 4-biphenylylnitrenium ion (**137**), and the *N,N*-diphenylnitrenium ion (**66**, Fig. 13.67). The former two nitrenium ions were generated through azide photolysis in MeCN–H$_2$O mixtures, and the latter nitrenium ion was generated via the *N*-aminopyridinium ion route. Some of the major Raman bands measured for these species appear in Figure 13.67. As with the TRIR experiments, excellent agreement is obtained between measured and DFT-calculated values.

5.4. Direct Detection of Intermediates in Nitrenium Ion Reactions

The LFP studies of the reaction of the *N*-methyl-*N*-4-biphenylylnitrenium ion with a series of arenes showed that no detectable intermediate formed in these reactions.[162] The rate constants of these reactions correlated neither with the oxidation potentials of the traps (as would be expected were the initial step electron transfer) nor with the basicity of these traps (a proxy for their susceptibility toward direct formation of the sigma complex). Instead, a good correlation of these rate constants was found with the ability of the traps to form π complexes with picric acid (Fig. 13.68). On this basis, it was concluded the initial step in these reactions was the rapid formation of a π complex (**140**) between the nitrenium ion (**138**) and the arene (**139**). This was followed by σ-complex formation and tautomerization to give adducts, or a relatively slow homolytic dissociation to give (ultimately) the parent amine.

In a subsequent study using diphenylnitrenium ion, several intermediates were detected. With 1,3,5-trimethoxybenzene or 1,3-dimethoxybenzene, the decay of the nitrenium ion occurred concurrently with the appearance of sigma adducts (**141**, Fig. 13.69).[168] These were characterized on the basis of their absorption maxima and their behavior toward pyridine bases. On the other hand, when readily oxidized arenes, such as *N,N*-dimethylaniline were employed, the characteristic ion radicals were detected (Fig. 13.70).[189]

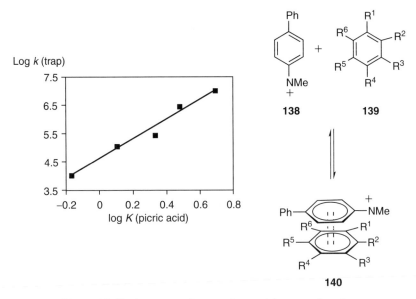

Figure 13.68. Arene trapping correlates with π complexation.

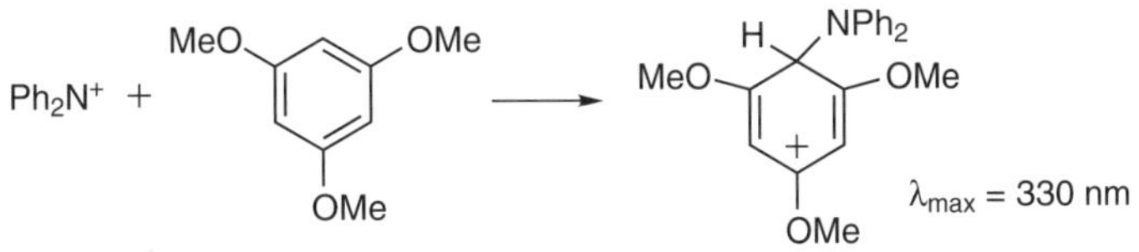

Figure 13.69. Sigma complex detected with diphenylnitrenium ion.

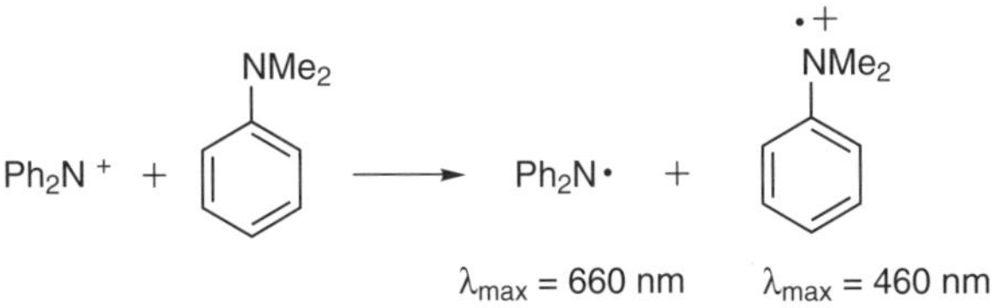

Figure 13.70. Radicals detected from electron-transfer reaction.

These results lead to a general mechanism for the reaction nitrenium ions with aromatic compounds (Fig. 13.71). Initial encounter leads to a π-complex (**141**). The latter is converted into isomeric σ complexes (**142–144**) which, in turn, either tautomerize to give stable adducts (**145–147**) or else dissociate to give radicals. The relative rates of these processes depend on the reactivities of the nitrenium ion and the arene. With less reactive nitrenium ions the π-complex is relatively long lived. With more reactive nitrenium ions the π complex forms in a low steady-state

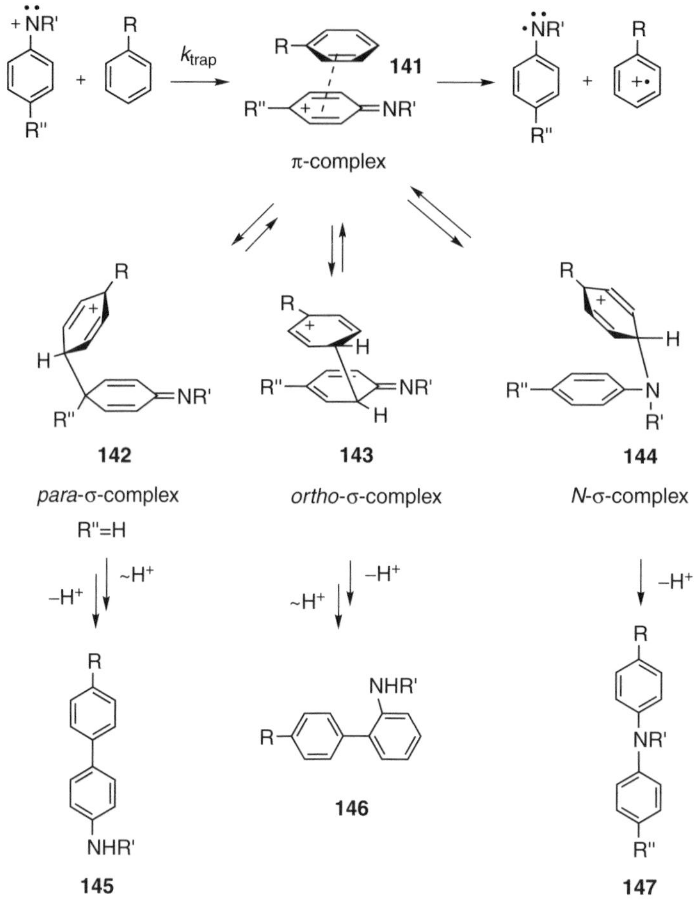

Figure 13.71. Mechanism for arylnitrenium ion reactions with arenes.

concentration and proceeds to give radicals in cases where the arenes are readily oxidized, or to σ complexes when they are not.

6. THE ROLE OF ARYLNITRENIUM IONS IN DNA DAMAGING REACTIONS

Arylnitrenium ions are considered to be key intermediates in carcinogenic DNA damage by metabolically activated amines. The enzymatic activation of amines and the propogation of DNA–amine adducts into mutations and cellular transformations is beyond the specific scope of this chapter. Presented here is an outline of the key mechanistic issues relevant to the interaction of DNA components with various arylnitrenium ions.

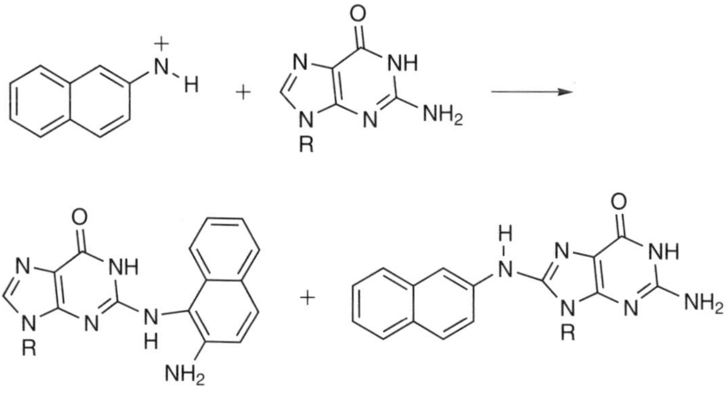

Figure 13.72. Reaction of naphthylnitrenium ion with guanosine.

Boche and co-worker[190] carried out detailed studies of the decomposition reactions of various N-hydroxylamine esters in the presence of DNA and deoxyguanosine, and characterized the adducts that resulted from these reactions. Thus, the nitrenium ion derived from O-acetyl-N-(2-aminofluorene) added to DNA to give an adduct joining the C8 position of the base with the nitrenium ion nitrogen. Similar experiments carried out with the precursor of 2-naphthylnitrenium ion gave N2-ortho **148** and C8–N adducts **149** as shown in Figure 13.72.[96]

With regard to DNA damage, the most extensively studied arylnitrenium ions are the 2-fluorenenylnitrenium ion and the 4-biphenylylnitrenium ion. The toxicity of these species is explained by the fact that they are relatively long lived in water (>1 µs) yet react with DNA components at nearly the diffusion limit. Novak and Kennedy[191] carried out extensive competitive trapping studies using N-acetyl-4-biphenylylnitrenium ion with various nucleosides. The pyrimidines, cytidine, uracidine, and thymidine did not show any reaction with this nitrenium ion (i.e., reaction with water was faster). On the other hand, gaunosine and its derivatives trapped this nitrenium ion with rate constants in excess of $10^9 M^{-1} s^{-1}$. Other purines, adenine, and inosine showed about a tenth of the reactivity of guanosine.

McClelland used LFP methods to measure rate constants for the reactions of various substituted 4-biphenylylnitrenium ions with deoxyguanosine and found that they fell mostly in the narrow range of $1.0–2.2 \times 10^9 M^{-1} s^{-1}$.[129,131,183] An exception was the highly stabilized 4'-methoxy derivative that was trapped at $3.6 \times 10^7 M^{-1} s^{-1}$. The 2-fluorenyl derivatives also showed somewhat attenuated reactivity, with rate constants with deoxyguanosine of $2–9 \times 10^8 M^{-1} s^{-1}$.[131]

With free bases or nucleosides in solution, the main adduct arises from a net coupling of the C8 on gaunosine with the nitrenium ion nitrogen. This result is interesting for two reasons. First, most reactive nucleophiles tend to add to the ring carbons of the arylnitrenium ion rather than the nitrogen. In some cases, highly reactive arenes have been observed to give a mixture of regioisomers, including N adducts;

only with guanine is the N adduct predominant. Second, whereas guanine bases in DNA react with a variety of electrophiles, the primary sites for addition seem to be N7 and O6. Adducts at C8 are apparently very rare. The reason for this divergent behavior with guanine has not been adequately explained.

A particularly important result from the point of view of chemical toxicology was the detection of an intermediate in the reaction of N-acetyl-2-fluorenylnitrenium ion (**150**) with guanosine (Fig. 13.73).[133] Concurrent with the decay of the nitrenium ion, a new absorbance band with $\lambda_{max} = 330$ nm appeared. On the basis of kinetic isotope effects, and the pH dependence of its decay, this intermediate was assigned to the σ adduct generated by joining the nitrenium ion's nitrogen to the C8 position of guanine (**152**). Thus it was concluded that the key DNA damaging reaction proceeded via a relatively straightforward electrophilic aromatic substitution mechanism. This contrasts with a proposal by Humphreys et al.,[192] which held that the initial adduct was between the N7 of guanine and the N of the nitrenium ion (**153**). The latter was supported by experiments on guanine derivatives wherein the C8 position was substituted. It was argued that **153** was converted into the stable adduct via deprotonation to give zwitterion **154**, followed by a 1,2-shift to give final product **151**.

Novak and Kennedy[193] studied the reaction of N-acetyl-N-2-fluorenylnitrenium ion (**150**) with single and double-stranded oligomeric DNA (6-mer), as well as

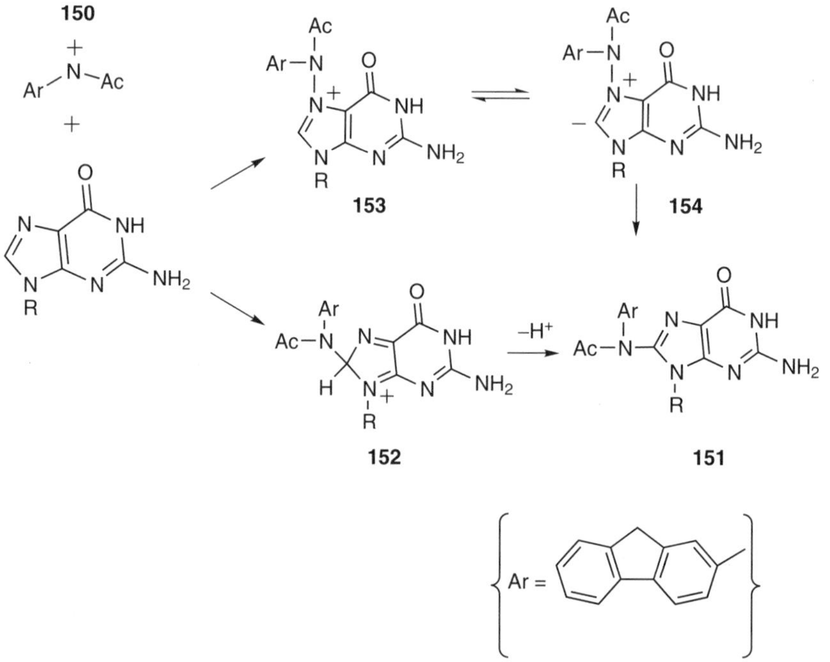

Figure 13.73. Mechanisms for addition to N-acetyl-N-2-fluorenenyl guanosine.

plasmid DNA. With the oligomeric DNA, virtually all of the trapping could be attributed to the fraction of the 6-mer considered to be in the single-stranded form. With double stranded plasmid DNA the reactivity with nitrenium ion was significantly attenuated, being ~2% of the reactivity of the free nucleoside. Single-stranded DNA, on the other hand showed reactivity that was slightly lower (~30%) than the monomeric nucleoside. Thus it was suggested that rapidly replicating cells, which have a higher fraction of single-stranded DNA, should be particularly susceptible to damage by activated mutagens.

More recently attention has turned to the reactions of heteroarylnitrenium ions,[166,173,194] These intermediates are derived from the activation of heterocyclic amines such as **155**, shown in Figure 13.74.[166] These amines are in turn created in cooked foodstuffs through the high-temperature thermolysis of amino acids or

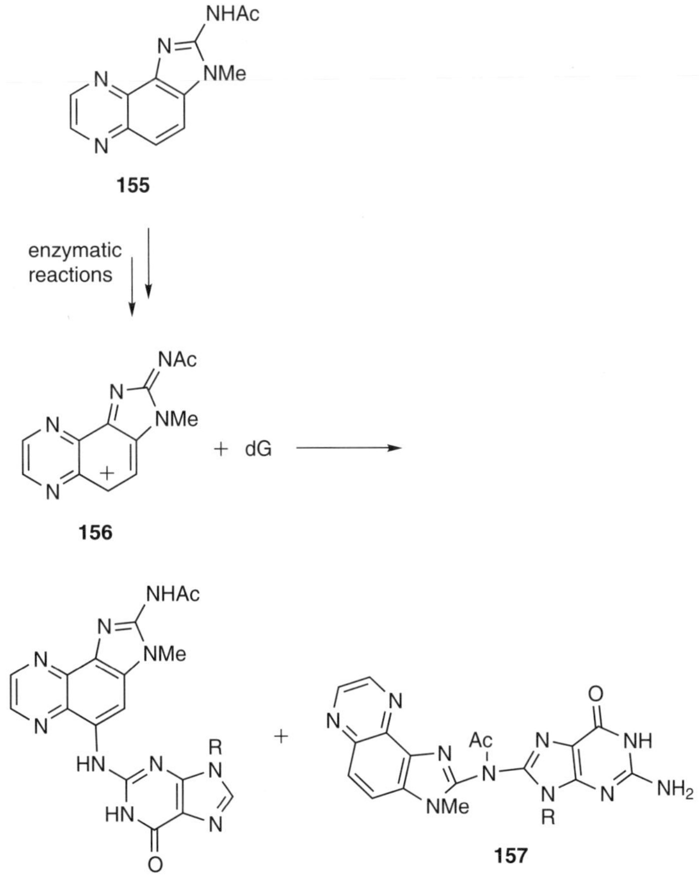

Figure 13.74. Reactions of heteroarylnitrenium ions with guanosine.

proteins. The nitrenium ion **156** is trapped by deoxyguanosine to give a C8 adduct (**157**) analogous to those observed for the carbocyclic nitrenium ions. An adduct connecting the guanine's N2 with the ring carbon of the heterocycle (**158**) is also observed. Interestingly, this nitrenium ion also decays, presumably through hydride abstraction, to give the **155**, along with some dimeric products. It was also found that similar reactions could occur through the less reactive conjugate base of this nitrenium ion.

7. CONCLUSION AND OUTLOOK

The past two decades have witnessed remarkable progress in the understanding of nitrenium ions. It is now widely agreed that these intermediates have a discrete existence. Several examples have been detected and characterized by direct means, including LFP. Numerous decay pathways for singlet-state nitrenium ions have been established both through product studies as well as by direct detection. The vast majority of studies have focused on the behavior of singlet arynitrenium ions. In a few cases, the chemical reactions of triplet arylnitrenium ions have been characterized, although few triplets have been studied by direct means. More information about the reactivity of alkyl and other non-arylnitrenium ions would be desirable.

Density functional-based theoretical methods have been demonstrated to give highly accurate predictions of nitrenium ions structures, singlet–triplet energy gaps, and vibrational spectra. Future challenges in the area of theoretical predictions include the accurate modeling of nitrenium ion reactions and reaction rates.

More recently, time-resolved vibrational spectroscopy has been applied to nitrenium ions. These techniques hold much promise for future research as they can provide more detailed structural information about nitrenium ions. In addition, the vibrational frequencies measured in this way can be used to assess the accuracy of theoretical calculations.

Nitrenium ions have been applied to the synthesis of macrocycles and other medicinally interesting compounds. The most successful reactions have been in cases where the nitrenium ion is covalently tethered to its intended target. Further efforts aimed at modulating the selectivity of these intermediates would increase their synthetic utility.

ADDITIONAL READING

R. A. Abramovitch and R. Jeyaraman, "Nitrenium Ions," in *Azides and Nitrenes*, E. F. V. Scriven, Ed., Academic Press, New York **1984**, p. 297.

M. Novak and S. Rajagopal, "*N*-Arylnitrenium Ions," *Adv. Phys. Org. Chem.* **2001**, *136*, 167.

R. A. McClelland, "Flash Photolysis Generation and Reactivities of Carbenium Ions and Nitrenium Ions," *Tetrahedron*, **1996**, *52*, 6823.

D. E. Falvey, "Photochemical Generation and Studies of Nitrenium Ions," in *Organic, Physical and Materials Photochemistry*, V. Ramamurthy and K. S. Schanze, Eds., Marcel Dekker, New York, **2000**, pp. 249–284.

F. F. Kadlubar, "DNA Adducts of Carcinogenic Amines," in *DNA Adducts Identification and Biological Significance*, K. Hemmink, A. Dipple, D. E. G. Shugar, F. F. Kadlubar, D. Segerback, and H. Bartsch, Eds., University Press, Oxford, UK, **1994**, pp. 199–216.

P. G. Gassman, "Nitrenium Ions," *Acc. Chem. Res.* **1970**, *3*, 26.

REFERENCES

1. C. J. Cramer, F. J. Dulles, and D. E. Falvey, *J. Am. Chem. Soc.* **1994**, *116*, 9787.

2. A. P. Dicks, A. Ahmad, R. D'Sa, and R. A. McClelland, *J. Chem. Soc. Perkin Trans. 2* **1999**, 1.

3. R. H. Wiley and J. Moffat, *J. Am. Chem. Soc.* **1955**, *77*, 1703.

4. G. Boche, P. Andrews, K. Harms, M. Marsch, K. S. Rangappa, M. Schimeczek, and C. Willecke, *J. Am. Chem. Soc.* **1996**, *118*, 4925.

5. M. Robert, A. Neudeck, G. Boche, C. Willecke, K. S. Rangappa, and P. Andrews, *New J. Chem.* **1998**, 1437.

6. H.-W. Wanzlick and E. Schikora, *Chem. Ber.* **1961**, *94*, 2389.

7. A. J. Arduengo, III, R. L. Harlow, and M. Kline, *J. Am. Chem. Soc.* **1991**, *113*, 361.

8. R. A. Abramovitch and B. A. Davies, *Chem. Rev.* **1964**, *64*, 149.

9. P. T. Lansbury, in *Nitrenes*, W. Lwowski, Ed., Wiley-Interscience, New York, **1970**, pp. 405–418.

10. P. G. Gassman, *Acc. Chem. Res.* **1970**, *3*, 26.

11. R. A. McClelland, *Tetrahedron* **1996**, *52*, 6823.

12. D. E. Falvey, *J. Phys. Org. Chem.* **1999**, *12*, 589.

13. D. E. Falvey, in *Organic, Physical, and Materials Photochemistry*, V. Ramamurthy and K. Schanze, Eds., Marcel Dekker, New York, **2000**, pp. 249–284.

14. M. Novak and S. Rajagopal, *Adv. Phys. Org. Chem.* **2001**, *36*, 167.

15. R. A. Abramovitch and R. Jeyaraman, in *Azides and Nitrenes: Reactivity and Utility*, E. F. V. Scriven, Ed., Academic Press, Orlando, FL, **1984**, pp. 297–357.

16. F. F. Kadlubar, in *DNA Adducts Identification and Biological Significance*, K. Hemmink, A. Dipple, D. E. G. Shugar, F. F. Kadlubar, D. Segerback, and H. Bartsch, Eds., University Press, Oxford, UK, **1994**, pp. 199–216.

17. R. C. Garner, C. N. Martin, and D. B. Clayson, in *Chemical Carcinogens*, Vol. 1, 2nd ed., C. E. Searle, Ed., American Chemical Society, Washington, DC, **1984**, pp. 175–276.

18. J. A. Miller, *Cancer Res.* **1970**, *30*, 559.

19. G. R. Hoffman and R. P. P. Fuchs, *Chem. Res. Toxicol.* **1997**, *10*, 347.

20. H. A. J. Schut and E. G. Snyderwine, *Carcinogenesis* **1999**, *20*, 353.

21. R. A. Abramovitch, H. H., Gibson, Jr., T. Nguyen, S. Olivella, and A. Solé, *Tetrahedron Lett.* **1994**, *35*, 2321.

22. R. A. Abramovitch, J. M. Beckert, P. Chinnasamy, H. Xiaohua, W. Pennington, and A. R. V. Sanjivamurthy, *Heterocycles* **1989**, *28*, 623.

23. Y. Kikagawa and M. Kawase, *J. Am. Chem. Soc.* **1984**, *106*, 5728.

24. R. A. Abramovitch, X. Ye, W. T. Pennington, G. Schimek, and D. Bogdal, *J. Org. Chem.* **2000**, *65*, 343.

25. D. J. Wardrop and A. Basak, *Org. Lett.* **2001**, *3*, 1053.

26. D. J. Wardrop and W. Zhang, *Org. Lett.* **2001**, *3*, 2353.

27. E. M. Genies and M. Lapowski, *J. Electroanal. Chem.* **1987**, *236*, 189.

28. G. Zotti, N. Comisso, G. D'Aprano, and M. Leclerc, *Adv. Mater.* **1992**, *4*, 749.

29. R. Holze, *Collect. Czech. Chem. Commun.* **2000**, *65*, 899.

30. Y. Wei, R. Hariharan, and S. Patel, *Macromolecules* **1990**, *23*, 758.

31. W. Liu, J. Kumar, S. Tripathy, K. J. Senecal, and L. Samuelson, *J. Am. Chem. Soc.* **1999**, *121*, 71.

32. W. Liu, A. L. Cholli, R. Nagarajan, J. Kumar, S. Tripathy, F. F. Bruno, and L. Samuelson, *J. Am. Chem. Soc.* **1999**, *121*, 11345.

33. Y. Ding, A. B. Padias, and H. K., Hall, Jr., *J. Polym. Sci. Part A: Chem.* **1999**, *37*, 2569.

34. F. Lux, *Polymer Rev.* **1995**, *35*, 2915.

35. Y. Wei, X. Tang, and Y. Sun, *J. Polym. Sci. Part A: Chem.* **1989**, *27*, 2385.

36. E. Bamberger, *Chem. Ber.* **1894**, *27*, 1548.

37. E. Bamberger and J. Lagutt, *Chem. Ber.* **1898**, *31*, 1500.

38. E. Bamberger, *Liebigs Ann. Chem.* **1925**, *441*, 297.

39. H. E. Heller, E. D. Hughes, and C. K. Ingold, *Nature (London)* **1951**, *168*, 909.

40. P. G. Gassman and R. L. Cryberg, *J. Am. Chem. Soc.* **1969**, *91*, 2047.

41. P. G. Gassman and R. L. Cryberg, *J. Am. Chem. Soc.* **1969**, *91*, 5176.

42. P. G. Gassman and G. A. Campbell, *J. Am. Chem. Soc.* **1971**, *93*, 2567.

43. P. G. Gassman, G. A. Campbell, and R. C. Frederick, *J. Am. Chem. Soc.* **1972**, *94*, 3884.

44. P. G. Gassman and G. A. Campbell, *J. Am. Chem. Soc.* **1972**, *94*, 3891.

45. C. H. Gudmundsen and W. E. McEwen, *J. Chem. Soc.* **1957**, *79*, 329.

46. P. G. Gassman, K. Uneyama, and J. L. Hahnfeld, *J. Am. Chem. Soc.* **1977**, *99*, 647.

47. J. C. Koziar and D. O. Cowan, *Acc. Chem. Res.* **1978**, *11*, 334.

48. N. J. Turro, *Modern Molecular Photochemistry*, Benjamin/Cummings: Menlo Park, CA, **1978**.

49. R. V. Hoffman and D. J. Poelker, *J. Org. Chem.* **1979**, *44*, 2364.

50. R. V. Hoffman, A. Kumar, and G. A. Buntain, *J. Am. Chem. Soc.* **1985**, *107*, 4731.

51. R. V. Hoffman and N. B. Christophe, *J. Org. Chem.* **1988**, *53*, 4769.

52. U. Svanholm and V. D. Parker, *J. Am. Chem. Soc.* **1974**, *96*, 1234.

53. D. Serve, *J. Am. Chem. Soc.* **1975**, *97*, 432.

54. P. Haberfield and M. DeRosa, *J. Chem. Soc. Chem. Commun.* **1985**, 1716.

55. V. J. Barclay, I. P. Hamilton, and P. Jensen, *J. Chem. Phys.* **1993**, *99*, 9709.

56. C. J. Cramer, F. J. Dulles, J. W. Storer, and S. E. Worthington, *Chem. Phys. Lett.* **1994**, *218*, 387.

57. O. Dopfer, D. Roth, and J. P. Maier, *Chem. Phys. Lett.* **1999**, *310*, 201.

58. C. Gonzales, A. Restrepo-Cossio, M. Márquez, K. B. Wiberg, and M. D. Rosa, *J. Phys. Chem. A.* **1998**, *102*, 2732.

59. Y. Kabbadj, T. R. Huet, D. Uy, and T. Oka, *J. Mol. Spectrosc.* **1996**, *175*, 277.

60. G. Osman, P. R. Bunker, P. Jensen, and W. P. Kraemer, *J. Mol. Spectroscopy* **1997**, *186*, 319.

61. S. D. Peyerimhoff and R. J. Buenker, *Chem. Phys.* **1979**, *42*, 167.

62. T. J. van-Huis, M. L. Leininger, C. D. Sherrill, and H. F. Schaefer, III, *J. Comput. Chem.* **1998**, *63*, 1107.

63. T. G. Wright and T. A. Miller, *J. Phys. Chem.* **1996**, *100*, 4408.

64. S. J. Dunlavey, J. M. Dyke, N. Jonathan, and A. Morris, *Mol. Phys.* **1980**, *39*, 1121.

65. S. T. Gibson, J. P. Greene, and J. Berkowitz, *J. Chem. Phys.* **1985**, *83*, 4319.

66. M. H. Lim, S. E. Worthington, F. E. Dulles, and C. J. Cramer, in *Chemical Applications of Density Functional Theory. ACS Symposium Series 629*, B. B. Laird, R. B. Ross, and T. Ziegler, Eds., American Chemical Society, Washington, DC, 1996, p. 402.

67. C. Qiao, G. Hong, and D. Wang, *J. Phys. Chem. A.* **1999**, *103*, 1972.

68. A. Gobbi and G. Frenking, *J. Chem. Soc. Chem. Commun.* **1993**, 1162.

69. M. Marquez, F. Marí, and C. A. Gonzales, *J. Phys. Chem. A.* **1999**, *103*, 6191.

70. C. J. Cramer and S. E. Worthington, *J. Phys. Chem.* **1995**, *99*, 1492.

71. G. P. Ford and J. D. Scribner, *J. Am. Chem. Soc.* **1981**, *103*, 4281.

72. S. A. Glover and A. P. Scott, *Tetrahedron* **1989**, *45*, 1763.

73. D. E. Falvey and C. J. Cramer, *Tetrahedron Lett.* **1992**, *33*, 1705.

74. M. B. Sullivan, K. Brown, C. J. Cramer, and D. G. Truhlar, *J. Am. Chem. Soc.* **1998**, *120*, 11778.

75. C. J. Cramer and D. E. Falvey, *Tetrahedron Lett.* **1997**, *38*, 1515.

76. C. J. Cramer, D. J. Truhlar, and D. E. Falvey, *J. Am. Chem. Soc.* **1997**, *119*, 12338.

77. M. B. Sullivan and C. J. Cramer, *J. Am. Chem. Soc.* **2000**, *122*, 5588.

78. M. Novak and J. Lin, *J. Org. Chem.* **1999**, *64*, 6032.

79. G. P. Ford and J. W. Thompson, *Chem. Res. Toxicol.* **1999**, *12*, 53.

80. J. M. Parks, G. P. Ford, and C. J. Cramer, *J. Org. Chem.* **2001**, *66*, 8997.

81. S. McIlroy, C. J. Cramer, and D. E. Falvey, *Org. Lett.* **2000**, *2*, 2451.

82. K. T. Potts, A. A. Kutz, and F. C. Nachrod, *Tetrahedron* **1975**, *31*, 2171.

83. H. Takeuchi, T. Taniguchi, and T. Ueda, *J. Chem. Soc. Perkin Trans. 2* **2000**, 295.

84. J. C. Fishbein and R. A. McClelland, *J. Am. Chem. Soc.* **1987**, *109*, 2824.

85. G. Kohnstam, W. A. Petch, and D. L. H. William, *J. Chem. Soc. Perkin Trans. 2* **1984**, 298.

86. H. Takeuchi, T. Adachi, H. Nishiguchi, K. Itou, and K. Koyama, *J. Chem. Soc. Perkin Trans I* **1993**, 867.

87. H. Takeuchi and K. Takano, *J. Chem. Soc. Perkin Trans. 1* **1986**, 611.

88. H. Takeuchi, M. Maeda, M. Mitani, and K. Koyama, *J. Chem. Soc. Perkin Trans 1* **1987**, 57.

89. P. G. Gassman and G. A. Campbell, *J. Chem. Soc. Chem. Commun.* **1970**, 427.

90. P. G. Gassman and G. D. Hartman, *J. Am. Chem. Soc.* **1973**, *95*, 449.

91. F. M. Schell and R. N. Ganguly, *J. Org. Chem.* **1980**, *45*, 4069.

92. M. Novak, K. A. Martin, and J. L. Heinrich, *J. Org. Chem.* **1989**, *54*, 5430.

93. M. Novak, M. J. Kahley, J. Lin, S. A. Kennedy, and T. G. James, *J. Org. Chem.* **1995**, *60*, 8294.

94. M. Novak, A. J. VandeWater, A. J. Brown, S. A. Sanzebacher, L. A. Hunt, B. A. Kolb, and M. E. Brooks, *J. Org. Chem.* **1999**, *64*, 6023.

95. F. Bosold and G. Boche, *Angew. Chem. Int. Ed. Engl.* **1990**, *29*, 63.

96. M. Famulok, F. Bosold, and G. Boche, *Angew. Chem. Int. Ed. Engl.* **1989**, *28*, 337.

97. C. Meier and G. Boche, *Tetrahedron Lett.* **1990**, *31*, 1693.

98. P. G. Gassman and J. E. Granrud, *J. Am. Chem. Soc.* **1984**, *106*, 1498.

99. J. D. Scribner and N. K. Naimy, *Cancer Res.* **1973**, *33*, 1159.

100. J. D. Scribner and N. K. Naimy, *Cancer Res.* **1975**, *35*, 1416.

101. J. D. Scribner, *J. Org. Chem.* **1976**, *41*, 3320.

102. J. D. Scribner, *J. Org. Chem.* **1977**, *42*, 7383.

103. J. S. Helmick and M. Novak, *J. Org. Chem.* **1991**, *56*, 2925.

104. M. Novak and A. K. Roy, *J. Org. Chem.* **1985**, *50*, 571.

105. M. Novak, M. Pelecanou, and L. Pollack, *J. Am. Chem. Soc.* **1986**, *108*, 112.

106. Y. Kikugawa and M. Kawase, *Chem. Lett.* **1990**, 581.

107. Y. Kikugawa and M. Kawase, *J. Chem. Soc. Chem. Commun.* **1991**, 1354.

108. A. R. Katritzky and M. Szajda, *J. Chem. Soc. Perkin Trans 1* **1985**, 2155.

109. A. R. Katriztky, J. Lewis, and P.-L. Nie, *J. Chem. Soc. Perkin Trans 1* **1979**, 446.

110. R. A. Abramovitch, K. Evertz, G. Huttner, H. H. Gibson, and H. G. Weems, *J. Chem. Soc. Chem. Commun.* **1988**, 325.

111. H. Takeuchi and K. Koyama, *J. Chem. Soc. Perkin Trans 1* **1988**, 2277.

112. N. F. Haley, *J. Org. Chem.* **1977**, *42*, 3929.

113. E. Giovaninni and B. F. S. E. deSousa, *Helv. Chim. Acta* **1979**, *62*, 185.

114. E. Giovaninni and B. F. S. E. deSousa, *Helv. Chim. Acta* **1979**, *62*, 198.

115. T. Doppler, H. Schmid, and H.-J. Hansen, *Helv. Chim. Acta* **1979**, *62*, 304.

116. T. Doppler, H. Schmid, and H.-J. Hansen, *Helv. Chim. Acta* **1979**, *62*, 271.

117. G. B. Anderson, L. L.-N. Yang, and D. E. Falvey, *J. Am. Chem. Soc.* **1993**, *115*, 7254.

118. R. J. Robbins and D. E. Falvey, *Tetrahedron Lett.* **1994**, *35*, 4943.

119. S. Srivastava and D. E. Falvey, *J. Am. Chem. Soc.* **1995**, *117*, 10186.

120. R. A. McClelland, P. A. Davidse, and G. Hadzialic, *J. Am. Chem. Soc.* **1995**, *117*, 4173.

121. R. A. McClelland, M. J. Kahley, P. A. Davidse, and G. Hadzialic, *J. Am. Chem. Soc.* **1996**, *118*, 4794.

122. S. T. Belt, C. Bohne, G. Charette, S. E. Sugamori, and J. C. Scaiano, *J. Am. Chem. Soc.* **1993**, *115*, 2200.

123. J. E. Chateauneuf, *Res. Chem. Intermed.* **1994**, *20*, 249.

124. W. Kirmse, J. Killian, and S. Steenken, *J. Am. Chem. Soc.* **1990**, *112*, 6399.

125. R. Born, C. Burda, P. Senn, and J. Wirz, *J. Am. Chem. Soc.* **1997**, *119*, 5061.

126. E. Leyva, M. S. Platz, G. Persy, and J. Wirz, *J. Am. Chem. Soc.* **1986**, *108*, 3783.

127. M. S. Platz, *Acc. Chem. Res.* **1995**, *28*, 487.

128. J. Michalak, H. B. Zhai, and M. S. Platz, *J. Phys. Chem.* **1996**, *100*, 14028.

129. R. A. McClelland, M. J. Kahley, and P. A. Davidse, *J. Phys. Org. Chem.* **1996**, *9*, 355.

130. P. Sukhai and R. A. McClelland, *J. Chem. Soc. Perkin Trans. 2* **1996**, 1529.

131. D. Ren and R. A. McClelland, *Can. J. Chem.* **1998**, *76*, 78.

132. T. A. Gadosy and R. A. McClelland, *J. Am. Chem. Soc.* **1999**, *121*, 1459.

133. R. A. McClelland, A. Ahmad, A. P. Dicks, and V. Licence, *J. Am. Chem. Soc.* **1999**, *121*, 3303.

134. P. Ramlall and R. A. McClelland, *J. Chem. Soc. Perkin Trans. 2* **1999**, 225.

135. H. Sawanishi, T. Hirai, and T. Tsuchiya, *Heterocycles* **1982**, *19*, 1043.

136. H. Sawanishi, T. Hirai, and T. Tsuchiya, *Heterocycles* **1982**, *19*, 2071.

137. H. Takeuchi and K. Watanabe, *J. Phys. Org. Chem.* **1998**, *11*, 478.

138. R. A. Abramovitch, J. M. Beckert, and W. T. Pennington, *J. Chem. Soc. Perkin Trans 1* **1991**, 1761.

139. R. A. Abramovitch and Q. Shi, *Heterocycles* **1994**, *37*, 1463.

140. H. Takeuchi, S. Hayakawa, and H. Murai, *J. Chem. Soc. Chem. Commun.* **1988**, 1287.

141. H. Takeuchi, S. Hayakawa, T. Tanahashi, A. Kobayashi, T. Adachi, and D. Higuchi, *J. Chem. Soc. Perkin Trans. 2* **1991**, 847.

142. H. Takeuchi, D. Higuchi, and T. Adachi, *J. Chem. Soc. Perkin Trans. 2* **1991**, 1525.

143. D. Chiapperino and D. E. Falvey, *J. Phys. Org. Chem.* **1997**, *10*, 917.

144. S. McIlroy, R. J. Moran, and D. E. Falvey, *J. Phys. Chem. A.* **2000**, *104*, 11154.

145. R. J. Moran and D. E. Falvey, *J. Am. Chem. Soc.* **1996**, *118*, 8965.

146. R. J. Moran, Ph.D. Thesis, University of Maryland, College Park, MD, 1997.

147. D. Bogdal, *Heterocycles* **2000**, *53*, 2679.

148. P. A. Davidse, M. J. Kahley, R. A. McClelland, and M. Novak, *J. Am. Chem. Soc.* **1994**, *116*, 4513.

149. A. Rieker and B. Speiser, *Tetrahedron Lett.* **1990**, *35*, 5013.

150. A. Rieker and B. Speiser, *J. Org. Chem.* **1991**, *56*, 4664.

151. B. Speiser, A. Rieker, and S. Pons, *J. Electroanal. Chem.* **1983**, *159*, 63.

152. B. Speiser, A. Rieker, and S. Pons, *J. Electroanal. Chem.* **1983**, *147*, 205.

153. D. Chinn, J. Dubow, M. Liess, M. Josowicz, and J. Janata, *Chem. Mater.* **1995**, *7*, 1504.

154. V. D. Nefedov, M. A. Toropova, T. P. Simonova, V. E. Zhuravlev, and A. M. Vorontsov, *Russ. J. Org. Chem.* **1989**, *25*, 141.

155. G. P. Ford and P. S. Herman, *J. Am. Chem. Soc.* **1989**, *111*, 3987.

156. D. Chiapperino, G. B. Anderson, R. J. Robbins, and D. E. Falvey, *J. Org. Chem.* **1996**, *61*, 3195.

157. R. J. Robbins, D. M. Laman, and D. E. Falvey, *J. Am. Chem. Soc.* **1996**, *118*, 8127.

158. S. Srivastava, M. Kercher, and D. E. Falvey, *J. Org. Chem.* **1999**, 64, 5853.

159. M. Novak, M. Pelecanou, A. K. Roy, A. F. Andronico, F. Plourde, T. M. Olefirowicz, and T. J. Curtin, *J. Am. Chem. Soc.* **1984**, *106*, 5623.

160. M. Novak, M. J. Kahley, E. Eiger, J. S. Helmick, and H. E. Peters, *J. Am. Chem. Soc.* **1993**, *115*, 9453.

161. M. Novak, M. J. Kahley, J. Lin, S. A. Kennedy, and L. A. Swanegan, *J. Am. Chem. Soc.* **1994**, *116*, 11626.

162. D. Chiapperino, S. McIlroy, and D. Falvey, *J. Am. Chem. Soc.* **2002**, *124*, 3567.

163. S. Srivastava, P. H. Ruane, J. P. Toscano, M. B. Sullivan, C. J. Cramer, D. Chiapperino, E. C. Reed, and D. E. Falvey, *J. Am. Chem. Soc.* **2000**, *122*, 8271.

164. M. Pelecanou and M. Novak, *J. Am. Chem. Soc.* **1985**, *107*, 4499.

165. M. Novak and J. Lin, *J. Am. Chem. Soc.* **1996**, *118*, 1302.

166. M. Novak, K. Toth, S. Rajagopal, M. Brooks, L. L. Hott, and M. Moslener, *J. Am. Chem. Soc.* **2002**, *124*, 7972.

167. R. Bose, A. R. Ahmad, A. K. Dicks, M. Novak, K. J. Kayser, and R. A. McClelland, *J. Chem. Soc. Perkin Trans. 2* **1999**, 1591.

168. S. McIlroy and D. E. Falvey, *J. Am. Chem. Soc.* **2001**, *123*, 11329.

169. M. Novak, K. S. Rangappa, and R. K. Manitsas, *J. Org. Chem.* **1993**, *58*, 7813.

170. R. A. Abramovitch and X. Ye, *J. Org. Chem.* **1999**, *64*, 5904.

171. P. Dalidowicz and J. S. Swenton, *J. Org. Chem.* **1993**, *58*, 4802.

172. R. J. Moran, C. J. Cramer, and D. E. Falvey, *J. Org. Chem.* **1996**, *61*, 3195.

173. M. Novak and S. Kazerani, *J. Am. Chem. Soc.* **2000**, *122*, 3606.

174. G. B. Anderson and D. E. Falvey, *J. Am. Chem. Soc.* **1993**, *115*, 9870.

175. R. J. Robbins, L. L.-N. Yang, G. B. Anderson, and D. E. Falvey, *J. Am. Chem. Soc.* **1995**, *117*, 6544.

176. G. A. Olah and D. J. Donovan, *J. Org. Chem.* **1978**, *43*, 1743.

177. V. V. Krishnamurthy, G. K. S. Prakash, P. S. Iyer, and G. A. Olah, *J. Am. Chem. Soc.* **1986**, *108*, 1575.

178. G. A. Olah, G. K. S. Prakash, and M. Arvanaghi, *J. Am. Chem. Soc.* **1980**, *102*, 6640.

179. G. A. Olah, M. Arvanaghi, and G. K. S. Prakash, *J. Am. Chem. Soc.* **1982**, *104*, 1628.

180. D. J. Raber, J. M. Harris, R. E. Hall, and P. v. R. Schleyer, *J. Am. Chem. Soc.* **1971**, *93*, 4821.

181. R. Ta-Shma and Z. Rappoport, *J. Am. Chem. Soc.* **1983**, *105*, 6082.

182. J. P. Richard, T. L. Aymes, and T. Vontor, *J. Am. Chem. Soc.* **1991**, *113*, 5871.

183. R. A. McClelland, T. A. Gadosy, and D. Ren, *Can. J. Chem.* **1998**, *76*, 1327.

184. J. C. Fishbein and R. A. McClelland, *J. Chem. Soc. Perkin Trans. 2* **1995**, 653.

185. J. C. Fishbein and R. A. McClelland, *J. Chem. Soc. Perkin Trans. 2* **1995**, 663.

186. R. C. Baetzold and L. K. J. Tong, *J. Am. Chem. Soc.* **1971**, *93*, 1347.

187. S. Srivastava, J. P. Toscano, R. J. Moran, and D. E. Falvey, *J. Am. Chem. Soc.* **1997**, *119*, 11552.

188. P. Zhu, S. Y. Ong, P. Y. Chan, Y. F. Poon, K. H. Leung, and D. L. Phillps, *Chem. Eur. J.* **2001**, *7*, 4928.

189. S. McIlroy and D. E. Falvey, 2002, unpublished results.

190. M. Famulok and G. Boche, *Angew. Chem. Int. Ed. Engl.* **1989**, *28*, 468.

191. M. Novak and S. A. Kennedy, *J. Am. Chem. Soc.* **1995**, *117*, 574.

192. W. G. Humphreys, F. F. Kadlubar, and F. P. Guengerich, *Proc. Natl. Acad. Sci. U.S.A.* **1992**, *89*, 8278.

193. M. Novak and S. A. Kennedy, *J. Phys. Org. Chem.* **1998**, *11*, 71.

194. M. Novak, L. Xu, and R. A. Wolf, *J. Am. Chem. Soc.* **1998**, *120*, 1643.

CHAPTER 14

Silylenes (and Germylenes, Stannylenes, Plumbylenes)

NORIHIRO TOKITOH

Institute for Chemical Research, Kyoto University, Kyoto 611-0011, Japan

WATARU ANDO

Department of Chemistry, University of Tsukuba, Ibaraki 305-8571, Japan

Reactive Intermediate Chemistry, edited by Robert A. Moss, Matthew S. Platz, and Maitland Jones, Jr.
ISBN 0-471-23324-2   Copyright © 2004 John Wiley & Sons, Inc.

1. INTRODUCTION

Much attention has been directed toward the heavier carbene analogues, that is, silylenes ($:SiR_2$),[1,2] germylenes ($:GeR_2$),[1,3] and stannylenes ($:SnR_2$). In the early stages of the investigation of silylenes, most of the results were concerned with dihalosilylenes ($:SiX_2$).[4] During the last two decades, however, the chemistry of organosilylenes ($:SiR_2$)[5] has progressed rapidly, accompanied by the synthesis of suitable precursors leading to silylenes and germylenes under various conditions. There has been special progress in the synthesis and isolation of "electronically (thermodynamically) stabilized" silylenes. Recently, some "kinetically stabilized" silylenes have also been isolated by the introduction of bulky substituents on the silicon atom. The synthesis of electronically stabilized silylenes has been performed by introduction of heteroatom substituents or conjugated systems on the silicon atom, while that of kinetically stabilized silylenes has been achieved by full substitution with bulky ligands, which prevent reactive divalent species from further oligomerization and/or intermolecular reactions.

2. GENERATION OF SILYLENES BY THERMALLY INDUCED α-ELIMINATION AND PHOTOEXTRUSION FROM OLIGOSILANES

2.1. Thermolysis of Polysilanes and Oligosilanes

Pyrolysis of polysilanes played an important role in the discovery of silylene reactions. Through the pyrolysis of alkoxydisilanes in the presence of diphenylacetylene, Atwell and Weynberg[6] obtained a product regarded as a dimer of dimethylsilylene adduct. 1,2-Shift of a methoxy group in disilanes takes place under relatively mild conditions (Scheme 14.1).[6]

The reactions are interpreted in terms of concerted silylene extrusions in which a substituent migrates from the incipient divalent silicon atom during Si–Si bond cleavage. The migrating group X can be hydrogen, halogen, alkoxy, amino groups,

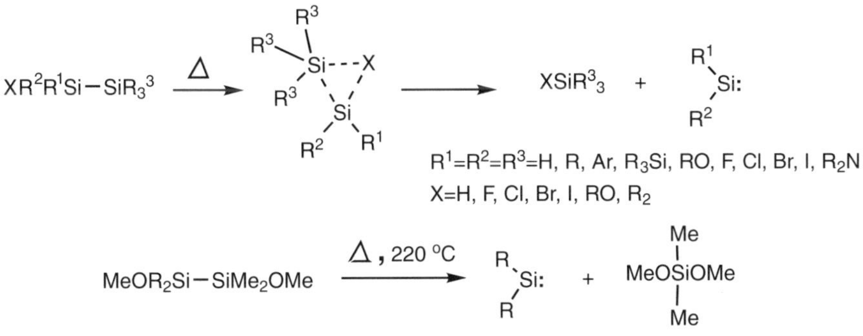

$XR^2R^1Si-SiR_3^3 \quad \xrightarrow{\Delta} \quad$ [structure] $\quad \longrightarrow \quad XSiR_3^3 \quad + \quad$ [structure] Si:

$R^1=R^2=R^3=$H, R, Ar, R_3Si, RO, F, Cl, Br, I, R_2N
X=H, F, Cl, Br, I, RO, R_2

$MeOR_2Si-SiMe_2OMe \quad \xrightarrow{\Delta,\ 220\ ^\circ C} \quad$ R Si: $\quad + \quad$ MeOSiOMe (Me, Me)

Scheme 14.1

and so on.[7] In 1964, Gilman et al.[8] studied the thermal elimination of dimethyl- and diphenylsilylenes from the corresponding 7-sila- and 7-germanorbornadienes at relatively high temperature. Many investigators used this method later with a variety of substituents on the basal carbon atoms (Scheme 14.2).

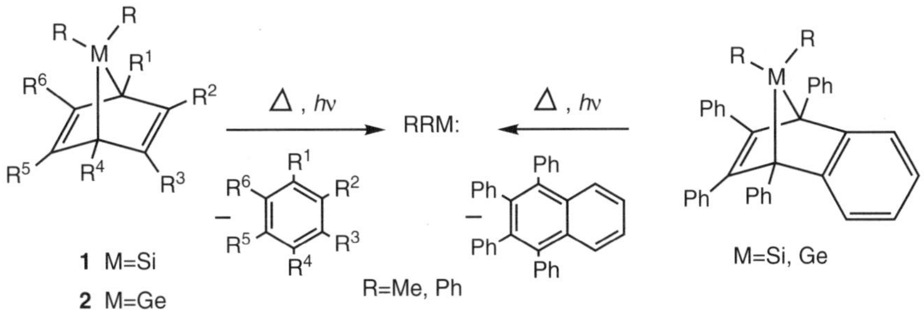

1 M=Si
2 M=Ge

R=Me, Ph

M=Si, Ge

Scheme 14.2

The larger the number of substituents on the basal carbon atoms, the more stable the 7-silanorbornadiene derivatives are. The reaction proceeds very slowly when the substituents are bulky and electron withdrawing.[2] On the other hand, the germanium analogues generate the corresponding germylenes both by photolysis with light of $\lambda = 254$ nm and thermolysis under relatively mild conditions (70–150 °C).[9,10] Pyrolysis of siliranes also has been employed as a mild route to generate silylenes since the pioneering work of Seyferth et al.[11,12] on hexamethylsilirane in 1976 (Scheme 14.3).

However, other siliranes are much more thermally stable than hexamethylsilirane, and do not serve as sources of :SiMe$_2$ at such low temperatures. Boudjouk et al.[13] made an important step when they found that bulky substituents such as the *tert*-butyl group on the silicon atom increased the stability of the silirane without preventing its thermolysis.[14,15]

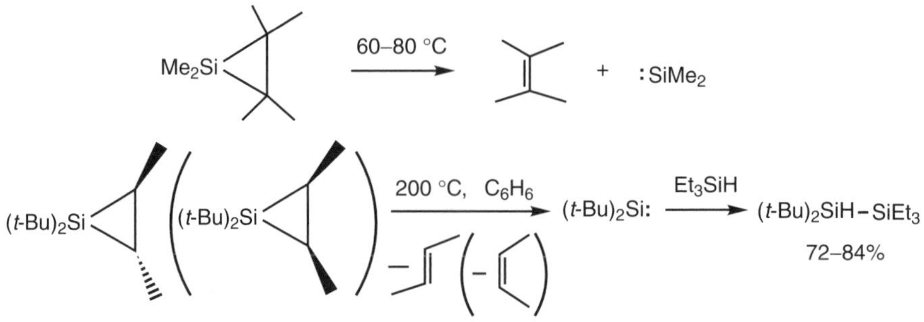

Scheme 14.3

The reversibility of most silylene addition reactions allows the cycloadduct of a silylene to 1,3-diene to be employed as a silylene source. Extrusion of a silylene from 1-silacyclopent-3-ene (**3**) has been achieved by thermolysis in the gas phase[16] and also by photolysis in solution (Scheme 14.4).

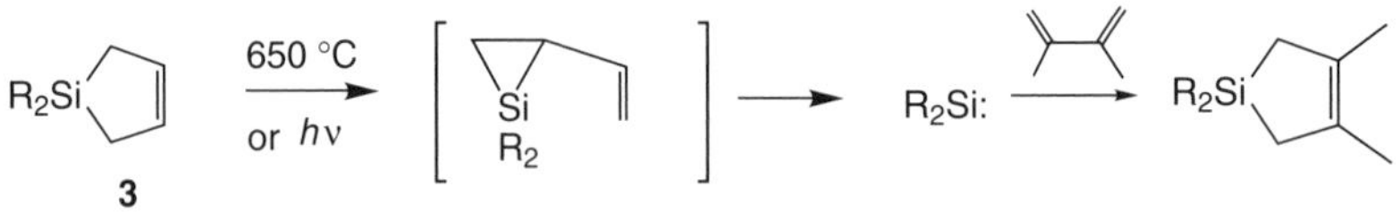

Scheme 14.4

In contrast to the somewhat complicated thermal behavior of siliranes, photolysis of silirene (**4**) readily leads to the loss of dimethylsilylene (Scheme 14.5).[17,18]

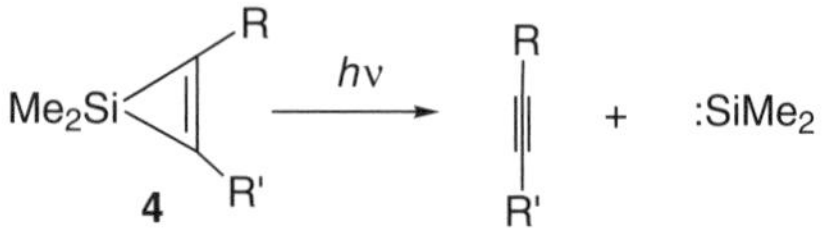

R=R'=Me;
R=Ph, R'=SiMe$_3$

Scheme 14.5

2.2. Generation of Dimethylsilylene from Polysilanes and Oligosilanes

In spite of its long-assumed intermediacy in several reactions, no carbon-substituted silylene was directly observed for many years. In 1979, however, Drahnak et al.[19] detected a broad ultraviolet (UV) absorption band (λ_{max} = 450 nm) after the photolysis of dodecamethylcyclohexasilane (**6**) in 3-methylpentane. This band was assigned to dimethylsilylene (**5**). Many different approaches to this intermediate, either photochemically or thermally, were examined (Scheme 14.6).[20]

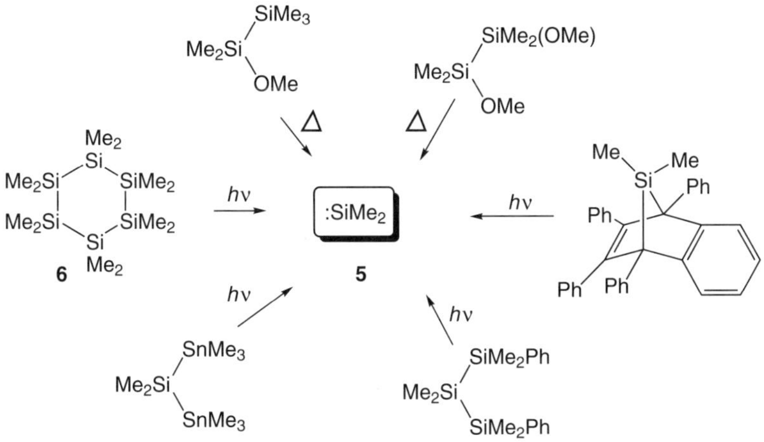

Scheme 14.6

2.3. Photolysis of Linear Polysilanes

There have been a number of routes to silylenes that are related to the simple photolysis of the precursors. The principal photoprocesses of alkylpolysilanes in solution are (a) chain abridgement through elimination of silylene and (b) chain scission by Si—Si bond homolysis. The photolysis of trisilanes [RR'Si(SiMe$_3$)$_2$] or cyclopolysilanes [(RR'Si)$_n$] is a well-established method for the generation of silylene (Scheme 14.7).[5,21]

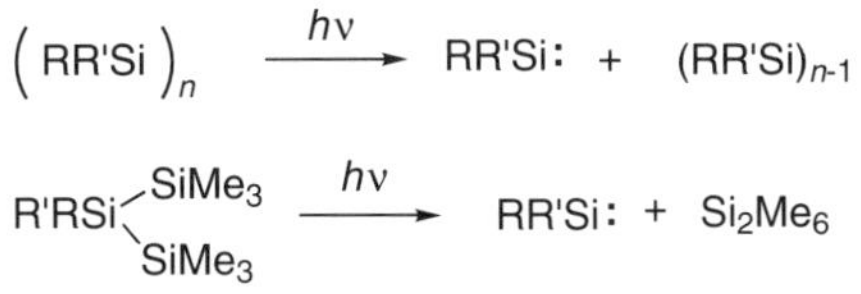

Scheme 14.7

In cyclohexane–Et$_3$SiH (1:1), the exhaustive irradiation of poly(di-n-hexasilane) [poly-(Hx$_2$Si)] (Hx = hexa) or poly(di-n-butylsilane) [poly(Bu$_2$Si)] with light of $\lambda = 248$ nm (pulsed) or 254 nm produced the silylene trapping product of the silylene [Et$_3$SiSiR$_2$H] and the homolytic cleavage product H(SiR$_2$)$_n$H (R = Hx or Bu), respectively.[22] The compound (Me$_3$Si)$_2$SiMes$_2$ is the precursor of choice for the generation of Mes$_2$Si:, which readily adds to alkenes to afford stable siliranes.[2]

Photolysis of 2,2-diaryl-substituted trigermanes generated the corresponding digermanes and diarylgermylenes. A similar reaction of bis(trimethylsilyl)diarylgermane gives hexamethyldisilane (HMD) and the expected diarylgermylene (Scheme 14.8).[10,23,24]

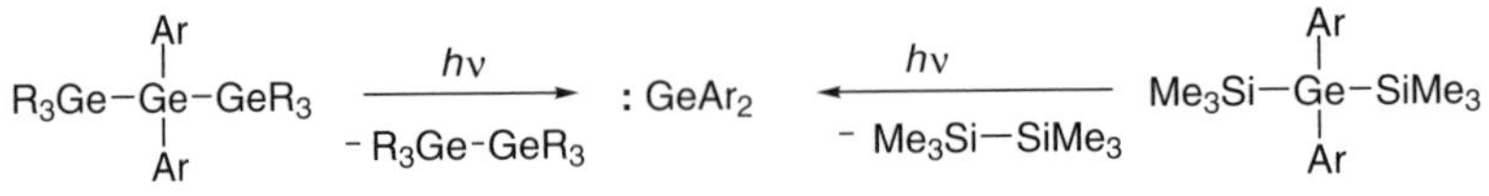

Scheme 14.8

2.4. Photolysis and Thermolysis of Cyclotrisilanes and Cyclotrigermanes

Strained cyclotrisilane derivatives, which are silicon analogues of cyclopropane, are prepared by the reductive coupling of overcrowded dichlorodiarylsilanes using lithium naphthalenide (Scheme 14.9).

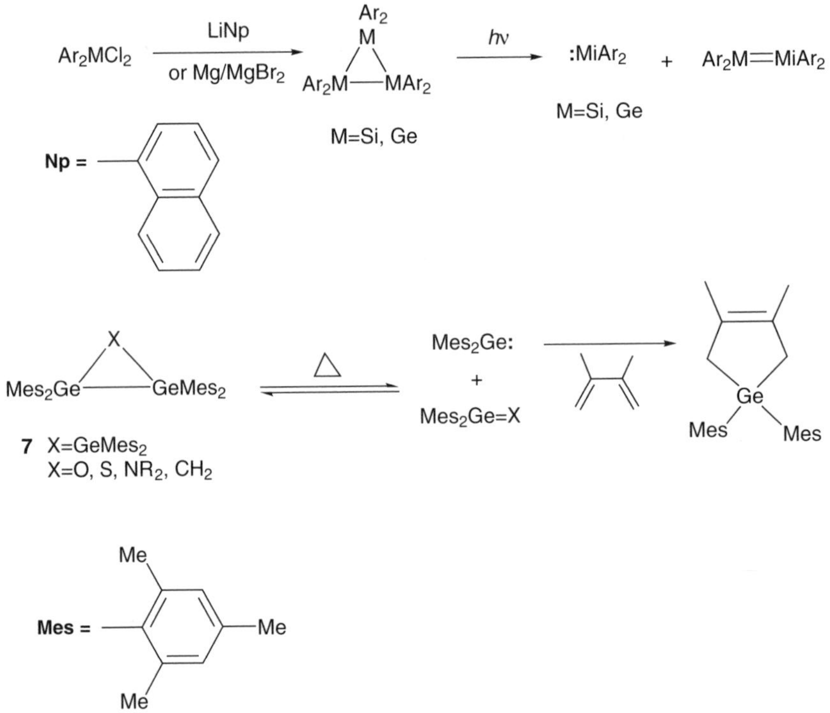

Scheme 14.9

The photolysis of cyclic polysilanes results in ring contraction with concomitant extrusion of a silylene fragment.[22,25] Although the formation of two reactive intermediates potentially complicates mechanisms for product formation, it has provided a useful method for the synthesis of both unstable and stable disilenes from photolysis of stable cyclotrisilanes $(Ar_2Si)_3$.[26] Cyclotrigermane derivatives are synthesized by the related reductive coupling of bulky dichlorodiarylgermanes using LiNp or Mg/MgBr$_2$. Although cyclotrigermane (**7**) is isolated as a stable compound at room temperature, thermolysis of **7** at 80 °C readily generates the expected dimesitylgermylene.[27,28]

2.5. Silylenes from Branched Cyclic Silylsilanes

Branched trisilacycloalkanes such as **8** undergo extrusion of a silylsilylene, for example, (Me$_3$Si)(Me)Si: (**9**), to produce the ring-contracted disilacycloalkane.[29–31] In addition to loss of silylene **9**, the five-membered ring gives a linear silylene (**10**) that is trapped by Et$_2$SiMeH via the competing migration of a Me$_3$Si group and a cleavage of the ring. Irradiation of the disilanyl-substituted 7-silanorbornene (**11**) with a 450-W high-pressure mercury lamp in *n*-hexane at 15 °C produces the disilanylsilylene (**12**), which is trapped by 2,3-dimethyl-1,3-butadiene to give adduct **13** in 31% yield (Scheme 14.10).

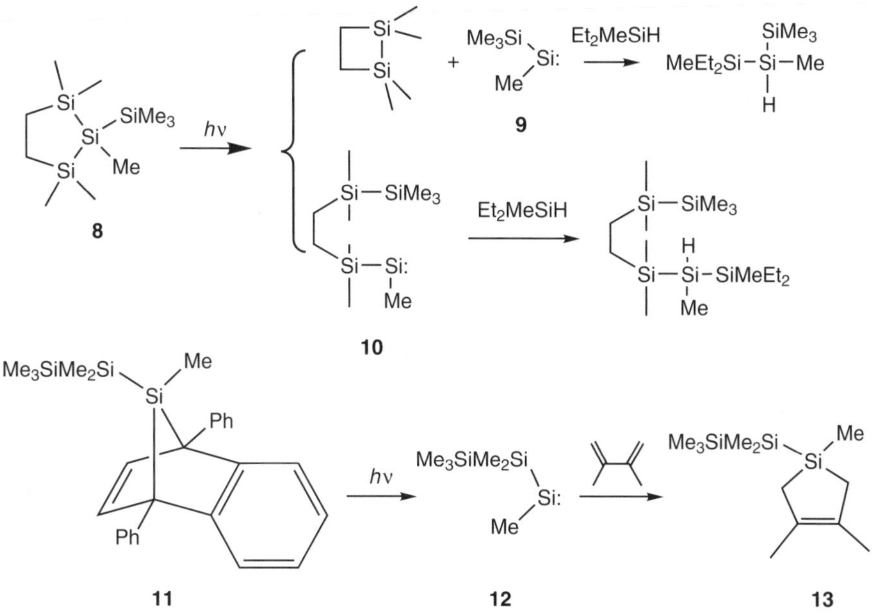

Scheme 14.10

Photochemical irradiation of (*i*-Pr$_3$Si)$_3$SiH (**14**) with light of 254 nm in either 2,2,4-trimethylpentane or pentane leads to the elimination of *i*-Pr$_3$SiH and the generation of bis(triisopropylsilyl)silylene (*i*-Pr$_3$Si)$_2$Si: (**15**). Silylene **15** can also be generated by the thermolysis of the same precursor **14** at 225 °C in 2,2,4-trimethylpentane (Scheme 14.11).[32] Reactions of **15** include the precedented insertion into an Si—H bond, and additions to the π bonds of olefins, alkynes, and dienes.

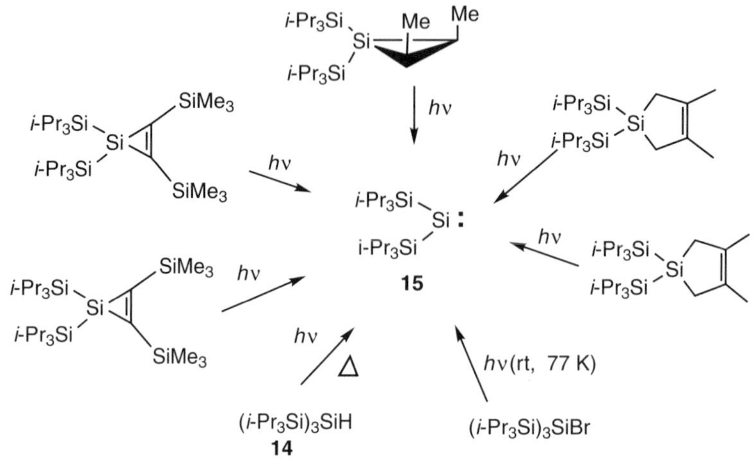

Scheme 14.11

2.6. Silylenes from Metal-Induced α Eliminations

There are also many routes to silylenes involving reductions of halo compounds with alkali metals. In many of these reactions, it is doubtful whether "free" silylenes are formed at all; the silylene may be complexed with metal or held in a solvent "cage" with a salt, or the intermediate may be an organometallic compound rather than a silylene, which is usually called a silylenoid; that is, a silylene-like intermediate. Silylenoids have been postulated as reaction intermediates in reduction of dihalosilanes[5h,13,33] with alkali metals, a reaction that affords polysilanes. With t-Bu$_2$SiBr$_2$ and t-Bu$_2$SiI$_2$, cyclotrisilanes were obtained, but with t-Bu$_2$SiCl$_2$, a disilane and a cyclotetrasilane were produced. The reactions of dichlorosilane, such as t-Bu$_2$SiCl$_2$, with lithium under ultrasound activation in the presence of Et$_3$SiH (Scheme 14.12) gave Et$_3$Si—SiH(t-Bu)$_2$ in ~60% yield. This result is rather persuasive evidence in favor of silylene formation, since Si—H bond insertion is characteristic for silylenes. It is difficult to see how the disilanes could arise from the reaction of Et$_3$SiH with silylenoids.[33b] Possibly, the halogen-containing silylenoids are in equilibrium with a very small amount of free silylene (**17**). The formation of silacyclopentanes (**18**) was reported in the reaction of dichlorosilanes and alkali metals under irradiation of ultrasound, in which the intermediary R$_2$Si: is intercepted by 2,3-dimethyl-1,3-butadiene. Although products are formed in low yields, the result was interpreted in terms of a silylene intermediate.[34] Reactions of diorganodihalosilanes with lithium in the presence of olefins gave silacyclopropanes, and decamethylsilicocene (**19**) was obtained by reduction of dichloro- or dibromo bis(η^1-pentamethylcyclopentadienyl)silane (Scheme 14.13).[33–36]

Reactions of t-Bu$_3$Si(i-Pr$_3$Si)SiBr$_2$ (**20**)[37,38] with activated magnesium or the reagents derived by the reduction of 1,3-dienes with magnesium gave rise to

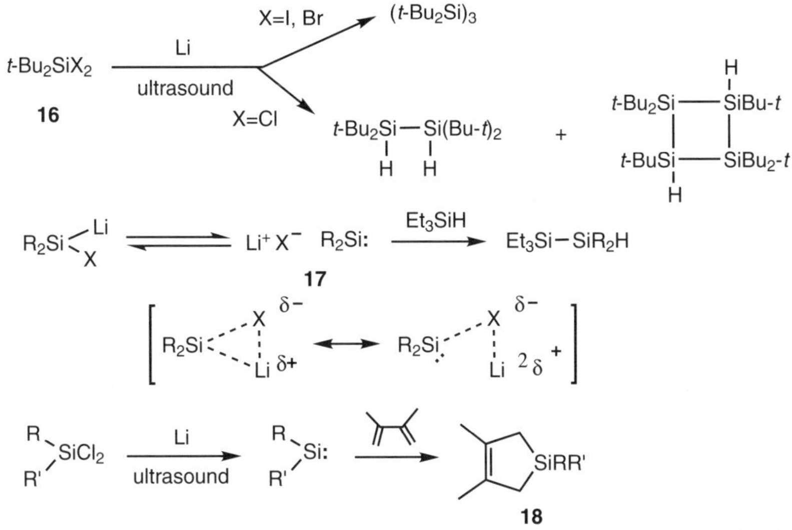

Scheme 14.12

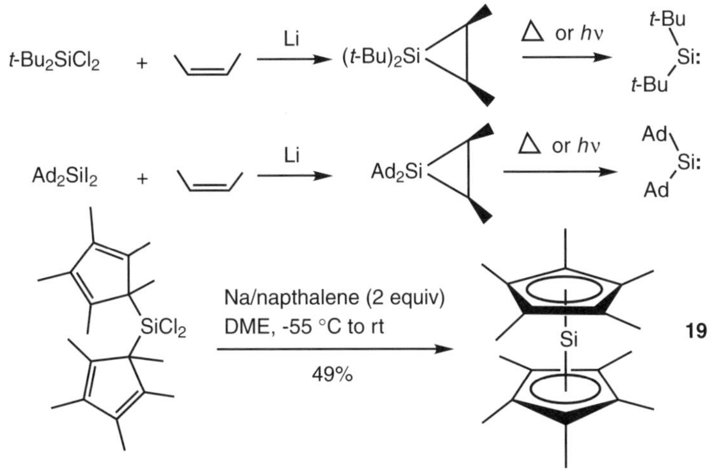

Scheme 14.13

products that look like silylene adducts. Selective inversions between metal-free and organometallic reaction systems for the generation of silylenes and their equivalents suggest that the silylenoid reactions are initiated by treatment of **20** with Mg* (Scheme 14.14).

A similar reaction of a bulky diaryl-substituted dihalogermane generates the corresponding diarylgermylene. Halogermylenes are also produced in the reaction of trihalogermanes with magnesium (Scheme 14.15).[39]

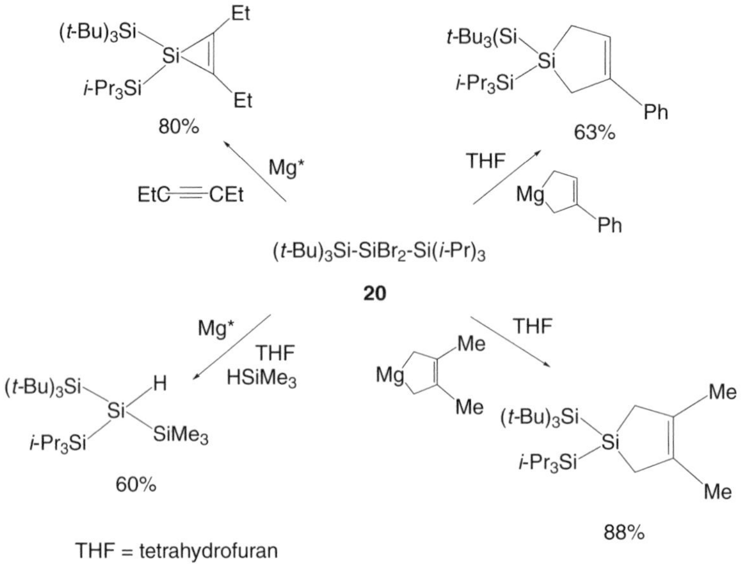

Scheme 14.14

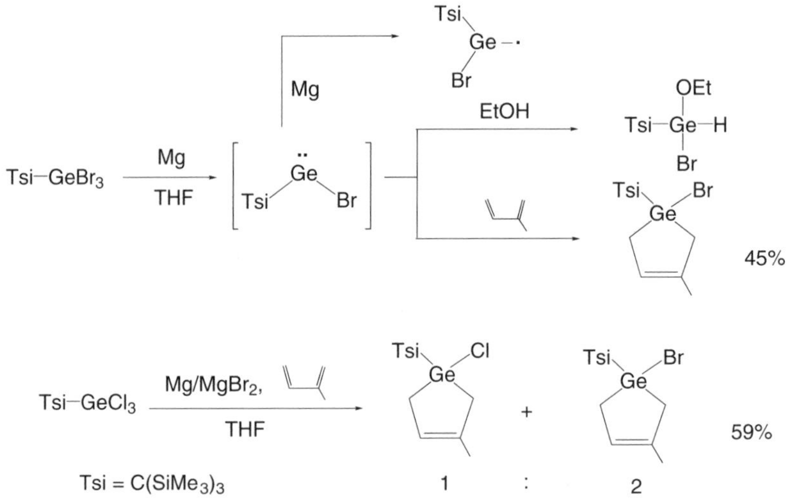

Scheme 14.15

3. STRUCTURES OF SILYLENES AND GERMYLENES

3.1. Singlet and Triplet States

While carbenes are found with both singlet and triplet ground electronic states, only singlet silylenes and germylenes are well documented.

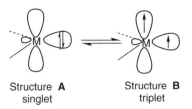

Structure **A**
singlet

Structure **B**
triplet

Figure 14.1

The ground states of silylenes and germylenes are the closed-shell singlet with the triplet state lying higher in energy, as evidenced by a variety of experimental and theoretical methods.[40–44] The different multiplicities of the ground state of the parent carbene (:CR$_2$) and other divalent species containing group 14 (IVA) elements (:MR$_2$) can be understood in terms of the energies of the highest occupied molecular orbital (HOMO) and lowest unoccupied molecular orbital (LUMO) in such species. The ground state has a pair of electrons in the orbital of σ symmetry as shown in structure **A** (Fig. 14.1).

In the triplet state shown in structure **B**, these electrons are not paired and one electron is in the orbital of π symmetry and the other remains in the orbital of σ symmetry. In the case of :SiH$_2$, the HOMO–LUMO energy gap is much larger than that of :CH$_2$ and the singlet electronic configuration is the ground state.[41]

Steric effects can also influence the singlet–triplet (S–T) gap, and that suggests different curvature of the potential curves for the singlet and triplet state as a function of the angle around the central metal.[44] In silylene chemistry, S–T crossing occurs at H–Si–H angle of $\sim$130°. This result indicates the possibility of formation of a silylene having triplet ground state with the R–Si–R angle >130°. According to calculations, the singlet ground state of the parent compound H$_2$Si: lies 21 kcal/ mol lower than the first triplet state. In the carbon analogues the situation is reversed. The first singlet state is calculated to lie 9 kcal/mol above its triplet ground state,[45] in agreement with experimental data.[43,46] In addition to the pairing energy and electrostatic effects, the small extent of s, p mixing in silicon seems to play an important role. The first factor is manifested not only in the large stabilization of the singlet state but also the bond angle of H$_2$Si:, which calculated to be as small as 93.4°,[41] which is a consequence of the relatively large difference between the sizes of the 3s and 3p orbitals compared with the difference between 2s and 2p.[47]

During the 1980s, several organosilylenes have been isolated and studied in argon or hydrocarbon matrices at very low temperatures such as 77 K, above which temperature they react rapidly with themselves or with solvent. By contrast, the divalent dicoordinate species of germanium, tin, and lead have been well characterized as stable molecules.

The enlargement of the bond angle caused by the triisopropylsilyl groups was encouraging with regard to the prospects of [(i-Pr)$_3$Si]$_2$Si: having a triplet ground state. It was found, however, that the addition of [(i-Pr)$_3$Si]$_2$Si: is totally stereospecific as seen for the reaction with *cis*-2-butene (Eq. 1).[38] One explanation is that the

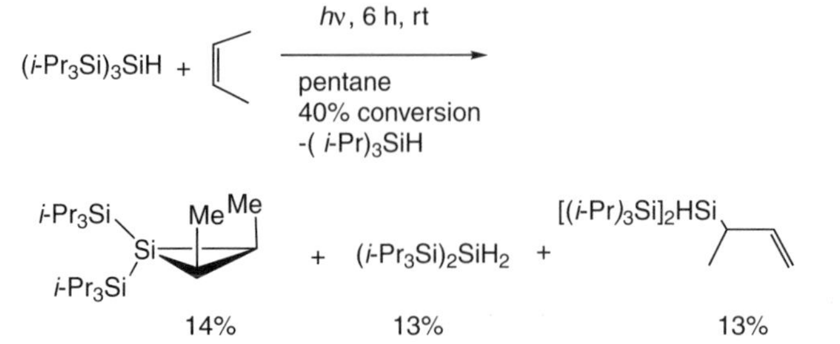

$$(1)$$

first excited singlet state is more reactive than the triplet, and lies so close in energy to the ground triplet that the triplet is siphoned off by reactions of the singlet. The larger size of the silicon orbitals leads to a decrease in the repulsion of the nonbonding electrons in the singlet state, and hence the energy lowering induced by their separation in the triplet state is less capable of compensating the attendant promotion energy. Treatment of $[(t\text{-Bu})_3\text{Si}]_2\text{SiBr}_2$ with $t\text{-Bu}_3\text{SiNa}$ led to products that were attributed to triplet $[(t\text{-Bu})_3\text{Si}]_2\text{Si:}$ (**21**).[48] The four-membered ring could arise by insertion of the divalent silicon of the silylene into a C—H bond of a methyl group. So, this result could be chemical evidence for a triplet state of $[(t\text{-Bu})_3\text{Si}]_2\text{Si:}$ (Scheme 14.16).[49] But for $[(t\text{-Bu})_3\text{Si}]_2\text{Si:}$, with a predicted energy difference of 7.1 kcal/mol between the excited singlet and ground-state triplet, the equilibrium constant is four orders of magnitude smaller, so even a 1000-fold more reactive singlet would account for <1% of the product.

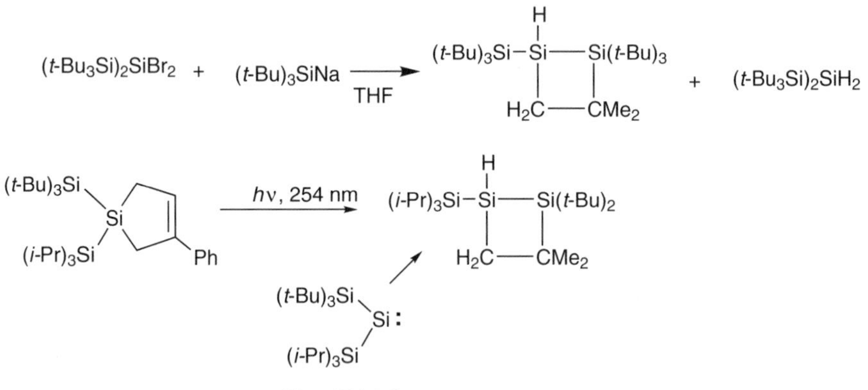

Scheme 14.16

3.2. Electronic Spectra of Silylenes and Germylenes

The electronic absorption of silylenes can be represented as a transition between $^1A'$ singlet S^0, with both electrons (mainly) in a $3s$ orbital on Si, to $^1A''$ excited S^1,

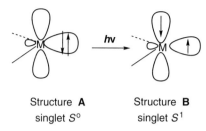

Structure **A**   Structure **B**
singlet S^0 singlet S^1

Figure 14.2

which has one electron in $3p$ orbital perpendicular to the molecular plane (Fig. 14.2). This transition usually appears in the visible region. Studies of the electronic spectra of organosilylenes have been reported in which photochemically generated silylenes were isolated in argon matrices at $\sim$10 K, or in hydrocarbon glasses at 77 K (Table 14.1).

There is much evidence for silylenes reacting as Lewis bases, but complexes of silylenes acting as a Lewis acid are now well established (Fig. 14.3, Table 14.2).[61–65] These complexes are also described as silaylides, $R_2Si^--X^+$. Formation of silylene complexes with Lewis bases is confirmed by a strong blue shift of the n–p transition. Matrix isolated dimesitylsilylene reacts with carbon monoxide to form the complex shown in Eq. 2.[65–67] The complex absorbs at $\sim$354 nm,

$$\text{Me}_2\overset{\text{Me}\quad\text{Me}}{\underset{}{Si}}\!\!\leftarrow\!\!CO \xleftarrow{CO} :SiMe_2 \xrightarrow{O_2} Me_2Si\overset{O}{\underset{O}{<}} \qquad (2)$$

thus showing a blue shift of $>$100 nm compared with the free silylene. These long-wavelength bands may possibly be the result of the silaketene

TABLE 14.1. Absorption Bands (nm) for Silylenes RR′Si

R	R′	λ_{max}	Reference	R	R′	λ_{max}	Reference
		(in Ar matrix $\sim$ 10 K)				(in 3-methylpentane (3MP) at 77 K)	
H	H	487	50	Me$_3$Si	Mes	760,776	55,56
H	Me	480	51,52	Me$_3$Si	Ph	660	55,56
H	NH$_2$	342	53	Mes	Mes	577	57
H	OMe	349	54	Mes	Ph	530	58
Me	Me	460	54	Mes	t-Bu	505	59
Me	Ph	495	51,52	Mes	Me	496	58
Me	Cl	407	51,52	Ph	Ph	495	58
Me	OMe	355	53	Ph	Me	490	58
				t-Bu	t-Bu	480	60
				Me	Me	453	20b

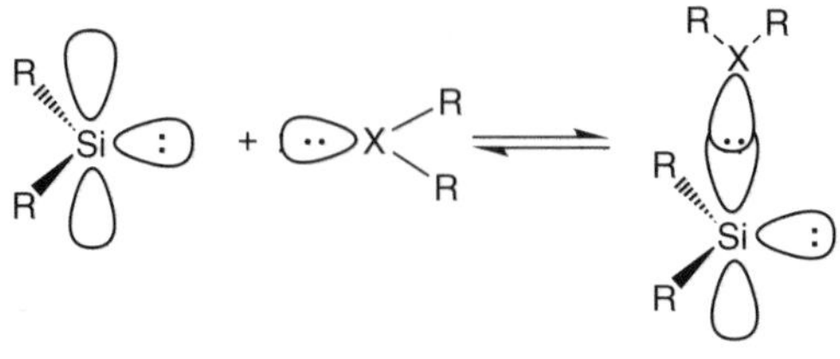

22

Figure 14.3

isomers.[68] Photochemically generated dimesitylsilylene in an oxygen matrix yielded silanone-*O*-oxide (**23**) and not the dioxasilirane (**24**) (Eq. 3).[69]

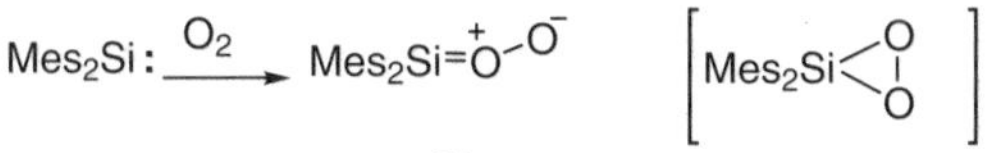

$$\text{Mes}_2\text{Si}: \xrightarrow{\text{O}_2} \text{Mes}_2\overset{+}{\text{Si}}{=}\overset{-}{\text{O}}{-}\overset{-}{\text{O}} \qquad \left[\text{Mes}_2\text{Si}\overset{O}{\underset{O}{\diagdown}} \right] \tag{3}$$

23   **24**

In germylene chemistry, ab initio valence calculations were performed on the three ground states of :GeH$_2$, :GeF$_2$, and :GeMe$_2$,[70] which are predicted to be singlets. Self-consistent field (SCF) S–T energy gaps are 10 kcal/mol for :GeH$_2$, 64 kcal/mol for :GeF$_2$, and 14 kcal/mol for :GeMe$_2$. The SCF ground state equilibrium geometries correspond to Ge–H, 1.60 Å, H–Ge–H, 93°; Ge–F, 1.76 Å; F–Ge–F, 97°, Ge–C, 2.20 Å; C–Ge–C, 98°. The most stable conformer of a germylene is a singlet, with the experimental UV–vis observations ($\lambda_{max} = 420 – 430$ nm) of :GeMe$_2$.[71a]

Diorganogermylenes were generated in hydrocarbon matrices at 77 K by the photolysis of 7-germanorbornadiene (**25**) or bis(trimethylsilyl)germanes (**26**). In the presence of an appropriate Lewis base, the germylenes thus generated show the electronic absorptions of the adducts, which have characteristic absorption bands at shorter wavelengths than those of free germylenes (Scheme 14.17, Table 14.3).[72–74]

TABLE 14.2. Absorption Band for Silylenes in 3MP at 77 K

Base	Me$_2$Si:	Mes$_2$Si:	Mes(ArO)Si:[a]
None	453	577	395
n-Bu$_3$N	287	325	346
Et$_2$O	299	320	332
i-PrOH		322	321
t-Bu$_2$S	322	316	350
n-Bu$_3$P	390	345	358
2-Me–THF	294	326	330
CO	345	354	328
Et$_3$P		336	349

[a] 2,6-Diisopropylphenyl = Ar.

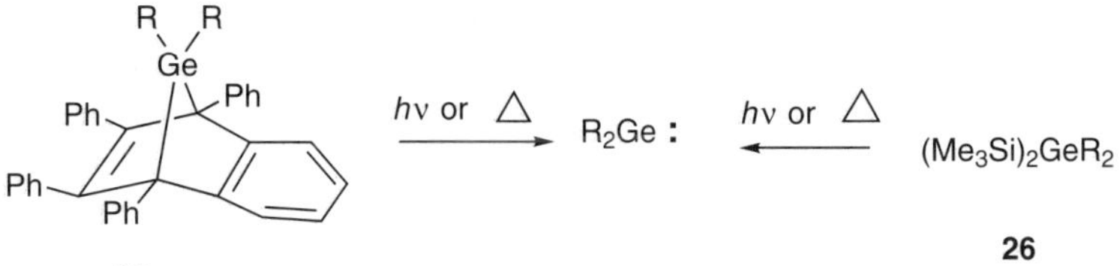

Scheme 14.17

3.3. Silylene Isomerization

Silene (**27**) can undergo a 1,2-shift to give either methylsilylene (**28**) or, less favorably, to silylcarbene (**29**). The thermochemistry and the kinetics of these reactions have been points of major disparity between theory and experiments

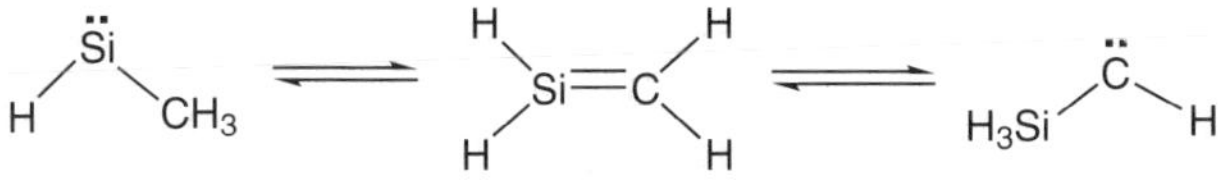

Scheme 14.18

(Scheme 14.18).[75,76] The silylene–silene rearrangement **27** → **28** is nearly thermoneutral, with the silene being slightly more stable. The photolysis of α-diazo compounds (**30**) is the only frequently used reaction path to silenes (**31**) via a carbene–silene rearrangement (Scheme 14.19).[77,78] Irradiation of the cyclopropenyltrisilane

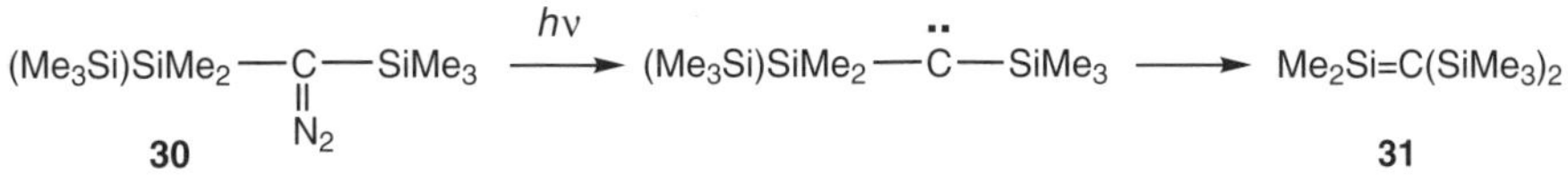

Scheme 14.19

TABLE 14.3. Absorption Bands (nm) for Germylenes in 3MP at 77 K[a]

	Me$_2$Ge:	Ph$_2$Ge:	Mes$_2$Ge:	Ar$_2$Ge:[a]	Ar$_2'$Ge:[b]
None	420	463	550	544	558
n-Bu$_3$P		306	314	334	
Me$_2$S		326	348	357	357
Thiophene		332	352	359	366
2-Me•THF		325	360	369	376
EtOH		320	333	332	

[a] 2,6-Diethylphenyl = Ar.
[b] 2,4,6-Triisipropylphenyl = Ar'.

(**32**) gives the relatively stable cyclopropenylsilylene (**33**), which can be efficiently converted into silacyclobutadiene (**34**) by irradiation into the visible absorption band of the silylene (Scheme 14.20).[79] The rearrangement of 1-silacyclopent-3-

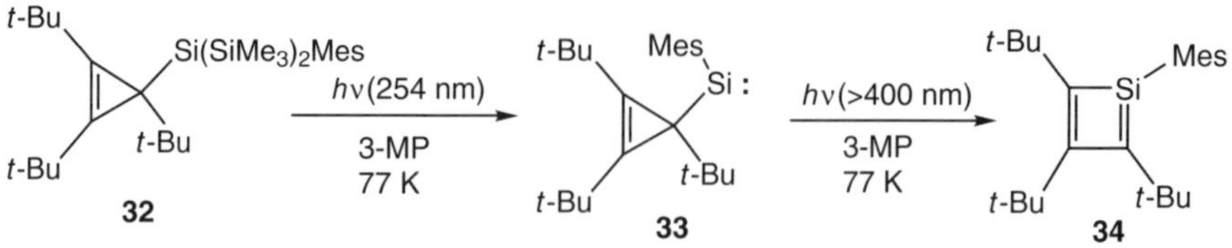

Scheme 14.20

ene-1,1-diyl (**35**) to 1-silacyclopenta-2,4-diene (**36**) has been confirmed by the UV–vis and infrared (IR) spectra of **35**, **36** and 1-silacyclopent-1,3-diene (**37**) in matrix isolation experiments (Scheme 14.21).[80]

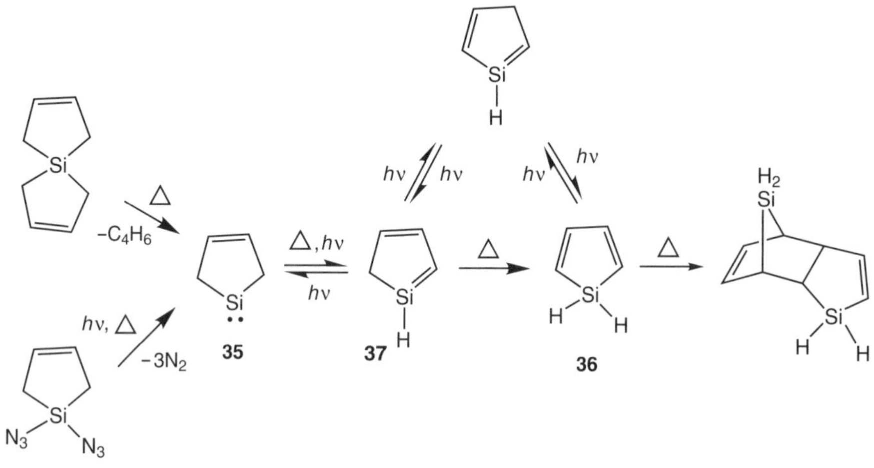

Scheme 14.21

The photochemical interconversion of silylenes and silenes is an important link between these two classes of compounds. It was first established that irradiation of the parent silene (**38**) with irradiation with light of $\lambda = 254$ nm resulted in the formation of methylsilylene (**39**) (Scheme 14.22). The reaction is reversible by using light of $\lambda = 320$ nm.[52] Ultraviolet absorption of silylene **39** strongly depends on the matrix

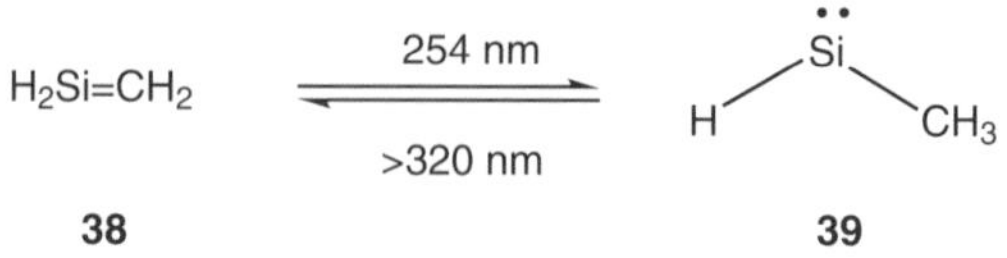

Scheme 14.22

material. Irradiation of **38** in solid argon results in the growing of an absorption with $\lambda_{max} = 480$ nm, whereas $\lambda_{max} = 330$ nm is found if nitrogen is used in the matrix instead.[51]

Rearrangement of a silene to a silylene (**40**) via migration of a Me$_3$Si group has been suggested as a step in the gas-phase silylene to silylene rearrangement. Labeling experiments, however, have indicated an alternative mechanism (Scheme 14.23).[81]

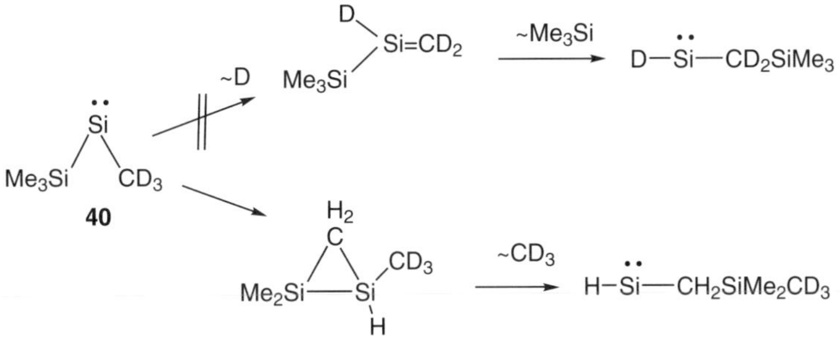

Scheme 14.23

Rearrangements of disilanes to α-silylsilenes are well established and are involved in the exchange of substituents between a silylene center and the adjacent silicon.[80b] Pulsed flash pyrolysis of acetylenic disilane (**41**) gave rise to the acetylenic silene (**42**), which subsequently rearranged to the cyclic silylene, 1-silacyclopropenylidene (**43**).[82] Irradiation of the cyclic silylene resulted in the isomerization to the isomeric **42**, which itself could be photochemically converted into the allenic silylene (**44**). Both **42** and **43** also were reported to isomerize on photolysis to the unusual (**45**), which was characterized spectroscopically (Scheme 14.24).

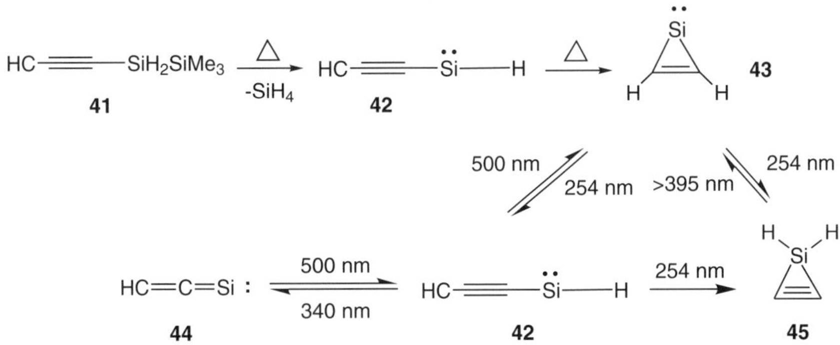

Scheme 14.24

Azidomethylsilane (**46**) eliminates nitrogen upon irradiation with the light of $\lambda = 254$ nm in an argon matrix to give silanimine (**47**) as the initially observable product, based on the comparison of the experimental and calculated IR wavenumbers (Scheme 14.25).[82,83b,c] On further irradiation, a second hydrogen-shift occurs yielding aminosilylene (**48**) ($\lambda_{max} = 330$ nm). Additional isomerization can be achieved after an exposure of **48** to light of longer wavelength to afford aminosilene (**49**) ($\lambda_{max} = 256$ nm), which regenerates **48** upon irradiation with light of shorter wavelengths.

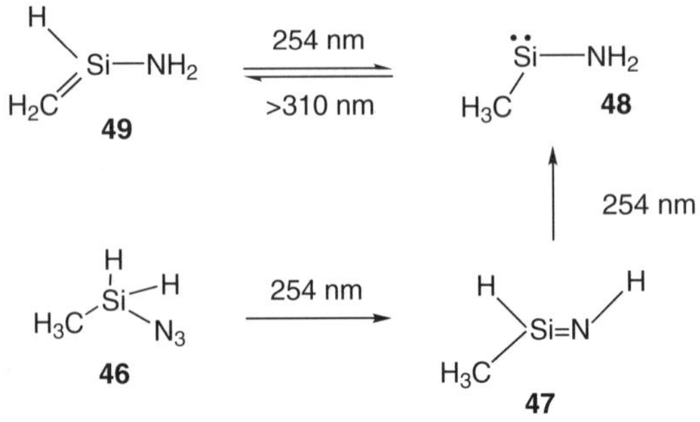

Scheme 14.25

4. REACTIONS OF SILYLENES

4.1. Insertion into Single Bonds

It has been well established that silylene is a ground-state singlet species with an electron pair in a hybrid orbital and a vacant orbital of π symmetry. A Lewis base can thus coordinate to silicon through the vacant orbital to form a silylene-base adduct. All the interactions involve the initial formation of a donor–acceptor complex, followed by a high-energy transition state leading to an insertion product.[84] The complex may also be represented as a zwitterionic species, that is, a silaylide, through the formation of the base-to-silicon σ bond, in which the silicon center would no longer be an electrophile, but instead have nucleophile character.

Insertion reactions of silylene into a number of single bonds have been observed. The bonds include Si–O, Si–N, Si–H, Si–halogen, strained C–O, Si–Si, O–H, N–H, and C–H (intramolecular only). Insertion into an X–H bond can be initiated by the formation of silaylide (**50**) with the donation of a pair of electrons from a heteroatom X to form a bond to a divalent silicon atom (Scheme 14.26).

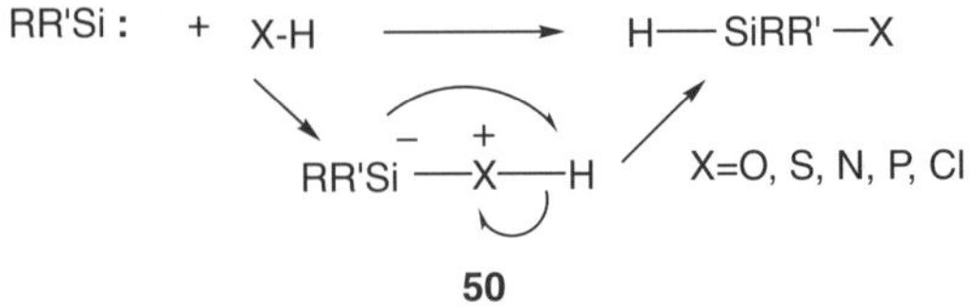

Scheme 14.26

4.1.1. *Insertion into O—H, N—H, and C—Halogen Bonds.*

Alcohols are bases toward silylenes. Complexes (**51**) of alcohols with Mes_2Si: and other hindered silylenes are observed spectroscopically when a 3-MP matrix containing the silylene and 5% of 2-propanol or 2-butanol is annealed.[85] Melting of the matrix results in a rapid reaction of the complex to give the O—H insertion product (Scheme 14.27).

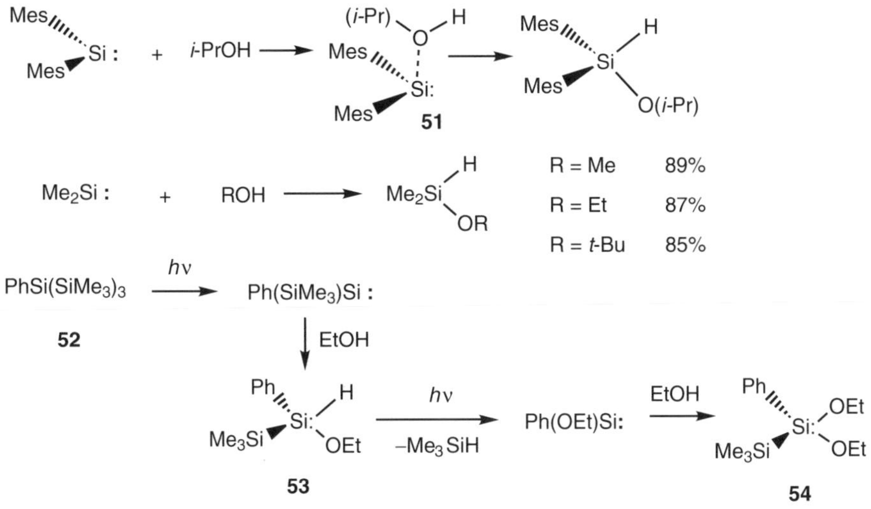

Scheme 14.27

The silylene-donor adducts of ethers, tertiary alcohols, tertiary amines, and sulfides were found to revert to silylenes upon warming, but the adducts of primary and secondary alcohols underwent rearrangement to the O—H insertion products. Dimethylsilylene inserts into the O—H bonds of water and alcohols to give bifunctional organosilanes. Interestingly, the photolysis of $PhSi(SiMe_3)_3$ (**52**) in the presence of excess ethanol gave the EtOH insertion product **53** as the initial product, but further photolysis gave $PhSi(H)(OEt)_2$ (**54**) by elimination of Me_3SiH (Scheme 14.27).[86] Dimethylsilylene inserted into the N—H bonds of primary and secondary amines to yield aminosilanes (**55**) (Eq. 4).[87]

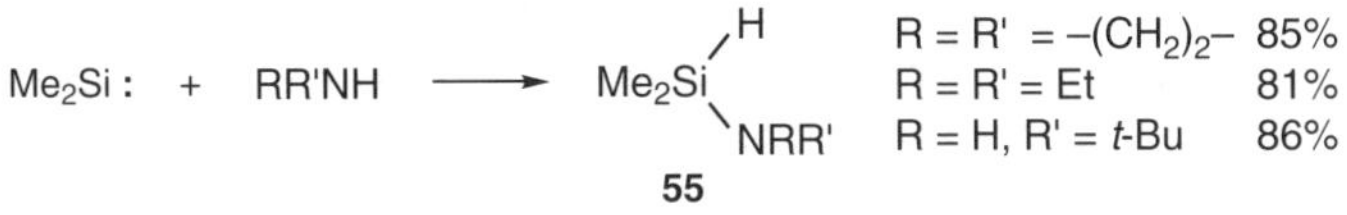

$$Me_2Si: + RR'NH \longrightarrow Me_2Si \overset{H}{\underset{NRR'}{}} \qquad \begin{array}{ll} R = R' = -(CH_2)_2- & 85\% \\ R = R' = Et & 81\% \\ R = H, R' = t\text{-}Bu & 86\% \end{array} \qquad (4)$$

The reaction of $Ph(Me_3Si)Si$: with CCl_4 in pentane produces $Ph(Me_3Si)SiCl_2$ and CCl_3-CCl_3, while chloroform and alkyl chlorides give the additional C—Cl insertion products. Hydrogen chloride abstraction products are also formed if a hydrogen is present vicinal to chlorine in the alkyl chlorides used as trapping agents. The C—Cl insertion and HCl abstraction might result from a collapse of an initial complex of the silylene with an alkyl halide (Scheme 14.28).[86,88]

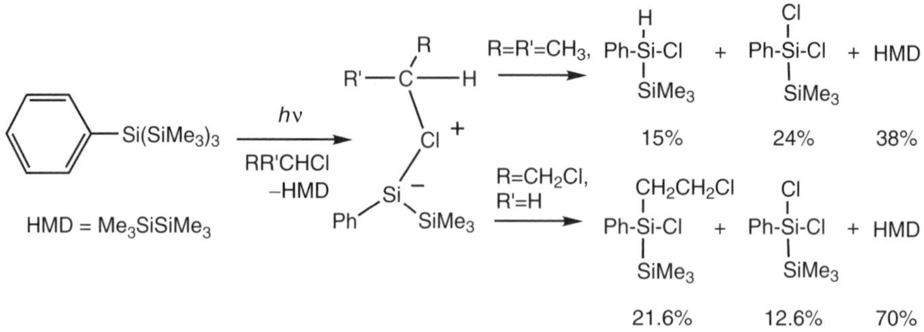

Scheme 14.28

The products observed in the reactions of $Me_2Ge:$ with CCl_3X ($X = Cl$, Br), $PhCH_2X$ ($X = Br$, I), and Ph_2CHCl, by proton chemically induced dynamic nuclear polarization (^1H CIDNP) are those of net insertion of $Me_2Ge:$ into the C—X bond and $Me_2Ge_2X_2$ ($X = Cl$, Br) (Scheme 14.29). A two-step radical reaction takes place

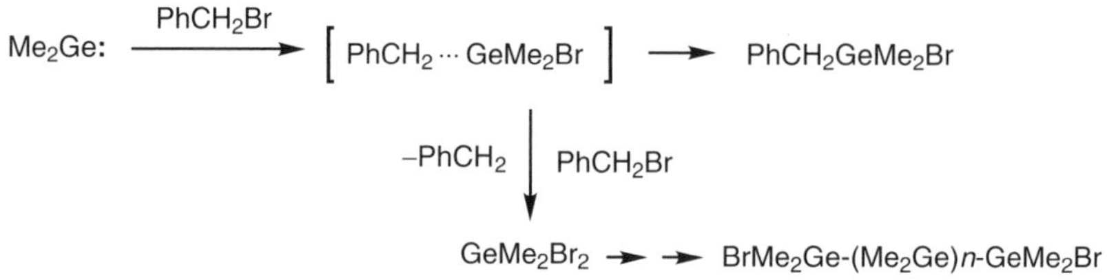

Scheme 14.29

by an abstraction–recombination mechanism, giving typical ^1H CIDNP effects as evidence for reaction of the singlet-state dimethylgermylene.[89] No reaction takes place with alkyl halides. The reactions of a silylene with crotyl chloride afford two types of C—Cl insertion products **56** and **57**, (ratio 72:28). The former product is formed by the direct silylene insertion into C—Cl, while the other results from a [2,3]-sigmatropic rearrangement of the chloronium ylide (**58**) (Scheme 14.30). Ylide formation and rearrangement have been reviewed in the carbene reaction with heteroatoms.[90] The photolysis of $(Me_3Si)_2Ge(Mes)_2$ in matrices containing crotyl chloride at 77 K produced a band at 515 nm, and upon annealing, chlorocrotyldimesitylgermane, the product of direct insertion of the germylene into a carbon–chlorine bond, was formed in quantitative yield. No 2,3-sigmatropic rearranged product was found. In the thermal reaction of dimethylgermylene with crotyl chloride, two products, direct insertion and rearranged product were obtained in the ratio 1:3.[73b]

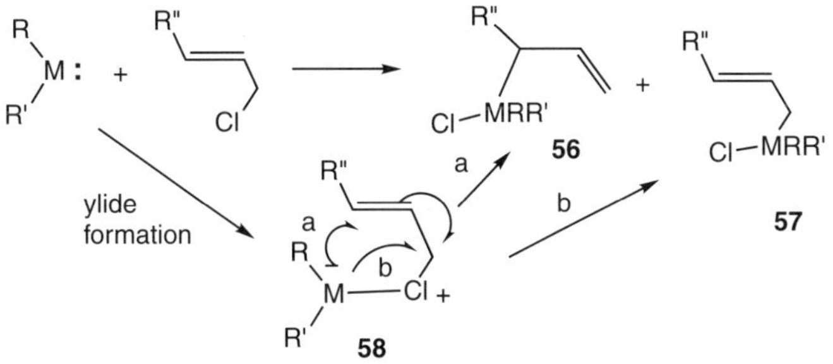

$$M = Si, R = Ph, R' = SiMe_3, R'' = Me$$
$$M = Ge, R = R' = Me, R'' = Me$$
$$M = Ge, R = R' = Mes, R'' = Me$$

UV Absorption of the ylide **58**

$$M = Ge, R = R' = Mes, R'' = H, \lambda_{max} = 530 \text{ nm}$$
$$M = Ge, R = R' = Mes, R'' = Me, \lambda_{max} = 515 \text{nm}$$

Scheme 14.30

Reaction of bis[(trimethylsilyl)methyl]germylene (**59**) with an excess of *trans*-1,2-dichloroethylene, and *cis*-1,2-dichloroethylene gave the corresponding vinyl chlorogermanes (**60** and **61**), in quantitative yields (Scheme 14.31).[91] The attack of a germylene on the chlorine atom appears to be faster than addition of the germylene to a double bond. The stereochemistry of the starting material was retained in the products, showing that these insertion reactions proceeded stereospecifically.

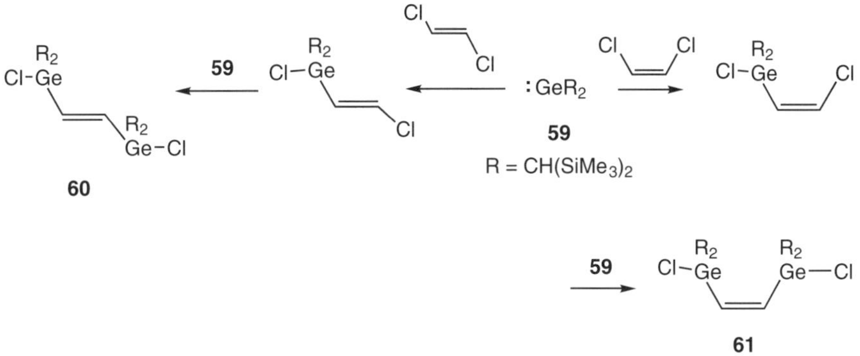

Scheme 14.31

4.1.2. Insertion into C–O and Si–O Bonds.

Photochemically generated dimethylsilylene reacts with oxetane at 0 °C to give allyloxydimethylsilane (**62**) and 2,2-dimethyl-1-oxa-2-silacyclopentane (**63**, Scheme 14.32).[92] No reaction of dimethylsilylene with unstrained aliphatic ethers such as tetrahydrofuran or diethyl ether could be observed. The deoxygenation of cyclooctene oxide with both thermally and photochemically generated dimethylsilylene are also thought to involve the formation of ylide **64**, which either extrudes dimethylsilanone (Me$_2$Si=O) or simply acts as a "silanone transfer agent" (Scheme 14.33).[93] Dimesitylsilylene

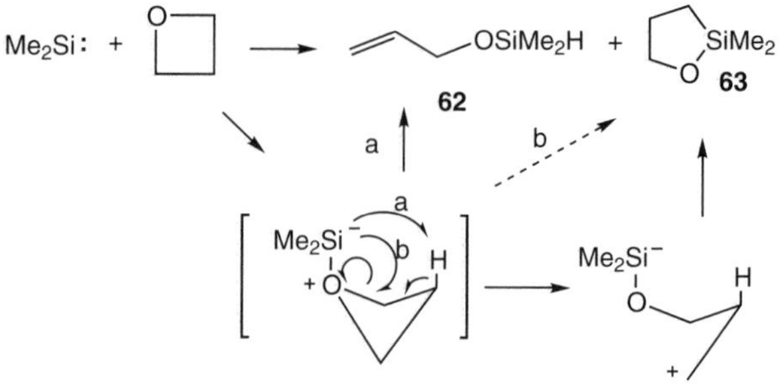

Scheme 14.32

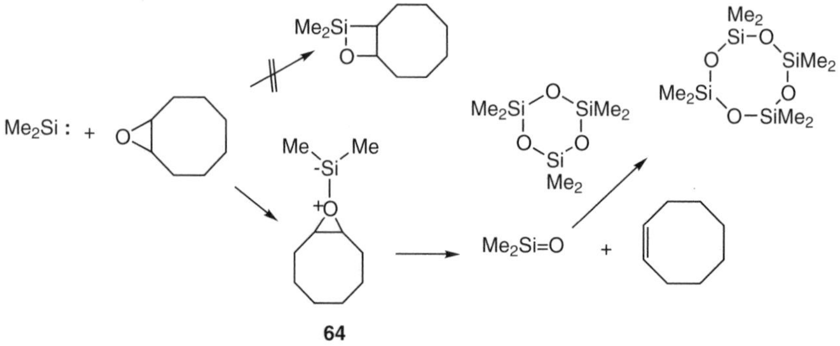

Scheme 14.33

generated by the photolysis of 2,2-dimesitylhexamethyltrisilane reacted with epoxides to give dimesitylsilanone-epoxide adducts (**66**, Scheme 14.34).[94]

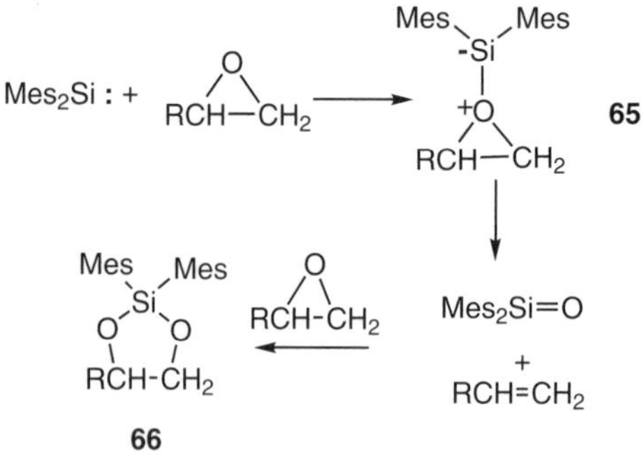

Scheme 14.34

Theoretical calculations suggest that these reactions are viable routes for the preparation of compounds with Si=O, Si=S, Si=Se double bonds via the barrierless formation of an encounter complex between a silylene and an oxirane or its heavier analogues.[95] Silylenes also insert into the silicon–oxygen bond of alkoxysilanes

$$\text{MeO(SiMe}_2)_2\text{OMe} \xrightarrow{225\,^\circ\text{C}} \text{Me}_2\text{Si(OMe)}_2 \; + \; \text{MeO(SiMe}_2)_n\text{OMe} \qquad (5)$$

$$n = 3\text{–}6$$

(Eq. 5).[96] The formation of the series $\text{MeO(SiMe}_2)_n\text{OMe}$ could be explained by the insertion of dimethylsilylene into either the Si–Si or Si–O bond. The results show that the reaction proceeds via the insertion into the Si–O bond, and alkoxypolysilanes are much more reactive than alkoxymonosilanes toward the silylene insertion. Photochemically generated dimethylsilylene inserts into Si–O single bonds of hexamethylcyclohexasiloxane (**67**) to yield 3,5,7-trioxa-1,2,4,6-tetrasilacycloheptane (**68**, Eq. 6).[97]

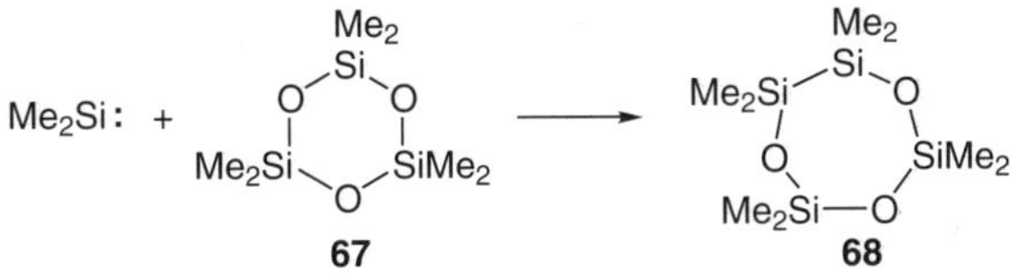

$$(6)$$

4.1.3. Insertion into Si–H and Si–Si Bonds.

Silylenes, generated by thermolysis of cyclotrisilanes, inserted into the Si–Cl or Si–H bonds of monosilane to yield a variety of disilanes, which could be further functionalized. In contrast to carbenes, the insertion of silylenes into C–H bonds has not been observed. However, the insertion into Si–H bonds has been studied extensively. The occurrence of direct insertion has been indicated by formation of nongeminate homocoupling products.[98,99]

Theoretical calculations suggested three steps for the insertion of silylene into a Si–H bond, that is (a) the formation of the complex with the interaction of the hydrogen atom of a silane and the empty orbital of a silylene; (b) the formation

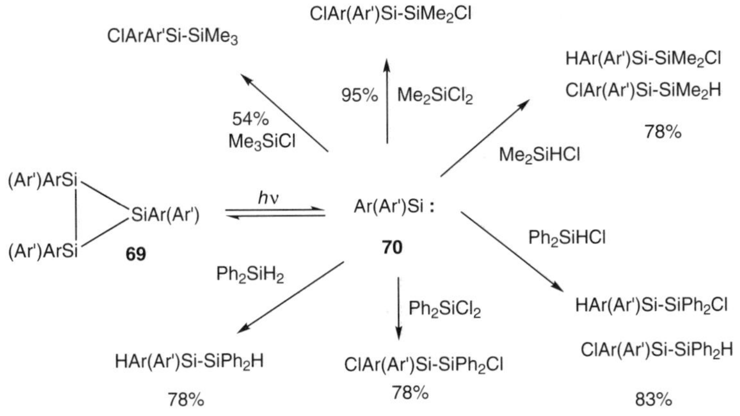

Ar = 2-(Me$_2$NCH$_2$)C$_6$H$_4$; R' = 2,4,6-Me$_3$C$_6$H$_2$

Scheme 14.35

of the complex between the silyl cation and silyl anion, from which follows the migration of a hydride ion from a silane to a silylene; and (c) a disilane formation through the inversion of the silyl anion.[100] The reaction of cyclotrisilane with diphenylsubstituted silanes (**70**) proceeded analogously; again the insertion of silylene (**70**) into the Si—H bond of Ph_2SiHCl being preferred to that into the Si—Cl bond (Scheme 14.35). The observation that photolysis of a precursor, silirene (**71**), to silylene $(t\text{-}Bu)_3Si\text{—}Si\text{—}Si(i\text{-}Pr)_3$ (**72**) leads to the formation of a product of the formal intramolecular C—H insertion, in addition to products from intermolecular Si—H insertion (Scheme 14.36).[37,38] Photolysis of $t\text{-}Bu_2Si(N_3)_2$ (**73**) was studied in

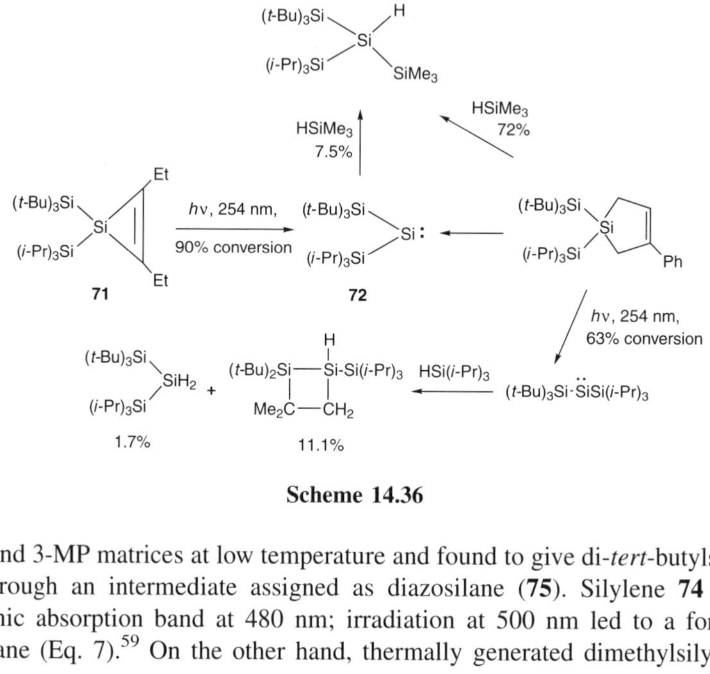

Scheme 14.36

argon and 3-MP matrices at low temperature and found to give di-*tert*-butylsilylene (**74**) through an intermediate assigned as diazosilane (**75**). Silylene **74** has an electronic absorption band at 480 nm; irradiation at 500 nm led to a formation of silirane (Eq. 7).[59] On the other hand, thermally generated dimethylsilylene or

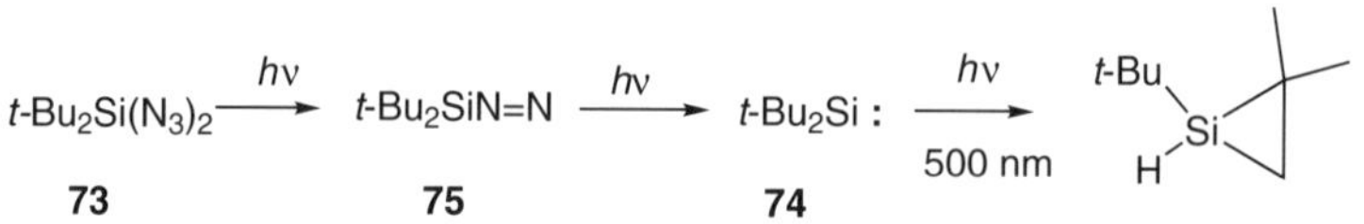

methylphenylsilylene reacts with 1,2-disilacyclobutenes to give trisilacyclopentenes (**76**, Scheme 14.37),[101] which demonstrated the insertion of a silylene into

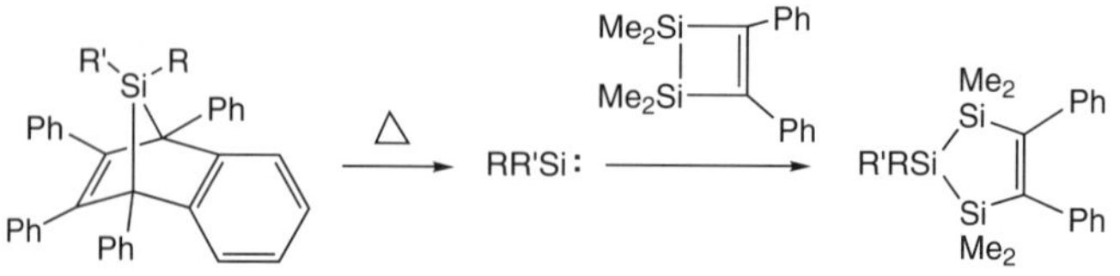

Scheme 14.37

the Si—Si bond. However, no insertion is observed with hexamethyldisilane. The difference between the reactivities is the result of the strain of the Si—Si bond.

4.2. Additions of Silylenes and Germylenes to Multiple Bonds

4.2.1. Addition to Acetylenes. The nature of the reactions of silylenes with acetylenes is a most perplexing problem. In 1976, Gaspar and Conlin[102] isolated tetramethylsilirene (**77**) from the flash pyrolysis of 1,2-dimethoxy-1,1,2,2-tetra-methyldisilane in the presence of dimethylacetylene (Scheme 14.38). The thermolysis of **77** at 105 °C gave only a polymer, and no disilacyclohexadiene was detected. However, thermolysis of 1,1-dimethyl-2-phenyl-3-trimethylsilyl-1-silacyclopropene at 250 °C gave 1,1,4,4-tetramethyl-2,5-diphenyl-3,6-bis(trimethylsilyl)-1,4-disilacyclohexa-2,5-diene and its isomer in 80% yields in the ratio of 3:1. In the presence of diphenylacetylene, disilacyclohexadiene was not detected.[103] The formation of disilacyclohexadiene might take place by direct dimerization of the silacyclopropene even in the presence of diphenylacetylene.

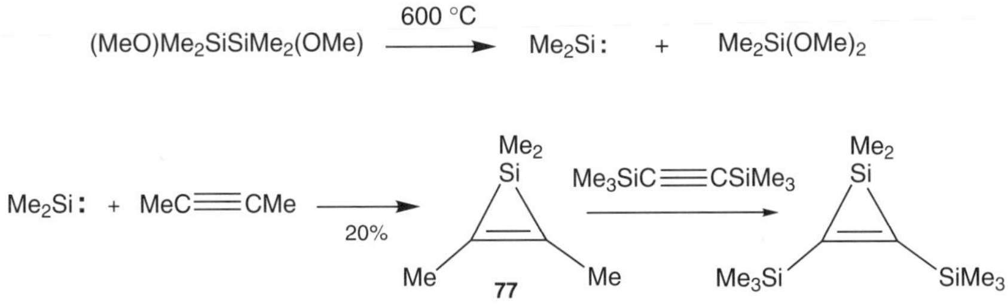

Scheme 14.38

Diadamantylsilylene Ad_2Si: formed from Ad_2SiI_2 upon treatment with lithium under ultrasonic irradiation was compared with that formed upon pyrolysis of a silirane (Scheme 14.39).[35]

Germirenes were also isolated by the reaction of dialkylgermylenes with a cyclo-heptyne derivative.[104] The addition of Me_2Si: (and Me_2Ge:) has been explained as the result of stepwise addition of triplet divalent species (Scheme 14.40).[105]

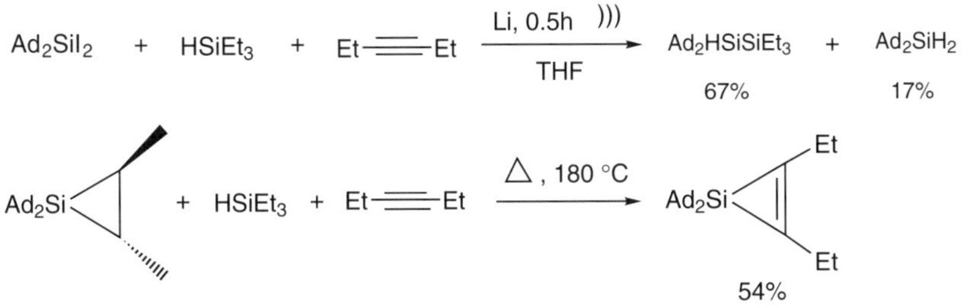

Scheme 14.39

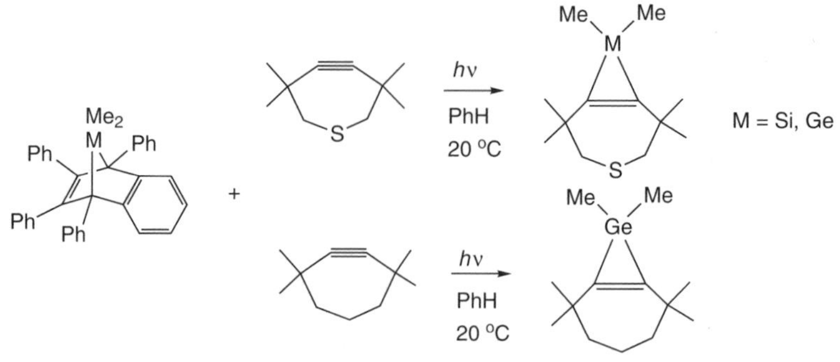

Scheme 14.40

Several polysilabicyclosilirenes (**79**) have been obtained in the reactions of polysilacyclooctynes (**78**) with dimesitylsilylene generated from the photolysis of 2,2-dimesityl-1,1,1,3,3,3-hexamethyltrisilane (Eq. 8).[106] Thermolysis of bis(silacyclopropane) (**80**) in the presence of bis(trimethylsilyl)acetylene at 60 °C affords bis-(silacyclopropene) (**81**) in 61% yield (Scheme 14.41).[107]

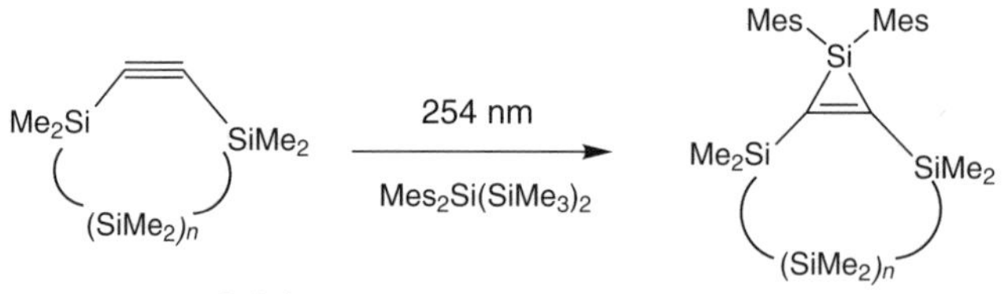

78 $n = 2,3,4$ (8)

79 $n = 2,3,4$

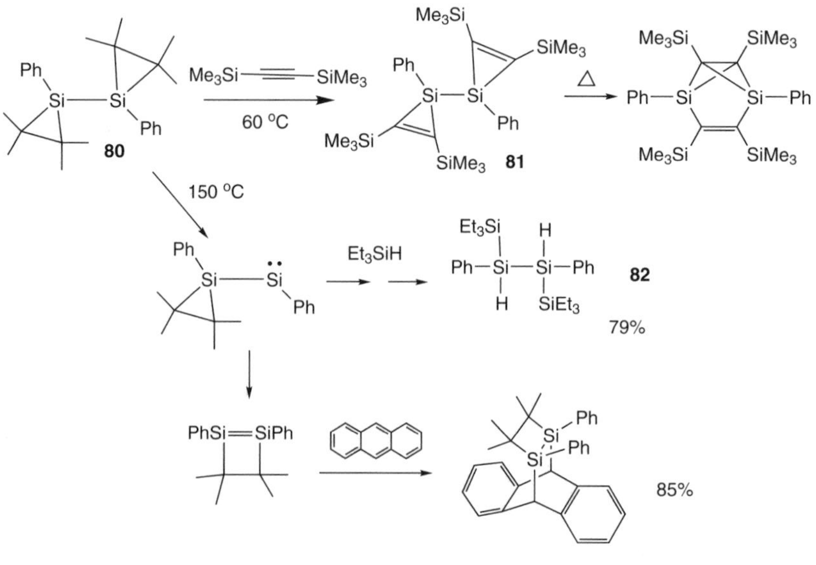

Scheme 14.41

Evidence for the generation of a silylene was obtained in the presence of triethylsilane, which afforded 2,3-diphenyltetrasilane (**82**) in 79% yield. The reactions of some 1,3-diynes, $R'C\equiv CC\equiv CR'$ (**83**), with silylenes afforded the silylene adduct. The course of this reaction strongly depends on the nature of the substituents R and R'. When R is an alkyl group, the bis(silirene) (**84**) is formed initially but undergoes rearrangement to a bicyclohexadiene derivative upon longer photolysis (Eq. 9).[108]

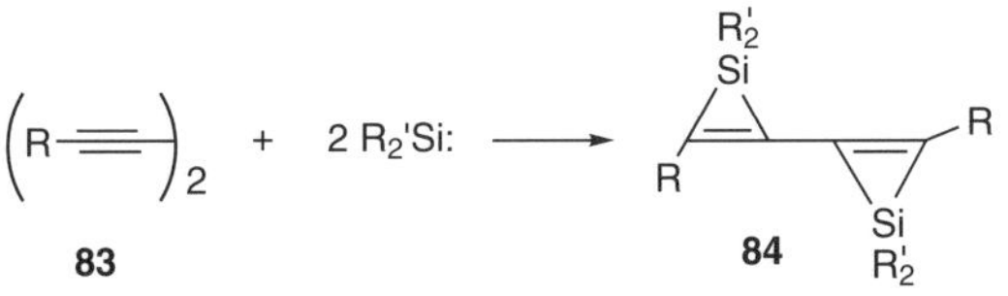

$$\left(R\text{---}\equiv\right)_2 \;+\; 2\,R_2'Si: \;\longrightarrow\; \text{[structure 84]} \tag{9}$$

83   **84**

Dimesityl- and dibutylgermylenes react with acetylene in the presence of a catalytic amount of palladium complexes to yield the corresponding C-unsubstituted germoles (**85**) and compound (**86**). When acetylene was bubbled into a toluene solution of hexamesitylcyclotrigermane in the presence of $Pd(PPh_3)_4$ at 80 °C, 1,1-dimesitylgermole was obtained in 85% yield (Eq. 10).[109]

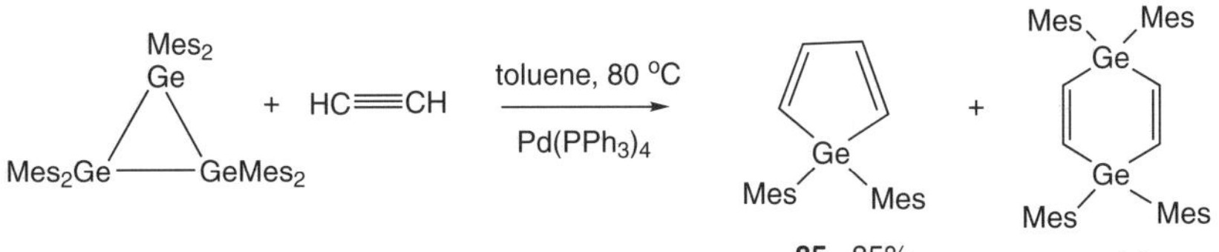

$$\text{[structure]} + HC\equiv CH \xrightarrow[\text{Pd(PPh}_3)_4]{\text{toluene, 80 °C}} \text{[structure 85]} + \text{[structure 86]} \tag{10}$$

85 85%   **86**

4.2.2. Addition to Olefins, Dienes, and Related Compounds.

Reaction of methylphenylsilylene with cyclohexene, followed by treatment with methanol gives the methanolysis product expected from a silirane.[110] In this reaction, attempts to isolate the siliranes from the reaction mixture were unsuccessful. Hexa-*tert*-butyl-cyclotrisilane photolyzes to yield both a silylene and a disilene.[111] Stable siliranes are obtained by the reactions with alkenes, as predicted (Scheme 14.42). Stereospecific addition of bis(triisopropylsilyl)silylene $(i\text{-}Pr_3Si)_2Si$: to *cis*-2-butene has been observed,[112] and a singlet ground state was deduced for diadamantylsilylene when it was found to undergo stereospecific addition to *cis*- and *trans*-2-butenes.[38] Dimethylsilylene reacted with 2,3-dimethyl-1,3-butadiene to afford a silacyclopentene via rearrangement of a vinylsilirane intermediate (**87**) (Eq. 11).[113]

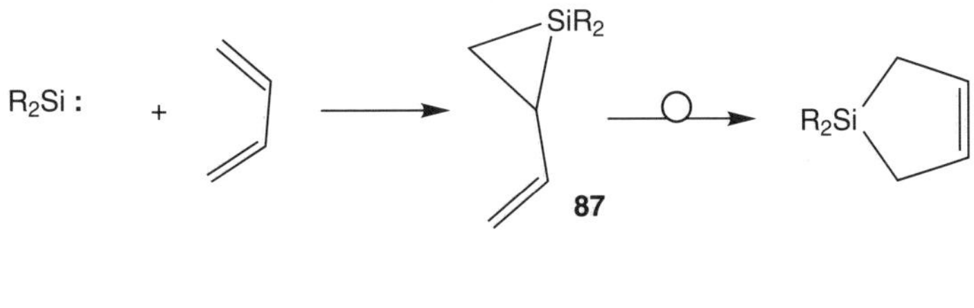

$$R_2Si: \;+\; \text{[structure]} \longrightarrow \text{[structure 87]} \longrightarrow R_2Si\text{[structure]} \tag{11}$$

87

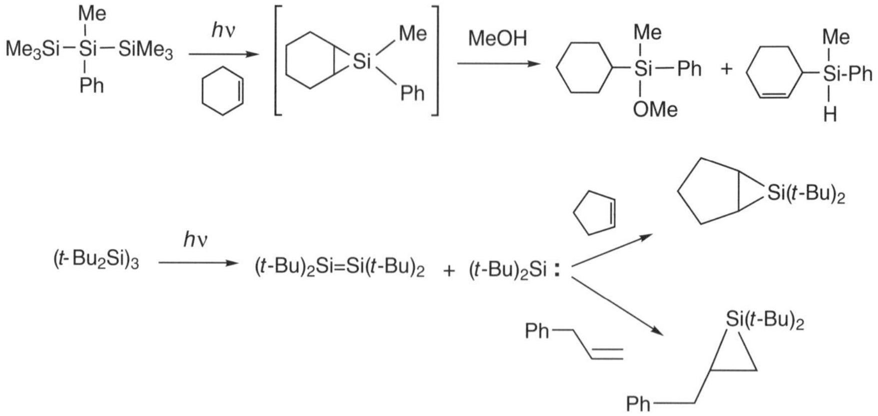

Scheme 14.42

Addition of Me$_2$Si: to cyclopentadiene gave both the conjugated 1-sila-2,4- and 2,5-cyclohexadienes. A vinylsilirane intermediate probably rearranges to the final product via a diradical intermediate (Scheme 14.43).[114]

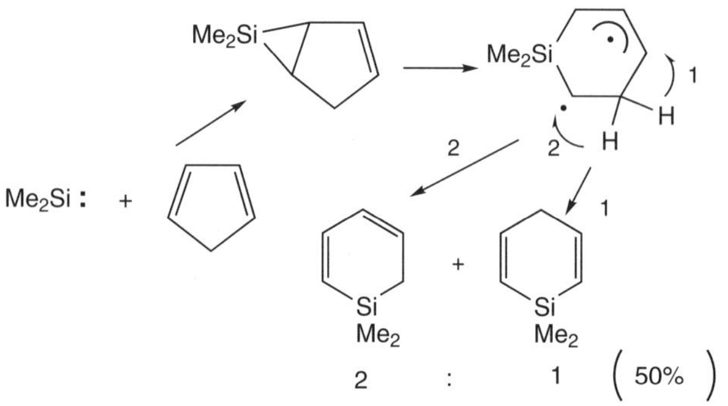

Scheme 14.43

Addition of dimesitylsilylene to *tert*-butylallene and 1,1,3,3-tetramethylbuta-triene led to the stable alkylidene siliranes (**88** and **89**) (Scheme 14.44).[115]

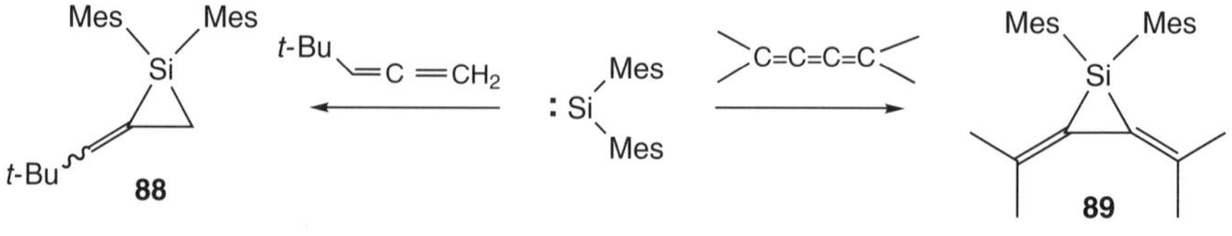

Scheme 14.44

When 2,2-bis(2,6-diisopropylphenyl)-1,1,1,3,3,3-hexamethyltrisilane (**90**), a silylene precursor of Dip$_2$Si:, was photolyzed with a low pressure mercury lamp in a toluene solution of C$_{60}$ thermally stable siliranes, (Dip$_2$Si)C$_{60}$ (**91**) and (Dip$_2$Si)$_2$C$_{60}$ (**92**) were formed, in 58 and 27% yields, respectively, together with other polyadducts, which evidently arose from addition of the silylene across the C=C double bond between the two six-membered rings of the fullerene (Scheme 14.45).[116]

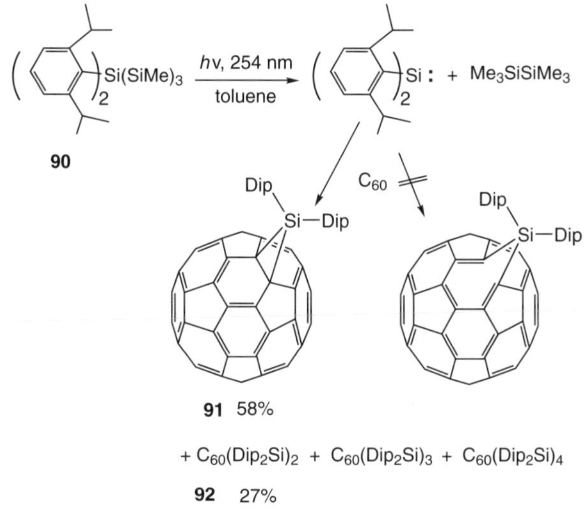

91 58%

$+ C_{60}(Dip_2Si)_2 + C_{60}(Dip_2Si)_3 + C_{60}(Dip_2Si)_4$

92   27%

Scheme 14.45

Although there are many examples of reactions of carbenes and silylenes with olefins, experiments concerning reactions of germylenes with olefins are rather limited. The reaction of :GeMe$_2$ with styrene gave 3,4-diphenylcyclopentane (**93**) via a germirane intermediate or a germylene–olefin π complex (**94**), and the reaction with a 1,3-butadiene derivative also gave germacyclopentane (**95**) together with the 2:1 cycloadduct (**96**) (Scheme 14.46).[117] On the other hand, the reaction of

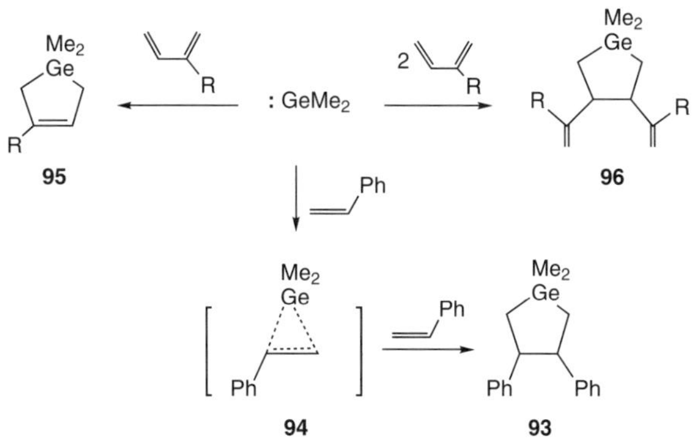

Scheme 14.46

stable germylene (**97**) with 2,5-dimethyl-2,3,4-hexatriene and *N*-phenylmaleimide provided the stable germirane derivatives, bis(alkylidene)germirane (**98**) and bicyclogermirane (**99**, Scheme 14.47).[118]

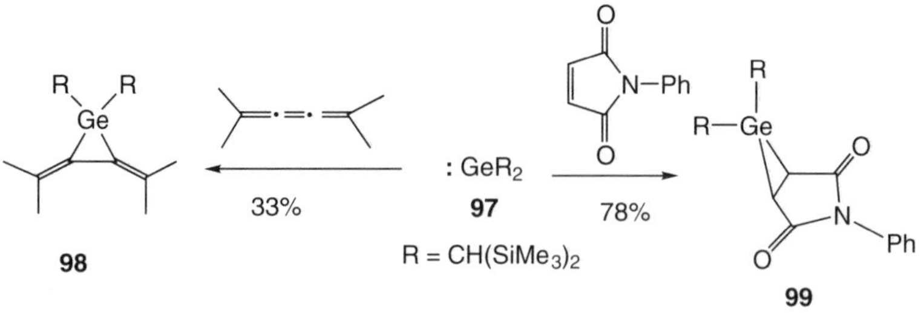

Scheme 14.47

The reactions of a stable germylene (:GeDis$_2$) with ethylene afforded the first 1,2-digermacyclobutane (**100**) via a germirane intermediate.[119] Interestingly, the photolysis of 1,2-digermacyclobutane derivatives in the presence of 2,3-dimethyl-1,3-butadiene gave the germacyclopentene (**101**) and in the presence of C$_{60}$ provided the germylene adduct of C$_{60}$ (**102**) as well as the germirane adducts (**103**) of C$_{60}$ (Scheme 14.48).[120]

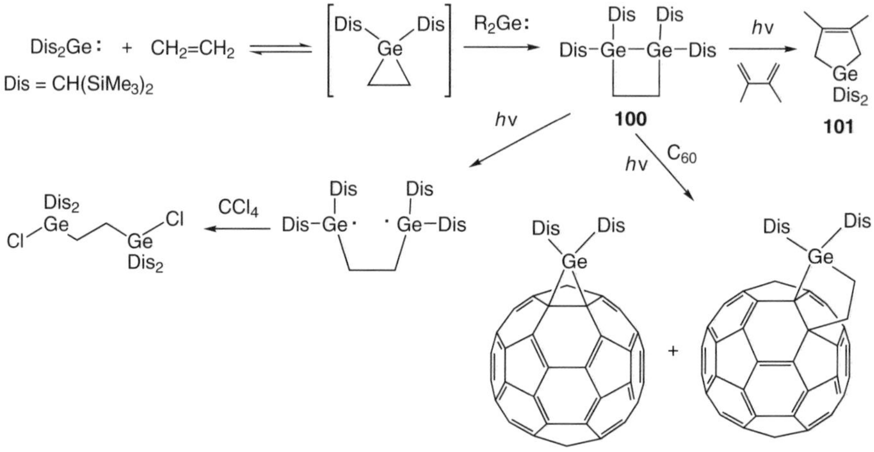

Scheme 14.48

4.2.3. Addition to Carbonyl and Thiocarbonyl Compounds.

Thermal generation of dimethylsilylene in the presence of carbonyl compounds leads to the addition of the silylene to the C=O bond with the formation of oxasiliranes

(**104**), followed by rearrangement to (**106**), probably through a biradical intermediate (**105**) (Scheme 14.49).[121] Under mild conditions, more direct evidence was found for the intermediacy of oxasiliranes. When 2-adamantanone and 7-norbornone were allowed to react with photochemically generated dimethylsilylene, the major products could be formulated as dimers (**108**) and the carbonyl adducts

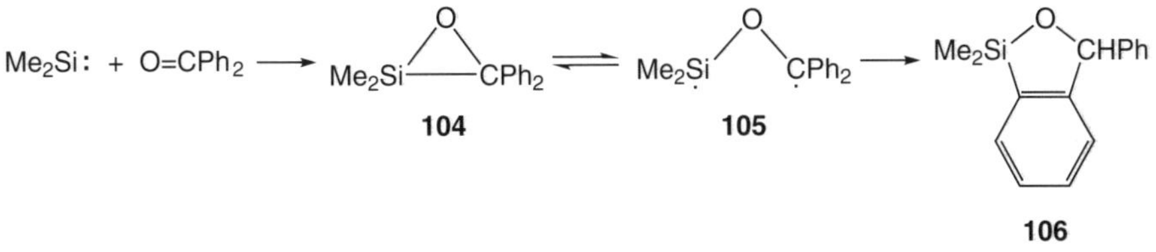

Scheme 14.49

(**110**) of oxasiliranes (**107**) (Scheme 14.50).[122] Stable oxasilirane (**111a**) was isolated by the reaction of dimesitylsilylene with 1,1,3,3-tetramethyl-2-indanone (**112a**).[123] The sulfur analogue thiasilirane (**111b**) was also isolated from the reaction of dimesitylsilylene with 1,1,3,3-tetramethyl-2-indanethione (**112b**). Although

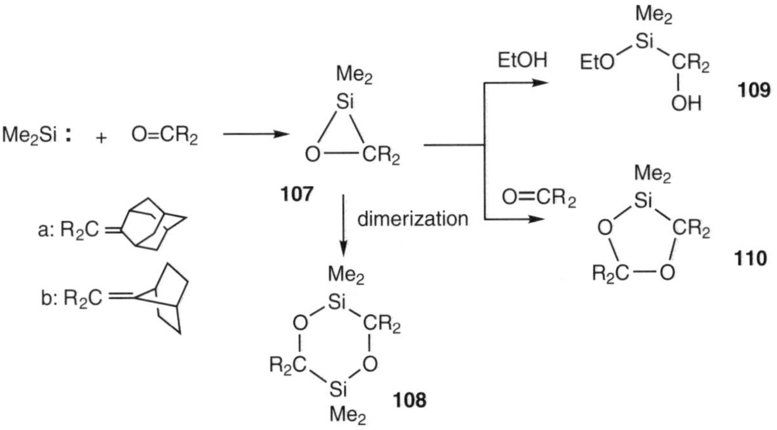

Scheme 14.50

adamantanone (**113a**) yielded the 2:1 adduct (**114**) instead of oxasilirane, adamantanethione (**113b**) gave the stable thiasilirane (**115**) (Scheme 14.51). Using matrix isolation techniques, spectroscopic evidence was obtained for the intermediacy of the silacarbonyl and silathiocarbonyl ylides (**117a** and **117b**) in low-temperature matrices (Scheme 14.52).[124] Irradiation of the oxasilirane (**111a**) in matrix (Ip/3-Mp) at 77 K with a low-pressure mercury lamp led to the appearance of a new

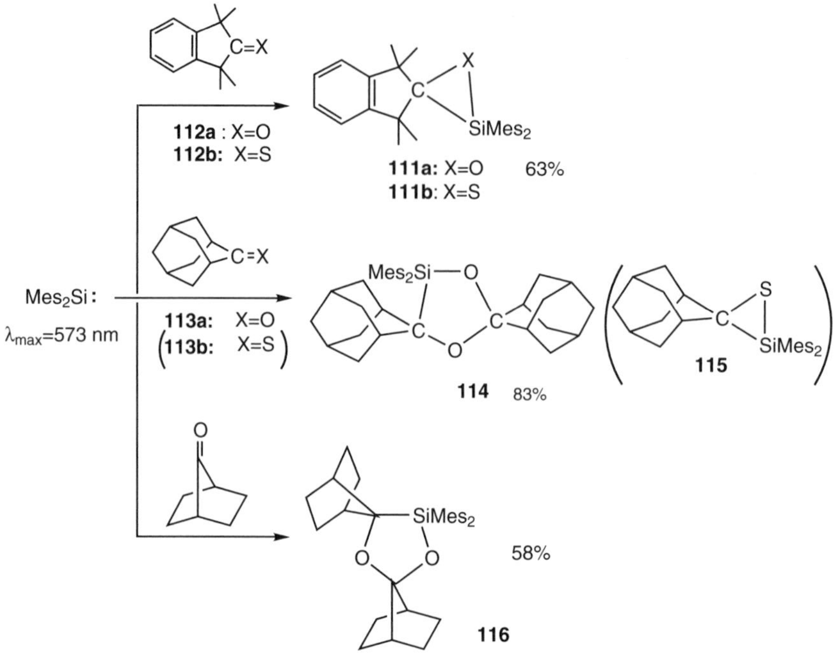

Scheme 14.51

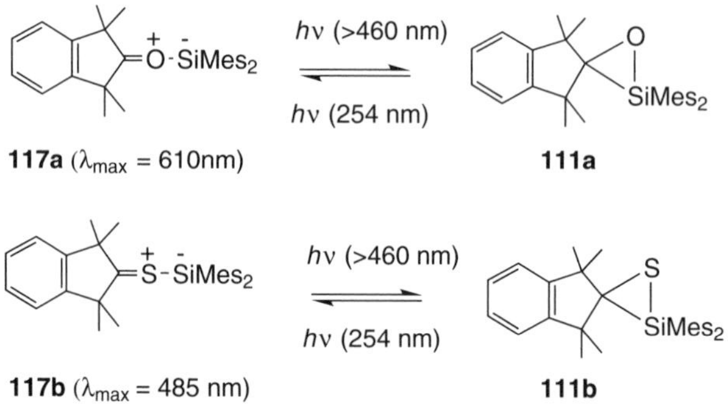

Scheme 14.52

band at 610 nm in the UV–vis spectrum and the matrix became interestingly blue. This absorption band was stable at 77 K on prolonged standing. However, it immediately disappeared on brief irradiation with a Xenon lamp ($\lambda > 460$ nm) or when the matrix was allowed to melt. This colored species was independently generated by the reaction of dimesitylsilylene with **113a**.[122c]

Dimethylsilylene reacts smoothly with α-diketones to yield 1,3-dioxa-2-sila-cylopentane-4-enes (**118** and **119**, Scheme 14.53).[125]

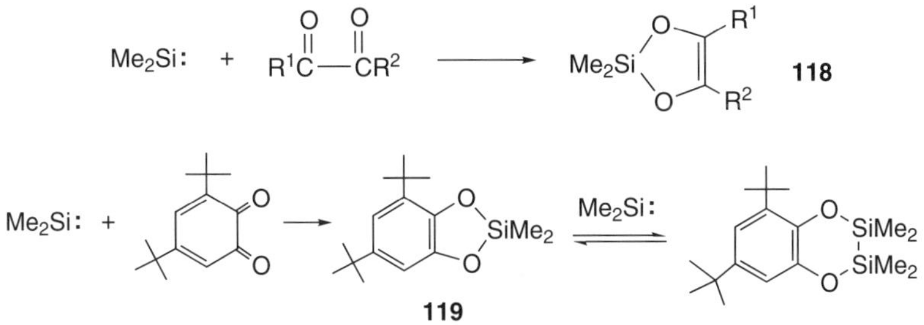

Scheme 14.53

Similarly, dimesitylgermylene reacts smoothly with thioketones and di-*tert*-butylthioketene to give the corresponding digermadithiolanes, respectively, via a germathiirane intermediate.[126]

Thiagermirane (**120**) and alkylidenethiagermirane (**121**) are isolated in the reactions of dimesitylgermylene, generated from hexamesitylcyclotrigermane, with adamantanethione and di-*tert*-butylthioketene, respectively (Scheme 14.54).[127]

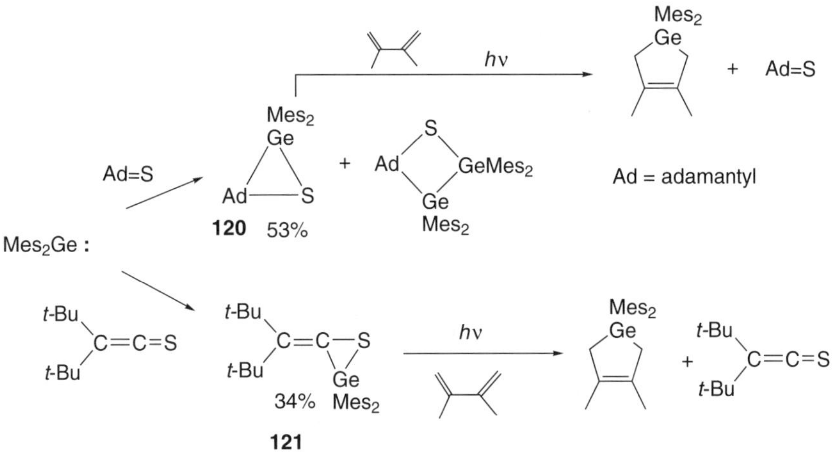

Scheme 14.54

Reactions of germylenes with 1,3-dienes, vinyl ketones, α-diketones, vinyl imines, and α-diimines gave corresponding [4 + 1] cycloadducts (Scheme 14.55).[79]

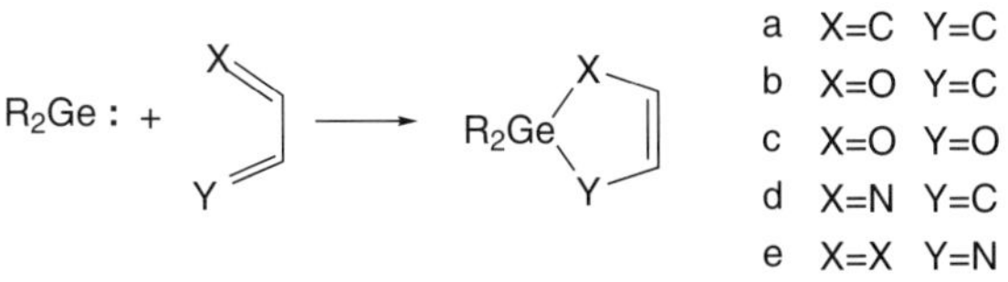

a	X=C Y=C
b	X=O Y=C
c	X=O Y=O
d	X=N Y=C
e	X=X Y=N

Scheme 14.55

5. SYNTHESIS, STRUCTURES, REACTIONS, AND DIMERIZATIONS OF STABLE SILYLENES

5.1. Synthesis and Isolation of a Stable Dialkylsilylene

Divalent silicon compounds (silylenes), one of the most interesting class of low-coordinated silicon compounds, had been known as highly reactive, short-lived transient species that resembled the carbon analogues (carbenes), until the recent success in the synthesis of the first stable silylene by West and co-workers.[128,129] Their structures had been much less explored than those of their heavier congeners such as germylenes and stannylenes. Although dodecamethylsilicocene (**19**), formally also a silylene, has been known since 1986,[36] compound **19** is stabilized by η^5-coordination of pentamethylcyclopentadienyl ligands and is not a congener of a carbene.

Recently, introduction of amino substituents on the silicon atom resulted in an epoch-making breakthrough in this field, that is, the synthesis of stable silylenes such as **122–125** (Scheme 14.56).[128,130] Although these silylenes are all well characterized by either X-ray crystallographic analysis or electron-diffraction analysis,

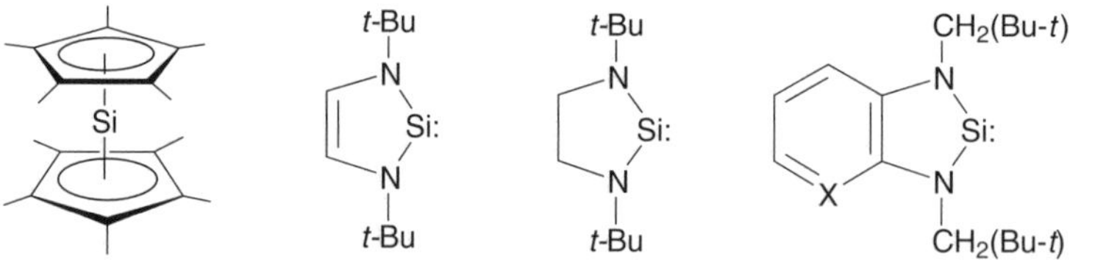

| 19 | 122 | 123 | 124 (X = CH), 125 (X = N) |

Scheme 14.56

the structural features of these special silylenes revealed that they are stabilized by strong interaction between the vacant *p*-orbital at the divalent silicon and filled π-type orbitals of the nitrogen atoms in the substituents. The structures and reactivities of these silylenes indicate that their frontier orbitals are considerably perturbed by their heteroatom substituents to such an extent that they are different from those of a true analogue of a singlet carbene. Several stable silylenes have been synthesized, and some of them in fact showed remarkable thermal stability (Scheme 14.56). West, Denk et al.[128] prepared silylene **122** according to Eq. 12. Dehalogenation of the dichloride was carried out using molten potassium metal in refluxing THF, and their rather vigorous conditions have become standard for the synthesis of stable silylenes of this type. Silylene **122** is a colorless crystalline

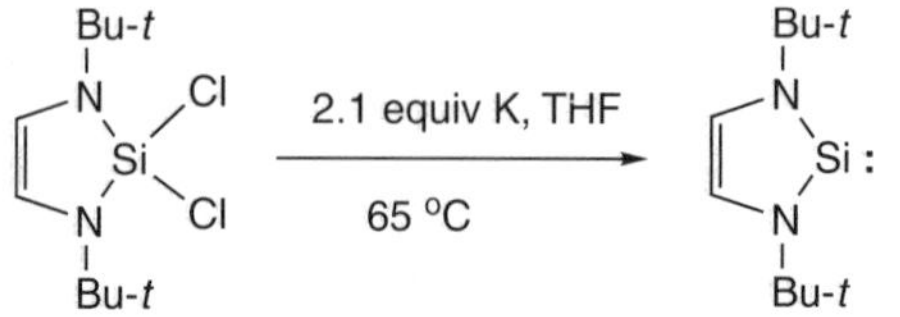

$$\text{2.1 equiv K, THF} \quad 65\ ^\circ\text{C} \tag{12}$$

substance having true thermal stability. Silylene **122** is not only indefinitely persistent at room temperature but also unchanged after heating at 150 °C in toluene for 4 months. Solid **122** finally decomposes at its melting point of 220 °C. The analogous compounds, **123**,[129a] **124**,[130b] and **125**,[130c] have also been synthesized. Two nitrogen atoms bonded to the divalent silicon stabilizes these compounds, although **123** is only marginally stable.

The five-membered rings in **122** and **124** are planar, while the ring in **123** is necessarily puckered. The N—C distances in the ring are shorter in **122** and **124** than in **123** by ∼10 pm. The Si—N bond lengths are similar to the normal Si—N single bond distances (∼172–174 pm). However, bonds to a divalent silicon atom are predicted to be longer, by ∼8 pm, than those to a tetravalent Si.[131] The bond lengths in the silylenes mentioned above are, therefore, consistent with some multiple bond character in the Si—N bonds.

Stable silylenes all react with alcohols via the expected insertion into the O—H bond to afford the corresponding alkoxyhydrosilanes (**126**).[130b,132] Reaction of silylene **122** with ketones afforded four-membered disilaoxetane compounds **127** in high yields (Scheme 14.57).[133] The mechanism for this transformation is best

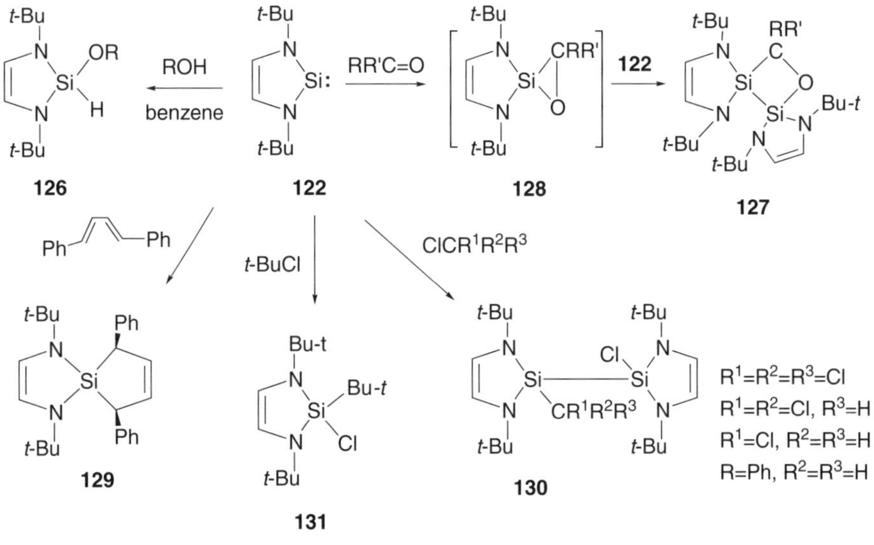

Scheme 14.57

interpreted in terms of a [2 + 1] cycloaddition to form an ephemeral oxasilacyclopropane intermediate (**128**), a transformation common for transient silylenes. This intermediate further reacts with the second silylene **122** to give the final product **127**.[129c] Stable silylenes undergo [1 + 4] cycloadditions with dienes. For example, silylene **122** reacts with 1,4-diphenylbutadiene to form the expected silacyclopentene product (**129**). When trichloromethane was added to a colorless solution of **122** in hexane, the yellow 2:1 adduct **130** was formed immediately in quantitative yield. The disilane was the exclusive product even in the presence of a 100-fold excess of

CHCl$_3$. Similar 2:1 adducts were formed quantitatively when **122** was treated with CCl$_4$, CH$_2$Cl$_2$, or benzyl chloride.[128,134] However, product **131** was obtained exclusively via silylene insertion into the C—Cl bond when **122** was allowed to react with *tert*-butyl chloride.[135] It is proposed that the formation of **130** is a result of the initial halophilic interaction between the silylene and halocarbon.

In 1999, Kira et al.[136] succeeded in the synthesis and isolation of the first stable dialkylsilylene (**132**) by taking advantage of their original cyclic ligand having four trimethylsilyl groups (Eq. 13). Thus, the reduction of the corresponding dibromosilane (**133**) with potassium graphite resulted in the formation of **132** as stable orange crystals, the crystallographic analysis of which revealed that the helmet-like bidentate ligand effectively protected its reactive silicon center and that the shortest distance between the divalent silicon atoms in the crystals of **132** is 7.210(1) Å. The relatively small C—Si—C angle [93.88(7)°] of **132** is indicative of the larger *p* character of the silicon hybrid orbitals used in the C—Si bonds of **132**.

$$\text{(13)}$$

Me$_3$Si SiMe$_3$ $\xrightarrow[\text{-50 °C}]{\text{KC}_8/\text{THF}}$ Me$_3$Si SiMe$_3$

SiBr$_2$ → Si:

Me$_3$Si SiMe$_3$ Me$_3$Si SiMe$_3$

133   **132**

Silylene **132** showed absorption maxima at 260 and 440 nm in its UV–vis spectra, the latter of which is assignable to the n(Si)–3p(Si) transition and close to those observed for dimethylsilylene (453 nm)[19] and 1-silacyclopentane-1,1-diyl (436 nm).[58,61,137] Thus, the electronic structure of silylene **132** is not much perturbed by the substituents. Silylene **132** showed a ^{29}Si NMR signal for the central divalent silicon at 567.4 ppm in C$_6$D$_6$, which is the lowest field ^{29}Si NMR resonance reported so far.[138] Theoretical GIAO calculations revealed that the parent silylene (SiH$_2$) and the closely related model compounds (silacyclopentane-1,1-diyl and 2,2,5,5-tetrakis(trihydrosilyl)silacyclopentane-1,1-diyl) should have ^{29}Si NMR resonances at 817, 754, and 602 ppm, respectively, again indicating that electronic perturbation by substituents is much smaller in **132** than that in **122–124**.

The most intriguing reactivity of **132** is the facile 1,2-migration of the trimethylsilyl group in the ligand to the divalent silicon atom giving the corresponding silaethene derivative **134** (Scheme 14.58). Such migratory insertion reaction has never been observed in the cases of stable germylene (**135**)[139] and stannylene (**136**)[140] bearing the same ligand even at 100 °C. It should be noted that this is the first experimental evidence for the isomerization of silylmethylsilylene to 1-silylsilaethene. Such a migration has already been predicted by theoretical calculations.[141]

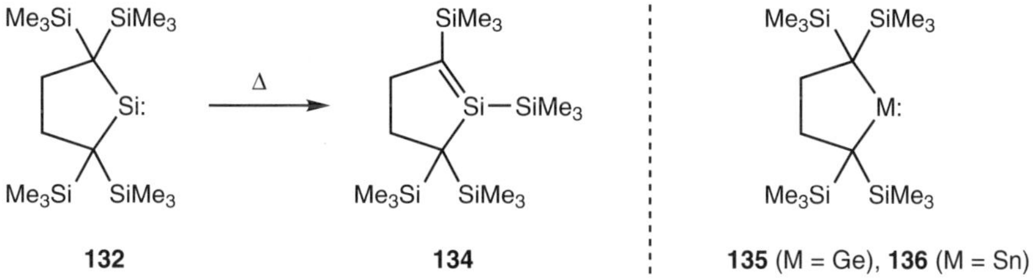

Me$_3$Si SiMe$_3$ / Si: / Me$_3$Si SiMe$_3$

132

$\Delta \longrightarrow$

SiMe$_3$ / Si—SiMe$_3$ / Me$_3$Si SiMe$_3$

134

Me$_3$Si SiMe$_3$ / M: / Me$_3$Si SiMe$_3$

135 (M = Ge), **136** (M = Sn)

Scheme 14.58

5.2. Generation and Reactions of an Overcrowded Diarylsilylene

Until recently, the intrinsic nature of silicon–silicon double bond has not been fully disclosed, and there had been no report concerning the thermal dissociation of disilenes into the corresponding silylenes, in contrast to the facile thermal dissociation for the germanium and tin analogues.[142] The high thermodynamic stability of the C=C and Si=Si double bonds to relative to those for the Ge=Ge and Sn=Sn double bonds is in good agreement with the computed dissociation energies for the process $H_2E=EH_2 \rightarrow 2H_2E$: (~140 kcal/mol for E = C,[143] 52–58 kcal/mol for E = Si,[144] 30–45 kcal/mol for E = Ge,[145] and 22–28 kcal/mol for E = Sn.[145a,146]

Meanwhile, Tokitoh et al.[147] reported the first example of thermal dissociation of extremely hindered disilenes [Tbt(Mes)Si=Si(Mes)Tbt; Tbt = 2,4,6-tris[bis(trimethylsilyl)methyl]phenyl, Mes = 2,4,6-trimethylphenyl (mesityl); (Z)-**137**: cis isomer, (E)-**137**: trans isomer] into the corresponding silylene [Tbt(Mes)Si:, **138**] under very mild conditions (~70 °C) as shown in Scheme 14.59.

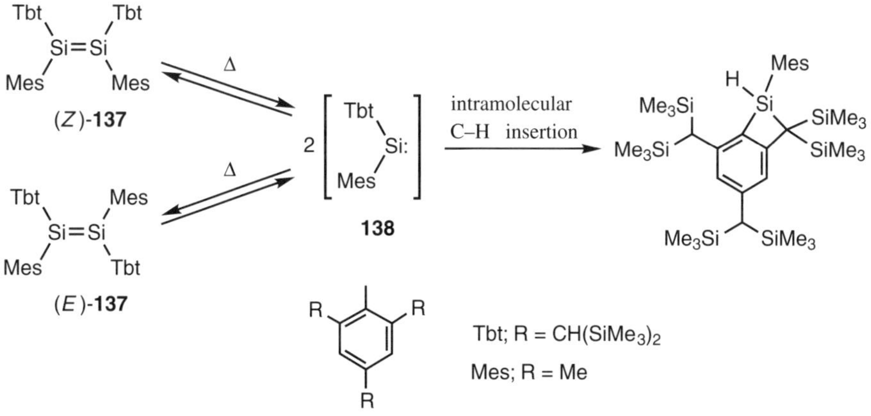

Tbt Tbt / Si=Si / Mes Mes

(Z)-**137**

Tbt Mes / Si=Si / Mes Tbt

(E)-**137**

Δ

$2\left[\begin{array}{c} \text{Tbt} \\ \text{Si:} \\ \text{Mes} \end{array}\right]$

138

intramolecular C–H insertion $\longrightarrow$

H Mes / Si / Me$_3$Si SiMe$_3$ / Me$_3$Si SiMe$_3$ / Me$_3$Si SiMe$_3$

R R / R

Tbt; R = CH(SiMe$_3$)$_2$
Mes; R = Me

Scheme 14.59

Kinetic studies on the thermal dissociation of **137** revealed that the unusual behavior of **137** in contrast to the stable disilenes obviously resulted from the presence of extremely bulky Tbt groups that weaken the Si=Si double bond.[147] The overcrowded diarylsilylene (**138**) thus generated was found to undergo facile reactions with methanol, triethylsilane, 2,3-dimethyl-1,3-butadiene, and elemental chalcogens such as sulfur and selenium (Scheme 14.60).[147] Silylene **138** readily reacted also with olefins and acetylenes to afford the corresponding [1 + 2]cycloaddition products (Scheme 14.60).[147] Furthermore, **138** was found to undergo addition reactions with aromatic hydrocarbons such as naphthalene and benzene to give the corresponding cycloadducts (Scheme 14.60), the formation of which should be noted

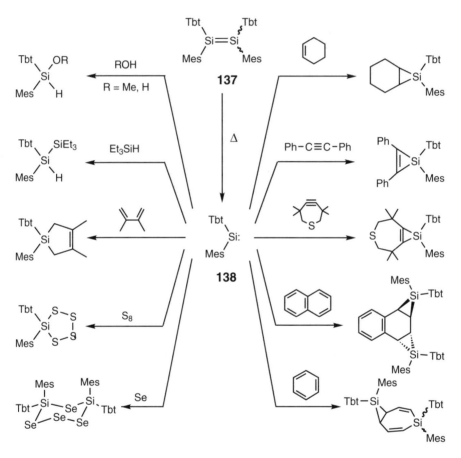

Scheme 14.60

as the first examples of [1 + 2]cycloaddition reactions of silylene to the carbon–carbon double bond in aromatic π-conjugated systems.[147,148] The successful isolation of these silylene adducts with aromatic hydrocarbons is obviously due to the much milder reaction conditions than those used in the conventional methods for generating silylenes (see above).

The very mild conditions of generation and the extreme bulkiness of the silylene **138** leads to a variety of important addition products with heteroatom-containing multiple bond compounds such as carbon disulfide,[149] nitriles,[150] phospha-alkynes,[150] and isonitriles (Scheme 14.61).[151] It is especially interesting that the reaction of **138** with isonitriles bearing a bulky aromatic substituent afforded the corresponding silylene–isonitrile complexes **139**, the first stable silylene–Lewis base complexes. The ^{13}C and ^{29}Si NMR spectroscopic data and theoretical calculations on the triphenyl-substituted model compound lead to the conclusion that the silylene–isonitrile adducts **139** do not have a cumulene structure (silaketenimine) but a zwitterionic structure (i.e., a silylene–isonitrile complex).

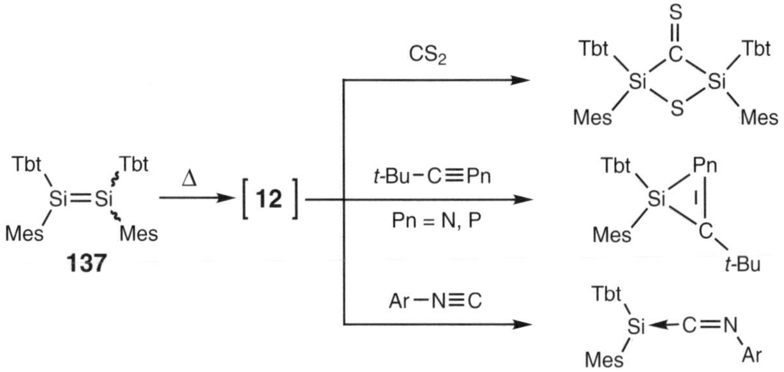

Scheme 14.61

Since Spitzer and West[152] reported a similar pyrolysis of tetraalkyldisilene $[(Me_3Si)_2CH]_2Si=Si[CH(SiMe_3)_2]_2$ leading to the generation of corresponding dialkylsilylene $[(Me_3Si)_2CH]_2Si$: at somewhat higher temperature ($\sim 100\,°C$), thermal dissociation of disilenes into silylenes may be a more general phenomenon than has previously been thought.

5.3. Reactions of a Silylene–Isonitrile Complex as a Masked Silylene

As mentioned above, there have been no examples of stable diarylsilylene although dialkyl- or diamino-substituted silylenes are now available as stable crystalline compounds. However, Tokitoh et al.[151] found that the overcrowded diarylsilylene–isonitrile complexes **139** act as a free silylene **138** in solution even at room temperature. Since there is a severe limitation in a study concerning the reactivities of silylenes under the conventional generating conditions, the reactivity of silylene–isonitrile complexes **139** as a masked silylene is of great interest.

The reactions of silylenes with 1,3-dienes giving the corresponding 3-silolenes are typical of the cycloaddition reactions of silylenes.[2,153] The mechanism of these reactions has been investigated in detail,[114,154] and it has been proposed that the reactions of silylenes with 1,3-dienes proceed via initial $[1 + 2]$ addition followed by the isomerization of the resulting 2-vinylsiliranes to the corresponding 3-silolenes. However, the observation and isolation of the intermediary 2-vinylsiliranes has been limited to only a few examples[155] because vinylsiliranes readily isomerize to 3-silolenes under the conditions of generation of the silylenes. Tokitoh and co-workers[156] attempted the synthesis and isolation of a 2-vinylsilirane intermediate by using silylene–isonitrile complex **139a** as a silylene source (Scheme 14.62).

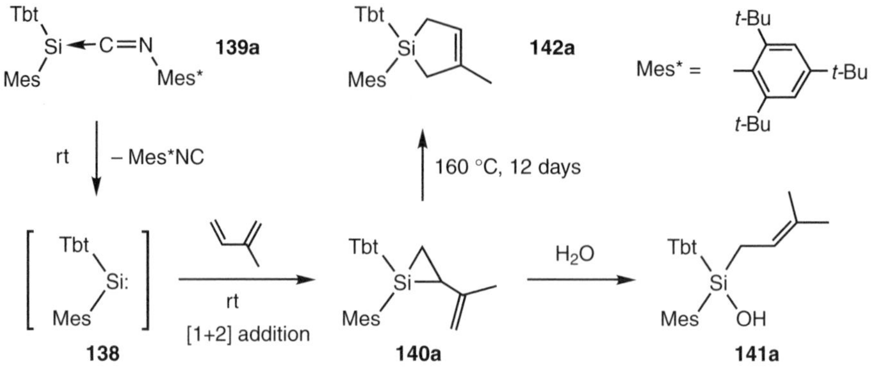

Scheme 14.62

Thus, complex **139a** reacted with isoprene at room temperature to give the corresponding 2-vinylsilirane (**140a**, 84%). 2-Vinylsilirane (**140a**) was also isolated in 89% yield by the reaction of disilene **137** with isoprene at 70 °C, which indicates that these reactions proceeded via silylene **138** as an intermediate. The exclusive formation of **140a** is probably the result of steric repulsion between bulky substituents on the silicon atom and the methyl group of isoprene. Although 2-vinylsilirane (**140a**) was stable in the solid state even on exposure to air, it underwent hydrolysis on silica gel to give the corresponding silanol **141a**. Thermolysis of **140a** in C_6D_6 at 160 °C for 12 days resulted in the formation of the formal [1 + 4] adduct **142a** in 88% yield. This reaction is the first isolation of the initial intermediate in the reaction of thermally generated silylene with a 1,3-diene, that is, the 2-vinylsilirane derivative, although the intermediacy of this type of [1 + 2] adduct in the reactions of photochemically generated silylenes has been already evidenced by NMR spectroscopy.[155a]

On the other hand, when complex **139a** was allowed to react with 2,3-dimethyl-1,3-butadiene in C_6D_6 in a sealed tube at room temperature, the formation of both [1 + 2] adducts **140b** and [1 + 4] adduct **142b** was observed by ^{29}Si NMR spectroscopy (Scheme 14.63; **14b**: $\delta_{Si} = -76.3, -72.9$, **16b**: $\delta_{Si} = -5.3$). Since heating the reaction mixture at 50 °C for 7 h resulted in no isomerization of **140b** into **142b**, silylene **138** seems to undergo [1 + 2] and [1 + 4] addition reactions with 2,3-dimethyl-1,3-butadiene competitively. This is the first clear evidence for the occurrence of direct [1 + 4] addition of a silylene to a 1,3-diene. This phenomenon is most likely due to the suppression of [1 + 2] addition by steric repulsion between the methyl groups of the butadiene and the bulky substituents on the silicon atom of **138**. This interpretation is also supported by the exclusive production of **140a** in the reaction with isoprene. Thus, the results obtained here suggest that silylenes bearing very bulky substituents may undergo direct [1 + 4] addition to dienes. Vinylsilirane

(**140b**) completely isomerized into **142b** on heating at 100 °C for 5 h and underwent hydrolysis on silica gel to give silanol **141b** as well as **140a** (Scheme 14.63).

Consequently, the reactivity of silylene–isonitrile complexes **139** as a masked silylene may provide us with not only a new and useful synthetic building block for a variety of organosilicon compounds but also a new research field in silylene chemistry.

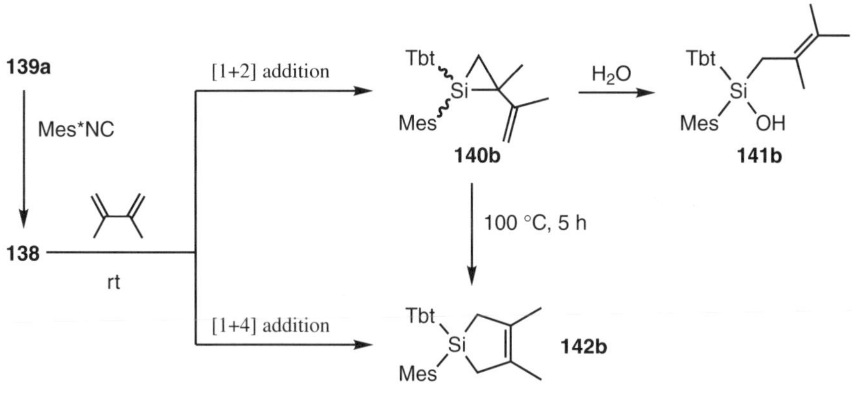

Scheme 14.63

6. SYNTHESIS, STRUCTURES, AND REACTIONS OF STABLE GERMYLENES

6.1. Synthesis of Stable Dialkylgermylenes

In 1976, Lappert and co-workers[157] reported the synthesis and characterization of the first stable germylene, $[(Me_3Si)_2CH]_2Ge$: (**143**). Germylene (**143**) was found to be a monomer in solution but exist as the corresponding dimer, $[(Me_3Si)_2CH]_2$ $Ge=Ge[CH(SiMe_3)_2]_2$ (**144**) in the solid state (Scheme 14.64). Jutzi et al.[158] synthesized the more sterically hindered dialkylgermylene, $[(Me_3Si)_3C][(Me_3Si)_2$ $CH]Ge$: (**145**) in 1991 (Scheme 14.64). In 1999, Kira et al.[139] applied their original helmet-like bidentate ligand to the kinetic stabilization of a germylene and

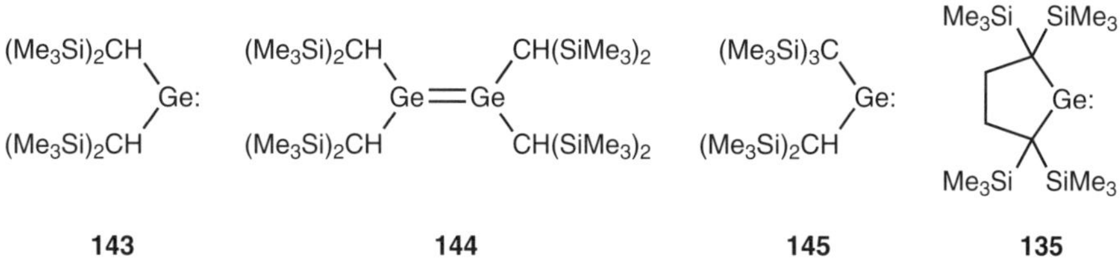

Scheme 14.64

succeeded in the synthesis and isolation of very stable dialkylgermylene (**135**) (Scheme 14.64). In this case, germylene **135** was found to show an identical UV–vis spectrum in THF or hexane, indicating that this cyclic germylene **135** does not form a complex with THF because of severe steric hindrance around the central germanium atom. This result is in sharp contrast to other germylenes that are known to form complexes with bases such as ethers and amines and to cause significant blue shift of the n–p transition of the germylenes.

6.2. Synthesis of Stable Diarylgermylenes

As for stable diarylgermylenes, three examples (**146**,[159] **147**,[160] and **148**,[161] Scheme 14.65) have been structurally analyzed by X-ray crystallographic analysis. The solid-state structure of germylene **146** indicated that its remarkable stability is mainly the result of intramolecular coordination of the fluorine atoms of the -CF_3

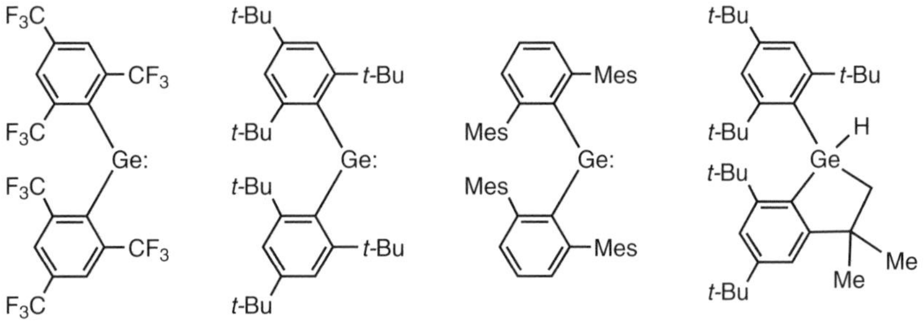

Scheme 14.65

substituents to its electron-poor germanium center, while the latter two germylenes (**147** and **148**) were found to be only kinetically stabilized by their extremely bulky aromatic substituents. However, the stability of germylene **148** was only marginal. In solution, **147** underwent intramolecular insertion of the germanium atom into one of the methyl groups in the substituents in the presence of a Lewis acid or at high temperature to give a germaindane derivative **149** (Scheme 14.65).[162]

Another stable diarylgermylene (**150**)[163] bearing extremely bulky aromatic substituents (Tbt group) together with 2,4,6-triisopropylphenyl group (Tip group) has been reported. That there is no color change on concentration of the blue solution of **150** in hexane ($\lambda_{max} = 581$ nm) strongly suggests that germylene **150** is stable as a monomeric species both in solution and in the solid state, although the solid-state structure of **150** has not been established yet. Germylene **150** was readily synthesized by either the reductive debromination of the corresponding dibromogermane, Tbt(Tip)GeBr$_2$,[164a–c] or the thermal retrocycloadditon of the 2,3-diphenylgermirene bearing Tbt and Tip groups on the germanium atom (Scheme 14.66).[164d] The latter reaction was found to be a useful synthetic method for germylene **150** under neutral conditions. In spite of the presence of such bulky substituents, germylene **150** showed versatile reactivity toward a variety of reagents such as alcohols, buta-dienes, acetylenes, hydrosilanes, and elemental chalcogens (sulfur and selenium) (Scheme 14.66). This effective combination of bulky aromatic ligands was success-fully applied to the synthesis and isolation of a series of stable germanium–chalco-gen double-bond species Tbt(Tip)Ge=X (X = S,[164a,165] Se,[165,166] Te,[166,167]).

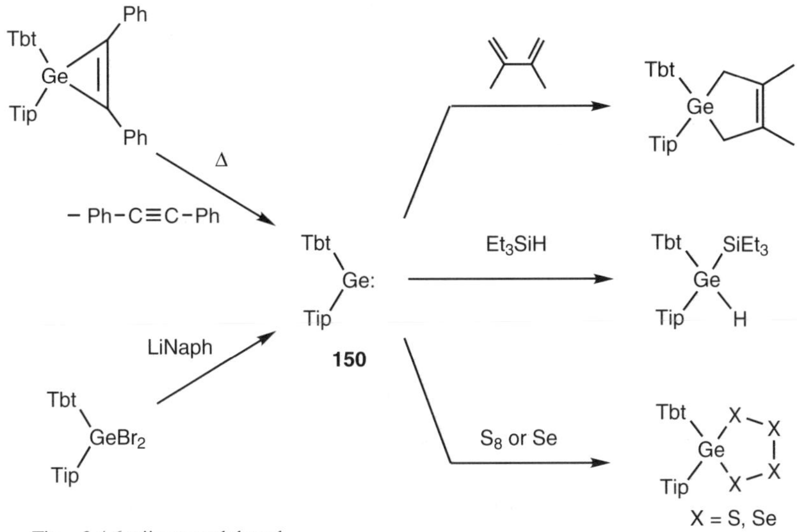

Tip = 2,4,6-triisopropylphenyl

Scheme 14.66

On the other hand, when a slight modification from Tip to Mes was made in the ligand attached to the germanium center in Tbt-substituted diarylgermylene, the stability of the germylene drastically changed.[168] Thus, the less hindered diarylger-mylene Tbt(Mes)Ge: (**151**) was found to be in equilibrium with its dimer, that is, digermene **152** (Scheme 14.67). Germylene **151** showed interesting thermochro-mism in hexane [blue at 295 K vs. yellow at 190 K], which is in sharp contrast to that of the temperature independence of the UV–vis spectra observed for **150**.[164a,165] The absorptions observed at 190 K ($\lambda_{max} = 439$ nm, $\varepsilon\,2.0 \times 10^4$) and 295 K ($\lambda_{max} = 575$ nm, $\varepsilon\,1.6 \times 10^3$) are assignable to the π–π^* transition of digermene **152** and the n–p transition of germylene **151**, respectively. The isosbestic points observed at 335, 390, and 509 nm indicate the quantitative interconversion between **152** and **151**.

The thermodynamic parameters ($\Delta H = 14.7 \pm 0.2$ kcal/mol and $\Delta S = 42.4 \pm 0.8$ cal/mol/K) for the dissociation of digermene **152** to germylene **151** were obtained from the temperature dependence of the absorptions. The bond dissocia-tion energy (BDE) (14.7 kcal/mol) of **152** into **151** is much smaller than the calcu-lated value (30–45 kcal/mol[145]) for the parent system ($H_2Ge=GeH_2$), indicating that the germanium–germanium double bond in **152** is considerably weakened

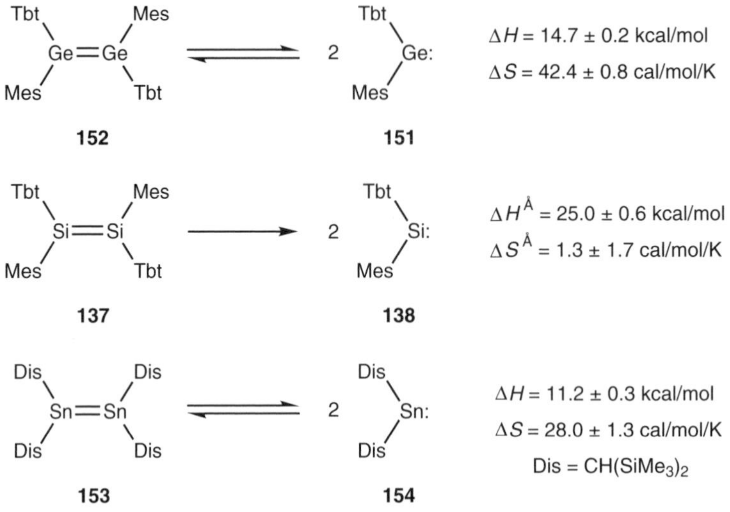

Tbt–Ge=Ge(Mes)(Tbt)(Mes) **152** ⇌ 2 Ge:(Tbt)(Mes) **151**

$\Delta H = 14.7 \pm 0.2$ kcal/mol
$\Delta S = 42.4 \pm 0.8$ cal/mol/K

Tbt–Si=Si(Mes)(Tbt)(Mes) **137** → 2 Si:(Tbt)(Mes) **138**

$\Delta H^{\text{Å}} = 25.0 \pm 0.6$ kcal/mol
$\Delta S^{\text{Å}} = 1.3 \pm 1.7$ cal/mol/K

Dis–Sn=Sn(Dis)(Dis)(Dis) **153** ⇌ 2 Sn:(Dis)(Dis) **154**

$\Delta H = 11.2 \pm 0.3$ kcal/mol
$\Delta S = 28.0 \pm 1.3$ cal/mol/K

Dis = CH(SiMe$_3$)$_2$

Scheme 14.67

because of the severe steric repulsion between the bulky substituents. Indeed, X-ray crystallographic analysis of the orange crystal of *(E)*-**152** revealed that it has a considerably elongated Ge=Ge double bond of 2.416(4) Å in the solid state,[169] which is in good agreement with the ready dissociation into germylene **151** in solution.

Meanwhile, when the orange solid of *(E)*-disilene (**137**) bearing the same substituents as **152** was dissolved in pentane, the generation of the corresponding silylene **138** was not observed as judged by UV–vis spectra.[147] The fact that the disilene **137** exists essentially as a dimer of **138** and dissociates into **138** in solution only to a minor extent at ambient temperature is also noteworthy, although a similar phenomenon is known for the bis(trimethylsilyl)methyl-substituted disilene–digermene pairs: [(Me$_3$Si)$_2$CH]$_2$M=M[CH(SiMe$_3$)$_2$]$_2$ (M=Si,[170] Ge[157]). On the other hand, *(E)*-disilene (**137**) dissociates into the corresponding silylene **138** upon heating with activation parameters ($\Delta H = 25.0 \pm 0.6$ kcal/mol and $\Delta S = 1.3 \pm 1.7$ cal/mol/K) the ΔH of the equilibrium between **137** and **138** was not determined because of the low concentration of **138** (Scheme 14.67). The most reasonable explanation for the difference between disilene **137** and digermene **152** is that the bond dissociation energy of **137** into **138** is larger than that of **152** into **151**. In the case of tin analogues, the equilibrium between distannene (**153**) and the corresponding stannylene (**154**) was reported and the thermodynamic parameters ($\Delta H = 11.2 \pm 0.3$ kcal/mol and $\Delta S = 28.0 \pm 1.3$ cal/mol/K) for the equilibrium were obtained from the temperature dependence of the NMR chemical shifts (Scheme 14.67).[142d] The BDE of distannene (**153**) is smaller than those of digermene (**152**) and disilene (**137**), and the BDE of the double-bond species of heavier group 14 (IVA) elements were found to decrease on going from Si to Sn as the theoretical calculations predicted.

6.3. Reactions of Kinetically Stabilized Germylenes

Although very little is known for the reactions of isolated germylenes **146**, **147**, and **148**, diarylgermylene (**150**) was found to undergo a variety of reactions with olefins, dienes, acetylenes, alcohols, isothiocyanates, elemental chalcogens, and hydrosilanes as in the cases of less hindered, transient germylenes previously reported.

Furthermore, germylene (**150**) showed a unique reactivity toward some heteroatom-containing compounds such as carbon disulfide,[163] nitrile oxide,[171] and gemdihalogenated olefins.[172] Treatment of **150** with carbon disulfide resulted in a formation of a novel heteroatom-substituted germene **155** (Scheme 14.68).[163] The germaketenedithioacetal structure of the core part of **155** was crystallographically established and the oxidation of **155** with molecular oxygen was found to lead to the diarylgermanone **156** (Scheme 14.68).[173]

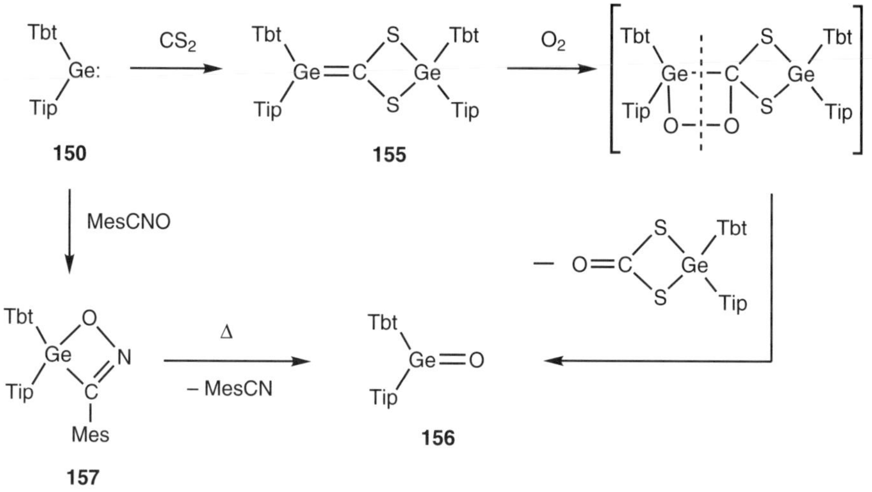

Scheme 14.68

Germylene (**150**) was allowed to react with a bulky nitrile oxide such as mesitonitrile oxide to give the corresponding [1 + 3] cycloaddition product **157**, which is the first stable example of an oxazagermete derivative.[171] Alternative formation of germanone (**156**) was achieved by thermolysis of this strained germaheterocycle **157** (Scheme 14.68).[171] The first stable alkylidenetelluragermirane **158** was synthesized by the reaction of **151** with 9-(dichloromethylene)fluorene, followed by addition of triphenylphosphine telluride (Scheme 14.69).[172] The formation of **158** is most likely interpreted in terms of the initial insertion of germylene **151** into the C—Cl bond of 9-(dichloromethylene)fluorene followed by dechlorination with more germylene **151** leading to 1-germaallene and subsequent telluration. The selenium and sulfur analogues of **158**, that is, selena- and thiagermiranes **159** and **160**,

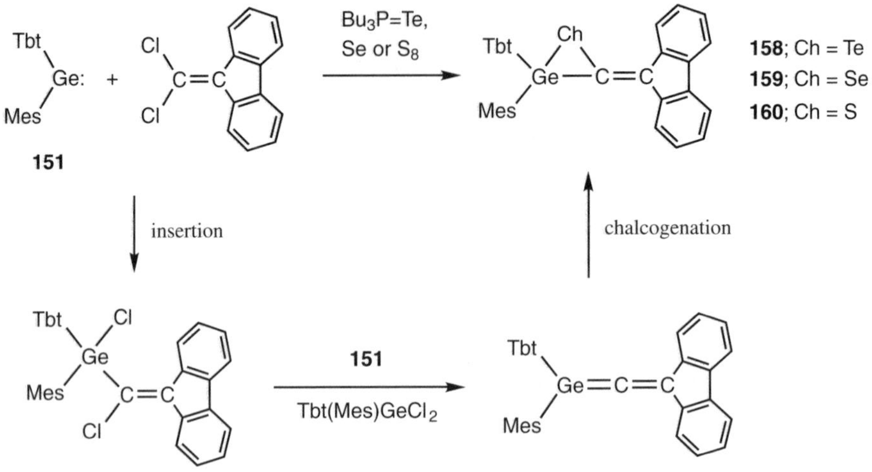

Scheme 14.69

were synthesized by a similar synthetic approach (Scheme 14.69).[172] Complexation of germylene (**150**) with some transition metal carbonyl complexes was also examined to give the first base-free examples of a germylene-group 6 (VIB) metal mononuclear complex, **161** and **162** (Scheme 14.70).[164b]

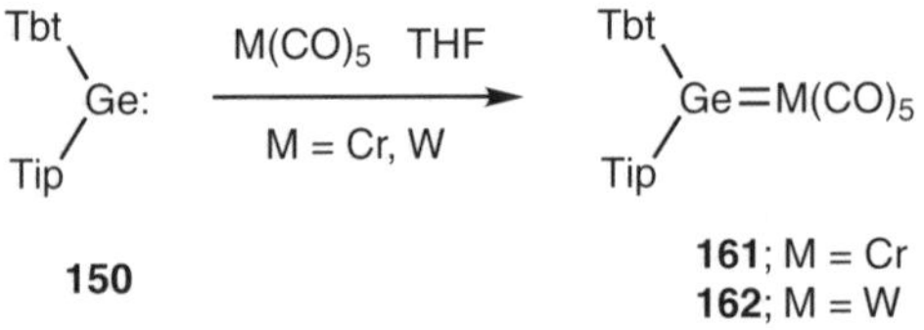

Scheme 14.70

Thus, Tbt-substituted germylenes **150** and **151** have enough space for functionalization, thus enabling them to be versatile synthetic building blocks for a variety of new organogermanium compounds.

7. SYNTHESIS, STRUCTURES, AND REACTIONS OF STABILIZED STANNYLENES

The structures of some stable stannylenes, such as several amino-,[174] alkoxy-,[175] and arylthio-substituted[176] intermediates, have been revealed by X-ray crystallography. They are monomeric crystals and the tin atom has the coordination number 2. The divalent tin in such compounds is stabilized by the effects of electronegativity of the ligand atoms and by the donation of the lone-pair electrons to the vacant $5p$ orbital of the tin. Although the first monomeric dialkyl- and diaryl- stannylenes in the solid state, bis[2-pyridyl-2,2-bis(trimethylsilyl)methyl]stannylene (**163**)[177] and bis[2,4,6-tris(trifluoromethyl)phenyl]stannylene (**164**)[178] were synthesized and characterized by X-ray crystallography. The stabilization of these stannylenes is the result of intramolecular contacts between the tin and neighboring nitrogen or fluorine atoms (Scheme 14.71). The chemical shift in ^{119}Sn NMR spectrum for

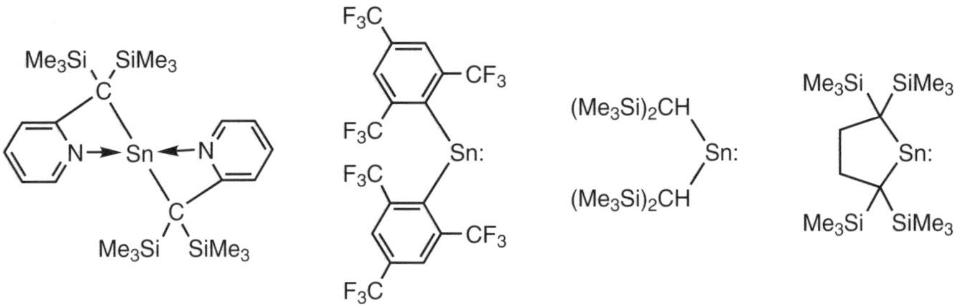

Scheme 14.71

163 was strongly temperature dependent, varying in the range of 120–150 ppm. The ^{119}Sn NMR resonance of **164** is split into 13 lines by coupling with the fluorine atoms of the trifluoromethyl groups in the ortho positions [J([119]Sn—F) = 240 Hz], clearly indicating the existence of the fluorine–tin contacts in solution as well.

In 1976, Lappert and co-workers[142c,157,179] reported the first stable dialkylstannylene (**154**, Scheme 14.71) in solution. They found that **154** exists as a monomer in the gas phase and as a dimer **153** in the solid state, whereas it exists as a monomer–dimer equilibrium mixture in solution. Extensive studies on the reactions of **154** were carried out, especially on oxidative addition and insertion reactions leading to a variety of new organotin compounds such as **165** and **166** (Scheme 14.72).[142c,157,179]

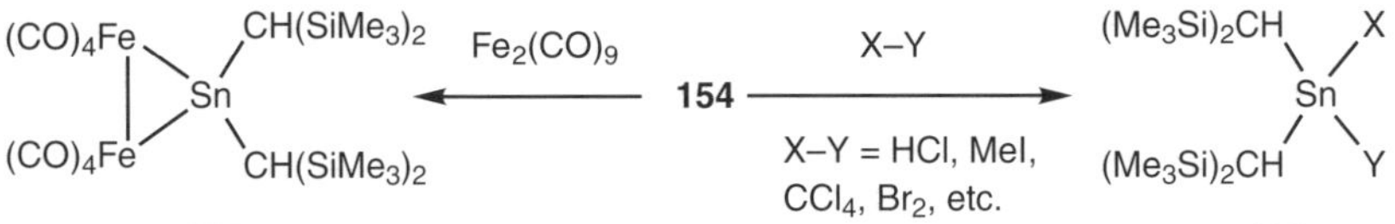

Scheme 14.72

On the other hand, the first stable dialkylstannylene (**136**) that is monomeric in the solid state was synthesized and crystallographically characterized by Kira et al. in 1991[140] (Scheme 14.71). Their helmet-like bidentate ligand was useful for the stabilization of the central tin atom as in the cases of its silicon and germanium analogues (see above), and the protecting ability is much higher than the two bis(trimethylsilyl)methyl groups in **154**. The signal for the central tin atom in the ^{119}Sn NMR spectrum of **136** appears at 2323 ppm as a sharp singlet that is the lowest[119] Sn chemical shift ever reported. In contrast to **154**, the bandwidth and the chemical shift were almost unchanged between -40 and $80\,°C$, clearly indicating no equilibration between monomer and dimer.

In 1994, Weidenbruch et al.[180] reported that introduction of two 2,4,6-tri-*tert*-butylphenyl groups onto the tin atom led to the successful isolation of **167**, which is the first example of kinetically stabilized monomeric diarylstannylene characterized in the solid state. Note that ^{119}Sn NMR spectrum of **167** shows two signals at 961 and 1105 ppm instead of a singlet expected at room temperature in solution. The striking feature of **167** is its partial isomerization into a sterically less encumbered stannylene **168** in solution (Scheme 14.73). The two signals of **167** coalesce

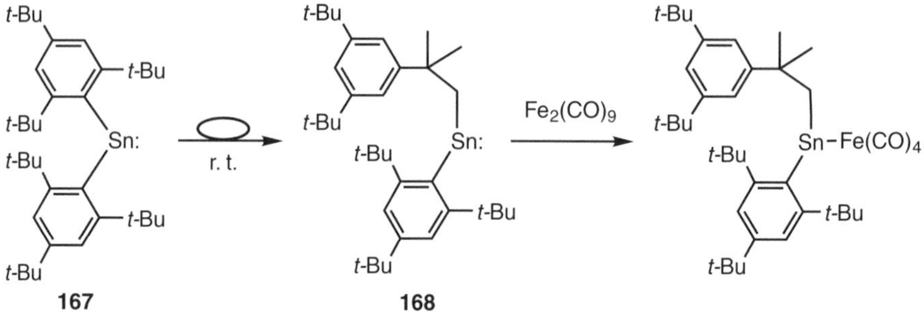

Scheme 14.73

upon heating to $50\,°C$ into a singlet at 980 ppm belonging to **168**, which persists even after cooling. The reason why **167** and **168** show such drastically shielded ^{119}Sn NMR signals, which are similar to those for heteroatom-substituted stannylenes, in view of the fact that the signal of **136** appears at 2323 ppm is still obscure. The newly obtained alkylaryl-substituted stannylene **168** formed a stannylene–iron complex with the stannylene unit occupying an axial position (Scheme 14.73).

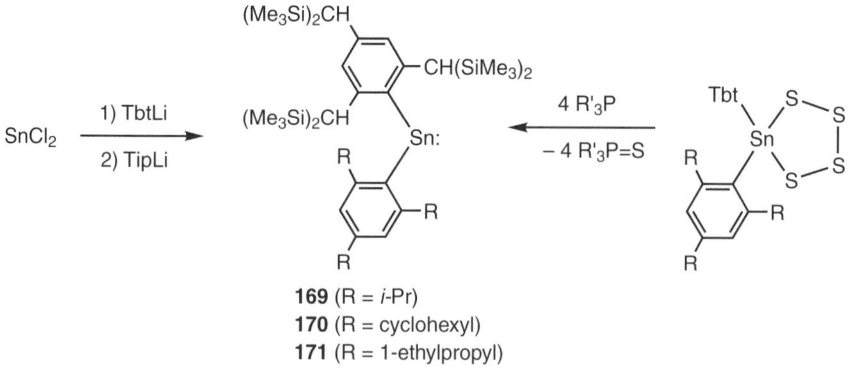

169 (R = *i*-Pr)
170 (R = cyclohexyl)
171 (R = 1-ethylpropyl)

Scheme 14.74

Soon after the isolation of **136**, Tokitoh et al.[181] described the synthesis of the first kinetically stabilized diarylstannylene stable in solution, that is, Tbt(Tip)Sn: (**169**), by treatment of TbtLi with stannous chloride followed by addition of TipLi (Scheme 14.74). Under an inert atmosphere, stannylene **169** was found to be quite stable even at 60 °C in solution, and it showed a deep purple color ($\lambda_{max} = 561$ nm) in hexane. The ^{119}Sn NMR spectrum of **169** showed only one signal at 2208 ppm, the chemical shift of which is characteristic of a divalent organotin compound as in the case of a monomeric dialkylstannylene (**136**). The bandwidth and the chemical shift of **169** were almost unchanged between -30 and 60 °C, indicating the absence of a monomer–dimer equilibrium.

Tokitoh and co-workers[182] further succeeded in the synthesis of overcrowded diarylstannylenes, Tbt(Tcp)Sn: (**170**; Tcp = 2,4,6-tricyclohexylphenyl) and Tbt (Tpp)Sn: [**171**; Tpp = 2,4,6-tris(1-ethylpropyl)phenyl], by the exhaustive desulfurization of the corresponding tetrathiastannolanes with a trivalent phosphine reagent (Scheme 14.74). Since only a few convenient precursors have been available for the generation of stannylenes, this new method should provide us with a useful synthetic route for a variety of overcrowded stannylenes. The successful synthesis of a series of Tbt-substituted diarylstannylenes enabled the systematic comparison of their electronic absorptions with those of the previously reported overcrowded diarylstannylenes, which led to the elucidation of the substituent effect on the $n \rightarrow p$ transition of stannylenes.

Power and co-workers[161] reported the synthesis and characterization of a new type of sterically crowded diarylstannylene (**172**), that bears two bulky m-terphenyl type ligands (2,6-Mes$_2$C$_6$H$_3$) and exists as a monomer even in the solid state (Scheme 14.75).

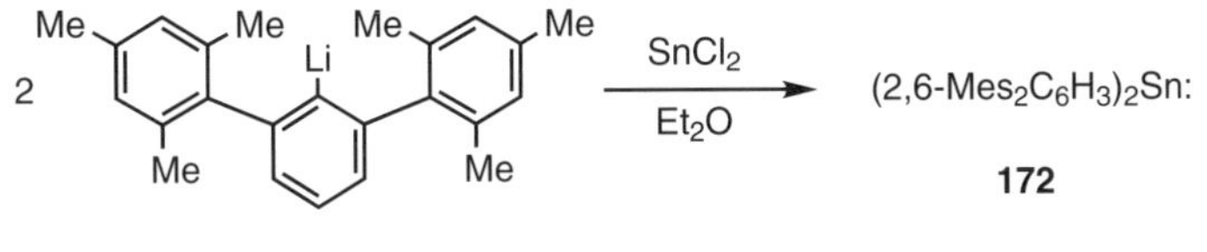

Scheme 14.75

8. SYNTHESIS AND REACTIONS OF STABILIZED PLUMBYLENES

In contrast to the remarkable progress in the chemistry of divalent organic compounds of silicon, germanium, and tin, the heaviest congener of this series, that is, divalent organolead compounds (plumbylenes), are less well known. They usually occur as reactive intermediates in the preparation of plumbanes (R$_4$Pb) and undergo polymerization and/or disproportionation in the absence of suitable stabilizing groups on the lead atom.[183]

Although dicyclopentadienyllead(II) compounds, formally called plumbylenes, have been known since 1956,[184] they are not the congeners of carbenes since they are stabilized by η^5-coordination of cyclopentadienyl ligands. In 1974, the first stable diaminoplumbylene [(Me$_3$Si)$_2$N]$_2$Pb: (**173**) was synthesized by Lappert and co-worker,[185] and since then, other stable plumbylenes with heteroatom substituents have also been reported. Recently, Tokitoh and co-workers[186] reported the synthesis and characterization of a stable aryl(arylthio)plumbylene (**174**), which is one of the rare examples of heteroleptic plumbylenes (Scheme 14.76).

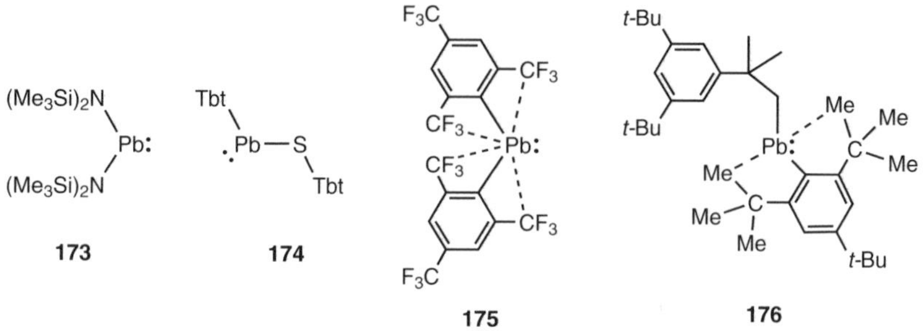

Scheme 14.76

A few plumbylenes bearing only carbon substituents have also been reported. Some of them, however, are stabilized by intramolecular coordination of the lone pair of a donor group in the organic substituent, thus giving the lead a coordination number >2. For example, an X-ray diffraction study of the first stable diarylplumbylene $R_2^f Pb$ (**175**; R^f = 2,4,6-tris(trifluoromethyl)phenyl) showed that four CF_3 fluorine atoms were coordinated to the lead atom (Scheme 14.76).[187] Four-coordinate divalent organolead compounds were also reported recently,[188] in which intramolecular coordination of nitrogen atoms is responsible for their stability. Similar intramolecular interactions were also observed in the first stable alkylarylplumbylene (**176**),[189] the X-ray structural analysis of which showed the short contacts between the lead atom and the carbon atoms of the methyl groups (in *ortho-tert*-butyl groups) indicating weak agostic interactions (Scheme 14.76).

On the other hand, kinetically stabilized plumbylenes bearing only organic substituents that do not contain donor groups are scarce, and their structures and reactivities are almost unexplored. The first dialkylplumbylene, Dis_2Pb: [**177**; Dis = bis(trimethylsilyl)methyl] was obtained in only 3% yield in 1973,[142c,157,190] and its structure was crystallographically determined recently.[191] Another dialkylplumbylene (**178**) with a lead atom in a seven-membered ring system was synthesized and characterized by X-ray diffraction (Scheme 14.77).[192] Two stable diarylplumbylenes, [2,6-Mes$_2$C$_6$H$_3$]$_2$Pb: (**179**)[173] and R_2Pb (**180**; R = 2-*tert*-butyl-4,5,6-Me$_3$C$_6$H), [189] were also structurally characterized (Scheme 14.77).

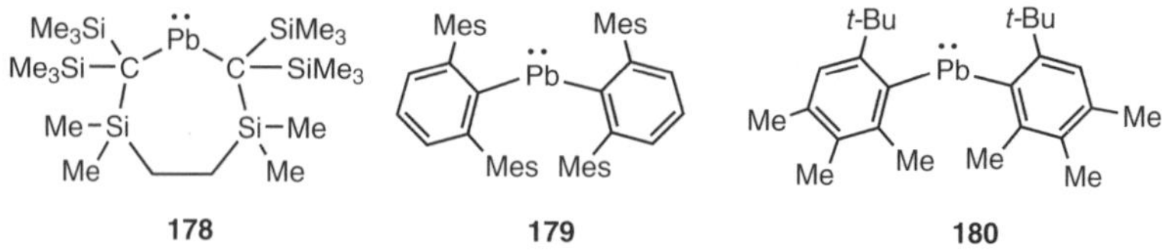

Scheme 14.77

Although all of the stable plumbylenes reported to date have been synthesized by nucleophilic substitution of a lead(II) compound, Tokitoh and co-workers[193] recently described two kinds of new synthetic routes to a plumbylene from tetravalent organolead compounds. The first one is the reductive debromination of the overcrowded dibromoplumbanes, Tbt(R)PbBr$_2$ {**181a** (R = 2,4,6-tris[(trimethyl-silyl)methyl]phenyl (Ttm)}, **181b** (R = Tip), and **181c** (R = Dis) with lithium naphthalenide leading to the formation of a series of hindered plumbylenes, Tbt(R)Pb: (**182a–c**) (Scheme 14.78). The other one is the exhaustive desulfuriza-

Scheme 14.78

tion of the corresponding overcrowded tetrathiaplumbolanes with a trivalent phosphorus reagent as in the cases of their tin analogues (see above). Although the more hindered diarylplumbylene, Tbt$_2$Pb: (**182d**) was not obtained by these synthetic methods, the conventional nucleophilic substitution reaction of [(Me$_3$Si)$_2$N]$_2$Pb: (**173**) with two molar amounts of TbtLi in ether resulted in the isolation of **182d** as stable blue crystals (Scheme 14.78).[194a] The molecular structure of **182d** was definitively determined by X-ray crystallographic analysis, which showed an extremely large C–Pb–C angle of 116.3(7)°. This value not only deviates greatly from the calculated one (97.3°) for Me$_2$Pb:,[195] but also from the averaged values of those for previously reported diaryl- and dialkylplumbylenes [88.2(2)–117.1(2)°; average = 103°], indicating the extremely large steric repulsion between the two Tbt groups. Plumbylenes **178** and **179** also showed a similar large widening of their C–Pb–C angles [117.1(2)° for **178** [62] and 114.5(6)° for **179**,[161] respectively]. The widening might be mainly owing to its seven-membered ring structure in the former case while it is again due to the huge steric hindrance between the bulky terphenyl ligands in the latter case.

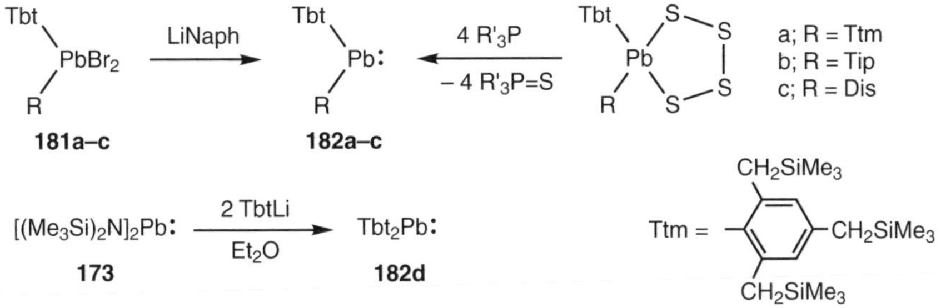

All the isolated plumbylenes mentioned above exists as V-shaped monomers in the solid state, Klinkhammer et al.[196] found a quite interesting feature for their new heteroleptic plumbylene R^f[(Me$_3$Si)$_3$Si]Pb: (**183**). Although plumbylene (**183**) was synthesized by a ligand disproportionation between the corresponding stannylene R$^f_2$Sn: (**164**) and plumbylene [(Me$_3$Si)$_3$Si]$_2$Pb:, the crystallographic analysis of the isolated lead product revealed that **183** has a dimeric form **183′** in the solid state with a considerably shorter Pb···Pb separation [3.537(1) Å] and a trans-bent angle of 40.8° (Scheme 14.79). Weidenbruch and co-workers[189] also reported the synthesis of

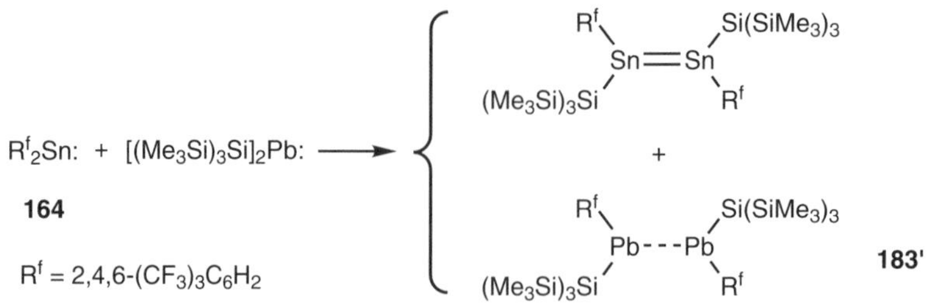

$R^f{}_2Sn$: + [(Me$_3$Si)$_3$Si]$_2$Pb: ⟶

164

$R^f = 2,4,6$-(CF$_3$)$_3$C$_6$H$_2$

Scheme 14.79

another heteroleptic plumbylene R[(Me$_3$Si)$_3$Si]Pb: (**184**; R = 2-*tert*-butyl-4,5,6-trimethylphenyl), the structural analysis of which showed a dimeric structure **184′** in the solid state with a similar short Pb···Pb separation [3.370(1) Å] and a trans-bent angle of 46.5° (Scheme 14.80).

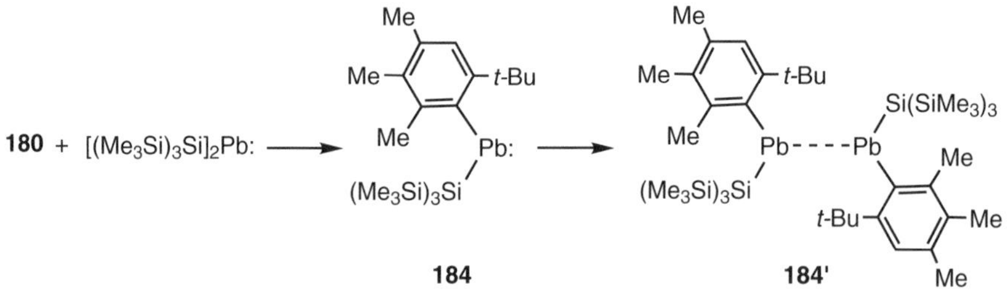

180 + [(Me$_3$Si)$_3$Si]$_2$Pb: ⟶

184   **184′**

Scheme 14.80

Although these two plumbylenes **183** and **184** were found to have close intramolecular contacts and exist as a dimer in the solid state, the lead–lead distances in their dimeric form are still much longer than the theoretically predicted values (2.95–3.00 Å) for the parent diplumbene, H$_2$Pb=PbH$_2$.[196,197] Quite recently, however, Weidenbruch and co-workers[18] succeeded in the synthesis and isolation of the dimer of a less hindered diarylplumbylene Tip$_2$Pb: (**185**), that is, Tip$_2$Pb=Pb Tip$_2$ (**186**, Scheme 14.81). Compound **186** showed a shorter Pb–Pb length [3.0515(3) Å] and much larger trans-bent angles (43.9° and 51.2°) than those observed for **183′** and **184′**. These data strongly indicate that **186** is the first molecule with a lead–lead double bond in the solid state, although **186** was found to dissociate into the monomeric plumbylene (**185**) in solution. Furthermore, they examined the synthesis of a heteroleptic plumbylene, Tip[(Me$_3$Si)$_3$Si]Pb: (**187**), by the treatment of diarylplumbylene Tip$_2$Pb: (**185**) and disilylplumbylene [(Me$_3$Si)$_3$Si]$_2$Pb:, and found that the product exists as the diplumbene Tip[[(Me$_3$Si)$_3$Si]Pb=Pb[Si(SiMe$_3$)$_3$]Tip (**188**) in the solid state (Scheme 14.81).[199] The X-ray structure analysis of **188** reveals the centrosymmetrical diplumbene structure with a trans-bent angle of 42.7° and a Pb–Pb bond length of 2.9899(5) Å. This bond is even shorter than that of **186** and very close to the theoretically predicted value for the parent diplumbene.

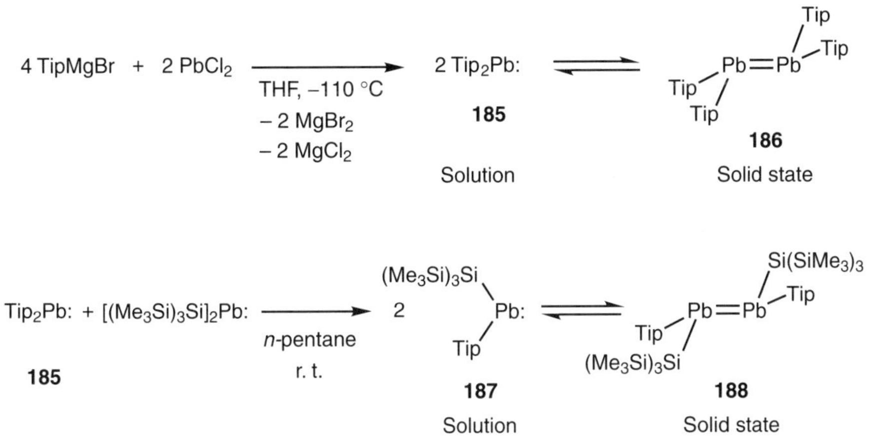

4 TipMgBr + 2 PbCl$_2$ $\xrightarrow[\substack{- 2\ MgBr_2 \\ - 2\ MgCl_2}]{THF,\ -110\ °C}$ 2 Tip$_2$Pb: $\rightleftharpoons$

185

Solution

Tip$_2$Pb=PbTip (structure **186**)

186

Solid state

Tip$_2$Pb: + [(Me$_3$Si)$_3$Si]$_2$Pb: $\xrightarrow[\substack{r.\ t.}]{n\text{-pentane}}$ 2 (Me$_3$Si)$_3$Si(Tip)Pb: $\rightleftharpoons$

185

187

Solution

(structure **188**)

188

Solid state

Scheme 14.81

In order to elucidate the relationship between the structure of plumbylene dimers and the bulkiness of substituents, Weidenbruch and co-workers[199] synthesized and characterized much less hindered diarylplumbylene Mes$_2$Pb: (**189**), which was isolated as a plumbylene dimer (**190**) stabilized with coexisting magnesium salt [MgBr$_2$(THF)$_4$] (Scheme 14.82). The large Pb$\cdots$Pb separation [3.3549(6) Å] and trans-bent angle (71.2°) of **190** suggest that the character of the lead–lead bonding interaction in plumbylene dimers is delicate and changeable.

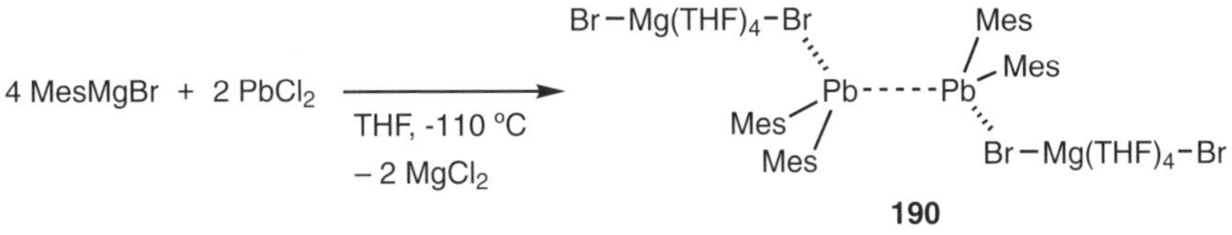

4 MesMgBr + 2 PbCl$_2$ $\xrightarrow[\substack{- 2\ MgCl_2}]{THF,\ -110\ °C}$

Br–Mg(THF)$_4$–Br ⋯ Pb–Pb ⋯ Br–Mg(THF)$_4$–Br, Mes groups

190

Scheme 14.82

The progress in the chemistry of plumbylenes is far behind those of its lighter congeners (silylenes, germylenes, and stannylenes). Most of the recent reports on plumbylenes are describe their synthesis and structural analysis; there have been very few descriptions of their reactivities.

Recently, however, Tokitoh and co-workers[186,193,194,200] reported a versatile reactivity of Tbt-substituted plumbylenes **182a–d**, that is, insertion reactions with alkyl halides, diphenyl dichalcogenides, bromine abstraction from carbon tetrabromide, and sulfurization with elemental sulfur as shown in Scheme 14.83.

These results indicate that the reactivity of a plumbylene, the heaviest congener of a carbene, is essentially the same as those observed for its lighter relatives. The successful isolation of the reaction products in the case of plumbylenes **182** might be the result of effective steric protection not only for the starting plumbylenes but

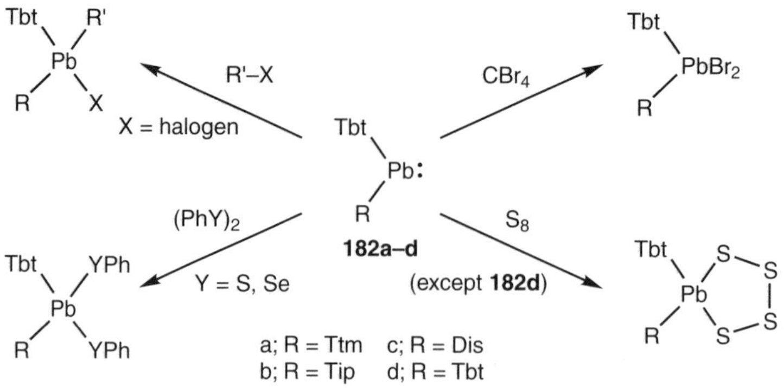

Scheme 14.83

also for the reaction products, which inevitably contain weak bonds between the lead and other atoms. Among the results obtained by Tokitoh and co-workers,[194b] the sulfurization of **182d** should be especially noted. Treatment of the overcrowded plumbylene **182d** with one atom equivalent of elemental sulfur at low temperature (−78 °C) resulted in the formation of plumbanethione Tbt$_2$Pb=S (**191**), the expected sulfurized product, while similar sulfurization of **182d** at 50 °C gave a new heteroleptic plumbylene Tbt(TbtS)Pb: (**174**), that is, the 1,2-aryl migration product of **191** (Scheme 14.84). These results demonstrate that R(RS)Pb: is more

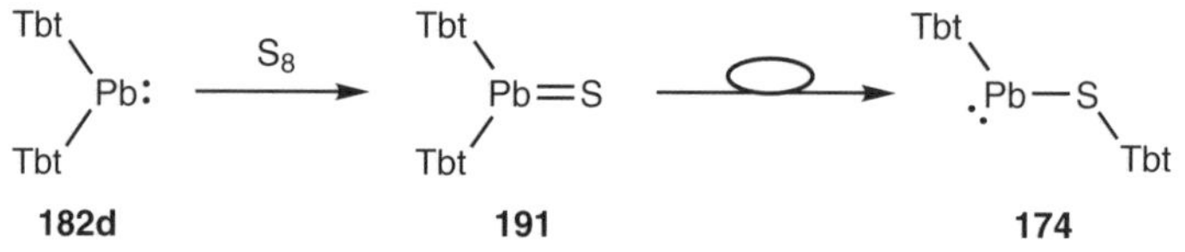

Scheme 14.84

stable than R$_2$Pb=S, and this fact is in sharp contrast to the relative stabilities of their lighter congeners. The kinetically stabilized metallanethiones Tbt(R)M=S (M = Si,[201] Ge,[164,165,166] and Sn[181,202]) are isolated as stable crystalline compounds and no isomerization to heteroleptic divalent species via 1,2-migration of the ligand is observed on heating. This unique relative stabilities found for plumbanethione and its plumbylene isomer was corroborated by ab initio calculations on a series of double-bond compounds, [H$_2$Pb=X] and their isomers [*trans*-H—Pb—S—H] and [*cis*-H—Pb—S—H] (X = O, S, Se, and Te).[203,204]

Note also the reactivity of the heteroleptic plumbylene **174** toward carbon disulfide. Treatment of **174** with excess amount of carbon disulfide resulted in a formation of unexpected product, lead(II) bis(aryl trithiocarbonate) (**192**), as a yellow, air- and moisture-stable solid (Scheme 14.85).[205] The reaction must involve not only the insertion of carbon disulfide into the Pb—S bond, but also the formal insertion of a sulfur atom into the Pb—C(Tbt) bond and subsequent insertion of another

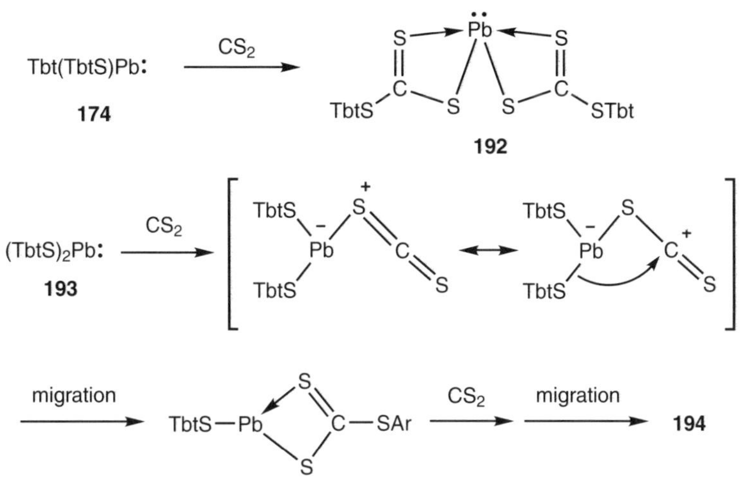

Scheme 14.85

carbon disulfide. Compound **192** was also obtained by the reaction of bis(arylthio) plumbylene (**193**) with carbon disulfide, indicating the double insertion of carbon disulfide into the two Pb—S bonds (Scheme 14.85). Although the migratory insertion of carbon disulfide into a metal–sulfur bond is known for some transition metal complexes, such a double insertion of carbon disulfide as suggested above has not been observed in the case of other divalent species of heavier group 14 (IVA) elements, that is, silylenes,[149] germylenes,[163] and stannylenes.[206]

9. CONCLUSION AND OUTLOOK

As can be seen in this chapter, the chemistry of divalent species of heavier group 14 (IVA) elements is still rapidly growing. The remarkable development of ligands for thermodynamic and kinetic stabilization has provided a number of stable examples of heavier homologues of carbenes, whose remarkable chemical and physical properties will stimulate the further progress in this field. The progress in the chemistry of stannylenes and plumbylenes is much slower than that in the areas of silylenes and germylenes. This relatively slow pace is probably the result of the weakness of bonds involving such heavy elements. A comparative study on the whole chemistry of divalent species of group 14 (IVA) elements including carbenes should be important to produce new concepts and applications for these low-coordinate species.

SUGGESTED READING

P. P. Gaspar, M. Xiao, D. H. Pae, D. J. Berger, T. Haile, T. Chen, D. Lei, W. R. Winchester, and P. Jiang, "The Quest for Triplet Ground State Silylenes," *J. Organometal. Chem.* **2002**, *646*, 68.

N. Tokitoh and R. Okazaki, "Recent Topics in the Chemistry of Heavier Congeners of Carbenes," *Coord. Chem. Rev.* **2000**, *210*, 251.

M. Haaf, T. A. Schmedake, and R. West, "Stable Silylenes," *Acc. Chem. Res.* **2000**, *33*, 704.

M. Weidenbruch, "Some Silicon, Germanium, Tin and Lead Analogues of Carbenes, Alkenes, and Dienes," *Eur. J. Inorg. Chem.* **1999**, 373.

P. P. Gaspar and R. West, "Silylenes," in *The Chemistry of Organosilicon Compounds*, Vol. 2, Z. Rappoport and Y. Apeloig, Eds., John Wiley & Sons, Inc., Chichester, UK, **1998**, Part 3, p. 2463.

W. Ando and Y. Kabe, "Highly Reactive Small-ring Monosilacycles and Medium-ring Oligosilacycles," in *The Chemistry of Organosilicon Compounds*, Vol. 2, Z. Rappoport and Y. Apeloig, Eds., John Wiley & Sons, Inc., Chichester, UK, **1998**, Part 3, p. 2401.

W. P. Neumann, "Germylenes and Stannylenes," *Chem. Rev.* **1991**, *91*, 311.

C.-H. Liu and T.-L. Hwang, "Inorganic Silylenes. Chemistry of Silylene, Dichlorosilylene and Difluorosilylene," in *Advances in Inorganic Chemistry and Radiochemistry*, Vol. 28, H. J. Emeleus and A. G. Sharpe, Eds., Academic Press, Orlando, FL, **1985**, p. 1.

REFERENCES

1. (a) P. P. Gaspar and B. J. Herold, in *Carbene Chemistry*, W. Kirmse, Ed., Academic Press, New York, **1971**. (b) P. P. Gaspar and G. S. Hammond, in *Carbenes*, R. A. Moss and M. Jones, Jr., Eds., Wiley-Interscience, New York, **1975**, 207. (c) P. P. Gaspar, in *Reactive Intermediates*, Vol. 1, M. Jones, Jr. and R. A. Moss, Eds., John Wiley & Sons, Inc., New York, **1978**, p. 229. (d) P. P. Gaspar, in *Reactive Intermediates*, Vol. 2, M. Jones, Jr. and R. A. Moss, Eds., John Wiley & Sons, Inc., New York, **1981**, p. 335.

2. P. P. Gaspar and R. West, in *The Chemistry of Organosilicon Compound*, Vol. 2, Z. Rappoport and Y. Apeloig, Eds., John Wiley & Sons, Inc., UK. Chichester, **1998**, Part 3, p. 2463.

3. (a) P. Rivière, M. Rivière-Baudet, J. Satgé, in *Comprehensive Organometallic Chemistry*, G. Wilkinson, F. G. Stone, and E. W. Abel, Eds., Pergamon, Oxford, **1982**, Chapter 10. (b) W. P. Neumann, *Chem. Rev.* **1991**, *91*, 311.

4. (a) C.-H. Liu and T.-L. Hwang, in *Advances in Inorganic Chemistry and Radiochemistry*, Vol. 28, H. J. Emeleus and A. G. Sharpe, Eds., Academic Press, Orlando, FL, **1985**, p. 1. (b) J. Satgé, *Pure Appl. Chem.* **1984**, *56*, 137.

5. For reviews on silylenes, see: (a) R. West, *Pure Appl. Chem.* **1984**, *56*, 1634. (b) G. Raabe and J. Michl, *Chem. Rev.* **1985**, *85*, 419. (c) R. West, *Angew. Chem., Int. Ed. Engl.* **1987**, *26*, 1201. (d) J. Barrau, J. Escudié, and J. Satgé, *Chem. Rev.*, **1990**, *90*, 283. (e) P. Jutzi, *J. Organometal. Chem.* **1990**, *400*, 1. (f) M. F. Lappert and R. S. Rowe, *Coord. Chem. Rev.* **1990**, *100*, 267. (g) J. Satgé, *J. Organometal. Chem.* **1990**, *400*, 121. (h) T. Tsumuraya, S. A. Batcheller, and S. Masamune, *Angew. Chem., Int. Ed. Engl.* **1991**, *30*, 902. (i) M. Weidenbruch, *Coord. Chem. Rev.* **1994**, *130*, 275. (j) A. G. Brook and M. Brook, *Adv. Organometal. Chem.*, **1996**, *39*, 71. (k) I. Hemme and U. Klingebiel, *Adv. Organometal. Chem.* **1996**, *39*, 159. (l) M. Driess, *Adv. Organometal. Chem.*, **1996**, *39*, 193. (m) R. Okazaki and R. West, *Adv. Organometal. Chem.* **1996**, *39*, 232. (n) K. M. Baines and W. G. Stibbs, *Adv. Organometal. Chem.* **1996**, *39*, 275. (o) M. Driess and H. Grützmacher, *Angew. Chem., Int. Ed. Engl.* **1996**, *35*, 827. (p) P. P. Power, *J. Chem.*

Soc., Dalton Trans. **1998**, 2939. (q) M. Weidenbruch, *Eur. J. Inorg. Chem.* **1999**, 373. (r) N. Tokitoh and R. Okazaki, *Coord. Chem. Rev.* **2000**, *210*, 251. (s) M. Haaf, T. A. Schmedake, and R. West, *Acc. Chem. Res.* **2000**, *33*, 704. (t) P. P. Gaspar, in *Science of Synthesis*, Vol. 4.4.3, Thieme, Stuttgart, **2002**, p. 135.

6. (a) W. H. Atwell and D. R. Weynberg, *J. Organometal. Chem.* **1966**, *5*, 594. (b) W. H. Atwell and D. R. Weynberg, *Angew. Chem., Int. Ed. Engl.* **1969**, *8*, 469.

7. I. M. T. Davidson, K. J. Huges, and S. Ijadi-Maghsoodi, *Organometallics* **1986**, *6*, 639.

8. J. H. Gilman, S. G. Cottis, and W. H. Atwell, *J. Am. Chem. Soc.* **1964**, *86*, 1596.

9. (a) W. P. Neumann and M. Schriewer, *Tetrahedron Lett.* **1980**, *21*, 3273. (b) M. Schriewer and W. P. Neumann, *Angew. Chem., Int. Ed. Engl.* **1981**, *20*, 1019. (c) M. Schriewer and W. P. Neumann, *J. Am. Chem. Soc.* **1983**, *105*, 897.

10. (a) W. Ando, T. Tsumuraya, and A. Sekiguchi, *Chem. Lett.* **1987**, 317. (b) Y. Kabe and W. Ando, *Adv. Strain Org. Chem.* **1993**, *3*, 59.

11. D. Seyferth, D. C. Annarelli, and S. C. Vick, *J. Am. Chem. Soc.* **1976**, *98*, 6382.

12. D. Seyferth and S. C. Vick, *J. Organometal. Chem.* **1977**, *125*, C11.

13. P. Boudjouk, R. Samaraweera, R. Sooriyakumaran, J. Chrusciel, and K. R. Anderson, *Angew. Chem., Int. Ed. Engl.* **1988**, *27*, 1355.

14. R. L. Lambert and D. Seyferth, *J. Am. Chem. Soc.* **1972**, *94*, 9246.

15. D. Seyferth and D. C. Annarelli, *J. Am. Chem. Soc.* **1975**, 97, 325.

16. (a) E. A. Chernyshev, N. G. Komalenkova, and S. A. Bashikirova, *J. Organometal. Chem.* **1984**, 271, 129. (b) D. Lei and P. P. Gaspar, *Organometallics* **1985**, *4*, 1471. (c) M. G. Steinmetz and C. Yu, *Organometallics* **1992**, *11*, 2686.

17. M. Ishikawa, T. Fuchigami, and M. Kumada, *J. Am. Chem. Soc.* **1977**, *99*, 245.

18. H. Sakurai, Y. Kamiyama, and Y. Nakadaira, *J. Am. Chem. Soc.* **1977**, *99*, 3879.

19. T. J. Drahnak, J. Michl, and R. West, *J. Am. Chem. Soc.* 1979, *101*, 5427.

20. (a) C. W. Carlson and R. West, *Organometallics* **1983**, *2*, 1792. (b) H. Vancik, G. Raabe, M. J. Michalczyk, R. West, and J. Michl, *J. Am. Chem. Soc.* **1985**, *107*, 4097.

21. (a) M. Ishikawa and M. Kumada, *J. Chem. Soc., Chem. Commun.* **1970**, 612. (b) M. Ishikawa and M. Kumada, *J. Chem. Soc., Chem. Commun.* **1971**, 489.

22. (a) P. Trefonas, R. West, and R. D. Miller, *J. Am. Chem. Soc.* **1985**, *107*, 2737. (b) G. Raabe and J. Michl, *Chem. Rev.* **1985**, *85*, 419. (c) A. G. Brook and K. M. Baines, *Adv. Organometal. Chem.* **1986**, *25*, 1.

23. (a) M. Wakasa, I. Yoneda, and K. Mochida, *J. Organometal. Chem.* **1989**, *366*, C1. (b) K. Mochida, I. Yoneda, and M. Wakasa, *J. Organometal. Chem.* **1990**, *399*, 53.

24. S. Konieczny, S. J. Jacobs, J. K. Braddock-Wilking, and P. P. Gaspar, *J. Organometal. Chem.* **1988**, *341*, C17.

25. (a) M. Ishikawa, T. Takaoka, and M. Kumada, *J. Organometal. Chem.* **1972**, *42*, 333. (b) M. Ishikawa and M. Kumada, *Adv. Organometal. Chem.* **1981**, 19, 51. (c) A. G. Brook, in *The Chemistry of Organosilicon Compounds*, S. Patai and Z. Rappoport, Eds., John Wiley & Sons, Inc., Chichester, UK, **1989**, Part 2, Chapter 15.

26. S. Masamune, Y. Hanzawa, S. Murakami, T. Bally, and F. J. Blount, *J. Am. Chem. Soc.* **1982**, *104*, 1150.

27. (a) S. Masamune, Y. Hanzawa, and D. J. Williams, *J. Am. Chem. Soc.*, **1982**, *104*, 6136. (b) J. T. Snow, S. Murakami, S. Masamune, and D. J. Williams, *Tetrahedron Lett.* **1984**, 25, 4191. (c) S. Collins, S. Murakami, J. T. Snow, and S. Masamune, *Tetrahedron Lett.*

1985, *26*, 1281. (d) M. Weidenbruch, F. T. Grim, M. Herrndorf, A. Schäfer, K. Peters, and S. H. Geroge, *J. Organometal. Chem.* **1988**, *341*, 335.

28. (a) W. Ando and T. Tsumuraya, *Tetrahedron Lett.* **1986**, *27*, 3251. (b) W. Ando and T. Tsumuraya, *J. Chem. Soc., Chem. Commun.* **1987**, 1514. (c) T. Tsumuraya, S. Sato, and W. Ando, *Organometallics* **1988**, *7*, 2015. (d) T. Tsumuraya, S. Sato, and W. Ando, *Organometallics* **1989**, *8*, 161. (e) T. Tsumuraya, S. Sato, and W. Ando, *Organometallics* **1990**, *9*, 2061. (f) Y. Kabe and W. Ando, *J. Organometal. Chem.* **1994**, *482*, 131.

29. M. Ishikawa, T. Yamanaka, and M. Kumada, *J. Organometal. Chem.* **1985**, *292*, 167.

30. H. Sakurai, Y. Nakadaira, and H. Sakabe, *Orgnometallics* **1983**, *2*, 1484.

31. (a) J. A. Hawari and D. Griller, *Orgnometallics* **1984**, *3*, 1123. (b) J. A. Hawari, M. Lesage, D. Griller, and W. P. Weber, *Orgnometallics* **1987**, *6*, 880.

32. (a) P. P. Gaspar, A. M. Beatty, T. Chen, T. Haile, D. Lei, W. R. Winchester, J. Braddock-Wilking, N. P. Rath, W. T. Klooster, T. F. Koetzle, S. A. Mason, and A. Albinati, *Organometallics* **1999**, *18*, 3921. (b) J. A. Hawari, D. Griller, W. P. Weber, and P. P. Gaspar, *J. Orgnometal. Chem.* **1987**, *326*, 335.

33. (a) R. West, *J. Organometal. Chem.* **1986**, *300*, 327. (b) R. J. P. Corriu, G. Lanneau, C. Priou, F. Soulairol, N. Auner, R. Probst, R. Conlin, and C. J. Tan, *J. Organometal. Chem.* **1994**, *466*, 55. (c) A. Kawachi and K. Tamao, *Bull. Chem. Soc. Jpn.* **1997**, *70*, 945.

34. J. Grobe and J. Scherholt, *Organosilicon Chemistry II*, N. Auner and J. Weis, Eds., VCH, Weinheim, **1996**, 317.

35. D. H. Pae, M. Xiao, M. Y. Chiang, and P. P. Gaspar, *J. Am. Chem. Soc.*, **1991**, *113*, 1281.

36. (a) P. Jutzi, D. Kanne, and C. Krüger, *Angew. Chem., Int. Ed. Engl.* **1986**, *25*, 164. (b) P. Jutzi, U. Holtmann, D. Kanne, C. Krüger, Gleiter, Blom. Hyla-Krispin, R. I, *Chem. Ber.* **1989**, *122*, 1629.

37. P. Jiang and P. P. Gaspar, *J. Am. Chem. Soc.* **2001**, *123*, 8622.

38. P. P. Gaspar, M. Xiao, D. H. Pae, D. H. Berger, T. Haile, T. Chen, D. Lei, W. R. Winchester, and P. Jiang, *J. Organometal. Chem.* **2002**, *646*, 68.

39. (a) H. Ohgaki, N. Fukaya, and W. Ando, *Organometallics* **1997**, *16*, 4956. (b) M. P. Egorov, O. M. Nefedov, T.-S. Lin, and P. P. Gaspar, *Organometallics* **1995**, *14*, 1539.

40. (a) S. Patai and Z. Rappoport, Eds., *The Chemistry of Organic Silicon Compounds, Part 1 and 2*, John Wiley & Sons, Inc., **1989**. (b) Y. Apeloig and M. Karni, in *The Chemistry of Organosilicon Compounds*, Vol. 2, Z. Rappoport and Y. Apeloig, Eds., John Wiley & Sons, Inc., Chichester, UK, **1998**, p. 1.

41. B. T. Luke, J. A. Pole, M. B. Krogh-Jespersen, Y. Apeloig, J. Chandrasekhar, and P. v. R. Schleyer, *J. Am. Chem. Soc.* **1986**, *108*, 260.

42. B. T. Luke, J. A. Pole, M. B. Krogh-Jespersen, Y. Apeloig, M. Karni, J. Chandrasekhar, and P. v. R. Schleyer, *J. Am. Chem. Soc.* **1986**, *108*, 270.

43. J. Berkowitz, J. P. Greene, H. Cho, and R. Ruscic, *J. Chem. Phys.* **1987**, *86*, 1235.

44. M. S. Gordon, *Chem. Phys. Lett.* **1985**, *114*, 348.

45. H. J. Werner and E. A. Reinsch, *J. Chem. Phys.* **1982**, *76*, 3144.

46. T. J. Sears and P. R. Bunker, *J. Chem. Phys.* **1983**, *79*, 5265.

47. R. Janoschek, *Chem. Unserer Z.* **1988**, *21*, 128.

48. N. Wiberg, *Coord. Chem. Rev.* **1997**, *163*, 217.

49. (a) M. S. Gordon and D. R. Gano, *J. Am. Chem. Soc.* **1985**, *106*, 5421. (b) I. M. T. Davidson and R. J. Scampton, *J. Organometal. Chem.* **1984**, *271*, 249.

50. I. Dubois, *Can. J. Phys.* **1968**, *46*, 2485.

51. H. P. Reisenauer, G. Mihm, and G. Maier, *Angew. Chem., Int. Ed. Engl.* **1982**, *21*, 854.

52. G. Maier, G. Mihm, H. P. Reisenauer, and D. L. Littman, *Chem. Ber.* **1984**, *117*, 2369.

53. G. Maier, S. Glatthaar, and H. P. Reisenauer, *Chem. Ber.* **1989**, *122*, 2403.

54. G. Maier, H. P. Reisenauer, K. Schttler, and U. Wessolek-Kraus, *J. Organometal. Chem.* **1989**, *366*, 25.

55. M. Kira, T. Maruyama, and H. Sakurai, *Chem. Lett.* **1993**, 1345.

56. S. G. Bott, P. Marshall, P. E. Wagenseller, Y. Wang, and R. T. Conlin, *J. Organometal. Chem.* **1995**, *499*, 11.

57. M. Weidenburch, P. A. Will, K. Peters, H. G. Schnering, and H. Marsmann, *J. Organometal. Chem.* **1996**, *521*, 355.

58. M. J. Michalczyk, M. J. Fink, D. J. D. Young, C. W. Carlson, K. M. Welsh, R. West, and J. Michl, *Silicon, Germanium Tin and Lead Compd.* **1986**, *9*, 75.

59. G. R. Gillette, G. H. Noren, and R. West, *Organometallics* **1987**, *6*, 2617.

60. K. M. Welsh, J. Michl, and R. West, *J. Am. Chem. Soc.* **1988**, *110*, 6689.

61. G. R. Gillette, G. H. Noren, and R. West, *Organometallics* **1990**, *9*, 2925.

62. G. Trinquier, *J. Chem. Soc., Faraday Trans.* **1993**, *89*, 775.

63. W. Ando, A. Sekiguchi, K. Hagiwara, A. Sakakibara, and H. Yoshida, *Organometallics* **1988**, *8*, 558.

64. T. Akasaka, W. Ando, S. Nagase, and A. Yabe, *Nippon Kagaku Kaishi* **1989**, 1440.

65. M. A. Pearsall and R. West, *J. Am. Chem. Soc.* **1988**, *110*, 6689.

66. C. A. Arrington, J. A. Petty, S. E. Payne, and W. C. K. Haskins, *J. Am. Chem. Soc.* **1988**, *110*, 6240.

67. H. Bornemann and W. Sander, *J. Organometal. Chem.* 2002, *641*, 156.

68. T. F. Hamilton and H. F. Schaefer, *J. Chem. Phys.* **1989**, *90*, 1031.

69. T. Akasaka, S. Nagase, A. Yabe, and W. Ando, *J. Am. Chem. Soc.* **1988**, *110*, 6270.

70. J. C. Barthelat, B. S. Roch, G. Trinquier, and J. Satgé, *J. Am. Chem. Soc.* **1980**, *102*, 4080.

71. (a) S. Tomoda, M. Shimoda, Y. Takeuchi, Y. Kaji, K. Obi, I. Tanaka, and K. Honda, *J. Chem. Soc., Chem. Commun.* **1988**, 910. (b) K. Mochida, N. Kanno, R. Kato, M. Kotani, S. Yamauchi, M. Wakasa, and H. Hayashi, *J. Organometal. Chem.* **1991**, *415*, 191.

72. W. Ando, H. Itoh, T. Tsumurya, and H. Yoshida, *Organometallics* **1988**, *7*, 1880.

73. (a) W. Ando, T. Tsumuraya, and A. Sekiguchi, *Chem. Lett.* **1987**, 317. (b) W. Ando, H. Ito, T. Tsumuraya, and H. Yoshida, *Organometallics* **1988**, *7*, 1880.

74. W. Ando, H. Itoh, and T. Tsumurya, *Organometallics* **1989**, *8*, 2759.

75. Y. Apeloig, in *The Chemistry of Organic Silicon Compounds*, S. Patai and Z. Rappoport, Eds., John Wiley & Sons, Inc., Chichester, **1989**.

76. H. F. Schaefer, *Acc. Chem. Res.* **1982**, *15*, 283.

77. A. Sekiguchi and W. Ando, *Organometallics* **1987**, *6*, 1857.

78. (a) A. G. Brook, in *The Chemistry of Organic Silicon Compounds*, S. Patai and Z. Rappoport, Eds., John Wiley & Sons, Inc., Chichester, **1989**, p. 965. (b) M. G. Steinmetz, *Chem. Rev.* **1995**, *95*, 1527.

79. (a) M. J. Fink, D. B. Puranik, and M. P. Johnson, *J. Am. Chem. Soc.* **1988**, *110*, 1315. (b) D. B. Puranik and M. J. Fink, *J. Am. Chem. Soc.* **1989**, *111*, 5951.

80. (a) V. N. Khabashesku, B. Balaji, S. A. Boganov, P. M. Matveichev, E. A. Chernyshev, O. M. Nefedov, and J. Michl, *Mendeleev Commun.* **1992**, 38. (b) V. N. Khabashesku, B. Balaji, S. A. Boganov, O. M. Nefedov, and J. Michl, *J. Am. Chem. Soc.* **1994**, *116*, 320.

81. (a) M. E. Lee, M. A. North, and P. P. Gaspar, *Phosphorus, Sulfur, and Silicon* **1991**, *56*, 203. (b) P. P. Gaspar, in *Reactive Intermediates*, Vol. 3, M. Jones, Jr., and R. A. Moss, John Wiley & Sons, Inc., New York, 1985, p. 297.

82. G. Maier and J. Glathaar, in *Organosilicon Chemistry-From Molecules to Materials*, N. Auner and J. Weiss, Eds., VCH, Weinhim, **1994**, 131.

83. (a) A. P. Dickenson, K. E. Nares, M. A. Ring, and H. E. O'Neal, *Organometallics* **1987**, *6*, 2596. (b) G. Maier and J. Glatthaar, *Angew. Chem., Int. Ed. Engl.* **1994**, *33*, 473. (c) G. Maier, H. Pacl, H. P. Reisenauer, A. Meudt, and R. Janoschek, *J. Am. Chem. Soc.* **1995**, *117*, 12712.

84. (a) D. Seyferth and T. F. O. Lim, *J. Am. Chem. Soc.* **1978**, *100*, 7074. (b) M.-D. Su and S.-Y. Chu, *J. Phys. Chem. A.* **1999**, *103*, 11011.

85. J. M. Jasinski, *J. Chem. Phys.* **1986**, *5*, 3057.

86. (a) K. Oka and R. Nakao, *Res. Chem. Intermed.* **1990**, *13*, 143. (b) K. Oka and R. Nakao, *J. Organometal. Chem.* **1990**, *390*, 7.

87. Y. T. Gu and W. P. Weber, *J. Organometal. Chem.* **1980**, *184*, 7.

88. (a) M. Ishikawa, K. Nakagawa, and M. Kumada, *J. Organometal. Chem.* **1981**, *214*, 277. (b) M. Ishikawa, M. Nakagawa, S. Katayama, and M. Kumada, *J. Organometal. Chem.* **1981**, *216*, C48.

89. (a) J. Koecher and M. Lehnig, *Organometallics* **1984**, *3*, 937. (b) J. Koecher, M. Lehnig, and W. P. Neumann, *Organometallics* 1988, *7*, 1201.

90. W. Ando, *Acc. Chem. Res.* **1977**, *10*, 179.

91. H. Ohgaki and W. Ando, *J. Orgamomet. Chem.* **1996**, *521*, 387.

92. T. Y. Gu and W. P. Weber, *J. Am. Chem. Soc.*, **1980**, *102*, 1641.

93. W. F. Goure and T. J. Barton, *J. Organometal. Chem.* **1980**, *199*, 33.

94. W. Ando, M. Ikeno, and Y. Hamada, *J. Chem. Soc., Chem. Commun.* **1981**, *621*.

95. Y. Apeloig and S. Sklenak, *Can. J. Chem.* **2000**, *78*, 1496.

96. W. H. Atwell, L. G. Mahone, S. F. Hayes, and J. G. Uhlmann, *J. Organometal. Chem.* **1969**, *18*, 69.

97. H. S. D. Soya, H. Okinoshima, and W. P. Weber, *J. Organometal. Chem.* **1977**, *133*, C-17.

98. (a) P. S. Skell and E. J. Goldstein, *J. Am. Chem. Soc.* **1964**, *86*, 1442. (b) B. Cox and H. Purnell, *J. Chem. Soc., Faraday Trans. I* 1975, *71*, 859.

99. D. P. Paquin and M. A. Ring, *J. Am. Chem. Soc.* **1977**, *99*, 1793.

100. S. Sakai and M. Nakamura, *J. Phys. Chem.* **1993**, *97*, 4960.

101. H. Sakurai, T. Kobayashi, and Y. Nakadira, *J. Organometal. Chem.* **1978**, *162*, C-43.

102. R. T. Conlin and P. P. Gaspar, *J. Am. Chem. Soc.*, **1976**, *98*, 3715.

103. M. Ishikawa, T. Fuchikami, and M. Kumada, *J. Organometal. Chem.* **1977**, *142*, C-45.

104. A. Krebs and J. Berndt, *Tetrahedron Lett.* **1983**, *24*, 4083.

105. M. P. Egorov, M. B. Ezova, M. S. P. Kolesnikov, O. M. Nefedov, M. B. Taraban, A. I. Kruppa, and T. V. Leshina, *Mendeleev Commun.* **1991**, *4*, 143.

106. F. Hojo, S. Sekigawa, N. Nakayama, T. Shimizu, and W. Ando, *Organometallics* **1993**, *12*, 803.

107. W. Ando, T. Shiba, T. Hidaka, K. Morihashi, and O. Kikuchi, *J. Am. Chem. Soc.* **1997**, *119*, 3629.

108. F. Meiners, W. Saak, and M. Weidenbruch, *Organometallics* **2000**, *19*, 2835

109. T. Tsumuraya and W. Ando, *Organometallics* **1990**, *9*, 869.

110. M. Ishikawa and M. Kumada, *J. Organometal. Chem.* **1974**, *81*, C-3.

111. S. Annemarie, M. Weidenbruch, and S. Pohl, *J. Organometal. Chem.* **1985**, *282*, 305.

112. W. R. Winchester and P. P. Gaspar, presented at *30th Symposium Organosilicon Chemistry*, London, Ontario, Canada, **1997**, p. A-16.

113. (a) W. H. Atwell and D. R. Weyenberg, *J. Am. Chem. Soc.* **1968**, *90*, 3438. (b) P. P. Gaspar, in *Reactive Intermediates*, Vol. 1, M. Jones, Jr. and R. A. Moss, Eds., John Wiley & Sons, New York, **1978**, 229.

114. R. J. Hwang, R. T. Conlin, and P. P. Gaspar, *J. Organometal. Chem.* **1975**, *94*, C-38.

115. (a) W. Ando and H. Saso, *Tetrahedron Lett.* **1986**, *27*, 5625. (b) H. Saso and W. Ando, *Chem. Lett.* **1988**, 1567. (c) T. Yamamoto, Y. Kabe, and W. Ando, *Organometallics* **1993**, *12*, 1996. (e) W. Ando, T. Yamamoto, H. Saso, and Y. Kabe, *J. Am. Chem. Soc.* **1991**, *113*, 2791.

116. (a) T. Akasaka, W. Ando, K. Kobayashi, and S. Nagase, *J. Am. Chem. Soc.* **1993**, *113*, 1605. (b) T. Akasaka, E. Mitsuhida, W. Ando, K. Kobayashi, and S. Nagase, *J. Chem. Soc., Chem. Commun.* **1995**, 1529.

117. (a) W. P. Neumann, *Chem. Rev.* **1991**, 91, 311; (b) J. Barrau, J. Excudié, and J. Satgé, *Chem. Rev.* **1990**, *90*, 283. (c) M.-D. Su and S.-Y. Chu, *J. Am. Chem. Soc.* **1999**, *121*, 11478.

118. W. Ando, H. Ohgaki, and Y. Kabe, *Angew. Chem., Int. Ed. Engl.* **1994**, *33*, 659.

119. H. Ohgaki, Y. Kabe, and W. Ando, *Organometallics* **1995**, *14*, 2139.

120. Y. Kabe, H. Ohgaki, Y. Yamagaki, H. Nakanishi, and W. Ando, *J. Organometal. Chem.* **2001**, *636*, 82.

121. (a) W. Ando, M. Ikeno, and A. Sekiguchi, *J. Am. Chem. Soc.* **1977**, *99*, 6447. (b) W. Ando and M. Ikeno, *Chem. Lett.* **1978**, 609.

122. W. Ando, M. Ikeno, and A. Sekiguchi, *J. Am. Chem. Soc.* **1978**, *100*, 3613.

123. (a) W. Ando, Y. Hamada, A. Sekiguchi, and K. Ueno, *Tetrahedron Lett.* **1982**, *23*, 5323. (b) W. Ando, Y. Hamada, A. Sekiguchi, and K. Ueno, *Tetrahedron Lett.* **1983**, *24*, 4033.

124. (a) W. Ando, Y. Hamada, and A. Sekiguchi, *J. Chem. Soc., Chem. Commun.* **1983**, 952. (b) W. Ando, Y. Hamada, and A. Sekiguchi, *Tetrahedron Lett.* **1984**, *25*, 5057. (c) W. Ando, K. Hagiwara, and A. Sekiguchi, *Organometallics* **1987**, *6*, 2270.

125. W. Ando and M. Ikeno, *J. Chem. Soc., Chem. Commun.* **1979**, 655.

126. (a) W. Ando, T. Tsumuraya, and A. Sekiguchi, *Tetrahedron Lett.* **1985**, *26*, 4523. (b) W. Ando and T. Tsumuraya, *Tetrahedron Lett.* **1986**, *27*, 3251. (c) W. Ando, T. Tsumuraya, and M. Goto, *Tetrahedron Lett.* **1986**, *27*, 5105.

127. (a) T. Tsumuraya, S. Sato, and W. Ando, *Organometallics* **1989**, *8*, 161. (b) W. Ando and T. Tsumuraya, *Organometallics* **1989**, *8*, 1467.

128. M. Denk, R. Lennon, R. Hayashi, R. West, A. V. Belyakov, H. P. Verne, A. Haaland, M. Wagner, and N. Metzler, *J. Am. Chem. Soc.* **1994**, *116*, 2691.

129. For reviews of stable diaminosilylenes, see (a) R. West and M. Denk, *Pure Appl. Chem.* **1996**, *68*, 785. (b) M. Denk, R. West, R. Hayashi, Y. Apeloig, R. Pauncz, and M. Karni, in *Organosilicon Chemistry II*, (Eds. N. Auner and J. Weis), VCH, Weinheim, 1996, 251. (c) B. Gehrhus and M. F. Lappert, *Phosphorus, Sulfur Silicon Relat. Elem.* **1997**, *124*, *125*, 537. See also, Ref. 1q.

130. (a) M. Denk, J. C. Green, N. Metzler, and M. Wagner, *J. Chem. Soc., Dalton Trans.* **1994**, 2405. (b) B. Gehrhus, M. F. Lappert, J. Heinicke, R. Boese, and D. Bláser, *J. Chem. Soc., Chem. Commun.* **1995**, 1931. (c) J. Heinicke, A. Opera, M. K. Kindermann, T. Karbati, and L. Nyulászi, *Chem. Eur. J.* **1998**, *4*, 541.

131. M. Haaf, A. Schmiedl, T. A. Schmedake, D. R. Powell, A. J. Millevolte, M. Denk, and R. West, *J. Am. Chem. Soc.* **1998**, *120*, 12714.

132. B. Gehrhus, P. B. Hitchcock, and M. F. Lappert, *Organometallics* **1997**, *16*, 4861

133. B. Gehrhus and M. F. Lappert, *Phosphorus, Sulfur, Silicon* **1997**, *125*, 537.

134. For a review, see: M. F. Lappert, *Main Group Metal Chem.* **1994**, *17*, 183.

135. D. F. Moser, T. Bosse, J. Olson, J. L. Moser, I. A. Guzei, and R. West, *J. Am. Chem. Soc.* **2002**, *124*, 4186.

136. M. Kira, S. Ishida, T. Iwamoto, and C. Kabuto, *J. Am. Chem. Soc.* **1999**, *121*, 9772.

137. J. Belzner and H. Ihmels, *Adv. Organometal. Chem.* **1998**, 43, 1.

138. E. A. Williams, in *The Chemistry of Organic Silicon Compounds*, S. Patai and Z. Rappoport, Eds., John Wiley & Sons, Inc., New York, **1989**, Part 1, Chapter 8, pp. 511–554.

139. M. Kira, S. Ishida, T. Iwamoto, M. Ichinohe, C. Kabuto, L. Ignatovich, and H. Sakurai, *Chem. Lett.* **1999**, 263.

140. M. Kira, R. Yauchibara, R. Hirano, C. Kabuto, and H. Sakurai, *J. Am. Chem. Soc.* **1991**, *113*, 7785.

141. S. Nagase and T. Kudo, *J. Chem. Soc., Chem. Commun.* **1984**, 1392.

142. (a) S. Masamune and L. R. Sita, *J. Am. Chem. Soc.* **1985**, *107*, 6390. (b) T. Fjeldberg, A. Haaland, M. F. Lappert, B. E. R. Schilling, R. Seip, and A. J. Thorne, *J. Chem. Soc., Chem. Commun.* **1982**, 1407. (c) J. D. Cotton, P. J. Davidson, and M. F. Lappert, *J. Chem. Soc., Dalton Trans.* **1976**, 2275. (d) K. W. Zilm, G. A. Lawless, R. M. Merrill, J. M. Millar, and G. G. Webb, *J. Am. Chem. Soc.* **1987**, *109*, 7236.

143. N. S. Issacs, in *Physical Organic Chemistry*, Longman Scientific & Technical, UK, 1986, pp. 1–76.

144. (a) K. Krogh-Jespersen, *J. Am. Chem. Soc.* **1985**, *107*, 537. (b) R. S. Grev, H. F. Schaefer, III, and K. M. Baines, *J. Am. Chem. Soc.* **1992**, *112*, 9458.

145. (a) D. E. Goldberg, P. B. Hitchcock, M. F. Lappert, K. M. Thomas, A. J. Thorne, T. Fjeldberg, A. Haaland, and B. E. R. Schilling, *J. Chem. Soc., Dalton Trans.* **1986**, 2387. (b) G. Trinquier, J.-P. Marliew, and P. Riviére, *J. Am. Chem. Soc.* **1982**, *104*, 4529.

146. A. Márquez, G. G. Gonzàlez, and J. F. Sanz, *Chem. Phys.* **1989**, *138*, 99.

147. (a) N. Tokitoh, H. Suzuki, R. Okazaki, and K. Ogawa, *J. Am. Chem. Soc.* **1993**, *115*, 10428. (b) H. Suzuki, N. Tokitoh, R. Okazaki, J. Harada, K. Ogawa, S. Tomoda, and M. Goto, *Organometallics* **1995**, *14*, 1016. (c) H. Suzuki, N. Tokitoh, and R. Okazaki, *Bull. Chem. Soc. Jpn.* **1995**, *68*, 2471.

148. H. Suzuki, N. Tokitoh, and R. Okazaki, *J. Am. Chem. Soc.* **1994**, *116*, 11572.

149. N. Tokitoh, H. Suzuki, and R. Okazaki, *Chem. Commun.* **1996**, 124.

150. N. Tokitoh, H. Suzuki, N. Takeda, T. Kajiwara, T. Sasamori, and R. Okazaki, *Silicon Chemistry*, in press.

151. N. Takeda, H. Suzuki, N. Tokitoh, R. Okazaki, and S. Nagase, *J. Am. Chem. Soc.* **1997**, *119*, 1456.

152. H. Spitzner and R. West, unpublished results.

153. J. Hermanns and B. Schmidt, *J. Chem. Soc., Perkin Trans. 1* **1998**, 2209.

154. (a) M. Ishikawa, F. Ohi, and M. Kumada, *J. Organometal. Chem.* **1975**, *86*, C23. (b) M. Ishikawa, K. Nakagawa, and M. Kumada, *J. Organometal. Chem.* **1979**, *178*, 105. (c) P. P. Gaspar and R.-J. Hwang, *J. Am. Chem. Soc.* 1974, *96*, 6198. (d) H. Sakurai, Y. Kobayashi, R. Sato, and Y. Nakadaira, *Chem. Lett.* **1983**, 1197. (e) D. Lei, R.-J. Hwang, and P. P. Gaspar, *J. Organometal. Chem.* **1984**, 271. (f) D. Lei and P. P. Gaspar, *Res. Chem. Intermed.* **1983**, *12*, 103. (g) K. L. Bobbitt and P. P. Gaspar, *J. Organometal. Chem.* **1995**, *499*, 17.

155. (a) S. Zhang and R. T. Conlin, *J. Am. Chem. Soc.* **1991**, *113*, 4272. (b) M. Weidenbruch, E. Kroke, H. Marsmann, S. Pohl, and W. Saak, *J. Chem. Soc., Chem. Commun.* **1994**, 1233. (c) E. Kroke, S. Willms, M. Weidenbruch, W. Saak, S. Pohl, and H. Marsmann, *Tetrahedron Lett.* **1996**, *37*, 3675.

156. (a) N. Takeda, N. Tokitoh, and R. Okazaki, *Chem. Lett.* **2000**, 622. (b) N. Takeda, T. Kajiwara, H. Suzuki, R. Okazaki, and N. Tokitoh, *Chem. Eur. J.*, in press.

157. P. J. Davidson, D. H. Harris, and M. F. Lappert, *J. Chem. Soc., Dalton Trans.* **1977**, 2268.

158. P. Jutzi, A. Becker, H. G. Stammler, and B. Neumann, *Organometallics* **1991**, *10*, 1647.

159. J. Bender, IV., M. M. B. Holl, and J. W. Kamp, *Organometallics* **1997**, *16*, 2743.

160. P. Jutzi, H. Schmidt, B. Neumann, and H. G. Stammler, *Organometallics* **1996**, *15*, 741.

161. R. S. Simons, L. Pu, M. M. Olmstead, and P. P. Power, *Organometallics* **1997**, *16*, 1920.

162. L. Lange, B. Meyer, and W.-W. du Mont, *J. Organometal. Chem.* **1987**, *329*, C17.

163. N. Tokitoh, K. Kishikawa, and R. Okazaki, *J. Chem. Soc., Chem. Commun.* **1995**, 1425.

164. (a) N. Tokitoh, T. Matsumoto, K. Manmaru, and R. Okazaki, *J. Am. Chem. Soc.* **1993**, *115*, 8855. (b) N. Tokitoh, K. Manmaru, and R. Okazaki, *Organometallics* **1994**, *13*, 167. (c) N. Tokitoh, K. Manmaru, and R. Okazaki, *Nippon Kagaku Kai Shi* **1994**, 240. (d) N. Tokitoh, K. Kishikawa, T. Matsumoto, and R. Okazaki, *Chem. Lett.* **1995**, 827.

165. T. Matsumoto, N. Tokitoh, and R. Okazaki, *J. Am. Chem. Soc.* **1999**, *121*, 8811.

166. N. Tokitoh, T. Matsumoto, and R. Okazaki, *Bull. Chem. Soc. Jpn.* **1999**, *72*, 1665.

167. N. Tokitoh, T. Matsumoto, and R. Okazaki, *J. Am. Chem. Soc.* **1997**, *119*, 2337.

168. K. Kishikawa, N. Tokitoh, and R. Okazaki, *Chem. Lett.* **1998**, 239.

169. N. Tokitoh, K. Kishikawa, R. Okazaki, T. Sasamori, N. Nakata, and N. Takeda, *Polyhedron* **2002**, *21*, 563.

170. H. Spitzner and R. West, unpublished results. For the synthesis of compound [(Me$_3$Si)$_2$CH]$_2$Si=Si[CH(SiMe$_3$)$_2$]$_2$, see: S. Masamune, Y. Eriyama, and T. Kawase, *Angew. Chem., Int. Ed. Engl.* **1987**, *26*, 584.

171. T. Matsumoto, N. Tokitoh, and R. Okazaki, *J. Chem. Soc., Chem. Commun.* **1997**, 1553.

172. K. Kishikawa, N. Tokitoh, and R. Okazaki, *Organometallics* **1997**, *16*, 5127.

173. K. Kishikawa, N. Tokitoh, and R. Okazaki, *Chem. Lett.* **1996**, 695.

174. T. Fjeldberg, H. Hope, M. F. Lappert, P. P. Power, and A. J. Thorne, *J. Chem. Soc., Chem. Commun.* **1983**, 639.

175. B. Çetinkaya, I. Gümrükçü, M. F. Lappert, J. L. Atwood, R. D. Rodgers, and M. J. Zatorotoko, *J. Am. Chem. Soc.* **1980**, *102*, 2088.

176. P. B. Hitchcock, M. F. Lappert, B. J. Samways, and E. L. Weinberg, *J. Chem. Soc., Chem. Commun.* **1983**, 1492.

177. L. M. Engelhardt, B. S. Jolly, M. F. Lappert, C. L. Laston, and A. H. White, *J. Chem. Soc., Chem. Commun.* **1988**, 336.

178. (a) H. Grützmacher, H. Pritzkow, and F. T. Edelmann, *Organometallics* **1991**, *10*, 23. (b) M. P. Bigwood, J. Corvan, and J. J. Zuckermann, *J. Am. Chem. Soc.* **1981**, *103*, 7643.

179. (a) D. E. Goldberg, D. H. Harris, M. F. Lappert, and K. M. Thomas, *J. Chem. Soc., Chem. Commun.* **1976**, 261. (b) J. D. Cotton, P. J. Davidson, M. F. Lappert, and J. D. Donaldson, *J. Chem. Soc., Dalton Trans.* **1976**, 2286.

180. M. Weidenbruch, J. Schlaefke, A. Schäfer, K. Peters, H. v. G. Schnering, and H. Marsmann, *Angew. Chem., Int. Ed. Engl.* **1994**, *33*, 1846.

181. (a) N. Tokitoh, M. Saito, and R. Okazaki, *J. Am. Chem. Soc.* **1993**, *115*, 2065. (b) M. Saito, N. Tokitoh, and R. Okazaki, *Organometallics* **1996**, *15*, 4531.

182. M. Saito, N. Tokitoh, and R. Okazaki, *Chem. Lett.* **1996**, 265.

183. For reviews, see: (a) P. Rivière, M. Riviére-Baudet, and J. Satgé, in *Comprehensive Organometallic Chemistry*, Vol. 2, Pergamon, New York, **1982**, p. 670. (b) P. G. Harison, in *Comprehensive Organometallic Chemistry II*, Vol. 2, Pergamon, New York, **1995**, p. 305. (c) P. G. Harrison, in *Comprehensive Coordination Chemistry*, Vol. 3, Pergamon, Oxford, UK, **1987**, p. 185. (d) E. W. Abel, in *Comprehensive Inorganic Chemistry*, Vol. 2, Pergamon, Oxford, UK, **1973**, p. 105.

184. E. O. Fischer and H. Grubert, *Z. Anorg. Allg. Chem.* **1956**, *286*, 237.

185. (a) D. H. Harris and M. F. Lappert, *J. Chem. Soc., Chem. Commun.* **1974**, 895. (b) M. J. S. Gyane, D. H. Harris, M. F. Lappert, P. P. Power, P. Riviére, M. Riviére-Baudet, *J. Chem. Soc., Dalton Trans.* **1977**, 2004.

186. (a) N. Kano, N. Tokitoh, and R. Okazaki, *Organometallics* **1997**, *16*, 4237. (b) K. Kano, N. Tokitoh, and R. Okazaki, *Phosphorus, Sulfur, Silicon, Related Elem.* **1997**, *124–125*, 517.

187. S. Brooker, J.-K. Buijink, and F. T. Edelmann, *Organometallics* **1991**, *10*, 25.

188. (a) C. Drost, P. B. Hitchcock, M. F. Lappert, and L. J.-M. Pierseens, *J. Chem. Soc., Chem. Commun.* **1997**, 1141. (b) P. B. Hitchcock, M. F. Lappert, and Z.-X. Wang, *J. Chem. Soc., Chem. Commun.* **1997**, 1113. (c) W.-P. Leung, W.-H. Kwok, L.-H. Weng, L. T. Low, Z. Y. Zhou, and T. C. Mak, *J. Chem. Soc., Dalton Trans.* **1997**, 4301.

189. M. Stürmann, M. Weidenbruch, K. W. Klinkhammer, F. Lissner, and H. Marsmann, *Organometallics* **1998**, *17*, 4425.

190. (a) P. J. Davidson and M. F. Lappert, *J. Chem. Soc., Chem. Commun.* **1973**, 317. (b) J. D. Cotton, P. J. Davidson, D. E. Goldberg, M. F. Lappert, and K. M. Thomas, *J. Chem. Soc., Chem. Commun.* **1974**, 893.

191. K. W. Klinkhammer, unpublished results.

192. C. Eaborn, T. Ganicz, P. B. Hitchcock, J. D. Smith, and S. E. Sözerli, *Organometallics* **1997**, *16*, 5621.

193. C. Kano, N. Tokitoh, and R. Okazaki, *Organometallics* **1997**, *16*, 2748.

194. (a) N. Kano, K. Shibata, N. Tokitoh, and R. Okazaki, *Organometallics* **1999**, *18*, 2999. (b) N. Kano, N. Tokitoh, and R. Okazaki, *J. Synth. Org. Chem. Jpn. (Yuki Gosei Kagaku Kyokai Shi)* 1998, *56*, 919.

195. C. Glidewell, *J. Organometal. Chem.* **1990**, *398*, 241.

196. K. W. Klinkhammer, T. F. Fässler, and H. Grützmacher, *Angew. Chem. Int. Ed. Engl.* **1998**, *37*, 124.

197. (a) G. Trinquier and J.-P. Marlieu, *J. Am. Chem. Soc.* **1987**, *109*, 5303. (b) G. Trinquier, *J. Am. Chem. Soc.* **1990**, 112, 2130. (c) H. Jacobsen and T. Ziegler, *J. Am. Chem. Soc.* **1994**, *116*, 3667.

198. M. Stürmann, W. Saak, H. Marsmann, and M. Weidenbruch, *Angew. Chem. Int. Ed. Engl.* **1999**, *38*, 187.

199. (a) M. Stürmann, W. Saak, and M. Weidenbruch, *Z. Anorg. Alleg. Chem.* **1999**, *625*, 705. (b) M. Stürmann, W. Saak, M. Weidenbruch, and K. W. Klinkhammer, *Eur. J. Inorg. Chem.* **1999**, 579.

200. (a) N. Tokitoh, N. Kano, K. Shibata, and R. Okazaki, *Organometallics* **1995**, *14*, 3121. (b) N. Kano, N. Tokitoh, and R. Okazaki, *Chem. Lett.* **1997**, 277.

201. (a) H. Suzuki, N. Tokitoh, S. Nagase, and R. Okazaki, *J. Am. Chem. Soc.* **1994**, *116*, 11578. (b) H. Suzuki, N. Tokitoh, R. Okazaki, S. Nagase, and M. Goto, *J. Am. Chem. Soc.* 1998, *120*, 11096.

202. M. Saito, N. Tokitoh, and R. Okazaki, *J. Am. Chem. Soc.* **1997**, *119*, 11124.

203. S. Nagase, unpublished results. See also, Ref. 194b.

204. For the calculation on $R_2M=O$ systems, see: J. Kapp, M. Remko, and P. v. R. Schleyer, *J. Am. Chem. Soc.* **1996**, *118*, 5745.

205. N. Kano, N. Tokitoh, and R. Okazaki, *Organometallics* **1998**, *17*, 1241.

206. M. Saito, N. Tokitoh, and R. Okazaki, *Organometallics* **1995**, *14*, 3620.

CHAPTER 15

Strained Hydrocarbons: Structures, Stability, and Reactivity

KENNETH B. WIBERG

Department of Chemistry, Yale University, New Haven, CT

Reactive Intermediate Chemistry, edited by Robert A. Moss, Matthew S. Platz, and Maitland Jones, Jr.
ISBN 0-471-23324-2   Copyright © 2004 John Wiley & Sons, Inc.

1. INTRODUCTION: BAEYER STRAIN THEORY; THE ORIGIN AND TYPES OF STRAIN

Molecules that are distorted from their normal geometry usually have an increased energy that may lead to increased reactivity. Such destabilization is commonly referred to as "strain." The concept of strain was first introduced by Adolph von Baeyer in 1874.[1] He noted that whereas there were many compounds with five- or six-membered rings, there were few examples of smaller or larger rings. He proposed that the other ring sizes would lead to deformed bond angles, and thus be "strained." We now know that this is correct for three- and four-membered rings. However, he thought all rings were planar, and it was subsequently shown that essentially all saturated rings, except cyclopropane, are nonplanar. Rings with more than six atoms have some strain, but they have components other than bond angle deformation. Here, torsional strain and steric strain become particularly important.

Strained hydrocarbons range from being quite stable thermally and relatively unreactive (cyclobutane) to being transient intermediates. The following will be concerned with the factors that control stability and reactivity.

2. STRAIN ENERGIES

The energies of molecules vary over a wide range making it difficult to compare them. Strain energies allow a direct comparison of deviations in energy from "normal" values, independent of molecular size. The following will summarize calculations of strain energies.

2.1. Experimental Data

The energy of a molecule is usually given as the heat (enthalpy) of formation from the elements in their normal state (graphite for carbon, diatomic gases for hydrogen, oxygen, nitrogen and fluorine, etc.). It may be derived from the heat of combustion, which is a standard calorimetric measurement. Consider cyclohexane. The heat of combustion in the gas phase is given by

$$C_6H_{12(l)} + 9O_{2(g)} \rightarrow 6CO_{2(g)} + 6H_2O_{(l)} \qquad \Delta H = -944.7 \text{ kcal/mol}$$

where the negative sign indicates that the reaction is exothermic. This result may be compared with the combustion of the elements

$$6C_{(s)} + 6H_{2(g)} \rightarrow 6CO_{2(g)} + 6H_2O_{(l)} \qquad \Delta H = -974.2 \text{ kcal/mol}$$

The difference between these quantities, -29.5 kcal/mol, is the heat of formation of cyclohexane and the negative sign indicates that it is more stable than the elements that it contains. These data are available for many strained hydrocarbons.[2] Unfortunately, the measurement of heats of combustion is no longer an active area of experimental thermochemistry, and energies of new compounds are now commonly estimated using theoretical calculations.

The heats of formation of alkenes may sometimes be obtained by combining heats of hydrogenation with the heat of formation of the saturated product. This procedure has been used to obtain heats of formation of many strained alkenes.[3] The heats of hydrogenation are usually measured in solution, and may differ somewhat from the value appropriate for the gas phase. Thus, the heats of formation of alkenes obtained in this way have a somewhat higher uncertainty than those obtained from heats of combustion.

2.2. Calculated Energies

Molecular mechanics (MM)[4] is useful for estimating heats of formation for many compounds that have modest bond angle deformation. Here, one starts with an approximate geometry, and the MM program calculates the bond lengths, angles, torsional angles, and nonbonded distances. The deviations from "standard" values are calculated, along with the energies associated with the deviations. Then, a systematic search is carried out for the lowest energy conformation that can be reached from the starting geometry. This procedure allows the energy of the molecule to be estimated, and for molecules that do not deviate markedly from normal geometrical parameters, this procedure is usually capable of providing satisfactory estimates of the heats of formation. In common with all methods for geometry optimization, only the nearest local minimum will be found, and a number of different starting points must be examined for flexible molecules such as cyclooctane in order to locate the global minimum.[5]

However, MM is not generally useful for compounds that have more severe angle deformation or other unusual structural features. Ab initio molecular orbital (MO) calculations are not subject to this limitation. Here, again, an approximate geometry is used as the starting point, and geometry optimization will lead to the nearest local minimum. Satisfactory results are usually obtained when moderately high level calculations are carried out. The simplest procedure is a Hartree–Fock (HF) calculation, but it has the disadvantage that the electron–electron repulsions that are generated in the course of the calculation are too large because they do not recognize that electrons will tend to stay out of each others way because of their mutual repulsion. A better calculation takes this into account, and is described as correcting for electron correlation.[6]

It is now practical to carry out these calculations for a wide variety of molecules, and here the use of density functional theory (DFT)[7] and particularly the hybrid functionals such as B3LYP has proven to be particularly useful. A basis set (equivalent to a set of atomic orbitals) such as 6-311+G* appears to be generally satisfactory. Greater accuracy may be obtained using more advanced methods such as the G2[8] or CBSQ[9] model chemistries. The calculations also provide estimates of the vibrational frequencies from which the zero-point energies may be obtained.

Correction for the differences in zero-point energies may be important when examining energy changes for reactions.

The heat of formation of a compound may be derived from the results of these calculations in several different ways. First, it is possible to calculate a set of group equivalents using the same level of theory as used for the compound in question.[10] Then,

$$\Delta H = 627.5^* (E_{calc} - n_{CH_3}E_{CH_3} + n_{CH_2}E_{CH_2} + n_{CH}E_{CH} + \cdots)$$

where E_{calc} is the calculated energy, n is the number of groups of a given type, E is the corresponding group equivalent, and 627.5 is the conversion factor from atomic units (used in the calculations) to kilocalories per mole (kcal/mol). This procedure often can give estimated heats of formation with an uncertainly of $\sim$1–2 kcal/mol. The heats of formation of some of the cycloalkanes in Figure 15.1 were estimated in this fashion using B3LYP/6-311+G* calculated energies.

A second procedure makes use of heats of atomization. The heat of atomization of carbon is the energy of converting graphite to carbon atoms. With diatomic molecules such as hydrogen, nitrogen, oxygen, and fluorine, it is the energy required to convert them into atoms. The heats of atomization for most elements in their normal form are known from experimental data.[11]

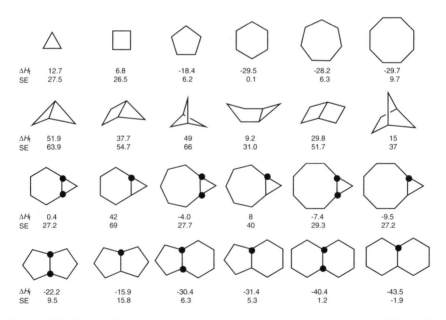

Figure 15.1. Heats of formation and strain energies of cyclic hydrocarbons in kilocalories per mole. The values given to 0.1 kcal/mol are experimental values (Refs. 2, 3, 17) and have an uncertainty of $\sim$0.5–1.0 kcal/mol. The values given as integers are calculated values (B3LYP/6-311+G*) and have an uncertainty of $\sim$3 kcal/mol.

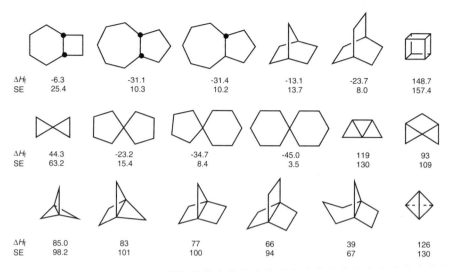

Figure 15.1   (*Continued*)

The heat of atomization of a compound may also be derived from the calculated energy and the calculated heats of atomization of the elements.[12] The difference between the heats of atomization of the compound and that of its constituent elements gives the heat of formation of the compound.

A third method makes use of an isodesmic reaction,[13] which is a balanced reaction in which the number and types of groups are as similar as possible on the two sides of the equation. Consider the following example:

The energy change for the reaction may be estimated using the calculated energies of the component molecules. The heats of formation of all of the molecules except bicyclo[2.2.0]hexane (**1**) are well known, and having the calculated energy change for the overall reaction, its heat of formation may be estimated. The calculated energy change is 47 kcal/mol, and combined with the experimental heats of formation of cyclohexane, ethane and methane, it leads to a heat of formation of **1** of 33 kcal/mol. This value may be compared with the value 30 kcal/mol obtained from the heat of hydrogenation of **1** and the heat of formation of cyclohexane.

2.3. Models for Strain Energy Calculations

The strain energy (SE) of a molecule is the difference between its heat of formation and the heat of formation of a "suitable unstrained model." Different choices of

TABLE 15.1. Franklin Group Equivalents[a]

Group	Group Value	Group	Group Value
CH_3	−10.12	cis-HC=CH	18.88
CH_2	−4.926	trans-HC=CH	17.83
CH	−1.09	$C=CH_2$	16.89
C	+0.8	C=C	24.57

[a] ΔH_f at 25 °C.

models may lead to different strain energies. Since the strain energies are not absolute quantities, all that is needed is a consistent, easily applied model. One of the simplest is derived from the Franklin group equivalents.[14] These are the heats of formation at 25 °C of various structural groups such as CH_3, CH_2, CH, C and olefinic groups (Table 15.1). By taking the appropriate sum of these equivalents, one obtains an unstrained model.

As an example, the heat of formation of cyclopentane in the gas phase is −18.4 kcal/mol. The Franklin group equivalent for CH_2 is −4.93 kcal/mol, and thus an unstrained model for cyclopentane is five times this quantity or −24.6 kcal/mol. The difference between the two values gives the strain energy as 6.2 kcal/mol. The strain energies given in Figure 15.1 were obtained in this way.

The strain energies may similarly be derived from the ab initio calculated energies by obtaining calculated group equivalents corresponding to the level of theory used. In the simplest approximation, the value for an unstrained CH_2 group might be taken as the difference in energy between butane and pentane, that for a CH_3 group would be one-half the energy of butane minus the CH_2 equivalent. The CH equivalent could be obtained by subtracting three times the CH_3 equivalent from the energy of isobutane. Other group equivalents may be obtained in a similar fashion.

A more detailed calculation would include the zero-point energies and the change in energy on going from 0 to 298 K. However, this usually has only a small effect on the calculated strain energies. Taking cyclopentane as an example, the B3LYP/6-311+G* energy is −196.59878 hartrees (1H = 627.51 kcal/mol), and the difference between butane and pentane is −39.32190 H. Subtracting five times the latter from the calculated energy of cyclopentane, and converting to kilocalories per mole gives 6.7 kcal/mol as the strain energy, in good agreement with the value derived from the experimental heat of formation.

The heats of formation and the strain energies of a group of experimentally known hydrocarbon systems are summarized in Figure 15.1. It can be seen that the strain energies cover a very wide range, with many of them greater than the strength of a carbon–carbon bond (90 kcal/mol).[15]

Strained alkenes may have a combination of the strain in the basic ring system, and the added strain that results from the introduction of the double bond. Thus, it is convenient to define olefinic strain (OS)[16] that may be given as the difference in heat of hydrogenation of the alkene in question and that of a similarly substituted unstrained alkene such as cyclohexene. The heats of hydrogenation of many alkenes are known,[17] and in other cases they may be obtained from the heats of formation of the alkene and the corresponding alkane.

However, in many interesting cases, the requisite experimental data are not available, and the olefinic strain for disubstituted alkenes may be estimated using the calculated energies and the following isodesmic reaction:

$$alkene + cyclohexane \rightarrow alkane + cyclohexene$$

The calculated energy change for this reaction gives the difference in olefinic strain between the alkene and cyclohexene, and the olefinic strain of the latter is negligible. With tri- and tetrasubstituted alkenes, it is more appropriate to use methylcyclohexane or 1,2-dimethylcyclohexane as the unstrained reactant since it is known that alkyl substitution stabilizes double bonds. The values of olefinic strain for a number of alkenes are summarized in Figure 15.2. Compounds with high OS are expected to be relatively reactive.

It is interesting to note that in some cases, the olefinic strain is negative indicating that some strain terms are lost on going from the alkane to the alkene. With cyclooctene, for example, some of the torsional interactions and H$\cdots$H nonbonded

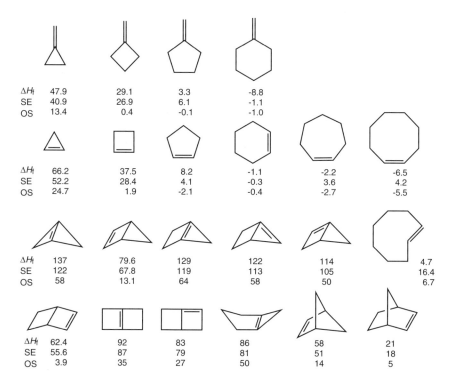

Figure 15.2. Heats of formation, strain energies, and olefinic strain for cycloalkenes.

repulsions in cyclooctane are removed by converting it into cyclooctene and the olefinic strain is −6 kcal/mol.

There is a group of alkenes in which the olefinic strain has relatively large negative values. These alkenes are termed hyperstable,[16] and are expected to be relatively unreactive. One example is bicyclo[4.4.4]tetradec-1-ene (**2**) with a predicted OS of −14 kcal/mol.

2

3. STRUCTURES OF STRAINED HYDROCARBONS

3.1. Angle Deformation

Bond angle deformation is one of the more common sources of strain. A C—C—C bond in an open-chain alkane has an angle of ~111°. For relatively small angular deviations, the increase in energy would approximately be given by

$$\Delta E = \tfrac{1}{2} k_\varphi \Delta\varphi^2 = 0.0175 \Delta\varphi^2$$

where k_φ is the bending force constant and $\Delta\varphi$ is the change in angle from the normal value. A typical value of k_φ for a C—C—C bond is 0.035 kcal/deg^2 (0.8 mdyn/Å^2). With cyclopentane, the average C—C—C bond angle is 105°, leading to $\Delta E = 3.2$ kcal/mol. This value is less than the observed strain energy because cyclopentane also has significant torsional strain (see below).

With cyclopropane, the conventional C—C—C bond angle is 60°, which is too small to be accommodated by any bonding model. Coulson and Moffitt[18] showed that the bonds in cyclopropane are bent and have an interorbital angle of ~104° (Fig. 15.3). This leads to high p-orbital character in the C—C bonds, and correspondingly higher s-orbital character in the C—H bonds. Clearly, a conventional description of bond angle deformation is not appropriate for compounds having three-membered rings. Cyclobutane is much closer to a conventional hydrocarbon.

The strain energies of cyclopropane (26.5 kcal/mol) and cyclobutane (26 kcal/mol) are nearly the same despite the apparent much greater bond angle deformation with the former. One reason is that the weak C—C bonds in cyclopropane are compensated in part by the stronger C—H bonds. The increased strength results from the greater s character, and it is known that C—H bond strengths increase with

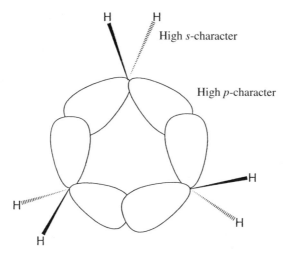

Figure 15.3. Bent bonds in cyclopropane.

increasing s character.[19] The C—H bond dissociation energy (BDE) of cyclopropane is 105 kcal/mol[20] as compared to 98 kcal/mol for an open-chain CH_2 group.[15]

3.2. Inverted Tetrahedral Geometries

The most remarkable structural feature of [1.1.1]propellane (**3**) is the geometry at the bridgehead carbons.[21] Since all four bonds point in the same direction, this might best be described as having an inverted tetrahedral geometry.

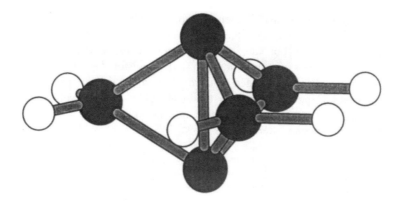

Although the bonding is unusual, the central bond in **A** [1.1.1]propellane has 0.8 times the electron density at the center of the C—C bond in butane. There is a significant amount of electron density near the bridgehead carbons, and this presumably is responsible for the high reactivity of the compound toward electrophiles and free radicals.[22] This type of geometry is found with many of the small ring propellanes. Not surprisingly, the strain energies of these compounds are relatively large. However, as indicated below, there is not a simple relationship between strain energy and reactivity, and one must also consider the strain energies of the products of the reaction, as well as possible differences in transition states for reaction.

3.3. Torsional Strain

Carbon–carbon single bonds prefer a staggered arrangement of the attached groups. Thus, eclipsed ethane has a 3 kcal/mol greater energy than staggered ethane.[23] The effects of torsional strain may be seen in many structures. Cyclobutane (**4**) is non-planar with conventional C–C–C bond angles of 88° as compared to 90° in planar cyclobutane.[24] Puckering will decrease the torsional strain, and the observed geometry is a compromise between torsional strain and bond angle strain. Similarly, cyclopentene (**5**) adopts a puckered arrangement in order to minimize torsional strain,[25] and this puckering can move around the ring leading to "pseudorotation." Torsional strain is one of the important components of the strain in medium sized rings.[26]

4   **5**

3.4. Steric Strain

Many molecules have groups close enough to each other to cause significant steric interactions. Cyclooctane is a simple example. The "crown conformation" (**6**) can be drawn with very good bond angles and torsional angles, but it also has hydrogens pointing to the center of the ring, leading to steric repulsion. Geometry optimization will relieve some of the nonbonded repulsion, but only at the expense of somewhat deformed C–C–C bond angles and torsional angles. As a result, cyclooctane adopts an alternate conformation (**7**) that is a better compromise between bond angle, torsional angle and steric strain.[27]

6   **7**

The paracyclophanes (**8**) have a strong repulsive steric interaction between the two parallel aromatic rings, which leads to considerable deformation of the rings.[28]

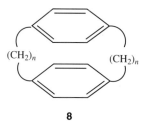

8

Compounds with n as small as 2 have been prepared. With a substituent on one of the rings they are chiral and have been resolved.[29] When the chains are longer, it is possible for one benzene ring to rotate, leading to racemization, and when $n = 4$ resolution was not possible indicating that rotation of a benzene ring is no longer strongly hindered.[30]

Although tri-*tert*-butylethylene has been prepared,[31] it has not as yet been possible to prepare tetra-*tert*-butylethylene, which should have very severe repulsive interactions between the *tert*-butyl groups. There are many other examples of intramolecular steric interactions that lead to increased strain energies.[32] Intermolecular steric interactions can lead to reduced reactivity.[26]

3.5. Twisted Double Bonds

It is relatively difficult to twist a double bond because this would lead to a disruption of the π bond. As a result, compounds with twisted double bonds are relatively rare. One notable series of such compounds are the *trans*-cycloalkenes. The smallest one that can exist at room temperature is *trans*-cyclooctene. The difference in energy between *cis*- and *trans*-cyclooctane was determined from the heats of hydrogenation and is 11 kcal/mol.[33] *trans*-Cycloheptene has been observed at a low temperature,[34] and there are indications that *trans*-cyclohexene is a transient intermediate in a reaction.[35] When the ring is larger, as with cyclodecene,[36] the normal tendency for the trans alkene to have the lower energy is again found. The OS values for these compounds are given in Table 15.2, and the examples with large positive values have been found to be relatively reactive toward electrophiles and in cycloaddition reactions.[16]

The structures of these compounds are interesting (Table 15.2). The cis compounds have both H–C=C–H and C–C=C–C torsional angles close to 0°. The trans compounds have calculated H–C=C–H torsional angles close to 180°, but the C–C=C–C torsional angles increasingly deviate from 180° as the size of the ring is decreased, indicating that the π bonds are becoming pyramidalized. The calculated H–C=C–H torsional angles are in agreement with the electron diffraction study of *trans*-cyclooctene[37] and with the NMR coupling constant for *trans*-cycloheptene.[34] It is difficult to determine the angle between the *p*-orbitals of the double bond in these cases[38] since bonds that are deformed are generally bent,[39] and so the simple geometric angles are not very useful.

TABLE 15.2. Calculated Energies and Structures of *cis*- and *trans*-Cycloalkenes

Compound	E_{rel}^a	OS^a	τ (HC=CH)^b	τ (CC=C–C)^b	Angle Deviation^c	
					σ	π
cis-Cyclohexene	0	0	2.2	1.5	2.3	0.2
trans-Cyclohexene	54	54	183.8	88.1	26.0	21.1
cis-Cycloheptene	0	−2.5	0.0	0.0	4.6	1.6
trans-Cycloheptene	30	27	184.3	116.1	19.3	15.4
cis-Cyclooctene	0	−5.5	−0.2	2.2	2.6	0.2
trans-Cyclooctene	11.5	6.7	182.4	136.8	13.1	10.0

^{a} In kilocalories per mole (kcal/mol).
^{b} Torsional angles in degrees.
^{c} The σ bond deviation in degrees from the line of centers, and the π bond deviation from perpendicular to the line of centers.

One way in which the π bonds in these compounds may be examined is by calculating natural bond orbitals (NBO) from the molecular wave function.[40] Here, the MOs are localized, and the nature of the localized orbitals may be examined. The p–π localized orbital is found to have decreased occupation as the carbon–carbon bond is distorted, indicating that the p orbitals are being used in part in forming other localized orbitals (i.e., rehybridized). It is also possible to calculate natural hybrid orbitals, and their deviation from the line of centers between the atoms involved. The deviations are included in Table 15.2, and are seen to increase markedly with increased twisting.

Another group of compounds that have a twisted double bond are the bicyclic compounds with bridgehead double bonds such as 1,2-norbornene (**9**) and 1,7-norbornene (**10**).[16,41] It has been found that many compounds, such as **11**, which is based on *trans*-cyclooctene, may be isolated whereas those based on smaller *trans*-cycloalkenes are usually quite unstable. Some evidence for the formation of **9** has been obtained by trapping the product of the dehalogenation of 1,2-dihalonorbornanes.[42] Here, the simplest view is that the two p orbitals that form the "double bond" in **9** and **10** are roughly perpendicular to each other. However, pyramidalization and rehybridization also are involved. One indication is the reduced localized π-orbital population found in the NBO analysis. Whereas normal alkenes have π populations of 1.96 e, for **9** with OS = 57 kcal/mol, it is 1.921, and for **10** with OS = 86 kcal/mol, it is 1.896. With **9**, the deviations of the σ and π orbitals from the line of centers are 24° and 19°, respectively, and with **10**, the deviations are 34° and 29°.

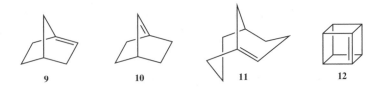

9   **10**   **11**   **12**

In addition to twisting, another mode of distortion for double bonds is pyramidalization. Cubene (**12**) provides a particularly notable example of pyramidalization. It is readily prepared by the dehalogenation of the corresponding diiodide, and although it has not been isolated in pure form, its reactions have been studied in some detail,[43] and its heat of formation has been determined using a novel gas-phase experiment.[44] The calculated OS is 51 kcal/mol, and the OS based on the experimental heat of formation is 61 ± 5 kcal/mol.

A number of other markedly pyramidalized alkenes have been prepared and studied both experimentally and theoretically.[45]

4. STABILITY–REACTIVITY IN THE ABSENCE OF EXTERNAL REAGENTS

Compounds are commonly considered to be stable if they can exist at room temperature. The strain energy is not the important quantity in determining stability, but rather the change in strain gives information on the driving force for the reaction. Then, there is normally a barrier to reaction that must be overcome. Unstable compounds have a relatively large driving force for reactions combined with low energy paths for reaction.

Cubane, with a strain energy of 156 kcal/mol, is one of the most highly strained hydrocarbons. It is also among the more stable of the highly strained small ring hydrocarbons because only a small part of the strain can be relieved in a reaction, and because there is no low-energy path for reaction. On the other hand, tricyclo[2.1.0.0^{1,3}]pentane (**13**) with a strain energy of 130 kcal/mol can only be observed as a transient species by ^{13}C NMR at -55 °C.[46] Here, a large part of the strain can be relieved by the cleavage of one of the internal carbon–carbon bonds.

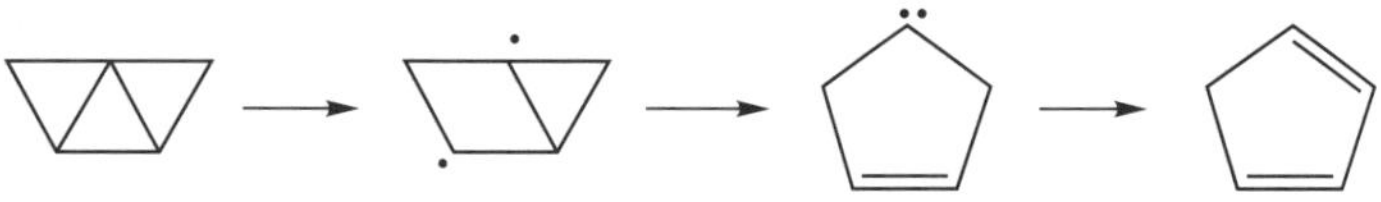

13

It is interesting to note that inversion of configuration at one of the radical centers is necessary in order for ring opening to the carbene to occur.

The role of strain relief in a reaction is nicely illustrated by the small ring propellanes. One way in which they may be destroyed in a condensed medium is via a free radical chain reaction, and this may be illustrated by the energies of adding a methyl radical across the bridgehead–bridgehead bonds. The estimated energies are shown at the top of the next page. The three smallest ring propellanes have close to the same strain energies, but the ring systems that are formed by adding to the ring differ markedly in energy. The resulting large difference in strain relief correlates well with the observation that **3** is quite stable at room temperature, whereas **14** and **15** undergo polymerization when the argon matrix in which they are formed is allowed to warm to 50 K.[47]

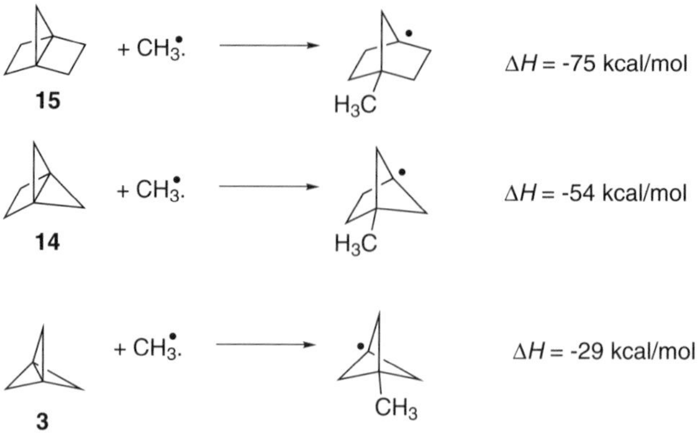

All three compounds are prepared by elimination of bromine from the corresponding bridgehead–bridgehead dibromides. In the case of **3**, this may be effected by butyllithium in a hydrocarbon solvent, but for **14** and **15** the elimination must be carried out by potassium vapor in the gas phase followed by trapping the product as an argon matrix at 10–15 K.

4.1. Thermal Reactivity

Small ring hydrocarbons have a wide range of thermal reactivity, with cyclopropane and cyclobutane being quite stable thermally. With these compounds, the thermolysis is known to proceed via initial cleavage of one C—C bond giving a diyl,[48] which has a relatively high energy.

There often is a correlation between the strain relief in a reaction and the rate of thermolysis, but other factors may also be of importance. The C_8 propellanes **16** and **17** have quite different reactivities. Whereas **16** undergoes thermal cleavage at 360 °C,[49] **17** undergoes cleavage at 25 °C.[50] Aside from the difference in reactivity, the two reactions are essentially the same.

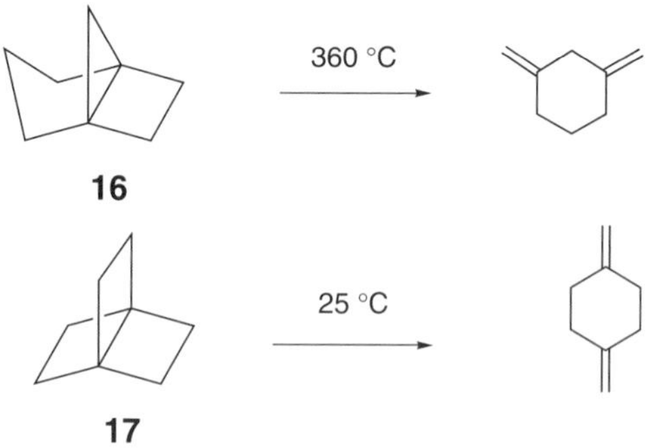

One factor that contributes to the higher reactivity of **17** is its strain energy (94 kcal/ mol) that is higher than that of **16** (67 kcal/mol). Here, the products have essentially the same strain and so the strain energies may be compared directly. There is also another possible factor that may contribute to the reactivity of **17**. It is relatively easy to move the bridgehead carbons apart in **17** due to the flexibility afforded by the four-membered rings. Hoffmann and Stohrer[51] suggested that there is a special mechanism that will aid ring cleavage if these carbons can be moved apart. Such a mechanism is not possible with **16** because the three-membered ring prevents the bridgehead carbons from moving apart a sufficient amount.

The thermolysis of a series of homologues of **17** has been examined and the free energies of activation were linearly related to the relief of strain in the cleavage of the central C–C bond forming a diyl. The compounds related to **16** fell on a different line.[49]

In some cases, steric interactions can prevent unimolecular reactions. Tetrahedrane (**18**) has been the subject of a number of studies, and the conclusion is that, if formed, it would rapidly decompose to form two molecules of acetylene.[52] However, tetra-*tert*-butyltetrahedrane (**19**) is a quite stable substance, and on heating rearranges to tetra-*tert*-butylcyclobutadiene.[53] An orbital symmetry[54] analysis of the cleavage of tetrahedrane to acetylene indicates that it involves a torsional motion that in the case of the *tert*-butyl substituted derivative would bring the *tert*-butyl groups very close to each other. As a result, this mode of reaction is not possible, and the compound is relatively stable.

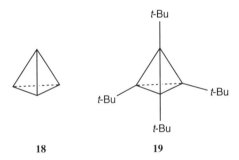

18   **19**

The thermolysis of cycloalkenes is often a more facile process than for the cycloalkenes. Cyclobutene undergoes thermolysis at 175 °C and yields butadiene in an orbital symmetry controlled reaction as shown by stereochemical studies of substituted cyclobutenes.[55]

4.2. Dimerization

Dimerization is an important mode of reaction of strained alkenes, and two different modes of reaction may be found. With cyclopropene, an ene reaction occurs to give a dimer,[56] and this in turn may further react in the same fashion to give a polymer.

The stereochemistry and regiochemistry of the dimerization of 2-*tert*-butylcyclopropenecarboxylic acid has been examined,[57] and other examples of this reaction have been reported.[58]

Another reaction of strained alkenes is cycloaddition to form a cyclobutane derivative. An example is[59]

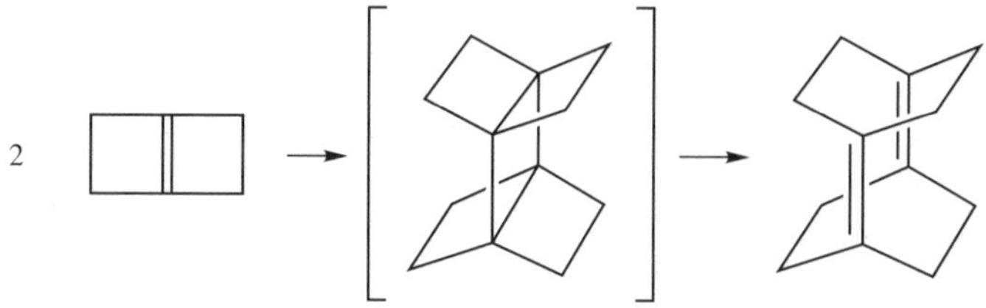

The initial dimerization presumably leads to a [2.2.2]propellane derivative that undergoes thermolysis to the diene that is the observed product. The dimerization occurs in dilute solutions, but at higher concentrations, polymers are found. This finding is in accord with the proposal that these reactions occur via initial formation of a 1,4-diyl, which closes to form the cyclobutane ring. In this way, the reaction avoids an orbital symmetry disallowed pathway. The diyl could react with the alkene at higher concentrations to give a polymer.

Cubene undergoes a similar reaction in which the initial cycloadduct undergoes further reactions.[43] The bridged cyclopropene, bicyclo[4.1.0]hept-1,6-ene gives a rapid ene reaction forming two diastereomeric cyclopropenes, which further dimerize leading to a tetramer.[56] This type of reaction appears to be common when the alkene has a high enough energy to make 1,4-diyl formation energetically possible.

Note that the formation of cyclobutane from ethylene has a favorable free energy change at room temperature, but the reaction will not occur because of the large reaction barrier. The free energy change becomes increasingly less favorable as the temperature is raised because of the negative entropy of reaction. If this can be avoided by having the two double bonds in the same molecule, the conversion to a cyclobutane ring can occur on heating.[49] All of these reactions probably occur via a 1,4-butane diradical intermediate, just as in the thermal decomposition of cyclobutane to ethylene at higher temperatures.[48]

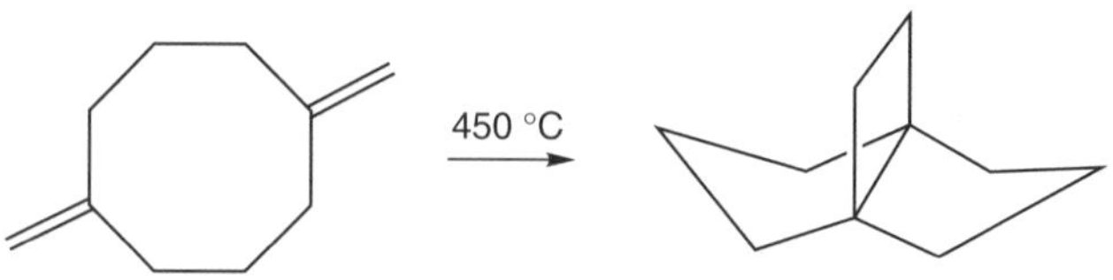

5. REACTIVITY OF STRAINED HYDROCARBONS TOWARD EXTERNAL REAGENTS

Some strained hydrocarbons are quite reactive, whereas others are not. Strain is one of the factors that control reactivity, but other factors are often important.

5.1. Cycloaddition Reactions

Cycloaddition reactions of C=C double bonds are often facilitated by increased olefinic strain. For example, phenyl azide does not react with unstrained alkenes, but does react readily with the smaller *trans*-cycloalkenes. Many interesting cycloalkenes are not sufficiently stable to be isolated, or even observed in solution. However, in many cases they can be trapped by reagents that lead to a {3+2} or [4+2] (Diels–Alder) reaction.

Most cyclopropenes are good dienophiles. Bicyclo[3.1.0]hex-1,5-ene and bicyclo[4.1.0]hept-1,6-ene are examples of strained alkenes that cannot be isolated, but can be trapped by diphenylisobenzofuran.[60] A number of bridged bicyclo[1.1.0]but-1,3-ene derivatives that cannot be isolated also have been trapped via Diels–Alder reactions.[61]

5.2. Reaction with Free Radicals

The strained saturated hydrocarbons are generally not reactive in free radical additions across C—C bonds, although hydrogens may be extracted from many of them by the more reactive free radicals. The propellanes are one group that frequently undergo free radical addition. Although [1.1.1]propellane is quite stable thermally, it does undergo facile free radical addition,[22] which frequently leads to oligomers[62] and may be related to the nonbonded electron density near the bridgehead carbons. It is interesting to note that this propellane readily adds iodine, and that the iodine may be removed and the propellane recovered via reactions with nucleophiles.[63]

The cycloalkenes are, of course, reactive in free radical additions just as the open chain alkenes.

5.3. Reactions of Cyclopropanes and Cyclobutanes with Electrophiles

Although cyclopropane and cyclobutane have similar strain energies, they differ markedly in their reactivity toward electrophiles. Thus, whereas cyclopropane reacts readily with bromine to give 1,3-dibromopropane, cyclobutane does not react with bromine.

The reaction of small ring hydrocarbons with acids has been extensively studied. Although cyclopropane does not react with acetic acid, bicyclo[2.1.0]pentane (**20**) reacts rapidly to give cyclopentyl acetate. On the other hand, bicyclo[2.2.0]hexane (**21**) does not react. Other small ring hydrocarbons that react rapidly with acetic acid are [3.2.1]propellane (**22**), bicyclo[1.1.0]butane, and [1.1.1]propellane (**3**). The compound [2.2.2]propellane appears to be relatively stable toward weak acids.

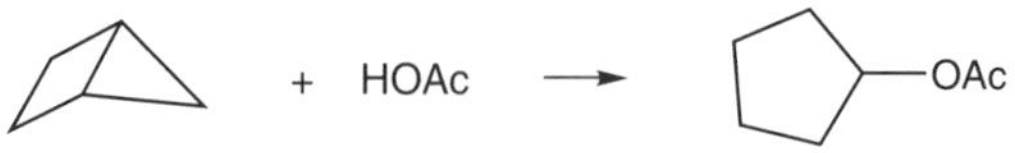

20   + HOAc ⟶ (cyclopentyl)—OAc

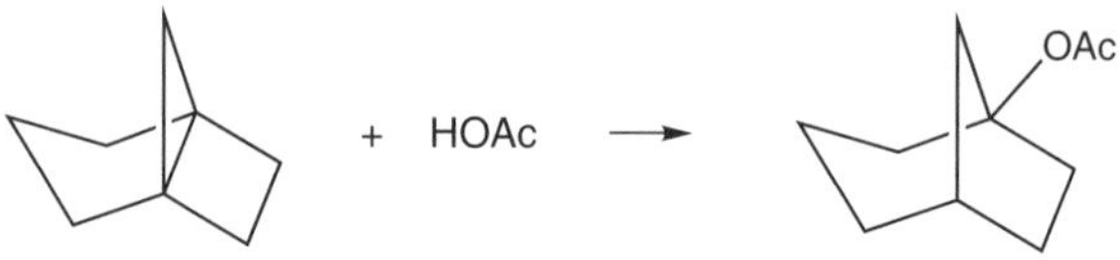

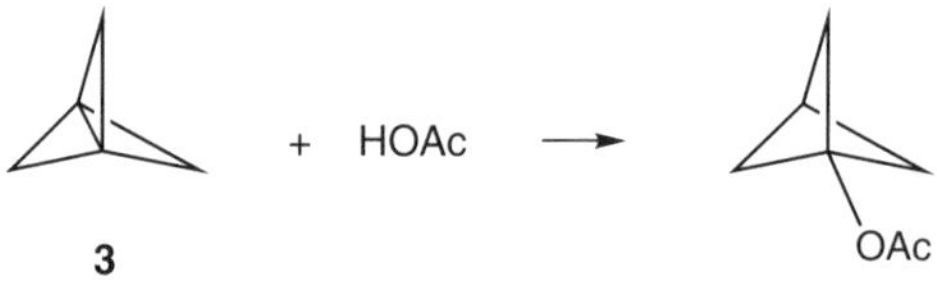

21   + HOAc ⟶ No reaction

22   + HOAc ⟶ (product with OAc)

3   + HOAc ⟶ (product with OAc)

In general, high reactivity toward electrophiles is found only with cyclopropane derivatives.

The origin of the difference between cyclopropanes and cyclobutanes in these reactions was suggested by a study of the cleavage of cyclopropane with D_2SO_4.[64] The product 1-propyl hydrogen sulfate had deuterium scrambled among the carbons, and it was found that H_2SO_4 reacted more rapidly than D_2SO_4. These data suggest that protonation is rate determining, and that a protonated cyclopropane intermediate is formed in which the deuterium can be scrambled before reacting with a nucleophile to give the product. There is now much evidence for protonated cyclopropane intermediates.[65]

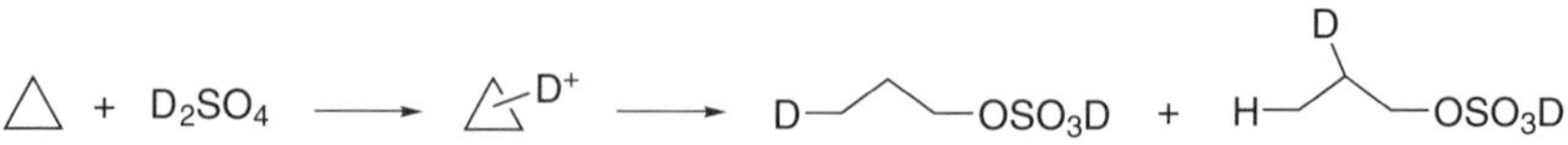

△ + D_2SO_4 ⟶ △—D^+ ⟶ D—⟍⟍—OSO_3D + H—⟍⟍(D)—OSO_3D

Calculations have shown that there are two ways in which cyclopropane may be protonated: at one of the carbons (corner protonated) and at one of the C—C bonds (edge protonated). They have about the same energy, and are involved in the movement of a proton around the ring. The calculations indicate that cyclopropane is unusually basic for a hydrocarbon. The reason can be seen in the structures of the protonated cyclopropanes (Fig. 15.4).[66] The bent bonds permit a proton to bond to a C—C bond without coming close to the positively charged carbon nuclei.

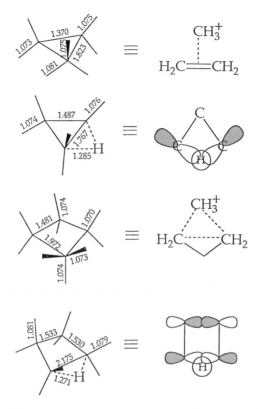

Figure 15.4. Calculated structures of protonated cyclopropane and cyclobutane.

Corner protonation leads to a species that could be described as a methyl cation (a known ion) coordinated with a carbon–carbon double bond.

In contrast to cyclopropane, the cyclobutane C–C bonds are only slightly bent, so that in order for the proton to form a bond, it must come close to the positively charged carbon nuclei, leading to increased Coulombic repulsion. Similarly, an attempt to bond one of the carbons does not lead to an ion with any apparent stabilization. This attempt to bond to one of the carbons leads to a relatively unstable species, and a slow rate of reaction.[66]

5.4. Carbon–Carbon Bond Cleavage by Transition Metal Species

The large difference in reactivity between cyclopropanes and cyclobutanes toward most electrophiles arises because there is little strain relief in the rate-determining step. The cleavage of C–C bonds by transition metal species leads to metallocycloalkanes, and if cleavage proceeds to a significant extent in the rate-determining step, the reactivities of cyclopropanes and cyclobutanes should become more comparable.

Cyclopropanes are cleaved by a variety of transition metal derivatives. The first observation was that $PtCl_4$ reacts to give a platinocyclobutane.[67] Cyclobutanes appear to be less reactive, but cubane reacts with $[Rh(CO)_2Cl]_2$ leading to C—C bond cleavage.[68] The subject of C—C bond cleavage has been reviewed.[69]

6. EFFECT OF STRAIN ON PHYSICAL PROPERTIES

6.1. NMR Spectra

Some of the more remarkable effects of strain are found in NMR chemical shifts. Cyclopropane derivatives usually have upfield proton chemical shifts with regard to the corresponding cyclohexane derivatives, whereas cyclobutanes commonly have downfield shifts.[70] The upfield shift for cyclopropane protons have sometimes been attributed to a ring current in the three-membered ring,[71] but there is little evidence for such a phenomenon. The unusual shift for these protons has proven valuable in demonstrating the presence of a three membered ring.

More information may be obtained from the carbon-13 nuclear magnetic resonance (^{13}C NMR) chemical shifts.[72] Cyclopropane has a shift of -2.8 ppm, whereas cyclobutane is found at 22.4 ppm, which is close to that of cyclohexane, 27.0 ppm. The bridgehead carbon of bicyclo[1.1.0]butane is at -3.0 ppm, whereas the methylene group is at 33.0 ppm. The annelation of a third cyclopropane ring, giving [1.1.1]propellane (**3**) has its bridgehead carbon resonance at 1.0 ppm. whereas the methylene carbon is found at 74 ppm. Thus, on going from cyclopropane to [1.1.1]propellane, the methylene carbon resonance increases by 77 ppm! However, in comparing cyclopropane with the bridgehead carbons of bicyclobutane and [1.1.1]propellane, one finds only a small change in chemical shift. Although the changes in these and other related compounds[71] are well documented, there is at present no comprehensive explanation for the changes in ^{13}C NMR chemical shifts.

6.2. Electronic Spectra

The vacuum ultraviolet (UV) spectrum of cyclopropane has been examined in some detail, and its first band is found at a significantly lower energy than for cyclobutane or cyclohexane.[73] This result has been attributed to the increase in strain that raises the ground-state energy. All of the transitions in the cycloalkanes appear to have Rydberg character in which one electron is promoted to a high-energy diffuse orbital, leaving an electron deficient core that resembles a radical cation. The spectrum of cyclobutane is similar to that of cyclohexane.

Bicyclo[1.1.0]butane has been studied in some detail.[74] Again, all of the states appear to involve Rydberg transitions. The transition energies are at lower energies than those for cyclopropane, which may be a result of the strain in the bicyclobutane ring that further increases the ground-state energy. The spectrum of [1.1.1]propellane has not as yet been subjected to a detailed analysis.

6.3. Vibrational Spectra

The vibrational spectra of several small ring compounds have been studied in sufficient detail so as to provide force constants for the vibrational modes.[75] Cyclopropane has shorter CH bonds with larger stretching force constants than for cyclobutane. The CH bonds in cyclopropane resemble those in ethylene, corresponding to their increased *s*-orbital character.

7. CONCLUSION AND OUTLOOK

Molecules may be "strained" by a variety of modes of distortion that include angle bending, torsional strain, and steric interactions. In some cases, strain can provide an important driving force for reaction. However, it is not the strain in the molecule that is important, but rather the change in strain in a reaction. Even this is not sufficient because there must also be a mechanism for the conversion of a compound to a lower energy product that leads to a relatively low barrier for reaction.

Strained molecules have provided much useful information about bonding in carbocyclic systems. There is still more to be learned about the differences between cyclopropanes and cyclobutanes, especially with regard to their NMR spectra and their interaction with other groups and rings. Their use as intermediates in synthetic studies will surely continue because strain relief provides significant driving force for reactions. Although the synthesis of new strained carbocyclic systems with high strain energies (such as compounds with near-planar carbons) will probably remain difficult, chemists will surely rise to the challenge, and devise new means for the preparation of these compounds.

SUGGESTED READING

A. Greenberg and J. Liebman, *Strained Organic Molecules*, Academic Press, New York, **1978**.

K. B. Wiberg, "The Concept of Strain in Organic Chemistry," *Angew. Chem. Int. Ed. Engl.* **1986**, *25*, 312.

A. de Meijere and S. Blechert, Ed., *Strain and its Implications in Organic Chemistry*, Kluwer Academic Publishers, Dordrecht, The Netherlands, **1989**.

U. Burkert and N. L. Allinger, *Molecular Mechanics*, American Chemical Society Washington, DC, **1982**.

P. Warner, "Strained Double Bonds," *Chem. Rev.* **1989**, *89*, 1067.

M. Saunders, P. Vogel, E. L. Hagen, and J. Rosenfeld, "Evidence for Protonated Cyclopropane Intermediates," *Acc. Chem. Res.* **1973**, *6*, 53.

"Strained Organic Compounds" (thematic issue) *Chem. Rev.* **1989**, *89*, 973.

E. L. Eliel and S. H. Wilen, *Stereochemistry of Organic Compounds*, John Wiley & Sons, Inc., New York, **1994**.

M. S. Newman, *Steric Effects in Organic Chemistry*, John Wiley & Sons, Inc., New York, **1963**.

REFERENCES

1. A. von Baeyer, *Chem. Ber.* **1885**, *13*, 2278.

2. J. B. Pedley, R. D. Naylor, and S. P. Kirby, *Thermochemical Data of Organic Compounds*, 2nd ed., Chapman and Hall, London, **1986**.

3. J. L. Jensen, *Prog. Phys. Org. Chem.* **1976**, *12*, 189.

4. U. Burkert and N. L. Allinger, *Molecular Mechanics*, American Chemical Society, Washington, DC, 1982. N. L. Allinger and P. v. R. Schleyer, Eds., *J. Comput. Chem.* **1996**, *17*, pp. 488ff.

5. M. Saunders, *J. Comput. Chem.* **1991**, *12*, 645.

6. W. J. Hehre, L. Radom, P. v. R. Schleyer, and J. A. Pople, *Ab Initio Molecular Orbital Theory*, John Wiley & Sons, Inc., New York, **1986**.

7. W. Koch and M. C. Holthausen, *A Chemist's Guide to Density Functional Theory*, 2nd ed., Wiley-VCH, Weinheim, **2001**.

8. L. A. Curtiss, K. Raghavachari, G. W. Trucks, and J. A. Pople, *J. Chem. Phys.* **1991**, *94*, 7221.

9. J. W. Ochterski, G. A. Petersson, and J. A. Montgomery, *J. Chem. Phys.* **1996**, *104*, 2578.

10. K. B. Wiberg, *J. Comput. Chem.* **1984**, *5*, 197. K. B. Wiberg, *J. Org. Chem.* **1985**, *50*, 5285. M. R. Ibrahim and P. v. R. Schleyer, *J. Comput. Chem.* **1985**, *6*, 157. L. R. Schmitz and Y. R. Chen, *J. Comput. Chem.* **1994**, *15*, 1337.

11. M. W. Chase, Jr., C. A. Davies, J. R. Downey, Jr., D. J. Frurip, R. A. MacDonald, and A. N. Syverud, *J. Phys. Chem. Ref. Data* **1985**, *14*, Suppl. 1.

12. For an example of the use of heats of atomization, see J. W. Ochterski, G. A. Petersson, and K. B. Wiberg, *J. Am. Chem. Soc.* **1995**, *117*, 11299.

13. Compare K. B. Wiberg and J. W. Ochterski, *J. Comput. Chem.* **1997**, *18*, 108.

14. J. L. Franklin, *Ind. Eng. Chem.* **1949**, *41*, 1070. Compare K. M. Engler, J. D. Andose, and P. v. R. Schleyer, *J. Am. Chem. Soc.* **1973**, *95*, 8005.

15. S. J. Blanksby and G. B. Ellison, *Acct. Chem. Res.* **2003**, *36*, 255.

16. W. F. Maier and P. v. R. Schleyer, *J. Am. Chem. Soc.* **1981**, *103*, 1891.

17. W. R. Roth, F.-G. Klarner, and H.-W. Lennartz, *Chem. Ber.* **1990**, *113*, 1818. W. Fang and D. W. Rogers, *J. Org. Chem.* **1992**, *57*, 2294.

18. C. A. Coulson and W. E. Moffitt, *Philos. Mag.* **1949**, *40*, 1.

19. C. A. Coulson, *Valence*, Clarendon Press, Oxford, **1952**, p. 200.

20. Unpublished calculations.

21. K. B. Wiberg and F. H. Walker, *J. Am. Chem. Soc.* **1982**, *104*, 5239. K. B. Wiberg, W. P. Dailey, F. H. Walker, S. T. Waddell, L. S. Crocker, and M. Newton, *J. Am. Chem. Soc.* **1985**, *107*, 7247.

22. K. B. Wiberg and S. T. Waddell, *J. Am. Chem. Soc.* **1990**, *112*, 2194.

23. J. D. Kemp and K. S. Pitzer, *J. Chem. Phys.* **1936**, *4*, 749.

24. R. M. Moriarty, *Top. Stereochem.* **1974**, *8*, 270.

25. M. K. Leong, V. S. Mastryukov, and J. E. Boggs, *J. Mol. Struct.* **1998**, *445*, 149.

26. E. L. Eliel and S. H. Wilen, *Stereochemistry of Organic Compounds*, John Wiley & Sons, Inc., New York, 1994, pp. 762ff.

27. F. A. L. Anet, *Top. Curr. Chem.* **1974**, *45*, 169.

28. F. Vogtle, *Cyclophane Chemistry: Synthesis, Structure and Reactions*, John Wiley & Sons, Inc., New York, **1993**.

29. D. J. Cram, R. B. Hornby, E. A. Truesdale, H. J. Heich, M. H. Dalton, J. M. Cram, *Tetrahedron*, **1974**, *30*, 1757.

30. D. J. Cram and R. A. Reeves, *J. Am. Chem. Soc.* **1958**, *80*, 3094.

31. G. J. Abruscato and T. T. Tidwell, *J. Org. Chem.* **1972**, *37*, 4151.

32. Compare M. S. Newman, *Steric Effects in Organic Chemistry*, John Wiley & Sons, Inc., New York, **1963**.

33. D. W. Rogers, H. von Voitherberg, and N. L. Allinger, *J. Org. Chem.* **1978**, *43*, 360.

34. M. Squillacote, A. Bergman, and J. De Felippis, *Tetrahedron Lett.* **1989**, *30*, 6805. G. M. Wallraff and J. Michl, *J. Org. Chem.* **1986**, *51*, 1794. Y. Inoue, T. Ueoka, T. Kuroda, and T. Hakushi, *J. Chem. Soc. Chem. Commun.* **1981**, 1031. E. J. Corey, F. A. Carey, R. A. E. Winter, *J. Am. Chem. Soc.* **1965**, *87*, 934.

35. R. G. Solomon, K. Folting, W. E. Streib, and J. K. Kochi, *J. Am. Chem. Soc.* **1974**, *96*, 1145. *trans*-1-Phenylcyclohexene has been observed spectroscopically (R. Bonneau, J. Joussot-Dubien, L. Salem, and A. J. Yarwood, *J. Am. Chem. Soc.* **1976**, *98*, 4329) and its strain energy has been measured via photoacoustic calorimetry (J. L. Goodman, K. S. Peters, H. Misawa, and R. A. Caldwell, *J. Am. Chem. Soc.* **1986**, *108*, 6803. It is 47 ± 3 kcal/mol less stable than the cis isomer.)

36. A. T. Blomquist, R. E. Burge, Jr., and A. C. Sucsy, *J. Am. Chem. Soc.* **1952**, *74*, 3636.

37. M. Traetteberg, *Acta. Chem. Scand.* **1975**, *B29*, 29.

38. R. C. Haddon, *J. Am. Chem. Soc.* **1987**, *109*, 1676.

39. K. B. Wiberg, *Acc. Chem. Res.* **1996**, *29*, 229.

40. A. E. Reed, L. A. Curtiss, and F. Weinhold, *Chem. Rev.* **1988**, *88*, 899.

41. P. M. Warner, *Chem. Rev.* **1989**, *89*, 1067.

42. R. Keese and E. P. Krebs, *Angew. Chem. Int. Ed. Engl.* **1971**, *10*, 262. Compare C. M. Geise and C. M. Hadad, *J. Am. Chem. Soc.* **2000**, *122*, 5861 and E. Kohl, T. Stroter, C. Siedschlag, K. Polborn, and G. Szeimies, *Eur. J. Org. Chem.* **1999**, 3057.

43. K. Lukin and P. E. Eaton, *J. Am. Chem. Soc.* **1995**, *117*, 7652.

44. M. Hare, T. Emrick, P. E. Eaton, and S. R. Kass, *J. Am. Chem. Soc.* **1997**, *119*, 237.

45. D. A. Hrovat and W. T. Borden, *J. Am. Chem. Soc.* **1988**, *110*, 4710. W. T. Borden, *Synlett* **1996**, 711.

46. K. B. Wiberg, N. McMurdie, J. V. McClusky, and C. M. Hadad, *J. Am. Chem. Soc.* **1993**, *115*, 10653.

47. F. H. Walker, K. B. Wiberg, and J. Michl, *J. Am. Chem. Soc.* **1982**, *104*, 2056. K. B. Wiberg, F. H. Walker, W. E. Pratt. and J. Michl, *J. Am. Chem. Soc.* **1983**, *105*, 3683.

48. C. J. Doubleday, *Am. Chem. Soc.* **1993**, *115*, 11968.

49. K. B. Wiberg, J. J. Caringi, M. G. Matturro, *J. Am. Chem. Soc.* **1990**, *112*, 5854.

50. P. Eaton, G. Temme, III, *J. Am. Chem. Soc.* **1973**, *95*, 7508. The compound that was studied had an *N,N*-dimethylcarboxamide group.

51. W.-D. Stohrer and R. Hoffmann, *J. Am. Chem. Soc.* **1972**, *94*, 779 (cf. Ref. 39).

52. P. B. Shevlin and A. P. Wolf, *J. Am. Chem. Soc.* **1970**, *92*, 406.

53. G. Maier, S. Pfriem, U. Schafer, and R. Matusch, *Angew. Chem. Int. Ed. Engl.* **1978**, *17*, 520. R. Notario, O. Castano, J. L. Andres, J. Elguero, G. Maier, and C. Hermann, *Chem.—A Eur. J.* **2001**, *7*, 342.

54. For a review of orbital symmetry, see R. B. Woodward and R. Hoffmann, *The Conservation of Orbital Symmetry*, Academic Press, New York, **1970**.

55. R. Criegee and K. Noll, *Ann. Chem.* **1959**, *627*, 1. R. E. K. Winter, *Tetrahedron Lett.* **1965**, 1207.

56. R. Breslow and P. Dowd, *J. Am. Chem. Soc.* **1963**, *85*, 2729.

57. M. S. Baird, H. Hussain, and W. Clegg, *J. Chem. Soc. Res. Synop.* **1988**, 110.

58. Compare W. E. Billups, W. Luo, G-A. Lee, J. Chee, B. E. Arney, Jr., K. B. Wiberg, and D. R. Artis, *J. Org. Chem.* **1996**, *61*, 764.

59. K. B. Wiberg, M. G. Matturo, P. J. Okarma, and M. E. Jason, *J. Am. Chem. Soc.* **1984**, *106*, 2194.

60. K. B. Wiberg and G. Bonneville, *Tetrahedron Lett.* **1982**, 5385.

61. G. Szeimies, in *Strain and its Implications in Organic Chemistry*, A. de Meijere and S. Blechert, Eds., Kluwer Academic Publishers, Dordrecht, The Netherlands, **1988**, p. 361.

62. M. D. Levin, P. Kaszynski, and J. Michl, *Chem. Rev.* **2000**, *100*, 169.

63. K. B. Wiberg and N. McMurdie, *J. Am. Chem. Soc.* **1991**, *113*, 8995.

64. R. Baird and A. A. Aboderin, *J. Am. Chem. Soc.* **1964**, *86*, 252, 2300.

65. M. Saunders, P. Vogel, E. L. Hagen, and J. Rosenfeld, *Acc. Chem. Res.* **1973**, *6*, 53.

66. K. B. Wiberg and S. R. Kass, *J. Am. Chem. Soc.* **1985**, *107*, 988.

67. D. M. Adams, J. Chatt, G. Guy, and N. Sheppard, *J. Chem. Soc.* **1961**, 738.

68. L. Cassar, P. E. Eaton, and J. Halpern, *J. Am. Chem. Soc.* **1970**, *92*, 3535.

69. B. Rybtchinski and D. Milstein, *Angew. Chem. Int. Ed. Engl.* **1999**, *38*, 871. K. C. Bishop, III, *Chem. Rev.* **1976**, *76*, 461.

70. J. J. Burke and P. C. Lauterbur, *J. Am. Chem. Soc.* **1964**, *86*, 1870.

71. D. J. Patel, M. E. H. Howden, and J. D. Roberts, *J. Am. Chem. Soc.* **1963**, *85*, 3218. C. D. Poulter, R. S. Boikess, J. I. Brauman, and S. Winstein, *J. Am. Chem. Soc.* **1972**, *94*, 1191.

72. H.-O. Kalinowsi, S. Berger, and S. Braun, *Carbon-13 NMR Spectroscopy*, John Wiley & Sons, Inc., Chichester, **1988**.

73. M. B. Robin, *Higher Excited States of Polyatomic Molecules*, Vol. 1, Academic Press, New York, **1974**, pp. 140 ff.

74. V. A. Walters, C. M. Hadad, Y. Thiel, S. D. Colson, K. B. Wiberg, P. M. Johnson, and J. B. Foresman, *J. Am. Chem. Soc.* **1991**, *113*, 4782.

75. K. B. Wiberg and R. E. Rosenberg, *J. Phys. Chem.* **1992**, *96*, 8282. K. B. Wiberg, R. E. Rosenberg, and S. T. Waddell, *J. Phys. Chem.* **1992**, *96*, 8293.

CHAPTER 16

Arynes

MICHAEL WINKLER, HANS HENNING WENK, and WOLFRAM SANDER

Lehrstul für Organische Chemie II der Ruhr-Universitat Bochum, 44780 Bochum, Germany

1. INTRODUCTION

Arynes were postulated as reactive intermediates more than 100 years ago.[1] First indications for the existence of such molecules were reported by Stoermer and Kahlert[2] in 1902. The isolation of 2-ethoxybenzofuran (**3**) after

Reactive Intermediate Chemistry, edited by Robert A. Moss, Matthew S. Platz, and Maitland Jones, Jr.
ISBN 0-471-23324-2   Copyright © 2004 John Wiley & Sons, Inc.

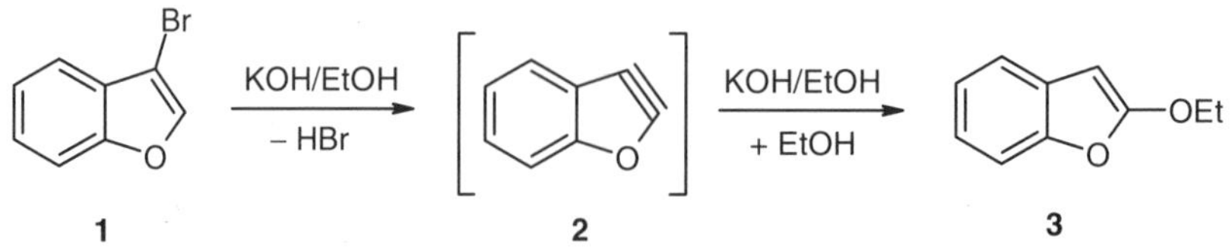

Scheme 16.1. The reaction **1** → **3** gave the first hint on the existence of an aryne, *o*-didehydrobenzofuran (**2**).[2]

treatment of 3-bromobenzofuran (**1**) with bases in ethanol suggests that *o*-dide-hydrobenzofurane (**2**) is formed in the course of this reaction (Scheme 16.1).[2]

Despite occasional speculation about arynes in the following years,[3] it took several decades until the groups around Roberts,[4] Huisgen,[5] and Wittig[6] could obtain definitive evidence for the existence of these reactive intermediates. The discovery of the endiyne cytostatics[7] has initiated a renaissance in aryne chemistry and many research efforts in this area have been reported during the last 20 years.[1,8] Furthermore, benzynes have been recognized as important reaction intermediates in combustion processes,[9] and they are expected to play a key role in the transformations of aromatics on transition metal surfaces.[10] This chapter will focus primarily on direct spectroscopic and theoretical investigations of arynes, whereas indirect evidence, as, for example, from trapping reactions, is only dealt with incidentally. Biological aspects and research directed toward the reactivity of endiynes with desoxyribonucleic acid (DNA), though very important, are not discussed in this chapter, because they have been reviewed elsewhere in more detail recently.[1]

2. THE PARENT BENZYNES

2.1. The *o*-Benzyne Story

o-Benzyne (**4**) was first suggested as a reactive intermediate in 1927,[11] and the existence of this species was further elaborated by Wittig in the following years.[6] In a classic investigation in 1953, Roberts et al.[4a] found compelling evidence for the intermediate formation of **4** in the reaction of ^{14}C labeled chlorobenzene with sodium amide (Scheme 16.2).

Formula 16.1

In 1956 Wittig was able to trap *o*-benzyne in various Diels–Alder reactions.[6c] Nonetheless, he remained skeptical about his own hypothesis of the existence of

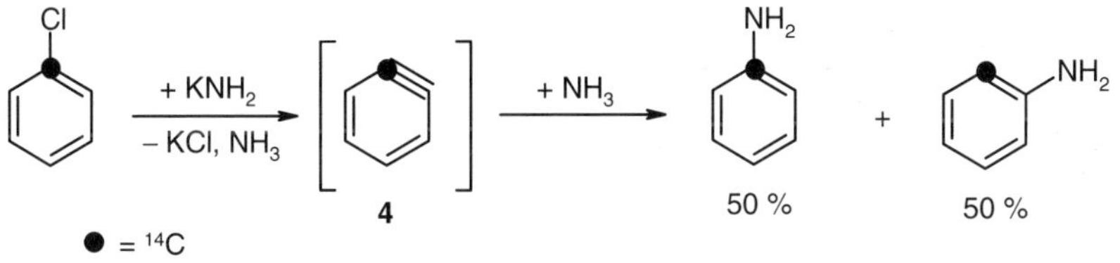

Scheme 16.2. In a classic experiment, Roberts et al.[4a] demonstrated the involvement of **4** in the nucleophilic aromatic substitution by isotopic labeling experiments. The product distribution can only be explained by the assumption of a symmetric intermediate.

4,[6d] until Huisgen and Knorr[5c] could demonstrate in an elegant work that the same intermediate is formed from four different precursors of **4** (Scheme 16.3). At approximately the same time, Fisher and Lossing[12] investigated the pyrolysis of the three isomeric diiodobenzenes using mass spectrometry (MS) and identified **4**

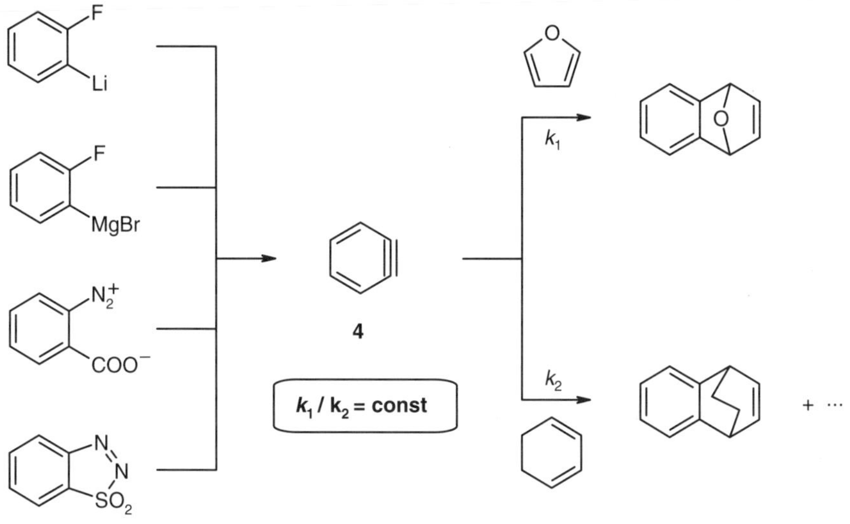

Scheme 16.3. In a very elegant trapping experiment, Huisgen and Knorr demonstrated that the same intermediate is formed from four different precursors.[5c] If **4** is trapped by a mixture of furan and cyclohexadiene, the same product ratio is observed in all four reactions.

on the basis of the measured ionization potential. Berry et al.[13] studied the photo-initiated decomposition of benzenediazonium carboxylates and characterized **4** by its ultraviolet (UV) and mass spectrum in the gas phase.

The first direct infrared (IR) spectroscopic detection of *o*-benzyne was accomplished by Chapman et al.,[14] using matrix isolation spectroscopy at very low temperatures to generate **4** starting from phthaloyl peroxide (**5**) and benzocyclo-

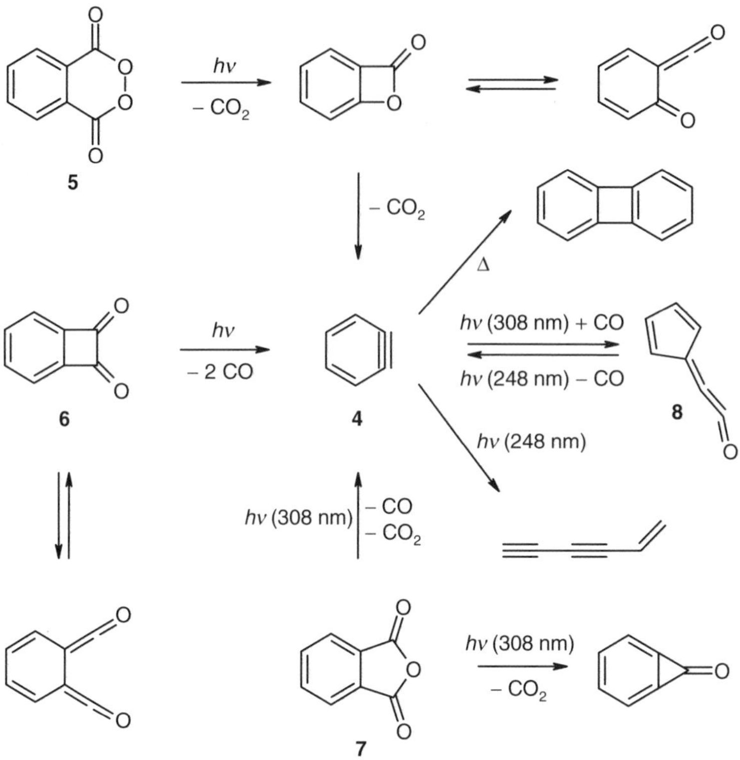

Scheme 16.4. Photochemistry of **4** and the corresponding precursors.

butenedione (**6**, Scheme 16.4). High yields of **4** were also obtained using phthalic anhydride (**7**) as a precursor. The comparatively complex reactions of the molecules involved in this photochemistry led to some controversy about the frequency of the $C\equiv C$ stretching vibration, surely the most interesting feature in the IR spectrum of **4**. Chapman et al.[14] assigned a band at 2085 cm^{-1} to the $C\equiv C$ stretching vibration and this finding was confirmed in a later investigation by Dunkin and MacDonald.[15] Wentrup et al. found this band at 2080 cm^{-1} in a phthalic anhydride matrix at 77 K. On annealing of the matrix a second overlapping absorption at the same position was observed, which was assigned to cyclopentadienylideneketene (**8**).[16] The assignment of the 2085-cm^{-1} trace to the $C\equiv C$ stretching vibration in **4** was further supported theoretically.[17] Doubts arose when Leopold et al.[18] inferred a frequency of 1860 cm^{-1} from the gas-phase PE spectrum. In order to clarify this discrepancy, Schaefer and co-workers[19] calculated the IR spectrum of **4** and proposed that the absorption should be found in the range from 1965 to 2010 cm^{-1}. In 1990 Schweig and co-workers[20] were able to demonstrate that an absorption at 2087 cm^{-1} belongs to cyclopentadienylideneketene rather than **4**. This issue was finally settled in 1992 by Radziszewski et al.,[21] who could definitively identify the frequency of the $C\equiv C$ stretching vibration at 1846 cm^{-1} by thorough analysis of the IR spectra of various isotopomers of **4**.

As expected, the formal C—C triple bond in benzyne is significantly weaker than in unstrained alkynes, the $C\equiv C$ stretching vibrations of which usually fall in the region $\sim$2150 cm^{-1}. Nevertheless, o-benzyne is better described as a strained alkyne rather than a biradical, which is evident from the large singlet–triplet splitting of 37.5 ± 0.3 kcal/mol[22] as well as the alkyne-like reactivity (e.g., in Diels–Alder reactions). The enthalpy of formation of **4** was determined to be 106.6 ± 3.0 kcal/mol by Wenthold and Squires.[23] For the $C\equiv C$ bond length a value of 124 ± 2 pm was found experimentally,[24] which comes closer to a typical C—C triple bond (120.3 pm in acetylene) rather than a C—C double bond (133.9 pm in ethylene).

In remarkable work, Warmuth et al.[25] were able to isolate o-benzyne in a hemi-carcerand—a "molecular container"—and to measure its NMR spectrum in solution. Whereas previous theoretical studies[26] concerning the structure of **4** had to rely more or less solely on IR spectra,[27] there were now additional experimental data available to judge the quality of the theoretical predictions.[27b] The ^{1}H and ^{13}C chemical shifts were most accurately reproduced using the B3LYP/6-311G** structure $[r(C_1C_2) = 124.5$ pm], while calculations on the basis of CASSCF(8,8)/DZP or CCSD(T)/6-31G** geometries led to poorer agreement with experiment.[27b] Furthermore, it was concluded that **4** more closely resembles a cyclic alkyne rather than a cumulene.

The important role of o-benzyne in combustion processes[28] has been realized only recently. In 1997, Lin and co-workers[9a] suggested that loss of a hydrogen atom from the phenyl radical, which is well known to be a key intermediate in the formation of polycyclic aromatic hydrocarbons (and soot) in flames, is a dominant process in the combustion of lead-free gasoline (which contains up to 20–30% of small aromatics). A deeper understanding of combustion processes thus requires a knowledge of the high-temperature chemistry of arynes. In this context, a series of elegant studies by R.F.C. Brown et al.[29–31] deserves special attention. Flash vacuum pyrolysis (FVP) of phthalic anhydride (**7**) ($T = 850$ °C), partly (5%) ^{13}C labeled in the 1 and 6 position, gave a pyrolysate containing biphenylene in which one ^{13}C label was distributed between two quaternary positions (Scheme 16.5).[29a] This rearrangement was explained in terms of a ring contraction of benzyne (**4**) leading to the exocyclic carbene (**9**), a reaction analogous to the acetylene–methylenecarbene (vinylidene) rearrangement. Different precursors (**6**, $T = 650$–830 °C; **11**, $T = 400$–850 °C) have been used, all of which contain a CO moiety in the leaving group, however.[29b,29c] The interpretation of the labeling experiments has been questioned by Wentrup et al., who proposed that the observed scrambling may better be explained by a Wolff type ring contraction of the intermediate ketenecarbene (**12**, Scheme 16.5).[32] Thus, the mechanism of the thermal decomposition of **7** has not been fully clarified, yet although the aryne contraction pathway has been established for some benzannellated derivatives of **4**.[29] A number of theoretical studies has been devoted to rearrangements on the C$_6$H$_4$ potential energy surface.[9b,33] At

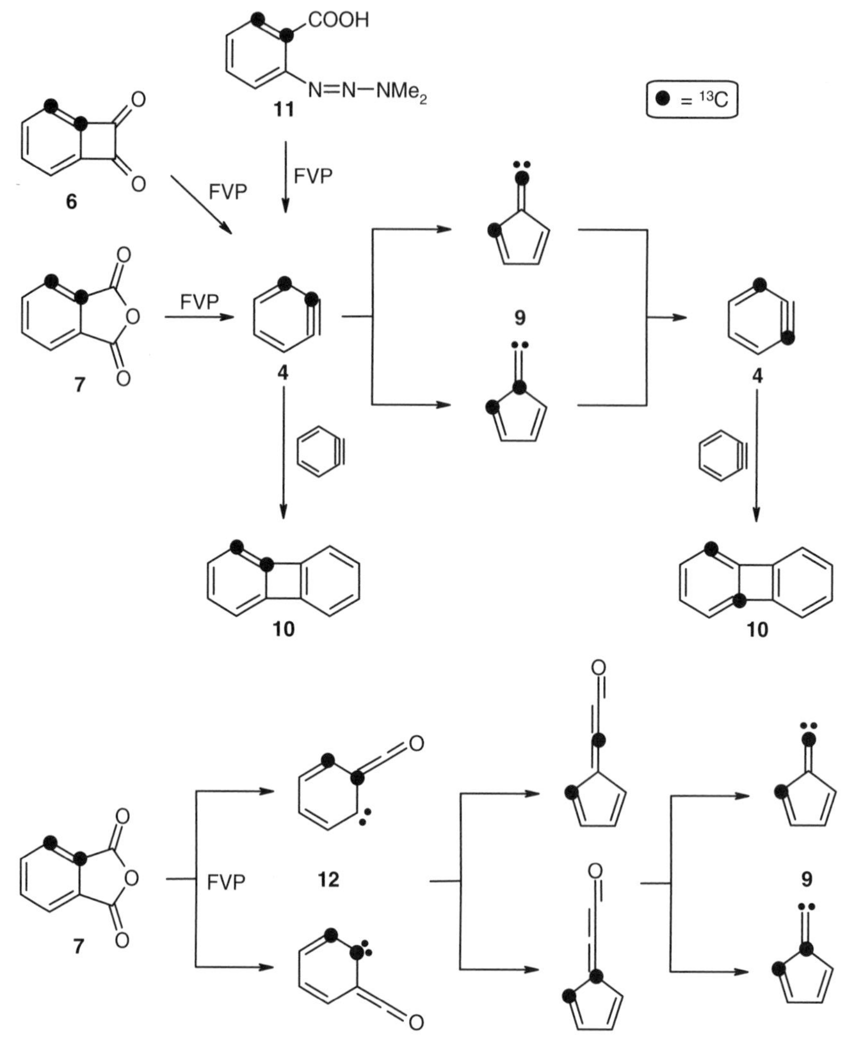

Scheme 16.5. The FVP of isotopically labeled phthalic anhydride (**7**) indicates a scrambling of carbon atoms. Yet, it is not clear whether a reversible ring contraction of **4** is responsible for this exchange or if it takes place already at the stage of ketenecarbene (**12**).[29,30,32] According to recent calculations, the barrier for the rearrangement **4** → **9** is only 32 kcal/mol, while the back-reaction proceeds with a very small or even vanishing barrier.[33] Thus, at 850 °C the ring contraction of **4** to **9** should at least energetically be possible.

the G2M level of theory the reaction **4** → **9** is endothermic by 31 kcal/mol and the barrier is only slightly larger (32 kcal/mol), thus rendering this reaction of *o*-benzyne possible at elevated temperatures.[33]

   o-Benzyne also plays an important role in benzene transformations and decomposition on transition metal surfaces and thus in heterogenous catalysis.[10] An ordered overlayer of adsorbed benzyne on Ir{100} has recently been prepared by thermal decomposition of adsorbed benzene, and characterized by low-energy electron diffraction (LEED). It has been demonstrated that **4** is di-σ-bonded to the surface with its ring plane at 47.2° to the surface normal.[10a] In a theoretical study, it has been shown that this tilt arises from back-bonding interactions of the aromatic π system with the surface *d* orbitals.[10b]

This short outline of the history of *o*-benzyne, one of the classic reaction intermediates, already shows how difficult apparently easy things (assigning IR absorptions or investigating elementary rearrangements) can be. Progress in aryne chemistry generally comes only slowly; despite the long history, many studies remain to be done in this challenging field of research.

2.2. The *meta*-Benzyne Story

Early attempts to generate and characterize *m*-benzyne (**13**) comprise the pyrolysis of 1,3-diiodobenzene (**14**)[12] and the flash photolysis of benzenediazonium-3-carboxylate (**15**).[34] While in the former case only enediynes (**16**) were in accordance with the ionization potentials determined (pyrolysis temperature: 960 °C), the results of the latter work remain unclear. A transient with $m/z = 76$ and a lifetime of 400 μs was observed, however, it was not possible to unambiguously identify this species as **13** due to the complexity of the product spectrum (Scheme 16.6). Indirect evidence for the existence of **13** stems from trapping experiments.[35–37]

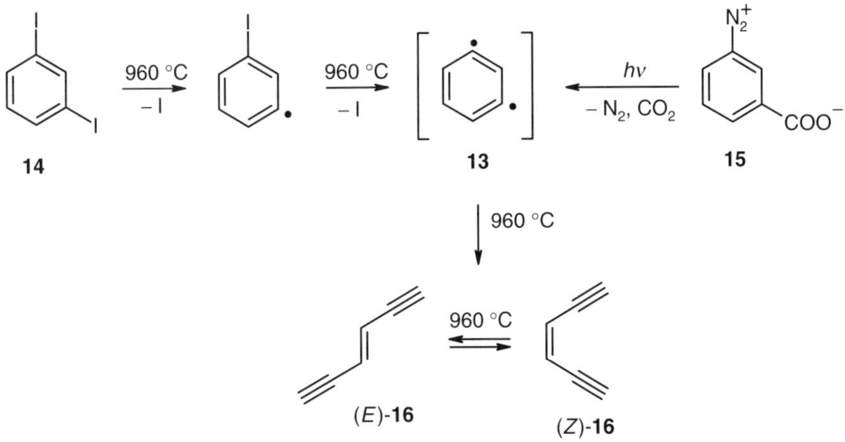

Scheme 16.6. Early attempts to generate *m*-benzyne did not give conclusive evidence for existence and structure of **13**.

In 1975, Washburn et al.[38] investigated the dehydrohalogenation of 2,6-dibromo-bicyclo[3.1.0]hex-3-ene (**17**) with bases and postulated the intermediate formation of bicyclo[3.1.0]hexatriene (**18**, Scheme 16.7).

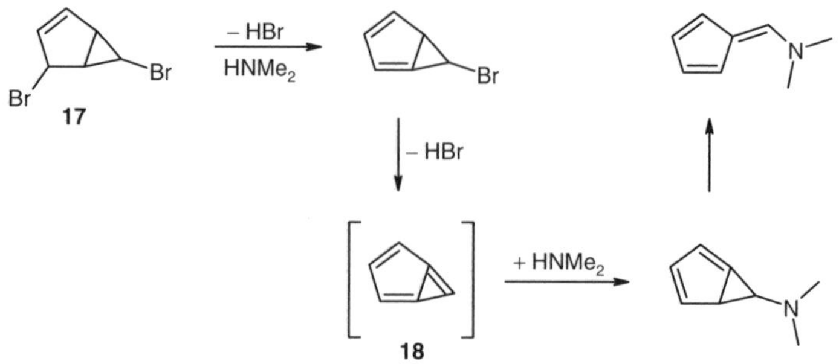

Scheme 16.7. From the observed trapping chemistry Washburn et al.[38] concluded that the bicyclic intermediate **18** is formed instead of biradical **13**. The question of whether **18** exists has been the subject of some controversy until recently.

The first spectroscopic evidence for a derivative of **13** was presented by Bucher et al.[39] in 1992, who were able to isolate 2,4-didehydrophenol (**19**) in cryogenic matrices (Scheme 16.8). Irradiation of matrix isolated quinone diazide (**20**) with light of wavelength $\lambda = 432 \pm 10$ nm yields carbene (**21**), which is

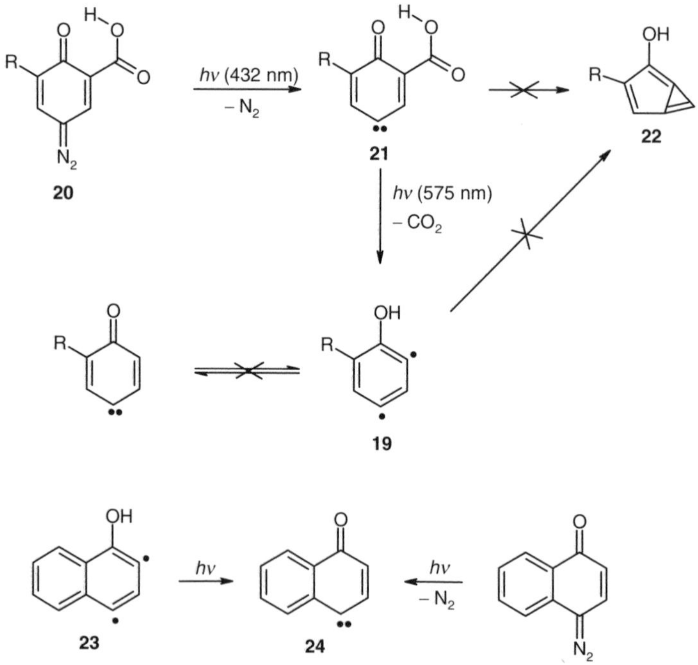

Scheme 16.8. Didehydrophenol (**19**) was the first derivative of **13** that could be isolated in an argon matrix at 8 K. Comparison of measured and calculated spectra supports the assignment to a monocyclic structure **19**. Evidence for bicyclic isomer **22** was not found. In contrast to **19**, derivative **23** can be photochemically converted into carbene **24**.[39]

photochemically converted into didehydrophenol (**19**) by long wavelength irradiation ($\lambda = 575 \pm 10$ nm). The existence of an OH group in **19** was confirmed by
isotopic labeling using $[D]_1$-**20**.[39] Evidence for a bicyclic isomer **22** was not
found.[40] Benzannellated didehydrophenol (**23**) tautomerizes photochemically to
carbene (**24**), in contrast to didehydrophenol (**19**), which yields ring-opened products of unknown constitution upon irradiation.[39]

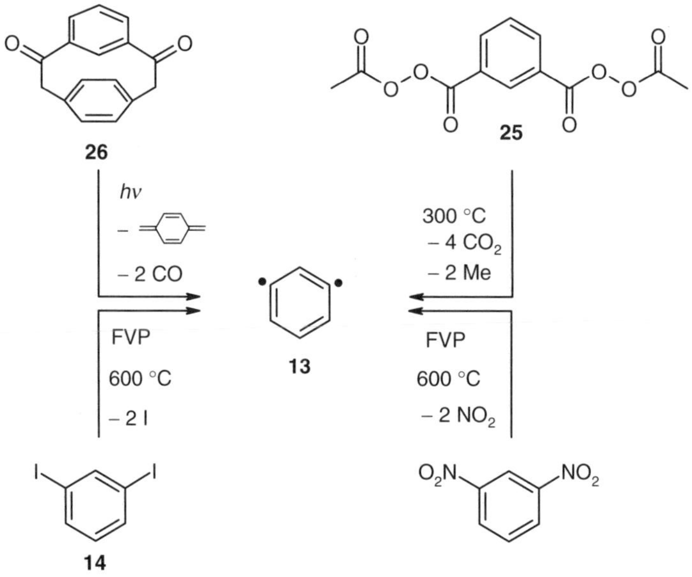

Scheme 16.9. The matrix isolation and IR spectroscopic characterization of **13** was achieved
starting from four different precursors.[41,42]

The isolation of the parent *m*-benzyne (**13**) in a cryogenic matrix was finally
achieved in 1996 starting from diacetyl peroxide (**25**) as a thermal source and
m-*p*-cyclophane-9,10-dione (**26**) as a photochemical precursor (Scheme 16.9).[41]
In recent work, the IR measurements were extended to the far infrared (FIR) region
(>200 cm^{-1}) and additional starting materials were used.[42] In particular, it
was shown that pyrolysis of 1,3-diiodobenzene at 600 °C gives good yields of *m*-
benzyne, together with (*Z*)- and (*E*)-hex-3-ene-1,5-diyne (**16**), the latter being
formed exclusively at temperatures >650 °C, in accordance with the earlier results
obtained by Fisher and Lossing.[12] The FVP of 1,3-dinitrobenzene also yields small
amounts of **13**.[42] By using these four different routes to *m*-benzyne, all of the more
intense IR absorptions of **13** were unambiguously identified. The IR spectrum of the
perdeuterated isotopomer $[D]_4$-**13** was also obtained.[42]

The vibrational spectrum of **13** is nicely reproduced by quantum chemical calculations at the CCSD(T) level of theory,[42] whereas density functional theory
(DFT) shows a variable performance for **13** depending on the functional
employed.[43–46] This caution holds especially for the distance between the radical

TABLE 16.1. Calculated Equilibrium Distances r(C1C3) of the Radical Centers in m-Benzyne[a]

Method	Basis Set		
	cc-pVDZ	cc-pVTZ	cc-pVQZ
HF	148.8	147.9	147.8
BHandHLYP	153.2	152.1	152.1
SPW91	155.0	154.0	154.0
MPW1PW91	155.9	154.9	154.9
B3PW91	157.3	156.2	156.2
B3P86	157.4	156.3	156.2
SVWN	158.1	156.9	157.0
SLYP	158.0	157.0	157.1
SVWN5	158.6	157.5	157.6
MPW1LYP	160.1	159.0	159.1
B3LYP	161.5	160.3	160.4
BPW91	187.8	184.3	182.2
BP86	193.3	190.4	189.9
BLYP	202.1	199.7	199.5
BVWN	203.4	200.7	200.4
BVWN5	204.1	201.4	201.2
BCCD(T)[b]	211 ± 1	204 ± 1	203 ± 1
RCCSD(T)[b]	211 ± 1	205 ± 1	203 ± 1
CAS(8,8)–RS3[b]	211 ± 1	205 ± 1	203 ± 1
CAS(8,8)–CISD+Q[b]	213 ± 1	207 ± 1	
CAS(8,8)–CISD[b]	215 ± 1	210 ± 1	
CAS(8,8)–RS2[b]	215 ± 1	210 ± 1	209 ± 1

[a] References 44 and 46; distances r(C1C3) in picometers.
[b] Lowest energy structure calculated for (V)B3LYP/cc-pVTZ optimized geometries with constrained distances r(C1C3).

centers and the question of whether the matrix isolated species can also be assigned to the bicyclic structure **18**. In more recent theoretical and experimental investigations, it was shown beyond any doubt that **13** has a monocyclic structure with a distance of 205 ± 5 pm between the radical centers (Table 16.1).[44,45] Nevertheless, the potential energy surface along the "cyclization coordinate" is very flat. Despite the significant distance of the radical centers the biradical character of **13** is low (19–32%, depending on the definition of this quantity).[26q,44] The electronic structure of **13** is most appropriately described as a σ-allylic system, in which primarily the σ*(C—H) orbital, located between the dehydrocarbons, participates in the interaction of the radical lobes. Negative-ion photoelectron spectroscopy (NIPES) experiments also support a monocyclic structure for m-benzyne, although the spectra of **13** are considerably more complex than in the case of the ortho and para-isomers.[22] The singlet–triplet energy splitting was determined to be 21.0 ± 0.3 kcal/mol, confirming the low biradical character of **13**.[22]

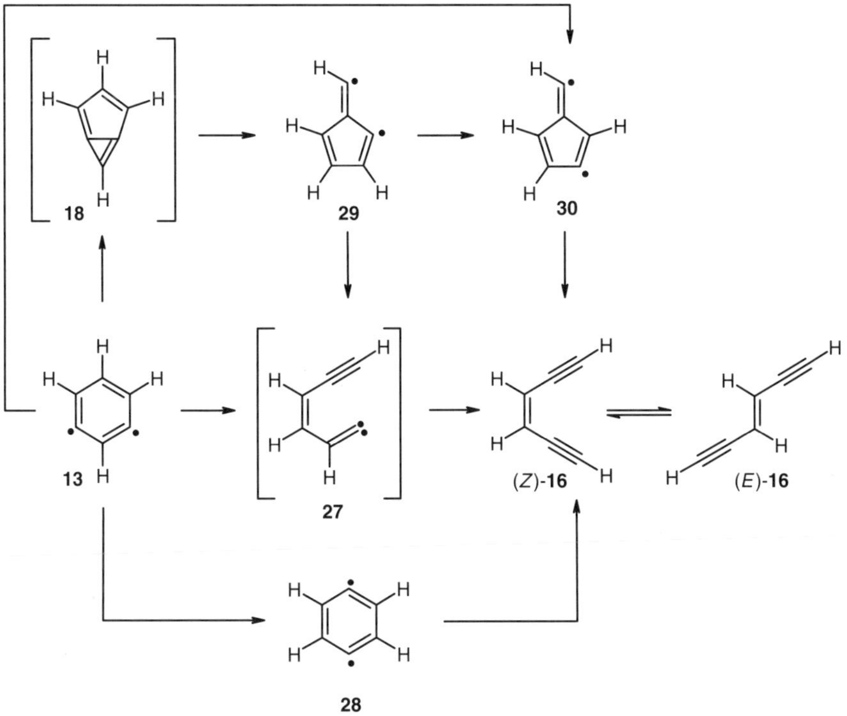

Scheme 16.10. The most probable pathways for the **13** → **16** rearrangement.[42]

The mechanism of the rearrangement of *m*-benzyne (**13**) to hex-3-ene-1,5-diyne (**16**) has been elucidated recently.[42] Possible isomerization pathways include ring opening to a vinylidene **27** with subsequent hydrogen-shift or rearrangement to the para-isomer **28** followed by *retro*-Bergman reaction (Scheme 16.10, and Section 2.3).[42,47] Experimental data obtained by Zewail and co-workers[33] using femtosecond time-resolved spectroscopy with mass spectrometric detection on the photolysis of 1,2-, 1,3-, and 1,4-dibromobenzene were interpreted by the authors in terms of a rapid equilibrium between *o*-, *m*-, and *p*-benzyne via quantum mechanical tunneling of hydrogen atoms, from which the para-isomer reacts to the entropically favored enediyne **16**. Investigations by Sander et al.,[42] however, favor an alternative mechanism of direct ring opening of **13**. According to DFT calculations the vinylidene (**27**) does not occupy a stationary point on the potential energy surface, and ring opening and hydrogen migration occur in one step. At the (U)B3LYP/6-31G(d,p) level the enthalpy of activation for the rearrangement **13** → **16** is 48.5 kcal/mol, whereas that for **13** → **28** is 64.6 kcal/mol. At the G2M level of theory, the latter reaction requires an activation enthalpy of 60–61 kcal/mol.[9b,33] The rearrangement of **13** to **30** requires 72 kcal/mol (B3LYP),[42] whereas the ring opening of **18** to give biradical **29** has not yet been discussed in the literature. Rearrangement of *m*- to *o*-benzyne (**4**) has an activation enthalpy of 56 kcal/mol at the G2M level, although it is not entirely clear, whether this reaction leads initially to intermediate **9** (cf. Scheme 16.5).[9b,33] Since no *o*-benzyne is observed under FVP conditions at 600 °C, hydrogen scrambling in *m*-benzyne seems to be less likely than an alternative lower energy pathway (e.g., **13** → [**27**] → **16** or **13** → [**18**] → **29** → **16**). The rearrangment **13** → **16** could also be induced photochemically.[42]

In summary, structure and vibrational spectrum of *m*-benzyne have been clarified during the last years. The (surprisingly) large variability of different quantum chemical methods in accounting for the structure and properties of **13** makes a more detailed exploration of intramolecular rearrangements of *m*-benzyne at more sophisticated (and coherent) levels of theory desireable. Even more complex appears to be the study of the intermolecular chemistry of **13**. Almost nothing is known about thermal or photochemical reactions of *m*-benzyne with atoms or small molecules. A method for the investigation of (charged) *m*-benzyne derivatives in the gas phase using FT ion cyclotron resonance MS has recently been described by Kenttämaa and co-workers[48] (distonic ion approach).[48] It could be shown that the reactivity of *m*-benzyne is reduced compared to that of the phenyl radical because of strong coupling of the formally unpaired electrons. In addition, it was demonstrated that charged derivatives of **13** are reactive towards electrophiles and nucleophiles.[48] Further insight into the intra- and intermolecular reactivity of *m*-arynes can be expected in the near future, although many technical problems (concerning both experiment and theory) have to be solved, before the chemistry of these compounds can be completely clarified.

2.3. The *p*-Benzyne Story

Experimental access to *p*-benzyne **28** turned out to be even more difficult than to the meta-isomer **13**. Pyrolysis of 1,4-diiodobenzene results in exclusive formation of hex-3-ene-1,5-diyne **16**.[12] Berry's attempt to generate **28** by photoinitiated decomposition of benzene diazonium-4-carboxylate in the gas phase gave dubious results.[34] Although the authors postulated the formation of **28**, a lifetime of >2 min was measured for the product observed. Such a lifetime is incompatible with a reactive biradical, in particular since the lifetime of ortho- and meta-isomers **4** and **13** are reported to be in the order of milliseconds and microseconds, respectively.[34]

Despite some earlier publications pointing out the intermediate formation of para-arynes,[49] **28** came into the focus of organic chemistry not before the classic experiments of Jones and Bergman in 1972.[50] The authors observed that 1,6-[D]$_2$-(Z)-hex-3-ene-1,5-diyne (**16**) rapidly reacts to a mixture of 3,4-[D]$_2$- and 1,6-[D]$_2$-**16** at 300 °C. The fact that no 1,3-[D]$_2$- or 1,4-[D]$_2$-**16** is formed can only be explained by presumption of a cyclic, symmetrical intermediate (Scheme 16.11).[50] Numerous trapping experiments supported the hypothesis of the thermal formation of *p*-benzyne from (Z)-**16**. In CCl$_4$, for example, thermolysis of the enediyne yields 1,4-dichlorobenzene.[50,51] The possibility of an alternative C1–C5 cyclization of enediyne **16** yielding fulvenediyl **30** was recently proposed by Schreiner and co-workers.[52] However, the reaction is endothermic by ~40 kcal/mol, and the cyclization barrier amounts to ~42 kcal/mol.[52] Still, an experimental

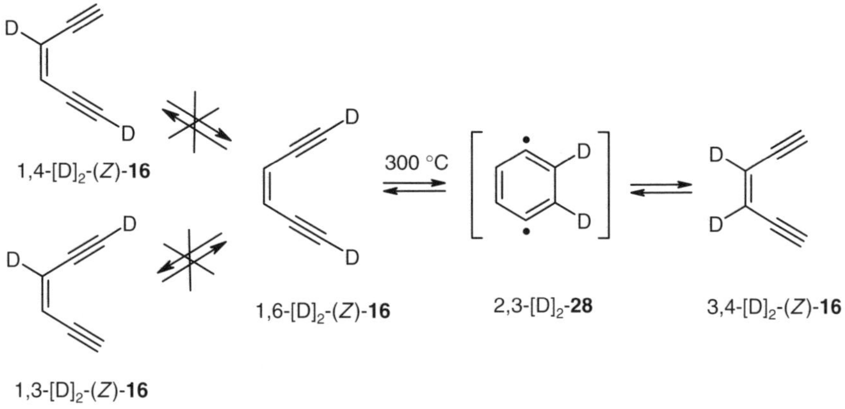

Scheme 16.11. The Bergman cyclization.[50]

realization of this cyclization might be facilitated by adding sterically demanding substituents to the alkyne termini.[53]

While the majority of trapping experiments favors a biradicaloid structure, evidence for a bicyclic intermediate (butalene, **31**) was also found.[54] Breslow et al.[54] reported the formation of **31** (the possible existence of which had been pointed out by Dewar before[55]) upon dehydrohalogenation of 3-chloro[2.2.0]bicyclohexadiene

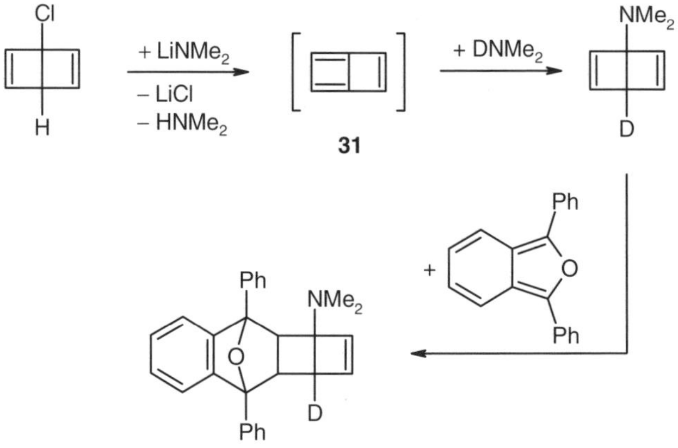

Scheme 16.12. Indirect evidence for the formation of butalene upon dehydrohalogenation of 3-chlorobicyclo[2.2.0]hexadiene.[52]

(Scheme 16.12). The bicyclic nature of the intermediate was inferred from the observed Diels–Alder type reactivity with activated dienes.

The energy profile of the reaction **28** ↔ **31** and the question whether butalene exists at all have been a matter of controversy until recently. According to DFT calculations, the biradical structure **28** is more stable by 36–39 kcal/mol than the highly strained butalene.[43a,56] On the basis of resonance energies[57] and isodesmic equations[56] it has been shown that **31** is considerably less antiaromatic than cyclo-butadiene. Ring current patterns in **31** and related bicyclic molecules (e.g., **18**) have recently been analyzed as well.[58] The barrier for ring opening of butalene (**31**) is ∼4–6 kcal/mol, which could be sufficient to allow its isolation in cryogenic matrices.[43a,56] Most interestingly, the **28–31** rearrangement proceeds through a transition state with only D_2 symmetry, according to DFT calculations.[43a,56]

One of the most important motivations for the investigation of *p*-benzyne and the Bergman reaction was the discovery of the enediyne cytostatics, namely, calichea-micins, esperamicins, and dynemicins, which are able to cleave the DNA double strand sequence specifically in vivo.[7] These substances have in common a (*Z*)-hex-3-ene-1,5-diyne moiety, usually embedded in a complex cyclic system, which, upon activation, forms a biradical that attacks DNA by abstracting two hydrogens from neighboring DNA strands (Fig. 16.1). Endiyne cytostatics, stereoelectronic effects on the Bergman reaction and on the hydrogen-abstraction reactivity of

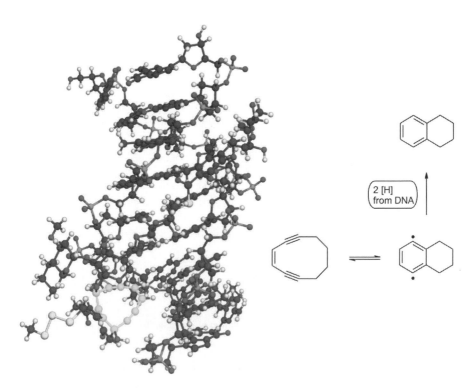

Figure 16.1. The structure of calicheamicin.

p-benzynes have been reviewed recently and the interested reader is refered to the cited literature for further information.[1,59]

Detailed thermochemical data for the Bergman rearrangement were determined by Roth et al.[60] from gas-phase NO trapping experiments. The activation barrier for ring opening of **28** to enediyne (*Z*)-**16** was reported as 19.8 kcal/mol, the enthalpy of formation of (*Z*)-**16** ($\Delta H^{\circ}_{f,298} = 129.5$ kcal/mol) is lower than that of *p*-benzyne ($\Delta H^{\circ}_{f,298} = 138 \pm 1$ kcal/mol) by 8.5 kcal/mol. The too low values for the enthalpy of formation of **28** determined in an earlier work by Squires and co-workers[23a] using collision induced dissociation (CID) measurements was corrected to 137.8 ± 2.9 kcal/mol in a later reinvestigation, now in perfect agreement with Roth's data.[22] Problems are encountered with an exact quantum chemical treatment of biradicals (and in particular their reactions), which requires a balanced treatment of dynamic and nondynamic (near-degeneracy) contributions to the correlation energy. On the other hand, precise experimental data are available for *p*-benzyne and the Bergman cyclization. Thus, this reaction has become a testing ground for ab initio and DFT methods.[61] Still, there is no general agreement about the performance and suitability of different quantum mechanical approximations, especially with regard to density functional theory.

In certain cases, the Bergman cyclization can be triggered photochemically, although the number of examples described in the literature is limited.[62] Recently, the electronic details of the photochemical enediyne cycloaromatization were investigated computationally.[63] The low-lying excited states of enediyne (**16**) can be described as a linear combination of the configurations of weakly interacting ethylene and acetylene units. According to this model, excitation of the ethylene fragment induces cis–trans isomerizations and hydrogen abstractions in addition to Bergman cyclization, and the photochemical cycloaromatization is most likely initiated by excitation of the acetylene unit. Moreover, the cyclization is more likely to occur on the $2^{1}A$ potential energy surface than from the $1^{3}B$ state. Although the reaction is more exothermic in the latter case (-42 kcal/mol compared to -18 kcal/mol at the CASMP2 level), the probability of cis–trans isomerization and hydrogen-abstraction is increased concurrently.[63] Consistently, Hopf et al.[64] observed photochemical cis–trans isomerization of **16** in the presence of a triplet sensitizer (Michler's ketone). Note, however, that systematic investigations of factors (substituents or sensitizers) facilitating the photochemical ring closure or inhibiting competing reaction pathways, have not been carried out yet, although this reaction is of potential interest for tumor therapy.

The first direct observation of a derivative of *p*-benzyne was reported by Chapman et al.[65] in 1976. Irradiation of anthraquinone bisketene (**32**) in organic glasses at 77 K produced a new species with UV absorptions in the range from 256 to 449 nm. After subsequent warming to room temperature, anthracene was isolated from the solution. An analogous experiment in the presence of 5% CCl_4 yielded 9,10-dichloroanthracene. In an argon matrix at 10 K, two IR absorptions at 710 and 760 cm^{-1} were observed for the new species. Based on the trapping reactions and the absence of an electron spin resonance (ESR) signal, the photolysis product was identified as 9,10-didehydroanthracene $^{1}A_{1}$-**33** (Scheme 16.13).[65] In

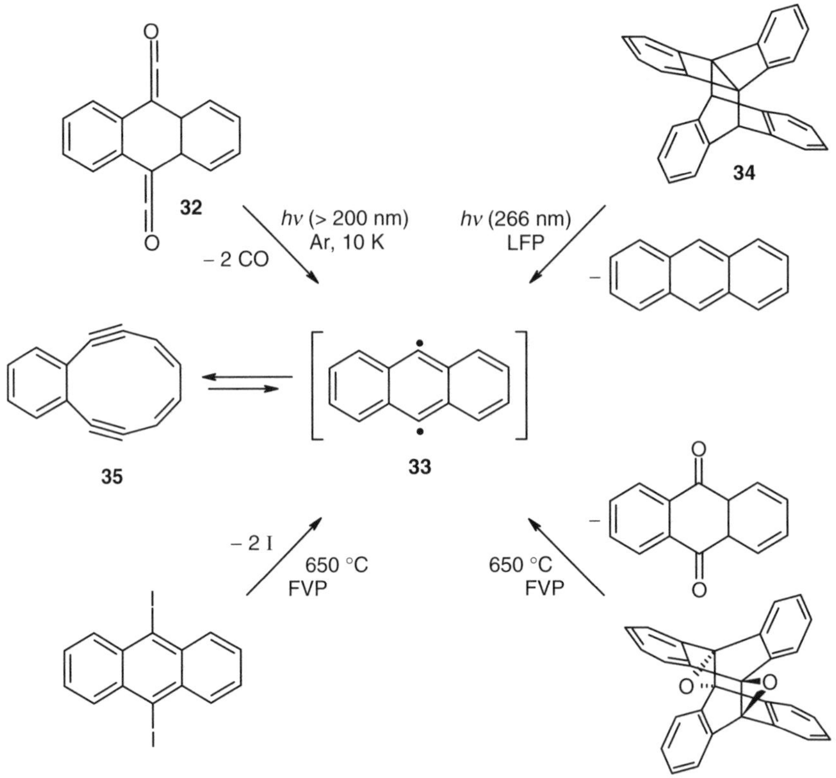

Scheme 16.13. 9,10-Didehydroanthracene (**33**) was generated from various precursors, but rapidly rearranges to cyclic diyne **35**.

1996, Chen and co-workers investigated the photochemistry of dehydrodianthracene (**34**) using laser flash photolysis (LFP) and were able to identify two transients: a short lived species with a lifetime of 2.1 μs and an absorption maximum at 295 nm, and a long-lived species (500 μs) with an absorption at 335 nm. The short-lived transient was assigned to 9,10-didehydroanthracene (**33**), the long-lived photolysis product was identified as 3,4-benzocyclododeca-3,7,9-triene-1,5-diyne (**35**).[66a] Since the UV spectroscopic data of **33** from the LFP study were in disaccord with Chapman's early study, Sander and co-workers[67] reinvestigated the photochemistry of bisketene (**32**) in argon matrices. By using alternative precursors and by comparison with DFT calculated IR spectra, it could be demonstrated that 9,10-didehydroanthracene (**33**) is labile even under matrix isolation conditions, and that the IR absorptions observed by Chapman et al.[65] belong to cyclic diyne **35**. Pyrolysis of various precursors produced only **35** as well (Scheme 16.13).[67,68] According to the DFT calculations and LFP measurements, the barrier for ring opening of **33** was estimated to be 10–12 kcal/mol, and diyne **35** is ∼6–8 kcal/mol more stable than didehydroanthracene (**33**).[67,68]

The comparatively low barrier for the *retro*-Bergman cyclization of **28** accounts for the numerous unsuccessful attempts to detect *p*-benzyne directly. Still, the barrier is significantly higher than that for didehydroanthracene (**33**). Photochemical cyclization of **16** could not be achieved in solution or under matrix isolation conditions. The breakthrough came in 1998 with experiments by Sander and co-workers.[8,69] Photolysis of diacetylterephthaloyldiperoxide (**36**) in argon matrices at 10 K yields (in addition to large amounts of CO_2) alkyl radicals and various other compounds (Scheme 16.14). Prominent IR absorptions were observed at 724.8 and 980.0 cm^{-1}, which disappeared along with the bands of the monoradicals upon annealing of the matrix at 40 K. Apart from these signals the IR spectrum of **28**, calculated at the UB3LYP level of theory, predicted only two other bands of comparable intensity: an absorption at 427 cm^{-1}, which could not be identified due to the low signal/noise ratio in this spectral region, and a band at 886 cm^{-1}, which was obscured by other intense signals.[69] The identification of **28** was confirmed by photolysis of partially deuterated derivatives of **36**. Independently, Radziszewski observed the same IR bands after 248-nm photolysis of 1,4-diiodobenzene **37** in solid neon at 6 K.[69] Despite the low overall intensity he was able to assign all IR active absorptions of **28** up to 2000 cm^{-1}. The experimental vibrational spectrum was further completed by NIPES measurements by Wenthold et al.,[22]

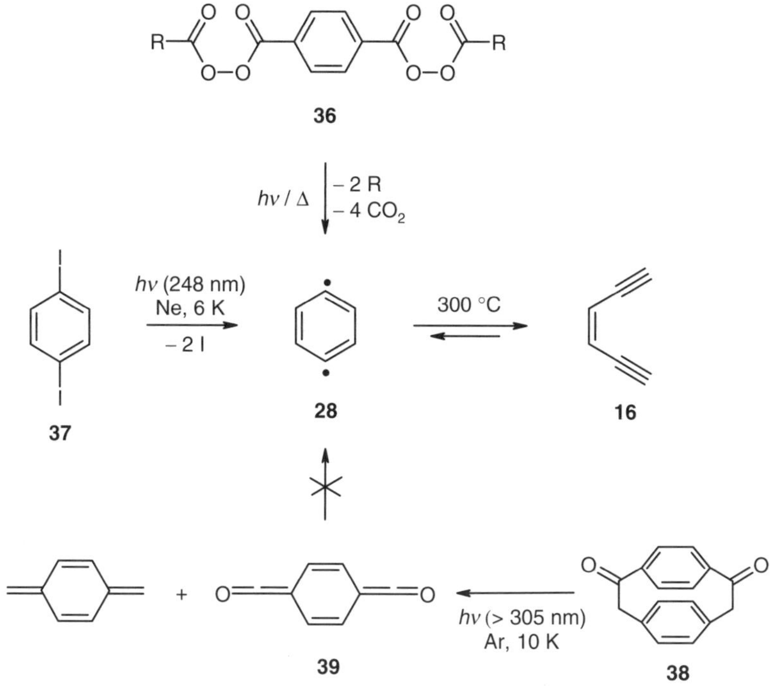

Scheme 16.14. Matrix isolation of *p*-benzyne **28**.[69]

which allowed identification of the frequencies of two totally symmetric (and therefore IR inactive) modes at ~600 and 1000 cm^{-1}, again in good agreement with the DFT-calculated spectrum.[8]

Recently, the assignment of the band at 980 cm^{-1} to **28** has been doubted based on new calculations (this band is shifted to 976 cm^{-1} if **28** is generated from 1,4-diiodobenzene (**37**), which is not unusual in the presence of iodine atoms. This shift may also be attributable to the change of the matrix host from argon to neon).[70] On the other hand, ab initio calculations of the IR spectrum of 28 are complicated by the existence of orbital instabilities,[71] the effect of which may (often) be negligible for first order properties (such as geometry and energy), but can result in severe deviations for second-order properties (vibrational frequencies, IR intensities).[70,71] In further matrix isolation studies, cyclophanedione (**38**), was envisioned as a precursor, which upon long wavelength irradiation fragments quantitatively to *p*-xylylene and bisketene (**39**) (Scheme 16.14).[72] In contrast to benzannellated derivative **32**, however, the latter turned out to be photochemically stable under matrix isolation conditions.[72]

In 1968, Hoffmann et al.[73] systematized the properties of the benzynes by formally separating the interaction of the radical centers into *through-space* and *through-bond* contributions. Specifically, this concept predicts that in *p*-benzyne the antisymmetric combination (b_{1u}) of the formally nonbonding orbitals at C1 and C4 is lowered beyond the symmetric combination (a_g). This idea has been confirmed by high-level calculations (Scheme 16.15).[26,70] The partial occupation of the antibonding C2—C3- and C4—C5-σ* orbitals results in elongation of these bonds, while the other C—C bonds are shortened.[70,73,74] This coupling of the electrons in *p*-benzyne stabilizes the singlet state relative to the triplet. Experimentally a singlet ground state with a singlet–triplet splitting of 3.8 ± 0.5 kcal/mol was determined by NIPES.[22]

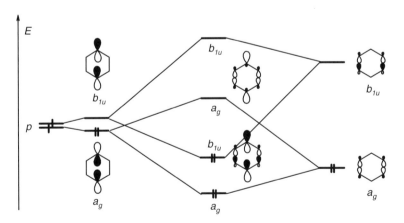

Scheme 16.15. Schematic MO representation of *through-space* and *through-bond* coupling in *p*-benzyne.[73,74]

This simple MO diagram forms the basis for understanding the structure and electronic properties of several derivatives of **28**, which will become clear in the following sections. As in the case of *m*-benzyne, however, many aspects of *p*-benzyne chemistry are still poorly understood. The stereoelectronic factors influencing the thermal Bergman cyclization are a subject of ongoing and intense research;[1,75] the photochemical cycloaromatization is even more difficult to rationalize. First attempts to investigate the intermolecular chemistry of **28** and its derivatives have been described only recently.[76] Thus, *p*-benzynes will continue to fascinate theorists and experimentalists alike.

3. SUBSTITUTED ARYNES

3.1. *o*-Benzynes

Numerous nucleophilic aromatic substitution reactions take place by an elimination–addition mechanism via intermediate formation of arynes (cf. Scheme 16.2).[77] In order to understand the regioselectivity of these reactions it is essential to take substituent effects into account.[78] Several substituted didehydrobenzenes have been investigated theoretically.[26w,79] A systematization of substituent effects with respect to structure and reactivity of **4** is anything but trivial, since the substituents act on the σ- as well as the π-system, resulting in a complex interplay of inductive and resonance effects.[79] Nevertheless, substituent effects can often be understood qualitatively in terms of an increase or decrease of the contribution of zwitterionic structures to the resonance hybrid.[26w,79] Generally, effects are significantly more pronounced in 2,3-arynes (polarization of the triple bond) than in the 3,4-didehydro isomers.[79] The interpretation of experimental reactivity data is often complicated by the fact that the attack of approaching nucleophiles is not solely charge controlled.[26w] It appears that more systematic (experimental and computational) studies are necessary for an extensive understanding of substituent effects on *o*-benzyne (**4**). Experimental data (such as C≡C stretching vibrational frequencies or gas-phase structures) to compare the theoretical results with are rare. Nonetheless, some examples have been described in the literature and will be discussed in this chapter.

The perfluorinated *o*-benzyne (**40**) could be isolated in cryogenic matrices by photolysis of the corresponding phthalic anhydride.[80] According to CASSCF calculations the singlet–triplet gap of tetrafluoro-*o*-benzyne is larger than that of **4** by several kilocalories per mole. The fluorine substituents only have a minor influence on the length of the C≡C bond. Interestingly, a nonplanar C_2 symmetrical structure is found with DFT methods (BLYP, B3LYP) for the triplet state, whereas CASSCF predicts a planar C_{2v} symmetrical geometry.[80] In a careful matrix isolation study, Radziszewski et al.[81] identified the C≡C stretching vibration of **40** at 1878 cm^{-1}. Despite the similar C≡C bond lengths, the chemistry of **40** differs significantly from that of **4**, the former being much more electrophilic and reactive. Thus, in contrast to **4**, compound **40** reacts readily with thiophene.[82]

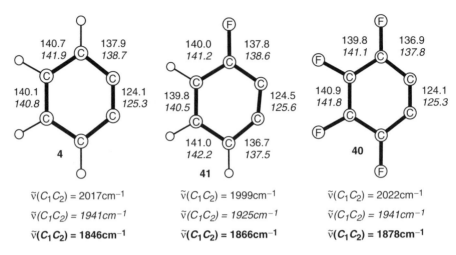

$\tilde{v}(C_1C_2) = 2017cm^{-1}$ $\tilde{v}(C_1C_2) = 1999cm^{-1}$ $\tilde{v}(C_1C_2) = 2022cm^{-1}$

$\tilde{v}(C_1C_2) = 1941cm^{-1}$ $\tilde{v}(C_1C_2) = 1925cm^{-1}$ $\tilde{v}(C_1C_2) = 1941cm^{-1}$

$\tilde{v}(\boldsymbol{C_1C_2}) = \boldsymbol{1846cm^{-1}}$ $\tilde{v}(\boldsymbol{C_1C_2}) = \boldsymbol{1866cm^{-1}}$ $\tilde{v}(\boldsymbol{C_1C_2}) = \boldsymbol{1878cm^{-1}}$

Figure 16.2. Calculated structures and C≡C stretching vibrational frequencies of **4**, **40**, and **41**. Normal print: B3LYP/cc-pVTZ data; in italics: BLYP/cc-pVTZ, bold print: experimental data.[81,83a] Bond lengths in picometers.

3-Fluoro-*o*-benzyne (**41**) has also been isolated in cryogenic matrices by photolysis of the corresponding phthalic anhydride. The C≡C stretching vibration of **41** has been observed at 1866 cm^{-1}.[81] Compared to the parent, the triple bond length is only modestly increased according to DFT calculations (Fig. 16.2).[81,83] Note, however, that the observed frequency shifts upon fluorine substitution in **4** are *not* reproduced at the B3LYP or BLYP level of theory (Fig. 16.2).[81,83] Thus, the performance of different DFT functionals to account for the properties of *o*-arynes still remains a matter of debate,[26y] and casts some doubts on the results of certain DFT studies.

3.2. *m*-Benzynes

Substituted *m*-arynes have been extensively studied computationally.[79] One important parameter for understanding substituent effects on the structure and reactivity of **13** is the distance between the radical centers. Despite the flat potential between **13** and **18** (Section 2.2),[44–46] the effect of substitution on the structure of *m*-benzyne seems to be rather small.[79] The data obtained with different theoretical methods could again be interpreted in terms of varying weights of mesomeric structures. Two resonance structures may be drawn for the bicyclic structure **18**; apart from the anti-Bredt form **18A**, a charge separated structure **18B** contributes to the resonance hybrid. Substituents, which stabilize the cyclopropenyl cation or allyl anion unit are expected to favor the bicyclic structure and to decrease the distance between the radical centers.[79] Accordingly, calculations show that donor substituents (NH$_2$, SiH$_3$) decrease, acceptor substituents (CN, NH$_3^+$) increase the C2–C6 distance in 2,6-didehydro isomers slightly. In 2,4-isomers the substituents interact with a negative partial charge, and therefore donor substituents result in increased distances, and acceptor substituents lead to decreased distances. For 3,5 isomers the shortest C3–C5 distances are predicted for σ-acceptor–π-donor substituents, since the central carbon atom (C1) of the allyl anion unit is positively polarized (significant negative charge density at the termini of the allyl anion).[79]

Formula 16.2

The influence of substituents on the stability of singlet *m*-benzynes is weak, and small variations have been rationalized on the basis of electrostatic arguments. The triplet states are even less affected by substitution.[79] Note that high structural flexibility is obviously characteristic for all *m*-benzynes. It can therefore not be taken for granted that calculated tendencies (and in particular those obtained with DFT methods) represent actual physical trends instead of just reflecting the deficiencies of the respective level of theory.

Photolysis of cyclophanediones was applied to the generation of several derivatives of **13**. Thus, 5-fluoro- and 5-methyl-1,3-didehydrobenzene were isolated in argon matrices and characterized by their IR spectra.[84] The generation of 2-substituted derivatives of **13** is not feasible with this method, since the corresponding cyclophanediones could not be synthesized.

The perfluorinated *m*-benzyne (**42**) was generated by photolysis of 1,3-diiodo-tetrafluorobenzene in solid neon (but not in argon) at very low temperature, whereas pyrolysis of this substance results in complete fragmentation of the ring system.[80] More recently, a number of different 5-substituted fluorinated *m*-benzynes could be prepared from the corresponding diiodo compounds in neon and were characterized IR spectroscopically.[85] Selected data for these molecules are shown in Table 16.2. In all cases, the calculated singlet–triplet energy splitting is larger than that of the unsubstituted system **13** and the C1—C3 internuclear distance is slightly reduced. Two characteristic vibrational modes are found for all 2-fluoro derivatives of **13**: an

TABLE 16.2. Calculated and Experimental Data for Some *m*-Benzyne Derivatives

	13	42	43	44	45	46
$\Delta E_{ST}{}^{a}$	19.5	26.3	22.4	24.8	20.5	23.7
$r(C_1C_3)^{b}$	200.4	189.5	192.0	187.1	190.4	193.5
$\tilde{v}_{calc.}^{b}$		1804.9	1788.3	1816.8	1789.3	1780.1
$\tilde{v}_{exp.}^{c}$		1823.8	1770.7	1805.3	1765.3	1773.5

[a] Singlet–triplet splitting in kilocalories per mole calculated at the BLYP/6-311G(d,p) level of theory.
[b] BLYP/6-311G(d,p); bond lengths *r* in picometers, vibrational frequencies in reciprocal centemeters.
[c] Observed position of the characteristic C—C/C—F stretching vibration in reciprocal centemeters.

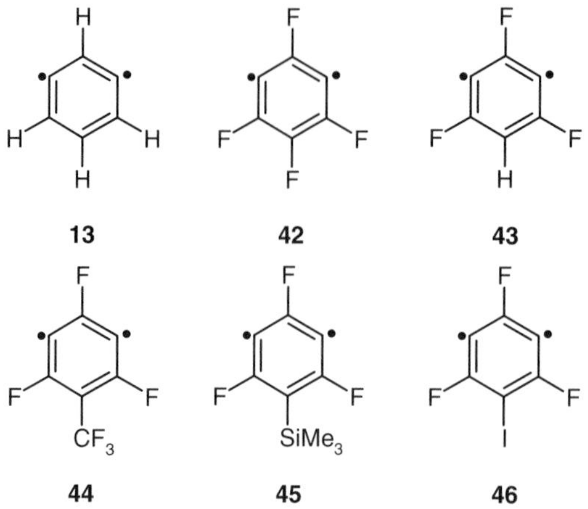

13   **42**   **43**

44   **45**   **46**

Formula 16.3

a_1 symmetrical vibration in the range 1765–1824 cm^{-1} and a b_2 symmetrical ring stretching vibration between 1551 and 1605 cm^{-1}. Substituents at C5 are hardly involved in these vibrations (as is evident, e.g., from the neglegible H/D shift between **43** and its 5-deuterated isotopomer). Therefore changes in the frequencies arise mainly from electronic effects of the substituents on the ring system. Thus, these vibrations may be used as a probe for substituent effects on 2-fluoro-*m*-benzynes.[85]

In summary, substituent effects on the structure and electronic properties of *m*-benzyne are rather small in many cases. One aspect of forthcoming investigations may be the search for derivatives of **13**, in which the distance of the radical centers is reduced drastically, which should lead to markedly different properties and reactivity.

3.3. *p*-Benzynes

Experimental data on substituted *p*-benzynes are rare. Only the perfluorinated derivative **47** has been the subject of a recent matrix isolation study. Irradiation of 1,4-diiodotetrafluorobenzene (**48**) in solid neon at 3 K generates *p*-didehydrotetrafluorobenzene (**47**) in good yield.[86] According to calculations, this derivative has a singlet ground state. However, the lowest triplet state is only 0.5 kcal/mol higher in energy (CASPT2/cc-pVDZ). Definitive experimental evidence for the ground state of **47** is not available yet, since the calculated IR spectra of the singlet and triplet states are essentially identical, and an ESR spectrum of the photolysis products in solid neon could not be measured so far.[86] Interestingly, longer wavelength irradiation (260–320 nm) converts **47** into the ring-opened 1,3,4,6-tetrafluorohex-3-ene-1,5-diyne (**49**, Scheme 16.16). This observation is particularly remarkable, since the high

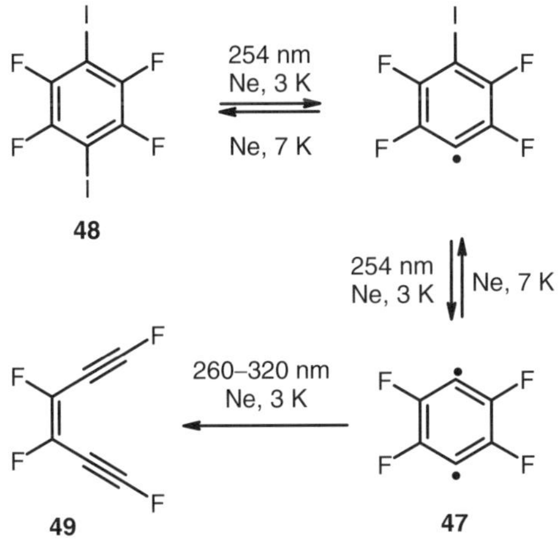

Scheme 16.16. Photochemistry of **48** in a neon matrix at 3 K.[86]

electronegativity of the fluorine substituents destabilizes the enediyne relative to the biradical, so that the ring opening of **47** is endothermic by 8 kcal/mol [UB3LYP/6-311++G(d,p)], and the barrier for *retro*-Bergman cyclization is significantly increased (37.5 kcal/mol) relative to that of the parent system **28**.[86]

Substituent effects on geometry and singlet–triplet splitting in *p*-didehydro-arenes were investigated in several theoretical studies and can in most cases be rationalized in terms of amplification or attenuation of *through-bond* coupling. Complete H/F exchange leads to an increase in ΔE_{ST} for the *o*- and *m*-benzyne, but decreases the energy gap for the para-derivative **47**. This effect has been explained by a stabilization of the bonding MO (symmetric combination of the non-bonding orbitals, S) by interaction with $\sigma^*(C-F)$ orbitals.[86,87] As shown in Figure 16.3, this interaction increases the gap between the highest occupied (HOMO) and lowest unoccupied molecular orbitals (LUMO) for **40** and **42**, but decreases the orbital energy difference in **47**. (Note, however, that the relationship between state energy differences and orbital energy differences is all but trivial.) A similar argumentation has been used by Cramer and Johnson[79] to rationalize the calculated singlet–triplet splittings of several singly substituted *p*-benzynes. For σ-withdrawing substituents ($X = NH_2$, CN, OH), it has been shown that the reduction of ΔE_{ST} is the result of a stabilization of the triplet state rather than desta-bilization of the singlet ground state according to the calculated biradical stabiliza-tion energies (BSE). The MO derived from symmetric combination of the nonbonding orbitals (S), which is empty in the singlet, but singly occupied in the triplet state, is stabilized by mixing with the empty $\sigma^*(C_{ipso}-X)$ orbital in all cases.[79]

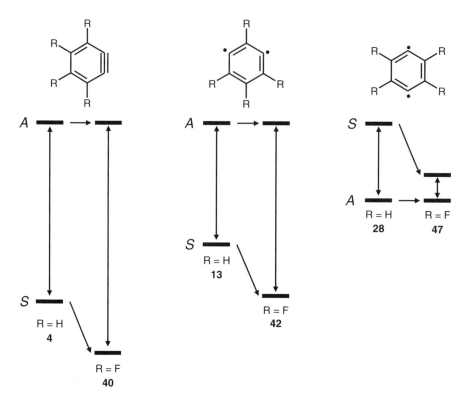

Figure 16.3. Schematic representation of the HOMO–LUMO energy changes upon perfluorination of benzynes.[86]

4. ANNELLATED ARYNES

4.1. Naphthynes

Naphthynes are among the most studied benzyne derivatives. Calculated relative energies of the lowest singlet and triplet states of all 10 isomers are summarized in Table 16.3.[88] 1,2-Naphthyne (**50**) is the most stable compound within this series, the second ortho-isomer **57** being slightly less stable. This effect has been interpreted in terms of the lower energy necessary for the distortion toward the formal triple bond in **50** than in **57**; the 1,2-bond in the parent naphthalene (**60**) is well known to be shorter than the 2,3-bond (cf. Fig. 16.9).[88,89] The same tendency holds for the o-naphthynes (see below), which is immediately obvious, considering the possible resonance structures of both isomers (Scheme 16.17). The preference for the singlet state in **50** is greater than that in **4** (30.4 kcal/mol at CASPT2/cc-pVDZ) by ~2 kcal/mol, while it is lower in the 2,3-isomer **57** by approximately the same amount.

TABLE 16.3. Calculated energies of the Naphthynes Relative to the Singlet Ground State of 1,2-Naphthyne (50) [CASPT2(12,12)/cc-pVDZ + ZPE(BPW91/cc-pVDZ)][a]

Isomer	No.	E(singlet)	E(triplet)	ΔE_{ST}
1,2	**50**	0.0	32.2	32.2
1,3	**51**	10.8	28.1	17.3
1,4	**52**	21.5	27.1	5.6
1,5	**53**	19.4	27.3	7.9
1,6	**54**	24.8	25.6	0.8
1,7	**55**	24.4	26.0	1.6
1,8	**56**	25.5	26.4	0.9
2,3	**57**	2.1	30.5	28.4
2,6	**58**	24.0	25.8	1.8
2,7	**59**	23.0	25.9	2.9

[a] See Ref. 88. All values are in kilocalories per mole.

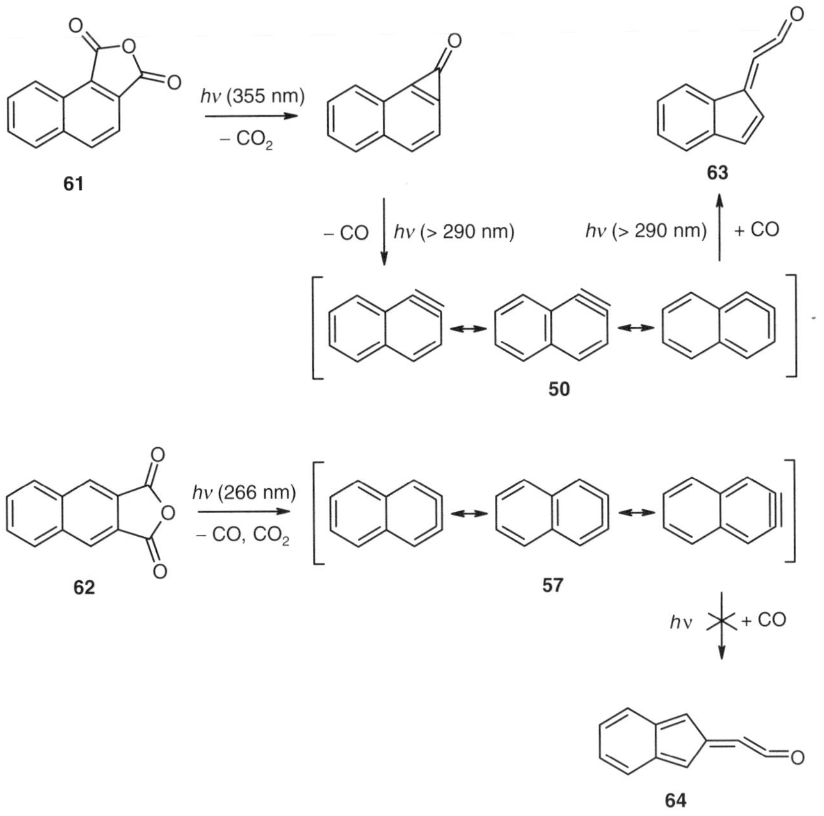

Scheme 16.17. Photochemical generation of **50** and **51** from anhydrides **61** and **62**. Compound **50** is photocarbonylated to cyclopentadienylideneketene (**63**). The analogous photoreaction of **57** to ketene **64** was not observed.[93]

Several naphthynes have already been investigated in the early 1970s. Grützmacher and Lohmann[90] generated some derivatives by pyrolysis of different precursors and measured the ionisation potentials of the pyrolysis products. At about the same time, Lohmann[91] investigated the photochemistry of the isomeric naphthalenedicarboxylic anhydrides **61** and **62** using LFP and found that dimerization of the 2,3-naphthyne (**57**) is significantly faster than that of **4**. The 1,2-isomer **50**, on the other hand, hardly dimerizes at all.

Matrix isolation of 2,3-didehydronaphthalene (**57**) was described for the first time by Weltner and co-workers[92] and has been the subject of a recent investigation by Sato et al.[93] Intermediates **50** and **57** are accessible photolytically from the anhydrides **61** and **62**, respectively, when the photolysis conditions are selected carefully.[93] Noteworthy again is the difference of both isomers compared with **4**. Whereas—in analogy to the parent system (cf. Scheme 16.4)–**50** is photochemically carbonylated to cyclopentadienylideneketene (**63**) (albeit irreversibly, in contrast to **4**), photocarbonylation of **57** to **64** is not observed at all (Scheme 16.17).[93] This finding was attributed to the different stability of the ketenes. The C≡C stretching vibrations of **50** and **57** could not be detected experimentally because of their low intensity. The calculated frequencies (**4**: 1953 cm^{-1}; **50**: 1994 cm^{-1}; **57**: 1922 cm^{-1}; B3LYP/6-31G**) follow the same trend as the bond lengths (**4**: 125.1 pm; **50**: 124.1 pm; **57**: 126.0 pm; B3LYP/6-31G**).

In their early contribution, Grützmacher and Lohmann[90b] identified diethynylbenzene (**65**) instead of 1,3-didehydronaphthalene (**51**) as the pyrolysis product of 1,3-dinitro- and 1,3-dibromonaphthalene, based on the high ionization potential measured (8.96 ± 0.02 eV). Obviously, a rearrangement analogous to the ring opening of m-benzyne takes place for the annellated derivative as well (Schemes 16.10 and 16.18).

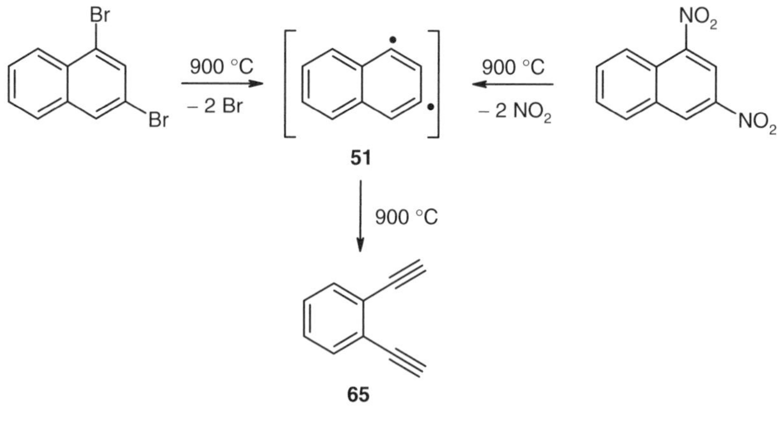

Scheme 16.18. Pyrolysis with mass spectrometric detection of 1,3-dibromo- and 1,3-dinitronaphthalene indicates that **51** is not thermally stable at 900 °C. The measured ionization potential suggests that the ring opened product **65** is formed under these conditions (cf. Scheme 16.10).[90b]

A derivative of 1,3-naphthyne was postulated by Billups et al.[94] as an intermediate of the dehydrohalogenation of the dichlorocarbene adduct of 2-bromoindene with potassium *tert*-butoxide; however, the complex product mixture did not allow the reaction mechanism to be conclusively elucidated (Scheme 16.19).

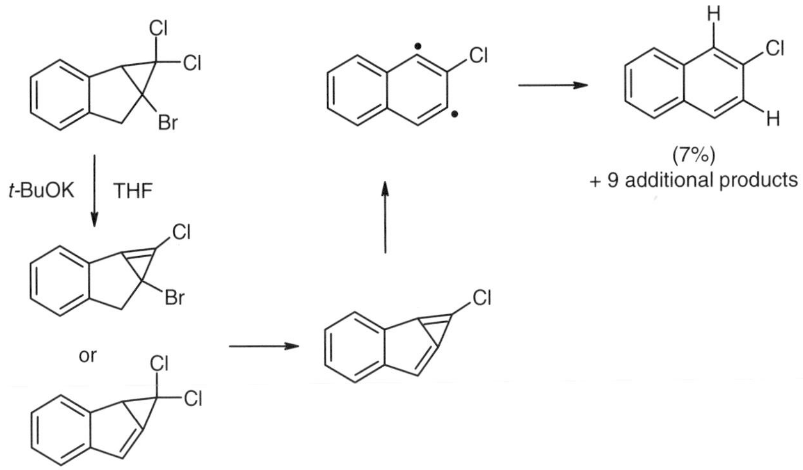

(7%)
+ 9 additional products

t-BuOK | THF

or

Scheme 16.19. 2-Chloro-1,3-didehydronaphthalene was postulated as an intermediate of a complex dehydrohalogenation reaction by Billups et al.[94] (Tetrahydrofuran = THF.)

In a broader sense, 1,8-naphthyne (**56**) may also be regarded as a derivative of *m*-benzyne, since the radical centers are separated by the same number of bonds as in **51**.

51   56

Formula 16.4

However, whereas 1,3-naphthyne (**51**) displays a strong preference for the singlet state (17.3 kcal/mol, comparable to that of *m*-benzyne at the same level of theory), singlet and triplet states are almost degenerate in the 1,8-isomer ($\Delta E_{ST} = 0.9$ kcal/mol).[88] The ionization potential (IP) of **56** (7.92 eV) is lower than that of naphthalene (8.26 eV), while the IPs of most other naphthynes are higher.[90] Attempts to generate **56** by irradiation of 1,8-naphthalenedicarboxylic anhydride in an argon matrix were unsuccessful.[95]

The properties of the benzannellated *p*-benzyne (**52**) are very similar to that of the parent system **28**. Both vicinal C—C bonds are elongated due to *through-bond* coupling of the unpaired electrons (see Fig. 16.9, Section 7 for a calculated

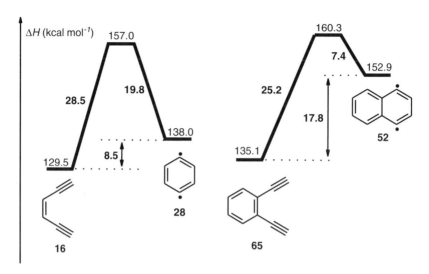

Figure 16.4. The energy profiles of the Bergman cyclization of **16** and **65** as measured by Roth et al.[60,96]

structure of **52**). The effect of benzannellation on the energetics of the Bergman cyclization (**65** → **52**) has been investigated in the gas phase by Roth et al. (Fig. 16.4).[96] The barrier of 25 kcal/mol for cyclization of **65** is lower than for (Z)-**16**. This value is in excellent agreement with the activation parameters determined by Grissom et al.[97] for the reactions **65** → **52** ($E_a = 25.1$ kcal/mol in solution). The reaction enthalpy for cyclization of the benzannellated system is higher, that is the reaction is more endothermic, which has been rationalized by the smaller amount of resonance energy released by formation of the second six-membered ring in naphthalene compared to benzene.[96] Accordingly, the barrier for the reverse (ring-opening) reaction **52** → **65** is lower than that for p-benzyne **28** (19.8 kcal/mol).[96]

In 1,5-naphthyne (**53**) the separation of A and S orbitals is larger than in **52**, since in the former there is no *through-space* interaction counteracting the *through-bond* coupling.[88] Accordingly, the singlet–triplet energy splitting is slightly enlarged in **53**.

52

53

Formula 16.5

Myers et al.[98] were able to trap 1,5-naphthyne (**53**), which is formed in solution from 1,2,6,7-tetradehydro[10]annulene (**66**) even at −40 °C.[98] The intermediate formation of the naphthyne was confirmed by formation of the 1,5-dideuterated product in deuterated solvents (Scheme 16.20). In a similar fashion, the 2,6-isomer **58** could be generated by a cyclization cascade from *(Z,Z)*-deca-3,7-diene-1,5,9-triene (**67**) by Bergman and co-workers.[99]

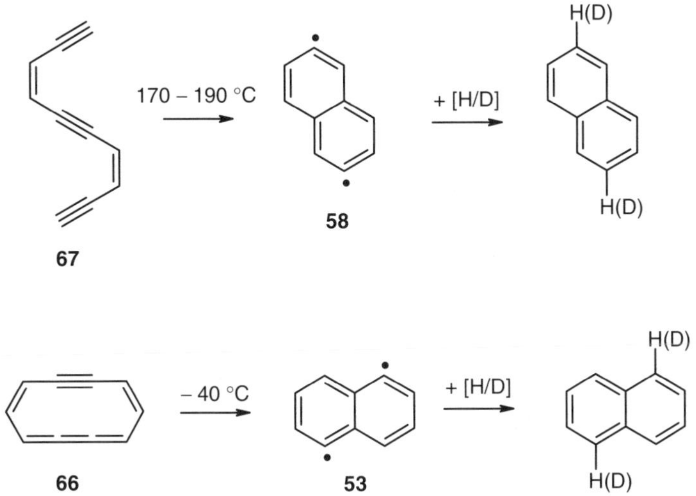

Scheme 16.20. Formation of naphthynes **53** and **58** by cycloaromatization of **66** and **67**.[98,99]

The remaining two isomers **54** and **59** have not yet been observed experimentally. According to calculations, the singlet–triplet splitting in **59** is comparably large (~3 kcal/mol), while both states are essentially degenerate in **54**.[88] Such "zig–zag couplings" for "W"-type arrangements of orbitals are well known in NMR[100] and ESR spectroscopy.[88,101]

54   **59**

Formula 16.6

4.2. Didehydroindenes

Interest in the didehydroindenes has been facilitated by the discovery of the neocarcinostatin cytostatics, that are closely related to the endiyne antitumor drugs (Scheme 16.21).[102]

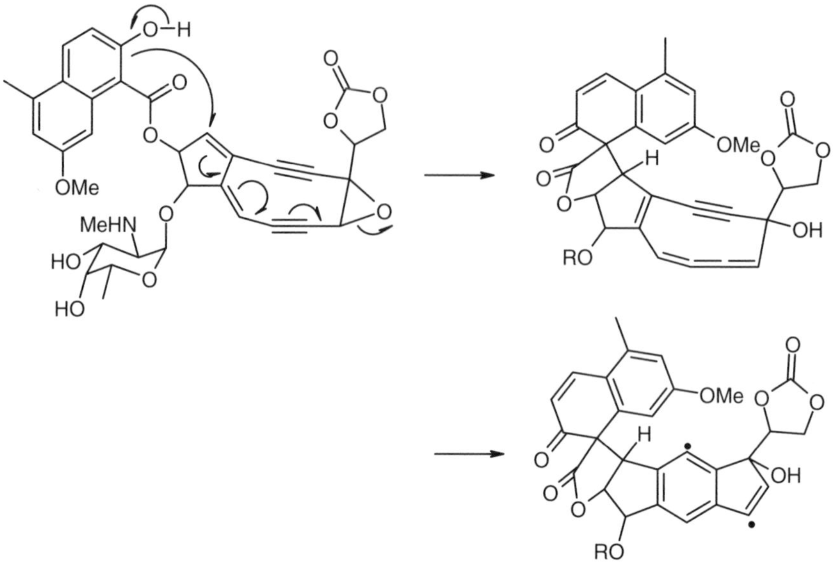

Scheme 16.21. Activation of neocarcinostatin (thiol independent pathway; see Ref. 7f for details).

Cycloaromatization of the cyclic eyneyneallene (**68**) produces the 1,5-didehydroin-dene biradical **69**, that abstracts hydrogen atoms from the DNA strand in analogy to the calicheamicins (cf. Section 3.3).

Formula 16.7

The prototype of this reaction is the Myers–Saito reaction, the rearrangement of eyneyneallene (Z)-hepta-1,2,4-triene-6-yne (**70**) to α,3-didehydrotoluene (**71**). This C2–C7 cyclization yields a benzylic π-conjugated σ,π-biradical and is therefore exothermic by 15 ± 3 kcal/mol (Fig. 16.5).[103] The reaction barrier was determined to be ∼22 kcal/mol.[103] Also known in literature is the alternative C2–C6-cyclization (Schmittel cyclization), which generates a less stable fulvenic structure **72**.[104] According to calculations, the Schmittel cyclization is endothermic by ∼10–19 kcal/mol with an activation barrier of ∼30–35 kcal/mol.[105] Replacing the acetylenic hydrogen atoms by sterically demanding substituents, however, favors the C2–C6 over C2–C7 cyclization.[104,105] Phenyl groups also reverse the energetics by benzylic stabilization in the transition state of the Schmittel

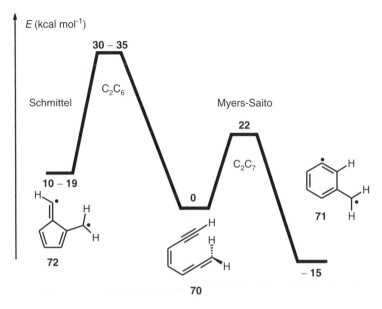

Figure 16.5. Myers–Saito versus Schmittel cyclization.[103–105]

cyclization.[104,105] The activation enthalpy for the cyclization **68** → **69** was calculated to be ~18 kcal/mol; the reaction is slightly exothermic by ~1 kcal/mol (in
contrast to acyclic system **71**, **69** does not profit from benzylic stabilization).[106]

The interesting α,3-didehydrotoluene biradical **71** has not yet been matrix isolated and structurally characterized. However, it could be shown experimentally
by CID measurements in the gas phase that **71** (in contrast to α,2- and α,4-didehy-
drotoluene) possesses a singlet ground state.[107] According to these experiments the
lowest triplet state is ~3 kcal/mol higher in energy,[107] in reasonable agreement with
photoelectron spectroscopic data of Chen and co-workers (5 kcal/mol)[108] and quantum chemical calculations.[109]

The first isolation of a derivative of **71** was reported by Sander et al. in 1998.[110]
Irradiation of quinone diazide (**73**) with λ > 475 nm in an argon matrix at 10 K
quantitatively yields 2,6-dimethylcyclohexa-2,5-diene-1-one-4-ylidene (**74**). Short
wavelength irradiation (λ > 360 nm) converts **74** into the substituted α,3-
didehydrotoluene derivative **75** (Scheme 16.22). Upon prolonged irradiation **76**
and (presumably) **77** are formed by Myers cycloreversion. No evidence was found
for the formation of bicyclic **78**.[110]

Didehydroindenes have been studied in detail computationally.[111] Relative
enthalpies of all possible isomers are summarized in Table 16.4. The three isomers
bearing the formal C≡C bond within the six-membered ring hardly differ in their
stability and singlet–triplet gaps. This finding is in accordance with the fact that
indene (in contrast to naphthalene) displays essentially no bond length alternation
within the six-membered ring.[111] The enthalpy of formation of the highly strained

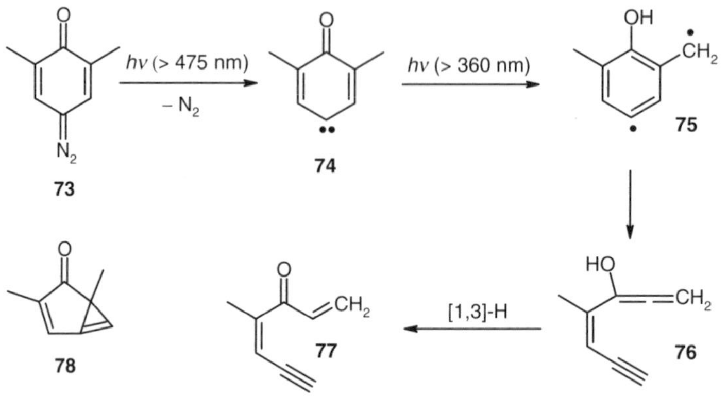

Scheme 16.22. The first direct detection of a derivative of α,3-didehydrotoluene.[110]

5,6-didehydroindene (**92**) is ~16 kcal/mol higher. In close analogy to the corresponding naphthynes **51** and **56**, the singlet–triplet gaps of **80** and **84** are significantly larger than in the 4,5-didehydroindene **90**.[111]

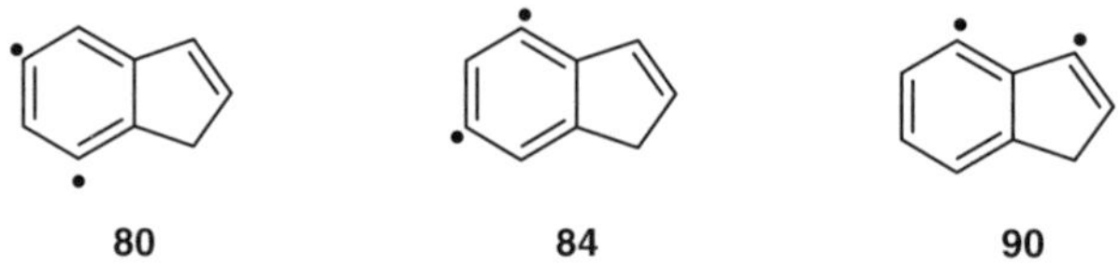

Formula 16.8

TABLE 16.4. Relative Enthalpies (H_{298}) and Singlet–Triplet Splittings of the Didehydroindenes

Isomer	No.	E(singlet)	E(triplet)	$\Delta E_{ST}{}^{a}$
1,2	**79**	0.1	30.2	30.1
1,3	**80**	13.8	26.6	13.0
1,4	**81**	18.3	26.2	7.9
1,5	**69**	21.4	29.0	7.7
1,6	**82**	27.0	27.5	0.4
2,3	**83**	0.5	30.2	29.8
2,4	**84**	14.2	26.9	12.9
2,5	**85**	28.1	27.7	0.3
2,6	**86**	26.0	27.9	1.9
3,4	**87**	0.0	30.6	30.6
3,5	**88**	28.0	27.8	0.2
3,6	**89**	26.3	27.5	1.2
4,5	**90**	27.1	28.6	1.5
4,6	**91**	27.1	28.0	0.8
5,6	**92**	15.7	34.9	19.3

[a] Calculated as difference of the singlet and triplet *BSE*s [CASPT2/cc-pVDZ + ZPE(BPW91/cc-pVDZ)]. See Ref. 111. All values are in kilocalories per mole.

Apart from the 1,5 isomer, there is experimental evidence only for the intermediate formation of 1,2-didehydroindane (**93**), the saturated analogue of **79**. The FVP of nona-1,3,8-triyne (**94**) yields indane (**95**) and indene (**96**), which is formed from **95** by loss of H_2.[112] Results from isotopic labeling suggest an unusual (1,3-diyne + alkyne)[2 + 4]-cycloaromatization resembling a Diels–Alder reaction, forming 1,2-didehydroindane (**93**, Scheme 16.23).[112]

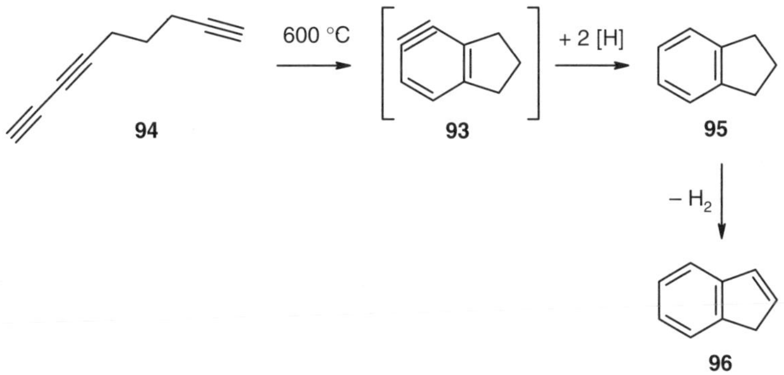

Scheme 16.23. Thermolysis of **94** results in formation of indane (**95**), presumably via intermediate formation of didehydroindane (**93**).[112]

5. HETEROCYCLIC ARYNES

Despite the fact that the history of hetarynes is older than that of the benzynes (cf. **2**), physical data on these compounds are scarce. Numerous trapping experiments furnished evidence for the formation of biradicaloid intermediates in the field of five-membered heterocycles (didehydrofurans, -thiophenes, and -pyrroles).[113] Direct spectroscopic data on these species, however, do not exist, which may be attributable to the increased ring strain in the five-membered *o*-arynes, associated with a strong tendency to undergo ring-opening reactions.

Investigations of six-membered heteroarynes concentrate on didehydropyridines, which will be dealt with exclusively in this section. The influence of the nitrogen lone pair on the structural and electronic properties of the benzynes was investigated by Cramer and Debbert.[114] Relative energies of the six isomeric pyridynes and their singlet–triplet splittings are summarized in Table 16.5.

In agreement with prior calculations[26u,115] at low levels of theory, it could be shown that 3,4-pyridyne (**101**) is significantly more stable than the 2,3-isomer (**97**).[114] The formal C—C triple bond length is similar in **4** and **101**, while it is markedly elongated in **97** (**4**: 124.1 pm; **101**: 124.8 pm; **97**: 126.7 pm; B3LYP/cc-pVTZ).[83,114] In addition, the singlet–triplet gap of **4** and **101** is comparable, whereas it is reduced by ~16 kcal/mol in **97**. This difference is caused by a

TABLE 16.5. Relative Energies and Singlet–Triplet Splittings of the Six Pyridynes

Isomer	No.	E(singlet)	E(triplet)	$\Delta E_{ST}{}^a$
2,3	**97**	7.1	21.7	14.6
2,4	**98**	1.2	24.5	23.3
2,5	**99**	11.3	24.0	13.0
2,6	**100**	17.5	14.9	−1.2
3,4	**101**	0.0	33.5	32.9
3,5	**102**	14.7	28.2	14.2

^{a} Calculated as the difference of the singlet and triplet *BSE*s [CASPT2(8,8)/cc-pVDZ//CASSCF(8,8)/cc-pVDZ)]. See Ref. 114.

cumulative effect of both, destabilization of the singlet ground state, and stabilization of the triplet state of **97**, as is evident from the calculated BSEs. The triplet state gains stability from the efficient delocalization to the third center. The calculated ^{14}N hyperfine coupling constant of 37 G in 2,3-pyridyne is significantly larger than in the 2,4 and 2,5 isomers with 32 and 31 G, respectively. On the other hand, the singlet state of **97** is destabilized by an unfavorable three-center, four-electron interaction, which makes the major contribution to the difference in ΔE_{ST} between **101** and **97**. In addition, an anomeric effect of the nitrogen lone pair on the vicinal C—C bonds is likely to be operative in **97**, which leads to an elongation of the formal triple bond and causes destabilization and an increase of the biradical character.[114]

In 1972 Berry and co-worker[116] detected 3,4-pyridyne (**101**) by MS. Trapping experiments also provided evidence for the existence of this intermediate, although the chemistry of **101** differs considerably from that of *o*-benzyne.[113] Thus, neither anthracene[113a] nor dimethylfulvene[117] form Diels–Alder adducts with **101**. Nam and Leroi[118] were able to generate **101** in nitrogen matrices at 13 K and characterized it by IR spectroscopy. Irradiation of 3,4-pyridinedicarboxylic anhydride (**103**) with $\lambda > 340$ nm results in formation of **101**, which upon short wavelength photolysis ($\lambda > 210$ nm) fragments to buta-1,3-diyne (**104**) and HCN, and to acetylene (**105**) and cyanoacetylene (**106**, Scheme 16.24). The assignment of an intense

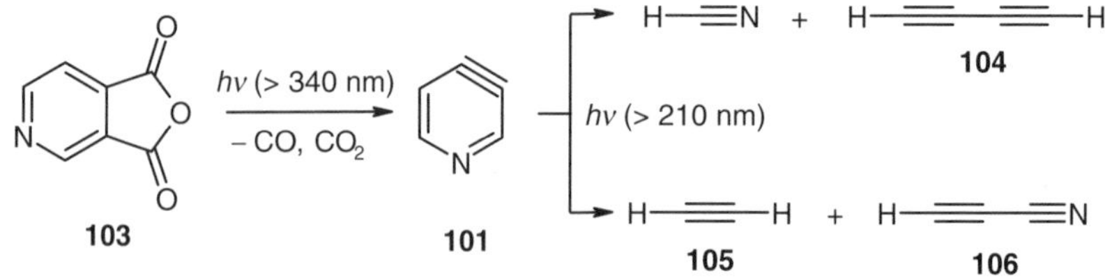

Scheme 16.24. Generation and photofragmentation of **101** in a nitrogen matrix at 13 K.[118]

absorption at 2085 cm^{-1} to the C≡C stretching mode in **101**[118] might have to be revised in view of Radziszewski's reinterpretation of the *o*-benzyne spectrum; ketene structures are apparently formed in this case as well (cf. Section 2.1).

2,3-Didehydropyridine (**97**) is less well investigated, and the existence of this intermediate is not yet established definitively. Aryne **97** has been postulated as an intermediate in the gas-phase pyrolysis of 2,3-pyridinedicarboxylic anhydride (**107**) at 600 °C, which finally yields β-ethinyl acrylonitrile (**108**).[119] Indirect evidence stems from trapping experiments with benzene and thiophene.[113e] Photolysis of matrix isolated **107** does not produce **97** (Scheme 16.25). Nam and Leroi[120]

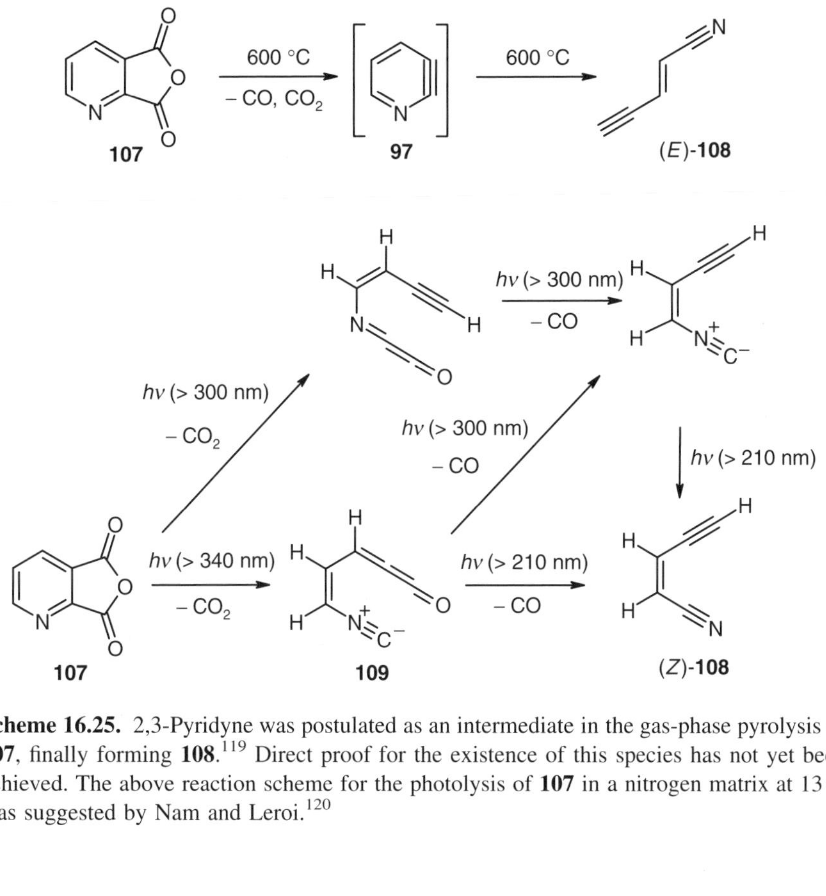

Scheme 16.25. 2,3-Pyridyne was postulated as an intermediate in the gas-phase pyrolysis of **107**, finally forming **108**.[119] Direct proof for the existence of this species has not yet been achieved. The above reaction scheme for the photolysis of **107** in a nitrogen matrix at 13 K was suggested by Nam and Leroi.[120]

suggested that loss of CO_2 leads initially to the 3-acyl-2-pyridyl biradical, which ring opens to **109** significantly faster than the second radical center at the aromatic ring is formed by CO release.

So far the three isomeric *m*-pyridynes have eluded experimental observation,[83a] although the properties of **98**, **100**, and **102** are of some interest from a theoretical point of view.[114]

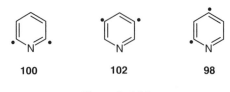

100   **102**   **98**

Formula 16.9

Isomer **100** is the least stable of all six pyridynes and its singlet and triplet state are essentially degenerate. It is less stable than the singlet ground state of ortho-isomer **101** by ~21 kcal/mol.[114] The structural flexibility of **100** is even more pronounced than in the case of the other *m*-arynes. In addition to mono- and bicyclic structures analogous to **13** and **18**, an allylic structure with a widened C2—N—C6 angle is found depending on the level of theory applied.[83,114,115d] The 3,5-isomer (**102**), which lies energetically above **101** by 15–18 kcal/mol, is the second least stable pyridyne. The singlet–triplet gap of **102** is ~18 kcal/mol, still below that of *m*-benzyne (**13**).[114] The unprecedented properties of the *m*-pyridynes, and in particular of **100**, result from a strong σ-allylic interaction, as schematically depicted in Scheme 16.26.[114]

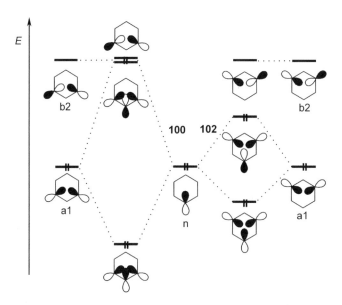

Scheme 16.26. σ-Allylic interaction of the lone pair (n) at the nitrogen atom with the bonding (*a*1) and antibonding (*b*2) combination of the *p* orbitals in **100** and **102**. Because of the stronger overlap, the HOMO and LUMO in **100** are essentially degenerate, which results in a very small singlet–triplet gap. In **102** ΔE_{ST} is decreased by only 3 kcal/mol compared to **13**, since the structure can more easily distort to avoid destabilizing interactions.[46,114] Note that antibonding orbital interactions are always stronger than bonding, so that both structures are destabilized relative to **13**.

2,4-Pyridyne (**98**) is the most stable of the three isomers, and, in contrast to **100** and **102**, has a higher preference for the singlet ground state than **13**. This change suggests the participation of nitrilium ion structure **98B** and possibly **98C** in the resonance hybrid, which is also evident from the shortening of the N—C2 bond compared to the N—C6 bond.

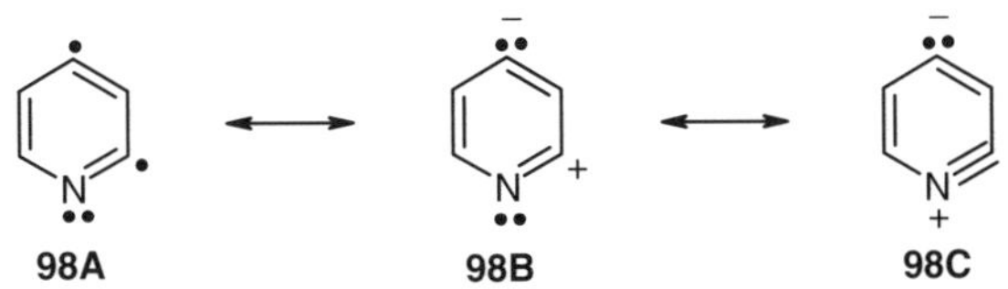

98A　　　　**98B**　　　　**98C**

Formula 16.10

An interesting route to **98** was suggested by Cramer et al.[114] cyclization of isonitrile **110** could lead to **98** in a rearrangement analogous to the Bergman cyclization. The activation enthalpy for the only slightly exothermic reaction was estimated at 18 kcal/mol. However, this novel access to *m*-arynes could not be realized experimentally so far.

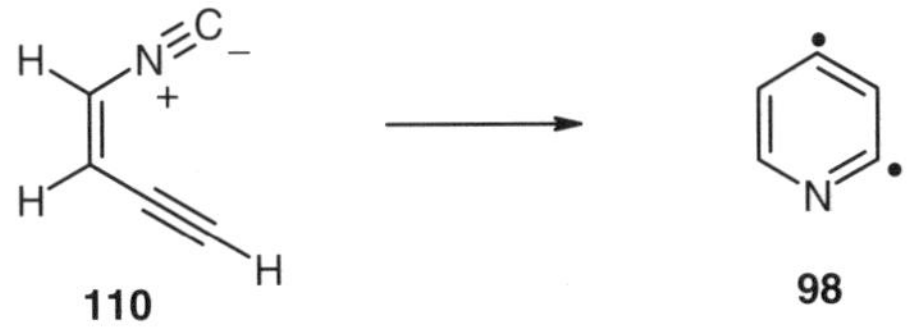

110　　　　　**98**

Formula 16.11

A three-center interaction similar to that in *m*-pyridynes **100** and **102** (Scheme 16.26) results in substantial stabilization in the two-electron case. 3,5-Didehydrophenyl cation (**111**), the classic example[121] for "in-plane aromaticity,"[122] is the global minimum of the $C_6H_3^+$ hypersurface.[123] Remarkably, the

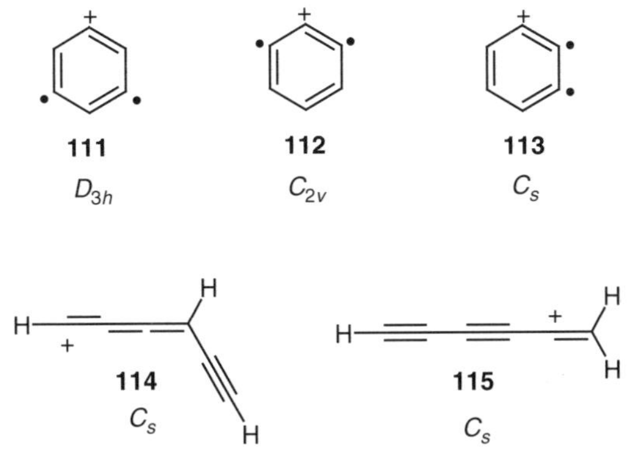

111　　　　**112**　　　　**113**
D_{3h}　　　　C_{2v}　　　　C_s

114　　　　　**115**
C_s　　　　　C_s

Formula 16.12

2,6-isomer **112** and 2,3-isomer **113** are less stable than **111** by 11.7 and 23.2 kcal/mol, respectively.[123] Obviously, for the two-electron case the 1,3,5-topology (comparable to that in the trishomocyclopropenium cation) is favored over the 1,2,3-topology, while for the four-electron, three-center systems the destabilization is most pronounced in the second case. In addition to its remarkable stability, **111** also fulfills magnetic aromaticity criteria. At the same level of theory [MP4(SDTQ)/6-31G*//MP2/6-31G*+ZPE] open-chain isomers **114** and **115** are less stable by 11.8 and 30.4 kcal/mol, respectively.[123]

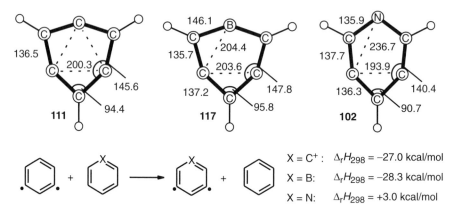

Figure 16.6. The structures of **111** and **117** reveal substantial two-electron, three-center interaction, accompanied by stabilization, whereas destabilization in **102** is minimized by distortion towards a larger C3—N/C5—N distance (BLYP/cc-pVTZ).[46] Bond lengths in picometers angles in degrees.

In contrast to the Jahn–Teller distorted, C_{2v} symmetric 1,3,5-tridehydrobenzene (**116**) (cf. Section 6.1), cation **111** has D_{3h} symmetry.[123] According to BLYP calculations, the isoelectronic 3,5-borabenzyne (**117**) is even more stabilized relative to **13** than is **111** (Fig. 16.6). Both systems are strongly distorted to maximize the stabilizing two-electron three-center interaction, whereas 3,5-pyridyne circumvents destabilization to some degree by maximizing the C3—N/C5—N distances (Fig. 16.6).[83a] In this context, it is interesting to mention the influence of protonation of the nitrogen atom in the pyridynes.[114b] Replacement of CH by N in benzynes has a pronounced impact on the singlet states (rather than on the triplets), which is largely attenuated by protonation (Tables 16.5 and 16.6).[114]

The properties of 2,5-pyridyne (**99**) can once more be rationalized within the *through-bond* coupling scheme. The singlet–triplet splitting in (**99**) is ~6 kcal/mol higher than in *p*-benzyne or in the protonated form (**99**—H$^+$).[114,124–126] A detailed interpretation of these energy differences has been given by Kraka and Cremer.[124] Anomeric delocalization of the lone pair at the nitrogen atom into the vicinal C—C bonds results in shortening of the N—C2 and N—C6 bonds

TABLE 16.6. Relative Energies and Singlet–Triplet Splittings of the Six Pyridinium Ions

Isomer	No.	E(singlet)	E(triplet)	$\Delta E_{ST}{}^{a}$
2,3	**97-H$^+$**	8.6	32.7	23.6
2,4	**98-H$^+$**	10.1	28.4	18.3
2,5	**99-H$^+$**	24.5	29.2	5.2
2,6	**100-H$^+$**	20.3	30.7	10.6
3,4	**101-H$^+$**	0.0	31.5	30.9
3,5	**102-H$^+$**	12.8	29.9	17.5

[a] Calculated as difference of the singlet and triplet *BSE*s [CASPT2(8,8)/cc-pVDZ//CASSCF(8,8)/cc-pVDZ]. See Ref. 114b. All values are in kilocalories per mole.

(Scheme 16.27); the shortened C2–N bond enhances *through-bond* coupling by increasing the overlap of the interacting orbitals. Additionally, the σ^*(C6–N) orbital is lower in energy than the σ^*(C–C) orbital (due to the higher electronegativity of nitrogen compared to carbon), which also increases the *through-bond* interaction.[124] Thus, the efficient coupling of the electrons in **99** takes place primarily via the C6–N bond.[124] The length of this bond is determined by two counteracting effects: shortening by anomeric delocalization of the lone pair and lengthening by efficient *through-bond* coupling. In **99**–H$^+$ the anomeric effect is switched off, which is also reflected in the calculated geometries (Fig. 16.7).[114b,124]

The aza-Bergman rearrangement of *C,N*-dialkynylimines to β-alkynyl acrylonitriles has also been studied intensively. In 1997, David and Kerwin investigated the thermochemistry of (*Z*)-**118** in solution and observed rearrangement to (*Z*)-**119**

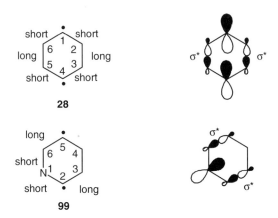

Scheme 16.27. Orbital interactions that determine the geometry and electronic structure of **28** and **99**.[124]

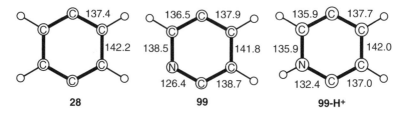

Figure 16.7. The structures of *p*-benzyne (**28**), 2,5-pyridyne (**99**), and (**99-H$^+$**), calculated at the UB3LYP/6-31G(d,p) level of theory.[124] Bond lengths in picometers.

(Scheme 16.28).[127] However, a 2,5-didehydropyridine intermediate could not be detected or trapped, which was attributed to the low barrier for ring opening of this species (a mechanism involving an aza-butalene structure was also not excluded).[127]

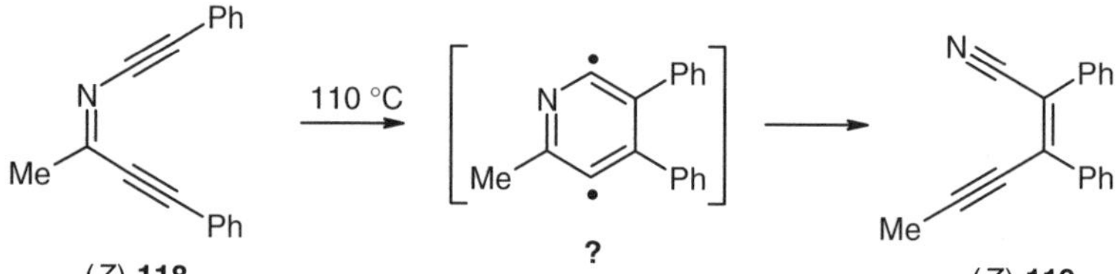

Scheme 16.28. Rearrangement of *C,N*-dialkynylimine **118** to β-alkynyl acrylonitrile **119**.[127]

One year later Chen and co-workers[126] reported trapping experiments of protonated 2,5-pyridyne. Detectable hydrogen-abstraction rates were only observed under low pH conditions, which suggest the existence of a protonated 2,5-pyridyne derivative.[126] Theoretical investigations show that—in contrast to *p*-benzyne **28** and enediyne (*Z*)-**16** (Fig. 16.4)—2,5-didehydropyridine (**99**) is stabilized relative to imine **120** (Fig. 16.8); the relative stability of nitrile **108** is also increased significantly.[114,125,126] However, the barrier for ring-opening of 2,5-pyridyne (**99–108**) was determined to be only 0.9 kcal/mol (cf. Ref. 125 for numerical details). In addition to the short lifetime, the large singlet–triplet separation accounts for the poor hydrogen-abstraction ability of **99**, as discussed below. Protonation does not only decrease the singlet–triplet gap significantly (see above), but also increases the barrier for *retro*-Bergman reaction of **99**—H$^+$ to **108**—H$^+$ to >10 kcal/mol.[114,124–126]

Interest in the aza-Bergman reaction and the influence of protonation on 2,5-pyridyne and its derivatives was (once more) directly stimulated by the search for less toxic antitumor drugs. One approach to increase the selectivity is to decrease the reactivity of the biradical intermediate in hydrogen-abstraction reactions.[59,66,124–126,128] In this regard, Chen has established a simple model that correlates the reactivity of singlet biradicals with their singlet–triplet gap ΔE_{ST}.[66,126,128] This model is based on the assumption that the reactivity of the triplet biradical is

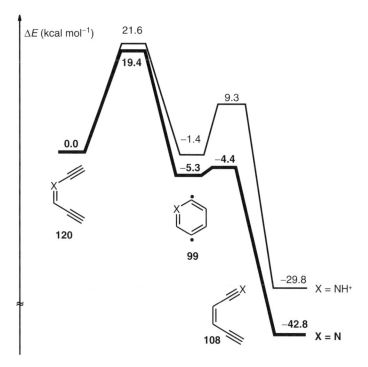

Figure 16.8. Calculated energy profile of the Bergman cyclization of imine **120** and its protonated form **120**–**H**$^+$ and of the subsequent *retro*-Bergman cyclization to nitrile **108** (**108**–**H**$^+$).[125]

comparable to that of the monoradical, while effects stabilizing the singlet state have to be overcome in the transition state of the hydrogen abstraction. Accordingly, an increased singlet–triplet gap is expected to raise the barrier for hydrogen abstraction and (possibly) the selectivity of the reaction. An interesting method to estimate ΔE_{ST} based on the isotropic hyperfine coupling constant of the radical center in the respective monoradical with the proton at the second center has been developed by Cramer and Squires.[129] Since this approach provides reliable results with DFT methods it can be applied to larger systems such as potential antitumor drugs.

In this context, it is noteworthy that the pH value of tumor cells (6.2–6.6) is lower than that of normal cells (7.5).[7,130] This difference could be used to control the Bergman cyclization of substituted and/or heterocyclic derivatives of **28** by protonation–deprotonation or to control the hydrogen abstraction depending on the pH value. In an exhaustive theoretical study Kraka and Cremer[59] investigated a large number of substituted and heterocyclic enediynes and the corresponding benzynes and proposed potential candidates for future anticancer drugs. While simply replacing CH by N allows pH-dependent control of the Bergman cyclization to some extent, calculations by these authors indicate, that the proton affinity of **99** is not sufficient to achieve efficient protonation under physiological conditions.[59]

Whether the investigation of *p*-arynes, as very briefly outlined in the foregoing sections, is going to facilitate the development of novel antitumor drugs is not foreseeable at the moment, and undue optimism is certainly not justified. Nevertheless, the concerted efforts of many research groups have contributed to the refinement of the experimental and theoretical tools of Physical-Organic chemistry, and it is no exaggeration to state that arynes are prototypical examples for the interplay of experiment and theory in modern chemistry.

6. RELATED SYSTEMS

6.1. Tridehydrobenzynes

In contrast to benzynes and benzdiynes (cf. Section 6.2), knowledge of tridehydroarenes is scarce. According to calculations by Bettinger et al.[26x] 1,2,4-tridehydrobenzene (**121**) and 1,3,5-tridehydrobenzene (**122**) are less stable than the 1,2,3-isomer **123** by 4.0 and 11.8 kcal/mol, respectively. In contrast to didehydrophenyl cation **111** (Section 5), **122** possesses C_{2v} symmetry; the calculated C1—C3 and C1—C5 distances are different (194 and 230 pm, respectively, at the CCSD/DZP level of theory).[26x] For the 1,2,3-isomer **123** a significantly shorter C1—C3-distance (~170 pm) is predicted by various methods.

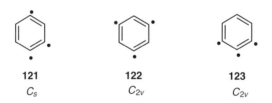

| 121 | 122 | 123 |
| C_s | C_{2v} | C_{2v} |

Formula 16.13

6.2. Benzdiynes

Polar carbon-rich species such as hexapentaeneylidene (**124**)[131] have been identified in interstellar clouds,[132] which has attracted some attention and interest in C_6H_2 compounds.[26x,133,134] The existence of **124** in interstellar space is remarkable, since this structure is 50 kcal/mol [CCSD(T)/cc-pVTZ] less stable than hexatriyne (**125**),[133] which, however, is nonpolar and therefore difficult to observe radioastronomically. At the same level of theory the three isomeric benzdiynes **126–128** are higher in energy than **125** by 39.4, 46.3, and 56.2 kcal/mol, respectively.[133]

The C1—C3 distance of 176 pm in **128**[26x,133,134] is rather low, which has been attributed to strong σ-allylic interaction as in the case of *m*-benzyne **13** (Section 2.2).[26x] Like *o*-benzyne (**4**), **127**, and **128** possess C—C bonds with

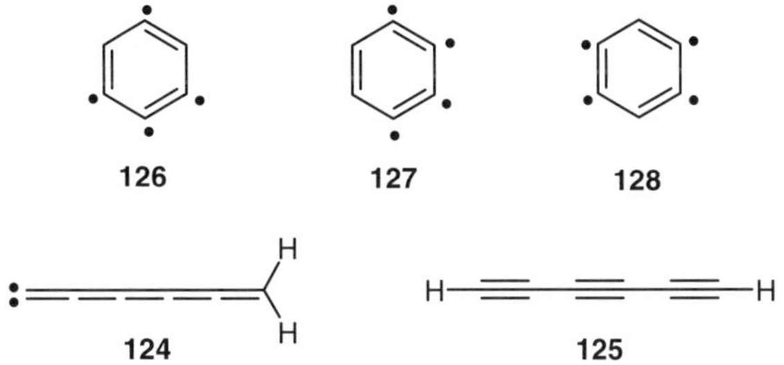

Formula 16.14

some triple-bond character. Due to the increased ring strain, which destabilizes the singlet ground states more than the triplet states, the singlet–triplet splittings in **127** and **128** (26.3 and 18.3 kcal/mol, respectively) are lower than that in **4** [35.8 kcal/mol at the CCSD(T)/TZ2P//CCSD/DZP level of theory].[16i] In contrast, the difference between both states is increased by 2 kcal/mol in **126** compared to o-benzyne. The LUMO (which is singly occupied in the triplet state) has not only a bonding interaction between C1 and C3, but also a strong antibonding interaction with C5, resulting in a large HOMO–LUMO gap and destabilization of the triplet state (BSE = −7 kcal/mol).[26x]

Despite early mention of benzdiynes as reactive intermediates,[135] the first direct evidence for the existence of a substituted derivative of **128** was not achieved until 1997.[136] Upon irradiation in cryogenic matrices, 1,4-bis(trifluoromethyl)-2,3,5,6-benzenetetracarboxylic dianhydride (**129**) loses CO_2 and CO in a stepwise manner. A photoproduct with a characteristic IR absorption at 1465 cm^{-1} was assigned the structure of 1,4-bis(trifluoromethyl)-2,5-benzdiyne (**130**) on the basis of its chemistry and comparison with the calculated spectrum (Scheme 16.29).[137] Prolonged

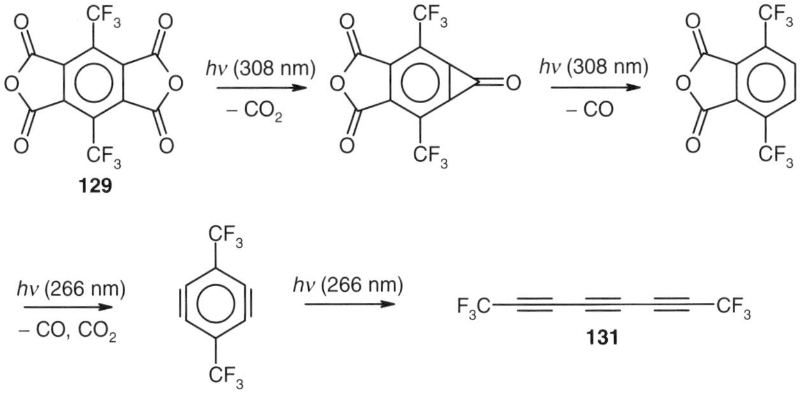

Scheme 16.29. First synthesis and photochemistry of a benzdiyne derivative.[137]

irradiation converts **130** into bis(trifluorormethyl)-hexatriyne (**131**). The difluoro derivative of **128** has been prepared and characterized in a similar fashion.[137] The overall reaction scheme is more complex in this case and two different ketene species of yet unknown constitution are involved in the photochemistry of the anhydride.[137]

The parent systems **127** and **128** are not accessible via this route.[137,138] Photolysis of 1,2,3,4- and 1,2,4,5-benzenetetracarboxylic dianhydride with different wavelengths invariably results in formation of hexatriyne **125**.[137] Attempts to isolate 1,2,5,6- and 2,3,6,7-tetradehydronaphthalenes (naphthdiynes) by photolysis of the corresponding naphthalenetetracarboxylic dianhydrides yield only acetylenic compounds, which suggests that naphthdiynes are formed, but rapidly decompose under photochemical conditions.[139]

6.3. Cyclo[6]-carbon

Aside from fullerene research, carbon clusters have always attracted much interest among scientists.[140] Linear cumulene-like C_6 ($D_{\infty h}$) **132** is well known from computational studies as well as experimentally.[141] Cyclic C_6, which will be considered as the extreme case of a dehydrobenzene here, has a long and varied history.[142] In particular, two cyclic isomers possessing D_{3h} and D_{6h} symmetry (**133** and **134**) have been discussed.[142] According to the most recent calculations, **133** ($^1A_1'$) is lowest in energy; **134** ($^1A_{1g}$) and **132** ($^3\Sigma_g^-$) being less stable by ~8 and 10 kcal/mol, respectively.[142e]

134	**133**	**132**
D_{6h}	D_{3h}	$D_{\infty h}$

Formula 16.15

The IR spectrum of **133** was obtained by laser vaporization of graphite and subsequent condensation of the reaction products in solid argon at 10 K. However, only the most intense mode at 1695 cm^{-1} could be detected.[143] The antisymmetric stretching vibration of the linear isomer **132** is observed at 1952 cm^{-1}.[141b,141e] The assignment could be corroborated by measuring the spectra of isotopically labeled compounds. In a more recent theoretical work, the UV spectra of **133** and **134** were calculated,[144] but experimental data are lacking so far.

7. CONCLUSIONS AND OUTLOOK

The interest in arynes has changed the status of these molecules from laboratory curiosities to well-established reactive intermediates. Yet, our knowledge of this

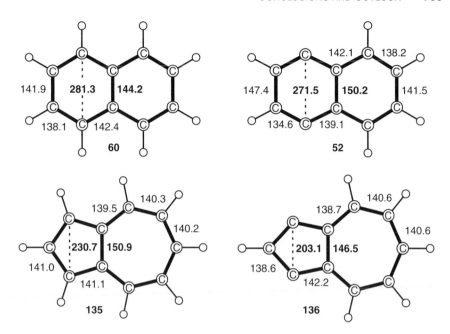

Figure 16.9. Calculated structures of azulene **135** and 1,3-azulyne **136** compared to naphthalene **60** and 1,4-naphthyne **52**.[46]

class of compounds is still incomplete and many aspects of their chemistry remain to be investigated in more detail. This lack pertains in particular to the mechanisms of intra- and intermolecular reactions and rationalizing the influences of hetero-atoms or substituents on the structure and reactivity of arynes. This knowledge will contribute to our understanding of basic processes in a large variety of fields, spanning synthesis, biological activity of cytostatics, heterogenous catalysis, combustion processes, and the chemistry of interstellar clouds.

Despite the long history, there are still many branches of aryne chemistry that are totaly unexploited. Not even one five-membered aryne has so far been characterized spectroscopically. Do we expect to find something new in this direction? Are the concepts that serve well to clarify the structures and electronic properties of six-membered arynes sufficient to understand the properties of the five-membered inter-mediates as well? The structures of azulene **135** and 1,3-azulyne **136**, calculated at the (U)BLYP/cc-pVTZ level of theory, are shown in Figure 16.9. Whereas the central bond in 1,4-naphthyne (**52**) is longer than that in naphthalene (**60**) because of *through-bond* coupling,[88] the opposite is found for the isomeric azulyne.[46] A detailed discussion of this (at first-glance counterintuitive) finding will be given elsewhere; here this example may suffice to demonstrate that surprises cannot be excluded, turning from six- to five-(or seven-)membered systems. Just as the concept of *through-bond–through-space* interactions, introduced by Hoffmann et al.[73,74] in 1968, had profound influence on almost all branches of chemistry, arynes will continue to be a source of inspiration for theorists and experimentalists alike.

SUGGESTED READING

H. H. Wenk, M. Winkler, and W. Sander, "One Century of Aryne Chemistry," *Angew. Chem. Int. Ed. Engl.*, **2003**, *42*, 502.

For a nice review on heterocyclic arynes, see M. G. Reinecke, "Hetarynes." *Tetrahedron*, **1982** *38*, 427.

H. C. van der Plas, "The Chemistry of Triple Bonded Groups," in Supplement C of *The Chemistry of Functional Groups*, S. Patai and Z. Rappoport, Eds., Wiley-Interscience, New York, **1982**.

For synthetic aspects of aryne chemistry, see C. Grundmann, "Arene und Arine," in Houben-Weyl, *Methoden der Organischen Chemie*, Vol. 5, C. Grundmann, Ed., Thieme, Stuttgart, **1981**.

H. Heaney, "Chemistry of Highly Halogenated Arynes," *Fortschr. Chem. Forsch.* **1970**, *16*, 35.

For a summary of early work, see R. W. Hoffmann, *Dehydrobenzene and Cycloalkynes*; Academic Press, New York, **1967**.

REFERENCES

1. H. H. Wenk, M. Winkler, and W. Sander, *Angew. Chem. Int. Ed. Engl.* **2003**, *42*, 502.

2. R. Stoermer and B. Kahlert, *Ber. Dtsch. Chem. Ges.* **1902**, *35*, 1633.

3. A. Lüttringhaus and G. Saaf, *Justus Liebigs Ann. Chem.* **1930**, *592*, 250.

4. (a) J. D. Roberts, H. E. Simmons, L. A. Carlsmith, and C. W. Vaughan, *J. Am. Chem. Soc.* **1953**, *75*, 3290; (b) J. D. Roberts, D. A. Semenow, H. E. Simmons, and L. A. Carlsmith, *J. Am. Chem. Soc.* **1956**, *78*, 601, 611.

5. (a) R. Huisgen and H. Rist, *Naturwissenschaften* **1954**, *41*, 358; (b) R. Huisgen and H. Rist, *Ann. Chem.* **1955**, *594*, 137; (c) R. Huisgen and R. Knorr, *Tetrahedron Lett.* **1963**, 1017.

6. (a) G. Wittig, *Naturwissenschaften* **1942**, *30*, 696; (b) G. Wittig and G. Harborth, *Ber. Dtsch. Chem. Ges.* **1944**, *77*, 306, 316; (c) G. Wittig and L. Pohmer, *Chem. Ber.* **1956**, *89*, 1334. (d) G. Wittig, *Pure Appl. Chem.* **1963**, *7*, 173. For a general overview about Wittig's achievements, see: (e) R. W. Hoffmann, *Angew. Chem. Int. Ed. Engl.* **2001**, *40*, 1411.

7. See, for example: (a) K. C. Nicolaou and W.-M. Dai, *Angew. Chem. Int. Ed. Engl.* **1991**, *30*, 1387; (b) K. C. Nicolaou and A. L. Smith, *Acc. Chem. Res.* **1992**, *25*, 497; (c) D. B. Borders and T. W. Doyle, Eds., *Endiyne Antibiotics as Antitumor Agents*, Marcel Dekker, New York, 1995; (d) W. K. Pogozelski and T. D. Tullius, *Chem. Rev.* **1998**, *98*, 1089; (e) Z. Xi and I. H. Goldberg, in *Comprehensive Natural Products Chemistry*, Vol. 7, D. H. R. Barton and K. Nakanishi, Eds., Pergamon, Oxford, **1999**, (f) S. E. Wolkenberg and D. L. Boger, *Chem. Rev.* **2002**, *102*, 2477.

8. W. Sander, *Acc. Chem. Res.* **1999**, *32*, 669.

9. (a) L. K. Madden, L. V. Moskaleva, S. Kristyan, and M. C. Lin, *J. Phys. Chem.* **1997**, *101*, 6790; (b) L. V. Moskaleva, L. K. Madden, and M. C. Lin, *Phys. Chem. Chem. Phys.* **1999**, *1*, 3967; and references cited therein.

10. (a) K. Johnson, B. Sauerhammer, S. Titmuss, and D. A. King, *J. Chem. Phys.* **2001**, *114*, 9539; (b) S. Yamagishi, S. J. Jenkins, and D. A. King, *J. Chem. Phys.* **2002**, *117*, 819; and literature cited therein.

11. W. E. Bachman and H. T. Clarke, *J. Am. Chem. Soc.* **1927**, *49*, 2089.

12. I. P. Fisher and F. P. Lossing, *J. Am. Chem. Soc.* **1963**, *85*, 1018.

13. (a) R. S. Berry, G. N. Spokes, and M. Stiles, *J. Am. Chem. Soc.* **1962**, *84*, 3570. (b) R. S. Berry, J. Clardy, and M. E. Schafer, *J. Am. Chem. Soc.* **1964**, *86*, 2738.

14. (a) O. L. Chapman, K. Mattes, C. L. McIntosh, J. Pacansky, G. V. Calder, and G. Orr, *J. Am. Chem. Soc.* **1973**, *95*, 6134; (b) O. L. Chapman, C. C. Chang, J. Kolc, N. R. Rosenquist, and H. Tomioka, *J. Am. Chem. Soc.* **1975**, *97*, 6586.

15. I. R. Dunkin and J. G. MacDonald, *J. Chem. Soc. Chem. Commun.* **1979**, 772.

16. C. Wentrup, R. Blanch, H. Briehl, and G. Gross, *J. Am. Chem. Soc.* **1988**, *110*, 1874.

17. (a) J. W. Laing and R. S. Berry, *J. Am. Chem. Soc.* **1976**, *98*, 660; (b) N.-H. Nam and G. E. Leroi, *Spectrochim. Acta A* **1985**, *41*, 67.

18. D. G. Leopold, A. E. Stevens-Miller, and W. C. Lineberger, *J. Am. Chem. Soc.* **1986**, *108*, 1379.

19. A. C. Scheiner, H. F. Schaefer, III, and B. Liu, *J. Am. Chem. Soc.* **1989**, *111*, 3118.

20. J. G. G. Simon, N. Muenzel, and A. Schweig, *Chem. Phys. Lett.* **1990**, *170*, 187.

21. J. G. Radziszewski, B. A. Hess, Jr., and R. Zahradnik, *J. Am. Chem. Soc.* **1992**, *114*, 52.

22. P. G. Wenthold, R. R. Squires, and W. C. Lineberger, *J. Am. Chem. Soc.* **1998**, *120*, 5279.

23. (a) P. G. Wenthold and R. R. Squires, *J. Am. Chem. Soc.* **1994**, *116*, 6401. For earlier work concerning the enthalpy of formation of **4**, see: (b) P. G. Wenthold, J. A. Paulino, and R. R. Squires, *J. Am. Chem. Soc.* **1991**, *113*, 7414 (106 ± 3 kcal/mol); (c) J. M. Riveros, S. Ingemann, and N. M. M. Nibbering, *J. Am. Chem. Soc.* **1991**, *113*, 1053 (105 ± 3 kcal/mol); (d) Y. Guo and J. J. Grabowski, *J. Am. Chem. Soc.* **1991**, *113*, 5923 (105 ± 5 kcal/mol).

24. A. M. Orendt, J. C. Facelli, J. G. Radziszewski, W. J. Horton, D. M. Grant, and J. Michl, *J. Am. Chem. Soc.* **1996**, *118*, 846.

25. (a) R. Warmuth, *Angew. Chem. Int. Ed. Engl.* **1997**, *36*, 1347; (b) R. Warmuth, *Eur. J. Org. Chem.* **2001**, 423; (c) R. Warmuth and J. Yoon, *Acc. Chem. Res.* **2001**, *34*, 95.

26. From the numerous studies up to 1998 the following are mentioned here as examples: (a) E. Haselbach, *Helv. Chim. Acta* **1971**, *54*, 1981; (b) M. J. S. Dewar and W.-K. Li, *J. Am. Chem. Soc.* **1974**, *96*, 5569; (c) W. Thiel, *J. Am. Chem. Soc.* **1981**, *103*, 1420; (d) M. J. S. Dewar, G. P. Ford, and C. H. Reynolds, *J. Am. Chem. Soc.* **1983**, *105*, 3162; (e) C. W. Bock, P. George, and M. Trachtman, *J. Phys. Chem.* **1984**, *88*, 1467; (f) M. J. S. Dewar and T.-P. Tien, *J. Chem. Soc. Chem. Commun.* **1985**, 1243; (g) L. Radom, R. H. Nobes, D. J. Underwood, and L. Wai-Kee, *Pure Appl. Chem.* **1986**, *58*, 75; (h) K. Rigby, I. H. Hillier, and M. A. Vincent, *J. Chem. Soc. Perkin Trans. 2* **1987**, 117; (i) I. H. Hillier, M. A. Vincent, M. F. Guest, and W. von Niessen, *Chem. Phys. Lett.* **1987**, *134*, 403; (j) A. C. Scheiner, and H. F. Schaefer, III, *Chem. Phys. Lett.* **1991**, *177*, 471; (k) R. Liu, Z. Xuenfeng, and P. Pulay, *J. Phys. Chem.* **1992**, *96*, 259; (l) H. U. Sutter and T.-K. Ha, *Chem. Phys. Lett.* **1992**, *198*, 259; (m) A. Nicolaides and W. T. Borden, *J. Am. Chem. Soc.* **1993**, *115*, 11951; (n) S. G. Wierschke, J. J. Nash, and R. R. Squires, *J. Am. Chem. Soc.* **1993**, *115*, 11958; (o) P. B. Karadahov, J. Gerrat, G. Raos, D. L. Cooper, and M. Raimondi, *Isr. J. Chem.* **1993**, *33*, 253; (p) E. Kraka and D. Cremer, *Chem. Phys. Lett.* **1993**, *216*, 333; (q) E. Kraka and D. Cremer, *J. Am. Chem. Soc.* **1994**, *116*, 4936; (r) R. Lindh and B. J. Persson, *J. Am. Chem. Soc.* **1994**, *116*, 4963; (s) R. Lindh, T. J. Lee, A. Bernhardsson, B. J. Persson, and G. Karlström, *J. Am. Chem. Soc.* **1995**, *117*, 7186; (t) J. J. Nash and R. R. Squires, *J. Am. Chem. Soc.* **1996**, *118*, 11872; (u) R. Liu,

D. R. Tate, J. A. Clark, P. R. Moody, A. S. Van Buren, and J. A. Krauser, *J. Phys. Chem.* **1996**, *100*, 3430; (v) C. J. Cramer, J. J. Nash, and R. R. Squires, *Chem. Phys. Lett.* **1997**, *277*, 311; (w) W. Langenaeker, F. De Proft, and P. Geerlings, *J. Phys. Chem. A* **1998**, *102*, 5944. For more recent computational studies, see: (x) H. F. Bettinger, P. v. R. Schleyer, and H. F. Schaefer, III, *J. Am. Chem. Soc.* **1999**, *121*, 2829; (y) B. A. Hess, Jr., and L. J. Schaad, *Mol. Phys.* **2000**, *98*, 1107; (z) F. De Proft, P. v. R. Schleyer, J. H. van Lenthe, F. Stahl, and P. Geerlings, *Chem. Eur. J.* **2002**, *8*, 3402; and references cited therein.

27. Additional spectroscopic data were available, which have, however, less frequently been used for comparison with theoretical calculations. These comprise the UV spectrum, Ref. 20, and the microwave spectrum of **4**: (a) R. D. Brown, P. D. Godfrey, and M. J. Rodler, *J. Am. Chem. Soc.* **1986**, *108*, 1296. The dipolar ^{13}C NMR spectrum of 1,2-$[^{13}C]_2$-**4** in an argon matrix was measured and analyzed in order to determine the C≡C distance., cf. Ref. 24. A detailed comparison of measured and calculated chemical shifts is given in (b) H. Jiao, P. v. R. Schleyer, B. R. Beno, K. N. Houk, and R. Warmuth, *Angew. Chem.* **1997**, *109*, 2929; *Angew. Chem. Int. Ed. Engl.* **1997**, *36*, 2761.

28. (a) J. A. Miller, R. J. Kee, and C. K. Westbrook, *Annu. Rev. Phys. Chem.* **1990**, *41*, 345; (b) I. Glassman, *Combustion*, Academic Press, San Diego, **1996**.

29. (a) R. F. C. Brown, D. V. Gardner, J. F. W. McOmie, and R. K. Solly, *Aust. J. Chem.* **1967**, *20*, 139; (b) M. Barry, R. F. C. Brown, F. W. Eastwood, D. A. Gunawardana, and C. Vogel, *Aust. J. Chem.* **1984**, *37*, 1643; (c) R. F. C. Brown, K. J. Coulston, F. W. Eastwood, and C. Vogel, *Aust. J. Chem.* **1988**, *41*, 1687; (d) R. F. C. Brown, *Eur. J. Org. Chem.* **1999**, 3211.

30. R. F. C. Brown and F. W. Eastwood, *Synlett*, **1993**, 9.

31. R. F. C. Brown, K. J. Coulston, F. W. Eastwood, and S. Saminathan, *Aust. J. Chem.* **1987**, *40*, 107.

32. C. Wentrup, R. Blanch, H. Briel, and G. J. Gross, *J. Am. Chem. Soc.* **1988**, *110*, 1874.

33. E. W.-G. Diau, J. Casanova, J. D. Roberts, and A. H. Zewail, *Proc Natl. Acad. Sci.* **2000**, *97*, 1376.

34. R. S. Berry, J. Clardy, and M. E. Schafer, *Tetrahedron Lett.* **1965**, *415*, 1003, 1011.

35. H. E. Bertorello, R. A. Rossi, and R. Hoyos de Rossi, *J. Org. Chem.* **1970**, *35*, 3332.

36. W. E. Billups, J. D. Buynak, and D. Butler, *J. Org. Chem.* **1979**, *44*, 4218.

37. S. V. Luis, F. Gavina, V. S. Safont, M. C. Torres, and M. I. Burguete, *Tetrahedron* **1989**, *45*, 6281.

38. (a) W. N. Washburn, *J. Am. Chem. Soc.* **1975**, *97*, 1615; (b) W. N. Washburn, and R. Zahler, *J. Am. Chem. Soc.* **1976**, *98*, 7827, 7828.

39. G. Bucher, W. Sander, E. Kraka, and D. Cremer, *Angew. Chem. Int. Ed. Engl.* **1992**, *31*, 1230; (b) W. Sander, G. Bucher, H. Wandel, E. Kraka, D. Cremer, and W. S. Sheldrick, *J. Am. Chem. Soc.* **1997**, *119*, 10660.

40. E. Kraka, D. Cremer, G. Bucher, H. Wandel, and W. Sander, *Chem. Phys. Lett.* **1997**, *268*, 313.

41. R. Marquardt, W. Sander, and E. Kraka, *Angew. Chem. Int. Ed. Engl.* **1996**, *35*, 746.

42. W. Sander, M. Exner, M. Winkler, A. Balster, A. Hjerpe, E. Kraka, and D. Cremer, *J. Am. Chem. Soc.* **2002**, *124*, 13072.

43. The IR spectrum of **13** has been reinterpreted in favor of a bicyclic structure **18** based on B3LYP calculations: (a) B. A. Hess, Jr., *Eur. J. Org. Chem.* **2001**, 2185; (b) B. A. Hess,

Jr., *Chem. Phys. Lett.* **2002**, *352*, 75. Regarding this interpretation, it should be mentioned that the measured IR spectrum, Ref. 41, has not been reproduced correctly. Two additional bands were assigned to *m*-benzyne, which are unequivocally due to other species (cf. Ref. 42 and 44–46).

44. M. Winkler and W. Sander, *J. Phys. Chem. A* **2001**, *105*, 10422.

45. E. Kraka, J. Anglada, A. Hjerpe, M. Filatov, and D. Cremer, *Chem. Phys. Lett.* **2001**, *348*, 115.

46. M. Winkler and W. Sander, in preparation.

47. W. Sander, R. Marquardt, G. Bucher, and H. Wandel, *Pure Appl. Chem.* **1996**, *68*, 353.

48. (a) K. K. Thoen and H.I. Kenttämaa, *J. Am. Chem. Soc.* **1997**, *119*, 3832; (b) K. K. Thoen and H. I. Kenttämaa, *J. Am. Chem. Soc.* **1999**, *121*, 800; (c) E. D. Nelson, A. Artau, J. M. Price, and H. I. Kenttämaa, *J. Am. Chem. Soc.* **2000**, *122*, 8781; (d) E. D. Nelson, A. Artau, J. M. Price, S. E. Tichy, L. Jing, and H. I. Kenttämaa, *J. Phys. Chem. A* **2001**, *105*, 10155. (e) J. M. Price, K. E. Nizzi, J. L. Campbell, H. I. Kenttämaa, M. Seierstad, and C. J. Cramer, *J. Am. Chem. Soc.* **2003**, *125*, 131.

49. (a) N. Darby, C. U. Kim, J. A. Salaün, K. W. Shelton, S. Takada, and S. Masamune, *J. Chem. Soc. Chem. Commun.* **1971**, 1516; (b) S. Masamune and N. Darby, *Acc. Chem. Res.* **1972**, *5*, 272.

50. (a) R. R. Jones and R. G. Bergman, *J. Am. Chem. Soc.* **1972**, *94*, 660; (b) R. G. Bergman, *Acc. Chem. Res.* **1973**, *6*, 25.

51. T. P. Lockhart, P. B. Comita, and R. G. Bergman, *J. Am. Chem. Soc.* **1981**, *103*, 4082, 4091.

52. M. Prall, A. Wittkopp, and P. R. Schreiner, *J. Phys. Chem. A* **2001**, *105*, 9265.

53. A similar reaction has been observed experimentally: (a) M. Chakraborty, C. A. Tessier, and W. J. Youngs, *J. Org. Chem.* **1999**, *64*, 2947. A photochemical C1C5 endiyne cyclization has recently been observed, but follows probably not a biradical mechanism via a fulvenediyl, but involves photoinduced electron transfer to the endiyne with subsequent cyclization of the radical anion: (b) I. V. Alabugin and S. V. Kovalenko, *J. Am. Chem. Soc.* **2002**, *124*, 9052.

54. (a) R. Breslow, J. Napierski, and T. C. Clarke, *J. Am. Chem. Soc.* **1975**, *97*, 6275. Compare: (b) R. Breslow and P. L. Khanna, *Tetrahedron Lett.* **1977**, 3429.

55. (a) M. J. S. Dewar and W.-K. Li, *J. Am. Chem. Soc.* **1974**, *96*, 5569. Compare (b) M. J. S. Dewar, G. P. Ford, and C. H. Reynolds, *J. Am. Chem. Soc.* **1983**, *105*, 3162.

56. P. M. Warner and G. B. Jones, *J. Am. Chem. Soc.* **2001**, *123*, 10322; and references cited therein.

57. B. A. Hess Jr., and L. J. Schaad, *Org. Lett.* **2002**, *4*, 735.

58. R. W. A. Havenith, F. Lugli, P. W. Fowler, and E. Steiner, *J. Phys. Chem. A* **2002**, *106*, 5703.

59. E. Kraka and D. Cremer, *J. Am. Chem. Soc.* **2000**, *122*, 8245; and references cited therein.

60. W. R. Roth, H. Hopf, and C. Horn, *Chem. Ber.* **1994**, *127*, 1765.

61. For example: (a) C. J. Cramer, *J. Am. Chem. Soc.* **1998**, *120*, 6261; (b) P. R. Schreiner, *J. Am. Chem. Soc.* **1998**, *120*, 4184; (c) R. Lindh, A. Bernhardsson, and M. Schütz, *J. Phys. Chem. A* **1999**, *103*, 9913; (d) J. Gräfenstein, A. Hjerpe, E. Kraka, and D. Cremer, *J. Phys. Chem. A* **2000**, *104*, 1748; (e) G. B. Jones and P. M. Warner, *J. Am. Chem. Soc.* **2001**, *123*, 2134; (f) M. Prall, A. Wittkopp, A. F. Fokin, and P. R. Schreiner, *J. Comp. Chem.* **2001**, *22*, 1605; and literature cited therein.

62. (a) I. D. Campbell and G. Eglington, *J. Chem. Soc. C* **1968**, 2120; (b) N. J. Turro, A. Evenzahav, and K. C. Nicolaou, *Tetrahedron Lett.* **1994**, *35*, 8089; (c) D. Ramkumar, M. Kalpana, B. Varghese, S. Sankararaman, M. N. Jagadeesh, and J. Chandrasekhar, *J. Org. Chem.* **1996**, *61*, 2247; (d) T. Kaneko, M. Takahashi, and M. Hirama, *Angew. Chem. Int. Ed. Engl.* **1999**, *38*, 1267; (e) N. Choy, B. Blanko, J. Wen, A. Krishan, and K. C. Russell, *Org. Lett.* **2000**, *2*, 3761.

63. (a) A. Evenzahav and N. J. Turro, *J. Am. Chem. Soc.* **1998**, *120*, 1835; (b) A. E. Clark, E. R. Davidson, and J. M. Zaleski, *J. Am. Chem. Soc.* **2001**, *123*, 2650.

64. L. Eisenhuth, H. Siegel, and H. Hopf, *Chem. Ber.* **1981**, *114*, 3772.

65. O. L. Chapman, C. C. Chang, and J. Kolc, *J. Am. Chem. Soc.* **1976**, *98*, 5703.

66. (a) M. J. Schottelius and P. Chen, *J. Am. Chem. Soc.* **1996**, *118*, 4896; (b) C. F. Logan and P. Chen, *J. Am. Chem. Soc.* **1996**, *118*, 2113.

67. H. H. Wenk and W. Sander, *Eur. J. Org. Chem.* **1999**, 57.

68. C. Koetting, W. Sander, S. Kammermeier, and R. Herges, *Eur. J. Org. Chem.* **1998**, 799.

69. R. Marquardt, A. Balster, W. Sander, E. Kraka, D. Cremer, and J. G. Radziszewski, *Angew. Chem. Int. Ed. Engl.* **1998**, *37*, 955.

70. T. D. Crawford, E. Kraka, J. F. Stanton, and D. Cremer, *J. Chem. Phys.* **2001**, *114*, 10638.

71. For a general treatment of orbital instabilities, see: T. D. Crawford, J. F. Stanton, W. D. Allen, H. F. Schaefer, III, *J. Chem. Phys.* **1997**, *107*, 10626.

72. R. Marquardt, W. Sander, T. Laue, and H. Hopf, *Liebigs Ann.* **1995**, 1643.

73. R. Hoffmann, A. Imamura, and W. J. Hehre, *J. Am. Chem. Soc.* **1968**, *90*, 1499.

74. For a general treatment of *through-bond* interactions, see: (a) R. Hoffmann, *Acc. Chem. Res.* **1971**, *4*, 1; (b) R. Gleiter, *Angew. Chem. Int. Ed. Engl.* **1974**, *13*, 696; (c) M. N. Paddon-Row, *Acc. Chem. Res.* **1982**, *15*, 245; (d) K. Jordan, *Theor. Chem. Acc.* **2000**, *103*, 286.

75. The reactivity of endiynes towards cyclization is, to some degree, controlled by the distance of the terminal acetylenic carbons, cf. (a) P. Magnus, R. T. Kewis, and J. C. Huffman, *J. Am. Chem. Soc.* **1988**, *110*, 1626; (b) P. A. Carter and P. Magnus, *J. Am. Chem. Soc.* **1988**, *110*, 6921; (c) J. P. Snyder, *J. Am. Chem. Soc.* **1989**, *111*, 7630; (d) J. P. Snyder, *J. Am. Chem. Soc.* **1990**, *112*, 5367; (e) P. Magnus, S. Fortt, T. Pitterna, and J. P. Snyder, *J. Am. Chem. Soc.* **1990**, *112*, 4986; (f) M. F. Semmelhack, T. Neu, and F. Foubelo, *Tetrahedron Lett.* **1992**, *33*, 3277; (g) K. C. Nicolaou, W.-D. Dai, Y. P. Hong, S.-C. Tsay, K. K. Baldrige, and J. S. Siegel, *J. Am. Chem. Soc.* **1993**, *115*, 7944; (h) M. F. Semmelhack, T. Neu, and F. Foubelo, *J. Org. Chem.* **1994**, *59*, 5038; (i) P. R. Schreiner, *Chem. Commun.* **1998**, 483; (j) P. R. Schreiner, *J. Am. Chem. Soc.* **1998**, *120*, 4184; (k) P. J. Benites, D. S. Rawat, and J. M. Zaleski, *J. Am. Chem. Soc.* **2000**, *122*, 7208; Ref. 7a; and literature cited therein. This distance can be adjusted by metal complexation, see, e.g., (l) B. P. Warner, S. P. Millar, R. D. Broene, and S. L. Buchwald, *Science* **1995**, *269*, 814; (m) B. König, W. Pitsch, and I. Thondorf, *J. Org. Chem.* **1996**, *61*, 5258; (n) A. Basak and J. C. Shain, *Tetrahedron Lett.* **1998**, *39*, 3029; (o) E. W. Schmitt, J. C. Huffman, and J. M. Zaleski, *Chem. Commun.* **2001**, 167; (p) D. S. Rawat, P. J. Benites, C. D. Incarvito, A. L. Rheingold, and J. M. Zaleski, *Inorg. Chem.* **2001**, *40*, 1846; and literature cited therein. Benzannellation and substituent effects on the rate of the Bergman cyclization are discussed in: (q) C.-S. Kim and K. C. Russell, *J. Org. Chem.* **1998**, *63*, 8229; (r) T. Kaneko, M. Takahashi, and M. Hirama, *Tetrahedron Lett.* **1999**, *40*, 2015; (s) S. Koseki, Y. Fujimura, and M. Hirama, *J. Phys. Chem. A* **1999**, *103*, 7672; (t) N. Choy,

B. Blanko, J. Wen, A. Krishan, and K. C. Russell, *Org. Lett.* **2000**, *2*, 3761; (u) G. B. Jones and G. W. Plourde, II, *Org. Lett.* **2000**, *2*, 1757; (v) A. Basak, J. C. Shain, U. K. Khamrai, K. R. Rudra, and A. Basak, *J. Chem. Soc. Perkin Trans. 1* **2000**, 1955; (w) B. König, W. Pitsch, M. Klein, R. Vasold, M. Prall, and P. R. Schreiner, *J. Org. Chem.* **2001**, *66*, 1742; (x) M. Prall, A. Wittkopp, A. F. Fokin, and P. R. Schreiner, *J. Comp. Chem.* **2001**, *22*, 1605; Ref. 61e; and literature cited therein.

76. F. S. Amegayibor, J. J. Nash, A. S. Lee, J. Thoen, C. J. Petzold, and H. I. Kenttämaa, *J. Am. Chem. Soc.* **2002**, *124*, 12066.

77. For example: (a) F. A. Carey and R. J. Sundberg, *Advanced Organic Chemistry*, 4th. ed., part A, Plenum Press, New York, 2000; (b) J. March, *Advanced Organic Chemistry*, 4th. ed., John Wiley & Sons, Inc., New York, **1992**; (c) J. Miller, *Aromatic Nucleophilic Substitution*, in *Reaction Mechanisms in Organic Chemistry*, C. Eaborn and N. B. Chapman, Eds., Elsevier, Amsterdam, The Netherlands, **1968**.

78. For example: (a) E. R. Biehl, E. Nieh, and K. C. Hsu, *J. Org. Chem.* **1969**, *34*, 3595; (b) M. Tielemans, V. Areschka, J. Colomer, R. Promel, W. Langenaeker, and P. Geerlings, *Tetrahedron* **1992**, *48*, 10575; (c) R. W. Hoffmann, *Dehydrobenzene and Cycloalkynes*, Academic Press, New York, **1967**; and literature cited therein.

79. (a) W. T. G. Johnson and C. J. Cramer, *J. Phys. Org. Chem.* **2001**, *14*, 597; (b) W. T. G. Johnson and C. J. Cramer, *J. Am. Chem. Soc.* **2001**, *123*, 923.

80. H. H. Wenk and W. Sander, *Chem. Eur. J.* **2001**, *7*, 1837.

81. R. J. Radziszewski, J. Waluk, P. Kaszynski, J. Spanget-Larsen, *J. Phys. Chem. A* **2002**, *106*, 6730.

82. (a) D. D. Callander, P. L. Coe, and J. C. Tatlow, *Chem. Commun.* **1966**, 143; (b) D. D. Callander, P. L. Coe, and J. C. Tatlow, *Tetrahedron* **1969**, *25*, 25; (c) H. Heaney, *Fortschr. Chem. Forsch.* **1970**, *16*, 35.

83. M. Winkler and W. Sander, unpublished results.

84. W. Sander and M. Exner, *J. Chem. Soc. Perkin Trans. 2* **1999**, 2285.

85. H. H. Wenk and W. Sander, *Eur. J. Org. Chem.* **2002**, 3927.

86. H. H. Wenk, A. Balster, W. Sander, D. A. Hrovat, and W. T. Borden, *Angew. Chem. Int. Ed. Engl.* **2001**, *40*, 2295.

87. W. T. Borden, *Chem. Commun.* **1998**, 1919.

88. R. R. Squires and C. J. Cramer, *J. Phys. Chem. A* **1998**, *102*, 9072.

89. G. P. Ford and E. R. Biel, *Tetrahedron Lett.* **1995**, *36*, 3663.

90. (a) H.-F. Grützmacher and J. Lohmann, *Liebigs Ann. Chem.* **1970**, *733*, 88; (b) H.-F. Grützmacher and J. Lohmann, *Liebigs Ann. Chem.* **1975**, 2023.

91. J. Lohmann, *J. Chem. Soc. Perkin Trans. 1* **1972**, *68*, 814.

92. H. A. Weimer, B. J. McFarland, S. Li, W. Weltner, Jr., *J. Phys. Chem.* **1995**, *99*, 1824.

93. (a) T. Sato, M. Moriyama, H. Niino, and A. Yabe, *Chem. Commun.* **1999**, 1089; (b) T. Sato, H. Niino, and A. Yabe, *J. Phys. Chem. A* **2001**, *105*, 7790.

94. W. E. Billups, J. D. Buynak, and D. Butler, *J. Org. Chem.* **1979**, *44*, 4218.

95. H. H. Wenk and W. Sander, unpublished results.

96. W. R. Roth, H. Hopf, T. Wasser, H. Zimmermann, and Ch. Werner, *Liebigs Ann.* **1996**, 1691.

97. J. W. Grissom, T. L. Calking, H. A. McMillen, and Y. Jiang, *J. Org. Chem.* **1994**, *59*, 5833.

98. (a) A. G. Myers and N. S. Finney, *J. Am. Chem. Soc.* **1992**, *114*, 10986; (b) A. G. Myers and P. S. Dragovich, *J. Am. Chem. Soc.* **1993**, *115*, 7021.

99. K. N. Bharucha, R. M. Marsh, R. E. Minto, and R. G. Bergman, *J. Am. Chem. Soc.* **1992**, *114*, 3120.

100. S. Sternhell, *Quart. Rev. Chem. Soc.* **1969**, *23*, 236.

101. F. W. King, *Chem. Rev.* **1976**, *76*, 157.

102. For reviews on neocarcinostatin cytostatics, see, for example: (a) H. Maeda, K. Edo, and N. Ishida, Eds., *Neocarzinostatin: The Past, the Present and Future of an Anticancer Drug*, Springer, Tokyo, **1997**; (b) A. G. Myers and C. A. Parrish, *Bioconjugate Chem.* **1996**, *7*, 322; see also Ref. 7.

103. (a) A. G. Myers, E. Y. Kuo, and N. S. Finney, *J. Am. Chem. Soc.* **1989**, *111*, 8057, 9130; (b) A. G. Myers, P. S. Dragovich, and E. Y. Kuo, *J. Am. Chem. Soc.* **1992**, *114*, 9369; (c) R. Nagata, H. Yamanaka, E. Okazaki, and I. Saito, *Tetrahedron Lett.* **1989**, *30*, 4995; (d) R. Nagata, Y. Hidenori, E. Murahashi, and I. Saito, *Tetrahedron Lett.* **1990**, *31*, 2907.

104. (a) M. Schmittel, M. Strittmatter, and S. Kiau, *Tetrahedron Lett.* **1995**, *36*, 4975; (b) M. Schmittel, S. Kiau, T. Siebert, and M. Strittmatter, *Tetrahedron Lett.* **1996**, *37*, 999, 7691; (c) M. Schmittel, M. Strittmatter, and S. Kiau, *Angew. Chem. Int. Ed. Engl.* **1996**, *35*, 1843; (d) M. Schmittel, M. Keller, S. Kiau, and M. Strittmatter, *Chem. Eur. J.* **1997**, *3*, 807; (e) T. Gillmann, T. Hülsen, W. Massa, and S. Wocadlo, *Synlett* **1995**, 1257; (f) J. G. Garcia, B. Ramos, L. M. Pratt, and A. Rodriguez, *Tetrahedron Lett.* **1995**, *36*, 7391.

105. (a) B. Engels and M. Hanrath, *J. Am. Chem. Soc.* **1998**, *120*, 6356; (b) P. R. Schreiner and M. Prall, *J. Am. Chem. Soc.* **1999**, *121*, 8615; (c) C. J. Cramer, B. L. Kormos, M. Seierstad, E. C. Sherer, and P. Winget, *Org. Lett.* **2001**, *3*, 1881; (d) S. P. de Visser, M. Filatov, and S. Shaik, *Phys. Chem. Chem. Phys.* **2001**, *3*, 1242; (e) P. W. Musch and B. Engels, *J. Am. Chem. Soc.* **2001**, *123*, 5557.

106. (a) C. J. Cramer and R. R. Squires, *Org. Lett.* **1999**, *1*, 215; (b) P. W. Musch and B. Engels, *Angew. Chem. Int. Ed. Engl.* **2001**, *40*, 3833.

107. (a) P. G. Wenthold, S. G. Wierschke, J. J. Nash, and R. R. Squires, *J. Am. Chem. Soc.* **1993**, *115*, 12611; (b) P. G. Wenthold, S. G. Wierschke, J. J. Nash, and R. R. Squires, *J. Am. Chem. Soc.* **1994**, *116*, 7378.

108. C. F. Logan, J. C. Ma, and P. Chen, *J. Am. Chem. Soc.* **1994**, *116*, 2137.

109. For quantum chemical studies on **71** and related open-shell singlet biradicals, see: (a) J. Cabrero, N. Ben-Amor, and R. Caballol, *J. Phys. Chem. A* **1999**, *103*, 6220; (b) J. Gräfenstein and D. Cremer, *Phys. Chem. Chem. Phys.* **2000**, *2*, 2091; cf. also Ref. 105.

110. W. Sander, H. Wandel, G. Bucher, J. Gräfenstein, E. Kraka, and D. Cremer, *J. Am. Chem. Soc.* **1998**, *120*, 8480.

111. C. J. Cramer and J. Thompson, *J. Phys. Chem. A* **2001**, *105*, 2091.

112. A. Z. Bradley and R. P. Johnson, *J. Am. Chem. Soc.* **1997**, *119*, 9917.

113. Reviews on heteroarynes, see, e.g., (a) T. Kauffmann, *Angew. Chem. Int. Ed. Engl.* **1965**, *4*, 543; (b) H. J. den Hertog and H. C. van der Plas, *Adv. Heterocycl. Chem.* **1965**, *4*, 121; (c) T. Kaufmann and R. Wirthwein, *Angew. Chem. Int. Ed. Engl.* **1971**, *10*, 20;

(d) M. G. Reinecke, in *Reactive Intermediates*, Vol. 2, R. A. Abramovitch, Ed., Chapter 5, Plenum Press, New York, **1982**; (e) M. G. Reinecke, *Tetrahedron* **1982**, *38*, 427.

114. (a) C. J. Cramer and S. Debbert, *Chem. Phys. Lett.* **1998**, *287*, 320; (b) S. L. Debbert and C. J. Cramer, *Int. J. Mass Spectrom.* **2000**, *201*, 1.

115. (a) T. Yonezawa, H. Konishi, and H. Kato, *Bull. Chem. Soc. Japan* **1969**, *42*, 933; (b) W. Adam, A. Grimison, and R. Hoffmann, *J. Am. Chem. Soc.* **1969**, *91*, 2590; (c) M. J. S. Dewar and G. P. Ford, *J. Chem. Soc. Chem. Commun.* **1977**, 539; (d) H. H. Nam, G. E. Leroi, and J. F. Harrison, *J. Phys. Chem.* **1991**, *95*, 6514.

116. J. Kramer and R. S. Berry, *J. Am. Chem. Soc.* **1972**, *94*, 8336.

117. T. Sasaki, K. Kanematsu, and M. Uchide, *Bull. Chem. Soc. Jpn.* **1971**, *44*, 858.

118. H.-H. Nam and G. E. Leroi, *J. Am. Chem. Soc.* **1988**, *110*, 4096.

119. (a) M. P. Cava, M. J. Mitchell, D. C. DeJongh, and R. Y. van Fossen, *Tetrahedron Lett.* **1966**, 2947; cf. also: (b) I. R. Dunkin and J. G. MacDonald, *Tetrahedron Lett.* **1982**, *23*, 4839.

120. H. H. Nam and G. E. Leroi, *Tetrahedron Lett.* **1990**, *31*, 4837.

121. J. Chandrasekhar, E. D. Jemmis, and P. v. R. Schleyer, *Tetrahedron Lett.* **1979**, 3707.

122. (a) M. N. Glukhovtsev, V. I. Minkin, and B. Y. Simkin, *Russ. Chem. Rev.* **1985**, *54*, 86; (b) M. Feyereisen, M. Gutowski, and J. Simons, *J. Chem. Phys.* **1992**, *96*, 2926. For a general discussion of the concept of aromaticity, see: (c) V. I. Minkin, M. N. Glukhovtsev, and B. Y. Simkin, *Aromaticity and Antiaromaticity*, John Wiley & Sons, Inc., New York **1994**; (d) P. v. R. Schleyer, Ed., *Chem. Rev.* **2001**, *101*, 1115 (Special Issue: *Aromaticity*).

123. P. v. R. Schleyer, H. Jiao, M. N. Glukhovtsev, J. Chandrasekhar, and E. Kraka, *J. Am. Chem. Soc.* **1994**, *116*, 10129.

124. E. Kraka and D. Cremer, *J. Comp. Chem.* **2001**, *22*, 216.

125. C. J. Cramer, *J. Am. Chem. Soc.* **1998**, *120*, 6261.

126. J. H. Hoffner, M. J. Schottelius, D. Feichtinger, and P. Chen, *J. Am. Chem. Soc.* **1998**, *120*, 376.

127. W. M. David and S. M. Kerwin, *J. Am. Chem. Soc.* **1997**, *119*, 1464.

128. (a) P. Chen, *Angew. Chem. Int. Ed. Engl.* **1996**, *35*, 1478. For a recent discussion of the Chen-model and detailed investigations of hydrogen-abstractions by benzynes, see: (b) A. E. Clark and E. R. Davidson, *J. Am. Chem. Soc.* **2001**, *123*, 10691.

129. C. J. Cramer and R. R. Squires, *J. Phys. Chem. A* **1997**, *101*, 9191.

130. (a) M. von Ardenne, *Adv. Pharmacol. Chemot. (San Diego)* **1972**, *10*, 339; (b) I. F. Tannock and D. Rotin, *Cancer Res.* **1989**, *49*, 4373; (c) E. M. Sevick and R. K. Jain, *Cancer Res.* **1988**, *48*, 1201. Compare also Ref. 126 and literature cited therein.

131. W. D. Langer, T. Velusamy, T. B. H. Kuiper, R. Peng, M. C. McCarthy, M. J. Travers, A. Kovacs, C. A. Gottlieb, and P. Thaddeus, *Astrophys. J.* **1997**, *480*, L63.

132. R. I. Kaiser, *Chem. Rev.* **2002**, *102*, 1309.

133. K. W. Sattelmeyer and J. F. Stanton, *J. Am. Chem. Soc.* **2000**, *122*, 8220.

134. S. Arulmozhiraja, T. Sato, and A. Yabe, *J. Comp. Chem.* **2001**, *22*, 923.

135. (a) E. K. Fields and S. Meyerson, *J. Org. Chem.* **1966**, *31*, 3307; (b) E. K. Fields and S. Meyerson, *Adv. Phys. Org. Chem.* **1968**, *6*, 1. For trapping studies of benzdyines, see (c) H. Hart, N. Raju, M. A. Meador, and D. L. Ward, *J. Org. Chem.* **1983**, *48*, 4357.

136. M. Moriyama, T. Ohana, and A. Yabe, *J. Am. Chem. Soc.* **1997**, *119*, 10229.

137. T. Sato, S. Arulmozhiraja, H. Niino, S. Sasaki, T. Matsuura, and A. Yabe, *J. Am. Chem. Soc.* **2002**, *124*, 4512; and literature cited therein.

138. (a) M. Moriyama, T. Ohana, and A. Yabe, *Chem. Lett.* **1995**, 557; (b) M. Moriyama, T. Sato, T. Uchimaru, and A. Yabe, *Phys. Chem. Chem. Phys.* **1999**, *1*, 2267.

139. T. Sato, H. Niino, and A. Yabe, *J. Photochem. Photobiol. A* **2001**, *145*, 3.

140. (a) W. Weltner, Jr., and R. J. van Zee, *Chem. Rev.* **1989**, *89*, 1713; (b) J. M. L. Martin, J. P. Franccois, and R. J. Gijbels, *Mol. Struct.* **1993**, *294*, 21.

141. (a) K. R. Thompson, R. L. DeKock, and W. Weltner, Jr., *J. Am. Chem. Soc.* **1971**, 93, 4688; (b) M. Vala, T. M. Chandrasekhar, J. Szcepanski, and R. Pellow, *High Temp. Sci.* **1990**, *27*, 19; (c) D. W. Arnold, S. E. Bradforth, T. N. Kitsopoulos, and D. N. Neumark, *J. Chem. Phys.* **1991**, *95*, 8753; (d) C. C. Arnold, Y. Zhao, T. N. Kitsopoulos, and D. N. Neumark, *J. Chem. Phys.* **1992**, *97*, 6121; (e) R. H. Kranze and W. R. M. Graham, *J. Chem. Phys.* **1993**, *98*, 71; (f) H. J. Hwang, A. van Orden, K. Tanaka, E. W. Kuo, J. R. Heath, and R. J. Saykally, *Mol. Phys.* **1993**, *79*, 769.

142. Relative stabilities of C_6 isomers have been calculated at different levels of theory: (a) K. Raghavachari, R. A. Whiteside, and J. A. Pople, *J. Chem. Phys.* **1986**, *85*, 6623; (b) K. Raghavachari and J. S. Binkley, *J. Chem. Phys.* **1987**, *87*, 2191; (c) J. Hutter and H. P. Lüthi, *J. Chem. Phys.* **1994**, *101*, 2213; (d) J. Hutter, H. P. Lüthi, and F. Diederich, *J. Am. Chem. Soc.* **1994**, *116*, 750. The historical developement of calculations on C_6 is outlined in (e) J. M. L. Martin and P. R. Taylor, *J. Phys. Chem.* **1996**, *100*, 6047. Concerning the aromaticity of **133** and **134**, see Ref. 123.

143. S. L. Wang, C. M. L. Rittby, and W. R. M. Graham, *J. Chem. Phys.* **1997**, *107*, 6032.

144. F. Grein, J. Franz, M. Hanrath, and S. D. Peyerimhoff, *Chem. Phys.* **2001**, *263*, 55.

PART 2

METHODS AND TEMPORAL REGIMES

CHAPTER 17

Matrix Isolation

THOMAS BALLY

Department of Chemistry, University of Fribourg, CH-1700, Fribourg, Switzerland

Reactive Intermediate Chemistry, edited by Robert A. Moss, Matthew S. Platz, and Maitland Jones, Jr.
ISBN 0-471-23324-2   Copyright © 2004 John Wiley & Sons, Inc.

1. WHAT IS MATRIX ISOLATION?

If one peruses the literature on reactive intermediates, or discusses the subject with colleagues interested in that field, one soon finds that the term "matrix isolation" means different things to different people, so some semantic clarification appears to be in order at the outset.

The term "matrix isolation" was coined by George Pimentel who pioneered this field[1,2] together with George Porter.[3] Pimentel intended this term to refer to a method whereby a substrate is mixed with a large excess of an (unusally unreactive) host gas and is condensed on a surface that is sufficiently cold to assure rapid solidification of the material. In this way, one ends up with a sample where (ideally) each substrate molecule is immobilized in a cavity surrounded by one or more layers of inert material and is thus "isolated" from the other substrate molecules in a "matrix" of the host gas.

In the course of time, the term matrix isolation came to be applied in a more general sense, encompassing a range of techniques where guest molecules are trapped in rigid host materials and are thereby prevented from undergoing diffusion. Such host materials may be, for example, crystals, zeolites or clays, polymers, boric acid glasses, or cryptands. However, most relevant in the present context are studies of reactive intermediates in *frozen solutions* that are often referred to under the heading of "matrix isolation," but should perhaps more appropriately be referred to as "low-temperature spectrosopy in rigid media."

Such experiments have provided (and continue to provide) much valuable spectrosopic information on many types of reactive intermediates discussed in this volume. In particular, solvents that provide transparent glasses on freezing, such as the ether–pentane–alcohol (EPA) mixture introduced by G.N. Lewis,[4] methyltetrahydrofuran (MTHF), or the mixture of $CFCl_3$ and $CF_2Br–CF_2Br$ discovered by Sandorfy[5] have proven to be very convenient media for obtaining ultraviolet–visible (UV–vis) absorption spectra of reactive intermediates that can be generated by photolysis or radiolysis at 77 K. On the other hand, many odd electron species (radicals, triplet biradicals and carbenes, radical ions) were characterized by electron spin resonance (ESR) methods in solvents that form polycrystalline matrices long before matrix isolation techniques became widely available.

However, to include this enormous body of work, and the techniques that stand behind it, into the present chapter would surpass its limits. Hence, the reader is encouraged to visit the contributions on carbocations, carbanions, radicals, radical ions, carbenes, sylilenes, nitrenes, and arynes where studies involving frozen solutions will be referred to in their topical context. Thus, this chapter will deal only with matrix isolation in the "proper" sense of the word, which can be summed up in a sketch (Scheme 17.1, adapted from the book of Almond and Downs[6]) that encompasses in schematic form most of what will be discussed in the following sections.

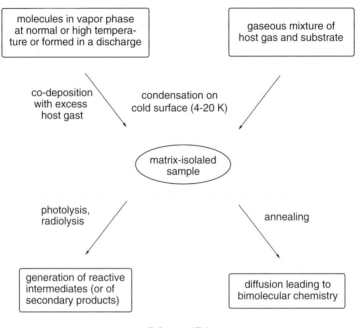

Scheme 17.1

2. WHY MATRIX ISOLATION?

When one wants to engage in the study of species that are only of fleeting existence under ambient conditions, one has basically two choices: either one looks at them very quickly, that is, immediately after their formation, which nowadays can be as short as a few femtoseconds, or one attempts to form or trap them under circumstances where they can be studied leisurly, using conventional spectroscopic tools. The two methods are complementary in that time-resolved techniques provide *kinetic information*, but often at the expense of spectroscopic detail, whereas investigations under "stable conditions" can yield much more detailed insight into the *electronic and* (often indirectly) the *molecular structure* of reactive intermediates.

Note that there is a price to pay for this, that is, the transfer of these species from their "natural" environment, where they occur as fleeting intermediates on the way from reactants to products (e.g., aqueous solution or the protein environment of an enzyme in the case of intermediates of biological processes), to an "artificial" environment. This transfer may impart different properties onto reactive intermediates and their chemistry, a feature that should be remembered when one uses insight gained from studies of chemically isolated species to draw conclusions about their behavior in their "native" environment.

With the above caveat in mind, one usually tries to create an environment that provides for minimal interaction with the target reactive intermediate, a goal that cannot always be achieved with frozen solvents. By this token, noble gases, especially neon and argon, are clearly the best choices, although in many cases the heavier noble gases, nitrogen, methane, or other inert molecular hosts, may also be used. On the other hand, one may want to deliberately choose a matrix material that is *reactive* toward the targeted intermerdiate, for example, CO or O_2, to trap carbenes or nitrenes, but that preserves the other advantages over polyatomic solvents (see below). Often, doping of an inert host material with an additive that fulfils a certain function (e.g., an alkyl chloride that can trap electrons when one wants to study radical cations), or can enter into diagnostic reactions on annealing (see below) is the strategy of choice.

Frozen solvents have another disadvantage, namely, that they are often opaque throughhout wide regions of the infrared (IR) and part of the UV spectral range. Since vibrational spectroscopy can provide (at least potentially) much more structural information than UV–vis spectroscopy, this creates an incentive to generate reactive intermediates in a medium that allows their observation in the IR. Since all atomic and homonuclear diatomic gases are entirely transparent in the IR (and in most cases also throughout the visible and UV), they also represent ideal choices as matrix hosts also from this spectroscopic point of view. Actually, these materials have the additional advantage that they offer generally much higher spectral resolution (i.e., narrower bands) than frozen solvents, which can be a decisive factor if it comes to disentangle complex spectral patterns. The spectra of the naphthalene radical cation in a frozen mixture of butyl chloride (BuCl) and isopentane (iP) at 77 K,[7] and in an Ar matrix at 12 K[8] shown in Figure 17.1 may serve to illustrate this feature, as well as the extended observational range in the UV.

A third advantage that matrix isolation has over frozen solvents is that the reactive intermediates must not necessarily be generated *in situ*, but can be made by flash vacuum pyrolysis or in plasma processes prior to their quenching with an excess of the host gas on the cold surface. Of course, this considerably widens the range of reactive intermediates that can be investigated, beyond those that require photolysis or some form of radiolysis for their formation.

Finally, because very low temperatures are needed to solidify the above host gases, reactive intermediates may be generated under conditions where unimolecular thermally activated processes are also largely suppressed, unless they involve quantum mechanical tunneling. Thus, species that rearrange or fragment spontaneously even at 77 K can be stabilized, say, at 10 K, provided that such processes are not driven by excess energy that is imparted onto the incipient intermediates on their formation by photolysis or radiolysis (see below).

Thus, there are many good reasons why someone who is interested in studying reactive intermedieates may want to purchase the necessary equipment and go to the trouble of familiarizing himself or herself with the technique of matrix isolation.

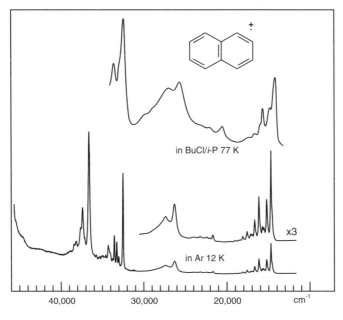

Figure 17.1

In fact, as will be shown below in Section 3, this has become much less daunting than it used to be in the early days.

3. LIMITATIONS OF MATRIX ISOLATION

It is of course important to realize that the technique of matrix isolation also has its limits, or that certain conditions must be fulfilled so that it can be applied. The first and most important one is that the precursor of the reactive intermediate to be studied must be an isolable substance and volatilizable without decomposition, which sets limits on the size of species that can be studied and/or on their thermal lability. Thus many interesting compounds (e.g. of biological relevance) are excluded, at least in their native forms. Also very nonvolatile substrates, such as metals, require special techniques such as Knudsen cells for controlled evaporation.

Second, rich bimolecular chemistry (attack by nucleophiles, electrophiles, oxidants, or reductants) that can be used to create reactive intermediates in solution is not generally available in the context of matrix isolation (exceptions to this rule will be discussed in the proper context below). Usually, reactive intermediates to be studied by matrix isolation must be accessible by means of unimolecular processes (fragmentations, rearrangements, ionization) induced by external sources of energy (light or other forms of radiation, discharges).

Third, even precursors that readily provide some reactive intermediate in solution or in the gas phase, will not necessarily "work" inside a matrix because of the so-called *cage effect*. With the exception of hydrogen and fluorine atoms, all fragments that might be created during the cleavage of a precursor will usually remain trapped in the same matrix cavity and may therefore undergo recombination, either in very low activation thermal processes or driven by (inadvertent) excitation of the target intermediate. Thus, a precursor that undergoes efficient photocleavage in solution or in the gas phase may appear to be entirely photostable under matrix conditions, simply because the fragments recombine immediately after their formation.

A fourth, often overlooked problem is most prominent in noble gas matrices which are notoriously poor heat sinks because only very low energy lattice phonons are available to accept molecular vibrational quanta. Hence, thermalization is very slow compared to solution, and the excess energy that may be imparted onto an incipient reactive intermediate in the process of its formation (e.g., from a precursor excited state) may therefore be dissipated in secondary chemical processes such as rearrangments or fragmentations, which may make it impossible to generate the primary reactive intermediate. Often, this problem can be alleviated by attaching alkyl groups that serve as "internal heat sinks," but sometimes this is not acceptable for other reasons.

Finally, one should pay attention to the fact that especially polar substrates have a tendency to form dimers or higher aggregates, even in the gas phase (although that tendency is much smaller than in frozen solutions). Thus, on radiolytic ionization of anthracene evaporated into a stream of Ar and trapped at 20 K a sizeable yield of the dimer cation is obtained, indicating that at least a part of the anthracene sample exists in the gas phase and later in Argon in the form of dimers. Another classical example is diazocyclopentadiene, which invariably yields a fulvene (the dimer of the targeted reactive intermediate, cyclopentadienylidene) on photolysis in matrices. Going to higher host/guest ratios may diminish or eliminate aggregation of precursors in the gas phase, but often this is not practical due to the ensuing loss of spectroscopic signals. A simple trick is to pass the stream of gas through a tube immersed into an ultrasound bath, thus providing energy to separate aggregates before deposition on the cold surface.

4. TECHNICAL ASPECTS, EQUIPMENT

When matrix isolation was invented in the 1950s, carrying out such experiments required a degree of adventurousness, because it involved the handling of liquid hydrogen or helium in open Dewar vessels, and stories of much destroyed laboratory equipment are recounted by people who were active in the field in its pioneer heydays. Luckily, matrix isolation is nowadays no riskier than any other standard laboratory activity. It requires mostly commercially available equipment that has undergone many cycles of improvement and is functioning reliably. Nevertheless, competent support from a machine shop and a glass blower is of great help to set up a laboratory for matrix isolation, as will become evident in the following sections that will be devoted to short descriptions of the different pieces of equipment required to carry out matrix isolation experiments. For more detailed information, and for many helpful practical hints, the reader is referred to the excellent guide by Dunkin.[9] Where it goes into detail, this chapter will focus on aspects that are not covered in the above book.

4.1. Cryostats and Associated Apparatus

Unless temperatures below 10 K are required for the condensation of the host gas or to prevent some low-activation thermal process from occurring, a *closed cycle cryostat* will invariably form the heart of each matrix isolation setup. These devices, which brought matrix isolation also to laboratories with no (affordable) access to liquid helium, consist basically of a compressor and an expander, connected by two lengths of high-pressure steel bellows tubing. Helium is used as the working medium in a so-called Gifford–McMahon cycle to generate temperatures down to 10 K at the cold end of the second stage of the expander (with special regenerator materials in the second stage of the expander down to 6 K may be attained, if needed).

If one intends to use Ne or H_2 as a host gas, one needs to take recourse to cryostats that are able to attain temperatures of 4 K or less. Until recently, *flow cryostats*, where liquid He is guided in a controlled fashion to the cold end, were usually required for such experiments. However, two-stage 4 K closed-cycle cryostats have recently become available that appear to obviate the need for liquid He.[10] The author knows of two prominent matrix-isolation research groups who have been using such devices for a little while now and have no complaints about them.

During sample deposition, or in annealing experiments, one does not want to operate a cryostat at the lowest temperature it can attain. Rather than reducing the cooling power of the cryostat, this is achieved by fitting the cold end with a resitive heating element and a thermocouple or a silicon diode. Higher temperatures are then attained by passing variable amounts of current through the resistive heater, a task that is usually handled by an electronic device that can be programmed to assure that a chosen temperature is maintained in the presence of varying heat loads without undue oscillations due to temporary under- or overheating. Most suppliers of cryostats will be able to provide expanders that are fully equipped and ready to cooperate with one or another temperature controller. For little additional cost, they will also mount an extra thermocouple or silicon diode that can be connected to a spare channel of the temperature controller and may be used, for example, to monitor the temperature directly in the sample, where it is usually a few degrees higher than directly on the cold end of the expander.

Of course all surfaces that are at such low temperatures must be kept out of contact with the ambient environment. This is achieved by a detachable and rotable *vacuum shroud* that surrounds the two expander stages and the sample, all of which must be kept under high vacuum while they are cold to avoid *collisional heat transfer*. By default, evacuation of the assembly occurs through vacuum ports mounted on the main body of the expander, but in some cases it is advantageous to have extra ports on the vacuum shroud itself. Furthermore, the first expansion stage of closed-cycle cryostats, where a temperature of $\sim$35–40 K is attained, is usually fitted with

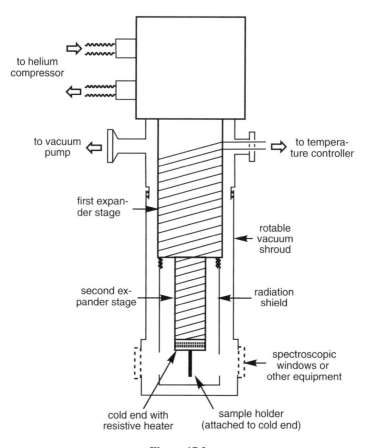

to helium
compressor

to vacuum
pump

to tempera-
ture controller

first expan-
der stage

rotable
vacuum
shroud

second ex-
pander stage

radiation
shield

spectroscopic
windows or
other equipment

cold end with
resistive heater

sample holder
(attached to cold end)

Figure 17.2

a detachable *radiation shield* that extends over the length of the second stage and
the sample to protect both from *radiative heat transfer.*

The part of the vacuum shroud that surrounds the sample has several openings
fitted with O-rings and threaded holes. These openings can be equipped with spectroscopic windows, different inlet systems, sources of high-energy radiation such as
open discharge lamps, or other devices. Some suppliers will furnish window
holders, and ARS even equips their vacuum shrouds with small side ports that
can be used as sample inlet systems, which are not very practical in many cases.
Most users will, however, want to customize the equipment to be mounted on the
shroud openings such as to optimally serve their particular uses. Figure 17.2 shows
a schematic drawing of a typical closed-cycle cryostat (minus the compressor).

4.2. Sample Holders, Spectroscopic Windows

The sample holder designates the body on which the matrix is deposited, and its
interface to the cold end of the cryostat. The choice of that body depends on the

type of spectroscopy that one wants to use. If one is only interested in UV–vis/NIR (near-infrared) absorption spectra, then a sapphire or well-polished CaF_2 disk are most appropriate.[11] Since one of the major advantages of matrix isolation (as compared to frozen solvent studies) is that IR spectra of reactive intermediates may be recorded, one will usually want to chose a material that is also transparent in this spectral range. Even if the full range of its IR transparency (down to 200 cm^{-1}) is not exploited, *cesium iodide* is the best choice, mainly because of its advantageous mechanical properties: NaCl and KBr crystals are so brittle that the mechanical stress during their cooling and heating over a range of nearly 300 K, while being fitted tightly into a copper ring of different expansivity, will often lead them to crack or cleave. In contrast, CsI behaves like a soft glass and weathers most conditions that occur in matrix isolation experiments. However, CsI is also very hygroscopic, and therefore one has to wait after every experiment until the sample holder has returned to ambient temperature before exposing it to atmosphere, to prevent condensed water from dissolving the precious material.

In contrast to popular belief, nourished by poorly maintained solution IR cells, well-polished alkali halide crystals are transparent down to 200 nm, so it is usually no problem to take UV/vis and IR spectra from the same samples deposited on CsI windows. One should, however, request the slightly more expensive "UV-quality" material that is free of color centers in this spectral range. Buying polished crystals is not worth the considerable additional expense, because one has to set up a facility for polishing them after each use anyway. In our experience, a soft suede leather cloth stretched over a glass slab and a bottle of isopropyl alcohol is all that is needed for this purpose, although crystals with coarsely ground or very foggy surfaces may need some pretreatment with very fine wetted sandpaper, followed by some wiping with water and ethanol, before polishing with isopropyl alcohol becomes effective.

Similarly, fused silica windows are not required except for the most demanding applications in UV spectroscopy, so fitting the vacuum shroud with a pair of well-polished KBr windows is usually appropriate,[12] unless one needs IR transparency at 500–200 cm^{-1}, when one has to use CsI windows. If one needs to probe the far-IR range (<200 cm^{-1}) one has to use polyethylene windows (the covers that come with KF flanges have proven adequate), but this makes it of course impossible to probe the same samples in the mid-IR or UV–vis range.

One of the most frequent reasons for failures of matrix isolation experiments is inefficient heat transfer between the sample and the cold end of the cryostat. Thus, it is very important to assure good thermal contact between the window and its holder (which must be made from a material of high heat conductivity), and between the latter and the cold end. This good thermal contact is effected by sandwiching the window between two copper plates fitted with indium O-rings that are flattened when the screws that hold the plates together are tightened.[13] Similarly, indium washers placed between the window holder and the cold end of the cryostat make sure that heat flow across this junction is not impeded. Figure 17.3 shows such a sample holder for optical absorption studies.

If one wants to measure spectra in *reflection*, the task of building a sample holder is much simpler, because it can be machined in one piece from solid copper that

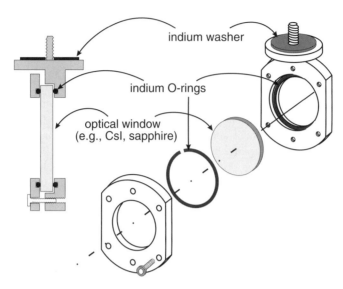

Figure 17.3

eliminates the above mentioned heat-transfer problems. For UV–vis studies the surface should be plated with nickel or silver, whereas in the IR a coating of gold assures optimal reflectivity. Similar sample holders may also be used for Raman scattering studies.

An interesting variant of a reflective sample holder that can be used both for absorption and emission studies was proposed by Rossetti and Brus.[14] They mounted two ∼100-μm slits on the edges of a mirrored plate and deposited a matrix to cover these slits. When one focuses a beam of exciting or probe light onto one of these slits, the mirrored surface and the matrix serve as a *waveguide* by internal reflection, and the light emanates from the opposite slit with very little loss. This technique allows absorption measurements on very dilute samples because the optical pathlengh (∼2 cm) is two orders of magnitude longer than the thickness of a typical matrix (∼200 μm). The same setup can be used to excite a sample (e.g., by a laser) and collect emitted light perpendicularly.

A wholly different setup is required for ESR studies of matrix isolated open-shell species, because the cavities of ESR spectrometers are too narrow to accept the standard vacuum shrouds. Therefore, these shrouds have to be fitted with a small extension tube at the bottom into which a sapphire rod can be lowered after the gas sample mixture is deposited on that rod. This modification requires in turn a special shroud that can be raised and lowered without breaking the vacuum. Such equipment is available commercially and is described in various books and reviews on matrix isolation.

4.3. Inlet Systems

If all that is required is the passage of a mixture of the host gas and the substrate from the sample preparation line onto a cold window, the inlet system can consist of a simple glass tube extending from the high vacuum side of a teflon valve through a Cajon ultratorr fitting that is welded onto a plate covering one of the free openings of the vacuum shroud. The ambient pressure side of the valve should provide for a connection to the sample prepration vacuum line (an example of such a setup is given in Fig. 17.4).

Substrates that are not sufficiently volatile to be premixed stoichiometrically with the host gas have to be placed in a U-tube that can be immersed into a temperature-controlled bath. A stream of host gas passing through this tube will then sweep the substrate along and form a mixture whose ratio depends (at a given geometry and substrate surface) on the temperature of the bath.[15a]

For substances that require more than ∼50 °C for slow evaporation, a furnace must be attached directly to the vacuum shroud. A very simple solution, proposed by Tomioka,[15b] consists of a double-walled glass tube that fits into a steel tube

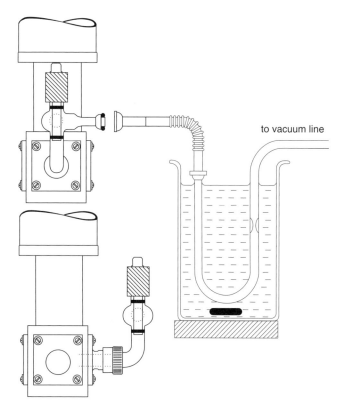

Figure 17.4

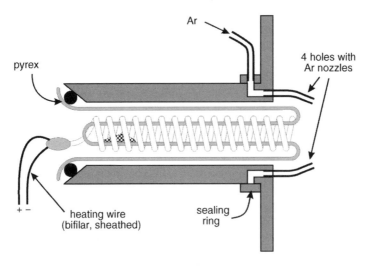

Figure 17.5

attached to a plate (cf. Fig. 17.5). The substrate is placed in the inner part of the glass tube that opens toward the sample window and that can be heated resistively up to ~300 °C. The host gas must be introduced separately, in a way that allows its efficient mixing with the substrate emanating from the heated tube. In the author's laboratory, a design with four small nozzles placed symmetrically around opening of the hot tube has proven suitable for this purpose (see Fig. 17.5). If higher temperatures are needed to volatilize a substrate, one must resort to Knudsen cells.

More elaborate setups are required if the target reactive intermediate is generated by pyrolysis or by passing the precursor through a microwave or electric discharge. The most clean and efficient pyrolysis can be achieved by a technique of pulsed deposition of the substrate–host gas mixture through a very hot ceramic tube. The design of such a system has been described by Chen and co-workers,[16] and variants thereof have found application in several matrix isolation laboratories. Many designs of such "functional" inlet systems that allow substrates to be deposited at high temperature or by sputtering, or pyrolysis, or after being subjected to discharges, or exposure to far-UV light or to beams of electrons are described in detail in the book by Moskovitz and Ozin,[17] so we will not go into more detail here, but refer to some of them in the context of the different techniques for the generation of reactive intermediates (Section 5).

A major problem with "functional" inlet systems is that they are often quite bulky, which makes it difficult to move the cryostat from the sample deposition line to different spectrometers, or from there to stationary sources of radiation. If only one form of spectroscopy is applied in a given study, then the vacuum shroud may be mounted with all parts attached to it in the sample chamber of the corresponding spectrometer. In this case the expander, on which the sample holder is attached, must be rotated within the vacuum shroud to switch the sample from matrix deposition to observation. Another solution is to design the inlet system in such a way that it can be detached from the vacuum shroud after deposition, thus restoring full mobility to the experiment, but this may be tricky because high vacuum must be maintained at all times inside the shroud. However, with the help of compact gate valves mounted on the one of the shroud's openings such solutions can be realized.

4.4. Sample Preparation Lines

The sample preparation vacuum line (often called "spray-on line") should allow for (a) controlled mixing of the host gas with the substrate (or with other components that are added to the matrix) by manometric techniques and (b) the controlled release of the gas (mixture) toward the inlet system of the cryostat. These conditions are met by a vacuum line that incoporates a storage bulb for the gas (mixture), inlets for attachment of evacuable containers that allow degassing of the substrate prior to its mixing with the host gas, pressure gauges that cover suitable ranges, a needle valve that allows the controlled release of the gas, possibly via a flowmeter, and interfaces to the bottles that contain the host gas(es), and to the inlet system that is attached to the vacuum shroud of the cryostat.

Suggestions on how to build and equip such vacuum lines are given in Dunkin's book,[9] so no details will be given here. Note that careful degassing of the sample prior to evaporation of the substrate, which needs to have a vapor pressure of at least 0.1 Torr to allow premixing with the host gas, is very important to assure good control over the host/guest ratio that should be kept below 1:500 to allow good isolation in the matrix.

4.5. Vacuum Equipment

High-vacuum pumps are needed, both for the evacuation of the cryostat and for the sample preparation line. In some laboratories, the same pumps are used to serve both purposes, but this entails several disadvantages, such as bellows tubes leading from the vacuum line to the cryostat that make for very inefficient pumping in the low-pressure regime. Since the volume of the cryostat is quite limited, even a very compact vacuum pump serves well for its efficient evacuation, provided that losses of pumping efficiency can be avoided. Several suppliers now feature small turbo-molecular pumps (pumping speeds of 50–100 l N_2/min suffice largely) that can be mounted horizontally, that is, flanged on in line with the expander port. Such pumps obviate the need for constructing complicated manifolds and at the same time minimize losses of pumping efficiency. Backing by a small rotary or membrane pump completes the vacuum system.

In order to avoid having to shut down the whole vacuum system each time the cryostat is vented after an experiment, it is advisable to place a small slide or butterfly valve between the turbopump and the expander port. When the system is reevacuated, the turbopump must be stopped before opening the valve, otherwise the ambient pressure shock wave that results might deform the delicate rotor and stator blades. However, one advantage of these small turbopumps is that they do not need to be bypasssed for roughing, which simplifies the design of the vacuum system.

Once the cryostat has attained about 40 K, the cold end of the expander is the most efficient vacuum pump for ambient gases in the system, and external evacuation is no longer required (i.e., the cryostat can in principle be detached from the vacuum system). Nevertheless, most practicioners of matrix isolation prefer to keep up the pumping, be it only to continue montioring the pressure inside the cryostat (Penning gauges may give erroneous readings in closed-off systems).

The choice of vacuum pumps for the sample preparation line will depend on the size of its volume and the budget. In general it is advantageous to have somewhat higher pumping speeds available for the vacuum line, but this can well be provided by a conventional oil diffusion pump, which is less expensive than a turbomolecular device, backed again by a rotary pump.

4.6. Trolleys

Unless a matrix isolation experiment is built into or around a spectrometer (which may be the only way to go if, say, a bulky inlet system has to be interfaced to the cryostat), the entire cryostat and the associated vacuum pumps (which together may easily weigh in at more than a hundred pounds) should be mobile in order to move them between different locations in a laboratory. The challenge then consists in constructing a device that will not only allow the whole setup to be wheeled around without too much effort, but permits also a precise and reproducible positioning of the sample next to the sample preparation line, inside different spectrometers, or in front of this or that external source of radiation. Ideally, one should be able to move the expander in all three directions of Cartesian space, at least within certain limits, without having to change the position of the cart or trolley on which the equipment is mounted.

When compact X-ray sources became available in the late 1970s, many dentists unburdened themselves of the somewhat unwieldy cranes that were required to position the older heavy devices next to their client's faces to obtain X-rays of their teeth. However, these cranes had precisely the features that are required for a good matrix isolation trolley, that is, the capability to position a rather heavy piece of equipment precisely in space without much physical effort. In the author's laboratory several such cranes have served (and continue to serve) splendidly for this purpose, and any potential practicioners of matrix isolation are well advised to keep their eyes open for such a device, some of which may still be had for free if they have not already been discarded.

Otherwise the construction of a trolley requires competent professional help. Once more, Dunkin's book[9] may serve as an excellent source of information to get started, although one may want to adapt the designs proposed therein to local circumstances. In any event, it pays to invest in the careful design of the trolley, because this can make life in the laboratory much easier afterwards!

4.7. Spectrometers

Advising the reader on the choice of spectrometers is of course beyond the scope of this chapter, but we would like to point out some features that are worth considering when purchasing or adapting such equipment for the purpose of matrix isolation experiments. In all cases, it is worthwhile to look for spectrometers that have spacious sample compartments, or are constructed in such a way that the probe beam can easily be deflected out of the spectrometer for passing through the sample, and then back toward the monochromator or detector. Usually one will find that the cover of the sample compartment must be dismounted to gain access with the cryostat, and often one will want to replace it with a customized cover to close off the compartment for purging, or to shield it from ambient light while the sample sits in there. Some spectrometers now come with fancifully shaped sample compartment covers that are not easily amenable to modification or replacement, so this should be kept in mind when considering the purchase of a spectrometer.

Whereas purging is usually not required for UV–vis spectrometry, it is inevitable in IR work, because the sharp bands of gaseous H_2O and CO_2 (whose shape is very sensitive to pressure and temperature), never cancel completely on ratioing sample and reference spectra, partly because the volume inside the vacuum shroud is completely free of these gases. The optimal solution to this problem is to evacuate the entire spectrometer, including the sample compartment, before taking IR spectra of matrix isolated species. Unfortuately, evacuable Fourier-transform infrared (FT–IR) spectrometers tend to be on the upscale side, but if one can afford such an instrument, then one never has to worry about dwindling nitrogen supplies on humid summer days or foggy beamsplitters. Instead, one has to design a special cover with a vacuum seal for the cryostat, but the author will gladly supply information on how to go about that.

One of the greatest advantages of matrix isolation IR (or Raman) spectroscopy is that vibrational bands are inherently very narrow, because rotations are largely suppressed, which means that much more detailed information can be obtained than in the liquid phase at room temperature. However, this additional information can only be obtained if the spectrometer offers high enough resolution, which in the case of interferometers translates into sufficient displacement of the movable mirror. For most practical purposes, a resolution of 1 cm^{-1} is adequate for matrix work, although it is good to have 0.5-cm^{-1} resoution available in case one needs it, say, for the elucidation of site structures.

Finally, in our age of powerful electronics, it is often forgotten that the most sophisticated signal conditioning circuitry or data handling software is no replacement for good optics and precise mechanics. Unfortunately, there is a price to be paid for the latter and the reason for the additional expense only becomes evident to a trained eye after opening the cover of the spectrometer, so the temptation is great to settle blindly for a more economical solution. Engaging in a news group of scientists who exchange technical information, and seeking the advice of colleagues who use this or that spectrometer for similar purposes, may help to avoid disappointments.

5. METHODS FOR GENERATING REACTIVE INTERMEDIATES FOR LOW-TEMPERATURE MATRIX STUDIES

Principally, there are three ways to generate reactive intermediates for matrix isolation studies, each of which has its own range of application, advantages, and limitations. They are

1. External generation followed by trapping in the matrix.
2. Cocondensation of a precursor with a reagent with which it will react in the matrix (usually during the condensation process).
3. Generation from a precursor isolated in the matrix (i.e., *in situ*).

These three methods will be highlighted briefly below.

5.1. External Generation

If the experiment is conducted properly, this method has the advantage that the reactive intermediate to be studied is truly isolated in the sense that it is (ideally) surrounded by nothing but the inert host material (doped perhaps with some deliberately added reagent). This feature can be very important, say, in the case of radicals which usually arise in pairs that have a propensity to recombine if trapped in the same matrix cavity, which often excludes method (3) above for the generation of such species.

On the other hand, external generation of reactive intermediates has two potential disadvantages, a chemical and a technical one. First, the target reactive intermediate must persist during the time that passes between its creation and its thermalization in the cold matrix environment. Thus, species that are prone to undergo thermal rearrangements or fragmentations at at the temperature that prevails in the environment where they are generated (say, a pyrolysis or discharge tube) will not appear in the form of the primary reactive intemediate in the matrix. For example, any attempt to study the radical cation of cyclobutene generated externally by matrix isolation is bound to fail because this species will have rearranged to the butadiene radical cation by the time it has reached the matrix. Similar considerations apply, for example, to carbenes carrying α-hydrogen atoms.

Second, contraptions used for external generation of reactive intermediates are often quite bulky, which may impede the mobility of the experiment. Such contraptions may render it difficult to investigate a sample by several different kinds of spectroscopy. In principle it is possible, with the aid of a gate valve mounted on the vacuum shroud, to construct devices that allow one to retract and detach the external source of a reactive intemediate after a matrix has been built, but the implementation of this strategy is technically quite challenging because high vacuum must be maintained at all times within the cryostat.

The methods for external generation of reactive intermediates are similar to those used in gas-phase experiments, that is, flash vacuum pyrolysis, passing a precursor through a microwave discharge, electron or proton radiolyis, or far-UV irradiation with open-tube discharge lamps during deposition (whereby most far-UV light is absorbed by gas-phase molecules excited into Rydberg states). Where necessary, details of these methods will be described in connection with their use to make different types of reactive intermediates in the following sections.

5.2. Cocondensation of Two Reagents

This method has been used very successfully by many groups, especially in the field of inorganic and organometallic chemistry. Typically, a beam of metal atoms, generated by evaporation from a high-temperature furnace (Knudsen cell), or by laser sputtering, is fed into a beam of the host material mixed with a compound whose adduct with the metal atom is the targeted reactive intermediate. For example, the group of Andrews has made quite a living over the past years from cocondensing a variety of transition metal atoms and clusters, obtained via laser sputtering, with H_2,[18] N_2,[19] O_2,[20] CO,[21] CO_2,[22] H_2O,[23] NO,[24] or S vapor[25] (only the most recent references are given). However, the same can be done with nonmetals, most importantly carbon (cf. Chapter 10 in this volume) or silicon.

The setup used in these experiments is remarkably simple: all that is required is a mechanical vacuum-feedthrough by which a rotatable rod, made of or topped by the desired material, can be introduced. Pulses of the fundamental of a Q-swiched, pulsed Nd:Yag laser (typically 20–80 mJ/pulse, 1–10 Hz) are focused on that target, either through a separate port,[26] or simply from behind the cold window.[27] In order to avoid formation of pits that lead to a greater propensity for cluster formation, the rod is rotated slowly such as to expose a fresh surface to successive laser pulses.[28]

On the other hand, alkali metal atoms can be obtained by simple evaporation at moderate temperatures, and these may serve to abstract bromine or iodine from organic halides, thus providing access to radicals,[29] biradicals,[30] or highly strained hydrocarbons, such as small-ring propellanes.[31,32] The technique has been described in some detail by Otteson and Michl.[33]

Matrix-isolated alkali atoms (or small clusters) also undergo easy photoionization, and the electrons released in this process may attach themselves to nearby substrates to form the corresponding radical anions.[34] However, one drawback of alkali metal atoms or clusters is that they tend to swamp the electronic absorption spectrum of the target reactive intermediate that can only thus be detected by IR.

An interesting variant of method (2) was discovered some time ago by Jacox who found that passing a mixture of NF_3 or CF_4 with Ar through a microwave discharge gave a high yield of fluorine atoms which could be used to abstract hydrogen atoms or add to multiple bonds or lone pairs.[35] However, since F atoms may diffuse through an Ar matrix, especially at a time when it is not yet completely solid, most of the products appeared in the matrix as complexes with HF, which often gave rise to rather substatial shifts of vibrational transitions compared to the free radicals.

5.3. Generation *In Situ*

Finally, *in situ* generation is by far the most popular method for the preparation of many types of organic reactive intermediates, because it is usually easiest to implement and does not impede the mobility of the experiment. Furthermore, the design of suitable precursor molecules often offers an interesting challenge to the synthetic chemist. Invariably, the energy required for the generation of a reactive intermediate from a precursor that is already embedded in a matrix comes in the form of *electromagnetic radiation*. In the case of neutral intermediates, this will usually be near UV light such as that provided by the widespread Hg, Xe, or Hg/Xe arc lamps. Often judicious filtering is required to avoid unwanted secondary reactions, or sometimes lasers are even employed to ensure monochromatic irradiation at a desired wavelength.

If direct *ionization* of a substate is to be effected, higher energy light is required, which can be generated by so-called resonance lamps where a suitable gas is enclosed in (or flowing through) a tube, capped with a LiF or MgF_2 window, where it is excited in a microwave discharge. Wavelengths between 122 nm (H_2) and 175 nm (N_2) can be obtained in this way, and tables for different gases have been compiled.[36] and are reproduced in some books on matrix isolation techniques.[17]

Although irradiation of a matrix will usually be effected at an energy where the targeted precursor shows some optical absorption, this is not invariably a prerequisite to achieve an effect on this precursor. For example, matrix-isolated organic radical cations can be generated by irradiation of Ar matrices by X-rays (to which molecules containing only second-row atoms are practically transparent) whereby the host material is ionized. Subsequent hole transfer to an added substrate leads to the formation of substrate radical cations[37] (see Section 5.6). Also, Maier and Lautz[38] recently found that 254-nm irradiation of Xe matrices doped with halogen atoms leads to bond cleavage in alkanes or olefins that do not absorb (and are otherwise photostable) at this wavelength. The mechanism whereby the energy that is primarily stored in Xe · Br complexes is transferred to the hydrocarbons has not been elucidated, but this method has allowed researchers to achieve some remarkable transformations.

If *in situ* generation of a reactive intermediate involves the formation of fragments, which is often the case, then one has to consider that these will be trapped in the same matrix cavity. With few exceptions (H and F atoms), the fragments will not be able to separate by diffusion through the matrix, so their recombination may occur more or less readily, depending on the activation energy for this process and geometric factors that may come into play. This so-called *cage effect* is especially obstructive in the formation of radicals which usually arise in pairs that may recombine with little, if any, activation energy. For example, certain binuclear metal carbonyls that readily undergo cleavage of the metal–metal bond on photolysis in solution, thus yielding reactive intermediates that undergo further interesting chemistry, appear as entirely photostable or undergo only loss of CO in matrices, due to the rapid recombination of the primary fragments. Further manifestations of the cage effect will be illustrated in the following sections.

Another potential limitation of *in situ* generation of reactive intermediates is that they must be reasonably photostable under the conditions of irradiation required for their generation. The higher the energy of the light that is needed to decompose a precursor, the more likely it is that the resulting intermediate is also affected by that light. Thus, phenol, which is in principle a very good precursor of the phenoxy radical because the hydrogen atom that is formed concomitantly can escape by diffuion through most matrices, gives the decomposition products of the phenoxy radical $(CO + C_5H_5)$ as major products on 254-nm irradiation in Ar matrices.[39]

However, the cage effect may allow an investiator to monitor processes that cannot be observed in the gas phase or in solution because fragmentation prevails under these conditions. A good example are radical cations which usually undergo dissociation on photoexcitation in the gas phase,[40] but that display a rich variety of photorearrangements in Ar matrices.[37]

6. PREPARATION OF DIFFERENT TYPES OF REACTIVE INTERMEDIATES

6.1. Carbenes, Nitrenes

The great majority of matrix isolation studies of carbenes and nitrenes have employed their formal adducts with molecular nitrogen, that is, diazo compounds or diazirines in the case of carbenes, azides in the case of nitrenes, as precursors for their *in situ* generation. Usually, these compounds will readily release N_2 on irradiation with a low-pressure mercury lamp (254 nm), and this fragment has the advantage that it will usually not react with or perturb the targeted reactive intermediate (see Scheme 17.2).

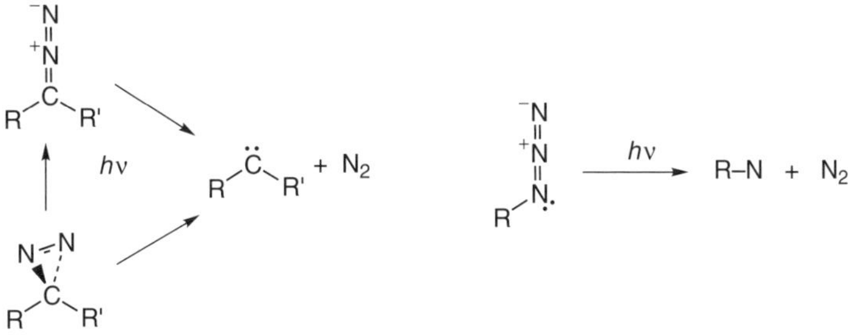

Scheme 17.2

A notable exception to this rule is the parent carbene (CH_2) whose characterization as a matrix isolated species was impeded for a long time because it recombines readily with N_2, even at 10 K, as demonstrated by the exchange of ^{15}N against ^{14}N when isotopically labeled diazomethane was photolyzed in a N_2 matrix.[41]

From the *singlet* state of CH_2, this recombination is exothermic by about 35 kcal/mol,[42] and, according to calculations, it should proceed with no barrier.[43] However, it would be surprising if all CH_2 would be trapped by N_2 before it undergoes intersystem crossing to the triplet ground state, which is thermally unreactive toward N_2 at matrix isolation temperatures. Perhaps the recombination is due to excitation of $^3CH_2$ under the conditions where diazomethane is decomposed. Photoinduced recapture of N_2 by carbenes has also been reported for cyclopentadienylidene and 2,5-diazacyclopentadienylidene (the latter is a singlet carbene which also adds N_2 thermally on annealing a matrix to 25 K).[44]

Diazo compounds are usually easy to synthesize and handle if the N_2 group is next to a carbonyl or an aryl group. In other cases, especially when electron-releasing substituents such as halogen or oxygen atoms are adjacent to the diazo moiety, diazirines prove to be more practical. Sometimes, diazirines undergo photo-rearrangement to diazo compounds prior to decomposition (or vice versa, as in the case of phenyldiazomethane[45]), but this does not impede the formation of the desired carbene.

A disadvantage of diazo compounds is that they are quite polar, which endows them with a propensity to form dimers in the gas phase, even at concentrations as small as 1:5000. This feature becomes evident after their decomposition, which sometimes leads predominantly to carbene dimers (e.g., pentafulvalene in the case of diazocyclopentadiene[46]), or adducts of the target carbene with its diazo precursor, as in the case of methylene, where the main products are $CH_2=NH$ and HCN.[47] The problem disappears at very high guest/host ratios, but it is often impractical to achieve these. Fortunately, diazirines are less prone to dimerization in the gas phase.

Other possible carbene precursors (vicinal dihalides, peresters, ketenes, carbene adducts with stable hydrocarbons, etc.) cannot generally be used for *in situ* generation of carbenes, because the fragments are likely to recombine. However, they can be used, for example, in experiments involving pyrolysis or other forms of external carbene generation where the fragments get a chance to separate in the gas phase and become trapped in distant matrix sites. All conceivable halo- and dihalocarbenes were made and studied in this way (see, e.g., the 1993 review by Sander et al.[48]) However, such methods can only be applied to carbenes which resist thermal rearrangement to more stable products.

Azides are virtually the only nitrene precursors that have been used in matrix isolation studies. They are usually easily accessible, but should only be made and handled in very small quantities because certain azides can be violently explosive.

6.2. Radicals

The cage effect can be a source of great frustration in matrix isolation studies of monoradicals, because such species are usually formed by homolysis from closed-shell compounds, and hence any radical generated *in situ* is invariably accompanied by another radical that will be trapped in the same matrix cage.

Hence, most photochemical precursors of radicals in solution (peroxides, azo compounds, most halides, etc.) are not generally suitable for *in situ* generation of these species in confined environments such as matrices (of course reagents that generate radicals by abstraction, such as the popular trialkyl tin hydrides, are also excluded from matrix isolation).

Two "chemical" strategies may be used to alleviate the cage problem: one consists in choosing precursors whose decomposition products include either a hydrogen atom, which can readily escape the matrix cage by diffusion, or a radical that will be unreactive toward the target radical. Cleaving C—H bonds photochemically is only possible in compounds which carry a chromophore that absorbs at (or above) the energy of the targeted bond. Unless far-UV light sources are available, this limits the strategy to precursors that produce resonance-stabilized radicals, such as benzyl or polyenyl radicals. On the other hand, it was found that atomic iodine or NO survive at least partially next to such reactive radicals as phenyl.[49] Thus, iodides and nitroso compounds should be considered as possible precursors for photolytic *in situ* generation of radicals, although they cannot be used generally, for example, to produce alkyl radicals.

The second strategy is to design precursors that interpose one or two inert molecules between two fragments that are prone to recombination (see Scheme 17.3). For example, diacylperoxides undergo clean photocleavage into two radicals R$^{\bullet}$ separated by a pair of CO_2 molecules that prevent at least some of the radicals enclosed in adjacent matrix cages from dimerizing.[50] The same strategy can also be employed to make matrix isolated biradicals (see, e.g., Chapter 16 in this volume). It was reported that a CO_2 and a CO molecule, such as they are produced on decomposition of acid anhydrides, R—CO—O—CO—R, may also suffice to keep two radicals at bay.[51]

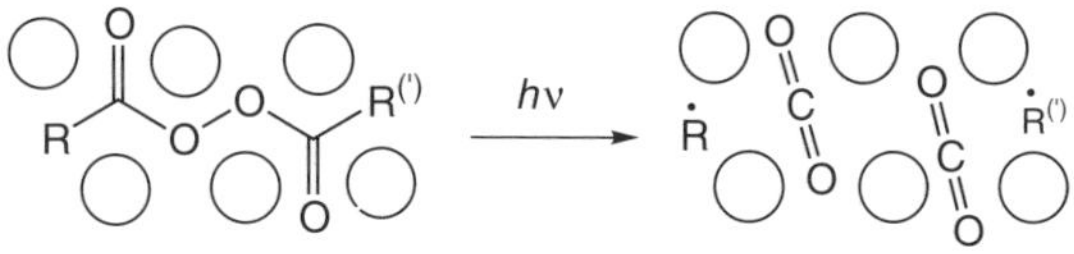

Scheme 17.3

Occasionally, steric factors may prevent radicals from recombining. Thus we have recently found that the radicals $PhCH_2^{\bullet}$ + $^{\bullet}COCl$ which are formed on photolysis of oxychlorocarbene $PhCH_2OCCl$ (which is itself a reactive intermediate!) do not recombine spontaneously to form phenacetyl chloride $PhCH_2COCl$, presumably because the unpaired electron of the OCCl radical "points" in the "wrong" direction to attack the benzyl radical. However, radical recombination *does* occur on annealing of the matrix, or on subsequent photoexcitation of the benzyl radical at >515 nm, which seems to provide the required energy its reorientation inside the rigid matrix cage (see Scheme 17.4).[52]

Of course all of the above problems are avoided if a radical is generated *externally* and subsequently trapped in a matrix. If one wants to prepare a radical in a

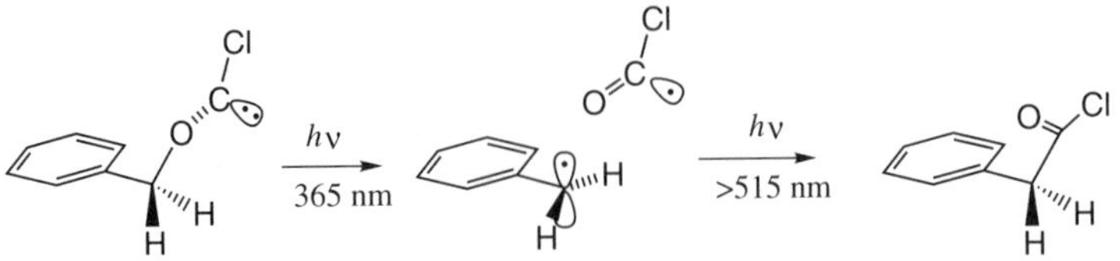

Scheme 17.4

planned and controlled fashion, some form of *pyrolysis* is usually the method of choice. Most of the precursors that may be used for *in situ* generation of radicals described above are also useful sources of thermally created radicals. In addition, azo compounds (R—N=N—R) are convenient thermal precursors, and precursors with weak C—C bonds, such as diphenylethane or 1,5-hexadiene, can also be employed as very clean sources of π-conjugated radicals. Also, bromides can be used instead of iodides, which are sometimes more difficult to synthesize.

In early studies, flash vacuum pyrolysis, a method that has proven very valuable in preparative studies of closed-shell compounds,[53] was regarded as the method of choice for the production of radicals for matrix isolation studies.[54] The disadvantage of this method, which is very well suited for preparative studies of closed-shell compounds, is that the reaction occurs on the walls of a hot tube whose surface may trap radicals (this problem may be alleviated by coating the inside of the tube with gold[55]). Also, unless a very low vacuum can be maintained in the pyrolysis tube, collisions between radicals may lead to gas-phase dimerization.

All of these problems are overcome if the precursors are pyrolyzed in an inert host gas at high pressure, because under these conditions energy transfer occurs mostly by collisions with the host gas, and radical recombination is largely suppressed. Of course, a constant stream of hot gas at high pressure is incompatible with the requirements of matrix isolation, so the experiment has to be carried out in a *pulsed* fashion. Chen and co-workers[16,56] were the first to propose what they called a "hyperthermal nozzle" for pulsed pyrolysis at very high temperatures, at that time for gas-phase studies. Several research groups have implemented variants of this design[49,57] for work in matrix isolation and have used it successfully for a variety of studies.

6.3. Biradicals

Whereas generation of monoradicals require cleavage of a *single* bond, formation of biradicals by fragmentation involves sequential or simultaneous cleavage of *two* bonds. If those two bonds involve the same moiety, one of the fragments is therefore a closed-shell species. By taking advantage of this fact, many matrix isolation studies on biradicals have employed precursors designed to yield these species on thermal and/or photochemical (*in situ*) extrusion of N_2 or CO. A few typical examples are given in Scheme 17.5.

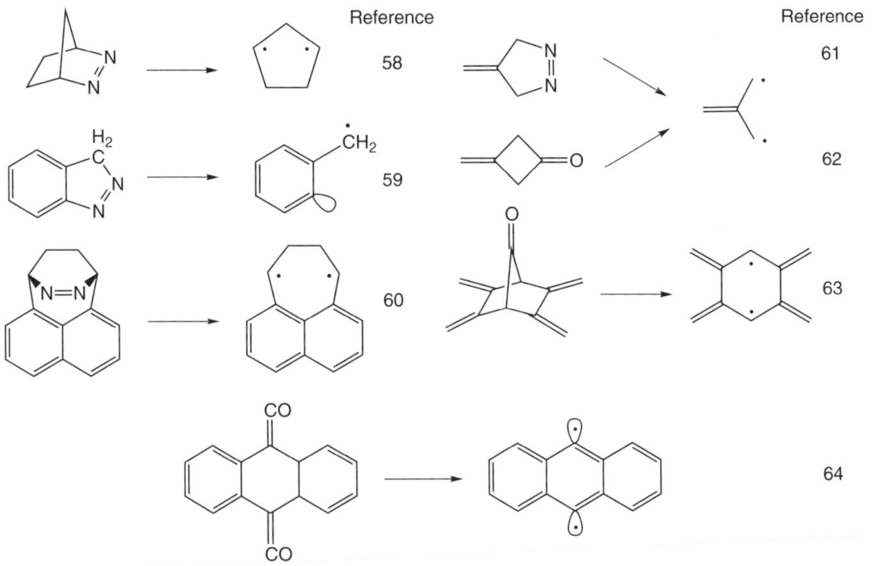

Scheme 17.5

Of course, biradicals can also be generated by a sequence of two cleavage reactions such as described above for monoradicals, for example, from diiodides or bis(acylperoxides). Several examples of this strategy can be found in Chapter 16 in this volume.

In biradicals, where the two unpaired electrons can recombine to form a covalent bond, the resulting species may in turn serve as a potential precursor for the biradical. This only works in cases where the resulting biradical has a triplet ground state, which creates a spin barrier for the re-formation of the original bond. Two typical examples where this strategy succeeded are illustrated in Scheme 17.6.[65,66]

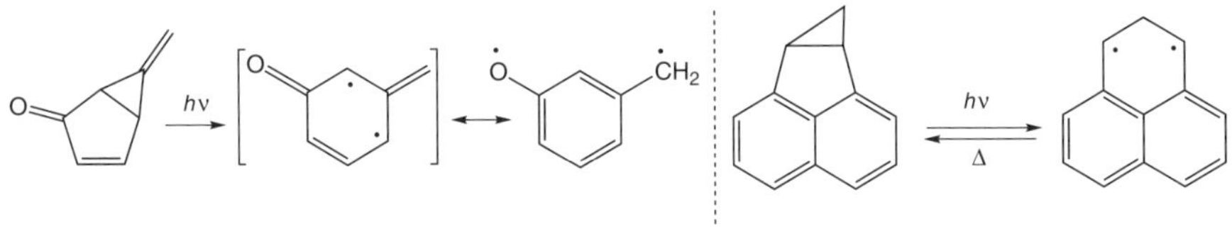

Scheme 17.6

Finally, some diradicals can be made *in situ* by an internal hydrogen-transfer reaction from a suitable hydrogen donor to a carbene or nitrene. In benzene derivatives, this is a well-tested route to *o*-quinoid compounds, which are not biradicals, although a biradical valence structure probably makes a significant contribution to their electronic structure. However, if the donor and the carbene or nitrene are attached at the 1 and 6 positions of a naphthyl moiety, then perinaphthadiyl diradicals result (see Scheme 17.7).[67,68]

X = CH$_2$, NH$_2$, OH
Y = CH, N

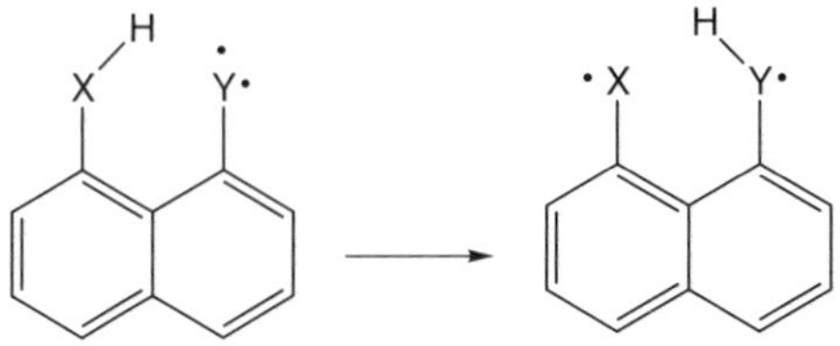

Scheme 17.7

The yield of triplet biradicals in all of the above processes is usually high enough to get very good ESR spectra (provided or course that the species has a triplet ground state, or that this state can be significantly populated from a singlet ground state at the temperature of the experiment). However, this says nothing about the relative or absolute yield of the target biradicals obtained from one or the other type of precursor. The samples studied by ESR spectroscopy may contain mainly ESR-silent byproducts that may predominate in UV–vis or IR absorption spectra. Several cases are known where ESR and UV–vis spectra gave conflicting evidence with regard to the identity of the species formed from a certain precursor.

An example is trimethylenemethane (TMM), one of the first hydrocarbon biradicals whose triplet ESR spectrum was recorded[61,62] (cf. Chapter 5 in this volume), but which eluded 25 years of efforts (among others by the present author) to obtain its UV–vis or IR spectrum, because the predominant product of photolysis or pyrolysis of different precursors was invariably methylenecyclopropane (MCP). Eventually, Maier et al.[69] successfully generated a sufficient quantity of TMM, by irradiation of MCP in halogen-atom doped Xe matrices, to record the IR spectrum of this elusive compound.

An intriguing aspect of matrix isolated biradicals is that they can in some cases persist in two different spin states, which may even be interconverted photochemically or thermally. The reader is referred to Chapter 5 in this volume for a more detailed discussion of this phenomenon.

6.4. Radical Ions

Although radical ions may appear as somewhat exotic species to many physical organic chemists, they are pivotal reactive intermediates in all processes involving *single electron transfer*, which occur in many important reactions of biological (e.g., enzyme catalyzed oxidations) and technical relevance (e.g., photoinduced electron transfer, imaging). More recently, radical cations have also moved into the focus of astrochemistry, because they may account for many of the absorptions observed in the diffuse interstellar clouds. Therefore, assessing the identity, structure, and properties of radical ions is of great interest, and matrix isolation constitutes at least potentially an excellent tool for their investigation, because it allows one to leisurely study the properties of these highly reactive species.

Actually, the history of matrix isolation studies of radical ions is long and rich and has been reviewed repeatedly,[26,28,70,71] but since most of these studies involve small molecules isolated in Ne matrices (where the interaction with charged species is minimal), not many of examples have found their way into the literature consulted by physical organic chemists. In these studies, radical ions were usually produced in open microwave or electric discharges, or by direct irradiation during deposition with far-UV light. More recently, several laboratories have succeeded in incorporating sufficient quantities of radical ions from mass selected ion beams into noble gas matrices to allow investigations by UV–vis and IR absorption spectroscopy.[28] In particular, Maier's group has applied this technique very successfully to study a variety of carbon chain molecules and ions by matrix isolation.[72]

Although many highly interesting species can be generated by the above techniques, these methods are not suitable for systematic investigations of organic radical ions because they involve either direct interaction of the precursors with high energy radiation or, in the case of mass selected ion beams, a very high collision energy when the ion hits the matrix. These processes may lead to unwanted and often uncontrollable fragmentations or other chemistry.

On the other hand, in the 1960s Hamill[73] pioneered a technique for *radiolytic* production of radical ions in selected frozen solvents, which was further developed and applied to the study of a great variety of organic species by Shida.[74] Although the samples investigated by this method are exposed to very high-energy radiation (usually ^{60}Co γ-rays of > 1 MeV), the main result of this radiolysis is ionization of the frozen solvent that subsequently oxidizes the substrate (which itself is largely transparent to this radiation) to form its radical cation. If a halide, such as a Freon, is used as a solvent, then the ejected electron gets trapped by dissociative attachment. If MTHF, which is a very good hole trap, is used as a solvent, then the added substrate is often a better electron scavenger than the solvent, and thus radical anions can be formed.[7]

In the 1980s, the group of the present author applied this method to noble gas matrix isolation, with the hope of recording the IR spectra of organic radical cations. We took advantage of the fact that Ar has a very high cross-section for the bremsstrahlung emanating from a conventional X-ray tube fitted with a tungsten target, and Ar is thus very efficiently ionized by such X-radiolysis. Subsequently, the hole and the ejected electron travel through the matrix by resonant charge transfer (or in the form of polarons) until the hole gets captured by an added substrate with a lower oxidation potential than Ar (in solid Ar). If no extra measures are taken, most of the electrons get scavenged by some impurities such as CO_2 that are always present in matrices, but we found that much cleaner results are obtained when an organic halide (CH_2Cl_2 has proven very useful) is added as an electron scavenger in roughly equimolar amount to the substrate to be oxidized.

The above technique, which is very easy to implement (all that is needed besides the usual matrix isolation equipment is a conventional X-ray source for crystallography), was used first to record UV–vis spectra[37] and later IR spectra[75,76] of numerous radical cations isolated in Ar matrices. The method seems to be quite generally applicable to any substrate that can be volatilized without decomposition.

A great advantage of the method is that the positive and negative charges reside in different cavities of the matrix, and therefore do not perturb each other.

Two main liminations have, however, become evident: the first is that molecules whose gas-phase ionization energy exceeds > 9.5 eV cannot be oxidized by ionized solid Ar. The reason for this limitation is unclear, because the process is exothermic (the ionization energy of solid Ar is 13.9 eV,[77] that of organic molecules in Ar is typically lowered by ~1 eV in solid Ar relative to the gas phase[78]). Perhaps the localization of the spin and charge onto the substrate entails a Frank–Condon barrier that cannot easily be surmounted at 12 K.

The second limitation arises from the fact that the excess energy that is imparted onto the incipient substrate radical cations (which can be several eV!) cannot be efficiently dissipated in an Ar matrix, and is therefore temporarily available to drive low activation rearrangements or fragmentations. As a consequence, metastable primary radical cations, such as those of compounds containing, for example, diazo groups or small rings, often cannot be observed in Ar matrices. In these cases, recourse must be taken to frozen glasses where energy dissipation is much more efficient.

An interesting aspect of this method of preparing radical cations is that, as the substrate radical cations accumulate in the matrix, they begin to compete very successfully with the added halide for the electrons that are liberated during the radiolysis of the Ar. Usually, the only effect of this process is that the yield of radical cations tapers off after a while (usually at 10–15% of the neutral substrate) due to continuous reneutralization. However, if ionization is accompanied by some spontaneous rearrangement or fragmentation, then what gets neutralized is the rearranged species, or the fragment that carries the charge. This permits the accumulation of *neutral* reactive intermediates that may be difficult to generate by other means.[79] An illustrative example are azirines which undergo spontaneous ring opening to yield radical cations of azo methine ylids on ionization. Reneutralization then leads to the neutral azomethine ylids that cannot be generated directly, for example, by photolysis of azirines, in quantity (see Scheme 17.8).[80]

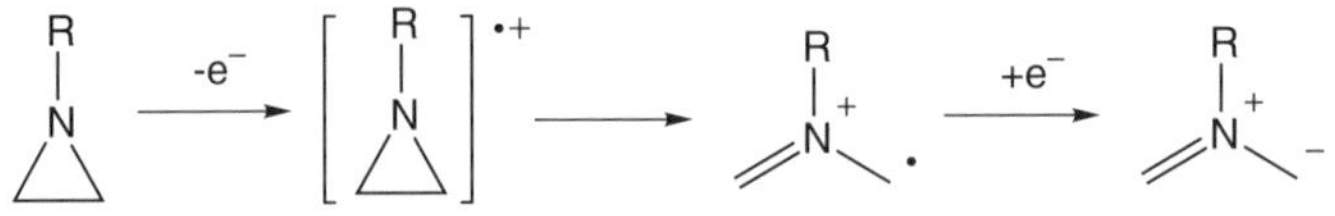

Scheme 17.8

So far, most of this discussion applies to the generation and study of *radical cations*. How about *radical anions?* These species are more difficult to generate in isolated form in cryogenic matrices because in solvents that do not significantly stabilize charged species (such as noble gases), isolated radical anions are prone to suffer detachment of the loosely bound extra electron, either spontanously or under the impact of low-energy radiation. This situation contrasts with that encountered in solvents such as methyltetrahydrofuran, which can be routinely used to generate and study radiolytically generated organic radical anions at 77 K.[7,74]

Due to this limitation, the literature on matrix isolated radical anions is much less plentiful than that on radical cations. Several small radical anions were obtained more or less coincidentally as products of plasma reactions in discharges, or on vacuum ultraviolet (VUV) irradiation during sample deposition, but their identification was usually difficult due to the presence of numerous species with overlapping transitions. Of course the technique of mass selection in ion beams allows one to isolate negatively as well as positively charged species. Thus, Maier and co-workers[81] fitted a source of negatively charged carbon chain radicals C_{n^-} (obtained by sputtering a graphite target with Cs^+) with a mass filter and interfaced the entire source to a 4 K matrix isolation apparatus, By virtue of their high electron affinities ($\sim$4 eV), these anions survived probing by UV/vis and IR absorption spectroscopy in Ne matrices, although they could be readily bleached by UV-photodetachment.[72] The only systematic way to generate radical anions for matrix isolation studies is by capturing electrons liberated by photolysis of codeposited alkali atoms, a technique that was pioneered by Kasai.[82] However, this technique has two disadvantages: First, in order to effect photoionization with visible light (UV light usually bleaches radical anions) the wave functions of the excited donor (alkali metal atom) and the acceptor must overlap to some degree. Therefore the resulting radical anions are not really isolated and their spectroscopic manifestations will be influenced by the nearby alkali cation. Second, matrix isolated alkali atoms show very strong absorptions throughout the visible spectrum,[83] which make it very difficult to probe radical anions in this spectral range. The pioneering studies of Kasai were done by ESR spectroscopy, which is not affected much by either of the above limitations.

Recently, the group of Salama has, however, begun to apply the above technique to the study of vibronic spectra of polyacenes that have the dual advantage that they have rather large electron affinities and that they absorb in the near-IR region where alkali atoms, in particular Na, do not absorb.[83,84]

Finally, we wish to mention an attempt to create radical anions by X-irradiation of Ar matrices containing trialkylamines as codopants,[85] which release electrons on oxidation by ionized Ar (cf. Section 6.4). However, only small amounts of the targeted radical anions could be obtained by this technique.

6.5. Closed-Shell Ions

Closed-shell ions are among the most important intermediates in solution chemistry, and no treatise on reactive intermediates (including the present one) would be complete without extensive sections on carbocations and carbanions, if not also on heteroanalogues of the above species. Nevertheless, closed-shell ions are conspicuously absent from matrix isolation studies, apart from a few cases where such species were coincidentally formed in discharges, or where charged species were deliberately "isolated" by mass spectrometry (cf. Section 6.4). The reason for this is that closed-shell ions are usually generated in *bimolecular* reactions, typically with Brønsted or Lewis acids or bases, that cannot easily be realized under the conditions of matrix isolation.

However, there are some exceptions. One of them is the possibility of *(photo)-protonation* or *-deprotonation*. If a matrix is doped with sufficient amounts of a proton donor or acceptor, chances are that the substrate will give up or accept a proton already on cocondensation or on subsequent photoexcitation. In fact, the higher noble gases (Ar, Kr, Xe) are themselves good proton acceptors, forming $(NG_n \cdot H)^+$ complexes that can be identified by their characteristic IR vibrations.[86] This feature allows occasionally to observe radicals formed by deprotonation of radical cations formed in noble gas matrices, for example, benzyl radical from ionized toluene.[87] However, we know of no examples where a carbanion was formed by deprotonation in matrices.

On the other hand, it is known that carbenes and nitrenes can be protonated to yield carbocations[88,89] or nitrenium ions, respectively.[90,91] Thus, if carbenes or nitrenes are generated in the presence of a protic acid in a matrix, at least some of them should be protonated. This concept has recently been tested in the case of arylnitrenes whereby substantial quantities of nitrenium cations were identified when the corresponding azides were photolyzed in Ar matrices containing 10–30% HCl.[92] However, the species formed in this way arise of course in the form of *ion pairs*, and therefore one cannot speak of truly matrix-isolated cations.

Another possibility to generate closed-shell cations in matrices is by ionization of the corresponding radicals, for example, by VUV photolysis in Ne or by X-radiolysis in Ar (cf. Section 6.4 on radical ions). What is required for such experiments is a clean source of radicals (to avoid the occurrence of multiple products) and a source of ionizing radiation. The only example where this experimental ansatz, which can potentially lead to truly matrix isolated closed-shell cations, has been put into practice was recently provided by Winkler and Sander. Among the compounds that they obtained on irradiation of iodobenzene with the 105–107-nm light from an Ar discharge lamp during deposition, they were able to identify the phenyl cation that was doubtlessly formed by ionization of the phenyl radical, which was also present in the matrix. However, a generalization of this procedure would require a separation of the radical forming from the ionization step, such that difference spectra for the latter step, which contain only signals from ionized products, could be obtained. Also, it would be desirable to generate radicals in the absence of nearby electron scavengers (such as iodine atoms) in order to avoid formation of ion pairs.

6.6. Other Reactive Intermediates

Of course, matrix isolation studies are not limited to the classes of reactive intermediates discussed in the above five sections. There are many other types of highly reactive species that can be stabilized (and were often observed for the first time) in cryogenic matrices. Only a comprehensive review could do justice to all those efforts, but this effort is beyond the scope of this chapter. So, we will limit ourselves to a few typical cases.

One highly reactive structural motif that has found great attention are bent triple bonds (cycloalkynes or -arynes), and twisted or strongly pyramidalized double bonds, such as anti-Bredt olefins or unsaturated cage compounds. Several such species have been immobilized and identified in cryogenic matrices (for *o*-benzynes, the reader is referred to Chapter 16 in this volume). In solution, such compounds are often generated by dehalogenation (e.g., with Na amalgam) or dehydrohalogenation (with strong bases). Dehalogenations can also be carried out by co-condensation of dihalides (usually diiodides) with alkali atoms.[33] This strategy has been followed to generate such species as the highly twisted adamantene[93] or the strongly pyramidalized tricyclo[3.3.1.0^{3,7}]nonene[94] shown in Scheme 17.9.

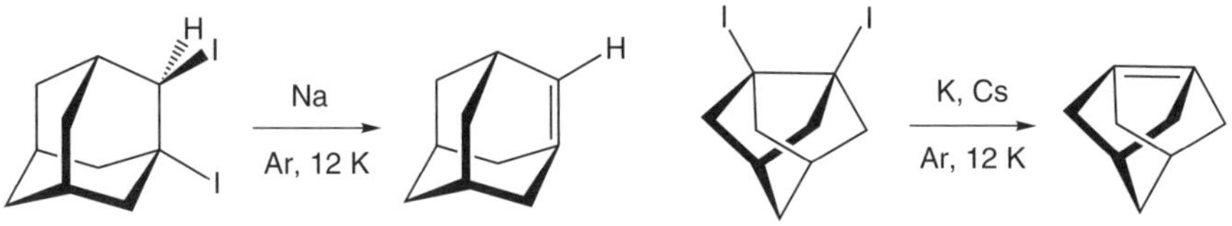

Scheme 17.9

Dehydrohalogenations (elimination of HX) can in turn be effected by passing the halide over a solid base such as vacuum-dried hot KOH[95] or *t*-BuOK[96] adsorbed on some carrier material. Recently, Billups et al.[97] showed that elimination of Me$_3$SiBr may even be effected at room temperature over solid CsF. Several very interesting strained alkenes have been generated by such techniques, but although these techniques would lend themselves very well for interfacing to matrix isolation we know of no example where this has been achieved.

In situ generation of strained alkenes or alkynes requires photoelimination of a leaving group toward which the targeted product is unreactive. Thus, cycloalkynes may be generated by elimination of CO from cyclopropenones, but these must in turn be generated *in situ* (e.g., from diazoketones). The generation of acenaphthyne[98] (Scheme 17.10) may serve as an illustration of this procedure that did, however, not work for the parent cyclopentyne, a compound that still awaits spectroscopic characterization.

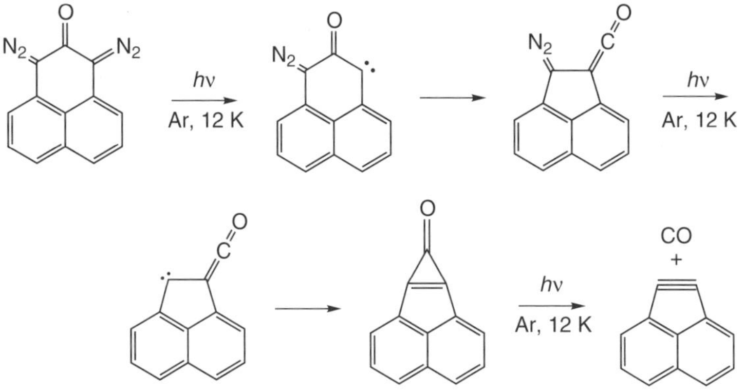

Scheme 17.10

Note that not all cyclopropenones eliminate CO photochemically. Thus, the first cyclohexynes had to be generated by *pyrolysis* of the precursor cyclopropenone.[99,100]

A very particular strained alkene, cyclobutadiene (CB), has a rich and instructive history that has been reviewed repeatedly.[101–103] As matrix isolation work by several groups played a pivotal role, and as some important lessons can be learned from this story, it merits closer attention in the present context. In trapping studies conducted in the 1960s, CB displayed a Janus-faced pattern of reactivity in that it underwent cycloaddition reactions, which are typical for dienes, but it also proved to be extremely reactive toward molecular oxygen, halogen atoms, methyl- and allyl radicals, as is typical for triplet biradicals. Much of the discussion centered therefore on the disposition of the singlet and triplet potential energy surfaces as depicted in Figure 17.6.

All calculations agreed that triplet CB would be square, whereas singlet CB would have a rectangular geometry, but the calculations disagreed in their predictions with regard to the relative energies of the two states (situations *a*, *b*, and *c* in Fig. 17.6), especially with respect to the ground state multiplicity (and hence the structure) of CB. As it was unclear in what way substituents would influence this situation, studies on the parent compound were called for to settle this question, and these could only be done in the gas phase or in matrices.

By 1973, the groups of Chapman et al.[104] and Krantz et al.[105] had obtained IR spectra of parent CB as well as of its mono- and dideuterated derivatives, which

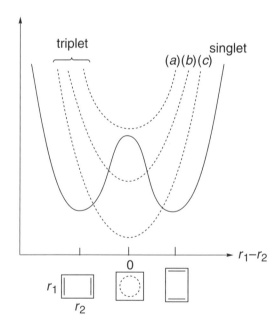

Figure 17.6

they generated from photo-α-pyrone (Corey's lactone), created *in situ* from α-pyrone. The spectra of CB-h_4 obtained by the two groups showed two strong bands at ~1240 and 570 cm^{-1} and a weaker one at ~650 cm^{-1}, close to the bending vibration of the coproduct, CO_2, at 662 cm^{-1} (Scheme 17.11).

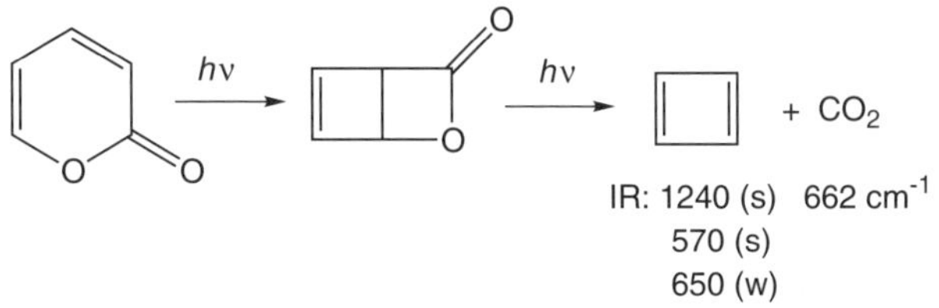

IR: 1240 (s)   662 cm^{-1}
570 (s)
650 (w)

Scheme 17.11

Group theory predicts that if parent CB is square, it should show *four* IR active fundamentals (of which one was a degenerate C—H stretch expected to be of very weak intensity). On the other hand, if parent CB were rectangular, it should show *seven* IR active fundamentals of which two are weak C—H stretches. Since only *three* (rather than five) bands were observed in the IR spectrum of CB below 2000 cm^{-1}, Krantz and Newton concluded that CB must be square. They supported this conclusion with a force-field calculation, based on a square geometry, that reproduced the observed frequencies, including those of CB-d_1, to within a few wavenumbers.[105] On the other hand, Chapman conjectured that the absence of evidence for two different vicinally dideuterated CB-d_2 (which would be expected for rectangular CB-d_2 and can exist as a 1,2- and 1,4-isomer) agrees also with a square structure.

The problem with the above assignment was that theory predicted unanimously that square CB should have a triplet ground state. However, all attempts to record an ESR spectrum of this putative triplet diradical under various conditions failed. Furthermore, theory also predicted that square triplet CB should show a strong transition at ~370 nm,[106] whereas rectangular singlet CB should be transparent down to 200 nm.[107,108] Although the exact position of an absorption that should be ascribed to CB was a matter of some debate, agreement prevailed with regard to the fact that it was weak and not above 300 nm, which posed an additional problem with respect to the assignment to square triplet CB.

All of this led the groups of Maier et al.[109] and Masamune et al.[110] to reinvestigate the IR spectrum of CB, in search of possible bands that might have escaped detection in the experiments of Chapman and Krantz. To this end, they employed different precursors of which the formal adduct of CB with phthalan proved to be the optimally suited.

To their great surprise, both groups started by finding *fewer* bands than had been found in the studies of Chapman and Krantz! In particular, the strong 660-cm^{-1} band appeared to be linked to the presence of CO_2. Maier concluded that it must have been due to a splitting of the degenerate bending vibration of CO_2 due to complexation with CB.[109] This conclusion was confirmed a year later through ^{13}C labeling experiments.[111]

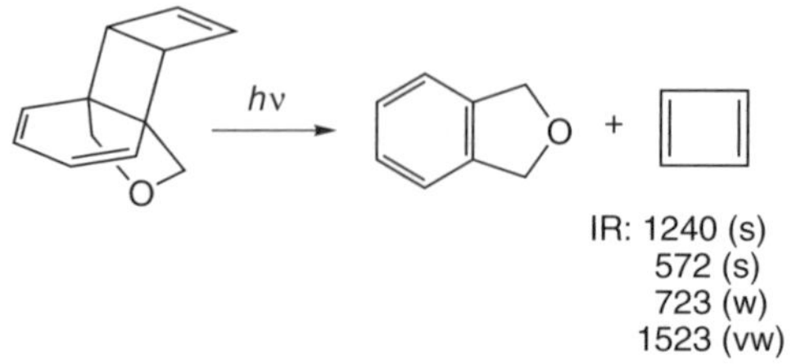

IR: 1240 (s)
572 (s)
723 (w)
1523 (vw)

Scheme 17.12

However, this finding did not solve the problem at hand. In fact, it took Masamune et al.[112,113] several years, involving studies with differently deuterated precursors, to finally confirm the presence of two additional weak bands at 723 and 1523 cm^{-1} (1456 and 609 cm^{-1} in CB-d_4), which they could unambiguously assign to CB (Scheme 17.12). With the proven presence of *four* IR bands below 2000 cm^{-1} the hypothesis that CB is square was definitely refuted.[114a]

The reason why we relate this particular story in some detail is because two very important lessons can be learned from it: first, *drawing conclusions on the basis of missing evidence is often misleading*.[114b] Second, it is always highly desirable to generate a reactive intermediate independently *from different precursors*. Only spectral features that appear in experiments involving at least two independent pathways of formation of a targeted species can be assigned with confidence to that species. If chemically different precursors are not available, one should at least try to play with isotopic substitution and make sure that the shifts of vibrational bands are in accord with expectations for the targeted species.

7. PROBING MATRIX-ISOLATED REACTIVE INTERMEDIATES

7.1. General Remarks

With few exceptions, all investigations of matrix-isolated reactive intermediates are done by *absorption spectroscopy*, in the UV–vis and/or in the IR spectral range, or, in the case of open-shell species, by *ESR*. Occasionally, one also finds studies where emission or Raman scattering of reactive intermediates is probed in matrices, but these studies are few and far between, so we will focus in this section on the first group of techniques that can be easily implemented with commercially available equipment.

A first point to consider is that thermal or photochemical decomposition of a precursor often does not lead to a single product, due to parallel or consecutive secondary reactions. Since absorption spectroscopy invariably probes all components of a mixture, the problem of how to distinguish between these components may arise in the context of studies on reactive intermediates. This problem can be addressed by searching conditions under which the concentration of only one or the other compound change, for example, by taking recourse to *trapping reactions* (Scheme 17.13). Thus, triplet carbenes are very reactive toward molecular oxygen

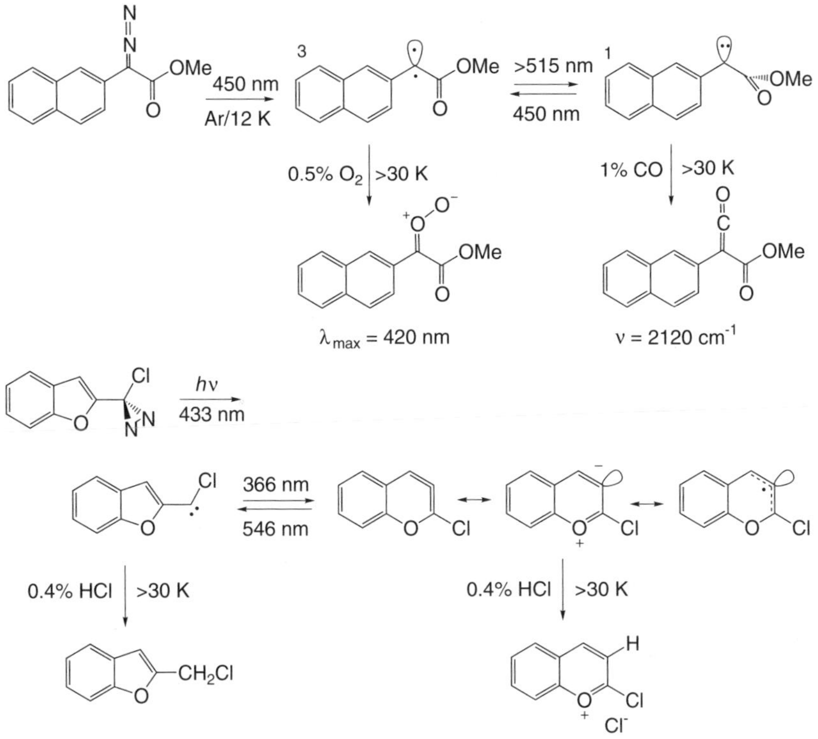

Scheme 17.13

and slight annealing of matrices doped with 1–2% O_2 usually leads to the rapid disappearance of triplet carbenes and the formation of carbonyl oxides,[115] which distinguish themselves by strong, broad UV bands peaking at 420–450 nm. On the other hand, singlet carbenes react readily with CO to yield ketenes that can easily be recognized by their strong IR peaks at ~2120 cm^{-1}. Singlet carbenes, nitrenes, and strained olefins can also be trapped in matrices by HCl, which leads either to protonation or to addition of HCl. Some representative examples of this are given in the above Scheme 17.12.[116,117]

Another strategy consists in irradiating a sample at judiciously selected wavelengths to effect *selective bleaching* of one or another component of a mixture. Thereby, it is usually advisable to start by irradiating at the longest wavelength where the targeted product mixture absorbs, in order to avoid simultaneous bleaching of several compounds or the establishment of photostationary equilibria. Unfortunately, irradiation into the first absorption band does not always provide sufficient energy to effect a photorearrangement or photocleavage reaction, and then one must move to shorter wavelengths which, of course, increases the risk that other components are bleached at the same time. However, by using different cutoff and interference filters, it is often possible to find a wavelength where at least one component is bleached much more efficiently than another, which allows the IR or UV–vis absorption bands of the two be distinguished by their different "photokinetic" behavior.

Such photochemical experiments have the additional benefit that they may lead to the discovery of (unexpected) new species that can be just as interesting as the originally targeted reactive intermediate. For example, attempts to selectively bleach the triplet napththyl carbomethoxy carbene shown in the Scheme 17.13 in view of its spectroscopic identification led to the serendipitous finding that the corresponding singlet carbene is also stable in an Ar matrix at 12 K, that is, the discovery of one of the rare cases of spin isomerism.[116] Similarly, Sheridan and co-workers[117,118] found recently that bleaching of 2-benzoylfurylchlorocarbenes (also shown in Scheme 17.13) leads to an entirely novel type of reactive intermediates, didehydropyrans, which have an interesting electronic structure.

Irrespective of how a targeted component of a mixture is affected, it is often easier to discern spectral changes by looking at *difference spectra*, that is, the results of subtraction of a spectrum taken after annealing or bleaching from that taken before. However, if meaningful difference spectra are to be obtained it is crucial to assure a highly *reproducible sample positioning* so that the beam of the spectrometer always probes the exact same part of the matrix. A reproducible sample positioning can be achieved by means of a template whose dimensions fit the bottom of the vacuum shround and which is mounted inside the spectrometer's sample chamber.

One is in an even better situation with regard to the identification of a reactive intermediate when one can examine *pairs* of difference spectra for the formation and the disappearance of a compound, because one can then readily discern the *mirror-image patterns* that must belong to the targeted species from other spectral changes that may accompany one or the other of these processes. Examples for this will be given in Sections 7.2–7.4 on IR and UV–vis spectroscopy.

7.2. Infrared Spectroscopy: Experiment

Infrared spectroscopy is by far the most popular tool for the inverstigation of matrix-isolated species. By virtue of the suppression of most rotations in solid matrices, IR spectra recorded under these conditions typically show patterns of very narrow peaks, compared to spectra obtained under normal laboratory conditions (solution, Nujol, or KBr pellets), where bands due to different vibrations often overlap to the extent that they cannot be separated. As a consequence, matrix isolation IR spectra are—at least potentially—are a very rich source of information on the species under investigation. Whether and how all this information can be used depends on the ability to assign the spectra, a subject to which we will return below.

However, when doing matrix isolation IR spectroscopy one should also be aware of some special features of this technique, and of some problems that one needs to address in order to be successful. One of these features is the so-called *site effect* that may lead to the splitting of bands that belong to a single normal mode. Site effects are a very well known phenomenon in the field of spectral hole burning or single molecule spectroscopy in solid media[119] (indeed they make it possible to observe single molecules in many cases). Such site effects result from the fact that a rigid host material may accommodate a guest molecule in different local environments (sites) where these molecules may experience slightly different external fields or geometrical constraints. Consequently, the frequencies of certain normal modes may be shifted, either through the influence of the varying external field on the electronic structure of a host molecule, or through an effective increase in a force constant when the atoms have to work against the rigid environment along some normal mode. Depending on the type and size of the molecule and on the host material, this can lead to multiple sharp lines or to a general broadening of the IR absorptions.

The extent to which site effects manifest themselves in the spectra depends also on the way a matrix is made. It has been reported that pulsed deposition leads to a simpler spectral site structure and sharper lines than slow, continuous deposition. But then this depends on the backing pressure and pulse duration, as well as on the temperature of the matrix gas and the speed with which extra gas is removed, so no general rules can be given. Every practicioner of matrix isolation has to find a combination of these above variables that leads to the best spectra under their laboratory conditions. The search for these conditions should, however, not be guided by purely aesthetic spectral criteria, but by the need to acquire a maximum of useful information with minimal effort.

In practice, one finds that site effects often affect high-frequency framework or functional group vibrations that occur in the $1700–2300$-cm^{-1} range most pronouncedly. Everyone who has worked, for example, with heterocumulenes, diazo compounds, or azides, is familiar with the fact that the very intense asymmetric stretching vibrations that are associated with such groups often appear in the form of multiple, sometimes overlapping peaks that may extend over a region of 50 cm^{-1} or more. Since all these peaks disappear simultaneously on photolysis, they must belong to the same species. To our best knowledge, the exact origin of these very strong site effects in the case of heterocumulenes is unknown, but they usually pose no assignment problems because no other vibrations occur in this spectral region.

When *pairs* of lines of different intensity occur in a region where only a single one is expected on the basis of normal mode calculations (cf. below), these need not indicate a site splitting, but are often indicative of *Fermi-resonances*, that is, a (near) coincidence of an IR active fundamental with a combination band that "borrows" intensity from the former. Such phenomena can often only be detected by virtue of the high resolution that is available in matrix isolation IR spectroscopy.

As mentioned above, it is often advantageous to look at *difference spectra*, and this is especially true in matrix isolation IR spectroscopy, because, due to the

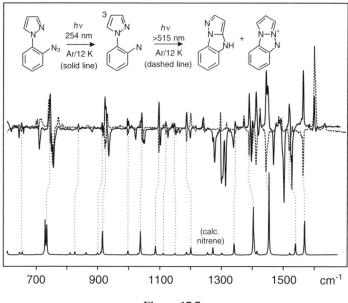

Figure 17.7

sharpness of bands, overlap between spectra of different compounds is often minimal. Fortunately, with the advent of interferometry, which eliminates the problem of hysteresis, precise wavenumber reproducibility, which, of course, is a prerequisite for obtaining meaningful difference spectra, is no longer a problem. Furthermore, the spectra automatically are present in digital form, which makes arithmetic operations trivial.

The case of pyrazolo-phenylnitrene (formed from the corresponding azide, solid line, and bleached selectively to yield a pyrazoloindole shown in Fig. 17.7) may serve as an illustration of this feature, which becomes even more evident when color is used. The trace shown at the bottom of this figure leads into the problem of spectroscopic assignments discussed in the Section 7.3.

7.3. Infrared Spectroscopy: Modeling

Up until about 1990, the main use of matrix isolation IR spectroscopy in connection with research on reactive intermediates was in the detection of signpost bands that allowed the detection and unambiguous assignment of certain functional groups. In many cases, where other spectroscopic evidence was lacking or inconclusive, such data were of course highly significant as a basis for mechanistic conclusions.

For example, the ring expansion reactions undergone by phenylnitrene and phenylcarbene manifested themselves for the first time by conspicuous IR bands at 1895 or 1815 cm^{-1} that were observed after photolysis of matrix-isolated phenyl azide[120] or phenyldiazomethane, respectively[121] and which Chapman and co-workers[122] assigned to the asymmetric stretching vibration of the cyclic allenes, (aza)cycloheptatetraene. Often, however, it was found necessary to generate an intermediate independently from several precursors, or to take recourse to isotopomers to substantiate such assignments.

Unfortunately, most of the structural information of IR spectra is contained in the often very crowded region of 500–1600 cm^{-1}, which was hardly exploited for diagnostic purposes except in the case of very small molecules with few vibrations, or for pattern matching of spectra of reactive intermediates obtained independently from different precursors. The reason for this was that the prediction of IR spectra was only possible on the basis of empirical valence force fields, and the unusual bonding situations that prevail in many reactive intermediates made it difficult to model the force fields of such species on the basis of force constants obtained from stable molecules.

On the other hand, quantum chemistry was of limited help, although techniques to compute force fields from second derivatives became widely available in the 1980s. However, the force constants computed at the Hartree–Fock level generally proved to be too inaccurate to permit a reliable modeling of experimental spectra. The situation improved a bit when the effects of dynamic correlation were included, usually at the MP2 level (for details of the methods discussed in this section, see Chapter 22 in this volume). However, especially the off-diagonal elements that describe the mixing of valence deformations were still often of the wrong magnitude (or even of the wrong sign!), which led to unreliable predictions of normal modes and their frequencies. These shortcomings could be improved upon by a judicious method of scaling that was proposed by Pulay et al.[123] in 1983, but in order to work reliably, this scaling required the correct assignment of vibrational spectra obtained from several isotopomers. Although this method was occasionally applied to reactive intermediates,[124] it never gained much poplularity in this field.

This situation changed drastically when it was discovered in the 1990s that *density functional* (DF) methods do a much better job of modeling force fields than (affordable) wave function based methods. Already within the local density approximation (LDA) of DF theory, vibrational frequencies were predicted with similar accuracy as at the MP2 level,[125,126] and the agreement between calculated and experimental vibrational spectra became notably better when gradient-corrected functionals such as BP86 or BLYP were employed. In the mid-1990s, several studies were undertaken to assess the performance of different exchange and correlation functionals for predicting vibrational spectra.

Of these studies, we only wish to mention the seminal 1996 work of Scott and Radom[127] who investigated the performance of a variety of gradient-corrected and hybrid functionals (BLYP, BP86, B3LYP, B3P86 and P3PW91) with the 6-31G* basis set for predicting over one thousand vibrational frequencies from a large suite of test molecules. By least-squares fitting, they also derived scaling factors for each method that bring the predicted (harmonic) vibrational frequencies into optimal agreement with experimental ones. Thereby, they found that frequencies calculated by BLYP or BP86 require hardly any scaling, but are in slightly worse overall agreement with experimental frequencies than those computed with the B3 hybrid exchange functional. Overall the best combination of functionals proved to be B3PW91 with a root-mean square (rms) deviation over the entire test set of 34 cm^{-1} and only 4% of frequencies that deviate by $>10\%$, all after scaling all frequencies by 0.9573. A close runner-up was B3LYP with the same rms deviation, but 6% of outliers, after scaling by 0.9614. Both DFT methods were much better than MP2/6-31G* (rms $= 63 \text{ cm}^{-1}$, 10% outliers) and the much more expensive QCISD method (rms $= 37 \text{ cm}^{-1}$, 6% outliers). Another very important finding of this study was that it usually does not pay to go to larger basis sets or finer quadrature grids, which is good news in terms of computational economy.

Thus it has now become possible for anyone with access to a modern PC to compute full vibrational spectra of any species that represents a minimum on its potential energy surface with sufficient accuracy to compare them directly with experimental ones. The *intensities* of IR transitions which are proportional to the changes of the dipole moment along the different normal coordinates are also an important element in this comparison. Although even relative intensities cannot be computed with similar accuracy as frequencies can, the assignment of an experimental spectrum to a certain molecule requires that no IR bands that are calculated to have high intensities are missing from the experimental spectrum, and that this spectrum does in turn not contain intense bands that are absent from the simulated spectrum. To summarize, there must be an evident and consistent relationship between the pattern of observed and the pattern of calculated IR bands for a structural assignment to be valid. It is very important to remember this because the temptation to succumb to a degree of wishful thinking in the process of pattern matching should not be neglected.

Amidst all the enthusiasm about this versatile new tool that quantum chemistry has put at the hands of practioners of IR spectroscopy in matrices, one should not forget its limitations. First, a valid prediction can only come from a calculation based on a correct structure. In the case of reactive intermediates, this is not always as evident as one might wish. A famous example is given in Chapter 16 in this volume: Much of the recent discussion on the correct assignment of the IR spectrum of *m*-benzyne was caused by the fact that different theoretical methods predict different structures, with more or less bonding between the radical centers, for this species. The DFT methods appear to overestimate this bonding, and hence are unsuitable for the prediction of the IR spectrum of *m*-benzyne.

A second instructive example that cautions us to watch out not only for molecular but also for the electronic structural features of species whose IR spectra we wish to assign comes from the recent work of Sheridan and Nikitina[118] on didehydrobenzoxazine, which they were able to generate from an acetylenic quinonimine: A normal (spin restricted) B3LYP calculation of this species yielded an IR spectrum that showed "only marginal similarity" to the experimental spectrum. However, on removing the requirement that pairs of electrons occupy the same spatial molecular orbitals, the calculation relaxed to a broken-spin structure describing a mixed singlet–triplet diradicaloid state that led to a satisfactory IR prediction.

Another limitation is inherent to the harmonic approximation on which standard quantum mechanical force-field calculations are invariably based. Due to a fortuitious (but surpisingly systematic) cancellation of errors, the harmonic frequencies calculated by modern density functional methods often match very well with the experimental ones, in spite of the fact that the latter involve necessarily more or less anharmonic potentials. Thus one is tempted to forget that the harmonic approximaton can become perilous when strong anharmonicity prevails along one or another molecular deformation coordinate.

An example of an otherwise fairly "normal" reactive intermediate where the harmonic approximation fails is phenylcarbene (see Fig. 17.8): Although the match beween calculated and observed IR bands is quite satisfactory across most of the spectrum, it breaks down completely in the region of 750–950 cm^{-1}, where there is not even a remote similarity between the pattern of three bands predicted by B3LYP calculations (dashed line) and those found experimentally (solid line, downward pointing peaks).

An analysis of the reasons for this discrepancy[45] revealed that the three normal modes that are excited in this frequency range comprise to varying degrees the Ph—C—H bending deformation which, by explicit calculation of the potential energy surface, turned out to be rather strongly anharmonic. A selective reduction of the Ph—C—H bending force constant in a Pulay-style scaled force-field calculation led to a prediction that is at least qualitatively in better accord than the unscaled B3LYP prediction in this frequency range (see inset in Fig. 17.8). However, acommodating anharmonicities of selected deformations in this way cannot be regarded

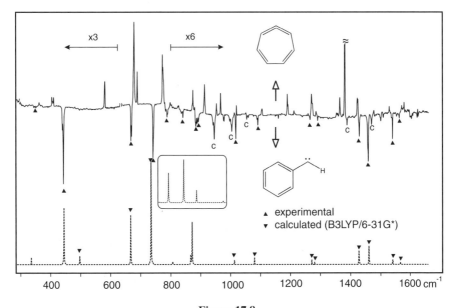

Figure 17.8

as a valid remedy, because anharmonicity affects also the mixing of valence deformations that interact to give normal modes.

7.4. Ultraviolet–Visible Spectroscopy: Experiment

Although IR spectra contain inherently more information than UV–vis absorption spectra, the information contained in the latter is often more directly useful. In particular, selective bleaching experiments require a knowledge of the electronic absorption spectra, and these spectra may also give valuable hints with regard to the presence of possible products. For example, triplet aryl carbenes are readily recognized by their characteristic sharp bands in the 400–500-nm range,[45] and polyenyl radicals or polyene radical cations by their pattern of a weak but sharp vis–NIR band accompanied by an intense near-UV transition.[128] Therefore UV–vis spectra should be taken routinely also in experiment targeted at recording IR spectra.

Fortunately, this requires no additional investment (apart from a UV–vis spectrometer that is, however, available anyway in most laboratories), because highly polished salt (KBr or CsI) windows are sufficiently transparent down to 200 nm so that high-quality UV–vis spectra can be obtained with the exact same sample configuration as is used for IR absorption spectroscopy. As in IR spectroscopy, it is often useful to look at difference spectra, but as bands due to different compounds have much more of a tendency to overlap in this case, which may lead to distortions in the difference spectra, one has to be cautious in assigning band maxima from difference spectra.

Although *site effects* are not as prevalent in UV–vis absorption as they are in IR spectra, they do exist and manifest themselves sometimes very clearly in band systems that comprise sharp peaks. An example is the radical cation of all-*trans*-octatetraene whose first absorption band consists of multiple peaks that can be selectively bleached by highly monochromatic light.[129] The site structure can become more evident in laser-induced fluorescence, where excitation of individual sites is possible down to the level of single molecules in favorable cases, but a discussion of this fascinating phenomenon is beyond the scope of this chapter.

7.5. Ultraviolet–Visible Spectroscopy: Modeling

Unfortunately, the situation with respect to the modeling of electronic absorption (or emission) spectra in view of their use in assigments of reactive intermediates is not nearly as comfortable as in the case of vibrational spectroscopy. No universal and affordable "magic bullet" such as the B3LYP/6-31G* method has been found, and thus reliable calculations of excited-state energies and transition moments (i.e., the constituents of electronic absorption spectra) require usually more effort and circumspection. However, that should not discourage interested users from venturing into this very exciting (albeit a bit slippery) field of computational chemistry, and these users may take comfort from the fact that many successes have been achieved, and that modeling of electronic absorption spectra has permitted us to clarify complicated mechanistic schemes in cases where IR spectroscopy could not be applied for one reason or another.

The difficulties are mainly caused by two problems: (1) the fact that even a qualitatively correct description of excited states often requires *multiconfigurational wave functions*; and (2) that dynamic electron correlation effects in excited states are often significantly greater than in the electronic ground states of molecules, and may also vary greatly between different excited states. For explanations of the concepts invoked in this section, see Section 3.2.3 of Chapter 22 in this volume. Therefore, an accurate modeling of electronic spectra requires methods that account for both effects simultaneously.

The multiconfigurational problem can be readily handled by a CI (configuration interaction) calculation limited to a restricted number of "active" orbitals and electrons and/or to a certain excitation level (simultaenous excitation of one, two, or three electrons), However, such calculations do not account sufficiently for dynamic electron correlation, which therefore has to be introduced "on top." Over the course of time, different strategies have been elaborated to achieve this. The simplest and most straightforward one is to introduce *parameters* into a model and to calibrate those on a test set of well-known and correctly assigned electronic transitions. This approach, which was implemented back in the 1950s at the Hückel level, and was subsequently applied to π- and later to all-valence-SCF–CI semiempirical methods, has a long and successful history. However, the application of these methods to species with unusual bonding situations (such as many reactive intermediates) is problematic because such compounds were usually not included in the parametrization. Nevertheless, methods like the popular INDO/S-CI model of Zerner and Ridley[130] have been and continue to be applied very successfully to the modeling of the electronic spectra of reactive intermediates.[116,131] Apart from the fact that these semiempirical methods are inexpensive in terms of computational resources, another advantage is that these are very transparent because the excited configurations are buildt from a set of Hartree–Fock MOs.

Accounting for dynamic correlation effects at the ab initio level requires that a multiconfigurational calculation be supplemented by a large CI (MR–CI) or by a perturbative treatment, for example, by the MP2 or coupled cluster methods described in Section 3.2.3.2 of Chapter 22 in this volume. The most widely applied of these is the CASPT2 method that was developed by Roos and co-workers[132] and is implemented in the MOLCAS suite of programs.[133] There an MP2-type calculation is carried out on the basis of a zero-order CAS (complete active space) self-consistent field (SCF) wave function (cf. section 3.2.3.3 of Chapter 22 in this volume). Conceptually, this is a very elegant approach, but in practice CASPT2 calculations on excited states are often fraught with problems. The reason is that excitations from or into MOs *outside* the active space, which are generated in the MP2 step, may be nearly degenerate with excitations *within* the active space, which leads to a breakdown of the perturbative *ansatz.* For small molecules, this problem can sometimes be overcome by increasing the active space, but for large molecules this is often impossible because the size of a CASSCF calculations scales factorially with the number of MOs in the active space. Then one is forced to take recourse to level-shifting techniques that lack a solid theoretical foundation and may lead to artifacts. Nevertheless, if applied with due circumspection, CASPT2 is a very reliable method and has been applied succesfully to the modeling of electronic spectra of a wide variety of compounds, including many matrix-isolated reactive intermediates in the author's laboratory.[8,134-142]

An entirely different approach to ab initio calculations of excited states is to look at the *response* of the ground-state electron distribution to oscillating electric fields of different frequencies. If the frequency of such an oscillation is in resonance with an electronic excitation, then the polarizability will increase discontinuously, that is, the *frequency-dependend polarizability* will show a *pole* whose intensity is proportional to the electronic transition moment. To probe this, one must solve a *time-dependent* Schrödinger equation. The corresponding formalism, which allows one also to calculate states where electrons are detached or added, was developed a long time ago for Hartree–Fock theory, but the lack of dynamic correlation at this level made it impossible to obtain accurate results. This situation changed when the above *ansatz* was carried over to density functional theory where dynamic correlation is accounted for by means of the exchange and correlation functionals.[143] After the implementation of this so-called time-dependent DFT (TD–DFT) method into the Gaussian series of programs,[144] the application of this method has gained a measure of popularity in recent years, mainly because it is considerably more economical in terms of computational and human resources than the above CASPT2 method. Since the excited-state electron density is modeled within the framework of a singles-CI, the results of TD–DFT calculations also lend themselves to a transparent interpretation in terms of electronic configurations. The TD-DFT theory also has recently been applied successfully to matrix-isolated reactive intermediates,[142,145,146] but the experiences made in our laboratory are mixed and the jury is perhaps still out on how reliable this promising new method really is.

Finally, we wish to mention that Grimme et al.[147] proposed a very economical multireference (MR) method based on the MOs obtained by DFT calculations (DFT–MRCI), whereby an overaccount for dynamic correlation is prevented by introducing empirical parameters. Although the introduction of these parameters gives this method a semiempirical flavor, its quantitative accuracy in the prediction of spectra appears to be comparable to that of CASPT2, at a fraction of the expense of computational and human effort.[142] In fact, DFT–MRCI has even yielded predictions in excellent agreement with experiment even in a recent case where CASPT2 failed entirely.[148]

8. CONCLUSION AND OUTLOOK

Research on reactive intermediates can be carried out by different approaches and techniques that are often complementary. The first, and the oldest of these approaches, involves *product studies* where conclusions about the properties of a reactive intermediate are drawn from the chemistry it undergoes. Translated into forensic terms, such studies amount to an investigation of the scene and the victims of the crime after the crime has been committed. Much closer to the action is *time-resolved* spectroscopy, which looks at the appearance or disappearance of some manifestation of the reactive intermediate or one of its "victims" (i.e., the proverbial "smoking gun"). Although, if judiciously designed and interpreted, both of the above methods make it possible to obtain much valuable insight into the crime, they do not allow Sherlock Holmes and his modern descendants to arrest the perpetrator. Here is exactly where *matrix isolation* steps in. This technique serves to incarcerate Dr. Moriarty and his cohorts who pulled the trigger and to investigate leisurely and in detail his demeanor, his weapons, his motives, and to see if these could eventually be used productively.

However, not all reactive intermediates are kind enough to provide spectroscopic signatures that allow their immediate and unambiguous identification, and it is therefore often necessary to compare those signatures to ones obtained by means of *modeling calculations* (the reader may note that with this we leave the realm of forensic analogy that we have perhaps already stretched too far). In fact, many recent matrix isolation studies owe their success to the tremendous advances in the field of computational chemistry, and to the increased availability of the hard- and software required to carry out such calculations. This situation provides an opportunity for much creative work in the field of reactive intermediates, but it also implies an obligation on the part of those who use such methods to apply them with due care and circumspection.

A productive exploitation of the synergy between experiment and theory requires that practitioners familiarize themselves with the scope and limitations of the methods they use, so they can avoid pitfalls due to *artifacts* that may occur both in experiment and in theory. It is, for example, disturbingly easy to "create" or "annihilate" bands by formation of suitably scaled difference spectra. On the other hand, the harmonic approximation that is at the basis of all practicable modeling calculations of vibrational spectra may lead to predictions that have no relation to experiment (as demonstrated above for the case of phenylcarbene).

Another malicious enemy of good science, which is looming over every spectrometer and computer, is *wishful thinking*. Burdened by expectations one has about the outcome of an experiment, it is often very tempting to pick, say, the peaks in a matrix isolation IR spectrum that appear to correlate with those calculated for an anticipated product (perhaps after a bit of squinting), and to use those peaks as a basis for a structural identification, sweeping others under the proverbial rug as spectroscopic garbage. Not only can this lead to false assignments, but one may miss very important hints that Nature provides on the chemistry that is taking place, and which is often outside the scope of what one originally set out to consider.

It is to be hoped (and, indeed, to be anticipated) that modeling calculations will become more reliable in the future, especially in the area of electronically excited states. However, this will not free us from using these powerful tools critically and circumspectly. With this caveat, a bright future lies ahead of matrix isolation spectroscopy, and there can be no doubt that it will continue to provide much valuable insight, as it has in the past.

ACKNOWLEDGMENTS

I am grateful to the many graduate students on whose hard work some of the examples that are used in this chapter are based, and to the Swiss National Science Foundation for continued support. I am very indebted to Professor Matthew Platz and to Professor Weston T. Borden for their contributions to improve this chapter.

SUGGESTED READING

I. R. Dunkin, *Matrix Isolation Techniques: A Practical Approach*, Oxford University Press, Oxford, **1998**.

M. J. Almond and A. J. Downs, *Spectroscopy of Matrix Isolated Species*; John Wiley & Sons, Inc., Chichester, **1989**.

L. Andrews and M. Moskovits, Eds., *Chemistry and Physics of Matrix Isolated Species*, North-Holland, Amsterdam, The Netherlands, **1988**.

A. J. Barnes, W. J. Orville-Thomas, A. Müller, and R. Gaufrès, Eds., *Matrix Isolation Spectroscopy*, Nato Advanced Study Institute Series C, D, Reidel, Dordrecht, The Netherlands, **1981**.

M. Moskovits and G. A. Ozin, *Cryochemistry*, Wiley-Interscience, New York, **1976**.

V. E. Bondybey, A. M. Smith, and J. Agreiter, *Chem. Rev.* **1996**, *96*, 2113.

REFERENCES

1. E. Whittle, D. A. Dows, and G. C. Pimentel, *J. Chem. Phys.* **1954**, *22*, 1943.

2. E. D. Becker and G. C. Pimentel, *J. Chem. Phys.* **1956**, *25*, 224.

3. I. Norman and G. Porter, *Nature (London)* **1954**, *174*, 58.

4. G. N. Lewis and D. Lipkin, *J. Am. Chem. Soc.* **1942**, *64*, 2801.

5. C. Sandorfy, *Can. J. Chem.* **1965**, *10*, 85.

6. M. J. Almond and A. J. Downs, *Spectroscopy of Matrix Isolated Species*, John Wiley & Sons, Inc., Chichester, **1989**.

7. T. Shida, E. Haselbach, and T. Bally, *Acc. Chem. Res.* **1984**, *17*, 180.

8. T. Bally, C. Carra, M. P. Fülscher, and Z. Zhu, *J. Chem. Soc. Perkin Trans. 2* **1998**, 1759.

9. I. R. Dunkin, *Matrix Isolation Techniques: A Practical Approach*, Oxford University Press, Oxford, **1998**.

10. Two providers currently offer closed cycle cryostats for matrix isolation with 0.5–1 W of cooling power at 4 K. The first is Janis Reserach (http://www.janis.com/p-a4k.html), which uses hardware from Sumitomo, the other is Cryomech (http://www.cryomech.com/cryostats.html).

11. Note that fused silica, which has an even higher UV transparency, is not a good choice because of its extremely poor heat conductivity that makes for very inefficient condensation of the host gas.

12. Here the mechanical properties are not so important because the external windows are not subjected to big temperature variations, and they do not need to be clamped down hard because the vacuum inside the cryostat holds them in place.

13. Good tightening of the screws improves the thermal contact between the window and the sample holder. To avoid slippage of the screwdriver into the CsI window it is advantageous to use screws with hexagonal holes.

14. R. Rossetti and L. E. Brus, *Rev. Sci. Instrum.* **1980**, *51*, 467.

15. (a) The appropriate temperature is one where a few milligrams of the substrate sublime over about an hour. It usually has to be determined by trial and error. If it is higher than room temperature, the parts of the inlet system that lead from the sample tube to the cryostat have to be wrapped with heating tape and warmed to avoid condensation of the substrate on its way to the sample window. (b) H. Tomioka, personal communication.

16. H. Clauberg, D. W. Minsek, and P. Chen, *J. Am. Chem. Soc.* **1992**, *114*, 99.

17. M. Moskovits and G. A. Ozin, *Cryochemistry*; Wiley Interscience, New York, **1976**.

18. X. Wang and L. Andrews, *J. Phys. Chem. A* **2002**, *106*, 3706, 3744, 6720.

19. X. Wang and L. Andrews, *J. Phys. Chem. A* **2002**, *106*, 2457.

20. X. Wang and L. Andrews, *J. Phys. Chem. A* **2001**, *105*, 5812.

21. M. Zhou, L. Andrews, and C. W. Bauschlicher, *Chem. Rev.* **2001**, *101*, 1931.

22. B. Liang and L. Andrews, *J. Phys. Chem. A* **2002**, *106*, 595, 4042.

23. B. Liang, L. Andrews, J. Li, and B. E. Bursten, *J. Am. Chem. Soc.* **2002**, *124*, 6723.

24. L. Andrews and A. Citra, *Chem. Rev.* **2002**, *102*, 885.

25. B. Liang and L. Andrews, *J. Phys. Chem. A* **2002**, *106*, 3738, 4038, 6945.

26. L. B. Knight, *Acc. Chem. Res.* **1986**, *19*, 313.

27. P. Hassanzadeh and L. Andrews, *J. Phys. Chem.* **1992**, *96*, 9177.

28. V. E. Bondybey, A. M. Smith, and J. Agreiter, *Chem. Rev.* **1996**, *96*, 2113.

29. L. Andrews and G. Pimentel, *J. Chem. Phys.* **1967**, *47*, 3637.

30. K. Hassenrück, J. G. Radziszewski, V. Balaji, G. S. Murthy, A. J. McKinley, D. E. David, V. M. Lynch, H.-D. Martin, and J. Michl, *J. Am. Chem. Soc.* **1990**, *112*, 873.

31. F. H. Walker, K. B. Wiberg, and J. Michl, *J. Am. Chem. Soc.* **1982**, *104*, 2056.

32. K. B. Wiberg, F. H. Walker, W. E. Pratt, and J. Michl, *J. Am. Chem. Soc.* **1983**, *105*, 3638.

33. D. Otteson and J. Michl, *J. Org. Chem.* **1984**, *49*, 866.

34. P. Kasai, *Phys. Rev. Lett.* **1968**, *21*, 67.

35. M. Jacox, *Rev. Chem. Intermed.* **1985**, *6*, 77.

36. D. Davis and W. Brain, *Appl. Opt.* **1968**, *7*, 2071.

37. T. Bally, in *Radical Ionic Systems*, A. Lund and M. Shiotani, Eds., Kluwer, Dordrecht, The Netherlands, **1991**, pp. 3–54.

38. G. Maier and C. Lautz, *Angew. Chem. Int. Ed. Engl.* **1999**, *38*, 2038.

39. J. Sparget-Larsen, M. Gil, A. Gorki, D. M. Blake, J. Waluk, and J. G. Radziszewski, *J. Am. Chem. Soc.* **2001**, *123*, 11253.

40. An entire analytical technique (ion photodissociation spectroscopy) relies on this phenomenon (see, e.g., R. C. Dunbar, in *Gas Phase Ion Chemistry*, Vol 2 and 3, M. T. Bowers, Ed., Academic Press, New York, **1979**, p. 1984).

41. C. B. Moore and G. C. Pimentel, *J. Chem. Phys.* **1964**, *41*, 3504.

42. The enthalpy of formation of singlet methylene is known experimentally (102.6 kcal/mol: C. C. Hayden, D. M. Neumark, K. Shabatake, R. K. Sparks, and Y. T. Lee, *J. Chem. Phys.* **1982**, *76*, 3607), wheras the best current estimate of the enthalpy of formation of diazomethane comes from a G2 calculation (67.1 kcal/mol: M. S. Gordon and S. R. Kass, *J. Phys. Chem.* **1995**, *99*, 6548).

43. S. A. Walch, *J. Chem. Phys.* **1995**, *103*, 4930.

44. G. Maier and H. P. Reisenauer, in *Advances in Carbene Chemistry*, Vol. 3, U. H. Brinker, Ed., Elsevier, Amsterdam, The Netherlands, **2001**, p. 115.

45. S. Matzinger and T. Bally, *J. Phys. Chem. A* **2000**, *104*, 3544.

46. M. S. Baird, I. R. Dunkin, N. Hacker, M. Poliakoff, and J. J. Turner, *J. Am. Chem. Soc.* **1981**, *103*, 5190.

47. C. B. Moore, G. C. Pimentel, and T. D. Goldfarb, *J. Chem. Phys.* **1965**, *43*, 63.

48. W. Sander, G. Bucher, and S. Wierlacher, *Chem. Rev.* **1993**, *93*, 1583.

49. A. V. Friderichsen, J. G. Radziszeweski, M. R. Nimlos, P. R. Winter, D. C. Dayton, D. E. David, and G. B. Ellison, *J. Am. Chem. Soc.* **2001**, *123*, 1977.

50. J. Pacansky and M. Dupuis, *J. Am. Chem. Soc.* **1982**, *104*, 415, and references cited therein.

51. J. G. Radziszewski, M. F. Nimlos, P. R. Winter, and G. B. Ellison, *J. Am. Chem. Soc.* **1996**, *118*, 7400.

52. R. A. Moss, Y. Ma, F. Zheng, R. R. Sauers, T. Bally, A. Maltsev, J. P. Toscano, and B. M. Showalter, *Phys. Chem., Chem. Phys.* **2002**, *5*, 1010.

53. See, for example, G. Seybold, *Angew. Chem. Int. Ed. Engl.* **1977**, *16*, 365.

54. E. Hedaya, *Acc. Chem. Res.* **1969**, *2*, 367.

55. J. Pacansky and J. S. Chang, *J. Chem. Phys.* **1981**, *74*, 5539.

56. D. W. Kohn and P. Chen, *Rev. Sci. Instrum.* **1992**, *63*, 4003.

57. G. Maier, T. Preiss, and H. P. Reisenauer, *Chem. Ber.* **1994**, *127*, 779.

58. S. L. Buchwalter and G. L. Closs, *J. Am. Chem. Soc.* **1973**, *97*, 3857.

59. G. C. Closs, L. P. Kaplan, and V. I. Budall, *J. Am. Chem. Soc.* **1967**, *89*, 3376.

60. C. R. Watson, R. M. Pagni, J. R. Dodd, and J. E. Bloor, *J. Am. Chem. Soc.* **1976**, *98*, 2551.

61. P. Dowd, A. Gold, and K. Sachdev, *J. Am. Chem. Soc.* **1968**, *90*, 2715.

62. P. Dowd and K. Sachdev, *J. Am. Chem. Soc.* **1967**, *89*, 715.

63. W. Roth, R. Langer, M. Bartmann, B. Stevermann, G. Maier, H. P. Reisenauer, R. Sustmann, and W. Müller, *Angew. Chem. Int. Ed. Engl.* **1987**, *26*, 256.

64. O. L. Chapman, C. C. Chang, and J. Kolc, *J. Am. Chem. Soc.* **1976**, *98*, 5703.

65. M. Rule, A. R. Matlin, E. F. Hilinski, D. A. Dougherty, and J. R. Berson, *J. Am. Chem. Soc.* **1979**, *101*, 5098.

66. J.-F. Muller, D. Muller, H. J. Dewey, and J. Michl, *J. Am. Chem. Soc.* **1978**, *100*, 1629.

67. M. S. Platz and J. R. Burns, *J. Am. Chem. Soc.* **1979**, *101*, 4425.

68. M. S. Platz, G. Carroll, F. Pierrat, J. Zayas, and S. Auster, *Tetrahedron* **1982**, *38*, 777.

69. G. Maier, H.-P. Reisenauer, K. Lanz, R. Tross, D. Jürgen, B. A. Hess, and L. J. Schaad, *Angew. Chem.* **1993**, *105*, 119.

70. T. A. Miller and V. Bondybey, Eds., *Molecular Ions: Spectroscopy, Stucture, and Chemistry*, North-Holland: Amsterdam, The Netherlands, 1980.

71. M. E. Jacox and W. E. Thompson, *Res. Chem. Intermed.* **1989**, *12*, 33.

72. J. P. Maier, *J. Phys. Chem. A* **1998**, *102*, 3462.

73. W. H. Hamill, in *Radical Ions*, E. T. Kaiser and L. Kevan, Eds., Wiley-Interscience, New York, **1968**.

74. T. Shida, *Electronic Absorption Spectra of Organic Radical Ions*, Elsevier, Amsterdam, The Netherlands, **1988**.

75. W. Tang, X.-L. Zhang, and T. Bally, *J. Phys. Chem.* **1993**, *97*, 4373.

76. L. Truttmann, K. R. Asmis, and T. Bally, *J. Phys. Chem.* **1995**, *99*, 17844.

77. B. Raz and J. Jortner, *Chem. Phys. Lett.* **1969**, *4*, 155.

78. A. Gedanken, B. Raz, and J. Jortner, *J. Chem. Phys.* **1973**, *58*, 1178.

79. T. Bally and J. Michalak, *J. Photochem. Photobiol. A: Chem.* **1992**, *69*, 185.

80. B. Müller, Ph. D. Thesis No. 1299, University of Fribourg, **2000**.

81. D. Forney, J. Fulara, P. Freivolgel, M. Jakobi, D. Lessen, and J. P. Maier, *J. Chem. Phys.* **1995**, *103*, 48.

82. P. Kasai, *Acc. Chem. Res.* **1971**, *4*, 329.

83. T. M. Halasinski, D. M. Hudgins, F. Salama, L. J. Allamandola, and T. Bally, *J. Phys. Chem. A* **2000**, *104*, 7484.

84. T. M. Halasinski, J. L. Weisman, R. Ruiterkamp, T. J. Lee, F. Salama, and M. Head-Gordon, *J. Chem. Phys.* **2003**, in print.

85. J. Gebicki and J. Michl, *J. Phys. Chem.* **1988**, *92*, 6452.

86. M. Beyer, A. Lammers, E. V. Savchenko, G. Niedner-Schatteburg, and V. E. Bondybey, *Phys. Chem. Chem. Phys.* **1999**, *1*, 2213, and references cited therein.

87. L. Andrews, J. H. Miller, and B. W. Keelan, *Chem. Phys. Lett.* **1980**, *71*, 207.

88. W. Kirmse, *Adv. Carbene Chem.* **2001**, *3*, 1.

89. S. T. Belt, C. Bohne, G. Charette, S. E. Sugamori, and J. C. Scaiano, *J. Am. Chem. Soc.* **1993**, *115*, 2200.

90. R. A. McClelland, M. J. Kahley, P. A. Davidse, and G. Hadzialic, *J. Am. Chem. Soc.* **1996**, *118*, 4794.

91. J. Michalak, H. B. Zhai, and M. S. Platz, *J. Phys. Chem.* **1996**, *100*, 14028.

92. C. Carra, T. Bally, Z. Zhu, and M. S. Platz, manuscript in preparation.

93. R. T. Coulin, R. D. Miller, and J. Michl, *J. Am. Chem. Soc.* **1979**, *101*, 7637.

94. J. G. Radziszwesky, T.-K. Yin, G. E. Renzoni, D. A. Hrovat, W. T. Borden, and J. Michl, *J. Am. Chem. Soc.* **1993**, *115*, 1454.

95. T. Bally and E. Haselbach, *Helv. Chim. Acta* **1978**, *61*, 754.

96. W. E. Billups and L.-J. Lin, *Tetrahedron* **1986**, *42*, 1575.

97. W. E. Billups, C. Giesenberg, and R. Cole, *Tetrahedron Lett.* **1997**, *38*, 1115.

98. O. L. Chapman, J. Gano, P. R. West, M. Regitz, and G. Maas, *J. Am. Chem. Soc.* **1981**, *103*, 7033.

99. W. Sander and O. Chapman, *Angew. Chem.* **1988**, *100*, 402.

100. C. Wentrup, R. Blanch, H. Briehl, and G. Gross, *J. Am. Chem. Soc.* **1988**, *110*, 1874.

101. G. Maier, *Angew. Chem. Int. Ed. Engl.* **1974**, *13*, 425.

102. T. Bally and S. Masamune, *Tetrahedron* **1980**, *36*, 343.

103. B. R. Arnold and J. Michl, in *Kinetics and Spectroscopy of Carbenes and Biradicals*, M. S. Platz, Ed., Plenum Press, New York, **1990**, p. 1.

104. O. L. Chapman, P. De la Cruz, R. Roth, and J. Pacansky, *J. Am. Chem. Soc.* **1973**, *95*, 1337.

105. A. Krantz, C. Y. Lin, and M. D. Newton, *J. Am. Chem. Soc.* **1973**, *95*, 2744.

106. N. Allinger, C. Gilardeau, and L. W. Chow, *Tetrahedron* **1968**, *24*, 2401.

107. R. C. Haddon and G. R. J. Williams, *J. Am. Chem. Soc.* **1975**, *97*, 6582.

108. M. Nakayama, M. Nisihira, and Y. I'Haya, *Bull. Chem. Soc. Jpn.* **1975**, *49*, 1502.

109. G. Maier, H.-G. Hartan, and T. Sayrac, *Angew. Chem. Int. Ed. Engl.* **1976**, *15*, 226.

110. S. Masamune, Y. Sugihara, K. Morio, and J. E. Bertie, *Can. J. Chem.* **1976**, *54*, 2679.

111. R. G. S. Pong, B.-S. Huang, J. Laureni, and A. Krantz, *J. Am. Chem. Soc.* **1977**, *99*, 4153.

112. S. Masamune, F. A. Souto-BAchiller, T. Machiguchi, and J. E. Bertie, *J. Am. Chem. Soc.* **1978**, *100*, 4889.

113. It is noteworthy that this was one of the first published chemical studies where IR spectra were taken with a Michelson-Interferometer rather than with a dispersive instrument. In fact, Masamune claims that it was only by virtue of this new technology that he and his co-workers were able to pinpoint the missing bands of CB (S. Masamune, personal communication).

114. (a) This discovery did not put an end to research on cyclobutadiene. For developments that occurred after 1978, the reader is encouraged to consult the review by Arnold and Michl.[103] (b) The same error is unlikely to happen nowadays because calculations have evolved to a level where a mere comparison of observed and calculated IR spectra would permit a clear distinction between the two hypotheses that were entertained in the 1970s.

115. W. Sander, *Angew. Chem.* **1990**, *102*, 362.

116. Z. Zhu, T. Bally, L. L. Stracener, and R. J. McMahon, *J. Am. Chem. Soc.* **1999**, *121*, 2863.

117. T. Khasanova and R. S. Sheridan, *J. Am. Chem. Soc.* **2000**, *122*, 8585.

118. A. Nikitina and R. S. Sheridan, *J. Am. Chem. Soc.* **2002**, *124*, 7670.

119. See, for example, T. Plakhotnik, E. A. Donley, and U. P. Wild, *Ann. Rev. Phys. Chem.* **1997**, *48*, 181.

120. O. L. Chapman and J.-P. LeRoux, *J. Am. Chem. Soc.* **1978**, *100*, 282.

121. P. R. West, O. L. Chapman, and J.-P. LeRoux, *J. Am. Chem. Soc.* **1982**, *104*, 1779.

122. R. J. McMahon, C. J. ABelt, O. L. Chapman, J. W. Johnson, C. L. Kreil, J.-P. LeRoux, A. M. Mooring, and P. L. West, *J. Am. Chem. Soc.* **1987**, *109*, 2456.

123. P. Pulay, G. Forarasi, G. Pongor, J. E. Boggs, and A. Varga, *J. Am. Chem. Soc.* **1983**, *105*, 7037.

124. See, for example, W. Tang, X.-L. Zhang, and T. Bally, *J. Phys. Chem.* **1993**, *97*, 4373.

125. C. W. Murray, G. J. Laming, N. C. Handy, and R. D. Amos, *Chem. Phys. Lett.* **1992**, *199*, 551.

126. J. Andzelm and E. Wimmer, *J. Chem. Phys.* **1992**, *96*, 1280.

127. A. P. Scott and L. Radom, *J. Phys. Chem.* **1996**, *100*, 16502.

128. T. Bally, K. Roth, W. Tang, R. R. Schrock, K. Knoll, and L. Y. Park, *J. Am. Chem. Soc.* **1992**, *114*, 2440.

129. T. Bally, S. Nitsche, and K. Roth, *J. Chem. Phys.* **1986**, *84*, 2577.

130. M. C. Zerner and J. E. Ridley, *Theor. Chim. Acta* **1973**, *32*, 111.

131. T. Bally, L. Truttmann, J.-T. Wang, and F. Williams, *J. Am. Chem. Soc.* **1995**, *117*, 7923.

132. K. Andersson and B. O. Roos, in *Modern Electronic Structure Theory*, Vol. 1, Part 1, Vol. 2, World Scientific Publishers Co., Singapore, **1995**, p. 55.

133. K. Andersson, M. R. A. Blomberg, M. P. Fülscher, V. Kellö, R. Lindh, P.-Å. Malmqvist, J. Noga, J. Olson, B. O. Roos, A. Sadlej, P. E. M. Siegbahn, M. Urban, and P.-O. Widmark, MOLCAS, Versions 3 and 4, University of Lund, Sweden, 1994.

134. M. P. Fülscher, S. Matzinger, and T. Bally, *Chem. Phys. Lett* **1995**, *236*, 167.

135. T. Bally, L. Truttmann, S. Dai, and F. Williams, *J. Am. Chem. Soc.* **1995**, *117*, 7916. Using Smart Source Parsing.

136. L. Truttmann, K. R. Asmis, and T. Bally, *J. Phys. Chem.* **1995**, *99*, 17844.

137. K. Huben, Z. Zhu, T. Bally, and J. Gebicki, *J. Am. Chem. Soc.* **1997**, *119*, 2825.

138. A. Marcinek, J. Adamus, K. Huben, J. Gebicki, T. J. Bartczak, P. Bednarek, and T. Bally, *J. Am. Chem. Soc.* **2000**, *122*, 437.

139. T. Bally, S. Bernhard, S. Matzinger, J.-L. Roulin, G. N. Sastry, L. Truttmann, Z. Zhu, A. Marcinek, J. Adamus, R. Kaminski, J. Gebicki, F. Williams, G.-F. Chen, and M. P. Fülscher, *Chem. Eur. J.* **2000**, *6*, 858.

140. T. Bally, Z. Zhu, J. Wirz, M. Fülscher, and J.-Y. Hasegawa, *J. Chem. Soc. Perkin Trans. 2* **2000**, 2311.

141. P. Bednarek, Z. Zhu, T. Bally, T. Filipiak, A. Marcinek, and J. Gebicki, *J. Am. Chem. Soc.* **2001**, *123*, 2377.

142. E. Haselbach, M. Allan, T. Bally, P. Bednarek, A.-C. Sergenton, A. d. Meijere, S. Kozhushkov, M. Piacenza, and S. Grimme, *Helv. Chim. Acta* **2001**, *84*, 1670.

143. M. E. Casida, in *Recent Advances in Density Functional Methods, Part I*, D. P. Chong, Ed.; World Scientific, Singapore, **1995**, p. 155.

144. R. E. Stratmann, G. E. Scuseria, and M. J. Frisch, *J. Chem. Phys.* **1998**, *109*, 8218.

145. K. Exner, B. Grossmann, G. Gescheidt, J. Heinze, M. Keller, T. Bally, P. Bednarek, and H. Prinzbach, *Angew. Chem. Int. Ed. Engl.* **2000**, *39*, 1455.

146. P. Brodard, A. Sarbach, J.-C. Gumy, T. Bally, and E. Vauthey, *J. Phys. Chem. A* **2001**, *105*, 6594.

147. S. Grimme and M. Waletzke, *J. Chem. Phys.* **1999**, *111*, 5645.

148. C. Carra, T. Bally, T. A. Jenny, and A. Albini, *Photochem. Photobiol. Sci.* **2002**, *1*, 38.

CHAPTER 18

Nanosecond Laser Flash Photolysis: A Tool for Physical Organic Chemistry

J. C. SCAIANO

Department of Chemistry, University of Ottawa, Ontario, K1N 6N5 Canada

1. INTRODUCTION

The understanding of chemical reactions taking place in the "normal" laboratory time scale, seconds-to-days, requires insight into much faster processes, with ultimate consequences in the "normal" time scale. For example, free radicals frequently have lifetimes in the microsecond or millisecond time scale. There are normally two approaches to study short-lived intermediates. In one, the experimental conditions are adjusted, so as to lengthen the intermediate lifetime to the point

Reactive Intermediate Chemistry, edited by Robert A. Moss, Matthew S. Platz, and Maitland Jones, Jr.
ISBN 0-471-23324-2   Copyright © 2004 John Wiley & Sons, Inc.

where its detection becomes possible with standard spectroscopic techniques. Cryogenic techniques, frequently combined with matrix isolation, can easily achieve this for reactive intermediates with activated decay (i.e., showing temperature dependence), or where their decay involves bimolecular reactions (e.g., free radical dimerization); in this case the lengthening of the lifetime is largely the result of restrictions to diffusion. This approach can yield excellent spectroscopic and structural data on the intermediate, but its key shortcoming is that is cannot provide any information about the reactivity of the species under conditions where its chemistry is of interest.

This brings us to the second approach: Allow the reaction to proceed under normal laboratory conditions for reaction, but accelerate the detection technique so that the time evolution of the reactive intermediate can be adequately monitored. In fact, having a fast detection technique is not sufficient, one also requires a fast generation methodology for the species under study.

The detection of reactive intermediates can be achieved with numerous techniques, such as magnetic and optical spectroscopies. This chapter deals only with optical spectroscopy in the ultraviolet (UV) and visible (vis) regions.

The 1967 Nobel Prize was shared by Eigen, Norrish, and Porter. The half-prize shared by Porter and Norrish was awarded for the development of the flash photolysis method, as "this provided a powerful tool for the study of the various states of molecules and the transfer of energy between them"[1] The work that led to this award was carried out largely in the preceding two decades. In the flash photolysis method the reactive intermediates were generated following excitation by the light pulse produced by the flash lamp. With this methodology millisecond studies are easy, and multimicrosecond resolution can also be achieved. It was clear, however, that better time resolution would be desirable. Interestingly, the detection system was capable of faster resolution, the real limitation being the duration of the light pulse.

The invention of the laser in the early 1960s provided the opportunity for a faster methodology. Many scientists were involved in the quest for better time resolution; Lindqvist should be credited with the first report of laser flash photolysis (LFP).[2] By taking advantage of the short pulse from a nitrogen laser (337 nm), Lindqvist was able to detect the triplet state of acridine, with the now classic instrumental design of Figure 18.1. Nanosecond LFP is today a common tool for kinetic studies, and while each of the components used by Lindqvist over 35 years ago has improved, the basic system remains much the same. By the time Porter and Norrish were receiving their Nobel Prize for "conventional" flash photolysis, the bar had already been raised, and Porter, in his Nobel lecture, referred to the potential of nanosecond techniques being developed.[3] The next major step was the introduction of computer control and data acquistion, with the first such system built at the University of Notre Dame in the late 1970s.[4]

Nanosecond techniques have now been superceded by picosecond and femtosecond techniques, allowing detection in time domains as short as 10^{-15} s. Yet, nanosecond techniques remain powerful tools in the arsenal of the physical organic chemist; quite simply, many radical, carbene,[5,6] carbocation,[7] carbanion[8] reactions take place in the nano- and microsecond time scales.

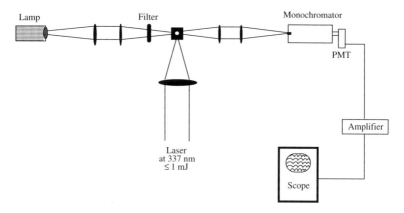

Figure 18.1. Laser flash photolysis system developed by Lindqvist. (Adapted from Ref. 2.)

2. TIME-RESOLVED ABSORPTION TECHNIQUES

It is important to realize that millisecond, microsecond, and nanosecond techniques all reflect the same technology as far as detection is concerned. In all cases, modern electronics are adequate to monitor the evolution of signal with time (remember that nanoseconds in time scale reflects gigaherz in terms of frequency). In all these techniques, the speed of light is sufficiently fast compared with the experiment, so that one can regard the propagation of light as an essentially instantaneous process; as a benchmark, light travels ~30 cm (1 ft) in 1 ns. In contrast in the case of picoseconds, the light travels distances that are comparable, or shorter than typical experimental dimensions. Thus, light travels ~0.3 mm in 1 ps. At the same time "picoseconds" are too fast for most electronic detection devices. Thus, in picosecond techniques timing is effectively done by moving optical components (frequently mirrors) so as to change the optical path.[9] Thus, picosecond techniques can rarely be stretched beyond 10 ns, since this would require changes in optical path >3 m (10 ft), and this is impractical in most laboratories.

Thus, nanosecond techniques are usually limited to a few nanoseconds on the short end, presently related to the pulse duration (2–20 ns) of typical nanosecond lasers. Long times are frequently limited by either the stability of electronic–optical components, or by the fact that in the absence of other processes many reaction intermediates (e.g., free radicals and carbenes) can undergo rapid self-reactions. "Nanosecond" techniques rarely extend beyond a few tens of milliseconds.

2.1. Nanosecond Laser Flash Photolysis

Nitrogen and ruby lasers played a key role in the early development of nanosecond laser flash photolysis (nLFP). Their role has gradually been taken over by the more convenient excimer and Nd/YAG lasers. Table 18.1 gives typical wavelengths

TABLE 18.1. Common Lasers Used for LFP

Laser	Fundamental (nm)	Harmonics (nm)
Nitrogen	337	
Excimer	157	
	193	
	248	
	308	
	351	
Ruby	694	347
Nd/YAG	1064	532, 355, 266

available from various lasers. Note that in the case of solid-state lasers (such as Nd/YAG) most of the useful wavelengths are obtained by generation of harmonics from the fundamental wavelengths. Excimer lasers offer great versatility by changing the gas mixture, and most of them are also suitable for operation with nitrogen at 337 nm.

A xenon lamp is usually the choice for the monitoring beam. For short time scales, the xenon lamp is frequently "pulsed"; that is, its output intensity is enhanced by a factor of 5–100 for a few milliseconds.[10] This reduces the lamp lifetime but greatly improves the signal to noise. Modern high-intensity ceramic lamps can be used without pulsing due to their high efficiency, and the fact that they incorporate internal high-reflectance mirrors.[11] An outline of a modern nLFP system is shown in Figure 18.2.

The detection system consists of a monochromator and photomultiplier, the latter wired for fast response, and able to support relatively high output currents

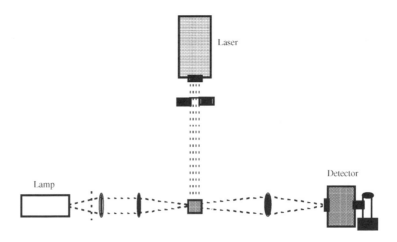

Figure 18.2. Typical configuration for a LFPs system.

(typically ≥ 2 mA) for a short time. Detection from ~ 200 nm and into the near-infrared (NIR) is readily achieved with modern photomultipliers, although the same detector may not be suitable for both wavelength extremes. The use of fiber optic technologies and compact photomultipliers has greatly reduced the size of nLFP systems.

The system, as described above, is capable of signal detection as a function of time at a selected wavelength. Spectra are constructed by monitoring time-resolved traces at many wavelengths, and then extracting absorbance data at a given time from each trace.

An alternate approach consists of using a gated spectrally resolved detector to capture spectral data in a well-defined, and frequently narrow, time window. Image capture is achieved with either a diode array detector or a CCD camera. An intensifier located between the output window of the spectrograph and the detector provides the gating capabilities. In this approach, time resolution can be achieved by collecting spectral data at different times and then reconstructing the time profile at one or more wavelengths.

Ultimately, the two approaches mentioned above collect data in a three-dimensional (3D) matrix in which the vertical axis is signal intensity (usually transient absorbance), and the other two axes are time and wavelength. The difference is simply which direction in the matrix is collected for each laser shot.

2.2. Data Acquisition and Processing

The signal from a nLFP system operating in the time-resolved mode (see above) is the voltage across a resistor (terminator) at the output of the photomultiplier. This terminator (typically 50 or 93 Ω) needs to be matched by the type of signal cables used. The output has the shape shown in Figure 18.3 for a continuous wave

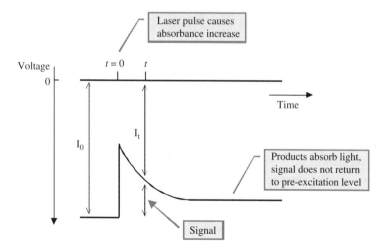

Figure 18.3. Time evolution of the photomultiplier output as an absorbing transient is generated and decays.

(CW) (i.e., not pulsed) monitoring source. Note that the voltage signals from photomultipliers are negative.

The laser fires at $t = 0$ and causes an increase in absorbance in the sample; as a consequence the intensity of light reaching the detector decreases. While laser photolysis systems are normally single-beam spectrometers, in fact they behave as dual-beam instruments. The reference beam is separated from the sample beam in time, rather than space. Thus, the reference signal is acquired before laser excitation and leads to I_0. The absorbance at time t in Figure 18.3 is given by Eq. 1:

$$\left(\begin{array}{c} \text{Absorbance} \\ \text{change} \end{array} \right) = \Delta\text{OD} = -\log\left(\frac{I_0}{I_t}\right) = \log\left(1 - \frac{\text{Signal}}{I_0}\right) \tag{1}$$

Since the reference beam is the same signal prior to laser excitation, the technique measures changes in absorbance, normally abbreviated as ΔOD. The abbreviation is derived from "change in optical density"; however, the preferred IUPAC term is absorbance. It is therefore possible to acquire inverted signals, when the laser pulse causes bleaching of the sample. Naturally, bleaching can only be observed where the starting materials can absorb the monitoring beam; it is easy to confuse emission from the sample with bleaching, since both lead to inverted signals. However, bleaching requires the monitoring beam on, while fluorescence (or scattered light, phosphorescence, etc.) can also be observed with the monitoring beam off.

2.3. Transient Spectroscopy

The output from a nLFP is normally a plot of ΔOD versus time at a given wavelength, following processing of the data as indicated in Section 2.2. The signal intensity obtainable, and indeed desirable, depends on many parameters. Some of the key parameters are outlined below.

- The signal intensity depends on the energy-per-pulse delivered by the laser, however, more is not always better. A range of 5–25 mJ/pulse delivered to the sample is convenient frequently achieving reasonable signals without the problems that can be encountered at very high laser doses (see below).
- The signal is proportional to the product of the quantum yield of transient formation times the extinction coefficient difference between the transient and its precursor.
- Many reactive intermediates can decay via self-reactions, giving dimers or disproportionation products, as is the case of free radicals and carbenes. When these self-reactions are not the ones under study, it is desirable to keep the transient concentration low enough to minimize this type of interference. For example, for a radical that dimerizes with $k_t = 3 \times 10^9\ M^{-1}\,s^{-1}$ and a concentration "c" of $10^{-5}M$, its first half-life ($t_{1/2} = 1/kc$) would be 33 μs. Note that excited triplet states also undergo bimolecular decay by triplet–triplet

annihilation (TTA). Rate constants for this process usually approach or exceed $10^{10}\ M^{-1}\,s^{-1}$.

- The absorbance, ΔOD, is independent of the value of I_0, the monitoring light intensity. However, the signal/noise ratio is strongly dependant on the I_0 value. Very high I_0 values can lead to distorted signals and nonlinear response. The I_0 value that a given nLFP can tolerate depends on the photomultiplier used and on the wiring of the dynod chain. Typically, ~2 mA are acceptable for short periods of time; with a 93 Ω terminator this corresponds to ~200 mV.[10]

2.4. Kinetic Studies

The simplest situation is that in which the reactive intermediate of interest can be directly observed and decays with clean first-order kinetics, with the signal after decay returning to the preexcitation level. There are many cases in which this situation is encountered, such as the case of many short-lived excited triplets. The decay follows simple first order kinetics, that is,

$$\ln\left(\frac{[R]_0}{[R]_t}\right) = \ln\frac{(\Delta OD)_0}{(\Delta OD)_t} = -kt \tag{2}$$

where the subscript 0 indicates initial conditions, and t after time t. Note that it makes no difference whether the transient concentration $[R]$ or the absorbance is used for the calculation–plot.

The lifetime of the intermediate is given by $\tau = 1/k$ and the half-life by Eq. 3,

$$t_{1/2} = 0.69\tau = \frac{0.69}{k} = \frac{\ln 2}{k} \tag{3}$$

where 0.69 is ln 2. The k obtained from this analysis is seldom the reflection of a single elementary process. Rather, it is more frequently than not, a composite of more than one parallel mode of decay. We will usually refer to it as the experimental rate constant or k_{exp}.

It is common for many second-order reactions to exhibit pseudo-first-order behavior under conditions of nLFP. This is due to the fact that, while reactive intermediates are present in micromolar concentrations, (typically 10–50 μM), the molecules with which they react are present in concentrations several orders of magnitude larger. As a result, the concentration of these reagents remains essentially constant during the decay of the transient species. An example is shown in Figure 18.4, where triplet benzophenone is quenched by melatonin.[12]

The decay of the benzophenone triplet follows pseudo-first-order kinetics, and the experimental rate constant is related to the parameters of interest by Eq. 4,

$$k_{exp} = k_0 + k_X[X] \tag{4}$$

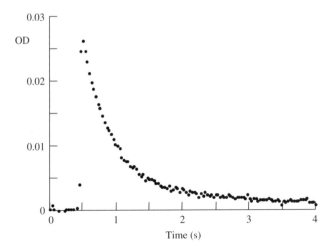

Figure 18.4. Quenching of benzophenone triplet by 0.24-mM melatonin in acetonitrile, monitored at 600 nm following 355-nm laser excitation. The spectrum was recorded at room temperature, under a nitrogen atmosphere.[12]

where k_0 is the rate constant for triplet decay in the absence of quencher ($k_0 = \tau_0^{-1}$) and k_X is the rate constant for triplet quenching, in this example by melatonin. By plotting the values of k_{exp} for different concentrations of quencher, as shown in Figure 18.5, it is possible to separate k_0 (interpret) from k_X (slope).

The trace in Figure 18.5 returns approximately to the original, preexcitation baseline. This trace was recorded at 600 nm (away from the 525-nm triplet

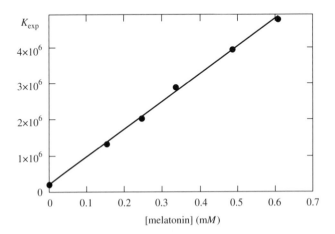

Figure 18.5. Quenching of benzophenone triplet by 0.24-mM melatonin in acetonitrile, monitored at 600 nm following 355-nm laser excitation. Each point is derived from a trace such as that shown in Figure 18.4. The spectrum was recorded at room temperature, under a nitrogen atmosphere.[12]

maximum) in order to minimize interference by the benzophenone ketyl radical, that has an absorption maximum ~545 nm. It is not uncommon to observe traces that do not return to the same baseline, but rather to a higher ΔOD level, showing that the products of reaction have absorbance at the monitoring wavelength.

A case in point was described in detail by Encinas et al. in 1980.[13] This example of a biradical is derived from γ-methylvalerophenone, which can be scavenged by octanethiol according to the mechanism of Scheme 18.1. In this case, the triplet state is very short lived (≤ 2 ns) and for practical purposes laser excitation leads to the biradical (BR) as the first detectable intermediate.

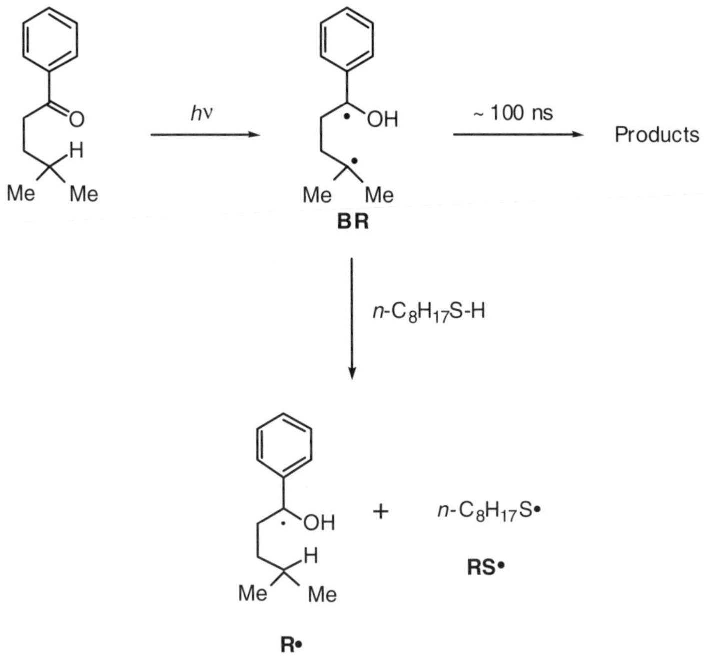

Scheme 18.1

The absorbance from the biradical is known to be exclusively due to the ketyl radical center,[14] and therefore trapping by the thiol to produce R$^\bullet$ not only preserves the chromophore, but in fact it makes it longer lived. Since the biradical lives ≤ 100 ns, and the ketyl radical (R$^\bullet$) many microseconds, the latter can be regarded as a stable product in the time scale of the experiment; that is, the residual absorbance following biradical decay appears flat, as shown in Figure 18.6. In this example we have a biradical decaying completely with pseudo-first-order kinetics. The radical R$^\bullet$ forms concurrently with biradical decay, and since both processes are directly linked, the decay (BR) and growth (R$^\bullet$) follow the same kinetics. Figure 18.6 shows the calculated traces for BR and R$^\bullet$, and the sum of both signals, corresponding to the experimental observable.

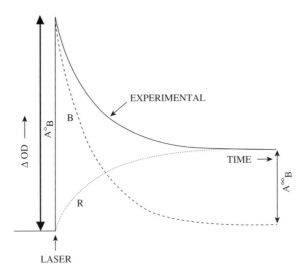

Figure 18.6. Calculated trace (solid line) resulting from the addition of transient absorbances due to biradical (dashed line) and free radical (dotted line).[13] [Reproduced with permission from M. V. Encinas, P. J. Wagner, and J. C. Scaiano, *J. Am. Chem. Soc.* **1980**, *102*, 1357. Copyright © 1980 American Chemical Society.]

Thus, Eq. 5 corresponds to the decay of the signal, while Eq. 6 is the normal expression for a first-order growth.

$$\ln \frac{(\Delta OD)_0^{BR}}{(\Delta OD)_t^{BR}} = -k_{\text{decay}}t \tag{5}$$

$$\ln \left(\frac{(\Delta OD)_\infty^{R}}{(\Delta OD)_\infty^{R} - (\Delta OD)_t^{R}} \right) = k_{\text{growth}}t \tag{6}$$

The superscripts in Eqs. 5 and 6 indicate the species, radical (R) or biradical (BR), while the subscripts indicate the time. The ∞ sign indicates the plateau region following the decay (experimental signal) or growth (R).

Combining Eqs. 5 and 6 leads to Eq. 7.

$$\ln \left(\frac{(\Delta OD)_t - (\Delta OD)_\infty^{R}}{(\Delta OD)_0^{BR} - (\Delta OD)_\infty^{R}} \right) = -k_{\text{exp}}t \tag{7}$$

The fact that the absorbance in the flat, postdecay region, is subtracted from both the numerator and denominator implies that in order to fit to correct value of k_{exp} all that is required is to shift the baseline to the "infinite" level, rather than the preexcitation level.

While it is rather intuitive that the decay lifetime of a species is shortened by any quencher or scavenger, this simple concept is less than intuitive in the case of growing signals. Growth signals are observed whenever the product of reaction (even a transient product) is more strongly absorbing than its precursor. A typical case is the reaction of alkoxy radicals with hydrogen donors.[4,15] For example, *tert*-butoxyl can be readily generated by photodecomposition of the peroxide; the radical is essentially transparent in the spectral region of interest. Scheme 18.2 shows a representative example, the reaction with melatonin.[12]

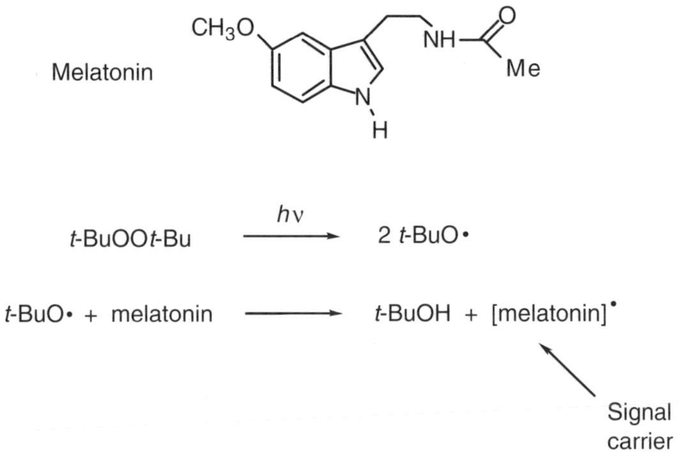

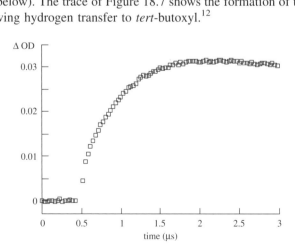

Scheme 18.2

Numerous examples of hydrogen abstraction by *tert*-butoxyl radicals have been reported.[16] The unusual choice of melatonin is not accidental; note that the radical has simply been identified as (melatonin)• and seems to emphasize that the site of reaction (in fact, rather well established in this case) may not necessarily be derived from the laser experiment. The laser technique observes all sites and forms of reaction regardless of which species is monitored and how well characterized it may be (see below). The trace of Figure 18.7 shows the formation of the melatonin radical following hydrogen transfer to *tert*-butoxyl.[12]

Figure 18.7. Kinetic trace recorded at 490 nm following 337-nm laser excitation of a sample containing 52-m*M* melatonin in 2:3 di-*tert*-butyl peroxide: acetonitrile.[12]

A first-order rate constant for the growth can be derived from the same analysis of Eq. 5; the values of k_{growth} are then related to the rate constants of interest according to Eq. 9.

$$k_{growth} = k_0 + k_X[XH] \tag{8}$$

Thus, a plot of k_{growth} versus [X] yields k_0 from the intercept and k_x from the slope.

2.5. The Probe Technique

The problem with the methodologies discussed above is that they require that either the reaction intermediate under study, or the product of its reaction with the substrate, give a signal in the spectral region accessible. Could one employ nLFP to determine rate constants where all the reagents and *all* the products are invisible to the detection technique employed? The answer to this question is affirmative, and the technique that we introduced in the late 1970s is now known as "the probe method." Interestingly, when we first used this technique we did not realize that it was a new method and no claim was made of its originality; it was simply a straightforward outcome of the kinetic expressions involved.[4,15,17]

There are numerous reactions of interest that fit the description mentioned above. For example, the reaction of *tert*-butoxyl with aliphatic alcohols is essentially invisible or "silent" in nLFP.[15] A practical example would be the reaciton of t-BuO$^\bullet$ with 2-propanol, in which the reaction with benzhydrol (diphenyl methanol) could be used as a probe. The mechanism is shown in Scheme 18.3 for this example.

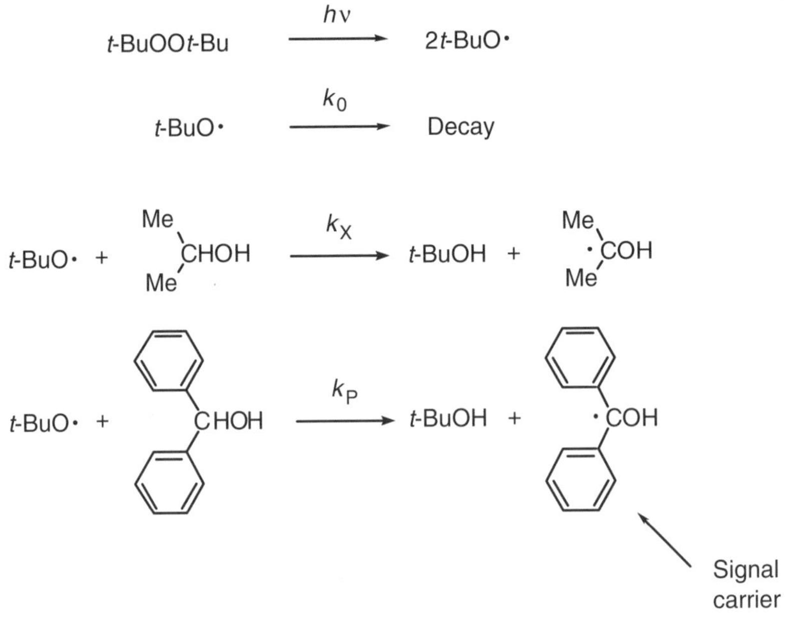

Scheme 18.3

The reaction can be generalized, as illustrated in Scheme 18.4, for the case of *tert*-butoxyl radicals.

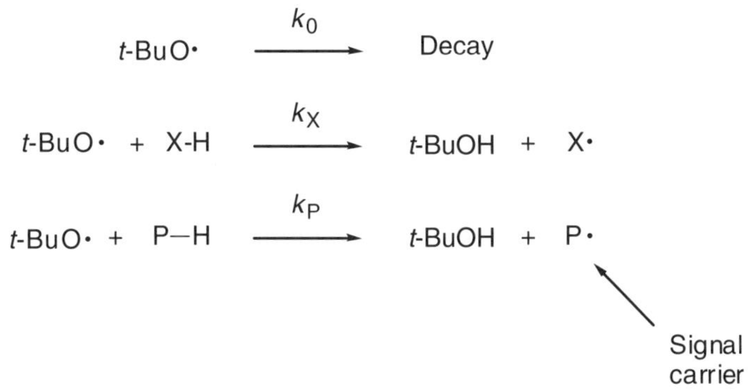

$$t\text{-BuO}\cdot \xrightarrow{\ k_0\ } \text{Decay}$$

$$t\text{-BuO}\cdot \ + \ \text{X-H} \xrightarrow{\ k_X\ } t\text{-BuOH} \ + \ \text{X}\cdot$$

$$t\text{-BuO}\cdot \ + \ \text{P-H} \xrightarrow{\ k_P\ } t\text{-BuOH} \ + \ \text{P}\cdot$$

Signal
carrier

Scheme 18.4

Thus, according to Scheme 18.4 the rates of consumption of alkoxyl (RO•) and of formation of the probe radical (P•) will be given by Eqs. 9 and 10.

$$-\frac{d[\text{RO}\bullet]}{dt} = (k_0 + k_P[\text{PH}] + k_X[\text{XH}])[\text{RO}\bullet] \tag{9}$$

$$\frac{d[\text{P}\bullet]}{dt} = k_P[\text{RO}\bullet][\text{PH}] \tag{10}$$

It is important to understand the relationship between $d[\text{P}\bullet]$ and $d[\text{RO}\bullet]$, given by Eq. 11.

$$d[\text{P}\bullet] = -d[\text{RO}\bullet]\left[\frac{k_P[\text{PH}]}{k_0 + k_P[\text{PH}] + k_X[\text{XH}]}\right] \tag{11}$$

The term in square brackets in Eq. 11 determines which fraction of the alkoxyl radicals will actually react with the probe PH. Thus, the increase in [P•] equals the decrease in [RO•], corrected by the fraction of reaction with PH, relative to all other processes in which the alkoxyl radicals participate.

When the reaction is complete the concentration of P• in the plateau region, $[\text{P}\bullet]_\infty$, will be given by Eq. 12:

$$[\text{P}\bullet]_\infty = [\text{RO}\bullet]_0\left[\frac{k_P[\text{PH}]}{k_0 + k_P[\text{PH}] + k_X[\text{XH}]}\right] \tag{12}$$

and thus, at any intermediate time, the concentration of alkoxyl will be given by

$$[RO\cdot]_t = ([P\cdot]_\infty - [P\cdot]_t)\left[\frac{k_0 + k_P[PH] + k_X[XH]}{k_P[PH]}\right] \tag{13}$$

In other words, the correction factor between formation of $P\cdot$ and consumption of $RO\cdot$ is always the reactivity ratio in the square brackets in Eqs. 11–13. This allows substitution of the actual alkoxyl concentration, Eq. 13, into Eq. 10, leading to Eq. 14.

$$\frac{d[P\cdot]}{dt} = k_P[PH]([P\cdot]_\infty - [P\cdot]_t)\left[\frac{k_0 + k_P[PH] + k_X[XH]}{k_P[PH]}\right] \tag{14}$$

$$\frac{d[P\cdot]}{[P\cdot]_\infty - [P\cdot]_t} = (k_0 + k_P[PH] + k_X[XH])\,dt \tag{15}$$

Integration of Eq. 15 is straightforward and yields Eq. 16.

$$\ln\left(\frac{[P\cdot]_\infty}{[P\cdot]_\infty - [P\cdot]_t}\right) = (k_0 + k_P[PH] + k_X[XH])t \tag{16}$$

or

$$\ln\left(\frac{[P\cdot]_\infty}{[P\cdot]_\infty - [P\cdot]_t}\right) = k_{growth}\cdot t \tag{17}$$

where

$$k_{growth} = k_0 + k_P[PH] + k_X[XH] \tag{18}$$

Analysis of Eqs. 17 and 18 tells us that if we monitor the formation of signal derived from PH, and determine the pseudo-first-order rate constant for the growth of signal due to $P\cdot$, (the probe-derived species) the value of k_{growth} is determined not only by k_P, but also by k_0 and by k_X. Thus, a plot of k_{growth} against [XH] at constant concentration of the probe, PH, will yield k_X from the slope, in spite of the fact that all reagents *and* all products in the reaction between $RO\cdot$ and XH do not yield a detectable signal.

A classic example of the use of the probe method is shown in Figures 18.8 and 18.9, reproduced from the literature.[15]

In the examples of Figures 18.8 and 18.9 the probe molecule is diphenyl-methanol, and it reacts with *tert*-butoxyl radicals as shown in Scheme 18.3. Usual probe concentrations were between 50 and 200 mM. Figure 18.8 shows a representative trace for the formation of the ketyl radical from diphenylmethanol (i.e., the same formed by photoreduction of benzophenone), the only detectable species in the system. Figure 18.8 shows how the value of k_{growth}, given by the slope of the plots, changes with substrate (1,7-octadiene) concentration, as predicted by Eq. 18.

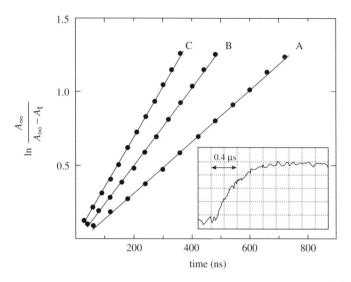

Figure 18.8. Evaluation of k_{growth} for 1,7-octadiene. (A) 0.067 M; (B) 0.54 M, and (C) 0.94 M. Concentration of diphenylmethanol (used as a probe) 0.133 M. Only 12 representative points out of 50 used for the calculations have been plotted in each case. *Insert:* Oscilloscope trace for 1,7-octadiene 0.94 M.[15] [Reproduced with permission from H. Paul, R. D. Small, Jr., and J. C. Scaiano, *J. Am. Chem. Soc.* **1978**, *100*, 4520. Copyright © 1978 American Chemical Society.]

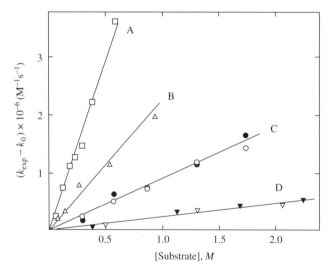

Figure 18.9. Evaluation of k_{XH} using equation 18 for cyclohexane (A), 1,7-octadiene (B), cumene (C), and toluene (D).[15] [Reproduced with permission from H. Paul, R. D. Small, Jr., and J. C. Scaiano, *J. Am. Chem. Soc.* **1978**, *100*, 4520. Copyright © 1978 American Chemical Society.]

The value of $k_0 + k_P[PH]$ can be obtained independently by a determination of k_{growth} in the absence of the substrate XH. Thus, it is possible to determine k_X using different probe concentrations if each trace is corrected by the value of k_{growth} in the absence of XH, labeled k_{growth}^0.

$$k_{\text{growth}}^0 = k_0 + k_P[PH] \tag{19}$$

$$k_{\text{growth}} - k_{\text{growth}}^0 = k_X[XH] \tag{20}$$

Figure 18.9 illustrates several plots according to Eq. 20, including one for 1,7-octadiene, the substrate of Figure 18.8. Note that while labeled differently, the ordinate in Figure 18.9 corresponds to the left term in Eq. 20.

We may be left to wonder how the rate of formation for the ketyl radical from benzophenone could get faster by adding 1,7-octadiene, or any other "invisible" substrate. The answer is simple, the "rate" is not faster, but rather, the rate constant for the signal growth is larger. In fact, the actual initial rate (see Eq. 10) is the same, as long as the same concentration of RO$^\bullet$ is generated initially in the presence or absence of XH. At any other time the rate will be lower in the presence of XH than in its absence, since XH will cause a decrease in RO$^\bullet$ concentration at all times except at zero time.

The signal in the plateau region (see Fig. 18.7 and the inset in Fig. 18.8) will obey the proportionality of Eq. 21.

$$\text{Signal due to P}^\bullet \propto \frac{k_P[PH]}{k_0 + k_P[PH] + k_X[XH]} \tag{21}$$

Thus, the plateau signal will be smaller in the presence of XH, as the denominator in Eq. 21 will increase as XH is added. Thus, we can easily see that if the initial rate for P$^\bullet$ is the same, but the infinite level is lower, the observable lifetime will be shorter; that is, the rate constant observed will be larger.

Interestingly, Eq. 21 can be converted into a Stern–Volmer type of expression, as shown in Eq. 22,

$$\frac{[P^\bullet]_\infty^0}{[P^\bullet]_\infty} = 1 + \frac{k_X[XH]}{k_0 + k_P[PH]} \tag{22}$$

where the superscript "o" indicates the value in the absence of XH.

Clearly there must be a price to pay for determining rate constants by a method where all the participants (reagents and products) are invisible to the technique. The reaction of Figure 18.8 provides an excellent example. In principle, 1,7-octadiene can react with *tert*-butoxyl radicals by all the mechanisms of Scheme 18.5. The probe method cannot distinguish them, the rate constant obtained $(2.3\times 10^6\, M^{-1}\, s^{-1})$ includes all possible modes and sites of reaction.[15] It is simply the rate constant with which 1,7-octadiene removes *tert*-butoxyl from the system. In this case the first reaction in Scheme 18.5 accounts for the reactivity of *tert*-butoxyl, perhaps with a minor contribution from the second reaction. In all cases, the site and form of reaction must be known independently. Chemical intuition, product studies, and other methods such as EPR spectroscopy assist in assigning the rate constants to specific reaction paths.

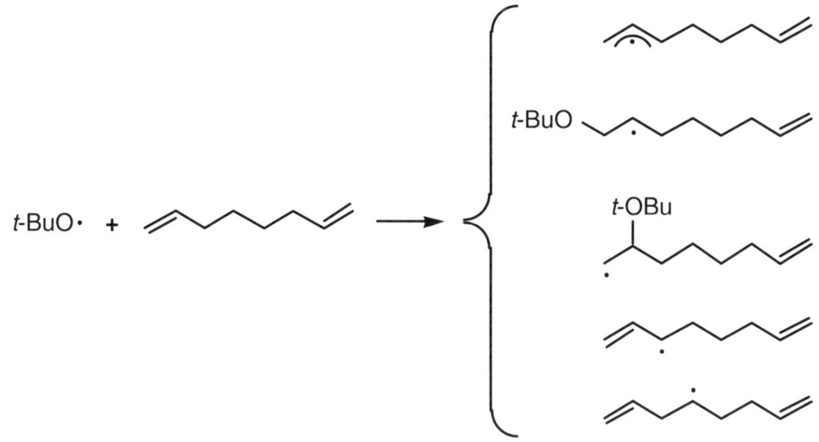

Scheme 18.5

Note that detecting a well-characterized intermediate as the product of a given reaction does not prove that the rate constant determined corresponds *exclusively* to that reaction. Let us take the hypothetical reaction of Scheme 18.6, to give radicals A• and B•. While benzylic radical B• will be readily detectable, we expect A• to be "silent." However, even if the rate constant is determined by monitoring only B•, the value obtained will correspond to both reaction paths, that is, $k_A + k_B$.

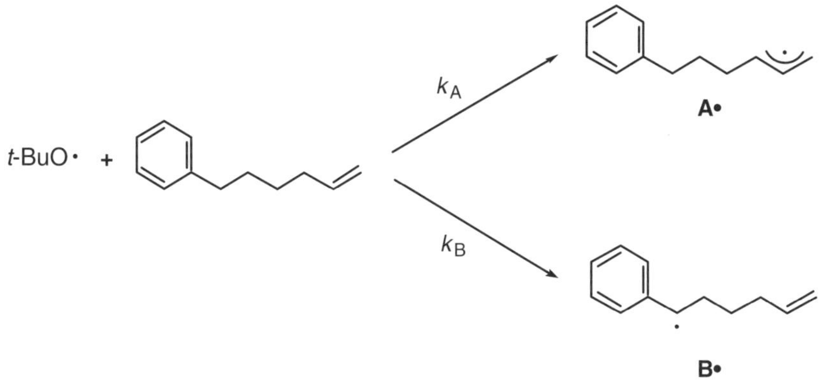

Scheme 18.6

The underlying reason for all these observations is that *the growth of any product of reaction always reflects the lifetime of the precursor species, tert*-butoxyl radical in our examples. Interestingly, this is the same concept that applies when one measures properties, such as fluorescence; that is, the observable fluorescence lifetime reflects the lifetime of the singlet state and not its radiative lifetime.

There are two limitations to the probe technique: first the lack of information on the mechanism or site of reaction just discussed. The second limitation is evident from Eq. 21: as XH is added the process occurs faster and the signal gets smaller. Under some conditions the signal-to-noise may deteriorate significantly at high XH concentrations.

The probe technique has been used for many other radicals, atomic species, carbenes, and carbocations; some of these are discussed in other chapters in this volume.

2.6. Second-Order Processes

Many of the reactions discussed in the preceding pages are in fact bimolecular processes, which would normally follow second-order kinetics. However, as already discussed, under the regime of LFP they behave as pseudo-first-order reactions. The corresponding rate constants and lifetimes are independent of the initial concentration of transient, and therefore knowledge of extinction coefficients and quantum yields is not needed. Further, it is not important to have a homogenous transient concentration.

In this section, we will deal with processes that are second order in transient concentration. A few examples are illustrated in Scheme 18.7.

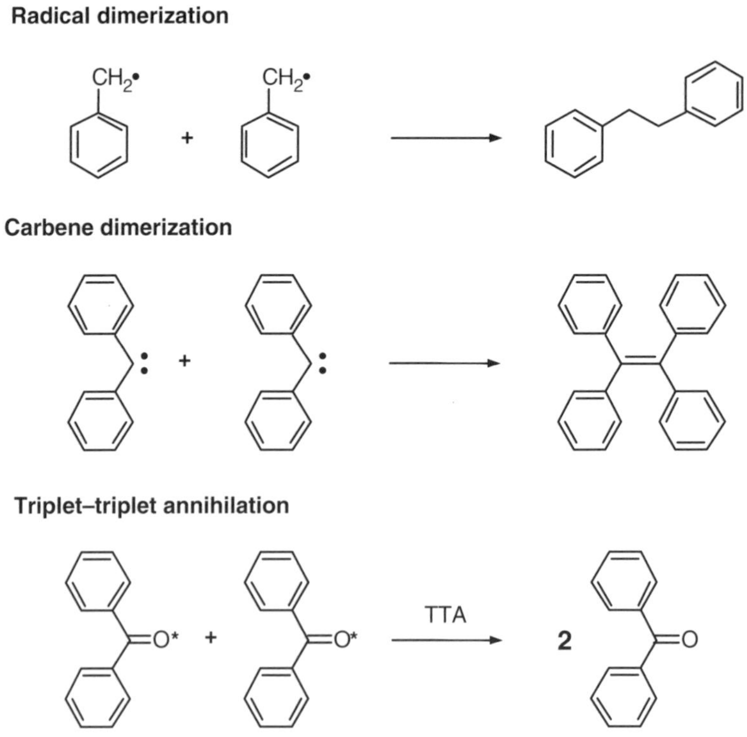

Radical dimerization

Carbene dimerization

Triplet–triplet annihilation

Scheme 18.7

In the absence of other reactions or reversibility, all the processes of Scheme 18.7 can be represented by Eq. 23 and follow the kinetic law of Eq. 24.

$$R\cdot + R\cdot \xrightarrow{k_t} Products \tag{23}$$

$$-\frac{d[R\cdot]}{dt} = 2k_t[R\cdot]^2 \tag{24}$$

where the factor of 2 reflects the stoichiometry of the reaction. Integration of this expression leads to Eq. 25.

$$\left[\frac{1}{[R\cdot]_0} - \frac{1}{[R\cdot]_t}\right] = -2k_t t \tag{25}$$

Therefore, a plot of $[R\cdot]_\tau^{-1}$ versus time will yield $2k_\tau$ from the slope. Simple enough, except that the nLFP data correspond to ΔOD, rather than $[R\cdot]$. Of course, where only the transient absorbs, $[R\cdot]$ and ΔOD are related by Beer's law, that is,

$$\Delta OD = \varepsilon\ell[R\cdot] \tag{26}$$

where ε is the molar extinction coefficient for the transient, and ℓ the optical path. The latter refers to the optical path for the monitoring light beam; this is straightforward only if the laser pulse covers the complete monitoring beam, with no gradients along the monitoring beam path, or in the penetration depth of the laser, something that requires low sample absorbance.

The extinction coefficient for the reaction intermediate can be determined (see below), but it is not uncommon for the error to be 10–20%.

Substitution of Eq. 26 into Eq. 25 leads to

$$\frac{1}{\Delta OD_t} = \frac{1}{\Delta OD_0} + \frac{2k_t}{\varepsilon\ell}t \tag{27}$$

Thus, a plot of $(\Delta OD_\tau)^{-1}$ against time yields $2k_\tau/\varepsilon\ell$ from the slope. It is not unusual for kinetic data from nLFP work to be reported in this manner, but it is essential to specify ℓ and the wavelength, since $2k_\tau/\varepsilon\ell$ will include the dependence of ε with the wavelength.

2.7. Quantum Yields and Extinction Coefficients

The signals obtained in LFPs experiments are proportional to the product of the extinction coefficient (ε) and the quantum yield (Φ) of transient generation. As a result, these two parameters are intimately linked and all methods for obtaining either one are based on protocols that allow the evaluation of the other one.

While beyond the coverage of this chapter, it is worth mentioning that in the technique of photoacoustic spectroscopy,[18,19] signals are related to Φ, but not ε. It therefore represents one of the possible tools to separate the two parameters.

Extinction coefficients are usually determined by competitive techniques, employing a well-documented transient as a reference. Many extinction coefficients can be traced to the ketyl radical from benzophenone, for which the data are available from pulse radiolysis studies. This value is in turn referred to the yields of the solvated electron, $HO^\bullet$ and $H^\bullet$ produced in the radiolysis of water. The value for $Ph_2C^\bullet OH$ is 3220 $M^{-1}cm^{-1}$ in water at 537 nm, at room temperature.[20]

As an example, let us assume we would like to determine the extinction coefficient for the naphthalene triplet at 425 nm in a polar solvent (e.g., acetonitrile) using as a reference the triplet state of benzophenone, with a reported extinction coefficient of 7220 $M^{-1}cm^{-1}$ at 530 nm.[21] We will select a laser that can excite both benzophenone and naphthalene, such as an excimer laser operating at 308 nm (see Table 18.1). We will then prepare two samples with matched absorbance at 308 nm, probably selecting an absorbance value <0.30 in the sample cuvette. Now we will perform sequential experiments with the two samples, ensuring that the laser dose is constant; recording the signal amplitude (ΔOD) at 425 nm for the naphthalene sample, A_N, and at 530 nm for the benzophenone sample, A_B.

Since both samples receive and absorb the same photon dose, the respective signals will be given by

$$A_N = a \times I_a \times \Phi_N^T \times \varepsilon_N^{425} \tag{28}$$

$$A_B = a \times I_a \times \Phi_B^T \times \varepsilon_B^{530} \tag{29}$$

where a is a proportionality constant, the same for both samples, I_a the laser dose absorbed, Φ_B^T, and Φ_N^T the intersystem crossing yields for benzophenone and naphthalene, respectively, and ε_B^{530} and ε_N^{425} the corresponding extinction coefficients at the monitoring wavelengths. In this particular case, we know that intersystem crossing for benzophenone is generally accepted to be one, that is, $\Phi_B^T = 1.0$, thus:

$$\frac{A_N}{\Phi_N^T \times \varepsilon_N^{425}} = \frac{A_B}{\Phi_B^T \times \varepsilon_B^{530}} \tag{30}$$

$$\Phi_N^T \times \varepsilon_N^{425} = \Phi_B^T \times \varepsilon_B^{530} \frac{A_N}{A_B} \tag{31}$$

Equation 31 yields the product $\Phi_N^T \varepsilon_N^{425}$, but the two values cannot be separated in this experiment. A further refinement of the method would determine both A_B and A_N as a function of laser dose and substitute the ratio A_N/A_B by the ratio of slopes in Eq. 31. This refinement seems to improve the data and to ensure that there are no nonlinear effects, or to correct for them if this is the case.

An alternate method can be used to obtain ε_N^{425} separate from the intersystem crossing quantum yield. In this case, we will also use benzophenone (BP) and naphthalene (N) as our example. Now one of our samples will contain both compounds, and we will choose the concentration of naphthalene so that it will quench at least 98% of the benzophenone triplets. This will require knowledge of the lifetime of the benzophenone triplet ($^3BP^*$) lifetime, τ_T, that will be readily available from the control experiment, which is the same as the benzophenone sample in the previous example (see below). We will also need the value of the triplet quenching rate constant, k_q, although in this particular case an estimate of $\sim 10^{10}\ M^{-1}s^{-1}$ is a good approximation in fluid solvents. Scheme 18.8 shows the process involved.

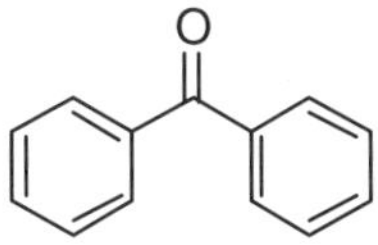

BP 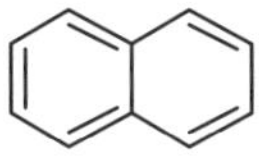   **N**

$$BP \xrightarrow{h\nu} {}^1BP^* \xrightarrow{ISC} {}^3BP^*$$

$$^3BP^* \xrightarrow{\tau_T} {}^3BP$$

$$^3BP^* + N \xrightarrow{k_q} BP + {}^3N^*$$

Scheme 18.8

Thus, the fraction of benzophenone triplets quenched by naphthalene (F_q) is given by

$$F_q = \frac{k_q[N]}{\tau_T^{-1} + k_q[N]} \tag{32}$$

If we set $F_q > 0.98$, and we assume $\tau_T = 4\ \mu s$ then the minimum concentration of naphthalene is 1.23 mM.

Next, we need to select a laser wavelength where benzophenone absorbs, but naphthalene does not. Among those shown in Table 18.1, 337, 347, and 355 nm are good options. Our control sample will be identical to the actual sample, except that it will not contain naphthalene. We will then measure A_B in the control and A_N^{et} in the sample; in both cases the measurement needs to be done before significant decay occurs.

$$A_B = a \times I_a \times \Phi_B^T \times \varepsilon_B^{530} \tag{29}$$

$$A_N^{et} = a \times I_a \times \Phi_B^T \times F_q \times \varepsilon_N^{425} \tag{33}$$

where the superscript "et" indicates that the naphthalene triplet was formed by energy transfer.

Note the important difference between Eq. 33 and 28. The term Φ_N^T has been replaced by $\Phi_B^T F_q$, since in the case of energy-transfer intersystem crossing in naphthalene is not involved (see Scheme 18.8); thus, the yield of triplet formation is determined by the properties of benzophenone, not naphthalene; thus, the extinction coefficient is given by Eq. 34.

$$\varepsilon_N^{425} = \frac{\varepsilon_B^{530}}{F_q} \times \frac{A_N^{et}}{A_B} \qquad (34)$$

Note that choosing concentrations of quencher leading to high quenching yields (>98% in our example) minimizes errors introduced by F_q, that for practical purposes can be assumed to be one.

Naturally, both methodologies can be expanded beyond the example discussed, and apply to other reaction intermediates in addition to excited states. The key problems are usually obtaining a good reference substrate, and separating the quantum yield from the extinction coefficient.

3. EXPANDING THE CAPABILITIES OF NANOSECOND LASER FLASH PHOTOLYSIS

There are at least three directions in which the capabilities of nLFP can be expanded; all have been developed and are useful tools for the physical organic chemist.

- Expansion of the time scale to monitor phenomena in other time scales. Picosecond and femtosecond techniques are well established. Use of nLFP for very long time scales (10 ms to s) requires very stable light sources, and frequently, long optical path cells.
- Application of nLFP techniques in the IR region has been available for well over a decade. In one approach, laser diodes are used to generate the monitoring beam and a fast IR detector employed. Alternatively, a step-scan spectrometer uses the same methodologies employed by normal Fourier transform infrared (FTIR) spectrometers, but spectral capture is much faster.
- nLFP can be employed with opaque (but reflective) samples by using time-resolved diffuse reflectance as the detection method. The method developed by Wilkinson and Kelly[22] has found applications with powders (silica, zeolites, etc.), self-supporting materials (paper, cloth), and even scattering suspensions.

While the methods outlined above are not part of our coverage, awareness of their basic capabilities is important.

4.   DOs AND DON'Ts OF nLFP

This section deals with a few aspects where awareness of potential problems is important. The list below is in no way exhaustive.

- *Two-photon processes* caused by absorption of photons by reaction inter-mediates and excited states are common under condition of high-power laser excitation. The consequence of two-photon excitation can include the forma-tion of new reaction intermediates (electron photoejection is common) and the partial depletion of intermediates formed in monophotonic processes.[23] To minimize this problem, do not use higher laser power then required to obtain a good signal/noise ratio, and do not focus the laser too tightly. There are in fact techniques used to obtain a more diffuse and homogenous laser beam (see below).

- *Shock waves* or acoustic waves lead to repetitive but attenuated signals every few microseconds. Their frequency is determined by the size of the sample cuvette, and the speed of sound in the medium. A detectable acoustic wave is generated when a significant fraction of the laser pulse energy is absorbed in a small volume. This can be caused by a variety of problems, including tight focusing of the laser beam, high absorption by the sample at the laser wavelength, high absorption by reaction intermediates at the laser wavelength, and absorbing defects (spots) on the surface of the sample cuvette.

- *The kinetics of second-order processes* (see above) present several problems, since kinetic values depend directly on the absolute concentration of transient. Two ways of making transient concentration fairly homogeneous are to work in dilute solutions (low absorbance) and to place a scattering plate in from of the sample cuvette. A scattering (or diffusing) plate is easily made by sandblasting a fused silica plate. Of course, these two suggestions (dilute solutions and light scattering) lead to a significant decrease in signal intensity.

- *Complex decay kinetics analysis*, common and acceptable in fluorescence spectroscopy, presents some difficulties in the case of transient absorption experiments. There are two main reasons for this. (1) The signal/noise in nLFP is such that rarely can a decay be monitored for over two decades in signal intensity, and (2) in the case of fluorescence it is generally a good assumption that after the process is complete the signal (emission) will be zero; such is not the case with absorption, where the signal (absorption) after completion of a process may be higher, lower or the same as before laser excitation. In general, a simplification of the chemistry in order to "clean-up" the kinetics so as to achieve simpler decay modes should be preferred over complex kinetic analysis.

- *Flow systems* are frequently required when samples are exposed to many laser pulses, when the substrate concentrations are very low (certainly $<10^{-4}\,M$), or when the products of photolysis are efficient quenchers or strongly absorbing.

5. CONCLUSION AND OUTLOOK

After 35 years, nLFPs has become a standard tool in photochemistry and physical organic chemistry. The choice of the right tools in terms of laser and detection systems, and a proper understanding to the kinetic methodologies available make it useful whether or not the intermediates of interest have a useful absorption in the spectral region accessible.

6. ACKNOWLEDGMENT

Thanks are due to the Natural Sciences and Engineering Research Council of Canada for generous support.

SUGGESTED READING

The reader is specifically referred to the following references: In addition, we recommend:

J. C. Scaiano, Ed., *Handbook of Photochemistry*, Vols. 1 and 2, CRC Press, Boca Raton, FL, **1989**.

G. Bucher, J. C. Scaiano, and M. S. Platz, "Kinetics of Carbene Reactions in Solution," in *Radical Reaction Rates in Liquids*, H., Fischer, Ed., Landolt-Börnstein, New Series II/18 Subvolume E2, Springer-Verlag, Berlin, **1998**, Chapter 14, pp. 141–349.

REFERENCES

1. Nobel Prize presentation speech, 1967.

2. L. Lindqvist, *Hebd. Seances Acad. Sci., Ser. C* **1966**, *263*, 852.

3. G. Porter, "Flash Photolysis and Primary Processes in the Excited State," in *Fast Reactions and Primary Processes in Chemical Kinetics*, Proc. 5th Nobel Symposium, S. Claesson, Ed., Almqvist and Wiksell, Stockholm, **1967**, pp. 141–161.

4. R. D. Small, Jr., and J. C. Scaiano, *J. Am. Chem. Soc.* **1978**, *100*, 296.

5. M. S. Platz and V. M. Maloney, "Laser Flash Photolysis Studies of Triplet Carbenes," in *Kinetics and Spectroscopy of Carbenes and Biradicals*, M. S. Platz, Ed., Plenum Press, New York, **1990**, p. 239.

6. R. A. Moss and N. J. Turro, "Laser Flash Photolysis Studies of Arylhalocarbenes," in *Kinetics and Spectroscopy of Carbenes and Biradicals*, M. S. Platz, Ed., Plenum Press, New York, **1990**, p. 213.

7. R. A. McClelland, C. Chan, F. L. Cozens, A. Modro, and S. Steenken, *Angew. Chem. Int. Ed. Engl.* **1991**, *30*, 1337.

8. G. Cosa, L. Llauger, J. C. Scaiano, and M. A. Miranda, *Org. Lett.* **2002**, *4*, 3083.

9. J. A. Schmidt and E. F. Hilinski, *Rev. Sci. Instrum.* **1989**, *60*, 2902.

10. J. C. Scaiano, *J. Am. Chem. Soc.* **1980**, *102*, 7747.

11. Luzchem Research, Inc. has developed a miniaturized laser system that takes advantage of these ceramic lamps: http : //www.luzchem.com.

12. J. C. Scaiano, *J. Pineal Res.* **1995**, *19*, 189.

13. M. V. Encinas, P. J. Wagner, and J. C. Scaiano, *J. Am. Chem. Soc.* **1980**, *102*, 1357.

14. R. D. Small, Jr., and J. C. Scaiano, *Chem. Phys. Lett.* **1977**, *50*.

15. H. Paul, R. D. Small, Jr., and J. C. Scaiano, *J. Am. Chem. Soc.* **1978**, *100*, 4520.

16. J. A. Howard and J. C. Scaiano, "Oxyl-, Peroxyl-, and Related Radicals," in *Rate and Equilibrium Constants for Reactions of Polyatomic Free Radicals, Biradicals and Radical Ions in Liquids*, H. Fischer, Ed., Landolt-Börnstein, New Series II/13, Berlin, **1984**, chapter 8.

17. R. D. Small, Jr., and J. C. Scaiano, *Chem. Phys. Lett.* **1977**, *48*, 354.

18. D. E. Falvey, *Photochem. Photobiol.* **1997**, *65*, 4.

19. S. E. Braslavsky and G. E. Heibel, *Chem. Rev.* **1992**, *92*, 1381.

20. R. V. Bensasson, E. J. Land, and T. G. Truscott, *Flash Photolysis and Pulse Radiolysis*; Pergamon Press, **1983**.

21. I. Carmichael and G. L. Hug, *J. Phys. Chem. Ref. Data* **1986**, *15*, 1.

22. F. Wilkinson and G. Kelly, "Diffuse Reflectance Flash Photolysis," in *Handbook of Organic Photochemistry*, Vol. I, J. C. Scaiano, Ed., CRC Press, Boca Raton, FL, **1989**, pp. 293f.

23. J. C. Scaiano, L. J. Johnston, W. G. McGimpsey, and D. Weir, *Acc. Chem. Res.* **1988**, *21*, 22.

CHAPTER 19

The Picosecond Realm

EDWIN F. HILINSKI

Department of Chemistry and Biochemistry, The Florida State University, Tallahassee, FL

Reactive Intermediate Chemistry, edited by Robert A. Moss, Matthew S. Platz, and Maitland Jones, Jr.
ISBN 0-471-23324-2 Copyright © 2004 John Wiley & Sons, Inc.

1. OVERVIEW OF PICOSECOND-RESOLVED METHODS

The nanosecond and femtosecond realms are bridged by the picosecond regime not only in time but also in the level of complexity associated with the experiments. One must be prepared to evaluate critically the spectroscopic information that is obtained in the picosecond domain—more so than the nanosecond regime, but less so than the femtosecond time scale. Cautions will be noted and explained at appropriate times.

Think of a reactive intermediate that you want to study. Design a precursor that photochemically can produce your chosen intermediate. However, be careful! Chemical intuition should guide you in your preparation of a precursor. Previous reports of spectra for the reactive intermediate recorded in slower time domains or via matrix isolation should be used for comparison with picosecond spectra when attempting to make a firm assignment of a spectrum to a structure. If it is the first time that a spectrum has been assigned to a structure, use several chemically reasonable precursors in the experiments to learn if they all give the same— within experimental error—spectrum for the intermediate. There can be pitfalls. Unexpected chemical turns in the path from precursor to intermediate can lead to incorrect assignments. Beyond questions about chemistry lie important considerations about the specific spectroscopic technique that is employed in the investigation. While the picosecond realm differs less from the use of traditional spectrometers, such as an ultraviolet–visible (UV–vis) absorption spectrometer, than femtosecond-resolved methods, there are additional factors not considered in the nanosecond regime or with the use of continuous-wave spectrometers that can lead to the misinterpretation of data. It should be emphasized that the fundamental measurements need to be reproducible. The interpretations are subject to change as more information about a particular reactive intermediate is gathered.

In this chapter, we will look at several different picosecond-resolved spectroscopies and their applications to studies of organic reactive intermediates. Our focus will be on the use of time-resolved spectrometers whose first generated laser pulses have picosecond time durations. It is not meant to be an extensive review, but to provide examples of methods that are usable by organic chemists to study intermediates. These spectroscopic methods developed by chemical physicists have evolved to states that are welcome in the laboratories of physical organic chemists. Some of these methods may be more recognizable to organic chemists than others. However, the benefits of less common approaches will be described in sufficient detail to demonstrate the power of the broad range of laser-based systems that are currently in the arsenal of interrogating tools available for developing a greater mechanistic understanding of the roles of reactive intermediates in organic chemical reactions. The descriptions of the several systems herein are not intended to provide all the details, but enough of them to gain an appreciation of what is involved in the application of each spectroscopy covered in this chapter. Some of the picosecond-resolved spectroscopies, such as transient electronic absorption spectroscopy, have been established for quite some time whereas others have been developed more recently. Henceforth, any specific time-resolved spectroscopy that probes the time regime from 1 to 999 ps will be referred to as a picosecond spectroscopy.

1.1. Ultraviolet–Visible Absorption Spectroscopy

One of the most commonly applied types of spectroscopy in the picosecond realm is pump–probe electronic absorption spectroscopy. The absorption spectra of reactive intermediates are usually just as featureless as those of the other two time domains described in this volume. It is simply the inherent nature of these spectra in condensed phases, most typically in solution. Spectroscopic studies in solution most closely mimic reaction conditions that reactive intermediates may find themselves involved in when they are formed and consumed during the course of an organic chemical reaction.

The key considerations of this experiment are the generation of the laser pulse of useful energy, the splitting of the pulse into the pump and the probe pulses, the conversion of the pump pulse into a wavelength suitable for excitation of the precursor to the reactive intermediate, the production of a probe pulse for interrogation of the evolution of the excited precursor, the timing of the arrival of the probe pulse relative to the pump pulse, and the measurement of how the intensity versus wavelength profile of the probe pulse changes as a function of the time that it passes through the sample relative to the excitation pulse (Fig. 19.1). The laser and detection systems may be of several types. Picosecond absorption spectrometers based on a solid-state laser and on a dye laser will be described.

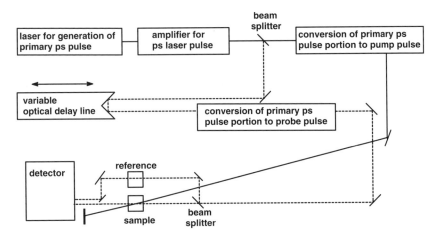

Figure 19.1. Schematic diagram of a general pump–probe-detect laser spectrometer suitable for picosecond electronic absorption, infrared (IR) absorption, Raman, optical calorimetry, and dichroism measurements. For picosecond fluorescence—a pump-detect method, no probe pulse needs to be generated.

1.1.1. Solid-State Laser Systems. The workhorse picosecond spectrometers are solid-state lasers and are based on an actively–passively mode-locked Nd:YAG oscillator, an yttrium aluminum garnet rod doped with neodymium(III) ions. What is described here is a typical spectrometer,[1–3] but note that details may be different depending on the specific system that is under consideration. The Nd:YAG rod is pumped with microsecond-pulsed flashlamps at a rate of 10 Hz. Each flashlamp excitation produces a train of linearly polarized, 1064-nm pulses that are 30 ps at one-half maximum. A pulse selector, consisting of two crossed polarizers set on either side of a Pockels cell, senses the passage of a pulse located early in the pulse train and uses this pulse to trigger the Pockels cell to rotate the polarization of one of the more intense pulses in the train. This single, selected pulse is amplified in energy by passage through another optically pumped Nd:YAG rod—an amplifier— and then split into two pulses of approximately equal energy. Each of these two pulses needs to be amplified to sufficiently high energy to be converted from 1064 nm into 30-ps pulses to be used for excitation and interrogation of the sample.

A pulse suitable for excitation of an organic precursor to a reactive intermediate typically needs to be of a wavelength shorter than 1064 nm. To accomplish this without the use of a dye laser, which complicates matters relative to the use of a solid-state laser alone, other solid-state optical components—harmonic-generating crystals, also known as sum frequency generating crystals—are used to convert 1064-nm pulses into laser pulses of the second harmonic at 532 nm, to mix 532- and 1064-nm pulses to generate the third harmonic at 355 nm, or to convert 532-nm second harmonic pulses into those of the fourth harmonic at 266 nm. These harmonics do not provide the broader excitation wavelength tunability exhibited via the use of a dye laser but do provide three wavelengths that have proved useful for a range of experiments. To achieve a greater palette of excitation wavelengths, one may use stimulated Raman scattering of the three harmonics, which involves the focusing of the selected, sufficiently energetic harmonic pulse through a Raman shifter—a cell fitted with entrance and exit windows and containing a gas under pressure. For example, if the gas were methane and the second harmonic were passed through the Raman shifter, $\sim$30-ps laser pulses shifted by the energy of the C–H stretch to higher and lower energy from 532 nm along with residual 532 pulses can be selected through the use of filters or a prism and directed to the sample for excitation at a wavelength other than one of the harmonics. This is how the pump or excitation pulse is generated.

For picosecond, pump–probe transient absorption spectroscopy, the detector is not responsible for the time resolution. The recording of spectra as a function of time after sample excitation is achieved by optically delaying the arrival of the probe pulse relative to the pump pulse. These two pulses are born from a single, amplified 1064-nm pulse. If each one travels the same distance, corrected for delays induced as a result of passage through optical components, the pump and probe pulses will arrive at the sample at the same time. However, if the probe pulse travels a greater distance to reach the sample than the pump pulse, the probe pulse will interrogate the sample at a time after the pump pulse given by the distance–time relationship provided by the speed of light.

Before going further into the transient absorption experiment, let us turn our attention to the generation of the probe or interrogation pulse. One-half of the original 1064-nm pulse that is not used for making the pump pulse is focused into a cell that contains a 1:1 mixture of H_2O and D_2O to which $\sim$1% w/w P_4O_{10} is added. An $\sim$30-ps, white-light continuum pulse, resulting from the nonlinear optical event of superbroadening of the 1064-nm pulse, exits the cell. For best continuum generation, 1064-nm pulses whose energies are in the range of 7–12 mJ are used. Competing with continuum generation is the occurrence of a longer time scale flash resulting from dielectric breakdown in the solution. It is important for the continuum-generating solution to be as pure as possible and free from particulate matter. The former is accomplished through the use of high-quality H_2O, D_2O, and P_4O_{10} while the latter can be insured by means of filtration through a membrane with 0.5-μm pores. Usable continuum intensity from $\sim$370 to $\sim$850 nm can be realized. The intensity versus wavelength profile of the continuum is shaped with filters so that the detector is not saturated for bands of wavelengths over the range of wavelengths that are monitored.

Now that we have the probe continuum and pump pulses generated, they are directed to the sample and the reference cells in a dual beam spectrometer configuration. The pump pulse is manipulated with reflectors and simple lenses. Prior to its passage through the sample cell, $\sim$8% is split from the pump pulse and sent to the probe of an energy meter so that pump pulse energies are monitored and can be maintained between 0.05 and 0.20 mJ/pulse. The continuum pulse is collected from the continuum generation cell with an achromatic lens that focuses it onto a pinhole covered with a diffuser. This pinhole target helps to separate the optics prior to continuum generation from the continuum direction optics. The image of the pinhole is magnified using an achromatic pair of lenses so that an $\sim$1.5-mm diameter spot exists at the sample cell and that the polychromator used to disperse the white-light continuum into a spectrum is suitably matched optically. The pump pulse spot is adjusted to be 1–2 mm in diameter.

The liquid sample is flowed through either a 1-, 2-, or 5-mm pathlength flow-cell at a rate sufficiently fast to replace the volume subjected to excitation by each pump pulse. Large sample volumes are used when possible to minimize the fraction converted to products. The sample solution is replaced when the product concentration becomes unacceptably high. If the sample emits light upon excitation, spatial filtering is used to maximize the rejection of the emission while permitting the confined diameter of the probe pulse to be collected optimally.

Prior to reaching the sample, the probe pulse is split into two pulses. One of the two is directed through the sample cell, and the other is passed through the reference cell. The reference cell is not flowed and contains sample solution that does not undergo excitation. The sample and reference probe pulses are dispersed and imaged onto two tracks of a multichannel detector. Types of detectors that are used include two-dimensional (2D) silicon intensified targets,[1] double diode arrays,[3–5] and charge-coupled devices (CCDs).[6] Data collection and calculation of difference absorption spectra are performed with computer control. For the recording of a difference absorption spectrum, the probe pulse at a set optical delay relative to the pump pulse and the pump pulse are allowed to pass through the sample; one-half of the probe pulse is directed through the reference cell. The computer is used to collect and average the continuum intensity versus wavelength profiles that are transmitted through the sample and reference cells. This is done for a sufficient number of laser pulses to obtain an acceptable signal/noise ratio. Then shutters are closed via computer control to block the probe and pump pulses from the sample. For an amount of time equal to that of the previous acquisition of probe pulses, background is recorded and subtracted from the previously recorded probe pulses. After this, the same data-acquisition routine is followed except that the pump pulse is blocked from the sample. The absorbance change (ΔA) at a wavelength for a selected time postexcitation is given by Eq. 1.

$$\Delta A(\lambda, t) = -\log\{[I_s(\lambda, t)I_{0r}(\lambda, t)]/[I_{0s}(\lambda, t)I_r(\lambda, t)]\} \tag{1}$$

The parameters $I_s(\lambda, t)$ and $I_r(\lambda, t)$ are the continuum intensities of the sample and reference tracks, respectively, when the sample is excited. The parameters $I_{0s}(\lambda, t)$ and $I_{0r}(\lambda, t)$ are the continuum intensities of the sample and reference tracks, respectively, when the sample is not excited. The continuum versus wavelength profile is dependent on the pulse energy used to generate it, and because it is a nonlinear optical phenomenon, changes markedly with relatively small changes in pulse energy. To correct for these changes, the ratio of $I_{0r}(\lambda, t)/I_r(\lambda, t)$ is taken. Typically, 400 excitation–no excitation pairs are used to get good signal/noise ratios.

1.1.2. Dye Lasers.

An example of a dye laser system used for transient absorption spectroscopy is that described by Wirz and co-workers.[4] They based their system on the use of a dual-cavity excimer laser; one cavity is used to pump a cascade of dye lasers. At the core of the dye laser subsystem is a feedback dye laser that generates 0.5-ps, 496-nm pulses and whose output is amplified, frequency doubled (second harmonic generation), and used to seed a second, synchronously pumped cavity of the excimer laser. This process yields a 0.5-ps, 248-nm, 3.5-mJ pulse with a spot size of $\sim$1.5 cm^2. The residual 0.5-ps, 496-nm pulse with an energy of $\sim$180 μJ is used for the generation of the continuum pulse. The continuum pulse is collected with a spherical mirror, which is inherently achromatic, and the central portion containing the residual 496-nm pulse is blocked. The rest of the principles of the pump–probe transient absorption experiment are analogous to those described above for the Nd:YAG system. The dye laser in this instrument is not tuned to provide more excitation wavelengths. Additional details for the synchronously pumped dye laser spectrometer are given by Wirz and co-workers[4] who also refer to a subpicosecond transient spectrometer described by Ernsting and Kaschke.[5] This instrument used by Wirz and co-workers[4] is noteworthy because its useful continuum spans from 310 to 700 nm. The ability to record spectra to shorter wavelengths is useful for studying organic reactive intermediates. However, one must keep in mind that dye laser systems are more challenging than solid-state lasers to operate and to keep tuned during the course of recording a series of time-resolved absorption spectra.

Before moving to our next picosecond spectroscopy, there are three points of caution. First, be as confident as possible that the intensities of absorption bands change linearly with changes in pump pulse energy. This criterion does not insure that one-photon changes are induced upon excitation, however, because with the photon fluxes that can be generated with these lasers, the one-photon transition may be saturated and therefore give a linear response that reflects changes induced after the absorption of a second photon. Second, spectroscopists who use, without the recording of spectra, one of the harmonics, Raman-shifted, or a narrow band of the continuum to probe the kinetics of formation or decay of a reactive intermediate are cautioned in the interpretation of the data. This warning is particularly important in the study of the shorter end of tens of picoseconds after excitation. Without the benefit of viewing spectra, measurement of absorbance change for narrow wavelength bands can lead to conclusions about kinetics that are solely due to the shifting of the absorption band exhibited by an intermediate that is the result either of intramolecular vibrational relaxation or of stabilization via the reorientation of solvent molecules, and not a measure of the formation or decay of the reactive intermediate. The third path to incorrect interpretation of data is manifest at very short times after excitation, when the pump and probe pulses simultaneously occupy portions of the same space and therefore time. The generation of continuum, like any nonlinear optical process, gives rise to a pulse whose temporal profile is slightly less than the pulse used to generate it. Because the continuum spans a range of wavelengths, the shorter wavelength end is separated from the longer wavelengths through the effects of group velocity dispersion, commonly known as chirp. When chirp is not taken into account, an absorption that spans more than $\sim$100 nm and truly grows with a single time constant will appear to have its shorter wavelength end appear first, with the longer wavelength end appearing several or tens of picoseconds later depending on the width of wavelengths spanned by the absorption band. This problem arises when the continuum pulse is chirped after passing through optical components in which the shorter wavelengths travel more slowly than the longer wavelengths. At early times, the leading edge, richer in longer wavelengths, has already passed through the sample before the tailing portion of the continuum pulse, which is richer in longer wavelengths, overlaps with the leading edge of the pump pulse. Only the portion of the probe pulse richer in shorter wavelengths interrogates the excited volume of the sample, so only the shorter wavelength region of the time-resolved spectrum exhibits a positive absorption. The longer wavelength region of the spectrum lags in its increase in absorption as the probe pulse is moved to longer times after excitation that still overlap with the pump pulse. If the decay of this band is equally fast, the shorter wavelengths will appear to decay before the longer wavelength region of the absorption band decays. This artifact, resulting from the chromatography of light in the continuum pulse, will lead one to conclude that a shorter wavelength band feeds the longer wavelength band. If the decay of the shorter wavelength portion of the absorption band is not so fast, then one will erroneously conclude that there is a lag between the growth of the shorter wavelength portion relative to the longer wavelength portion when, in reality, both are appearing with the same rate constant.

1.2. Fluorescence Spectroscopy

Picosecond-resolved measurements of fluorescence intensities have been performed through the use of a streak camera[1] and by means of time-correlated single photon counting (TCSPC)[3]. The former is a less instrumentationally involved approach provided that one has access to a streak camera. The TCSPC does allow a greater dynamic range for signal detection than streak camera measurements; however, the dynamic range for streak camera based detection has been a focus for improvement. Both of these approaches will be described here.

1.2.1. Streak Camera Detection. When a streak camera is used to measure the intensity versus time profiles of fluorescence in the picosecond regime,[1,7] the components that are used are based on those used for transient absorption described in Section 1.1, with the exception of the generation of a continuum probe pulse and the associated optics used for collecting reference and sample continuum light intensities. The streak camera coupled with a multichannel 2D detector capture the streak image of the fluorescence rise and decay. Both the quality of the streaking of the fluorescence with time and the correction of the response of the 2D detector over its area need to be known in order to obtain physically correct profiles of measured light intensity versus time. The sample is flowed through an optical cell or a 3-mm diameter quartz tube and is excited with a pump pulse. At 90° to the path of the excitation pulse, the fluorescence is collected with the use of an aspheric lens that can be precisely focused from the UV to the visible as needed for each specific sample.

A fluorescence event is recorded by splitting a portion of the 1064-nm pulse and directing it to a photodiode to trigger the electronics of the streak camera. The 1064-nm pulse is directed over a suitable distance to be used within the trigger window and converted to the wavelength desired for excitation of the sample. A portion of this excitation pulse is directed into the slit of the streak camera as a timing marker. The remainder of the pulse is passed to the sample. The light emitted from the sample can be filtered as needed to obtain information about the fluorescence behavior of the sample. The collected light is stored in a computer and treated with the time vector and 2D detector corrections. The rise times and decays are fitted in the usual manners that take into account the instrument response function.

1.2.2. Time-Correlated Single-Photon Counting. For the application of TCSPC in the picosecond time domain, lasers with pulses whose half-widths are 20 ps or less are used.[3] For better time resolution, the combination of a microchannel plate photomultiplier tube (MCP-PMT) and a fast constant fraction discriminator (CFD) are used instead of a conventional photomultiplier tube (PMT). A TCSPC system with a time response as short as 40 ps has at its core a Nd:YLF (neodymium: yttrium lithium fluoride) laser generating 70-ps, 1053-nm pulses at a repetition rate of 75 Hz with an average power of 10 W. These pulses are converted to slightly shorter pulses of the second harmonic at 532 nm with an average power of 1 W that are used to pump a dye laser. This dye laser is cavity dumped to reduce the repetition rate to 1 MHz for the output 700-nm, 6-ps, 20-mW pulses. The pulses from the dye laser are mixed with residual 1053-nm pulses to give, when pyridine 1 dye is used, 420- nm pulses for sample excitation. The fluorescence is collected at 90° to the excitation path, directed through a monochromator, and measured with an MCP-PMT whose output is amplified with a CFD. A portion of the 700-nm pulse is sent to a photodiode connected to another channel of the CFD. The output of the CFD start emission and laser stop pulses are used by a time-to-amplitude converter whose output in turn is sent to a computer containing a multichannel analyzer card and software for hardware control and data acquisition and analysis.

1.3. Raman Spectroscopy

In the picosecond regime, Raman spectroscopy takes the kinetic capabilities of transient absorption and fluorescence methods and combines them with more featureful vibrational spectra. The potentially greater detail provided by the bands in a Raman spectrum is offset with the need to assign a group of bands firmly to the structure of a reactive intermediate. Libraries of assignments of vibrational frequencies to functional groups and structural features have been compiled for stable, ground-state molecules. Similar libraries need to be compiled for the new manifestations of bonding and structure that exist within reactive intermediates. Interpretations of the shifts of bands from the ground to an excited state or from one reactive intermediate to a different, but structurally related, reactive intermediate help one to understand the structure of the transient species.

Several types of time-resolved Raman spectroscopies have been reported and reviewed by Hamaguchi and co-workers[8] and Hamaguchi and Gustafson.[9] These include pump–probe spontaneous and time-resolved coherent Raman spectroscopy of the anti-Stokes and Stokes varieties [coherent anti-Stokes Raman spectroscopy (CARS) and coherent Stokes Raman spectroscopy (CSRS)], respectively. Here we will focus on pump–probe time-resolved spontaneous Raman spectroscopy.

The laser used to generate the pump and probe pulses must have appropriate characteristics in both the time and the frequency domains as well as suitable pulse power and repetition rates. The time and frequency domains are related through the Fourier transform relationship that limits the shortness of the laser pulse time duration and the spectral resolution in reciprocal centimeters. The limitation has its basis in the Heisenberg uncertainty principle. The shorter pulse that has better time resolution has a broader band of wavelengths associated with it, and therefore a poorer spectral resolution. For a 1-ps, sech^2-shaped pulse, the minimum spectral width is 10.5 cm^{-1}.[9] The pulse width cannot be <10 ps for a spectral resolution of 1 cm^{-1}. An optimal choice of time duration and spectral bandwidth are 3.2 ps and 3.5 cm^{-1}.[8] The pump pulse typically is in the UV region. The probe pulse may also be in the UV region if the signal/noise enhancements of resonance Raman

(RR) scattering are exploited; otherwise the probe pulse may be in the visible to avoid complicating fluorescence. The probe pulse, which is scattered by the sample into a Raman spectrum recorded over the time duration and at the selected time that the probe pulse passes through the sample, has a narrow band of wavelengths associated with it, which differs from the white-light continuum typically used for picosecond electronic absorption spectroscopy and from the need to generate probe pulses over a desired wavenumber range to record a picosecond-resolved IR spectrum.

The pump pulse energy is controlled to minimize two-photon phenomena and to maximize the concentration of the desired excited-state or other reactive intermediate. The optimal average power of the probe pulse changes with a specific experiment but is often maintained at $\sim$10 mW; peak powers in the range of 0.1–10 MW with repetition rates of 1 kHz–1 MHZ are best for picosecond spontaneous Raman spectroscopy.

The system described by Iwata et al.[8] uses a continuous wave (CW) mode-locked Nd:YAG laser to generate 65-ps, 1064-nm laser pulses at a repetition rate of 82 MHz with an average power of 11 W. These pulses are subjected to temporal compression with an optical fiber and a diffraction grating and to second harmonic generation whose average power is kept constant with a feedback loop driven by an acousto-optic modulator (AOM). The 532-nm second-harmonic laser pulses, exhibiting a half-width of 5 ps, a repetition rate of 82 MHz, and an average power of 0.8 W, synchronously pump a rhodamine 6G dye laser that has a three-plate birefringent filter for tuning the wavenumber of the 588-nm output pulses generated at a repetition rate of 82 MHz with an average power of 130 mW. These pulses are amplified with a two-stage dye amplifier pumped with the 532-nm second harmonic from a CW Nd:YAG regenerative amplifier operated at a repetition rate of 2 kHz and seeded with a portion of the 1064-nm output from the CW mode-locked Nd:YAG laser. The use of Kiton Red 620 in the dye amplifier achieves a gain of 10,000 for the 588-nm output pulses, created at a repetition rate of 2 kHz—500 µs between pulses—and with an average power of 25 mW, that are subsequently frequency doubled to 294-nm, 6-mW pump pulses. The residual 588-nm pulses transmitted through the last frequency doubling crystal and separated from the 294-nm pump pulses with a dichroic mirror are used as probe pulses. The probe pulse is optically delayed (0.3 mm = 1 ps) from its partner pump pulse. Then the two pulses are colinearly reunited with the use of a dichroic mirror so that the probe pulse interrogates the flowing, thin jet of sample solution at a selected time after the pump pulse has excited the sample. The light scattered at right angles to the direction of travel of the pump and probe pulses is collected and focused with two lenses, separated by a narrow band rejection filter, also known as a line-rejection or notch filter, onto the entrance slit of a single-grating spectrograph. The dispersed Raman scattering is measured with the use of a liquid nitrogen cooled CCD detector for typical data acquisition conditions of the following: a 200-µm entrance slit width that corresponds to 5.9 cm^{-1} at 1400 cm^{-1} from the 3.5 cm^{-1} half-width pump pulse; a 0.1–10-mW probe pulse average power; a 5-mW pump pulse average power; and exposure time of 10 min.

1.4. Infrared Absorption Spectroscopy

Picosecond IR spectroscopy complements the vibrational information provided via picosecond Raman spectroscopy. While picosecond Raman spectroscopy has a longer history and has been more widely applied, picosecond IR spectroscopy has been undergoing development and has been applied to an increasing number of chemical systems. Although the components comprising a picosecond Raman spectrometer are simpler and fewer in number than those that make up a picosecond IR spectrometer, picosecond IR spectroscopy does have its advantages including the general absence of complications from fluorescence and the ability to perform polarized IR measurements in order to obtain more information about molecular structure.[10] From the perspective of experimental design, two different picosecond IR spectrometers will be described herein—one that primarily covers the fingerprint region and another that covers the remainder of the IR region.

1.4.1. The >1500-cm^{-1} Region. The >1500-cm^{-1} spectral region is less demanding instrumentationally than the fingerprint region to be considered next and has been studied by several groups (e.g., see Refs. 11–14 and 15a). Developments in IR spectroscopy have lagged far behind those for Raman spectroscopy in the picosecond regime.

Here we will focus in detail on a UV pump–IR probe spectrometer described by Ernsting and co-workers[15] the system is based on an excimer laser and a dye laser operating with a pulse repetition rate ranging from 5 to 10 Hz. Pump pulses at 308 nm excite the sample and are followed at a selected time by probe IR pulses that range from 1950 to 4300 cm^{-1}. Absorbance changes can be recorded with a time resolution of 1.8 ps and with an accuracy in absorbance (A) of 0.001.

Initial laser pulse generation is achieved with the use of a twin-tube excimer laser in which one channel is a XeCl* laser oscillator delivering 15-ns, 308-nm, 80-mJ pulses for the driving of two dye lasers needed for difference frequency generation. The second channel is used for amplification of subpicosecond 308-nm pulses that become pump pulses.

The first dye laser generates 10-ps, 2-μJ, 365-nm pulses that pump a microscopic distributed feedback dye laser (DFDL) producing Fourier transform limited 0.7-ps pulses at 616 nm. These DFDL pulses are amplified in two stages to give 100-μJ, diffraction-limited pulses.

The UV pump pulses are generated from the 616-nm pulses passed through a BBO (β-barium borate) crystal for frequency doubling to give 1-μJ, 308-nm seed pulses that are separated from the 616-nm fundamental with the use of a dichroic mirror. These 308-nm pulses are amplified in two passes through the second channel XeCl* amplifier to give 3-mJ, 1-ps, 308-nm pulses. The 616-nm pulse remaining after second harminc generation is split into two unequal portions; 20% is used to pump the second dye laser while 80% is used for difference frequency generation. Tunable red-shifted pulses are needed that overlap in time with the 80% portion of the 0.7-ps, 616-nm pulses in order to generate IR pulses in the 2-5 μm region (4300–1950 cm^{-1}, respectively). This is accomplished through the use of a widely tunable prismatic DFDL. The red-shifted spectral region from 700 to 840 nm can be covered with four different dye solutions. The output is amplified to give 0.5-μJ, 0.4-nm bandwidth, 4-ps pulses. The difference frequency is generated with a $LiIO_3$ crystal that yields IR pulses of several nanojoules that are continuously tunable from 1950 to 4300 cm^{-1} by means of computer controlled adjustment of the phase-matching angle of the crystal, simultaneously with the wavelength of the prismatic DFDL.

After this difference frequency probe pulse generation, the residual visible pulses are separated from the IR pulses by transmission through 3-mm silicon plates, after which the IR pulse is focused with a CaF_2 lens behind a 1-mm Ge plate that splits the probe pulse into sample and reference pulses with approximately equal pulse energies. Each probe pulse is directed through its own respective sample and reference flow cells. Ultimately, the sample and reference probe pulses are focused with CaF_2 lenses onto liquid nitrogen cooled InSb detectors. A nearly collinear path for the UV pump pulse is used. The time delay between the UV pump and IR probe pulses is selected through the use of a computer-controlled translation stage. One data point for one delay position is typically the combination of 100 laser shots. Difference IR absorption spectra are recorded.

The temporal widths of the IR pulses and the time resolution of this spectrometer are tested with the use of a Ge sample that, when exposed to the pump pulses, results in transient IR absorption at 2290 cm^{-1}. Modeling the risetime of this absorption gives a cross-correlation width (full width at half-maximum, fwhm) of 1.8 ps.

1.4.2. The Fingerprint Region.[16-18] The pump–probe approach has been used to study the more difficult, but very informative, fingerprint region of the IR spectrum of a transient species. The 527-nm second harmonic of a mode-locked Nd:YLF (Nd^{3+}:yttrium lithium fluoride) laser is used to pump synchronously two dye lasers so that tunable pulses between 545 and 578 nm are generated when the laser dye rhodamine 560 is used. Pulses of the fundamental 1053-nm output of the Nd:YLF laser are amplified via passage through a CW Nd:YLF amplifier, are converted to the second harmonic, and are used to pump two rhodamine 590 dye amplifiers that are used to boost the power of the pulses produced by the two synchronously pumped dye lasers to the level of microjoules. To generate an IR pulse in the 1640–940 cm^{-1}—the mid-IR or fingerprint—region, an amplified pulse from one of the dye lasers and a 1053-nm pulse from the Nd:YLF regenerative amplifier are tunably difference-frequency mixed in a β-barium borate (BBO) crystal to give a pulse that can be tuned in the range from 1.15 to 1.30 μm in the near-IR. The residual dye laser pulses are removed with the use of a cut-off filter, and the near-IR pulses are difference-frequency mixed in an $AgGaS_2$ (AGS) crystal to give tunable mid-IR pulses at a repetition rate of 1 kHz. The wavenumber of the IR probe pulse is tuned by changing the wavelength of the dye laser and by phase-matching the angles of the BBO and AGS crystals. Unconverted portions of the near-IR pulses are filtered from the IR pulses with the use of a germanium plate. The IR probe pulses are sent to the sample.

Approximately 10% of a mid-IR pulse is reflected, by means of a BaF_2 plate, for use as a probe pulse that is focused into the sample cell with an aluminum-plated concave mirror. The mid-IR probe pulse transmitted through the sample is reflected and focused with a concave Al-plated mirror onto a liquid-nitrogen cooled MCT (HgCdTe) detector. The portion of the mid-IR pulse transmitted through the first BaF_2 plate is again sent through a second BaF_2; the reflected portion is used as a reference pulse that is focused with a CaF_2 lens onto another MCT detector. The signals from the two MCT detectors are amplified, and the pulsed components are gated with a boxcar system. The probe signal is electronically divided by the reference signal in order to reduce the noise resulting from instability of the IR pulses. The processed signals are converted from analogue to digital, sent to a computer, and stored.

The pump pulse for sample excitation comes from the output of a second dye amplifier. Its wavelength depends on the dye that is used; for rhodamine 560, the range is from 545 to 578 nm. The polarization angle of this pulse is selected with the use of a Fresnel rhomb prism. An optical delay line is used to change the path length traveled by the pump pulse relative to its corresponding probe pulse. The pump pulse is noncollinearly focused onto as much as possible of the same volume of the sample as the probe pulse. The train of pump pulses is modulated with a mechanical chopper that is synchronized to the regenerative amplifier at one-half of the 1- kHz repetition rate of the probe pulses. This process gives rise to a set of pump–probe–no pump–probe pulses analogous to those described for the picosecond difference electronic absorption spectrum. The detected mid-IR probe pulses for the sample and the reference with the pump pulse on and off are separately stored in the computer, and the difference IR absorption spectrum is calculated. The temporal and spatial overlaps of the pump and probe pulses are checked with a standard such as detecting the rise of a broad transient IR absorption resulting from pump–pulse generated free carriers in silicon.[19] The timing jitter between the pulses from the two synchronously pumped dye lasers can be as much as $\sim$10 ps.

With this spectrometer, a difference mid-IR spectrum at a selected time after sample excitation is recorded by sweeping from 1640 to 940 cm^{-1} in steps that may be as short as approximately equal to the spectral resolution of the spectrometer—in this case,[16] $\sim$8 cm^{-1}. The sample solution is pumped through a flow cell that has IR-transmitting CaF_2 windows set with a 0.1-mm optical pathlength. The BaF_2 windows have also been used for the sample cell.[18]

1.5. Optical Calorimetry

Picosecond-resolved thermochemical information can be extracted from the evolution of a transient grating produced by the crossing of two laser pulses and interrogated with a third short pulse of light. Several groups have applied this method to thermodynamic questions about the decay of excited states and the evolution of excited states into reactive intermediates.[20–25]

Pulses of one of the harmonics of an actively–passively mode-locked Nd:YAG such as the third harmonic at 355 nm are split into two parts, and after traveling different paths are simultaneously recombined in the sample noncollinearly.[22] An optical delay is used to insure temporal coincidence in the sample. These two pulses traveling along crossed paths generate a transient grating in the sample cell. A third pulse (e.g., at 532 nm) serving as the probe pulse is sent along a variable optical delay and passes through the sample along a counterpropagating path, relative to the two crossed 355-nm pulses, at an appropriate angle from Bragg diffraction. The diffracted pulse is filtered with a 500-nm cutoff filter and reflected onto a vacuum photodiode. The output signal of the photodiode is amplified and fed into a multichannel analyzer board residing in a personal computer. The signals from 20 laser pulses are averaged and analyzed appropriately. Those interested in the details of the data analysis are referred to the cited literature.

2. STUDIES ON REACTIVE INTERMEDIATES

The examples that are presented in this section are not rigorous reviews of reactive intermediates but highlight instead the application of picosecond-resolved methods to selected problems. The symbols and abbreviations used herein may be different from those in the original publications, but if they have been changed, it is to achieve a uniformity of meaning for which they are used in this chapter.

2.1. Excited States of Conjugated Pi Systems

2.1.1. 1,2-Diphenylethene (Stilbene). This molecule has been the subject of many photophysical and photochemical investigations and the subject of several reviews.[26–28] It is the prototypical alkene for studies of photoisomerization. Transient spectroscopic measurements in the picosecond time domain have been performed on electronically excited *trans*-stilbene in a wide range of environments. Selections from these studies are described here.

An absorption band with a maximum near 585 nm resulted from 265-nm, 7-ps pulsed excitation of a 1-mM solution of *trans*-stilbene (tS) in hexane at room temperature.[29,30] The absorption band was assigned to an $S_n < S_1$ absorption of tS in a geometry close to that of ground-state tS. Narrowing of this absorption band occurred for ~50 ps after excitation. This narrowing has been attributed to intramolecular vibrational relaxation of vibrationally hot tS in its S_1 state generated via 265-nm excitation. Information from Raman spectroscopic measurements supports the case for vibrational relaxation.[31–37] Pulsed excitation at 306 nm created the tS S_1 state with less excess vibrational energy and as a consequence narrowing of the $S_n < S_1$ absorption band was not observed. The lifetime of the tS S_1 state is 95 ps in room temperature hexane as measured by means of transient absorption spectroscopy[29] and is consistent with streak camera based measurements of the decay fluorescence that competes with the major pathway of twisting about what was the ethylenic carbon–carbon double bond in the ground state.[30]

Three possible paths have been discussed for the unimolecular decay of the near vertical geometry of the tS S_1 state.[27,38,39] The commonly accepted, dominant mechanism has the original ethylenic bond undergo a twist on the excited singlet state surface to a perpendicular geometry—the twisted excited singlet state or phantom state—that partitions upon return to the ground state to a mixture of *trans*- and *cis*-stilbene. A second pathway, taken by some substituted stilbenes after direct excitation, proceeds via intersystem crossing to a triplet surface upon which isomerization occurs followed by return to the ground state. A third mechanism, involving internal conversion to high vibrational levels of the ground state and subsequent trans–cis isomerization, has been found to be unimportant.

The picosecond IR absorption spectrum of the tS S_1 state in the fingerprint region is different in *n*-heptane and in acetonitrile.[13] The spectrum recorded for S_1 tS in the nonpolar solvent *n*-heptane is consistent with a species that has a center of symmetry. In acetonitrile, the spectrum exhibits additional weak bands near 1570, 1250, and 1180 cm^{-1}, which are approximately at the same frequencies as strong Raman bands assigned to in-plane vinylic vibrational modes in S_1. This result was taken to suggest a molecular structure for S_1 that lacks a center of symmetry in acetonitrile. However, because the intensities of these three bands are weak, it was concluded that either the polarization of S_1 or the contribution from polarized S_1 structures to all of the S_1 structures in acetonitrile may be small.

tS

Picosecond Raman measurements have led to the proposal of a dynamic polarization model.[40,41] In this model, S_1 tS undergoes reversible changes in vibrational frequencies that are induced by solvent fluctuations. The mixing of a perturbing state with S_1 interconverts carbon–carbon double bonds with single bonds that leads S_1 tS near the vertical geometry to proceed along the pathway for isomerization.

While much has been learned about S_1 near the geometry of tS in solution, there have been no reports on the spectroscopic detection of the twisted excited singlet state of stilbene. Information about the twisted excited singlet state of tetraphenylethene will be described in this chapter in Section 2.3. Interesting photochemistry also occurs following the excitation of *cis*-stilbene; however, because of the much shorter lifetime of S_1 *cis*-stilbene, femtosecond-pulsed lasers must be used.[42–45]

At higher concentrations in solution, the photodimerization of tS has been studied by means of picosecond electronic absorption spectroscopy.[6] The S_1 state of tS in benzene at 22°C is quenched with a diffusion-controlled rate constant of 2.03×10^{10} M^{-1} s^{-1} to give a new reactive intermediate exhibiting an absorption maximum at 480 nm. This new species decays unimolecularly with a rate constant of $(2.40 \pm 0.37) \times 10^9$ s^{-1}. It has tentatively been assigned to either the excimer or a biradicaloid species located at the pericyclic minimum.

2.1.2. 1,4-Diphenyl-1,3-butadiene. The excited-state behavior of this diene differs significantly from stilbene and is the subject of a review.[26] Unlike tS in which the lowest vertical excited singlet state is the 1^1B_u state and S_2 is the 2^1A_g state in solution, these two excited states lie very close to each other in *all-trans*-1,4-diphenyl-1,3-butadiene (DPB). The additional carbon–carbon double bond introduces a new conformational equilibrium involving the s-trans and s-cis rotamers. Most spectroscopic studies in solution have concluded that the 1^1B_u state is S_1. The DPB compound has a low quantum yield for photoisomerization, so the use of DPB in time-resolved spectroscopic studies on photoisomerization, especially those that monitor only fluorescence decay, needs to be considered cautiously and critically.

Femtosecond electronic absorption spectroscopy was used to study DPB and the following four structurally related butadienes: 1,4-diphenyl-1,3-cyclopentadiene (DPCP); (*E,E*)-diindanylidenylethane (stiff-5-DPB); 1,1,4,4-tetraphenyl-1,3-butadiene (TPB); and 1,2,3,4-tetraphenyl-1,3-cyclopentadiene (TPCP).[46] The related compounds were used to infer additional information about DPB through interpretations of the spectroscopic observations that were made. The absorbance changes of S_1 DPB in hexane at 620, 660, and 690 nm within the 640-nm absorption band were measured following excitation at 310 nm. At each of these three wavelengths, the data were fitted well to an instrument limited rise time of 120 fs and a exponential decay time of 465 ± 10 ps. Although only these three wavelengths were

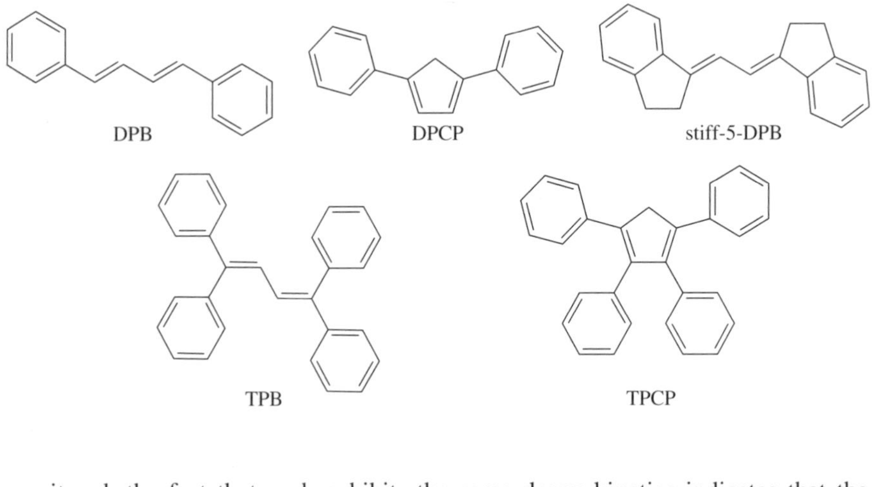

monitored, the fact that each exhibits the same decay kinetics indicates that the entire 640-nm absorption band evolved uniformly. This procedure is not as good as monitoring the entire absorption band, but it is clearly more reliable than merely monitoring the kinetics at a single narrow band of wavelengths. Kinetics studied at a single narrow band of wavelengths, often reported as a single wavelength, must be carefully evaluated because, at short times after excitation, absorbance changes may be the result of intramolecular vibrational relaxation or changes in solvation in addition to an overall increase or decrease in the concentration of a species. In this femtosecond study,[46] no evidence exists for 1^1B_u–2^1A_g interconversion as was described in an earlier picosecond spectroscopic study.[47]

Picosecond Raman spectroscopy provided information about the excited states of DPB.[48,49] Raman spectra of the S_1 state of DPB in several solvents including pentane, hexane, heptane, octane, decane, dodecane, and tetrahydrofuran (THF) exhibit bands assignable to both the 1^1B_u and 2^1A_g states. Because of a small energy gap of ~500 cm^{-1} between these two states, it was concluded that the lowest excited singlet state in solution is of mixed character. The changes in the Raman bands of S_1 with solvent and probe wavelength are consistent with a distribution of s-trans conformers in the S_1 state that controls the degree of mixing of the 1^1B_u and 2^1A_g states. Changes in the position and bandwidth of the peak assigned to the phenyl–vinyl stretch were attributed to vibrational relaxation within S_1.

The excited-state ordering is expected to play an important role in governing the photophysical and photochemical modes of deactivation of DPB following electronic excitation in solution.[26]

2.1.3. 1,6-Diphenyl-1,3,5-hexatriene.
The quantum yield for photoisomerization of *all-trans*-1,6-diphenyl-1,3,5-hexatriene (DPH) is much lower than that of *trans*-stilbene.[50–53] The DPH compound is the first in the series of vinylogous stilbenes for which the 2^1A_g state is lower than the 1^1B_u state. Early picosecond fluorescence[54] and electronic absorption[55] measurements were interpreted as providing

DPH

evidence for the equilibration of the 1^1B_u state, initially populated via one-photon absorption, with the lower lying 2^1A_g state. A subsequent picosecond study[56] demonstrated that there was no evidence from either picosecond fluorescence or electronic absorption measurements to support equilibration of the 1^1B_u and 2^1A_g states. The fluorescence decay recorded with a streak camera[56] did not exhibit a <30-ps decay component after excitation of DPH in cyclohexane of the type previously recorded for DPH in hexane.[54] It was proposed[56] that the previously reported fast decay was actually due to a solvent Raman band induced by the excitation pulse. Transient absorption spectra in the later work[56] did exhibit spectral changes reported earlier.[55] However, these changes were attributed[56] to the effects of chirp and not to equilibration of excited states. Additional complications due to fluorescence from the sample and to multiphoton absorption leading to the ionization of DPH to give the DPH radical cation were also considered[56] as factors affecting the appearance of the transient absorption spectra previously.[55] Also, the absorbance ratio of the two maxima near 460 and 650 nm were found to be temperature independent[56] contrary to the earlier report.[55] It was concluded[56] that the relaxation of the initially formed 1^1B_u state to the equilibrium mixture is complete within ~10 ps, much faster than previously reported.[54,55] This faster relaxation time is supported by more recent femtosecond absorption[57] and picosecond Raman investigations.[58,59]

Picosecond-resolved CARS has been used to study the excited states of DPH in cyclohexane and methanol solutions[58,59] in order to learn about the vibrational spectroscopy of the excited singlet state, the time scale for equilibration of the two lowest lying singlet excited states, and the structure of the excited state. The vibrational spectrum was assigned to an excited singlet state of both 1^1B_u and 1^1A_g character that is produced from the intially populated 1^1B_u Franck–Condon state on a subpicosecond time scale. In methanol solution, shifts of high-frequency chain modes are observed that indicate changes in bond orders in the chain. It was noted[58] that these changes should be linked to altered yields for competing biradicaloid and zwitterionic pathways of DPH photoisomerization reported for solvents of different polarity.[53]

2.1.4. Diphenylacetylene. The photophysics and photochemistry of diphenylacetylene (DPA), also known as tolan, and several derivatives have been reviewed.[60] The unusual photophysical properties, such as a very slow rate of conversion of S_2 to S_1, a dependence of the S_2 lifetime on temperature, and a strong dependence of excitation wavelength on the fluorescence quantum yield of the lowest excited singlet states of DPA were correlated with the S_2–S_1 energy gap.

Picosecond electronic absorption and fluorescence measurements were performed on DPA in a series of solvents, including isopentane, *n*-hexane, cyclohexane, methylcyclohexane, *trans*-decalin, ethanol, and acetonitrile.[61] The picosecond fluorescence measurements were performed by means of time-correlated single-photon counting. Picosecond absorption spectra recorded 295-nm excitation with a 4.2-ps dye laser pulse and corrected for the effects of group velocity dispersion (chirp) associated with the white-light continuum probe pulse. In hexane, a 500-nm absorption band appears immediately after excitation, has a lifetime of 8 ps, and was assigned to an absorption from the S_2 state. The fluorescence lifetime agrees well with the 8-ps lifetime of S_2. Note that a shorter lifetime is indicated as a result of a femstosecond-resolved study.[62] The rapid decay of the 500-nm band gives rise to two absorption bands with maxima at 436 and 700 nm. These two bands have a lifetime of $\sim$200 ps and were assigned to absorptions from S_1. The decay of the S_1 absorption bands are concomitant with the rise of an absorption band at 415 nm assigned to a $T_n < T_1$ transition. No fluorescence with a lifetime of 200 ps is observed; the fluorescent state of DPA in solution is the S_2 state, which is an exception to Kasha's rule.[63] The S_2, S_1, and T_1 states have been assigned to the 1^1B_{1u}, 1^1A_u, and 1^3B_{1u} states, respectively.[60,64] The temperature dependence of the fluorescence lifetime gives an activation energy (E_a) of $\sim$890 cm^{-1} or 2.5 kcal/mol; this is the activation barrier that must be traversed on going from S_2 to S_1.[61] An investigation of the temperature dependence of the fluorescence quantum yield gives an E_a of 14 kJ/mol or 3.3 kcal/mol.[64] The difference between these two activation energies was explained[64] by the presence of a temperature-independent contribution that was not detected in the earlier investigation.[61]

Several vibrational modes in the excited singlet states of DPA have been identified from picosecond CARS measurements.[65,66] The central C—C bond of DPA
retains much of its triple-bond-like character in the S_2 state; however, the central
C—C bond has double-bond-like character in the S_1 state. By analogy with the
trans-bent form of S_1 acetylene,[67] it was proposed[65,66] that S_1 DPA has a trans-
or cis-form bent structure that is consistent a previously proposed structure.[64] It
has been cautioned that, because of the conjugation between the phenyl groups
and the central C—C bond, the S_1 state may assume a structure different from the
bent form.[60] The interpretation of recent picosecond IR absorption measurements
provide support for the trans-bent planar structure for S_1 of DPA. The diphenylace-
tylenic fluorophore has recently been incorporated into several chemosensor structures whose fluorescent signaling properties are controlled by the relative
flexibilities of the molecules.[68]

2.2. Photolysis of *tert*-Butyl 9-Methyl-9-fluoreneperoxycarboxylate

Picosecond electronic absorption spectroscopy was used to investigate the formation of the 9-methyl-9-fluorenyl radical following 266-nm, 18-ps pulsed excitation
of the peroxyester in either cyclohexane or acetonitrile.[69] The absorption of the 9-
methylfluorenyl radical (**3**) grows with a rise time of 55 ± 15 ps as monitored at a
single, unreported wavelength. The proposed events following 266-nm, pulsed excitation is outlined in Scheme 19.1 in terms of a pathway in which the O—O bond is

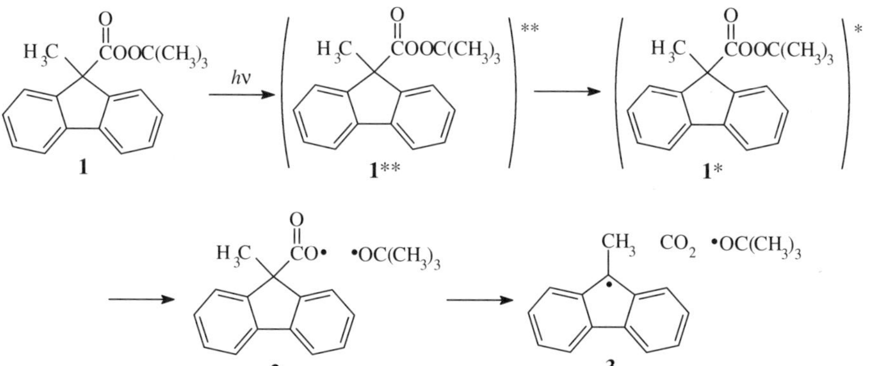

Scheme 19.1

first broken and then decarboxylation occurs. The electronic absorption spectrum of
1 was described as nearly superimposable on the spectrum of 9-methyl-9-fluorene-
carboxylic acid, methyl ester, except for the broad, weak absorption band, extending to
wavelengths >400 nm, that is characteristic of peroxides such as **1**. Excitation at
266 nm was thought to generate a higher excited singlet state (**1****) in which the

excitation is mainly localized in the fluorenyl moiety and that subsequently relaxes to the lowest singlet excited state ($1*$) in which the excitation resides on the peroxy-ester group. From the inability to detect fluorescence or transient absorption from $1**$ or $1*$, it was concluded that the slow step giving rise to the 55-ps rise time is due to the decarboxylation of acyloxy radical **2** to give radical **3** as the only reactive intermediate detected spectroscopically.

More information about the sequential pathway versus one in which the O—O and C—C bond homolyses occur concertedly was obtained by means of picosecond IR absorption spectroscopy.[15b] While IR spectroscopy, in principle, can provide more structural information, limitations stemming from the weakness of some IR intensities can be detrimental to the detection of a reactive intermediate. In this case, the CO_2 photoproduct was monitored (near 2335 cm^{-1}) instead of radical intermediates **2** or **3**. It was argued that neither **1**, **2**, nor **3** absorbed in this region. Excitation of **1** in CCl_4 was achieved with a 308-nm, 1-ps pulse. The 1.3-ps probe pulse may be tuned to cover a range from 2000 to 4300 cm^{-1} with the spectrometer that was used, but in these specific experiments the interrogation window was from 2000 to 2400 cm^{-1}.

2.3. Pi-Bond Heterolysis versus Homolysis

The excited-state behavior of 1,1,2,2-tetraphenylethene (TPE) has been studied by means of picosecond fluorescence,[70,71] absorption,[72,73] and Raman[74] spectroscopies and picosecond optical calorimetry.[71,75–77] It has been shown that, like stilbene, TPE derivatives substituted with minimally perturbing stereochemical labels such as methyl groups undergo efficient photoisomerization.[78] However, unlike stilbene, strong spectroscopic evidence exists for the direct detection of the twisted excited singlet state, S_{1P} herein but traditionally designated as $^1p*$, of TPE.

Excitation of TPE in hexane with a 305-nm, 0.5-ps laser pulse led to the appearance of absorption bands at 423 and 650 nm during the time duration of the excitation pulse.[79] The 630-nm band decays and the 423-nm band shifts to 417 nm with a lifetime of 5 ± 1 ps, and the 423-nm band shifts to 417 nm on this time scale. The 417-nm absorption then decays with a lifetime of 3.0 ± 0.5 ns. The 630- and 417-nm bands were assigned to an electronic transition from the vertical singlet excited state (S_{1V}) and to an electronic transition from the nonfluorescent S_{1P}, respectively.[79]

Picosecond fluorescence measurements[70] made at several temperatures ranging from 4 to 293 K with the use of a streak camera are consistent with results obtained from transient absorption spectroscopy. These fluoresence studies performed on TPE in several solvents, including 3-methylpentane, isopentane, decalin, phenol, ethylene glycol, and triacetin, at several temperatures provided more information about the excited-state behavior of TPE. After excitation with a 355-nm, 5-ps laser pulse to a vibrationally hot TPE vertical excited singlet state designated here as $(S_{1V})*$, intramolecular vibrational relaxation (ivr) gives the relaxed vertical singlet excited state S_{1V}, which undergoes further relaxation, proposed to be torsional, to another geometry on the excited singlet state surface to S_{1R}.

A subsequent picosecond electronic absorption spectroscopic study[72] of TPE excited with 266- or 355-nm, 30-ps laser pulses in cyclohexane found what was reported previously.[79] However, in addition to the nonpolar solvent cyclohexane, more polar solvents such as THF, methylene chloride, acetonitrile, and methanol were employed. Importantly, the lifetime of S_{1P} becomes shorter as the polarity is increased; this was taken to be evidence of the zwitterionic, polar nature of TPE S_{1P} and the stabilization of S_{1P} relative to what is considered to be a nonpolar S_{0P}, namely, the transition state structure for the thermal cis–trans isomerization. Although perhaps counterintuitive to the role of a solvent in the stabilization of a polar species, the decrease in the S_{1P} lifetime with an increase in solvent polarity is understood in terms of internal conversion from S_1 to S_0, which should increase in rate as the S_1–S_0 energy gap decreases with increasing solvent polarity. Along with the solvent-dependent lifetime of S_{1P}, it was noted that the TPE S_{1P} absorption band near 425 nm is located where the two subchromophores—the diphenylmethyl cation and the diphenylmethyl anion—of a zwitterionic S_{1P} should be expected to absorb light. A picosecond transient absorption study[73] on TPE in supercritical fluids with cosolvents provided additional evidence for charge separation in S_{1P}.

Picosecond optical calorimetry[71,75–77] was used to learn about changes in the S_{1P} energy when the solvent is changed. For the calorimetry studies, 634-nm, ~50-ps pulses from a synchronously pumped dye laser were used. From an analysis[75] of the heat released after excitation of TPE in cyclohexane, TPE S_{1P} was estimated to lie 73.1 kcal/mol above the ground state. This put S_{1P} 39.4 kcal/mol above the transition state for thermal isomerization. This value was approximated by analogy to that of 1,2-di(4-methylphenyl)-1,2-diphenylethene, which is known to exist 36.0 kcal/mol above the ground state.[80] Additional picosecond optical calorimetric experiments[76] with several solvents refined the initial estimate of the energy of S_{1P} above the ground state to 67.0 ± 1.3, 66.2 ± 1.4, and 65.3 ± 1.8 in alkane (pentane, hexane, and nonane), diethyl ether, and THF, respectively, as solvents. Within the context of the dielectric continuum model for solvation, the dipole moment of S_{1P} was estimated to be 6.3 D, which compares well with a dipole moment resulting from the separation of a unit charge by the length of a C—C single bond.[76] Cautions have been raised concerning the ability to separate contributions from enthalpy and volume changes in these photothermal measurements.[76,81] A somewhat larger dipole moment of at least 7.5 D was measured by means of time-resolved microwave conductivity that used 7-ns, 308-nm pulsed excitation.[82]

A combined picosecond fluorescence, which employed time-correlated single photon counting, and picosecond optical calorimetry investigation[71] followed two nanosecond fluorescence studies[77,82] and gives a mapping of the TPE S_1 surface. The two nanosecond studies demonstrated that S_{1P} is not formed irreversibly from the fluorescent states reported previously[70] but is in equilibrium with the fluorescent S_{1R} state in hydrocarbon solvents. In S_1,[71] S_{1V} is 77 kcal/mol above the ground state and traverses a barrier of 2.6 kcal/mol to get to S_{1R} at an energy of 69 kcal/mol. The S_{1R} state passes over a barrier of 2.4 kcal/mol to arrive at S_{1P} whose energy of 67 kcal/mol lies 2 kcal/mol below S_{1R}.

To this point, twisting to give S_{1P} has been inferred from the time-resolved absorption, fluorescence, and photothermal studies described herein along with product studies and applications of steady-state spectroscopies. Nanosecond Raman spectroscopy provides the needed evidence for the twisting. Both TPE and its analogue with ^{13}C present at the ethylenic positions in heptane were excited at 316 nm with a 10-ns laser pulse and probed with a 417-nm, 10-ns pulse. In S_0, strong coupling exists between the vinyl C—C stretch and the phenyl C—C stretch. In S_1, two Raman bands are assigned to the phenyl C—C stretch, but none are observed for the vinyl C—C stretch. This absence indicates that, in S_1, coupling between the vinyl C—C stretch and the phenyl C—C stretch is absent, which was interpreted as being consistent with the decrease in the bond order of the ethylenic C—C bond in S_0 on going to S_{1P}. Unfortunately, no Raman band for this C—C stretch could be assigned in the spectrum of S_1. Further interpretation of the Raman spectra led to following conclusions: (a) that the large difference in the frequencies of the two phenyl C—C stretch bands is consistent with the presence of cationic and anionic moieties in S_{1P} and may be considered as manifestations of its zwitterionic nature and (b) that the resonance enhancement of the phenyl Raman bands is consistent with localized electronic transitions from cation and anion subchromophores in S_{1P} as was described previously.[72]

More recent femtosecond spectroscopic investigations provide interesting information about shorter time scale phenomena associated with excitation of TPE.[83,84]

3. CONCLUSION AND OUTLOOK

Much useful kinetic and structural information can be provided by means of picosecond spectroscopies. The future may be considered in terms of the development of new picosecond spectroscopic methods and investigations of new chemical systems including the excited states of reactive intermediates. As picosecond electronic absorption and fluorescence methods have evolved to more user-friendly implementations in greater number, one can anticipate the same for the more instrumentally challenging infrared absorption and Raman spectroscopies and the more challenging analyses associated with optical calorimetric and dichroism measurements. For the measurement of transient absorption spectra, the importance of probing wavelengths <400 nm will make necessary an increase in the number of spectrometers capable of measuring further into the UV region.

The outlook is good for applications of these picosecond methods to an increasing number of studies on reactive intermediates because of the limitations imposed by the time resolution of nanosecond methods and the generally greater challenges of the use of a femtosecond spectrometer. The pump–probe technique will be augmented in more widespread applications of the preparation–pump–probe method that permits the photophysics and photochemistry of reactive intermediates to be studied.

SUGGESTED READING

W. G. Herkstroeter and I. R. Gould, "Absorption Spectroscopy of Transient Species," in *Determination of Electronic and Optical Properties*, Vol. VIII, 2nd ed., B. W. Rossiter and R. C. Baetzold, Eds., John Wiley & Sons, Inc., New York, 1993, pp. 225–319.

J. C. Scaiano, Ed., *Handbook of Organic Photochemistry*, Vol. I, CRC Press, Boca Raton, FL, **1989**.

J. C. Scaiano, Ed., *Handbook of Organic Photochemistry*, Vol. II, CRC Press, Boca Raton, FL, **1989**.

J. R. Lakowicz, *Principles of Fluorescence Spectroscopy*, 2nd ed., Kluwer Academic/Plenum Publishers, New York, **1999**.

M. D. Fayer, Ed., *Ultrafast Infrared and Raman Spectroscopy*, Vol. 26, Marcel-Dekker, New York, **2001**.

J. R. Schoonover and G. F. Strouse, "Time-Resolved Vibrational Spectroscopy of Electronically Excited Inorganic Complexes in Solution," *Chem. Rev.* **1998**, *98*, 1335.

H.-O. Hamaguchi and K. Iwata, "Physical chemistry of the lowest excited singlet state of trans-stilbene in solution as studied by time-resolved Raman spectroscopy," *Bull. Chem. Soc. Jpn.* **2002**, *75*(5), 883.

REFERENCES

1. J. A. Schmidt and E. F. Hilinski, *Rev. Sci. Instrum.* **1989**, *60*, 2902.

2. W. G. Herkstroeter and I. R. Gould, in *Determination of Electronic and Optical Properties*, 2nd ed., B. W. Rossiter and R. C. Baetzold, Eds., John Wiley & Sons, Inc., New York, 1993, Vol. VIII, pp. 225–319.

3. B. R. Arnold, S. J. Atherton, S. Farid, J. L. Goodman, and I. R. Gould, *Photochem. Photobiol.* **1997**, *65*, 15.

4. E. Hasler, A. Hormann, G. Persy, H. Platsch, and J. Wirz, *J. Am. Chem. Soc.* **1993**, *115*, 5400.

5. N. P. Ernsting and M. Kaschke, *Rev. Sci. Instrum.* **1991**, *62*, 600.

6. K. S. Peters, S. C. Freilich, and J. Lee, *J. Phys. Chem.* **1993**, *97*, 5482.

7. T. Tahara and H.-o. Hamaguchi, *Chem. Phys. Lett.* **1995**, *234*, 275.

8. K. Iwata, S. Yamaguchi, and H.-o. Hamaguchi, *Rev. Sci. Instrum.* **1993**, *64*, 2140.

9. H.-o. Hamaguchi and T. L. Gustafson, *Annu. Rev. Phys. Chem.* **1994**, *45*, 593.

10. H. Okamoto, *J. Phys. Chem. A* **1999**, *103*, 5852.

11. J. N. Moore, P. A. Hansen, and R. M. Hochstrasser, *Proc. Natl. Acad. Sci. U.S.A.* **1988**, *85*, 5062.

12. H. Graener, T. Q. Ye, A. J. Laubereau, *J. Phys. Chem.* **1989**, *93*, 7044.

13. J. R. Sprague, S. M. Arrivo, and K. G. J. Spears, *J. Phys. Chem.* **1991**, *95*, 10528.

14. P. O. Stoutland, R. B. Dyer, and W. H. Woodruff, *Science* **1992**, *257*, 1913.

15. (a) J. Aschenbrücker, U. Steegmüller, N. P. Ernsting, M. Buback, J. Schroeder, and J. Jasny, *Appl. Phys. B* **1997**, *65*, 441. (b) J. Aschenbrücker, M. Buback, N. P. Ernsting, J. Schroeder, and U. Steegmüller, *J. Phys. Chem. B* **1998**, *102*, 5552.

16. H. Okamoto and M. Tasumi, *Chem. Phys. Lett.* **1996**, *256*, 502.

17. H. Okamoto, *Chem. Phys. Lett.* **1998**, *283*, 33.

18. T.-a. Ishibashi, H. Okamoto, and H.-o. Hamaguchi, *Chem. Phys. Lett.* **2000**, *325*, 212.

19. T. M. Jedju and L. Rothberg, *Appl. Opt.* **1988**, *27*, 615.

20. J. Morais, J. Ma, and M. B. Zimmt, *J. Phys. Chem.* **1991**, *95*, 3885.

21. C. Högemann, M. Pauchard, and E. Vauthey, *Rev. Sci. Instrum.* **1996**, *67*, 3449.

22. E. Vauthey and A. Hensler, *J. Phys. Chem.* **1995**, *99*, 8652.

23. L. Genberg, Q. Bao, S. Gracewski, and R. J. D. Miller, *Chem. Phys.* **1989**, *131*, 81.

24. M. Takezaki, N. Hirota, and M. Terazima, *J. Phys. Chem.* **1996**, *100*, 10015.

25. J. T. Fourkas and M. D. Fayer, *Acc. Chem. Res.* **1992**, *25*, 227.

26. J. Saltiel and Y.-P. Sun, in *Photochromism—Molecules and Systems*, H. Dürr and H. Bouas-Laurent, Eds., Elsevier, Amsterdam, The Netherlands, 1990, p. 64.

27. D. H. Waldeck, *Chem. Rev.* **1991**, *91*, 415.

28. H. Görner and H. J. Kuhn, in *Advances in Photochemistry*, Vol. 19, D. C. Neckers, D. H. Volman, and G. von Bünau, Eds., John Wiley & Sons, New York, 1995, p. 1.

29. B. I. Greene, R. M. Hochstrasser, and R. B. Weisman, *Chem. Phys. Lett.* **1979**, *62*, 427.

30. R. M. Hochstrasser, *Pure Appl. Chem.* **1980**, *52*, 2683.

31. H.-o. Hamaguchi, *Chem. Phys. Lett.* **1987**, *126*, 185.

32. H.-o. Hamaguchi, *J. Chem. Phys.* **1988**, *89*, 2587.

33. T. L. Gustafson, D. M. Roberts, and D. A. Chernoff, *J. Chem.Phys.* **1983**, *79*, 1559.

34. T. L. Gustafson, D. M. Roberts, and D. A. Chernoff, *J. Chem. Phys.* **1984**, *81*, 3438.

35. A. B. Myers, M. O. Trulson, and R. A. Mathies, *J. Chem. Phys.* **1985**, *83*, 5000.

36. K. Iwata and H.-o. Hamaguchi, *J. Phys. Chem. A* **1997**, *101*, 632.

37. K. Iwata and H.-o. Hamaguchi, *J. Mol. Liq.* **1995**, *65/66*, 417.

38. J. Saltiel, J. D'Agostino, E. D. Megarity, L. Metts, K. R. Neuberger, M. Wrighton, and O. C. Zafiriou, in *Org. Photochem.*, Vol. 3, O. L. Chapman, Ed., Marcel Dekker, New York, **1973**, p. 1.

39. J. Saltiel and J. L. Charlton, in *Rearrangements in Ground and Excited States*, P. de Mayo, Ed., Academic Press, New York, **1980**, Vol. 3, p. 25.

40. H.-o. Hamaguchi, *Mol. Phys.* **1996**, *89*, 463.

41. K. Iwata, R. Ozawa, and H.-o. Hamaguchi, *J. Phys. Chem. A* **2002**, *106*, 3614.

42. S. Abrash, S. Repinec, and R. M. Hochstrasser, *J. Chem. Phys.* **1990**, *93*, 1041.

43. R. J. Sension, A. Z. Szarka, and R. M. Hochstrasser, *J. Chem. Phys.* **1992**, *97*, 5239.

44. D. C. Todd and G. R. Fleming, *J. Chem. Phys.* **1993**, *98*, 269.

45. P. Matousek, A. W. Parker, D. Phillips, G. D. Scholes, W. T. Toner, and M. Towrie, *Chem. Phys. Lett.* **1997**, *278*, 56.

46. S. E. Wallace-Williams, B. J. Schwartz, S. Moeller, R. A. Goldbeck, W. A. Yee, M. A. El-Bayoumi, and D. S. Kliger, *J. Phys. Chem.* **1994**, *98*, 60.

47. C. Rullière, A. Declémy, and P. Kottis, *Laser Chem.* **1985**, *5*, 185.

48. D. L. Morris, Jr. and T. L. Gustafson, *J. Phys. Chem.* **1994**, *98*, 6725.

49. D. L. Morris, Jr., and T. L. Gustafson, *Appl. Phys. B* **1994**, *59*, 389.

50. J. Saltiel, G. Krishnamoorthy, Z. Huang, D.-H. Ko, and S. Wang, *J. Phys. Chem. A* **2003**, *107*, 3178.

51. J. Saltiel, S. Wang, L. P. Watkins, and D.-H. Ko, *J. Phys. Chem. A* **2000**, *104*, 11443.

52. J. Saltiel and S. Wang, *J. Am. Chem. Soc.* **1995**, *117*, 10761.

53. J. Saltiel, D.-H. Ko, and S. A. Fleming, *J. Am. Chem. Soc.* **1994**, *116*, 4099.

54. T. C. Felder, K.-J. Choi, and M. R. Topp, *Chem. Phys.* **1982**, *64*, 175.

55. C. Rullière and A. Declémy, *Chem. Phys. Lett.* **1987**, *135*, 213.

56. E. F. Hilinski, W. M. McGowan, D. F. Sears, Jr., and J. Saltiel, *J. Phys. Chem.* **1996**, *100*, 3308.

57. W. A. Yee, R. H. O'Neil, J. W. Lewis, J. Z. Zhang, and D. S. Kliger, *Chem. Phys. Lett.* **1997**, *276*, 430.

58. S. Hogiu, W. Werncke, M. Pfeiffer, A. Lau, and T. Steinke, *Chem. Phys. Lett.* **1998**, *187*, 8.

59. W. Werncke, S. Hogiu, M. Pfeiffer, A. Lau, and A. Kummrow, *J. Phys. Chem. A* **2000**, *104*, 4211.

60. Y. Hirata, *Bull. Chem. Soc. Jpn.* **1999**, *72*, 1647.

61. Y. Hirata, T. Okada, N. Mataga, and T. Nomoto, *J. Phys. Chem.* **1992**, *96*, 6559.

62. D. Zimdars, R. S. Francis, C. Ferrante, and M. D. Fayer, *J. Chem. Phys.* **1997**, *106*, 7498.

63. M. Kasha, *Disc. Faraday Soc.* **1950**, *9*, 14.

64. C. Ferrante, U. Kensy, and B. Dick, *J. Phys. Chem.* **1993**, *97*, 13457.

65. T.-a. Ishibashi and H.-o. Hamaguchi, *Chem. Phys. Lett.* **1997**, *264*, 551.

66. T.-a. Ishibashi and H.-o. Hamaguchi, *J. Phys. Chem. A* **1998**, *102*, 2263.

67. C. K. Ingold and G. W. King, *J. Chem. Soc.* **1953**, 2725.

68. S. A. McFarland and N. S. Finney, *J. Am. Chem. Soc.* **2002**, *124*, 1178.

69. D. E. Falvey and G. B. Schuster, *J. Am. Chem. Soc.* **1986**, *108*, 7419.

70. P. F. Barbara, S. D. Rand, and P. M. Rentzepis, *J. Am. Chem. Soc.* **1981**, *103*, 2156.

71. J. Ma, G. B. Dutt, D. H. Waldeck, and M. B. Zimmt, *J. Am. Chem. Soc.* **1994**, *116*, 10619.

72. C. L. Schilling and E. F. Hilinski, *J. Am. Chem. Soc.* **1988**, *110*, 2296.

73. Y.-P. Sun and M. A. Fox, *J. Am. Chem. Soc.* **1993**, *115*, 747.

74. T. Tahara and H.-o. Hamaguchi, *Chem. Phys. Lett.* **1994**, *217*, 369.

75. M. B. Zimmt, *Chem. Phys. Lett.* **1989**, *160*, 564.

76. J. Morais, J. Ma, and M. B. Zimmt, *J. Phys. Chem.* **1991**, *95*, 3885.

77. J. Ma and M. B. Zimmt, *J. Am. Chem. Soc.* **1992**, *114*, 9723.

78. W. J. Leigh and D. R. Arnold, *Can. J. Chem.* **1981**, *59*, 3061.

79. B. I. Greene, *Chem. Phys. Lett.* **1981**, *79*, 51.

80. W. J. Leigh and D. R. Arnold, *Can. J. Chem.* **1981**, *59*, 609.

81. M. B. Zimmt and P. A. Vath, *Photochem. Photobiol.* **1997**, *65*, 10.

82. W. Schuddeboom, S. A. Jonker, J. M. Warman, M. P. deHaas, M. J. W. Vermeulen, W. F. Jager, B. de Lange, B. L. Ferigna, and R. W. Fessenden, *J. Am. Chem. Soc.* **1993**, *115*, 3286.

83. E. Lenderink, K. Duppen, and D. A. Wiersma, *J. Phys. Chem.* **1995**, *99*, 8972.

84. R. W. J. Zijlstra, P. T. Duijnen, B. L. Feringa, T. Steffen, K. Duppen, and D. A. Wiersma, *J. Phys. Chem. A* **1997**, *101*, 9828.

███████ CHAPTER 20

Reactions on the Femtosecond Time Scale

JOHN E. BALDWIN

Department of Chemistry, Syracuse University, Syracuse, New York

1. INTRODUCTION

By 1960, the evolving conceptual framework of physical organic chemistry clearly recognized the significance of electronic and thermochemical interpretations of

Reactive Intermediate Chemistry, edited by Robert A. Moss, Matthew S. Platz, and Maitland Jones, Jr.
ISBN 0-471-23324-2   Copyright © 2004 John Wiley & Sons, Inc.

structure and reactivity. It espoused the ideas associated with "reactive intermediates." These conceptual insights were scarcely glimpsed by organic chemists in 1920; today these mutually supportive intellectual foundations are generally appreciated as durable and powerful supports for the impressive advances in chemical understanding achieved over the past 40 years.

The modes of thinking about structures and reactions and intermediates facilitated or even demanded by these conceptual innovations meshed productively with new tools of a different sort to drive the progress of recent decades. Advances in electronics and computers, all sorts of spectroscopy, laser optics and physics, chromatography, mass spectrometry, quantum theory and computational strategies, molecular beam experiments, and so on, radically expanded the limits of experimental and theoretical investigations.

By now many sorts of reactive intermediates, such as free radicals, carbocations, carbenes, benzynes, and Bredt's rule violating olefins, have been generated and isolated under conditions permitting full structural characterizations. Theory-based structural parameters are generally in nearly perfect agreement with experimentally determined values.

Experimental work providing information on reaction kinetics—the time dependence of reactants and products under defined conditions—served indispensably to correlate structure–reactivity data and to provide estimates of transition state energies. Theory-based definitions of transition structures gave some clues as to how reactions might actually take place. But the dynamic aspects of chemical reactions remained inaccessible, or only poorly accessible.

2. REACTION DYNAMICS

The dream or vision of being able to see atomic motions as a chemical reaction takes place has long been an aspiration, one approached with understandable diffidence and with reliance on theory-based modeling. Since the 1930s physical chemists concerned with dynamics have explored through computations and experiments on very simple reactions, involving few atoms, just how reactions take place. But the dynamics of more complicated organic reactions remained unapproachable using the experimental tools at hand before the advent of experimental femtochemistry.

With better electronic structure theory and faster computers, theory-based limitations to gaining deeper insights on reaction dynamics began to recede in the 1970s. In a landmark contribution, Salem and co-workers[1,2] calculated just how the thermal cis, trans interconversions of $1,2\text{-}d_2$-cyclopropanes take place—how one C—C bond lengthens, how methylene groups rotate to form a diradical intermediate, and how further exquisitely choreographed coupled motions of terminal methylene groups and C—C—C bond angle variations lead to formation of an isomeric labeled cyclopropane.

The sequence of three-dimensional (3D) structures for this thermal epimerization, a series of virtual photographs in timed sequence (Fig. 20.1), constituted

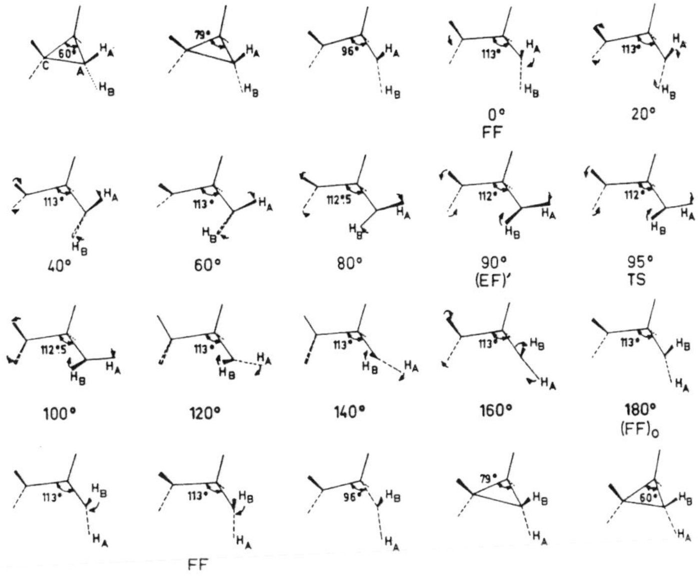

Figure 20.1. A computationally inferred lapsed-time sequence of structures representing reaction dynamics for a thermal stereomutation of cyclopropane.[1,2] [Reproduced with permission from J. A. Horsley, Y. Jean, C. Moser, L. Salem, R. M. Stevens, and J. S. Wright, *J. Am. Chem. Soc.* **1972**, *94*, 279. Copyright © 1972 American Chemical Society.] Reproduced with permission from Y. Jean, L. Salem, J. S. Wright, J. A. Horsley, C. Moser, and R. M. Stevens, *Pure Appl. Chem. Suppl.* (23rd Congr., Boston) **1971**, *1*, 197.

a chemical equivalent of stop-motion or lapsed-time photography as pioneered by Eadweard Muybridge, Etienne-Jules Marey, and Harold Edgerton.[3]

The visual and conceptual impact of seeing the timed sequence of structures, a full representation of atomic-scale events as a complex chemical reaction took place, was powerful. This achievement, the product of state-of-the-art calculations applied to an ambitious objective as well as excellent presentation graphics, was not diminished through a repressed awareness that it all depended on theory. Nothing experimentally based provided an anchor for the visually compelling rendition of the reacting system as a cyclopropane cleaved a C–C bond, formed a trimethylene diradical intermediate, and executed a net one-center epimerization before reverting to the cyclopropane structure.

As better and better methods for following fast reactions with precision were introduced and exploited, characteristic reaction times faster than a second—times measured in milliseconds (ms, 10^{-3} s), or microseconds (μs, 10^{-6} s), or nanoseconds (ns, 10^{-9} s) and then in picoseconds (ps, 10^{-12} s)—were measured through stopped-flow techniques (Chance, 1940), flash photolysis (Norrish and Porter, 1949), temperature-jump and related relaxation methods (Eigen, 1954), and then with pulsed lasers. But even with the temporal resolutions on the picosecond time scale reached in the 1970s, the time-dependent parameters observable experimentally related to changing populations of reactants and products—to reaction kinetics. Dynamic details remained hidden; the atomic-scale events transpiring at the transition state, or within the transition state region, continued to be objects of speculation, theoretical conjecture, or computation-based modeling. Direct evidence was not accessible.

Reaction dynamics as opposed to reaction kinetics strives to unravel the fundamentals of reactions—just how they transpire, how intramolecular vibrational energy redistributions provide energy to the modes most involved along the reaction coordinate, how specific reaction states progress to specific product states, why product energy distributions and ratios of alternative products are as they are, and, of course, how fast the basic processes on an atomic scale and relevant timeframe occur.

Dynamics deals with the nature of transition regions, treating them today as far more substantial ranges of configurations in time and space than one might infer from a simple, rigid calculated transition structure. The transition state concept enlarged to a transition region perspective is just "the full family of configurations through which the reacting particles evolve en route from reagents to products" at or close to the minimum energy required for the reaction to occur.[4]

Molecular hydrogen has a vibrational period of only 7.6 fs, while vibrational modes involving heavier atoms are slower. The motions of atoms in activated molecules, the various vibrations and rotations leading to chemical reactions, will generally take place over time periods of 10 to 100 or 1000 fs, the femtosecond time scale. One femtosecond is 10^{-15} s; "femton" is Swedish for 15, making the femtosecond unit, the next beyond the nicely classical milli-, micro-, nano-, picosecond sequence, easy to remember.

Intramolecular vibrational energy relaxation typically redistributes energy throughout all vibration modes in a molecule in ~ 1 ps, so the observational window for following the contributions of specific vibrations to overall reactions is narrow. One may have significant reaction events taking place much faster than the time required for complete vibrational energy redistribution—a basic presumption of transition state theory. Or a reaction may involve transition regions spacious enough to allow transition structures to exhibit the characteristics of reactive intermediates and persist over times much longer than typical vibrational periods.

To secure detailed descriptions of chemical transformations as they occur within the transition state region based on direct, real-time experimental observations is the ambitious goal of femtochemistry.[5] It is chemistry on the femtosecond time scale, the time scale appropriate to the dynamic details responsible for transformations taking place over the transition state region as bond-breaking and bond-making chemical events evolve. Sometimes the methods used are termed femtosecond transition state spectroscopy (FTS), a descriptor that emphasizes the intent to observe manifestations of chemical transformations in detail as reactants are converted to products over the femtosecond time domain while traversing the transition state region of the potential energy surface.

3. TIME-RESOLVED FEMTOSECOND DYNAMICS

Today, ultrafast pulsed-laser techniques, high-speed computers, and other sophisticated instrumentation make it possible to measure the time evolutions of reactants, intermediates, transition structures, and products following an abrupt photoactivation of a starting material. Detailed theoretical calculations, experienced judgments based on the literature, and newly accessible femtosecond-domain experimental data providing observed intensities of chemical species versus time can provide insights on the atomic-scale events responsible for overall reaction outcomes.

Such experiments depend on a pump–probe–detect strategy.[3,6] In a typical femtosecond study, a narrowly defined laser pulse is amplified and split into two components. The pump and probe pulses may each be manipulated by various nonlinear optical interactions and frequency conversion schemes to afford pulses centered at different wavelengths and having defined energies, shapes, durations, and phases. The pump component that excites the sample is typically of higher energy and intensity. The probe component is sent along a variable length path route, then directed along with the pump pulse for delivery to the sample. The probe pulse will arrive at the sample with a small time delay, as defined by the speed of light and a longer path length (Fig. 20.2).

The adjustable optical path length for the probe pulse is defined by a computer controlled translation stage for the time delay desired; given the speed of light, small variations in path length correspond to appropriate changes in the time delay. An optical path increment of 1 μm corresponds to a time delay of 3.3 fs: The translation stage does not have to be adjusted long distances to give suitable pulse delays!

The combined pump and probe beams are focused onto a supersonic beam of reactant in helium within a vacuum chamber containing a time-of-flight mass spectrometer. The pump pulse promotes the reactant to some excited state; the probe

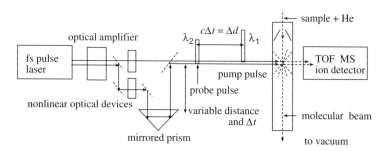

Figure 20.2. Schematic outline of typical pump–probe–detect experiments with femtosecond pulses, a molecular beam source, and mass spectrometric detection of transient species. Computer control and data processing instruments, as well as various optical components, are not shown. The time separation Δt between pump and probe pulses is dictated by the difference in optical path lengths, Δd, traversed by the two components of the original pulse.

pulse, arriving shortly thereafter, ionizes an excited neutral species that is then detected by mass spectrometry. Variations in the pulse delay time and time-dependent ion intensities reveal the rise and fall of excited species and neutral fragments.

The pump and probe pulses employed may be subjected to a variety of nonlinear optical mixing processes; they may be prepared and characterized by intensity, duration, spectral band width, and polarization. They may arrive in the reaction chamber at a desired time difference, or none. The probe pulse may lead to ionizations followed by detections of ions by mass spectrometry, but many alternatives for probing and detection have been used, such as laser-induced fluorescence, photoelectron spectroscopic detection, absorption spectroscopy, and the like.

Another alternative involves focusing the probe pulse onto the silver photocathode of an electron gun, providing a well-defined pulse of electrons through the photoelectric effect so that time-dependent electron diffraction data for short-lived intermediates may be obtained.

4. COHERENCE

The success of femtosecond time-resolved experiments depends on having lasers able to provide sufficiently short and intense pulses, together with an effective integration of all the other instrumentation, optical components, and computers necessary for the enterprise, and on the susceptibility of starting material to photoexcitation. These requirements are readily understood. Success also depends on coherent formation of the excited system through the pump pulse, a consideration not so readily grasped.

For a time there was concern over just how the uncertainty principle would limit what could be achieved with ultrafast laser pulses. The potential problem turned out to be chimerical, not a problem, once coherence came to be appreciated.[7-9] This conceptual advance was essential to the successful development of experimental femtosecond chemistry; it did not progress through better lasers and nonlinear optical tricks and faster computers alone.

The concepts and applications of coherent light within the physics and optics literatures were developed vigorously in parallel with advances in lasers and nonlinear optics. They were introduced within the context of molecular systems through nuclear magnetic resonance (NMR) spectroscopy, especially over the past 30 years. The importance of coherence in studies of chemical dynamics came to be appreciated only more recently.[7]

If the pump pulse is ultrashort, and $\Delta t \Delta E \approx h/2\pi$, where $\Delta t = \sigma_t$ and $\Delta E = \sigma_E$, then won't the uncertainty in E be too large for useful experiments? This anticipation or anxiety turned out not to be the case. A 60-fs width pulse implies an uncertainty in energy, ΔE, of $\sim$0.7 kcal/mol, a very large uncertainty when compared with the precision routinely achieved in various spectroscopic measurements—but small in relation to typical bond dissociation energies!

Femtosecond laser pulses are *not* monochromatic. The pulse duration is very well defined and very short, which means that its frequency in not precisely known, given the uncertainly principle. A femtosecond laser pulse is a coherent combination over a range of frequency components: the in-phase generation of the components defines the pulse shape, energy, and intensity. It excites a molecule to a number of close-lying energy levels, to give a coherent preparation of a superposition of states having very nearly identical molecular geometrical parameters. The femtosecond laser pulse delivers excitation energy to the molecule under study, sets the time reference for following subsequent chemical events, and, just as importantly, excites many molecular states coherently. The atoms of the excited molecules are initially localized in phase space; all vibrational states formed start to progress toward products in phase.

The energy landscape relevant to a molecular reaction is given by the potential energy surface, a function of electron distributions and nuclear coordinates. The total energy of an isolated molecule is constant and equal to the sum of the potential energy (of the molecule) and the kinetic energy (of the molecule and its constituent atoms). The kinetic energy cannot be negative, so an excited molecule may look for reaction options at energies at or below the energy imparted upon excitation and at or above the potential energy surface.

Ultrafast laser excitation gives excited systems prepared coherently, as a coherent superposition of states. The state wave function (a probability wave) is a coherent sum of matter wave functions for each molecule excited. The exponential terms in the relevant time-dependent equation, the phase factors, define phase relationships between constituent wave functions in the summation.

$$\Psi(r,t) = \sum_j c_j \Psi_j(r)\exp(-iE_j t/\hbar)$$

The matter wave function is formed as a coherent superposition of states or a state ensemble, a "wave packet." As the phase relationships change the wave packet moves, and spreads, not necessarily in only one direction; the localized launch configuration disperses or propagates with the wave packet. The initially localized wave packets evolve like single-molecule trajectories.

The initial ground-state equilibrium geometry of reactant molecules is largely preserved in the initially formed electronically excited species, with uncertainties at atomic positions much smaller than the range of changes of distances between atoms characteristic of the reaction being followed. This remarkable outcome is encompassed by quantum mechanics and the wave-particle duality of matter recognized by de Broglie (1924) and expressed as $\lambda = h/p$: the wave length (λ) associated with a particle is equal to Planck's constant (h) divided by the momentum of the particle (p). Coherent laser-pulse excitation ensures that the structural evolution of the excited molecules will follow single molecular trajectories, until the excited molecules and ensemble coherences atrophy through intramolecular or possibly intermolecular perturbations.

When Schrödinger introduced matter wave groups or wave packets in 1926 they were strictly theoretical, quantum mechanical formalisms. There were no experimentally accessible ways to prepare matter wave packets from molecules. Now they can be generated with excellent spatial resolution and temporal definition, coherently, over times much shorter than the periods associated with molecular vibrations and rotations. Once created, the matter wave packets evolve over time and reflect chemical dynamics events on the femtosecond time scale.

The photochemical excitation delivered by a narrowly defined pump laser pulse achieves three indispensable things: it sets time = 0, energizes the reactant molecules, and localizes them in space. It induces molecular coherence as excitation of each of the individual molecules involved leads to a coherent superposition of separate wave packets, a highly localized, geometrically well-defined and moving packet—analogous to a classical system, one that can be described using classical concepts of atomic positions and momentum.

5. REPRESENTATIVE FEMTOSECOND DYNAMICS STUDIES

As recently as 15 years ago direct experimental scrutiny of atomic-scale events in the femtosecond time domain was not possible. That long dreamed of goal, to follow in real time the processes involved as molecules are energized and proceed rapidly through transition regions to give reaction products, has now been attained. Exactly what molecules do and how they do it, and how fast they do it on this extremely fast time scale, is a very young but rapidly maturing branch of chemistry: femtochemistry.

The examples gathered here deal primarily with relatively simple organic reactions, though the photodissociation of one inorganic diatomic molecule is included as an historically important paradigmatic process. The examples cover but a fraction of the field, but may serve to illustrate the powers and limitations of the area at its present state of development.

5.1. Sodium Iodide

Excitation of NaI to an energy well above the exit channel for formation of Na and I atoms provides a wave packet for bound but highly excited molecules on a steep repulsive potential wall. It moves quickly to a constant energy space and vibrates within the transition state region. The probe pulse induced fluorescence signal indicative of a short, covalent species within the transition state region shows a periodic rise and fall, and an overall damping. The characteristic time for the oscillation depends on the pump pulse wavelenth: When it is centered at 311 nm τ_o is ~1.2 ps.[10] The oscillation is damped, not exponentially but in steps, for conversion to Na and I atoms occurs most efficiently when the $[Na–I]^{\ddagger}$ internuclear distance matches the distance at the surface crossing. The signals proportional to covalent $[Na–I]^{\ddagger}$ become broader with time, as the wave packet suffers some dephasing, but the coherent vibrational motion is clearly recorded over some 10 periods (Fig. 20.3).[10–13]

The excited-state potential energy curve for $[Na–I]^{\ddagger}$ has a covalent inter part and, at longer bond distances, an ionic part, a consequence of a crossing of covalent

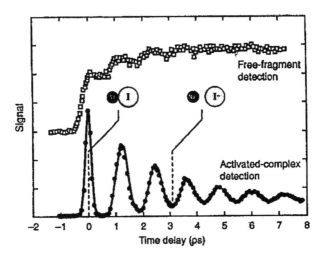

Figure 20.3. The time evolution of photoexcited sodium iodide. The signal intensity proportional to covalent ([Na–I]‡) structures oscillates following the femtosecond pump pulse; the [Na–I]‡ vibrates between covalent and ionic structures before crossing to the lower potential energy surface and dissociating to give Na and I atoms.[13] [Reproduced with permission from A. H. Zewail, *J. Phys. Chem. A.* **2000**, *104*, 5660.]

and ionic potential energy curves. The oscillation is thus between covalent bonding at short [Na–I]‡ distances and ionic bonding at longer distances. The crossing from the potential heading toward Na$^+$ and I$^-$ down onto the surface leading to Na and I atoms is rather inefficient. The probability of crossover leading to homolytic dissociation is at a maximum at the shorter [Na–I]‡ bond lengths, and vice versa.

$$Na{-}I^* \; \rightleftarrows \; Na^+ \; \cdots \cdots \cdots \; I^-$$

$$Na{-}I^* \; \xrightarrow{\text{crossing}} \; \longrightarrow \; Na + I$$

Monitoring the formation of Na atoms by laser induced fluorescence spectroscopy (at another wavelength) as a function of time shows a stepwise growth in the Na atom intensity, following the implications of the periodic vibration of the [Na–I]‡ species exactly. Thus product formation depends on both a vibrational mode and a curve crossing; both play a role along the reaction coordinate.

This experimental work on the dissociation of excited NaI clearly demonstrated behavior one could describe with the vocabulary and concepts of classical motions.[10–13] The incoherent ensemble of molecules just before photoexcitation with a femtosecond laser pump pulse was transformed through the excitation into a coherent superposition of states, a wave packet that evolved as though it represented a single vibrationally activated molecule.

5.2. Methyl Iodide

The dissociation of methyl iodide to give an iodine atom and a methyl radical has been studied extensively by a variety of experimental and theoretical techniques. Femtosecond resolved studies augmented the accumulated store of information and insight on this reaction by defining the essential two-dimensional (2D) nature of the reactive vibrational modes leading to products.[14,15]

Two-photon photoexcitation of CH_3I in a molecular beam with a narrow laser pulse at $\lambda = 315$ nm provides 184 kcal/mol of energy and generates coherent wave packets in two Rydberg states, repulsive states that may lead to C–I bond cleavage. The states are well recognized and characterized by spectroscopists; they involve promotions of a $5p$ π nonbonding electron from iodine to a σ^* orbital, giving specific $6p$ or $7s$ Rydberg states in specific vibronic levels. The signals for the two CH_3I^* states decayed very quickly, with lifetimes of 166 ± 6 and 131 ± 6 fs. Much longer decays were seen following excitations to the respective excited states of CD_3I: the lifetimes were 331 ± 26 and 415 ± 20 fs.

The pronounced isotope effects on excited-state lifetimes imply some involvement of one or more vibration modes in addition to the C–I bond distance lengthening along in the reaction coordinate. Quantum molecular dynamics calculations solving the time-dependent Schrödinger equation based on a 2D reaction coordinate model, with the CH_3 symmetrical stretching along one dimension and the C–I bond length along the other, provided a persuasive rationale for the experimental observations. The dissociation takes place initially along the symmetric CH_3 vibrational coordinate of the Rydberg states; the wave packets tunnel through a barrier across a seam to access the exit channel leading through elongation of C–I to the methyl radical and iodine atom products. The lifetime ratios for CH_3I and CD_3I Rydberg excited states stem from tunneling and the large differences in reduced mass and zero-point energies of the reactants.

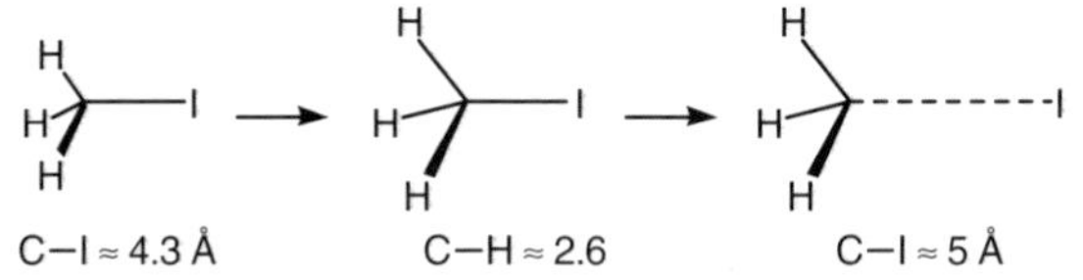

C–I ≈ 4.3 Å   C–H ≈ 2.6   C–I ≈ 5 Å

5.3. 1,3-Dibromopropane

A femtosecond time-resolved study of reactions following excitation of 1,3-dibromopropane to a $(n,5p)$ Rydberg state uncovered a striking instance of coherent dynamics in a relatively large molecular environment, one characterized by 27 internal degrees of freedom.[16,17]

The multiphoton ionization mass spectrum of 1,3-dibromopropane shows molecular ions at 200/202/204 amu, thanks to the bromine-79 and bromine-81 isotopes, weak ion intensities for $BrCH_2CH_2CH_2$ at 121/123, and a base peak at 41,

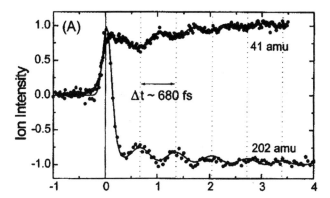

Figure 20.4. The time evolution of one isotopic version of $BrCH_2CH_2CH_2Br^*$ (202 amu) and of the allyl cation (41 amu) formed from the $BrCH_2CH_2CH_2$ radical upon ionization by the probe pulse. Both intensity versus time profiles show constant periods of 680 fs, with a phase shift of exactly one-half a period.[16] [Reproduced with permission from C. Kötting, E. W. G. Diau, J. E. Baldwin, and A. H. Zewail, *J. Phys. Chem. A* **2001**, *105*, 1677. Copyright © 2001 American Chemical Society.]

corresponding to the allyl cation. The 2,2-d_2-labeled system gives the corresponding ions at 202/204/206, 123/125, and at both 42 and 43.

The excitation of the unlabeled dibromide followed by a time-delayed probe pulse gives a time-dependent intensity profile for the 202 amu signal. It shows a rapid decay component near time zero (time constant τ_0) followed by a slower decay (τ_1). The slower decay exhibits a periodic coherent modulation (τ_c) and a gradual dephasing (Fig. 20.4).

The two species of 202 amu are taken to be the initially formed Franck–Condon structure and the parent species giving coherent resonance motion; loss of a bromine atom gives $BrCH_2CH_2CH_2$, which is detected after ionization by the probe pulse as C_3H_5, at 41 amu. The decay of $BrCH_2CH_2CH_2$ leads to cyclopropane, a product not ionized by the probe pulse, and hence not seen through mass spectrometry.

Data analysis of the transient signals leads to a deconvolution providing values for the constants $\tau_0 = 50$ fs, $\tau_1 = 2.5$ ps, and $\tau_c = 680$ fs.

The signal for the 41 amu transient, a measure of the time-dependent rise and fall of $BrCH_2CH_2CH_2$, rises ($\tau_1 = 2.5$ ps) and then decays ($\tau_2 = 7.5$ ps), and it shows the same periodic coherent modulation, with a characteristic oscillation time $\tau_c = 680$ fs, phased shifted by π radians! The local peaks of signal intensity proportionate to the $BrCH_2CH_2CH_2$ radical concentration match the local troughs of signal decay for the 202 amu periodic modulation; they are 180° out of phase.

$$BrCH_2CH_2CH_2Br \xrightarrow{h\nu} BrCH_2CH_2CH_2Br^* \longrightarrow BrCH_2CH_2\dot{C}H_2$$
$$\tau \approx 2.5 \text{ ps} \qquad\qquad \tau \approx 7.5 \text{ ps}$$

The most stable ground-state conformation of 1,3-dibromopropane is a C_2-symmetric gauche-gauche structure, shown in two perspectives below.

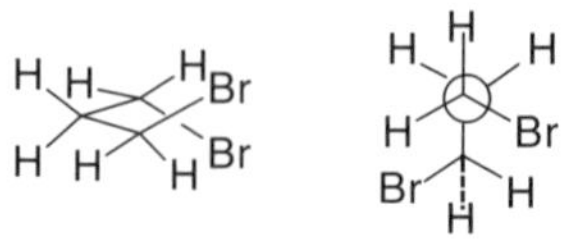

It has only one vibrational mode of low frequency, with a wavenumber $\tilde{v} \approx$ 50 cm^{-1}. This mode preserves C_2 symmetry as the BrCCC dihedral angle oscillates about an equilibrium value of 60°. A wavenumber of ~50 cm^{-1} is equivalent to a wavelength of ~2.0×10^{-2} cm, or a frequency of oscillation v of 1.5×10^{12} s^{-1}; the reciprocal gives the period of the oscillation, ~670 fs, a value quite comparable to the observed coherence ($\tau_c = 680$ fs).

The excited dibromide executes this vibrational mode and the vibrational coherence persists as some trajectories find the C—Br bond-cleavage exit channel, a reaction favored when the dihedral angle is close to 60°. The proper phasing of two vibrations, the BrCCC dihedral-angle-modifying torsional mode and the C—Br stretching vibration, leads to the cleavage of the C—Br bond.

This reading of the situation is supported by parallel studies of BrCH$_2$CD$_2$CH$_2$Br. The fact that time-dependent decays of 42 and 43 amu signals—stand-ins for the BrCH$_2$CD$_2$CH$_2$ radical—both show $\tau_c = 7.5$ ps, that is, no isotope effect, implies that the decay process does not involve rupture of a C—D bond. The loss of HBr or DBr occurs after the monobromo radical is ionized by the probe pulse.

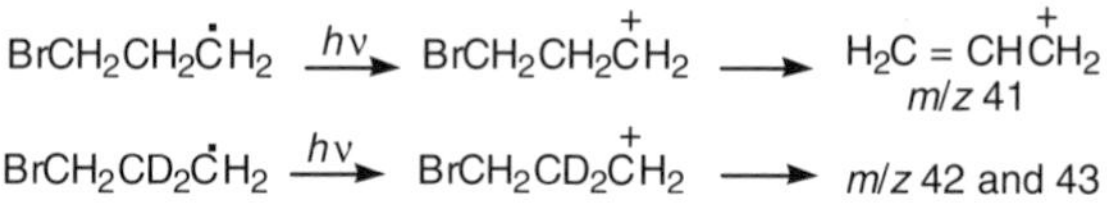

5.4. 1,4-Dehydrobenzene and the Bergman Rearrangement

Photoexciting 1,4-dibromobenzene in a molecular beam with 307-nm pump pulse photons leads to sequential loss of both bromine atoms within 100 fs; the excited dibromide has a lifetime of ~50 fs and the monobromide of ~80 fs. The p-benzyne generated decays relatively slowly, showing a lifetime of ~400 ps.[18] It presumably leads to (Z)-3-hexene-1,5-diyne through a valence isomerization of the sort associated with the Bergman rearrangement.

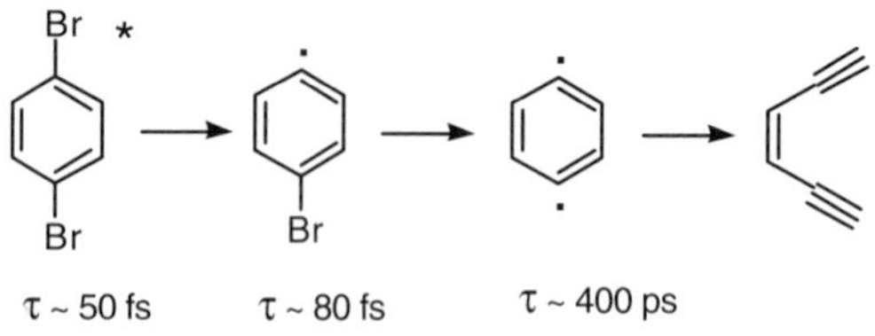

Similar femtosecond observations of the dynamics of 1,2- and 1,3-dehydrobenzene molecules starting from 1,2- and 1,3-dibromobenzene show that each dibromide gives rise to 76 amu intermediates within 100 fs, and that the 76 amu species have lifetimes of $\sim$400 fs! Whether o- and m-benzyne intermediates isomerize to p-benzyne and then decay, or just how such isomerizations might occur, remain open to further inquiry.

5.5. 1,4-Cyclohexadiene to Benzene

1,4-Cyclohexadiene may be converted to molecular hydrogen and benzene either thermally or photochemically. The thermal reaction on the ground-state potential energy surface is orbital symmetry "allowed" and is generally considered a concerted pericyclic process. Studies using femtosecond resolved mass spectrometry detected another pathway: A pump pulse excitation of the diene at $\Delta E = 186$ kcal/mol provides an electronically excited state well above the energies required for concerted loss of molecular hydrogen or for loss of a hydrogen atom with formation of a cyclohexadienyl radical.[19] The excited starting material of 80 amu decays in 95 ± 10 fs. A transient of 79 amu corresponding to the C_6H_7 radical is seen: it has a rise time of 50 ± 30 fs and a decay time of 110 ± 30 fs.

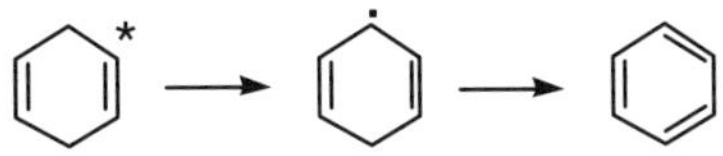

80 amu   79 amu
τ ~ 95 fs   τ ~ 110 fs

A transient at 78 amu is also observed ; since it has the same dynamic characteristics seen for the 79 transient species it may be ascribed to fragmentation of the $m/z = 79$ cation formed by ionization of the cyclohexadienyl radical induced by the probe pulse.

Careful theoretical analysis indicates a large difference in energy between the concerted pathway and the stepwise, consecutive release of two hydrogen atoms by the excited diene. The concerted elimination may well be strongly favored, and yet the nonconcerted pathway for the elimination to form benzene does contribute. The mechanism for the reaction under the given reaction conditions must include it along with the dominant concerted hydrogen molecular elimination. Further, theoretical assessments of the excited diene at the conical intersection leading rapidly to the ground-state surface suggest that it is highly asymmetrical, and that concerted loss of molecular hydrogen from the ground-state species well above the minimum energy required for the concerted elimination of molecular hydrogen will be controlled by vibrational energy redistribution into the reaction coordinate.

5.6. Norrrish Type-I Cleavages

Photoexcitation of an aliphatic ketone leads to cleavage of a C—C bond linking the carbonyl function and an alkyl residue. This α-cleavage process, known as a

Norrish type-I reaction, has been studied over the years in extreme detail, with every imaginable physical and theoretical method at hand. Data gathered through studying such reactions on the femtosecond time scale, together with new theoretical work prompted by the dynamics observed, have provided a detailed picture of the processes involved and a fresh perspective on nonconcerted α-cleavage events.[20–24]

Excitation of acetone with two photons at $\lambda = 307$ nm delivers 186 kcal/mol of energy, more than enough to break both C—C bonds and give carbon monoxide and two methyl radicals. Following the process with femtosecond mass spectrometry shows that the excited state of acetone of 58 amu rises and falls very quickly, in a spike-like fashion. It is formed and decays with a time constant of ∼50 fs. For acetone-d_6, a similar time dependence is seen: The rise and fall of the 64 amu excited species take place in ≪80 and 80 fs (Fig. 20.5).

By using a full panoply of theory—time-dependent density functional theory and CASSCF and CASMP2 ab initio methods—the states and potential-energy surface features associated with the dynamics were uncovered. The high-energy excitation of acetone gives several excited Rydberg states that reach the $S_2(n, 3s)$ surface in

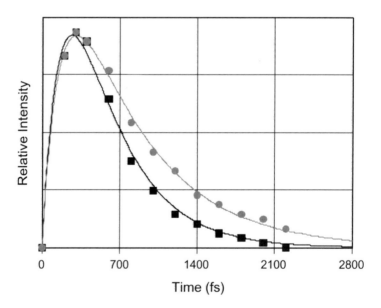

Figure 20.5. A graphical representation of the time evolution of transients for the Norrish type-I α-cleavage 43 and 46 amu fragments from acetone and from acetone-d_6. The representative sets of "data points" (■ for 43 amu, ● for 46 amu fragments) are modeled with simple buildup and decay response functions, $I(t) = A[\exp(-t/\tau_2) - \exp(-t/\tau_1)]$; the time constants of buildup and decay are τ_1 and τ_2, respectively. A modest isotope effect on the characteristic time for formation of these acyl radicals (60 and 80 fs, respectively) and a more prominent —CH_3/—CD_3 effect on decays through loss of CO (420 and 670 fs, respectively) were recorded.[24]

50 fs or less, and proceed through cleavage of a C—C bond in ~50 fs. Larger ketones decay in slightly longer times (e.g., 2-butanone, in ~100 fs; 3-pentanone, in ~160 fs), but the trend is much less pronounced than would be expected if there were full vibrational energy redistribution prior to the first cleavage reaction. The α-cleavage is highly nonstatistical. The energy redistribution from the C—O stretching mode to C—C stretching and C—C—O bending modes leading to bond cleavage is quite efficient.

Acetone in the S_2 state gives an excited-state linear acetyl radical along with the methyl radical; the radical exists in a double-well potential about the CCO angle of $180°$.

The fragment radical of 43 amu from acetone, the acetyl intermediate, builds up in ~60 fs and decays in ~420 fs. The first α-cleavage takes place in a time comparable to the vibrational period of the C—C bond, ~43 fs, while the second is much slower. The intermediate of 46 amu from acetone-d_6 rises in 80 fs and decays in ~670 fs.

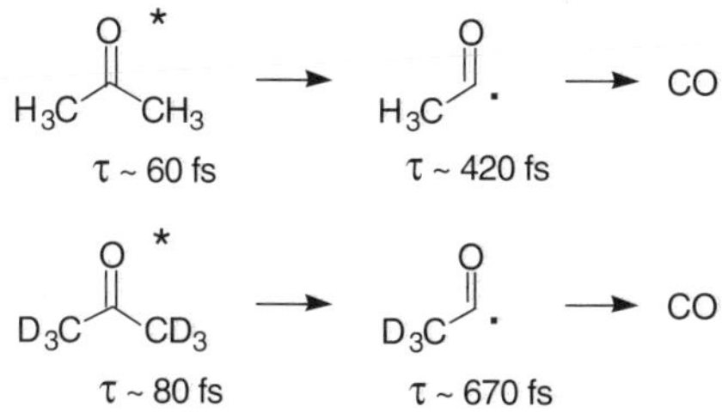

The second cleavage depends on moving vibrational energy into the C—C bond stretching coordinate while trajectories along the C—C—O bending coordinate are nonreactive. This partial vibrational energy redistribution takes in ~420 fs, substantially longer than required to break the first C—C bond. The CH_3/CD_3 isotope effect reflects nuclear motions involving the $-CH_3$ and $-CD_3$ substructures, an effect seen previously in the decomposition of methyl iodide.

When acetone is photoactivated at lower energies, with one photon at $\lambda = 307$ nm (93 kcal/mol), quite different photochemical events lead to the α-cleavage.[25] The initial femtosecond motion leads to very fast dephasing of the wave packet from the Franck–Condon region on the $S_1(n,\pi^*)$ surface, with a pulse-limited time constant of <50 fs. The photoexcited acetone on the S_1 surface then persists on a nanosecond time scale, eventually giving methyl and acetyl radicals through an α-cleavage. This fragmentation does not occur directly from S_1; it takes place indirectly, through intersystem crossing from S_1 to T_1, a state that has available a much less energetically demanding α-cleavage option.

When cyclic ketones are pumped with two photons of a $\lambda = 307$-nm femtosecond pulse, imparting some 186 kcal/mol of excitation energy, they are converted rapidly, within ~50 fs, through a Norrish type-I C—C cleavage to acylalkyl diradicals of varying size and number of degrees of freedom, N_v. From cyclobutanone to cyclodecanone, N_v ranges from 27 to 81. The decays of the acylalkyl diradicals through loss of CO (and perhaps other, minor paths) occur with characteristic times of 100 ± 20, 125 ± 10, 180 ± 10, and 180 ± 20 fs, for the diradicals formed from cyclobutanone, cyclopentanone, cyclohexanone, and cyclodecanone, respectively.[26,27]

$\tau \approx 100$ fs $\tau \approx 125$ fs $\tau \approx 180$ fs $\tau \approx 180$ fs

Statistical theories would require decreases in reaction rates that are orders of magnitude larger than the modest differences noted. The key vibrational energy redistribution leading to the second α-cleavage is restricted to modes near the acyl function and involved importantly in the reaction coordinate. These acylalkyl diradical intermediates do not achieve complete statistical redistributions of vibrational energy throughout all vibrational modes, and only then lose CO. The experimental results indicate non-RRKM behavior.

5.7. Photolysis of Cyclobutanone

The photochemistry of cyclobutanone presents a special case since the Norrish type-I cleavage to give an acylalkyl diradical intermediate releases ring-strain energy. Thus the energy available for subsequent reactions is reduced correspondingly, compared to the energy retained in an acyl radical from an acyclic ketone, or less strained cyclic ketones.

Very thorough femtosecond studies of cyclobutanone and parallel theoretical work have served to clarify the complex events following photoexcitation of cyclobutanone with a one-photon laser pulse at $\lambda = 307$ nm (excitation energy 93 kcal/mol) to the $S_1(n, \pi^*)$ state.[28] The excitation activates three vibrational modes, those for C—O stretching, C—O out-of-plane wagging, and ring puckering, and they facilitate very rapid dephasing, in <50 fs, as the initially formed wave packet propagates away from the Franck–Condon region. The S_1 intermediate is formed with ~6 kcal/mol of excess vibrational energy, and the nearby α-C—C bond dissociation barrier is only ~2 kcal/mol high. As soon as intramolecular vibrational-energy redistribution moves energy from the three initially activated vibrations into the appropriate α-C—C bond-stretching mode, an efficient conversion to the acyclic acylalkyl diradical occurs. The lifetime of the S_1 intermediate is only 5 ps; it decays some $1000 \times$ faster than S_1 states for typical aliphatic ketones, thanks to the very low energy barrier. The lifetime is limited by the α-cleavage, not by intersystem crossing to the T_1 state.

Most interestingly, the S_1 lifetimes for 3,3-d_2-cyclobutanone and 2,2,4,4-d_4-cyclobutanone are substantially longer, 9.0 ± 1.5 ps and 6.8 ± 1.0 ps, respectively. The large isotope effect on the lifetime of the d_2-system stems from vibrational energy redistribution requirements, not the sorts of factors associated with standard Bigeleisen and Mayer theory.[29,30]

5.8. Trimethylene and Tetramethylene Diradicals

The postulation of trimethylene and tetramethylene diradicals as reactive intermediates involved in many thermal isomerization and fragmentation reactions has a long history,[31] but not until 1994 had they ever been detected in real time. The validity of the "diradical hypothesis" was tested through femtosecond studies, and the tests provided dramatic evidence confirming that these short-lives species are indeed real, directly experimentally accessible chemical entities.[27,32]

The pump–probe–detect arrangements for the femtosecond experiments was similar to those described above. When cyclobutanone was pumped with two photons of a $\lambda = 307$-nm femtosecond pulse, two consecutive C—CO bond cleavages led to the formation of the trimethylene diradical, detected as an easily ionized transient at 42 amu, with buildup and decay times of 120 ± 20 fs. The decay presumably involves isomerizations to cyclopropane and to propylene—structures not ionized by the probe pulse and thus undetected during the experiment.

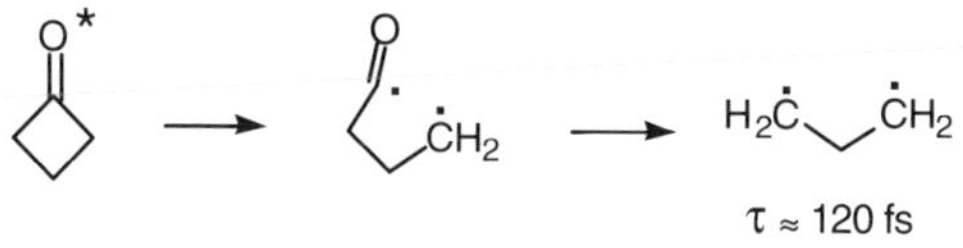

$$\tau \approx 120 \text{ fs}$$

Starting with cyclopentanone and a pump pulse delivering two photons at 310 nm, the parent mass at 84 amu grows in intensity and then decays. So does a species at 56 amu, the parent minus CO, with a buildup time of 150 ± 30 fs and a decay time of 700 ± 40 fs.[27] It attains a peak intensity in ~300 fs.

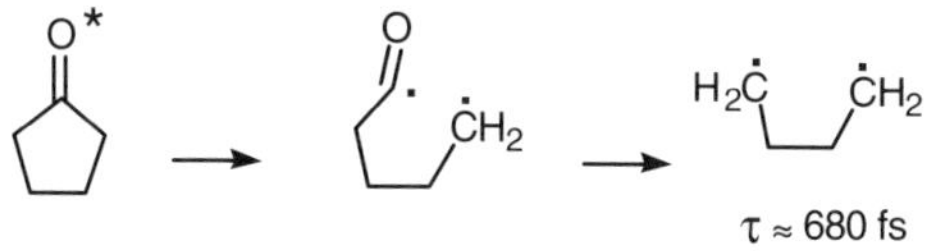

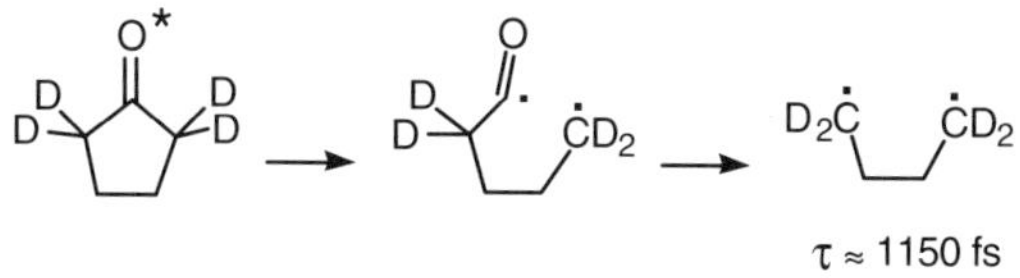

$$\tau \approx 680 \text{ fs}$$

$$\tau \approx 1150 \text{ fs}$$

The decay of the tetramethylene diradical derived from 2,2,5,5-d_4-cyclopentanone is much slower than seen for the C_4H_8 diradical.[24] Both principal decay modes, fragmentation to two ethylenes and ring-closure to cyclobutane, may be dependent dynamically on torsional motions of the terminal methylene groups.

These studies raised many interesting questions and set an ambitious agenda for further work, served admirably to establish the time scales appropriate to tetramethylene and trimethylene diradicals, and confirmed them as distinct molecular species.

Unpublished work by Herek and Zewail on the dynamics of photodecarbonylations of 2,2,4,4-d_4-cyclobutanone and 3,3-d_2-cyclobutanone uncovered that the

$CD_2CH_2CD_2$ and $CH_2CD_2CH_2$ diradicals decay at somewhat longer lifetimes, in ~183 and ~129 fs, respectively. Torsional motions of the terminal methylene groups are obviously critical to the reaction coordinate leading to vibrationally excited cyclopropane-d_i products.[33]

Calculational estimates of the lifetimes of the trimethylene diradical[34,35] based on microcanonical variational unimolecular rate theory and direct dynamics simulations have been reported. The lifetimes derived from theory, 91 and 118 fs, are comparable to the experimental estimate, 120 ± 20 fs. Similar lifetime estimates from theory for tetramethylene are comparable, or slightly below, the experimental value.[36]

Dynamics calculations have also provided new approaches to the stereochemical modes through which cyclopropanes and trimethylene intermediates may be related.[37–39] Full quantum dynamics calculations for the trimethylene diradical based on a reduced dimensionality model that followed wave packet densities and time constants for formation of products led to the conclusion that conrotatory and disrotatory "double" rotations of both terminal methylene groups are favored over a "single" rotation of just one by a 2.2:1 ratio.[40]

A related study sought the lifetime of the 1,3-cyclopentanediyl diradical, generated photochemically from the 2,3-diazabicyclo[2.2.1]hept-2-ene.[41] This diradical may be viewed as a trimethylene diradical constrained geometrically by an ethano tether, or as a tetramethylene diradical bridged by a methylene function. In either view, the 1,3-cyclopentanediyl diradical might possibly decay through cyclization to bicyclo[2.1.0]pentane, or a 1,2-hydrogen shift to form cyclopentene, or with C—C bond cleavage to give 1,4-pentadiene. Theory suggests that the ring closure is by far the preferred alternative.

The mass spectrum of diazabicyclo[2.2.1]hept-2-ene shows only a weak molecular ion and a very strong fragment at 68 amu. The femtosecond studies found that the 68 amu easily ionized transient profile could be modeled with a rise time of 30 ± 10 fs and decay time of 190 ± 10 fs, a value comparable to the decay time of trimethylene.

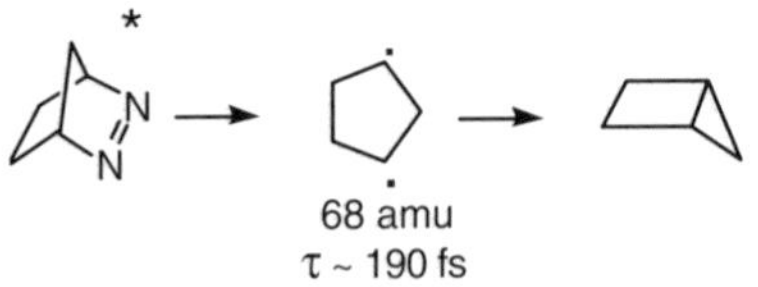

68 amu
$\tau \sim 190$ fs

Now the 1,3-cyclopentanediyl diradical is constrained to cyclize in a disrotatory fashion while the trimethylene species might well close in both disrotatory and conrotatory ways. Were all other factors constant one could infer that the geometrical restrictions imposed on the 1,3-cyclopentanediyl diradical entailed no significant deduction in rate of cyclization, and thus that conrotatory cyclization of the trimethylene diradical is not strongly preferred under the given reaction conditions and circumstances.

5.9. Oxyalkyl Diradical and Formylalkyl Radical Intermediates

Two-photon excitation at $\lambda = 307$ nm provides 186 kcal/mol of energy to tetra-hydrofuran (THF); the excited molecule breaks a C—O bond with a characteristic time of 55 ± 15 fs to form an oxytetramethylene diradical. This species, now with 114 kcal/mol of available energy, has a lifetime of 65 ± 15 fs; it decays primarily through a C—C β-cleavage reaction, giving the trimethylene diradical (42 amu, $t \sim 120$ fs). The cleavage to trimethylene is the dominant reaction.[42]

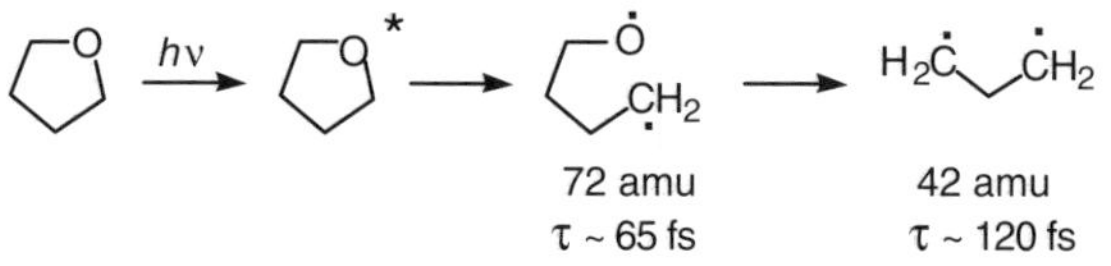

72 amu
τ ~ 65 fs

42 amu
τ ~ 120 fs

An alternative β-cleavage process contributes to a lesser extent: Loss of a hydrogen gives a formylalkyl radical of 71 amu, which loses ethylene to provide another formylalkyl radical of 43 amu. The transient species at 71 and 43 amu have decay times of ~120 fs, similar to the time seen for the trimethylene diradical.

72 amu   71 amu   43 amu

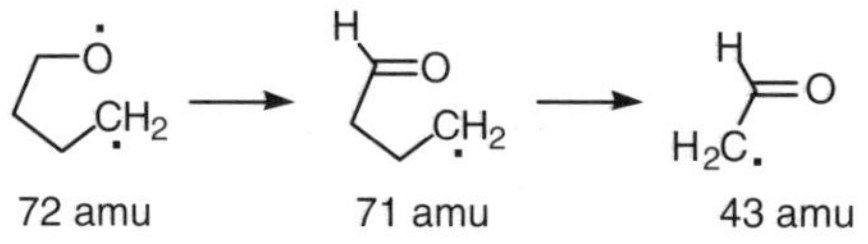

This femtosecond study confirmed the involvement of the oxytetramethylene diradical as a reactive intermediate, and found that the trimethylene formed from it had the same lifetime as the trimethylene generated through the photodecarbonyl-ation of cyclobutanone. For tetrahydropyran, the oxypentamethylene diradical (86 amu) is formed readily and the 85 amu transient, from the β-cleavage of a C—H bond, is the dominant fragmentation product.

5.10. Retro-Diels–Alder Reactions

Retro-Diels–Alder reactions have long been studied and discussed with an emphasis on whether they should be considered "concerted" or "step-wise" processes. Femtosecond real time studies of representative retro-Diels–Alder reactions of simple hydrocarbons have helped to provide an answer and to sharpen the nature of the question.[43,44]

Cyclohexene, upon excitation through a two-photon process providing 186 kcal/mol, gives two species detected through ionization by a probe pulse and mass spectrometry: a species at 82 amu, the parent structure or the diradical species formed through β-cleavage, and at 54 amu, a mass corresponding to butadiene. An ion at M-15, at 67, is also recorded. The femtosecond transients show that the 82 amu species rises in <10 fs and decays in 225 ± 20 fs; the amu 54 signal rises in 15 ± 10 fs and decays in 150 ± 15 fs.

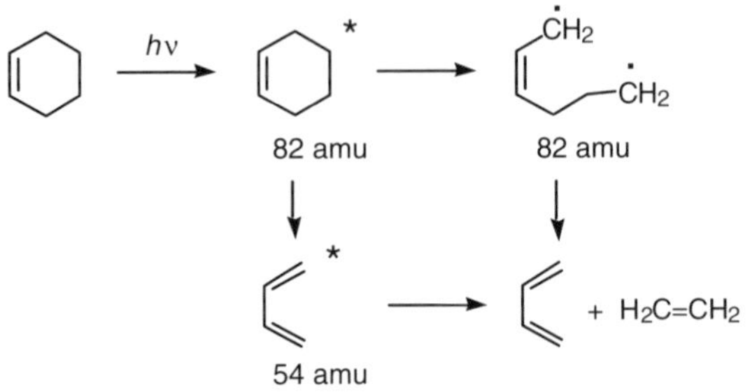

Clearly, the 54 amu species is not formed from the 82 amu entity; rather both derive from the very rapid decay of a common precursor, the photoexcited cyclohexene. The 82 amu species may then be associated with the diradical intermediate on the nonconcerted pathway, and the 54 signal with an excited form of butadiene, one vulnerable to ionization when probed with a $\lambda = 615$-nm photon (46 kcal/mol).

For norbornene, a similar situation with respect to dynamics obtains. The 94 amu species corresponding to a C_7H_{10} entity rises in <10 fs and decays in 190 ± 10 fs; the fragment species at 66 amu rises in 30 ± 15 fs and decays in 230 ± 30 fs.

For bicyclo[2.2.2]oct-2-ene, the same pattern is observed: a C_8H_{12} species at 108 forms very rapidly, rising in <10 fs and then decays in 190 ± 10 fs; the fragment at 80 amu rises in 15 ± 10 fs and decays in 185 ± 10 fs.

All three of these retro-Diels–Alder reactions give excited diene intermediates that decay in comparable times: the τ values range from 150 to 230 fs. The exact structural characteristics of these intermediates is currently unclear. Perhaps this issue could be addressed using femtosecond spectroscopic studies applying laser-induced fluorescence techniques, or through theory-based approaches.

In all cases, the diradical intermediate formed along the nonconcerted reaction pathway persists for characteristic times ranging from 190 ± 10 to 230 ± 20 fs—many times longer than the time associated with a typical C–C vibration, ~30 fs. The diradical intermediates are real enough, though the lifetimes themselves give no direct information on the relative importance of concerted versus nonconcerted alternatives on ground-state potential energy surfaces. That matter may be contingent on system-dependent couplings between symmetric and antisymmetry C–C bond stretching modes leading toward concerted and nonconcerted transition regions in cooperation with other modes contributing to the reaction coordinates.

The retro-Diels–Alder reactions studied may well involve *many* transient configurations appropriate to the potential energy surface and vibrational possibilities approaching and traversing transition regions.[45,46]

Comprehensive theoretical investigations of such retro-Diels–Alder reactions, with particular attention devoted to the very rapid transitions between photoexcited and ground-state potential energy surfaces at conical intersections, have appeared.[47]

6. FEMTOSECOND STRUCTURAL DETERMINATIONS

Several approaches to modifying femtosecond experiments are being developed so that structures, or at least structural information, as functions of time may be secured. One tactic implements an ultrafast electron diffraction strategy.

Electron diffraction structural definitions for transient species may be attained by modifying the pump–probe–detect approach through suitable instrumental modifications. The third-generation ultrafast diffraction setup at Caltech uses a femtosecond laser pulse split to provide a reaction-initiating pump pulse and a second time-delayed pulse directed to a silver photocathode of an electron gun.[48] Bursts of electrons generated through the photoelectric effect, as many as $\sim$25,000 per pulse at a pulse width of $\sim$4 ps, have been generated. Pulse widths as narrow as 1.07 ± 0.27 ps have been attained. The electron pulses are accelerated, collimated, and focused on the scattering volume containing the molecular species from pump–pulse initiated chemical events. Diffraction images are obtained using a 2D single-electron detection system based on a CCD detector, a charge-coupled device, as a function of reaction time.

Earlier work had demonstrated the concept: using CF_2I_2 as precursor, excitation at $\lambda = 307$ nm using a laser pulse gives primarily CF_2 within 4 ps, and its structure was determined by analyzing diffraction-difference data as functions of delay time. The radial distribution and molecular scattering functions led to structures for CF_2I_2 and CF_2 in excellent agreement with earlier determinations and theory. In CF_2, the C—F bond length is 1.30 Å and the FCF bond angle is 104.9°.[49]

Ultrafast electron diffraction was used to define the structure of the cyclopentadienyl radical formed through photodissociation of $CpCo(CO)_2$. The structure obtained reflected the Jahn–Teller distortion from D_{5h} symmetry, a dynamic structure thanks to pseudorotations converting dienylic and elongated conformations.[50,51]

Photoexcited IF_2C—CF_2I gives the IF_2C—CF_2 radical in $\sim$200 fs, which subsequently loses an iodine atom at a much slower rate, in $\sim$32 ps. Thus the IF_2C—CF_2 species lasts long enough to be structurally defined by electron diffraction. The time-resolved diffraction data for species within the scattering volume were analyzed to learn whether the intermediate had a bridged structure or a classical form. Excellent matches with theory-based structural parameters were realized when the IF_2C—CF_2 radical was taken to be a mixture of anti and gauche conformations of classical, nonbridged species (Fig. 20.6).[48,52–54] The spatial and temporal resolutions for these structures were estimated to be $\sim$0.01 Å and $\sim$1 ps.

anti   gauche

A still more demanding ultrafast electron diffraction study of the electrocyclic ring-opening isomerization of 1,3-cyclohexadiene to 1,3(Z),5-hexatriene was

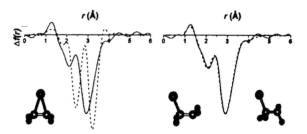

Figure 20.6. Comparisons of experimental and theoretical electron diffraction radial distribution curves based on ab initio geometries for the bridged C_2F_4I radical structure (on left) and a sum of classical anti and gauche structures (on right).[48] The intermediate present 5 ps after the pump pulse was defined structurally through 2D diffraction difference images. [Reproduced with permission from H. Ihee, V. A. Lobastov, U. M. Gomez, B. M. Goodson, R. Srinivasan, C.-Y. Ruan, and A. H. Zewail, *Science* **2001**, *291*, 458. Copyright © 2001 American Association for the Advancement of Science.]

undertaken. The time-dependent diffraction data recorded by the CCD detector provided structural data for the cyclic diene in fine agreement with the literature. Difference curves at various delay times were consistent with depletion of covalent and next-nearest-neighbor C—C pairs in the starting diene; a number of new positive contributions at distances ranging from $\sim$3.5 to 5.8 Å were evident, indicative of new C—C interactions at separations longer than those found in the cyclic diene.[48] Given the conformational flexibility of the product triene, and the lack of heavy atoms in the system, the agreement between experimentally recorded and theory-based diffraction-difference-defined radial distribution functions may be considered remarkable. The electrocyclic ring opening was observed directly following femtosecond excitation: Still greater sensitivity and more narrowly defined electron pulses may bring the day of full structural visualizations of reactions in progress closer.

and conformers

Another approach would use X-ray crystallography; promising preliminary examples have been reported, but much more needs to be done before the prospect is realized in any general sense.[55]

7. MORE COMPLEX REACTIONS

This introduction to experiments following chemical reactions in the femtosecond time domain could be extended to considerations of other gas-phase reactions involving molecular structures of similar complexity. Or one could extend this chapter extensively and attempt to survey femtosecond chemical investigations of far more complex systems and dynamic issues, including studies involving other phases of matter—gases, liquids, solids, clusters, surfaces, and biological systems. There have been reports on the femtosecond dynamics of DNA assemblies, and bacteriorhodopsin photochemistry, and electron-transfer processes in proteins, and the photodissociation of carbon monoxide from its complex with myoglobin: one might say, "The sky's the limit!"

There are technical challenges still to be overcome before every thought experiment in this brave new realm of molecular science can be realized in practice, but there is good reason for optimism. While "simple" gas-phase reactions of comparatively small molecules will continue to attract serious attention, the forefronts of femtochemistry today encompass far wider perspectives.

8. CONCLUSION AND OUTLOOK

The experimental and theoretical strategies of femtochemistry have provided telling insights on chemical dynamics over the past 15 years. The breakthrough examples and many of the prototypical organic reactions that have been reported already permit some important generalizations.

Experimentally, it has become possible to follow chemical reactions on a femtosecond time scale with excellent time resolution—with precise definitions of $t = 0$, time separations between pulses, and with very narrow pulse widths.

Each pump–probe sequence may involve on the order of a million or a billion independent molecules; the excitations give a spatially well-defined coherent matter wave packet. An initially randomly oriented ensemble of molecules is launched through the narrow pump laser pulse to give molecules with nuclear–nuclear distances defined to $\sim$0.1 Å. The sequence typically is repeated many, many times, over the full range of relevant times, so that the relatively weak signals detected and the intrinsic scatter of the experimental data may be signal averaged. Hence, the absolute imperative of precisely defined timing for the pulses, to initiate a reaction with a pump pulse and to monitor progress along the reaction coordinate using a probe pulse and some detection strategy such as mass spectrometry. Without excellent synchronization, the computer implemented signal averaging would give blurred composites of the individual runs, and little useful information.

Excitations of molecules with femtosecond laser pulses lead to excited-state matter wave packets coherently, launching them with such well-defined spatial resolution and coherence in nuclear motions that they evolve like single-molecule trajectories. Both electronically excited and vibrationally excited ground-state species may be studied. The structural change versus time profile of a reaction turns out to be compatible with classical modes of thinking.

Reaction dynamics may be controlled by the cooperative conjunction of two or more vibrational variables: A rate may be dependent on the correct phasing of such modes, rather than simply upon achieving or exceeding an energy barrier.

A similar understanding has emerged from recent semiclassical trajectory calculations for organic reactions: Each molecule gives a product defined by the potential energy surface, the specific distribution of vibrational energy among the available modes, and the relative phases among these modes. The reaction outcome is not dependent on the potential energy surface alone. The overall distribution of products and energies observed represents the composite of all of the individual vibrational-mode-and-phase distinctive reactant molecules as each follows its trajectory.

In a variety of ways, deuterium-substituted reactants can provide vital information on dynamics: Substantial isotope effects point to vibrational modes playing a key role in dynamics, and vice versa.

The once rather ephemeral transition state construct derived from logic and statistical mechanics, a virtual entity, has emerged as an experimental reality. Structural changes associated with specific nuclear vibrations in energized molecules in the transition region may be correlated with reaction dynamics.

The transition state concept, once understood in static terms only, as the saddle point separating reactants and products, may be fruitfully expanded to encompass the transition region, a landscape in several significant dimensions, one providing space for a family of trajectories and for a significant "transition state lifetime." The line between a traditional transition structure and a reactive intermediate thus is blurred: The latter has an experimentally definable lifetime comparable to or longer than some of its vibrational periods.

SUGGESTED READING

D. L. Smith, "Coherent Thinking," *Eng. Sci.* **1999**, LXII (4), 6.

O. M. Sarkisov and S. Ya. Umanskii, "Femtochemistry," *Russ. Chem. Rev.* **2001**, *70*, 449.

M. Chergui, "Femtochemistry," *Chimia* **2000**, *54*, 83.

G. Roberts, "Femtosecond Chemical Reactions," *Philos. Trans. R. Soc. London A* **2000**, *358*, 345.

R. Hoffmann, "Pulse, Pump & Probe," *Am. Sci.* **1999**, *87*, 308. See also R. Hoffmann, *Am. Sci.* **1998**, *86*, 326; **1999**, *87*, 21; **2000**, *88*, 14.

A. H. Zewail, "Femtochemistry," *J. Phys. Chem.* **1993**, *97*, 12427.

A. H. Zewail, "Transient Species at Femtosecond Resolution," *Proc. Robert A. Welch Found. Conf. Chem. Res.* **1994**, *38*, 129.

A. H. Zewail, "Femtochemistry: Recent Progress in Studies of Dynamics and Control of Reactions and Their Transition States," *J. Phys. Chem.* **1996**, *100*, 12701.

A. H. Zewail, "Femtochemistry: Chemical Reaction Dynamics and Their Control," *Adv. Chem. Phys.* **1997**, *101*, 3, 892.

A. H. Zewail, "Femtochemistry. Atomic-scale Resolution of Physical, Chemical and Biological Dynamics," *Proc. Robert A. Welch Foundation Conf. Chem. Res.* **1997**, *41*, 323.

A. H. Zewail, "Femtochemistry: Atomic-Scale Dynamics of the Chemical Bond Using Ultrafast Lasers (Nobel lecture)," *Angew. Chem., Int. Ed. Engl.* **2000**, *39*, 2586.

A. H. Zewail, "Femtochemistry: Atomic-Scale Dynamics of the Chemical Bond," *J. Phys. Chem. A* **2000**, *104*, 5660.

A. H. Zewail, "Femtochemistry. Past, Present, and Future," *Pure Appl. Chem.* **2000**, *72*, 2219.

REFERENCES

The references cited here depend heavily on investigations reported by the Professor Ahmed H. Zewail, the 1999 Nobel Laureate in Chemistry, and his collaborators at the California Institute of Technology. In them, one will find extensive citations of work leading up to recent advances in femtochemistry as well as to contemporary studies from other laboratories.

1. Y. Jean, L. Salem, J. S. Wright, J. A. Horsley, C. Moser, and R. M. Stevens, *Pure Appl. Chem., Suppl. (23rd Congr., Boston)* **1971**, *1*, 197.

2. J. A. Horsley, Y. Jean, C. Moser, L. Salem, R. M. Stevens, and J. S. Wright, *J. Am. Chem. Soc.* **1972**, *94*, 279.

3. S. J. Baskin and A. H. Zewail, *J. Chem. Educ.* **2001**, *78*, 737.

4. J. C. Polanyi and A. H. Zewail, *Acc. Chem. Res.* **1995**, *28*, 119.

5. A. H. Zewail and R. B. Bernstein, *Chem. Eng. News* **1988**, *Nov. 7*, 24.

6. R. Hoffmann, *Am. Sci.* **1999**, *87*, 308.

7. A. H. Zewail, *Laser Phys.* **1995**, *5*, 417.

8. A. H. Zewail, *Angew. Chem., Int. Ed. Engl.* **2001**, *40*, 4371.

9. A. H. Zewail, *Nature (London)* **2001**, *412*, 279.

10. M. J. Rosker, T. S. Rose, and A. H. Zewail, *Chem. Phys. Lett.* **1988**, *146*, 175.

11. T. S. Rose, M. J. Rosker, and A. H. Zewail, *J. Chem. Phys.* **1988**, *88*, 6672.

12. V. Engel, H. Metiu, R. Almeida, R. A. Marcus, and A. H. Zewail, *Chem. Phys. Lett.* **1988**, *152*, 1.

13. A. H. Zewail, *J. Phys. Chem. A* **2000**, *104*, 5660.

14. H. Guo and A. H. Zewail, *Can. J. Chem.* **1994**, *72*, 947.

15. M. H. M. Janssen, M. Dantus, H. Guo, and A. H. Zewail, *Chem. Phys. Lett.* **1993**, *214*, 281.

16. C. Kötting, E. W. G. Diau, J. E. Baldwin, and A. H. Zewail, *J. Phys. Chem. A* **2001**, *105*, 1677.

17. C. Kötting, E. W. G. Diau, T. I. Sølling, and A. H. Zewail, *J. Phys. Chem. A* **2002**, *106*, 7530.

18. E. W. D. Diau, J. Casanova, J. D. Roberts, and A. H. Zewail, *Proc. Nat. Acad. Sci. U. S. A.* **2000**, *97*, 1376.

19. S. De Feyter, E. W. G. Diau, and A. H. Zewail, *Phys. Chem. Chem. Phys.* **2000**, *2*, 877.

20. S. K. Kim, S. Pedersen, and A. H. Zewail, *J. Chem. Phys.* **1995**, *103*, 477.

21. S. K. Kim and A. H. Zewail, *Chem. Phys. Lett.* **1996**, *250*, 279.

22. S. K. Kim, J. Guo, J. S. Baskin, and A. H. Zewail, *J. Phys. Chem.* **1996**, *100*, 9202.

23. E. W. G. Diau, C. Kötting, T. I. Sølling, and A. H. Zewail, *ChemPhysChem* **2002**, *3*, 57.

24. T. I. Sølling, E. W. G. Diau, C. Kötting, S. De Feyter, and A. H. Zewail, *ChemPhysChem* **2002**, *3*, 79.

25. E. W. G. Diau, C. Kötting, and A. H. Zewail, *ChemPhysChem* **2001**, *2*, 273.

26. E. W. G. Diau, J. Herek, Z. H. Kim, and A. H. Zewail, *Science* **1998**, *279*, 847.

27. S. Pedersen, J. L. Herek, and A. H. Zewail, *Science* **1994**, *266*, 1359.

28. E. W. G. Diau, C. Kötting, and A. H. Zewail, *ChemPhysChem* **2001**, *2*, 294.

29. J. Bigeleisen and M. G. Mayer, *J. Chem. Phys.* **1947**, *15*, 261.

30. J. Bigeleisen, *J. Chem. Phys.* **1949**, *17*, 675.

31. J. E. Baldwin, in *The Chemistry of the Cyclopropyl Group*, Vol. 2, Z. Rappoport, Ed., John Wiley & Sons, Inc., Chichester, **1995**, pp. 469–494.

32. R. Baum, *Chem. Eng. News* **1994**, *Nov. 28*, 6.

33. J. E. Baldwin, T. B. Freedman, Y. Yamaguchi, and H. F. Schaefer, *J. Am. Chem. Soc.* **1996**, *118*, 10934.

34. C. Doubleday, Jr., *J. Phys. Chem.* **1996**, *100*, 3520.

35. C. Doubleday, Jr., K. Bolton, G. H. Peslherbe, and W. L. Hase, *J. Am. Chem. Soc.* **1996**, *118*, 9922.

36. C. Doubleday, Jr., *Chem. Phys. Lett.* **1995**, *233*, 509.

37. C. Doubleday, Jr., K. Bolton, and W. L. Hase, *J. Phys. Chem. A* **1998**, *102*, 3648.

38. C. Doubleday, Jr., K. Bolton, and W. L. Hase, *J. Am. Chem. Soc.* **1997**, *119*, 5251.

39. K. Bolton, W. L. Hase, and C. Doubleday, Jr., *Ber. Bunsen-Ges. Phys. Chem.* **1997**, *101*, 414.

40. E. M. Goldfield, *Faraday Discuss.* **1998**, *110*, 185.

41. S. De Feyter, E. W. G. Diau, A. A. Scala, and A. H. Zewail, *Chem. Phys. Lett.* **1999**, *303*, 249.

42. A. A. Scala, E. W. D. Diau, Z. H. Kim, and A. H. Zewail. *J. Chem. Phys.* **1998**, *108*, 7933.

43. E. W. G. Diau, S. De Feyter, and A. H. Zewail, *Chem. Phys. Lett.* **1999**, *304*, 134.

44. B. A. Horn, J. L. Herek, and A. H. Zewail, *J. Am. Chem. Soc.* **1996**, *118*, 8755.

45. D. K. Lewis, B. Brandt, L. Crockford, D. A. Glenar, G. Rauscher, J. Rodriquez, and J. E. Baldwin, *J. Am. Chem. Soc.* **1993**, *115*, 11728.

46. D. K. Lewis, D. A. Glenar, S. Hughes, B. L. Kalra, J. Schlier, R. Shukla, and J. E. Baldwin, *J. Am. Chem. Soc.* **2001**, *123*, 996.

47. S. Wilsey, K. N. Houk, and A. H. Zewail, *J. Am. Chem. Soc.* **1999**, *121*, 5772.

48. H. Ihee, V. A. Lobastov, U. M. Gomez, B. M. Goodson, R. Srinivasan, C.-Y. Ruan, and A. H. Zewail, *Science* **2001**, *291*, 458.

49. J. Cao, H. Ihee, and A. H. Zewail, *Chem. Phys. Lett.* **1998**, *290*, 1.

50. H. Ihee, J. S. Feenstra, J. Cao, and A. H. Zewail, *Chem. Phys. Lett.* **2002**, *353*, 325.

51. S. Zilberg, and Y. Haas, *J. Am. Chem. Soc.* **2002**, *124*, 10683.

52. H. Ihee, J. Kua, W. A. Goddard, III, and A. H. Zewail, *J. Phys. Chem. A* **2001**, *105*, 3623.

53. H. Ihee, A. H. Zewail, and W. A. Goddard, III, *J. Phys. Chem. A* **1999**, *103*, 6638.

54. H. Ihee, B. M. Goodson, R. Srinivasan, V. A. Lobastov, and A. H. Zewail, *J. Phys. Chem. A* **2002**, *106*, 4087.

55. Y. Tanimura, K. Yamashita, and P. A. Anfinrud, *Proc. Natl. Acad. Sci. U. S. A.* **1999**, *96*, 8823.

CHAPTER 21

Potential Energy Surfaces and Reaction Dynamics

BARRY K. CARPENTER

Department of Chemistry and Chemical Biology, Cornell University, Ithaca, NY

Reactive Intermediate Chemistry, edited by Robert A. Moss, Matthew S. Platz, and Maitland Jones, Jr.
ISBN 0-471-23324-2   Copyright © 2004 John Wiley & Sons, Inc.

1. TOPOLOGY OF POTENTIAL ENERGY HYPERSURFACES

1.1. Introduction: Beyond the Potential Energy Profile

Generations of organic chemists have learned to think about the reactions they study in terms of potential energy (PE) profiles of the kind shown in Figure 21.1. While such graphs have been helpful in guiding our thinking about the thermo-chemistry and kinetics of reactions, there is growing recognition that they may also have been misleading, at least in some circumstances.

The principal problem arises from compressing the structural changes that occur during a reaction into a single "reaction coordinate." In reality, a unimolecular reaction of a nonlinear, N-atom molecule requires $3N - 6$ geometrical coordinates to describe fully the structural changes that occur between reactant and product. Thus a graph that seeks to depict potential energy as a function of molecular struc-ture should be a $(3N - 5)$ dimensional hypersurface, often just called a potential energy surface (PES). For most people, this object is unimaginable, and so some projection onto a smaller number of geometrical coordinates is necessary if the plot is going to be useful for understanding a reaction. It is the limit of that projection—reducing $3N - 6$ down to just one geometrical coordinate—that creates the familiar PE profile, but there is good reason to think that sometimes this may be taking things too far. Addition back of even one more geometrical coordinate reveals properties of PESs that cannot be seen in the usual PE profile. Sec-tions 1.2–1.4 describe some of the more commonly occurring of these topological features.

1.2. The Stepwise versus Concerted Controversy

In a reaction that involves the formation or scission of more than one covalent bond, there is inevitably a question of timing. Do the bonding changes all occur at once (concertedly) or not?[1] The question threatens to become ill-defined when one

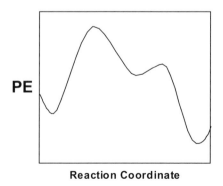

Reaction Coordinate

Figure 21.1. A typical PE profile.

recognizes that there might be a continuum of possibilities from a synchronous concerted process, in which all of the bonding changes occur to the same extent at the same time, through varying degrees of asynchronous transformation, to the final limit of a stepwise process in which, say, one bond is fully broken before a second one begins to break. In order to try to formulate a better defined mechanistic inquiry, one might therefore recast the question to ask whether the reaction involves an intermediate. At first sight, it would seem that the stepwise mechanistic limit would require the creation of an intermediate, but, as we will see, even this attempted clarification cannot always produce a black-or-white, experimentally testable question.

The issue is conveniently illustrated with examples from nominally pericyclic reactions, where the controversy about concertedness has been the most heated. The first example is the vinylcyclopropane rearrangement,[2] depicted (somewhat unrealistically as we will see) in Scheme 21.1. The overall transformation is clear: One must break the bond between C1 and C2 of the reactant, and then reconnect C2 to C3′. The controversy resides in the details of the stereochemistry and timing of these two events.[3] First the stereochemistry: The bond formation at C2 can occur with retention or inversion of its original configuration. In addition, the bond formation at C3′ can occur to the same face (suprafacial) of the allylic unit (C1−C2′−C3′) to which C2 was originally attached, or to the opposite face (antarafacial). These two stereochemical variables combine to create four possible stereoisomeric products, for which the two-letter designations are shown in Scheme 21.1. Experimental study of the stereochemistry would appear to allow one to differentiate between two extremes of the timing question. If the breaking of the C1−C2

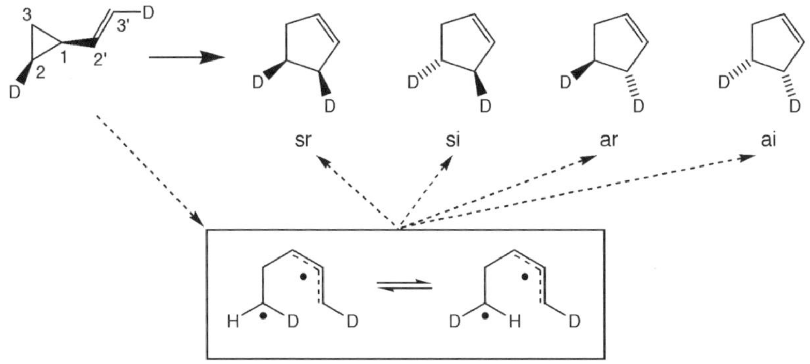

Scheme 21.1. The four possible stereoisomeric cyclopentenes-d_2 that could be formed by C1−C2 scission of the indicated single enantiomer of vinylcyclopropane-d_2. One possible mechanism, involving interconverting achiral biradicals, is also depicted. The pairs of letters under each product isomer indicate the stereochemical changes between reactant and product. Note that in an actual experiment, the reaction would be complicated by competitive C1−C3 scission. In this scheme s = superfacial, a = antarafacial, r = retention, and i = inversion.

bond and the formation of the C2–C3′ bond occurred concertedly, the reaction would be pericyclic (a [1,3] sigmatropic rearrangement) and, hence, subject to the Woodward–Hoffmann rules of orbital symmetry conservation.[4] These rules indicate that, for a thermal reaction, the products designated si and ar would be "allowed" whereas the other pair, sr and ai, would be "forbidden." The alternative, stepwise mechanism would presumably involve the singlet-state biradical depicted in Scheme 21.1. On average, this biradical might be expected to be achiral (in the same way that butane is achiral despite the fact that some of its low-energy conformations are chiral). In addition, it might be expected to undergo facile internal rotations about the remaining C–C bond to C2. Since the intermediate is achiral, and since achiral intermediates are supposed to give only achiral or racemic products, this mechanism would seem to predict that the product isomers related to each other as enantiomers should be formed in equal amounts, namely, [sr] = [ai] and [si] = [ar].

Before discussing the experimental results, let us make the connection to the topic of this section—PE profiles. The representation of concerted or stepwise mechanisms on a PE profile seems clear cut. If the reaction is stepwise it involves an intermediate, appearing as a local minimum on the PE profile—such as that near the center of the profile in Figure 21.1. On the other hand, a concerted reaction can have no intermediate, and so its PE profile should have just a reactant minimum and a product minimum connected by a single transition state.

In reality, the reaction could not be persuaded to go exactly as shown in Scheme 21.1, because the C1–C3 bond would certainly break at very nearly the same rate as C1–C2. In the experiments actually conducted by Baldwin et al.,[3c] this problem was resolved by deuterium labeling *both* C2 and C3—creating diastereomerically pure, but achiral molecules. Even then, there remained a large number of technical difficulties, which in the end the researchers were able to overcome. Their results indicated that the four stereochemical courses for the reaction run at 300 °C were sr 23%, si 40%, ar 13%, and ai 24%. These numbers do not fit the expectations from either mechanism. Clearly, the Woodward–Hoffmann "forbidden" and "allowed" products are formed in nearly equal amounts ([sr] + [ai] = 47%; [si] + [ar] = 53%)—hardly what one would expect for a pericyclic reaction. On the other hand, the stereochemical paths do not show the pairwise equalities expected from the stepwise mechanism.

The current explanation of these results comes from a combination of high-level ab initio electronic-structure calculations and molecular-dynamics simulations (see Section 3).[5] The picture that emerges from the electronic-structure calculations is that breaking the C1–C2 bond of the reactant creates a biradical that sits on a kind of plateau on the PES. This means that the biradical can undergo a variety of quite substantial geometrical changes that are accompanied by very little (<1 kcal/mol) change in its PE. There are exits to all four possible products from the plateau region, with little or no barrier to the formation of any of them. This situation cannot be depicted in a traditional PE profile because of its overly compressed dimensionality. Furthermore, the kinetics of such reactions cannot be properly described by the traditional statistical models, such as Rice–Ramsperger–Kassel–Marcus

(RRKM) or transition state theory (TST) (see Section 2). The product ratios are determined by the detailed dynamics of the reaction, as described in Section 3.

There is computational and/or experimental evidence that a structurally diverse range of singlet biradicals occupy plateau-like regions of their PESs.[6] A selection of them, including the reactions that they mediate, are shown in Scheme 21.2. The occurrence of a plateau at a mechanistically crucial region of the PE hypersurface guarantees that there can be no clear distinction between "concerted" and "step-wise" mechanisms, and also strongly indicates that the reaction may be one whose course will not be describable by any simple kinetic model.

A plateau on the PE hypersurface is not the only feature that can serve to muddy the distinction between concerted and stepwise reactions. Another occurs when an intermediate, even one in a quite deep PE "well," is located off the direct path from reactant to product. An example comes from the nominal [3,3] sigmatropic rearrangement of 1,2,6-heptatriene. If heated alone, in the gas phase, the reactant rearranges cleanly to the [3,3] product, 3-methylene-1,5-hexadiene. However, Roth et al.[7] showed that an intermediate could be trapped when the reaction was conducted in high-pressure oxygen (!). They postulated that the intermediate was the

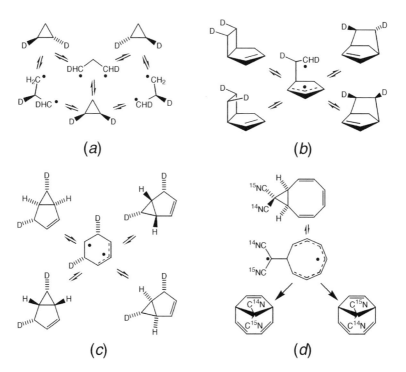

Scheme 21.2. Some of the reactions for which there is evidence of mediation by a biradical on an energy plateau. See Ref. (6a–e).

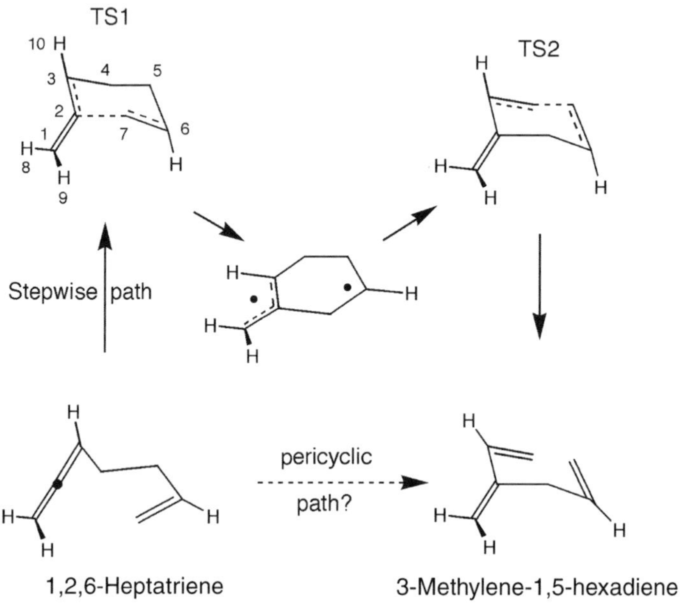

Scheme 21.3. Possible mechanisms for the rearrangement of 1,2,6-heptatriene.

biradical shown in Scheme 21.3. Curiously, when the trapping efficiency was studied as a function of O_2 pressure, it was discovered that roughly one-half of the rearrangement occurred without an interceptible intermediate. This observation led the authors to suggest that there must be competitive stepwise and concerted mechanisms for the reaction.[7] However, subsequent high-level ab initio calculations by Hrovat et al.[8] found no evidence for a pericyclic transition state; only the stepwise pathway could be located. This apparent disagreement between theory and experiment could be rationalized when the detailed geometries of the biradical and the transition states for its formation and conversion to 3-methylene-1,5-hexadiene were examined. Of particular interest was the dihedral angle between the H8–C1–H9 plane and the C4–C3–H10 plane. In the reactant this dihedral angle is 90°, since the planes in question are at the ends of an allene. Interestingly, in both of the transition states found in the calculations, the corresponding dihedral angle was still near 90°. However, in the biradical intermediate the dihedral angle is 0° in order to achieve allylic stabilization for one of the radical sites. This dihedral-angle difference means that the biradical can be thought of as being displaced from the two transition states along a direction that is roughly at right angles to the reaction coordinate. Figure 21.2 illustrates the point. As a consequence of the PES topology, one can contemplate a mechanism in which some molecules would follow a path (red arrow in Fig. 21.2) from the first transition state directly to and over the second, lower energy one. Other molecules may follow routes (or trajectories, as we will discuss later) that miss the second transition state and, instead fall into the biradical "trap." The oxygen-trappable intermediate would be generated only by molecules

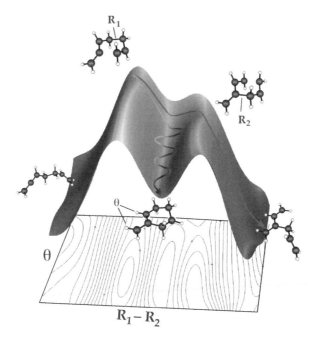

Figure 21.2. Schematic PE surface for the rearrangement of 1,2,6-heptatriene. The geometrical coordinates, θ and R_1-R_2 are defined as follows: θ is the dihedral angle between the H8–C1–H9 and C4–C3–H10 planes. R_1 is the C4–C5 distance and R_2 is the C2–C7 distance. See Scheme 21.3 for atom numbering (see color insert).

following the general direction of the blue arrow in Figure 21.2.[8] Molecular dynamics calculations (see Section 3) have provided support for this picture.[9] For the present discussion, the important conclusion is that the mechanistic distinction between stepwise and concerted mechanisms becomes moot on PESs of this kind.

In summary, one can recognize that the familiar PE profile may sometimes hide features of a PE hypersurface that become apparent when even a single additional geometrical coordinate is added to the graph. While the profile may encourage us to think that any reaction involving making or breaking of several covalent bonds should be describable as either stepwise or concerted, the higher dimensional representation reveals that there are probably many reactions for which that distinction is not meaningful.

1.3. Bifurcations and the Nature of Transition States

Most—although, as we will see, not all—of the chemically interesting places on a PES correspond to stationary points—that is, places where the partial first derivatives of the PE with respect to all geometrical coordinates are simultaneously zero. Two familiar classes of features on the PES meet this criterion: minima—corresponding to reactants, products, or intermediates—and transition states. These classes are distinguished by the partial *second* derivatives of the PE with respect to the geometrical coordinates. The second derivatives describe the direction of curvature of the hypersurface. For a minimum, all of the partial second derivatives are positive (or, more accurately, all of the eigenvalues of the Hessian or second-derivative matrix are positive) because the PES is heading upward in all directions. For a transition state, there is one (and only one) direction in which the PES is curved downward. Computing partial second derivatives is an important part of the calculation of the vibrational frequencies for a molecule or transition state, although the calculation is usually done in so-called mass weighted coordinates, which allow convenient separation of kinetic and potential energy terms in the equations of motion. The curvature of the PES at a transition state means that it will always have a single normal mode of imaginary frequency. The atomic displacements of this normal mode serve to define the reaction coordinate.

The geometrical properties of the PES in the vicinity of a transition state mean that the steepest descent path down from the transition state (also generally calculated in mass-weighted coordinates, and called the intrinsic reaction coordinate or IRC) will usually lead only to a single reactant in one direction and a single product (or intermediate) in the other. However, a transition state can sometimes be "shared" by more than one reactant and/or product. One of these cases arises when the PES possesses a so-called valley-ridge inflection point (VRI).[10]

It is easy to construct a simple mathematical function that has such a point. One begins by plotting out a simple cubic equation of the form shown in Eq. 1.

$$z = c_0 + c_1 y + c_2 y^2 + c_3 y^3 \tag{1}$$

Then, at each of the turning points of the curve one constructs a parabola in the xz plane. The parabola at the higher (larger z value) of the turning points is chosen to curve upward while the one at the lower turning point is chosen to curve downward. Now suppose that the z direction represents the PE of a molecule, while the x and y directions represent two geometrical coordinates (q_1 and q_2). The resulting PE surface is shown in Figure 21.3.

Because of our construction choices, both of the stationary points on this surface are transition states. However, one of them (TS1) has its reaction coordinate along the y direction (q_1), whereas the other (TS2) has it along the x direction (q_2). Imagine now trying to follow the IRC down from TS1 toward TS2. At first, the IRC must be along the q_1 direction because that is the reaction coordinate for TS1. However, at some point, as we begin to approach TS2 the reaction coordinate direction has to change from q_1 to q_2. The point where this occurs is the VRI point. It is located, in this simple case, by calculating the second derivative of the PE with respect to q_2 at each point along the IRC. At TS1 this quantity must be positive, since q_2 is the upward-curved direction. However, as we proceed along the IRC it must decrease in magnitude, on its way to becoming negative at TS2. The

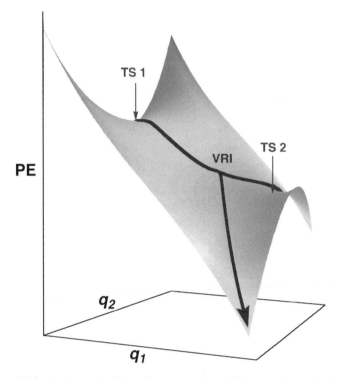

Figure 21.3. A schematic PE surface possessing a VRI point (see color insert).

only way to get smoothly from a positive quantity to a negative one is to go through zero. The place where the second derivative goes to zero is the VRI.

The chemical significance of this rather abstract idea is that the IRC bifurcates at the VRI, which is shown by the blue line in Figure 21.3. Consequently, the reactant and product that are linked by TS2 are either both products for TS1 or both reactants for TS1, depending on which way the reaction proceeds. In principle, there could be a VRI on each side of TS1, meaning that two reactants and two products would be interconverted via a single transition state. There could even be additional VRIs on either side of TS2, which would increase the number of reactants and products that could share TS1.

The VRIs are chemically relevant features on a PE hypersurface even though they do not happen to be stationary points. They represent perplexing places for traditional kinetic models, such as TST, because these models have no way of predicting what fraction of molecules will choose one path or the other at the bifurcation. In other words, TST cannot tell you what the product ratio will be in a reaction that occurs via a VRI. Several examples of such reactions are now known.[11]

The VRIs are not the only place where bifurcations can occur. The mechanism of the 1,2,6-heptatriene rearrangement that is summarized in Figure 21.2 also depicts a bifurcation (the red arrow or the blue arrow) occurring at TS1. At first sight, this would seem to contravene the requirement that the PE hypersurface be downward curved in only one direction at a transition state. However, it really does not. It is true that the steepest descent path is uniquely defined at a transition state, but that fact does not stop one from taking a nonsteepest descent path that is still all downhill but not along the IRC. For the surface depicted in Figure 21.2, the IRC begins along the direction of the blue arrow. The red arrow is a nonsteepest descent path. Why would one ever follow anything but the IRC down from a transition state? That is also a question to which transition state theory can provide no insight because the answer depends on details of reaction dynamics to be described in Section 3.

1.4. Conical Intersections

So far our discussion has presumed that, for a given reaction, a single PES is sufficient to describe all of the interesting chemistry. For thermal reactions, this is usually (but not always) a reasonable approximation. For photochemistry it never is. By its very nature, photochemistry involves the generation of one or more electronic excited states, each of which has its own PES. How excited-state surfaces are related to each other and to the ground-state surface is an issue of crucial importance to understanding photochemistry. It is an interesting historical fact that the discussion of this topic in almost all of the organic photochemistry literature was based on erroneous understanding, until about the mid-1980s. In order to grasp the somewhat challenging concept of conical intersections, which turn out to be at the root of the problem, it is perhaps useful to repeat this error and then to discuss its origins and consequences. The photochemistry of butadiene serves as a convenient example.

The photochemical disrotatory closure of butadiene to cyclobutene has been described with a state-correlation diagram, like that shown in Figure 21.4.[12] It is based on the familiar orbital-correlation diagram of Woodward and Hoffmann,[4] from which the intended correlations indicated by the dashed lines can readily be deduced. The solid lines indicate that there is an avoided crossing, which is put in as a result of the quantum mechanical noncrossing rule. It says that two states of the same total symmetry cannot cross. Instead, as they approach each other in energy, they will mix and separate, as the solid lines indicate.

The description of the photochemical ring closure of butadiene that derives from this picture is as follows. Absorption of an ultraviolet (UV) photon by the butadiene generates the ^1A state. The ^1A state evolves along the correlation line to the ^1A state of cyclobutene until it encounters an allowed crossing with the excited ^1S state. It hops over to that state and rolls down to the local minimum on the upper surface. From there it drops down to the maximum on the ground-state surface, allowing it either to return to the reactant or proceed on to ground-state cyclobutene.

So what's the error in this description? It turns out to be the invocation of the noncrossing rule. It has been known for many decades that PE hypersurfaces for states of the same total symmetry actually *can* cross.[13] Quite why this information

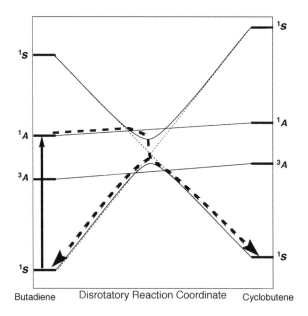

Figure 21.4. A state-correlation diagram depicting the avoided-crossing representation of butadiene photochemistry.

did not make it from the chemical physics community to the organic chemistry community is not clear, but its consequences are significant.

For an N-atom molecule, the PES of each state has $3N - 5$ dimensions ($3N - 6$ geometrical coordinates plus the energy coordinate). Two hypersurfaces for electronic states of the same symmetry are allowed to cross in $3N - 7$ dimensions. Potential energy hypersurfaces for electronic states of different spin and/or spatial symmetry can cross in $3N - 6$ dimensions. Note that diatomic molecules ($N = 2$) represent a special case. For them, $3N - 7 = -1$ and so in this case the noncrossing rule *is* rigorously obeyed. But for all molecules of three or more atoms, crossing between PE hypersurfaces for states of the same symmetry *is* permitted.[13] The crossing is usually depicted in the two special coordinates (i.e., the difference between $3N - 5$ and $3N - 7$) along which the degeneracy of the two states is lifted. These coordinates are usually given the somewhat forbidding names of the nona-diabatic coupling vector and the gradient difference vector.[14] When projected onto these two coordinates, the crossing between the two surfaces takes on the geometry of a pair of cones touching at a point—hence the name "conical intersection." It is shown schematically in Figure 21.5.

The chemical significance of conical intersections is that they provide sites of unit efficiency for return from an excited electronic state to the ground state. It turns out that the probability of (nonradiative) hopping between two electronic states is inversely dependent on the energy gap between them. So the return from the excited-state minimum to the ground-state maximum in Figure 21.4 would be a

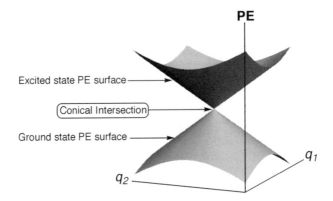

Figure 21.5. A schematic representation of a conical intersection between two electronic states of a molecule. Coordinates q_1 and q_2 are the nonadiabatic coupling vector and the gradient difference vector, along which the degeneracy between the states is lifted (see color insert).

low-probability event. By contrast, since the energy gap between states at a conical intersection is zero, the probability of returning from the excited to the ground state is unity.

Since the crossing between electronic states is itself describable as a hypersurface, there is no single molecular geometry at which this can be said to occur. Instead, it is usually assumed that the most probable geometries at which return to the ground state will occur correspond to energy minima on the crossing hypersurface. Modern computer programs capable of doing ab initio electronic structure calculations allow one to search for such structures.

The significance of conical intersections for organic photochemistry is twofold. First, their molecular geometries are frequently quite unlike any of the minima or transition states on the ground-state surface.[14] Consequently, the return to the ground state can occur at structures and energies that are essentially never accessed in any thermal reaction. Second, the initial direction of the trajectories across the ground-state surface is restricted to the plane defined by the nonadiabatic coupling vector and the gradient difference vector. Inspection of Figure 21.5 may suggest that this latter requirement hardly represents much of a restriction since it would appear to allow motion in any direction at all. However, it is important to remember that Figure 21.5 is a schematic projection of a hypersurface onto just the two unique coordinates. The restriction of the initial trajectories to this plane is actually quite severe.

In the case of butadiene photocyclization, detailed computational studies have revealed that some features of the avoided-crossing model were correct, but others were not.[15] As the older mechanism had suggested, there is a crossing from the initially accessed excited state to a nominally doubly excited state that is dropping in energy along a disrotatory coordinate. However, this crossing occurs at a geometry of C_1 symmetry, and is therefore necessarily a conical intersection. The disrotatory motion evolves on the new excited-state surface until a conical intersection between it and the ground state is encountered. Examination of the nonadiabatic coupling and gradient difference vectors shows that the disrotatory ring closure to cyclobutene is completed on the ground-state surface.

2. STATISTICAL KINETIC MODELS

2.1. Introduction

Construction of at least a partial PE hypersurface for a reaction is obviously an important step in the elucidation of its mechanism. Of special interest in the context of this volume is investigation of the part of the PES in the vicinity of a putative intermediate. Typically, one would like to know whether there actually is an intermediate formed during a reaction of interest, and if so what its properties are. Modern ab initio electronic-structure calculations can be very helpful in providing some of that information, but so far there are few who would be willing to accept their results as substitutes for experimental facts. The principal historical challenge for the experimental study of reactive intermediates has been that many have lifetimes that are too short to permit direct detection. The development of lasers with pulse widths in the nanosecond to subpicosecond range has gone a long way to overcoming that problem, but laser-flash experiments are still largely limited to reactions that can be initiated photochemically. Even today, there exist few good ways to detect transient intermediates directly formed in thermal reactions, because the steady-state concentrations of such species are too low, especially given the fact that they have to be distinguished from structurally related reactants and products present in concentrations that are orders of magnitude higher.

The paucity of generally applicable direct methods for observing thermally generated reactive intermediates has led over the years to the development of a variety of techniques for indirectly detecting their presence and deducing their properties. Many of these techniques depend, explicitly or implicitly, on kinetic models, particularly TST. Since the later discussions in this chapter will question the general applicability of these models, at least as they have been typically employed by organic chemists, it seems appropriate to begin by reviewing their basic assumptions.

2.2. Concepts and Approximations

2.2.1. Phase Space.[16] It will be useful here to anticipate a formulation that we will use in more detail in Section 3, namely, the solution of the classical equations of motion for the atoms of a molecule undergoing a chemical reaction. One starts with a molecule of defined geometry (say, in Cartesian coordinates) and with defined velocities for each of its atoms (expressible as components in the x, y, and z directions). The problem then is to solve Newton's second law of motion, $\mathbf{F} = m\mathbf{A}$, for each atom. The force, $\mathbf{F}$, can be calculated as the first derivative of the PE with respect to the x, y, and z coordinates of each atom, the mass, m of each atom is known, and so Newton's law allows one to deduce the acceleration, **A**, on each atom. Together with its known velocity, the acceleration permits the calculation of a new position for each atom (i.e., a new geometry for the molecule) after some time step δt. At the new geometry one starts the calculation all over again. When these time steps are strung together, they define the dynamics of the molecule—the way that its shape and PE change with time. The change in shape could be traced out as a path across the PE hypersurface, but this would not be a complete record of the calculation, because at each point we need to know not only the molecular geometry but also the velocity components on each atom. A complete specification of these quantities for a N-atom molecule thus requires $6N$ variables (or $6N - 12$ if we are not interested in the translational and rotational motion of the molecule). The time evolution of the coordinates and velocities can be described as the path followed by a point in a $(6N - 12)$ dimensional space, called the phase space of the system. The path for the molecule is called a trajectory. In order for the reader to make a connection between this very sketchy description of classical trajectory calculations and the much more detailed ones in the cited references, it is necessary to mention one technical issue. It turns out that Newton's second law can be transformed from a set of second-order differential equations in positions and velocities to a set of first-order differential equations in positions and momenta. This reformulation, which is due to Hamilton, permits an easier numerical solution, and so is the basis for most descriptions of trajectory calculations that one finds in textbooks. Similarly, the concept of phase space is often cast in terms of the positions and momentum components of each atom rather than their positions and velocities, but the two formulations are equivalent.

For reasons described in Section 3, solving the equations of motion is usually time consuming, even with modern computers. In the 1930s, when many of the theories of chemical kinetics were being developed, electronic computers did not exist, and so there was no hope of taking this approach for systems of any complexity. Instead, theories were developed that sought to avoid the calculation of individual molecular trajectories, while still being able to describe how an *ensemble* of trajectories, for a large number of molecules simultaneously, would behave *on average*. For the later discussion, it will be useful to describe here two different kinds of ensembles that appear frequently in the discussion of kinetic models.[17] In a *microcanonical* ensemble, every molecule has the same total energy, but a different position in phase space. In a *canonical* ensemble, the molecules have a range of energies corresponding to a Boltzmann distribution at a defined temperature, and again they have different positions in phase space.

2.2.2. The Transition State Hypothesis.

The general idea that a transition state is located at a saddle point on the PES, as detailed in Section 1.3, is familiar to most organic chemists. However, the original concept of a transition state started out as something rather different. In the development of both transition state and RRKM theory, the transition state was defined as the location of a plane (actually a hyperplane) in phase space, perpendicular to the reaction coordinate.[18]

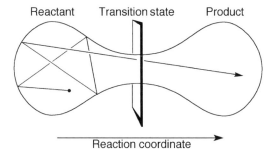

Figure 21.6. Schematic representation of the relative phase-space volumes available to reactant, transition state, and product. A plane located at the most constricted place has the highest probability of being crossed only once by a molecular trajectory, which is the location of the transition state.

The location was chosen so that all trajectories starting on the reactant side of the phase space would proceed on to the product side with unit probability once they had crossed this plane. In other words, *no trajectory should recross the plane and return to the reactant side*. The proposed existence of such a plane was called the transition state hypothesis. Its location is shown schematically in Figure 21.6. The peanut-shaped object in this figure is supposed to be a symbolic representation of the volume of phase space for a microcanonical ensemble of molecules in the reactant region, product region, and between the two. For passage to occur between reactant and product regions, the total energy of our ensemble must be at least slightly higher than the value of the maximum PE along the reaction coordinate. In the reactant and product regions, where the minimum PEs are low, our molecules will consequently have considerable amounts of excess energy. That allows them to have large velocities for their atoms and/or large numbers of structures that deviate significantly from the minimum-energy ones. In other words, the volume of phase-space accessible to the ensemble is large. However, in the vicinity of the PE maximum the amount of excess energy is greatly reduced, and the phase-space consequently becomes constricted. Hence, the peanut shape. If one now imagines the trajectories of our molecules, rattling more or less randomly around in the available phase space, it is apparent that the constriction in the vicinity of the potential energy maximum should force most of the trajectories to flow roughly parallel to the reaction coordinate. Consequently, this is the location for the plane that should minimize the chance of trajectory recrossing.

The recrossing question is important because, if none occurs, the number of trajectories traversing the plane per unit time defines the rate of product formation, which is one of the fundamental quantities one wants to get from any kinetic theory. It also turns out that a famous approximation of TST—the supposed thermal equilibrium between reactant and transition state molecules—arises as a direct consequence of the nonrecrossing hypothesis.[18c]

2.2.3. The Statistical Approximation. This approximation gives the statistical kinetic models their name; it is the one that permits estimations of rate constants to be made without the need for any trajectory calculations, and it is also the one whose careless application is going to be most heavily criticized in the following pages of this chapter. There are a variety of equivalent statements of the statistical approximation.[18] One is to say that, for a microcanonical ensemble of reactant molecules, all states of the same total energy are equally likely to be populated throughout the reaction. Another is to say that reactant–molecule trajectories are *ergodic* (i.e., they explore all of the available phase space in the reactant region before passage through the transition state). A third is to say that the lifetime (time prior to reaction) of a molecule in the reactant region is random, and therefore that the population in this region decays (by reaction) exponentially. In the end, though, what these statements all imply is that the rate of intramolecular vibrational energy redistribution (IVR) is much faster than the rate of passage through the transition state.

Some classic experiments[19] that have contributed to the acceptance of the statistical approximation have led to the conclusion that the effective "rate constant" for IVR in a typical polyatomic molecule is of the order of $10^{12}\,\text{s}^{-1}$. In other words, if an ensemble of molecules could be prepared in a defined vibrational state, they would be roughly one-half way toward having a statistical distribution of their vibrational energy (in the absence of collisions) within about a picosecond. No serious disagreement with this general picture arises from the dynamical simulations and experiments cited here. Proponents of a statistical model go on to say, correctly, that if the lifetime of an intermediate is very much longer than a picosecond, then IVR will be largely complete before it reacts, and therefore the statistical approximation will be valid by definition. It is the final step in the argument that is brought into question by the recent studies. The assumption is made that an intermediate facing a PE (or, more accurately, standard free energy) barrier to product formation of more than a few $k_B T$, where k_B is the Boltzmann constant and T is the temperature, will have a lifetime significantly longer than a picosecond, and by the preceding arguments, will consequently have suffered near-complete IVR. The problem is that the argument is circular. By associating a lifetime (or equivalently a rate constant) with a barrier height one is accepting a statistical kinetic model whose validity the exercise was supposed to be testing. In molecular dynamics simulations, one often finds that there exists no well-defined rate constant for the reaction of a thermally generated reactive intermediate—its concentration is not describable by a single-exponential decay. In particular, there can be components of the population that react much faster than would have been expected from any statistical kinetic model because the dynamics of their formation place them on more-or-less direct trajectories over the barrier(s) to formation of one (or sometimes more) of the products. At this stage in the research, it is not clear whether one can ever say that a barrier to product formation of a given magnitude will be sufficient to ensure statistical behavior of any reactive intermediate.

If the statistical approximation were correct, one could estimate the rate constant for a microcanonical ensemble of reactant molecules by estimating the volume of phase space available to the reactant and the transition state or, in a quantized version, the number of rotational and vibrational states available in each region. It is this latter, counting approach that turns the estimation of rate constants into a statistical calculation.

2.3. RRKM Theory[18]

The RRKM theory is the most widely used of the microcanonical, statistical kinetic models.[20] It seeks to predict the rate constant with which a microcanonical ensemble of molecules, of energy E (which is greater than E_0, the energy of the barrier to reaction) will be converted to products. The theory explicitly invokes both the transition state hypothesis and the statistical approximation described above. Its result is summarized in Eq. 2

$$k(E) = \frac{\sigma N^{\ddagger}(E - E_0)}{h\rho(E)} \tag{2}$$

where $k(E)$ is the rate constant, σ is a statistical factor that counts reaction path degeneracies arising from possibly different symmetries of the reactant and transition state, $N^{\ddagger}(E - E_0)$ is the number of vibrational states of the transition state between E_0 and E, h is Planck's constant, and $\rho(E)$ is the density of vibrational states in the reactant at energy E. This expression needs some modification in order to include rotational states properly (and to ensure conservation of angular momentum), but that will not be discussed further here.[18d]

Of course, in a thermal reaction, molecules of the reactant do not all have the same energy, and so application of RRKM theory to the evaluation of the overall unimolecular rate constant, k_{uni}, requires that one specify the distribution of energies. This distribution is usually derived from the Lindemann–Hinshelwood model, in which molecules **A** become activated to vibrationally and rotationally excited states **A*** by collision with some other molecules in the system, **M**. In this picture, collisions between **M** and **A*** are assumed to transfer energy in the other direction, that is, returning **A*** to **A**:

$$\mathbf{A} + \mathbf{M} \underset{k_{-1}}{\overset{k_1}{\rightleftharpoons}} \mathbf{A}^* + \mathbf{M}$$

$$\mathbf{A}^* \xrightarrow{k_2} \text{Products}$$

Application of the usual steady-state approximation to this mechanism reveals that

$$k_{uni} = \frac{k_1 k_2 [\mathbf{M}]}{k_{-1}[\mathbf{M}] + k_2} \tag{3}$$

Under the conditions used for most organic reactions, it is reasonable to assume that the concentration of collision partners, [**M**], is sufficiently high that

$k_{-1}[\mathbf{M}] \gg k_2$, in which case $k_{uni} = k_2(k_1/k_{-1})$. This is condition called the high-pressure limit. The trick then is to note that k_1/k_{-1} defines the equilibrium ratio $[\mathbf{A}^*]/[\mathbf{A}]$, which can be determined from the Boltzmann distribution. Thus, if one calculates $k_2(E)$ for a range of energies above E_0 (the higher the value of E and the larger the number of steps between E_0 and E, the more accurate the calculation, but the longer it takes) from the RRKM Eq. (2), multiplies each of these microcanonical rate constants by the Boltzmann factor $\exp(-E/k_B T)$, and then adds up the results, the result is the RRKM value for k_{uni}. Importantly, provided the corrections for rotational effects are treated properly, the RRKM expression for k_{uni} in the high-pressure limit is the same as the transition state theory result, which is discussed next.

2.4. Transition State Theory[21]

Transition state theory, as worked out by Evans and Polanyi, and by Eyring, uses the techniques of equilibrium statistical mechanics to derive an expression for k_{uni}. It depends explicitly (and not surprisingly) on the transition state hypothesis and implicitly on the statistical approximation. The original papers on TST do not discuss any of the versions of the statistical approximation listed in Section 2.2.3, but that the theory does nonetheless make that approximation is clear given that the TST result can be derived as the high-pressure-limit version of RRKM theory where the invocation of the statistical approximation is unambiguous. Alternatively, one can recognize that the assumed maintenance of a canonical ensemble of reactant molecules at a well-defined temperature throughout the reaction only makes sense if the repopulation of all energetically accessible parts of the reactant phase space occurs more rapidly than the passage through the transition state.

One version of the TST result is given by Eq. 4:

$$k_{uni} = \frac{k_B T}{h} \frac{Q^{\ddagger}}{Q} e^{-E_0/k_B T} \tag{4}$$

where Q and $Q^{\ddagger}$ are the partition functions of the reactant and transition state, respectively. Alternatively, one can recognize that the ratio of partition functions and the exponential term together define the hypothetical equilibrium constant between reactant and transition state. This transformation leads to the familiar thermodynamic formulation of TST (Eq. 5) that most organic chemists will encounter:

$$k_{uni} = \frac{k_B T}{h} e^{-\Delta G_0^{\ddagger}/RT} \tag{5}$$

where $\Delta G_0^{\ddagger}$ is defined as the standard free energy of activation (which is itself decomposable into enthalpy and entropy components in the usual way).

It is perhaps worth pointing out that several texts claim that the TST result is *exact* provided that quantum mechanical effects such as tunneling are negligible. However, for the purposes of the discussion in Section 3, one needs to be clear what "exact" means. An expression for a rate constant can only be exact if the rate constant itself is well defined. That will be true only if the reactant population decays exponentially with time (the unimolecular rate constant then being the proportionality constant in the exponent), and that in turn will be generally true only if the statistical approximation is correct. Thus the TST result is exact if quantum mechanical effects are negligible, *and* if the statistical approximation is correct.

2.5. Variational Transition State Theory[22]

We have seen that TST depends on the nonrecrossing of the transition state plane for the accuracy of its predicted rate constant. To the extent that trajectories *do* recross the transition state, the real rate constant will be reduced. One could consequently seek to find the optimum location for the transition state plane by looking for the position along the reaction coordinate that *minimizes* the computed rate constant. This idea is behind variational transition state theory (VTST). It comes in two varieties that differ in how the idea is implemented. In canonical VTST, one admits the possibility that the ideal transition state may not be located at the maximum in the potential energy profile, but wherever the ideal position is, one assumes that it is fixed and independent of the energies of individual molecules. Thus the calculation is to find the transition state position along the reaction coordinate, $q^{\ddagger}$, that makes $dk_{\mathrm{uni}}(T)/dq^{\ddagger} = 0$. This condition will always locate the transition state at a maximum on the *standard free energy surface* rather than a maximum on the PE surface. Obviously, this search for the optimum transition state is more work than simply assuming that it occurs at the PE maximum as conventional TST does. For reactions with large barriers, the improvement in k_{uni} by using the VTST approach may not be worth the extra effort because the optimum transition state is likely to be very close to the conventional one. However, for reactions with small or zero conventional barriers, the VTST approach is essential.

The most sophisticated and computationally demanding of the variational models is microcanonical VTST. In this approach one allows the optimum location of the transition state to be energy dependent. So for each $k(E)$ one finds the position of the transition state that makes $dk(E)/dq^{\ddagger} = 0$. Then one Boltzmann weights each of these microcanonical rate constants and sums the result to find k_{uni}. There is general agreement that this is the most reliable of the statistical kinetic models, but it is also the one that is most computationally intensive. It is most frequently necessary for calculations on reactions with small barriers occurring at very high temperatures, for example, in combustion reactions.

3. NONSTATISTICAL DYNAMICS

3.1. Introduction: Elements of Molecular Dynamics[23]

The basic idea behind molecular dynamics (MD) simulation was presented in Section 2.2.1. From a specified set of initial conditions one wishes to solve Newton's second law of motion or, equivalently, Hamilton's equations, in order to predict how the system under study will evolve in time. This task has three prerequisites: (a) some way of selecting the initial conditions; (b) some way of calculating the potential energy of the system, as well its first derivatives with respect to nuclear positions, at all steps of the trajectory; (c) some way of numerically integrating the equations of motion. It will be convenient to discuss these requirements in reverse order.

The reason that one needs a numerical integration algorithm is easy to understand. The argument will be framed in terms of Newton's equations, but a similar analysis applies to the Hamiltonian formulation. In the first step of a MD simulation, Newton's second law allows one to calculate the accelerations experienced by each atom of a system, given their masses and the forces acting on them. What one then needs to do is to use the computed accelerations, plus the velocities and positions specified in the initial conditions to predict some new set of coordinates for the atoms. However, since velocities are first-time derivatives and accelerations are second-time derivatives of position, this task is going to require some sort of integration. If the forces (and hence accelerations) remained constant, or varied in some simple way with time one might be able to do the integration analytically. Unfortunately, for all systems of chemical interest the forces acting on the atoms change in very complicated ways as the system moves around on the PES. It is for that reason that one needs some numerical integration scheme. Inevitably, the trajectory will be calculated as a sequence of small time steps. How small depends on the rate of change of the forces with time, and the sophistication of the integration algorithm, but for systems of interest to the organic chemist time steps will typically be 0.1 to 1 fs (i.e., 10^{-16}–10^{-15} s). Whether the steps have been small enough is easily checked by making sure that total energy conservation is obeyed (i.e., the sum of the potential and kinetic energies must be constant). It is also prudent to use the more rigorous (but costly) check of time reversal. Since the system being simulated is purely classical, the equations of motion are deterministic and so at any point of a trajectory one should be able to reverse the arrow of time (by changing the signs of the velocity or momentum components on the atoms) and follow the trajectory backward to its starting point.

For the systems discussed in this chapter, a typical duration of a complete trajectory might be of the order of a picosecond (10^{-12} s), for other kinds of problems it could be several orders of magnitude longer. Even a 1-ps trajectory is likely to require 10^3–10^4 steps, and at each step one must calculate the PE of the system plus the $3N$ first derivatives with respect to nuclear coordinates. In principle, there are two general ways of approaching this task. One is to have calculated ahead of time a complete PE hypersurface for the system, and to have expressed it as some sort of algebraic function of the nuclear coordinates. This task turns out to be almost impossible for systems of a size interesting to organic chemists. The alternative is to use so-called direct dynamics in which some sort of computation of the PE and its derivatives is done on the fly as the trajectory is evolving. The problem with this approach is easily seen. Suppose one used a sophisticated post-Hartree–Fock electronic structure calculation, as one might want to do in order to get an accurate estimate of the energy. With current computational resources and algorithms, such a calculation might take on the order of 1 min for a medium-sized organic molecule. That means a 1-ps trajectory might take as long as a week of CPU time, at the end of which you would know how one single molecule had reacted. A typical simulation requires several thousand complete trajectories with different initial conditions in order to get a sample of how the whole system will behave.

An obvious solution to this problem would be to compromise on the quality of method used to estimate the PE at each point. Indeed, if one goes as far as using molecular mechanics (MM) for the purpose, it becomes possible to run MD simulations on full-size proteins.[24] The drawback is that MM methods are not capable of describing the making and breaking of covalent bonds, and so such techniques have not been useful for the kinds of reactions discussed in this chapter. There have been efforts to hybridize molecular mechanics and quantum mechanics (QM) so that the reactive centers in a molecule can be treated by the QM part and the nonreactive ones by the MM part.[25] Some considerable successes have been reported with these techniques, but they are not yet easy to use for the nonspecialist because the division of molecules into reactive and nonreactive components, and then the correct blending of the MM and QM descriptions of these two parts, requires careful thought and a fair amount of experience. One might expect that the recent great success of density functional theory (DFT) for electronic-structure calculations would carry over into MD simulation, and indeed there has been available for some time a direct dynamics technique due to Car and Parrinello that relies on DFT methods. The Car–Parrinello approach has also enjoyed some considerable success, but has not been widely used in reactions involving covalent bonding changes, perhaps because DFT with a plane-wave basis set is not well suited to describing such processes.[26]

Probably the most widely used method for carrying out MD simulations on reactions of the kind considered here is the AM1–SRP technique introduced by Truhlar and co-workers.[27] In this approach, one carries out high-level ab initio calculations on the stationary points for a reaction of interest and then reparametrizes a semiempirical molecular orbital model, such as AM1, to fit the results. The resulting model is capable of the very fast calculation of PE and its derivatives required for direct dynamics, but has parameters that are specific for the reaction of interest, and have to be readjusted for a new reaction. The SRP part of the acronym stands for specific reaction parameters. Even this technique is not without its problems. The fitting procedure requires that one try to reproduce not only the relative potential energies of the key stationary points but also their molecular geometries and vibrational frequencies (which give information about the local curvature of the PE hypersurface). Unfortunately, energy, geometry, and frequency are measured in different units, and so there is no algorithmic way to specify an optimum fit to them all simultaneously. The reparametrization requires some subjective assessments of the relative weights to be given to various properties of the stationary points, and so it could well be that two people, given the same set of ab initio results, would end up with two different "best fit" AM1 parameters. Despite these difficulties, the AM1–SRP approach seems to be the most generally applicable one at the moment, at least for reactions involving covalent bonding changes in medium-sized organic molecules.

The selection of initial conditions for a trajectory needs to be done carefully. A single trajectory, even if completely and accurately representative of reality, would still tell one only about the behavior of a single molecule (assuming the reaction is unimolecular). The behavior of that molecule would almost certainly depend heavily on the initial conditions (its geometry and the velocity or momentum components assigned to its atoms) that were selected to start the calculation. In order to assess how a collection of molecules will behave it is therefore necessary to take some sort of statistical sample of the possible initial conditions, and to run trajectory calculations for each of them. What that sample looks like depends on a number of things. First, one needs to decide what the population of starting molecules would look like in the experiment to be simulated. If the reaction is thermal, then a canonical ensemble might be expected; if it is a state-to-state laser study, then a microcanonical ensemble might be a better representation. Next, one needs to decide where on the PE hypersurface to begin the simulation. This choice may seem trivial—why would it be anywhere but in the reactant region? The answer is that, for a thermal, unimolecular reaction, this is rarely feasible. As discussed in Section 2.3, all current kinetic theories rely on some version of the Lindemann–Hinshelwood model for thermal reactions. In this picture, the promotion of a molecule to an energy high enough for reaction is a very rare event if the barrier is much bigger than $k_B T$. Even when a molecule makes it to this elevated position, there is, in the high-pressure limit, a much higher chance of its being deactivated by the next collision than there is of sufficient vibrational energy happening to accumulate in the reaction coordinate to permit passage through the transition state region. This reason is why thermal reactions take minutes or hours rather than picoseconds. The upshot is that attempted simulation of a thermal reaction of a typical organic molecule, starting from the reactant region is not computationally feasible, by ~12–14 orders of magnitude! Consequently, all of the MD simulations described in this chapter were initiated in the vicinity of the transition state for formation of the reactive intermediate of interest. Typically, the assumption is made that molecules in this region form a microcanonical or canonical ensemble. To the extent that assumption is wrong, it would presumably tend to bias the results in favor of a statistical model for the overall kinetics, and so any finding of nonstatistical behavior probably cannot be explained away on that basis. Finally, one has to decide how to apportion kinetic energy to the molecules in one's sample. There are many schemes for doing this, each with their own virtues and liabilities. Most common for the kinds of study described here is so-called quasiclassical normal-mode sampling.[28] In this procedure, kinetic energy is supplied in quantized fashion to the vibrational normal modes of the molecule, with randomization of the vibrational phases. This last stipulation means that different trajectories will start from different molecular structures in the vicinity of the stationary point; if all of the vibrational oscillators were at the minima of their potential curves, the relative phase angles between them could be only 0 or 180°. In order to start with phase angles other than these two values, oscillators have to be "caught" at some position other than that corresponding to the minimum in PE. This requirment implies that the molecule itself will start with a distorted geometry. Usually, zero-point energy (0.5 $h\nu$) is supplied to each normal mode. In addition, some of the modes may be excited to levels above $v = 0$, in a way that preserves the energetic requirements for the ensemble as a whole. Quantized initial conditions are used, despite the fact that the subsequent trajectory calculations employ only the equations of classical mechanics, because quantization ensures a proper distribution of energy among modes of different frequency. For reactions run under typical thermal conditions, the low-frequency modes should have average energies near the classical limit of $k_B T$ (above the zero-point level), whereas high-frequency modes will generally have average energies below this limit.

3.2. The Dynamics of Reactive Intermediates

3.2.1. Influence of the PE Profile.
Although we saw at the beginning of this chapter that PE profiles should be treated with extreme caution when one is analyzing mechanistic possibilities, they are nevertheless useful devices for classifying the kinds dynamic of behavior one might expect for reactive intermediates.

For reactions whose PE profile is like that in of Figure 21.7 (*a*), there is neither experimental nor theoretical reason to doubt that the statistical kinetic models will do a good job of describing both absolute and, in the case where there are alternative rate-determining exits from the intermediate minimum, relative rates of product formation. However, the same cannot be said for any of the other three profiles in Figure 7. In Figure 21.7 (*b*) the intermediate is formed with kinetic energy (largely in the form of molecular vibrations) that far exceeds $k_B T$, and hence also far exceeds the energy needed to traverse the barrier to products. Such reactions have a high probability of exhibiting large nonstatistical dynamical effects.

Figure 21.7 (*c*) is a representation of the kind of PE profile found for many reactions involving the opening of a strained hydrocarbon ring. As described in Section 1.2, the species residing on the energy plateau is a singlet biradical. In this kind of reaction (of which eight examples have been studied at the time of writing) there is not an intermediate with a single well-defined equilibrium geometry. Rather, the biradical is highly conformationally flexible and usually capable of forming a number of stereoisomeric products with little or no barrier. The statistical kinetic models are ill equipped to deal with such situations. When trajectory calculations are run, one typically finds that all stereoisomeric products can be formed, but not with equal probability. A combination of the entrance dynamics (as defined roughly by the direction of the reaction-coordinate eigenvector at the transition state for formation of the biradical) and the presence of very small barriers ($\sim k_B T$) appear to determine the favored products. Thus far it has proven difficult to predict how these factors will combine, and hence which product(s) will be favored before doing the molecular dynamics simulation. It would be a significant contribution to this kind of chemistry if qualitative or semiquantitative rules

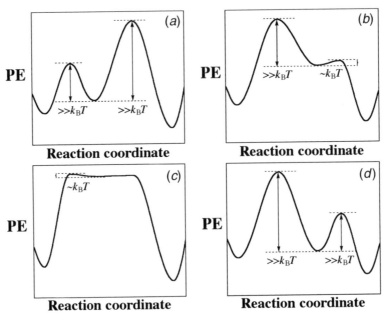

Figure 21.7. Schematic PE profiles. All except the one in panel (a) represent reactions that are potentially susceptible to nonstatistical dynamical effects. See text for further discussion.

for making such predictions could be developed, and that is an area of active research.

Figure 21.7 (d) represents the kind of profile for which it has typically been thought that the statistical models will work. However, in two examples cited below, this turns out not to be the case. Even when the intermediate sits in a relatively deep potential energy minimum, the kinetic energy acquired during its formation can be an important factor controlling its fate. Of particular importance is the fact that this kinetic energy will not be statistically deposited into all vibrational modes, but rather will selectively excite certain modes of the intermediate. Which ones can be identified by running a single trajectory from the rate-determining TS toward the intermediate, with just enough kinetic energy in the reaction coordinate (say 0.1 kcal/mol) to get things moving and with no kinetic energy (even zero-point motion) in any other mode. Such a trajectory will almost never pass exactly through the minimum-energy geometry of the intermediate, because that would require the highly unlikely event that its vibrational modes be excited with relative phase angles of exactly 0 or 180°. Instead, the trajectory is likely to follow a path that explores various regions of the PE minimum corresponding to the intermediate. If the statistical approximation were correct, this motion would be chaotic (i.e., there would be no discernible periodicity). However, in reality one typically finds quite clear periodicity for at least a picosecond or so. Whether this has consequences for the subsequent chemistry of the intermediate is a question that is still under investigation, but in two cases that have been studied in detail, it seems that it does. They are described below.

3.2.2. The Subtle Role of Symmetry.

An appeal to symmetry seems like one of the least controversial and most rigorous bases for making an argument about the behavior of a physical system. And indeed it is. However, it is possible for arguments that appear to be based solely on symmetry actually to depend on some additional ancillary assumptions that may be less obviously valid. An important example for the present discussion concerns prediction of product ratios from reactive intermediates.

Almost every textbook on introductory organic chemistry will tell its readers that a reaction occurring via an achiral intermediate in an achiral medium can lead only to achiral and/or racemic products, even if the reactant was chiral and optically pure. The argument seems unassailable: If there were a chiral product available from an achiral intermediate, then the transition structures leading to the two enantiomers of the product would themselves have to be related as enantiomers. Since enantiomers have identical thermochemical properties in an achiral medium, the rate constants for conversion of the intermediate to the enantiomeric products would have to be identical by symmetry. Hence, the product would have to be racemic. However, there is a covert assumption in the argument. We have treated the intermediate as if it were the reactant for the product-forming step of the process. By so doing, we have implicitly ignored the history of formation of the intermediate. This decision is equivalent to making the statistical approximation, because we are saying that the dynamics of formation of the intermediate are not relevant to its subsequent chemistry. In reality, there are several cases that show such an assumption is not generally valid. In effect, the symmetry of the phase space in the vicinity of an intermediate can be lower than the symmetry of the PE surface; the momentum components describing the dynamics of the system can carry chiral information even if the static minimum-energy geometry of the intermediate appears to be achiral. A similar argument applies to the formation of isotopic isomers of products formed from a labeled intermediate, under circumstances where kinetic isotope effects are negligible.

The analysis summarized in Section 3.2.1 suggests that the reactions of an intermediate *can* depend on how it was made, because its formation from the preceding TS imprints it with dynamic information that can influence its subsequent selection of exit channels to the products. This claim flies in the face of some distinguished physical organic history. Some of the classic experiments for probing the existence of a reactive intermediate have relied on the product ratios from the intermediate both reflecting its static symmetry and being *independent* of its mode of generation. And, in fact, such behavior has been demonstrated in some celebrated cases.[29] However, recognize that in almost all of these cases the product-forming step was a bimolecular trapping event. Even if the PE barrier is very small, a bimolecular reaction will usually be relatively slow because it can occur only when the trapping reagent happens to be near to and correctly oriented to the intermediate. There are few cases in which independent generation of a reactive intermediate from different sources has led to the same ratio of products when the product-forming step is unimolecular.[30]

3.2.3. Examples.
There are by now several reactions for which the best available levels of ab initio electronic structure theory find a plateau on the PE hypersurface, in the vicinity of a singlet-state biradical.[6] Several of these have been studied by MD simulation and/or experiment, and in each case the conclusion is that application of statistical kinetic models will give a misleading description of how the reaction really occurs. There is not room to describe each of these studies, and so just one is chosen as a representative, and described in Section 3.2.3.1.

The demonstration that reactions with more traditional PE profiles, similar to those in Figure 21.7 (b) and (d), may also exhibit significant nonstatistical dynamics has broadened the range of processes for which this effect may need to be considered. For that reason, three examples are described below.

3.2.3.1. The Vinylcyclopropane Rearrangement.
In Section 1.2, we saw that the vinylcyclopropane rearrangement is a reaction for which the standard classification of concerted versus stepwise mechanism seems inadequate. It turns out that, in large measure, this is because the statistical kinetic models are also inadequate for describing the reaction. Electronic structure calculations at various levels of theory agree that the reaction occurs with the involvement of a biradical that sits on a plateau on the potential-energy hypersurface.[5] From this plateau region more-or-less isoenergetic paths lead to the four stereochemically distinct products (see Scheme 21.1). Doubleday, et al.[5a] carried out extensive quasiclassical trajectory studies on this reaction using an AM1–SRP model (see Section 3.1) fit to the highest level ab initio results—those from a multireference configuration interaction calculation. Their simulations matched the experimental outcome very well: the computed product ratios were 42:30:10:18 (si/sr/ar/ai), whereas the experimental ratios[3c] were 40:23:13:24. Importantly, the simulations revealed that the reaction was dominated by nonstatistical dynamics. One symptom of the nonstatistical behavior was a highly time-dependent product ratio. Thus, trajectories that completed the passage to products in <200 fs gave a ratio of 53:43:0:4, whereas those taking >600 fs gave 20:22:30:28. Clearly, the reaction becomes more stereorandom the longer the trajectories last. This picture is just what one would expect if IVR were occurring on a time scale comparable to that for product formation. The simulations also revealed that the product ratio could be strongly influenced by the distribution of kinetic energy among the vibrational modes of the TS from which the trajectories were initiated. This behavior is also clearly nonstatistical.

There is every reason to believe that the results found for the vinylcyclopropane rearrangement are typical of those to be expected for reactions involving biradicals (and presumably other reactive intermediates) on energetic plateaus. Such reactions simply cannot be understood within the context of any statistical kinetic model.

3.2.3.2. Generation and Dissociation of the Acetone Radical Cation.
It has been known for some time, and verified by a number of research groups, that the

generation of the acetone radical cation by [1,3] hydrogen shift from its enol isomer is followed by a fragmentation for which the branching ratio does not reflect the symmetry of the intermediate.[31] The reaction is summarized in Figure 21.8.[32] Quasiclassical trajectory studies initiated in the vicinity of the [1,3] migration TS have provided an explanation for this behavior. The results are summarized in the graph

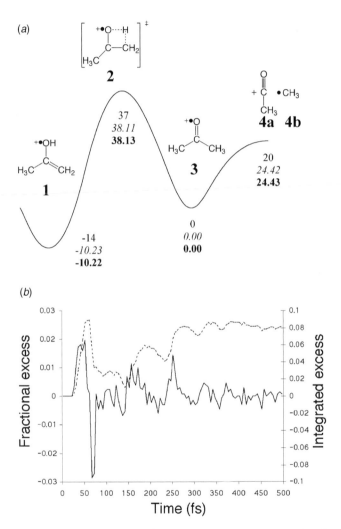

Figure 21.8. (*a*) Potential energy profile for the acetone radical cation formation and dissociation. Relative energies are in kilocalories per mole. Plain figures are experimental enthalpies, italicized figures are UB3LYP/cc-pVTZ potential energies, and bold figures are AM1–SRP potential energies. [Reproduced with permission from J. A. Nummela and B. K. Carpenter, *J. Am. Chem. Soc.* **2002**, *124*, 8512. Copyright © 2002 American Chemical Society.] (*b*) Graph of the fractional excess of newly formed methyl group loss versus time. The dashed line is the integral of the curve.[32]

of Figure 21.8 (*b*). It shows the time dependence of the methyl group dissociation from the acetone radical cation. One sees evidence of oscillatory behavior in which the first methyl to be lost is the newly formed one. A short while later there is a preference for loss of the existing methyl, and then the graph shows a damped oscillatory behavior between the two. The integral of the branching-fraction curve does not converge on zero at long times (i.e., there is an asymmetry in the final product ratio that favors loss of the newly formed methyl). This result is consistent with the experimental observations.

The simulation suggests that the geometries of the [1,3] migration TS and the methyl-dissociation asymptote (there is not a maximum in PE along the dissociation pathway) are crucial to understanding this behavior. The methyl dissociation occurs preferentially from a geometry in which one OCC angle is small (near 90°) and the other large (>130°). This results because the product acylium ion has an OCC angle of 180°. The [1,3] migration TS has a similar asymmetry in the OCC bond angles, caused by the need to get the oxygen and methylene carbon sufficiently close to transfer a hydrogen atom. Thus the migration TS starts out closer in geometry to one of the dissociation exits than to the other. The simulation reveals that some trajectories simply leave the [1,3] migration TS and take the nearer dissociation exit without ever accessing the acetone radical ion minimum. These trajectories constitute the first peak in Figure 21.8 (*b*). They necessarily lose the newly formed methyl. Those trajectories that fail to take the nearer exit do enter the acetone radical ion minimum, and they pick up roughly 40 kcal/mol of kinetic energy as they do so. This kinetic energy appears largely in a vibrational motion consisting of an in-plane bend of the carbonyl oxygen, first toward one methyl and then toward the other. Importantly, this vibration is excited with a particular phase; as trajectories leave the [1,3] migration TS, the OCC angle to the newly formed methyl is increasing and the other is decreasing. This means that the existing methyl becomes the first to experience a geometry of the acetone radical cation appropriate for dissociation. This behavior is responsible for the large downward peak in Figure 21.8 (*b*). After that the carbonyl oxygen swings backward and forward for several hundred femtoseconds, encouraging one methyl and then the other to dissociate. The dissociation peaks are damped both because of IVR, which is spreading the kinetic energy into other vibrational modes, and because of loss of population due to dissociation.

This case is clearly one in which nonstatistical dynamics (indicated both by the time-dependent branching ratio and the lack of correspondence between the static symmetry of the intermediate and its branching ratio) plays an important role, despite the depth of the PE minimum in which the acetone radical cation resides.

3.2.3.3. The Rearrangement of 1,2,6-Heptatriene. The experimental facts[7] and basic mechanistic idea[8] behind this reaction were outlined in section 1.2. The molecular dynamics study[9] began with a single CASSCF(8,8)/6-31G(d) trajectory started from TS1 (see Fig. 21.2) with no kinetic energy (not even ZPE) in any of the real-frequency normal modes. The purpose of such an unphysical trajectory calculation is to see what is the steepest descent path down from the transition state and which modes of the product or intermediate receive most of the kinetic energy picked up along the way. In this case, the trajectory led to a local minimum conformation of the biradical intermediate. Its vibrational motions turned out to be very interesting. As soon as it was formed, the biradical started to exhibit a large-amplitude stretching motion of the C4—C5 bond (see Scheme 21.3). No other bond in the molecule showed such behavior. It was of interest because scission of this very bond would be necessary to complete the conversion of the intermediate to the product. The mechanism by which kinetic energy could be "channeled" into this motion was revealed when a second trajectory was run. It was identical to the first save for the fact that H10 (see Scheme 21.3) was replaced by a deuterium. Now the amplitude of the C4—C5 stretch induced in the biradical was much reduced. The picture that arises from these observations is as follows. In TS1, the C3—H10 bond is almost normal to the C1—C2—C3 plane, reflecting the allenic origin of this moiety. However, in the biradical, the C3—H10 bond is in the C1—C2—C3 plane, as required for the intermediate to enjoy allylic stabilization. Thus, an important motion taking one from TS1 to the biradical is the bending of the C3—H10 bond into the C1—C2—C3 plane. Consequently, the out-of-plane bend of C3—H10 receives a large fraction of the kinetic energy released in the formation of the intermediate. It then turns out that the C3—H10 bend is in near resonance with the C4—C5 stretch, which leads to very efficient energy transfer between the two. The evidence for this picture comes from the deuterium substitution, which significantly influences the frequency of the bend but hardly changes the frequency of the stretch, and therefore "detunes" the resonance. The significance of the excitation of the C3—H10 bend and C4—C5 stretch is that the combination of these two geometrical changes is precisely what is required to take the biradical over the second transition state and on to the product. And, indeed, AM1–SRP simulations reveal a significant number of trajectories that enter the biradical minimum but then exit to product in <500 fs. This reaction is much faster than would be expected on the basis of a statistical model kinetic analysis.

The trajectory calculations reveal another important piece of information about the reaction. When CASSCF(8,8)/6-31G(d) trajectories are started from TS1 with ZPE in the vibrational modes, it is found that they can either go to the biradical or they can cross the second transition state directly, without ever accessing the biradical minimum. The difference between the two sets is only the relative phases of the vibrational motions. This result is also seen in the AM1–SRP simulation. Thus, while the chemistry is very different, the general picture that emerges for the 1,2,6-heptatriene rearrangement is strikingly similar to that found for the acetone radical cation dissociation. In both cases, a significant fraction of the trajectories leaving the rate-limiting TS fail to follow the steepest descent path, and thereby miss the subsequent intermediate minimum altogether. Those that do enter the minimum experience nonstatistical dynamical effects that have an important influence on how the intermediate proceeds on to the products.

3.2.3.4. Thermal Deazetization of 2,3-Diazabicyclo[2.2.1]hept-2-ene.[33] In the previous two examples, the reactions involved intermediates residing in potential

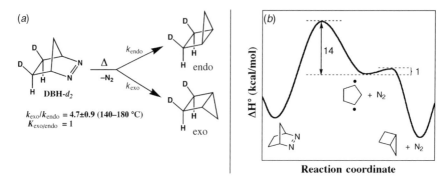

Figure 21.9. (*a*) Experimental data on the thermal deazetization of DBH-d_2. (*b*) Schematic enthalpy profile for the reaction. The cyclopentane-1,3-dyl intermediate is calculated to have a C_2-symmetry equilibrium structure.

energy wells that were 10–20-kcal/mol deep. The nonstatistical dynamics observed for such intermediates, while chemically significant, amounted to no more than a 60% deviation in branching ratio from that expected on the basis of the statistical approximation (ignoring contributions from trajectories that bypassed the local intermediate minimum). In the thermal deazetization of 2,3-diazabicyclo[2.2.1]-hept-2-ene-*exo,exo-d$_2$*, (DBH-d_2, see Fig. 21.9) the deviation from the statistical prediction is nearly 500%. The reason is that the PE profile for the reaction looks like that in Figure 21.9(*b*): The intermediate has a barrier-to-product formation of only ∼1 kcal/mol, but is formed with excess kinetic energy of roughly 14 kcal/mol. Reactions of this kind can be expected to show large effects from nonstatistical dynamics, because the PE surface in the region of the intermediate is relatively flat, and so it is possible to take non-IRC paths to the products with an energy penalty that is much smaller than the excess energy available to the system. This means that the dynamics for formation of the intermediate do not need to be perfectly coupled to those for passage of the intermediate over its product-forming transition state, because direct trajectories can take slightly higher energy routes to the products. In the case of DBH-d_2, CASSCF(6,6) calculations show a C_s symmetry transition structure for loss of N$_2$. The resulting cyclopentane-1,3-diyl has a C_2 symmetry equilibrium geometry. However, a CASSCF trajectory run from the rate-determining TS maintains C_s symmetry throughout—it does not ever access the minimum energy structure of the biradical. Nor does it pass through the C_1 symmetry transition structure for ring closure. Instead, it takes a higher barrier (∼3 kcal/mol) C_s symmetry path, because this is the direction that the dynamics of N$_2$ loss have imposed on it. Most important is the fact that the trajectory shows formation of the *inverted* product bicyclo[2.1.0]pentane-*exo,exo-2,3-d$_2$*. This product is in fact favored experimentally (ratio 4.7 ± 0.9:1 *exo/endo* in the gas phase). The calculations suggest that its preferential formation is due to the Newtonian mechanics of N$_2$ loss from the reactant. As the N$_2$ departs in one direction, the carbons to which it was attached move in the opposite direction to conserve momentum. This phenomenon drives an out-of-plane bending motion in the cyclopentane-1,3-diyl with sufficient energy to carry it over the barrier (although not via the transition state) to formation of the inverted product.

3.3. Experimental Tests for Effects Due to Nonstatistical Dynamics

The most direct evidence for nonstatistical behavior comes from state-to-state laser-induced reactions conducted in molecular beams. In such studies, the tracking of energy flow during chemical reactions is sometimes possible, and in those cases the validity of the statistical approximation can be assessed directly.[34] However, for reactions conducted under conditions used by most organic chemists, the unambiguous detection of nonstatistical dynamics is more challenging. Thus far only two techniques have been developed, and both really need the support of high-level electronic-structure theory to make a convincing case.

The first technique is relatively simple: It is the measurement of the temperature dependence of product ratios, particularly in the case where products are related to each other as optical isomers or isotopic label isomers.[6b,35] For reactions of this kind the thermodynamic product ratio can usually be deduced by symmetry, provided any equilibrium isotope effects are small. If the observed product ratio differs significantly from the thermodynamic value, one can be confident that the products are formed under kinetic control. In a mechanistic interpretation that relied on the validity of the statistical approximation, the product ratio would then be ascribable to competitive reaction paths with barriers of different heights to formation of the products. Generally, one would expect these barriers to have different heights on the *potential* energy hypersurface, since the supposed competitive mechanisms will almost always have transition states of different geometry, and, barring coincidence, different geometry usually means different energy. If indeed there are PE barriers of different heights to formation of the products, then any of the statistical kinetic models will predict that there should be a temperature dependence to the product ratio. Often that turns out to be the case, and then one can be satisfied with the conventional mechanistic description. However, in several instances studies have found temperature-independent kinetic product ratios.[6b,35] These experiments present a problem for the standard mechanistic description, because in each case such a result requires that the barriers to product formation be identical on the PE surface but different on the standard free energy surface. In other words, the transition states would have to have coincidentally identical heats of formation but very different entropies. There could be odd circumstances where such a thing might happen occasionally, but the repeated observation of this behavior in reactions of structurally diverse molecules stretches the limits of credulity for the "coincidence" explanation. Actually, models based on nonstatistical dynamics also fail to predict absolute rigorous temperature independence to product ratios. However, in those cases where the question has been studied, the energy or temperature dependence is found to be very weak, and well within the error of a typical experiment.[5b,36]

Somewhat more direct evidence for nonstatistical behavior has come from studies in supercritical fluids.[33] In these experiments, the idea is to try to intercept by collision any molecules on direct, nonstatistical trajectories to the products. The expectation is that collisions should promote IVR, and that increasing collision frequency should therefore lead to increasing conformance of the product ratio to that predicted from a statistical kinetic model. The reason that supercritical fluids are useful in such studies is that their high compressibility means that the density of the medium, and hence the collision frequency, is controllable by changing the pressure. This finding is also true for gases, of course, but in the gas phase it is difficult to attain the collision frequency necessary to intercept an intermediate whose lifetime might be only $\sim$100 fs. At least in one case, quantitative fitting of the pressure dependence of the product ratio has permitted an experimental estimate of the lifetime of the nonstatistical population component for a reactive intermediate.[33] This number can be compared directly with results from MD simulation.

4. CONCLUSION AND OUTLOOK

Two principal points serve to encapsulate the discussion in this chapter. The first is that it is necessary to think about the PE hypersurface for a chemical reaction in a higher dimensionality projection than has typically been employed by mechanistic organic chemists to date. Even the three-dimensional projections on which most of the discussions in this chapter have been based are undoubtedly concealing important topological features that a still higher dimensional view would reveal, but for now the features summarized in Section 1.1 provide more than enough material for research. The second point is that kinetic models employing the statistical approximation may not have served us as well as we had thought. Increasing application of MD simulation, made possible by rapidly increasing computer power, is continuing to show the shortcomings of the TST and RRKM models, and hence of the mechanistic descriptions that were based on them.

There are (at least) two major opportunities for research by those interested in this topic. On the computational side, there is definite room for improvement in simulation methods. Right now none of the simulation approaches has the "user friendliness" that has brought electronic-structure calculation into the realm of routine applicability by nonspecialists. Nor has the field seen the development of the qualitative or semiquantitative models that did so much to make the results of molecular orbital calculations useful to organic chemists. On the experimental side, it will be obvious to the reader that the techniques for detecting the effects of nonstatistical dynamics are still very rudimentary and indirect. There is clearly room for creative scientists to come up with techniques whose results can give us more direct insight into these issues.

Looking a little further out, it is already apparent that the dynamical models outlined in this chapter are themselves inadequate. There is growing evidence that some fundamentally quantum mechanical phenomena, such as tunneling, can have important roles to play in everyday reactions.[37] These phenomena are not describable by models based on classical mechanics. Progress is being made on the incorporation of quantum mechanical components into MD simulations, although the full quantum mechanical description of dynamics remains far from computationally feasible for systems of more than a few degrees of freedom.[38] Similarly, nonadiabatic dynamic simulations, in which trajectories (or, better, wave packets) are propagated simultaneously on more than one PES, are becoming increasingly sophisticated and are starting to prove their value in analyzing photochemical (and some kinds of thermal) reactions.[39]

Even for purely adiabatic reactions, the inadequacies of classical MD simulations are well known. The inability to keep zero-point energy in all of the oscillators of a molecule leads to unphysical behavior of classical trajectories after more than about a picosecond of their time evolution.[40] It also means that some important physical organic phenomena, such as isotope effects, which are easily explained in a TST model, cannot be reproduced with classical molecular dynamics. So it is clear that there is much room for improvement of both the computational and experimental methods currently employed by those of us interested in reaction dynamics of organic molecules. Perhaps some of the readers of this book will be provide some of the solutions to these problems.

SUGGESTED READING

L. Sun, K. Song, and W. L. Hase, "A S_N2 Reaction that Avoids its Deep Potential Energy Minimum," *Science*, **2002**, *296*, 875.

C. Doubleday, "Mechanism of the Vinylcyclopropane–Cyclopentene Rearrangement Studied by Quasiclassical Direct Dynamics," *J. Phys. Chem. A* **2001**, *105*, 6333.

B. K. Carpenter, "Dynamic Behavior of Organic Reactive Intermediates," *Angew. Chem., Int. Ed. Engl.* **1998**, *37*, 3341.

R. D. Levine, "Molecular Reaction Dynamics Looks Toward the Next Century: Understanding Complex Systems," *Pure Appl. Chem.* **1997**, *69*, 83.

R. Q. Topper, "Visualizing Molecular Phase Space: Nonstatistical Effects in Reaction Dynamics," *Rev. Comput. Chem.* **1997**, *10*, 101.

R. D. Levine and R. B. Bernstein, *Molecular Reaction Dynamics and Chemical Reactivity*, Oxford University Press, New York, **1987**.

REFERENCES

1. (a) K. N. Houk, S. L. Wilsey, B. R. Beno, A. Kless, M. Nendel, and J. Tian, *Pure Appl. Chem.* **1998**, *70*, 1947. (b) K. N. Houk, Y. Li, J. Storer, L. Raimondi and B. Beno, *J. Chem. Soc., Faraday Trans.* **1994**, *90*, 1599. (c) F.-G. Klärner, B. Krawczyk, V. Ruster, and U. K. Deiters, *J. Am. Chem. Soc.* **1994**, *116*, 7646. (d) M. J. S. Dewar and S. Kirschner, *J. Am. Chem. Soc.* **1974**, *96*, 5244. (e) J. A. Berson, *Acc. Chem. Res.* **1972**, *5*, 406. (f) J. E. Baldwin, A. H. Andrist, and R. K. Pinschmidt, Jr., *Acc. Chem. Res.* **1972**, *5*, 402. (g) W. von E. Doering, M. Franck-Neumann, D. Hasselmann, and R. L. Kaye, *J. Am. Chem. Soc.* **1972**, *94*, 3833. (h) P. D. Bartlett, *Nucleus* **1966**, *43*, 251.

2. H. M. Frey and D. C. Marshall, *J. Chem. Soc.* **1962**, 3981.

3. (a) J. E. Baldwin and S. J. Bonacorsi, Jr., *J. Am. Chem. Soc.* **1996**, *118*, 8258. (b) J. J. Gajewski, L. P. Olson, and M. R. Willcott, III, *J. Am. Chem. Soc.* **1996**, *118*, 299. (c) J. E. Baldwin, K. A. Villarica, D. I. Freedberg, and F. A. L. Anet, *J. Am. Chem. Soc.* **1994**, *116*, 10845. (d) J. E. Baldwin and S. Bonacorsi, Jr., *J. Am. Chem. Soc.* **1993**, *115*, 10621. (e) J. J. Gajewski and L. P. Olson, *J. Am. Chem. Soc.* **1991**, *113*, 7432. (f) J. E. Baldwin and N. D. Ghatlia, *J. Am. Chem. Soc.* **1991**, *113*, 6273. (g) J. J. Gajewski and M. P. Squicciarini, *J. Am. Chem. Soc.* **1989**, *111*, 6717. (h) G. D. Andrews and J. E. Baldwin, *J. Am. Chem. Soc.* **1976**, *98*, 6705. (i) P. H. Mazzocchi and H. J. Tamburin, *J. Am. Chem. Soc.* **1970**, *92*, 7220. (j) M. R. Willcott III, and V. H. Cargle, *J. Am. Chem. Soc.* **1969**, *91*, 4310. (k) H. E. O'Neal and S. W. Benson, *J. Phys. Chem.* **1968**, *72*, 1866. (l) M. R. Willcott, III, and V. H. Cargle, *J. Am. Chem. Soc.* **1967**, *89*, 723.

4. R. B. Woodward and R. Hoffmann, *The Conservation of Orbital Symmetry*, Verlag Chemie Academic Press, Weinheim, **1970**.

5. (a) C. Doubleday, G. Li, and W. L. Hase, *Phys. Chem. Chem. Phys.* **2002**, *4*, 304. (b) C. Doubleday, *J. Phys. Chem. A* **2001**, *105*, 6333. (c) C. Doubleday, M. Nendel, K. N. Houk, D. Thweatt, and M. Page, *J. Am. Chem. Soc.* **1999**, *121*, 4720. (d) K. N. Houk, M. Nendel, O. Wiest, and J. W. Storer, *J. Am. Chem. Soc.* **1997**, *119*, 10545. (e) E. R. Davidson and J. J. Gajewski, *J. Am. Chem. Soc.* **1997**, *119*, 10543.

6. (a) C. P. Suhrada and K. N. Houk, *J. Am. Chem. Soc.* **2002**, *124*, 8796. (b) M. B. Reyes, E. B. Lobkovski, and B. K. Carpenter, *J. Am. Chem. Soc.* **2002**, *124*, 641. (c) B. R. Beno, S. Wilsey, and K. N. Houk, *J. Am. Chem. Soc.* **1999**, *121*, 4816. (d) D. A. Hrovat, S. Fang, W. T. Borden, and B. K. Carpenter, *J. Am. Chem. Soc.* **1997**, *119*, 5253. (e) C. Doubleday, Jr., K. Bolton, and W. L. Hase, *J. Am. Chem. Soc.* **1997**, *119*, 5251.

7. W. R. Roth, D. Wollweber, R. Offerhaus, V. Rekowski, H. W. Lennartz, R. Sustmann, and W. Mueller, *Chem. Ber.* **1993**, *126*, 2701.

8. D. A. Hrovat, J. A. Duncan, and W. T. Borden, *J. Am. Chem. Soc.* **1999**, *121*, 169.

9. S. L. Debbert, B. K. Carpenter, D. A. Hrovat, and W. T. Borden, *J. Am. Chem. Soc.* **2002**, *124*, 7896.

10. H. Metiu, J. Ross, R. Silbey, and T. F. George, *J. Chem. Phys.* **1974**, *61*, 3200.

11. (a) C. Zhou and D. M. Birney, *Org. Lett.* **2002**, *4*, 3279. (b) P. Caramella, P. Quadrelli, and L. Toma, *J. Am. Chem. Soc.* **2002**, *124*, 1130. (c) Y. Kumeda and T. Taketsugu, *J. Chem. Phys.* **2000**, *113*, 477. (d) M. Hirsch, W. Quapp, and D. Heidrich, *Phys. Chem. Chem. Phys.* **1999**, *1*, 5291. (e) P. Valtazanos, S. F. Elbert, and K. Ruedenberg, *J. Am. Chem. Soc.* **1986**, *108*, 3147.

12. W. T. A. M. Van der Lugt and L. J. Oosterhoff, *J. Am. Chem. Soc.* **1969**, *91*, 6042.

13. J. Von Neumann and E. Wigner, *Physik. Z.* **1929**, *30*, 467.

14. (a) D. R. Yarkony, *Acc. Chem. Res.* **1998**, *31*, 511. (b) M. Klessinger, *Pure Appl. Chem.* **1997**, *69*, 773. (c) F. Bernardi, M. Olivucci, and M. A. Robb, *Chem. Soc. Rev.* **1996**, 321. (d) M. A. Robb, F. Bernardi, and M. Olivucci, *Pure Appl. Chem.* **1995**, *67*, 783.

15. M. Olivucci, I. N. Ragazos, F. Bernardi, and M. A. Robb, *J. Am. Chem. Soc.* **1993**, *115*, 3710.

16. R. Q. Topper, *Rev. Comput. Chem.* **1997**, *10*, 101.

17. D. A. McQuarrie, *Statistical Mechanics*, Harper and Row, New York, **1975**.

18. (a) T. Baer and W. L. Hase, *Unimolecular Reaction Dynamics*, Oxford University Press, New York, **1996**. (b) R. C. Gilbert and S. C. Smith, *Theory of Unimolecular and*

Recombination Reactions, Blackwell Scientific, Oxford, **1990**. (c) J. I. Steinfeld, J. S. Francisco, and W. L. Hase, *Chemical Kinetics and Dynamics*, Prentice Hall, Englewood Cliffs, **1989**. (d) P. J. Robinson and K. A. Holbrook, *Unimolecular Reactions*, Wiley-Interscience, Bristol, **1972**.

19. (a) J. N. Butler and G. B. Kistiakowsky, *J. Am. Chem. Soc.* **1960**, *82*, 759. (b) J. D. Rynbrandt and B. S. Rabinovitch, *J. Phys. Chem.* **1970**, *74*, 4175. (c) J. D. Rynbrandt and B. S. Rabinovitch, *J. Phys. Chem.* **1971**, *75*, 2164.

20. (a) O. K. Rice and H. C. Ramsperger, *J. Am. Chem. Soc.* **1927**, *49*, 1617. (b) L. S. Kassel, *Chem. Rev.* **1932**, *10*, 11. (c) R. A. Marcus, *J. Chem. Phys.* **1952**, *20*, 352. (d) R. A. Marcus, *J. Chem. Phys.* **1952**, *20*, 355.

21. (a) M. G. Evans and M. Polanyi, *Trans. Faraday Soc.* **1935**, *31*, 875. (b) H. Eyring, *J. Chem. Phys.* **1935**, *3*, 107.

22. (a) D. G. Truhlar and B. C. Garrett, *Annu. Rev. Phys. Chem.* **1984**, *35*, 159. (b) P. Pechukas, *Annu. Rev. Phys. Chem.* **1981**, *32*, 159. (c) D. G. Truhlar and B. C. Garrett, *Acc. Chem. Res.* **1980**, *13*, 440. (d) W. H. Miller, *J. Chem. Phys.* **1976**, *65*, 2216.

23. (a) J. M. Haile, *Molecular Dynamics Simulation: Elementary Methods*, Wiley-Interscience, New York, 1992. (b) D. L. Bunker, *Methods Comput. Phys.* **1971**, *10*, 287.

24. (a) W. Wang, O. Donini, C. M. Reyes, and P. A. Kollman, *Annu. Rev. Biophys. Biomol. Struct.* **2001**, *30*, 211.

25. (a) P. Carloni, U. Rothlisberger, and M. Parrinello, *Acc. Chem. Res.* **2002**, *35*, 455. (b) P. Amara, M. J. Field, C. Alhambra, and J. Gao, *Theor. Chem. Acc.* **2000**, *104*, 336. (c) M. J. Bearpark, F. Bernardi, S. Clifford, M. Olivucci, M. A. Robb, B. R. Smith, and T. Vreven, *J. Am. Chem. Soc.*, **1996**, *118*, 169.

26. R. Car and M. Parrinello, *Phys. Rev. Lett.* **1985**, *55*, 2471.

27. A. Gonzalez-Lafont, T. N. Truong, and D. G. Truhlar, *J. Phys. Chem.* **1991**, *95*, 4618.

28. G. H. Peslherbe, H. Wang, and W. L. Hase, *Adv. Chem. Phys.* **1999**, *105*, 171.

29. (a) J. D. Roberts, H. E. Simmons, Jr., L. A. Carlsmith, and C. Vaughan Wheaton, *J. Am. Chem. Soc.* **1953**, *75*, 3290. (b) P. S. Skell and M. S. Cholod, *J. Am. Chem. Soc.* **1969**, *91*, 6035.

30. P. B. Dervan, T. Uyehara, and D. S. Santilli, *J. Am. Chem. Soc.* **1979**, *101*, 2069.

31. (a) F. W. McLafferty, D. J. McAdoo, J. S. Smith, and R. Kornfeld, *J. Am. Chem. Soc.* **1971**, *93*, 3720. (b) D. J. McAdoo, F. W. McLafferty, and J. S. Smith, *J. Am. Chem. Soc.* **1970**, *92*, 6343. (c) G. Depke, C. Lifshitz, H. Schwarz, and E. Tzidony, *Angew. Chem.* **1981**, *93*, 824. (d) F. Turecek and F. W. McLafferty, *J. Am. Chem. Soc.* **1984**, *106*, 2525. (e) F. Turecek and V. Hanus, *Org. Mass Spectrom.* **1984**, *19*, 631. (f) T. H. Osterheld and J. I. Brauman, *J. Am. Chem. Soc.* **1993**, *115*, 10311.

32. J. A. Nummela and B. K. Carpenter, *J. Am. Chem. Soc.* **2002**, *124*, 8512.

33. M. B. Reyes and B. K. Carpenter, *J. Am. Chem. Soc.* **2000**, *122*, 10163.

34. For example, see: (a) A. H. H. Chang, D. W. Hwang, X.-M. Yang, A. M. Mebel, S. H. Lin, and Y. T. Lee, *J. Chem. Phys.* **1999**, *110*, 10810. (b) H. Choi, D. H. Mordaunt, R. T. Bise, T. R. Taylor, and D. M. Neumark, *J. Chem. Phys.* **1998**, *108*, 4070. (c) E. W.-G. Diau, J. Herek, Z. H. Kim, and A. H. Zewail, *Science* **1998**, *279*, 847.

35. (a) B. A. Lyons, J. Pfeifer, T. H. Peterson, and B. K. Carpenter, *J. Am. Chem. Soc.* **1993**, *115*, 2427. (b) R. H. Newman-Evans, R. J. Simon, and B. K. Carpenter, *J. Org. Chem.* **1990**, *55*, 695. (c) R. H. Newman-Evans and B. K. Carpenter, *J. Am. Chem. Soc.* **1984**, *106*, 7994.

36. (a) D. C. Sorescu, D. L. Thompson, and L. M. Raff, *J. Chem. Phys.* **1995**, *102*, 7910. (b) B. A. Lyons, J. Pfeifer, and B. K. Carpenter, *J. Am. Chem. Soc.* **1991**, *113*, 9006.

37. M. J. Knapp and J. P. Klinman, *Eur. J. Biochem.* **2002**, *269*, 3113.

38. (a) M. Ben-Nun and T. J. Martinez, *Adv. Chem. Phys.* **2002**, *121*, 439. (b) S. Hammes-Schiffer and S. R. Billeter, *Int. Rev. Phys. Chem.* **2001**, *20*, 591. (c) A. Donoso, D. Kohen, and C. C. Martens, *J. Chem. Phys.* **2000**, *112*, 7345.

39. (a) J. L. Liao and G. A. Voth, *J. Chem. Phys.* **2002**, *116*, 9174. (b) E. A. Coronado, J. Xing, and W. H. Miller, *Chem. Phys. Lett.* **2001**, *349*, 521. (c) M. Garavelli, F. Bernardi, M. Olivucci, M. J. Bearpark, S. Klein, and M. A. Robb, *J. Phys. Chem. A* **2001**, *105*, 11496. (d) M. D. Hack, A. M. Wensmann, D. G. Truhlar, M. Ben-Nun, and T. J. Martinez, *J. Chem. Phys.* **2001**, *115*, 1172. (e) J. C. Burant and J. C. Tully, *J. Chem. Phys.* **2000**, *112*, 6097.

40. (a) G. Stock and U. Muller, *J. Chem. Phys.* **1999**, *111*, 65. (b) M. Ben-Nun and R. D. Levine, *J. Chem. Phys.* **1996**, *105*, 8136. (c) Y. Guo, D. L. Thompson, and T. D. Sewell, *J. Chem. Phys.* **1996**, *104*, 576. (d) G. H. Peslherbe and W. L. Hase, *J. Chem. Phys.* **1994**, *100*, 1179.

CHAPTER 22

The Partnership between Electronic Structure Calculations and Experiments in the Study of Reactive Intermediates

WESTON THATCHER BORDEN

Department of Chemistry, University of Washington, Seattle, WA

Reactive Intermediate Chemistry, edited by Robert A. Moss, Matthew S. Platz, and Maitland Jones, Jr.
ISBN 0-471-23324-2 Copyright © 2004 John Wiley & Sons, Inc.

1. INTRODUCTION

Because reactive intermediates (RIs) undergo facile intra- and/or intermolecular reactions, they are, by definition, short lived. As described in the other chapters in this volume, the transient nature of RIs provides a challenge to their direct spectroscopic observation. When the spectra of a putative RI can be obtained, it may not be certain that the observed spectra actually belong to the RI.

What makes RIs short lived is, of course, their unusual electronic structures. The effect of their unusual electronic structures on the geometries and relative energies of RIs is often of tremendous interest. Unfortunately, the short lifetimes of RIs usually precludes their study by X-ray crystallography; and measuring the energies of RIs, especially relative to those of stable molecules, can be very challenging.

The unusual electronic structures of RIs frequently also make their chemistries very unusual. Understanding the electronic structures of RIs well enough to make both qualitative and quantitative predictions about their reactivities is an important goal of research in this area of chemistry.

It is easy to appreciate why electronic structure calculations that are accurate enough to compute reliably the geometries, spectra, and relative energies of RIs and to predict their reactivities, would be a very important adjunct to experiments. Fortunately, during the last three decades of the twentieth century, advances in both computer hardware and software began to make it possible to perform electronic structure calculations on a wide variety of RIs with the required accuracy.

This fact was first brought to the attention of experimentalists, interested in RIs, by two different computational predictions, made in the 1970s, about methylene.[1] Triplet methylene was computed to have a bent geometry, with an H—C—H bond angle of ∼135°; and the best calculations predicted the triplet to lie ∼10 kcal/mol below the lowest singlet state. Both of these computational results, when published, were in apparent conflict with experiments, which had been interpreted as indicating that triplet methylene is linear and that it lies >20 kcal/mol below the lowest singlet state. However, subsequent experiments proved the calculations to be correct.

At the same time that experimentalists were beginning to realize that electronic structure calculations were capable of making predictions, accurate enough to be useful, the development by John Pople (who shared the Nobel Prize in Chemistry in 1998) and his co-workers of the Gaussian series of programs made performing such calculations relatively easy, even for experimentalists. Experimentalists began to stop relying on their theoretician friends to perform electronic structure calculations for them and started carrying out calculations themselves.

In many, if not most, groups currently engaged in experimental studies of RIs, electronic structure calculations play an important role; and theoretical chemists continue to find RIs fascinating subjects for computational studies. The partnership between electronic structure calculations and experiments in the study of RIs is the subject of this chapter.

Section 2 describes in general terms the types of information, useful to experimentalists, about RIs that can be obtained from electronic structure calculations. References are given to various applications of electronic structure calculations that are discussed in other chapters of this book.

In both Sections 2 and 3, terms that are frequently used in the literature in discussing the results of electronic structure calculations, are given in italics. It is hoped that, by having their attention directed to these terms, the reader will become familiar with them and will learn what each term means.

Section 3 provides a brief introduction to the methods used in performing electronic structure calculations. Methods based on both wave functions and on density functional theory (DFT) are discussed. The assumptions that underlie each of these two different types of calculation are described, and a comparison between these methods is provided.

Section 4 discusses the application of electronic structure calculations to understanding and predicting the outcome of experiments on three different types of RI. The examples in this section are taken from research, performed in collaborations between the author's group and the groups of experimentalists. The three RIs discussed in Section 4 are phenylnitrene, cubyl cation, and propane-1,3-diyls.

Section 5 provides some final observations about the uses of electronic structure calculations in the experimental studies of RIs; and Section 6 gives suggestions for further reading, both about electronic structure theory and about its application to specific experimental studies of RIs. A brief description of each of the suggestions for further reading is provided.

2.   WHAT ROLES CAN ELECTRONIC STRUCTURE CALCULATIONS PLAY IN EXPERIMENTAL STUDIES OF REACTIVE INTERMEDIATES?

2.1. Predictions of Spectra

Because of the fleeting existence of RIs under ordinary conditions, if RIs are to be observed directly, they must be observed either by detection during their very short lifetimes or by extending their lifetimes. The latter strategy is sometimes pursued by studying RIs in the gas phase, under conditions where bimolecular reactions can be minimized. However, immobilizing RIs in glasses or rare gas matrices at low temperatures has increasingly become the technique of choice for obtaining detailed spectroscopic information about RIs.

Whether laser flash photolysis (LFP) is used to detect RIs before they react, or matrix isolation at very low temperatures is employed to slow down or quench these reactions, spectroscopic characterization of RIs is frequently limited to infrared (IR) and/or ultraviolet–visible (UV–vis) spectroscopy. Nuclear magnetic resonance (NMR) spectroscopy, which is generally the most useful spectroscopic technique for unequivocally assigning structures to stable organic molecules, is inapplicable to many types of RI.

It is often the case that the putative structure assigned to an RI is so unusual that vibrational and electronic spectra of analogous molecules are unavailable. How then can an experimentalist know whether an observed IR and/or UV–vis spectrum actually does belong to the putative structure assigned to the RI? As exemplified in the sections on matrix-isolated benzynes in Chapter 16 in this volume and also discussed in Chapter 17 in this volume, suitable electronic structure calculations can provide accurate predictions of the IR spectra and UV–vis excitation energies for RIs. The IR and UV–vis spectra, predicted for a particular structure, can then be compared with those obtained experimentally.

The ability of calculations, especially those based on DFT, to predict IR spectra accurately is generally very good. Currently, most, if not all of the papers that report the IR spectrum of a matrix-isolated RI, contain a figure that shows graphically how well the observed spectrum and the spectrum computed for the structure assigned to the RI correspond.

As described in Chapter 1 in this volume, electronic structure theory calculations are also very useful for computing ^{13}C chemical shifts in carbocations. The predicted ^{13}C NMR spectra and the experimental spectra, obtained in superacids under conditions where the carbocations are stable, can be compared. These comparisons are particularly helpful in differentiating between classical and nonclassical structures for these electron-deficient species.

2.2. Predictions of Geometries

One of the most desirable and experimentally least accessible pieces of information about an RI is its molecular structure. Unusual features of the electronic structures of RIs (e.g., nonclassical bonding in carbocations) often result in unusual bond lengths and bond angles. Unfortunately, although the geometries of a few carbocations have been obtained by X-ray crystallography, most RIs are too reactive to allow their structures to be determined experimentally.

However, except in pathological cases (e.g., that of *m*-benzyne, which is discussed in Chapter 16 in this volume), even fairly low-level calculations can produce rather accurate structural parameters. Therefore, it is almost always much easier to compute the structural parameters for an RI than to measure them. If there is good agreement between the spectra observed for an RI and those computed for the structure assigned to it, then it is usually assumed that the bond lengths and bond angles in the calculated structure are very close to those in the RI.

2.3. Predictions of Enthalpies

As discussed in Section 3, it is usually more challenging to compute accurately the energies than the geometries of RIs. In contrast, although geometries of most RIs are hard to obtain experimentally, it is often possible to measure the enthalpy of activation for the formation and/or disappearance of an RI accurately. This experimental value can then be compared with the values predicted by different levels of electronic structure calculations, in order to test the ability of a particular level of theory to mirror accurately the experimental energetics.

In computing an activation enthalpy for a reaction, it is first necessary to optimize the geometry of the reactant and to locate the transition structure (TS). Both geometries are *stationary points* (i.e., points at which all the first derivatives of the energy are zero) on the *potential energy surface* (PES); but a TS is a *saddle point* (i.e., a stationary point with one and only one negative second derivative). Finding a TS is computationally more demanding than optimizing the geometry of an energy minimum. Nevertheless, in many cases, a level of electronic structure theory that is sufficient to provide an accurate geometry for an RI also suffices to find with comparable accuracy the geometry of the TS by which the RI is formed and/or reacts.

The adequacy of calculations, performed at a particular level of electronic structure theory, for finding equilibrium and TS geometries, can be tested, even if comparisons with experimental enthalpy differences cannot be made. The relative energies of two geometries, calculated at the level of theory that is used to find these stationary points, can be recomputed by *single-point* calculations, performed at the same two geometries, but with a higher level of electronic structure theory (usually one that is too demanding of computational resources to be used routinely for finding geometries). If the two different levels of theory give similar energy differences, this finding provides evidence that the lower level of theory is probably as good as the higher level would be for finding the geometries of these two stationary points on the PES.

In order to compute the enthalpy difference between two stationary points, the difference between their electronic energies must be corrected for the difference between their vibrational energies at 0 K and also for the difference between the amounts of thermal energy that each absorbs between 0 K and the temperature of interest. These *zero-point* and *heat capacity* corrections are available from *vibrational analyses*, which require calculation of the second derivatives of the energy with respect to all internal coordinates at each stationary point. Vibrational analyses must also be performed in order to confirm that the geometry of a putative energy minimum has only real vibrational frequencies and that the geometry of a stationary point, which is thought to be a TS, has one and only one vibrational frequency that is imaginary.

Good agreement between a measured enthalpy of activation and that computed at a particular level of theory, furnishes evidence that calculations at this level of theory are accurate enough to provide reliable information about the enthalpy differences between the reactant, the TS, and other stationary points on the PES for an RI. As already noted, differences between the heats of formation of an RI and other energy minima on a PES (i.e., stable molecules and other RIs, which are either formed from an RI or from which an RI is formed) are usually harder to measure experimentally than the activation enthalpy for appearance or disappearance of an RI. Therefore, being able to compute accurately the enthalpy differences between an RI and other energy minima on a PES can provide very valuable information that is usually not easy to obtain experimentally.

2.4. Predictions of Free Energies and Isotope Effects

Rotational and vibrational partition functions can be computed from the geometry and vibrational frequencies that are calculated for a molecule or TS. The entropy can then be obtained from these partition functions. Thus, electronic structure calculations can be used to compute not only the enthalpy difference between two stationary points but also the entropy and free energy differences.

The ability to compute the free energy difference between two minima on a PES allows the effect of isotopic substitution on the equilibrium constant between them to be calculated. Similarly, a kinetic isotope effect (KIE) can be predicted from the effect of isotopic substitution on the free energy difference between a minimum and a TS (i.e., a free energy of activation).

KIEs are often very sensitive to the lengths of the forming and breaking bonds in the TS especially when rehybridization occurs at the site of isotopic substitution in a TS. By comparing measured KIEs with those computed for different TS geometries, Houk co-workers[2] obtained valuable information about the geometries and lengths of the forming and breaking bonds in the TSs for a wide variety of reactions. The use of KIE effect calculations in an attempt to reconcile the results of two experiments on an RI, propane-1,3-diyl, is described in Section 4.3.1.

2.5. Development of Qualitative Models

Perhaps more valuable over time than the quantitative predictions of spectra, structural parameters, and relative enthalpies and entropies of RIs, which can be obtained from electronic structure calculations, are the qualitative models of the electronic structures and reactivities of RIs that emerge from the computational results. Any model, to be successful, must do two things.

First, a successful model must provide explanations of all the experimental and computational results that are known at the time that the model is formulated. Since the electronic structures of RIs are usually rather different from those of stable molecules, a new model, or a modification of an old one, is frequently necessary in order to explain the puzzling chemistry and/or spectroscopy of an RI.

Second, a successful model makes predictions that can be tested experimentally. The formulation of a model that stimulates experiments, whose results may lead to more calculations, is probably the most important type of synergism between calculations and experiments. Section 4 provides some specific examples of this type of partnership in the study of RIs and of the reactions that they undergo.

3. METHODS FOR PERFORMING ELECTRONIC STRUCTURE CALCULATIONS

This section provides a "beginners' guide" to electronic structure calculations. A much more detailed introduction than is possible here can be found in the authoritative and very readable monograph by Cramer.[3a]

Until the last decade, for all but a few theoreticians, electronic structure calculations meant calculations based on approximate solution of the time-independent Schrödinger equation,

$$\mathcal{H}\Psi = E\Psi \qquad (1)$$

In Eq. 1, $\mathcal{H}$ is the *Hamiltonian operator*, Ψ is one of the possible wave functions for all of the electrons in a molecule, and E is the energy associated with each of these possible *many-electron wave functions*.

However, within the last 10 years, calculations based on density functional theory (*DFT*) have come into wide use. The DFT calculations are usually much less computationally demanding than *wave function based calculations* that provide results of comparable accuracy. Hence, DFT calculations are particularly useful for very large molecules, for which wave function based calculations of sufficient accuracy may be too computer intensive to perform.

3.1. The Born–Oppenheimer Approximation

Since a molecule consists of both nuclei and electrons, it may not be at all obvious why calculations can be performed on just the electrons. However, the huge differences in masses between nuclei and electrons justify the *Born–Oppenheimer approximation*, which allows the motions of the electrons to be separated from the motions of the nuclei.

Using the Born–Oppenheimer approximation, electronic structure calculations are performed at a fixed set of nuclear coordinates, from which the electronic wave functions and energies at that geometry can be obtained. The first and second derivatives of the electronic energies at a series of molecular geometries can be computed and used to find energy minima and to locate TSs on a PES.

The wave functions for the periodic motions (i.e., vibrations) of the nuclei and the associated vibrational energy levels can then be obtained from the geometries and the second derivatives of the energy at the stationary points. As described in Sections 2.3 and 2.4, the vibrational energies are necessary to correct the difference between the electronic energies of two stationary points into differences between their enthalpies and free energies.

3.2. Wave Function Based Calculations

The Hamiltonian operator in Eq. 1 contains sums of different types of quantum mechanical operators. One type of operator in $\mathcal{H}$ gives the kinetic energy of each electron in Ψ, by computing the second derivative of the electron's wave function with respect to all three Cartesian coordinates axes. There are also terms in $\mathcal{H}$ that use Coulomb's law to compute the potential energy due to (a) the attraction between each nucleus and each electron, (b) the repulsion between each pair of electrons, and (c) the repulsion between each pair of nuclei.

The last set of terms is independent of the electrons and depends only on the coordinates of the nuclei. At a particular geometry, the sum of these terms provides the total nuclear repulsion energy, which is just an additive constant to the electronic energy. Consequently, the terms for the nuclear–nuclear repulsion energy in $\mathcal{H}$ can be neglected when attempting to solve Eq. 1 for the electronic wave functions and their associated energies.

Since Ψ in Eq. 1 is the wave function for all of the electrons in the molecule, it is simplest to begin trying to find Ψ by assuming that it can be approximated as the product of *one-electron wave functions*, one wave function for each of the electrons in a molecule. These one-electron wave functions are called *orbitals*, and they are distinguished from the many-electron wave function by using a lower case psi (ψ) for the former and an upper case psi (Ψ) for the latter.

In order for Ψ to embody the *Pauli exclusion principle*, it must be an *antisymmetrized wave function*. Antisymmetrization requires that exchange of any two electrons between orbitals or exchange of the spins between electrons in the same orbital causes Ψ to change sign.

Antisymmetrization results in Ψ vanishing, not only if two electrons with the same spin occupy the same orbital, but also if electrons with the same spin have the same set of spatial coordinates. Thus, antisymmetrization of Ψ results not only in the Pauli exclusion principle but also in the *correlation* of electrons of the same spin.

To put it crudely, this correlation ensures that electrons of the same spin cannot be in the same place at the same time. Therefore this type of correlation makes the Coulombic repulsion energy between electrons of the same spin smaller than that between electrons of opposite spin. This is the reason why *Hund's rule* states that, an electronic state in which two electrons occupy different orbitals with the same spin is lower in energy than an electronic state in which the electrons occupy the orbitals, but with opposite spins.

The simplest expression for Ψ consists of a single *configuration*, in which each electron is assigned to an orbital with spin up (α) or spin down (β). If two electrons with opposite spin occupy the same orbital, ψ_i, it is notationally more convenient to denote this orbital occupancy by replacing $\psi_i^{\alpha}\psi_i^{\beta}$ with Ψ_i^2.

Unfortunately, if a single configuration is used to approximate the many-electron wave function, electrons of opposite spin remain uncorrelated. The tacit assumption that electrons of opposite spin move independently of each other is, of course, physically incorrect, because, in order to minimize their mutual Coulombic repulsion energy, electrons of opposite spin do certainly tend to avoid each other. Therefore, a wave function, Ψ, that consists of only one configuration will overestimate the Coulombic repulsion energy between electrons of opposite spin.

The major difficulty in wave function based calculations is that, starting from an *independent-particle model*, correlation between electrons of opposite spin must somehow be introduced into Ψ. Inclusion of this type of electron correlation is essential if energies are to be computed with any degree of accuracy. How, through the use of multiconfigurational wave functions, correlation between electrons of opposite spin is incorporated into Ψ, is the subject of Section 3.2.3.

3.2.1. Hartree–Fock Theory.

In the lowest electronic state of most stable molecules the n orbitals of lowest energy are all doubly occupied, thus forming a *closed shell*. If, at least as a first approximation, such an electronic state is described by a single configuration, the wave function for this state can be written as

$$\Psi = |\psi_1^2\psi_2^2 \ldots \psi_m^2\psi_n^2\rangle \qquad (2)$$

where the Dirac "ket" ($|~\rangle$) is used to symbolize that Ψ has been antisymmetrized.

Minimizing the energy, E, in Eq. 1 with respect to variations in the filled orbitals, $\psi_1, \psi_2, \ldots, \psi_m, \psi_n$, in Ψ Eq. 2, leads to the finding that the optimal orbitals must satisfy

$$\mathcal{F}\psi_i = \varepsilon_i\psi_i \qquad (3)$$

The operator $\mathcal{F}$ in this equation is called the *Fock operator*, and ε_i is the energy of orbital ψ_i. According to *Koopmans' theorem*, $-\varepsilon_i$ is approximately equal to the energy required to ionize a molecule by removing an electron from ψ_i.

Two of the terms in $\mathcal{F}$ are the same *one-electron operators* that appear in $\mathcal{H}$. One of these operators gives the kinetic energy of an electron in ψ_i, and the other computes the attractive Coulombic energy between an electron in ψ_i and each of the nuclei in the molecule.

However, there are two more types of one-electron operators in $\mathcal{F}$. One of them, $\mathcal{J}_j$, is called the *Coulomb operator*. The other, $\mathcal{K}_j$, is called the *exchange operator*. Together, they replace the *two-electron operators*, e^2/r_{kl}, in $\mathcal{H}$, which give the Coulombic repulsion energy between each pair of electrons, k and l.

$2\mathcal{J}_j - \mathcal{K}_j$ operating on ψ_i gives the expression in *Hartree–Fock (HF) theory* for the effective Coulombic repulsion energy between an electron ψ_i and the pair of electrons in ψ_j. In the Fock operator, $2\mathcal{J}_j - \mathcal{K}_j$ is summed over the n doubly occupied orbitals; so the sum, $\Sigma(2\mathcal{J}_j - \mathcal{K}_j)$ in $\mathcal{F}$, gives the HF expression for the effective Coulombic repulsion energy between an electron in ψ_i and all the other electrons in a molecule.

The operator $2\mathcal{J}_j$ computes the Coulombic repulsion energy between an electron in ψ_i and the pair of electrons in ψ_j, assuming that the electrons in ψ_i and in ψ_j move independently of each other. The operator $\mathcal{K}_j$ corrects the Coulombic repulsion energy, computed from $2\mathcal{J}_j$, for the fact that antisymmetrization of Ψ results in correlation between an electron in ψ_i and the electron of the same spin in ψ_j.

The operators $\mathcal{J}_i$ and $\mathcal{K}_i$ give the same result when they operate on ψ_i. Thus, $2\mathcal{J}_i - \mathcal{K}_i$ operating on ψ_i has the same effect as just $\mathcal{J}_i$ operating on ψ_i. This cancellation is physically correct, because an electron in ψ_i only feels the field from the one other electron in ψ_i, not from itself. Thus, the exchange operator, $\mathcal{K}_i$, in $\mathcal{F}$ also has the effect of canceling the fictional repulsion between an electron in ψ_i and itself, which is included in $2\mathcal{J}_i$.

The n doubly occupied orbitals in the closed-shell wave function in Eq. 2 are required in order to construct (i.e., compute) the Coulomb and exchange operators in $\mathcal{F}$. Thus, it might seem impossible to use Eq. 3 to find the optimal orbitals for a molecule, because these orbitals must already be known in order to construct the Fock operator in Eq. 3. However, this seemingly impossible task is made possible by using an iterative process.

Starting with a guess as to what the optimal orbitals are, this guess is used to construct the Coulomb and exchange operators in the Fock operator, and Eq. 3 is then used to find an improved set of orbitals. This set of orbitals is used to construct the Coulomb and exchange operators for a new Fock operator; and, by using this new $\mathcal{F}$ operator in Eq. 3, another set of ψ_i is obtained. The process is repeated until the set of ψ_i that are used to construct $\mathcal{F}$ is essentially the same as the set of ψ_i obtained by using this Fock operator in Eq. 3. These converged orbitals are said to have reached self-consistency, and they are the orbitals that give the lowest possible energy for the HF wave function in Eq. 2.

Since the electric field, computed from the filled ψ_i, is used to construct the Coulomb operator in $\mathcal{F}$, the electric field that is used to construct $\mathcal{F}$ from the converged orbitals is the same as the electric field that is computed from the orbitals that solve Eq. 3 for this Fock operator. Therefore, at convergence, both the orbitals and the electric field computed from them are self-consistent. Consequently, HF theory is also known as *self-consistent field (SCF)* theory.

3.2.2. Basis Sets of Atomic Orbitals and the LCAO–MO Approximation. It

seems reasonable to assume that, like isolated atoms, atoms in molecules have orbitals. Therefore, it makes physical sense to try to write the orbitals, ψ_i, for a molecule as linear combinations of the *atomic orbitals (LCAOs)*, ϕ_r, of the atoms in the molecule.

$$\psi_i = \Sigma c_{ir} \phi_r \tag{4}$$

Equation 4, in which *molecular orbitals (MOs)* are expressed as LCAOs, is known as the *LCAO–MO approximation*.

Within the LCAO–MO approximation, the process for finding the MOs that minimize the total HF energy becomes clearer. If the mathematical form of each of the AOs, ϕ_r, in a molecule is assumed to be known, it is the coefficient, c_{ir}, of each AO in each filled MO that is optimized variationally. This is accomplished by solving Eq. 3 until self-consistency is reached.

However, it is not safe to assume that the optimal AOs, with which to build the MOs for a particular molecule, are the same as the optimal AOs in other molecules in which a particular atom appears. In fact, it is not even safe to assume the optimal AOs for an atom are the same in all the MOs in the same molecule. In finding the *variationally correct* LCAO–MOs (i.e., the LCAO–MOs that give the lowest possible HF energy), it is important to allow both the sizes of the AOs, as well as their coefficients, to vary in each MO.

This variability in AO sizes is most easily introduced by using more than one independent mathematical function to represent each AO. The contribution of each function to each AO in each MO can then be optimized variationally, by performing a HF calculation. The set of mathematical functions used to represent each of the AOs in a molecule is called the *basis set*.

The valence AOs in a molecule might be expected to vary much more with their environment than the inner (*core*) AOs. Thus, an economical way to construct a basis set is to use a single *basis function* to represent the core AOs, but to use two independent basis functions to represent the valence AOs. This type of basis set is called a *split-valence* basis set.

In such a basis set one of the basis functions for the valence AOs might have its maximum value close to the nucleus, while the second might have its maximum value much farther from the nucleus. Thus, as illustrated graphically in Figure 22.1(*a*), a linear combination of the two basis functions can be used to represent an orbital that has a size which is intermediate between that of the two basis functions. By allowing the contribution of these two basis functions to each MO to be determined by solving Eq. 3 to self-consistency, the effective size of each AO, ϕ_r, in each MO, ψ_i, in Eq. 4 will be optimal for the HF wave function in Eq. 2.

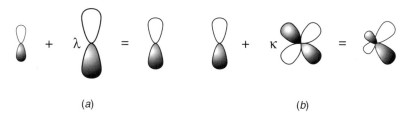

(*a*)   (*b*)

Figure 22.1. Schematic depiction of (*a*) the mixing of two-$2p$ basis functions, to give a $2p$ AO of intermediate size. (*b*) The mixing of a $3d$ basis function into a $2p$ basis function, to give a polarized $2p$ AO, which is no longer centered on the nucleus.

Further improvements in the flexibility with which the AOs in Eq. 4 are described mathematically can be obtained by adding a third independent basis function to a split valence basis set. In an anion, electrons are likely to be spread over a greater volume than in a neutral molecule, so adding very *diffuse basis functions* to the basis set for a negatively charged molecule is usually important. A further improvement in the basis set for a molecule would be to use two or three independent basis functions to describe, not only the valence AOs, but also the core AOs. Such basis sets are called, respectively, *double-zeta* or *triple-zeta* basis sets.

Another very important way to improve the basis sets used for LCAO–MO calculations is to abandon the tacit assumption that the AOs in a molecule are centered exactly on the nuclei of the atoms. This improvement is most easily accomplished by incorporating *polarization functions* into the basis sets. Polarization functions have one more angular node than the basis functions that they are polarizing; so mixing in polarization functions desymmetrizes basis functions by turning them into hybrid orbitals. Thus, $2p$ basis functions serve as polarization functions for the $1s$ basis functions on hydrogen; and, as shown graphically in Figure $1(b)$, $3d$ basis functions serve as polarization functions for the $2p$ basis functions on first-row elements.

A final point about basis functions concerns the way in which their radial parts are represented mathematically. The AOs, obtained from solutions of the Schrödinger equation for one-electron atoms, fall-off exponentially with distance. Unfortunately, if exponentials are used as basis functions, computing the integrals that are required for obtaining electron repulsion energies between electrons is mathematically very cumbersome. Perhaps the most important software development in wave function based calculations came from the realization by Frank Boys that it would be much easier and faster to compute electron repulsion integrals if Gaussian-type functions, rather than exponential functions, were used to represent AOs.

However, especially for core AOs, where an exponential function has a very sharp cusp at the nucleus, several Gaussian functions are required even to begin to represent an exponential function accurately. For example, a split-valence basis set, developed by John Pople's group and widely employed in current calculations, uses six Gaussian functions to represent the $1s$ core orbitals on first row atoms; whereas, a set of only three Gaussians, plus one more independent Gaussian, is used to represent the valence $2s$ and $2p$ AOs. With $3d$ functions included, to serve as polarization functions for the valence orbitals on the heavy atoms, Pople denotes this *Gaussian basis set* as 6-31G(d) or, more commonly, 6-31G*.[4a]

A better Gaussian basis set, created by Pople and co-workers[4b,c] for first-row atoms, is 6-311G. It is *valence triple zeta*, in that it uses three independent basis functions to represent each valence AO. With one $4f$ and two independent $3d$ polarization functions on non-hydrogen atoms and one $3d$ and two independent sets of $2p$ polarization functions on hydrogen, in Pople's notation the resulting basis set is called 6-311G(2d1f,2p1d).

Not all the basis sets that are in common use were created by Pople's group. Also frequently employed and built into many programs for performing electronic structure calculations are Dunning's *correlation-consistent (cc)-basis sets*.[5] As their name implies, the cc-basis sets are especially useful for calculations that include electron correlation. These basis sets provide a means for systematically increasing the amount of *correlation energy* that is recovered by such calculations.

3.2.3. Inclusion of Electron Correlation.

HF calculations, performed with basis sets so large that the calculations approach the *HF limit* for a particular molecule, still calculate total energies rather poorly. The reason is that, as already discussed, HF wave functions include no correlation between electrons of opposite spins. In order to include this type of correlation, *multiconfigurational (MC) wave functions*, like that in Eq. 5, must be used.

$$\Psi = c_1|\psi_1^2\psi_2^2\ldots\psi_m^2\psi_n^2\rangle + c_2|\psi_1^2\psi_2^2\ldots\psi_m^2\psi_v^2\rangle + c_3|\psi_1^2\psi_2^2\ldots\psi_v^2\psi_n^2\rangle$$
$$+ c_4\left(|\psi_1^2\psi_2^2\ldots\psi_m^\alpha\psi_n^\beta\psi_v^2\rangle - |\psi_1^2\psi_2^2\ldots\psi_m^\beta\psi_n^\alpha\psi_v^2\rangle\right)\ldots \quad (5)$$

In an MC wave function, the HF wave function in Eq. 2 is augmented by additional configurations, only four of which are actually given in Eq. 5. As shown pictorially in Figure 22.2, in these four additional configurations different pairs of electrons have been excited from two of the orbitals (ψ_m and ψ_n) that are doubly occupied in the HF wave function. The electrons have been excited into one of the many *virtual orbitals* (ψ_v), which, if empty, are the HF wave function. It is the

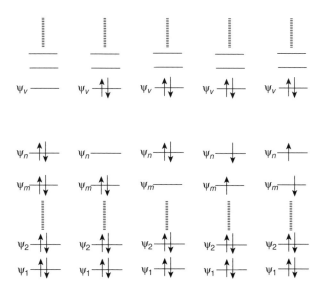

Figure 22.2. Pictorial depiction of the HF and the four excited configurations that are given in Eq. 5. Many more configurations can be generated by excitations of electrons from other orbitals that are filled in the HF configuration into the many virtual orbitals that are unoccupied in this configuration.

mixing of the excited configurations into the HF configuration that provides correlation between electrons of opposite spin, which is absent from the HF wave function.

For example, mixing of the second configuration in Eq. 5 into the HF configuration with a minus sign results in the mixing of ψ_v into ψ_n with a plus sign for one of the electrons in ψ_n and with a minus sign for the other. Provided that ψ_n and ψ_v overlap, the resulting orbitals, $\sqrt{c_1}\psi_n + \sqrt{c_2}\psi_v$ and $\sqrt{c_1}\psi_n - \sqrt{c_2}\psi_v$, no longer span exactly the same region of space. Therefore, the Coulombic repulsion energy in the HF wave function in Eq. 2, between the electrons of opposite spin in ψ_n, is reduced in the MC wave function in Eq. 5.

3.2.3.1. Configuration Interaction Calculations.

For the ground state of a molecule, the optimal coefficients for the MC wave function in Eq. 5 are those that cause Ψ to satisfy the Schrödinger equation with the lowest energy. Other sets of coefficients that cause Ψ to satisfy Eq. 1 give the wave functions for excited states of the molecule.

The different sets of coefficients that cause Ψ to satisfy Eq. 1, and the energy that corresponds to each wave function can be obtained by computing the energies of the interactions between each pair of configurations in Eq. 5, due to the Hamiltonian operator, $\mathcal{H}$. If these interaction energies are displayed as a matrix, the coefficients that result in the MC wave function in Eq. 5 satisfying Eq. 1 are those that diagonalize this *Hamiltonian matrix*.

If good basis sets are used, the Hamiltonian matrices that are generated in *configuration interaction* (CI) calculations on even comparatively small molecules are very large. Fortunately, efficient computer algorithms have been developed for computing the elements of these matrices and finding the wave functions for the lowest energy states, even for MR wave functions that consist of 10^6 configurations. For such a wave function, the effect of the Hamiltonian operator on Ψ is represented by a matrix that contains $10^6 \times 10^6 = 10^{12}$ elements.

If all possible excitations are included in a CI wave function, except for very small molecules, many more than 10^6 configurations are generated. Thus, full CI calculations are almost always impractical or impossible.

One of the ways to reduce the size of CI calculations is by limiting the number of electrons that are simultaneously excited. For example, including only *single and double excitations* of electrons from the HF configuration gives what is called an SD–CI (or CISD) calculation. Sometimes the effects of including higher levels of excitation (e.g., triples and/or quadruples) are estimated, rather than obtained by actually performing an SDTQ–CI calculation. An SD–CI calculation in which the effect of including quadruple excitations is estimated would be denoted SD(Q)–CI.

3.2.3.2. Calculations Based on Perturbation Theory.

An alternative to performing a CI calculation is to use *perturbation theory*. As in the expansion of a mathematical function in a power series, in perturbation theory the coefficients for and the energy of an MC wave function are expanded in a series. The more terms in the series that are actually included, the more closely the coefficients and energies that are obtained by perturbation theory approximate the exact values that would be computed by a CI calculation with the same configurations.

If only the leading terms in the perturbation theory expansions are included, the computer resources required for such calculations are much more modest than those that would be needed for a CI calculation with the same set of configurations. This is the biggest advantage of calculations that are based on perturbation theory over CI calculations.

The lowest level of perturbation theory that can be used to calculate energies which include the effects of electron correlation is called second-order *Möller–Plesset (MP2)* perturbation theory.[6] MP2 calculations include the interactions of the excited configurations with just the HF configuration and ignore the effects of the interactions of the excited configurations with each other. For closed-shell molecules, *Brillouin's theorem* shows that singly excited configurations do not interact directly with the HF configuration. Therefore, only doubly excited configurations are included in an MP2 calculation.

Higher level calculations, based on higher orders of Möller–Plesset perturbation theory, can also be performed, albeit with the consumption of much more computer time. For example, an MP4SDTQ calculation uses fourth-order Möller–Plesset perturbation theory, includes excitations through quadruples, and gives better energies than MP2 does.

3.2.3.3. Size Consistency in CI Calculations.

Not only are MPn calculations less demanding of computer resources than CI calculations that include the same levels of excitations, but MPn calculations are *size-consistent*; whereas, CISD calculations are not. A computational method is size consistent if the energy, obtained in a calculation on two identical molecules at infinity, is exactly twice the energy that is obtained in a calculation on just one of these molecules. The reason why CISD calculations are not size consistent is easy to understand.

If the CISD wave functions for two identical molecules are multiplied, to give the wave function for the pair of molecules, there are terms in the resulting wave function in which both molecules are doubly excited. Since these terms represent quadruple excitations from the HF configuration, they are not included in the CISD wave function for two identical molecules at infinity. Consequently, the CISD energy for a pair of identical molecules is higher than twice the CISD energy for an individual molecule.

Clearly, the way to remedy the lack of size consistency in a CISD calculation is to include the missing terms that contain quadruple excitations. Because these terms are products of simultaneous double excitations, their sizes can be determined from information that is available from CISD. Two methods that include quadruple excitations in this manner and, hence, are size consistent are discussed in Section 3.2.3.4.

3.2.3.4. Coupled-Cluster and Quadratic CI Calculations.

The most powerful method for including electron correlation that can be used in calculations on medium-sized molecules is based on *coupled-cluster (CC) theory*. This method, which was developed by Cizek and successsfully implemented by Bartlett et al.,[7] expresses the effects of quadruple and higher excitations as products of the effects of double excitations.

The CCSD calculations are similar, both in methodology and in the accuracy of the results obtained, to calculations performed with Pople's *quadratic CI* method.[8] Like CCSD calculations, QCISD calculations also explicitly include single and double excitations; and the effects of quadruple excitation in QCISD are obtained from quadrature of the effects of double excitations. However, CCSD does contain terms for the effects of excitations beyond quadruples, which are absent from QCISD.

In both methods the effects of triple excitations can be estimated. These effects can be significant, if quantitative accuracy is the goal of the calculations. However, performing a calculation at either the CCSD(T) or QCISD(T) level of theory comes at the cost of substantially increasing the computer time required, beyond that consumed by a CCSD or QCISD calculation.

3.2.3.5. G2, G3, and CBS Calculations.

The Gaussian-2 (G2)[9a] and Gaussian-3 (G3)[9b] methods, developed by Pople and co-workers, and the complete basis set (CBS)[10] method, developed by Petersson and co-worker, all attempt to provide very accurate thermochemical predictions. The energies computed by very large calculations with very large basis sets are extrapolated from the results of high-level calculations with moderate-sized basis sets and lower level calculations with very large basis sets. Like the MPn, CCSD(T), and QCISD(T) methods, the G2, G3, and CBS methods are available in the current Gaussian package of programs for performing electronic structure calculations.[11]

3.2.3.6. CASSCF and CASPT2 Calculations.

For many molecules and reactions, a relatively small number of electrons (e.g., those in π orbitals) can be identified as being important. For such molecules and such reactions, variational calculations can be performed with configurations that provide correlation between just those *active electrons*. For each bonding orbital that is occupied by active electrons, a virtual orbital is also usually included in the *orbital active space*; and all possible configurations that arise from distributing the active electrons among the active orbitals are included in the resulting MC wave function.

In an *MCSCF calculation*, not only the coefficients of the multiple configurations in the MC wave function, but also the orbitals in them, are simultaneously optimized. An (n/m)MCSCF calculation, in which the n active electrons and m active orbitals are chosen in the manner described in the preceding paragraph, is called a *complete active space (CAS)SCF calculation*.[12]

It is imperative to use CASSCF wave functions for singlet diradicals and other open-shell molecules for which a single configuration provides an inadequate description of the wave function.[13] However, perhaps surprisingly, CASSCF calculations often perform rather poorly in calculations on molecules and TSs with closed shells of electrons, if the active electrons are delocalized. An example is provided by (6/6)CASSCF calculations with the 6-31G* basis set on the chair Cope rearrangement of 1,5-hexadiene, which gave a calculated activation enthalpy that was >12 kcal/mol above the experimental value.[14]

In order to get good agreement between CASSCF and experimental results, it is often necessary to include correlation between the active electrons and the other valence electrons.[15] This type of correlation, which is called *dynamic correlation*, can be added to a CASSCF wave function by using the equivalent of MP2 perturbation theory. However, in an MP2 calculation on an MC wave function, rather than including double excitations from a single HF *reference configuration*, excitations from all of the configurations must be included.

Several different versions of second-order perturbation theory for *multireference* wave functions have been implemented; but the one currently in widest use is probably the CASPT2 method.[16] This method was developed by Roos and co-workers[17] in Lund, Sweden, and it is available in their MOLCAS package of computer programs.

A "CASMP2" module is also available in the Gaussian package of programs. However, at least as implemented in Gaussian 98, CASMP2 gives much poorer results than CASPT2. For example, Dr. David Hrovat in my group found (6/6) CASMP2/6-31G* calculations gave an enthalpy of activation for the Cope rearrangement that is in even worse agreement with experiment than the (6/6)CASSCF/ 6-31G* value.[18]

On the other hand, when the (6/6)CASSCF geometry for the Cope TS was partially reoptimized at the (6/6)CASPT2 level, (6/6)CASPT2/6-31G* calculations gave an enthalpy of activation for the Cope rearrangement that was ∼3 kcal/mol lower than the experimental value.[19] Moreover, (6/6)CASPT2 calculations with the 6-311G* basis set gave an enthalpy of activation that was only ∼1 kcal/mol lower than the measured value.

CASPT2 calculations give excellent results for not only the enthalpic barriers to pericyclic reactions, such as the Cope rearrangement, but also for the excitation energies in the UV–vis spectra of conjugated molecules. Examples of the use of CASPT2 calculations of UV–vis spectra for the identification of RIs are provided by singlet 1,2,4,5-tetramethylbenzeneene diradical[20] and each of the rearrangement products, formed sequentially from the radical cation of the syn dimer of cyclobutadiene.[21]

3.3. Calculations Based on Density Functional Theory

As discussed in Section 3.2.3.6, the major difficulty in wave function based calculations is that correlation between electrons of opposite spin must somehow be introduced into a theory that starts with the physically unrealistic premise that electrons of opposite spin move independently of each other. However, this tacit assumption not only provides a mathematically tractable starting point (i.e., HF theory) for wave function based calculations, but this assumption also underpins the entire concept of orbitals (i.e., wave functions for single electrons). The existence of MOs may be a construct, but it is a construct that has proven to be very useful in interpreting the results of both calculations and experiments.

An attractive way to maintain the conceptual simplicity that arises from positing the existence of MOs, while improving upon the accuracy of HF calculations, would be to augment the Coulomb and exchange operators in the Fock operator with an additional operator that somehow accounts for correlation between electrons of opposite spin. This is essentially what is done in calculations based on DFT. Moreover, the general strategy behind modifying the Fock operator, so that it includes the effects of both exchange between electrons of the same spin and correlation between electrons of opposite spin, can be rigorously justified by two theorems. These theorems were proved by Walter Kohn, who shared the 1998 Nobel Prize in Chemistry with John Pople.

3.3.1. The Hohenberg–Kohn and Kohn–Sham Theorems. The first of these two theorems is called the *Hohenberg–Kohn theorem*. It proves that the energy of a molecule can, *in principle*, be computed exactly from the exact electron density. Since the electron density in a molecule is a mathematical function of the three spatial coordinates, and since functions of functions are called *functionals*, the Hohenberg–Kohn theorem can be formulated as stating that a functional exists from which the energy of a molecule can be computed exactly from the exact electron density. Unfortunately, the Hohenberg–Kohn theorem does not provide any information on what that functional is or how to go about finding it.

The second theorem is called the *Kohn–Sham theorem*. It proves that, if the exact electron density in a molecule can be represented by a single configuration, the energy, computed from the density, can be minimized by variationally optimizing the orbitals in that configuration. Therefore, a one-electron, Fock-like, operator can be used in the DFT analogue of Eq. 3, to find the optimal *Kohn–Sham (KS) orbitals* with which to compute the density. The same type of iterative process, employed in HF theory, can be used to find the KS orbitals that give the electron density of lowest energy for a given functional.

3.3.2. Functionals. The difference between the Fock operator, $\mathcal{F}$, in wave function based calculations and the analogous operator in DFT calculations is that the Coulomb and exchange operators in $\mathcal{F}$ are replaced in DFT by a functional of the electron density. In principle, this functional should provide an exact formula for computing the Coulombic interactions between an electron in a KS orbital and all the other electrons in a molecule. To be exact, this functional must include corrections to the Coulombic repulsion energy, computed directly from the electron density, for exchange between electrons of the same spin and correlation between electrons of opposite spin.

In addition, the functional must somehow cancel the fictitious repulsion energy between an electron and itself, which arises if the electron density, due to all the electrons, is used to compute the Coulombic energy of a single electron. As discussed in Section 3.2.1, in HF theory cancellation of the *self-repulsion energy* results from the presence of the exchange operator in $\mathcal{F}$. If this effect of $\mathcal{K}_j$, in the Fock operator is not mirrored exactly by the functional chosen, the cancellation of the self-repulsion energy will not occur.

During the last decade of the twentieth century, creation and testing of functionals provided gainful employment for the many theoretical chemists who were interested in the development of DFT as an accurate method for use in electronic structure calculations. The chapter on DFT in Cramer's book[3] provides not only descriptions of different types of functionals but also information on how well each performs in calculating energies, bond lengths, and dipole moments for several different test sets of molecules.

Here it is sufficient to note that the functional, currently in use, which usually provides the most accurate results for organic molecules, is B3LYP. It consists of a three-parameter functional that was developed by Becke,[22] combined with a correlation functional due to Lee, Yang, and Parr[23] "Be three lip" is not an injunction in pidgin English to effect an impossible change in human physiology, but the acronym for the functional that is currently the one most widely used by chemists doing DFT calculations on organic and organometallic compounds.

3.4. Comparisons of Calculations Based on Wave Functions and on DFT

Wave function based calculations are often called "ab initio" (literally, "from the beginning") calculations, because energies, geometries, and properties are all computed in terms of fundamental quantities (i.e., Planck's constant and the charge and mass of an electron). In contrast, no one has yet figured out how to find the holy grail of DFT—the functional that computes energies exactly from exact densities—"from the beginning". The successes and failures of each functional have to be evaluated by comparisons of the energies, bond lengths, etc. computed using the functional, with experimental values.

Some functionals, in particular B3LYP, incorporate parameters that are optimized by comparisons between DFT and experimental results. Thus, B3LYP calculations are, like the MINDO, MNDO, and AM1 methods that were developed by the late Michael Dewar in the 1970s and 1980s, semiempirical in nature. In this sense, B3LYP could be called "the AM1 of the twenty-first century".

However, most wave function based calculations also contain a semiempirical component. For example, the *primitive* Gaussian functions in all commonly used basis sets (e.g., the six Gaussian functions used to represent a $1s$ orbital on each first row atom in the 6-31G* basis set) are *contracted* into sums of Gaussians with fixed coefficients; and each of these linear combinations of Gaussians is used to represent one of the independent basis functions that contribute to each AO. The sizes of the primitive Gaussians (compact versus diffuse) and the coefficient of each Gaussian in the contracted basis functions, are obtained by optimizing the basis set in calculations on free atoms or on small molecules.[4d]

If the optimizations are performed in calculations on atoms, the basis functions for at least some of the atoms are usually scaled, because AOs in molecules are generally more compact than AOs on atoms. Thus, a set of scaling factors must be chosen. If the basis set optimizations are performed in calculations on small molecules, the molecules to use in the optimizations must be chosen. These types of choices in basis set construction are usually made by assaying how well different versions of a basis set perform in test calculations.[4d] Since basis set performance is usually judged by comparing the results of these test calculations to the results of experiments, this comparison introduces a semiempirical component into most of the contracted basis sets that are commonly used for electronic structure calculations.

The incorporation of experimental results into wave function based methods can be much less subtle. For example, G2 and G3 are both very high-level "ab initio" methods that have been designed to provide results that accurately mirror the results of experiments.[9] However, part of the way close agreement is achieved between the enthalpies computed by the G2 or G3 methods and those found by experiments, is by carrying out large electronic structure calculations, extrapolating the energies obtained from them, and then applying semiempirical corrections to these energies.

Even if most wave function based calculations cannot claim superiority to DFT by virtue of being purely ab initio, wave function based calculations do have an advantage over those based on DFT. Wave function based calculations can be *systematically* improved; whereas DFT calculations cannot. For example, increasing the size of the basis set used and the amount of electron correlation energy recovered is not only guaranteed to provide a lower total energy for a molecule; but also, performing a better calculation usually (though not always) results in energy differences between molecules being computed more accurately.

In contrast, increasing the size of the basis set in B3LYP calculations or using a different functional will certainly affect the energy difference computed between two stationary points on a PES, but such changes may or may not increase the accuracy with which the energy difference is computed. Since, B3LYP is known to give poor results for energy differences between certain types of isomers,[24] there is no assurance that one can always depend on B3LYP to be the best functional to use; but there is usually no way of knowing a priori whether a different functional will give better results.

On the other hand, DFT calculations, especially with the B3LYP functional, frequently do give results that are as good as those obtained from CCSD(T). The latter is currently the "gold standard level of theory"[3b] for wave function based calculations that employ a single reference configuration.

However, a DFT calculation consumes much less computer time than a CCSD(T) calculation on the same molecule. The time it takes to perform DFT calculations scales no worse than the cube of the number of basis functions used; and efforts are being made by computational chemists to reduce this scaling toward being more nearly linear. In contrast, the time it takes to perform CCSD(T) calculations scales as the number of basis functions to the seventh power.

Since DFT calculations take much less time to perform than CCSD(T) calculations, DFT calculations can be carried out on *much* larger molecules than CCSD(T) calculations. In addition, with the same amount of computer resources, many more problems in chemistry can be addressed by DFT calculations than by CCSD(T) calculations.

A particular advantage of B3LYP calculations is that the IR frequencies, computed with this functional, are usually more accurate than those obtained from wave function based calculations. The frequencies obtained from wave function based calculations have to be adjusted, using semiempirical scaling factors; whereas B3LYP frequencies usually require little or no scaling. Therefore, B3LYP calculations, rather than wave function based methods, are usually used in identifying RIs in matrix isolation and in time-resolved IR studies, by comparing calculated and observed IR frequencies.

UV–vis spectra can also be computed, using time-dependent (TD)DFT;[25] and it is now possible to perform such calculations with relative ease, using the Gaussian suite of electronic structure codes.[11,26] Using CASPT2 to compute UV–vis spectra, usually requires more computational expertise than using TDDFT.

On the other hand, since DFT calculations are usually based on a single electronic configuration, DFT has problems in dealing properly with some open-shell RIs. For example, B3LYP works fairly well for small radicals but gives poor spin distributions for radicals with extended conjugation.[27] Inadequate solution of the self-repulsion problem causes B3LYP to fail very badly for radical cations.[28] Finally, since DFT is based on a single electronic configuration, it does not properly describe singlet diradicals, for which an adequate wave function must contain at least two configurations.[13]

For calculations on many, if not most RIs, there is no reason to choose between DFT (e.g., B3LYP) and wave function based methods, if both types of calculations can be carried out. For example, since DFT calculations can be used *much* more economically than CCSD(T) calculations to optimize geometries, locate TSs, and perform vibrational analyses at these stationary points, performing single-point CCSD(T) calculations at B3LYP optimized geometries (denoted CCSD(T)// B3LYP) is an attractive option and a way of checking the accuracy of energy differences computed by B3LYP. To the extent that the results from B3LYP and CCSD(T) calculations agree, one can be confident that the answers provided by these two types of calculations are correct.

If the results of DFT and wave function based calculations disagree, then one can try to carry out wave function based calculations with larger basis sets and/or with inclusion of more electron correlation. If wave function based calculations at several different levels of electronic structure theory give similar results, it is likely that these calculations are correct and that the DFT calculations are in error.

4. APPLICATIONS OF ELECTRONIC STRUCTURE CALCULATIONS TO EXPLAINING AND PREDICTING THE CHEMISTRY OF THREE REACTIVE INTERMEDIATES—PHENYLNITRENE, CUBYL CATION, AND PROPANE-1,3-DIYL

The three examples in this section are taken largely from the work of our research group. Other, perhaps better, examples of the application of electronic structure calculations to the explanation and prediction of the behavior of RIs could have been selected; and some of them are provided in the Suggested Reading Section.

The research projects described here were chosen for two reasons. First, they are examples with which the author of this chapter is very familiar. Second, in each case, calculations led not only to the solution of the problem that motivated them, but the calculations also made predictions that inspired additional experiments.

4.1. Differences between the Ring Expansion Reactions of Phenylcarbene, Phenylnitrene, and Phenylphosphinidene

Few areas of research illustrate the partnership between calculations and experiments as well as the investigation of the chemistry of phenylnitrene and why its chemistry is very different from that of phenylcarbene and phenylphosphinidene. This topic has been the subject of two recent reviews,[29] and it is also discussed by Platz in Chapter 11 of this volume. Therefore, only a very brief description will be given here of the synergy between the experiments done in Platz's lab and the calculations performed in our lab.

Singlet phenylcarbene (**1a**) undergoes ring expansion to 1,2,4,6-cycloheptate-traene (**3a**), and singlet phenylnitrene (**1b**) undergoes the analogous reaction to form 1-aza-1,2,4,6-cycloheptatetraene (**3a**) (Fig. 22.3). However, the reactions differ in several important ways, one of them being a very large difference in rate. Singlet **1b** reacts much faster than singlet **1a**. In fact, the ring expansion of **1a** only takes place at elevated temperatures; so **1a** has a rich intermolecular chemistry. In contrast, the ring-expansion reaction of **1b** is so fast that, even at very low temperatures, the rate can only be measured by LFP.[29]

This large difference between the rates of ring expansion of **1a** and **1b** is surprising in view of the apparently similar electronic structures of nitrenes and carbenes. Both types of RIs have two nonbonding electrons and two nonbonding orbitals. However, there is one important difference between the nonbonding orbitals in **1a** and **1b**.

As shown in Figure 22.4, the nonbonding orbital on the carbenic carbon that lies in the plane of the benzene ring in **1a** is a hybrid, containing large amounts of $2s$ character. Consequently, this orbital is lower in energy than the $2p$–π AO on the carbenic carbon, which is conjugated with the benzene ring. As a result, the dominant electronic configuration in the wave function for the lowest singlet state of **1a**

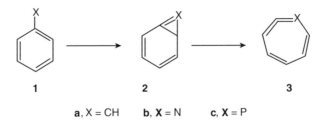

a, X = CH   b, X = N   c, X = P

Figure 22.3. Pathway for the ring expansion reactions of phenylcarbene (**1a**), phenylnitrene (**1b**) and phenylphosphinidene (**1c**).

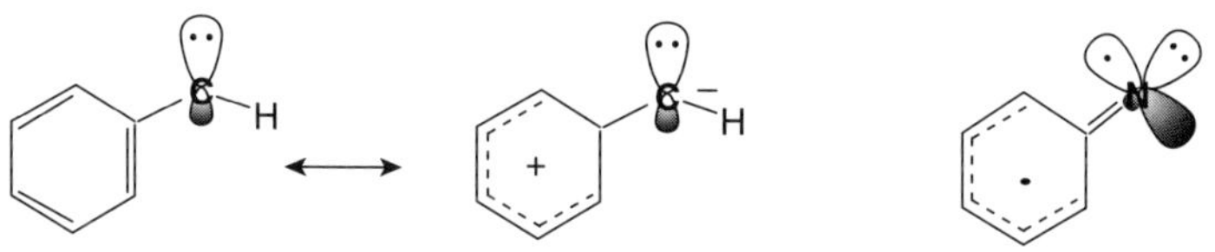

1a **1b**

Figure 22.4. Schematic depiction of the electronic structures of the lowest singlet states of phenylcarbene (**1a**) and phenylnitrene (**1b**).

places both nonbonding electrons in the hybridized σ orbital. As depicted in the second resonance structure for **1a** in Figure 22.4, the empty $2p–\pi$ AO on the carbenic carbon can accept π electrons from the benzene ring in **1a**.

In contrast to the case in **1a**, in **1b** both of the nonbonding orbitals on nitrogen are pure $2p$ AOs. Electronic structure calculations[30] revealed that in the lowest singlet state of **1b**, the $2p–\sigma$ AO on nitrogen, which lies in the plane of the benzene ring, is occupied by one nonbonding electron. The other nonbonding electron formally occupies a $2p–\pi$ orbital on nitrogen, but this electron is largely delocalized into the benzene ring, The equilibrium geometry calculated for this state indicates that its electronic structure can be depicted schematically as in Figure 22.4, with the σ and π nonbonding electrons occupying very different regions of space.

Confining these two electrons of opposite spin to different regions of space minimizes the Coulombic repulsion between these electrons.[30b] Consequently, the singlet–triplet energy gap in **1b** is computed to be only about one-half the size of those measured in N–H[31] and in methylnitrene.[32] Experimental evidence that the electronic structure calculations are correct came from the excellent agreement between two independent measurements of the singlet–triplet energy splitting in **1b**[33] and the values predicted by the calculations.[30]

The difference between the reactivities of **1a** and **1b** toward ring expansion is nicely explained by the difference between the electronic structures of their lowest singlet states.[29] With incluson of dynamic electron correlation, calculations of the PESs connecting **1a** to **3a**[34] and **1b** to **3b**[35] found that in both reactions electrocyclic ring opening of **2** to **3** has a low barrier, so that the rate-determining step is ring closure of **1** to **2**. In agreement with experiment, the barrier-to-ring closure, computed by (8/8)CASPT2 calculations, was found to be considerably lower for **1b** → **2b** than for **1a** → **2a**.[35]

The greater reactivity of **1b** than **1a** is easily rationalized by the depictions of the lowest singlet states in Figure 22.4. Just bending the nitrogen out of the plane of the benzene ring in **1b** is sufficient to allow bonding between the σ nonbonding electron on nitrogen and the π nonbonding electron of opposite spin in the benzene ring. On the other hand, cyclization of **1a** not only requires bending the carbenic carbon out of the plane of the benzene ring but also an increased contribution from the ionic resonance structure in the lowest singlet state.

Without the calculations that showed that there is a fundamental difference between the lowest singlet states of **1a** and **1b**,[30] and that this difference is responsible for the difference between their reactivities,[35] it is not clear how the correct explanation for the much faster ring expansion of **1b** than of **1a** would ever have been found.

From the understanding, provided by the calculations, of the mechanism by which **1b** cyclizes, what can be predicted about how the rate of this reaction might be affected by substituents on the benzene ring? The substituent effects would, in fact, be expected to be small, except for possible steric effects due to substituents in the ortho positions. If both ortho positions are substituted, one would expect to see a decrease in rate, relative to unsubstituted **1b**. On the other hand, if only one ortho position is substituted, cyclization should be about as fast as in unsubstituted **1b**; but cyclization should preferentially occur at the unsubstituted ortho carbon. Additional (8/8)CASPT2/6-31G* calculations by Bill Karney in our group[36] and subsequent experiments by the Platz group[37] confirmed these qualitative predictions about the effects of ortho substituents.

Although electronic effects of substituents on the rate of cyclization of **1b** should be small, it does seem possible that radical-stabilizing substituents on the benzene ring might have at least a modest effect on the rate of cyclization. If the unpaired π electron in **1b** tends to localize at a ring carbon to which a radical-stabilizing substituent is attached, this effect should tend to favor cyclization at an ortho carbon that bears such a substituent. In contrast, if a radical-stabilizing substituent were attached to the para carbon, the rate of cyclization should be retarded. Confirmation of these qualitative expectations for radical-stabilizing cyano substituents came from a combined computational–experimental study by our group and Platz's.[38]

Unlike the case with phenylcarbene (**1a**) and phenylnitrene (**1b**), ring expansion of phenylphosphinidene (**1c**) to 1-phospha-1,2,4,6-cycloheptatetraene (**3c**) has not been observed; and **1c** lives long enough to undergo intermolecular reactions.[39] Explaining why chemical reactions do not occur can be as important as explaining why they do take place; and recent (8/8)CASPT2/6-31G* calculations offer a very simple explanation of why the reaction **1c** → **3c** has never been observed[40]—the calculations predict that this reaction is quite endothermic. The difference between the thermodynamics of this reaction and the ring expansions of **1a** and **1b** is that the bonds which would be formed to phosphorus in the ring expansion of **1c** are much weaker than the corresponding bonds formed to carbon and to nitrogen in the ring expansions of, respectively, **1a** and **1b**.

Interestingly, the calculations find the difference in the bond strengths to the heteroatoms in **3b** and **3c** to be largely compensated for in **2b** and **2c** by the greater ability of phosphorus, compared to nitrogen, to accommodate the small bond angle at the heteroatom in the cyclopropene ring. As a result, closure of **1c** to **2c** is predicted to be considerably less endothermic than the next step in the ring expansion, the electrocyclic ring opening of **2c** to **3c**. Therefore, the CASPT2 calculations offer some hope that it may be possible to trap the small amount of **2c** that is predicted to be in equilibrium with **1c**, using a reagent that reacts with **2c** but not with **1c**.

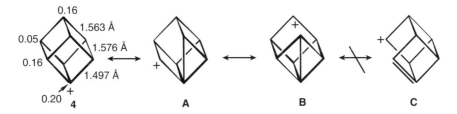

Figure 22.5. The MP2/6-31G* bond lengths and net Mulliken charges (in italics) at C–H groups, computed for cubyl cation **4**. Resonance structures **A** and **B** are consistent with the calculated bond lengths and charges; whereas, structure **C** is not.

4.2. Why Is Cubyl Cation Formed So Easily?

Another example of the power of a partnership between calculations and experiments is provided by cubyl cation (**4** in Fig. 22.5). As pointed out by Eaton, "Everything about the cubyl cation seems unfavorable: (1) the geometry about the positively charged carbon is very far from flat; (2) the exocyclic orbitals in cubane are . . . . rich [in 2*s* character, making the bonds to potential leaving groups unusually strong], and hyperconjugative stabilization would require high-energy, cubene-type structures."[41] Nevertheless, Eaton et al.[42] and Moriarity et al.[43] both found that this cation can be formed by solvolysis under surprisingly mild conditions.

For many polycyclic carbocations there is a good correlation between the changes in strain energies on carbocation formation, predicted by molecular mechanics calculations, and the solvolysis rates. However, at 70° cubyl cation is formed 10^{15} times faster than expected from this correlation.[41,44] What is responsible for this huge rate acceleration?

4.2.1. Calculations and Additional Experiments on Cubyl Cations. In order to answer the question, posed by the experiments of Eaton and Moriarity, we undertook electronic structure calculations on **4**. Calculations at the HF level did not predict the special stability, found experimentally for **4**, but MP2 calculations did.[45] It was known that inclusion of dynamic electron correlation is necessary for the proper description of "nonclassical" cations, in which delocalization of electrons in σ bonds provides stabilization.[46] Therefore, the contrast between the HF and MP2 results suggested that **4** was stabilized by such delocalization. But what was the nature of this delocalization?

The optimized geometry of **4**, obtained at the MP2/6-31G* level and shown in Figure 22.5, offered more clues about the type of delocalization that is *not* responsible for the stabilization of **4** than about the type of delocalization that is. If delocalization of the strained bonds between the carbons, α and β to the cationic center, were involved and occurred in the manner depicted in resonance structure **C**, the bonds between the ipso and α carbons should have considerable double-bond character. One would then expect these bonds to be much shorter than they are in the MP2/6-31G* optimized structure of **4** in Figure 22.5. One would also expect to see lengthening of the bonds between the α and β carbons. These latter bonds are longer than those in the MP2/6-31G* optimized geometry for cubane, but only by 0.011 Å.

The Mulliken charges at the C—H groups, which are shown in Figure 22.5, are also inconsistent with an appreciable contribution from resonance structures like **C**. Such structures place positive charges on the β C—H groups, but these carbons are computed to be the least positively charged in **4**. The most positively charged carbons are predicted to be those α to the cationic carbon and, surprisingly, the carbon γ to it. The population analyses suggest that resonance structures such as **A** and **B** in Figure 22.2 best describe the delocalized bonding in **4**.

Bonding between the β carbons and the ipso carbon, as depicted in structures **A** and **B**, is suggested by the positive bond orders that are computed between the relevant orbitals on these carbons. The distance between these carbons (2.034 Å) in the MP2/6-31G* optimized geometry, although longer than in cyclobutyl cations where there is only one β carbon,[47] is about the same as some of the distances, found by calculations, in the transition structures for many pericyclic reactions (e.g., the chair Cope rearrangement)[19,48] Thus the distance between the β carbons and the ipso carbon in the optimized geometry of **4** is short enough to allow significant bonding to occur between these carbons.

The most intriguing result of the electronic structure calculations is the prediction of substantial positive charge on the γ carbon in **4**. Indeed, it was known from experiments, which had already been completed at the time that our calculations were performed, that halogens and other inductively withdrawing substituents at the γ carbon depress the rate of formation of **4**.[42,43] However, a methyl substituent, although rate accelerating when attached to an α carbon, at the γ carbon had also been found to be slightly rate retarding. Since methyl substituents usually stabilize carbocations, Moriarty argued that development of positive charge at the γ carbon of **4** seemed highly unlikely.[43]

We pointed out[45] that the lowest unoccupied (LU)MO of **4** must have the same cylindrical symmetry (a_1 in the C_{3v} point group) at the γ carbon that it has at the ipso carbon. Since methyl groups stabilize carbocations principally through hyperconjugative donation of electrons from π-like combinations of C—H bonds, and since the C—H orbitals of a methyl group at the γ carbon with π symmetry (e in C_{3v}) are orthogonal to the LUMO, a methyl substituent at the γ carbon of **4** should provide much less stabilization than a methyl at an α carbon.

Our calculations did predict that a γ methyl group should provide a small amount of stabilization for **4** in the gas phase.[18] However, subsequent calculations that included solvation effects did find, in agreement with experiment, that a γ methyl group on **4** should slightly depress the rate of carbocation formation.[49]

After the publication of our article on **4**,[45] Eaton showed that a methoxyl substituent is also rate retarding when attached to the γ carbon.[44] Since a methoxyl group is a strongly π-electron-donating but σ-electron-withdrawing substituent, this finding makes perfect sense in terms of the symmetry of the LUMO of **4**.

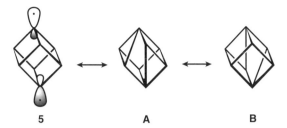

Figure 22.6. 1,4-Dehydrocubane (**5**), showing the nonbonding orbital that is essentially doubly occupied in the lowest singlet. Both **A** and **B** are two of the six, equivalent resonance structures that contribute to the low energy of this state, relative to the triplet.

Because silyl and stannyl groups are good σ-electron donors, their attachment at the γ carbon of **4** should be stabilizing and thus accelerate the rate of formation of **4**. In agreement with this conjecture, our calculations predicted that stabilization of **4** by a γ substituent should increase in the order $Me < SiR_3, < SnR_3$.[18] Eaton's experiments confirmed these predictions.[44]

4.2.2. Why Is Singlet 1,4-Dehydrocubane Predicted to Lie Far Below the Triplet in Energy?

Our calculations on **4** led to the solution of two more puzzles, one posed by the results of calculations and the other by the results of experiments. (2/2)CASSCF and MR–CI calculations by Michl and co-workers[50] and (2/2)CASSCF calculations by Hrovat[51] had previously found that the singlet state of 1,4-dehydrocubane (**5**) lies 10–13 kcal/mol below the lowest triplet. Not only are the singly occupied AOs at C1 and C4 in **5** too far apart to interact significantly through space; but, as indicated in Figure 22.6, the calculations also found that the nonbonding MO with the largest occupation number is comprised of the out-of-phase combination of these two AOs. Clearly, the interaction between these AOs in **5** must be through the bonds of the cube, not through space.

The results of our calculations on cubyl carbocation **4**, which showed that stabilization of **4** comes from interactions between the ipso carbon and the β carbons (as represented by resonance structures **A** and **B** in Fig. 22.5), suggested that stabilization of diradical **5** comes from the same types of interactions. These too can be represented by resonance structures, two of which are shown in Figure 22.6. Delocalization of electrons from all six of the bonds between the α and β carbons in **5** requires that the in-phase combination of nonbonding AOs be left empty, and that is why the pair of nonbonding electrons preferentially occupies the out-of-phase combination of nonbonding AOs at C1 and C4.

4.2.3. Why Are the Cubyl Hydrogens of Methylcubane More Reactive than the Methyl Hydrogens Toward Abstraction by tert-Butoxyl Radicals?

The experimental puzzle, which our calculations on **4** helped to solve, was posed by Della's finding that *tert*-butoxyl radicals preferentially abstract hydrogens from the carbons of the cube in methylcubane (**6**), rather than from the methyl group.[52] Although the cubyl hydrogens are attached to tertiary carbons, both the large amount of $2s$ character in these C—H bonds and the inability of the radical center in cubyl radical to planarize, make the C—H bonds in cubane unusually strong. As Della noted, the C—H bonds of the cube are, in fact, stronger than the methyl C—H bonds in **6**. Therefore, why are the C—H bonds of the cube in **6** more reactive than the weaker C—H bonds of the methyl group toward *tert*-butoxyl radicals?

Polar effects in the TSs for hydrogen atom abstraction reactions by radicals—the development of opposite charges at the hydrogen-donor and hydrogen-acceptor atoms, are known to have an accelerating effect on these reactions.[53] The Borden group reasoned that the stability of the cubyl cation would favor the separation of charge that is depicted in the bottom resonance structure in Figure 22.7, thus accelerating the rate of abstraction of cubyl hydrogens by *tert*-butoxyl radical. In order to test this explanation, we performed unrestricted[13] (U)HF and UMP2/6-31G* calculations on this reaction; but, for the sake of computational convenience, we replaced *tert*-butoxyl radical with methoxyl radical.

The results of the UMP2 calculations, which included dynamic electron correlation, confirmed our hypothesis.[54] The UMP2 calculations found, in agreement with the results of Della's experiments,[52] that abstraction of a cubyl hydrogen has a lower energy barrier than abstraction of a methyl hydrogen in the reaction of **6** with an alkoxyl radical. Moreover, when population analyses were performed on the TSs for both reactions, the TS for abstraction of a cubyl hydrogen was found to be much more polar and to have much more positive charge on the hydrocarbon moiety than the TS for abstraction of a methyl hydrogen.

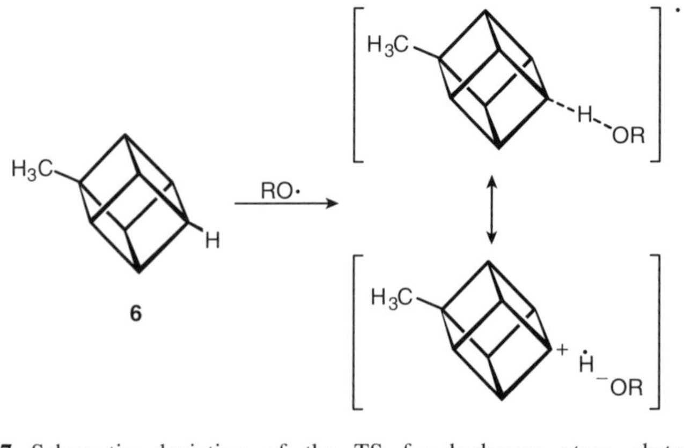

Figure 22.7. Schematic depiction of the TS for hydrogen atom abstraction from methylcubane (**6**) by an alkoxyl radical. The polarity of the TS, depicted in the bottom resonance structure, was confirmed by the results of population analyses.[54]

4.3. Can Hyperconjugation in a 1,3-Diradical Control the Stereochemistry of Cyclopropane Ring Opening and Make a Singlet the Electronic Ground State of the Diradical?

We began our calculations of the PES for the ring opening of cyclopropane in an attempt to resolve an apparent conflict between the results of experiments by Berson and co-workers[55] and by Baldwin and co-workers.[56] However, graduate students in our research group were among the experimentalists whose research ultimately benefited the most from the predictions made by our calculations.

4.3.1. Calculations and Experiments on the Stereomutation of Cyclopropane.
In 1965, Hoffmann published a seminal paper on trimethylene, another name for propane-1,3-diyl (**8**).[57] He used extended hückel (EH) calculations and an orbital interaction diagram to show that hyperconjugative electron donation from the central methylene group destabilizes the symmetric combination of $2p$–π AOs on the terminal carbons in the "(0,0)" conformation of this diradical. Hoffmann's calculations predicted that the resulting occupancy of the antisymmetric combination of $2p$–π AOs in **8** should favor conrotatory opening of cyclopropane (**7**), as depicted in Figure 22.8.

Numerous experiments with alkyl-substituted cyclopropanes failed to detect any sign of the preference predicted by Hoffmann's EH calculations.[58] In retrospect, this failure was due to shortcomings in both the calculations and in the design of the experiments. The EH calculations used a single configuration to represent the wave function for **8**; whereas, as noted in Section 3.2.3.6, the wave function for such a singlet diradical should include a second configuration, in which the symmetric combination of $2p$–π AOs is doubly occupied.[13] Therefore, the experimental preference for conrotatory opening of **7** is expected to be much weaker than was predicted by Hoffmann's EH calculations.

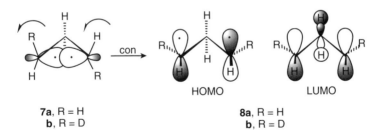

7a, R = H
b, R = D

8a, R = H
b, R = D

HOMO   LUMO

Figure 22.8. Conrotatory ring opening of cyclopropane (**7**) to what Hoffmann called the "(0,0)" conformation of propane-1,3-diyl (**8**).[57] The in-phase combination of $2p$–π AOs in the LUMO is destabilized by an antibonding interaction with the "π" combination of C–H bonding orbitals at C2. A lower energy MO, which is not shown, is stabilized by hyperconjugative electron donation; that is, the bonding version of this interaction. The out-of-phase combination of $2p$–π AOs in the HOMO has a node at C2; hence, it does not mix with the C–H orbitals at this carbon.

Alkyl substituents are not just stereochemical markers. At C1 and C3 of **8** they can hyperconjugatively donate electrons to both the symmetric and antisymmetric combinations of $2p-\pi$ AOs at these carbons. As a result, hyperconjugative destabilization of both these combinations of $2p-\pi$ AOs by electron donation from alkyl substituents at C1 and C3 should reduce the preference for the configuration in which the antisymmetric combination of $2p-\pi$ AOs is doubly occupied. In fact, our (2/2)CASSCF calculations on **8a** and on its 1,3-dimethyl derivative indicated that the methyl substituents should wipe out almost completely the weak preference for conrotation over monorotation that we computed for the stereomutation of unsubstituted **7**.[59]

In 1975, Berson and co-workers[55] disclosed the results of elegant experiments on the stereomutation of **7b**, using deuterium atoms, instead of alkyl groups, as the stereochemical markers. From measurements of the relative rates of enantiomerization and trans $\rightarrow$ cis isomerization of optically active **7b**, Berson and co-workers[55] found that there was, indeed, a preference for double over single methylene rotations.

However, quantitative evaluation of the size of this preference depended on knowing the size of the secondary deuterium isotope effect on which C—C bond in **7b** cleaves. With the seemingly reasonable assumption of a secondary isotope effect of 1.10 on bond cleavage, the experimental data led to the conclusion that double methylene rotation was favored over single methylene rotation by a factor of $\sim$50 in the stereomutation of **7b**. Although the error limits on the measurements were large enough to allow the actual ratio to be much smaller, Berson wrote, "There is no doubt that the double rotation mechanism predominates by a considerable factor."[55]

Sixteen years later Baldwin and co-workers[56] published the results of even more elegant experiments in which the stereomutation of optically active **7**-1-^{13}C-1,2,3-d_3 was studied. Because a deuterium atom is attached to each of the carbons in this compound, it was unnecessary for Baldwin to assume the size of a secondary deuterium isotope effect on which bond cleaves, in interpreting his kinetic data, The results of his experiments led him to conclude, "the double rotation mechanism does *not* predominate by a substantial factor."[56]

In discussing the contrast between his findings on the stereomutation of **7**-1-^{13}C-1,2,3-d_3 and Berson's results on the stereomutation of **7**-1,2-d_2, Baldwin attributed the difference between them to Berson's having made "some reasonable but nevertheless erroneous assumptions about kinetic isotope effects...."[56] In order to test this conjecture, Steve Getty (in our group) calculated the size of the secondary kinetic isotope effects that electronic structure theory predicts for Berson's experiments.[60]

Getty performed (2/2)CASSCF/6-31G* geometry optimizations and single-point calculations, in which dynamic correlation was added by allowing SD excitations from the (2/2)CASSCF wave functions. He then used the vibrational frequencies from the (2/2)CASSCF calculations in the Biegeleisen equation to compute the isotope effects for all of the possible reaction pathways—conrotation, disrotation, and monorotation—in the ring opening of **7** to **8**.

The calculations found that the secondary isotope effect predicted for Berson's experiments was 1.13, close to the value of 1.10 assumed by Berson. This computational result led us to write, "On the basis of these computational findings, we conclude that there is no way to reconcile the experimental results [obtained] by Berson and by Baldwin."[60] Subsequent isotope effect calculations by Baldwin et al.[61] led to the same conclusion.

What did Getty's calculations and those of Baldwin and co-workers on the PES for ring opening of **7a** predict about whether a strong preference for double rotation should be expected in the experiments of Berson and Baldwin? At the temperatures at which these experiments were performed, both sets of calculations did predict conrotatory ring opening to be faster than either disrotatory and monorotatory ring opening, but by factors of only 2 − 4. However, whether these computational results mean that double methylene rotation is actually predicted to be preferred to single methylene rotation in the automerization of **7a** depends on what is assumed about the behavior of diradical **8a**, formed by crossing the TSs for conrotatory and disrotatory ring opening of **7a**.

Suppose that, as Getty calculated,[60] conrotation is favored over disrotation by a factor of 3 in the ring opening of **7a** to **8a**. Then, if all branching ratios are assumed to be predictable from transition state theory (TST), without explicit inclusion of any possible dynamical effects, conrotation must be favored over disrotation by exactly the same factor of 3 in the closure of **8a** to **7a**. Thus, a TST model predicts that three-quarters of the molecules of **8a**, whether formed by conrotatory or disrotatory ring opening, will undergo conrotatory ring closure to **7a**. Since **7a** is three times more likely to undergo conrotatory than disrotatory ring opening to **8a**, a TST model then predicts the ratios of molecules undergoing con/con, dis/dis, con/dis, and dis/con ring opening and ring closure should be, respectively: $3 \times 3/4 : 1 \times 1/4 : 3 \times 1/4 : 1 \times 3/4$, or $2.25 : 0.25 : 0.75 : 0.75$.

Both our lab and Baldwin et al. noted that, for **7b**, con/dis and dis/con ring opening and closure both produce the same net stereochemical outcome as monorotation. In addition, in the experiments of Berson and Baldwin, con/con and dis/dis opening and closure are operationally indistinguishable, and both contribute to the net amount of double methylene rotation observed. Thus, if conrotation is preferred to disrotation by a factor of 3, a TST model predicts that, of the molecules of **7b** which undergo stereomutation by passage through the (0,0) geometry, the fraction expected to have undergone double methylene rotations would be $(2.25 + 0.25)/(2.25 + 0.25 + 2 \times 0.75) = 2.5/4 = 62.5\%$. The remaining 37.5% would be expected to have undergone *net* rotation of a single methylene group, which is operationally indistinguishable from cleavage of the bond between C1 and C2 in **7b** and passage over a monorotatory TS. Consequently, if a TST model were correct in predicting the partitioning of the (0,0) diradical, ring opening of **7b** by coupled rotations of the C1 and C2 methylene groups would be expected to make a substantial contribution to the formation of product, found to have undergone *net* rotation of just one methylene group in Berson's experiments.[61]

On the other hand, it is conceivable that, if **7a** undergoes disrotatory ring opening to **8a**, energy remains in the disrotatory mode of methylene rotations for the short time that **8a** takes to cross the TS for disrotatory closure to **7a**. In this dynamical model, disrotatory ring opening to **8a**, rather than resulting in preferential closure to **7a** via conrotation, would always lead to disrotatory ring closure. Similarly, in this model, molecules of **7a** that opened to **8a** via conrotation would always reclose to **7a** by this mode of coupled methylene rotations. Thus, if this dynamical model were correct, both conrotatory and disrotatory ring opening of **7b** to the (0,0) diradical would contribute only to the fraction of product found to have undergone double methylene rotations in Berson's experiments.

Consequently, whether the PES calculations by Getty and by Baldwin and co-workers predict that a preference for double methylene rotations should have been observed in the experiments of Berson (and Baldwin), depends on whether the fate of **8b** (and the corresponding diradical in Baldwin's eperiments on 7-1-^{13}C-1,2,3-d_3) is better described by a TST model or by the dynamical model, discussed in the previous paragraph. If the fate of **8b** is better described by a TST model, ring opening of **7b** to **8b** would be expected to make a substantial contribution to the fraction of product found to have undergone net rotation of just one methylene group. On the other hand, if the fate of **8b** is dictated by a dynamical effect, which favors ring closure of **8b** to **7b** by the same mode of coupled methylene rotation by which **8b** is formed from **7b**, ring opening of **7b** to **8b** would contribute only to the formation of product observed to have undergone double methylene rotations.

In order to investigate which model for the behavior of **8** is closer to being correct, our group provided Carpenter with an analytical expression, fit to our PES, so that Carpenter could perform semiclassical *trajectory calculations* on our PES. At the same time, Doubleday and Hase undertook *direct dynamics calculations*. As discussed in Chapter 21 in this volume, in the latter type of trajectory calculation the forces acting on a molecule at different points on a PES are found by performing electronic structure calculations. For this purpose, Doubleday and Hase used a reparameterized version of AM1 that provided a PES, similar to those calculated by Getty and by Baldwin, Yamaguchi, and Schaefer.

Both types of dynamics calculations gave the same type of results. The calculations found that the ratio of double-to-single methylene rotations is higher than the ~1 : 1 ratio predicted by application of a TST model to the PES. Ratios of double-to-single rotations between 2.9 and 3.5 were obtained from the direct dynamics calculations of Doubleday et al.,[62] and a ratio of 4.7 was computed from the trajectory calculations performed by Carpenter and co-workers.[63]

Both sets of calculations found that ring closure of **8** preferentially occurs by the same mode of coupled methylene rotations as ring opening of **7**. Crudely put, the dynamical behavior of **8** can be predicted by, what would be called in classical mechanics, "conservation of angular momentum." Chapter 21 in this volume provides examples of other reactions in which dynamical effects cause statistical models, such as TST, to fail to make correct predictions.

4.3.2. Calculations and Experiments on the Stereomutation of 1,1-Difluorocyclopropanes. C—F bonds have low-lying unfilled, σ* orbitals;[64] therefore, C—F bonds are strong, hyperconjugative electron acceptors. Consequently, as shown schematically in Figure 22.9, hyperconjugative interaction with

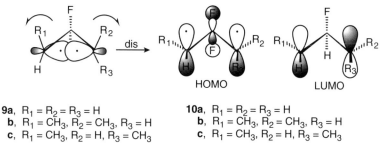

9a, $R_1 = R_2 = R_3 = H$
  b, $R_1 = CH_3$, $R_2 = CH_3$, $R_3 = H$
  c, $R_1 = CH_3$, $R_2 = H$, $R_3 = CH_3$

10a, $R_1 = R_2 = R_3 = H$
  b, $R_1 = CH_3$, $R_2 = CH_3$, $R_3 = H$
  c, $R_1 = CH_3$, $R_2 = H$, $R_3 = CH_3$

Figure 22.9. Disrotatory ring opening of 1,1-difluorocyclopropane (**9**) to 2,2-difluorocyclo-propane-1,3-diyl (**10**). The in-phase combination of $2p$–π AOs in the highest occupied molecular orbital (HOMO) is stabilized by a bonding interaction with the $2p$ AO at C2 in the "π" combination of low-lying, C—F, antibonding orbitals.

the low-lying, unfilled orbitals of the C—F bonds at C2 in 2,2-difluoropropane-1,3-diyl (**10a**), stabilizes the symmetric combination of $2p$–π AOs on the terminal carbons. Therefore, disrotation should be preferred to conrotation in the steromutation of 1,1-difluorocyclopropane (**9a**).

SD(Q)-CI/6-31G*//(2/2)CASSCF/6-31G* calculations confirmed the qualitative expectation that coupled disrotation should be preferred to both conrotation and monorotation in the stereomutation of **9a**.[65] In addition, unlike the case in **8a**, the calculations found that the preference for coupled rotation was enhanced by the addition of methyl groups to the terminal carbons of **10a**. Hyperconjugative electron donation from the methyl groups in **10b** and **10c** enhances electron donation from the symmetric combination of $2p$–π AOs on the terminal carbons into the antibonding orbitals of the C—F bonds at C2.

It is true that, in general, experiments are more likely than calculations to produce totally unanticipated findings. However, our calculations on **10b** and **10c** gave a wholly unexpected result—the preference for having both methyl groups on the "outside" in the transoid,transoid conformation (as in **10b**), rather than having one on the "inside" in the transoid,cisoid conformation (as in **10c**), was computed to be much larger in diradical **10** than in diradical **8**. This difference between the two diradicals was shown to have an electronic origin.[66]

Long-range interaction between the pseudo-π combination of methyl C—H bonds on one terminal carbon and the $2p$–π AO on the other terminal carbon was found to be bonding in diradicals, such as **8**, with hyperconjugatively electron-donating bonds at C2. In contrast, the hyperconjugatively electron-accepting C—F bonds in **10** were found to make the long-range interaction between the pseudo-π combination of methyl C—H bonds on one terminal carbon and the $2p$–π AO on the other terminal carbon strongly antibonding in **10c**. Since **9b** can undergo disrotatory ring opening to **10b**; whereas, **9c** must undergo disrotatory ring opening to **10c**, disrotatory ring opening was predicted to be much faster for **9b** that for **9c**.[65,66]

These computational findings lead to the expectations that, if optically active *cis*- and *trans*-1,1-difluoro-2-ethyl-3-methylcyclopropanes were both pyrolyzed, they each should undergo racemization faster than epimerization; but the cis isomer should racemize much faster than the trans isomer. In order to test these predictions, Scott Lewis, then a graduate student in our group, carried out the syntheses of the two, optically active, 1,1-difluorocyclopropanes, and Feng Tian in Dolbier's group, improved the syntheses and measured the kinetics of racemization and epimerization.[67]

For the cis isomer, the ratio of racemization to epimerization was found to be 107 at 274.5 °C, and the cis isomer was found to undergo racemization 16.2 times faster than the trans isomer. The ratio of racemization to epimerization for the latter was found to be only 6.6 at 274.5 °C. Thus, these experiments confirmed the computational predictions that in **9b** and **9c**, double methylene rotations are faster than any process that leads to net single methylene rotation, and the preferred mode of double rotation is disrotation, rather than conrotation.[65,66]

4.3.3. Calculations on the Stereomutation of 1,1-Disilylcyclopropanes.

It is to be hoped that future experiments on the stereochemistry of the ring-opening reactions of 1,1-bis(trimethylsilyl)cyclopropanes (**11**) will test the computational prediction by Anne Skancke (in our group) that the hyperconjugatively electron-donating C—Si bonds at C2 in 2,2-bis(trimethylsilyl)propane-1,3-diyl (**12a**) will result in a substantial preference for ring opening and ring closure by coupled conrotation.[68]

In agreement with the results of experiments on pyrolysis of 1,1-bis(trimethylsilyl)cyclopropane,[69] additional (2/2)CASPT2/6-31G* calculations predict that a rapid 1,2-silyl shift will occur in **12a**, forming **13a**.[18] However, if *cis*- and *trans*-1,1-bis(trimethylsilyl)-2,3-dimethylcyclopropanes (**11b** and **11c**) were pyrolyzed, then, as shown in Figure 22.10, the stereochemistry of ring opening could, presumably, still be inferred from the stereochemistry of the double bonds in the expected rearrangement products (**13b** and **13c**).

Interestingly, the calculations predict that, in the pyrolysis of **11c**, formation of cisoid,cisoid diradical **12c** should be favored over formation of the transoid,transoid stereoisomer.[18] (Since the latter conformer is predicted not to be formed, it is not shown in Fig. 22.10.) The C—Si bonds at C2 in **12** are strong hyperconjugative electron donors; so, as discussed in Section 4.3.2, there is a long-range, bonding

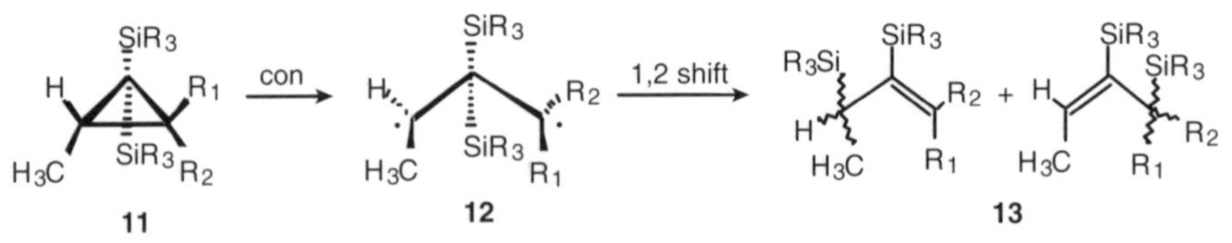

a, R = R$_1$ = R$_2$ = H; **b**, R$_1$ = H, R = R$_2$ = CH$_3$; **c**, R = R$_1$ = CH$_3$, R$_2$ = H

Figure 22.10. Rearrangement products (**13**), predicted to be formed from 1,1-disilylcyclopropanes (**11**) by conrotatory ring opening, followed by a 1,2-shift of a trimethylsilyl group in the 1,3-diradicals (**12**) generated. Pyrolysis of **11b** should lead to a mixture of (*E/Z*) isomers (**13b**), but pyrolysis of **11c** is predicted to form only the *E* stereoisomer (**13c**).[18]

interaction in **12** between the pseudo-π combination of methyl C—H bonds on one terminal carbon and the $2p$–π AO on the other terminal carbon.[66] There are two such interactions in **12c**, and they undoubtedly play a major role in the computational finding that the cisoid,cisoid conformation in **12c** is lower in energy than the transoid,transoid conformation.

A 1,2-shift of a trimethysilyl group in **12c** can lead to formation of only **13c**; so, if the calculations are correct and pyrolysis of **11c** results in ring opening to just **12c**, **13c** should be the only product formed in the pyrolysis of **11c**. In contrast, although, conrotatory ring opening of **11b** can afford only transoid,cisoid diradical **12b**, a 1,2-trimethylsilyl shift in **12b** can lead to formation of either **13b** or **13c**, depending on which radical center in **12b** is the migration terminus. Therefore, unlike the pyrolysis of **11c**, which is predicted to afford only **13c**, pyrolysis of **11b** should produce a mixture of **13b** and **13c**.

4.3.4. Calculations on the Stereomutation of 1,2-Dimethylspiropentanes and Reinterpretation of the Experimental Results.

Unlike the case in the experiments on the stereomutations of alkyl-substituted cyclopropanes,[58] experiments by Gajewski et al. [70] did find a preference for coupled over single methylene rotations in the stereomutations of *cis*- and *trans*-1,2-dimethylspiropentanes (**14a** and **14b**). This difference in outcomes is not really surprising, because the strained C—C bonds of cyclopropane rings in diradicals **15a** and **15b** are much stronger hyperconjugative electron donors than the C—H bonds in propane-1,3-diyl (**8**) (Fig. 22.11).

The strongly electron-donating cyclopropane bonds in **15** should result in conrotation being the dominant mode of coupled methylene rotation. However,

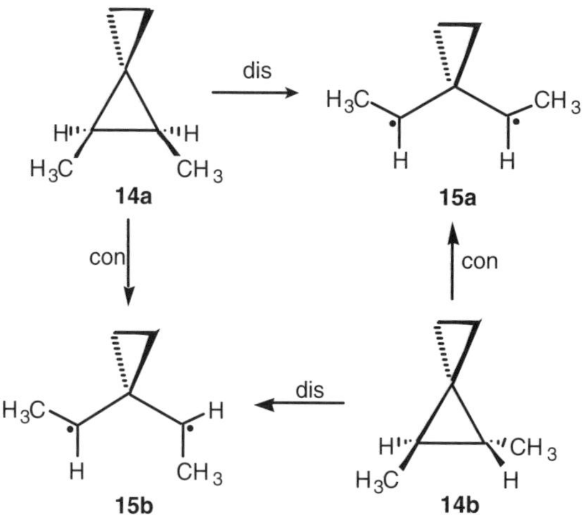

Figure 22.11. 1,3-Diradicals (**15a** and **15b**) formed by conrotatory and disrotatory ring opening of *cis*- and *trans*-1,2-dimethylspiropentanes (**14a** and **14b**).

Gajewski's experiments found that the cis isomer (**14a**) underwent ring opening by coupled rotation faster than the trans isomer (**14b**). Gajewski quite reasonably supposed that the transoid,transoid diradical (**15a**) would be lower in energy than the transoid,cisoid diradical (**15b**). Since a disrotatory pathway connects **14a** with **15a**, Gajewski came to the logical, but very surprising, conclusion that the preferred mode of coupled methylene rotations in the stereomutation of **14** must be disrotation, rather than the expected conrotation.

(2,2)CASPT2/6-31G* calculations on the energies of diradicals **15a** and **15b**, performed by Bill Johnson, then a graduate student in our group, led to a reinterpretation of Gajewski's experimental results.[71] The calculations showed that, although Gajewski's assumption that **15a** is lower in energy than **15b** is perfectly reasonable, it turns out to be incorrect.

As discussed in Section 4.3.2, when, in propane-1,3-diyls, the bonds to the substituents at C2 are hyperconjugative electron donors there is a long-range, bonding interaction between the pseudo-π combination of methyl C—H bonds on one terminal carbon and the $2p$–π AO on the other terminal carbon.[66] The cyclopropyl bonds in **15** are such strong electron donors that this effect results in the transoid, cisoid conformer (**15b**) being computed to be lower in energy than the transoid, transoid conformer (**15a**).

However, Johnson's calculations did find that Gajwski was correct in interpreting the faster rate of reaction of **14a**, relative to **14b**, as meaning that **14a** undergoes ring opening to the more stable diradical. However, since the more stable diradical is **15b**, rather than **15a**, and since **14a** is connected to **15b** by a conrotatory pathway, Gajewski's experiments actually confirm the qualitative prediction that **14** should preferentially undergo stereomutation by conrotation.

4.3.5. Calculations and Experiments on the Ground States of Cyclopentane-1,3-diyls.

Our calculations on 2,2-difluoropropane-1,3-diyl (**10**)[65,66] led to another prediction[72] that was subsequently tested by us experimentally and confirmed.[73] Both (2,2)CASSCF and SD-CI calculations predict the triplet to be the ground state of propane-1,3-diyl (**8**)[60,66] and cyclopentane-1,3-diyl (**16a**),[74] a cyclic derivative of **8**. Prior experiments by Closs had shown that the ground state of **16a** is, in fact, a triplet;[75] and Dougherty, Adam, Wirz, and their co-workers[76] found the same to be true for the 1,3-diphenyl derivative (**16b**) (Fig. 22.12).[76]

In contrast, our (2,2)CASSCF and (2,2)CASPT2/6-31G* calculations predicted that hyperconjugation with the fluorines is strong enough in both diradical **10**[65,66] and in the cyclic derivative **16c**[72] to make a singlet the ground state of each diradical. In the lowest singlet state both nonbonding electrons can occupy the MO, formed from the in-phase combination of $2p$–π AOs on the terminal atoms and stabilized by mixing with the C—F antibonding orbitals. However, in the triplet only one nonbonding electron can occupy this MO.

In collaboration with Adam and Wirz, our group confirmed this prediction experimentally, by synthesizing a derivative of **16d** and showing that it does have a singlet ground state.[73] Subsequently, a derivative of **16e**, in which the geminal fluorines at C2 in **16d** were replaced by ethoxy groups, was also found to have

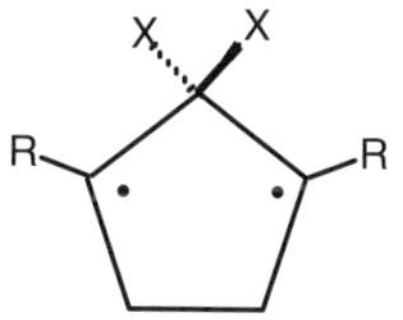

16a, X = R = H; **b**, X = H, R = Ph;
c, X = F, R = H; **d**, X = F, R = Ph;
e, X = OCH$_2$CH$_3$, R = Ph;
f, X = OCH$_3$, R = Ph;
g, X = SiH$_3$, R = H

Figure 22.12. Cyclopentane-1,3-diyl (**16a**) has been both calculated[74] and found[75] to have a triplet ground state, and the 1,3-diphenyl derivative (**16b**) has also been found to have a triplet ground state.[76] In contrast, calculations on **16c**[72] and **16g**[79] have predicted both diradicals should have singlet ground states, and experiments on derivatives of **16d**,[73] **16e**,[77] and **16f**[78] found these three diradicals do, in fact, have singlet ground states.

a singlet ground state.[77] Most recently, Abe et al.[78] measured the lifetimes of derivatives of **16f**, with different substituents in the para positions of the phenyl rings. The longest lived singlet diradical was, as might have been expected, found to be the one with the substituents in the aromatic ring that are the best electron donors. Donation of electrons to the $2p-\pi$ AOs on C1 and C3 increases the amount of electron donation from the in-phase combination of these AOs into the low-lying antibonding orbitals of the C—O bonds at C2.

We also predicted that hyperconjugatively electron-donating, as well as electron-accepting bonds at C2, can confer a singlet ground state on derivatives of **16**. Our (2,2)CASSCF and CASPT2/6-31G* calculations found that the silyl substituents at C2 in **16g** should make this diradical a ground-state singlet too.[79]

However, in contrast to the case for singlet diradical **16c**, singlet **16g** was calculated to have a high barrier-to-ring closure. Unlike the fluorines in **16c**, the silyl groups in **16g** cause conrotatory closure to be strongly preferred; but disrotation is required for closure of diradical **16**. Unfortunately, the predictions of a singlet ground state and a high barrier-to-ring closure in diradical **16g** are unlikely to be testable experimentally, because, as in the case of singlet diradical **12**, a 1,2-silyl shift is calculated to occur rapidly in singlet **16g**.[18]

5. CONCLUSION AND OUTLOOK

Section 2 described some of the properties of RIs that can be computed, using the methodology discussed in Section 3. Section 4 provided several examples of how electronic structure calculations can be used to obtain information about RIs that is crucial to interpreting the results of experiments, but that would be either difficult or impossible to obtain experimentally, Examples of such information, discussed in

Section 4, are the nature of the lowest singlet state of phenylnitrene (**1b**),[30] the charge distribution in cubyl cation **4**,[45] and the existence of a long-range, bonding interaction that counterintuitively makes diradical **15b** lower in energy than diradical **15a**.[71] Electronic structure calculations have obviously become very useful tools for studying, understanding, and predicting the behavior of RIs.

Future advances in computer hardware and software will surely make it possible to perform electronic structure calculations at higher and higher levels of theory on larger and larger molecules. Suppose that, at some time in the not too distant future, the properties of RIs can be computed at a level of theory so high that, as shown by repeated comparisons with experiments, the results of calculations at this level can be *guaranteed* to be as accurate as experimental measurements. If the cost of calculating the geometries, energies, and spectra of RIs at this level of theory were significantly less than the cost of measuring these properties, would there be any point in doing experiments to measure them?

It might be argued that theoretical predictions, no matter how good the record of success of the method used to make them, must always be verified experimentally. However, putting aside this philosophical issue, there would still be a very good reason to do experiments. Nature has demonstrated an amazing ability to surprise experimentalists, by providing not only unexpected solutions to puzzles that experimentalists have sought to solve, but also by furnishing new and previously unimagined puzzles. Without experiments, these puzzles would not be discovered; and, without these puzzles, electronic structure calculations would never be performed, in order to solve them.

Of course, calculations done "in silico," like experiments done in silica, have the potential for yielding unexpected results. Several examples of theoretical results, which were very surprising when first obtained, were provided in Section 4. However, experimental research has an even greater propensity than computational studies for making unexpected discoveries, far removed from the questions that initially motivated the research.

For example, if, in studying the reaction of methylcubane (**6**) with *tert*-butoxyl radicals, Della et al.[52] were only trying to see whether this reaction could be used to generate cubylcarbinyl radical, their finding that *tert*-butoxyl radical preferentially abstracts a hydrogen from the cube, rather than from the methyl group, was only indirectly related to the question they were actually trying to answer. However, without the stimulus provided by Della's experimental results, we would never have been motivated to try to solve the puzzle of why the stronger C–H bond is preferentially broken in this reaction.

In fact, all of the computational projects described in Section 4, were motivated by the results of experiments. Without the puzzles provided by the experimental results, the calculations would never have been performed. Therefore, if calculations ever did replace experiments, computational chemists would very likely become the victims of the success of the theoretical methods they helped to develop.

The focus in this chapter, on how electronic structure calculations can be used by experimentalists who are studying RIs, is due to the expectation that most of the readers will be experimentalists. However, computational chemists might interpret this chapter (especially Section 4) rather differently. They might read this chapter as a treatise on how experiments on RIs can serve computational chemists, by providing them with problems that can be addressed by performing calculations.

Neither perspective is entirely correct. Calculations can be very useful in designing, executing, and interpreting experiments; *and* experimental results motivate many, if not most, electronic structure calculations. As indicated by the title of this chapter, electronic structure calculations and experiments really have become partners in the study of RIs.

ACKNOWLEDGMENT

My research was generously supported by the National Science Foundation. I am grateful to Professor Thomas Bally, Professor Barry Carpenter, Professor Chris Cramer, Professor Ernest Davidson, Professor Matt Platz, and Professor Daisy Zhang for reading and commenting on various drafts of this chapter. I am also very grateful to Dr. David Hrovat for proofreading the penultimate draft and for being the major contributor to much of the research from my group during the past 18 years.

SUGGESTED READING

C. J. Cramer, *Essentials of Computational Chemistry*, John Wiley & Sons, Inc., New York, **2002**. This monograph provides an authoritative, detailed, and very readable treatise on the methods that are currently used for performing electronic structure calculations, which are described briefly in Section 3.

W. J. Hehre, L. Radom, P. von R. Schleyer, and J. A. Pople, *Ab Initio Molecular Orbital Theory*, Wiley-Interscience, New York, **1986**. For an excellent discussion of basis sets and how they are constructed see, pages 65–88.

The following is a small collection of papers that use the results of electronic structure calculations to explain and/or predict the results of experiments on different types of RIs:

M. Mauksch, V. Gogonea, H. Jiao, and Paul. von R. Schleyer, "The Remarkable Nature of the Monocyclic $(CH)_9^+$ Cation. A Heilbronner Möbius Aromatic System Revealed," *Angew. Chem. Int. Ed. Engl.* **1998**, *37*, 2395. B3LYP calculations provide an interpretation of puzzling experimental results, obtained nearly three decades earlier.

Y. Osamura, H. F. Schaefer, III, S. K. Gray, and W. H. Miller, "Vinylidene: A Very Shallow Minimum on the C_2H_2 Potential Energy Surface. Static and Dynamic Considerations." *J. Am. Chem. Soc.* **1981**, *103*, 1904. At the time these calculations were published, there was no convincing experimental evidence that $H_2C=C$: is an energy minimum, rather than a TS. The prediction that it is an energy minimum was subsequently confirmed experimentally by S. M. Burnett, A. E. Stevens, C. S. Feigerle, and W. C. Lineberger, "Observation of Vinylidene by Photoelectron Spectroscopy of the Vinylidene $(C_2H_2^-)$ Ion," *Chem. Phys. Lett.* **1983**, *100*, 124.

K. N. Houk, N. G. Rondan, and J. Mareda, "Theoretical Studies of Halocarbene Cycloaddition Selectivities. A New Interpretation of Negative Activation Energies and Entropy Control of Selectivities," *Tetrahedron* **1985**, *41*, 1555. Calculations on carbene addition reactions led to a general explanation of why it is possible for very exothermic, bimolecular reactions to have negative activation enthalpies.

D. M. Smith, S. D. Wetmore, and L. Radom, "Theoretical Studies of Coenzyme-B_{12}-Dependent Carbon-Skeleton Rearrangements," in *Theoretical Biochemistry—Processes and Properties of Biological Systems*, L. A. Ericksson, Ed., Elsevier, Amsterdam, The Netherlands, **2001**, pp. 183–214. Electronic structure calculations are applied to the understanding and prediction of how enzymes can lower the barriers to the 1,2-shifts in radicals that occur in reactions catalyzed by B_{12}.

F. Ogliaro, N. Harris, S. Cohen, M. Filatov, S. P. de Visser, and S. Shaik, "A Model 'Rebound' Mechanism of Hydroxylation by Cytochrome P450. Stepwise and Effectively Concerted Pathways, and Their Reactivity Patterns," *J. Am. Chem. Soc.* **2000**, *122*, 8977. Calculations explain puzzling aspects of cytochrome P450 hydroxylation reactions in terms of two, different, reactive spin states of the enzyme.

T. Bally, S. Bernhard, S. Matzinger, J.-L. Roulin, G. N. Sastry, L. Truttmann, Z. Zhu, A. Marcinek, J. Adamus, R. Kaminski, J. Gebicki, F. Williams, G.-F. Chen, and M. P. Fülscher, "The Radical Cation of *syn*-Tricyclooctadiene and Its Rearrangement Products," *Chem. Eur. J.* **2000**, *6*, 858. A combination of CASPT2 calculations of UV spectra and B3LYP and CCSD(T) calculations of PESs is used to identify the intermediates formed sequentially in the rearrangement of the radical cation of the syn dimer of cyclobutadiene and to understand why they differ from those formed from the anti dimer.

D. A. Hrovat, J. H. Hammons, C. D. Stevenson, and W. T. Borden, "Calculations of the Equilibrium Isotope Effects on the Reductions of Benzene-d_6 and Cyclooctatetraene-d_8," *J. Am. Chem. Soc.* **1997**, *119*, 9523. B3LYP/6-31+G* calculations on the title compounds and on the radical anions formed from them show that the very large difference between the equilibrium isotope effects, found by Stevenson, is due to an inverse isotope effect on the planarization of the COT ring. This explanation was subsequently confirmed by KIE measurements, carried out by C. D. Stevenson, E. C. Brown, D. A. Hrovat, and W. T. Borden, "Isotope Effects on the Ring Inversion of Cyclooctatetraene," *J. Am. Chem. Soc.* **1998**, *120*, 8864.

REFERENCES

1. Review: I. Shavitt, *Tetrahedron* **1985**, *41*, 1531.

2. See, inter alia: (a) B. R. Beno, K. N. Houk, and D. A. Singleton, *J. Am. Chem. Soc.* **1996**, *118*, 9984; (b) A. J. DelMonte, J. Haller, K. N. Houk, K. B. Sharpless, D. A. Singleton, T. Strassner, and A. A. Thomas, *J. Am. Chem. Soc.* **1997**, *119*, 9907; (c) A. E. Keating, S. R. Merrigan, D. A. Singleton, and K. N. Houk, *J. Am. Chem. Soc.* **1999**, *121*, 3933.

3. (a) C. J. Cramer, *Essentials of Computational Chemistry*, John Wiley & Sons, Inc., New York, 2002. (b) The quotation is from page 212.

4. (a) P. C. Hariharan and J. A. Pople, *Theor. Chim. Acta* **1973**, *28*, 213; (b) R. Krishnan, J. S. Binkley, R. Seeger, and J. A. Pople, *J. Chem. Phys.* **1980**, *72*, 650; (c) M. J. Frisch, J. A. Pople, and J. S. Binkley, *J. Chem. Phys.* **1984**, *80*, 3265. (d) For an excellent discussion

of basis sets and how they are constructed see, W. J. Hehre, L. Radom, P. von R. Schleyer, and J. A. Pople, *Ab Initio Molecular Orbital Theory*, Wiley-Interscience, New York, 1986, pp. 65–88.

5. T. H. Dunning, Jr., *J. Chem. Phys.* **1989**, *90*, 1007

6. (a) C. Möller and M. S. Plesset, *Phys. Rev.* **1934**, *46*, 618; (b) J. A. Pople, J. S. Binkley, and R. Seeger, *Int. J. Quantum Chem.* **1976**, *S10*, 1.

7. (a) R. J. Bartlett and G. D. Purvis, *Int. J. Quant. Chem.* **1978**, *14*, 516; (b) R. J. Bartlett and J. F. Stanton, in *Reviews in Computational Chemisty*, Vol. 5, K. B. Lipkowitz and D. B. Boyd, Eds., VCH Publishers, New York, 1994, pp. 65–169.

8. J. A. Pople, M. Head-Gordon, and K. Raghavachari, *J. Chem. Phys.* **1987**, *87*, 3700.

9. (a) L. A. Curtiss, K. Raghavachari, G. W. Trucks, and J. A. Pople, *J. Chem. Phys.* **1991**, *94*, 7221; (b) L. A. Curtiss, P. C. Redfern, Raghavachari, V. Rassolov, and J. A. Pople, *J. Chem. Phys.* **1998**, *109*, 7764.

10. M. R. Nyden and G. A. Petersson, *J. Chem. Phys.* **1981**, *75*, 1843 (1981).

11. GAUSSIAN98, Revision A.7, M. J. Frisch, G. W. Trucks, H. B. Schlegel, G. E. Scuseria, M. A. Robb, J. R. Cheeseman, V. G. Zakrzewski, J. A. Montgomery, R. E. Stratmann, J. C. Burant, S. Dapprich, J. M. Millam, A. D. Daniels, K. N. Kudin, M. C. Strain, O. Farkas, J. Tomasi, V. Barone, M. Cossi, R. Cammi, B. Mennucci, C. Pomelli, C. Adamo, S. Clifford, J. Ochterski, G. A. Petersson, P. Y. Ayala, Q. Cui, K. Morokuma, D. K. Malick, A. D. Rabuck, K. Raghavachari, J. B. Foresman, J. Cioslowski, J. V. Ortiz, A. G. Baboul, B. B. Stefanov, G. Liu, A. Liashenko, P. Piskorz, I. Komaromi, R. Gomperts, R. L. Martin, D. J. Fox, T. Keith, M. A. Al-Laham, C. Y. Peng, A. Nanayakkara, C. Gonzalez, M. Challacombe, P. M. W. Gill, B. Johnson, W. Chen, M. W. Wong, J. L. Andres, C. Gonzalez, M. Head-Gordon, E. S. Replogle, and J. A. Pople, Gaussian, Inc., Pittsburgh PA, 1998.

12. B. O. Roos, *Adv. Chem. Phys.* **1987**, *69*, 339.

13. T. Bally and W. T. Borden, in *Reviews in Computational Chemisty*, Vol. 13, K. B. Lipkowitz and D. B. Boyd, Eds., Wiley-VCH Publishers, New York, 1999, pp. 1–97.

14. M. DuPuis, C. Murray, and E. R. Davidson, *J. Am. Chem. Soc.* **1991**, *113*, 9756.

15. W. T. Borden and E. R. Davidson, *Acc. Chem. Res.* **1996**, *29*, 67.

16. (a) K. Andersson, P. Å. Malmqvist, B. O. Roos, A. J. Sadlej, and K. Wolinski, *J. Phys. Chem.* **1990**, *94*, 5483. (b) K. Andersson, P.-Å. Malmqvist, and B. O. Roos, *J. Chem. Phys.* **1992**, *96*, 1218.

17. K. Andersson, M. R. A. Blomberg, M. P. Fülscher, G. Karlström, V. Kellö, R. Lindh, P.-Å. Malmqvist, J. Noga, J. Olsen, B. O. Roos, A. J. Sadlej, P. E. M. Siegbahn, M. Urban, and P.-O. Widmark, *MOLCAS-4*, University of Lund, Sweden.

18. D. A. Hrovat and W. T. Borden, unpublished results.

19. D. A. Hrovat, K. Morokuma, and W. T. Borden, *J. Am. Chem. Soc.* **1994**, *116*, 1072.

20. D. A. Hrovat and W. T. Borden, *J. Am. Chem. Soc.* **1994**, *116*, 6327.

21. T. Bally, S. Bernhard, S. Matzinger, J.-L. Roulin, G. N. Sastry, L. Truttmann, Z. Zhu, A. Marcinek, J. Adamus, R. Kaminski, J. Gebicki, F. Williams, G.-F. Chen, and M. P. Fülscher, *Chem. Eur. J.* **2000**, *6*, 858.

22. A. D. Becke, *J. Chem. Phys.* **1993**, *98*, 5648.

23. C. Lee, W. Yang, and R. G. Parr, *Phys. Rev. B* **1988**, *37*, 785.

24. For examples see (a) D. A. Plattner and K. N. Houk, *J. Am. Chem. Soc.* **1995**, *117*, 4405; (b) A. Karpfen, C. H. Choi, and M. Kertesz, *J. Phys. Chem. A* **1997**, *101*, 7426; and (c) Footnote 25 in D. A. Hrovat, J. Chen, K. N. Houk, and W. T. Borden, *J. Am. Chem. Soc.* **2000**, *122*, 7456.

25. (a) R. E. Stratmann, G. E. Scuseria, and M. J. Frisch, *J. Chem. Phys.* **1998**, *109*, 8218; (b) M. E. Casida, C. Jamorksi, K. C. Casida, and D. R. Salahub, *J. Chem. Phys.* **1998**, *108*, 4439.

26. For a recent application of TDDFT to the identification of an RI see A. C. Goren, D. A. Hrovat, M. Seefelder, H. Quast, and W. T. Borden, *J. Am. Chem. Soc.* **2002**, *124*, 2469.

27. T. Bally, D. A. Hrovat, and W. T. Borden, *Phys. Chem. Chem. Phys.* **2000**, *2*, 3363.

28. T. Bally and G. N. Sastry, *J. Phys. Chem. A* **1997**, *101*, 7923.

29. (a) W. T. Borden, N. P. Gritsan, C. M. Hadad, W. L. Karney, C. R. Kemnitz, and M. S. Platz, *Acc. Chem. Res.* **2000**, *33*, 765; (b) W. R. Karney and W. T. Borden, in U. H. Brinker, Ed., *Advances in Carbene Chemistry*, Vol. 3, Elsevier, Amsterdam, The Netherlands, **2001**, pp. 206–251.

30. (a) S.-J. Kim, T. P. Hamilton, and H. F. Schaefer, *J. Am. Chem. Soc.* **1992**, *114*, 5349; (b) D. A. Hrovat, E. E. Waali, and W. T. Borden, *J. Am. Chem. Soc.* **1992**, *114*, 8698.

31. P. C. Engelking and W. C. Lineberger, *J. Chem. Phys.* **1976**, *65*, 4323.

32. M. J. Travers, D. C. Cowles, E. P. Clifford, G. B. Ellison, and P. C. Engelking, *J. Phys. Chem.* **1999**, *111*, 5349.

33. (a) M. J. Travers, D. C. Cowles, E. P. Clifford, and G. B. Ellison, *J. Am. Chem. Soc.* **1992**, *114*, 8699. (b) R. N. McDonald and S. J. Davidson, *J. Am. Chem. Soc.* **1993**, *115*, 10857.

34. (a) S. Matzinger, T. Bally, E. V. Patterson, and R. J. McMahon, *J. Am. Chem. Soc.* **1996**, *118*, 1535. (b) M. W. Wong and C. Wentrup, *J. Org. Chem.* **1996**, *61*, 7022. (c) P. R. Schreiner, W. L. Karney, P. v. R. Schleyer, W. T. Borden, T. P. Hamilton, and H. F. Schaefer, *J. Org. Chem.* **1996**, *61*, 7030.

35. W. L. Karney and W. T. Borden, *J. Am. Chem. Soc.* **1997**, *119*, 1378.

36. W. L. Karney and W. T. Borden, *J. Am. Chem. Soc.* **1997**, *119*, 3347.

37. (a) N. P. Gritsan, A. D. Gudmundsdóttir, D. Tigelaar, and M. S. Platz, *J. Phys. Chem. A* **1999**, *103*, 3458. (b) N. P. Gritsan, D. Tigelaar, and M. S. Platz, *J. Phys. Chem. A* **1999**, *103*, 4465; (c) N. P. Gritsan, A. D. Gudmundsdóttir, D. Tigelaar, Z. Zhu, W. L. Karney, C. M. Hadad, and M. S. Platz, *J. Am. Chem. Soc.* **2001**, *123*, 1951, and references cited therein.

38. (a) N. P. Gritsan, I. Likhotvorik, M.-L. Tsao, N. Çelebi, M. S. Platz, W. L. Karney, C. R. Kemnitz, and W. T. Borden, *J. Am. Chem. Soc.* **2001**, *123*, 1425. (b) For a combined computational-experimental study by our group and Platz's of the rearrangement of *ortho*-biphenylnitrene, see M.-L. Tsao, N. Gritsan, T. R. James, M. S. Platz, D. A. Hrovat, and W. T. Borden, *J. Am. Chem. Soc.* **2003**, *125*, 9343.

39. X. Li, D. L. Lei, M. Y. Chiang, and P. P. Gaspar, *J. Am. Chem. Soc.* **1992**, 8526.

40. J. M. Galbraith, P. P. Gaspar, and W. T. Borden, *Am. Chem. Soc.* **2002**, *1234*, 11669.

41. P. E. Eaton, *Angew. Chem. Int. Ed. Engl.* **1992**, *31*, 1421.

42. P. E. Eaton, C.-X. Yang, and Y. Xiong, *J. Am. Chem. Soc.* **1990**, *112*, 3225.

43. (a) R. M. Moriarity, M. T. Sudersan, R. Penmasta, and A. K. Awasthi, *J. Am. Chem. Soc.* **1990**, *112*, 3225; (b) D. N. Kevill, M. J. D'Souza, R. M. Moriarity, M. T. Sudersan, R. Penmasta, and A. K. Awasthi, *J. Chem. Soc. Chem. Commun.* **1990**, 623.

44. P. E. Eaton and J. P. Zhou, *J. Am. Chem. Soc.* **1992**, *114*, 3118.

45. D. A. Hrovat and W. T. Borden, *J. Am. Chem. Soc.* **1990**, *112*, 3227.

46. See, for example, W. J. Hehre, L. Radom, P. von R. Schleyer, and J. A. Pople, "Ab Initio Molecular Orbital Theory," Wiley-Interscience, New York, 1986, pp. 379–396.

47. For examples see: (a) M. L. McKee, *J. Phys. Chem.* **1986**, *90*, 4908; (b) W. Koch, B. Liu, and D. J. DeFrees, *J. Am. Chem. Soc.* **1988**, *110*, 7325; (c) M. Saunders, K. Laidig, K. B. Wiberg, and P. von R. Schleyer, *J. Am. Chem. Soc.* **1988**, *110*, 7652. (d) E. W. Della, P. M. W. Gill, and C. H. Schiesser, *J. Org. Chem.* **1988**, *53*, 4354.

48. (a) O. Weist, K. A. Black, and K. N. Houk, *J. Am. Chem Soc.* **1994**, *116*, 10336; (b) H. Jiao and P. Von R. Schleyer, *Angew. Chem. Int. Ed. Engl.* **1995**, *34*, 334.

49. A. G. Martinez, E. T. Vilar, J. O. Barcina, and S. M. Cerero, *J. Am. Chem Soc.* **2002**, *124*, 6676.

50. K. Hassenrüuck, J. G. Radziszewski, V. Balaji, G. S. Murthy, A. J. McKinley, D. E. David, V. M. Lynch, H.-D. Martin, and J. Michl, *J. Am. Chem. Soc.* **1990**, *112*, 873.

51. D. A. Hrovat and W. T. Borden, *J. Am. Chem. Soc.* **1990**, *112*, 875.

52. E. W. Della, N. J. Head, P. Mallon, and J. C. Walton, *J. Am. Chem. Soc.* **1992**, *114*, 10730.

53. W. A. Pryor, W. H. Davis, Jr., and J. P. Stanley, *J. Am. Chem. Soc.* **1973**, *95*, 4754; and the many book chapters and reviews referenced in this paper.

54. D. A. Hrovat and W. T. Borden, *J. Am. Chem. Soc.* **1994**, *116*, 6459.

55. (a) L. D. Pedersen and J. A. Berson, *J. Am. Chem. Soc.* **1975**, *97*, 238; (b) J. A. Berson, L. D. Pedersen, and B. K. Carpenter, *J. Am. Chem. Soc.* **1976**, *98*, 122. See also, (c) S. J. Cianciosi, N. Ragunathan, T. R. Freedman, L. A. Nafie, and J. E. Baldwin, *J. Am. Chem. Soc.* **1990**, *112*, 8204.

56. S. J. Cianciosi, N. Ragunathan, T. R. Freedman, L. A. Nafie, D. K. Lewis, D. A. Glenar, and J. E. Baldwin, *J. Am. Chem. Soc.* **1991**, *113*, 1864.

57. R. Hoffmann, *J. Am. Chem. Soc.* **1968**, *90*, 11475.

58. Reviews: (a) J. J. Gajewski, *Hydrocarbon Thermal Isomerizations*, Academic Press: New York, 1981, pp. 27–35; (b) J. A. Berson, in *Rearrangements in Ground And Excited States*, Vol. 1, P. de Mayo, Ed., Academic Press, New York, 1980, pp. 311–390.

59. See footnote 28 of Ref. 60.

60. S. J. Getty, E. R. Davidson, and W. T. Borden, *J. Am. Chem. Soc.* **1992**, *114*, 2085.

61. J. E. Baldwin, Y. Yamaguchi, and H. F. Schaefer, III, *J. Phys. Chem.* **1994**, *98*, 7513.

62. C. Doubleday, Jr., K. Bolton, and W. L. Hase, *J. Am. Chem. Soc.* **1997**, *119*, 5251.

63. D. A. Hrovat, S. Fang, W. T. Borden, and B. K. Carpenter, *J. Am. Chem. Soc.* **1997**, *119*, 5253.

64. Review: W. T. Borden, *Chem. Commun.* **1998**, 1919.

65. S. J. Getty, D. A. Hrovat, and W. T. Borden, *J. Am. Chem. Soc.* **1994**, *116*, 1521.

66. S. J. Getty, D. A. Hrovat, J. D. Xu, S. A. Barker, and W. T. Borden, *J. Chem. Soc. Faraday Trans.* **1994**, *90*, 1689.

67. F. Tian, S. B. Lewis, M. D. Bartberger, W. R. Dolbier, Jr., and W. T. Borden, *J. Am. Chem. Soc.* **1998**, *120*, 6187.

68. A. Skancke, D. A. Hrovat, and W. T. Borden, *J. Am. Chem. Soc.* **1998**, *120*, 7079.

69. V. F. Mironove, V. V. Shcherbinin, N. A. Viktorov, and V. D. Sheludyakov, *Zh. Obshch. Khim.* **1975**, *45*, 1908.

70. (a) J. J. Gajewski and M. J. Chang, *J. Am. Chem. Soc.* **1980**, *102*, 7542; (b) J. J. Gajewski, R. J. Weber, and M. J. Chang, *J. Am. Chem. Soc.* **1979**, *101*, 2100.

71. W. T. G. Johnson, D. A. Hrovat, and W. T. Borden, *J. Am. Chem. Soc.* **1999**, *121* 7766.

72. J. D. Xu, D. A. Hrovat, and W. T. Borden, *J. Am. Chem. Soc.* **1994**, *116*, 4888.

73. W. Adam, W. T. Borden, C. Burda, H. Foster, T. Heidenfelder, M. Heubes, D. A. Hrovat, F. Kita, S. B. Lewis, D. Scheutzow, and J. Wirz, *J. Am. Chem. Soc.* **1998**, *120*, 593

74. (a) M. P. Conrad, R. M. Pitzer, and H. F. Schaefer, *J. Am. Chem. Soc.* **1979**, *101*, 2245; (b) C. D. Sherrill, E. T. Seidl, and H. F. Schaefer, *J. Phys. Chem.* **1992**, *96*, 3712.

75. (a) S. L. Buchwalter and G. L. Closs, *J. Am. Chem. Soc.* **1975**, *97*, 3857. (b) S. L. Buchwalter and G. L. Closs, *J. Am. Chem. Soc.* **1979**, *101*, 4688.

76. (a) F. D. Coms and D. A. Dougherty, *Tetrahedron Lett.* **1988**, *29*, 3753; (b) W. Adam, S. Grabowski, H. Platsch, K. Hannemann, J. Wirz, and M. Wilson, *J. Am. Chem. Soc.* **1989**, *111*, 751; (c) F. D. Coms and D. A. Dougherty, *J. Am. Chem. Soc.* **1989**, *111*, 6894; (d) W. Adam, H. Platsch, and J. Wirz, *J. Am. Chem. Soc.* **1989**, *111*, 6896.

77. M. Abe, W. Adam, T. Heidenfelder, W. M. Nau, and X. Zhang, *J. Am. Chem. Soc.* **2000**, *122*, 2019.

78. M. Abe, W. Adam, M. Hara, M. Hattori, T. Majima, M. Nojima, K. Tachibana, and S. Tojo, *J. Am. Chem. Soc.* **2002**, *124*, 6450.

79. W. T. G. Johnson, D. A. Hrovat, A. Skancke, and W. T. Borden, *Theor. Chem. Acc.* **199**, *102*, 207.

Reactive Intermediate Chemistry, edited by Robert A. Moss, Matthew S. Platz, and Maitland Jones, Jr.
ISBN 0-471-23324-2   Copyright © 2004 John Wiley & Sons, Inc.

1016 INDEX